T0178382

Springer Finance

Textbooks

Springer Finance Textbooks

Springer Finance is a programme of books addressing students, academics and practitioners working on increasingly technical approaches to the analysis of financial markets. It aims to cover a variety of topics, not only mathematical finance but foreign exchanges, term structure, risk management, portfolio theory, equity derivatives, and financial economics.

This subseries of Springer Finance consists of graduate textbooks.

More information about this series at http://www.springer.com/series/3674

Emilio Barucci • Claudio Fontana

Financial Markets Theory

Equilibrium, Efficiency and Information

Second Edition

 Springer

Emilio Barucci
Dipartimento di Matematica
Politecnico di Milano
Milano, Italy

Claudio Fontana
Laboratoire de Probabilités et Modèles
    Aléatoires
Université Paris Diderot (Paris 7)
Paris, France

ISSN 1616-0533   ISSN 2195-0687   (electronic)
Springer Finance
ISBN 978-1-4471-7404-2   ISBN 978-1-4471-7322-9   (eBook)
DOI 10.1007/978-1-4471-7322-9

Mathematics Subject Classification (2010): 91B06, 91B08, 91B16, 91B24, 91B25, 91B30, 91B50, 91G10, 91G20

Printed on acid-free paper

This Springer imprint is published by Springer Nature
The registered company is Springer-Verlag London Ltd.
The registered company address is: 236 Gray's Inn Road, London WC1X 8HB, United Kingdom

To Piero, Maria, Elena, Teresa and Maria
  E.B.

To Anna, Angelo and Liliana
  C.F.

Preface to the Second Edition (2017)

Already at the time of the first edition of this book, financial markets theory represented, and still represents nowadays, an extremely rich field of research, characterized by a huge and growing literature. Since the publication of the first edition, the literature has grown even more significantly, to the extent that it is simply impossible to give a truly comprehensive account of financial markets theory. In this second edition, we aim at providing a broad overview of the fundamental aspects of the theory, balancing economic intuition and mathematical detail. Moreover, following the philosophy of the first edition, the presentation of theoretical results is always accompanied by a discussion of the empirical evidence reported in the financial literature.

In comparison with the first edition, the book has been thoroughly revised and significantly expanded. While the overall structure of the book is similar to that of the first edition, most of the chapters have been entirely rewritten. In particular, besides presenting a large amount of additional material, we aim at providing a self-contained derivation of almost all the results and models discussed in the book, with detailed proofs, thus making the book accessible for self-study. Furthermore, this second edition contains more than 200 exercises. The solutions to most of the exercises are given at the end of the book, while the solutions to the remaining exercises can be downloaded from the publisher's website. Some exercises contain proofs of theoretical results which are not proven in the text, and other exercises develop interesting side results or explore some models not discussed in the text, while other exercises introduce simple applications of the results presented in the text, with the aim of helping the reader to develop a better understanding of the theory. Our presentation is partly influenced by several excellent books which have appeared after the publication of the first edition. In particular, without any pretence to be exhaustive, these include the books by Back [98], Cvitanić & Zapatero [511], De Jong & Rindi [540], Demange & Laroque [551], Eeckhoudt et al. [631], Munk [1363], Hens & Rieger [938], Lengwiler [1182], Pennacchi [1411] and Vives [1630].

The book is organized as follows. The first seven chapters contain a presentation of classical asset pricing theory, first in a single period setting (Chaps. 1–5) and then in a dynamic setting (Chaps. 6–7), together with a discussion of the empirical evidence on the implications of the theory. The last three chapters deal with more advanced topics and present a number of extensions of classical asset pricing theory as well as models going beyond classical asset pricing theory, in order to address several asset pricing puzzles. Each chapter ends with a section dedicated to notes and further readings and a collection of exercises. We now give a more detailed outline of the contents of each chapter.

Chapter 1 presents a quick overview of some fundamental results of decision theory and equilibrium theory under certainty (classical microeconomic theory). The chapter starts with a discussion of preference relations and utility functions and then introduces the concepts of general equilibrium, Pareto optimality and representative agent. The aim of this chapter is to make the book as self-contained as possible as far as classical economic analysis is concerned.

Chapter 2 lays the foundations of decision theory in the presence of risk. The chapter starts with the presentation of expected utility theory and then analyses the concepts of risk aversion, certainty equivalent, stochastic dominance and mean-variance preferences, also explaining the diversification and the insurance principle.

Chapter 3, in a simple two-period setting, analyses the optimal portfolio problem, first in the case of a single risky asset and then in the more general case of multiple risky assets. The chapter contains several comparative statics results as well as closed-form solutions under suitable assumptions on the utility function and on the distribution of asset returns. The second part of the chapter contains a presentation of the classical Markowitz's mean-variance theory, exploring the properties of the mean-variance portfolio frontier. We also establish several mutual fund separation results. The last part of the chapter deals with optimal insurance problems and optimal consumption-saving problems in the context of a two-period economy.

Chapter 4 deals with general equilibrium theory in a risky environment, where agents interact in a financial market. The chapter starts by presenting the notion of Pareto optimality and its implications in terms of risk sharing. The concept of rational expectations equilibrium is introduced and characterized in the context of a two-period economy. In this chapter, different financial market structures are considered, with a particular attention to the important case of complete markets. The last part of the chapter is devoted to the fundamental theorem of asset pricing, which links the absence of arbitrage opportunities to the existence of a strictly positive linear pricing functional. The relation of this important result to the existence of an equilibrium of an economy and its implications for the valuation of financial assets are also discussed.

Chapter 5, on the basis of the general equilibrium theory developed in Chap. 4, presents some of the most important asset pricing models, including the consumption capital asset pricing model (CCAPM), the capital asset pricing model (CAPM) and the arbitrage pricing theory (APT). The relations of these asset pricing models with the absence of arbitrage opportunities are also discussed. In this chapter, the theoretical presentation of the models and of their implications is accompanied by an overview of the empirical evidence reported in the literature. In particular, several asset pricing anomalies and puzzles are described and discussed.

Chapter 6 extends the analysis developed in the previous chapters to the case of dynamic multiperiod economies. The chapter starts by studying the optimal investment-consumption problem of an individual agent in a multiperiod setting, by relying on the dynamic programming approach. Under suitable assumptions on the utility function, closed-form solutions are derived. The chapter then proceeds by extending the general equilibrium theory established in Chap. 4 to a dynamic setting, introducing the notion of dynamic market completeness and analysing the aggregation property of the economy. The fundamental theorem of asset pricing is then established in a multiperiod setting, and its relation to the equilibrium of an economy is discussed. Later in the chapter, the asset pricing relations presented in Chap. 5 are extended to a dynamic context and specialized for several classes of utility functions. The chapter ends by considering multiperiod economies with an infinite time horizon and the possibility of asset price bubbles.

Chapter 7 is devoted to an extensive overview of the empirical evidence on classical asset pricing theory. In particular, the attention is focused on the empirical properties of the observed prices and returns and on several anomalies reported in the literature, including the excess volatility phenomenon, the predictability of asset returns, the equity premium puzzle, the risk free rate puzzle and other related asset pricing puzzles.

Chapter 8 concerns the role of information in financial markets. The chapter starts by studying the value of information from the point of view of an individual agent and then proceeds by analysing the impact of information in the economy. This leads to the introduction of the notion of Green-Lucas equilibrium and, in particular, to the question of whether equilibrium prices transmit, aggregate and reveal the private information of the agents. This concept is intrinsically linked to the informational efficiency of financial markets and provides a microfoundation to the efficient markets theory. In the chapter, the possibility or the impossibility of informationally efficient markets is discussed and established in the context of several models. The chapter closes by briefly examining the case where agents exhibit heterogeneous opinions and surveying the empirical evidence on the informational efficiency of financial markets.

Chapter 9 presents a critical analysis of several fundamental hypotheses underlying the theory developed in the previous chapters of the book. More specifically, the attention is focused on the notions of risk and uncertainty, on the classical assumptions of expected utility theory, on the agents' rationality and on the imperfections of financial markets. By relaxing the assumptions made so far in the book, several alternative paradigms are proposed and their implications are discussed, also on the basis of the empirical evidence reported in the literature. Some basic aspects of behavioural finance are explored and compared to the implications of classical asset pricing theory.

Finally, Chap. 10 deals with financial markets microstructure. Relaxing the assumption of perfect competition adopted so far in the book, this chapter explores the real functioning of financial markets and the role of different categories of market participants. In particular, the central themes are represented by the role of information under imperfect competition and market liquidity. The chapter closes with an overview of insider trading, market manipulation and the different institutional features of financial markets.

In our presentation, we do not aim at presenting results in their most general formulations with full mathematical details, but rather we aim at explaining the central themes of financial markets theory in a transparent setting. For this reason, we have chosen to keep the mathematical technicalities at a minimum level, only requiring a basic knowledge of algebra, differential calculus and elementary probability theory. While we present the theoretical results with complete and rigorous proofs, we always try to emphasize the economic intuition over the technical aspects. Even though some familiarity with basic microeconomics is helpful, the book does not require any previous knowledge of economic theory, since all the relevant results are reviewed in Chap. 1.

This book can be used in several ways. The first five chapters can form the basis for a first course on the economics of uncertainty, focused on expected utility theory, portfolio selection, equilibrium and no-arbitrage. The remaining chapters contain more advanced material and can be used for more specialized courses or as a starting point for the self-study of these topics. The extensive bibliography contained in the book should be useful to every researcher in the field as a guide to the wide literature on financial markets theory. An updated pdf file containing the errata of the book will be available for download on the webpage http://sites.google.com/site/fontanaclaud/fmt.

We are thankful to Gaetano La Bua who has checked the exercises of the book and made several useful comments. We are also thankful to Gaetano for his help in the preparation of the figures appearing in the text. We are grateful to Rémi Lodh and Catriona Byrne from Springer for their precious assistance in the several stages of the preparation of this second edition.

Milano, Italy   Emilio Barucci
Paris, France   Claudio Fontana
February 2017

From the Preface to the First Edition (2003)

This is just another book on financial markets theory. Why another book?

Most of the time an author will answer that there was no book covering the same topics with the same approach. This is also my answer. Organizing my lectures for an advanced financial markets theory course, I tried to make my students understand how financial markets theory was a body in continuous transformation animated both by a rich theoretical debate and by a strict interaction with real financial markets. I did not find a book with such an approach and so I decided to write this one. I hope my students were and will be fascinated by the picture.

The book is driven by two perspectives: on the one side the theoretical debate about financial markets and agent's behaviour under risk and on the other the comparison of theoretical results with the empirical evidence. I intend to highlight how financial markets theory has developed during the last 50 years along these two perspectives. The first one has been a driving force for a long time; my personal reconstruction of the development of the theory includes Bachelier (1900), Arrow (1953), Modigliani and Miller (1958), Debreu (1959), Sharpe (1964), Lintner (1965), Black and Scholes (1973), LeRoy (1973), Merton (1973), Jensen and Meckling (1976), Rubinstein (1976), Ross (1977), Lucas (1978), Harrison and Kreps (1979) and Grossman (1981). As a matter of fact, the classical asset pricing theory was almost fully developed by the mid-1980s; the debate then turned to the empirical evidence and hence the second perspective came into play. Results are mixed with some puzzles still alive after 20 years. These puzzles generated a debate with developments inside the classical asset pricing theory as well as outside with an attempt to build alternative paradigms (e.g. behavioural finance). Classical asset pricing theory tries to explain asset pricing anomalies by changing agent's preferences, probability distributions for fundamentals, economic environment maintaining its traditional pillars, i.e. agents' rationality, equilibrium or no-arbitrage arguments, and rational expectations. According to the classical asset pricing theory, asset prices are explained through risk factors related to asset fundamentals. On the other hand, behavioural finance tries to build an alternative paradigm by relaxing some hypotheses on agents' rationality and introducing market frictions. The debate on asset pricing puzzles is one of the main topics of the book. On this point we follow the Kuhn (1970) perspective: a paradigm has never been rejected by falsification through direct comparison with the real world; a new paradigm should be accepted.

The book is an adventure in modern financial markets theory with two plots: theoretical developments and real financial markets. There are two actors: classical asset pricing theory and heretics. Conclusions on the status of the theory are left to the reader. My personal belief is that classical asset pricing theory is a strong and flexible paradigm. Many anomalies can be reinterpreted inside the paradigm, but some of them are hard to kill. On the other hand, behavioural finance and alternative approaches provide useful insights to understand real financial markets day by day, but they are not an alternative paradigm. Agent's rationality is a simplification of human behaviour and therefore some anomalies are expected.

The book can be used in several ways. It is an advanced financial markets theory textbook, and it provides a handbook on recent developments of the literature. The book covers a wide spectrum of topics. We are not going to deal with mathematical finance topics (option pricing, term structure, interest rate derivatives) because in our view they are mainly an application of the fundamental asset pricing theorem; there are many interesting and hard to solve problems in this field, but they are mainly technical problems. We handle the topics in the simplest setting (finite states-discrete time). Some parts of the book have been heavily inspired by Barucci (2000). The book owes a debt to Huang and Litzenberger (1988), which introduced me to the economic analysis of financial markets. I strongly believe that the advancement of the theory will be driven in the future by economic analysis, empirical evidence and the incorporation of the institutional setting in the picture. This book testifies this belief. I hope the reader will find interesting hints in reading the book.

It took 5 years to complete this book. During these years, the book has been a living companion for me with frustrations, worries and many other emotions; now it is only printed paper. I have to thank many for their encouragement and help. First of all I would like to thank all my coauthors for having contributed a lot to this book. I owe enormous intellectual debt to them. Special thanks to Maria Elvira Mancino for her encouragement. I would like to thank my colleagues at the University of Pisa, Bruce Marshall and Claudia Neri, for carefully reading the final manuscript. The book is first of all dedicated to my students in the past and in the future, to their enthusiasm which is a strong motivation for a teacher. The book is dedicated to all who sympathized, sympathize or will sympathize with me, in particular my parents, Teresa and those who are not with me anymore.

Firenze/Pisa, Italy   Emilio Barucci
September 2002

Contents

Chapter 1
Prerequisites

*All theory depends on assumptions which are not quite true.
That is what makes it theory. The art of successful theorizing is
to make the inevitable simplifying assumptions in such a way
that the final results are not very sensitive. A "crucial"
assumption is one on which the conclusions do depend
sensitively, and it is important that crucial assumptions be
reasonably realistic. When the results of a theory seem to flow
specifically from a special crucial assumption, then if the
assumption is dubious, the results are suspect.*

Solow (1956)

*Once it has achieved the status of paradigm, a scientific theory
is declared invalid only if an alternative candidate is available
to take its place. No process yet disclosed by the historical study
of scientific development at all resembles the methodological
stereotype of falsification by direct comparison with nature.
[...] The act of judgement that leads scientists to reject a
previously accepted theory is always based upon more than a
comparison of that theory with the world. The decision to reject
one paradigm is always simultaneously the decision to accept
another, and the judgement leading to that decision involves the
comparison of both paradigms with nature and with each other.*

Kuhn (1970)

In this chapter, we provide a brief overview of the fundamental results on pure exchange economies under certainty. In Sect. 1.1, we present the classical axiomatic framework to decision making under certainty, discussing the axioms of preference relations and introducing the notion of utility function. In Sects. 1.2 and 1.3, we provide the main results on general equilibrium theory and we introduce the notions of Pareto optimality and representative agent together with the two Welfare Theorems.

© Springer-Verlag London Ltd. 2017
E. Barucci, C. Fontana, *Financial Markets Theory*,
Springer Finance, DOI 10.1007/978-1-4471-7322-9_1

1.1 Choices Under Certainty

Let us consider an agent who makes choices under certainty. In an environment which is not affected by any source of randomness, choices can be analysed by means of two ingredients:

- a set of choices X;
- a weak preference relation $\mathscr{R}$ representing the agent's preferences over couples of elements of X.

In the classical consumption problem, choices concern bundles of goods and, therefore, X will be a subset of $\mathbb{R}^L$ (e.g., $X = \mathbb{R}^L_+$), where $L \in \mathbb{N}$ is the number of goods.[1] A vector $x \in X$ identifies a bundle of goods.

A (weak) *preference relation* $\mathscr{R}$ is defined on X. Given $x, y \in X$, if $x \mathscr{R} y$ we say that *the bundle of goods x is at least as preferred as the bundle of goods y*. The preference relation $\mathscr{R}$ allows us to deduce two relations:

- the strong preference relation $\mathscr{P}$ defined as

$$x \mathscr{P} y \iff x \mathscr{R} y \text{ but not } y \mathscr{R} x,$$

  meaning that *the bundle of goods x is strictly preferred to the bundle of goods y*;
- the indifference relation $\mathscr{I}$ defined as

$$x \mathscr{I} y \iff x \mathscr{R} y \text{ and } y \mathscr{R} x,$$

  meaning that *the bundle of goods x and the bundle of goods y are equally preferred*.

The couple $(X, \mathscr{R})$ characterizes completely the agent's choice problem: the choice set X and the preference relation $\mathscr{R}$. To yield a meaningful analysis, some hypotheses on the preference relation are needed. We will assume that the preference relation $\mathscr{R}$ satisfies the following *rationality* axiom.

Assumption 1.1 (Rationality) The preference relation $\mathscr{R}$ on the set of choices X is said to be *rational* if it satisfies the following properties:

- *Reflexivity*: for every $x \in X$, it holds that $x \mathscr{R} x$;
- *Completeness*: for every $x, y \in X$, it holds that $x \mathscr{R} y$ or $y \mathscr{R} x$;
- *Transitivity*: for every $x, y, z \in X$, if $x \mathscr{R} y$ and $y \mathscr{R} z$, it then holds that $x \mathscr{R} z$.

---

[1]The set $\mathbb{R}^L_+$ denotes the set of non-negative vectors and $\mathbb{R}^L_{++}$ the set of strictly positive vectors with L elements.

According to the above assumption, an agent is rational if he is able to rank any couple of alternatives and if he is consistent over different choices. The transitivity of the preference relation represents the hypothesis that the agent is able to "think big", not only with respect to a single couple of alternatives.

In order to mathematically describe decision making problems, it is useful to introduce a *utility function*. To this end, we require the preference relation $\mathscr{R}$ to satisfy an additional technical hypothesis called the *continuity* hypothesis.

Assumption 1.2 (Continuity) The preference relation $\mathscr{R}$ on the set of choices X is said to be *continuous* if, for every $x \in X$, the sets $\{y \in X : y \mathscr{R} x\}$ and $\{y \in X : x \mathscr{R} y\}$ are closed.

We say that a *utility function* $u : X \to \mathbb{R}$ represents the preference relation $\mathscr{R}$ if

$$x \mathscr{R} y \iff u(x) \geq u(y), \quad \text{for all } x, y \in X.$$

The following theorem holds (see Mas-Colell et al. [1310, Proposition 3.C.1] for a proof).

Theorem 1.1 *Let $\mathscr{R}$ be a preference relation on the set of choices X satisfying Assumptions 1.1 and 1.2. Then $\mathscr{R}$ can be represented through a continuous utility function $u : X \to \mathbb{R}$.*

If Assumptions 1.1 and 1.2 hold, then the decision making problem of an agent characterized by the preference relation $\mathscr{R}$ on the set of choices X will be completely described by the couple (X, u). Note that Theorem 1.1 does not assert the uniqueness of the utility function u representing the preference relation $\mathscr{R}$. Indeed, if the utility function u represents the preference relation $\mathscr{R}$, then any strictly increasing transformation of u will also represent the same preference relation $\mathscr{R}$. For a given utility function $u : X \to \mathbb{R}$, the *indifference sets* (or *indifference curves*) of u are defined as the level sets of the function u, i.e., the sets $\{x \in X : u(x) = c\}$, for $c \in \mathbb{R}$.

Some important properties of a preference relation $\mathscr{R}$ are the following[2]:

- $\mathscr{R}$ is *strictly monotone* if, for every $x, y \in X$ such that $y > x$, then $y \mathscr{P} x$;
- $\mathscr{R}$ is *strictly convex* if, for every $x, y, z \in X$ with $x \neq y$ such that $x \mathscr{R} z$ and $y \mathscr{R} z$, then $\alpha x + (1 - \alpha) y \mathscr{P} z$, for every $\alpha \in (0, 1)$.

If the preference relation $\mathscr{R}$ satisfies the two above properties, then the indifference sets of the utility function u are ordered in an increasing sense (equivalently, u is strictly increasing) and $u(\alpha x + (1 - \alpha) y) > \min(u(x), u(y))$, for all bundles of goods $x, y \in X$ with $x \neq y$ and $\alpha \in (0, 1)$. In the following, we shall generally assume utility functions to be twice continuously differentiable. In that case, if the above properties hold, then the first partial derivative of a utility function

---

[2] Given two vectors $x, y \in \mathbb{R}^L$, the inequality $x \geq y$ means that $x_n \geq y_n$ ($\forall n = 1, \ldots, L$), while $x > y$ means that $x_n \geq y_n$ ($\forall n = 1, \ldots, L$) with $x \neq y$, and $x >> y$ means that $x_n > y_n$ ($\forall n = 1, \ldots, L$).

(*marginal utility*) with respect to a consumption good is positive and the second partial derivative is negative. In the following, Assumptions 1.1 and 1.2 will be always supposed to hold unless otherwise specified.

We dip the agent in a perfectly competitive market. Following the classical approach, the interaction of multiple agents in the market can be addressed in two subsequent steps. The first part of the problem is addressed in this section, the second part in the next one. In a perfectly competitive market, agents do not affect market prices (i.e., agents are *price takers*) and, therefore, we can parametrize their behavior with respect to market prices. In the following, we identify market prices by a vector $p \in \mathbb{R}_+^L$.

We assume that every agent is characterized by *substantial rationality*, i.e., an agent pursues his goals in the most appropriate way under the constraints imposed by the environment. In a perfectly competitive market, the constraints are given by the *budget constraint* (determined by market prices and wealth): given the wealth $w \in \mathbb{R}_+$ and the price vector $p \in \mathbb{R}_+^L$, the set on which an agent makes his choices is

$$B := \{x \in X : p^\top x \leq w\},$$

for some closed set $X \subseteq \mathbb{R}_+^L$ and with the superscript $^\top$ denoting transposition. According to the substantial rationality paradigm, the most appropriate bundle of goods $x^* \in X$ solving the agent's optimal choice problem will be the one resulting from the maximization of the utility function representing the agent's preferences subject to the budget constraint. Hence, a vector $x^* \in X$ is obtained as the solution of the following optimization problem:

$$\max_{x \in B} u(x). \tag{MP0}$$

Note that, if the utility function u is continuous (as follows from Theorem 1.1), then there always exists a solution to Problem (MP0) if $p \in \mathbb{R}_{++}^L$. This is a direct consequence of the fact that any continuous function admits a maximum over a compact set (see Mas-Colell et al. [1310], Proposition 3.D.1). Moreover, by the strict monotonicity of the utility function u, it is never optimal for an agent not to use all his wealth w, so that the budget constraint can be written as an equality constraint and the optimum problem can be handled via the Lagrange multipliers method. For simplicity of presentation, we shall suppose from now on that $X = \mathbb{R}_+^L$. Given the Lagrangian

$$L(x, \lambda) = u(x) + \lambda(w - p^\top x),$$

where λ represents the Lagrange multiplier, the first order necessary conditions for a strictly positive optimal consumption vector $x^* \in \mathbb{R}^L_{++}$ become[3]

$$u_{x_k}(x^*) = \lambda^* p_k, \quad \text{for all } k = 1, \ldots, L;$$
$$w = p^\top x^*, \tag{1.1}$$

where u_{x_k} denotes the first partial derivative of the utility function u with respect to its k-th argument. The quantities x_k and p_k denote the amount of good k and its price, respectively. Condition (1.1) shows that the gradient of the utility function u in correspondence of the optimal choice x^* must be equal to the price vector p multiplied by the Lagrange multiplier λ^*.

Associated with the utility maximization problem (MP0), we define the *indirect utility function* $v : \mathbb{R}_+ \times \mathbb{R}^L_{++} \to \mathbb{R}$ by $v(w, p) := u(x^*)$, where $x^* \in \mathbb{R}^L_{++}$ is the solution to problem (MP0) for wealth w and market prices p. As can be checked from (1.1), the Lagrange multiplier λ^* represents the derivative of the indirect utility function with respect to wealth.

A strictly positive optimal solution $x^* \in \mathbb{R}^L_{++}$ is obtained if the marginal utility for every good goes to $+\infty$ when the amount of the good tends to zero and if it goes to zero when the amount of the good increases to $+\infty$ (*Inada conditions*), i.e.,

$$\lim_{x \to +\infty} u'(x) = 0 \quad \text{and} \quad \lim_{x \searrow 0} u'(x) = +\infty.$$

Moreover, when the utility function is strictly concave, the optimal solution x^* is unique. Due to condition (1.1), the optimality conditions for a vector $x^* \in \mathbb{R}^L_{++}$ can be synthesized in the following classical requirement (provided that $p \in \mathbb{R}^L_{++}$):

$$\frac{u_{x_k}(x^*)}{u_{x_\ell}(x^*)} = \frac{p_k}{p_\ell}, \quad \text{for all } k, \ell = 1, \ldots, L. \tag{1.2}$$

We call *marginal rate of substitution* between good k and good ℓ the ratio of the partial derivatives of the utility function u with respect to x_k and x_ℓ, respectively. Condition (1.2) means that, in correspondence of the optimal solution x^*, marginal rates of substitution are equal to price ratios between different goods. The optimal consumption vector x^* associated with the price vector p will be denoted by the function $x(p)$. The demand of agent i, endowed with the vector of goods $e^i \in \mathbb{R}^L_+$ (which corresponds to the wealth endowment $w = p^\top e^i$), is given by the function $z^i : \mathbb{R}^L_+ \to \mathbb{R}^L$ defined as $z^i(p) := x^i(p) - e^i$. The inequality $z^i_k(p) > 0$ means that agent i has a positive demand of good k, while $z^i_k(p) < 0$ means that agent i has a positive supply (negative demand) of good k, in correspondence of the vector of prices p.

---

[3]Given a function $f : \mathbb{R}^N \to \mathbb{R}$ of N real variables, the partial derivative with respect to the variable x_i will be denoted by f_{x_i}, while the partial derivative evaluated in $x \in \mathbb{R}^N$ will be denoted by $f_{x_i}(x)$.

The maximization of the utility function under the budget constraint in a perfectly competitive market represents one of the pillars of economic analysis. In what follows, we will refer to it as the *internal consistency requirement*.

In the present context, we can also analyse saving and optimal consumption decisions in an intertemporal setting under certainty. For simplicity, let us consider a one good economy ($L = 1$), with the representative good being generically interpreted as wealth, and two distinct time periods $t = 0$ (today) and $t = 1$ (tomorrow). Let e_0 denote the wealth at time $t = 0$ and e_1 the wealth (income) at time $t = 1$. We suppose that at $t = 0$ the agent can choose to save an amount s of his initial wealth e_0 and invest that amount in a risk free asset (i.e., bank account) yielding $r_f > 0$ at date $t = 1$ per unit of wealth invested at date $t = 0$. If the agent decides to save the amount s, then the consumption c_0 and c_1 at $t = 0$ and $t = 1$, respectively, will be given by

$$c_0 = e_0 - s \quad \text{and} \quad c_1 = e_1 + s\, r_f,$$

thus leading to the budget constraint

$$c_0 + \frac{c_1}{r_f} = e_0 + \frac{e_1}{r_f}. \tag{1.3}$$

Observe that r_f can be interpreted as the ratio between the price of one unit of wealth obtained tomorrow (at $t = 1$) and the price of one unit of wealth today (at $t = 0$). We assume that the agent's preferences can be represented by a utility function additive across time, i.e., there exists a utility function $u : \mathbb{R}_+ \to \mathbb{R}$ such that the agent's preferences can be represented by

$$u(c_0) + \delta\, u(c_1), \tag{1.4}$$

where $\delta \in (0, 1)$ represents a (subjective) discount factor, accounting for the natural fact that wealth obtained in the future is less attractive than wealth obtained today. Assuming that u is strictly increasing, the first order condition for the maximization of the utility function (1.4) with respect to c_0 and c_1 subject to the budget constraint (1.3) yields

$$\frac{u'(c_0^*)}{u'(c_1^*)} = r_f\, \delta. \tag{1.5}$$

In particular, assuming that the function u is strictly concave, this optimality condition implies that

$$c_0^* < c_1^* \iff r_f\, \delta > 1,$$

meaning that the agent consumes more in $t = 1$ than in $t = 0$ if and only if the return r_f on the risk free asset is greater than the reciprocal of the discount factor

δ. Furthermore, it can be easily shown from (1.5) that the optimal saving rate s^* is decreasing with respect to the income at $t = 1$, while it is increasing with respect to the wealth at $t = 0$. As far as the risk free rate is concerned, the sensitivity analysis depends on the substitution and income effect (compare with the analysis in Sect. 3.4 under risk). Note that if $r_f \delta = 1$, the optimality condition yields $c_0^* = c_1^*$, showing a preference for *consumption smoothing* over time.

1.2 General Equilibrium Theory

In the previous section we have studied the decision making problem of a single agent. Let us now consider an economy with $I > 1$ agents and $L > 1$ goods. The economy is fully described by I couples $(\mathscr{R}^i, e^i)$, for $i = 1, \ldots, I$. The vector $e^i \in \mathbb{R}_+^L$ represents the endowment of agent i in terms of the L goods and $\mathscr{R}^i$ denotes his preference relation.

We assume that every agent i has solved his optimal consumption problem (MP0) in correspondence of the wealth $w^i = p^\top e^i$, so that his choices are represented by the demand function $z^i(p)$. By the function $z : \mathbb{R}_+^L \rightarrow \mathbb{R}^L$ defined as $z(p) := \sum_{i=1}^{I} z^i(p)$ we denote the aggregate excess demand in the economy.

In a perfectly competitive economy, the market represents an anonymous fictitious place of centralized trades where agents communicate exclusively through prices. Economic theory has addressed market interaction favoring the equilibrium analysis, i.e., the analysis of the economy in correspondence of a price vector p^* such that the aggregate demand is equal to the aggregate supply:

$$z(p^*) = 0. \tag{1.6}$$

Equation (1.6) actually consists of L equations in L unknowns (prices), one for each good. A price vector p^* solving system (1.6) is called an *equilibrium price vector*. A couple $(p^*, x^*) \in \mathbb{R}_+^L \times \mathbb{R}_+^{L \times I}$, where $x^* = (x^1(p^*), \ldots, x^I(p^*))$ is the optimal demand of the I agents in the economy, is called *competitive equilibrium* and the matrix $x^* \in \mathbb{R}_+^{LI}$ is called *equilibrium allocation*. We will refer to (1.6) as the *external consistency requirement*. A competitive equilibrium is thus represented by a couple of price vector and allocation of goods satisfying the *internal* as well as the *external* consistency requirements. In other words, in a competitive market equilibrium, every agent chooses an optimal consumption bundle, coherently with his budget constraint (internal consistency), and all goods are fully allocated among the agents (external consistency). Condition (1.6) is required with strictly increasing preferences: under weaker assumptions (local non-satiation), it can be replaced by $z(p^*) \leq 0$.

The question of the existence of a vector of prices p^* solving system (1.6) is addressed in the following result (see Mas-Colell et al. [1310, Propositions 17.B.2 and 17.C.1]).

Theorem 1.2 *Let the preference relation $\mathscr{R}^i$ be rational, continuous, strictly monotone and strictly convex, for all $i = 1, \ldots, I$, and suppose that $\sum_{i=1}^{I} e^i \in \mathbb{R}_{++}^L$. Then there exists an equilibrium price vector $p^* \in \mathbb{R}_{++}^L$ solving (1.6).*

This result does not represent the end of the story. Indeed, it might well happen that the equilibrium is not unique, in which case coordination problems arise: on which equilibrium will agents coordinate? There are coordination failure and selection of equilibria problems. Provided that agents have the capacity to carry out the above steps, which equilibrium price will they choose among those compatible with the economy? We just mention that, in order to establish the uniqueness of the equilibrium, additional properties on the excess demand function are required (see also the notes at the end of this chapter).

In correspondence of a competitive equilibrium $(p^*, x^*) \in \mathbb{R}_{++}^L \times \mathbb{R}_{++}^{LI}$ with a strictly positive allocation of goods, it holds that

$$\frac{u_{x_k}^i(x^{*i})}{u_{x_\ell}^i(x^{*i})} = \frac{p_k^*}{p_\ell^*}, \qquad \text{for all } k, \ell = 1, \ldots, L \text{ and } i = 1, \ldots, I, \tag{1.7}$$

where x^{*i} denotes the equilibrium consumption bundle of agent i, for $i = 1, \ldots, I$. The above condition means that the marginal rate of substitution of every agent i for any couple of goods (k, ℓ) is equal to the equilibrium price ratio p_k^*/p_ℓ^*. As a consequence, marginal rates of substitution are equal among all agents. Condition (1.7) can also be expressed in the following equivalent way: given an equilibrium $(p^*, x^*) \in \mathbb{R}_{++}^L \times \mathbb{R}_{++}^{L \times I}$, there exists a vector of positive real numbers $(\lambda^1, \ldots, \lambda^I)$ such that $\nabla u^i(x^{*i}) = \lambda^i p^*$, for all $i = 1, \ldots, I$. The term λ^i is the Lagrange multiplier of the optimal consumption problem (MP0) of agent i in correspondence of the equilibrium prices p^*. Therefore, in correspondence of the equilibrium prices p^*, the gradients of the utility functions of all the agents are proportional to the equilibrium price vector.

1.3   Pareto Optimality and Aggregation

Equilibrium theory provides us with a tool for the analysis of the market. As economics is a social science, we need tools for evaluating the outcome of the market's coordination. The privileged tool is represented by the *Pareto optimality criterion*.

Given an economy described by I couples $(\mathscr{R}^i, e^i)$, with $e^i \in \mathbb{R}_+^L$, for $i = 1, \ldots, I$, we denote an *allocation* of goods by the tuple $x = (x^1, \ldots, x^I)$, with $x^i \in \mathbb{R}_+^L$ for all $i = 1, \ldots, I$. We say that the allocation x is *feasible* if $\sum_{i=1}^{I} x^i \leq e$, where $e := \sum_{i=1}^{I} e^i$ is the aggregate endowment. We can now introduce the following Pareto optimality criterion to evaluate the efficiency of an allocation: given two feasible allocations x, y, we say that y *Pareto dominates* x if $y^i \mathscr{R}^i x^i$, for all

$i = 1, \ldots, I$, and $y^j \mathscr{P}^j x^j$ for at least one agent $j \in \{1, \ldots, I\}$. We can then give the following definition.

Definition 1.3 A feasible allocation x is an *efficient allocation* (or *Pareto optimal allocation*) if there exists no other feasible allocation y that Pareto dominates x.

Using the representation of the preference relations $(\mathscr{R}^1, \ldots, \mathscr{R}^I)$ via utility functions $(u^1, \ldots, u^I)$ (see Theorem 1.1), an allocation $x^* \in \mathbb{R}_+^{L \times I}$ is Pareto optimal if and only if there does not exist an allocation $x \in \mathbb{R}_+^{L \times I}$ such that

$$\sum_{i=1}^{I} x_k^i \leq \sum_{i=1}^{I} e_k^i, \qquad \text{for all } k = 1, \ldots, L,$$

and

$$u^i(x^i) \geq u^i(x^{*i}), \qquad \text{for all } i = 1, \ldots, I,$$

with strict inequality holding for some $j \in \{1, \ldots, I\}$.

The set of Pareto optimal allocations constitutes the *contract curve* which can be described by means of the *Edgeworth box* in a simple economy with two agents and two goods (see Sect. 4.1).

The following two *Welfare Theorems* establish a connection between equilibrium and Pareto optimality (see Mas-Colell et al. [1310, Chapter 16] for more details).

Theorem 1.4 (First Welfare Theorem) *If (p^*, x^*) is a competitive equilibrium and the preference relations $\mathscr{R}^i$ are strictly monotone, for all $i = 1, \ldots, I$, then the equilibrium allocation x^* is a Pareto optimal allocation.*

Theorem 1.5 (Second Welfare Theorem) *Let x^* be a Pareto optimal allocation, with $x^{*i} \in \mathbb{R}_{++}^L$ for all $i = 1, \ldots, I$. If $\mathscr{R}^i$ is continuous, strictly monotone and convex, for all $i = 1, \ldots, I$, then x^* is a competitive equilibrium allocation once the initial endowment $e^i = x^{*i}$, $i = 1, \ldots, I$, is allocated to the agents.*

The first theorem provides a formal proof of the presence of an "invisible hand" in the market: the agents, maximizing their utility in a perfectly competitive market, reach an allocation which is socially optimal. The two theorems can be read in opposite perspectives. The first Welfare Theorem is the crowning of the dreams of the supporters of the free market: a perfectly competitive market with agents maximizing their utilities leads to an outcome which is optimal for the society (according to the Pareto optimality criterion). On the other side, there are many optimal allocations and the Pareto criterion does not induce a complete order among feasible allocations. Therefore, there is room for wealth redistribution, i.e., for choosing among Pareto optimal allocations. The second Welfare Theorem establishes that we can reach in equilibrium any Pareto optimal allocation if the resources are redistributed ex-ante in an appropriate way (lump sum transfers). In particular, if the initial allocation is already a Pareto optimal allocation then it also represents an equilibrium allocation and the agents have no incentive to trade.

The identification of Pareto optimal allocations as equilibrium allocations provides an interesting tool for the analysis of the economy at the aggregate level and, in particular, of the equilibrium price vector p^*. Indeed, under some conditions, the equilibrium price vector of the economy can be determined by the choices of a single *representative agent* or *consumer*.

Let us consider an economy described by $\{(u^i, e^i); i = 1, \ldots, I\}$ and a corresponding competitive equilibrium (p^*, x^*). We want to characterize the representative agent/consumer through a utility function such that the optimal choice for that utility function in correspondence of p^* will be the aggregate endowment $e = \sum_{i=1}^{I} e^i$. This means that, if the representative agent is endowed with the aggregate endowment e, then his excess demand will be equal to zero in correspondence of the equilibrium prices p^* of the economy with I agents. Note that, in an economy with a single agent, the external consistency requirement amounts to a no-trade condition and, therefore, the couple (p^*, e) is an equilibrium for the economy with one representative agent (*no-trade equilibrium*).

We endow our representative agent with a utility function $\mathbf{u} : \mathbb{R}_+^L \to \mathbb{R}$ defined on the set $\mathbb{R}_+^L$ of aggregate consumption vectors x. For any $x \in \mathbb{R}_+^L$, the function $\mathbf{u}(x)$ is obtained by maximizing a weighted sum of the utilities u^i of the I agents among all allocations $(x^1, \ldots, x^I)$ which are feasible with respect to a given aggregate endowment x. More precisely, the function $\mathbf{u}(x)$ is called *social welfare function* and is defined as follows, for any $x \in \mathbb{R}_+^L$:

$$\mathbf{u}(x) := \max_{x^i \in \mathbb{R}_+^L, \, i=1,\ldots,I} \sum_{i=1}^{I} a^i u^i(x^i), \tag{1.8}$$

where the maximization is subject to the constraint

$$\sum_{i=1}^{I} x_k^i \leq x_k, \quad \text{for all } k = 1, \ldots, L. \tag{1.9}$$

The coefficients $a^i \geq 0$, for $i = 1, \ldots, I$, represent the weights assigned to the utility functions of the individual agents. For a fixed $x \in \mathbb{R}_+^L$, an allocation $(x^{*1}, \ldots, x^{*I}) \in \mathbb{R}_+^{L \times I}$ satisfying the feasibility constraint (1.9) is said to *define* the function $\mathbf{u}(x)$ if it holds that $\mathbf{u}(x) = \sum_{i=1}^{I} a^i u^i(x^{*i})$. The following proposition describes the relation between a Pareto optimal allocation and the social welfare function $\mathbf{u}$ introduced in (1.8)–(1.9) (see Varian [1610] and Mas-Colell et al. [1310, Proposition 16.E.2] for a proof).

Proposition 1.6 *Consider an economy with aggregate endowment $e \in \mathbb{R}_{++}^L$ and continuous, strictly increasing and concave utility functions u^i, for all $i = 1, \ldots, I$. Let $\mathbf{u}(e)$ be defined as in (1.8)–(1.9), with $x = e$. Then the following hold:*

  (i) *the allocation $(x^{*1}, \ldots, x^{*I}) \in \mathbb{R}_+^{L \times I}$ which defines $\mathbf{u}(e)$ with $a^i > 0$, for all $i = 1, \ldots, I$, is Pareto optimal;*

(ii) let $(x^{*1}, \ldots, x^{*I})$ be a Pareto optimal allocation such that $x^{*i} \in \mathbb{R}^L_{++}$ for all $i = 1, \ldots, I$. Then there exists a set of weights $\{a^{*i} \geq 0; i = 1, \ldots, I\}$ such that $(x^{*1}, \ldots, x^{*I})$ defines $\mathbf{u}(e)$ and not all weights are null;

(iii) let $\{a^{*i} > 0; i = 1, \ldots, I\}$ be the set of weights associated to a Pareto optimal allocation $(x^{*1}, \ldots, x^{*I}) \in \mathbb{R}^{L \times I}_{++}$. Then, for every $i = 1, \ldots, I$, it holds that $a^{*i} = 1/\lambda^{*i}$, i.e., the weight a^{i*} is equal to the reciprocal of the Lagrange multiplier λ^{*i} of the optimum problem (MP0) of agent i in correspondence of the vector of equilibrium prices p^* and the allocation $(x^{*1}, \ldots, x^{*I})$.

Thanks to the above proposition, one can choose a set of strictly positive weights $a^i, i = 1, \ldots, I$. Then, with such a choice of weights, the problem

$$\max_{x^i \in \mathbb{R}^L_+, i=1,\ldots,I} \sum_{i=1}^{I} a^i u^i(x^i),$$

under the constraint $\sum_{i=1}^{I} x^i_k \leq e_k$, for all $k = 1, \ldots, L$, yields as solution an allocation $(x^{*1}, \ldots, x^{*I})$ which is Pareto optimal and, therefore, of equilibrium for an economy with I agents. Furthermore, the Lagrange multiplier λ^{*i} of the optimal consumption problem of agent i in correspondence of the equilibrium price vector p^* associated to the allocation $(x^{*1}, \ldots, x^{*I})$ is given by $\lambda^{*i} = 1/a^i$.

The above result allows us to establish that the representative agent endowed with the aggregate resources $e \in \mathbb{R}^L_+$ of the economy and with a utility function given by the social welfare function will not trade in correspondence of the equilibrium prices of the original economy with I agents. Let p^* be an equilibrium price vector supported by the equilibrium allocation x^* in the economy with I agents. By Theorem 1.4, the allocation x^* is Pareto optimal and, in view of Proposition 1.6, it defines the social welfare function $\mathbf{u}$ with respect to the aggregate endowment e, see (1.8)–(1.9). Consider then the problem of maximizing the function $\mathbf{u}$:

$$\max_{x \in \mathbb{R}^L_+} \mathbf{u}(x), \tag{1.10}$$

under the constraint $(p^*)^\top (x - e) \leq 0$. Note that, for all $k = 1, \ldots, L$,

$$\mathbf{u}_{x_k}(e) = \sum_{i=1}^{I} a^i \, \nabla u^i(x^{*i})^\top \frac{dx^{*i}}{dx_k}(e) = \sum_{i=1}^{I} \frac{1}{\lambda^{*i}} \lambda^{*i} p^{*\top} \frac{dx^{*i}}{dx_k}(e) = p^*_k, \tag{1.11}$$

where $\nabla u^i(x^{*i})$ denotes the gradient of the utility function u^i in correspondence of the optimal consumption vector x^{*i} of agent i. In (1.11), the second equality follows from the fact that x^{*i} represents the optimal consumption vector for agent i, for all $i = 1, \ldots, I$, in correspondence of the price vector p^*, together with the first order condition (1.1) and part *(iii)* of Proposition 1.6. The last equality in (1.11) is due to the fact that the allocation $(x^{*1}, \ldots, x^{*I})$ is feasible, together with the strict monotonicity of the utility functions u^i, for $i = 1, \ldots, I$. Condition (1.11) implies that the representative agent will not trade in correspondence of the prices p^* (i.e., $x^* = e$), thereby defining a competitive equilibrium (p^*, e) for an economy with a unique agent (*no-trade equilibrium*). Hence, the equilibrium prices of the economy with I agents are obtained as equilibrium prices of an economy composed by a single agent with a specific utility function (social welfare function) and endowed with the aggregate resources of the entire economy.

The function $\mathbf{u}(x)$ defined in (1.8)–(1.9) depends on the weights a^i, $i = 1, \ldots, I$, and, therefore, on the Pareto optimal allocation under consideration. Indeed, the utility function of the representative agent associates a Pareto optimal allocation $(x^{*1}, \ldots, x^{*I})$ with a vector of weights $(a^1, \ldots, a^I)$ and, through the no-trade equilibrium condition, with the corresponding vector of equilibrium prices p^*. The vector p^* coincides with the equilibrium prices of the economy with I agents where each agent i is endowed with x^{*i}, for $i = 1, \ldots, I$. As the weights change, the Pareto optimal allocation and the social welfare function $\mathbf{u}$ change: what happens to the equilibrium prices?

In general, equilibrium prices also change as the weights change and, therefore, the prices of the economy with I agents depend on the wealth distribution. We will say that an economy enjoys the *aggregation property* (or that there exists a *representative agent in a strong sense*) if equilibrium prices only depend on the resources of the whole economy and not on the weights or on the initial allocation. A sufficient condition for an economy to satisfy this property is that the preferences of all the agents are *quasi-homothetic*, i.e., characterized by indirect utility functions of the Gorman form $v_i(w_i, p) = \alpha_i(p) + \beta(p) w_i$, for all $i = 1, \ldots, I$, with the coefficient $\beta(p)$ being the same for all the agents (see e.g. Mas-Colell et al. [1310, Chapters 4 and 16]). In that case, the aggregate demand is independent of the initial allocation and the equilibrium prices p^* defined through the no-trade equilibrium condition depend neither on the weights $(a^i, \ldots, a^I)$ nor on the allocation of the goods, but only on the resources of the economy as a whole. Three classes of utility functions satisfy these requirements: logarithmic utility, power utility and exponential utility. If the aggregation property holds, then the equilibrium prices of the economy with I agents can be identified through the optimum problem of the representative agent endowed with the resources of the economy and the no-trade condition, regardless of the wealth distribution. In this sense, the economy with I agents and the economy with the representative agent are observationally equivalent.

1.4 Notes and Further Readings

In economic theory, the focus on equilibrium analysis is due to several reasons. In what follows we only propose some remarks, referring to Arrow [77], Hahn [874], Ingrao & Israel [1001] for a more comprehensive analysis. In the first place, equilibrium analysis has a methodological-historical motivation: in comparison with other sciences, economic theory is a young science. As a consequence, economics has borrowed methodologies and instruments from more mature sciences (e.g., physics) where equilibrium analysis represents a classical tool.

The definition of an equilibrium state implies the identification of dynamics in the phenomenon under consideration, dynamics which seem to be completely outside the picture described so far. Actually, we have not proposed a dynamic system, but we can identify a dynamic process in our context by referring to the internal and to the external consistency requirements. If $p \neq p^*$, then one of the two consistency requirements is violated and one of the following situations is verified: (a) given the prices, not all agents maximize their utility; (b) there are no goods to satisfy the demand of all agents or some goods do not have a buyer. In both cases, it seems plausible that some forces inside the market will move the prices. The main problem is how to describe these forces. On the other hand, if the economy stands exactly in p^*, then there are no endogenous forces moving prices: if agents trade in a perfectly competitive market and communicate exclusively through prices, then only p^* can represent a stationary state of the economy.

This interpretation appears rather weak, since nothing is said about the ability of the market to converge to p^* starting from a different price vector $p \neq p^*$. Stability analysis of the equilibrium requires introducing the time dimension in the above picture and to formalize the dynamics outside the equilibrium. Pursuing this goal, general equilibrium theory shows all its limits: it is difficult in fact to represent internal and external consistency requirements through a dynamical system. The two consistency requirements, which represent a strong point to prove the existence of the equilibrium and to characterize it analytically, represent a serious obstacle to introducing dynamic elements in the context of general equilibrium theory. It is difficult to model disequilibrium dynamics with consumption-exchange of goods. These problems have been avoided in the literature by extending the time for trade and admitting consumption only at the final time when an equilibrium is reached. In this way, the internal consistency requirement is maintained, every agent during the bargaining declares his theoretical demand, while at that time there is no trade and no consumption. Trade and consumption only occur in equilibrium. Prices change over time when demand and supply differ. The dynamics is modelled by assuming a *tâtonnement process*: the price dynamics is described by a differential-difference equation defined through a function which preserves the sign of the excess demand and assumes value zero if the excess demand is equal to zero. If the excess demand function satisfies some additional hypothesis (typically, goods are gross substitutes: two goods i and j are said to be gross substitutes if for any price vector p it holds that $\partial z^i / \partial p_j > 0$ with $i \neq j$), then the competitive equilibrium will be stable with respect to the tâtonnement process, see Mas-Colell et al. [1310, Proposition 17.H.9]. The dynamic process described above formally represents the so-called *demand and supply law*.

There is another reason for using equilibrium analysis. The existence of a vector of equilibrium prices means that the decisions of the agents pursuing their self-interest are compatible in a perfectly competitive market. This motivation goes back to the axiomatic approach of general equilibrium theory developed in the '30s as an answer to the following question: under what conditions are agents' decisions compatible? The question circulated for a long time in the environment of social scientists (since Adam Smith, if not before) and it was motivated by the fact that the market seemed able to coordinate the demand of all agents. It is obvious that in this perspective the accent is on equilibrium existence rather than on its dynamics. To this end, consistency requirements simplify the analysis in a substantial way and allow to bring the problem of the compatibility of agents' decisions back to the solution of a (non-linear) system of L equations in L variables. In this perspective, a positive answer to the problem of the existence of a vector of equilibrium prices appears a complete success for general equilibrium theory.

In this perspective, the existence of a general economic equilibrium leads to interesting philosophical-moral-theoretical implications, but little implications on the real world. On this side, an implication is the following: if agents have a thorough knowledge of the economy and know all the couples $(\mathscr{R}^i, e^i)$, for $i = 1, \ldots, I$, then they can cover the road described in this chapter to identify the vector of equilibrium prices p^* and determine their demand consistently. This interpretation of general equilibrium theory requires very strong hypotheses: the two consistency conditions are satisfied, agents pursue their self-interest, they know the economic model and they are endowed with a remarkable computational capacity.

Concerning the uniqueness of the equilibrium vector $p^* \in \mathbb{R}^L_{++}$ in Theorem 1.2, the fact that all goods are gross substitutes ensures that the equilibrium price vector $p^* \in \mathbb{R}^L_{++}$ is unique, see Mas-Colell et al. [1310, Proposition 17.F.3]. However, this assumption implies strong hypotheses on the initial allocation of goods as well as on the preference relations of the agents.

Chapter 2
Choices Under Risk

> *Behaviour is substantively rational when it is appropriate to the achievement of given goals within the limits imposed by given conditions and constraints. [...] behaviour is procedurally rational when it is the outcome of appropriate deliberation. is the outcome of appropriate deliberation.*
>
> Simon (1976)

The optimal consumption problem introduced in the previous chapter is formulated in a very simple context: agents act in a single time period and choices concern bundles of goods. The characteristics of the goods are perfectly known ex-ante and, hence, the optimal consumption problem is a purely deterministic optimization problem. The introduction of risk requires at least two time periods $t \in \{0, 1\}$. This is due to the necessity of describing the agents' imperfect knowledge: at the initial date $t = 0$ agents do not have a full knowledge of the world, they only have some beliefs about it, while at $t = 1$ the uncertainty is fully resolved and revealed. The agents' ignorance at the initial date $t = 0$ with respect to the state of the world at $t = 1$ may concern many different features (economic conditions, meteorological conditions, quality of the goods, dividends, losses, etc.) and several examples will be discussed in the following.

If an agent only makes his choices at $t = 1$, once the state of the world has been revealed, then we relapse into the analysis developed in the previous chapter. In this chapter, we aim at analysing the choices of an agent at the initial date $t = 0$, when he does not have a complete knowledge of the state of the world and, therefore, of the future consequences of the choices made at $t = 0$. We assume that the true state of the world can be only observed at $t = 1$. In this setting, we first have to describe the status of ignorance of the agent and then to analyse his choices through something similar to the couples $(X, \mathscr{R})$ or, equivalently, (X, u) used in the previous chapter under certainty. To this end, we introduce two crucial elements: a *probability space* and a *utility function u*.

The beliefs of the agents at $t = 0$ about the state of the world at the future date $t = 1$ are described by means of a *probability space* $(\Omega, \Theta, \mathbb{P})$. The set Ω is the set of all possible states of the world at time $t = 1$: a *state of the world* (or *state of nature*) is represented by an element $\omega \in \Omega$ and provides a complete and exhaustive description of the world at $t = 1$ (an element $\omega \in \Omega$ is also called

© Springer-Verlag London Ltd. 2017
E. Barucci, C. Fontana, *Financial Markets Theory*,
Springer Finance, DOI 10.1007/978-1-4471-7322-9_2

elementary event). The states of the world are mutually exclusive, i.e., only one elementary event is realized at $t = 1$. The collection Θ is a σ-algebra on Ω.[1] If Ω is a finite set, then a natural choice for Θ is given by the set of all the subsets of Ω. An *event* is an element $E \in \Theta$ and, therefore, a subset of Ω. The application $\mathbb{P} : \Theta \to [0, 1]$ is a *probability measure*, with $\mathbb{P}(E)$ denoting the probability of the event $E \in \Theta$. Intuitively, $\mathbb{P}(E)$ represents the confidence an agent places on the realization of the event E. The probability of an event can have an objective or a subjective interpretation. The distinction is however subtle and it also depends on the situation faced by the agent (see Chap. 9). In what follows, we will typically assume that the probability has an objective interpretation.[2]

At the initial date $t = 0$, the agent is assumed to know the probability space and, therefore, the set Ω of all possible elementary events as well as the probability measure $\mathbb{P}$. At $t = 1$, the state of the world is fully revealed and the agent discovers the realized elementary event $\omega \in \Omega$. When an agent cannot observe at $t = 0$ the state of the world but has a complete knowledge of the probability space, we say that he faces a *risky* situation. This situation differs from the case of an agent who does not perfectly know the probability of the events: in this case, we will say that the agent faces an *uncertain* situation. In our analysis, the word *risk* refers to a

---

[1]Θ is a collection of subsets of Ω that includes the empty set and satisfies the following conditions:

– if $E \in \Theta$, then its complement E^c also belongs to Θ;
– if $\{E_n\}_{n \in \mathbb{N}} \subseteq \Theta$ then $\bigcup_{n=1}^{\infty} E_n \in \Theta$.

[2]In the case where Ω is a finite set, the probability measure $\mathbb{P} : \Theta \to [0, 1]$ is assumed to satisfy the following axioms:

1. for every $E_1, E_2 \in \Theta$ such that $E_1 \cap E_2 = \emptyset$ we have $\mathbb{P}(E_1 \cup E_2) = \mathbb{P}(E_1) + \mathbb{P}(E_2)$;
2. $\mathbb{P}(\Omega) = 1$.

These axioms lead to some fundamental properties of the probability measure $\mathbb{P}$:

– for every $E \in \Theta$, it holds that $\mathbb{P}(E) \leq 1$ and $\mathbb{P}(E) + \mathbb{P}(E^c) = 1$;
– for every $E_1, E_2 \in \Theta$ with $E_1 \subset E_2$, it holds that $\mathbb{P}(E_1) \leq \mathbb{P}(E_2)$ and $\mathbb{P}(E_2 \setminus E_1) = \mathbb{P}(E_2) - \mathbb{P}(E_1)$;
– for every $E_1, E_2, \ldots, E_n \in \Theta$, $n \in \mathbb{N}$, it holds that

$$\mathbb{P}(E_1 \cup E_2 \cup \ldots \cup E_n) \leq \mathbb{P}(E_1) + \mathbb{P}(E_2) + \ldots + \mathbb{P}(E_n),$$

with equality holding for disjoint events (*finite additivity*).

In the more general case where Ω is not a finite set, then finite additivity has to be replaced by σ-*additivity*: for any countable collection of disjoint events $\{E_k\}_{k \in \mathbb{N}} \subseteq \Theta$ it holds that

$$\mathbb{P}\left(\bigcup_{k=1}^{\infty} E_k \right) = \sum_{k=1}^{\infty} \mathbb{P}(E_k).$$

We refer to Chung [448] for a classical account of probability theory.

setting with *known* probabilities, while the word *uncertainty* refers to a setting with *unknown* probabilities. We will return to this distinction in Sect. 9.1.

The probability space $(\Omega, \Theta, \mathbb{P})$ describes an agent's knowledge/beliefs about the world: it remains to describe his choices in a risky environment. Following Savage [1499], we identify choices as real-valued *random variables* (or *acts*). An act is an application $\tilde{x} : \Omega \to \mathbb{R}$, where $\mathbb{R}$ (the set of real numbers) represents the *space of consequences*. From the point of view of an agent, this application describes the "consequence" $\tilde{x}(\omega)$ associated by the act $\tilde{x}$ to the state of the world $\omega \in \Omega$. In the case of financial choices we shall generally think of the space of consequences $\mathbb{R}$ (or $\mathbb{R}_+$) as representing wealth (or returns).

We have to understand how an act introduces risk in the space of consequences of an agent. Let $\mathscr{B}(\mathbb{R})$ denote the Borel σ-algebra of $\mathbb{R}$ (i.e., the σ-algebra generated by all the intervals in $\mathbb{R}$). Given that an agent has chosen the act $\tilde{x}$, what is the probability of its consequences belonging to a Borel set $B \in \mathscr{B}(\mathbb{R})$? Intuitively, this corresponds to asking what is the probability of a set B of consequences associated to a given act $\tilde{x}$ of the agent. The natural answer is to evaluate the probability of B as the probability of the event in Θ representing the pre-image $\tilde{x}^{-1}(B)$ of B according to the act $\tilde{x}$. As a consequence, the probability measure induced by $\tilde{x}$ on $(\mathbb{R}, \mathscr{B}(\mathbb{R}))$ is defined as $\pi(B) := \mathbb{P}(\tilde{x}^{-1}(B))$. Note that this construction is possible only if for every $B \in \mathscr{B}(\mathbb{R})$ it holds that $\tilde{x}^{-1}(B) \in \Theta$, i.e., only if the application $\tilde{x} : \Omega \to \mathbb{R}$ is $(\Theta, \mathscr{B}(\mathbb{R}))$-measurable. In that case, $\tilde{x}$ is said to be a *random variable*. Hence, given the probability space $(\Omega, \Theta, \mathbb{P})$, a random variable $\tilde{x}$ induces a probability measure π on $(\mathbb{R}, \mathscr{B}(\mathbb{R}))$. Each act-random variable induces a specific probability measure. Given a probability space, in our analysis we will directly refer to acts as random variables (or *gambles*) and to their probability measures on $(\mathbb{R}, \mathscr{B}(\mathbb{R}))$.

Representing risk in this way is only apparently innocuous. Indeed, an agent does not possess a complete knowledge of the future state of the world, but he is assumed to be able to describe in an exhaustive way his ignorance status through a probability space (and, therefore, he is assumed to know all possible future states of the world). Furthermore, he is able to quantify his ignorance through a probability measure. This means that the agent is able to quantify his confidence on the realization of any event, where that quantity can be the outcome of objective and/or subjective elements. According to our representation of choices in a risky environment, the probability assumes an objective interpretation, in contrast with the subjectivist approach proposed by De Finetti and Savage amongst others (see Chap. 9 for more details).

In this chapter, we investigate decision making problems in a risky setting. More specifically, we shall be concerned with the optimal choice of an agent among several possible acts (or gambles-random variables). The most natural criterion to evaluate a gamble-random variable $\tilde{x}$ seems to be provided by its expected value $\mathbb{E}[\tilde{x}]$. However, this evaluation rule poses some problems, as illustrated by the classical *Saint Petersburg paradox* proposed by Daniel Bernoulli in 1738. Let us consider a game consisting of repeated tosses of a coin: if at the first toss a head appears then the player receives the payoff $1 = 2^0$, otherwise the game continues with another toss. If at that second toss a head appears then the player receives the payoff $2 = 2^1$, otherwise the game continues with another toss, at which the payoff in the case of a head appearing will be 2^2, and so on. At the k-th toss, for any $k \in \mathbb{N}$, if no head appeared on the first $k-1$ tosses, the payoff in the case of a head appearing will be 2^{k-1}. We suppose that the coin is fair, so that the probability of a head appearing at any toss is equal to $1/2$. Since the probability of no head appearing in the k first tosses is $1/2^k$, for any $k \in \mathbb{N}$, the expected value of this game is given by

$$\mathbb{E}[\tilde{x}] = \sum_{k=1}^{\infty} \left(\frac{1}{2} \right)^k 2^{k-1} = \frac{1}{2} \sum_{k=1}^{\infty} 1 = +\infty.$$

Hence, if an agent evaluates this gamble through its expectation, then the gamble would appear infinitely attractive. In reality, when confronted with such a game, no one would be prepared to pay an infinite amount of money in order to take part to the gamble, the reason being that significant payoffs are only obtained with very small probabilities. In order to resolve this apparent paradox, Daniel Bernoulli proposed to evaluate gambles via the expected value of their logarithm, i.e., by computing $\mathbb{E}[\log(\tilde{x})]$. In that case:

$$\mathbb{E}[\log(\tilde{x})] = \sum_{k=1}^{\infty} \left(\frac{1}{2} \right)^k \log(2^{k-1}) = \frac{\log 2}{2} \sum_{k=1}^{\infty} \frac{k}{2^k} = \log 2 < +\infty.$$

Therefore, if an agent evaluates a random variable by taking the expectation of a logarithmic transformation, then the possibility of playing the game $\tilde{x}$ and the possibility of receiving the sure payoff 2 should be equally attractive. Equivalently, to participate to the game an agent would be willing to pay up to 2 and not an arbitrarily large amount of wealth as suggested by the expected value criterion.

In synthesis, Bernoulli proposed that an agent should evaluate a random variable through the expected value of an increasing non-linear transformation of the random variable. This hypothesis did not find a theoretical justification for a long time until the expected utility theory was proposed by Von Neumann and Morgenstern in 1947, as we are now going to explain.

This chapter is structured as follows. In Sect. 2.1, we prove that a preference relation can be represented in terms of expected utility if it satisfies a certain set of axioms. In Sect. 2.2, we introduce the notions of risk aversion and certainty equivalent, while in Sect. 2.3 we study stochastic dominance criteria. Section 2.4 deals with mean-variance preferences and introduces the diversification and insurance principles. At the end of the chapter, we provide a guide to further readings as well as a series of exercises.

2.1 Expected Utility Theory

For simplicity, we consider the case of random variables with a finite number $S \in \mathbb{N}$ of values. Specifying a priori a finite set of values $\{x_1, \ldots, x_S\}$, with $x_i \in \mathbb{R}$ for all $i = 1, \ldots, S$ and $x_i < x_{i+1}$ for all $i = 1, \ldots, S - 1$, we have a set $\mathcal{M}$ of random variables (or gambles) $\tilde{x}$ valued on the finite support $\{x_1, \ldots, x_S\}$. A random variable $\tilde{x} \in \mathcal{M}$ will be identified by a probability distribution $\{\pi_1, \ldots, \pi_S\}$ associated with the S possible realizations:

$$\tilde{x} \in \mathcal{M} \iff \tilde{x} \equiv \{x_1, \ldots, x_S; \pi_1, \ldots, \pi_S\} \text{ with } \pi_s \geq 0, \forall s = 1, \ldots, S, \text{ and } \sum_{s=1}^{S} \pi_s = 1.$$
(2.1)

In the risky environment under consideration, the set $\mathcal{M}$ becomes the new set of choices instead of the set X considered in the previous chapter under certainty. Therefore, from now on an agent will be identified by a pair $(\mathcal{M}, \mathcal{R})$, where the preference relation $\mathcal{R}$ is defined over the gambles in $\mathcal{M}$.

Introducing a probability space and a set of random variables allows to mathematically represent the decision making problem of an agent in a risky environment. In this setting, we would like to parallel the theory established under certainty, i.e., to represent the preference relation $\mathcal{R}$ through a function defined on the space of random variables $\mathcal{M}$. For clarity of presentation, we limit our analysis to a single good which we generically interpret as wealth (the analysis can be easily extended to multiple goods).

Due to the presence of risk, the optimal choice problem is more complex than the one considered in the previous chapter. The approach we present below refers to the axiomatic *Von Neumann-Morgenstern expected utility theory*, see Von Neumann & Morgenstern [1632]. According to this approach, one first identifies a set of hypotheses on the preference relation $\mathcal{R}$ which, if satisfied, allows to represent the relation $\mathcal{R}$ through a function defined over $\mathcal{M}$. Such a function will be a linear combination of the utilities associated with wealth/consumption in each elementary state of the world, where the coefficients of the linear combination are given by the probabilities of realization of the different states.

In order to represent $\mathcal{R}$ by means of an expected utility function, we require $\mathcal{R}$ to satisfy the following axioms (compare with Assumption 1.1).

Assumption 2.1 (Rationality) The preference relation $\mathcal{R}$ is *rational* if it satisfies the following properties:

- *(Reflexivity)*: for every $\tilde{x} \in \mathcal{M}$, it holds that $\tilde{x} \mathcal{R} \tilde{x}$;
- *(Completeness)*: for every $\tilde{x}_1, \tilde{x}_2 \in \mathcal{M}$, it holds that $\tilde{x}_1 \mathcal{R} \tilde{x}_2$ or $\tilde{x}_2 \mathcal{R} \tilde{x}_1$;
- *(Transitivity)*: for every $\tilde{x}_1, \tilde{x}_2, \tilde{x}_3 \in \mathcal{M}$ such that $\tilde{x}_1 \mathcal{R} \tilde{x}_2$ and $\tilde{x}_2 \mathcal{R} \tilde{x}_3$, it holds that $\tilde{x}_1 \mathcal{R} \tilde{x}_3$.

The rationality hypothesis alone is not enough for our purposes. Two further hypotheses are needed, one analogous to the continuity Assumption 1.2 and a crucial one called the *independence axiom*.

Assumption 2.2 (Continuity) The preference relation $\mathcal{R}$ is *continuous* if, for every $\tilde{x}_1, \tilde{x}_2, \tilde{x}_3 \in \mathcal{M}$, such that $\tilde{x}_1 \, \mathcal{R} \, \tilde{x}_2$ and $\tilde{x}_2 \, \mathcal{R} \, \tilde{x}_3$, there exists a scalar $\alpha \in [0, 1]$ such that $\alpha \, \tilde{x}_1 + (1 - \alpha)\tilde{x}_3 \, \mathcal{I} \, \tilde{x}_2$.

In Assumption 2.2, the random variable $\alpha \, \tilde{x}_1 + (1 - \alpha)\tilde{x}_3$ represents the gamble $\{x_1, \ldots, x_S; \alpha \, \pi_1^1 + (1 - \alpha)\pi_1^3, \ldots, \alpha \, \pi_S^1 + (1 - \alpha)\pi_S^3\}$, i.e., the gamble obtained as a convex linear combination with weights α and $1 - \alpha$ of the probabilities associated by the two gambles $\tilde{x}_1$ and $\tilde{x}_2$ to the values $\{x_1, \ldots, x_S\}$. This hypothesis allows us to represent the order induced by $\mathcal{R}$ over $\mathcal{M}$ through a function from $\mathcal{M}$ to $\mathbb{R}$. The function will be of the expected utility form if the preference relation $\mathcal{R}$ satisfies the following additional hypothesis.

Assumption 2.3 (Independence) The preference relation $\mathcal{R}$ satisfies the *independence* assumption if, for all $\tilde{x}_1, \tilde{x}_2, \tilde{x}_3 \in \mathcal{M}$ and $\alpha \in [0, 1]$, we have that $\tilde{x}_1 \, \mathcal{R} \, \tilde{x}_2$ if and only if $\alpha \, \tilde{x}_1 + (1 - \alpha)\tilde{x}_3 \, \mathcal{R} \, \alpha \, \tilde{x}_2 + (1 - \alpha)\tilde{x}_3$.

As can be readily checked, if the preference relation $\mathcal{R}$ satisfies Assumption 2.3, then also $\mathcal{I}$ and $\mathcal{P}$ do (this fact will be used in the proof of the next theorem). Expected utility theory states that the preference relation $\mathcal{R}$ over the set of gambles $\mathcal{M}$ can be represented through a function $U : \mathcal{M} \to \mathbb{R}$ that assigns to a gamble $\tilde{x} = \{x_1, \ldots, x_S; \pi_1, \ldots, \pi_S\} \in \mathcal{M}$ the value

$$U(\tilde{x}) = \sum_{s=1}^{S} \pi_s \, u(x_s), \tag{2.2}$$

where u is the utility function that represents $\mathcal{R}$ for wealth obtained with certainty. In other words, if $\mathcal{R}$ satisfies Assumptions 2.1–2.3, then there exists a utility function u such that the preference relation $\mathcal{R}$ can be represented as in (2.2). Note that the function u does not depend on the state $s = 1, \ldots, S$ (*state independent utility*). The function U is a linear combination of the utilities $u(x_s)$ that the agent obtains with certainty in correspondence of the events $s = 1, \ldots, S$ and the weights of the linear combination are given by the probabilities π_s. The function U is called *expected utility function*.

Theorem 2.1 *Given the pair $(\mathcal{M}, \mathcal{R})$, if the preference relation $\mathcal{R}$ satisfies Assumptions 2.1–2.3, then there exist S scalars $u(x_s) \in \mathbb{R}$, $s = 1, \ldots, S$, such that, for every $\tilde{x}_1, \tilde{x}_2 \in \mathcal{M}$,*

$$\tilde{x}_1 \, \mathcal{R} \, \tilde{x}_2 \iff U(\tilde{x}_1) \geq U(\tilde{x}_2), \tag{2.3}$$

where $U(\cdot)$ is defined as in (2.2).

Proof Following Mas-Colell et al. [1310, Proposition 6.B.3], we divide the proof into seven steps.

Step 1. If $\tilde{x} \mathcal{R} \tilde{x}'$ and $\alpha \in (0, 1)$, then $\tilde{x} \mathcal{R} \alpha \tilde{x} + (1-\alpha)\tilde{x}'$ and $\alpha \tilde{x} + (1-\alpha)\tilde{x}' \mathcal{R} \tilde{x}'$. In view of Assumption 2.3, this claim can be readily verified. Indeed:

$$\tilde{x} = \alpha \tilde{x} + (1-\alpha)\tilde{x} \; \mathcal{R} \; \alpha \tilde{x} + (1-\alpha)\tilde{x}' \; \mathcal{R} \; \alpha \tilde{x}' + (1-\alpha)\tilde{x}' = \tilde{x}'.$$

Note that, as can be easily checked, the same property also holds for $\mathcal{P}$.

Step 2. There exist two gambles $\tilde{x}^*, \tilde{x}_* \in \mathcal{M}$ which represent respectively the best and the worst gamble in $\mathcal{M}$ with respect to $\mathcal{R}$, in the sense that $\tilde{x}^* \mathcal{R} \tilde{x}$ and $\tilde{x} \mathcal{R} \tilde{x}_*$ for every $\tilde{x} \in \mathcal{M}$ (see also Mas-Colell et al. [1310, Exercise 6.B.3]). Indeed, without loss of generality, suppose that the preference relation $\mathcal{R}$ for wealth obtained with certainty is monotone, meaning that $\tilde{x}_{x_S} \mathcal{R} \tilde{x}_{x_{S-1}} \mathcal{R} \ldots \mathcal{R} \tilde{x}_{x_1}$, where $\tilde{x}_{x_s}$ denotes the gamble that takes the value x_s with probability one, for $s = 1, \ldots, S$. Due to Assumptions 2.1–2.3 and applying repeatedly the result established in the first step, one can show that $\tilde{x}^* = \tilde{x}_{x_S}$ and $\tilde{x}_* = \tilde{x}_{x_1}$. When $\mathcal{R}$ is not monotone, $\tilde{x}_*$ and $\tilde{x}^*$ will be two gambles which assign an amount of money different from x_1 and x_S with probability one.

Now, if $\tilde{x}^* \mathcal{I} \tilde{x}_*$, then all gambles in $\mathcal{M}$ are indifferent for the agent and, therefore, the claim of the theorem follows in a trivial way (it is enough to choose a constant utility function). So, let us assume that $\tilde{x}^* \mathcal{P} \tilde{x}_*$. We now have to show that there exists a function $F : \mathcal{M} \to \mathbb{R}$ which represents $\mathcal{R}$ and is a linear function with weights defined by the probabilities π_s of the S events, thus yielding the expected utility function U.

Step 3. For every $\alpha, \beta \in [0, 1]$ the following holds:

$$\beta \tilde{x}^* + (1-\beta)\tilde{x}_* \; \mathcal{P} \; \alpha \tilde{x}^* + (1-\alpha)\tilde{x}_* \iff \beta > \alpha.$$

Indeed, by step 1 (for the strict preference relation $\mathcal{P}$) we know that

$$\tilde{x}^* \; \mathcal{P} \; \alpha \tilde{x}^* + (1-\alpha)\tilde{x}_*$$

and that, for any $\phi \in (0, 1)$,

$$\phi \tilde{x}^* + (1-\phi)(\alpha \tilde{x}^* + (1-\alpha)\tilde{x}_*) \; \mathcal{P} \; \alpha \tilde{x}^* + (1-\alpha)\tilde{x}_*.$$

To prove the $\Leftarrow$ implication, it is enough to choose $\phi = (\beta - \alpha)/(1-\alpha)$, with $\beta > \alpha$. To show the converse implication, suppose that $\beta \leq \alpha$. In the case $\beta = \alpha$ we have that $\beta \tilde{x}^* + (1-\beta)\tilde{x}_* \; \mathcal{I} \; \alpha \tilde{x}^* + (1-\alpha)\tilde{x}_*$ and, therefore, a contradiction is obtained. If $\beta < \alpha$, then a contradiction with the hypothesis is easily reached by relying on the same arguments used above.

Step 4. For every $\tilde{x} \in \mathcal{M}$, there exists a unique $\alpha_{\tilde{x}} \in [0, 1]$ such that

$$\alpha_{\tilde{x}} \tilde{x}^* + (1-\alpha_{\tilde{x}})\tilde{x}_* \; \mathcal{I} \; \tilde{x}.$$

Indeed, the existence comes from Assumption 2.2, while the uniqueness can be easily deduced from Step 3.

Step 5. The function $F : \mathcal{M} \to \mathbb{R}$ defined as $F(\tilde{x}) = \alpha_{\tilde{x}}$ represents the relation $\mathcal{R}$ in the sense of (2.3), where $\alpha_{\tilde{x}} \in [0, 1]$ is as in Step 4.

Indeed, given two gambles $\tilde{x}, \tilde{x}' \in \mathcal{M}$, Step 4 implies that:

$$\tilde{x} \mathcal{R} \tilde{x}' \iff \alpha_{\tilde{x}} \tilde{x}^* + (1 - \alpha_{\tilde{x}}) \tilde{x}_* \ \mathcal{R} \ \alpha_{\tilde{x}'} \tilde{x}^* + (1 - \alpha_{\tilde{x}'}) \tilde{x}_*.$$

The claim then follows from the double implication established in Step 3.

Step 6. The function F is linear: for every $\tilde{x}, \tilde{x}' \in \mathcal{M}$ and $\beta \in [0, 1]$, it holds that

$$F\big(\beta \tilde{x} + (1 - \beta) \tilde{x}'\big) = \beta F(\tilde{x}) + (1 - \beta) F(\tilde{x}').$$

Indeed, by the definition of F, we have

$$\tilde{x} \ \mathcal{I} \ F(\tilde{x}) \tilde{x}^* + \big(1 - F(\tilde{x})\big) \tilde{x}_* \quad \text{and} \quad \tilde{x}' \ \mathcal{I} \ F(\tilde{x}') \tilde{x}^* + \big(1 - F(\tilde{x}')\big) \tilde{x}_*.$$

By Assumption 2.3 (applied to the indifference relation $\mathcal{I}$), we obtain

$$\beta \tilde{x} + (1 - \beta) \tilde{x}' \ \mathcal{I} \ \beta \big(F(\tilde{x}) \tilde{x}^* + \big(1 - F(\tilde{x})\big) \tilde{x}_*\big) + (1 - \beta) \tilde{x}'$$
$$\mathcal{I} \ \beta \big(F(\tilde{x}) \tilde{x}^* + \big(1 - F(\tilde{x})\big) \tilde{x}_*\big) + (1 - \beta) \big(F(\tilde{x}') \tilde{x}^* + \big(1 - F(\tilde{x}')\big) \tilde{x}_*\big).$$

Rearranging the terms we obtain

$$\beta \tilde{x} + (1 - \beta) \tilde{x}' \ \mathcal{I} \ \big(\beta F(\tilde{x}) + (1 - \beta) F(\tilde{x}')\big) \tilde{x}^* + \big(1 - \beta F(\tilde{x}) - (1 - \beta) F(\tilde{x}')\big) \tilde{x}_*.$$

By the definition of the function F in Step 5, we thus get

$$F\big(\beta \tilde{x} + (1 - \beta) \tilde{x}'\big) = \beta F(\tilde{x}) + (1 - \beta) F(\tilde{x}').$$

Step 7. The function F coincides with the function U defined in (2.2).

Indeed, let us define $u(x_s) := F(\tilde{x}_{x_s})$, for all $s = 1, \ldots, S$. Every gamble $\tilde{x} \in \mathcal{M}$, identified through the vector of probabilities $\{\pi_1, \ldots, \pi_S\}$ associated with the realizations $\{x_1, \ldots, x_S\}$, can be rewritten as $\tilde{x} = \sum_{s=1}^{S} \pi_s \tilde{x}_{x_s}$. Using the linearity of the function F established in Step 6, this allows us to write

$$F(\tilde{x}) = F\left(\sum_{s=1}^{S} \pi_s \tilde{x}_{x_s}\right) = \sum_{s=1}^{S} \pi_s u(x_s).$$

We have thus shown that the function F coincides with the expected utility function U introduced in (2.2), thus proving the theorem. $\square$

Note that, given a preference relation $\mathscr{R}$ satisfying the assumptions introduced in the present section, the expected utility function U representing $\mathscr{R}$ is unique up to strictly increasing linear transformations (see Exercise 2.1). If a gamble-random variable $\tilde{x}$ takes an infinite number of values in $\mathbb{R}$, letting $f_{\tilde{x}}$ be its probability density function, then the expected utility function can be written as follows, extending representation (2.2),

$$U(\tilde{x}) = \int_{\mathbb{R}} u(z) f_{\tilde{x}}(z) \mathrm{d}z = \mathbb{E}\left[u(\tilde{x})\right],$$

provided that the integral exists. In the following, $\tilde{x}$ will denote a gamble-random variable and $U(\tilde{x})$ and $\mathbb{E}[u(\tilde{x})]$ will equivalently denote the expected utility function. In the case of general probability spaces, the preference relation $\mathscr{R}$ admits a representation through an expected utility function if Assumptions 2.1–2.3 hold together with an additional assumption named the *sure thing principle* (if a gamble is concentrated on a subset A of $\mathbb{R}$ and each amount of money in A is preferred to another gamble, then the first gamble is preferred to the second) as well as other hypotheses of technical nature (see Fishburn [706], Kreps [1136]).

The representation of the preference relation $\mathscr{R}$ in a risky environment via an expected utility function can also be established in the case where the agent is allowed to consume in correspondence of two dates $t \in \{0, 1\}$. As explained at the beginning of this chapter, we assume that the state of the world is fully revealed only at $t = 1$, thus introducing a risk for an agent who has to take decisions at $t = 0$. In the case of a finite probability space, as in (2.1), the expected utility function $U : \mathbb{R} \times \mathscr{M} \to \mathbb{R}$ takes the form

$$U(x_0, \tilde{x}) = \sum_{s=1}^{S} \pi_s \, u(x_0, x_s), \qquad \text{for all } x_0 \in \mathbb{R} \text{ and } \tilde{x} \in \mathscr{M},$$

where the first argument x_0 denotes consumption at the initial date $t = 0$ and $\tilde{x} \in \mathscr{M}$ is the gamble which will assume one of the realizations $\{x_1, \ldots, x_S\}$ at the future date $t = 1$, representing consumption at $t = 1$. The expected utility is said to be *additively separable over time* if, for each $t \in \{0, 1\}$, the utility at date t depends only on consumption/wealth at t, i.e.,

$$U(x_0, \tilde{x}) = u_0(x_0) + \sum_{s=1}^{S} \pi_s \, u_1(x_s), \qquad \text{for all } x_0 \in \mathbb{R} \text{ and } \tilde{x} \in \mathscr{M}.$$

A particular case is the *time invariant* utility, where $u_1(\cdot) = \delta u_0(\cdot)$:

$$U(x_0, \tilde{x}) = u(x_0) + \sum_{s=1}^{S} \delta \, \pi_s \, u(x_s), \qquad \text{for all } x_0 \in \mathbb{R} \text{ and } \tilde{x} \in \mathscr{M},$$

for some utility function $u : \mathbb{R} \to \mathbb{R}$ and $0 < \delta \leq 1$. The term δ is a discount factor representing the natural fact that an agent prefers to consume today (at $t = 0$) rather than in the future (at $t = 1$).

The expected utility framework and the probability space represent two fundamental ingredients of our analysis. A discussion of their role is postponed to Chap. 9, after having acquired a complete picture of the results obtained assuming that the agents' preferences can be represented by means of expected utility functions.

2.2 Risk Aversion

In the setting considered in this chapter, an agent faces an optimal choice problem in a risky environment: loosely speaking, this means that an agent will not only consider the utility associated to given amounts of wealth, but he will also take into account the probabilities of receiving those amounts of wealth.

An agent's attitude towards risk is described by his preference relation $\mathscr{R}$ on the space of random variables $\mathscr{M}$. If $\mathscr{R}$ can be represented through an expected utility function U, then the attitude towards risk can be analysed through the properties of the function u, where u is the utility function representing the preference relation $\mathscr{R}$ with respect to wealth obtained with certainty, as in Theorem 2.1. In the following, we shall always assume that the preference relation $\mathscr{R}$ satisfies Assumptions 2.1–2.3.

The propensity-aversion of an agent towards risk can be evaluated by considering *actuarially fair* gambles-random variables. We say that a gamble-random variable $\tilde{x}$ is actuarially fair if $\mathbb{E}[\tilde{x}] = 0$. In the case of a finite probability space, as in (2.1), a gamble $\tilde{x} = \{x_1, \ldots, x_S; \pi_1, \ldots, \pi_S\}$ is actuarially fair if $\sum_{s=1}^{S} x_s \pi_s = 0$.

We say that an agent is *risk averse* if he does not accept (or is at most indifferent) to any actuarially fair gamble at all wealth levels, while an agent is *risk neutral* if he is indifferent to any actuarially fair gamble and is *risk lover* if he accepts any actuarially fair gamble, for every level of wealth. We shall also say that an agent is *strictly risk averse* if he rejects any actuarially fair gamble for every level of wealth. Note that these definitions are "global" in the sense that they refer to an indeterminate level of wealth and to any actuarially fair gamble. In our analysis, we shall mostly restrict our attention to risk neutral and risk averse agents, due to the fact that risk loving agents are considered to be a negligible minority whose behavior does not seem to significantly affect financial markets.

In order to illustrate the definition of risk aversion, suppose that the wealth of an agent is represented by a random variable $\tilde{x}$ with expected value $\mathbb{E}[\tilde{x}]$. If the agent is risk averse, then he will prefer to obtain the sure amount $\mathbb{E}[\tilde{x}]$ with certainty rather than being exposed to the random amount of wealth $\tilde{x}$. Equivalently, a risk averse agent will be willing to pay a positive price to move from a random variable $\tilde{x}$ describing his wealth to the expected value $\mathbb{E}[\tilde{x}]$ of that random variable obtained with certainty. Such a price is called the *risk premium*.

Definition 2.2 The *risk premium* $\rho_u(\tilde{x})$ of a gamble $\tilde{x}$ for an agent characterized by a utility function u is the maximum amount that the agent is willing to pay in order to receive, instead of the random variable $\tilde{x}$, its expected value $\mathbb{E}[\tilde{x}]$ with certainty, i.e., $\rho_u(\tilde{x})$ is such that

$$u\big(\mathbb{E}[\tilde{x}] - \rho_u(\tilde{x})\big) = \mathbb{E}\left[u(\tilde{x})\right]. \tag{2.4}$$

According to Definition 2.2, for a risk averse agent characterized by the utility function u and exposed to the gamble $\tilde{x}$, paying the price $\rho_u(\tilde{x})$ and receiving the sure amount $\mathbb{E}[\tilde{x}]$ is equally attractive to receiving the random realization $\tilde{x}$ of the gamble. Note that, in this representation, the payoff of the gamble $\tilde{x}$ includes the initial wealth of the agent. The definition of risk premium directly leads to the concept of *certainty equivalent* $CE_u(\tilde{x})$ of the gamble $\tilde{x}$ for an agent with utility function u.

Definition 2.3 The *certainty equivalent* of the gamble $\tilde{x}$ for an agent characterized by the utility function u is the amount of wealth $CE_u(\tilde{x})$ which makes the agent indifferent to the gamble, i.e., $CE_u(\tilde{x})$ is such that

$$u\big(CE_u(\tilde{x})\big) = \mathbb{E}\left[u(\tilde{x})\right].$$

Note that the certainty equivalent and the risk premium of a gamble are unique if the function u is strictly increasing. In this case, exploiting the two above definitions we get the equality

$$CE_u(\tilde{x}) = \mathbb{E}[\tilde{x}] - \rho_u(\tilde{x}).$$

If the risk premium $\rho_u(\tilde{x})$ can be interpreted as the maximum price that an agent is willing to pay in order to receive the sure amount $\mathbb{E}[\tilde{x}]$ instead of the gamble $\tilde{x}$, then the certainty equivalent $CE_u(\tilde{x})$ admits the interpretation of the smallest amount of wealth required by the agent to give up the gamble $\tilde{x}$ and accept a sure amount of wealth.

Exploiting concavity and Jensen's inequality,[3] we can establish that an agent is (strictly) risk averse if and only if u is (strictly) concave. More precisely, we have the following proposition (see Fig. 2.1 for a graphical illustration).

Proposition 2.4 *Given an agent characterized by an increasing utility function u, the following conditions are equivalent:*

(i) the agent is risk averse;
(ii) the utility function u is concave;
(iii) $CE_u(\tilde{x}) \leq \mathbb{E}[\tilde{x}]$, for every random variable $\tilde{x}$ with $\mathbb{E}[|\tilde{x}|] < \infty$;
(iv) $\rho_u(\tilde{x}) \geq 0$, for every random variable $\tilde{x}$ with $\mathbb{E}[|\tilde{x}|] < \infty$.

---

[3]Let $\tilde{x}$ be a random variable with finite expected value $\mathbb{E}[\tilde{x}]$. Jensen's inequality establishes that, for any concave function $g : \mathbb{R} \to \mathbb{R}$, it holds that $\mathbb{E}\left[g(\tilde{x})\right] \leq g(\mathbb{E}[\tilde{x}])$.

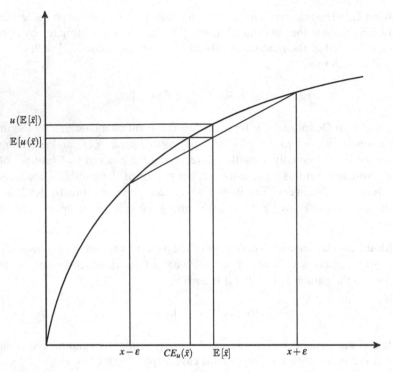

Fig. 2.1 Expected utility and certainty equivalent, $\{x - \epsilon, x + \epsilon; 1/2, 1/2\}$

Proof $(i) \Rightarrow (ii)$: Let us consider a risk averse agent characterized by the utility function u. For any $w, w' \in \mathbb{R}$ and $\lambda \in (0, 1)$, define $\bar{w} := \lambda w + (1 - \lambda)w'$ and consider the gamble $\tilde{\epsilon}$ defined as follows:

$$\tilde{\epsilon} = \begin{cases} (1 - \lambda)(w - w') & \text{with probability } \lambda; \\ \lambda(w' - w) & \text{with probability } 1 - \lambda. \end{cases}$$

Since $\mathbb{E}[\tilde{\epsilon}] = \lambda(1 - \lambda)(w - w') + (1 - \lambda)\lambda(w' - w) = 0$, the gamble $\tilde{\epsilon}$ is actuarially fair. Hence, due to the definition of risk aversion,

$$u\big(\lambda w + (1 - \lambda)w'\big) = u(\bar{w}) \geq \mathbb{E}[u(\bar{w} + \tilde{\epsilon})] = \lambda u(w) + (1 - \lambda)u(w').$$

Since this holds for every $w, w' \in \mathbb{R}$ and $\lambda \in (0, 1)$, this implies that the utility function $u : \mathbb{R} \to \mathbb{R}$ is concave.

$(ii) \Rightarrow (i)$: Suppose that u is concave. Then, for any $w \in \mathbb{R}$ and for any gamble $\tilde{\epsilon}$ such that $\mathbb{E}[\tilde{\epsilon}] = 0$, Jensen's inequality implies that

$$\mathbb{E}[u(w + \tilde{\epsilon})] \leq u\big(w + \mathbb{E}[\tilde{\epsilon}]\big) = u(w),$$

thus showing that the agent is risk averse, since he does not accept (or is at most indifferent to) any actuarially fair gamble.

(ii) ⇔ (iii) ⇔ (iv): Since the function u is assumed to be increasing, these equivalences follow directly from Definitions 2.2–2.3.   □

Note that the equivalence between the first two claims of Proposition 2.4 also holds for a non-increasing utility function u.

As can be seen from Definitions 2.2–2.3, the risk premium and the certainty equivalent of a gamble $\tilde{x}$ are defined with respect to a specific utility function u. We now intend to put the risk premium $\rho_u(\tilde{x})$ in relation with the random variable $\tilde{x}$ and the utility function u. To this end, we consider two particular cases: a gamble described by a random variable with additive noise and a gamble described by a random variable with multiplicative noise. This will lead to the introduction of the coefficients of *absolute* and *relative risk aversion*.

Let $\tilde{x} = x + \tilde{\epsilon}$, where $\tilde{\epsilon}$ is a random variable with zero mean and variance σ^2, so that $\mathbb{E}[\tilde{x}] = x$. We assume that the function u is twice differentiable and that the noise component $\tilde{\epsilon}$ is "small" with respect to x. By Definition 2.2, we have that

$$u(x - \rho_u(\tilde{x})) = U(\tilde{x}). \tag{2.5}$$

For every realization ϵ of $\tilde{\epsilon}$, we can approximate $u(x + \epsilon)$ by a second order Taylor expansion of the function u centered in x, thus obtaining

$$u(x + \epsilon) \approx u(x) + \epsilon\, u'(x) + \frac{\epsilon^2}{2} u''(x).$$

Using this expression to evaluate the expected utility, we obtain the approximation

$$U(\tilde{x}) = \mathbb{E}[u(x + \tilde{\epsilon})] \approx u(x) + \frac{\sigma^2}{2} u''(x). \tag{2.6}$$

If the noise component $\tilde{\epsilon}$ is "small" with respect to x, then also the risk premium $\rho_u(\tilde{x})$ will be "small" with respect to x and, therefore, we can use a Taylor expansion (centered in x and up to the first order) of the utility function u evaluated at the certainty equivalent $CE_u(\tilde{x}) = x - \rho_u(\tilde{x})$, thus obtaining

$$u(x - \rho_u(\tilde{x})) \approx u(x) - \rho_u(\tilde{x})u'(x). \tag{2.7}$$

Using the two approximations (2.6)–(2.7) and assuming that the function u is strictly increasing, we obtain the following estimate for the risk premium $\rho_u(\tilde{x})$:

$$\rho_u(\tilde{x}) \approx -\frac{1}{2} \frac{u''(x)}{u'(x)} \sigma^2, \tag{2.8}$$

which can be rewritten as $\rho_u(\tilde{x}) \approx \frac{1}{2} r_u^a(x) \sigma^2$, where we define

$$r_u^a(x) := -\frac{u''(x)}{u'(x)}.$$

The quantity $r_u^a(x)$ is called the *coefficient of absolute risk aversion* of the utility function u in correspondence of wealth x. In view of the approximation (2.8), the risk premium is decomposed into two factors: the variance of the gamble and the absolute risk aversion coefficient. Note that the risk premium is increasing in both factors. The reciprocal of the absolute risk aversion coefficient is called the *absolute risk tolerance* and is defined as (provided that $u'' > 0$)

$$t_u(x) := \frac{1}{r_u^a(x)} = -\frac{u'(x)}{u''(x)}.$$

Let us now consider the case where $\tilde{x} = x(1 + \tilde{\epsilon})$, where $\tilde{\epsilon}$ is a random variable with zero mean and variance σ^2. We assume that u is twice differentiable and strictly increasing. By relying on a reasoning similar to the one employed above, we can provide the following estimate of the relative risk premium $\rho_u^r(\tilde{x})$ defined by the relation $x\rho_u^r(\tilde{x}) := \rho_u(\tilde{x})$:

$$\rho_u^r(\tilde{x}) \approx -\frac{1}{2}\frac{u''(x)}{u'(x)}x\sigma^2,$$

which can be rewritten as $\rho_u^r(\tilde{x}) \approx \frac{1}{2} r_u^r(x)\sigma^2$, where

$$r_u^r(x) := -x\frac{u''(x)}{u'(x)}$$

is the *coefficient of relative risk aversion* of the utility function u in correspondence of the wealth x. By definition, it holds that $r_u^r(x) = r_u^a(x)x$, for every $x \in \mathbb{R}$.

The following classification of the utility function u can be given according to the behavior of the coefficient of absolute risk aversion with respect to wealth changes:

- u is characterized by decreasing absolute risk aversion (*DARA*) if $x \mapsto r_u^a(x)$ is a decreasing function;
- u is characterized by constant absolute risk aversion (*CARA*) if $x \mapsto r_u^a(x)$ is a constant function;
- u is characterized by increasing absolute risk aversion (*IARA*) if $x \mapsto r_u^a(x)$ is an increasing function.

Of course, a similar classification can be given on the basis of the behavior of the coefficient of relative risk aversion with respect to wealth changes.

Risk Aversion Comparison

Having characterized the risk aversion of an agent in terms of the properties of his utility function u, we now aim at comparing the behavior of different agents with respect to risky choices.

Let us consider two agents a and b characterized by utility functions u^a and u^b, respectively, and the same initial wealth. We say that agent a is *more risk averse* than agent b if agent b always accepts a gamble if agent a does. Equivalently (assuming that the utility functions u^a and u^b are increasing), we can affirm that, for each gamble $\tilde{x}$, agent a is more risk averse than agent b if and only if the risk premium $\rho_{u^a}(\tilde{x})$ of agent a is greater or equal than the risk premium $\rho_{u^b}(\tilde{x})$ of agent b. As far as the certainty equivalent is concerned, the opposite holds, due to the equivalence between claims (*iii*) and (*iv*) in Proposition 2.4. Intuitively, the fact that agent a is more risk averse than agent b amounts to saying that agent a requires a higher compensation than agent b in order to accept the riskiness of a gamble $\tilde{x}$.

The following result, first established in De Finetti [537], Pratt [1432], Arrow [78], gives three equivalent conditions for agent a to be more risk averse than agent b (for simplicity, we limit our attention to twice differentiable utility functions).

Proposition 2.5 *Given two strictly increasing and strictly concave utility functions u^a and u^b, the following conditions are equivalent:*

(*i*) $r_{u^a}^a(x) \geq r_{u^b}^a(x)$, *for every $x \in \mathbb{R}$;*
(*ii*) *there exists an increasing and concave function $g : \mathbb{R} \to \mathbb{R}$ such that, for every $x \in \mathbb{R}$, it holds that $u^a(x) = g(u^b(x))$;*
(*iii*) *u^a is more risk averse than u^b, i.e., $\rho_{u^a}(x + \tilde{\epsilon}) \geq \rho_{u^b}(x + \tilde{\epsilon})$, for every $x \in \mathbb{R}$ and for every random variable $\tilde{\epsilon}$ such that $\mathbb{E}[\tilde{\epsilon}] = 0$.*

Proof (*i*) $\Leftrightarrow$ (*ii*): Since u^a and u^b are both increasing, concave and twice differentiable, there exists a twice differentiable increasing function $g : \mathbb{R} \to \mathbb{R}$ such that $u^a(x) = g(u^b(x))$, for all $x \in \mathbb{R}$. Indeed, let $g := u^a \circ (u^b)^{-1}$, which can be verified to be increasing and twice differentiable. Differentiating u^a once and twice we obtain

$$u^{a'}(x) = g'\big(u^b(x)\big)u^{b'}(x) \quad \text{and} \quad u^{a''}(x) = g'\big(u^b(x)\big)u^{b''}(x) + g''\big(u^b(x)\big)\big(u^{b'}(x)\big)^2.$$

Dividing the second expression by $u^{a'}(x)$ and using the relation between u^a and u^b, we obtain, for every $x \in \mathbb{R}$,

$$r_{u^a}^a(x) = r_{u^b}^a(x) - \frac{g''\big(u^b(x)\big)}{g'\big(u^b(x)\big)}u^{b'}(x). \tag{2.9}$$

Therefore, $r_{u^a}^a(x) \geq r_{u^b}^a(x) \iff g''(u^b(x)) \leq 0$, for every $x \in \mathbb{R}$.

($ii $) $\Rightarrow$ (iii): Suppose that agents a and b are endowed with the same wealth $x \in \mathbb{R}$. For an arbitrary random variable $\tilde{\epsilon}$ with $\mathbb{E}[\tilde{\epsilon}] = 0$ we have that

$$u^a\big(x - \rho_{u^a}(x + \tilde{\epsilon})\big) = \mathbb{E}[u^a(x + \tilde{\epsilon})] = \mathbb{E}\big[g\big(u^b(x + \tilde{\epsilon})\big)\big] \leq g\big(U^b(x + \tilde{\epsilon})\big)$$
$$= g\big(u^b(x - \rho_{u^b}(x + \tilde{\epsilon}))\big) = u^a\big(x - \rho_{u^b}(x + \tilde{\epsilon})\big),$$
$$(2.10)$$

where the inequality is due to Jensen's inequality, since g is concave. Since u^a is increasing, we then have $\rho_{u^a}(x + \tilde{\epsilon}) \geq \rho_{u^b}(x + \tilde{\epsilon})$.

(iii) $\Rightarrow$ (ii): As in the first part of the proof, since u^a and u^b are strictly increasing, there exists an increasing function $g : \mathbb{R} \to \mathbb{R}$ such that $u^a(x) = g(u^b(x))$. We shall now prove that the function g is concave (in the range of u^b). Similarly as in the proof of Proposition 2.4, let $x, x' \in \mathbb{R}$ and $\lambda \in [0, 1]$, define $\bar{x} := \lambda x + (1 - \lambda)x'$ and consider the gamble $\tilde{\epsilon}$ defined as follows:

$$\tilde{\epsilon} = \begin{cases} (1 - \lambda)(x - x') & \text{with probability } \lambda; \\ \lambda(x' - x) & \text{with probability } 1 - \lambda. \end{cases}$$

Note that $\mathbb{E}[\tilde{\epsilon}] = 0$. Then, by the definition of risk premium,

$$\lambda u^b(x) + (1 - \lambda)u^b(x') = \mathbb{E}[u^b(\bar{x} + \tilde{\epsilon})] = u^b\big(\bar{x} - \rho_{u^b}(\bar{x} + \tilde{\epsilon})\big),$$

so that, since $u^a(x) = g(u^b(x))$,

$$g\big(\lambda u^b(x) + (1 - \lambda)u^b(x')\big) = u^a\big(\bar{x} - \rho_{u^b}(\bar{x} + \tilde{\epsilon})\big). \tag{2.11}$$

On the other hand,

$$\lambda g\big(u^b(x)\big) + (1 - \lambda)g\big(u^b(x')\big) = \lambda u^a(x) + (1 - \lambda)u^a(x') = \mathbb{E}[u^a(\bar{x} + \tilde{\epsilon})]$$
$$= u^a\big(\bar{x} - \rho_{u^a}(\bar{x} + \tilde{\epsilon})\big).$$
$$(2.12)$$

By (2.11)–(2.12), together with the increasingness of u^a and property (iii), we get

$$g\big(\lambda u^b(x) + (1 - \lambda)u^b(x')\big) \geq \lambda g\big(u^b(x)\big) + (1 - \lambda)g\big(u^b(x')\big).$$

Since $x, x' \in \mathbb{R}$ and $\lambda \in [0, 1]$ are arbitrary, this proves the concavity of g. $\qquad\square$

The above proposition bridges the gap between the coefficient of absolute risk aversion, which is a local measure of risk aversion (being computed in relation to a specific level of wealth), and a global characterization of risk aversion (which refers to an indeterminate level of wealth, as explained at the beginning of this section). If the coefficient of absolute risk aversion of an agent is larger than the coefficient of another agent for every level of wealth, then the first agent is more risk averse than the second agent. Note that, on the basis of this result, $u(\cdot)$ and $a + bu(\cdot)$ (with $b > 0$) are equivalent as far as risk aversion is concerned, since the coefficient of absolute risk aversion is invariant under positive linear transformations.

Proposition 2.5 can also be used to analyse the attitude of an agent towards risk with respect to changes in wealth, as shown in the next result.

Proposition 2.6 *If an agent is characterized by an increasing DARA utility function u and $\tilde{\epsilon}$ is a random variable such that $\mathbb{E}[\tilde{\epsilon}] = 0$, then the risk premium $\rho_u(x+\tilde{\epsilon})$ is a decreasing function of x, i.e., the agent becomes less risk averse as wealth increases.*

Proof We need to show that, for $x' \geq x$, we have

$$\rho_u(x + \tilde{\epsilon}) \geq \rho_u(x' + \tilde{\epsilon}).$$

Set $x' = x + k$, with $k := x' - x \geq 0$, and define the utility functions u^a and u^b by $u^a(x) := u(x)$ and $u^b(x) := u(k + x)$, respectively, for $x \in \mathbb{R}$. Since u is DARA, we have that

$$r_{u^a}^a(x) \geq r_{u^b}^a(x).$$

As a consequence, since $x \in \mathbb{R}$ is arbitrary, using the implication $(i) \Rightarrow (iii)$ of Proposition 2.5, it holds that

$$\rho_u(x + \tilde{\epsilon}) = \rho_{u^a}(x + \tilde{\epsilon}) \geq \rho_{u^b}(x + \tilde{\epsilon}) = \rho_u(x' + \tilde{\epsilon}),$$

thus proving the claim.   □

In other words, Proposition 2.6 says that, for a risk averse agent characterized by an increasing DARA utility function, any desirable gamble cannot become undesirable as wealth increases. Equivalently, in view of Proposition 2.4, the certainty equivalent of a gamble is increasing with respect to wealth. In view of this result, a DARA utility function seems a reasonable assumption to represent an agent's behavior with respect to risk.

Classes of Utility Functions

Four classes of utility functions play a particularly important role. Further properties of these utility functions will be derived in the exercises proposed at the end of the chapter.

Quadratic Utility Function

$$u(x) = x - \frac{b}{2}x^2, \qquad \text{with } b > 0.$$

The function $x \mapsto u(x)$ is increasing only in the domain $[0, 1/b]$. For $x \in [0, 1/b)$, the coefficient of absolute risk aversion r_u^a is positive and is given by

$$r_u^a(x) = \frac{b}{1 - bx}.$$

Since $r_u^{a'}(x) = \frac{b^2}{(1-bx)^2} > 0$, for $x \neq 1/b$, the function $x \mapsto r_u^a(x)$ is increasing in wealth, meaning that the quadratic utility function is IARA. The coefficient of relative risk aversion $r_u^r(x) = \frac{bx}{1-bx}$ is increasing, i.e., $r_u^{r'}(x) = \frac{b}{(1-bx)^2} > 0$.

Exponential Utility Function

$$u(x) = -\frac{1}{a} e^{-ax}, \qquad \text{with } a > 0.$$

The coefficient of absolute risk aversion r_u^a is positive and is given by

$$r_u^a(x) = a.$$

Therefore, the exponential utility function is CARA and the coefficient of relative risk aversion $r_u^r(x) = ax$ is increasing.

Power Utility Function

$$u(x) = \frac{b}{b-1} x^{1-\frac{1}{b}}, \qquad \text{with } b > 0.$$

The coefficient of absolute risk aversion r_u^a is positive and is given by

$$r_u^a(x) = \frac{1}{bx}.$$

Since $r_u^{a'}(x) = -\frac{1}{bx^2} < 0$, the map $x \mapsto r_u^a(x)$ is decreasing, meaning that the power utility function is DARA. Moreover, $r_u^r(x) = 1/b$ and, hence, the coefficient of relative risk aversion is constant (CRRA utility function).

Logarithmic Utility Function

$$u(x) = \log(bx), \qquad \text{with } b > 0.$$

The coefficient of absolute risk aversion r_u^a is positive and given by

$$r_u^a(x) = \frac{1}{x}.$$

Clearly, $r_u^a(x)$ is decreasing in wealth, so that the logarithmic utility function is DARA. Moreover, $r_u^r(x) = 1$ and, hence, the coefficient of relative risk aversion is constant (CRRA utility function). Note that, as shown in Exercise 2.2, the logarithmic utility function can be obtained as the limit of the power utility function as $b \to 1$.

In our analysis, we will often consider the large class of utility functions with *hyperbolic absolute risk aversion (HARA)*, i.e., the class of utility functions such that the absolute risk tolerance t_u is linear with respect to wealth:

$$t_u(x) = a + bx, \qquad \text{for some } a, b \in \mathbb{R}.$$

The HARA class embeds generalized versions of the above utility functions. Indeed, the following types of utility functions can be obtained (see Fig. 2.2 and Exercise 2.10):

1. exponential utility function $u(x) = -ae^{-x/a}$, for $a > 0$ and $b = 0$;
2. generalized power utility function $u(x) = \frac{b}{b-1}(a + bx)^{\frac{b-1}{b}}$, for $a \in \mathbb{R}$ and $b > 0$ with $b \neq 1$;
3. generalized logarithmic utility function $u(x) = \log(x+a)$, for $x > -a$ and $b = 1$.

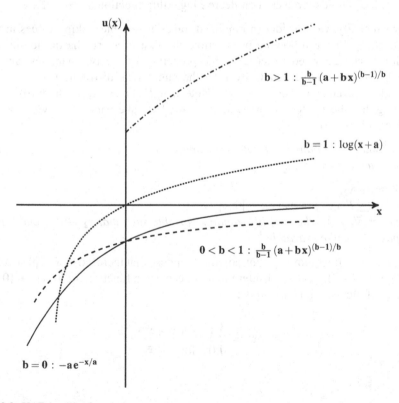

Fig. 2.2 HARA utility functions

2.3 Stochastic Dominance

Stochastic dominance criteria introduce an order among gambles-random variables which is shared by all agents characterized by utility functions satisfying some common properties. The interesting feature of stochastic dominance criteria is that they exclusively refer to some basic features of the probability distribution of the gamble-random variables, without referring to specific preference relations or utility functions. In this section, we present three different stochastic dominance criteria: *first order stochastic dominance*, *second order stochastic dominance* and *second order stochastic monotonic dominance*. The mean-variance criterion, partly related to stochastic dominance, will be discussed in Sect. 2.4.

In a nutshell, we aim at answering the following question: is it possible to rank gambles-random variables through the order induced by expected utility just by assuming that the utility function satisfies some basic properties?

The first criterion that we present is the *first order stochastic dominance*.

Definition 2.7 Let $\tilde{x}_1$ and $\tilde{x}_2$ be two random variables. We say that $\tilde{x}_1$ dominates $\tilde{x}_2$ according to the *first order stochastic dominance* criterion (written as $\tilde{x}_1 \succeq_{\text{FSD}} \tilde{x}_2$) if $U(\tilde{x}_1) \geq U(\tilde{x}_2)$ holds for every non-decreasing utility function $u : \mathbb{R} \to \mathbb{R}$.

For simplicity, we consider (normalized) random variables taking values in the interval $[0, 1]$. In the following proposition, the first order stochastic dominance criterion is characterized in terms of the properties of the two gambles-random variables $\tilde{x}_1$ and $\tilde{x}_2$. We denote by $F_i(\cdot)$ the cumulative distribution function of the random variable $\tilde{x}_i$, for $i = 1, 2$. Note that $F_i(1) = 1$, while $F_i(0) \geq 0$ (meaning that the random variable $\tilde{x}_i$ is allowed to take value zero with strictly positive probability).

Proposition 2.8 *For any two random variables $\tilde{x}_1$ and $\tilde{x}_2$ taking values in $[0, 1]$, the following are equivalent:*

(i) $\tilde{x}_1 \succeq_{\text{FSD}} \tilde{x}_2$;
(ii) $F_1(x) \leq F_2(x)$, *for every* $x \in [0, 1]$;
(iii) $\tilde{x}_1 \equiv^d \tilde{x}_2' + \tilde{\epsilon}$, *where $\tilde{x}_2'$ is a random variable such that $\tilde{x}_2 \equiv^d \tilde{x}_2'$ and $\tilde{\epsilon}$ is a positive random variable.*[4]

Proof (i) $\Rightarrow$ *(ii)*: Arguing by contradiction, suppose that there exists $\bar{x} \in [0, 1]$ such that $F_1(\bar{x}) > F_2(\bar{x})$. Let us consider the non-decreasing function $u : [0, 1] \to \{0, 1\}$ defined as follows, for all $x \in [0, 1]$:

$$u(x) := \begin{cases} 1 & \text{for } x > \bar{x}; \\ 0 & \text{for } x \leq \bar{x}. \end{cases}$$

---

[4]The notation $\equiv^d$ stands for equality in law (or in distribution).

Let $U(\tilde{x}_i) = \mathbb{E}[u(\tilde{x}_i)]$ be the expected utility associated with the random variable $\tilde{x}_i$, for $i = 1, 2$. Evaluating the difference $U(\tilde{x}_1) - U(\tilde{x}_2)$ of the expected utility functions of the two random variables $\tilde{x}_1$ and $\tilde{x}_2$, we get

$$U(\tilde{x}_1) - U(\tilde{x}_2) = \mathbb{E}[u(\tilde{x}_1)] - \mathbb{E}[u(\tilde{x}_2)] = \mathbb{P}(\tilde{x}_1 > \bar{x}) - \mathbb{P}(\tilde{x}_2 > \bar{x}) = F_2(\bar{x}) - F_1(\bar{x}) < 0.$$

This contradicts the assumption that $\tilde{x}_1 \succeq_{FSD} \tilde{x}_2$, thus proving that $(i) \Rightarrow (ii)$.

$(ii) \Rightarrow (iii)$: Let $\tilde{u}$ be a uniformly distributed random variable on $[0, 1]$, independent of $\tilde{x}_1$ and $\tilde{x}_2$ (such a random variable always exists, up to an enlargement of the original probability space), and define the auxiliary random variables $\tilde{x}_1'$ and $\tilde{x}_2'$ as

$$\tilde{x}_1' := \inf\{y \in [0, 1] : F_1(y) \geq \tilde{u}\} \quad \text{and} \quad \tilde{x}_2' := \inf\{y \in [0, 1] : F_2(y) \geq \tilde{u}\}.$$

Since $F_1(x) \leq F_2(x)$ for every $x \in [0, 1]$, it holds that $\mathbb{P}(\tilde{x}_1' \geq \tilde{x}_2') = 1$. Moreover, denoting by F_i' the distribution function of $\tilde{x}_i'$, for $i = 1, 2$, it holds that, for all $x \in [0, 1]$,

$$F_i'(x) = \mathbb{P}(\tilde{x}_i' \leq x) = \mathbb{P}(F_i(x) \geq \tilde{u}) = F_i(x),$$

thus showing that $\tilde{x}_i' \equiv^d \tilde{x}_i$, for $i = 1, 2$. The claim then follows by letting $\tilde{\epsilon} := \tilde{x}_1' - \tilde{x}_2'$, so that $\tilde{x}_1 \equiv^d \tilde{x}_1' = \tilde{x}_2' + \tilde{\epsilon}$ and $\tilde{x}_2' \equiv^d \tilde{x}_2$.

$(iii) \Rightarrow (i)$: If $\tilde{x}_1 \equiv^d \tilde{x}_2' + \tilde{\epsilon}$, where $\tilde{x}_2' \equiv^d \tilde{x}_2$ and $\tilde{\epsilon}$ is a positive random variable, then for every non-decreasing utility function u we have that

$$U(\tilde{x}_1) = \mathbb{E}[u(\tilde{x}_2' + \tilde{\epsilon})] \geq \mathbb{E}[u(\tilde{x}_2')] = U(\tilde{x}_2),$$

thus proving the implication. □

This first order stochastic dominance criterion is rather weak. It only induces a partial order on gambles-random variables. Furthermore, since the class of utility functions considered is very large, the order induced among random variables simply establishes that a gamble-random variable $\tilde{x}_1$ is preferred to a gamble-random variable $\tilde{x}_2$ if and only if the distribution function of the first random variable is always smaller than the one of the second random variable. Note that

$$\tilde{x}_1 \succeq_{FSD} \tilde{x}_2 \quad \Rightarrow \quad \mathbb{E}[\tilde{x}_1] \geq \mathbb{E}[\tilde{x}_2].$$

However, the converse implication does not necessarily hold (a counterexample is provided in Exercise 2.19).

A different order among random variables can be established considering risk averse agents, thus leading to the *second order stochastic dominance* criterion.

Definition 2.9 Let $\tilde{x}_1$ and $\tilde{x}_2$ be two random variables. We say that $\tilde{x}_1$ dominates $\tilde{x}_2$ according to the *second order stochastic dominance* criterion (written as $\tilde{x}_1 \succeq_{SSD} \tilde{x}_2$) if $U(\tilde{x}_1) \geq U(\tilde{x}_2)$ holds for every concave utility function $u : \mathbb{R} \to \mathbb{R}$.

Similarly to Proposition 2.8, the following result characterizes the second order stochastic dominance criterion between two random variables $\tilde{x}_1$ and $\tilde{x}_2$ in terms of the properties of their probability distributions (for simplicity, in the following proposition we restrict our attention to continuously differentiable utility functions).

Proposition 2.10 *For any two random variables $\tilde{x}_1$ and $\tilde{x}_2$ taking values in $[0, 1]$, the following are equivalent:*

(i) $\tilde{x}_1 \succeq_{\text{SSD}} \tilde{x}_2$;
(ii) $\mathbb{E}[\tilde{x}_1] = \mathbb{E}[\tilde{x}_2]$ and $G(y) := \int_0^y (F_1(z) - F_2(z))\mathrm{d}z \leq 0$, for every $y \in [0, 1]$;
(iii) $\tilde{x}_2 \equiv^d \tilde{x}_1 + \tilde{\epsilon}$, where $\tilde{\epsilon}$ is a random variable such that $\mathbb{E}[\tilde{\epsilon}|\tilde{x}_1] = 0$.

Proof $(i) \Rightarrow (ii)$: The function $\bar{u} : \mathbb{R} \to \mathbb{R}$ defined by $\bar{u}(x) := x$, for $x \in [0, 1]$, is of course concave. Hence, the assumption $\tilde{x}_1 \succeq_{\text{SSD}} \tilde{x}_2$ implies that

$$\mathbb{E}[\tilde{x}_1] - \mathbb{E}[\tilde{x}_2] = U(\tilde{x}_1) - U(\tilde{x}_2) \geq 0.$$

Applying the same reasoning to the function $x \mapsto -x$ yields the reverse equality, thus showing that $\mathbb{E}[\tilde{x}_1] = \mathbb{E}[\tilde{x}_2]$. Recall that, for any random variable $\tilde{x}$ with distribution function $F_{\tilde{x}} : [0, 1] \to [0, 1]$, the expected value can be computed as

$$\mathbb{E}[\tilde{x}] = \int_0^1 \left(1 - F_{\tilde{x}}(z)\right)\mathrm{d}z.$$

Together with the equality $\mathbb{E}[\tilde{x}_1] = \mathbb{E}[\tilde{x}_2]$, this implies that $G(0) = G(1) = 0$. Let us consider any twice differentiable concave utility function $u : \mathbb{R} \to \mathbb{R}$. Then, using twice the integration by parts formula and the assumption that $\tilde{x}_1 \succeq_{\text{SSD}} \tilde{x}_2$, we obtain

$$0 \leq U(\tilde{x}_1) - U(\tilde{x}_2) = \int_{[0,1]} u(z)\mathrm{d}\big(F_1(z) - F_2(z)\big)$$

$$= u(1)\big(F_1(1) - F_2(1)\big) - \int_{(0,1]} u'(z)\big(F_1(z) - F_2(z)\big)\mathrm{d}z$$

$$= -u'(1)G(1) + u'(0)G(0) + \int_{(0,1]} G(z)\, u''(z)\mathrm{d}z = \int_{(0,1]} G(z)\, u''(z)\mathrm{d}z.$$

$$(2.13)$$

Since the function u is arbitrary and concave, meaning that $u''(z) \leq 0$ for every $z \in [0, 1]$, this implies that $G(z) \leq 0$ for all $z \in [0, 1]$.

$(ii) \Rightarrow (iii)$: The proof of this implication is more technical and we refer the reader to the original paper Rothschild & Stiglitz [1472], where the implication is proved first for discrete random variables and then extended to general random variables.

$(iii) \Rightarrow (i)$: For any concave function u, Jensen's inequality (applied to the conditional expectation with respect to the random variable $\tilde{x}_1$) together with the tower property of the conditional expectation implies that

$$U(\tilde{x}_2) = \mathbb{E}[u(\tilde{x}_1 + \tilde{\epsilon})] = \mathbb{E}\big[\mathbb{E}[u(\tilde{x}_1 + \tilde{\epsilon})|\tilde{x}_1]\big] \le \mathbb{E}\big[u\big(\mathbb{E}[\tilde{x}_1 + \tilde{\epsilon}|\tilde{x}_1]\big)\big] = \mathbb{E}[u(\tilde{x}_1)],$$

thus proving the implication.   □

We want to emphasize that Proposition 2.10 makes no assumptions on the monotonicity of the utility function. Property (*iii*) is particularly interesting, since it implies that if a gamble $\tilde{x}_1$ is preferred to another gamble $\tilde{x}_2$ by every risk averse agent then the two gambles have the same expected value $\mathbb{E}[\tilde{x}_1] = \mathbb{E}[\tilde{x}_2]$ while the first gamble has a smaller variance. Indeed, if $\tilde{x}_1 \succeq_{SSD} \tilde{x}_2$, then property (*iii*) implies that

$$\sigma^2(\tilde{x}_2) = \sigma^2(\tilde{x}_1 + \tilde{\epsilon}) = \sigma^2(\tilde{x}_1) + \sigma^2(\tilde{\epsilon}) + 2\operatorname{Cov}(\tilde{x}_1, \tilde{\epsilon}) \ge \sigma^2(\tilde{x}_1),$$

where we have used the fact that

$$\operatorname{Cov}(\tilde{x}_1, \tilde{\epsilon}) = \mathbb{E}[\tilde{x}_1 \tilde{\epsilon}] - \mathbb{E}[\tilde{x}_1]\mathbb{E}[\tilde{\epsilon}] = \mathbb{E}\big[\tilde{x}_1 \mathbb{E}[\tilde{\epsilon}|\tilde{x}_1]\big] = 0.$$

Summing up, we have shown that:

$$\tilde{x}_1 \succeq_{SSD} \tilde{x}_2 \quad \Rightarrow \quad \mathbb{E}[\tilde{x}_1] = \mathbb{E}[\tilde{x}_2] \quad \text{and} \quad \sigma^2(\tilde{x}_1) \le \sigma^2(\tilde{x}_2).$$

However, the converse implication does not hold. A counterexample can be easily constructed. Indeed, let us consider the two gambles

$$\tilde{x}_1 = \{0, 4, 8, 12; \ 1/4, 1/4, 1/4, 1/4\} \qquad \text{and} \qquad \tilde{x}_2 = \{0.2, 6, 11.8; \ 1/3, 1/3, 1/3\}.$$

Elementary computations give us that

$$\mathbb{E}[\tilde{x}_1] = 6, \ \sigma^2(\tilde{x}_1) = 20 \qquad \text{and} \qquad \mathbb{E}[\tilde{x}_2] = 6, \ \sigma^2(\tilde{x}_2) \approx 22.43.$$

Consider the concave utility function

$$u(x) = \begin{cases} 2x & \text{for } x \in [0, 6]; \\ 6 + x & \text{for } x > 6. \end{cases}$$

Then, one can compute

$$\mathbb{E}[u(\tilde{x}_1)] = 10 \qquad \text{and} \qquad \mathbb{E}[u(\tilde{x}_2)] \approx 10.07.$$

Even though the two gambles $\tilde{x}_1$ and $\tilde{x}_2$ have the same expectation and the variance of $\tilde{x}_1$ is lower than the variance of $\tilde{x}_2$, an agent with the above utility function will prefer gamble $\tilde{x}_2$ to gamble $\tilde{x}_1$. Note that the order induced by the criterion of second order stochastic dominance is obviously partial, since it only allows to compare gambles-random variables with the same expected value.

As already remarked, the second order stochastic dominance criterion is defined without any hypothesis on the monotonicity of the utility functions. The *second order stochastic monotonic dominance* criterion is obtained by restricting the class of utility functions to non-decreasing concave functions.

Definition 2.11 Let $\tilde{x}_1$ and $\tilde{x}_2$ be two random variables. We say that $\tilde{x}_1$ dominates $\tilde{x}_2$ according to the *second order stochastic monotonic dominance* criterion (written as $\tilde{x}_1 \succeq_{\text{SSD}}^M \tilde{x}_2$) if $U(\tilde{x}_1) \geq U(\tilde{x}_2)$ holds for every non-decreasing concave utility function $u : \mathbb{R} \to \mathbb{R}$.

In the spirit of Proposition 2.10, the following result gives three equivalent characterizations of second order stochastic monotonic dominance (we omit the proof, which is similar to those of the previous propositions in this section).

Proposition 2.12 *For any two random variables $\tilde{x}_1$ and $\tilde{x}_2$ taking values in $[0, 1]$, the following are equivalent:*

(i) $\tilde{x}_1 \succeq_{\text{SSD}}^M \tilde{x}_2$;
(ii) $G(y) = \int_0^y \big(F_1(z) - F_2(z)\big)\mathrm{d}z \leq 0$, *for every $y \in [0, 1]$;*
(iii) $\tilde{x}_2 \equiv^d \tilde{x}_1 + \tilde{\epsilon}$, *where $\tilde{\epsilon}$ is a random variable such that $\mathbb{E}[\tilde{\epsilon}|\tilde{x}_1] \leq 0$;*
(iv) $\tilde{x}_2 \equiv^d \tilde{x}_1 + \tilde{\xi} + \tilde{v}$, *where $\tilde{\xi}$ is a non-positive random variable and $\tilde{v}$ is a random variable such that $\mathbb{E}[\tilde{v}|\tilde{x}_1 + \tilde{\xi}] = 0$.*

Note that part *(ii)* of Proposition 2.12 implies that $\mathbb{E}[\tilde{x}_1] \geq \mathbb{E}[\tilde{x}_2]$. Indeed, since the random variables $\tilde{x}_1$ and $\tilde{x}_2$ take values in $[0, 1]$, this simply follows from the fact that

$$\mathbb{E}[\tilde{x}_1] - \mathbb{E}[\tilde{x}_2] = \int_0^1 \big(1 - F_1(z)\big)\mathrm{d}z - \int_0^1 \big(1 - F_2(z)\big)\mathrm{d}z = -G(1) \geq 0.$$

By definition, second order stochastic dominance implies second order stochastic monotonic dominance, in the sense that, if a random variable $\tilde{x}_1$ dominates another random variable $\tilde{x}_2$ according to $\succeq_{\text{SSD}}$, then $\tilde{x}_1$ also dominates $\tilde{x}_2$ according to $\succeq_{\text{SSD}}^M$.

2.4 Mean-Variance Analysis

As we have seen at the beginning of this chapter by means of the Saint Petersburg paradox, the expected value fails to provide an adequate representation of a preference relation in a risky environment. In order to resolve the paradox, expected utility theory has been introduced under the assumption that the preference relation $\mathcal{R}$ satisfies Assumptions 2.1–2.3. Evaluating a gamble-random variable $\tilde{x}$ via the expected utility paradigm consists in taking the expected value $\mathbb{E}[u(\tilde{x})]$ of a utility function $u : \mathbb{R} \to \mathbb{R}$ of the gamble-random variable $\tilde{x}$. If the agent is risk averse, then the function u is concave. In general, the expected utility function of a gamble-random variable depends on the whole distribution of the latter.

In this section, we aim at answering the question of when a preference relation $\mathscr{R}$ can be fully described in terms of the first two moments (the mean and the variance) of a gamble-random variable. According to this paradigm and motivated also by the approximation (2.8) of the risk premium of a random variable $\tilde{x}$, the variance is taken as the most natural candidate in order to measure the riskiness of a gamble $\tilde{x}$. This leads us to introduce the *mean-variance* criterion: an agent prefers random variables with higher expected values and dislikes random variables with higher variances. More precisely:

$$\tilde{x}_1 \succeq_{MV} \tilde{x}_2 \iff \mathbb{E}[\tilde{x}_1] \geq \mathbb{E}[\tilde{x}_2] \quad \text{and} \quad \sigma^2(\tilde{x}_1) \leq \sigma^2(\tilde{x}_2).$$

We also say that an agent is characterized by *mean-variance preferences* if his expected utility $\mathbb{E}[u(\tilde{x})]$ of a gamble $\tilde{x}$ can be represented as

$$\mathbb{E}[u(\tilde{x})] = V\big(\mathbb{E}[\tilde{x}], \sigma^2(\tilde{x})\big)$$

for some function $V : \mathbb{R} \times \mathbb{R}_+ \to \mathbb{R}$ increasing with respect to the first argument and decreasing with respect to the second argument. Clearly, if $\tilde{x}_1 \succeq_{MV} \tilde{x}_2$, then the gamble $\tilde{x}_1$ is preferred to the gamble $\tilde{x}_2$ by any agent with mean-variance preferences. Note that the mean-variance criterion only yields a partial order among random variables, since it does not allow to compare random variables characterized by differences with the same sign in both the expected value and the variance.

Not every risk averse agent exhibits a preference structure compatible with the mean-variance criterion, as shown by the following counterexample. Let us consider the couple of random variables $\tilde{x}_1$ and $\tilde{x}_2$ described by

$$\tilde{x}_1 = \{0.5, 3, 4, 7; 1/4, 1/4, 1/4, 1/4\}, \qquad \tilde{x}_2 = \{1.5, 3, 8; 1/2, 1/4, 1/4\},$$

and a logarithmic utility function $u(x) = \log(x)$. Elementary computations lead to

$$\mathbb{E}[\tilde{x}_1] = 3.625, \ \sigma^2(\tilde{x}_1) = 5.422 \quad \text{and} \quad \mathbb{E}[\tilde{x}_2] = 3.5, \ \sigma^2(\tilde{x}_2) = 7.125,$$

while

$$\mathbb{E}[u(\tilde{x}_1)] \approx 0.406 \quad \text{and} \quad \mathbb{E}[u(\tilde{x}_2)] \approx 0.433.$$

Therefore, the gamble $\tilde{x}_1$ has a higher mean and a lower variance than the gamble $\tilde{x}_2$, but an agent with a logarithmic utility function will prefer $\tilde{x}_2$ to $\tilde{x}_1$. Therefore, such an agent does not act according to the mean-variance criterion.

This example shows that, in general, considering the class of risk averse agents and a random variable $\tilde{x}$ with mean $\mathbb{E}[\tilde{x}]$ and variance $\sigma^2(\tilde{x})$, it is not possible to express the expected utility $U(\tilde{x})$ in the form $V(\mathbb{E}[\tilde{x}], \sigma^2(\tilde{x}))$. However, the mean-variance representation of preferences holds true under additional assumptions on the utility function u or on the random variable involved. We now discuss some of the most important cases.

a) *Quadratic utility function.*

If the utility function u is quadratic, i.e., $u(x) = x - \frac{b}{2}x^2$ (for $b > 0$), then the expected utility $U(\tilde{x})$ of $\tilde{x}$ becomes

$$U(\tilde{x}) = \mathbb{E}[\tilde{x}] - \frac{b}{2}\mathbb{E}[\tilde{x}^2] = \left(1 - \frac{b}{2}\mathbb{E}[\tilde{x}]\right)\mathbb{E}[\tilde{x}] - \frac{b}{2}\sigma^2(\tilde{x}) =: V\left(\mathbb{E}[\tilde{x}], \sigma^2(\tilde{x})\right).$$

Since $b > 0$, the expected utility is decreasing in the variance $\sigma^2(\tilde{x})$ and, moreover, for $\mathbb{E}[\tilde{x}] \leq 1/b$, it is increasing in the expected value. However, representing preferences by means of a quadratic utility function is somewhat problematic. In particular, a quadratic utility function exhibits increasing absolute risk aversion (IARA, see Sect. 2.2) and, for $\mathbb{E}[\tilde{x}] > 1/b$, the expected utility is decreasing in the expected value.

b) *Second order approximation.*

Let us consider a random variable $\tilde{x}$ and a twice differentiable increasing and concave utility function $u : \mathbb{R} \to \mathbb{R}$. The expected utility function $U(\tilde{x}) = \mathbb{E}[u(\tilde{x})]$ can be approximated through its Taylor expansion up to the second order centered on the expected value $\mathbb{E}[\tilde{x}]$:

$$U(\tilde{x}) \approx u\left(\mathbb{E}[\tilde{x}]\right) + \frac{1}{2}u''\left(\mathbb{E}[\tilde{x}]\right)\sigma^2 =: V\left(\mathbb{E}[\tilde{x}], \sigma^2(\tilde{x})\right).$$

Since the function u is increasing, this approximation implies that, in the case of a random variable $\tilde{x}$ with a small dispersion around its mean (so that $u(\mathbb{E}[\tilde{x}])$ dominates) or in the case of a utility function with positive third derivative, the expected utility function is increasing with respect to the expected value and decreasing with respect to the variance, due to the concavity of u. However, it should be noted that the higher order terms in the Taylor expansion depend on the higher order moments of $\tilde{x}$.

c) *Normal distribution.*

The mean-variance criterion finds a justification within the expected utility paradigm if a random variable $\tilde{x}$ is normally distributed. In that case, moments of arbitrary order of $\tilde{x}$ can be expressed in terms of the mean $\mathbb{E}[\tilde{x}]$ and of the variance $\sigma^2(\tilde{x})$. As a consequence, the expected utility function $\mathbb{E}[u(\tilde{x})]$ can be written as $V(\mathbb{E}[\tilde{x}], \sigma^2(\tilde{x}))$ for some function $V : \mathbb{R} \times \mathbb{R}_+ \to \mathbb{R}$. If the utility function u is increasing and concave, then the function V will be increasing with respect to its first argument and decreasing with respect to the second. More precisely, we have the following proposition.

Proposition 2.13 *Let* $u : \mathbb{R} \to \mathbb{R}$ *be an increasing and concave utility function and let* $\tilde{x}$ *be a normally distributed random variable with mean* μ *and variance* σ^2. *Then there exists a function* $V : \mathbb{R} \times \mathbb{R}_+ \to \mathbb{R}$, *increasing with respect to the first argument and decreasing with respect to the second, such that* $\mathbb{E}[u(\tilde{x})] = V(\mu, \sigma^2)$.

Proof Let $\tilde{z}$ denote a normally distributed random variable with zero mean and unitary variance and define the function $V : \mathbb{R} \times \mathbb{R}_+ \to \mathbb{R}$ by

$$V(x, y) := \mathbb{E}[u(x + \sqrt{y}\,\tilde{z})], \quad \text{for } x \in \mathbb{R} \text{ and } y \in \mathbb{R}_+.$$

The representation $\mathbb{E}[u(\tilde{x})] = V(\mu, \sigma^2)$ follows by noting that $\mu + \sigma\tilde{z} \equiv^d \tilde{x}$. The sign of the two partial derivatives of V is established in Exercise 2.14. □

As a particular case, let us consider the case of an exponential utility function $u(x) = -\frac{1}{a}\exp(-ax)$, for $a > 0$, and a normally distributed random variable $\tilde{x}$ with mean μ and variance σ^2. Recall that the random variable $\exp(a\tilde{x})$ is distributed according to a log-normal distribution with expected value $\exp(a\mu + a^2\sigma^2/2)$. This implies that the expected utility admits the following representation:

$$U(\tilde{x}) = -\frac{1}{a}\mathbb{E}[\exp(-a\tilde{x})] = -\frac{1}{a}\exp\left(-a\mu + \frac{a^2}{2}\sigma^2\right) =: V(\mu, \sigma^2).$$

In particular, the function V is increasing with respect to the expected value μ and decreasing with respect to the variance σ^2 (mean-variance preferences). The coefficient of absolute risk aversion a represents the *mean-variance trade-off*.

Proposition 2.13 implies an interesting property: the indifference curves of the expected utility function in the expected return-standard deviation plane are ordered in an increasing sense, have a positive slope and are convex, see Fig. 2.3. Let us consider the case of a normally distributed random variable $\tilde{x}$ with mean μ and variance σ^2. The slope of the indifference curves of the expected utility function can be determined by setting equal to zero the total differential of the function V : $\mathbb{R} \times \mathbb{R}_+ \to \mathbb{R}$ defined in the proof of Proposition 2.13:

$$dV(\mu, \sigma^2) = \frac{\partial V(\mu, \sigma^2)}{\partial \mu}d\mu + \frac{\partial V(\mu, \sigma^2)}{\partial \sigma}d\sigma$$

$$= \mathbb{E}[u'(\tilde{x})]d\mu + \mathbb{E}\left[u'(\tilde{x})\frac{\tilde{x} - \mu}{\sigma}\right]d\sigma = 0.$$

The indifference curves in the expected return-standard deviation plane have a positive slope. Indeed:

$$\frac{d\mu}{d\sigma} = -\frac{\mathbb{E}\left[u'(\tilde{x})\frac{\tilde{x}-\mu}{\sigma}\right]}{\mathbb{E}[u'(\tilde{x})]} \geq 0, \tag{2.14}$$

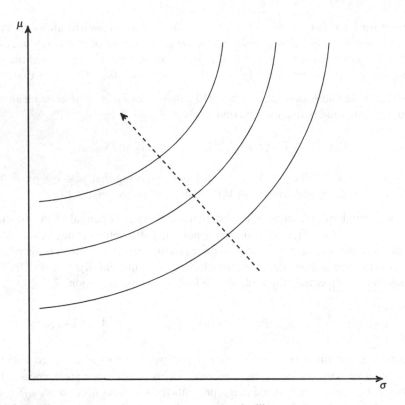

Fig. 2.3 Indifference curves of a mean-variance expected utility

where the inequality can be shown as in Exercise 2.14. The convexity of the indifference curves can be shown by a similar argument, see Ingersoll [1000, Chapter 4].

As an example, if the utility function is exponential, i.e., $u(x) = -\frac{1}{a}\exp(-ax)$, for $a > 0$, then we have (see Exercise 2.15)

$$\frac{\mathrm{d}\mu}{\mathrm{d}\sigma} = a\sigma \qquad (2.15)$$

and the convexity of the indifference curves is ensured since $a > 0$. Similarly, in the case of a quadratic utility function $u(x) = x - \frac{b}{2}x^2$, with $\mu < 1/b$, we have that the slope of the indifference curves is (see Exercise 2.15)

$$\frac{\mathrm{d}\mu}{\mathrm{d}\sigma} = \frac{b\sigma}{1 - b\mu} > 0. \qquad (2.16)$$

As long as $\mu < 1/b$, the slope is increasing in σ and, therefore, the indifference curves are convex in the expected return-standard deviation plane.

The Diversification and Insurance Principles

If an agent's preferences satisfy the mean-variance criterion, then the variance becomes the sole risk indicator. In this context, we can introduce two key principles in financial decision making problems: the *diversification* and the *insurance* principles.

Consider an agent who holds one unit of wealth and wants to determine his optimal exposure to two random variables $\tilde{x}_1$ and $\tilde{x}_2$. In other words, the agent needs to choose the optimal $w^* \in \mathbb{R}$ such that the linear combination $w^*\tilde{x}_1 + (1 - w^*)\tilde{x}_2$ will be preferred to any other linear combination of the two gambles $\tilde{x}_1$ and $\tilde{x}_2$, i.e.,

$$w^*\tilde{x}_1 + (1 - w^*)\tilde{x}_2 \; \mathscr{R} \; w\tilde{x}_1 + (1 - w)\tilde{x}_2, \quad \text{for every } w \in \mathbb{R}.$$

Equivalently, we can interpret this optimal choice problem as the problem of determining the optimal proportion w^* of wealth invested in the two lotteries $\tilde{x}_1$ and $\tilde{x}_2$. According to this interpretation, we can distinguish three cases:

a) $0 < w^* < 1$: the agent invests a positive amount of wealth in both gambles;
b) $w^* > 1$: the agent invests an amount larger than his endowment of wealth in the first gamble and goes short on the second gamble;
c) $w^* < 0$: the agent invests an amount larger than his endowment of wealth in the second gamble and goes short on the first gamble.

The expected utility function of our agent with mean-variance preferences in correspondence of a generic gamble $\tilde{x}$ satisfies $U(\tilde{x}) = V(\mathbb{E}[\tilde{x}], \sigma^2(\tilde{x}))$, for a function $V : \mathbb{R} \times \mathbb{R}_+ \to \mathbb{R}$ increasing with respect to the first argument and decreasing with respect to the second. In the present context, this means that our agent will choose the optimal value w^* achieving the optimal trade-off between the expected value and the variance of the linear combination $w\tilde{x}_1 + (1 - w)\tilde{x}_2$.

We denote $\mu_i := \mathbb{E}[\tilde{x}_i]$ and $\sigma_i^2 := \sigma^2(\tilde{x}_i)$, for $i = 1, 2$, and let $\rho \in [-1, 1]$ be the correlation coefficient between $\tilde{x}_1$ and $\tilde{x}_2$. For every $w \in \mathbb{R}$, we also define the random variable $\tilde{x}_w := w\tilde{x}_1 + (1 - w)\tilde{x}_2$, with expected value and variance

$$\mathbb{E}[\tilde{x}_w] = w\mu_1 + (1-w)\mu_2 \quad \text{and} \quad \sigma^2(\tilde{x}_w) = w^2\sigma_1^2 + (1-w)^2\sigma_2^2 + 2w(1-w)\rho\sigma_1\sigma_2.$$

Let us first consider the case where the two gambles have the same expected value and are uncorrelated, i.e., $\mu_1 = \mu_2 =: \mu$ and $\rho = 0$. In this case, since every linear combination of $\tilde{x}_1$ and $\tilde{x}_2$ has expected value μ, the agent's problem reduces to the problem of minimizing the variance of the linear combination $\tilde{x}_w$:

$$\min_{w\in\mathbb{R}} w^2\sigma_1^2 + (1 - w)^2\sigma_2^2. \tag{2.17}$$

Since $\sigma_1^2, \sigma_2^2 > 0$, the first order condition for the minimization problem (2.17) is both necessary and sufficient, yielding the optimal value

$$w^* = \frac{\sigma_2^2}{\sigma_1^2 + \sigma_2^2} .$$

In this case, the optimal choice w^* of the agent belongs to the interval $(0, 1)$, meaning that a positive amount of wealth is invested in both gambles $\tilde{x}_1$ and $\tilde{x}_2$. This result highlights the *diversification principle*: in order to minimize the risk (the variance), the agent will choose to diversify his exposure between the two gambles.

Note also that $w^* \geq (1 - w^*) \iff \sigma_1^2 \leq \sigma_2^2$, so that the agent will choose a greater exposure to the gamble with lower variance. In the particular case where $\sigma_1^2 = \sigma_2^2 =: \sigma^2$, the agent will achieve a *perfect diversification* between the two gambles, i.e., the optimal choice will be $w^* = 1/2$. In this case, the minimal variance $\sigma^2(\tilde{x}_{w^*})$ is equal to one half of the original variance $(\sigma^2/2)$.

Having discussed the case of two uncorrelated gambles $\tilde{x}_1$ and $\tilde{x}_2$, let us study the optimal choice problem when $\rho \neq 0$. Let us first consider the case of perfectly correlated gambles, i.e., $|\rho| = 1$. The random variables $\tilde{x}_1$ and $\tilde{x}_2$ are perfectly correlated if and only if there exist $a, b \in \mathbb{R}$ such that

$$\tilde{x}_1 = a + b\tilde{x}_2$$

and, moreover, we have $\rho = 1$ if and only if $b > 0$ (and, analogously, $\rho = -1$ if and only if $b < 0$). Let us start with the case $\rho = 1$, always assuming that $\mu_1 = \mu_2 =: \mu$. As above, in this case the agent's optimal choice problem reduces to the problem of finding the optimal $w^* \in \mathbb{R}$ which minimizes the variance of $\tilde{x}_w$:

$$\min_{w \in \mathbb{R}} w^2 \sigma_1^2 + (1 - w)^2 \sigma_2^2 + 2 w(1 - w)\sigma_1\sigma_2 . \qquad (2.18)$$

Note that the first two terms are positive, the third one instead becomes negative when $w < 0$ or $w > 1$. Since $\sigma_1^2 + \sigma_2^2 \geq 2\sigma_1\sigma_2$, the first order condition of the above minimization is both necessary and sufficient and yields the optimal value

$$w^* = \frac{\sigma_2}{\sigma_2 - \sigma_1} .$$

In correspondence of the optimal choice w^*, it holds that $\sigma^2(\tilde{x}_{w^*}) = 0$, meaning that the agent can completely eliminate the risk. It is interesting to observe that, in the case $\rho = 1$, the optimal choice w^* does not belong to the interval $(0, 1)$. Indeed, if $\sigma_1^2 < \sigma_2^2$, we have $w^* > 1$, meaning that the agent does not diversify his exposure between the two sources of risk $\tilde{x}_1$ and $\tilde{x}_2$, but invests an amount of wealth greater than his endowment of wealth in the first gamble $\tilde{x}_1$ (the one with smaller variance) and sells short the gamble $\tilde{x}_2$. A similar argument applies to the case $\sigma_1^2 > \sigma_2^2$.

Let us now consider the case of perfect negative correlation, i.e., $\rho = -1$. In that case, the minimization problem corresponding to (2.18) is

$$\min_{w \in \mathbb{R}} w^2 \sigma_1^2 + (1 - w)^2 \sigma_2^2 - 2 w(1 - w)\sigma_1\sigma_2 . \tag{2.19}$$

In this case, the third term of (2.19) becomes negative when $0 < w < 1$. It is easy to check that the optimal solution is given by

$$w^* = \frac{\sigma_2}{\sigma_2 + \sigma_1} .$$

As in the case of perfect positive correlation, we have $\sigma^2(\tilde{x}_{w^*}) = 0$, i.e., the agent can completely eliminate the risk of his exposure to the two gambles $\tilde{x}_1$ and $\tilde{x}_2$. However, unlike in the case of perfect positive correlation, here $w^* \in (0, 1)$, meaning that the agent will diversify his exposure between the two gambles, putting more weight on the gamble with smaller variance.

The analysis of the two cases of perfect positive and negative correlation allows us to introduce the *insurance principle*. Indeed, in the case of perfect negative correlation, the agent interested in minimizing the variance chooses to invest a positive amount of wealth in both gambles $\tilde{x}_1$ and $\tilde{x}_2$. Intuitively, since the two gambles are perfectly negatively correlated, values of $\tilde{x}_1$ above its mean μ will be compensated by values of $\tilde{x}_2$ below its mean μ and vice-versa. Hence, by combining the two gambles with positive weights, the agent can reduce to zero the riskiness of his overall portfolio. Analogously, in the case of perfect positive correlation, the agent chooses to invest a positive amount of wealth in the less risky gamble and a negative amount in the riskier gamble, in such a way to create a compensation between the two sources of randomness. This argument takes the name of *insurance principle*, since the exposure on the gamble $\tilde{x}_1$ insures the exposure on the gamble $\tilde{x}_2$.

Let us now consider the case where $|\rho| \neq 1$. In this case, the variance minimization problem corresponding to problems (2.17)–(2.18) is

$$\min_{w \in \mathbb{R}} w^2 \sigma_1^2 + (1 - w)^2 \sigma_2^2 + 2 \rho w(1 - w)\sigma_1\sigma_2 .$$

It can be readily checked that the optimal choice is given by

$$w^* = \frac{\sigma_2^2 - \rho\sigma_1\sigma_2}{\sigma_1^2 + \sigma_2^2 - 2\rho\sigma_1\sigma_2} .$$

Suppose that $\sigma_1^2 > \sigma_2^2$. In this case, since $\sigma_1^2 > \rho\sigma_1\sigma_2$, it holds that

$$\sigma_2^2 - \rho\sigma_1\sigma_2 = \sigma_1^2 + \sigma_2^2 - 2\rho\sigma_1\sigma_2 - (\sigma_1^2 - \rho\sigma_1\sigma_2) < \sigma_1^2 + \sigma_2^2 - 2\rho\sigma_1\sigma_2,$$

thus showing that $w^* < 1$. This means that, if the agent is interested in minimizing the variance, then he will never invest more than his endowment of wealth in the riskier gamble. The sign of w^* depends on the numerator $\sigma_2^2 - \rho\sigma_1\sigma_2$. Indeed, if $-1 < \rho \leq \sigma_2/\sigma_1$ then we have $w^* \in [0, 1]$, while if $\sigma_2/\sigma_1 < \rho < 1$ then we have $w^* < 0$. This means that, if the correlation is negative or positive but not too high, then the agent will choose to diversify his exposure between the two gambles $\tilde{x}_1$ and $\tilde{x}_2$, while, if the correlation is close to 1, then he will adopt a strategy similar to the one obtained in the case of perfect positive correlation, investing in the less risky gamble more than his endowment of wealth and selling short the riskier gamble. Note that, unlike the cases of perfect positive and negative correlation, it is not possible to reduce to zero the total variance if $|\rho| \neq 1$.

The above discussion can be illustrated by means of Fig. 2.4, which considers the case of two gambles $\tilde{x}_1$ and $\tilde{x}_2$ with means μ_1 and μ_2 and variances σ_1^2 and σ_2^2 (with $\mu_1 > \mu_2$ and $\sigma_1^2 > \sigma_2^2$). The figure represents the possible combinations of expected value and standard deviation obtained in the cases $\rho = 1$, $\rho = -1$ and

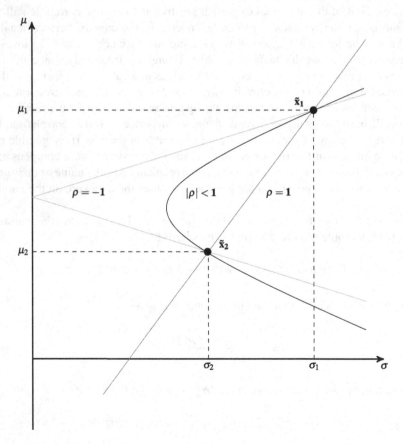

Fig. 2.4 Mean-variance of portfolios

$-1 < \rho < 1$. We can observe that, in the case of perfect negative correlation, it is possible to eliminate the risk with a diversified investment in $\tilde{x}_1$ and $\tilde{x}_2$, while this is possible by choosing $w^* < 0$ in the case of perfect positive correlation. In the case where $-1 < \rho < 1$, the risk can never be perfectly eliminated.

Let us conclude this section by generalizing the above problem to the case of $N > 2$ gambles-random variables $\tilde{x}_n$, for $n = 1, \ldots, N$. Let us assume that an agent needs to choose his exposure/investment $w_n \in \mathbb{R}$ in the n-th gamble, for each $n = 1, \ldots, N$, subject to the constraint $\sum_{n=1}^{N} w_n = 1$ (i.e., the agent must allocate all his wealth among the N gambles). Let us denote $\tilde{x}_w = \sum_{n=1}^{N} w_n \tilde{x}_n$. Then

$$\mathbb{E}[\tilde{x}_w] = \sum_{n=1}^{N} w_n \mu_n \quad \text{and} \quad \sigma^2(\tilde{x}_w) = \sum_{n=1}^{N} \sum_{m=1}^{N} w_n w_m \, \mathrm{Cov}(\tilde{x}_n, \tilde{x}_m).$$

Let us consider the special case where the N gambles are uncorrelated, meaning that $\mathrm{Cov}(\tilde{x}_i, \tilde{x}_j) = 0$ for all $i, j \in \{1, \ldots, N\}$ with $i \neq j$. Then, the problem of minimizing the variance of the linear combination $\tilde{x}_w$ becomes

$$\min_{\substack{w \in \mathbb{R}^N \\ \text{s.t. } \sum_{n=1}^{N} w_n = 1}} \sum_{n=1}^{N} w_n^2 \sigma^2(\tilde{x}_n).$$

We introduce the Lagrangian

$$L(w_1, \ldots, w_N, \lambda) = \sum_{n=1}^{N} w_n^2 \sigma^2(\tilde{x}_n) + \lambda \left(1 - \sum_{n=1}^{N} w_n \right). \tag{2.20}$$

As shown in Exercise 2.21, the optimal solution is given by

$$w_n^* = \frac{1/\sigma^2(\tilde{x}_n)}{\sum_{n=1}^{N} 1/\sigma^2(\tilde{x}_n)}, \quad \text{for all } n = 1, \ldots, N. \tag{2.21}$$

In other words, in order to minimize the overall variance, the exposure to each gamble $\tilde{x}_n$ should be proportional to the (normalized) reciprocal of the variance of the gamble $\tilde{x}_n$, for $n = 1, \ldots, N$. In particular, if the N gambles have the same variance σ^2, then the agent will equally distribute his wealth among the N gambles, achieving a *perfect diversification*, i.e.,

$$w_n^* = \frac{1}{N}, \quad \text{for all } n = 1, \ldots, N.$$

Note that, with a perfectly diversified portfolio, the overall variance amounts to σ^2/N. Therefore, thanks to the diversification principle, in the limit the variance will tend to zero for $N \to \infty$ (perfect diversification).

2.5 Notes and Further Readings

We refer the reader to Kreps [1136] for a comprehensive account of decision theory in a risky environment. Assumptions 2.1–2.3 play a key role in representing preferences via an expected utility function. In particular, referring to Chap. 9 for a more detailed discussion, the independence axiom (Assumption 2.3) turns out to be rather controversial on an empirical and theoretical ground. The most famous argument goes back to the *Allais Paradox* (see Allais [41]). Consider four gambles with monetary values $\{0, 1000, 5000\}$ and corresponding probabilities

$$A = \{0, 1, 0\}, \qquad B = \{0.01, 0.89, 0.1\},$$
$$C = \{0.9, 0, 0.1\}, \qquad D = \{0.89, 0.11, 0\}.$$

It is empirically observed that people express the following preference order: $A \mathscr{P} B$ and $C \mathscr{P} D$. It is easy to show that this preference relation does not satisfy the independence axiom. Indeed, let X, Y, Z be the gambles paying the amounts $0, 1000, 5000$ with certainty, respectively. A contradiction to the independence axiom can be easily obtained as follows:

$$A \mathscr{I} \frac{11}{100} Y + \frac{89}{100} Y \quad \text{and} \quad B \mathscr{I} \frac{11}{100} \left(\frac{1}{11} X + \frac{10}{11} Z \right) + \frac{89}{100} Y.$$

The transitivity of the preference relation $\mathscr{P}$ implies that

$$\frac{11}{100} Y + \frac{89}{100} Y \mathscr{P} \frac{11}{100} \left(\frac{1}{11} X + \frac{10}{11} Z \right) + \frac{89}{100} Y.$$

By the independence axiom, it holds that

$$A \mathscr{P} \frac{1}{11} X + \frac{10}{11} Z,$$

and

$$\frac{11}{100} Y + \frac{89}{100} X \mathscr{P} \frac{11}{100} \left(\frac{1}{11} X + \frac{10}{11} Z \right) + \frac{89}{100} X.$$

This means that $D \mathscr{P} C$, and, therefore, a contradiction is obtained. One can also exhibit other situations analogous to the Allais paradox: in general, they require agents to evaluate gambles with small/large probabilities. These paradoxes highlight the key role of the independence axiom, which amounts to assume that an agent is able to identify the common part of two gambles and to evaluate them only by considering what is different (i.e., an agent is able to "decontextualize" his choices). However, this is not observed in reality: an agent typically takes different decisions in relation with the context surrounding him, since the context itself affects the preferences.

A version of the expected utility representation theorem according to a subjectivist interpretation of probability has been provided in Savage [1499]. In this setting, agents do not know the true probability of the events. Savage [1499] starts from a complete preorder defined on $\mathcal{M}$ without pre-specified probabilities and shows that, if the preorder satisfies some assumptions, then there exists both a subjective probability distribution and a utility function such that the preorder can be represented by means of the expected value of a utility function, where the expectation is taken with respect to the subjective probability distribution. Agents behave as if those probabilities were assigned to the events (on this approach, see also Sect. 9.1 and Anscombe & Aumann [72]). Another interesting extension of the expected utility theory is represented by the situation where an agent exhibits state dependent preferences, i.e., his utility function u depends on the state of the world $\omega \in \Omega$. A generalization of the expected utility representation result (Theorem 2.1) can be provided in this setting, relaxing the independence axiom.

In Sect. 2.1, we have identified a set of conditions on a preference relation $\mathcal{R}$ so that the latter can be represented by an expected utility function. In applications, instead of describing the behavior of the agent through his preference relation, it is often directly assumed that the agent is characterized by a given utility function. The drawback of this approach is that it can lead to consider a utility function and a class of gambles which are not compatible with a representation through an expected utility function, because the utility function is unbounded and/or the expected value of the gamble is not finite. This drawback is avoided in the following cases: a bounded utility function; gambles taking values on a finite set; gambles with a finite expected value together with an increasing and concave utility function. The last case deserves to be verified. Given an increasing, concave and differentiable utility function $u : \mathbb{R} \to \mathbb{R}$, we have that $u(x) \le u(z) + u'(z)(x - z)$ for every $x, z \in \mathbb{R}$. If the expected value of the gamble $\tilde{x}$ is finite, then

$$\mathbb{E}[u(\tilde{x})] \le u(z) + u'(z)(\mathbb{E}[\tilde{x}] - x) < +\infty$$

and, hence, the preference relation admits a representation through the expected utility.

Note that the risk aversion hypothesis is not consistent with the observation that many agents stipulate insurance contracts and, at the same time, accept gambles which are far from being actuarially fair (think of casino games, national lotteries, etc.). This behavior can be explained by assuming that agents are characterized by a concave utility function for small levels of wealth, then convex and then again concave for large values of wealth, as proposed in Friedman & Savage [744]. Kahneman & Tversky [1060] instead conclude that investors maximize a function which is convex for positive outcomes and concave for negative outcomes, i.e., agents are *loss averse*. We will return to this type of utility functions in Chap. 9.

In Ross [1468], a different measure of risk aversion is proposed: an agent a is *strongly more risk averse* than an agent b if

$$\inf_{z \in \mathbb{R}} \frac{u^{a''}(z)}{u^{b''}(z)} \geq \sup_{z \in \mathbb{R}} \frac{u^{a'}(z)}{u^{b'}(z)}.$$

This condition implies that agent a is more risk averse than agent b (in the sense of Sect. 2.2), but the converse implication does not hold. It can be shown that the above condition is verified if and only if there exists a strictly positive constant λ and a non-increasing and concave function $g : \mathbb{R} \to \mathbb{R}$ such that

$$u^a(z) = \lambda u^b(z) + g(z), \qquad \text{for every } z \in \mathbb{R}.$$

Let agent a be strongly more risk averse than agent b. If the random variable $\tilde{x}$ represents the initial random endowment of an agent (background risk) then the risk premium of a gamble $\tilde{x} + \tilde{\epsilon}$ such that $\mathbb{E}[\tilde{\epsilon}|\tilde{x}] = 0$ for agent a is larger than that for agent b, see Ross [1468]. Note that risk aversion alone does not suffice for this property to hold true. In a multiple goods setting, risk aversion has been characterized in Kihlstrom & Mirman [1089, 1090] (see also Gollier [800]). More recently, new characterizations of risk aversion have been proposed referring to the utility that an agent gets from two gambles. For instance, the definition of *proper risk aversion* (see Pratt & Zeckauser [1433]) can be summarized as follows: if two independent gambles are individually unattractive, then the compound lottery offering both together is less attractive than either alone. The definition of *standard risk aversion* (see Kimball [1101]) extends the above definition to "loss aggravating" independent risks (see also Gollier & Pratt [803]). For a survey of the recent literature on risk aversion we refer to Gollier [800] (see also Eeckhoudt & Gollier [630] on decision making problems in a risky setting). We refer the interested reader to Levy [1205], Sriboonchitta et al. [1560] for two recent accounts on stochastic dominance and its implications for the purposes of financial modeling.

The reconciliation of the mean-variance criterion with the expected utility approach represented a challenging problem in the past. Assuming an increasing and concave utility function, what are the probability distributions compatible with the mean-variance criterion? In Tobin [1596], besides verifying that the normal multivariate distribution satisfies this property, the author claimed that every probability distribution characterized by two parameters makes the mean-variance criterion compatible with the expected utility. This conjecture turns out to be incorrect: the missing point is that a linear combination of random variables with probability distributions belonging to a given common family does not necessarily have a probability distribution belonging to the same family. The normal multivariate distribution obviously enjoys this property. More generally, the distributions which are compatible with the mean-variance are the *elliptical distributions*, see Ingersoll [1000], Owen & Rabinovitch [1389] and also Meyer [1334] for the characterization of the class of linear distribution functions. In Chamberlain [395], the class of probability distributions such that the expected utility is a function only of the mean and of the variance is characterized. An important property turns out to be that of *spherical symmetry*: a vector of random variables is spherically distributed (around the origin) if its distribution is invariant with respect to orthogonal linear transformations that leave the origin fixed. Note however that for this class of distributions the expected utility function for an increasing and concave utility function u is not necessarily increasing in the expected value and decreasing in the variance. Some counterexamples concerning mean-variance preferences are presented in Exercises 2.22 and 2.23.

2.6 Exercises

Exercise 2.1 In the context of a finite probability space, as in Theorem 2.1, show that the expected utility representation is determined up to a strictly positive linear transformation.

Exercise 2.2

(i) Show that the logarithmic utility function $u(x) = \log(a + x)$ is the limiting case of a power utility function $u(x) = \frac{1}{b-1}((a + bx)^{(b-1)/b} - 1)$ for $b \to 1$.

(ii) Show that the exponential utility function $u(x) = -\exp(-x/a)$ is the limiting case of a power utility function $u(x) = \frac{1}{b-1}(1 + \frac{bx}{a})^{\frac{b-1}{b}}$ for $b \to 0$.

(iii) Show that the utility function $u(x) = \frac{1-e^{ax}}{a}$ tends to a linear utility function as $a \to 0$.

Exercise 2.3 Let $\tilde{x}$ be a random variable with expected value $\mathbb{E}[\tilde{x}]$ and variance $k^2\sigma^2$, admitting the representation $\tilde{x} = \mathbb{E}[\tilde{x}] + k\tilde{\epsilon}$, where $\tilde{\epsilon}$ is a random variable with zero mean and variance σ^2. Show that, for any strictly increasing and differentiable utility function $u : \mathbb{R} \to \mathbb{R}$, the risk premium $\rho_u(\tilde{x}, k)$ parameterized by k satisfies $\rho_u(\tilde{x}, 0) = 0$ and $\frac{\partial \rho_u(\tilde{x},k)}{\partial k}|_{k=0} = 0$.

Exercise 2.4 Given a quadratic utility function $u(x) = x - \frac{b}{2}x^2$, for $b > 0$, and a random variable $\tilde{x}$ with expected value μ and variance σ^2, determine the risk premium $\rho_u(\tilde{x})$.

Exercise 2.5 Given the exponential utility function $u(x) = -\frac{1}{a}e^{-ax}$, $a > 0$, and a normally distributed random variable $\tilde{x} \sim \mathcal{N}(\mu, \sigma^2)$, determine the risk premium.

Exercise 2.6 Given a power utility function $u(x) = \frac{x^{1-a}}{1-a}$, for $a > 0$, and a random variable $\tilde{x}$ such that $\log \tilde{x}$ is distributed as a Normal $\mathcal{N}(\mu, \sigma^2)$, determine the risk premium.

Exercise 2.7 Consider an agent a with piecewise linear utility function $u^a(x) = x$ for $x \le x_0$ and $u^a(x) = x_0 + a(x - x_0)$ otherwise, for some $a \in (0, 1)$, and an agent b with utility function $u^b(x) = x$ for $x \le x_0$ and $u^b(x) = x_0 + b(x - x_0)$ otherwise, for some $b \in (0, 1)$. Show that agent b is more risk averse than agent a if and only if $b \leq a$.

Exercise 2.8 Let us consider a three times differentiable utility function $u :$ $\mathbb{R} \to \mathbb{R}$.

(i) Prove that, if u is increasing and DARA, then $u'''(x) > 0$ for every $x \in \mathbb{R}$.
(ii) For an increasing and concave utility function $u : \mathbb{R} \to \mathbb{R}$, define the degree of absolute prudence as $p_u^a(x) := -u'''(x)/u''(x)$, for $x \in \mathbb{R}_+$. Show that the utility function u is DARA if and only if $p_u^a(x) > r_u^a(x)$, for every $x \in \mathbb{R}$.
(iii) Show that $p_u^a(x) > 0$, for all $x \in \mathbb{R}$, is a necessary condition to have a DARA utility function.

Exercise 2.9 Let $u : \mathbb{R} \to \mathbb{R}$ be a twice differentiable utility function and denote by $r_u^a(x)$ the coefficient of absolute risk aversion. Show that, for suitable $c, k \in \mathbb{R}$:

$$u'(x) = ke^{-\int_c^x r_u^a(y)dy}, \qquad \text{for every } x \in \mathbb{R}.$$

Exercise 2.10 Consider the HARA class of utility functions introduced in Sect. 2.2 ($r_u^a(x) = 1/(a + bx)$, for some $a, b \in \mathbb{R}$). Verify that:

(i) a HARA utility function shows a decreasing coefficient of absolute risk aversion if and only if $b > 0$ and a decreasing coefficient of relative risk aversion if and only if $a < 0$;
(ii) the three specifications of utility function $\log(x + a)$ for $b = 1$, $-a\exp(-x/a)$ for $b = 0$ and $\frac{1}{b-1}(a + bx)^{\frac{b-1}{b}}$ in the remaining cases provide an exhaustive representation of the HARA class of utility functions. Determine the domain of these functions.

Exercise 2.11 An agent has to make a choice between two different gambles: a) receive 25000 euro with certainty; b) receive 32000 euro with probability 0.2, 10000 euro with probability 0.7 and 1000 euro with probability 0.1. The agent is characterized by a power utility function $u(x) = x^\gamma$, with $\gamma \in (0, 1)$, and he is indifferent between two gambles: c) receive 20.25 euro with certainty; d) receive 16 euro with probability $1/2$ and 25 euro with probability $1/2$. Determine γ and which of the two gambles a) and b) will be preferred by the agent.

Exercise 2.12

(i) Show that a utility function $u : \mathbb{R} \to \mathbb{R}$ is characterized by a constant relative risk aversion coefficient different from 1 if and only if $u(x) = \alpha x^{1-b} + \beta$ for $\alpha > 0$ and $\beta \in \mathbb{R}$, with $b \neq 1$.
(ii) Show that a utility function $u : \mathbb{R} \to \mathbb{R}$ is characterized by a constant relative risk aversion coefficient equal to one if and only if $u(x) = \alpha \log(x) + \beta$ for $\alpha > 0$ and $\beta \in \mathbb{R}$.

Exercise 2.13 Show that if agent a is strongly more risk averse than agent b, in the sense of Sect. 2.5, then agent a is also more risk averse than agent b, but the converse is not true.

Exercise 2.14 In the context of Proposition 2.13, prove that the mean-variance function $V : \mathbb{R} \times \mathbb{R}_+ \to \mathbb{R}$ is increasing with respect to the first argument and decreasing with respect to the second one.

Exercise 2.15 Prove that, in the case of a normally distributed random variable $\tilde{x}$, the slope $\mathrm{d}\mu/\mathrm{d}\sigma$ of the indifference curves in the expected return-standard deviation plane for an exponential utility function $u(x) = -\frac{1}{a}\exp(-ax)$, for $a > 0$, is given by formula (2.15) and, in the case of a quadratic utility function $u(x) = x - \frac{b}{2}x^2$, for $\mu < 1/b$, by formula (2.16).

Exercise 2.16 Prove that, if two random variables $\tilde{x}_1$ and $\tilde{x}_2$ are normally distributed and have the same mean μ, then $\tilde{x}_1 \succeq_{\mathrm{SSD}} \tilde{x}_2$ if and only if $\sigma^2(\tilde{x}_1) \leq \sigma^2(\tilde{x}_2)$.

Exercise 2.17 Consider two gambles $\tilde{x}_1 = \{0, 1; 0.3, 0.7\}$ and $\tilde{x}_2 = \{0, x; \pi, 1-\pi\}$, with $x \in [0, 1]$. Provide necessary and sufficient conditions on x and π so that

 (i) $\tilde{x}_1 \succeq_{\mathrm{FSD}} \tilde{x}_2$;
 (ii) $\tilde{x}_1 \succeq_{\mathrm{SSD}} \tilde{x}_2$;
 (iii) $\tilde{x}_1 \succeq_{\mathrm{SSD}}^M \tilde{x}_2$;
 (iv) $\tilde{x}_1 \succeq_{\mathrm{MV}} \tilde{x}_2$.

Exercise 2.18 Let $\tilde{x}_1$ be a random variable taking values $\{-1, 1, 0\}$ with probabilities $\{\pi, \pi, 1 - 2\pi\}$, respectively, and $\tilde{x}_2$ a random variable taking values $\{-1, 1, 0\}$ with probabilities $\{\gamma, \gamma, 1 - 2\gamma\}$, respectively. Show that $\tilde{x}_1 \succeq_{\mathrm{SSD}} \tilde{x}_2$ if and only if $\gamma \geq \pi$.

Exercise 2.19 Show that if $\tilde{x}_1 \succeq_{\mathrm{FSD}} \tilde{x}_2$ then $\mathbb{E}[\tilde{x}_1] \geq \mathbb{E}[\tilde{x}_2]$ and provide a counterexample showing that the converse implication does not hold.

Exercise 2.20

 (i) Does $\tilde{x}_1 \succeq_{\mathrm{SSD}}^M \tilde{x}_2$ imply $\sigma^2(\tilde{x}_1) \leq \sigma^2(\tilde{x}_2)$?
 (ii) Does $\tilde{x}_1 \succeq_{\mathrm{SSD}}^M \tilde{x}_2$ imply $\tilde{x}_1 \succeq_{\mathrm{MV}} \tilde{x}_2$?

Exercise 2.21 In the context of Sect. 2.4, prove that the solution to the variance minimization problem (2.20) is given by (2.21).

Exercise 2.22 (Hens & Rieger [938], Theorem 2.30) Suppose that the expected utility function $\mathbb{E}[u(\tilde{x})]$ of a risk averse agent can be represented through a continuous mean-variance function $V(\mathbb{E}[\tilde{x}], \sigma^2(\tilde{x}))$, strictly increasing with respect to the mean and strictly decreasing with respect to the variance. Then, there exist two random variables $\tilde{x}_1$, $\tilde{x}_2$ such that $\mathbb{P}(\tilde{x}_1 \geq \tilde{x}_2) = 1$ and $\mathbb{P}(\tilde{x}_1 > \tilde{x}_2) > 0$, but $\mathbb{E}[u(\tilde{x}_1)] < \mathbb{E}[u(\tilde{x}_2)]$.

Exercise 2.23 (Hens & Rieger [938], Corollary 2.31)   Suppose that the expected utility function $\mathbb{E}[u(\tilde{x})]$ of a risk averse agent can be represented through a continuous mean-variance function $V(\mathbb{E}[\tilde{x}], \sigma^2(\tilde{x}))$, strictly increasing in the first argument, for any $\sigma^2(\tilde{x})$. Then the preference relation represented by $\mathbb{E}[u(\tilde{x})]$ does not satisfy the independence axiom.

Chapter 3
Portfolio, Insurance and Saving Decisions

*The most important questions of life are, for the most part,
really only problems of probability.*
Pierre Simon de Laplace (Théorie Analytique des Probabilités)

*To withdraw is not to run away, and to stay is no wise action
when there is more reason to fear than to hope. 'Tis the part of a
wise man to keep himself today for tomorrow, and not venture
all his eggs in one basket.*
Miguel de Cervantes (Don Quixote)

In the previous chapter, we have presented a general framework for decision making problems under risk, based on expected utility theory. In that context, decisions concerned abstract random variables. In the present chapter, we will consider random variables representing returns of assets/insurance contracts. This allows us to analyse optimal portfolio, insurance and saving problems in a risky setting and, in particular, to characterize the optimal choice of a risk averse agent.

This chapter is structured as follows. In Sect. 3.1, we study the optimal portfolio problem, first in the case of a single risky asset and then in the presence of multiple risky assets. In Sect. 3.2, we develop the theory of mean-variance portfolio selection and study the properties of the mean-variance portfolio frontier. Section 3.3 deals with optimal insurance problems, while Sect. 3.4 lays the foundations of the analysis of optimal saving-consumption problems. At the end of the chapter, we provide a guide to further readings as well as a series of exercises.

3.1 Portfolio Theory

In this section, we consider the problem of an agent aiming at optimally investing his wealth by trading on a set of risk free and risky assets, where the optimality criterion is represented by the maximization of expected utility. As in the previous chapter, we consider a two-period model, where at the initial date $t = 0$ the agent has to make his investment choices without knowing the state of the world $\omega \in \Omega$, which will only become known at the future date $t = 1$. In this context, the agent's choices concern random variables representing the random returns of a set of traded assets, where the return of an asset is defined as the ratio between the (random)

© Springer-Verlag London Ltd. 2017
E. Barucci, C. Fontana, *Financial Markets Theory*,
Springer Finance, DOI 10.1007/978-1-4471-7322-9_3

dividend yielded by the asset at $t = 1$ and its current price at $t = 0$. The optimal choice problem of an agent consists in determining the *portfolio* of the traded assets which maximizes the expected utility of the random wealth obtained at $t = 1$.

We assume the existence of a financial market where each asset can be bought and sold without liquidity constraints, transaction costs or other market frictions. More specifically, we assume the existence of a risk free asset (denoted as asset 0 and generically interpreted as a bank account), the return of which is constant over all states of the world and denoted by $r_f > 0$. This asset represents the possibility of investing-borrowing at the risk free rate r_f. Besides the risk free asset, the market contains $N \in \mathbb{N}$ risky assets. The return of asset n at $t = 1$ is denoted by the random variable $\tilde{r}_n$, where $\tilde{r}_n := \tilde{d}_n/p_n$, with $\tilde{d}_n$ denoting the random dividend of asset n delivered at date $t = 1$ and $p_n > 0$ denoting the price of asset n at date $t = 0$, for $n = 1, \ldots, N$. According to this notation, $\tilde{r}_n$ represents the random wealth obtained at $t = 1$ by investing one unit of wealth in the n-th asset at the initial date $t = 0$.

Let us denote by w_0 the initial endowment/wealth of an agent at date $t = 0$. The agent's problem consists in allocating the initial wealth w_0 among the $N + 1$ available assets (i.e., the risk free asset together with the N risky assets) in order to maximize the expected utility of the random wealth obtained at $t = 1$. We denote by w_n, for $n = 1, \ldots, N$, the amount of wealth invested at time $t = 0$ in the n-th asset. For each $n \in \{1, \ldots, N\}$, the inequality $w_n > 0$ means that the agent is investing a positive amount of wealth in the n-th asset, while $w_n < 0$ means that the agent is selling short the n-th asset for an amount of wealth equal to $|w_n|$ (i.e., he sells the amount $|w_n|$ of the asset without owning it). Since the budget constraint has to be satisfied, $w_0 - \sum_{n=1}^{N} w_n$ represents the amount of initial wealth invested in the risk free asset. As before, the inequality $w_0 - \sum_{n=1}^{N} w_n > 0$ means that the agent is investing a positive amount of wealth in the bank account, while the converse inequality means that the agent is borrowing a positive amount of wealth from the bank account. Starting from initial wealth w_0, a vector $w = (w_1, \ldots, w_N)^\top \in \mathbb{R}^N$ represents a *portfolio*, i.e., an investment strategy in the $N + 1$ assets.

The random returns of the N risky assets are represented by different random variables. Without loss of generality, we assume that the random variables $(r_f, \tilde{r}_1, \ldots, \tilde{r}_N)$ are almost surely linearly independent, meaning that no random return can be expressed as the linear combination of the other $N - 1$ random returns and of the risk free return. We denote by $e = (e_1, \ldots, e_N)^\top \in \mathbb{R}^N$ the vector of the expected returns of the risky assets, i.e., $e_n = \mathbb{E}[\tilde{r}_n]$, for all $n = 1, \ldots, N$, and by $V \in \mathbb{R}^{N \times N}$ the variance-covariance matrix of the N random returns of the risky assets:

$$V = \begin{bmatrix} \sigma^2(\tilde{r}_1) & \mathrm{Cov}(\tilde{r}_1, \tilde{r}_2) \ldots \mathrm{Cov}(\tilde{r}_1, \tilde{r}_N) \\ \mathrm{Cov}(\tilde{r}_2, \tilde{r}_1) & \sigma^2(\tilde{r}_2) & \ldots \mathrm{Cov}(\tilde{r}_2, \tilde{r}_N) \\ \vdots & \ddots & \vdots & \vdots \\ \mathrm{Cov}(\tilde{r}_N, \tilde{r}_1) & \ldots & \ldots & \sigma^2(\tilde{r}_N) \end{bmatrix}.$$

The matrix V is symmetric and, since asset returns are assumed to be linearly independent, it is also positive definite.

Since the returns of the N risky assets are random, the wealth at time $t = 1$ generated by a portfolio $w = (w_1, \ldots, w_N)^\top \in \mathbb{R}^N$ will also be random. We denote by $\widetilde{W}$ the random variable representing the wealth obtained at $t = 1$ from an investment strategy $w \in \mathbb{R}^N$, starting from the initial wealth $w_0 \in \mathbb{R}$. More precisely:

$$\widetilde{W} = \left(w_0 - \sum_{n=1}^{N} w_n\right) r_f + \sum_{n=1}^{N} w_n \tilde{r}_n = w_0 \, r_f + \sum_{n=1}^{N} w_n \left(\tilde{r}_n - r_f\right). \tag{3.1}$$

The second equality in (3.1) shows that the portfolio $w = 0 \in \mathbb{R}^N$ (i.e., the strategy investing the total amount of the initial wealth w_0 in the risk free asset) yields a portfolio return equal to r_f (this represents the equivalent to the status quo in Chap. 2). On the contrary, if the agent decides to invest an amount of wealth equal to $w_n > 0$ on the n-th asset, for $n = 1, \ldots, N$, then his wealth at $t = 1$ will increase or decrease depending on whether the random return $\tilde{r}_n$ of the n-th asset will be greater or smaller than the risk free rate r_f.

The agent's *optimal portfolio problem* consists in solving the following maximization problem, starting from a given initial wealth $w_0 \in \mathbb{R}$:

$$\max_{w \in \mathbb{R}^N} \mathbb{E}\left[u(\widetilde{W})\right], \tag{3.2}$$

with $\widetilde{W}$ being given as in (3.1) and where we tacitly assume that the expected value $\mathbb{E}[u(\widetilde{W})]$ is finite for all $w \in \mathbb{R}^N$. Note that, if the agent is risk averse, then his expected utility function is concave in $\widetilde{W}$. Indeed, risk aversion implies that the utility function $u : \mathbb{R} \to \mathbb{R}$ is concave (see Proposition 2.4) and, hence, given two portfolios $w^1, w^2 \in \mathbb{R}^N$ and two initial wealths $w_0^1, w_0^2 \in \mathbb{R}$, by the linearity of the expectation it holds that

$$\mathbb{E}\left[u\left(\lambda \widetilde{W}^1 + (1-\lambda)\widetilde{W}^2\right)\right] \geq \lambda \mathbb{E}\left[u(\widetilde{W}^1)\right] + (1-\lambda)\mathbb{E}\left[u(\widetilde{W}^2)\right], \quad \forall \lambda \in [0, 1], \tag{3.3}$$

where $\widetilde{W}^i$ denotes the random wealth at $t = 1$ associated to the portfolio w^i and starting from the initial wealth w_0^i, for $i = 1, 2$. Moreover, if u is strictly concave, then the expected utility function is strictly concave in $\widetilde{W}$. Similarly, the function $w_0 \mapsto \mathbb{E}[u(\widetilde{W})]$ is also concave.

By relying on representation (3.1), the optimal portfolio problem can be rewritten as follows, for a given initial wealth $w_0 \in \mathbb{R}$:

$$\max_{w \in \mathbb{R}^N} \mathbb{E}\left[u\left(w_0 \, r_f + \sum_{n=1}^{N} w_n \left(\tilde{r}_n - r_f\right)\right)\right].$$

Assuming an increasing utility function u, a necessary condition for the existence of a solution to problem (3.2) is that the random return $\tilde{r}_n$ is lower than the risk free rate r_f in some states of the world and higher than r_f in some other states, for all $n = 1, \ldots, N$. In other words, if $\mathbb{P}(\tilde{r}_n \neq r_f) > 0$, we must have $\mathbb{P}(\tilde{r}_n < r_f) > 0$ and $\mathbb{P}(\tilde{r}_n > r_f) > 0$, for all $n = 1, \ldots, N$. Indeed, if this was not the case, then in the absence of trading constraints it would be possible to attain arbitrarily large values of expected utility. This condition is related to the absence of arbitrage opportunities in the market. Moreover, it turns out that, under suitable assumptions, the same condition is also sufficient for the existence of a solution to the portfolio problem (3.2) (see Sect. 4.4 for more details). We call $\mathbb{E}[\tilde{r}_n] - r_f$ the *return risk premium* of asset n, for $n = 1, \ldots, N$. Concerning the uniqueness of the optimal solution to problem (3.2), it is easy to show that, if the utility function u is strictly concave, then the solution in terms of wealth is necessarily unique. Indeed, if there were two optimal wealths $\widetilde{W}^{*1}, \widetilde{W}^{*2}$, any strictly convex combination $\lambda \widetilde{W}^{*1} + (1 - \lambda)\widetilde{W}^{*2}$ would yield a strictly higher expected utility, as shown in (3.3), thus contradicting the optimality of $\widetilde{W}^{*1}, \widetilde{W}^{*2}$. Moreover, under the assumption of linearly independent returns, it can be shown that the optimal portfolio $w^* \in \mathbb{R}^N$ is unique (see Exercise 3.3).

If the utility function u is differentiable, then the first order necessary conditions (i.e., the gradient of the expected utility function has to be equal to zero) for an optimal portfolio $w^* \in \mathbb{R}^N$ implies that

$$\mathbb{E}\big[u'(\widetilde{W}^*)(\tilde{r}_n - r_f)\big] = 0, \qquad \text{for all } n = 1, \ldots, N, \tag{3.4}$$

where u' denotes the first derivative of the utility function u and $\widetilde{W}^*$ denotes the optimal wealth resulting from the solution of problem (3.2). If the utility function u is strictly concave and twice differentiable, so that $u'' < 0$, then condition (3.4) is also sufficient for the optimality of the portfolio $w^* \in \mathbb{R}^N$. We denote by $\tilde{r}$ the random return of a generic portfolio w (so that $\widetilde{W} = w_0 \tilde{r}$) and by $\tilde{r}^*$ the optimal portfolio return, so that $\widetilde{W}^* = w_0 \tilde{r}^*$.

The analysis of optimal portfolio problems changes depending on the number of risky assets available in the market. In the following, we shall first consider the simple case of a single risky asset ($N = 1$) and then the more general case of multiple risky assets ($N > 1$). The key difference between the two cases is that, in the presence of multiple risky assets, each individual asset is evaluated not only for its own riskiness but also for its contribution to the risk of a portfolio as a whole. In the latter case, the diversification and the insurance principles introduced in the previous chapter will play a crucial role.

The Case of a Single Risky Asset

In the case of a single risky asset ($N = 1$), the optimal portfolio problem can be analysed in an explicit way, establishing an important relationship between the return risk premium of the risky asset and the optimal portfolio, as shown in the following proposition. We always assume that utility functions are twice differentiable and denote by $\tilde{r}$ the random return of the single risky asset.

Proposition 3.1 *If an agent is strictly risk averse and characterized by a strictly increasing utility function u and initial wealth w_0, then the following hold:*

(i) $w^* = 0$ *if and only if* $\mathbb{E}[\tilde{r}] - r_f = 0$;
(ii) $w^* > 0$ *if and only if* $\mathbb{E}[\tilde{r}] - r_f > 0$;
(iii) $w^* < 0$ *if and only if* $\mathbb{E}[\tilde{r}] - r_f < 0$.

Proof If $w^* = 0$, condition (3.4) directly implies that $\mathbb{E}[\tilde{r}] - r_f = 0$, since u is assumed to be strictly increasing and, hence, $u'(w_0 r_f) > 0$. Conversely, suppose that $\mathbb{E}[\tilde{r}] - r_f = 0$. Recall that strict risk aversion implies strict concavity of the utility function u. Then, by the optimality of $\widetilde{W}^*$, it holds that

$$u(w_0 r_f) = u(w_0 r_f) + u'(w_0 r_f)\left(\mathbb{E}[\tilde{r}] - r_f\right) w^* \geq \mathbb{E}\left[u\left(\widetilde{W}^*\right)\right] \geq u(w_0 r_f),$$

thus showing that $\mathbb{E}[u(\widetilde{W}^*)] = u(w_0 r_f)$. In turn, this implies that $w^* = 0$, since the optimal portfolio w^* is unique, by the strict concavity of u and the assumption of linearly independent returns. We have thus proved claim (*i*). Moreover, since $u'(w_0 r_f) > 0$, the sign of the first derivative of the expected utility function with respect to w in correspondence of $w = 0$ is strictly negative or strictly positive depending on whether $\mathbb{E}[\tilde{r}] - r_f > 0$ or $\mathbb{E}[\tilde{r}] - r_f < 0$, respectively. Since the utility function u is concave, this implies claims (*ii*)-(*iii*). $\qquad\square$

According to Proposition 3.1, a risk averse agent invests a strictly positive amount of wealth in the risky asset if and only if its return risk premium is strictly positive. Observe that the riskiness of the asset itself does not matter for the sign of the optimal portfolio w^*. Furthermore, if the return risk premium of the risky asset is zero, then any investment is less attractive than investing the total amount of wealth in the risk free asset.

We now proceed to derive several comparative statics results. Comparative statics results aim at understanding the impact on the optimal portfolio w^* of changes of the coefficient of risk aversion, of the initial wealth and of the distribution of the return of the risky asset. We start with the following proposition, which compares the optimal portfolios of two risk averse agents characterized by different degrees of risk aversion.

Proposition 3.2 *Let a and b denote two risk averse agents characterized by strictly increasing and strictly concave utility functions u^a and u^b, respectively, with the same initial wealth $w_0 > 0$. Suppose that $\mathbb{E}[\tilde{r}] - r_f > 0$. If agent a is more risk averse than agent b, then the demand of the risky asset by agent a is smaller than that by agent b, i.e., $w_a^* \leq w_b^*$.*

Proof Let w_a^* and w_b^* denote the optimal portfolios for agents a and b, respectively. Since $\mathbb{E}[\tilde{r}] - r_f > 0$, Proposition 3.1 implies that $w_a^* > 0$ and $w_b^* > 0$. By Proposition 2.5, there exists an increasing, concave and twice differentiable function $g(\cdot)$ such that $u^a(x) = g(u^b(x))$. The optimality condition (3.4) for agent a is

$$\mathbb{E}\big[u^{a'}\big(w_0\, r_f + w_a^*(\tilde{r} - r_f)\big)(\tilde{r} - r_f)\big] = 0.$$

Let

$$f(w) := \mathbb{E}\big[u^{a'}\big(w_0\, r_f + w(\tilde{r} - r_f)\big)(\tilde{r} - r_f)\big].$$

By the concavity of u^a, it can be readily verified that $w \mapsto f(w)$ is decreasing. To prove that $w_a^* \leq w_b^*$ it is enough to show that $f(w_b^*) \leq 0$. Observe that

$$f(w_b^*) = \mathbb{E}\Big[g'\big(u^b\big(w_0\, r_f + w_b^*(\tilde{r} - r_f)\big)\big)\, u^{b'}\big(w_0\, r_f + w_b^*(\tilde{r} - r_f)\big)(\tilde{r} - r_f)\Big] \qquad (3.5)$$

and

$$\mathbb{E}\big[u^{b'}\big(w_0\, r_f + w_b^*(\tilde{r} - r_f)\big)(\tilde{r} - r_f)\big] = 0. \qquad (3.6)$$

The difference between $f(w_b^*)$ in (3.5) and the expression in (3.6) is due to the positive and decreasing function $g'(\cdot)$ evaluated at $u^b(w_0\, r_f + w_b^*(\tilde{r} - r_f))$. It is easy to see that

$$g'\Big(u^b\big(w_0\, r_f + w_b^*(\tilde{r} - r_f)\big)\Big)(\tilde{r} - r_f) \leq g'\big(u^b(w_0\, r_f)\big)(\tilde{r} - r_f),$$

which in turn implies that

$$f(w_b^*) \leq g'\big(u^b(w_0\, r_f)\big)\mathbb{E}\big[u^{b'}\big(w_0\, r_f + w_b^*(\tilde{r} - r_f)\big)(\tilde{r} - r_f)\big] = 0,$$

thus proving the claim.   □

The result of Proposition 3.2 is in line with economic intuition. Indeed, it shows that a more risk averse agent will invest a larger amount of wealth in the risk free asset, in order to reduce the riskiness of his overall portfolio. In Proposition 2.5, we have shown that an agent's risk premium increases as his absolute risk aversion coefficient increases. Proposition 3.2 is coherent with this result, since it shows that the optimal investment in the risky asset decreases as the risk aversion increases.

In a setting with a single risky asset, it is also possible to derive comparative statics results on the risky asset demand as the initial wealth, the risk premium or the risk free rate change. We start by considering the effects of a change in the initial wealth on the risky asset demand by a risk averse agent. We denote by $w^*(w_0)$ the risky asset demand (i.e., the optimal portfolio) of an agent endowed at date $t = 0$ with the initial wealth $w_0 \in \mathbb{R}_+$.

Proposition 3.3 *Let u be a strictly increasing and strictly concave utility function. If $w^*(w_0) > 0$ for all $w_0 \in \mathbb{R}_+$ and $r_f > 0$, then:*

(i) if $r_u^a : \mathbb{R}_+ \to \mathbb{R}_+$ is decreasing then $w^* : \mathbb{R}_+ \to \mathbb{R}_+$ is increasing;
(ii) if $r_u^a : \mathbb{R}_+ \to \mathbb{R}_+$ is increasing then $w^* : \mathbb{R}_+ \to \mathbb{R}_+$ is decreasing;
(iii) if $r_u^a : \mathbb{R}_+ \to \mathbb{R}_+$ is constant then $w^* : \mathbb{R}_+ \to \mathbb{R}_+$ is constant.

Proof We only prove the first claim, the proof of (ii) and (iii) being similar. The optimal portfolio $w^*(w_0)$ for an agent endowed with wealth w_0 at $t = 0$ is implicitly defined by the optimality condition (3.4). By the implicit function theorem,

$$\frac{dw^*}{dw_0} = -\frac{\mathbb{E}[u''(\widetilde{W}^*)(\tilde{r} - r_f)]r_f}{\mathbb{E}[u''(\widetilde{W}^*)(\tilde{r} - r_f)^2]}.$$

Due to the strict concavity of u, the denominator of the above expression is negative and, therefore, the sign of dw^*/dw_0 depends on the sign of the numerator. If $w^*(w_0) > 0$ and u is DARA, then it can be easily shown that

$$u''(\widetilde{W}^*)(\tilde{r} - r_f) \geq -r_u^a(w_0\,r_f)u'(\widetilde{W}^*)(\tilde{r} - r_f). \tag{3.7}$$

Indeed, since $w^*(w_0) > 0$, we observe that $\widetilde{W}^* \geq w_0\,r_f$ on the event $\{\tilde{r} \geq r_f\}$ and therefore the fact that u is DARA implies that $u''(\widetilde{W}^*)/u'(\widetilde{W}^*) \geq -r_u^a(w_0\,r_f)$ on $\{\tilde{r} \geq r_f\}$. On the complementary event $\{\tilde{r} < r_f\}$, the converse inequality holds. By multiplying by $\tilde{r} - r_f$, we obtain inequality (3.7). In turn, inequality (3.7) implies that

$$\mathbb{E}\big[u''(\widetilde{W}^*)(\tilde{r} - r_f)\big] \geq -r_u^a(w_0\,r_f)\mathbb{E}\big[u'(\widetilde{W}^*)(\tilde{r} - r_f)\big] = 0,$$

where the last equality follows from the optimality condition (3.4). Therefore, if $r_u^a : \mathbb{R}_+ \to \mathbb{R}_+$ is decreasing (i.e., the utility function u is DARA) and $w^*(w_0) > 0$, then $dw^*/dw_0 \geq 0$. □

Proposition 3.3 shows that, if the utility function is DARA, then the risky asset is a *normal good*, in the sense that the agent's optimal demand is an increasing function of wealth. On the contrary, if the utility function is IARA, then the risky asset is an *inferior good*, meaning that an agent's optimal demand is a decreasing function of wealth. Finally, if the utility function is CARA, then the optimal investment in the risky asset is independent of wealth. This result suggests that a DARA utility function is the most plausible hypothesis for the majority of the agents in an economy (see Arrow [78]).

According to Proposition 3.3, the coefficient of absolute risk aversion gives information about changes in the risky asset demand. However, it does not give information on changes in the proportion of wealth invested in the risky asset. In this regard, the coefficient of relative risk aversion matters and we have the following proposition.

Proposition 3.4 *Let u be a strictly increasing and strictly concave utility function. If $w^*(w_0) > 0$ for all $w_0 \in \mathbb{R}_+$ and $r_f > 0$, then:*

(i) if $r_u^r : \mathbb{R}_+ \to \mathbb{R}_+$ is decreasing then $\frac{dw^*(w_0)}{dw_0} \frac{w_0}{w^*(w_0)} \geq 1$ and, in particular, the function $w_0 \mapsto w^*(w_0)/w_0$ is increasing;

(ii) if $r_u^r : \mathbb{R}_+ \to \mathbb{R}_+$ is increasing then $\frac{dw^*(w_0)}{dw_0} \frac{w_0}{w^*(w_0)} \leq 1$ and, in particular, the function $w_0 \mapsto w^*(w_0)/w_0$ is decreasing;

(iii) if $r_u^r : \mathbb{R}_+ \to \mathbb{R}_+$ is constant then $\frac{dw^*(w_0)}{dw_0} \frac{w_0}{w^*(w_0)} = 1$ and, in particular, the function $w_0 \mapsto w^*(w_0)/w_0$ is constant.

Proof We will only prove part (*i*), the proof of parts (*ii*) and (*iii*) being similar. By the implicit function theorem applied to the optimality condition (3.4), we have that the elasticity of the optimal demand of the risky asset is given by

$$
\frac{dw^*(w_0)}{dw_0} \frac{w_0}{w^*(w_0)} = -\frac{\mathbb{E}\left[u''(\widetilde{W}^*)(\tilde{r} - r_f)\right]r_f}{\mathbb{E}\left[u''(\widetilde{W}^*)(\tilde{r} - r_f)^2\right]} \frac{w_0}{w^*(w_0)}
$$

$$
= 1 - \frac{\mathbb{E}\left[u''(\widetilde{W}^*)\widetilde{W}^*(\tilde{r} - r_f)\right]}{w^*(w_0)\mathbb{E}\left[u''(\widetilde{W}^*)(\tilde{r} - r_f)^2\right]},
$$

where the second equality follows from $\widetilde{W}^* = w_0 r_f + w^*(\tilde{r} - r_f)$, see (3.1). Therefore, since it is assumed that $w^*(w_0) > 0$ for all $w_0 \in \mathbb{R}_+$ and the function u is strictly concave, it holds that

$$
\text{sign}\left(\frac{dw^*(w_0)}{dw_0} \frac{w_0}{w^*(w_0)} - 1\right) = \text{sign}\left(\mathbb{E}\left[u''(\widetilde{W}^*)\widetilde{W}^*(\tilde{r} - r_f)\right]\right).
$$

Then, a reasoning analogous to that used in the proof of Proposition 3.3 yields that $\mathbb{E}[u''(\widetilde{W}^*)\widetilde{W}^*(\tilde{r} - r_f)] \geq 0$ (see Exercise 3.7) if the mapping $x \mapsto r_u^r(x)$ is decreasing. To complete the proof, it suffices to note that

$$
\frac{d(w^*/w_0)}{dw_0} = \frac{w^*(w_0)}{w_0^2}\left(\frac{dw^*(w_0)}{dw_0} \frac{w_0}{w^*(w_0)} - 1\right).
$$

$\square$

In particular, if an agent is characterized by a constant relative risk aversion coefficient (CRRA utility function), then the proportion of wealth invested in the risky asset is independent of the initial wealth.

We now consider comparative statics results concerning the behavior of the risky asset demand with respect to changes in the risk free rate r_f and in the expected return $\mathbb{E}[\tilde{r}]$. In general, when r_f and/or $\mathbb{E}[\tilde{r}]$ change, there is an interplay between a *substitution* and a *income* effect on the optimal demand of the risky asset.[1] Which one of the two effects will prevail depends on further assumptions on the utility function, as shown in the following proposition (see Fishburn & Porter [707]).

Proposition 3.5 *Let u be a strictly increasing and strictly concave utility function. If $w^*(w_0) > 0$ for all $w_0 \in \mathbb{R}_+$, then the following hold:*

(i) *if u is DARA and the distribution of $\tilde{r} - r_f$ remains unchanged as r_f changes, then $\partial w^*/\partial r_f \geq 0$;*

(ii) *if u is IARA and $0 \leq w^*(w_0) \leq w_0$, then $\partial w^*/\partial r_f \leq 0$;*

(iii) *if $r_u^r(x) \leq 1$, for all $x \in \mathbb{R}_+$, and $\tilde{r}$ is non-negative, then $\partial w^*/\partial r_f \leq 0$;*

(iv) *if u is DARA and the distribution of $\tilde{r} - \mathbb{E}[\tilde{r}]$ does not change as $\mathbb{E}[\tilde{r}]$ changes, then $\partial w^*/\partial \mathbb{E}[\tilde{r}] \geq 0$.*

Proof For brevity of notation, we denote $\tilde{z} := \tilde{r} - r_f$.

(*i*): Since the distribution of $\tilde{z}$ is assumed not to change when r_f changes, the implicit function theorem applied to the optimality condition (3.4) gives (omitting the dependence of the optimal demand on the initial wealth w_0)

$$\frac{\partial w^*}{\partial r_f} = -w_0 \frac{\mathbb{E}[u''(w_0 r_f + w^*\tilde{z})\tilde{z}]}{\mathbb{E}[u''(w_0 r_f + w^*\tilde{z})\tilde{z}^2]}.$$

Since u is strictly concave, the numerator of the above expression is non-negative if u is DARA, as shown in the proof of Proposition 3.3.

(*ii*): The implicit function theorem applied to the optimality condition (3.4) allows us to establish that

$$\begin{aligned}
\frac{\partial w^*}{\partial r_f} &= -\frac{\mathbb{E}[u''(w_0 r_f + w^*\tilde{z})\tilde{z}](w_0 - w^*) - \mathbb{E}[u'(w_0 r_f + w^*\tilde{z})]}{\mathbb{E}[u''(w_0 r_f + w^*\tilde{z})\tilde{z}^2]} \\
&= \frac{\mathbb{E}[r_u^a(w_0 r_f + w^*\tilde{z})u'(w_0 r_f + w^*\tilde{z})\tilde{z}](w_0 - w^*)}{\mathbb{E}[u''(w_0 r_f + w^*\tilde{z})\tilde{z}^2]} \\
&\quad + \frac{\mathbb{E}[u'(w_0 r_f + w^*\tilde{z})]}{\mathbb{E}[u''(w_0 r_f + w^*\tilde{z})\tilde{z}^2]}.
\end{aligned} \qquad (3.8)$$

The denominator of the above expression is negative, since u is strictly concave. The last term of (3.8) is therefore also negative, since u is strictly increasing, and is related to the substitution effect on the demand of the risky asset: if the risk free rate increases, then the risk free asset becomes more

---

[1]In general terms, the substitution effect is related to changes in the relative prices of goods, while the income effect is due to changes in purchasing power. Technically, the substitution effect consists in a price change that alters the slope of the budget constraint but leaves the consumer on the same indifference curve, while the income effect represents the change in purchasing power due to price movements (a parallel shift of the budget constraint).

attractive and, hence, the optimal demand of the risky asset decreases. On the contrary, the numerator of the term on the second line of (3.8) is related to the income effect associated with a change in the risk free rate. Similarly as in the proof of Proposition 3.3, if the utility function u is IARA, then it can be shown that

$$r_u^a(\widetilde{W}^*)\tilde{z} \geq r_u^a(w_0\,r_f)\tilde{z},$$

and, hence,

$$\mathbb{E}\big[r_u^a\big(w_0\,r_f + w^*\tilde{z}\big)u'\big(w_0\,r_f + w^*\tilde{z}\big)\tilde{z}\big] \geq r_u^a(w_0\,r_f)\mathbb{E}\big[u'\big(w_0\,r_f + w^*\tilde{z}\big)\tilde{z}\big] = 0,$$

where the last equality follows from the optimality condition (3.4). Since we assume that $w^* \in [0, w_0]$, part (ii) is established.

(iii): Note that the numerator in (3.8) can be rewritten as

$$\mathbb{E}\big[u'(\widetilde{W}^*)\big(1 + r_u^a(\widetilde{W}^*)\tilde{z}(w_0 - w^*))\big)\big]$$
$$= \mathbb{E}\big[u'(\widetilde{W}^*)\big(1 - r_u^a(\widetilde{W}^*)\widetilde{W}^* + r_u^a(\widetilde{W}^*)w_0\,\tilde{r}\big)\big]$$
$$= \mathbb{E}\big[u'(\widetilde{W}^*)(1 - r_u^r(\widetilde{W}^*) + r_u^a(\widetilde{W}^*)w_0\,\tilde{r})\big].$$

If $r^r(\widetilde{W}^*) \leq 1$ and $\tilde{r}$ is non-negative, then the above quantity is also non-negative. Since u is strictly concave and, hence, the denominator in (3.8) is negative, part (iii) then follows.

(iv): Let us denote $\tilde{x} := \tilde{r} - \mathbb{E}[\tilde{r}]$, so that $\tilde{r} - r_f = \tilde{x} + \mathbb{E}[\tilde{r}] - r_f$. Then, the implicit function theorem applied to the optimality condition (3.4) gives

$$\frac{\partial w^*}{\partial \mathbb{E}[\tilde{r}]} = -\frac{\mathbb{E}\big[u''(\widetilde{W}^*)w^*(\tilde{x} + \mathbb{E}[\tilde{r}] - r_f) + u'(\widetilde{W}^*)\big]}{\mathbb{E}\big[u''(\widetilde{W}^*)(\tilde{x} + \mathbb{E}[\tilde{r}] - r_f)^2\big]}.$$

If the utility function u is DARA, then part (iv) follows by analogous arguments as for part (ii), making use of the fact that $w^*(w_0) > 0$ for all $w_0 \in \mathbb{R}_+$, together with the strict concavity of u. □

In view of Proposition 3.1, the assumption $w^*(w_0) > 0$, for all $w_0 \in \mathbb{R}_+$, appearing in the statement of Proposition 3.5 is satisfied if and only if $\mathbb{E}[\tilde{r}] - r_f > 0$. In all the cases considered in Proposition 3.5, the impact of a change in the return risk premium $\mathbb{E}[\tilde{r}] - r_f$ on the optimal demand of the risky asset is determined by the interplay of two effects: the substitution and the income effect. While the substitution effect has a clear sign, the income effect depends on the coefficient of absolute risk aversion. For instance, in part (i) of Proposition 3.5, the fact that u is DARA implies that the demand of the risky asset increases as r_f increases because, under the assumption of part (i), the distribution of $\tilde{r} - r_f$ remains unchanged with no substitution effect. On the contrary, in the case considered in part (ii), both the income and the substitution effects play a role. In particular, if the utility function

u is IARA, then both effects go in the same direction and lead to a decrease of the demand of the risky asset in correspondence of an increase of the risk free rate r_f. If in part (*ii*) of Proposition 3.5 we consider a DARA utility function, then the income effect would go in the opposite direction of the substitution effect and the overall impact on the demand of the risky asset could be positive or negative. The situation considered in part (*iv*) is also similar and shows that, if u is DARA, then both the income and the substitution effects go in the same direction, leading to an increase of the investment in the risky asset as long as the expected return $\mathbb{E}[\tilde{r}]$ increases, while leaving unchanged the distribution of $\tilde{r} - \mathbb{E}[\tilde{r}]$.

The Case of Multiple Risky Assets

In general, the comparative statics results obtained in a single risky asset setting cannot be easily extended to the case of $N > 1$ assets. Indeed, without further assumptions on the distribution of asset returns or on the utility function, general results on optimal portfolio choices with multiple risky assets are less informative. In particular, it is difficult to obtain an explicit relation between the optimal portfolio and the assets' risk premia.

We start by providing a simple necessary and sufficient condition for an arbitrary risk averse agent to invest the total amount of wealth in the risk free asset. We denote by $w^* = (w_1^*, \ldots, w_N^*)^\top \in \mathbb{R}^N$ the optimal portfolio.

Proposition 3.6 *Let u be a strictly increasing and strictly concave utility function. Then $w_n^* = 0$ for all $n = 1, \ldots, N$ if and only if $\mathbb{E}[\tilde{r}_n] = r_f$ for all $n = 1, \ldots, N$.*

Proof The optimality condition (3.4) gives that

$$\mathbb{E}\left[u'\left(w_0\, r_f + \sum_{k=1}^{N} w_k^* (\tilde{r}_k - r_f) \right)(\tilde{r}_n - r_f) \right] = 0, \qquad \text{for all } n = 1, \ldots, N.$$

If $\mathbb{E}[\tilde{r}_n] = r_f$ holds for all $n = 1, \ldots, N$, then the portfolio $w^* = 0 \in \mathbb{R}^N$ satisfies the above optimality condition. As a matter of fact, for $w^* = 0$, it becomes

$$u'(w_0\, r_f)\big(\mathbb{E}[\tilde{r}_n] - r_f\big) = 0, \qquad \text{for all } n = 1, \ldots, N$$

which is always satisfied if $\mathbb{E}[\tilde{r}_n] = r_f$, for all $n = 1, \ldots, N$. By the strict concavity of u, the portfolio w^* is then the unique optimal solution to the portfolio choice problem (compare with Exercise 3.3). Conversely, if $w_n^* = 0$ for all $n = 1, \ldots, N$, then the same optimality condition, together with the strict increasingness of u, shows that we must have $\mathbb{E}[\tilde{r}_n] = r_f$ for all $n = 1, \ldots, N$.   □

According to the above proposition, any risk averse agent will invest the total amount of wealth in the risk free asset if and only if every risky asset has a null return risk premium. However, this result is not really interesting for the sufficiency part, since it is indeed uncommon for every risky asset traded in the market to have a null risk premium. Instead, it is interesting for its converse implication: in reality, it is observed that only a small fraction of market participants invests in stocks, while many investors hold all their wealth in the bank account. According to the above proposition, it is difficult to reconcile this behavior with the hypothesis of perfect rationality.

Let us consider a risk averse agent. It is reasonable to guess that the expected return of his optimal portfolio will be greater or equal than the risk free rate r_f, since the agent can always choose to invest all his wealth in the risk free asset. This intuition is formalized in the following proposition. We recall that $\widetilde{W}^*$ denotes the optimal wealth resulting from the solution of problem (3.2).

Proposition 3.7 *Let u be a strictly increasing and concave utility function and denote by $\tilde{r}^*$ the random return of the optimal portfolio w^*, i.e., $\tilde{r}^* = \widetilde{W}^*/w_0$, for $w_0 > 0$. Then $\mathbb{E}[\tilde{r}^*] \geq r_f$.*

Proof We argue by contradiction. Suppose that $\mathbb{E}[\tilde{r}^*] < r_f$. Then:

$$U(\widetilde{W}^*) = \mathbb{E}[u(w_0\,\tilde{r}^*)] \leq u\big(\mathbb{E}[w_0\,\tilde{r}^*]\big) < u(w_0\,r_f),$$

where the first inequality follows from Jensen's inequality and the second from the strict increasingness of the utility function u together with the fact that $w_0 > 0$. This contradicts the optimality of w^*. $\square$

Note also that, if in Proposition 3.7 we assume that the utility function u is strictly concave, then the same result holds true with a strict inequality, i.e., $\mathbb{E}[\tilde{r}^*] > r_f$.

As shown in Proposition 3.1, in the context of simple financial markets with a single risky asset the demand of the risky asset is positive, negative or null according to whether the return risk premium of the asset if positive, negative or null, respectively. In the presence of multiple risky assets, we no longer have such a general and clear relation between risky assets' demand and risk premia.

In the absence of trading restrictions, if an agent is risk neutral (i.e., with a linear utility function), then he will invest in an unlimited way on the asset with the highest return risk premium, thus excluding the existence of an optimal portfolio. However, if a risk neutral agent is subject to a short sale constraint and cannot borrow from the bank account, then his optimal portfolio is well-defined and will consist of investing the total amount of wealth in the asset with the highest risk premium. The rationale of this result is very simple: a risk neutral agent only cares about the expected wealth and the latter is linear in the assets' expected returns. Therefore, a risk neutral agent will invest only in the asset with the highest risk premium.

On the contrary, if an agent is risk averse, then he aims at maximizing the expected value of an increasing and concave transformation (i.e., the utility function) of the random wealth obtained at $t = 1$. In this case, the agent will not only consider the risk premia of the assets, but also the distribution of the random wealth. Note that only in some specific cases (see Sect. 2.4) the expected utility is increasing in the expected value of the wealth and decreasing in its variance. In a general context, it may happen that a risk averse agent will choose to invest a positive amount of wealth in a risky asset with a negative return risk premium if that asset helps to reduce the riskiness of the overall portfolio. This suggests that, in a general setting, the diversification and insurance principles play a key role in portfolio selection.

Proposition 3.8 *Let u be a strictly increasing and concave utility function. Then, for any $w_0 > 0$, the following hold, for all $n = 1, \ldots, N$:*

(i) $\mathrm{Cov}\left(\tilde{r}_n, u'(\widetilde{W}^*)\right) \leq 0$ *if and only if* $\mathbb{E}[\tilde{r}_n] - r_f \geq 0$;
(ii) $\mathrm{Cov}\left(\tilde{r}_n, u'(\widetilde{W}^*)\right) \geq 0$ *if and only if* $\mathbb{E}[\tilde{r}_n] - r_f \leq 0$.

Proof By the optimality condition (3.4) and recalling that for two random variables $\tilde{x}_1, \tilde{x}_2$ it holds that $\mathrm{Cov}(\tilde{x}_1, \tilde{x}_2) = \mathbb{E}[\tilde{x}_1 \tilde{x}_2] - \mathbb{E}[\tilde{x}_1]\mathbb{E}[\tilde{x}_2]$, we obtain

$$\mathrm{Cov}\left(u'(\widetilde{W}^*), \tilde{r}_n\right) + \mathbb{E}[\tilde{r}_n]\mathbb{E}[u'(\widetilde{W}^*)] = r_f \, \mathbb{E}[u'(\widetilde{W}^*)], \qquad \text{for all } n = 1, \ldots, N.$$

This optimality condition implies the following return risk premium implicit in the optimal portfolio choice:

$$\mathbb{E}[\tilde{r}_n] - r_f = -\frac{\mathrm{Cov}\left(u'(\widetilde{W}^*), \tilde{r}_n\right)}{\mathbb{E}[u'(\widetilde{W}^*)]}, \qquad \text{for all } n = 1, \ldots, N.$$

The denominator on the right-hand side is positive, due to the fact that u is increasing, and the claim follows. □

While it is not possible to derive general relations between asset risk premia and optimal portfolios, Proposition 3.8 establishes a relation between the expected risk premium of an asset and the covariance of the random return of that asset with the marginal utility of the optimal wealth. Note that this result relates risk premia to optimal wealth and not to optimal portfolios. Moreover, the relation is not fully explicit due to the presence of the marginal utility function u'. The main message of Proposition 3.8 is that a risk averse agent will invest in such a way that the covariance of the marginal utility of the optimal wealth and the return of an asset is negative if and only if the risk premium of that asset is positive. In this sense, the result yields an implicit relation between asset returns and optimal wealth.

In some special cases, it is possible to obtain more explicit results on the relation between risk premia and optimal wealth (see Exercises 3.4 and 3.5). In particular, this is the case if the utility function is quadratic or if asset returns are distributed as a multivariate normal random variable. Let us first consider the case of a quadratic utility function $u(x) = x - \frac{b}{2}x^2$, with $b > 0$. It holds that:

$$\mathrm{Cov}\left(\tilde{r}_n, u'(w_0 \tilde{r}^*)\right) = -bw_0 \, \mathrm{Cov}\left(\tilde{r}_n, \tilde{r}^*\right).$$

Assuming that the expected marginal utility of optimal wealth is strictly positive, i.e.
$\mathbb{E}[u'(w_0 \tilde{r}^*)] = 1 - b w_0 \mathbb{E}[\tilde{r}^*] > 0$, and $w_0 > 0$, we have that, for all $n = 1, \ldots, N$,

(i) $\mathbb{E}[\tilde{r}_n] - r_f \geq 0$ if and only if $\text{Cov}(\tilde{r}_n, \tilde{r}^*) \geq 0$;
(ii) $\mathbb{E}[\tilde{r}_n] - r_f \leq 0$ if and only if $\text{Cov}(\tilde{r}_n, \tilde{r}^*) \leq 0$.

In the case where asset returns are distributed as a multivariate normal random variable, we can rely on the following result, known as *Stein's lemma*.

Lemma 3.9 *Let $(\tilde{x}_1, \tilde{x}_2)$ be a bivariate normal random vector and let $g : \mathbb{R} \to \mathbb{R}$ be a differentiable function. Then, as soon as the expectations are well-defined, it holds that*

$$\text{Cov}\big(g(\tilde{x}_1), \tilde{x}_2\big) = \mathbb{E}[g'(\tilde{x}_1)] \, \text{Cov}(\tilde{x}_1, \tilde{x}_2).$$

Letting $g(\cdot)$ be the marginal utility function $u'(\cdot)$, we can then obtain a result analogous to that obtained above in the case of a quadratic utility function.

These results directly relate the risk premium of an asset to the covariance between its return and the optimal wealth. The key feature that allows us to obtain such a relation is represented by the fact that, both in the case of a quadratic utility and of normally distributed returns, the expected utility function is an increasing function of the expected wealth and a decreasing function of its variance (provided that $\mathbb{E}[\widetilde{W}^*] \leq 1/b$ in the case of a quadratic utility function). This observation allows us to explain the role of the insurance and diversification principles in the context of portfolio choice problems. Indeed, an asset whose return is positively correlated with the optimal wealth does not contribute to reduce the variability of the optimal wealth and, therefore, a risk averse agent will require a positive risk premium for investing in such an asset. Conversely, an asset whose return is negatively correlated with the optimal wealth contributes to reduce the variability of the optimal wealth, so that a risk averse agent may accept a negative risk premium because such an asset provides insurance against wealth variability. In other words, an agent aiming to reduce the overall variance will choose a portfolio such that wealth is positively correlated with assets with a positive risk premium and negatively correlated with assets with a negative risk premium. We remark that all these results only concern the optimal wealth and not the optimal portfolio itself. Indeed, there is no direct and general relation between optimal portfolios and asset risk premia.

Concerning the diversification principle, risk aversion implies that an agent will diversify his wealth among several sources of risk, instead of investing all his wealth in a single risky asset. However, perfect diversification is a rare situation. In a setting without a risk free asset, the optimal portfolio is perfectly diversified (i.e., the agent will invest the same amount of wealth in each asset) if the risky asset returns are independent and identically distributed (i.i.d.), as shown in the next result (see Samuelson [1492]).

Proposition 3.10 *Suppose that the random returns $(\tilde{r}_1, \ldots, \tilde{r}_N)$ of N risky assets are independent and identically distributed and suppose that there exists no risk free asset. Then, the optimal portfolio of every risk averse agent is given by $w_n^* = w_0/N$, for all $n = 1, \ldots, N$.*

Proof The proof is based on a stochastic dominance argument. Denote by

$$\widetilde{W}^* = w_0 \frac{1}{N} \sum_{n=1}^{N} \tilde{r}_n$$

the wealth associated with the fully diversified portfolio, starting from a given initial wealth w_0. We show that the fully diversified portfolio dominates any other portfolio according to the second order stochastic dominance criterion. Indeed, let $w \in \mathbb{R}^N$ be any other portfolio investing in the N assets, so that $\sum_{n=1}^{N} w_n = w_0$ and denote by $\widetilde{W}'$ the random wealth associated to such a portfolio. The random variable $\widetilde{W}'$ can be equivalently written as follows:

$$\widetilde{W}' = \widetilde{W}^* + \sum_{n=1}^{N} \left(w_n - \frac{w_0}{N} \right) \tilde{r}_n .$$

In view of Proposition 2.10, in order to show that $\widetilde{W}^* \succeq_{\mathrm{SSD}} \widetilde{W}'$, it suffices to prove that $\mathbb{E}\big[\sum_{n=1}^{N} \left(w_n - \frac{w_0}{N} \right) \tilde{r}_n | \widetilde{W}^* \big] = 0$. This can be easily shown as follows:

$$\mathbb{E}\left[\sum_{n=1}^{N} \left(w_n - \frac{w_0}{N} \right) \tilde{r}_n \bigg| \widetilde{W}^* \right] = \sum_{n=1}^{N} \left(w_n - \frac{w_0}{N} \right) \mathbb{E}[\tilde{r}_n | \widetilde{W}^*]$$

$$= \mathbb{E}[\tilde{r}_1 | \widetilde{W}^*] \sum_{n=1}^{N} \left(w_n - \frac{w_0}{N} \right) = 0,$$

where we have used the fact that the random returns $(\tilde{r}_1, \ldots, \tilde{r}_N)$ are i.i.d. and, hence, $\mathbb{E}[\tilde{r}_n | \widetilde{W}^*] = \mathbb{E}[\tilde{r}_1 | \widetilde{W}^*]$, for every $n = 1, \ldots, N$. □

Closed Form Solutions and Mutual Fund Separation

As we have mentioned in the previous section, in the context of a financial market with multiple risky assets, it is difficult to establish general and explicit results on optimal portfolio choices. In what follows, we aim at investigating some special cases where, under additional assumptions on the utility function or on the distribution of the asset returns, explicit results can be obtained. In particular, we first focus on the two classical cases of quadratic and exponential utility functions with multivariate normal returns. Then, in a more general setting, we investigate the class of utility functions for which only the overall amount invested in the risky assets, and not the composition of the optimal portfolio, changes as the initial wealth changes (*mutual fund separation*).

Quadratic Utility Function

Let us consider the case of a quadratic utility function $u(x) = x - \frac{b}{2}x^2$, for some parameter $b > 0$. In this case, the optimality condition (3.4) becomes

$$\mathbb{E}\big[(1 - b\widetilde{W}^*)(\tilde{r}_n - r_f)\big] = 0, \qquad \text{for all } n = 1, \ldots, N.$$

Using the definition of covariance, the above optimality condition implies that, for all $n = 1, \ldots, N$,

$$\big(1 - b\, w_0\, \mathbb{E}[\tilde{r}^*]\big)\mathbb{E}[\tilde{r}_n - r_f] = b\, \mathrm{Cov}(\widetilde{W}^*, \tilde{r}_n) = b\sum_{k=1}^{N} w_k^*\, \mathrm{Cov}(\tilde{r}_k, \tilde{r}_n).$$

Equivalently, in vector notation, the optimal portfolio $w^* \in \mathbb{R}^N$ satisfies

$$w^* = \frac{1 - b\, w_0\, \mathbb{E}[\tilde{r}^*]}{b} V^{-1}(e - r_f \mathbf{1}), \tag{3.9}$$

where we have used the assumption that the risky assets are non-redundant (i.e., the variance-covariance matrix V is positive definite) and $\mathbf{1}$ denotes the column vector $(1, \ldots, 1)^\top \in \mathbb{R}^N$. Note that, up to a multiplicative factor, the optimal portfolio w^* is given by the inverse of the variance-covariance matrix multiplied by the vector of the risky assets' return risk premia. If $w_0\, \mathbb{E}[\tilde{r}^*] < 1/b$, then the scale factor $(1 - b\, w_0\, \mathbb{E}[\tilde{r}^*])/b$ is strictly positive and corresponds to the reciprocal of the coefficient of absolute risk aversion evaluated at the expected optimal wealth $(\widetilde{W}^* = w_0\, \tilde{r}^*)$. Note that (3.9) characterizes the optimal portfolio only in a semi-explicit form.

In the simpler case of a single risky asset ($N = 1$), with expected return $\mathbb{E}[\tilde{r}]$ and variance $\sigma^2(\tilde{r})$, the optimal portfolio can be written in a fully explicit form:

$$w^* = \frac{(1 - b\, w_0\, r_f)\big(\mathbb{E}[\tilde{r}] - r_f\big)}{b\big(\sigma^2(\tilde{r}) + \big(\mathbb{E}[\tilde{r}] - r_f\big)^2\big)}. \tag{3.10}$$

Under the assumption that $w_0\, r_f < 1/b$, the optimal investment in the risky asset $|w^*|$ is decreasing with respect to the risk-aversion parameter b, to the initial wealth w_0 and to the variance $\sigma^2(\tilde{r})$.

Exponential Utility Function and Normal Returns

Let us now suppose that the risky asset returns $\tilde{r}$ are distributed as a multivariate random variable with mean vector e and variance-covariance matrix V and, as in Sect. 2.4, let us consider an exponential utility function $u(x) = -\frac{1}{a}\exp(-ax)$, for a given risk aversion parameter $a > 0$. Recall that, for any vector $z \in \mathbb{R}^N$, the random variable $\exp(z^\top \tilde{r})$ is log-normally distributed with mean $\exp(z^\top e + z^\top Vz/2)$. The random wealth $\widetilde{W}$ obtained at $t = 1$ by investing in a portfolio $w \in \mathbb{R}^N$ is given by $\widetilde{W} = w_0 \, r_f + w^\top (\tilde{r} - r_f \mathbf{1})$ and, since the normal distribution is stable with respect to linear transformations, $\widetilde{W}$ is also normally distributed, with mean and variance given by

$$\mathbb{E}[\widetilde{W}] = w_0 \, r_f + w^\top (e - r_f \mathbf{1}) \qquad \text{and} \qquad \sigma^2(\widetilde{W}) = w^\top Vw.$$

Due to the above observations, maximizing the expected exponential utility is equivalent to the following optimization problem:

$$\max_{w \in \mathbb{R}^N} \left(w^\top (e - r_f \mathbf{1}) - \frac{a}{2} w^\top Vw \right).$$

Since the matrix V is assumed to be positive definite, the optimal portfolio w^* is then explicitly determined as

$$w^* = \frac{1}{a} V^{-1} (e - r_f \mathbf{1}). \tag{3.11}$$

Note that, since the exponential utility function is CARA (see section "Classes of Utility Functions"), the optimal investment in the risky assets does not depend on the initial wealth w_0 (compare with Proposition 3.3). As can be seen from (3.11), the investment in the risky assets is decreasing in the coefficient of absolute risk aversion a.

The appearance of the matrix V^{-1} makes the relation between the optimal portfolio w^* and the assets' risk premia $e - r_f$ rather complex, due to the presence of correlation effects. Indeed, if the random returns of the risky assets are uncorrelated (and, hence, due to the normality assumption, independent), then the matrix V reduces to a diagonal matrix whose elements are given by the variances of the individual returns $(\sigma^2(\tilde{r}_1), \ldots, \sigma^2(\tilde{r}_N))$. Hence, the inverse matrix V^{-1} has elements $V_{n,n}^{-1} = 1/\sigma^2(\tilde{r}_n)$, for $n = 1, \ldots, N$. In this special case, we recover the result obtained in the case of a single risky asset in Proposition 3.1: $w_n^* > 0$ if and only if $\mathbb{E}[\tilde{r}_n] > r_f$, for every $n = 1, \ldots, N$.

In the simpler case of a single risky asset with return $\tilde{r}$, it holds that

$$w^* = \frac{1}{a\sigma^2(\tilde{r})} (\mathbb{E}[\tilde{r}] - r_f). \tag{3.12}$$

We see that the amount of wealth invested in the risky asset is increasing in the risk premium $\mathbb{E}[\tilde{r}] - r_f$, decreasing in the coefficient of absolute risk aversion a (compare with Proposition 3.3) and its absolute value is decreasing in the variance $\sigma^2(\tilde{r})$. Moreover, the optimal portfolio does not depend on the initial wealth w_0.

In the present context, the diversification and insurance principles drive the optimal portfolio decisions. To illustrate the effects of these two principles in a simple setting, let us consider the case of two risky assets with returns $\tilde{r}_1, \tilde{r}_2$ jointly distributed as normal random variables, with expected returns e_1, e_2, variances σ_1^2, σ_2^2 and correlation coefficient $\rho \in (0, 1)$. Without loss of generality, we suppose that $\sigma_1^2 \geq \sigma_2^2$. The variance-covariance matrix V and its inverse V^{-1} then become

$$V = \begin{bmatrix} \sigma_1^2 & \rho\sigma_1\sigma_2 \\ \rho\sigma_1\sigma_2 & \sigma_2^2 \end{bmatrix} \quad \text{and} \quad V^{-1} = \begin{bmatrix} \frac{1}{\sigma_1^2(1-\rho^2)} & -\frac{\rho}{\sigma_1\sigma_2(1-\rho^2)} \\ -\frac{\rho}{\sigma_1\sigma_2(1-\rho^2)} & \frac{1}{\sigma_2^2(1-\rho^2)} \end{bmatrix}.$$

By formula (3.11), up to a scale factor of $1/a$, the optimal portfolio is given by

$$w_1^* = \frac{1}{\sigma_1^2(1-\rho^2)}(e_1 - r_f) - \frac{\rho}{\sigma_1\sigma_2(1-\rho^2)}(e_2 - r_f),$$

$$w_2^* = -\frac{\rho}{\sigma_1\sigma_2(1-\rho^2)}(e_1 - r_f) + \frac{1}{\sigma_2^2(1-\rho^2)}(e_2 - r_f).$$

From the above expressions, it is evident that the correlation coefficient ρ, in particular its sign, plays a crucial role in the determination of the optimal portfolio, see Fig. 3.1. Let us first consider the case when $\rho > 0$. Then, it is easy to see that:

$$w_1^* > 0 \iff e_1 - r_f > \frac{\rho\sigma_1}{\sigma_2}(e_2 - r_f),$$

$$w_2^* > 0 \iff e_1 - r_f < \frac{\sigma_1}{\rho\sigma_2}(e_2 - r_f).$$

As a consequence, the optimal portfolio is diversified (i.e., $w_1^*, w_2^* > 0$) if and only if

$$\frac{\rho\sigma_1}{\sigma_2}(e_2 - r_f) < e_1 - r_f < \frac{\sigma_1}{\rho\sigma_2}(e_2 - r_f). \tag{3.13}$$

If at least one of the two assets has a negative risk premium, then at least one asset will be sold short and the optimal portfolio will not be diversified. In other words, having positive risk premia represents a necessary condition in order to have a diversified optimal portfolio. The optimal portfolio is diversified if both assets are characterized by a positive risk premium and if the difference between the two risk premia is not too large, as follows from (3.13).

Let us now consider the case of negative correlation, i.e., $\rho < 0$. In this case, it holds that

$$w_1^* > 0 \iff e_1 - r_f > \frac{\rho\sigma_1}{\sigma_2}(e_2 - r_f),$$

$$w_2^* > 0 \iff e_1 - r_f > \frac{\sigma_1}{\rho\sigma_2}(e_2 - r_f).$$

In this case, the fact that both assets have a positive risk premium is a sufficient condition to have a diversified optimal portfolio. However, the optimal portfolio

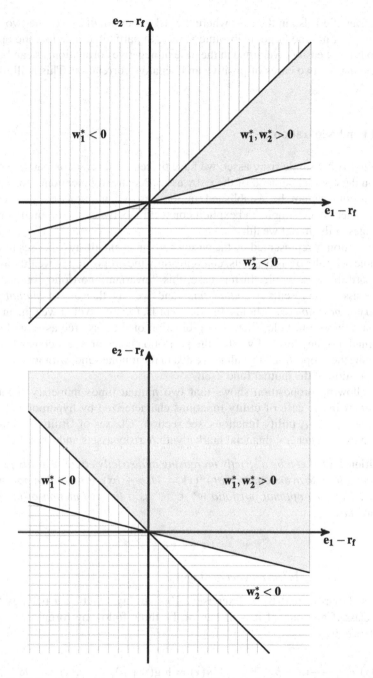

Fig. 3.1 Portfolios in the risk premium space (positive correlation, top; negative correlation, bottom)

may be diversified also in the case where the risk premium of one of the two assets is negative (but not too large in absolute value). In particular, note that the optimal portfolio is more easily diversified in the case of negative correlation, as can be seen by comparing the two cases of positive and negative correlation. This is illustrated in Fig. 3.1.

Mutual Fund Separation

In a setting with a single risky asset, we have presented several comparative statics results on the optimal demand of the risky asset. As already mentioned, analogous general results cannot be established in the case of multiple risky assets. In particular, there is no simple and explicit connection between the optimal portfolio and changes in the initial wealth.

An exception is represented by the case where an agent always chooses to invest in the same portfolio of risky assets, changing only the proportion of wealth invested in that portfolio as a whole. In this case, this "invariant" portfolio composed by the risky assets represents a *mutual fund* and we say that a *two mutual funds (monetary) separation* result holds. In other words, for any level of wealth, an agent will always choose to hold a linear combination of the risk free asset and of the mutual fund. For any level of wealth, the portfolio choice problem consists only in determining the proportion of wealth invested in the mutual fund, without modifying the composition of the mutual fund itself.

The following proposition shows that two mutual funds monetary separation always holds in the case of utility functions characterized by hyperbolic absolute risk aversion (HARA utility functions, see section "Classes of Utility Functions"), in the context of a general financial market with N risky assets and a risk free asset.

Proposition 3.11 *Let u be a strictly increasing and strictly concave utility function with hyperbolic risk aversion, so that $r_u^a(x) = 1/(a + bx)$, for suitable parameters $a, b \in \mathbb{R}$. Then the optimal portfolio $w^* \in \mathbb{R}^N$ is a linear function of the initial wealth w_0, i.e.,*

$$w^* = (a + b w_0 r_f)\gamma,$$

where the vector $\gamma \in \mathbb{R}^N$ is independent of w_0.

Proof As shown in part b) of Exercise 2.10, a utility function u belongs to the HARA class if and only if it is of one of the three following forms, for suitable parameters a, b:

$$u(x) = \frac{1}{b-1}(a + bx)^{\frac{b-1}{b}}, \qquad u(x) = \log(x + a), \qquad u(x) = -a e^{-\frac{x}{a}}.$$

This implies that, if u belongs to the HARA class, then the marginal utility u' takes one of the two following forms (with $b = 1$ in the case of logarithmic utility):

$$u'(x) = (a + bx)^{-\frac{1}{b}}, \qquad u'(x) = e^{-\frac{x}{a}}.$$

For wealth w_0, let us denote by $w^*(w_0)$ the associated optimal portfolio. In the remaining part of the proof, following LeRoy & Werner [1191, Theorem 13.6.1], we fix an arbitrary wealth $\bar{w}_0$ and aim at showing that the optimal portfolio $w^*(w_0)$ associated to any other value of wealth w_0 is given by

$$w^*(w_0) = \frac{a + b\,w_0\,r_f}{a + b\,\bar{w}_0\,r_f}\, w^*(\bar{w}_0).$$

The result will then follow by defining $\gamma := w^*(\bar{w}_0)/(a + b\,\bar{w}_0\,r_f)$.

Let us first consider the case where $u'(x) = (a + bx)^{-1/b}$, for some parameter $b > 0$. In this case, for wealth $\bar{w}_0$, the optimality condition (3.4) takes the form, for all $n = 1, \dots, N$,

$$\mathbb{E}\left[\left(a + b\,\bar{w}_0\,r_f + b\sum_{n=1}^{N} w_n^*(\bar{w}_0)(\tilde{r}_n - r_f)\right)^{-\frac{1}{b}}(\tilde{r}_n - r_f)\right] = 0.$$

Dividing the above equation by $(a + b\,\bar{w}_0\,r_f)^{-1/b}$ and multiplying by $(a + b\,w_0\,r_f)^{-1/b}$, we get, for all $n = 1, \dots, N$,

$$\mathbb{E}\left[\left(a + b\,w_0\,r_f + b\sum_{n=1}^{N}\frac{a + b\,w_0\,r_f}{a + b\,\bar{w}_0\,r_f}w_n^*(\bar{w}_0)(\tilde{r}_n - r_f)\right)^{-\frac{1}{b}}(\tilde{r}_n - r_f)\right] = 0,$$

thus showing that the portfolio $w^*(w_0) = \frac{a + b\,w_0\,r_f}{a + b\,\bar{w}_0\,r_f}w^*(\bar{w}_0)$ satisfies the optimality condition for an arbitrary wealth w_0, thus proving the claim.

In the case where $u'(x) = \exp(-x/a)$, the reasoning is analogous. Indeed, the optimality condition (3.4) gives, for the fixed level $\bar{w}_0$ of wealth,

$$\mathbb{E}\left[\exp\left(-\frac{1}{a}\bar{w}_0\,r_f - \frac{1}{a}\sum_{n=1}^{N}w_n^*(\bar{w}_0)(\tilde{r}_n - r_f)\right)(\tilde{r}_n - r_f)\right] = 0, \quad \text{for all } n = 1, \dots, N.$$

Then, multiplying the above expression by $\exp\left(-(w_0 - \bar{w}_0)r_f/a\right)$ gives

$$\mathbb{E}\left[\exp\left(-\frac{1}{a}w_0\,r_f - \frac{1}{a}\sum_{n=1}^{N}w_n^*(\bar{w}_0)(\tilde{r}_n - r_f)\right)(\tilde{r}_n - r_f)\right] = 0, \quad \text{for all } n = 1, \dots, N.$$

This shows that the portfolio $w^*(w_0) = w^*(\bar{w}_0)$ solves the optimality condition for an arbitrary wealth w_0, thus proving the claim.   □

The above result implies that ratios of the optimal demands of the risky assets do not depend on the level of initial wealth. Indeed, for any w_0, it holds that

$$\frac{w_n^*(w_0)}{w_m^*(w_0)} = \frac{\gamma_n}{\gamma_m}, \qquad \text{for every } n, m \in \{1, \ldots, N\}.$$

In other words, for different levels of initial wealth, the optimal portfolios differ only by the amount of wealth invested in the risky assets as a whole and not by the composition of the portfolio of risky assets. Indeed, the optimal portfolio corresponding to an initial wealth w_0 is given by

$$w^*(w_0) = \big((a + b\, w_0\, r_f)\gamma_1, \ldots, (a + b\, w_0\, r_f)\gamma_N\big)^\top \in \mathbb{R}^N,$$

while the amount of wealth invested in the risk free asset is given by

$$w_0 - (a + b\, w_0\, r_f) \sum_{n=1}^{N} \gamma_n.$$

As a further consequence, for the HARA class of utility functions, the comparative statics results obtained in the case of a single risky asset can be extended to the case of N risky assets by considering as a single risky asset the *mutual fund*, whose composition does not change as wealth changes. In particular, if $a = 0$ in the representation $r_u^a(x) = 1/(a + bx)$, then the relative risk aversion coefficient is constant (CRRA utility), thus implying that the proportion of wealth invested in the mutual fund of risky assets does not change with wealth, due to Proposition 3.4. On the other hand, the exponential utility function is characterized by an increasing coefficient of relative risk aversion, so that the proportion of wealth invested in the mutual fund will decrease as wealth increases.

3.2 Mean-Variance Portfolio Selection

In this section, we present one of the most popular and historically important portfolio selection approach, consisting in the determination of the portfolio which achieves the minimum variance among all portfolios with a given expected return. This problem has been originally introduced by Markovitz [1303] and represents one of the cornerstones of modern portfolio theory. Furthermore, it is also at the basis of the CAPM and is related to classical asset pricing models, as will be discussed in Chap. 5.

Some of the main reasons for the popularity of the mean-variance approach are represented by its analytical tractability and by the possibility of deriving clear empirical implications. However, as we have already discussed in Sect. 2.4, mean-variance preferences are consistent with expected utility maximization only in special cases, notably in the case of a quadratic utility function (which however suffers from some drawbacks) and in the case of normally distributed returns.

Mean-variance portfolio selection is also related to second order stochastic dominance. Indeed, in view of Proposition 2.10, if two portfolios w_1 and w_2 have the same expected return $\mathbb{E}[\tilde{r}]$ and portfolio w_1 dominates portfolio w_2 according to the second order stochastic dominance criterion, then portfolio w_1 must have a lower variance than w_2. In this sense, if there exists a portfolio w^* with expected return $\mathbb{E}[\tilde{r}]$ that dominates according to the second order stochastic dominance criterion any other portfolio with the same expected return, then the portfolio w^* also achieves the minimum variance among all portfolios with expected return $\mathbb{E}[\tilde{r}]$.

In this section, we shall first consider the case of an economy with $N \in \mathbb{N}$ risky assets with random returns $(\tilde{r}_1, \ldots, \tilde{r}_N)$ and then the case of an economy with N risky assets together with a risk free asset with deterministic return r_f.

The Case of N Risky Assets

Let us consider an economy with $N > 1$ risky assets with returns $(\tilde{r}_1, \ldots, \tilde{r}_N)$. As in the previous sections, we denote by $e = (e_1, \ldots, e_N)^\top \in \mathbb{R}^N$ the vector of expected returns and by $V \in \mathbb{R}^{N \times N}$ the variance-covariance matrix of the random vector $\tilde{r}$, assumed to be positive definite. In the present context, a *portfolio* is represented by a vector $w \in \mathbb{R}^N$ such that $\sum_{n=1}^N w_n = 1$, where w_n represents the *proportion of wealth* invested in the n-th asset (equivalently, the initial wealth is normalized to 1), for $n = 1, \ldots, N$. We denote by Δ_N the set of all portfolios. For any portfolio $w \in \Delta_N$, we denote by $\tilde{r}_w$ its random return, given by $\tilde{r}_w = \sum_{n=1}^N w_n \tilde{r}_n$, with expected value and variance given by

$$\mathbb{E}[\tilde{r}_w] = w^\top e = \sum_{n=1}^N w_n e_n \quad \text{and} \quad \sigma^2(\tilde{r}_w) = w^\top V w = \sum_{n=1}^N \sum_{m=1}^N w_n w_m \, \mathrm{Cov}(\tilde{r}_n, \tilde{r}_m).$$

Note also that the condition $w \in \Delta_N$ can be simply expressed as $w^\top \mathbf{1} = 1$, where $\mathbf{1} := (1, \ldots, 1)^\top \in \mathbb{R}^N$. We base our presentation on the derivation of the portfolio frontier proposed in Merton [1329] (compare also with Huang & Litzenberger [971, Chapter 3]).

We are interested in determining the portfolio $w^* \in \Delta_N$ which achieves the minimum variance among all possible portfolios with a given expected return $\mu \in \mathbb{R}$. The optimal portfolio w^* is determined as the solution to the following problem:

$$\min_{w \in \mathbb{R}^N} w^\top V w, \qquad (3.14)$$

under the constraints

$$w^\top e = \mu \quad \text{and} \quad w^\top \mathbf{1} = 1. \qquad (3.15)$$

Note that the two constraints simply amount to require that the vector w^* is a portfolio (i.e., $w^* \in \Delta_N$) and yields the required expected return μ. The solution to the above problem is given in the following theorem.

Theorem 3.12 *Let $(\tilde{r}_1, \dots, \tilde{r}_N)$ be random returns with expected value $e \in \mathbb{R}^N$ and positive definite variance-covariance matrix $V \in \mathbb{R}^{N \times N}$. Then, for any $\mu \in \mathbb{R}$, the solution to the constrained variance minimization problem (3.14) is given by*

$$w^* = g + h\mu, \qquad (3.16)$$

where

$$g = \frac{BV^{-1}\mathbf{1} - AV^{-1}e}{D}, \qquad h = \frac{CV^{-1}e - AV^{-1}\mathbf{1}}{D},$$

$$A = \mathbf{1}^\top V^{-1}e, \qquad B = e^\top V^{-1}e, \qquad C = \mathbf{1}^\top V^{-1}\mathbf{1}, \qquad D = BC - A^2.$$

Proof Clearly, problem (3.14) is equivalent to the minimization of $w^\top V w / 2$ over all $w \in \mathbb{R}^N$ subject to the constraints (3.15). To this minimization problem, we can associate the Lagrangian

$$L(w, \lambda, \gamma) := \frac{1}{2}w^\top V w + \lambda(\mu - w^\top e) + \gamma(1 - w^\top \mathbf{1}).$$

The necessary conditions for the optimality of w^* are given by

$$\frac{\partial L(w^*, \lambda, \gamma)}{\partial w} = V w^* - \lambda e - \gamma \mathbf{1} = 0, \qquad (3.17)$$

$$\frac{\partial L(w^*, \lambda, \gamma)}{\partial \lambda} = \mu - w^{*\top}e = 0, \qquad (3.18)$$

$$\frac{\partial L(w^*, \lambda, \gamma)}{\partial \gamma} = 1 - w^{*\top}\mathbf{1} = 0. \qquad (3.19)$$

In (3.17), $\partial L(w^*, \lambda, \gamma)/\partial w$ denotes the gradient of the function L evaluated at (w^*, λ, γ). Note that, since the matrix V is assumed to be positive definite, then the first order conditions (3.17)–(3.19) are also sufficient for the solution to the variance minimization problem (3.14). We now determine explicitly the optimal portfolio $w^* \in \Delta_N$. First, by (3.17),

$$w^* = \lambda V^{-1} e + \gamma V^{-1} \mathbf{1}. \tag{3.20}$$

Replacing this expression in (3.18) and in (3.19), we obtain

$$\lambda = \frac{C\mu - A}{D}, \tag{3.21}$$

$$\gamma = \frac{B - A\mu}{D}, \tag{3.22}$$

$$w^* = g + h\mu, \tag{3.23}$$

where A, B, C, D and g, h are defined as in the statement of the theorem. Observe that, since the inverse of a positive definite matrix is also positive definite, it holds that $B > 0$ and $C > 0$. Moreover, $D > 0$, since $BD = (Ae - B\mathbf{1})^\top V^{-1} (Ae - B\mathbf{1}) > 0$ and $B > 0$. $\qquad\qquad\qquad\qquad\qquad\qquad\qquad\qquad\qquad\qquad\qquad\qquad\qquad\qquad\qquad\Box$

Theorem 3.12 explicitly characterizes the optimal portfolio w^* which minimizes the variance among all portfolios with a given expected return μ. Note that the optimal portfolio is unique and is explicitly given by (3.16), where the terms A, B, C, D, g, h are derived by the data of the problem. Theorem 3.12 solves the problem of an agent who is interested in minimizing the risk (as measured by the variance) of his portfolio while ensuring a given level μ of expected return. We call *portfolio frontier* (denoted by PF) the set of all portfolios which solve problem (3.14) as the required expected return μ varies. Since the optimality conditions employed in the proof of Theorem 3.12 are both necessary and sufficient, it follows that a portfolio belongs to the portfolio frontier if and only if it admits the representation (3.16) for some expected return $\mu \in \mathbb{R}$. Moreover, Theorem 3.12 also provides the basis for the solution of several related mean-variance optimization problems. For instance, as shown in Exercise 3.19, it can be used for determining the portfolio maximizing the Sharpe ratio with respect to a given reference rate of return.

Note that, as a consequence of Proposition 2.10, portfolios belonging to the portfolio frontier cannot be dominated by other portfolios in the sense of second order stochastic dominance. However, it is not generally true that a portfolio belonging to the portfolio frontier dominates all other portfolios with the same expected return according to the second order stochastic dominance criterion (see Sect. 2.3).

In the following, we present several fundamental properties of the portfolio frontier.

Property 1: Linear Combinations of Frontier Portfolios

Given two arbitrary portfolios w^{*1}, w^{*2} belonging to the portfolio frontier, with $\mathbb{E}[\tilde{r}_{w*1}] \neq \mathbb{E}[\tilde{r}_{w*2}]$, any other portfolio w^* belonging to the portfolio frontier can be represented as a linear affine combination of w^{*1} and w^{*2}.

Indeed, let us consider two frontier portfolios w^{*1}, w^{*2} with expected returns equal to $\mu_1 = 0$ and $\mu_2 = 1$, respectively. Then, due to Theorem 3.12, it holds that $w^{*1} = g$ and $w^{*2} = g + h$. Then, by Theorem 3.12, given any other frontier portfolio w^* with expected return $\mathbb{E}[\tilde{r}_{w*}]$, we can write

$$w^* = g + h\,\mathbb{E}[\tilde{r}_{w*}] = \big(1 - \mathbb{E}[\tilde{r}_{w*}]\big)g + \mathbb{E}[\tilde{r}_{w*}](g+h) = \big(1 - \mathbb{E}[\tilde{r}_{w*}]\big)w^{*1} + \mathbb{E}[\tilde{r}_{w*}]w^{*2},$$

thus proving that w^* is an linear affine combination of w^{*1} and w^{*2}. The same reasoning can be extended to two arbitrary portfolios w^{*1}, w^{*2} belonging to the portfolio frontier with $\mathbb{E}[\tilde{r}_{w*1}] \neq \mathbb{E}[\tilde{r}_{w*2}]$.

Moreover, for any $M \in \mathbb{N}$, if the portfolios $w^{*1}, \dots, w^{*M}$ all belong to the portfolio frontier, then any linear combination of $w^{*1}, \dots, w^{*M}$ with weights summing up to one will also belong to the portfolio frontier, with expected return given by the linear combination of the expected returns associated to the portfolios $w^{*1}, \dots, w^{*M}$. Indeed, if $w^* = \sum_{i=1}^{M} \alpha_i w^{*i}$ with $\sum_{i=1}^{M} \alpha_i = 1$, then it holds that

$$w^* = \sum_{i=1}^{M} \alpha_i w^{*i} = \sum_{i=1}^{M} \alpha_i \big(g + h\,\mathbb{E}[\tilde{r}_{w*i}]\big) = g + h \sum_{i=1}^{M} \alpha_i \,\mathbb{E}[\tilde{r}_{w*i}].$$

Hence, by Theorem 3.12, the portfolio $w^* = \sum_{i=1}^{M} \alpha_i w^{*i}$ solves problem (3.14) with respect to the expected return $\mu = \sum_{i=1}^{M} \alpha_i \,\mathbb{E}[\tilde{r}_{w*i}]$.

Property 2: Covariance and Variance of Frontier Portfolios

For any two portfolios w^{*1} and w^{*2} belonging to the portfolio frontier, with expected returns $\mathbb{E}[\tilde{r}_{w*1}]$ and $\mathbb{E}[\tilde{r}_{w*2}]$, respectively, it holds that

$$\mathrm{Cov}\big(\tilde{r}_{w*1}, \tilde{r}_{w*2}\big) = (w^{*1})^\top V w^{*2} = \frac{C}{D}\left(\mathbb{E}[\tilde{r}_{w*1}] - \frac{A}{C}\right)\left(\mathbb{E}[\tilde{r}_{w*2}] - \frac{A}{C}\right) + \frac{1}{C} \tag{3.24}$$

and, in particular, for any portfolio w^* belonging to the portfolio frontier,

$$\sigma^2(\tilde{r}_{w*}) = \frac{C}{D}\left(\mathbb{E}[\tilde{r}_{w*}] - \frac{A}{C}\right)^2 + \frac{1}{C} = \frac{1}{D}\big(C\,\mathbb{E}[\tilde{r}_{w*}]^2 - 2A\,\mathbb{E}[\tilde{r}_{w*}] + B\big), \tag{3.25}$$

where A, B, C, D are defined as in Theorem 3.12.

The expressions (3.24)–(3.25) can be easily obtained by means of elementary computations, see Exercise 3.10. In particular, by (3.25), we can obtain an explicit characterization of the portfolio $w^{\mathrm{MVP}} \in \Delta_N$ which achieves the minimum variance among all portfolios, without any constraint on its expected return. Indeed, expression (3.25) is minimized by $\mathbb{E}[\tilde{r}_{w^*}] = A/C$. Hence, letting $\mu = A/C$ (recall that $C > 0$) in (3.16), the portfolio

$$w^{\mathrm{MVP}} = g + h\frac{A}{C} = \frac{V^{-1}\mathbf{1}}{C} \tag{3.26}$$

minimizes the variance, so that $\sigma^2(\tilde{r}_{w^{\mathrm{MVP}}}) = 1/C$, with a corresponding expected return of $\mathbb{E}[\tilde{r}_{w^{\mathrm{MVP}}}] = A/C$. The second equality in (3.26) easily follows by substituting the explicit expressions for g, h, A, C given in Theorem 3.12.

Moreover, the covariance between a frontier portfolio w^{*1} and an *arbitrary* portfolio w^2 (i.e., a portfolio $w^2 \in \Delta_N$ not necessarily belonging to the portfolio frontier) can be expressed as follows, making use of the representation (3.20) of a frontier portfolio:

$$\mathrm{Cov}\left(\tilde{r}_{w^{*1}}, \tilde{r}_{w^2}\right) = (w^{*1})^{\top}Vw^2 = \lambda_1 e^{\top}V^{-1}Vw^2 + \gamma_1 \mathbf{1}^{\top}V^{-1}Vw^2 = \lambda_1\mathbb{E}[\tilde{r}_{w^2}] + \gamma_1, \tag{3.27}$$

where λ_1 and γ_1 are the optimal Lagrange multipliers associated to the frontier portfolio w^{*1} and where $\mathbb{E}[\tilde{r}_{w^2}]$ is the expected return of the portfolio w^2.

Formula (3.27) can also be used to prove the following property of the minimum variance portfolio w^{MVP}: for any portfolio $w \in \Delta_N$ (not necessarily belonging to the portfolio frontier), it holds that

$$\mathrm{Cov}(\tilde{r}_{w^{\mathrm{MVP}}}, \tilde{r}_w) = \sigma^2(\tilde{r}_{w^{\mathrm{MVP}}}).$$

Indeed, w^{MVP} is the solution to Problem (3.14) for the expected return $\mu = A/C$, so that, according to the notation used in Theorem 3.12, we have $\lambda_{\mathrm{MVP}} = 0$ and $\gamma_{\mathrm{MVP}} = 1/C$. Hence, applying expression (3.27) with $w^{*1} = w^{\mathrm{MVP}}$ gives that

$$\mathrm{Cov}\left(\tilde{r}_{w^{\mathrm{MVP}}}, \tilde{r}_w\right) = \gamma_{\mathrm{MVP}} = \frac{1}{C} = \sigma^2(\tilde{r}_{w^{\mathrm{MVP}}}). \tag{3.28}$$

Property 3: Shape of the Portfolio Frontier

In the standard deviation - expected return plane $(\sigma, \mu) \in (0, +\infty) \times \mathbb{R}$, the portfolio frontier is represented by a hyperbola centered in the point $(0, A/C)$ and with asymptotes $\mu = A/C \pm \sqrt{D/C}\,\sigma$. In the variance - expected return plane $(\sigma^2, \mu) \in (0, +\infty) \times \mathbb{R}$, the portfolio frontier is represented by a parabola with vertex $(1/C, A/C)$.

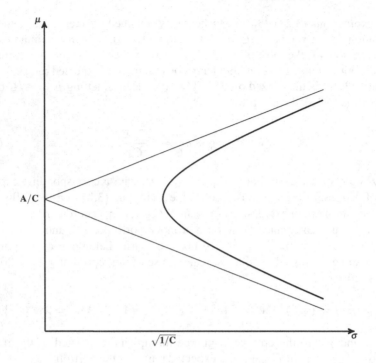

Fig. 3.2 Portfolio frontier in the standard deviation-expected return plane

Indeed, the fact that the portfolio frontier can be represented by a hyperbola in the standard deviation - expected return plane follows from equation (3.25), which can be equivalently rewritten in the following form, for any portfolio $w^* \in PF$:

$$\frac{\sigma^2(\tilde{r}_{w*})}{1/C} - \frac{\left(\mathbb{E}[\tilde{r}_{w*}] - A/C\right)^2}{D/C^2} = 1.$$

The above equation defines a hyperbola in the standard deviation - expected return plane, centered in $(0, A/C)$ and with asymptotes $\mu = A/C + \sqrt{D/C}\,\sigma$, see Fig. 3.2.

In the variance - expected return plane, the last part of equation (3.25) clearly shows that the portfolio frontier is represented by a parabola, see Fig. 3.3. In particular, the vertex of the parabola coincides with the minimum variance portfolio w^{MVP}, which is characterized by expected return $\mathbb{E}[\tilde{r}_{w,MVP}] = A/C$ and variance $\sigma^2(\tilde{r}_{w,MVP}) = 1/C$, as shown in Property 2 above. In particular, this graphical representation of the portfolio frontier makes clear that, in order to achieve a higher expected return, the investor needs to tolerate a greater risk (as measured by the variance) as there is a trade-off between variance and expected return.

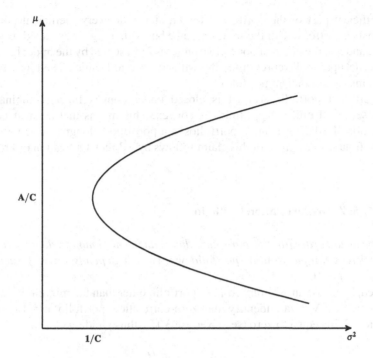

Fig. 3.3 Portfolio frontier in the variance-expected return plane

Property 4: Efficient Portfolio Frontier

All portfolios not dominated according to the mean-variance criterion belong to the following subset of the portfolio frontier:

$$\text{EPF} = \{w^* \in \text{PF } such\ that\ \mathbb{E}[\tilde{r}_{w^*}] \geq A/C\}. \tag{3.29}$$

As we have seen in Theorem 3.12, given any expected return $\mu \in \mathbb{R}$, there exists a unique portfolio $w^* \in \Delta_N$ with expected return $\mathbb{E}[\tilde{r}_{w^*}] = \mu$ which minimizes the variance. On the contrary, as can be deduced from equation (3.25), for a fixed value $\bar{\sigma}^2$ of the variance such that $\bar{\sigma}^2 > 1/C = \sigma^2(\tilde{r}_{w^{\text{MVP}}})$, we can find two frontier portfolios w_1^*, w_2^* with expected returns $\mathbb{E}[\tilde{r}_{w_1^*}] \neq \mathbb{E}[\tilde{r}_{w_2^*}]$ such that $\sigma^2(\tilde{r}_{w_1^*}) = \sigma^2(\tilde{r}_{w_2^*}) = \bar{\sigma}^2$. More specifically, we have $\mathbb{E}[\tilde{r}_{w_1^*}] > A/C > \mathbb{E}[\tilde{r}_{w_2^*}]$, where we recall that A/C is the expected return corresponding to the minimum variance portfolio w^{MVP}. We call *efficient portfolio frontier (EPF)* the subset of the portfolio frontier defined in (3.29). In this case, $w_1^* \in \text{EPF}$ and $w_2^* \notin \text{EPF}$. The efficient portfolio frontier is composed of those frontier portfolios that are not dominated according to the mean-variance criterion (see Sect. 2.4). We say that a frontier portfolio is *inefficient* if it does not belong to EPF. In the case of agents with mean-variance preferences (see Sect. 2.4), we can restrict our attention

to the efficient part of the portfolio frontier, since for every inefficient portfolio
there exists a portfolio with the same variance but with a higher expected return. In
Figs. 3.2 and 3.3, the efficient portfolio frontier is represented by the upper branch of
the hyperbola/parabola representing the whole portfolio frontier. The lower branch
contains instead inefficient portfolios.

The efficient portfolio frontier is closed under convex linear combinations,
i.e., the set of all efficient portfolios is convex. This means that a linear convex
combination of efficient frontier portfolios is a portfolio belonging to the efficient
portfolio frontier. The proof of this claim follows by a direct application of Property
1.

Property 5: Zero-Correlation Portfolio

*Let w^* be an arbitrary frontier portfolio other than the minimum variance portfolio.
There exists a unique frontier portfolio w^{zc} (which depends on w^*) such that
$\mathrm{Cov}(\tilde{r}_{w^*}, \tilde{r}_{w^{zc}}) = 0$.*

Indeed, let w^* be an arbitrary frontier portfolio other than the minimum variance
portfolio w^{MVP}. We can identify the zero-correlation portfolio w^{zc} by setting
expression (3.24) equal to zero (see Exercise 3.12), thus yielding

$$\mathbb{E}[\tilde{r}_{w^{zc}}] = \frac{A}{C} - \frac{D/C^2}{\mathbb{E}[\tilde{r}_{w^*}] - A/C}. \tag{3.30}$$

By Theorem 3.12, the portfolio w^{zc} is identified as the unique portfolio which
minimizes the variance given an expected return equal to (3.30). The portfolio w^{zc} is
the unique frontier portfolio that achieves zero correlation with the given portfolio
w^*. Moreover, as can be easily deduced from (3.30) together with Property 4 above,
the portfolio w^{zc} is efficient if and only if the portfolio w^* is inefficient. Indeed,
since $D > 0$, it holds that $\mathbb{E}[\tilde{r}_{w^{zc}}] > A/C$ if and only if $\mathbb{E}[\tilde{r}_{w^*}] < A/C$.

The zero-correlation portfolio w^{zc} admits an interesting geometrical interpretation. Indeed, in the standard deviation - expected return plane, the expected return
of the portfolio w^{zc} is identified by the intersection with the vertical axis of the
tangent to the portfolio frontier at the point $(\sigma(\tilde{r}_{w^*}), \mathbb{E}[\tilde{r}_{w^*}])$. In the variance -
expected return plane, the expected return of the portfolio w^{zc} is identified by the
intersection with the vertical axis of the line connecting $(\sigma^2(\tilde{r}_{w^*}), \mathbb{E}[\tilde{r}_{w^*}])$ to the
point $(1/C, A/C)$. We refer the reader to Exercise 3.14 for a proof of these claims.

The properties established above allow us to prove the following proposition
which represents one of the key results on the portfolio frontier.

Proposition 3.13 *Let $w^q \in \Delta_N$ be an arbitrary portfolio (with corresponding
return $\tilde{r}_{w^q}$) and $w^p \in \Delta_N$ a portfolio belonging to the portfolio frontier (with
corresponding return $\tilde{r}_{w^p}$) other than the minimum variance portfolio w^{MVP}. Denote*

by $w^{zc(p)}$ *the zero-correlation portfolio with respect to* w^p. *Then the following hold:*

(i) if $w^q \in$ PF *then*

$$\tilde{r}_{w^q} = (1 - \beta_{qp})\tilde{r}_{w^{zc(p)}} + \beta_{qp}\,\tilde{r}_{w^p};$$

(ii) if $w^q \notin$ PF *then*

$$\tilde{r}_{w^q} = (1 - \beta_{qp})\tilde{r}_{w^{zc(p)}} + \beta_{qp}\,\tilde{r}_{w^p} + \tilde{\epsilon}_{qp}, \tag{3.31}$$

where $\mathrm{Cov}\left(\tilde{r}_{w^p}, \tilde{\epsilon}_{qp}\right) = \mathrm{Cov}\left(\tilde{r}_{w^{zc(p)}}, \tilde{\epsilon}_{qp}\right) = \mathbb{E}[\tilde{\epsilon}_{qp}] = 0$ *and*

$$\beta_{qp} = \frac{\mathrm{Cov}(\tilde{r}_{w^q}, \tilde{r}_{w^p})}{\sigma^2(\tilde{r}_{w^p})}.$$

Moreover, for any portfolio $w^q \in \Delta_N$, *it holds that*

$$\mathbb{E}[\tilde{r}_{w^q}] = (1 - \beta_{qp})\mathbb{E}[\tilde{r}_{w^{zc(p)}}] + \beta_{qp}\,\mathbb{E}[\tilde{r}_{w^p}]. \tag{3.32}$$

Proof We start by proving the last part of the proposition. Let w^p and w^q be two arbitrary portfolios, with w^p belonging to the portfolio frontier and different from the minimum variance portfolio w^{MVP}. By (3.21)–(3.22) together with property (3.27), which holds for any portfolio w^q, we obtain

$$
\begin{aligned}
\mathbb{E}[\tilde{r}_{w^q}] &= \frac{A\,\mathbb{E}[\tilde{r}_{w^p}] - B}{C\,\mathbb{E}[\tilde{r}_{w^p}] - A} + \mathrm{Cov}(\tilde{r}_{w^q}, \tilde{r}_{w^p})\frac{D}{C\,\mathbb{E}[\tilde{r}_{w^p}] - A} \\
&= \frac{A}{C} - \frac{D/C^2}{\mathbb{E}[\tilde{r}_{w^p}] - A/C} + \frac{\mathrm{Cov}(\tilde{r}_{w^q}, \tilde{r}_{w^p})}{\sigma^2(\tilde{r}_{w^p})}\left(\frac{1}{C} + \frac{(\mathbb{E}[\tilde{r}_{w^p}] - A/C)^2}{D/C}\right)\frac{D}{C\,\mathbb{E}[\tilde{r}_{w^p}] - A} \\
&= \mathbb{E}[\tilde{r}_{w^{zc(p)}}] + \beta_{qp}\left(\mathbb{E}[\tilde{r}_{w^p}] - \mathbb{E}[\tilde{r}_{w^{zc(p)}}]\right),
\end{aligned}
$$

where β_{qp} is defined as in the statement of the proposition. This proves (3.32). In order to prove part *(ii)* of the proposition, observe that the relationship between the random variables $\tilde{r}_{w^q}$, $\tilde{r}_{w^p}$ and $\tilde{r}_{w^{zc(p)}}$ can always be written as

$$\tilde{r}_{w^q} = \alpha + \beta_1\tilde{r}_{w^{zc(p)}} + \beta_2\tilde{r}_{w^p} + \tilde{\epsilon}_{qp},$$

where β_1 and β_2 are obtained as the regression coefficients of $\tilde{r}_{w^q}$ on $(\tilde{r}_{w^{zc(p)}}, \tilde{r}_{w^p})$ and $\tilde{\epsilon}_{qp}$ is a random variable satisfying $\mathrm{Cov}\left(\tilde{r}_{w^p}, \tilde{\epsilon}_{qp}\right) = \mathrm{Cov}\left(\tilde{r}_{w^{zc(p)}}, \tilde{\epsilon}_{qp}\right) = \mathbb{E}[\tilde{\epsilon}_{qp}] = 0$. Note that, since the two random variables $\tilde{r}_{w^p}$ and $\tilde{r}_{w^{zc(p)}}$ are uncorrelated (Property 5), we have

$$\beta_2 = \mathrm{Cov}(\tilde{r}_{w^q}, \tilde{r}_{w^p})/\sigma^2(\tilde{r}_{w^p}) = \beta_{qp} \qquad \text{and} \qquad \beta_1 = \beta_{q\,zc(p)}.$$

Moreover, note that relation (3.32) can also be written taking $w^{\mathrm{zc(p)}}$ as the reference portfolio, instead of the portfolio w^{p}, thus yielding

$$\mathbb{E}[\tilde{r}_{w^{\mathrm{q}}}] = (1 - \beta_{\mathrm{q\,zc(p)}})\mathbb{E}[\tilde{r}_{w^{\mathrm{p}}}] + \beta_{\mathrm{q\,zc(p)}}\,\mathbb{E}[\tilde{r}_{w^{\mathrm{zc(p)}}}], \qquad (3.33)$$

where we have used the fact that the zero-correlation portfolio with respect to $w^{\mathrm{zc(p)}}$ is the portfolio w^{p}. Comparing (3.32) with (3.33), we obtain $\beta_{\mathrm{q\,zc(p)}} = 1 - \beta_{\mathrm{qp}}$, so that $\beta_1 = 1 - \beta_{\mathrm{qp}}$ and

$$\tilde{r}_{w^{\mathrm{q}}} = \alpha + (1 - \beta_{\mathrm{qp}})\tilde{r}_{w^{\mathrm{zc(p)}}} + \beta_{\mathrm{qp}}\tilde{r}_{w^{\mathrm{p}}} + \tilde{\epsilon}_{\mathrm{qp}}.$$

Finally, taking the expectation of the last expression and comparing to (3.32), we obtain that $\alpha = 0$. We have thus proved part (ii) of the proposition. Part (i) then follows from part (ii) together with Property 1, since any frontier portfolio w^{q} can be written as the linear affine combination of two arbitrary frontier portfolios, in this case w^{p} and $w^{\mathrm{zc(p)}}$. This implies that, if $w^{\mathrm{q}} \in \mathrm{PF}$, then the random variable $\tilde{\epsilon}_{\mathrm{qp}}$ is identically equal to zero. $\square$

In view of the above proposition, we have established the following properties:

- The return of any portfolio w^{q} (not necessarily belonging to the portfolio frontier) can be written as a linear combination of the return of an arbitrary portfolio w^{p} belonging to the portfolio frontier and the return of the frontier portfolio $w^{\mathrm{zc(p)}}$ uncorrelated with w^{p} plus a random variable $\tilde{\epsilon}_{\mathrm{qp}}$ with zero mean, uncorrelated with the two frontier portfolios w^{p} and $w^{\mathrm{zc(p)}}$. Moreover, if (and only if) the portfolio w^{q} belongs to the frontier, the last component is null.
- The coefficients of the linear combination (3.31) can be interpreted as the coefficients of the linear regression of $\tilde{r}_{w^{\mathrm{q}}}$ on $\tilde{r}_{w^{\mathrm{p}}}$ and $\tilde{r}_{w^{\mathrm{zc(p)}}}$.
- The expected return of any portfolio w^{q} (not necessarily belonging to the portfolio frontier) is given by the linear affine combination of the expected return of an arbitrary portfolio w^{p} belonging to the portfolio frontier and of the expected return of the frontier portfolio $w^{\mathrm{zc(p)}}$ uncorrelated with w^{p}, with weights given by β_{qp} and $1 - \beta_{\mathrm{qp}}$, respectively.
- The coefficient β_{qp} is a linear combination of the β coefficients associated to the individual N risky assets with respect to the portfolio w^{p}. Indeed, for any portfolio $w^{\mathrm{q}} = (w_1^{\mathrm{q}}, \ldots, w_N^{\mathrm{q}})^{\top}$, due to the bilinearity of the covariance operator, it holds that $\beta_{\mathrm{qp}} = \sum_{n=1}^{N} w_n^{\mathrm{q}} \beta_{\mathrm{np}}$, where $\beta_{\mathrm{np}} := \mathrm{Cov}(\tilde{r}_n, \tilde{r}_{w^{\mathrm{p}}})/\sigma^2(\tilde{r}_{w^{\mathrm{p}}})$.

We close this section by illustrating the concepts presented so far in the simple case of two risky assets with random returns $\tilde{r}_1$ and $\tilde{r}_2$. We denote by e_1, e_2 and σ_1^2, σ_2^2 the expected values and the variances of the random variables $\tilde{r}_1$ and $\tilde{r}_2$, respectively. In this context, a portfolio is simply identified by the couple $(w, 1-w)$, with $w \in \mathbb{R}$. The expected return and the variance of the return associated to a portfolio $(w, 1 - w)$, denoted by $\mathbb{E}[\tilde{r}_w]$ and $\sigma^2(\tilde{r}_w)$, respectively, are given by

$$\mathbb{E}[\tilde{r}_w] = w\,e_1 + (1 - w)e_2, \tag{3.34}$$

$$\sigma^2(\tilde{r}_w) = w^2\sigma_1^2 + (1 - w)^2\sigma_2^2 + 2w(1 - w)\rho\sigma_1\sigma_2,$$

where ρ denotes the correlation coefficient between $\tilde{r}_1$ and $\tilde{r}_2$.

In this simple situation, given a required expected return $\mu \in \mathbb{R}$, the identification of the portfolio with expected return μ achieving the minimum variance becomes trivial, since there will be a unique portfolio $(w, 1 - w)$ with expected return $\mathbb{E}[\tilde{r}_w] = \mu$. Indeed, due to (3.34), it holds that $w = (\mu - e_2)/(e_1 - e_2)$, provided that $e_1 \neq e_2$. In this case, the portfolio frontier can be easily described in the standard deviation - expected return plane. The shape of the portfolio frontier crucially depends on the correlation coefficient ρ and we can distinguish three different cases (see also Fig. 2.4):

a) If $\rho = 1$ (perfect positive correlation), then we have, for all $w \in \mathbb{R}$,

$$\mathbb{E}[\tilde{r}_w] = w\,e_1 + (1 - w)e_2,$$

$$\sigma(\tilde{r}_w) = w\,\sigma_1 + (1 - w)\sigma_2.$$

The above equations imply that, in the standard deviation - expected return plane (σ, μ), the portfolio frontier is described by the linear equation

$$\mu = e_2 + \frac{\sigma - \sigma_2}{\sigma_1 - \sigma_2}(e_1 - e_2).$$

This means that the portfolio frontier is represented by the straight line connecting the point (σ_1, e_1), which represents the first risky asset, to the point (σ_2, e_2) representing the second risky asset.

b) If $\rho = -1$ (perfect negative correlation), then we have, for all $w \in \mathbb{R}$,

$$\mathbb{E}[\tilde{r}_w] = w\,e_1 + (1 - w)e_2,$$

$$\sigma(\tilde{r}_w) = w\,\sigma_1 - (1 - w)\sigma_2, \quad \text{if } w \geq \frac{\sigma_2}{\sigma_1 + \sigma_2},$$

$$\sigma(\tilde{r}_w) = -w\,\sigma_1 + (1 - w)\sigma_2, \quad \text{if } w < \frac{\sigma_2}{\sigma_1 + \sigma_2}.$$

In this case, the portfolio frontier is identified on the standard deviation - expected return plane by the two half-lines originating from the point $(0, e_2 + \frac{\sigma_2}{\sigma_1 + \sigma_2}(e_1 - e_2))$ with slopes $\pm (e_1 - e_2)/(\sigma_1 + \sigma_2)$.

c) If $-1 < \rho < 1$, then the portfolio frontier is represented on the standard deviation - expected return plane as a hyperbola in an intermediate position with respect to the two cases considered above. For $0 \leq w \leq 1$, the portfolio frontier belongs to the interior of the region identified by the straight line connecting the points (σ_1, e_1) and (σ_2, e_2) and the half-lines described in case **b**).

The Case of N Risky Assets and a Risk Free Asset

So far we have considered an economy with N risky assets with random returns $(\tilde{r}_1, \ldots, \tilde{r}_N)$. We now extend the analysis to the case of $N + 1$ assets, adding to the original economy with N risky assets a risk free asset with deterministic return r_f. As before, we are interested in determining the portfolio composed of the $N + 1$ assets achieving the minimum variance for a given expected return. In the present context, a portfolio is simply identified by a vector $w = (w_1, \ldots, w_N)^\top \in \mathbb{R}^N$, where w_n determines the proportion of wealth invested in the n-th risky asset, for each $n = 1, \ldots, N$. The quantity $1 - \sum_{n=1}^N w_n$ represents the proportion of wealth invested in the risk free asset. According to whether $1 - \sum_{n=1}^N w_n$ is greater or less than zero, this represents the possibility of investing or borrowing, respectively, in the bank account. We denote by PF* the portfolio frontier obtained with N risky assets plus a risk free asset. To exclude trivial cases, we assume that $\mathbb{E}[\tilde{r}_n] \neq r_f > 0$ for at least one $n \in \{1, \ldots, N\}$.

To identify the portfolio frontier in the presence of a risk free asset, we need to solve the following problem:

$$\min_{w \in \mathbb{R}^N} w^\top V w, \tag{3.35}$$

under the constraint

$$w^\top e + (1 - w^\top \mathbf{1}) r_f = \mu,$$

where $\mu \in \mathbb{R}$ represents the required expected return of the portfolio. Since we now have an additional asset, namely the risk free asset with return r_f, the set of investment possibilities is larger than the one considered in the case of N risky assets. Therefore, the minimum variance that can be achieved in Problem (3.35) for a given expected return $\mu \in \mathbb{R}$ will always be less or equal than the minimum variance that can be achieved in Problem (3.14) for the same expected return μ. Actually, in the present context, it is always possible to reduce to zero the variance of a portfolio by simply investing the total amount of wealth in the risk free asset. Similarly as in the case of N risky assets, we can establish the following proposition.

Proposition 3.14 *Let $(\tilde{r}_1, \ldots, \tilde{r}_N)$ be N random returns with expected value $e \in \mathbb{R}^N$ and positive definite covariance matrix $V \in \mathbb{R}^{N \times N}$ and let $r_f > 0$ be the risk free return. Then, for any $\mu \in \mathbb{R}$, the solution to the constrained variance minimization problem (3.35) is given by*

$$w^* = \frac{\mu - r_f}{K} V^{-1}(e - r_f \mathbf{1}), \qquad (3.36)$$

where $K = (e - \mathbf{1}r_f)^{\mathsf{T}} V^{-1}(e - \mathbf{1}r_f) = B - 2Ar_f + Cr_f^2 > 0$.

Proof As in the case of Theorem 3.12, the problem can be solved through the associated Lagrangian

$$L(w, \lambda) := \frac{1}{2} w^{\mathsf{T}} V w + \lambda \big(\mu - w^{\mathsf{T}} e - (1 - w^{\mathsf{T}} \mathbf{1}) r_f\big).$$

The first order conditions amount to

$$\frac{\partial L(w^*, \lambda)}{\partial w} = V w^* - \lambda(e - \mathbf{1}r_f) = 0,$$

$$\frac{\partial L(w^*, \lambda)}{\partial \lambda} = \mu - w^{*\mathsf{T}} e - (1 - w^{*\mathsf{T}} \mathbf{1}) r_f = 0.$$

Solving for w^* we obtain $w^* = \lambda V^{-1}(e - \mathbf{1}r_f)$, so that, solving for λ, we get $\lambda = (\mu - r_f)/K$, with $K > 0$ defined as in the statement of the proposition. Since the matrix V is assumed to be positive definite, the first order conditions are both necessary and sufficient for the optimality of w^*.   □

In view of (3.36), the variance and the standard deviation of a portfolio w^* belonging to the portfolio frontier composed by the N risky assets together with the risk free asset can be explicitly computed as

$$\sigma^2(\tilde{r}_{w*}) = w^{*\mathsf{T}} V w^* = \frac{\big(\mathbb{E}[\tilde{r}_{w*}] - r_f\big)^2}{K} \qquad (3.37)$$

and

$$\sigma(\tilde{r}_{w*}) = \frac{\mathbb{E}[\tilde{r}_{w*}] - r_f}{\sqrt{K}}, \qquad \text{if } \mathbb{E}[\tilde{r}_{w*}] \geq r_f;$$

$$\sigma(\tilde{r}_{w*}) = -\frac{\mathbb{E}[\tilde{r}_{w*}] - r_f}{\sqrt{K}}, \qquad \text{if } \mathbb{E}[\tilde{r}_{w*}] \leq r_f. \qquad (3.38)$$

The portfolio frontier PF* can be represented in the standard deviation - expected return plane by two half-lines with slope $\sqrt{K}$ and $-\sqrt{K}$, respectively, originating from the point $(0, r_f)$. Analogously to the discussion in Property 4 above, the half-line with positive slope represents the *efficient* part of the portfolio frontier

(EPF*). Moreover, as in the case of Property 1, it can be easily shown that two arbitrary portfolios belonging to the portfolio frontier generate through linear affine combinations all portfolios belonging to the portfolio frontier and linear affine combinations of frontier portfolios belong to the portfolio frontier.

It is interesting to study the relation between the two portfolio frontiers PF* and PF obtained with and without, respectively, the risk free asset. Such a relation crucially depends on r_f and A/C, using the notation introduced in Theorem 3.12, where we recall that A/C is the expected return associated to the minimum variance portfolio constructed from the N risky assets.

Proposition 3.15 *Let denote by* PF* *and* PF *the portfolio frontiers obtained, respectively, with and without the risk free asset with return* r_f *and let A and C be defined as in Theorem 3.12. Then, if* $r_f \neq A/C$, *there exists a unique portfolio* w^e *which belongs to both* PF *and* PF*. *Moreover,* $w^e \in$ EPF *if and only if* $r_f < A/C$.

Proof Let us suppose that $r_f \neq A/C$ and consider the portfolio $w^e \in$ PF having zero correlation with the portfolio belonging to PF with expected return equal to r_f. Then, in view of (3.30), it holds that

$$\mathbb{E}[\tilde{r}_{w^e}] = \frac{A}{C} - \frac{BC - A^2}{C(C r_f - A)} = \frac{A r_f - B}{C r_f - A}. \tag{3.39}$$

Therefore, due to Theorem 3.12,

$$w^e = g + h\,\mathbb{E}[\tilde{r}_{w^e}] = \frac{BV^{-1}\mathbf{1} - AV^{-1}e}{BC - A^2} + \frac{CV^{-1}e - AV^{-1}\mathbf{1}}{BC - A^2}\frac{A r_f - B}{C r_f - A}$$

$$= V^{-1}\frac{e - r_f\mathbf{1}}{\mathbf{1}^\top V^{-1}(e - r_f\mathbf{1})}.$$

It can be easily checked that this expression coincides with (3.36) for an expected return μ equal to $\mu = (A\,r_f - B)/(C\,r_f - A)$. We have thus shown that the portfolio w^e belongs to both PF and PF*. We now show that the intersection of the two portfolio frontiers PF and PF* contains at most one element, thus yielding the uniqueness of w^e. Indeed, arguing by contradiction, suppose that there exists another portfolio w' belonging to both PF and PF*. Note first that $\mathbb{E}[\tilde{r}_{w^e}] \neq \mathbb{E}[\tilde{r}_{w'}]$, since each expected return identifies a unique frontier portfolio. Let $\alpha := (r_f - \mathbb{E}[\tilde{r}_{w'}])/(\mathbb{E}[\tilde{r}_{w^e} - \tilde{r}_{w'}])$ and define the portfolio $\bar{w} := \alpha w^e + (1 - \alpha)w'$. Then, since PF and PF* are both closed under linear affine combinations, the portfolio $\bar{w}$ belongs to both PF and PF*. Moreover, due to the definition of α, we have that $\mathbb{E}[\tilde{r}_{\bar{w}}] = r_f$ and, by formula (3.37), it also holds that $\sigma^2(\tilde{r}_{\bar{w}}) = 0 < 1/C$, thus contradicting the minimality property of the minimum variance portfolio w^{MVP}. We have thus shown that w^e is unique. Finally, recalling that $BC - A^2 > 0$ (see Theorem 3.12), it follows from (3.39) that $\mathbb{E}[\tilde{r}_{w^e}] > A/C$ if and only if $r_f < A/C$. In view of Property 4, this implies that w^e belongs to EPF if and only if $r_f < A/C$. $\qquad\qquad\square$

We can also explicitly compute the variance of the portfolio w^e as follows:

$$\sigma^2(\tilde{r}_{w^e}) = w^{e\top} V w^e = \frac{(e - r_f \mathbf{1})^\top V^{-1}(e - r_f \mathbf{1})}{\left(\mathbf{1}^\top V^{-1}(e - r_f \mathbf{1})\right)^2} = \frac{K}{(A - C r_f)^2}. \qquad (3.40)$$

In the context of Proposition 3.15, if $r_f \neq A/C$, then the portfolio w^e admits an interesting geometric characterization: in the standard deviation-expected return plane, the portfolio w^e is identified by the tangency condition between one of the two half-lines (3.38) and the portfolio frontier PF composed by the risky assets only. In view of expression (3.38), in order to confirm this claim, it suffices to show that the slope of the tangent line to PF in correspondence of $(\sigma(\tilde{r}^{w^e}), \mathbb{E}[\tilde{r}_{w^e}])$ is equal to $\pm\sqrt{K}$ (compare also with Exercise 3.14). By Property 3, the line tangent to PF at the point $(\sigma(\tilde{r}^{w^e}), \mathbb{E}[\tilde{r}_{w^e}])$ has a slope equal to

$$\frac{d\mu}{d\sigma} = \pm \frac{\sigma(\tilde{r}_{w^e})D}{C\mathbb{E}[\tilde{r}_{w^e}] - A}.$$

Substituting $\mathbb{E}[\tilde{r}_{w^e}]$ and $\sigma^2(\tilde{r}_{w^e})$ with the explicit expressions given in (3.39) and (3.40), respectively, we then obtain that $d\mu/d\sigma = \pm\sqrt{K}$. In view of this observation, we call the portfolio w^e (when it exists, i.e., when $r_f \neq A/C$) the *tangent portfolio*.

Thanks to Property 1, which also holds for the portfolio frontier PF^*, any frontier portfolio can be obtained as the linear affine combination of the tangent portfolio w^e and of the portfolio w^0 investing in the risk free asset alone. Note that the tangent portfolio satisfies $w^{e\top}\mathbf{1} = 1$, while the portfolio investing only in the risk free asset satisfies $w^{0\top}\mathbf{1} = 0$. Depending on the relation between r_f and A/C (where the latter quantity is the expected return of the minimum variance portfolio w^{MVP}), we can distinguish three cases (the first one is illustrated in Fig. 3.4):

a) The PF^* is composed of two half-lines, see (3.38). If $r_f < A/C$, due to Proposition 3.15, the tangent portfolio w^e is efficient and the frontier is divided into three regions. The first region corresponds to the segment connecting the two points that correspond to w^0 and w^e. This region contains all portfolios that are linear convex combinations of the two portfolios w^0 and w^e, i.e., all portfolios characterized by a positive investment in the risk free asset as well as in the tangent portfolio w^e, so that $0 \leq w^\top\mathbf{1} \leq 1$. The second region corresponds to the part at the right of the point representing w^e: any portfolio w belonging to this second region involves short selling the risk free asset and investing in the tangent portfolio w^e more than the total amount of wealth, i.e., $w^\top\mathbf{1} > 1$. Finally, the third region is located below the risk free rate r_f and contains all inefficient portfolios. Such portfolios involve short selling the tangent portfolio w^e and investing more than the total amount of wealth in the risk free asset w^0, i.e., $w^\top\mathbf{1} < 0$.

b) If $r_f > A/C$, due to Proposition 3.15, the tangent portfolio w^e is inefficient and, as in case a), the frontier is divided into three regions. The first region corresponds to the segment connecting the two points that correspond to w^0 and w^e. This region

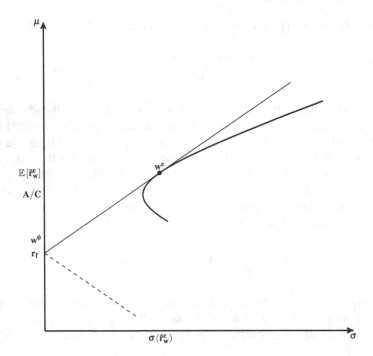

Fig. 3.4 Portfolio frontier with a risk free asset $(A/C > r_f)$

contains all portfolios that are linear convex combinations of the two portfolios w^0 and w^e, i.e., all portfolios characterized by a positive investment in the risk free asset as well as in the tangent portfolio w^e, so that $0 \leq w^\top 1 \leq 1$. Any such portfolio is inefficient. The second region corresponds to the part at the right of the point that represents w^e. Any portfolio w belonging to this second region is inefficient and involves short selling the risk free asset and investing in the tangent portfolio w^e more than the total amount of wealth, i.e., $w^\top 1 > 1$. Finally, the third region contains all efficient portfolios and is located above r_f. All portfolios belonging to this region involve short selling the tangent portfolio w^e and investing an amount of money larger than the total amount of wealth in the risk free asset w^0, i.e., $w^\top 1 < 0$.

c) Finally, it may happen that $r_f = A/C$. In this case $K = D/C > 0$ and the portfolio frontier PF* is represented in the standard deviation-expected return plane by the asymptotes of the portfolio frontier PF. In this case, there does not exist a portfolio belonging to both frontiers PF and PF* and every portfolio belonging to PF* involves a net investment in the risk free asset equal to the total amount of wealth and, consequently, the overall net amount of wealth invested in the risky assets is equal to zero. Indeed, let w^* be an arbitrary portfolio belonging to PF*, thus admitting the representation (3.36). Then, setting $r_f = A/C$, we get

$$\mathbf{1}^\top w^* = \mathbf{1}^\top V^{-1}\left(e - \frac{A}{C}\mathbf{1}\right)\frac{\mathbb{E}[\tilde{r}_{w^*}] - A/C}{K} = 0.$$

The previous discussion on the efficiency of portfolios belonging to the portfolio frontier PF* leads us to introduce the *Sharpe ratio*, an index frequently used to evaluate the performance of an asset or portfolio. Given a portfolio $w \in \mathbb{R}^N$, the corresponding Sharpe ratio SR(w) is defined as follows:

$$SR(w) := \frac{\mathbb{E}[\tilde{r}_w] - r_f}{\sigma(\tilde{r}_w)},$$

i.e., the ratio between the return risk premium of a portfolio and the standard deviation of the associated random return. In the standard deviation - expected return plane, the Sharpe ratio represents the slope of the straight line which connects the point $(0, r_f)$ to the point $(\sigma(\tilde{r}_w), \mathbb{E}[\tilde{r}_w])$ representing the portfolio w. Observe that, due to expression (3.38), the Sharpe ratio of all portfolios belonging to EPF* is constant and equal to $\sqrt{K}$. Moreover, since efficient frontier portfolios are not dominated according to the mean-variance criterion, there does not exist a portfolio with a Sharpe ratio greater than $\sqrt{K}$. On the Sharpe ratio, see also Exercise 3.19.

We can obtain an explicit expression for the covariance of an arbitrary portfolio $w \in \mathbb{R}^N$ (investing in the $N + 1$ assets and not necessarily belonging to the portfolio frontier) and a portfolio w^* belonging to the portfolio frontier PF*. Indeed, by relying on representation (3.36) and on the fact that $\mathbb{E}[\tilde{r}_w] = w^\top e + (1 - w^\top \mathbf{1})r_f$, we get

$$\text{Cov}(\tilde{r}_w, \tilde{r}_{w^*}) = w^\top V w^* = \frac{\mathbb{E}[\tilde{r}_{w^*}] - r_f}{K}w^\top(e - \mathbf{1}r_f) = \frac{(\mathbb{E}[\tilde{r}_{w^*}] - r_f)(\mathbb{E}[\tilde{r}_w] - r_f)}{K}.$$

In the case of an economy with N risky assets and a risk free asset with return r_f, we can establish the following proposition (we omit the proof, which is analogous to that of Proposition 3.13, noting that the riskless portfolio w^0 has zero correlation with any other portfolio).

Proposition 3.16 *Let $w^q \in \mathbb{R}^N$ be an arbitrary portfolio and $w^p \in \mathbb{R}^N$ a portfolio belonging to the portfolio frontier* PF* *such that $\mathbb{E}[\tilde{r}_{w^p}] \neq r_f$. Then the following hold:*

(i) if $w^q \in$ PF then*

$$\tilde{r}_{w^q} = (1 - \beta_{qp})r_f + \beta_{qp}\,\tilde{r}_{w^p};$$

(ii) if $w^q \notin$ PF then*

$$\tilde{r}_{w^q} = (1 - \beta_{qp})r_f + \beta_{qp}\,\tilde{r}_{w^p} + \tilde{\epsilon}_{qp},$$

where $\text{Cov}\left(\tilde{r}_{w^p}, \tilde{\epsilon}_{qp}\right) = \mathbb{E}[\tilde{\epsilon}_{qp}] = 0$ *and* $\beta_{qp} = \text{Cov}(\tilde{r}_{w^q}, \tilde{r}_{w^p})/\sigma^2(\tilde{r}_{w^p})$.
Moreover, for any portfolio $w^q \in \mathbb{R}^N$, *it holds that*

$$\mathbb{E}[\tilde{r}_{w^q}] - r_f = \beta_{qp}\big(\mathbb{E}[\tilde{r}_{w^p}] - r_f\big). \tag{3.41}$$

Relation (3.41) expresses the risk premium of any portfolio w^q (not necessarily belonging to the portfolio frontier PF*) in terms of the risk premium of an arbitrary portfolio $w^p \in$ PF*. The two risk premia have the same sign if and only if the coefficient β_{qp} is positive, where β_{qp} is the regression coefficient of the random return $\tilde{r}_{w^q}$ on $\tilde{r}_{w^p}$.

Extensions: Constraints on Borrowing

The portfolio frontier can also be constructed in the presence of additional constraints on the financial market. In what follows, we limit our attention to two possible situations: the case where agents are not allowed to borrow at the risk free rate r_f and the case where there are transaction costs on the risk free asset (i.e., there are different interest rates for lending and for borrowing).

Let us first consider the case where agents cannot borrow at the risk free rate r_f, i.e., all portfolios must satisfy the additional requirement $w^\top \mathbf{1} \leq 1$. As we have discussed above, if $r_f < A/C$, then the efficient portfolio frontier EPF* (obtained by including the risk free asset) consists of the tangent line to the portfolio frontier PF (with no risk free asset) in correspondence of an efficient portfolio w^e. Moreover, any portfolio belonging to the portfolio frontier can be represented as a linear combination of w^e and of the risk free asset (i.e., of the portfolio $w^0 = 0$). According to the analysis developed above (see Fig. 3.4), on the efficient part of the portfolio frontier, i.e., on the straight line with positive slope originating from the point $(0, r_f)$, borrowing occurs only at the right of the point $(\sigma(\tilde{r}_{w^e}), \mathbb{E}[\tilde{r}_{w^e}])$, since along the segment that connects the point $(0, r_f)$ to the point $(\sigma(\tilde{r}_{w^e}), \mathbb{E}[\tilde{r}_{w^e}])$ an agent will invest a positive amount of wealth both in the risk free asset and in the tangent portfolio (i.e., $0 \leq w^\top \mathbf{1} \leq 1$), while along the half-line with negative slope he will sell short the tangent portfolio (i.e., $w^\top \mathbf{1} \leq 0$) and will invest in the risk free asset. As a consequence, the portfolio frontier in the presence of the additional no-borrowing constraint coincides with the unconstrained portfolio frontier PF* up to the point $(\sigma(\tilde{r}_{w^e}), \mathbb{E}[\tilde{r}_{w^e}])$, which corresponds to the tangent portfolio w^e. However, at the right of that point the constrained portfolio frontier will coincide with the portfolio frontier PF obtained without the risk free asset. In the latter part of the portfolio frontier, due to the no-borrowing constraint, the agent will only invest in the N risky assets. Of course, a similar analysis can be developed in the symmetric case where $r_f > A/C$.

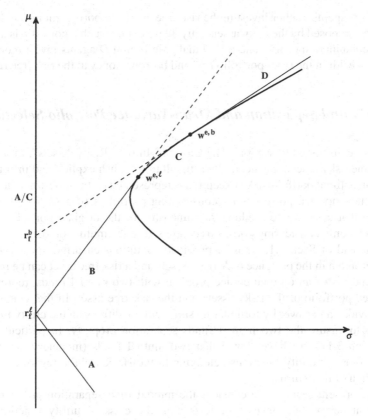

Fig. 3.5 Portfolio frontier with transaction costs

Let us now suppose that agents face two different risk free rates depending on whether they borrow or invest in the risk free asset. More specifically, we assume the existence of a risk free rate r_f^b (r_f^ℓ, resp.) which represents the rate at which it is possible to borrow from (to lend to, resp.) the bank account. We suppose that $r_f^b > r_f^\ell$ and that $A/C > r_f^b$. Considering two risk free assets with returns r_f^b and r_f^ℓ, respectively, we can build two portfolio frontiers as in Proposition 3.14. We denote by $w^{e,b}$ the tangent portfolio of the portfolio frontier obtained with a risk free asset with return r_f^b and, analogously, by $w^{e,\ell}$ the tangent portfolio of the frontier obtained with risk free rate r_f^ℓ. Note that, as can be seen from (3.38), the absolute value of the slope of the two half-lines defining the portfolio frontier with a risk free asset is decreasing in the risk free rate. In the present context, the portfolio frontier is illustrated in Fig. 3.5. In particular, the portfolio frontier with different borrowing and lending rates is composed of four regions. In region A, agents sell short the portfolio of the risky assets $w^{e,\ell}$ (i.e., $w^\top \mathbf{1} \leq 0$) and invest in the risk free asset (with a corresponding rate r_f^ℓ). In region B agents invest a positive amount of wealth both in the risk free asset (with corresponding rate r_f^ℓ) and in the risky portfolio $w^{e,\ell}$.

In region C agents neither invest in the risk free asset nor borrow money, holding a portfolio composed by the risky assets only. In particular, such a portfolio is a linear convex combination of $w^{e,b}$ and $w^{e,\ell}$. Finally, in region D agents invest more than the total wealth in the risky portfolio $w^{e,b}$ and borrow money at the risk free rate r_f^b.

Mutual Fund Separation and Mean-Variance Portfolio Selection

As we have discussed in section "The Case of Multiple Risky Assets", in the case of multiple risky assets, it is generally difficult to establish explicit and informative results on optimal portfolios. An exception is represented by the case when investors choose their optimal portfolios by constructing portfolios of a family of *mutual funds*. In that case, we can reduce the dimension of the original portfolio choice problem by only considering portfolios composed of the mutual funds.

At the end of Sect. 3.1, we have provided conditions ensuring that a portfolio choice problem in the presence of N risky assets and a risk free asset can be reduced to the analysis of an optimal choice problem with two mutual funds, represented by a fixed portfolio of the risky assets and the risk free asset. In that context, we have provided an answer by formulating sufficient conditions on the utility function in order to ensure the two mutual funds separation property. In particular (see Proposition 3.11), we have shown that two mutual funds (monetary) separation always holds for utility functions belonging to the HARA class, regardless of the distribution of the returns.

In the present section, we consider the mutual fund separation problem from a different perspective. Instead of restricting the class of utility functions, we shall consider restrictions on the probability distribution of the returns, aiming at establishing necessary and sufficient conditions for the validity of the mutual fund separation property for every risk averse agent, following Ingersoll [1000] and Ross [1467]. We aim at answering the following question: given a set of N assets, is it possible to identify $K < N$ portfolios (*mutual funds*) of the N assets such that, for any portfolio composed of the N original assets and for any utility function satisfying some basic properties, there exists a portfolio composed of the K mutual funds that is preferred to the first portfolio? More precisely, let us formulate the following definition.

Definition 3.17 We say that a set of N assets with returns $\tilde{\mathbf{r}} = (\tilde{r}_1, \ldots, \tilde{r}_N)$ satisfies the *separation property through K mutual funds* if there exist K portfolios (*mutual funds*) $(w^1, \ldots, w^K)$, with $w^k \in \Delta_N$ and return $\tilde{r}^k$, for all $k = 1, \ldots, K$, such that, for any portfolio $w \in \Delta_N$ of the N assets with return $\tilde{r}_w$, there exists a vector of weights $(\lambda_1, \ldots, \lambda_K)$, with $\sum_{k=1}^K \lambda_k = 1$, such that, for every concave utility function u,

$$\mathbb{E}\left[u\left(\sum_{k=1}^K \lambda_k \tilde{r}^k\right)\right] \geq \mathbb{E}[u(\tilde{r}_w)]. \tag{3.42}$$

According to the above definition, a set of N assets satisfies the separation property through K mutual funds if, for any portfolio $w \in \Delta_N$ of the original N assets, there exists a linear combination $\lambda_1 w^1 + \ldots + \lambda_K w^K$, with weights adding up to one, of the K mutual funds which dominates the portfolio w in the sense of second order stochastic dominance (see Sect. 2.3). Definition 3.17 is in a *strong* sense because the weights $\lambda_1, \ldots, \lambda_K$ are assumed not to depend on the utility function. If the weights are allowed to depend on the utility function u, then one can define the separation property in a *weak* sense (however, the two definitions coincide for $K \leq 2$, see Ross [1467]). We also mention that an alternative definition of separation property can be obtained by formulating Definition 3.17 with respect to all concave *non-decreasing* functions (see Ross [1467]).

In what follows, we shall focus our attention on the case $K = 2$ (see Exercise 3.24 for the simple case $K = 1$). As in the case of portfolio choice problems, the conditions which ensure the validity of the two mutual funds separation property depend on whether a risk free asset is present in the economy or not.

Let us first consider an economy with N risky assets with returns $\tilde{\mathbf{r}} = (\tilde{r}_1, \ldots, \tilde{r}_N)$ together with a risk free asset with return r_f. In order to establish the two mutual funds separation property, we can assume that the risk free asset acts as one of the two mutual funds (*monetary separation*). Indeed, since Definition 3.17 is formulated with respect to any concave utility function, with possibly unbounded risk aversion, we can consider risk averse agents for whom investing only in the risk free asset is the optimal choice. A necessary and sufficient condition on the probability distribution of the random vector $\tilde{\mathbf{r}}$ in order to satisfy the two mutual funds (monetary) separation property is given in the following proposition (see Ingersoll [1000, Chapter 6] for a proof based on the second order stochastic dominance criterion).

Proposition 3.18 *The two mutual funds (monetary) separation property holds if and only if the random vector $\tilde{\mathbf{r}}$ admits the representation*

$$\tilde{\mathbf{r}} = r_f \mathbf{1} + b\tilde{r}^* + \tilde{\epsilon},$$

for some random variable $\tilde{r}^*$, where $b \in \mathbb{R}^N$ and $\tilde{\epsilon}$ is an N-dimensional random vector such that $\mathbb{E}[\tilde{\epsilon} | \tilde{r}^*] = 0$, and there exist weights $(a_1, \ldots, a_N)$ such that $\sum_{i=1}^{N} a_i \tilde{\epsilon}_i = 0$ and $\sum_{i=1}^{N} a_i = 1$.

In the setting of the above proposition, the two separating mutual funds are represented by the risk free asset and the portfolio $(a_1, \ldots, a_N)^\top \in \Delta_N$ composed by the N risky assets.

If we instead consider an economy with N risky assets and without a risk free asset, then a necessary and sufficient condition in order to satisfy the two mutual funds separation property can be formulated as follows (see Ross [1467, Corollary 1]).

Proposition 3.19 *Suppose that $\mathbb{E}[\tilde{r}_i] \neq \mathbb{E}[\tilde{r}_1]$, for at least one $i \in \{2, \ldots, N\}$. Then the two mutual funds separation property holds if and only if the random vector $\tilde{\mathbf{r}}$ admits the representation*

$$\tilde{\mathbf{r}} = \tilde{r}^*\mathbf{1} + b\tilde{r}^{**} + \tilde{\epsilon},$$

for some random variables $\tilde{r}^*$ and $\tilde{r}^{**}$, where $b \in \mathbb{R}^N$ and $\tilde{\epsilon}$ is an N-dimensional random vector such that $\mathbb{E}[\tilde{\epsilon}|\tilde{r}^* + \gamma \tilde{r}^{**}] = 0$, for every $\gamma \in \mathbb{R}$, and there exist two vectors of weights $a, c \in \mathbb{R}^N$ such that $a^\top \mathbf{1} = c^\top \mathbf{1} = 1$ and $a^\top \tilde{\epsilon} = c^\top \tilde{\epsilon} = 0$, with $a^\top b \neq c^\top b$.

If the asset returns are distributed according to a distribution belonging to the elliptical class, as well as any distribution implying a mean-variance utility function (see Sect. 2.4), then the two mutual funds separation property holds, see Owen & Rabinovitch [1389] and Chamberlain [395]. In particular, the normal multivariate distribution satisfies the conditions for the two mutual funds separation property, see Ross [1467].

Let us now explore the relations between the mutual funds separation property and the mean-variance portfolio frontier. Let us suppose that the economy consists of N risky assets, without a risk free asset. It is easy to see that, if the two funds separation property holds, then the two separating portfolios w^1 and w^2 must belong to the portfolio frontier. Indeed, suppose on the contrary that one of the two portfolios does not belong to the portfolio frontier. Then, for any linear affine combination $\lambda w^1 + (1 - \lambda)w^2$ of the two portfolios w^1, w^2, there exists a portfolio w' such that

$$\mathbb{E}[\tilde{r}_{w'}] = \mathbb{E}[\lambda \tilde{r}_{w^1} + (1 - \lambda)\tilde{r}_{w^2}] \quad \text{and} \quad \sigma^2(\tilde{r}_{w'}) < \sigma^2(\lambda \tilde{r}_{w^1} + (1 - \lambda)\tilde{r}_{w^2}).$$

This implies that the portfolio w' is not dominated according to the second order stochastic dominance criterion by the linear combination of the two portfolios w^1, w^2. Clearly, this contradicts the assumption that the two funds separation property holds with respect to the couple w^1, w^2.

Due to the properties of the portfolio frontier, if the N assets satisfy the mutual funds separation property, then any pair of two portfolios belonging to the portfolio frontier can be chosen as a pair of mutual funds. In particular, as mutual funds we may take an arbitrary frontier portfolio w^p (different from the minimum variance portfolio w^{MVP}) and the associated zero-covariance portfolio $w^{zc(p)}$. Recall that, in view of Proposition 3.13, the return of any portfolio $w^q \in \Delta_N$ can be written as

$$\tilde{r}_{w^q} = \beta_{qp} \tilde{r}_{w^p} + (1 - \beta_{qp})\tilde{r}_{w^{zc(p)}} + \tilde{\epsilon}_{qp}, \tag{3.43}$$

where $\text{Cov}(\tilde{r}_{w^p}, \tilde{\epsilon}_{qp}) = \text{Cov}(\tilde{r}_{w^{zc(p)}}, \tilde{\epsilon}_{qp}) = \mathbb{E}[\tilde{\epsilon}_{qp}] = 0$. Exploiting this representation, we can establish the following result (see Huang & Litzenberger [971] and Litzenberger & Ramaswamy [1222]).

Proposition 3.20 *Consider N risky assets with returns $\tilde{\mathbf{r}} = (\tilde{r}_1, \ldots, \tilde{r}_N)$ and let w^p be a frontier portfolio, with corresponding zero-correlation portfolio $w^{zc(p)}$. The two funds separation property holds with respect to the pair of mutual funds $(w^p, w^{zc(p)})$ if and only if*

$$\mathbb{E}[\tilde{\epsilon}_{qp} \mid \beta_{qp}\tilde{r}_{w^p} + (1 - \beta_{qp})\tilde{r}_{w^{zc(p)}}] = 0,$$

for any portfolio $w^q \in \Delta_N$.

Proof We show the sufficiency of the above condition (we refer to Huang & Litzenberger [971, Chapter 4] for the proof of the necessity part). Due to the tower property of the conditional expectation together with Jensen's inequality, it holds that, for any portfolio $w^q \in \Delta_N$,

$$\mathbb{E}[u(\tilde{r}_{w^q})] = \mathbb{E}\big[\mathbb{E}[u(\tilde{r}_{w^q})|\beta_{qp}\tilde{r}_{w^p} + (1 - \beta_{qp})\tilde{r}_{w^{zc(p)}}]\big]$$
$$\leq \mathbb{E}\big[u\big(\mathbb{E}[\tilde{r}_{w^q}|\beta_{qp}\tilde{r}_{w^p} + (1 - \beta_{qp})\tilde{r}_{w^{zc(p)}}]\big)\big]$$
$$= \mathbb{E}\big[u\big(\beta_{qp}\tilde{r}_{w^p} + w_0(1 - \beta_{qp})\tilde{r}_{w^{zc(p)}}\big)\big],$$

for any concave utility function u. Hence, by Definition 3.17, the two portfolios w_p and $w_{w^{zc(p)}}$ act as separating mutual funds. $\qquad\square$

By relying on the above proposition, we can also establish the following result.

Corollary 3.21 *Consider N assets with normally distributed returns $\tilde{\mathbf{r}} = (\tilde{r}_1, \ldots, \tilde{r}_N)$, with $\mathbb{E}[\tilde{r}_i] \neq \mathbb{E}[\tilde{r}_j]$ for some $i, j \in \{1, \ldots, N\}$, and let w^p be a frontier portfolio, with $w^{zc(p)}$ being the corresponding zero-correlation portfolio. Then the two funds separation property holds with respect to the pair of mutual funds $(w^p, w^{zc(p)})$.*

Proof Since the normal distribution is stable with respect to linear transformations, $(\tilde{r}_{w^p}, \tilde{r}_{w^{zc(p)}})$ is jointly normally distributed. Moreover, due to the properties of the normal distribution (and noting that $\tilde{\epsilon}_{qp}$ is also normally distributed), the fact that $\mathrm{Cov}(\tilde{r}_{w^p}, \tilde{\epsilon}_{qp}) = \mathrm{Cov}(\tilde{r}_{w^{zc(p)}}, \tilde{\epsilon}_{qp}) = 0$ implies that $\tilde{r}_{w^p}$, $\tilde{r}_{w^{zc(p)}}$ and $\tilde{\epsilon}_{qp}$ are mutually independent. In turn, this implies that $\mathbb{E}[\tilde{\epsilon}_{qp} \mid \beta_{qp}\tilde{r}_{w^p} + (1 - \beta_{qp})\tilde{r}_{w^{zc(p)}}] = \mathbb{E}[\tilde{\epsilon}_{qp}] = 0$. Proposition 3.20 then implies that the two funds separation property holds. $\qquad\square$

The above results are in line with Proposition 2.13, where we have shown that, if asset returns are normally distributed, then the expected utility of a risk averse agent with a strictly increasing utility function can be written as a function of the mean and of the variance, increasing in the first argument and decreasing in the second argument. These features imply that the agent's optimal portfolio belongs to the efficient part of the portfolio frontier. A result analogous to Corollary 3.21 can be established in the case where the risky assets have normally distributed returns with equal expectations. In this case, as shown in Exercise 3.26, the one fund separation property holds with respect to the minimum variance portfolio w^{MVP}.

Let us now consider an economy with a risk free asset with return $r_f > 0$ together with N risky assets with returns $\tilde{\mathbf{r}} = (\tilde{r}_1, \ldots, \tilde{r}_N)$. In this setting, by Proposition 3.16, the random return $\tilde{r}_w$ associated to a portfolio w composed of the N risky assets plus the risk free asset can be represented as

$$\tilde{r}_w = \beta_{qe}\tilde{r}_{w^e} + (1 - \beta_{qe})r_f + \tilde{\epsilon}_{qe},$$

where $\tilde{r}_{w^e} \in PF^*$ and $Cov(\tilde{r}_{w^e}, \tilde{\epsilon}_{qe}) = \mathbb{E}[\tilde{\epsilon}_{qe}] = 0$. We have the following version of Proposition 3.20 (see also Huang & Litzenberger [971, Chapter 4]).

Proposition 3.22 *Consider N risky assets with returns $\tilde{\mathbf{r}} = (\tilde{r}_1, \ldots, \tilde{r}_N)$ and a risk free asset with return $r_f \neq A/C$ and let w^e be the tangent portfolio (see Proposition 3.15). The two funds (monetary) separation property holds with respect to the risk free asset and the tangent portfolio w^e if and only if*

$$\mathbb{E}[\tilde{\epsilon}_{qe} \mid \tilde{r}_{w^e}] = 0,$$

for any portfolio $w^q \in \Delta_N$, using the notation of (3.43).

Similarly to the analysis performed in the case of N risky assets, if asset returns are normally distributed and have different expected returns then the two funds separation property holds true.

We close this section with a simple example, which shows that, in the case of N risky assets with normally distributed returns and a risk free asset with return $r_f < A/C$, the optimal portfolio of an agent characterized by an exponential utility function is composed of the risk free asset and of the tangent portfolio w^e. Of course, such a result is not surprising, in view of the above discussion and of the link between exponential utility maximization and mean-variance portfolio selection in the case of normally distributed returns (see section "Closed Form Solutions and Mutual Fund Separation"). Indeed, for a given initial wealth w_0 and a risk aversion parameter $a > 0$, letting $\widetilde{W} = w_0 \, r_f + \sum_{n=1}^{N} w_n(\tilde{r}_n - r_f)$, we consider the optimal portfolio choice problem

$$\max_{w \in \mathbb{R}^N} -\frac{1}{a}\mathbb{E}\big[\exp(-a\widetilde{W})\big].$$

Due to expression (3.11), the optimal portfolio $w^* \in \mathbb{R}^N$ is explicitly given in terms of wealth invested in the N risky assets by

$$w^* = \frac{1}{a}V^{-1}(e - r_f\mathbf{1}).$$

On the other hand, due to Proposition 3.14, the frontier portfolio w_μ^* associated to the expected return μ can be written in terms of proportions of wealth invested in the N risky assets as

$$w_\mu^* = V^{-1}(e - r_f\mathbf{1})\frac{\mu - r_f}{K},$$

where K is defined as in Proposition 3.14. By comparing the two above portfolios, we see that the optimal portfolio w^* for the exponential utility corresponds to the frontier portfolio w_μ^* for $\mu = r_f + K/(aw_0)$. Recall also that, by Proposition 3.15, the tangent portfolio w^e is given by

$$w^e = V^{-1} \frac{(e - r_f \mathbf{1})}{\mathbf{1}^\top V^{-1}(e - r_f \mathbf{1})},$$

so that

$$w^* = \frac{\mathbf{1}^\top V^{-1}(e - r_f \mathbf{1})}{a} w^e.$$

In particular, note that the proportion of wealth invested in the tangent portfolio w^e is decreasing in the coefficient of absolute risk aversion a. In Fig. 3.6 the indifference curves of an exponential utility function are depicted together with the optimal portfolios, which are shown to belong to the efficient part of the portfolio frontier, given by a linear combination of the risk free asset and the tangent portfolio w^e.

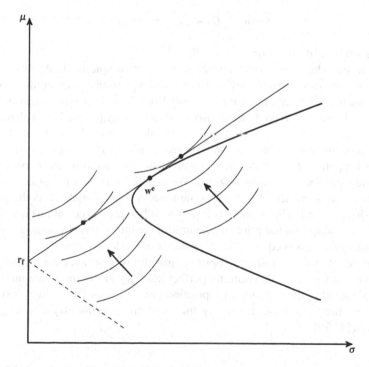

Fig. 3.6 Mean-variance utility and portfolio frontier

3.3 Insurance Demand

In this section, we shall consider the optimal choice problem of a risk averse agent
facing a risk (i.e., a potential loss) and having the possibility of buying an insurance
contract in order to mitigate the risk exposure. Insurance problems can be addressed
with tools analogous to those used in the context of portfolio selection problems,
since the insurance contract can be thought of as a risky asset whose payoff depends
on the realization of the loss event. Indeed, while in portfolio choice problems agents
need to decide on the optimal allocation of wealth, in insurance problems agents
need to determine the optimal demand of insurance contracts in order to hedge the
risk.

Let us first study an insurance problem in a very simple setting, where the
randomness is reduced to two elementary states of the world. Consider a risk averse
agent endowed with initial wealth $w_0 > 0$ at date $t = 0$, facing the possibility of
a loss event at the future date $t = 1$. The loss event is represented by a random
variable taking value $D > 0$ with probability $\pi \in (0, 1)$ and 0 with probability
$1 - \pi$ (i.e., at $t = 1$ there is the risk of losing an amount equal to D). If there does
not exist any insurance market, then the agent's expected utility is simply given by

$$\pi u(w_0 - D) + (1 - \pi)u(w_0),$$

for some given utility function $u : \mathbb{R} \to \mathbb{R}$.

We now introduce an insurance market, in a very simple form. At time $t = 0$,
the agent has the possibility of buying a particular type of insurance contract, which
delivers one unit of wealth at time $t = 1$ only if the loss occurs. Such an insurance
contract can be viewed as an *Arrow security*, i.e., a security paying 1 if an elementary
event is realized at $t = 1$ and 0 otherwise. In other words, the agent can trade in
the insurance market and buy one unit of wealth at $t = 1$ contingent on the event
"the loss happens" (for simplicity, we assume that the interest rate between date
$t = 0$ and date $t = 1$ is equal to zero). We denote by $p > 0$ the price at $t = 0$
of this insurance contract. Observe that the latter can be interpreted as the gamble
$[1 - p, -p; \pi, 1 - \pi]$. This gamble is *fair* (i.e., it has zero expectation) if and only
if $p = \pi$, meaning that the price of one unit of wealth at $t = 1$ contingent on the
occurrence of the loss is equal to the probability of the loss occurring.

Let us denote by $w^* \in \mathbb{R}_+$ the optimal quantity of insurance bought by a risk
averse agent at time $t = 0$, assuming perfect liquidity in the insurance market. In
this simple setting, the following proposition characterizes the optimal insurance
demand (we tacitly assume that utility functions are continuously differentiable),
see Mossin [1358].

Proposition 3.23 *Let u be strictly increasing and strictly concave utility function and let $w_0 > 0$. Then the following hold:*

(i) *if $p = \pi$ then $w^* = D$;*
(ii) *if $p > \pi$ then $w^* < D$;*
(iii) *if $p < \pi$ then $w^* > D$.*

Proof The agent's optimization problem amounts to

$$\max_{w \in \mathbb{R}_+} \left(\pi u(w_0 - D - wp + w) + (1 - \pi)u(w_0 - wp) \right).$$

By the Kuhn-Tucker conditions, w^* is a solution to the above problem if and only if

$$\pi(1 - p)u'(w_0 - D - w^*p + w^*) - p(1 - \pi)u'(w_0 - w^*p) \leq 0, \qquad (3.44)$$

with equality holding in the case of an interior solution (i.e., $w^* > 0$). If $p = \pi$, then condition (3.44) becomes

$$u'\left(w_0 - D + w^*(1 - p)\right) - u'(w_0 - w^*p) \leq 0.$$

Since the function u is assumed to be strictly concave (so that u' is strictly decreasing), this condition can only be verified for $w^* > 0$, meaning that the optimal solution w^* must satisfy

$$u'\left(w_0 - D + w^*(1 - p)\right) = u'(w_0 - w^*p).$$

Again, the strict concavity of u implies that the unique solution is given by $w^* = D$, thus proving part *(i)* of the proposition. If $p > \pi$, then the optimality condition (3.44) gives, for an optimal choice $w^* > 0$,

$$\frac{u'(w_0 - D - w^*p + w^*)}{u'(w_0 - w^*p)} = \frac{p(1 - \pi)}{\pi(1 - p)} > 1.$$

By the strict concavity of u, this implies that $w^* < D$, thus proving part *(ii)*. The proof of part *(iii)* is analogous.   □

In view of the above proposition, if the insurance contract is fairly priced (i.e., $p = \pi$), then a risk averse agent will insure himself completely, thus reducing to zero the variability of his wealth at the future time $t = 1$. On the contrary, if the price p of the insurance contract is higher than the probability π of the loss occurring, then the agent will only partially insure himself ($w^* < D$). Finally, in the case $p < \pi$, the agent will over-insure himself, buying an insurance coverage w^* greater than the potential loss D.

Let us now extend the above analysis to the case of several possible events at time $t = 1$. For simplicity, we shall restrict our attention to the case of two possible events, removing the risk free status where the initial wealth remains unaltered. This means that at the initial time $t = 0$ an agent faces a gamble $\tilde{x} = [x_1, x_2; \pi, 1 - \pi]$. The expected utility associated to such a gamble is

$$U(\tilde{x}) = \pi u(x_1) + (1 - \pi)u(x_2).$$

For instance, we can think of a farmer endowed with a field of wheat, which provides a crop x_1 if the weather conditions are favorable and a crop x_2 if the weather conditions are unfavorable.

We first characterize the indifference curves associated to the expected utility U in the space of wealth in correspondence of the first event and wealth in correspondence of the second event, keeping the probability π fixed. In this context, we denote a generic gamble $\tilde{x}$ through the pair (x_1, x_2) and the space $\mathbb{R}_+^2$ is called the *state space*. The marginal rate of substitution $\text{SMS}(x_1, x_2)$ of an agent with utility function u in correspondence of the gamble $\tilde{x} = (x_1, x_2)$ (given by the ratio between the marginal utilities in the two states of the world weighted by the corresponding probabilities of occurrence) is equal to minus the slope of the tangent line to the indifference curve at the point (x_1, x_2), i.e.,

$$\text{SMS}(x_1, x_2) = -\frac{dx_2}{dx_1} = \frac{\pi u'(x_1)}{(1 - \pi)u'(x_2)}. \tag{3.45}$$

The bisectrix of the state space $\mathbb{R}_+^2$ admits an interesting interpretation, see Fig. 3.7. Indeed, it represents the *certainty line*, i.e., the locus of all the gambles characterized by the same amount of wealth in correspondence of both elementary events. In other words, along the certainty line the agent faces no risk and the following equality holds:

$$\text{SMS}(x_1, x_2) = \frac{\pi}{1 - \pi}.$$

This means that, along the certainty line, the marginal rate of substitution is equal to the ratio of the probabilities of the two elementary events. This allows us to show an important property of utility functions of risk averse agents: if u is concave, then the indifference curves in the state space $\mathbb{R}_+^2$ are convex. To show this property, it suffices to prove that

$$\frac{d\text{SMS}(x_1, x_2)}{dx_1} \leq 0, \qquad \text{for all } (x_1, x_2) \in \mathbb{R}_+^2.$$

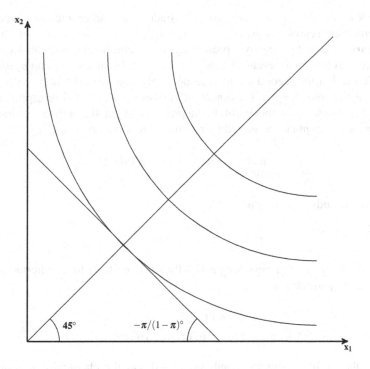

Fig. 3.7 Expected utility in the state space

Since $SMS(x_1, x_2) > 0$, due to the strict increasingness of u, the above inequality holds if and only if $d \log SMS(x_1, x_2)/dx_1 \leq 0$. Differentiating the logarithm of the marginal rate of substitution and using (3.45), we obtain

$$\frac{d \log(SMS(x_1, x_2))}{dx_1} = \frac{d}{dx_1} \left(\log u'(x_1) - \log u'(x_2) \right) = \frac{u''(x_1)}{u'(x_1)} - \frac{u''(x_2)}{u'(x_2)} \frac{dx_2}{dx_1}$$

$$= \frac{u''(x_1)}{u'(x_1)} + \frac{u''(x_2)}{u'(x_2)} \frac{\pi}{(1-\pi)} \frac{u'(x_1)}{u'(x_2)} \leq 0,$$

where the last inequality comes from the risk aversion hypothesis (i.e., the concavity of u). Analogously, one can show that if an agent is risk lover then his indifference curves in the state space $\mathbb{R}_+^2$ are concave.

As illustrated by Fig. 3.7, risk aversion implies that agents aim at diversifying risks. Indeed, since the indifference curves are convex in the state space $\mathbb{R}_+^2$, the convex linear combination of two gambles will always be preferred (or, at most, indifferent) to the worse of the two gambles, while if two gambles are indifferent among themselves then their convex linear combination will be preferred (or, at most, indifferent) to both of them. Of course, the opposite happens in the case of risk loving agents: they will concentrate their wealth in only one of the two gambles.

Let us now consider an agent who can trade in a market with two securities: a security that represents wealth contingent upon the occurrence of the first elementary event and a security representing wealth contingent upon the occurrence of the second elementary event (*Arrow securities*). We denote by p_1 and p_2 the price of the first and of the second security, respectively, and assume that $p_1, p_2 > 0$. Let also (e_1, e_2) denote an agent's endowment in terms of wealth in correspondence of the two possible states of the world. In the present context, the optimal consumption problem for an agent characterized by an utility function u becomes

$$\max_{(x_1, x_2) \in \mathbb{R}^2_+} \left(\pi u(x_1) + (1 - \pi)u(x_2) \right),$$

subject to the budget constraint

$$p_1 x_1 + p_2 x_2 \leq p_1 e_1 + p_2 e_2.$$

As can be readily verified, necessary and sufficient conditions for an interior solution $(x_1^*, x_2^*) \in \mathbb{R}^2_{++}$ amount to

$$\frac{\pi u'(x_1^*)}{(1 - \pi)u'(x_2^*)} = \frac{p_1}{p_2}. \tag{3.46}$$

Observe the analogy between condition (3.46) and the classical condition (1.1) obtained for the optimal consumption problem under certainty. The main difference between these two conditions is that, under uncertainty, the marginal utilities of optimal consumption in correspondence of the two states of the world are weighted by the respective probabilities of occurrence.

If the two Arrow securities are priced fairly in relative terms, meaning that $p_1/p_2 = \pi/(1 - \pi)$, then the optimal consumption will be the same in both states of the world (i.e., $x_1^* = x_2^*$), analogously to the case of a fair insurance contract described at the beginning of this section. Hence, if the two Arrow securities are priced fairly, then the agent will choose an optimal consumption vector belonging to the certainty line, thus reducing to zero his overall risk exposure. Note also that, if the Arrow securities are priced fairly, then the budget constraint (when satisfied as an equality) identifies consumption vectors (x_1, x_2) with an expected value equal to that of the initial endowment (e_1, e_2). Hence, choosing among lotteries with the same expected value, a risk averse agent will optimally achieve a risk free consumption plan by suitably investing in the two Arrow securities.

It can be easily shown that if $p_2/p_2 \neq \pi/(1 - \pi)$ (i.e., the two securities are not fairly priced in relative terms), then an agent's optimal consumption plan (x_1^*, x_2^*) does not belong to the certainty line. Indeed, as can be deduced from (3.46), if wealth contingent on the first elementary event has a relative price with respect to wealth contingent on the second elementary event higher than the fair price $\pi/(1 - \pi)$, then the optimal consumption x_1^* in correspondence of the first event will be smaller than the optimal consumption x_2^* in correspondence of the second event:

$$\frac{p_1}{p_2} > \frac{\pi}{1-\pi} \iff x_1^* < x_2^*.$$

So far, we have considered losses represented by random variables taking a finite number of possible values. However, the above results can be extended to general random variables. We shall think of a random variable $\tilde{x}$ with $\mathbb{E}[\tilde{x}] > 0$ as the potential loss incurred by an agent and assume that the agent can buy an insurance contract to offset the loss $\tilde{x}$. The price of one unit of insurance is assumed to be given by $(1+\lambda)\mathbb{E}[\tilde{x}]$, where $\lambda \geq 0$ can be interpreted as an *insurance premium*, and the payoff of one unit of the insurance contract is exactly $\tilde{x}$. The agent can choose the optimal coverage w^* (i.e., the number of units of the insurance contract bought at $t = 0$): he pays $w^*(1+\lambda)\mathbb{E}[\tilde{x}]$ at time $t = 0$ and he will receive $w^*\tilde{x}$ at time $t = 1$. In the current setting, the optimal insurance problem of an agent endowed with initial wealth $w_0 > 0$ can be formulated as follows:

$$\max_{w \in \mathbb{R}_+} \mathbb{E}\big[u(w_0 - (1-w)\tilde{x} - w(1+\lambda)\mathbb{E}[\tilde{x}])\big]. \tag{3.47}$$

The solution to problem (3.47) is given in the following proposition, where we implicitly assume enough regularity in order to ensure the existence of an interior solution (compare with Proposition 3.23).

Proposition 3.24 *Let u be a strictly increasing and strictly concave utility function and let $w_0 > 0$. Then the following hold:*

 (i) *if $\lambda = 0$ then $w^* = 1$;*
 (ii) *if $\lambda > 0$ then $w^* < 1$;*
 (iii) *if $\lambda < 0$ then $w^* > 1$.*

Proof The first order condition for an interior solution $w^* > 0$ to problem (3.47) is given by

$$\mathbb{E}\big[u'\big(w_0 - (1-w^*)\tilde{x} - w^*(1+\lambda)\mathbb{E}[\tilde{x}]\big)\big(\tilde{x} - (1+\lambda)\mathbb{E}[\tilde{x}]\big)\big] = 0.$$

Since the function u is assumed to be strictly concave, this condition is necessary and sufficient for a solution $w^* > 0$ to problem (3.47). If $\lambda = 0$, then the above condition becomes

$$0 = \mathbb{E}\big[u'\big(w_0 - (1-w^*)\tilde{x} - w^*\mathbb{E}[\tilde{x}]\big)\big(\tilde{x} - \mathbb{E}[\tilde{x}]\big)\big]$$

$$= \mathbb{E}[u'\big(w_0 - (1-w^*)\tilde{x} - w^*\mathbb{E}[\tilde{x}]\big)]\mathbb{E}\big[\tilde{x} - \mathbb{E}[\tilde{x}]\big]$$

$$\quad + \text{Cov}\,\big(u'\big(w_0 - (1-w^*)\tilde{x} - w^*\mathbb{E}[\tilde{x}]\big), \tilde{x} - \mathbb{E}[\tilde{x}]\big)$$

$$= \text{Cov}\,\big(u'\big(w_0 - (1-w^*)\tilde{x} - w^*\mathbb{E}[\tilde{x}]\big), \tilde{x} - \mathbb{E}[\tilde{x}]\big).$$

Clearly, $w^* = 1$ satisfies the latter condition, since the covariance between a random variable and a deterministic quantity is always null. The strict concavity of u ensures that $w^* = 1$ is the unique solution. Let us now consider the case $\lambda > 0$. In this case, the derivative of the expected utility with respect to w evaluated at $w = 1$ is negative:

$$\mathbb{E}\big[u'\big(w_0 - (1+\lambda)\mathbb{E}[\tilde{x}]\big)\big(\tilde{x} - (1+\lambda)\mathbb{E}[\tilde{x}]\big)\big] = -\lambda u'\big(w_0 - (1+\lambda)\mathbb{E}[\tilde{x}]\big)\mathbb{E}[\tilde{x}] < 0,$$

since $\mathbb{E}[\tilde{x}] > 0$. As a consequence, it holds that $w^* < 1$, thus proving part *(ii)* of the proposition. The proof of part *(iii)* is analogous.   □

Note that, similarly to Proposition 3.23, it the insurance contract is priced fairly (i.e., $\lambda = 0$), then a risk averse agent will choose a full insurance coverage ($w^* = 1$), thus eliminating the exposure to the risk generated by the random loss $\tilde{x}$. On the contrary, it can be shown that, for a sufficiently large insurance premium λ, the agent will choose not to buy insurance at all (see Exercise 3.29).

As in the case of portfolio choice problems with a single risky asset (see Sect. 3.1), several comparative statics results can be established on the optimal insurance demand with respect to changes in the initial wealth w_0, in the insurance premium λ and in the degree of risk aversion, as shown in the following proposition (where we always assume that the utility function u is sufficiently differentiable).

Proposition 3.25 *Let u be a strictly increasing and strictly concave utility function. Then the following hold:*

(i) *given the same initial wealth w_0, if agent a is more risk averse than agent b, then the optimal insurance demand of agent a is higher than that of agent b, i.e., $w_a^* \geq w_b^*$;*

(ii) *if the utility function u is DARA (IARA, resp.), then the optimal insurance demand w^* is decreasing (increasing, resp.) with respect to the initial wealth w_0;*

(iii) *if the utility function u is CARA or IARA, then the optimal insurance demand w^* is decreasing with respect to the insurance premium λ.*

Proof The proposition can be proved by the same arguments used to establish Propositions 3.2, 3.3 and 3.5 in the context of portfolio choice problems with a single risky asset (see Sect. 3.1), by identifying $w_0 - (1+\lambda)\mathbb{E}[\tilde{x}]$ with $w_0 r_f$, $(1-w)$ with w and $-(\tilde{x} - (1+\lambda)\mathbb{E}[\tilde{x}])$ with $\tilde{r} - r_f$ in terms of the notation used in Sect. 3.1. Part *(i)* of the proposition then follows from Proposition 3.2, while part *(ii)* follows from Proposition 3.3. Finally, part *(iii)* can be proved by relying on arguments analogous to those used in the proof of Proposition 3.5 (see Eeckhoudt et al. [631], Proposition 3.4 for more details).   □

Note that, in the case of a DARA utility function, the effect on the optimal insurance demand of an increase in the insurance premium λ is ambiguous (see Eeckhoudt et al. [631, Chapter 3] for more details and compare also with Proposition 3.5).

Finally, let us extend the above analysis by introducing an additional non-insurable risk $\tilde{y}$ (*background risk*), which is supposed to be independent of $\tilde{x}$ and with zero mean. In this case, the optimal insurance problem for a risk averse agent becomes

$$\max_{w \in \mathbb{R}_+} \left(\mathbb{E}\big[u\big(w_0 + \tilde{y} - (1-w)\tilde{x} - w(1+\lambda)\mathbb{E}[\tilde{x}]\big)\big] \right). \qquad (3.48)$$

To analyse the solution to the insurance problem (3.48) in the presence of background risk, we need to introduce the *coefficient of absolute prudence* p_u^a, for a given utility function u (tacitly assumed to be three times differentiable)[2]:

$$p_u^a(x) = -\frac{u'''(x)}{u''(x)}. \qquad (3.49)$$

We say that a risk averse agent is *prudent* if his marginal utility u' is convex, i.e., if he exhibits a positive coefficient of absolute prudence. Let us denote by w^{**} the optimal insurance demand for a risk averse agent with utility function u in the presence of the background risk $\tilde{y}$. As in the case without background risk, it can be easily shown that $w^{**} = 1$ if $\lambda = 0$. However, in the case where there exists a strictly positive insurance premium $\lambda > 0$, it holds that $w^* < w^{**} < 1$, provided that $p_u^a > 0$ and the coefficients of absolute prudence and of absolute risk aversion are both decreasing in wealth (see Eeckhoudt & Kimball [632]). In other words, under these assumptions, a prudent and risk averse agent will buy more insurance when exposed to the additional source of uncertainty represented by the background risk $\tilde{y}$, in line with economic intuition.

A similar analysis can be developed for the optimal portfolio choice problem with a single risky asset in the presence of a background risk $\tilde{y}$, independent of the random return $\tilde{r}$ of the risky asset and with $\mathbb{E}[\tilde{y}] = 0$. In this case, the optimal portfolio choice problem becomes

$$\max_{w \in \mathbb{R}} \mathbb{E}\big[u\big(w_0 r_f + \tilde{y} + w(\tilde{r} - r_f)\big)\big]. \qquad (3.50)$$

If the utility function u is increasing and concave and $\mathbb{E}[\tilde{r}] > r_f$, then the optimal portfolio w^{**} consists in a positive investment in the risky asset, thus confirming the result obtained without background risk (see Proposition 3.1). Moreover, in the presence of background risk, the optimal investment w^{**} in the risky asset is smaller than the optimal investment w^* obtained without background risk if the

---

[2]The coefficient of absolute prudence provides an approximation of the *precautionary premium* $\psi_u(\tilde{x})$ for the random variable $\tilde{x}$, defined as follows: $u'\,(\mathbb{E}[\tilde{x}] - \psi_u(\tilde{x})) = \mathbb{E}[u'(\tilde{x})]$. See also Eeckhoudt & Gollier [630] and the following section for more results on the coefficient of absolute prudence.

coefficient of absolute risk aversion is decreasing and convex with respect to wealth. A necessary condition for this to hold is that the coefficient of absolute prudence is larger than the coefficient of absolute risk aversion (see Exercise 3.32).

3.4   Optimal Saving and Consumption

In the analysis developed so far agents were only interested in maximizing the expected utility of wealth/consumption at the future date $t = 1$, while wealth/consumption at the initial date $t = 0$ did not enter explicitly into the agents' utility functions. In the present section, we shall consider portfolio problems where agents also take into account consumption at the initial date $t = 0$. In that context, an agent needs to decide the proportion of the initial wealth to be consumed at $t = 0$ and the proportion of wealth to be saved for future consumption at $t = 1$. In other words, an agent needs to optimally allocate the initial wealth between current consumption and saving for future consumption (the general multi-period case will be treated in Chap. 6).

Let us first recall the optimal saving and consumption problem under certainty, as discussed at the end of Sect. 1.1. We consider an agent endowed with wealth w_0 at time $t = 0$ and wealth w_1 at time $t = 1$ and we denote by c_0 and c_1 the consumption at $t = 0$ and $t = 1$, respectively. We denote by s the part of the initial wealth w_0 saved for future consumption and invested in a risk free asset yielding the risk free rate r_f. The optimal saving-consumption problem can then be formalized as follows, for a given utility function u, assumed to be twice differentiable, strictly increasing and strictly concave:

$$\max_{s \in \mathbb{R}} \left(u(w_0 - s) + u(w_1 + s r_f) \right). \tag{3.51}$$

The first order necessary condition for the optimality of s^* amounts to the following equality:

$$u'(w_0 - s^*) = u'(w_1 + s^* r_f) r_f. \tag{3.52}$$

Moreover, since the function u is assumed to be strictly concave, condition (3.52) is also sufficient for the optimality of s^*.

By relying on the optimality condition (3.52) and applying the implicit function theorem, we can obtain several comparative statics results on the optimal saving-consumption problem, analogously to the case of portfolio choice problems discussed in section "The Case of a Single Risky Asset". In particular, we can establish that the optimal saving s^* is increasing in w_0 and decreasing in w_1. The first claim follows by noting that

$$\frac{\partial s^*}{\partial w_0} = \frac{u''(w_0 - s^*)}{u''(w_0 - s^*) + u''(w_1 + s^* r_f) r_f^2} > 0,$$

due to the strict concavity of u. The fact that s^* is decreasing with respect to w_1 can be established similarly. This shows the preference of risk averse agents for *consumption smoothing*, i.e., risk averse agents prefer to consume a similar amount of wealth in both time periods $t = 0$ and $t = 1$.

With a similar reasoning, we can also study the effect on the optimal saving s^* of a change in the risk free rate r_f. Indeed, by the implicit function theorem,

$$\frac{\partial s^*}{\partial r_f} = \frac{-u''(w_1 + s^* r_f) s^* r_f - u'(w_1 + s^* r_f)}{u''(w_0 - s^*) + u''(w_1 + s^* r_f) r_f^2}. \tag{3.53}$$

Observe that, as in the case of portfolio choice problems (see Proposition 3.5), the overall effect of a change in the risk free rate r_f is undetermined, since it depends both on the income and on the substitution effect. However, if $s^* > 0$, a sufficient condition for the optimal saving s^* to be increasing with respect to r_f is that the coefficient of relative risk aversion is smaller than one (see Exercise 3.33).

Let us now consider a risky setting where at date $t = 1$ agents face a risk represented by a random variable $\tilde{y}$. Without loss of generality, we assume that $\mathbb{E}[\tilde{y}] = 0$. For example, such a random variable can represent the uncertain labor income received at time $t = 1$. Intuitively, for a risk averse agent, this additional uncertainty should lead to an increase in the optimal saving. In the present risky context, the optimal saving-consumption problem becomes

$$\max_{s \in \mathbb{R}} \left(u(w_0 - s) + \mathbb{E}[u(w_1 + s r_f + \tilde{y})] \right). \tag{3.54}$$

We denote by s^{**} the optimal saving in the presence of the additional risk $\tilde{y}$ in order to distinguish it from the optimal saving s^* obtained under certainty as a solution to problem (3.51). The optimality condition for the optimal saving s^{**} gives

$$u'(w_0 - s^{**}) = \mathbb{E}[u'(w_1 + s^{**} r_f + \tilde{y}) r_f].$$

By comparing the above optimality condition with the optimality condition (3.52) under certainty, we see that there exists an "extra saving" term $s^{**} - s^*$ induced by the future uncertainty represented by the random variable $\tilde{y}$. We call *precautionary saving* the difference $s^{**} - s^*$. As shown in the following proposition (see also Kimball [1100]), the precautionary saving is linked to the coefficient of absolute prudence introduced in (3.49). As before, we always assume that $r_f > 0$.

Proposition 3.26 *In problem* (3.54), *if* $p_u^a(x) \geq 0$ *for all* $x \in \mathbb{R}_+$, *then* $s^{**} - s^* \geq 0$.

Proof Let the function $H : s \mapsto H(s)$, denote the expected utility in correspondence of a saving level $s \in \mathbb{R}$, i.e.,

$$H(s) := u(w_0 - s) + \mathbb{E}[u(w_1 + s r_f + \tilde{y})].$$

Note that, since $u'' < 0$, the function H is concave. As a consequence, in order to prove that $s^{**} - s^* \geq 0$ it is enough to show that $H'(s^*) \geq 0$. By (3.52), it holds that, under the standing assumption $\mathbb{E}[\tilde{y}] = 0$,

$$H'(s^*) = -u'(w_0 - s^*) + r_f \, \mathbb{E}[u'(w_1 + s^* r_f + \tilde{y})]$$
$$= -u'(w_1 + s^* r_f + \mathbb{E}[\tilde{y}])r_f + r_f \, \mathbb{E}[u'(w_1 + s^* r_f + \tilde{y})].$$

Therefore,

$$H'(s^*) \geq 0 \quad \Longleftrightarrow \quad \mathbb{E}[u'(w_1 + s^* r_f + \tilde{y})] \geq u'(w_1 + s^* r_f + \mathbb{E}[\tilde{y}]) = u'(w_1 + s^* r_f).$$

Since $p_u^a(x) \geq 0$ for all $x \in \mathbb{R}_+$, the function u' is convex, so that the inequality on the right-hand side holds true, as a consequence of Jensen's inequality. In turn, this implies that $H'(s^*) \geq 0$, thus proving the claim.   □

Results similar to those on risk aversion (see, e.g., Propositions 2.5 and 2.6) can be also established with respect to prudence. In particular, analogously to the coefficient of risk aversion, the coefficient of absolute prudence can be increasing or decreasing in wealth. Decreasing absolute prudence implies that wealthier agents are less inclined towards precautionary saving. Note that a DARA utility function exhibits prudence, since the fact that the function $x \mapsto r_u^a(x)$ is decreasing implies that $p_u^a(x) \geq r_u^a(x)$, for all $x \in \mathbb{R}_+$, as can be easily checked provided that u''' exists, see Exercise 2.8.

Concerning the behavior of the optimal saving s^{**} with respect to changes in the risk free rate r_f, in the presence of the additional risk $\tilde{y}$, if the coefficient of relative risk aversion is smaller than one then the optimal saving s^{**} is increasing with respect to the risk free rate r_f (compare with Exercise 3.33).

Let us now consider what happens if the riskiness of the labor income represented by the random variable $\tilde{y}$ increases. To this end, we replace the random variable $\tilde{y}$ with a random variable $\tilde{z}$ such that $\tilde{z} =^d \tilde{y} + \tilde{\epsilon}$, where $\tilde{\epsilon}$ is a random variable with $\mathbb{E}[\tilde{\epsilon}|\tilde{y}] = 0$. In this case, the same arguments used in the proof of Proposition 3.26 allow us to show that if the agent is prudent then the optimal saving will increase.

We conclude the present section by considering the optimal saving-consumption problem in the presence of a risky asset with random return $\tilde{r}$. This means that the agent has the possibility of saving at time $t = 0$ an amount s and investing that amount into the risky asset, yielding the random wealth $s\tilde{r}$ at the future time $t = 1$. The optimal saving-consumption problem then becomes

$$\max_{s \in \mathbb{R}} \left(u(w_0 - s) + \mathbb{E}[u(w_1 + s\tilde{r})] \right).$$

Let s^{***} denote the optimal solution to this problem. As before, the optimality condition for s^{***} gives

$$u'(w_0 - s^{***}) = \mathbb{E}[u'(w_1 + s^{***}\tilde{r})\tilde{r}]$$

and an application of the implicit function theorem confirms that the optimal saving s^{***} is increasing with respect to the initial wealth w_0:

$$\frac{\partial s^{***}}{\partial w_0} = \frac{u''(w_0 - s^{***})}{u''(w_0 - s^{***}) + \mathbb{E}[u''(w_1 + s^{***}\tilde{r})\tilde{r}^2]} \geq 0,$$

similarly as above. Concerning the effect on the optimal saving of an increase in the riskiness of the random return of the risky asset, it can be shown that the optimal saving s^{***} increases if $2u''(x) + u'''(x)x > 0$ for all $x \in \mathbb{R}_+$, i.e., $xp_u(x) > 2$. If the coefficient of relative risk aversion is greater than one and decreasing, then s^{***} will increase as the riskiness of the risky asset return increases (see Exercise 3.34). In particular, this is the case of a power utility function with relative risk aversion coefficient greater than one (compare also with Eeckhoudt et al. [631, Section 6.3]).

In a multi-period setting, we will often consider a separable power utility function of the following form, for $b > 0$:

$$u(x_0, x_1) = \frac{b}{b-1}x_0^{1-\frac{1}{b}} + \frac{b}{b-1}x_1^{1-\frac{1}{b}}.$$

In this case, the absolute risk aversion coefficient is given by $1/(bx)$ and the relative risk aversion coefficient is $1/b$. The marginal rate of substitution between consumption at date $t = 0$ and consumption at date $t = 1$ is $(x_1/x_0)^{-1/b}$ and the elasticity of intertemporal substitution of consumption is b. Indeed:

$$\frac{\partial_{x_1} u(x_0, x_1)}{\partial_{x_0} u(x_0, x_1)} = \left(\frac{x_1}{x_0}\right)^{-\frac{1}{b}},$$

and

$$-\frac{d \log(x_1/x_0)}{d \log\left(\frac{\partial_{x_1} u(x_0,x_1)}{\partial_{x_0} u(x_0,x_1)}\right)} = -\left(\frac{x_1}{x_0}\right)^{-\frac{1}{b}} \frac{x_0}{x_1} \frac{d(x_1/x_0)}{d(x_1/x_0)^{-\frac{1}{b}}} = b.$$

Observe also that the elasticity of intertemporal substitution of consumption b is the reciprocal of the coefficient of relative risk aversion $1/b$. This fact represents a severe limitation of the time-additive power utility function, as we are going to discuss in more detail in Chaps. 6 and 9.

3.5 Notes and Further Readings

We refer the interested reader to Leland [1180] concerning the existence of an optimal solution to problem (3.2). In Chap. 4, we will also relate the existence of an optimal portfolio to the absence of arbitrage opportunities.

In the case of a single risky asset, as considered in the first part of Sect. 3.1, it is rather difficult to provide further comparative statics results beyond Proposition 3.5. In particular, it is difficult to analyse the behavior of the optimal demand of the risky asset as the riskiness of the risky asset increases or, similarly, when the initial wealth becomes random. Intuitively, one would expect that the optimal demand of a risk averse agent decreases as the riskiness of the asset increases. However, this is not always the case. Indeed, let us consider the portfolio problem (3.2) with two assets: a risk free asset with return r_f and a risky asset with random return $\tilde{r}_1$ and a positive risk premium. Let us now consider the same problem with a risky return $\tilde{r}_2$ such that $\tilde{r}_1 \succeq_{\text{SSD}} \tilde{r}_2$, i.e., $\tilde{r}_2 \equiv^d \tilde{r}_1 + \tilde{\epsilon}$, where $\mathbb{E}[\tilde{\epsilon}|\tilde{r}_1] = 0$ (the return $\tilde{r}_2$ is equal in distribution to $\tilde{r}_1$ plus a noise component which increases the riskiness). In Rothschild & Stiglitz [1473] and Hadar & Seo [872], it has been shown that the demand of the risky asset with return $\tilde{r}_1$ by a risk averse agent can be greater or smaller than that of the risky asset $\tilde{r}_2$. A sufficient condition for the (positive) demand of the asset with return $\tilde{r}_1$ to be larger than that of the risky asset with return $\tilde{r}_2$ is that the coefficient of relative risk aversion is increasing and less or equal than one and that the coefficient of absolute risk aversion is decreasing (see Rothschild & Stiglitz [1473]). The same effect is observed if relative prudence is positive and less than two (see Hadar & Seo [872]). Furthermore, if the change in the riskiness of the risky asset leads to a random return which is dominated according to first order stochastic dominance, then the optimal demand will decrease if the coefficient of relative risk aversion is less than one (see Hadar & Seo [872]).

As a related situation, let us consider the optimal portfolio problem (3.2) in the case where the initial wealth becomes $w_0 + \tilde{\epsilon}$, where $\tilde{\epsilon}$ is a random variable with zero mean, independent of $\tilde{r}$. We call $\tilde{\epsilon}$ a *background risk* (which may be interpreted as labor income) and denote by w^{**} the optimal portfolio in the case of random initial wealth and by w^* the optimal portfolio without background risk. One would guess that, if $\mathbb{E}[\tilde{r}] - r_f > 0$, a risk averse agent with a strictly increasing utility function would choose an optimal portfolio w^{**} such that $0 \leq w^{**} \leq w^*$, since the random initial wealth increases the overall risk and, hence, the optimal choice would be to invest less in the risky asset. Actually, this does not always hold true for a risk averse agent. It can be shown (Eeckhoudt et al. [631, Proposition 4.3]) that the above conjecture is true under the additional assumption that the coefficient of absolute risk aversion is decreasing and convex (for instance, the coefficient of absolute risk aversion of a power utility function is decreasing and convex).

As mentioned above, the second order stochastic dominance criterion does not allow us to establish that all risk averse agents reduce the demand after a shift in riskiness dominated according to this criterion. Gollier [799] provides a criterion (*central dominance*) that allows to establish that all risk averse agents reduce the demand of a risky asset after a shift in its distribution. According to this criterion, if the random return of the risky asset changes from the distribution function F_1 to the distribution function F_2, then every risk averse agent reduces his optimal demand of the risky asset if and only if there exists γ such that $\int_0^y (F_2(z) - \gamma F_1(z))dz \geq 0$, for all $y \in [0, 1]$ (assuming that the distribution is normalized onto $[0, 1]$). The central dominance and the second order stochastic dominance criterion cannot be compared (see also Gollier [800, Chapter 6]).

Further stochastic dominance results have been established by several authors (we refer to Levy [1204, 1205], Sriboonchitta et al. [1560] for surveys). If two asset returns are independently distributed with positive variance and the same mean (in particular, if they are identically distributed), then both assets must enter the optimal portfolio with a strictly positive weight, see Hadar & Russell [869], Hadar et al. [871]. The result has been extended in Hadar & Russell [870] to the case of two dependent risky assets provided that their marginal distributions are identical.

Ross [1468] establishes the following interesting result, making use of the notion of strong risk aversion (see Sect. 2.5). Consider two risky assets, the first one with a higher expected return and riskier than the second one (i.e., two random returns $\tilde{r}^1$ and $\tilde{r}^2$ such that $\mathbb{E}[\tilde{r}^1 - \tilde{r}^2 | \tilde{r}^2] \geq 0$). If agent a is strongly more risk averse than agent b, then the optimal demand of agent a of the first asset will be less than that of agent b. Agent a, being strongly more risk averse, will choose a less risky portfolio with a lower expected return. When the initial wealth is stochastic and the asset has a non-negative risk premium conditional on wealth, the strongly more risk averse agent's demand will be smaller than that of a less risk averse agent (see Ross [1468]).

An abstract and general framework allowing for a unified treatment of several mean-variance optimization problems has been recently proposed in Fontana & Schweizer [725] by relying on Hilbert space methods. The portfolio frontier has been derived allowing for short sales (see e.g. Dybvig [606]): imposing a no short sale constraint on risky assets (i.e., $w_n \geq 0$, for all $n = 1, \ldots, N$), the portfolio frontier of course changes. Indeed, if short sale constraints are introduced, then we can observe kinks in the portfolio frontier.

In Proposition 3.11 we have provided a general two mutual funds separation result in the case of HARA utility functions, in the presence of a risk free asset (*monetary separation*). In the absence of a risk free asset, a two mutual funds separation result still holds with respect to two risky mutual funds (*two risky mutual funds separation*) for any utility function belonging to the HARA class, provided that a complete set of Arrow securities is traded (i.e., markets are complete), as shown in Cass & Stiglitz [376, Theorem 4.1]. For a general set of traded securities, in a possibly incomplete market, the class of utility functions such that the two funds separation result holds is restricted to the class of utility functions satisfying

$$u'(x) = bx^c \quad \text{or} \quad u'(x) = a + bx,$$

see Cass & Stiglitz [376, Theorem 5.1]. Quadratic utility functions as well as CRRA utility functions (power and logarithmic utility functions) satisfy this condition.

Mutual fund separation results have been tested empirically. In particular, two mutual funds (monetary) separation results have been tested empirically with negative results, see Lo & Wang [1240] for an analysis based on trading volume and Canner et al. [360] for an analysis based on portfolio allocations proposed by financial advisors (for a reconciliation see Bajeux-Besnainou et al. [182]). Kroll et al. [1137] present experimental evidence showing that people often choose inefficient portfolios and violate the two mutual funds separation.

3.6 Exercises

Exercise 3.1 Consider a strictly risk averse agent endowed with initial wealth w_0 and with a strictly increasing and twice differentiable utility function. Let r_f and $\tilde{r}$ denote the return of the risk free asset and of the risky asset, respectively. Show that the minimum risk premium $\mathbb{E}[\tilde{r} - r_f]$ of the risky asset required by the agent to invest the totality of his wealth in the risky asset approximatively satisfies the following inequality

$$\mathbb{E}[\tilde{r} - r_f] \geq r_u^a(w_0 \, r_f) w_0 \, \mathbb{E}[(\tilde{r} - r_f)^2].$$

Exercise 3.2 Consider a quadratic utility function $u(x) = x - \frac{b}{2}x^2$, an initial wealth $w_0 = 100$, a risk free rate $r_f = 1.1$ and a risky asset with expected return $\mathbb{E}[\tilde{r}] = 1.3$ and variance $\sigma^2(\tilde{r}) = 1.5$.

  (i) Determine the optimal portfolio w^* for $b = 0.006$.
 (ii) Determine for which values of b we have $w^* < 0$.
(iii) How does w^* change if the risk free rate increases to $r_f' = 1.2$?
(iv) How does w^* change if the initial wealth increases to $w_0' = 150$?
 (v) How does w^* change if its expected return decreases to $\mathbb{E}[\tilde{r}'] = 1.2$?

Exercise 3.3 Consider the optimal portfolio choice problem in the presence of N risky assets with returns $(\tilde{r}_1, \ldots, \tilde{r}_N)$ and of a risk free asset with return $r_f > 0$. Suppose that there are no redundant assets in the economy in the sense that the random variables $(r_f, \tilde{r}_1, \ldots, \tilde{r}_N)$ are linearly independent. Show that the optimal portfolio $w^* \in \mathbb{R}^N$ solving problem (3.2), for a strictly increasing and strictly concave utility function u, is unique.

Exercise 3.4 (LeRoy & Werner [1191], Theorem 13.5.1) Consider a strictly increasing and strictly concave utility function u and suppose that there exist N risky assets whose returns $(\tilde{r}_1, \ldots, \tilde{r}_N)$ admit the representation

$$\tilde{r}_n = \sum_{\substack{k=1 \\ k \neq n}}^{N} \lambda_k \tilde{r}_k + \tilde{\epsilon}_n, \qquad \text{for all } n = 1, \ldots, N,$$

where $\sum_{k=1, k \neq n}^{N} \lambda_k = 1$, for all $n = 1, \ldots, N$, and $\tilde{\epsilon}_n$ is a random variable satisfying

$$\mathbb{E}[\tilde{\epsilon}_n | \tilde{r}_1, \ldots, \tilde{r}_{n-1}, \tilde{r}_{n+1}, \ldots, \tilde{r}_N] = \mathbb{E}[\tilde{\epsilon}_n], \tag{3.55}$$

for all $n = 1, \ldots, N$. Show that $w_n^* > 0$ if and only if $\mathbb{E}[\tilde{\epsilon}_n] > 0$, for all $n = 1, \ldots, N$.

Exercise 3.5 (LeRoy & Werner [1191], Corollary 13.5.2) Consider a strictly increasing and strictly concave utility function u and suppose that there exist a risk free asset with return $r_f > 0$ and N risky assets whose returns $(\tilde{r}_1, \ldots, \tilde{r}_N)$ satisfy the condition

$$\mathbb{E}[\tilde{r}_n | \tilde{r}_1, \ldots, \tilde{r}_{n-1}, \tilde{r}_{n+1}, \ldots, \tilde{r}_N] = \mathbb{E}[\tilde{r}_n],$$

for all $n = 1, \ldots, N$. Show that $w_n^* > 0$ if and only if $\mathbb{E}[\tilde{r}_n] > r_f$, for all $n = 1, \ldots, N$.

Exercise 3.6 Consider two risky assets with returns $\tilde{r}_1, \tilde{r}_2$, with corresponding expected returns e_1, e_2, variances σ_1^2, σ_2^2 and correlation ρ. Let $(w, 1 - w)$ denote a portfolio of the two risky assets, with corresponding expected return $\mathbb{E}[\tilde{r}]$ and variance $\sigma^2(\tilde{r})$. Verify the following claims:

(i) if $0 \le w \le 1$ then $\sigma^2(\tilde{r}) \le \max\{\sigma_1^2; \sigma_2^2\}$;
(ii) if $\rho = 1$ then $\sigma^2(\tilde{r}) = 0$ for $w = -\sigma_2/(\sigma_1 - \sigma_2)$;
(iii) if $\rho = -1$ then $\sigma^2(\tilde{r}) = 0$ for $w = \sigma_2/(\sigma_1 + \sigma_2)$;

Exercise 3.7 In the context of Proposition 3.4, show that $\mathbb{E}[u''(\widetilde{W}^*)\widetilde{W}^*(\tilde{r} - r_f)] \ge 0$ if the utility function u exhibits decreasing relative risk aversion.

Exercise 3.8 Consider an economy with a risk free asset with return r_f and a risky asset whose random return $\tilde{r}$ can take two possible values $\{d, u\}$ with probabilities $\{\pi, 1 - \pi\}$, respectively. Assume that $d < r_f < u$. Determine the optimal demand of the risky asset of an agent endowed with initial wealth w_0 according to the following utility functions:

(i) $u(x) = \sqrt{x}$;
(ii) $u(x) = \log(x)$;
(iii) $u(x) = x^\gamma/\gamma$, with $\gamma \ne 1$.

Exercise 3.9 Consider an exponential utility function $u(x) = -\frac{1}{a}\exp(-ax)$, with $a > 0$, and an economy with a risk free asset with return $r_f = 1.1$ and two risky assets with random returns $\tilde{r}_1, \tilde{r}_2$ distributed as normal random variables with means $e_1 = 1.2$ and $e_2 = 1.3$ and variances $\sigma_1^2 = 4$ and $\sigma_2^2 = 9$, respectively, with correlation coefficient $\rho \in (-1, 1)$.

(i) Determine when the optimal portfolio is diversified, i.e., when $w_1^*, w_2^* > 0$.
(ii) Suppose that ρ is such that the optimal portfolio is diversified (take for instance $\rho = 0.5$). If ρ increases, how should σ_1 and σ_2 vary in order for the optimal portfolio to remain diversified?
(iii) Suppose that $\rho \in \{-0.5, 0.5\}$. Give conditions on a such that the optimal portfolio invests more than one unit of wealth in the risky assets.

Exercise 3.10 Show that the covariance between the returns $\tilde{r}_{w*1}$ and $\tilde{r}_{w*2}$ of two frontier portfolios w^{*1} and w^{*2} is given by formula (3.24).

Exercise 3.11 Let w^{MVP} denote the minimum variance portfolio. Show that, for any frontier portfolio w^*, it holds that $\mathrm{Cov}(\tilde{r}_{w^{\mathrm{MVP}}}, \tilde{r}_{w^*}) = 1/C = \sigma^2(\tilde{r}_{w^{\mathrm{MVP}}})$.

Exercise 3.12 Given an arbitrary frontier portfolio w^*, with $w^* \neq w^{\mathrm{MVP}}$, there exists a unique frontier portfolio w^{zc} such that $\mathrm{Cov}(\tilde{r}_{w^*}, \tilde{r}_{w^{zc}}) = 0$.

Exercise 3.13 Determine the frontier portfolio w^* such that its variance is equal to the variance of its zero correlation portfolio, i.e., determine the frontier portfolio w^* such that $\sigma^2(\tilde{r}_{w^*}) = \sigma^2(\tilde{r}_{w^{zc}})$, where w^{zc} is the zero correlation portfolio with respect to w^*.

Exercise 3.14 Given a frontier portfolio w^*, show that the expected return $\mathbb{E}[\tilde{r}_{w^{zc}}]$ of its zero correlation portfolio w^{zc} is identified, in the variance-expected return plane, by the intersection of the line connecting the points $(\sigma^2(\tilde{r}_{w^*}), \mathbb{E}[\tilde{r}_{w^*}])$ and $(\sigma^2(\tilde{r}_{w^{\mathrm{MVP}}}), \mathbb{E}[\tilde{r}_{w^{\mathrm{MVP}}}])$ with the vertical axis. Similarly, show that $\mathbb{E}[\tilde{r}_{w^{zc}}]$ is identified, in the standard deviation-expected return plane, by the intersection of the tangent to the portfolio frontier at the point $(\sigma^2(\tilde{r}_{w^*}), \mathbb{E}[\tilde{r}_{w^*}])$ with the vertical axis.

Exercise 3.15 Consider a portfolio $w^p \notin PF$. Show that, in the variance - expected return plane, the line which connects the two points $(\sigma^2(\tilde{r}_{w^p}), \mathbb{E}[\tilde{r}_{w^p}])$ and $(\sigma^2(\tilde{r}_{w^{\mathrm{MVP}}}), \mathbb{E}[\tilde{r}_{w^{\mathrm{MVP}}}])$ intercepts the expected return axis at the level $\mathbb{E}[\tilde{r}_{w^q}]$, where w^q is the portfolio such that $\mathrm{Cov}(\tilde{r}_{w^q}, \tilde{r}_{w^p}) = 0$ and with the minimum variance among all portfolios with zero correlation with w^p.

Exercise 3.16 Let us consider an economy with two risky assets with returns $\tilde{r}_1$ and $\tilde{r}_2$. Show that the explicit formula $w = (\mu - \mathbb{E}[\tilde{r}_2])/(\mathbb{E}[\tilde{r}_1] - \mathbb{E}[\tilde{r}_2])$ is a special case of formula (3.16) in the case $N = 2$.

Exercise 3.17 Let w^p be a frontier portfolio and w^q be an arbitrary portfolio (i.e., not necessarily belonging to the portfolio frontier) such that $\mathbb{E}[\tilde{r}_{w^q}] = \mathbb{E}[\tilde{r}_{w^p}]$. Show that $\mathrm{Cov}(\tilde{r}_{w^q}, \tilde{r}_{w^p}) = \sigma^2(\tilde{r}_{w^p})$.

Exercise 3.18 Consider an economy with N risky assets with random returns $(\tilde{r}_1, \ldots, \tilde{r}_N)$ and a risk free asset with return r_f, as in section "The Case of N Risky Assets and a Risk Free Asset", and suppose that $r_f \neq A/C$. Show that any portfolio w^* belonging to the portfolio frontier PF* can be expressed as the linear combination of the risk free asset and the tangent portfolio w^e, so that

$$w^* = \alpha w^e = \alpha \, \frac{e - r_f \mathbf{1}}{\mathbf{1}^{\top} V^{-1}(e - r_f \mathbf{1})},$$

for some α, where the second equality follows from the proof of Proposition 3.15

Exercise 3.19 Let $\mu > A/C$ and consider the following optimization problem:

$$\max_{w \in \Delta_N} \frac{w^{\top} e - \mu}{\sqrt{w^{\top} V w}}, \tag{3.56}$$

corresponding to the maximization of the *Sharpe ratio* with respect to the reference rate of return μ over the set of all portfolios investing in the N risky assets. Prove that, for any $\mu > A/C$, the solution w^* to problem (3.56) is given by

$$w^* = g + h\frac{\mu A - B}{\mu C - A} = \frac{V^{-1}(e - 1\mu)}{1^\mathsf{T} V^{-1}(e - 1\mu)}. \tag{3.57}$$

Exercise 3.20 Consider an economy with N risky assets with returns $(\tilde{r}_1, \ldots, \tilde{r}_N)$ and a risk free asset with return r_f, with $r_f < \mathbb{E}[\tilde{r}_n]$ for all $n = 1, \ldots, N$. Show that the tangent portfolio w^e is diversified if the risky asset returns are uncorrelated.

Exercise 3.21 Consider a risk free asset with return $r_f = 1.1$ and two risky assets with normally distributed returns $(\tilde{r}_1, \tilde{r}_2)$ with $\mathbb{E}[\tilde{r}_1] = 1.2$, $\mathbb{E}[\tilde{r}_2] = 1.3$ and $\sigma^2(\tilde{r}_1) = 4$, $\sigma^2(\tilde{r}_2) = 9$ and correlation ρ.

 (i) Determine the portfolio frontier composed by the two risky assets only.
 (ii) In the case where $\rho = 0.5$, determine the minimum variance portfolio w^{MVP} and the corresponding variance $\sigma^2(\tilde{r}_{w\mathrm{MVP}})$.
 (iii) In the case where $\rho = 0.5$, determine the tangent portfolio w^e.
 (iv) Consider the problem of an agent maximizing the expected exponential utility (with $u(x) = -\exp(-ax)$). Show that the optimal portfolio $w^*(a)$ for such an agent is given by a multiple of the tangent portfolio w^e, as shown at the end of Sect. 3.2. Give conditions on the risk aversion parameter a so that the investment in the portfolio w^e is greater than one.
 (v) Determine the optimal portfolio that minimizes the variance, with and without the risk free asset, for the given expected return $\mu = 1.25$.
 (vi) Verify that the portfolio $w' = (0.6, 0.5)$ (with the remaining proportion of wealth being invested in the risk free asset) does not belong to the portfolio frontier.

Exercise 3.22 Consider an economy with a risk free asset with return $r_f = 1.13$ and two risky assets with normally distributed returns with expected values $\mathbb{E}[\tilde{r}_1] = 1.16$, $\mathbb{E}[\tilde{r}_2] = 1.25$ and variances $\sigma^2(\tilde{r}_1) = 2$, $\sigma^2(\tilde{r}_2) = 4$ and correlation ρ.

 (i) Determine the portfolio frontier composed by the risky assets in the two cases $\rho = 0.5$ and $\rho = -0.5$.
 (ii) Determine the minimum variance portfolio w^{MVP} in the two cases $\rho = 0.5$ and $\rho = -0.5$.
 (iii) Determine the tangent portfolio in the two cases $\rho = 0.5$ and $\rho = -0.5$.
 (iv) For $\rho = -0.5$, consider an agent with quadratic utility function $u(x) = x - \frac{b}{2}x^2$ and determine his optimal portfolio. Give conditions on b in order that the optimal investment in the tangent portfolio w^e is greater than one.

Exercise 3.23 Consider two risky assets with random returns $\tilde{r}_1$ and $\tilde{r}_2$, with expected values e_1 and e_2, respectively, and variances σ_1^2 and σ_2^2 (with $\sigma_1^2 \leq \sigma_2^2$), respectively, and correlation ρ. For a portfolio $(w, 1-w)$, denote by $\mathbb{E}[\tilde{r}_w]$ and $\sigma^2(\tilde{r}_w)$ the expectation and the variance, respectively, of the corresponding random return $\tilde{r}_w$. Verify the following claims.

(i) If $\rho < \sigma_1/\sigma_2$ then there exists $w \in (0,1)$ such that $\sigma(\tilde{r}_w) < \sigma_1$ and, for all $w \notin [0,1]$, it holds that $\sigma(\tilde{r}_w) \geq \sigma_1$.

(ii) If $\rho = \sigma_1/\sigma_2$ then $\sigma(\tilde{r}_w) > \sigma_1$ for every $w \in \mathbb{R}$.

(iii) If $\rho > \sigma_1/\sigma_2$ then there exists $w \notin (0,1)$ such that $\sigma(\tilde{r}_w) < \sigma_1$ and, for all $w \in (0,1)$, it holds that $\sigma(\tilde{r}_w) > \sigma_1$.

(iv) Show that, if $\sigma_1 = \sigma_2 =: \sigma$, then the minimum variance portfolio is given by $w^{\mathrm{MVP}} = (1/2, 1/2)^\top$, independently of the value of the correlation coefficient ρ. What happens in the case $\rho = 1$?

Exercise 3.24 Show that, if the mutual fund separation property holds for $K = 1$, i.e., there exists a portfolio w^* such that $\mathbb{E}[u(\tilde{r}_{w^*})] \geq \mathbb{E}[u(\tilde{r}_w)]$ for any $w \in \Delta_N$ and for any concave utility function u, then w^* must coincide with the minimum variance portfolio.

Exercise 3.25 Consider an economy with a risk free asset with return r_f and N risky assets with i.i.d. random returns $\tilde{\mathbf{r}} = (\tilde{r}_1, \ldots, \tilde{r}_N)$. Show that the two mutual funds separation property holds with respect to the risk free asset and the equally weighted portfolio $(1/N, \ldots, 1/N)^\top$ of the risky assets.

Exercise 3.26 Let us consider an economy with N risky assets with normally distributed returns $\tilde{\mathbf{r}} = (\tilde{r}_1, \ldots, \tilde{r}_N)$ with $\mathbb{E}[\tilde{r}_n] = \mathbb{E}[\tilde{r}_1]$, for all $n = 2, \ldots, N$, and let w^{MVP} denote the minimum variance portfolio. Show that the one fund separation property holds with respect to w^{MVP}.

Exercise 3.27 Consider the optimal insurance problem of an agent with quadratic utility function $u(x) = x - \frac{b}{2}x^2$, with initial wealth w_0, exposed to the possibility of a loss $D > 0$ which can occur at time $t = 1$ with probability $\pi \in (0,1)$. Let p be the price of one unit of wealth contingent on the occurrence of the loss event and denote by w^* the agent's optimal insurance demand. Verify that $w^* = D$ if $p = \pi$ and that $w^* < D$ if $p > \pi$.

Exercise 3.28 Consider a risky setting with two possible states of the world at the time $t = 1$, with probabilities π and $1 - \pi$, and an agent with power utility function $u(x) = x^\gamma$, with $\gamma \in (0,1)$. Verify that, if the prices of wealth contingent on the two states of the world are not fair, i.e., $p_1/p_2 > \pi/(1 - \pi)$, then the optimal consumption in the two states of the world satisfies $x_1^* < x_2^*$.

Exercise 3.29 In the setting of Proposition 3.24, for a given utility function u, define λ^* by

$$\lambda^* = \frac{\mathrm{Cov}\left(\tilde{x}, u'(w_0 - \tilde{x})\right)}{\mathbb{E}[\tilde{x}]\mathbb{E}[u'(w_0 - \tilde{x})]}.$$

Show that, if the insurance premium λ is greater than λ^*, then the optimal insurance demand w^* consists in buying zero units of the insurance contract.

Exercise 3.30 Consider a normally distributed loss $\tilde{x} \sim \mathcal{N}(\mu, \sigma^2)$ with $\mu > 0$ and an agent with exponential utility function $u(x) = -\exp(-ax)/a$, for $a > 0$. Verify that, in the setting of Proposition 3.24, if $\lambda > 0$, then the optimal insurance demand w^* is less than one.

Exercise 3.31 (Eeckhoudt et al. [631], Proposition 3.5) Consider a random loss $\tilde{x}$ which can take N possible ordered values $0 < x_1 < \ldots < x_N$ with probabilities $\pi_1, \ldots, \pi_N$, with $\sum_{n=1}^{N} \pi_n = 1$. Consider an insurance contract paying a non-negative indemnity $I(\tilde{x})$ at time $t = 1$, with price $p = (1 + \lambda)\mathbb{E}[I(\tilde{x})]$ at time $t = 0$. Show that, for any risk averse agent, an insurance contract with an indemnity of the form $I^*(\tilde{x}) = \max\{0; \tilde{x} - K\}$, for some $K > 0$ is optimal among all insurance contracts with price p.

Exercise 3.32 (Eeckhoudt et al. [631], Proposition 4.3) In the same setting as at the end of Sect. 3.3, show that, in the presence of background risk, the optimal investment w^{**} in the risky asset is smaller than the optimal investment w^* obtained in the case without background risk if the coefficient of absolute risk aversion is decreasing and convex with respect to wealth. Show also that a necessary condition is that the coefficient of absolute prudence is larger than the coefficient of absolute risk aversion.

Exercise 3.33 Consider the optimal saving-consumption problem under certainty, as presented at the beginning of Sect. 3.4. Show that the optimal saving s^* is increasing with respect to the risk free rate r_f if the coefficient of relative risk aversion is less than one.

Exercise 3.34 In the context of the optimal saving-consumption problem in the presence of a risky asset, as described at the end of Sect. 3.4, show that if the coefficient of relative risk aversion is greater than one and decreasing, then the condition $xp_u(x) > 2$, for all $x \in \mathbb{R}$, is satisfied. In particular, the optimal saving s^{***} increases with respect to an increase in the riskiness of the risky asset's return.

Chapter 4
General Equilibrium Theory and No-Arbitrage

An economy is in equilibrium when it produces messages which do not induce the agents to modify the theories they believe in or the policies which they pursue.

Hahn (1973)

In a rational expectations equilibrium, not only are prices determined so as to equate supply and demand, but individual economic agents correctly perceive the true relationship between the non price information received by the market participants and the resulting equilibrium market price. This contrasts with the ordinary concept of equilibrium in which the agents respond to prices but do not attempt to infer other agents' non price information from actual market prices.

Radner (1982)

Relying on the perfect competition hypothesis, similarly to the situation considered in Chap. 1 under certainty, the analysis of an economy under risk can be decomposed into two steps: first, the agents' individual behavior is studied by taking prices/returns as given (*internal consistency*); in a second step, the interaction of the agents in the market is studied. In this second step, the focus is on the determination of an equilibrium price vector, i.e., a price vector such that the market demand equals the total supply and, therefore, agents' decisions are compatible among themselves (*external consistency*). In Chaps. 2 and 3, we have addressed only the first step of the problem, assuming that at the initial date $t = 0$ agents trade wealth contingent on the state of the world realized in $t = 1$ or assets with dividends described by generic random variables. In the present chapter, we will address the second step.

In order to illustrate the topics dealt with in this chapter, let us first present an elementary example. Consider a farmer who owns a field of corn and, at date $t = 0$, has to take a consumption-investment decision. At $t = 0$, the farmer must decide how to allocate among a set of assets the wealth obtained from the previous harvest (investment decision), deciding how much to consume today (at $t = 0$) and how much to consume at the future date $t = 1$ in correspondence of the possible states of the world (consumption decision). We will consider two different settings: investment decisions taken at $t = 0$ with consumption only at the future date $t = 1$ and investment decisions taken at $t = 0$ with consumption at both dates $t = 0$ and $t = 1$. In order to allocate his wealth, our farmer has the possibility of trading

© Springer-Verlag London Ltd. 2017
E. Barucci, C. Fontana, *Financial Markets Theory*,
Springer Finance, DOI 10.1007/978-1-4471-7322-9_4

corn in markets open at $t = 0$ for assets delivering goods at the future date $t = 1$ (*future markets*) and markets open at $t = 1$ for immediate delivery (*spot markets*). To simplify the analysis, we will consider two stylized settings: when agents can consume only at $t = 1$, there exist $L > 1$ goods; when agents can consume at $t = 0$ as well as at $t = 1$, then there exists a single good ($L = 1$), generically identified as wealth. As in the previous chapter, we shall restrict our attention to a finite probability space, meaning that we only consider a finite number $S > 1$ of elementary states of the world that can be realized at $t = 1$. For instance, in the case of our farmer, each elementary event fully describes the weather conditions prevailing in the time period between $t = 0$ and $t = 1$ (since the crop obtained at $t = 1$ depends on the weather conditions). Let us suppose that there are $I \geq 2$ agents (farmers) in the economy, where every agent is fully described by a preference relation and by an endowment. We shall always make the crucial assumption that the probability space describing the uncertainty of the economy as well as the agents' rationality are *common knowledge* among the agents.[1] Furthermore, we will assume that the preference relation of each agent satisfies the axioms ensuring the expected utility representation (see Sect. 2.1). Note that most of the results presented in this chapter can be also established under the assumption that agents have heterogeneous beliefs.

To develop our analysis, we introduce the concept of *contingent good*. Indeed, goods differ not only for their intrinsic features, but also for the state of the world in which they become available. For instance, in the case of a farmer owning a field of corn, a unit of corn at $t = 1$ in the case where weather conditions have been unfavorable and a unit of corn at $t = 1$ in the case where weather conditions have been favorable are two rather different goods from the point of view of a farmer who has to take decisions at $t = 0$. Hence, if there are S elementary events and the farmer cultivates $L > 1$ goods, then the economy comprises $L \times S$ contingent goods. When consumption is only allowed at $t = 1$, recalling that in this case we consider $L > 1$ goods, agent i (for $i = 1, \ldots, I$) is endowed at the initial date $t = 0$ with a basket $e^i \in \mathbb{R}^{L \times S}$ of contingent goods (in the case of a farmer, the latter represents the crop obtained from the field at $t = 1$). For $l = 1, \ldots, L$, row l of the basket of goods e^i represents the quantity of good l obtained in correspondence of the S possible states of the world $(\omega_1, \ldots, \omega_S)$, while column s represents the quantities of the L goods obtained in correspondence of the state of the world ω_s, for $s = 1, \ldots, S$. Given a basket of contingent goods $x \in \mathbb{R}^{L \times S}$, we denote by x_{ls} the amount of good l in state of the world ω_s and by $x_s \in \mathbb{R}^L$ the vector of L goods available in state of the world ω_s. When consumption is allowed both at $t = 0$ and at $t = 1$ (so that $L = 1$), a basket of goods is simply described by a vector in $\mathbb{R}^{S+1}$, where the first component represents

---

[1]Some elements of the economy are common knowledge among the agents if each agent knows them, knows that the other agents know them, knows that the other agents know that he knows and so on. In that context, agents cannot agree to disagree (see Fuydenberg & Tirole [746] for a rigorous definition of common knowledge).

wealth/consumption at time $t = 0$ and the remaining S components represent wealth/consumption at time $t = 1$ in correspondence of the S possible states of the world.

This simple setting captures two essential features of financial markets: the intertemporal dimension of agents' decisions and their intrinsic riskiness. The forward-looking feature of economic decisions is due to the fact that, at the initial date $t = 0$, an agent faces different possible states of the world which will be realized at the future date $t = 1$ and, hence, wealth obtained at $t = 1$ is risky. Concerning the possible realizations of the state of the world at $t = 1$, an agent describes his beliefs through a probability measure $\mathbb{P}$ on the finite probability space represented by $\Omega = (\omega_1, \ldots, \omega_S)$. However, from the point of view of an agent, a probability measure alone does not suffice in order to solve his decision making problem. Indeed, if other agents are present in the economy, then every agent needs to formulate forecasts on the behavior of the other agents. This happens exactly when there are both *future markets* at $t = 0$ (i.e., markets open at $t = 0$ for assets delivering wealth/goods at $t = 1$) and *spot markets* at $t = 1$. In this case, taking their decisions at $t = 0$, agents have to forecast spot market prices at $t = 1$, which in turn depend on the behavior of all the agents in the economy.

A crucial point of the analysis consists in understanding how agents form their beliefs/expectations at $t = 0$ about future prices at $t = 1$. In our analysis, we will make the *rational expectations hypothesis*, which amounts to the following: a) the economic model is common knowledge among the agents (where by economic model we generically refer to all those features that turn out to be relevant for the solution to the decision problem); b) agents formulate their expectations by fully exploiting all the available information. In addition, we suppose that these two assumptions are common knowledge among all the agents themselves.

An assumption similar to the rational expectations hypothesis has been already introduced in Chap. 1 to strengthen the capability of general equilibrium theory to describe the economic world under certainty. More specifically, we assumed that the economic model as well as agents' rationality are common knowledge among the agents themselves and that agents fully exploit this common knowledge when taking their decisions. Note that, even without this assumption, general equilibrium theory in a riskless environment is able to show that agents' choices are compatible in a perfectly competitive market when agents pursue their self-interest. However, in a risky environment, the rational expectations hypothesis plays a much more relevant role. Indeed, unlike in a riskless environment, in a risky economy the rational expectations hypothesis is necessary to define the agents' behavior and the equilibrium. In particular, the rational expectations hypothesis is needed to define the agents' expectations about future prices, which are in turn needed to determine an agent's optimal choice. Without this hypothesis, or a similar one, we would not be able to establish the compatibility of agents' decisions in a perfectly competitive market. The rational expectations hypothesis will represent the third consistency requirement (*informational consistency*) and we will show that, in a rational expectations equilibrium, forecasts are self-confirming (perfect forecasts). We point out that one could replace the rational expectations assumption with a more realistic assumption about the agents' knowledge of the economy, but in that case agents' expectations would not be self-confirming (thus leading to biased forecasts), meaning that there is space for learning. We will return to this topic in Chap. 9 and we refer the reader to Radner [1439] for a critical introduction to rational expectations equilibrium theory.

Our analysis of agents' interactions in the economy will be based on the notion of equilibrium and we shall evaluate the efficiency of an allocation by relying on the Pareto optimality criterion, as introduced in Sect. 1.3. In particular, we will study the characteristics of future and spot markets allowing us to extend to a risky environment the important relationship between equilibrium and Pareto optimality established in Sect. 1.3 under certainty.

Extending the general equilibrium analysis to a risky environment, we shall also address the valuation problem. Given the fundamentals of the economy (dividends' distribution, agents' preferences and endowments), the general valuation problem consists in determining the prices of financial assets. We address this problem from two different perspectives: *equilibrium analysis* and *no-arbitrage pricing*. In the first case, equilibrium analysis allows to determine asset prices by assuming that each agent maximizes his expected utility and markets clear (i.e., the total demand equals the total supply). In particular, the equilibrium-based approach to asset pricing allows to relate the price of an asset to the agents' preferences and to their endowments. On the other hand, adopting the no-arbitrage perspective, asset prices will be determined by assuming that the market does not allow for arbitrage opportunities (loosely speaking, the market does not allow for opportunities to make money out of nothing without incurring in any risk). In this case, an asset's price is determined without any reference to the agents' preferences and endowments. We shall see that, according to the equilibrium-based valuation, the price of an asset will be given by the expectation of the future dividend multiplied by the marginal rate of substitution of an agent (*pricing kernel*), while, according to the no-arbitrage paradigm, the price of an asset will be given as the expectation of the future dividend with respect to a *risk neutral* probability measure, discounted by the risk free rate. In particular, the risk neutral probability measure will be different from the probability measure that describes the economy in the real world (historical/statistical probability) and will be derived from the assumption that the financial market does not allow for arbitrage opportunities. Moreover, we shall also discuss the relation between these two valuation paradigms.

This chapter is structured as follows. In Sect. 4.1, we study the notion of Pareto optimality in the presence of risk and explore its implications in terms of risk sharing. In Sect. 4.2, considering different market settings, we introduce the notion of rational expectations equilibrium. In Sect. 4.3, we deal with the intertemporal consumption problem in a simple two-period setting and we study the relations between equilibrium, market completeness, Pareto optimality and aggregation. Section 4.4 introduces the notion of arbitrage opportunity and presents the fundamental theorem of asset pricing and its implications for the valuation of financial securities. At the end of the chapter, we provide a guide to further readings as well as a series of exercises.

4.1   Pareto Optimality and Risk Sharing

Let us consider a two-period economy (i.e., $t \in \{0, 1\}$) with I agents ($i = 1, \dots, I$), L goods ($l = 1, \dots, L$) and S elementary states of the world ($s = 1, \dots, S$) realized at date $t = 1$, with $I \geq 2$, $L \geq 1$ and $S \geq 2$. As discussed in the introduction to the present chapter, all agents share homogeneous beliefs, meaning that the probability space (Ω, Θ) as well as the probability measure $\mathbb{P}$ are common knowledge among the agents. We furthermore assume that each elementary event ω_s occurs with a strictly positive probability $\pi_s > 0$, for all $s = 1, \dots, S$, with $\sum_{s=1}^{S} \pi_s = 1$. In the present section, we shall consider two different situations: consumption only at $t = 1$ and consumption at both dates $t = 0$ and $t = 1$. We start from the first case, assuming that there are $L > 1$ goods.

Assuming that consumption is only allowed at the future date $t = 1$, we consider I risk averse agents characterized by preference relations admitting the expected utility representation with state independent utility functions defined on the basket of consumption goods in $t = 1$, meaning that $u^i : \mathbb{R}_+^L \to \mathbb{R}$, for all $i = 1, \dots, I$. Each agent i is characterized by the couple (e^i, u^i), where u^i denotes his utility function and $e^i \in \mathbb{R}_+^{L \times S}$ represents his endowment in terms of quantities of the L goods available in the S states of the world at $t = 1$. We shall always tacitly assume that utility functions are differentiable, strictly increasing and strictly concave, unless otherwise mentioned.

Since the state of the world realized at $t = 1$ is unknown at the initial date $t = 0$ when decisions have to be taken, we can formulate two different notions of Pareto optimality. In the following, an allocation $\{x^i \in \mathbb{R}_+^{L \times S}; i = 1, \dots, I\}$ is said to be *feasible* if the feasibility constraint is satisfied in every possible state of the world: $\sum_{i=1}^{I} x_s^i \leq \sum_{i=1}^{I} e_s^i$ for all $s = 1, \dots, S$. Note also that, for each possible state of the world, this is a set of L constraints (one for each good).

Definition 4.1 A feasible allocation $\{x^i; i = 1, \dots, I\}$ is *ex-ante Pareto optimal* if there is no feasible allocation $\{x^{i'}; i = 1, \dots, I\}$ such that

$$\sum_{s=1}^{S} \pi_s \, u^i(x_s^{i'}) \geq \sum_{s=1}^{S} \pi_s \, u^i(x_s^i), \qquad \text{for all } i = 1, \dots, I,$$

with at least one strict inequality, for some $i \in \{1, \dots, I\}$.

Definition 4.2 A feasible allocation $\{x^i; i = 1, \dots, I\}$ is *ex-post Pareto optimal* if there is no feasible allocation $\{x^{i'}; i = 1, \dots, I\}$ such that

$$u^i(x_s^{i'}) \geq u^i(x_s^i), \qquad \text{for all } i = 1, \dots, I \text{ and } s = 1, \dots, S,$$

with at least one strict inequality, for some $i \in \{1, \dots, I\}$ and $s \in \{1, \dots, S\}$.

The two above definitions of Pareto optimality evaluate the optimality of an allocation $\{x^i; i = 1, \dots, I\}$ by referring to the different information available to the agents in correspondence of the two dates $t = 0$ and $t = 1$. Indeed, at $t = 0$ (*ex-ante*) the state of the world is still unknown and, therefore, utilities associated to consumption in the different states of the world are weighted by the corresponding probabilities, while at $t = 1$ (*ex-post*) the state of the world is fully revealed.

In Sect. 1.3, we have characterized Pareto optimal allocations under certainty in two distinct ways: a) maximizing the utility function of each single agent given the utility levels of all other agents as well as the feasibility constraint imposed by the available resources; b) maximizing a linear combination with positive weights of the agents' utility functions (social welfare function) subject to the feasibility constraint. These two approaches can also be employed in the present risky setting.

A Pareto optimal allocation $x^* \in \mathbb{R}_+^{I \times L \times S}$ can be characterized similarly as in Chap. 1 by requiring that the marginal rate of substitution between any couple of goods is the same for all the agents of the economy. Note that, in the present risky setting, we have $L \times S$ different goods, since the L consumption goods are contingent on the S possible states of the world, so that the total number of goods is $L \times S$ (contingent goods). Hence, considering an interior ex-ante Pareto optimal allocation $\{x^{i*} \in \mathbb{R}_{++}^{L \times S}; i = 1, \ldots, I\}$, the following condition must hold:

$$\frac{\pi_s u^i_{x_{ls}}(x^{i*}_s)}{\pi_r u^i_{x_{kr}}(x^{i*}_r)} = \frac{\pi_s u^j_{x_{ls}}(x^{j*}_s)}{\pi_r u^j_{x_{kr}}(x^{j*}_r)}, \quad \text{for all } i,j = 1,\ldots,I, \ s,r = 1,\ldots,S \text{ and } k,l = 1,\ldots,L,$$

$$(4.1)$$

where $u^i_{x_{ls}}$ denotes the partial derivative of the utility function u^i with respect to x_{ls}, for $l = 1,\ldots,L$, $s = 1,\ldots,S$ and $i = 1,\ldots,I$. In the case of an ex-post Pareto optimal allocation, it holds that

$$\frac{u^i_{x_{ls}}(x^{i*}_s)}{u^i_{x_{ks}}(x^{i*}_s)} = \frac{u^j_{x_{ls}}(x^{j*}_s)}{u^j_{x_{ks}}(x^{j*}_s)}, \quad \text{for all } i,j = 1,\ldots,I, \ s, = 1,\ldots,S, \text{ and } k,l = 1,\ldots,L.$$

$$(4.2)$$

It is easy to see that an ex-ante Pareto optimal allocation is also ex-post Pareto optimal. Indeed, this follows directly from Definitions 4.1 and 4.2 and can also be verified from conditions (4.1)–(4.2) (it suffices to take $s = r$ in (4.1)).

The set of Pareto optimal allocations defines the *contract curve* of the economy, as we are now going to illustrate by means of the *Edgeworth box* in the simple case of two agents ($I = 2$), two possible states of the world ($S = 2$) and a single consumption good ($L = 1$). We consider two agents ($i = a,b$), two states of the world ($s = 1,2$) with strictly positive probabilities ($\pi, 1 - \pi$) and a unique good, which we can think of as corn, recalling the example given in the introduction to this chapter. Suppose that the field of farmer a produces the crop $e^a = (e^a_1, e^a_2) \in \mathbb{R}_+^2$ in the two states of the world $\{1,2\}$, while the field of farmer b produces the crop $e^b = (e^b_1, e^b_2) \in \mathbb{R}_+^2$. The expected utility of agent $i \in \{a,b\}$ is given by

$$\pi u^i(x^i_1) + (1 - \pi)u^i(x^i_2)$$

and, in view of condition (4.1), an allocation $x^* \in \mathbb{R}^{2\times2}_{++}$ is ex-ante Pareto optimal if the following condition holds:

$$\frac{\pi \, u^{a'}(x_1^{a*})}{(1-\pi)u^{a'}(x_2^{a*})} = \frac{\pi \, u^{b'}(x_1^{b*})}{(1-\pi)u^{b'}(x_2^{b*})}. \tag{4.3}$$

Condition (4.3) identifies the allocations belonging to the contract curve (i.e., ex-ante Pareto optimal allocations) as those allocations in correspondence of which the indifference curves of the two agents a and b are tangent. In order to develop some intuition on the concept of (ex-ante) Pareto optimality, let us characterize the contract curve in three particular cases:

a) $e_1^a + e_1^b = e_2^a + e_2^b$ (*no aggregate risk*). In this case, the aggregate endowment of the economy (i.e., the aggregate crop of the two farmers) is constant in the two states of the world, so that the Edgeworth box is a square, see Fig. 4.1. In particular, the certainty line of agent a coincides with that of agent b. By the analysis developed in Sect. 3.3, in correspondence of the certainty line both agents have a marginal rate of substitution for consumption in the two states of the world equal to the ratio $\pi/(1-\pi)$ of the probabilities of the two states of the world. Therefore, in view of condition (4.3), the certainty line of the two farmers also coincides with the contract

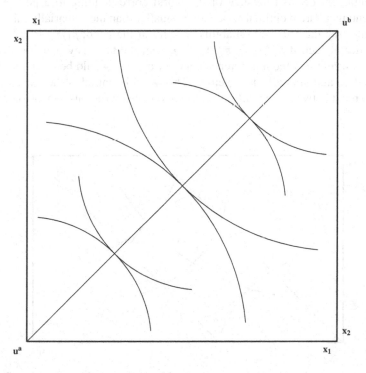

Fig. 4.1 Contract curve with no aggregate risk

curve. This implies that, if there is no aggregate risk in the economy, then Pareto optimal allocations are characterized by full mutual insurance between the two agents. Indeed, in correspondence of a Pareto optimal allocation, the consumption of each agent at $t = 1$ is the same in both possible states of the world. By the analysis developed in Sect. 3.3, the prices (p_1, p_2) at $t = 0$ for the consumption good at $t = 1$ contingent on the two states of the world (*Arrow securities*) implementing a Pareto optimal allocation are actuarially fair, in the sense that the relative price is equal to the ratio of the probabilities: $p_1/p_2 = \pi/(1 - \pi)$. In particular, it can be easily verified that in equilibrium the consumption of both agents in both states of the world is equal to the expectation of their respective endowments, i.e., $x_1^{i*} = x_2^{i*} = \pi e_1^i + (1 - \pi)e_2^i$, for $i \in \{a, b\}$.

 b) $e_1^a + e_1^b > e_2^a + e_2^b$ (*aggregate risk*). In this case, the aggregate endowment of the economy in the state of the world ω_1 is larger than that in the state of the world ω_2, meaning that there is aggregate risk. In the presence of aggregate risk, the Edgeworth box is a rectangle, see Fig. 4.2. The certainty lines of the two agents a and b do not coincide and the contract curve lies in the region between the certainty lines of the two agents. As a consequence, the prices of the Arrow securities implementing a Pareto optimal allocation are not actuarially fair and reflect the aggregate risk. The ratio between the price of the consumption good contingent on ω_1 (the state of the world corresponding to a rich harvest) and the price of the consumption good contingent on ω_2 (the state of the world corresponding to a poor harvest) implementing a Pareto optimal allocation is smaller than the actuarially fair relative price, reflecting the relative availability/scarcity of the good: $p_1/p_2 < \pi/(1 - \pi)$. As a matter of fact, if $p_1/p_2 \geq \pi/(1 - \pi)$, then the tangency point between the budget constraint and the indifference curve of agent a would be on the left of his certainty line and similarly for agent b. This would contradict the existence of a tangency point between the indifference curves of the two agents. On the other hand,

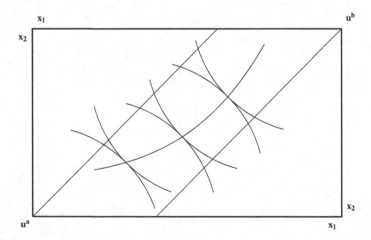

Fig. 4.2 Contract curve with aggregate risk

if $p_1/p_2 < \pi/(1-\pi)$, the tangency condition is verified inside the certainty lines of the two agents (see Exercise 4.1).

c) *One of the two agents is risk neutral.* Suppose that agent a is risk neutral. In this case, the indifference curves of agent a are straight lines with slope $\pi/(1-\pi)$. The contract curve is identified by the certainty line of agent b and, in correspondence of a Pareto optimal allocation, the risk neutral agent provides complete insurance to the risk averse agent against the risk coming from the uncertainty of the state of the world.

On the basis of the above observations, in the case of a simple economy with a single good, two agents and two possible states of the world, we can conclude that in correspondence of a Pareto optimal allocation all idiosyncratic risk is diversified by mutual insurance among the agents (*mutuality principle*, see Wilson [1661]). If there is no aggregate risk, then the agents will not bear any idiosyncratic risk because the latter will be completely diversified. Agents will bear a risk only in the case of aggregate risk. It should be noted, however, that this mutuality principle does not necessarily hold if agents have heterogeneous beliefs. Moreover, the results of this section rely on the assumption that the agents' utility functions are state independent.

The present analysis can be extended to an economy with $I \geq 2$ agents and $S \geq 2$ possible states of the world. When there is a single good and agents have homogeneous beliefs, condition (4.1) can be equivalently rewritten as follows (see Exercise 4.2): for an interior ex-ante Pareto optimal allocation $\{x^{i*}; i = 1, \ldots, I\}$ there exists a vector of strictly positive constants $(\lambda_1, \ldots, \lambda_I)$ such that

$$\lambda_i u^{i'}(x_s^{i*}) = \lambda_j u^{j'}(x_s^{j*}), \qquad \text{for all } i,j = 1, \ldots, I \text{ and } s = 1, \ldots, S. \qquad (4.4)$$

Condition (4.4) is often called the *Borch condition*, see Borch [269].

In the present setting, by relying on condition (4.4), we are able to give a general and easy proof of the mutuality principle in the presence of a single good. Let us first establish the *co-monotonicity property* of Pareto optimal allocations in the presence of a single consumption good. We define the aggregate endowment in correspondence of the state of the world ω_s as $e_s := \sum_{i=1}^{I} e_s^i$, for each $s = 1, \ldots, S$ (for simplicity, we assume that the aggregate endowment is strictly positive).

Theorem 4.3 *Let* $(x^{1*}, \ldots, x^{I*}) \in \mathbb{R}_{++}^{I \times S}$ *be an ex-ante Pareto optimal allocation. Then, for any $s, r \in \{1, \ldots, S\}$, the following properties are equivalent:*

(i) $x_s^{i*} \geq x_r^{i*}$ *for some* $i \in \{1, \ldots, I\}$;
(ii) $x_s^{i*} \geq x_r^{i*}$ *for all* $i \in \{1, \ldots, I\}$;
(iii) $e_s \geq e_r$.

Proof (i) $\Rightarrow$ (ii): suppose that $x_s^{i*} \geq x_r^{i*}$ for some $i \in \{1, \ldots, I\}$. Then, due to the concavity of the utility function u^i, it holds that $u^{i'}(x_s^{i*}) \leq u^{i'}(x_r^{i*})$. Due to condition (4.4), this implies that $u^{j'}(x_s^{j*}) \leq u^{j'}(x_r^{j*})$ for all $j \in \{1, \ldots, I\}$. By the concavity of the utility functions u^j, for all $j \in \{1, \ldots, I\}$, this implies property (ii).

(*ii*) $\Rightarrow$ (*iii*): suppose that $x_s^{i*} \geq x_r^{i*}$ for all $i \in \{1,\ldots,I\}$. Then, since the optimal allocation $\{x^{i*}; i = 1,\ldots,I\}$ is feasible and the utility functions are strictly increasing, it holds that $e_s = \sum_{i=1}^{I} x_s^{i*} \geq \sum_{i=1}^{I} x_r^{i*} = e_r$.

(*iii*) $\Rightarrow$ (*i*): suppose that $x_s^{i*} < x_r^{i*}$ holds for all $i \in \{1,\ldots,I\}$. Then, similarly as in the previous step, this would imply that $e_s < e_r$, thus yielding a contradiction. $\square$

The above proposition illustrates the co-monotonicity property of Pareto optimal allocations. Indeed, if a given agent $i \in \{1,\ldots,I\}$ consumes more in the state of the world ω_s than in the state of the world ω_r, then any other agent will also consume more in ω_s than in ω_r. Furthermore, since the aggregate consumption equals the aggregate endowment, this is also equivalent to $e_s \geq e_r$. In the case of two possible states of the world, two agents and a single good, the co-monotonicity property corresponds in the Edgeworth box to the fact that the contract curve lies between the certainty lines of the agents, see Fig. 4.2.

The result of Theorem 4.3 is of particular interest since it implies that agents do not bear risk (i.e., they are fully insured, in the sense that their optimal consumption plan is constant across all possible states of the world) if there is no aggregate risk (i.e., if the aggregate endowment is independent of the state of the world). As a matter of fact, as a consequence of Theorem 4.3, if the aggregate endowment is state independent for a subset $\bar{\Omega} \subset \Omega = \{\omega_1,\ldots,\omega_S\}$ of states of the world (i.e., there is no aggregate risk on $\bar{\Omega}$) then, in correspondence of an interior ex-ante Pareto optimal allocation, the optimal consumption plan of each agent is constant over all $\omega \in \bar{\Omega}$. Summing up, in correspondence of a Pareto optimal allocation only aggregate risk matters (*mutuality principle*).

Theorem 4.3 has another fundamental implication: in correspondence of a Pareto optimal allocation, the optimal consumption of any agent i in the state of the world ω_s does not depend on the agent's individual endowment but only on the aggregate endowment in the state ω_s, for all $s = 1,\ldots,S$. In other words, there is a one-to-one correspondence between the aggregate endowment in state ω_s and the Pareto optimal consumption plan of any agent in correspondence of ω_s. Since utility functions are assumed to be state independent, this correspondence is also state independent. Therefore, given a Pareto optimal allocation $(x^{1*},\ldots,x^{I*})$ we can without loss of generality assume that, for every $i = 1,\ldots,I$, there exists an increasing function $y^i : \mathbb{R}_+ \to \mathbb{R}_+$ such that $x_s^{i*} = y^i(e_s)$, for all $s = 1,\ldots,S$. Such functions are called *sharing rules* (see later in this section for more details and see also Gollier [800, Proposition 79]).

Condition (4.4) confirms that, in correspondence of a Pareto optimal allocation, a risk neutral agent will provide complete insurance to every risk averse agent. Indeed, if agent i is risk neutral (meaning that the utility function u^i is linear), it holds that

$$\lambda_j u^{j'}(x_s^{j*}) = \lambda_i, \qquad \text{for all } j = 1,\ldots,I \text{ with } j \neq i \text{ and } s = 1,\ldots,S,$$

thus implying that x_s^{j*} is constant across all possible states of the world, for all $j = 1,\ldots,I$ with $j \neq i$.

Let us now consider an economy with a single good (wealth) and consumption at the two dates $t = 0$ and $t = 1$. Similarly as before, agent i (for $i = 1, \ldots, I$) is represented by the couple (e^i, u^i), where the utility function $u^i : \mathbb{R}^2_+ \to \mathbb{R}$ is strictly increasing and concave and $e^i \in \mathbb{R}^{S+1}_+$ represents the agent's endowment in terms of wealth at $t = 0$ and contingent wealth at $t = 1$. A consumption plan of agent i is described by a non-negative $(S + 1)$-dimensional vector $x^i = (x^i_0, x^i_1, \ldots, x^i_S)^\top$, where x^i_0 denotes consumption at the initial date $t = 0$ and x^i_s denotes consumption at date $t = 1$ in correspondence of the state of the world ω_s, for $s = 1, \ldots, S$. In this context, in order to characterize Pareto optimal allocations, we rely on Proposition 1.6, which can also be established in a risky setting. We assume that agents have homogeneous beliefs. Given a vector of strictly positive weights $(a^1, \ldots, a^I)$, the maximization problem determining an interior (ex-ante) Pareto optimal allocation is given by

$$\max_{\substack{(x^i_0, x^i_1, \ldots, x^i_S) \in \mathbb{R}^{S+1}_+ \\ i=1, \ldots, I}} \sum_{i=1}^{I} a^i \sum_{s=1}^{S} \pi_s u^i(x^i_0, x^i_s),$$

subject to the feasibility constraint

$$\sum_{i=1}^{I} x^i_s \leq \sum_{i=1}^{I} e^i_s, \qquad \text{for all } s = 0, 1, \ldots, S.$$

Note that, since we always assume that utility functions are strictly increasing, the feasibility constraint is binding and can be expressed as an equality constraint.

The above problem can be dealt with by the Lagrange multiplier method introducing a vector $(\lambda_0, \lambda_1, \ldots, \lambda_S) \in \mathbb{R}^{S+1}$. Necessary and sufficient conditions (due to the strict concavity of the utility functions) for a strictly positive (ex-ante) Pareto optimal allocation $\{x^{i*} \in \mathbb{R}^{S+1}_{++}; i = 1, \ldots, I\}$ are given by the complete allocation of resources and by

$$a^i \sum_{s=1}^{S} \pi_s u^i_{x_0}(x^{i*}_0, x^{i*}_s) = \lambda_0, \qquad \text{for all } i = 1, \ldots, I, \tag{4.5}$$

$$a^i \pi_s u^i_{x_s}(x^{i*}_0, x^{i*}_s) = \lambda_s, \qquad \text{for all } i = 1, \ldots, I \text{ and } s = 1, \ldots, S. \tag{4.6}$$

where $a^i > 0$, for all $i = 1, \ldots, I$, and $u^i_{x_0}(\cdot)$ and $u^i_{x_s}(\cdot)$ denote the derivatives of the function $u^i(\cdot)$ with respect to its first and second arguments, respectively. The above conditions can be rewritten in terms of marginal rates of substitution as follows:

$$\frac{\pi_s\, u^i_{x_s}(x_0^{i*}, x_s^{i*})}{\sum_{r=1}^{S} \pi_r\, u^i_{x_0}(x_0^{i*}, x_r^{i*})} = \frac{\lambda_s}{\lambda_0}, \qquad \text{for all } s = 1, \dots, S \text{ and } i = 1, \dots, I, \qquad (4.7)$$

$$\frac{\pi_s\, u^i_{x_s}(x_0^{i*}, x_s^{i*})}{\pi_r\, u^i_{x_r}(x_0^{i*}, x_r^{i*})} = \frac{\lambda_s}{\lambda_r}, \qquad \text{for all } s, r = 1, \dots, S \text{ and } i = 1, \dots, I. \qquad (4.8)$$

Condition (4.8) establishes that for any couple of states of the world the marginal rate of substitution between wealth in correspondence of the two states is the same for all the agents. Condition (4.7) establishes the same result for the marginal rate of substitution between wealth in any state of the world in $t = 1$ and wealth in $t = 0$. Observe that the marginal utility at $t = 0$ is affected by consumption at $t = 1$. Condition (4.8) corresponds to the ex-ante Pareto optimality condition (4.1) with $L = 1$. As a consequence, also in this setting with consumption at both dates $t \in \{0, 1\}$, we can prove the analogue of Theorem 4.3 showing that the mutuality principle and the co-monotonicity property of a Pareto optimal allocation hold.

As a special case, if agents are characterized by time additive and state independent utility functions of the form

$$u^i(x_0, x_s) = u^i_0(x_0) + \delta_i\, u^i_1(x_s), \qquad \text{for all } i = 1, \dots, I,$$

where $0 < \delta_i \le 1$ represents the discount factor of agent i, then conditions (4.7)–(4.8) characterizing an ex-ante Pareto optimal allocation reduce to

$$\frac{\pi_s \delta_i\, u^{i\prime}_1(x_s^{i*})}{u^{i\prime}_0(x_0^{i*})} = \frac{\lambda_s}{\lambda_0}, \qquad \text{for all } s = 1, \dots, S \text{ and } i = 1, \dots, I, \qquad (4.9)$$

$$\frac{\pi_s\, u^{i\prime}_1(x_s^{i*})}{\pi_r\, u^{i\prime}_1(x_r^{i*})} = \frac{\lambda_s}{\lambda_r}, \qquad \text{for all } s, r = 1, \dots, S \text{ and } i = 1, \dots, I. \qquad (4.10)$$

As discussed in case of consumption only at date $t = 1$, if agents have homogeneous beliefs, time additive and state independent utility functions, then Theorem 4.3 implies that each Pareto optimal allocation is associated with a Pareto optimal *sharing rule*, see Wilson [1661], Rubinstein [1485], Breeden & Litzenberger [287], Constantinides [488]. A Pareto optimal sharing rule is a family of functions $\{y^i : \mathbb{R}_+ \to \mathbb{R}_+; i = 1, \dots, I\}$, depending only on the aggregate endowment and not on the initial allocation of resources nor on the state of the world, such that $x_s^{i*} = y^i(e_s)$, for all $i = 1, \dots, I$ and $s = 0, 1, \dots, S$. In other words, a Pareto optimal sharing rule describes a Pareto optimal allocation as a function of the aggregate endowment. The next proposition provides a more precise result on Pareto optimal sharing rules $\{y^i : \mathbb{R}_+ \to \mathbb{R}_+; i = 1, \dots, I\}$, showing in particular how marginal increases in the aggregate endowment are shared among the agents.

Proposition 4.4 *Consider an economy populated by I agents with homogeneous beliefs and time additive and state independent utility functions u^i, $i = 1, \ldots, I$. Let $e_s = \sum_{i=1}^{I} e_s^i$ be the aggregate endowment, for $s = 0, 1, \ldots, S$. The optimal sharing rule $\{y^i : \mathbb{R}_+ \to \mathbb{R}_+ ; i = 1, \ldots, I\}$ associated to a Pareto optimal allocation $\{x^{i*}; i = 1, \ldots, I\}$ satisfies*

$$y^{i'}(e_s) := \frac{dy^i(e_s)}{de_s} = \frac{t_{u^i}\big(y^i(e_s)\big)}{\sum_{j=1}^{I} t_{u^j}\big(y^j(e_s)\big)}, \tag{4.11}$$

for all $i = 1, \ldots, I$ and $s = 0, 1, \ldots, S$, where t_{u^i} denotes the risk tolerance associated to the utility function u^i, i.e., $t_{u^i}(x) = -u^{i'}(x)/u^{i''}(x)$.

Proof Given a vector of strictly positive weights a^i, $i = 1, \ldots, I$, consider the utility function of the representative agent defined as

$$\mathbf{u}(e_0, e) := \max_{\substack{(x_0^i, x_1^i, \ldots, x_S^i) \in \mathbb{R}_+^{S+1} \\ i=1,\ldots,I}} \sum_{i=1}^{I} a^i \left(u_0^i(x_0^i) + \delta_i \sum_{s=1}^{S} \pi_s u_1^i(x_s^i) \right),$$

with $e = (e_1, \ldots, e_S)$ and subject to the feasibility constraint $\sum_{i=1}^{I} x_s^i \leq \sum_{i=1}^{I} e_s^i$, for all $s = 0, 1, \ldots, S$. The above objective function is additively separable with respect to time and consumption in different states of the world and the feasibility constraints are formulated separately with respect to each state of the world. Hence, the representative agent's utility function admits the representation

$$\mathbf{u}(e_0, e) = \mathbf{u}_0(e_0) + \sum_{s=1}^{S} \pi_s \mathbf{u}_1(e_s),$$

where, for all $s = 1, \ldots, S$,

$$\mathbf{u}_0(e_0) := \max_{\substack{(x_0^1, \ldots, x_0^I) \in \mathbb{R}_+^I \\ \sum_{i=1}^{I} x_0^i \leq e_0}} \sum_{i=1}^{I} a^i u_0^i(x_0^i) \qquad \text{and} \qquad \mathbf{u}_1(e_s) := \max_{\substack{(x_s^1, \ldots, x_s^I) \in \mathbb{R}_+^I \\ \sum_{i=1}^{I} x_s^i \leq e_s}} \sum_{i=1}^{I} a^i \delta_i u_1^i(x_s^i).$$

By the envelope theorem, the first order conditions yield

$$\delta_i a^i u_1^{i'}(x_s^{i*}) = \mu_s = \mathbf{u}_1'(e_s), \qquad \text{for all } i = 1, \ldots, I \text{ and } s = 1, \ldots, S, \tag{4.12}$$

where μ_s denotes the Lagrange multiplier associated to the feasibility constraint for state s and the second equality amounts to the fact that μ_s represents the marginal increase of $\mathbf{u}_1$ as the endowment e_s is marginally increased. Differentiating with respect to e_s we obtain

$$\delta_i \, a^i u_1^{i''}(x_s^{i*}) \frac{dx_s^{i*}}{de_s} = \mathbf{u}_1''(e_s), \qquad \text{for all } i = 1, \ldots, I \text{ and } s = 1, \ldots, S. \tag{4.13}$$

Due to the mutuality principle (see Theorem 4.3 and the following discussion), it holds that $x_s^{i*} = y^i(e_s)$, for some function $y^i : \mathbb{R}_+ \to \mathbb{R}_+$, for all $i = 1, \ldots, I$ and $s = 1, \ldots, S$. Hence, equation (4.13) can be rewritten as

$$\delta_i \, a^i u_1^{i''}(x_s^{i*}) y^{i'}(e_s) = \mathbf{u}_1''(e_s), \qquad \text{for all } i = 1, \ldots, I \text{ and } s = 1, \ldots, S,$$

which in combination with (4.12) yields, for all $i = 1, \ldots, I$ and $s = 1, \ldots, S$,

$$y^{i'}(e_s) = \frac{\mathbf{u}_1''(e_s)}{\mathbf{u}_1'(e_s)} \frac{u_1^{i'}(x_s^{i*})}{u_1^{i''}(x_s^{i*})}. \tag{4.14}$$

Recalling that $\sum_{i=1}^{I} y^i(e_s) = e_s$, so that $\sum_{i=1}^{I} y^{i'}(e_s) = 1$, for all $s = 1, \ldots, S$, we get

$$1 = \sum_{i=1}^{I} y^{i'}(e_s) = \sum_{i=1}^{I} \frac{\mathbf{u}_1''(e_s)}{\mathbf{u}_1'(e_s)} \frac{u_1^{i'}(x_s^{i*})}{u_1^{i''}(x_s^{i*})} = -\frac{\mathbf{u}_1''(e_s)}{\mathbf{u}_1'(e_s)} \sum_{i=1}^{I} t_{u^i}\big(y^i(e_s)\big),$$

so that equation (4.14) can be rewritten as

$$y^{i'}(e_s) = \frac{t_{u^i}\big(y^i(e_s)\big)}{\sum_{j=1}^{I} t_{u^j}\big(y^j(e_s)\big)},$$

for all $i = 1, \ldots, I$ and $s = 1, \ldots, S$, where $t_{u^i}(x) = -u^{i'}(x)/u^{i''}(x)$ is the risk tolerance of agent i. A similar argument allows to prove the claim for consumption at the initial date $t = 0$ (i.e., for $s = 0$).   □

According to the above proposition, in correspondence of a Pareto optimal allocation, any increment in the aggregate resources of the economy is shared among all the agents in proportion to the individual risk tolerance. More specifically, the marginal increase in the aggregate endowment of the economy is shared among the agents in proportion to the ratio between their individual risk tolerance and the average risk tolerance in the economy. Therefore, in correspondence of a Pareto optimal allocation, a marginal increase of the aggregate endowment is allocated to agents who are less risk averse and, therefore, to agents who are more willing to bear risk. Note also that, due to the strict monotonicity and concavity of the utility functions, the derivative $y^{i'}$ is strictly positive. In turn, this directly implies that all the functions y^i are increasing:

$$y^i(e_s) \geq y^i(e_r) \text{ for all } i = 1, \ldots, I \quad \Longleftrightarrow \quad e_s \geq e_r,$$

for all $s, r \in \{1, \ldots, S\}$, thus confirming once more the co-monotonicity property. As we already mentioned, every agent consumes more in correspondence of state ω_s than of state ω_r if and only if $e_s \geq e_r$ (compare with Theorem 4.3).

In general, Pareto optimal sharing rules are non-linear. However, if agents' beliefs are homogeneous, the following proposition shows that the sharing rule is linear if and only if the agents' utility functions belong to the HARA class with a common parameter b (to this effect, see also Exercise 4.5).

Proposition 4.5 *Consider an economy populated by I agents with homogeneous beliefs and time additive and state independent utility functions u^i, $i = 1, \ldots, I$. A sharing rule $\{y^i : \mathbb{R}_+ \to \mathbb{R}_+; i = 1, \ldots, I\}$ is linear in correspondence of every Pareto optimal allocation if and only if all the agents have utility functions with linear risk tolerance with a common slope $b \geq 0$, i.e., if and only if*

$$t_{u^i}(x) = a^i + bx, \qquad \text{for all } i = 1, \ldots, I, \tag{4.15}$$

for some $(a^1, \ldots, a^I) \in \mathbb{R}_+^I$.

Proof We prove the sufficiency part, referring the reader to Exercise 4.4 for the necessity part (see also Wilson [1661] and Amershi & Stoeckenius [56]). In the case of linear risk tolerance, condition (4.11) can be rewritten as

$$y^{i'}(e_s) = \frac{a^i + by^i(e_s)}{\sum_{j=1}^I a^j + be_s}, \qquad \text{for all } i = 1, \ldots, I \text{ and } s = 0, 1, \ldots, S,$$

using the fact that $\sum_{j=1}^I y^j(e_s) = e_s$. It can be checked that, for a given vector $(y^1(0), \ldots, y^I(0))$ such that $\sum_{i=1}^I y^i(0) = 0$, the solution to the above differential equation is given by

$$y^i(e_s) = y^i(0) + \frac{a^i + by^i(0)}{\sum_{j=1}^I a^j} e_s, \qquad \text{for all } i = 1, \ldots, I \text{ and } s = 0, 1, \ldots, S.$$

Indeed, if the sharing rule admits this representation, then

$$y^{i'}(e_s) = \frac{a^i + by^i(0)}{\sum_{j=1}^I a^j} = \frac{a^i + by^i(0)}{\sum_{j=1}^I a^j} \left(\frac{\sum_{j=1}^I a^j}{\sum_{j=1}^I a^j + be_s} + \frac{b}{\sum_{j=1}^I a^j + be_s} e_s \right)$$

$$= \frac{a^i + by^i(0)}{\sum_{j=1}^I a^j + be_s} + \frac{b}{\sum_{j=1}^I a^j} \frac{a^i + by^i(0)}{\sum_{j=1}^I a^j + be_s} e_s = \frac{a^i + by^i(e_s)}{\sum_{j=1}^I a^j + be_s}.$$

$$\square$$

In view of the above proposition, a Pareto optimal sharing rule is linear if and only if the agents' utility functions belong to the HARA class with a common *cautiousness* coefficient (b). Recall that the generalized power, logarithmic and

exponential utility functions all belong to the HARA class (see Sect. 2.2 and
Exercise 2.10). In particular, if the utility functions belong to the CARA class, so
that $t_{u^i}(x) = a^i$, then the sharing rule is linear and is given by

$$y^i(e_s) = y^i(0) + \frac{a^i}{\sum_{j=1}^{I} a^j} e_s, \qquad \text{for all } i = 1, \ldots, I \text{ and } s = 0, 1, \ldots, S.$$

Similarly, if the utility functions are of the CRRA type (i.e., $t_{u^i}(x) = bx$), then
Proposition 4.4 implies that $y^{i'}(e_s) = y^i(e_s)/e_s$, so that

$$y^i(e_s) = k_i e_s, \qquad \text{for all } i = 1, \ldots, I \text{ and } s = 0, 1, \ldots, S,$$

where the vector $(k_1, \ldots, k_I)$ satisfies $\sum_{i=1}^{I} k_i = 1$.

The result of Proposition 4.5 is important. Indeed, in the case of agents with
utility functions characterized by linear risk tolerance with the same cautiousness
coefficient, any Pareto optimal allocation can be represented as an affine function
of the aggregate endowment. In particular, this feature implies that a Pareto optimal
allocation can be achieved by providing to the agents claims on a risk free asset and
on the aggregate endowment. In this context, a *two mutual funds* separation result
holds true: Pareto optimal allocations lie in the span of the risk free asset and of the
aggregate endowment (we will return on this point in Sect. 4.3).

4.2 Asset Markets and Equilibria

In this section, we shall be concerned with the relation between market equilibrium
and Pareto optimality, as discussed in Sect. 1.3 in the case of a riskless environment,
by making some assumptions on the structure of the economy and, in particular,
on the existence of *spot* and *future* markets for contingent goods. For the moment,
we consider a two-period economy, with $L \geq 1$ consumption goods and where
consumption is only allowed at $t = 1$. Agents have the possibility of trading at the
initial date $t = 0$ in *future markets* (markets open at $t = 0$ for goods delivered at
the future date $t = 1$) and also at $t = 1$ in *spot markets* (markets open at $t = 1$
for immediate delivery). We can consider three different types of assets. We call
real asset an asset delivering at $t = 1$ a bundle of consumption goods, while we
call *numéraire asset* an asset delivering a single good which directly enters the
agents' utility functions. Finally, we call *nominal asset* an asset delivering a good
which does not enter directly the agents' utility function (in that sense, money could
be considered as a nominal asset). In the following, we shall mostly consider an
economy with real or numéraire assets.

In the presence of spot and future markets, as will become clear from the
following analysis, a crucial role is played by the notion of *market completeness*.
In a nutshell, we say that markets are *complete* if every contingent consumption
plan at $t = 1$ can be *replicated* (or *spanned/reached*) via a suitable portfolio

composed of the available assets and a sufficient initial endowment, in the sense that the payoff delivered by such a replicating portfolio equals the consumption plan in correspondence of every possible state of the world $(\omega_1, \ldots, \omega_S)$ at $t = 1$. As will be shown below, a fundamental consequence of market completeness is that every equilibrium allocation is Pareto optimal.

Arrow-Debreu Equilibrium

Let us introduce a first notion of equilibrium, which represents the most natural extension of the concept of equilibrium introduced in Sect. 1.3 under certainty. Following Debreu [535], general equilibrium theory can be extended to a risky environment by assuming that at the initial date $t = 0$ there are $L \times S$ financial markets, namely one market for each of the L consumption goods in correspondence of each of the S possible states of the world. This means that agents have the possibility of investing in $L \times S$ contingent goods, where each contingent good (l, s) delivers one unit of the l-th consumption good in correspondence of the state of the world ω_s and zero otherwise, for $l = 1, \ldots, L$ and $s = 1, \ldots, S$. In particular, these $L \times S$ contingent goods are *real assets*, since each asset delivers a specific consumption good contingent on the state of the world. We denote by $q \in \mathbb{R}^{L \times S}$ the matrix of the prices of the $L \times S$ contingent goods, where $q_{ls} \in \mathbb{R}$ denotes the price at date $t = 0$ of the contingent good delivering at date $t = 1$ one unit of the l-th good only in correspondence of the state of the world ω_s. Note that the $L \times S$ markets can be considered *future* markets, since they are open at the initial date $t = 0$ for delivery at the future date $t = 1$. It is important to observe that, in the presence of $L \times S$ markets, the economy is *complete* by construction. In fact, in order to replicate an arbitrary consumption plan $c \in \mathbb{R}^{L \times S}$, it suffices to invest at time $t = 0$ in a suitable combination of the $L \times S$ basic assets.

Formally, this economy with $L \times S$ assets is not too different from the economy considered in Chap. 1, since there is a financial market for every possible contingent good (l, s), for all $l = 1, \ldots, L$ and $s = 1, \ldots, S$. Therefore, we can easily adapt to the present setting the general equilibrium theory introduced in Sect. 1.3. Given the matrix of prices $q \in \mathbb{R}^{L \times S}$, the optimal consumption problem of agent $i \in \{1, \ldots, I\}$, characterized by a utility function $U^i : \mathbb{R}^{L \times S} \to \mathbb{R}$ and endowed with a basket of contingent goods $e^i \in \mathbb{R}^{L \times S}$, becomes

$$\max_{x \in \mathbb{R}^{L \times S}} U^i(x), \tag{PO1}$$

subject to the budget constraint

$$\sum_{s=1}^{S} \sum_{l=1}^{L} q_{ls} x_{ls} \leq \sum_{s=1}^{S} \sum_{l=1}^{L} q_{ls} e_{ls}^i. \tag{4.16}$$

In the present setting, we can formulate the following definition of equilibrium.

Definition 4.6 The price-allocation couple $(q^*; x^{1*}, \ldots, x^{I*})$, with $q^* \in \mathbb{R}^{L \times S}$ and $x^{i*} \in \mathbb{R}^{L \times S}$, for all $i = 1, \ldots, I$, constitutes an *Arrow-Debreu equilibrium* if

(i) for every $i = 1, \ldots, I$, given the matrix of prices $q^* \in \mathbb{R}^{L \times S}$, the consumption plan $x^{i*} \in \mathbb{R}^{L \times S}$ is a solution of the optimum problem (PO1) of agent i;

(ii) $\sum_{i=1}^{I} x_{ls}^{i*} \leq \sum_{i=1}^{I} e_{ls}^{i}$, for all $l = 1, \ldots, L$ and $s = 1, \ldots, S$.

Of course, if agents' preferences are strictly increasing, then the inequality in part (ii) of the above definition can be replaced by an equality. Concerning the existence of an Arrow-Debreu equilibrium, Theorem 1.2 can be applied to the present context. If the agents' utility functions satisfy appropriate hypotheses (strict monotonicity, strict concavity and continuity) and are of the form $U^i(x) = \sum_{s=1}^{S} \pi_s u^i(x_s)$ and $\sum_{i=1}^{I} e^i \in \mathbb{R}_{++}^{L \times S}$, then the equilibrium prices are strictly positive ($q^* \in \mathbb{R}_{++}^{L \times S}$) and the strictly positive equilibrium allocation $(x^{1*}, \ldots, x^{I*})$ can be characterized by the following condition, for all $i = 1, \ldots, I$:

$$\frac{U_{x_{ls}}^{i}(x^{i*})}{U_{x_{kr}}^{i}(x^{i*})} = \frac{\pi_s u_{x_l}^{i}(x_s^{i*})}{\pi_r u_{x_k}^{i}(x_r^{i*})} = \frac{q_{ls}^*}{q_{kr}^*}, \qquad \text{for all } s, r = 1, \ldots, S \text{ and } k, l = 1, \ldots, L,$$

where $u_{x_l}^{i}$ denotes the first derivative of the utility function u^i with respect to its l-th argument, for $i = 1, \ldots, I$ and $l = 1, \ldots, L$.

In particular, in view of condition (4.1), this implies that an equilibrium allocation (in the sense of Definition 4.6) is ex-ante Pareto optimal, thus providing a version of the First Welfare Theorem in the context of Arrow-Debreu equilibria. If the above condition is satisfied, we say that the matrix of prices $q^* \in \mathbb{R}_{++}^{L \times S}$ *implements* the Pareto optimal allocation $(x^{1*}, \ldots, x^{I*})$ starting from the initial allocation $(e^1, \ldots, e^I)$. Moreover, it can be easily verified that also the Second Welfare Theorem (see Theorem 1.5) can be extended to a risky setting in terms of Arrow-Debreu equilibria.

The market structure adopted in the context of Arrow-Debreu equilibria allows for trade only at the initial date $t = 0$, before the resolution of uncertainty, with no trading taking place at $t = 1$. Note that, since ex-ante Pareto optimal allocations are also ex-post Pareto optimal, if an Arrow-Debreu equilibrium allocation is reached at the initial date $t = 0$ and the possibility of trading at the future date $t = 1$ is then introduced, then there will be no incentive to trade at $t = 1$. Indeed, if there were incentives to trade at $t = 1$, after the state of the world is revealed, then this would contradict the existence of an Arrow-Debreu equilibrium at the initial date $t = 0$ (see Exercise 4.6). Trading could occur at $t = 1$ only in the case of multiple equilibria if agents do not coordinate among themselves on the reached equilibrium.

Radner Equilibrium

As we have seen, the notion of Arrow-Debreu equilibrium allows for a straight-forward application of classical equilibrium theory to a risky economy. However, Pareto optimality is reached at equilibrium under the assumption of the existence of $L \times S$ future markets open at time $t = 0$, with all the trading activity taking place simultaneously before the resolution of uncertainty. Clearly, this setting is hardly realistic. However, it is possible to reach Pareto optimal allocations in equilibrium by considering a smaller number of open markets, assuming that there are both *future markets* open at $t = 0$ and *spot markets* open at $t = 1$. In particular, when some of the $L \times S$ contingent goods are not available for trade in future markets open at $t = 0$, spot markets open at $t = 1$ play an important role, since trading after the resolution of the uncertainty can allow the agents to reach Pareto optimality. This idea goes back to Arrow [76].

Let us consider the case of N future markets open at $t = 0$ for real assets. Asset n (for $n = 1, \ldots, N$) delivers at $t = 1$ a basket of contingent goods $\tilde{a}^n$ described by

$$
\tilde{a}^n =
\begin{bmatrix}
a^n_{11} & a^n_{12} & \cdots & a^n_{1S} \\
a^n_{21} & a^n_{22} & \cdots & a^n_{2S} \\
\vdots & \ddots & \ddots & \vdots \\
a^n_{L1} & \cdots & \cdots & a^n_{LS}
\end{bmatrix}
\in \mathbb{R}^{L \times S}.
$$

For $n = 1, \ldots, N$, $l = 1, \ldots, L$ and $s = 1, \ldots, S$, the quantity a^n_{ls} represents the amount of good l delivered by asset n at time $t = 1$ in correspondence of the state of the world ω_s. Equivalently, we can think of $\tilde{a}^n$ as an $\mathbb{R}^L$-valued random variable with S possible realizations. We denote by $p \in \mathbb{R}^N$ the vector of the prices of the N real assets. For the sake of convenience, we assume that the aggregate supply of all N assets is normalized to zero. Besides the N future markets open at time $t = 0$ for real assets, we assume that at $t = 1$ (i.e., after the state of the world is revealed) there are L spot markets, one for each of the L goods. Summing up, in the economy there are $N + L$ markets. We also denote by $q \in \mathbb{R}^{L \times S}$ the matrix collecting the prices of the L goods in the spot markets open at $t = 1$, so that the column q_s of the matrix q represents the prices of the L goods in correspondence of the state ω_s.

Similarly as above, equilibrium analysis proceeds by first determining the individual behavior of the agents, taking prices as given, and then by aggregating the agents' decisions, requiring that equilibrium prices make agents' decisions compatible among themselves. In the present setting, since trading is allowed to take place at both dates $t = 0$ and $t = 1$, agents have to take decisions at the initial date $t = 0$ without knowing the spot prices prevailing at the future date $t = 1$, since the latter will depend on the state of the world (which will only be revealed at $t = 1$). As a consequence, in order to solve their decision problem, agents have to formulate expectations/forecasts about future spot prices.

Agents' forecasts are crucial to determine first the agents' individual behavior and, in a second step, the market equilibrium. Moreover, an agent cannot make forecasts on future prices without making some hypotheses on the information available to the other agents, on their behavior and on their forecasts as well (the so-called *forecast the forecasts of the others* phenomenon). We assume that all agents formulate *rational expectations* and that this fact is common knowledge among all the agents. Therefore, every agent knows the economic model, fully exploits all the available information and knows that all the other agents will do the same (this corresponds to the *informational consistency* requirement mentioned in the introduction to the present chapter). In this context, unlike in the economy under certainty described in Chap. 1, this hypothesis turns out to be necessary to characterize agents' decisions and to establish their compatibility in a risky environment under perfect competition.

If the agents' preferences and (homogeneous) beliefs, the initial endowments, the probability space and the matrix $\tilde{a}^n$, for all $n = 1, \ldots, N$, are common knowledge among the agents, then, under the rational expectations hypothesis and thanks to the consistency requirements, the definition of spot market equilibrium prices in correspondence of each state of the world $(\omega_1, \ldots, \omega_S)$ reduces to a deterministic problem. From the point of view of time $t = 0$, spot market equilibrium prices are represented by random variables. This type of equilibrium is known as the *perfect foresight, rational expectations* equilibrium (or *Radner equilibrium*, see Radner [1437]). In correspondence of an equilibrium, agents have common expectations and correctly anticipate future prices: market uncertainty completely vanishes and is reduced to uncertainty about the state of the world realized at $t = 1$. In this context, the rational expectations hypothesis allows us to determine the agents' behavior and the equilibrium prices and, therefore, the assumptions on the agents' rationality play a more relevant role than under certainty. Without such hypotheses, one would not be able to define an agent's optimal behavior and to establish the compatibility of agents' decisions through a price system. Any other assumption on agents' forecasts does not lead to perfect foresight, meaning that forecasts will be biased, inducing agents to learn over time. In this sense, a perfect foresight equilibrium represents the conclusion of a learning process: agents have nothing more to learn about the world. Coherently with the interpretation of an equilibrium status provided in Chap. 1, a rational expectations equilibrium can be thought of as a rest point in the agents' learning process.

On the basis of the above discussion, assuming rational expectations, the optimal consumption problem of agent i, for $i = 1, \ldots, I$, can be written as

$$\max_{x \in \mathbb{R}^{L \times S}} U^i(x), \qquad\qquad (\text{PO2})$$

subject to the two following budget constraints (at time $t = 0$ and at time $t = 1$, respectively), for some $z \in \mathbb{R}^N$:

$$p^\top z \leq 0 \quad \text{and} \quad q_s^\top x_s \leq q_s^\top \left(\sum_{n=1}^N \tilde{a}_s^n z_n + e_s^i \right), \quad \text{for all } s = 1, \ldots, S,$$

(4.17)

where the vector $p \in \mathbb{R}^N$ denotes the prices of the N assets available for trade at the initial date $t = 0$, $z \in \mathbb{R}^N$ denotes the vector of trades in the N future markets, a_s^n denotes the s-th column of the matrix $\tilde{a}^n$, q_s is the s-th column of the matrix q and represents the prices of the L goods in state ω_s forecasted at $t = 0$ and, as usual, $e_s^i \in \mathbb{R}^L$ denotes the endowment of agent i in state ω_s. In the above problem, it is implicitly assumed that the initial endowment in the N future assets is null.

Problem (PO2) actually consists of two problems: a consumption problem and an investment problem. The consumption problem consists in choosing, for each possible state of the world $(\omega_1, \ldots, \omega_S)$, the consumption plan at $t = 1$ given the spot prices forecasted at $t = 0$, while the investment problem consists in choosing the vector z of trades in the N future assets available at time $t = 0$. In particular, the investment problem amounts to ensure that at time $t = 1$ the agent is endowed with the wealth needed to finance his consumption plan in each state of the world. Note also that the vector z is allowed to take values in $\mathbb{R}^N$, meaning in particular that *short sales* are allowed (i.e., the components of z can take negative values).

We are now in a position to formulate the following definition.

Definition 4.7 The tuple $\{p^*, q^*; (z^{1*}, x^{1*}), \ldots, (z^{I*}, x^{I*})\}$, with $p^* \in \mathbb{R}^N$, $q^* \in \mathbb{R}^{L \times S}$, $x^{i*} \in \mathbb{R}^{L \times S}$ and $z^{i*} \in \mathbb{R}^N$, for all $i = 1, \ldots, I$, constitutes a *Radner equilibrium* if

(i) for every $i = 1, \ldots, I$, given $(p^*, q^*) \in \mathbb{R}^N \times \mathbb{R}^{L \times S}$, the trade vector - consumption plan (z^{i*}, x^{i*}) is a solution of Problem (PO2) for agent i;

(ii) $\sum_{i=1}^I z_n^{i*} \leq 0$, for all $n = 1, \ldots, N$;

(iii) $\sum_{i=1}^I x_{ls}^{i*} \leq \sum_{i=1}^I e_{ls}^i$, for all $l = 1, \ldots, L$ and $s = 1, \ldots, S$.

Note that, in the above definition, the agents' expectations made at the initial date $t = 0$ on the spot prices prevailing at date $t = 1$ are self-fulfilling, in the sense that they will clear all the spot markets for every possible state of the world once the state of the world is revealed at date $t = 1$.

The Case of S Arrow Securities (Complete Market)

Let us now specialize the structure of the economy. We assume that, at the initial date $t = 0$, there exist S future markets for S assets (i.e., $N = S$), with the s-th asset delivering one unit of the first good in correspondence of the state of the world ω_s and zero otherwise, for each $s = 1, \ldots, S$. In other words, in the economy there are S numéraire *Arrow securities* available for trade at date $t = 0$ (see also Arrow [76]).

In this context, one can prove the existence of a Radner equilibrium and also the Pareto optimality of the equilibrium allocation.

If S future markets are open at time $t = 0$ for the S Arrow securities, then it can be easily verified that every consumption plan in $t = 1$ can be obtained by a suitable portfolio of the S Arrow securities. Indeed, given the forecasted spot prices q, a bundle of goods $c \in \mathbb{R}^{L \times S}$ can be simply obtained by holding a sufficient amount of the first good in each state of the world $(\omega_1, \ldots, \omega_S)$ and then trading in the spot markets open at time $t = 1$. The ex-ante Pareto optimality of a Radner equilibrium is proved in the following proposition, which also establishes a one-to-one correspondence between Arrow-Debreu equilibria and Radner equilibria in an economy with S markets for Arrow securities open at $t = 0$ (compare also with Mas-Colell et al. [1310, Proposition 19.D.1]).

Proposition 4.8 *Given an economy (e^i, U^i), with $e^i \in \mathbb{R}_+^{L \times S}$ and $U^i : \mathbb{R}_+^{L \times S} \to \mathbb{R}$, for all $i = 1, \ldots, I$, the following hold:*

(i) *if the tuple $(q^*; x^{1*}, \ldots, x^{I*})$, with $q^* \in \mathbb{R}_{++}^{L \times S}$, is an Arrow-Debreu equilibrium, then there exist a vector of prices of the Arrow securities $p^* \in \mathbb{R}_{++}^S$ and a set of trade vectors $(z^{1*}, \ldots, z^{I*}) \in \mathbb{R}^{S \times I}$ such that $\{p^*, q^*; (z^{1*}, x^{1*}), \ldots, (z^{I*}, x^{I*})\}$ is a Radner equilibrium;*

(ii) *if the tuple $\{p^*, q^*; (z^{1*}, x^{1*}), \ldots, (z^{I*}, x^{I*})\}$, with $p^* \in \mathbb{R}_{++}^S$ and $q^* \in \mathbb{R}_{++}^{L \times S}$, is a Radner equilibrium, then there exists a vector $\mu \in \mathbb{R}_{++}^S$ such that $(\mu_1 q_1^*, \ldots, \mu_S q_S^*; x^{1*}, \ldots, x^{I*})$ is an Arrow-Debreu equilibrium.*

Proof (i): let $p_s^* := q_{1s}^*$, for all $s = 1, \ldots, S$. For each $i = 1, \ldots, I$, if $x^i \in \mathbb{R}^{L \times S}$ satisfies the budget constraint (4.16), then there exists a vector $z^i \in \mathbb{R}^N$ such that the budget constraint (4.17) is satisfied (recall that $N = S$). Indeed, let

$$z_s^i := \frac{1}{q_{1s}^*} q_s^{*\top} (x_s^i - e_s^i), \qquad \text{for all } s = 1, \ldots, S.$$

By (4.16), it holds that

$$\sum_{s=1}^S p_s^* z_s^i = \sum_{s=1}^S q_s^{*\top} (x_s^i - e_s^i) \leq 0,$$

so that the first requirement of the budget constraint (4.17) is met. Furthermore, we also have $q_s^{*\top} (x_s^i - e_s^i) = q_{1s}^* z_s^i$, for all $s = 1, \ldots, S$ and, therefore, x^i also satisfies the second requirement of the budget constraint (4.17), recalling that at time $t = 0$ the available assets are the S Arrow securities for the first good. Conversely, for all $i = 1, \ldots, I$, let $x^i \in \mathbb{R}^{L \times S}$ and $(z_1^i, \ldots, z_S^i)$ be such that the budget constraint (4.17) is satisfied, so that $\sum_{s=1}^S p_s^* z_s^i \leq 0$ and $q_s^{*\top} (x_s^i - e_s^i) \leq p_s^* z_s^i$, recalling that $p_s^* = q_{1s}^*$ (for all $s = 1, \ldots, S$). By summing the constraints over s we obtain

$$\sum_{s=1}^S q_s^{*\top} (x_s^i - e_s^i) \leq \sum_{s=1}^S p_s^* z_s^i \leq 0.$$

Therefore, x^i satisfies the budget constraint (4.16), for all $i = 1, \ldots, I$. We have thus shown that, in correspondence of prices $p^* = (q_{11}^*, \ldots, q_{1S}^*)$, the budget sets associated to an Arrow-Debreu equilibrium and to a Radner equilibrium coincide. This implies that the allocation $(x^{1*}, \ldots, x^{I*})$ of an Arrow-Debreu equilibrium implemented by prices $q^* \in \mathbb{R}_{++}^{L \times S}$ is also an allocation of a Radner equilibrium in correspondence of Arrow security prices $p^* \in \mathbb{R}_{++}^S$, spot prices $q^* \in \mathbb{R}_{++}^{L \times S}$ and of the vector of trades on future markets $z^{i*} = (z_1^{i*}, \ldots, z_S^{i*}) \in \mathbb{R}^S$ defined by $z_s^{i*} := \frac{1}{q_{1s}^*} q_s^{*\top}(x_s^{i*} - e_s^i)$, for $i = 1, \ldots, I$. Indeed, $(x^{i*}, z^{i*}) \in \mathbb{R}^{L \times S} \times \mathbb{R}^S$ solves problem (PO2) for agent i given the price vectors (p^*, q^*) and it holds that $\sum_{i=1}^I z_s^{i*} = \sum_{i=1}^I \frac{1}{q_{1s}^*} q_s^{*\top}(x_s^i - e_s^i) \leq 0$, for all $s = 1, \ldots, S$, while the third requirement in Definition 4.7 comes directly from Definition 4.6.

(ii): set $\mu_s > 0$ so that $\mu_s q_{1s}^* = p_s^*$, for all $s = 1, \ldots, S$. By relying on arguments analogous to those used in the first part of the proof, it can be easily verified that the elements $x^i \in \mathbb{R}^{L \times S}$ satisfying the budget constraint (4.17) for a vector of trades $z \in \mathbb{R}^S$ also satisfy (4.16) with respect to prices $(\mu_1 q_1^*, \ldots, \mu_S q_S^*) \in \mathbb{R}_{++}^{L \times S}$. Therefore, $x^{i*} \in \mathbb{R}^{L \times S}$ solves problem (PO1) for agent i in correspondence of prices $(\mu_1 q_1^*, \ldots, \mu_S q_S^*)$, for every $i = 1, \ldots, I$, and contingent goods markets clear, as a consequence of part (iii) of Definition 4.7. By Definition 4.6, this implies that $(\mu_1 q_1^*, \ldots, \mu_S q_S^*; x^{1*}, \ldots, x^{I*})$ is an Arrow-Debreu equilibrium.   □

As can be seen from the proof of the above proposition, the quantity $\mu_s = p_s^*/q_{1s}^*$ represents the price at date $t = 0$ of one unit of wealth obtained at time $t = 1$ in correspondence of the state of the world ω_s, for all $s = 1, \ldots, S$. The strict monotonicity of the agents' utility functions and the fact that all the states of the world have a strictly positive probability of occurrence imply that the equilibrium prices of the S Arrow securities are strictly positive.

Proposition 4.8 shows that the set of Arrow-Debreu equilibria of an economy with $L \times S$ goods, as considered at the beginning of the present section, coincides with the set of Radner equilibria obtained with S Arrow securities available for trade at time $t = 0$. This result has two important consequences: first, the Radner equilibrium allocation is ex-ante Pareto optimal (and, hence, also ex-post Pareto optimal); second, in order to establish the existence of a Radner equilibrium with S Arrow securities, we can rely on classical equilibrium analysis to prove the existence of an Arrow-Debreu equilibrium.

A remarkable feature of the above result is represented by the fact that the presence of spot markets open at $t = 1$ allows for a reduction in the number of markets needed to achieve Pareto optimality in equilibrium. Indeed, when S Arrow securities are available for trade at time $t = 0$, only $L + S$ markets are needed to reach a Pareto optimal allocation through a Radner equilibrium instead of $L \times S$ markets as in the case of an Arrow-Debreu equilibrium. More specifically, since agents are allowed to trade goods in spot markets open at $t = 1$, at the initial date $t = 0$ agents are only interested in transferring wealth among different states of the world and this can be done through S assets each delivering one unit of a reference good in correspondence of a specific state of the world (Arrow securities). At time $t = 1$, in correspondence of the specific state of the world which will be realized, agents will trade in the L spot markets. We want to remark that, in order to analyse the economy, we had to introduce a rather strong hypothesis on the agents' ability to forecast future spot prices: agents' expectations are rational and, therefore, self-confirming in equilibrium.

The Case of Complex Securities

The results obtained so far on the existence and the properties of Radner equilibria require the existence of a number of Arrow securities equal to the number S of possible states of the world. This assumption is hardly realistic, because Arrow securities represent theoretical assets which are not traded in real financial markets. Indeed, assets typically traded in financial markets pay dividends in correspondence of more than one elementary state of the world, i.e., they are *complex securities*. In the remaining part of this section, we aim at extending the equilibrium analysis presented so far to this more realistic situation. To simplify the presentation, we restrict our attention to N numéraire assets delivering dividends expressed in terms of the first good,[2] so that the spot price of the first good at $t = 1$ is equal to one across all possible states of the world. The prices of the N future assets are denoted by $p \in \mathbb{R}^N$, the dividend of the n-th asset is denoted by $\tilde{d}_n \in \mathbb{R}^S$, for all $n = 1, \ldots, N$, and we let $D \in \mathbb{R}^{S \times N}$ be the matrix collecting the asset dividends:

$$D = \begin{bmatrix} d_{11} & d_{12} & \ldots & d_{1N} \\ d_{21} & d_{22} & \ldots & d_{2N} \\ \vdots & \ddots & \ddots & \vdots \\ d_{S1} & \ldots & \ldots & d_{SN} \end{bmatrix}.$$

The matrix D encodes significant information on the structure of the financial market. Indeed, the matrix D is associated with the linear map $z \mapsto Dz$ from $\mathbb{R}^N$ onto $\mathbb{R}^S$ which describes *contingent consumption plans* (i.e., vectors $c \in \mathbb{R}^S$) as a result of the investment decisions taken at the initial date $t = 0$ in the N assets (i.e., $z \in \mathbb{R}^N$). In other words, the map $z \mapsto Dz$ describes the consumption plan $c = Dz \in \mathbb{R}^S$ (at time $t = 1$) generated by an investment strategy $z \in \mathbb{R}^N$.

It is important to observe that the set of consumption plans attainable by trading in the N (future) markets open at time $t = 0$ is not necessarily the whole space $\mathbb{R}^S$. The linear subspace of $\mathbb{R}^S$ containing all consumption plans attainable by trading in the N markets is the image (or range) $I(D)$ of the linear map $z \mapsto Dz$ and the dimension of the linear subspace $I(D) \subseteq \mathbb{R}^S$ is less or equal than S. A consumption plan (or contingent wealth at $t = 1$) $c \in \mathbb{R}^S$ is said to be *reachable* (or *attainable*) by trading in the N assets if $c \in I(D)$, i.e., if it is possible to reach (or attain/replicate)

---

[2]Using the notation a_{ls}^n introduced before, this corresponds to $a_s^n = (d_{sn}, 0, \ldots, 0) \in \mathbb{R}^L$, where d_{sn} is the dividend paid by asset n in correspondence of the state of world ω_s, for all $n = 1, \ldots, N$ and $s = 1, \ldots, S$.

the consumption plan c by means of a portfolio $z^c \in \mathbb{R}^N$ of the N assets available for trade at time $t = 0$, starting from a suitable endowment, so that $c = Dz^c$. The portfolio $z^c \in \mathbb{R}^N$ which replicates the consumption plan c is called *replicating* portfolio.

The dividend matrix D allows for an easy characterization of *market completeness*. As already mentioned, markets are *complete* if every contingent consumption plan $c \in \mathbb{R}^S$ at $t = 1$ can be reached/replicated by a suitable portfolio $z^c \in \mathbb{R}^N$ of the N assets available for trade at the starting date $t = 0$. Hence, in the present setting, it is immediate to see that markets are complete if and only if $I(D) = \mathbb{R}^S$, i.e., the rank of the matrix D is equal to S (i.e., rank$(D) = S$). In other words, markets are complete if and only if the N traded assets allow to span the whole space $\mathbb{R}^S$. On the other hand, if markets are incomplete, then agents can only reach a subset of all possible consumption plans in $\mathbb{R}^S$ by investing in the N available assets.

The N assets available for trade at time $t = 0$ are said to be *non-redundant* if their dividends are linearly independent, i.e., if it is not possible to express the dividend of any asset as a linear combination of the dividends of all the other assets. This property is satisfied if $N \leq S$ and rank$(D) = N$ (i.e., the columns of the matrix D are linearly independent vectors). In particular, when $N = S$, meaning that the number of traded assets is equal to the number of possible states of the world, this is equivalent to the non-singularity of the square matrix D (i.e., det$(D) \neq 0$). We also say that the N assets enjoy the *uniqueness of representation* property if, for every $c \in I(D)$, there exists a unique vector $z^c \in \mathbb{R}^N$ such that $Dz^c = c$. The uniqueness of representation property holds if and only if the dimension of ker(D) is zero and, therefore, if and only if the N assets are non-redundant.

Without loss of generality, limiting our attention to a full rank matrix D, the following situations may arise:

- $N > S$: in this case, $I(D) = \mathbb{R}^S$ and, $\forall c \in \mathbb{R}^S$, dim$\{z \in \mathbb{R}^N : Dz = c\} = N - S$;
- $N = S$: in this case, $I(D) = \mathbb{R}^S$ and, $\forall c \in \mathbb{R}^S$, dim$\{z \in \mathbb{R}^N : Dz = c\} = 0$;
- $N < S$: in this case, $I(D) \subset \mathbb{R}^S$ and, $\forall c \in I(D)$, dim$\{z \in \mathbb{R}^N : Dz = c\} = 0$.

In the last two cases, the N assets enjoy the uniqueness of representation property, while in the first two cases markets are complete. A necessary condition for market completeness is that the number of traded assets is greater or equal than the number of possible states of the world, i.e., $N \geq S$. Loosely speaking, markets are complete when market participants can trade in a set of assets which span all possible risky scenarios of the economy. For an attainable contingent consumption plan $c \in I(D)$, we can define its *market value* as follows.

Definition 4.9 Let $c \in \mathbb{R}^S$ be a contingent consumption plan belonging to $I(D)$. The *market value* of c at $t = 0$, denoted by $V(c)$, is defined as the value at $t = 0$ of the replicating portfolio for c, i.e.,

$$V(c) := \{p^\top z^c : z^c \in \mathbb{R}^N \text{ and } Dz^c = c\}.$$

Note that, if the N assets do not enjoy the uniqueness of representation property, then there may exist two different portfolios $z^c, z^{c'} \in \mathbb{R}^N$ satisfying $Dz^c = c = Dz^{c'}$. In principle, the market values associated to the two portfolios z^c and $z^{c'}$ can be different. However, in correspondence of an equilibrium allocation, the market value of any consumption plan $c \in I(D)$ is uniquely defined (this corresponds to the so-called Law of One Price, see Proposition 4.18 and Exercise 4.22). This is closely connected to the absence of *arbitrage opportunities*. In Sect. 4.4 we shall investigate the relation between the absence of arbitrage opportunities and the existence of an equilibrium. In particular, this implies that Definition 4.9 is well posed.

In order to study the existence and the properties of equilibrium allocations, the notion of market completeness plays a particularly important role. Indeed, if markets are complete, then agents are unrestricted in their capacity to transfer wealth across different states of the world by forming suitable portfolios of the assets available for trade. In particular, this implies that agents can reach Arrow-Debreu equilibria, as shown in the following proposition. Note also that, as a consequence of the First Welfare Theorem, equilibrium allocations are Pareto optimal.

Proposition 4.10 *Given an economy* (e^i, U^i), *with* $e^i \in \mathbb{R}_+^{L \times S}$, *and* $U^i : \mathbb{R}_+^{L \times S} \to \mathbb{R}$, *for all* $i = 1, \dots, I$, *with* N *assets traded at* $t = 0$ *such that* $\mathrm{rank}(D) = S$, *then the following hold:*

(i) *if the tuple* $(q^*; x^{1*}, \dots, x^{I*})$, *with* $q^* \in \mathbb{R}_{++}^{L \times S}$, *constitutes an Arrow-Debreu equilibrium, then there exist a vector of prices* $p^* \in \mathbb{R}_{++}^N$ *and a set of trade vectors* $(z^{1*}, \dots, z^{I*}) \in \mathbb{R}^{N \times I}$ *such that* $\{p^*, q^*; (z^{1*}, x^{1*}), \dots, (z^{I*}, x^{I*})\}$ *constitutes a Radner equilibrium;*

(ii) *if the tuple* $\{p^*, q^*; (z^{1*}, x^{1*}), \dots, (z^{I*}, x^{I*})\}$, *with* $p^* \in \mathbb{R}_{++}^N$ *and* $q^* \in \mathbb{R}_{++}^{L \times S}$, *constitutes a Radner equilibrium, then there exists a vector* $\mu \in \mathbb{R}_{++}^S$ *such that* $(\mu_1 q_1^*, \dots, \mu_S q_S^*; x^{1*}, \dots, x^{I*})$ *constitutes an Arrow-Debreu equilibrium.*

The proof of the above proposition is similar to that of Proposition 4.8 (see Dana & Jeanblanc [514] and Mas-Colell et al. [1310, Proposition 19.E.2]). As in the case of Proposition 4.8, this result has two important implications: in the case of complete markets, the Radner equilibrium allocation is ex-ante Pareto optimal and, moreover, the existence of a Radner equilibrium can be established by relying on classical existence results for Arrow-Debreu equilibria.

When markets are incomplete, in the sense that not every contingent consumption plan can be spanned by the available assets (i.e., $I(D) \subset \mathbb{R}^S$), it is not possible to establish a general connection between Radner and Arrow-Debreu equilibria. While the existence of a Radner equilibrium can be established under suitable conditions, the equilibrium allocation will not necessarily be Pareto optimal. Indeed, unlike in the situation of complete markets, in the presence of incomplete market agents have a limited possibility of transferring wealth across different states of the world by trading on the available assets. As a consequence, the possibilities for risk sharing are limited.

It is worth pointing out that the allocation $(x^{1*}, \dots, x^{I*})$ obtained in correspondence of a Radner equilibrium with financial assets delivering dividends $D \in \mathbb{R}^{S \times N}$

is still an equilibrium allocation if the dividends' structure is modified in a way such that the new dividend matrix $D' \in \mathbb{R}^{S \times N'}$ satisfies $I(D) = I(D')$ (see Mas-Colell et al. [1310, Proposition 19.E.3]). In other words, the equilibrium allocation only depends on the linear subspace spanned by the financial assets and not on the specific structure of the financial markets open at $t = 0$. As a consequence of this property, equilibrium allocations are unaffected by the addition (or the deletion) of redundant assets.

Note that, for $L = 1$, the trade vector z (together with the initial endowment and the dividend matrix D) completely determines the amount of consumption good obtained at time $t = 1$, since $x_s = \sum_{n=1}^{N} d_{sn} z_n + e_s$, as a consequence of the constraint (4.17) together with the standing assumption of strictly increasing utility functions. In the case of incomplete markets, the following definition of *constrained ex-ante Pareto optimality* can be formulated, see Diamond [571] and Mas-Colell et al. [1310, Definition 19.F.1].

Definition 4.11 The asset allocation $(z^1, \ldots, z^I) \in \mathbb{R}^{N \times I}$ is *constrained ex-ante Pareto optimal* if it is feasible (i.e., $\sum_{i=1}^{I} z^i \leq 0$) and if there does not exist a feasible asset allocation $(z^{1'}, \ldots, z^{I'}) \in \mathbb{R}^{N \times I}$ such that the expected utility associated with the latter allocation is greater or equal than the expected utility associated with the allocation $\{z^1, \ldots, z^I\}$, for all the agents, and strictly greater for at least one agent $i \in \{1, \ldots, I\}$.

According to the above definition, an asset allocation is constrained ex-ante Pareto optimal if there is no asset redistribution which induces a Pareto improvement. Recalling that for $L = 1$ the asset allocation directly determines the consumption plan, it can be proved that any Radner equilibrium in incomplete markets is constrained ex-ante Pareto optimal in the sense of the above definition (see Mas-Colell et al. [1310, Proposition 19.F.1]). Note, however, that the situation $L = 1$ is rather particular, since with a single consumption good at $t = 1$ there are no possibilities for trade once the state of the world is revealed, so that the consumption plan is fully determined by the initial asset allocation. When $L \geq 2$, it is not possible to establish an analogous result.

4.3 Equilibrium with Intertemporal Consumption

In this section, we address the intertemporal consumption problem, by allowing utility functions to depend on consumption at both dates $t = 0$ and $t = 1$. In particular, we shall focus our attention on the relation between equilibrium allocations and Pareto optimality and on the possibility of characterizing the equilibrium allocation by means of a representative agent. We first consider the case of complete financial markets and then the more complex case of incomplete markets.

Complete Markets

Let us consider an economy with a single good (which we generically interpret as wealth) and S possible states of the world $(\omega_1, \ldots, \omega_S)$. In line with the previous section, we assume that at the initial date $t = 0$ agents can trade in S future markets, each of them corresponding to one of the S basic Arrow securities. The utility function of agent i (for each $i = 1, \ldots, I$) is defined with respect to wealth consumed at the initial date $t = 0$, denoted by x_0, and wealth consumed at the future date $t = 1$, denoted by x_s, in correspondence of the s-th state of the world ω_s, for each $s = 1, \ldots, S$. As before, the function $(x_0, x_s) \mapsto u(x_0, x_s)$ is assumed to be strictly increasing and concave. Moreover, we always assume that consumption is non-negative. For $s = 1, \ldots, S$, we denote by p_s the price of the s-th Arrow security paying one unit of wealth in correspondence of the state of the world ω_s and nothing in correspondence of all the other states. For $i = 1, \ldots, I$, we denote by $(e_0^i, e_1^i, \ldots, e_S^i)$ the endowment of agent i, with e_0^i denoting the initial wealth at time $t = 0$ and e_s^i the endowment in terms of wealth at time $t = 1$ contingent on the state of the world ω_s, for $s = 1, \ldots, S$. We assume that agents share homogeneous beliefs. The optimal consumption problem of agent i can be formulated as follows:

$$\max_{(x_0, x_1, \ldots, x_S) \in \mathbb{R}_+^{S+1}} \sum_{s=1}^{S} \pi_s \, u^i(x_0, x_s), \tag{PO3}$$

subject to the budget constraint

$$x_0 + \sum_{s=1}^{S} p_s x_s \leq e_0^i + \sum_{s=1}^{S} p_s e_s^i. \tag{4.18}$$

In the above formulation of the budget constraint, the price of wealth consumed at the initial date $t = 0$ has been conventionally set equal to one. As usual, since the utility functions are assumed to be strictly increasing, the budget constraint (4.18) can be formulated as an equality.

Problem (PO3) can be solved by means of the associated Lagrangian. Letting λ^i be the Lagrange multiplier associated with the optimal consumption problem of agent i, necessary and sufficient conditions (due to the strict concavity of the utility function) for a strictly positive optimal consumption plan $(x_0^{i*}, x_1^{i*}, \ldots, x_S^{i*})$ become

$$\sum_{s=1}^{S} \pi_s \, u_{x_0}^i(x_0^{i*}, x_s^{i*}) = \lambda^i, \tag{4.19}$$

$$\pi_s \, u_{x_s}^i(x_0^{i*}, x_s^{i*}) = \lambda^i p_s, \qquad \text{for all } s = 1, \ldots, S, \tag{4.20}$$

where, as before, $u^i_{x_0}$ and $u^i_{x_s}$ denote the derivatives of the utility function u^i with respect to its first and second argument, respectively. Since $\lambda^i > 0$, it holds that

$$\frac{\pi_s u^i_{x_s}(x^{i*}_0, x^{i*}_s)}{\sum^S_{r=1} \pi_r u^i_{x_0}(x^{i*}_0, x^{i*}_r)} = p^*_s, \qquad \text{for all } s = 1, \dots, S \text{ and } i = 1, \dots, I, \qquad (4.21)$$

in correspondence of the equilibrium price vector $(p^*_1, \dots, p^*_S) \in \mathbb{R}^S_{++}$. In particular, condition (4.21) means that, for all $s = 1, \dots, S$, the equilibrium price p^*_s of the s-th Arrow security corresponds to the ratio between the marginal utility of consumption at time $t = 1$ in the state of the world ω_s weighted by π_s and the (expected) marginal utility of consumption at the initial date $t = 0$. As a consequence, the marginal rates of substitution between consumption at the initial date $t = 0$ and consumption at $t = 1$ in correspondence of the state of the world ω_s, for $s = 1, \dots, S$, are equal among all the agents. By the analysis developed in Sect. 4.1, this implies the (ex-ante) Pareto optimality of the equilibrium allocation. This can also be seen by letting $a^i := 1/\lambda^i$, for all $i = 1, \dots, I$, and $\lambda_0 := 1$ and $\lambda_s := p^*_s$, for all $s = 1, \dots, S$, and comparing (4.19)–(4.20) with (4.5)–(4.6). We have thus shown that, if at date $t = 0$ agents can trade in future markets where all S basic Arrow securities are traded, then the equilibrium allocation is Pareto optimal. In other words, the possibility of spanning all possible future states of the world by trading in future markets open at $t = 0$ allows agents to reach a Pareto optimal allocation in equilibrium.

As a special case, when the agents' utility functions are time additive and state independent, i.e., $u^i(x_0, x_s) = u^i_0(x_s) + \delta^i u^i_1(x_s)$, for two strictly increasing and concave utility functions u^i_0 and u^i_1 and for some discount factor $\delta^i \in (0, 1)$, condition (4.21) becomes

$$\frac{\delta^i \pi_s u^{i'}_1(x^{i*}_s)}{u^{i'}_0(x^{i*}_0)} = p^*_s, \qquad \text{for all } s = 1, \dots, S \text{ and } i = 1, \dots, I. \qquad (4.22)$$

As pointed out in the previous section, Arrow securities are not traded in real financial markets: traded securities are in general complex and deliver a dividend described by a random variable, taking different values depending on the realization of the state of the world. Hence, let us now assume that N complex securities are traded at date $t = 0$, with security n delivering a random dividend $\tilde{d}_n$ with possible realizations $(d_{1n}, \dots, d_{Sn}) \in \mathbb{R}^S$, where d_{sn} denotes the amount of wealth delivered by security n in the state of the world ω_s, for $n = 1, \dots, N$ and $s = 1, \dots, S$.

In the presence of financial markets open at $t = 0$ with N complex securities, we assume that the initial endowment of agent i (for all $i = 1, \dots, I$) is given in terms of wealth and units of the N securities. More specifically, the endowment of agent i is represented by a vector $(e^i_0, e^i_1, \dots, e^i_N)$, where e^i_n, for $n = 1, \dots, N$, denotes the amount of security n initially held by agent i at $t = 0$ while e^i_0 denotes the endowment of wealth at $t = 0$. The optimal consumption problem of agent i can then be formulated as follows:

$$\max_{x_0 \in \mathbb{R}_+, (z_1, \dots, z_N) \in \mathbb{R}^N} \sum_{s=1}^{S} \pi_s u^i \left(x_0, \sum_{n=1}^{N} z_n d_{sn} \right), \tag{PO3'}$$

subject to the budget constraint

$$x_0 + \sum_{n=1}^{N} p_n z_n \leq e_0^i + \sum_{n=1}^{N} p_n e_n^i, \tag{4.23}$$

where $p_n \in \mathbb{R}_+$ denotes the price of security n at time $t = 0$, for $n = 1, \dots, N$, and z_n is the number of units of security n traded at $t = 0$ by agent i, for $n = 1, \dots, N$. In other words, the vector $z = (z_1, \dots, z_N) \in \mathbb{R}^N$ denotes the investment strategy of agent i in the N available securities. Note that allowing z to belong to $\mathbb{R}^N$ means in particular that short sales are allowed.

As can be easily checked, the conditions characterizing the optimal consumption choice in equilibrium are given by

$$\sum_{s=1}^{S} \frac{\pi_s u_{x_s}^i(x_0^{i*}, x_s^{i*})}{\sum_{r=1}^{S} \pi_r u_{x_0}^i(x_0^{i*}, x_r^{i*})} d_{sn} = p_n^*, \qquad \text{for all } n = 1, \dots, N \text{ and } i = 1, \dots, I, \tag{4.24}$$

where

$$x_0^{i*} = e_0^i - \sum_{n=1}^{N} p_n^*(z_n^{i*} - e_n^i) \qquad \text{and} \qquad x_s^{i*} = \sum_{n=1}^{N} z_n^{i*} d_{sn}, \qquad \text{for all } s = 1, \dots, S,$$

and $p^* \in \mathbb{R}^N$ denotes the vector of equilibrium prices. It is important to observe that condition (4.24) does not have the same implications of condition (4.21), because it does not imply that marginal rates of substitution between consumption at $t = 1$ in correspondence of the different states of the world and consumption at the initial date $t = 0$ are equal among all the agents. As a consequence, the equilibrium allocation associated to Problem (PO3') is not necessarily ex-ante Pareto optimal. However, in the special and important case of *complete* financial markets, we can prove the Pareto optimality of an equilibrium allocation, as shown in the following proposition (in line with the result of Proposition 4.10, see also Huang & Litzenberger [971, Section 5.7]).

Proposition 4.12 *If markets are complete (i.e., if* rank$(D) = S$, *using the notation introduced in Sect. 4.2), then the equilibrium allocation corresponding to Problem (PO3') is ex-ante Pareto optimal.*

Proof If markets are complete, ex-ante Pareto optimality can be obtained as a consequence of Proposition 4.10 together with the Pareto optimality of Arrow-Debreu equilibria (First Welfare Theorem). However, we shall prove the claim in a more direct and constructive way. The optimality conditions (4.24) defining an equilibrium allocation in correspondence of an equilibrium can be rewritten in matrix notation as follows, for all $i = 1, \ldots, I$:

$$
\begin{bmatrix}
d_{11} & d_{21} & \ldots & d_{S1} \\
d_{12} & d_{22} & \ldots & d_{S2} \\
\vdots & \ddots & \ddots & \vdots \\
d_{1N} & \ldots & \ldots & d_{SN}
\end{bmatrix}
\times
\begin{bmatrix}
\dfrac{\pi_1 u^i_{x_1}(x^{i*}_0, x^{i*}_1)}{\sum_{r=1}^{S} \pi_r u^i_{x_0}(x^{i*}_0, x^{i*}_r)} \\
\vdots \\
\dfrac{\pi_S u^i_{x_S}(x^{i*}_0, x^{i*}_S)}{\sum_{r=1}^{S} \pi_r u^i_{x_0}(x^{i*}_0, x^{i*}_r)}
\end{bmatrix}
=
\begin{bmatrix}
p^*_1 \\
p^*_2 \\
\vdots \\
p^*_N
\end{bmatrix}.
\tag{4.25}
$$

The first matrix appearing in the left-hand side of (4.25) is $D^\top$. If the matrix D has full rank and $N \geq S$, then we can build a submatrix $\underline{D} \in \mathbb{R}^{S \times S}$ corresponding to S linearly independent securities (i.e., S linearly independent rows of $D^\top$). We can then consider the following equation, where the matrix appearing on the left hand side is now $\underline{D}^\top$:

$$
\begin{bmatrix}
d_{11} & d_{21} & \ldots & d_{S1} \\
d_{12} & d_{22} & \ldots & d_{S2} \\
\vdots & \ddots & \ddots & \vdots \\
d_{1S} & \ldots & \ldots & d_{SS}
\end{bmatrix}
\times
\begin{bmatrix}
\dfrac{\pi_1 u^i_{x_1}(x^{i*}_0, x^{i*}_1)}{\sum_{r=1}^{S} \pi_r u^i_{x_0}(x^{i*}_0, x^{i*}_r)} \\
\vdots \\
\dfrac{\pi_S u^i_{x_S}(x^{i*}_0, x^{i*}_S)}{\sum_{r=1}^{S} \pi_r u^i_{x_0}(x^{i*}_0, x^{i*}_r)}
\end{bmatrix}
=
\begin{bmatrix}
p^*_1 \\
p^*_2 \\
\vdots \\
p^*_S
\end{bmatrix}.
$$

Multiplying by the inverse of the matrix $\underline{D}^\top$ (which exists since $\operatorname{rank}(\underline{D}) = S$), we obtain

$$
\begin{bmatrix}
\dfrac{\pi_1 u^i_{x_1}(x^{i*}_0, x^{i*}_1)}{\sum_{r=1}^{S} \pi_r u^i_{x_0}(x^{i*}_0, x^{i*}_r)} \\
\vdots \\
\dfrac{\pi_S u^i_{x_S}(x^{i*}_0, x^{i*}_S)}{\sum_{r=1}^{S} \pi_r u^i_{x_0}(x^{i*}_0, x^{i*}_r)}
\end{bmatrix}
=
\begin{bmatrix}
d_{11} & d_{21} & \ldots & d_{S1} \\
d_{12} & d_{22} & \ldots & d_{S2} \\
\vdots & \ddots & \ddots & \vdots \\
d_{1S} & \ldots & \ldots & d_{SS}
\end{bmatrix}^{-1}
\times
\begin{bmatrix}
p^*_1 \\
p^*_2 \\
\vdots \\
p^*_S
\end{bmatrix}
=:
\begin{bmatrix}
m^*_1 \\
m^*_2 \\
\vdots \\
m^*_S
\end{bmatrix}.
$$

In particular, observe that the right hand side of the last relation does not depend on i, so that

$$
\frac{\pi_s u^i_{x_s}(x^{i*}_0, x^{i*}_s)}{\sum_{r=1}^{S} \pi_r u^i_{x_0}(x^{i*}_0, x^{i*}_r)} = m^*_s, \qquad \text{for all } s = 1, \ldots, S \text{ and } i = 1, \ldots, I. \tag{4.26}
$$

We have thus shown that, in correspondence of an equilibrium allocation, marginal rates of substitution between consumption at $t = 1$ in the different states of the world and consumption at the initial date $t = 0$ are equal among all agents. In view of condition (4.7), this implies the ex-ante Pareto optimality of the equilibrium allocation. □

The strict monotonicity of the utility function u^i implies that the quantities m_s^* introduced in the proof of Proposition 4.12 are strictly positive. In particular, the vector $m^* \in \mathbb{R}_{++}^S$ is the unique solution to the equation (omitting possibly redundant assets)

$$D^\top m = p^*. \tag{4.27}$$

In Proposition 4.12, the equilibrium of the economy has been defined starting from the N securities paying the dividends represented by the matrix D. From the equilibrium prices of the original N securities we can derive the implicit equilibrium prices of other assets by means of replication arguments. As a consequence, under the assumption of market completeness, Problems (PO3) and (PO3') are equivalent, up to a suitable identification of the budget constraints.

Corollary 4.13 *If markets are complete (i.e., if* $\mathrm{rank}(D) = S$*), then the quantity m_s^* defined in (4.26) is the implicit equilibrium price (or market value) of the s-th Arrow security yielding* 1 *at time* $t = 1$ *in correspondence of the state of the world ω_s and* 0 *otherwise, for every* $s = 1, \ldots, S$.

Proof Since markets are complete, every contingent consumption plan $c \in \mathbb{R}^S$ can be replicated by investing in the N available securities. As in the proof of Proposition 4.12, let us define by $\underline{D}$ the submatrix composed of the linearly independent columns of D. Let us fix an arbitrary $s \in \{1, \ldots, S\}$ and consider the vector $\mathbf{1}_s := (0, \ldots, 0, 1, 0, \ldots, 0) \in \mathbb{R}^S$, where the 1 element is in the s-th position, representing the payoff of the s-th Arrow security. The payoff $\mathbf{1}_s$ can be replicated by investing in a portfolio $z(s) \in \mathbb{R}^S$ such that $\underline{D} z(s) = \mathbf{1}_s$. Multiplying by the inverse matrix $\underline{D}^{-1}$, we get $z(s) = \underline{D}^{-1} \mathbf{1}_s$. Hence, the implicit equilibrium price of the s-th Arrow security is given by the market value (see Definition 4.9) of the replicating portfolio $z(s)$, thus yielding $p^{*\top} z(s) = p^{*\top} \underline{D}^{-1} \mathbf{1}_s = m_s^*$. $\square$

In view of the above corollary, the quantity m_s^* associated with the state of the world ω_s is also called *state price*. Note that, if markets are complete and an asset paying c is introduced, then its equilibrium price is $m^{*\top} c$, as a consequence of the above corollary. The following proposition presents two important properties of asset prices in correspondence of an equilibrium allocation, namely that the market value $V(c)$ defined in Definition 4.9 is linear. Moreover, Proposition 4.14 implies that identical contingent consumption plans have identical market values at $t = 0$.

Proposition 4.14 *Suppose that markets are complete (i.e., $\mathrm{rank}(D) = S$). Then, in correspondence of an equilibrium allocation, the following hold:*

 (i) *if $c \in \mathbb{R}^S$ represents a contingent consumption plan, then its market value $V(c)$ at $t = 0$ (in the sense of Definition 4.9) is equal to $c^\top m^*$;*
(ii) *if $c, \tilde{c} \in \mathbb{R}^S$ represent two contingent consumption plans and $\hat{c} := c + \tilde{c}$, then it holds that*

$$V(\hat{c}) = V(c) + V(\tilde{c}).$$

Proof (i): market completeness implies the existence of a portfolio z^c of S linearly independent securities such that $\underline{D} z^c = c$, i.e., $z^c = \underline{D}^{-1} c$. Similarly as in the proof of Corollary 4.13, the market value at time $t = 0$ of such a portfolio is given by $m^{*\top} c$.

(ii): the claim follows from part *(i)*, since $m^{*\top} \hat{c} = m^{*\top}(c + \tilde{c}) = m^{*\top} c + m^{*\top} \tilde{c}$. $\square$

Summing up, if $N \geq S$ securities are traded in the market at time $t = 0$ and rank(D) $= S$, then the equilibrium prices of the N assets implicitly and uniquely define the prices of the S Arrow securities (state prices). Moreover, as shown in (4.26), the vector of the state prices coincides with the vector of the agents' marginal rates of substitution, thus implying the (ex-ante) Pareto optimality of the equilibrium allocation.

Representative Agent Analysis

In the context of a complete markets economy, equilibrium prices can also be characterized by means of a representative agent economy, in analogy to the riskless setting considered in Proposition 1.6. In particular, equilibrium prices can be determined in terms of the no-trade equilibrium of a single representative agent endowed with the resources of the whole economy, see Constantinides [488]. We denote by $e = (e_0, e_1, \ldots, e_S)$ the aggregate endowment of an economy with I agents, where $e_s := \sum_{i=1}^{I} e_s^i$, for all $s = 0, 1, \ldots, S$. In an economy with a single agent endowed with the aggregate resources of the original economy, the representative agent will not trade in correspondence of an equilibrium and his optimal consumption plan will be given by the aggregate endowment $(e_0, e_1, \ldots, e_S)$. We will develop the representative agent analysis in an intertemporal consumption setting, noting that the results presented below can also be established in an economy with consumption only at $t = 1$ and a single good.

Let us consider an economy populated by I agents with homogeneous beliefs, time additive and state independent utility functions $u^i(x_0, x_s) = u_0^i(x_0) + \delta u_1^i(x_s)$, with discount factor $\delta \in (0, 1)$, and with initial endowment $(e_0^i, e_1^i, \ldots, e_S^i)$, for all $i = 1, \ldots, I$. Note that the following analysis is also valid when agents exhibit different time preferences, i.e., when agents' preferences are characterized by different discount factors. In the present context, we construct the utility function of a representative agent, following the construction adopted in a riskless economy (see Proposition 1.6). Since markets are assumed to be complete, there is a unique vector of the prices of the Arrow securities $p^* = (p_1^*, \ldots, p_S^*)$ implicit in the equilibrium prices of the original N securities (see Corollary 4.13). As in Proposition 1.6, let $a^i := 1/\lambda^i$, where λ^i denotes the Lagrange multiplier associated with the optimal consumption plan $(x_0^{i*}, x_1^{i*}, \ldots, x_S^{i*})$ of agent i given as the solution to Problem (PO3) in correspondence of the equilibrium price vector p^* of the Arrow securities, for $i = 1, \ldots, I$. Similarly as in the proof of Proposition 4.4, we then define the utility function of the representative agent as

$$\mathbf{u}(e_0, e_1, \ldots, e_S) := \mathbf{u}_0(e_0) + \delta \sum_{s=1}^{S} \pi_s \, \mathbf{u}_1(e_s), \qquad (4.28)$$

where, for all $s = 1, \ldots, S$,

$$\mathbf{u}_0(e_0) := \max_{\substack{(x^1,\ldots,x^I) \in \mathbb{R}_+^I \\ \sum_{i=1}^{I} x^i \leq e_0}} \sum_{i=1}^{I} a^i u_0^i(x^i) \quad \text{and} \quad \mathbf{u}_1(e_s) := \max_{\substack{(x^1,\ldots,x^I) \in \mathbb{R}_+^I \\ \sum_{i=1}^{I} x^i \leq e_s}} \sum_{i=1}^{I} a^i u_1^i(x^i).$$

$$(4.29)$$

In other words, the representative agent's utility function is constructed as a weighted sum of the individual agents' utility functions, with the weights being given by the inverse of the Lagrange multipliers associated to each individual agent's optimal consumption problem, subject to the available resources constraints. Note that, since utility functions are assumed to be strictly increasing, the inequality constraints appearing in (4.29) can be equivalently formulated as equality constraints.

Similarly as in the proof of Proposition 4.4, the first order conditions yield, for all $i = 1, \ldots, I$ and $s = 1, \ldots, S$,

$$\mathbf{u}_0'(e_0) = \sum_{i=1}^{I} a^i u_0^{i'}(x_0^{i*}) \frac{\mathrm{d}x_0^{i*}}{\mathrm{d}e_0} = \sum_{i=1}^{I} \frac{\mathrm{d}x_0^{i*}}{\mathrm{d}e_0} = 1, \qquad (4.30)$$

$$\mathbf{u}_1'(e_s) = \sum_{i=1}^{I} a^i u_1^{i'}(x_s^{i*}) \frac{\mathrm{d}x_s^{i*}}{\mathrm{d}e_s} = \frac{p_s^*}{\delta \pi_s} \sum_{i=1}^{I} \frac{\mathrm{d}x_s^{i*}}{\mathrm{d}e_s} = \frac{p_s^*}{\delta \pi_s}, \qquad (4.31)$$

where we have used the optimality conditions (4.19)–(4.20) of the individual agents' optimal consumption problems together with the fact that $a^i = 1/\lambda^i$, for all $i = 1, \ldots, I$. Conditions (4.30)–(4.31) together imply that

$$\frac{\delta \pi_s \, \mathbf{u}_1'(e_s)}{\mathbf{u}_0'(e_0)} = p_s^*, \qquad \text{for all } s = 1, \ldots, S. \qquad (4.32)$$

Condition (4.32) shows that the equilibrium price vector $p^* = (p_1^*, \ldots, p_S^*)$ of the original economy corresponds to the vector of the equilibrium prices of the Arrow securities in an economy with a representative agent endowed with the resources of the whole economy and with the utility function $\mathbf{u}$ defined in (4.28)–(4.29). Indeed, by imposing that the optimal choice for the representative agent is obtained in correspondence of the aggregate endowment $(e_0, e_1, \ldots, e_S)$ (no-trade equilibrium) and by the optimality conditions (4.30)–(4.31), the marginal rate of substitution of the representative agent between consumption in the state of the world ω_s at time $t = 1$ and consumption at time $t = 0$ is equal to the Arrow security equilibrium price p_s^*, for all $s = 1, \ldots, S$. Condition (4.32) is of the same form of the optimality condition (4.22) obtained in the original economy with I agents with time additive and state independent preferences. Due to the assumption of complete markets, a no-trade equilibrium of a representative agent economy can equivalently be obtained by allowing the representative agent to trade in the N original assets instead of the S Arrow securities.

The representative agent's utility function defined in (4.28)–(4.29) depends on the weights $(a^1, \ldots, a^I)$. Since such weights are defined in terms of the Lagrange multipliers of Problem (PO3) of each individual agent $i \in \{1, \ldots, I\}$, they depend on the initial endowments of the I agents and, hence, on the initial allocation of resources. This means that the equilibrium price vector p^* in general depends on the distribution of resources among the agents. Therefore, the equivalence between the original economy with I agents and the representative agent's economy holds only in correspondence of equilibrium prices. In this sense, the representative agent is only *locally* representative of the original economy.

However, the results discussed at the end of Sect. 1.3 on the existence of a representative agent in a strong sense can be extended to the present context. Indeed, if agents have homogeneous beliefs, the same discount factor and time additive state independent utility functions with linear risk tolerance and with identical cautiousness (compare with Proposition 4.5), then equilibrium prices will only depend on the aggregate endowment, regardless of the distribution of resources among the I agents (*aggregation property*). In other words, equilibrium prices will not depend on the weights $(a^1, \ldots, a^I)$. The generalized power $u^i(x) = \frac{1}{b-1}(\gamma_i + bx)^{\frac{b-1}{b}}$, with $b \notin \{0, 1\}$, exponential $u^i(x) = -\gamma_i \exp(-x/\gamma_i)$ and logarithmic $u^i(x) = \log(\gamma_i + bx)$ utility functions, with the same coefficient b for all the agents, satisfy the above conditions and, hence, allow for aggregation in a strong sense, see Rubinstein [1485], Milne [1344]. More precisely, in the case of generalized power utility functions, we have the following result (compare with Huang & Litzenberger [971, Section 5.25]).

Proposition 4.15 *Suppose that markets are complete and let*

$$u_0^i(x) = u_1^i(x) = \frac{1}{b-1}(\gamma_i + bx)^{\frac{b-1}{b}},$$

with $b \notin \{0, 1\}$, for all $i = 1, \ldots, I$. Then the implicit equilibrium prices $p^* = (p_1^*, \ldots, p_S^*)$ of the S Arrow securities are given by

$$p_s^* = \frac{\delta \pi_s \left(\sum_{i=1}^I \gamma_i + be_s \right)^{-1/b}}{\left(\sum_{i=1}^I \gamma_i + be_0 \right)^{-1/b}}, \qquad \text{for all } s = 1, \ldots, S.$$

Proof Under the present assumptions, we can explicitly compute the functions $\mathbf{u}_0$ and $\mathbf{u}_1$ introduced in (4.29). Indeed, the generalized power utility form and the first order conditions involved in the constrained maximization problems of (4.29) give

$$a^i(\gamma_i + bx_s^{i*})^{-\frac{1}{b}} = \theta_s \qquad \text{and} \qquad \sum_{i=1}^{I} x_s^{i*} = e_s,$$

for all $s = 0, 1, \ldots, S$ and $i = 1, \ldots, I$, where θ_s is the Lagrange multiplier associated to the s-th state of the world. In particular, this implies that, for all $i = 1, \ldots, I$ and $s = 0, 1, \ldots, S$,

$$\left(\frac{\theta_s}{a^i}\right)^{-b} = \gamma_i + bx_s^{i*}.$$

Summing over all i, recalling that $\sum_{i=1}^{I} x_s^{i*} = e_s$, for all $s = 0, 1, \ldots, S$, and solving for θ_s, we get

$$\theta_s = \left(\sum_{i=1}^{I} \gamma_i + be_s\right)^{-\frac{1}{b}} \left(\sum_{i=1}^{I}(a^i)^b\right)^{\frac{1}{b}}.$$

In turn, from the above first order condition, this implies that

$$a^i(\gamma_i + bx_s^{i*})^{-\frac{1}{b}} = \left(\sum_{i=1}^{I} \gamma_i + be_s\right)^{-\frac{1}{b}} \left(\sum_{i=1}^{I}(a^i)^b\right)^{\frac{1}{b}},$$

so that

$$\mathbf{u}_0(e_0) = \sum_{i=1}^{I} \frac{a^i}{b-1}(\gamma_i + bx_0^{i*})^{\frac{b-1}{b}}$$

$$= \frac{1}{b-1} \left(\sum_{i=1}^{I} \gamma_i + be_0\right)^{-\frac{1}{b}} \left(\sum_{i=1}^{I}(a^i)^b\right)^{\frac{1}{b}} \left(\sum_{i=1}^{I} \gamma_i + be_0\right)$$

$$= \frac{1}{b-1} \left(\sum_{i=1}^{I} \gamma_i + be_0\right)^{\frac{b-1}{b}} \left(\sum_{i=1}^{I}(a^i)^b\right)^{\frac{1}{b}},$$

and similarly for all $\mathbf{u}_1(e_s)$, $s = 1, \ldots, S$. The equilibrium prices of the S Arrow securities are then given by condition (4.32), which in the present case reduces to

$$p_s^* = \frac{\delta\pi_s \left(\sum_{i=1}^I \gamma_i + be_s\right)^{-1/b}}{\left(\sum_{i=1}^I \gamma_i + be_0\right)^{-1/b}},$$

for all $s = 1, \ldots, S$, thus proving the claim. □

In particular, note that the equilibrium prices of the Arrow securities computed in Proposition 4.15 do not depend on the weights $(a^1, \ldots, a^I)$ but only on the aggregate endowment $(e_0, e_1, \ldots, e_S)$. As shown in Exercise 4.7, an analogous result holds true in the case of exponential utility functions.

Summing up, if the aggregation property holds in a strong sense, then the equilibrium prices of the original economy with I agents can be characterized in terms of the no-trade equilibrium for a representative agent endowed with the aggregate resources of the original economy, regardless of the initial resource distribution.

Incomplete Markets

In reality, financial markets are typically *incomplete*, in the sense that not every contingent consumption plan can be replicated by some portfolio of the securities available for trade at the initial date $t = 0$. This means that $S > N$, i.e., the number of (linearly independent) available securities is smaller than the number of possible states of the world. In other words, the available securities do not completely span the randomness of the economy.

In some cases, it is possible to make financial markets complete by introducing financial *derivatives*. In general terms, a financial derivative is a security whose dividend is defined in terms of the dividend of another security traded in the market (*underlying security*). As a canonical example, consider the case of a (*European*) *Call option*. A Call option is a financial derivative which gives to the holder the right, but not the obligation, to buy at the future date $t = 1$ the underlying security for some pre-specified price K (*strike price*), fixed at the initial date $t = 0$ of the writing of the option. Following Ross [1465] and Breeden & Litzenberger [287], we now show how the introduction of Call options can help completing the market.

Let us assume that there exists a portfolio (*state index portfolio*) $z^{\text{si}} \in \mathbb{R}^N$ such that its dividend d^{si} delivered at $t = 1$ takes distinct values in correspondence of different states of the world, i.e., the random variable $d^{\text{si}} : \Omega \to \mathbb{R}$ is such that $d^{\text{si}}(\omega_r) \neq d^{\text{si}}(\omega_s)$ if $r \neq s$, for all $r, s = 1, \ldots, S$. Without loss of generality, we suppose that $d^{\text{si}}(\omega_1) < \ldots < d^{\text{si}}(\omega_S)$. Let $d_s^{\text{si}} := d^{\text{si}}(\omega_s)$, for $s = 1, \ldots, S$. Let us also assume that agents have the possibility of investing in $S - 1$ Call options, having the state index portfolio as underlying security, with strike prices $K_s = d_s^{\text{si}}$, for $s = 1, \ldots, S - 1$. Since the holder of a Call option is not obliged to exercise the option at $t = 1$ (i.e., to buy the state index portfolio at the pre-specified price K_s), he will exercise the option only if the state index portfolio has at date $t = 1$ a dividend d^{si} greater than the exercise price. Hence, the payoff at time $t = 1$ of the Call option with strike price K_s is given by $\max\{d^{\text{si}} - K_s; 0\}$, so that the s-th Call option will be exercised only in correspondence of the states of the world $(\omega_{s+1}, \ldots, \omega_S)$. At the initial date $t = 0$, there are S open financial markets, one market for the state index portfolio and $S - 1$ markets for the $S - 1$ Call options. The dividend matrix D associated to this economy can be represented as

$$\begin{bmatrix} d_1^{\text{si}} & d_2^{\text{si}} & d_3^{\text{si}} & \cdots & d_S^{\text{si}} \\ 0 & d_2^{\text{si}} - d_1^{\text{si}} & d_3^{\text{si}} - d_1^{\text{si}} & \cdots & d_S^{\text{si}} - d_1^{\text{si}} \\ 0 & 0 & d_3^{\text{si}} - d_2^{\text{si}} & \cdots & d_S^{\text{si}} - d_2^{\text{si}} \\ 0 & \cdots & 0 & \ddots & \vdots \\ 0 & 0 & \cdots & 0 & d_S^{\text{si}} - d_{S-1}^{\text{si}} \end{bmatrix}.$$

Since we assumed that $d_1^{\text{si}} < \ldots < d_S^{\text{si}}$, it holds that $\text{rank}(D) = S$ and, therefore, markets are complete. As a consequence, if we allow agents to trade a sufficient number of Call options and the latter allow to span all possible states of the world, as it is the case under the present assumptions, then equilibrium allocations will be Pareto optimal. Note that the existence of a state index portfolio is equivalent to the existence of a security which allows to perfectly identify at time $t = 1$ the realization of the state of the world, so that any other security can be viewed as a derivative having the state index portfolio as underlying security. As such, the existence of a state index portfolio represents a rather strong assumption.

When markets are incomplete (i.e., when $\text{rank}(D) < S$), the available securities do not span all possible states of the world, meaning that agents have limited possibilities of transferring wealth across different states of the world by investing in the market at the initial date $t = 0$. As a consequence, in correspondence of an equilibrium allocation, agents will not have in general the same marginal rates of substitution, meaning that there are Pareto improving reallocations of wealth that cannot be implemented due to market incompleteness. This implies that, in the presence of market incompleteness, equilibrium allocations are not necessarily Pareto optimal. Moreover, when $\text{rank}(D) < S$, there exist infinitely many solutions to equation (4.27), so that the implicit prices of the Arrow securities are not uniquely defined by the equilibrium prices of the N traded securities. Indeed, all possible prices of the Arrow securities belong to a vector subspace of dimension $S - \text{rank}(D)$.

In a two-period economy with a single good (which we generically interpret as wealth) and not necessarily complete markets, it can be shown that every allocation associated with a Radner equilibrium (see Sect. 4.2) is *constrained ex-ante Pareto optimal*, in the sense of Definition 4.11, see Diamond [571], Stiglitz [1568], Geanakoplos & Polemarchakis [765]. We present this result in a rather general framework assuming that agents are endowed with traded assets as well as with exogenous state-contingent wealth. More specifically, we assume that, for every $i = 1, \ldots, I$, agent i is endowed with e_0^i units of wealth at the initial date $t = 0$ as well as with e_n^i units of the n-th security, for all $n = 1, \ldots, N$. Moreover, for every $i = 1, \ldots, I$, agent i is endowed with an additional state-contingent wealth at time $t = 1$ (which can be thought of as labor income or as an exogenous source of wealth), represented by the vector $(v_1^i, \ldots, v_S^i) \in \mathbb{R}^S$. In the present context, an allocation can be described by a tuple $(x_0^1, \ldots, x_0^I; z^1, \ldots, z^I) \in \mathbb{R}^I \times \mathbb{R}^{N \times I}$, where x_0^i denotes agent i's consumption at time $t = 0$ and z_n^i denotes the number of units of the n-th asset held in the portfolio by the i-th agent, for $i = 1, \ldots, I$ and $n = 1, \ldots, N$. Agent i's consumption at time $t = 1$ in correspondence of the state of the world ω_s is given by $x_s^i = v_s^i + \sum_{n=1}^{N} z_n^i d_{sn}$, for $s = 1, \ldots, S$. An allocation is said to be *feasible* if

$$\sum_{i=1}^{I} x_0^i \leq \sum_{i=1}^{I} e_0^i \quad \text{and} \quad \sum_{i=1}^{I} z_n^i \leq \sum_{i=1}^{I} e_n^i, \qquad \text{for all } n = 1, \ldots, N.$$

Proposition 4.16 *In the above setting, suppose that the agents' utility functions are strictly increasing and let $(x_0^{1*}, \ldots, x_0^{I*}; z^{1*}, \ldots, z^{I*}) \in \mathbb{R}^I \times \mathbb{R}^{N \times I}$ be an equilibrium allocation, with corresponding equilibrium prices $p^* \in \mathbb{R}_{++}^N$. Then the allocation $(x_0^{1*}, \ldots, x_0^{I*}; z^{1*}, \ldots, z^{I*})$ is constrained ex-ante Pareto optimal.*

Proof Arguing by contradiction, suppose that $(x_0^{1*}, \ldots, x_0^{I*}; z^{1*}, \ldots, z^{I*})$ is not constrained ex-ante Pareto optimal. In this case, there exists a feasible allocation $(\hat{x}_0^1, \ldots, \hat{x}_0^I; \hat{z}^1, \ldots, \hat{z}^I)$ such that

$$\sum_{s=1}^{S} \pi_s u^i(\hat{x}_0^i, \hat{x}_s^i) \geq \sum_{s=1}^{S} \pi_s u^i(x_0^{i*}, x_s^{i*}), \qquad \text{for all } i = 1, \ldots, I, \qquad (4.33)$$

with $\hat{x}_s^i = v_s^i + \sum_{n=1}^{N} \hat{z}_n^i d_{sn}$, for all $i = 1, \ldots, I$ and $s = 1, \ldots, S$, where the inequality in (4.33) is strict for at least one $i \in \{1, \ldots, I\}$. Since the utility function u^i is strictly increasing, the budget constraint is satisfied as an equality in correspondence of the optimal choice $(x_0^{i*}; z^{i*})$, for all $i = 1, \ldots, I$, so that

$$x_0^{i*} + \sum_{n=1}^{N} p_n^* z_n^{i*} = e_0^i + \sum_{n=1}^{N} p_n^* e_n^i, \qquad \text{for all } i = 1, \ldots, I. \qquad (4.34)$$

Hence, (4.33) and (4.34) together imply that

$$\hat{x}_0^i + \sum_{n=1}^{N} p_n^* \hat{z}_n^i \geq e_0^i + \sum_{n=1}^{N} p_n^* e_n^i, \qquad \text{for all } i = 1, \ldots, I,$$

with strict inequality for at least one $i \in \{1, \ldots, I\}$. Summing over all $i = 1, \ldots, I$, we get

$$\sum_{i=1}^{I} \left(\hat{x}_0^i + \sum_{n=1}^{N} p_n^* \hat{z}_n^i \right) > \sum_{i=1}^{I} \left(e_0^i + \sum_{n=1}^{N} p_n^* e_n^i \right).$$

Since $p^* \in \mathbb{R}_{++}^N$, this contradicts the feasibility of the allocation $(\hat{x}_0^1, \ldots, \hat{x}_0^I; \hat{z}^1, \ldots, \hat{z}^I)$, thus proving the claim.   □

As shown in Sect. 4.2, market completeness suffices to ensure the (ex-ante) Pareto optimality of equilibrium allocations. However, equilibrium allocations can be Pareto optimal even in incomplete markets. More specifically, we say that markets are *effectively complete* if every Pareto optimal allocation can be attained through security markets. It can be shown that, if the consumption sets of the agents are bounded from below and closed and markets are effectively complete, then every equilibrium allocation is Pareto optimal, thus providing a version of the First Welfare Theorem in the context of incomplete markets (see LeRoy & Werner [1191, Chapter 16]).

When agents have homogeneous beliefs and time additive state independent utility functions, a Pareto optimal allocation can be described in terms of a *sharing rule* $\{y^i : \mathbb{R}_+ \to \mathbb{R}_+; i = 1, \ldots, I\}$ such that $x_s^{i*} = y^i(e_s)$, for all $s = 0, 1, \ldots, S$ and $i = 1, \ldots, I$ (see Proposition 4.4). In particular, the existence of a sharing rule implies that Pareto optimal allocations only depend on the aggregate endowment $(e_1, \ldots, e_S)$ at time $t = 1$ in correspondence of the different states of the world $(\omega_1, \ldots, \omega_S)$ as well as on e_0. As a consequence, the financial market is *effectively complete* as long as it is complete with respect to the aggregate endowment states $(e_1, \ldots, e_S)$, meaning that, for all $s = 0, 1, \ldots, S$, there exists an Arrow security paying one unit of wealth if the aggregate endowment equals e_s and zero otherwise. If this is the case, in order to attain the Pareto optimal allocation, then every agent i will choose to hold $y^i(e_s)$ units of such Arrow security, for each $s = 0, 1, \ldots, S$. Note also that, as shown in Theorem 4.3, only the realization of the aggregate endowment matters and not the specific state of the world associated to the realization. In other words, in order to achieve effective completeness, it is not necessary to span different states of the world associated with identical realizations of the aggregate endowment. If markets are not complete with respect to the aggregate endowment states, then, similarly as above, Call options written on the aggregate endowment can be introduced in order to make the market effectively complete.

If the agents' endowments are given in terms of units of the N available securities, then the aggregate endowment portfolio (i.e., the portfolio delivering the aggregate endowment $\sum_{i=1}^{I} \sum_{n=1}^{N} e_n^i d_{sn}$ in correspondence of the state of the world ω_s, for each $s = 1, \ldots, S$) is always traded in the market. We call it the *market portfolio*. In this case, markets are effectively complete if and only if the agents' consumption plans in correspondence of every Pareto optimal allocation belong to the asset span (see LeRoy & Werner [1191, Section 16.3]). When the agents' endowments are not only defined in terms of the traded assets (for instance, agent i's endowment is composed of traded assets as well as of an exogenous vector $(v_1^i, \ldots, v_S^i)$, as considered before Proposition 4.16), markets are effectively complete only if the aggregate endowment $\sum_{i=1}^I \sum_{n=1}^N e_n^i d_{sn} + \sum_{i=1}^I v_s^i$, for $s = 1, \ldots, S$, lies in the asset span.

As a simple consequence of the above observations, in the absence of aggregate risk every equilibrium allocation will be Pareto optimal regardless of the completeness of the market, provided that the agents' endowments at $t = 1$ belong to the asset span. Indeed, suppose that the agents' endowments are defined in terms of units of the traded assets and the aggregate endowment does not depend on the state of the world (i.e., there is no aggregate risk). Then, as a consequence of Theorem 4.3, the consumption plan of every agent in correspondence of a Pareto optimal allocation is riskless and can be attained by simply holding a fraction of the market portfolio, whose dividend is constant across all possible states of the world. Hence, in the present situation, markets are effectively complete and every equilibrium allocation is Pareto optimal (see also Exercise 4.8 and compare with LeRoy & Werner [1191, Section 16.5] and Munk [1363, Section 7.4.2]).

In a similar way, Proposition 4.5 shows that when the agents' utility functions exhibit linear risk tolerance with the same cautiousness coefficient, then the Pareto optimal sharing rule is linear with respect to the aggregate endowment. This result implies that every equilibrium allocation will be Pareto optimal provided that a risk free asset is traded in the market and the agents' endowments lie in the asset span (for instance, agents' endowments are given in terms of units of the traded assets, as considered above). In this case, to reach a Pareto optimal allocation, every agent needs only to trade the risk free asset and the market portfolio, thus confirming that the agents' consumption plans in correspondence of Pareto optimal allocations exhibit a *two funds separation* property.

Concerning the possibility of characterizing equilibrium prices by means of the no-trade equilibrium of a single agent economy, an analysis based on a representative agent is possible whenever equilibrium allocations are Pareto optimal, as it is the case in complete markets (see Sect. 4.2). However, if the agents' utility functions are of the form considered in Proposition 4.5, then the construction of a representative agent in a strong sense and the aggregation property can also be established in incomplete markets provided that a risk free asset is traded and the aggregate endowment is spanned by the traded assets (or the agents' endowments are directly given in terms of units of the traded assets), see Milne [1344] and Detemple & Gottardi [563]. In this case, the representative agent's utility function will belong to the same class of the agents' utility functions (compare with the proof of Proposition 4.15). Equilibrium asset prices will be characterized as in (4.32) in terms of the marginal utility of the representative agent evaluated in correspondence of the aggregate endowment of the economy. Summing up, three properties are strictly related: effective completeness, demand aggregation and fund separation. In this regard, the HARA class of utility functions plays a crucial role allowing to obtain these three key properties, under suitable conditions on the economy's structure. We refer to Chap. 9 for a further analysis of the implications of market incompleteness.

4.4 The Fundamental Theorem of Asset Pricing

In this section, we study the relations between three cornerstones of modern finance theory, namely

(a) the absence of arbitrage opportunities,
(b) the existence of a positive linear pricing functional,
(c) the existence of an equilibrium of the economy.

Loosely speaking, the *Fundamental Theorem of Asset Pricing* asserts the equivalence of the above three properties, providing on the one hand a precise characterization of *viable* economies (i.e., economies for which there exists an equilibrium) and, on the other hand, opening the door to the *risk neutral valuation* of financial derivatives. Results on the fundamental theorem of asset pricing can be traced back to Ross [1464], Harrison & Kreps [904] and Harrison & Pliska [905] (see also the notes at the end of the chapter).

In the first part of this section, we introduce the notion of *arbitrage opportunity* and provide the precise connection between properties *(a)* and *(b)*, while, in the second part, we shall be concerned with the relation with property *(c)*. Finally, we illustrate the above concepts in the context of the classical binomial model, with a special focus on the valuation and hedging of financial derivatives, and we present the classical result of the Modigliani-Miller theorem.

As in Sect. 4.3, we always assume that agents live in a two-period economy (i.e., $t \in \{0, 1\}$) with a single consumption good and S possible states of the world, characterized by probabilities $(\pi_1, \ldots, \pi_S)$, with $\pi_s > 0$ for all $s = 1, \ldots, S$. As before, we assume that agents have homogeneous beliefs. However, we want to point out that the results presented below also hold in the case of heterogeneous beliefs $(\pi_1^i, \ldots, \pi_S^i)$, for $i = 1, \ldots, I$, provided that all the agents agree on null sets, i.e., $\pi_s^i > 0$ for all $i = 1, \ldots, I$ and $s = 1, \ldots, S$. In the economy, there are N securities available for trade at the initial date $t = 0$, with prices $p \in \mathbb{R}^N$. We denote by $D \in \mathbb{R}^{S \times N}$ the matrix representing the dividends of the N assets in correspondence of the S possible states of the world. In this context, as considered in Sect. 4.3, a *portfolio* of the N securities is represented by a vector $z \in \mathbb{R}^N$, where z_n denotes the number of units of security n held in the portfolio, for $n = 1, \ldots, N$. For a portfolio $z \in \mathbb{R}^N$, its price at date $t = 0$ is given by $p^\top z$ and the corresponding random dividend (or *payoff*) at $t = 1$ is represented by the vector $Dz \in \mathbb{R}^S$.

No-Arbitrage and Risk Neutral Probability Measures

In real financial markets, there should be no possibility of creating wealth out of nothing without risk. This intuitive concept is formalized by the notion of *arbitrage opportunity*, as presented in the following definition. Recall that, for a vector $x \in \mathbb{R}^S$, the inequality $x \geq 0$ means that $x_s \geq 0$ for all $s = 1, \ldots, S$, while $x > 0$ means that $x_s \geq 0$ for all $s = 1, \ldots, S$ and $x_s > 0$ for at least one $s \in \{1, \ldots, S\}$.

Definition 4.17 For a price-dividend couple $(p, D) \in \mathbb{R}^N \times \mathbb{R}^{S \times N}$, a portfolio $z \in \mathbb{R}^N$ is said to be

(i) an *arbitrage opportunity of the first kind* if $p^\top z \leq 0$ and $Dz > 0$;
(ii) an *arbitrage opportunity of the second kind* if $p^\top z < 0$ and $Dz \geq 0$.

A portfolio $z \in \mathbb{R}^N$ is simply said to be an *arbitrage opportunity* if it is an arbitrage opportunity of the first kind and/or an arbitrage opportunity of the second kind. We say that the market is *arbitrage free* if (p, D) does not admit arbitrage opportunities.

In view of the above definition, an arbitrage opportunity of the first kind represents an investment opportunity with zero (or even negative) cost at time $t = 0$ yielding a non-negative and non-zero payoff at time $t = 1$. Similarly, an arbitrage opportunity of the second kind represents an investment opportunity with a negative cost at time $t = 0$ yielding a non-negative (but possibly null) payoff at time $t = 1$. In general, the two concepts are distinct (see Exercise 4.15). Defining the matrix

$$\overline{D} := \begin{bmatrix} -p^\top \\ D \end{bmatrix} \in \mathbb{R}^{(S+1) \times N},$$

a portfolio $z \in \mathbb{R}^N$ is an arbitrage opportunity if and only if $\overline{D}z > 0$. The following proposition presents some simple but important consequences of the absence of arbitrage opportunities.

Proposition 4.18 *If there are no arbitrage opportunities, then the following hold:*

(i) *the* Law of One Price *holds, i.e., if $z, z' \in \mathbb{R}^N$ are two portfolios which satisfy $Dz = Dz'$, then $p^\top z = p^\top z'$;*
(ii) *if $z \in \mathbb{R}^N$ is a portfolio such that $Dz = 0$, then $p^\top z = 0$;*
(iii) *there is no riskless arbitrage opportunity, i.e., there does not exist a portfolio $z \in \mathbb{R}^N$ with $p^\top z \leq 0$ and $\sum_{n=1}^N d_{sn}z_n = c$, for all $s = 1, \ldots, S$, for some $c > 0$.*

Proof (i): Arguing by contradiction, let $z, z' \in \mathbb{R}^N$ satisfy $Dz = Dz'$ and suppose that $p^\top z < p^\top z'$. Define the portfolio $\bar{z} := z - z'$. Then $p^\top \bar{z} < 0$ and $D\bar{z} = 0$, thus showing that $\bar{z}$ is an arbitrage opportunity of the second kind. This contradicts the assumption of the absence of arbitrage opportunities (the case $p^\top z > p^\top z'$ can be treated in an analogous way).
(ii): It suffices to apply part *(i)* with $z' = 0 \in \mathbb{R}^N$.
(iii): This follows directly by Definition 4.17, since a riskless arbitrage opportunity is an arbitrage opportunity of the first kind.   □

As can be seen by inspecting the proof of the above proposition, statements *(i)*-*(ii)* are actually equivalent, in the sense that the Law of One Price holds if and only if $Dz = 0$ implies that $p^\top z = 0$. In particular, the Law of One Price implies that

portfolios with identical payoffs must also have identical prices, otherwise it would
be possible to realize an arbitrage opportunity. However, the implication does not
hold in the converse direction, meaning that it is possible that the Law of One Price
holds but, nevertheless, arbitrage opportunities (of the first kind) are present, as
shown in Exercise 4.16.

Having introduced the notion of arbitrage opportunity, we now aim at providing
a general characterization of an arbitrage free market. To this end, we have to define
the notion of *pricing functional*. As a preliminary, let $c \in \mathbb{R}^S$ be a payoff-contingent
consumption plan which can be attained by trading in the securities available in the
market, i.e., $c \in I(D)$, with $I(D)$ denoting the image of the dividend matrix D. This
means that there exists a portfolio $z^c \in \mathbb{R}^N$ satisfying $Dz^c = c$ (we say that z^c is the
replicating portfolio for the payoff c). As explained in Sect. 4.2, the market value
of the payoff c, defined in Definition 4.9 and denoted by $V(c)$, is given by the value
at time $t = 0$ of the replicating portfolio z^c, so that $V(c) = p^\top z^c$. In particular, if
there are no arbitrage opportunities, Proposition 4.18 implies that, if there exist two
distinct portfolios z and z' such that $c = Dz = Dz'$, then it holds that $p^\top z = p^\top z'$
(Law of One Price), so that $V : I(D) \to \mathbb{R}$ is indeed a function (meaning that, for all
$c \in I(D)$, the market value $V(c)$ is uniquely defined). Of course, if markets are not
complete (i.e., if $\text{rank}(D) < S$), then not every contingent consumption plan $c \in \mathbb{R}^S$
can be perfectly replicated. We introduce the following definition.

Definition 4.19 A mapping $Q : \mathbb{R}^S \to \mathbb{R}$ is said to be a *pricing functional* on $\mathbb{R}^S$ if
$Q(c) = V(c)$, for every $c \in I(D)$.

In other words, a pricing functional Q extends the market valuation functional
$V : I(D) \to \mathbb{R}$ to contingent consumption plans which do not necessarily lie in the
span of the traded securities. A pricing functional Q is said to be *linear* if, for every
$c, c' \in \mathbb{R}^S$ and $\alpha, \beta \in \mathbb{R}$, it holds that $Q(\alpha c + \beta c') = \alpha Q(c) + \beta Q(c')$. A pricing
functional Q is said to be *positive* (*strictly positive*, resp.) if, for every $c \geq 0$ ($c > 0$,
resp.), it holds that $Q(c) \geq 0$ ($Q(c) > 0$, resp.).

The following theorem contains the essence of the Fundamental Theorem
of Asset Pricing and, in particular, shows the equivalence between the absence
of arbitrage opportunities and the existence of a strictly positive linear pricing
functional.

Theorem 4.20 *For a price-dividend couple (p, D), the following are equivalent:*

  (i) *there are no arbitrage opportunities;*
 (ii) *there exists a strictly positive linear pricing functional $Q : \mathbb{R}^S \to \mathbb{R}$;*
(iii) *there exists a solution $m \in \mathbb{R}^S_{++}$ to the system $p = D^\top m$.*

Proof (i) $\Rightarrow$ (ii): let us define the linear space $M := \{\overline{D}z : z \in \mathbb{R}^N\}$ and the cone
$H := \mathbb{R}^{S+1}_+$. Both M and H are closed and convex subsets of $\mathbb{R}^{S+1}$. The assumption
that there are no arbitrage opportunities can be equivalently formulated as

$$H \cap M = \{0\}.$$

The separating hyperplane theorem[3] gives the existence of a non-zero linear functional $F : \mathbb{R}^{S+1} \to \mathbb{R}$ such that

$$F(x) < F(y), \qquad \text{for all } x \in M \text{ and } y \in H \setminus \{0\}.$$

Since M is a linear space, this implies that $F(x) = 0$, for all $x \in M$, and, hence, $F(y) > 0$, for all $y \in H \setminus \{0\}$. In turn, since F is a linear functional on $\mathbb{R}^{S+1}$, there exists a vector $u = (u_0, u_1, \ldots, u_S) \in \mathbb{R}^{S+1}_{++}$ such that $F(y) = u^\top y$, for all $y \in \mathbb{R}^{S+1}$. Define then the functional $Q : \mathbb{R}^S \to \mathbb{R}$ by $Q(c) := (u_1, \ldots, u_S)^\top c / u_0$, for all $c \in \mathbb{R}^S$. Clearly, Q is a strictly positive linear functional on $\mathbb{R}^S$. Moreover, if $c \in I(D)$, so that there exists $z^c \in \mathbb{R}^N$ satisfying $c = Dz^c$, it holds that

$$Q(c) = \frac{1}{u_0}(u_1, \ldots, u_S)^\top c = \frac{1}{u_0}(u_1, \ldots, u_S)^\top Dz^c = \frac{1}{u_0}\left(F(\overline{D}z^c) + u_0 p^\top z^c\right)$$

$$= p^\top z^c = V(c),$$

thus showing that $Q(c) = V(c)$, for all $c \in I(D)$. In view of Definition 4.19, we have thus proved that Q is a strictly positive linear pricing functional.

$(ii) \Rightarrow (iii)$: using the notation introduced in the previous step of the proof, define $m := (u_1, \ldots, u_S)/u_0 \in \mathbb{R}^S_{++}$. Since $u^\top x = 0$, for all $x \in M$ (equivalently, $u^\top \overline{D}z = 0$, for all $z \in \mathbb{R}^N$), we obtain, dividing by u_0 and recalling that $u_0 > 0$,

$$-p^\top z + m^\top Dz = 0, \qquad \text{for all } z \in \mathbb{R}^N.$$

This implies that $m \in \mathbb{R}^S_{++}$ solves the system $p = D^\top m$.

$(iii) \Rightarrow (i)$: arguing by contradiction, suppose that $z \in \mathbb{R}^N$ is an arbitrage opportunity, so that $\overline{D}z > 0$. Then, since $m \in \mathbb{R}^S_{++}$ and $p = D^\top m$, it holds that

$$0 < (1, m_1, \ldots, m_S)\overline{D}z = -p^\top z + m^\top Dz = 0,$$

thus yielding a contradiction. □

As can be seen be inspecting the proof of the above theorem, the pricing functional Q is completely characterized by the vector m, in the sense that $Q(c) = m^\top c$, for all $c \in \mathbb{R}^S$. Moreover, the quantities $(m_1, \ldots, m_S)$ admit the interpretation of *state prices*, i.e., m_s represents the market value at time $t = 0$ of one unit of wealth contingent on the realization of the state of the world ω_s, for $s \in \{1, \ldots, S\}$. Indeed, let $\mathbf{1}_s$ denote the vector $(0, \ldots, 0, 1, 0, \ldots, 0) \in \mathbb{R}^S$, with the element 1 being in the s-th position. Then, recalling that $Q(c) = m^\top c$, for all $c \in \mathbb{R}^S$, and applying the

---

[3]In the present context, the separating hyperplane can be stated as follows (see Duffie [593]). Let M and H be closed convex cones in $\mathbb{R}^{S+1}$ such that $M \cap H = \{0\}$. Then, if H does not contain a linear subspace other than $\{0\}$, than there is a non-zero linear functional $F : \mathbb{R}^{S+1} \to \mathbb{R}$ such that $F(x) < F(y)$, for all $x \in M$ and $y \in H \setminus \{0\}$.

pricing functional Q to the payoff represented by the vector $\mathbf{1}_s$, we get $Q(\mathbf{1}_s) = m_s$, for all $s = 1, \ldots, S$. In this sense, the state price vector m describes the arbitrage free prices of the Arrow securities implicit in the price-dividend couple (p, D). However, the state price vector m is not necessarily uniquely specified (equivalently, Theorem 4.20 does not assert the uniqueness of the pricing functional Q). As will be shown below in Proposition 4.26, m is unique as long as markets are complete.

Even though a risk free asset is not necessarily traded in the market, a pricing functional Q determines the risk free rate implicit in the price-dividend couple (p, D). Indeed, let us consider the riskless payoff $\mathbf{1} := (1, \ldots, 1) \in \mathbb{R}^S$. Then, in view of Theorem 4.20, we get $Q(\mathbf{1}) = m^\top \mathbf{1} = \sum_{s=1}^{S} m_s$. Hence, the risk free rate r_f implicitly determined by the pricing functional Q is given by $1/r_f = \sum_{s=1}^{S} m_s$. Note that the absence of arbitrage opportunities implies that there cannot exist two portfolios with riskless payoffs and different rate of returns (see Exercise 4.19 and compare also with Proposition 4.18).

The absence of arbitrage opportunities admits an alternative characterization in terms of the existence of a *risk neutral probability measure*, as defined below.

Definition 4.21 A vector $\pi^* = (\pi_1^*, \ldots, \pi_S^*) \in \mathbb{R}_{++}^S$ with $\sum_{s=1}^{S} \pi_s^* = 1$ represents a *risk neutral probability measure* for the price dividend couple (p, D) if

$$p_n = \frac{1}{r_f} \sum_{s=1}^{S} \pi_s^* d_{sn}, \qquad \text{for all } n = 1, \ldots, N,$$

where r_f denotes the risk free rate of return.

We have the following result, which relates the absence of arbitrage opportunities to the existence of a risk neutral probability measure.

Proposition 4.22 *For a price-dividend couple (p, D), the following are equivalent:*

(i) there are no arbitrage opportunities;
(ii) there exists a risk neutral probability measure $\pi^* = (\pi_1^*, \ldots, \pi_S^*)$.

Proof $(i) \Rightarrow (ii)$: if there are no arbitrage opportunities, then Theorem 4.20 gives the existence of a state price vector $m \in \mathbb{R}_{++}^S$. Let

$$m_0 := \sum_{s=1}^{S} m_s \qquad \text{and} \qquad \pi_s^* := \frac{m_s}{m_0}, \qquad \text{for all } s = 1, \ldots, S.$$

Then, recalling that $r_f = 1/m_0 > 0$ (see above) and that $p = D^\top m$, we get

$$p_n = \sum_{s=1}^{S} d_{sn} m_s = m_0 \sum_{s=1}^{S} d_{sn} \frac{m_s}{m_0} = \frac{1}{r_f} \sum_{s=1}^{S} \pi_s^* d_{sn}, \qquad \text{for all } n = 1, \ldots, N,$$

thus showing that $(\pi_1^*, \ldots, \pi_S^*) = (m_1/m_0, \ldots, m_S/m_0)$ is a risk neutral probability measure.

(ii) $\Rightarrow$ (i): let $(\pi_1^*, \ldots, \pi_S^*)$ be a risk neutral probability measure and let $z \in \mathbb{R}^N$ be an arbitrage opportunity. Then, since $p_n = \frac{1}{r_f} \sum_{s=1}^{S} \pi_s^* d_{sn}$, for all $n = 1, \ldots, N$, and $Dz \geq 0$, it holds that (recalling that $\pi^* \in \mathbb{R}_{++}^S$)

$$p^\top z = \frac{1}{r_f} \sum_{n=1}^{N} \sum_{s=1}^{S} \pi_s^* d_{sn} z_n \geq 0,$$

thus showing that z cannot be an arbitrage opportunity. $\square$

Denoting by $\mathbb{E}^*[\cdot]$ the expectation with respect to a risk neutral probability measure π^*, it then follows directly from Definition 4.21 that

$$p_n = \frac{1}{r_f} \mathbb{E}^*[\tilde{d}_n], \qquad \text{for all } n = 1, \ldots, N, \tag{4.35}$$

where $\tilde{d}_n$ denotes the random variable taking values $(d_{1n}, \ldots, d_{Sn})$.

The valuation formula (4.35) can also be extended to payoffs not traded in the market. Indeed, recalling the relation between a pricing functional Q, the corresponding state price vector m and the associated risk neutral probability measure π^* (see Proposition 4.22), it holds that[4]

$$Q(c) = m^\top c = \frac{1}{r_f} \sum_{s=1}^{S} \pi_s^* c_s = \frac{1}{r_f} \mathbb{E}^*[\tilde{c}], \qquad \text{for all } c \in \mathbb{R}^S. \tag{4.36}$$

Formulae (4.35)–(4.36) explain why a probability measure $\pi^* = (\pi_1^*, \ldots, \pi_S^*)$ as in Definition 4.21 is called *risk neutral*. Indeed, the arbitrage free price at time $t = 0$ of a random payoff is simply given by its expectation (computed with respect to the probability measure π^*), discounted by the risk free rate of return. Equivalently, letting $\tilde{r}_n := \tilde{d}_n/p_n$ denote the random return of security n (assuming that $p^n > 0$), for $n = 1, \ldots, N$, it holds that

$$\mathbb{E}^*[\tilde{r}_n] = \frac{1}{p_n} \mathbb{E}^*[\tilde{d}_n] = r_f, \qquad \text{for all } n = 1, \ldots, N.$$

This shows that, under a risk neutral probability measure π^*, the expected rate of return of every traded security coincides with the risk free rate of return. Moreover, for all $s = 1, \ldots, S$, the arbitrage free price of the s-th Arrow security $\mathbf{1}_s$ paying one

---

[4]In the following, with some abuse of notation, we equivalently denote by $\tilde{x}$ or $x = (x_1, \ldots, x_S) \in \mathbb{R}^S$ the random variable $\tilde{x}$ taking values $(x_1, \ldots, x_S)$ in the S states of the world.

unit of wealth in correspondence of the state of the world ω_s and zero otherwise is given by

$$Q(\mathbf{1}_s) = m_s = m_0\,\pi_s^* = \pi_s^*/r_f.$$

A risk neutral probability measure $\pi^* = (\pi_1^*, \ldots, \pi_S^*)$ is related to the original (also called physical/statistical/historical) probability measure $(\pi_1, \ldots, \pi_S)$ by the *likelihood ratio* $\ell = (\ell_1, \ldots, \ell_S) := (\pi_1^*/\pi_1, \ldots, \pi_S^*/\pi_S)$. In particular, it holds that $\ell_s > 0$, for all $s = 1, \ldots, S$. This shows that the two probability measures are *equivalent*, in the sense that they both assign a strictly positive probability to all states of the world $(\omega_1, \ldots, \omega_S)$. Moreover, we can derive an alternative representation of the risk neutral valuation formula (4.36) in terms of the original probability measure and of the likelihood ratio ℓ:

$$Q(c) = \frac{1}{r_f}\mathbb{E}[\tilde{\ell}\tilde{c}] = \frac{1}{r_f}\big(\mathbb{E}[\tilde{c}] + \mathrm{Cov}(\tilde{\ell}, \tilde{c})\big), \qquad (4.37)$$

where we have used the identity $\mathrm{Cov}(\tilde{\ell}, \tilde{c}) = \mathbb{E}[\tilde{\ell}\tilde{c}] - \mathbb{E}[\tilde{\ell}]\mathbb{E}[\tilde{c}]$, together with the fact that $\mathbb{E}[\tilde{\ell}] = 1$. Formula (4.37) shows that the arbitrage free price of a random payoff $\tilde{c}$ represented by a vector $c \in \mathbb{R}^S$ can be decomposed into two terms: a first term which is simply given by the expected payoff discounted by the risk free rate and a second term which depends on the covariance between the random payoff and the likelihood ratio.

As shown in Theorem 4.20, the absence of arbitrage opportunities implies the existence of a pricing functional Q, or, equivalently (see Proposition 4.22), the existence of a risk neutral probability measure π^*. For a random payoff represented by a vector $c = (c_1, \ldots, c_S) \in \mathbb{R}^S$, the value $Q(c)$ represents an *arbitrage free price* at time $t = 0$ in the sense that, if the original financial market composed of the N traded securities is extended to include the payoff c at the price $Q(c)$, then no arbitrage opportunity will be present in the extended market. To prove this claim, we shall first consider the case of a complete market and then the more general case of an incomplete market.

If markets are complete then $I(D) = \mathbb{R}^S$, so that all random payoffs $c \in \mathbb{R}^S$ can be attained by trading in the N available securities. As a consequence, the unique arbitrage free price of the payoff c is given by $Q(c) = V(c) = p^\top z^c$, where $z^c \in \mathbb{R}^N$ is a portfolio satisfying $Dz^c = c$. It is easy to check that any other price for the payoff c necessarily introduces arbitrage opportunities in the extended market composed of the N original securities together with the payoff c (see Exercise 4.20). This proves that, if markets are complete, then the arbitrage free price of any payoff-contingent consumption plan is uniquely determined by the pricing functional Q.

When markets are incomplete, things are not so simple and the arbitrage free price of a payoff-contingent consumption plan $c \in \mathbb{R}^S$ is not necessarily uniquely defined. Nevertheless, one can determine an interval to which the price of a given payoff $c \in \mathbb{R}^S$ must belong in order to exclude arbitrage opportunities in the extended market where the payoff c is traded together with the original N securities. As a preliminary, let us give the following definition.[5]

Definition 4.23 Let $c \in \mathbb{R}^S$ represent a random payoff. The *super-replication price* of c, denoted by $q_u(c)$, is defined as

$$q_u(c) := \min\{ p^\top z : z \in \mathbb{R}^N \text{ and } Dz \geq c \}.$$

Similarly, the *sub-replication price* of c, denoted by $q_l(c)$, is defined as

$$q_l(c) := \max\{ p^\top z : z \in \mathbb{R}^N \text{ and } Dz \leq c \}.$$

The super-replication price $q_u(c)$ of a payoff c represents the smallest amount of wealth needed to form a portfolio having a payoff always greater or equal than the payoff c, for every possible state of the world. An analogous interpretation can be given for the sub-replication price $q_l(c)$. The following result holds on $q_u(c)$ and $q_l(c)$ (the proof is given in Exercise 4.21).

Proposition 4.24 *If there are no arbitrage opportunities, then the following hold:*

  (i) for every $c \in \mathbb{R}^S$, it holds that $q_u(c) \geq q_l(c)$;
 (ii) if $c \in I(D)$, then $q_u(c) = q_l(c) = V(c)$;
(iii) if $c \notin I(D)$, then $q_u(c) > q_l(c)$.

In particular, statement *(ii)* of the above proposition shows that, if a payoff c is *attainable*, then its arbitrage free price is uniquely defined and is equal to the value of its replicating portfolio z^c. Equivalently, if a payoff c belongs to $I(D)$, we can say that c is *redundant*, in the sense that adding the payoff c to the original market does not alter the space of contingent consumption plans that can be spanned by the available securities. On the other hand, if $c \notin I(D)$, we say that c is *non-redundant*. If the payoff c is not attainable (if $c \notin I(D)$), then there exists a non-trivial interval of possible arbitrage free prices. This is illustrated in Fig. 4.3 in a case with two states and assets that span a line. For a payoff c which cannot be attained, any price outside of the interval $(q_l(c), q_u(c))$ leads to arbitrage opportunities in the extended financial market composed of the N original securities together with the payoff c (see Exercise 4.23).

As remarked after Theorem 4.20, the absence of arbitrage opportunities does not ensure the uniqueness of the pricing functional Q or, equivalently, of the risk neutral probability measure $\pi^* = (\pi_1^*, \ldots, \pi_S^*)$. In the following proposition, we present a dual characterization of the super-replication and sub-replication prices in terms of risk neutral probability measures. We denote by Π^* the set of all risk neutral

---

[5]In the absence of arbitrage opportunities, it can be shown that $\inf\{ p^\top z : z \in \mathbb{R}^N \text{ and } Dz \geq c \} = \min\{ p^\top z : z \in \mathbb{R}^N \text{ and } Dz \geq c \}$ and, similarly, $\sup\{ p^\top z : z \in \mathbb{R}^N \text{ and } Dz \leq c \} = \max\{ p^\top z : z \in \mathbb{R}^N \text{ and } Dz \leq c \}$, meaning that the infimum and the supremum are actually attained by some portfolios, see Föllmer & Schied [721, Theorem 1.32].

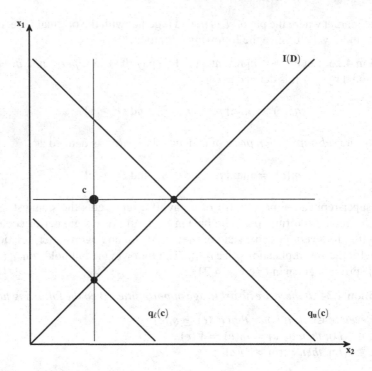

Fig. 4.3 Super and sub-replication in an incomplete market

probability measures associated to a given price-dividend couple (p, D), i.e.,

$$\Pi^* = \left\{ \pi^* \in \mathbb{R}^S_{++} : \sum_{s=1}^S \pi_s^* = 1 \text{ and } p_n = \sum_{s=1}^S \pi_s^* d_{sn}/r_f \text{ for all } n = 1, \ldots, N \right\},$$

with respect to a risk free return r_f.

Proposition 4.25 *Let $c \in \mathbb{R}^S$. If there are no arbitrage opportunities, then*

$$q_u(c) = \sup_{\pi^* \in \Pi^*} \frac{1}{r_f} \sum_{s=1}^S \pi_s^* c_s =: h_{\sup}(c), \qquad (4.38)$$

$$q_l(c) = \inf_{\pi^* \in \Pi^*} \frac{1}{r_f} \sum_{s=1}^S \pi_s^* c_s =: h_{\inf}(c). \qquad (4.39)$$

Proof We only present the proof of (4.38), the proof of (4.39) being analogous. Let $z \in \mathbb{R}^N$ be a portfolio such that $Dz \geq c$. Then, for any $\pi^* \in \Pi^*$, it holds that

$$\frac{1}{r_f} \sum_{s=1}^{S} \pi_s^* c_s \leq \frac{1}{r_f} \sum_{s=1}^{S} \sum_{n=1}^{N} \pi_s^* d_{sn} z_n = \sum_{n=1}^{N} p_n z_n = p^\top z.$$

Taking the infimum over all portfolios $z \in \mathbb{R}^N$ such that $Dz \geq c$ and the supremum over all risk neutral probability measures $\pi^* \in \Pi^*$, we get $h_{\sup}(c) \leq q_u(c)$. It remains to prove the converse inequality. Let $p^\varepsilon := h_{\sup}(c) + \varepsilon$, for some $\varepsilon > 0$, and consider the extended financial market given by the N original securities together with the payoff c traded at price p^ε. Since $p^\varepsilon > h_{\sup}(c)$, there exists no risk neutral probability measure for the extended market and, hence, the extended market is not arbitrage free (see Proposition 4.22). This means that there exist $z \in \mathbb{R}^N$ and $z' \in \mathbb{R} \setminus \{0\}$ such that

$$Dz + cz' \geq 0 \quad \text{and} \quad p^\top z + p^\varepsilon z' \leq 0, \tag{4.40}$$

with at least one the two inequalities being strict. By taking the expectation with respect to any risk neutral probability measure $\pi^* \in \Pi^*$ (and discounting by the risk free rate r_f) and then taking the supremum over all $\pi^* \in \Pi^*$, the first inequality in (4.40) gives $p^\top z + h_{\sup}(c) z' \geq 0$. Together with the second inequality in (4.40), this implies that $z' < 0$. But then, again the first inequality of (4.40) implies that $-Dz/z' \geq c$. In view of Definition 4.23, this means that

$$h_{\sup}(c) + \varepsilon = p^\varepsilon \geq -p^\top z/z' \geq q_u(c).$$

Since ε is arbitrary, this proves the desired inequality. □

In turn, Proposition 4.25 allows to derive a general criterion in order to establish when, in a general incomplete market, a given payoff can be attained by a portfolio of the N available securities. Indeed, summarizing the main results presented so far, we have that, if there are no arbitrage opportunities, then the following are equivalent, for any $c \in \mathbb{R}^S$:

(i) $c \in I(D)$, i.e., there exists a portfolio $z^c \in \mathbb{R}^N$ such that $Dz^c = c$;
(ii) the map $\Pi^* \ni \pi^* \mapsto \frac{1}{r_f} \mathbb{E}^*[\tilde{c}]$ is constant, i.e., the discounted risk neutral expectation of $\tilde{c}$ does not depend on the choice of the risk neutral probability measure;
(iii) $q_u(c) = q_l(c)$.

In incomplete markets, there exist infinitely many risk neutral probability measures. However, as shown above, the risk neutral expectation of any attainable payoff does not depend on the specific risk neutral probability measure chosen.

In general, financial markets are incomplete. It is therefore of interest to provide a precise characterization of market completeness. This is the content of the following theorem (which is usually called the *second Fundamental Theorem of Asset Pricing*), showing that, in the absence of arbitrage opportunities, the risk neutral probability measure is unique if and only if the market is complete.

Theorem 4.26 *If there are no arbitrage opportunities, then the following are equivalent:*

  (i) the market is complete, i.e., $I(D) = \mathbb{R}^S$;
  (ii) there exists a unique solution $m \in \mathbb{R}^S_{++}$ to the system $p = D^\top m$;
  (iii) there exists a unique risk neutral probability measure.

Proof In view of the proof of Proposition 4.22, it is clear that *(ii)* and *(iii)* are equivalent. The implication *(iii)* $\Rightarrow$ *(i)* follows from the above arguments. In order to show that *(i)* implies *(ii)*, suppose that there exist two vectors $m, m' \in \mathbb{R}^S_{++}$ such that $p = D^\top m = D^\top m'$, so that $D^\top (m - m') = 0$. Since $I(D) = \mathbb{R}^S$ is equivalent to rank$(D) = S$, this implies that $m = m'$, thus proving the claim. $\square$

We want to remark that, in general, there is no implication between the absence of arbitrage opportunities and the notion of market completeness, in the sense that markets can be complete even in the presence of arbitrage opportunities (an explicit example is given in Exercise 4.20).

No-Arbitrage and Equilibrium

Having analysed the relation between the absence of arbitrage opportunities and the existence of a strictly positive linear pricing functional (or, equivalently, of a risk neutral probability measure), we are now in the position to investigate the relation between the absence of arbitrage opportunities and the solvability of portfolio optimization problems and, in turn, the existence of an equilibrium.

We consider an agent's optimal portfolio problem in a general (complete or incomplete) market, with intertemporal consumption and S possible states of the world, as considered in Sect. 4.3. The general economy is populated by I agents, with state independent utility functions $u^i : \mathbb{R}^2 \to \mathbb{R}$ defined with respect to consumption at time $t = 0$ and consumption at time $t = 1$, for $i = 1, \ldots, I$. As before, we suppose that there are N securities traded at time $t = 0$ with dividend matrix $D \in \mathbb{R}^{S \times N}$ and price vector $p \in \mathbb{R}^N$. For all $i = 1, \ldots, I$, the endowment of agent i will be denoted by $(e_0^i, e_1^i, \ldots, e_S^i) \in \mathbb{R}^{S+1}$, where e_0^i represents the wealth of agent i in $t = 0$ and e_s^i the wealth of agent i in $t = 1$ contingent on state ω_s, for $s = 1, \ldots, S$. The endowment $(e_1^i, \ldots, e_S^i)$ can represent exogenous contingent wealth as well as the contingent wealth generated by an initial endowment in the N available securities. We generically denote by $z \in \mathbb{R}^N$ a trade vector (portfolio), expressed in terms of units bought/sold of the N securities. In this context, denoting by $(x_0^i, x_1^i, \ldots, x_S^i) \in \mathbb{R}^{S+1}$ the consumption plan of agent i in $t = 0$ and in $t = 1$ in correspondence of the S possible states of the world, the optimal portfolio problem can be written as

$$\max_{(x_0^i, x_1^i, \ldots, x_S^i) \in \mathbb{R}^{S+1}} \sum_{s=1}^{S} \pi_s u^i(x_0^i, x_s^i), \tag{PO4}$$

where the consumption plan $(x_0^i, x_1^i, \ldots, x_S^i) \in \mathbb{R}^{S+1}$ is subject to the budget constraint

$$x_0^i \leq e_0^i - p^\top z \quad \text{and} \quad x_s^i \leq e_s^i + \sum_{n=1}^{N} d_{sn} z_n, \qquad \text{for all } s = 1, \ldots, S, \tag{4.41}$$

for some trade vector $z \in \mathbb{R}^N$. We denote by $z^{i*} \in \mathbb{R}^N$ the optimal trade vector of agent i. Recall that, if the utility function u^i is strictly increasing in both of its arguments, then the budget constraint (4.41) can be written as an equality. In this case, given the endowment $(e_0^i, e_1^i, \ldots, e_S^i)$, the optimal consumption Problem (PO4) reduces to determining the optimal portfolio $z^{i*} \in \mathbb{R}^N$.

Proposition 4.27 *For any $i \in \{1, \ldots, I\}$, suppose that the utility function $u^i : \mathbb{R}^2 \to \mathbb{R}$ is strictly increasing in both of its arguments and that there exists an optimal portfolio $z^{i*} \in \mathbb{R}^N$ in correspondence of the price-dividend couple (p, D). Then the are no arbitrage opportunities.*

Proof Arguing by contradiction, let $z^{i*} \in \mathbb{R}^N$ denote the optimal portfolio associated to Problem (PO4) and suppose that $\bar{z} \in \mathbb{R}^N$ is an arbitrage opportunity. Consider then the following problem

$$\max_{\mu \in \mathbb{R}_+} \sum_{s=1}^{S} \pi_s u^i\left(x_0^{i*} - \mu p^\top \bar{z}, x_s^{i*} + \mu \sum_{n=1}^{N} d_{sn} \bar{z}_n\right), \tag{4.42}$$

where x_0^{i*} and x_s^{i*} denote the optimal consumption of agent i at date $t = 0$ and $t = 1$, respectively, given as the solution to Problem (PO4) subject to the budget constraint (4.41). Since the optimal portfolio z^{i*} satisfies the budget constraint, then the portfolio $z^{i*} + \mu \bar{z}$ also satisfies the budget constraint, for any $\mu \in \mathbb{R}_+$, because $\bar{z}$ is an arbitrage opportunity. Since at least one of the two inequalities $p^\top \bar{z} \leq 0$ and $D\bar{z} \geq 0$ is strict and the utility function u^i is strictly increasing in both of its arguments, the optimization problem (4.42) does not admit a solution, thus contradicting the existence of an optimal portfolio z^{i*}. □

Proposition 4.27 shows that, if agents' preferences are characterized by utility functions which are strictly increasing both in consumption at time $t = 0$ and in consumption at time $t = 1$, then the presence of arbitrage opportunities is incompatible with the existence of an optimal portfolio. Since arbitrage opportunities represent portfolios which "create wealth out of nothing", such a result is hardly surprising. In particular, Proposition 4.27 shows that the existence of an optimal portfolio z^{i*} excludes arbitrage opportunities of the first as well as of the second kind. If the utility function $u^i : \mathbb{R}^2 \to \mathbb{R}$ is only assumed to be strictly increasing in the first argument and non-decreasing in the second argument, then the result of Proposition 4.27 holds true with respect to arbitrage opportunities of the second kind (see Exercise 4.30 and compare also with LeRoy & Werner [1191, Section 3.6]).

Proposition 4.27 concerns the optimal portfolio problem of individual agents, but can be easily extended to a Radner equilibrium of the economy, as shown in the following corollary, the proof of which is an immediate consequence of the previous proposition. Indeed, in view of Definition 4.7, every agent's equilibrium portfolio has to be an optimal portfolio in the sense of Problem (PO4).

Corollary 4.28 *Suppose that $u^i : \mathbb{R}^2 \to \mathbb{R}$ is strictly increasing in both of its arguments, for some $i = 1, \ldots, I$, and let (p, D) be the price-dividend couple associated to a Radner equilibrium of the economy. Then there are no arbitrage opportunities.*

We now aim at studying the converse implication. More precisely, starting from a price-dividend couple (p, D) which does not admit arbitrage opportunities, we want to address the existence of an optimal portfolio for an individual agent. As a second step, we want to study the existence of a Radner equilibrium of an economy such that p corresponds to the price vector implementing the equilibrium allocation. We say that the price-dividend couple (p, D) represents a *viable* financial market if there exists an economy supporting the price vector p in correspondence of an equilibrium allocation. The following proposition provides a first and simple answer to such a question.

Proposition 4.29 *Let (p, D) be a price-dividend couple which does not admit arbitrage opportunities. Then there exists an economy populated by I risk neutral agents with arbitrary endowments $(e_0^i, e_1^i, \ldots, e_S^i) \in \mathbb{R}^{S+1}$, for all $i = 1, \ldots, I$, with beliefs given by risk neutral probabilities and a risk free discount factor $1/r_f$, such that the price vector p is associated to a Radner equilibrium of this economy.*

Proof If (p, D) does not admit arbitrage opportunities then, in view of Proposition 4.22, there exists a risk neutral probability measure $\pi^* = (\pi_1^*, \ldots, \pi_S^*)$ (note that, if the market is incomplete, there exist infinitely many risk neutral probability measures: for the purposes of this proof, it suffices to choose an arbitrary element $\pi^* \in \Pi^*$). For all $i = 1, \ldots, I$, let $(e_0^i, e_1^i, \ldots, e_S^i)$ be an arbitrary vector in $\mathbb{R}^{S+1}$ and consider the expected utility function

$$U^i(x_0^i, x_1^i, \ldots, x_S^i) = x_0^i + \frac{1}{r_f} \sum_{s=1}^{S} \pi_s^* x_s^i,$$

where $1/r_f$ is given by the sum of the state prices associated to π^* (compare with the discussion before Definition 4.21). Then, the portfolio optimization problem (PO4) subject to the budget constraint (4.41) can be rewritten as

$$\max_{z^i \in \mathbb{R}^N} \left(e_0^i - p^\top z^i + \frac{1}{r_f} \sum_{s=1}^{S} \pi_s^* \left(e_s^i + \sum_{n=1}^{N} d_{sn} z_n^i \right) \right).$$

Since π^* is a risk neutral probability measure, the equality $-p_n + \frac{1}{r_f} \sum_{s=1}^{S} \pi_s^* d_{sn} = 0$ is always satisfied, so that any portfolio which satisfies the budget constraint (4.41) as an equality is optimal. In particular, the portfolio $z^{i*} = 0 \in \mathbb{R}^N$ is an optimal choice, for all $i = 1, \ldots, I$ (i.e., each agent consumes the wealth generated by his initial endowment). In view of Definition 4.7, this proves the existence of a Radner equilibrium associated to the price vector $p \in \mathbb{R}^N$. $\qquad \square$

The above result shows that, if all the agents of the economy are risk neutral and their beliefs correspond to a risk neutral probability measure, then there exists a no-trade equilibrium associated with the arbitrage free price-dividend couple (p, D). In that sense, the couple (p, D) represents a viable financial market. As can be seen by inspecting the proof of Proposition 4.29, in the presence of an incomplete market, the I agents are also allowed to exhibit heterogeneous beliefs, as long as all the agents' beliefs correspond to risk neutral probabilities. Proposition 4.29 also explains why the probabilities π^* are called *risk neutral*. Indeed, a probability measure $\pi^* \in \Pi^*$ corresponds to the beliefs of a risk neutral agent supporting the price vector p in correspondence of a no-trade equilibrium. Note also that, in the setting of Proposition 4.29, nothing prevents us from choosing $I = 1$, so that the arbitrage free price-dividend couple (p, D) also corresponds to the no-trade equilibrium of a single risk neutral representative agent whose beliefs correspond to a risk neutral probability measure (in this regard, compare also with the discussion in Lengwiler [1182, Section 5.3.2]).

As we have seen in the previous chapters, economic agents are typically risk averse. We now aim at understanding the implications of the absence of arbitrage opportunities on the existence of an optimal portfolio for a risk averse agent and, consequently, on the existence of a Radner equilibrium for an economy populated by risk averse agents.

Let us consider the optimal consumption Problem (PO4) subject to the additional constraint that the optimal consumption plan $(x_0^{i*}, x_1^{i*}, \ldots, x_S^{i*})$ is non-negative, i.e.,

$$\max_{(x_0^i, x_1^i, \ldots, x_S^i) \in \mathbb{R}^{S+1}} \sum_{s=1}^{S} \pi_s u^i(x_0^i, x_s^i), \tag{PO4'}$$

subject to the budget constraint

$$0 \le x_0^i \le e_0^i - p^\top z \quad \text{and} \quad 0 \le x_s^i \le e_s^i + \sum_{n=1}^{N} d_{sn} z_n, \qquad \text{for all } s = 1, \ldots, S,$$

$$\tag{4.43}$$

for some portfolio $z \in \mathbb{R}^N$. The following proposition shows that the existence of an optimal portfolio for the above problem is actually equivalent to the absence of arbitrage opportunities, under the additional assumption that the utility function u^i is continuous.

Proposition 4.30 *For any $i \in \{1, \ldots, I\}$, suppose that $u^i : \mathbb{R}^2 \to \mathbb{R}$ is continuous and strictly increasing in both of its arguments. Then, there exists an optimal portfolio $z^{i*} \in \mathbb{R}^N$ for Problem (PO4') in correspondence of the price-dividend couple (p, D) if and only if (p, D) does not admit arbitrage opportunities.*

Proof As in Proposition 4.27, the existence of an optimal portfolio for Problem (PO4') subject to the budget constraint (4.43) implies that there are no arbitrage opportunities. Conversely, if the couple (p, D) does not admit arbitrage opportunities, it can be shown that agent i's budget set (identified by the budget constraint (4.43)) is closed and bounded, i.e., it is a compact set (see LeRoy & Werner [1191], Theorem 3.6.5). Since every continuous function has a maximum over a compact set (Weierstrass theorem), this implies the existence of an optimal portfolio $z^{i*} \in \mathbb{R}^N$.   □

If consumption is not restricted to be non-negative, then the absence of arbitrage opportunities does not suffice in general to ensure the existence of an optimal portfolio (see Exercise 4.33).

So far, in the case of concave utility functions, we have focused our attention on the relation between the absence of arbitrage opportunities and the existence of an optimal portfolio for an individual agent. Now, we aim at establishing a relation between the arbitrage free property of a price-dividend couple (p, D) and the existence of an economy populated by I risk averse agents such that p corresponds to the price vector implementing an equilibrium allocation. In other words, assuming that the couple (p, D) is arbitrage free, we want to prove the existence of an equilibrium of an economy such that $p = p^*$, where p^* denotes the equilibrium price vector of the economy, following the notation of Sect. 4.2. To this end, we rely on the representative agent technique, as introduced in Sect. 1.3 under certainty and in the previous section under risk.

Suppose that the price-dividend couple (p, D) does not admit arbitrage opportunities and consider an economy populated by I agents with strictly increasing and concave utility functions $u^i : \mathbb{R}^2 \to \mathbb{R}$, for $i = 1, \ldots, I$, separable with respect to time and with discount factor δ, i.e.,

$$u^i(x_0, x_s) = u_0^i(x_0) + \delta u_1^i(x_s), \qquad \text{for } i = 1, \ldots, I.$$

We assume that all the I agents have the same discount factor δ and believe in the same probability distribution. Furthermore, we assume that the agents' endowments lie in the asset span, i.e., $e^i \in I(D)$, for every $i = 1, \ldots, I$. For a given vector of weights $(a^1, \ldots, a^I) \in \mathbb{R}^I_{++}$, we define the candidate representative agent's utility function as in (4.28)–(4.29). Then, the utility functions $\mathbf{u}_0$ and $\mathbf{u}_1$ are strictly increasing and concave and we can consider the optimal consumption

Problem (PO4') for the representative agent's utility function **u**. Under a continuity assumption and in the absence of arbitrage opportunities, Proposition 4.30 can be applied to the representative agent's optimal consumption problem and we denote by $(x_0^*, x_1^*, \ldots, x_S^*)$ the optimal consumption plan in correspondence of the aggregate endowment $(e_0, e_1, \ldots, e_S)$ of the economy, with $e_s = \sum_{i=1}^{I} e_s^i$, for $s = 0, 1, \ldots, S$. Hence, if the representative agent is endowed with the optimal consumption plan $(x_0^*, x_1^*, \ldots, x_S^*)$, then the price vector p corresponds to the no-trade equilibrium of the single representative agent economy. The allocation $\{(x_0^{i*}, x_1^{i*}, \ldots, x_S^{i*}); i = 1, \ldots, I\}$ which defines the representative agent's utility function (4.28) will be a Pareto optimal allocation for the economy populated by the I agents. In turn, by the Second Welfare Theorem, the allocation $\{(x_0^{i*}, x_1^{i*}, \ldots, x_S^{i*}); i = 1, \ldots, I\}$ together with the price vector $p^* = p$ represents a Radner equilibrium of the economy with I agents endowed with $\{(x_0^{i*}, x_1^{i*}, \ldots, x_S^{i*}); i = 1, \ldots, I\}$ as initial allocation.

In the above argument, the utility functions u^i of the I agents were not assumed to be of a specific form. Indeed, the above construction of an economy yielding p as equilibrium price vector is based on taking the agents' utility functions as given and endowing the agents with a suitable allocation. As discussed in Sect. 4.3, in general equilibrium prices depend on the allocation of resources. However, if the agents' utility functions are of the generalized power, logarithmic or exponential form, then the aggregation property holds in a strong form and equilibrium prices only depend on the aggregate resources and not on the distribution of resources among the I agents. In this case, it suffices to endow the I agents with any feasible allocation such that $(e_0, e_1, \ldots, e_S) = (x_0^*, x_1^*, \ldots, x_S^*)$, with $(x_0^*, x_1^*, \ldots, x_S^*)$ denoting the optimal consumption plan in the representative agent economy supporting the price vector p. Note also that this result does not need market completeness, but only effective market completeness, as discussed at the end of Sect. 4.3.

Proposition 4.30 implies that there exists an economy with a single risk averse agent such that the price vector p is associated to a no-trade equilibrium, thus linking market viability to the absence of arbitrage opportunities. Moreover, there is a direct relation between the single agent's marginal utility of optimal consumption and a risk neutral probability measure. This is the content of the following proposition.

Proposition 4.31 *For any $i \in \{1, \ldots, I\}$, let $u^i : \mathbb{R}^2 \to \mathbb{R}$ be concave and strictly increasing in both of its arguments and denote by $(x_0^{i*}, x_1^{i*}, \ldots, x_S^{i*}) \in \mathbb{R}_+^{S+1}$ the optimal consumption plan of Problem (PO4') subject to the budget constraint (4.43) in correspondence of the price-dividend couple (p, D). Then the following hold:*

(i) *the price vector p corresponds to the no-trade equilibrium of the economy populated by the single agent i endowed with $(x_0^{i*}, x_1^{i*}, \ldots, x_S^{i*})$;*

(ii) *if $(x_0^{i*}, x_1^{i*}, \ldots, x_S^{i*}) \in \mathbb{R}_{++}^{S+1}$, then a risk neutral probability measure π^* can be defined by letting*

$$\pi_s^* := \frac{\pi_s \, u_{x_s}^i(x_0^{i*}, x_s^{i*})}{\sum_{r=1}^{S} \pi_r \, u_{x_r}^i(x_0^{i*}, x_r^{i*})}, \qquad \text{for all } s = 1, \ldots, S, \qquad (4.44)$$

where $u^i_{x_s}$ denotes the partial derivative of the function u^i with respect to the second argument evaluated in correspondence of state ω_s, for $s = 1, \ldots, S$ and $i = 1, \ldots, I$.

Proof (i): this follows directly from the optimality of $(x^{i*}_0, x^{i*}_1, \ldots, x^{i*}_S)$.
(ii): since $(x^{i*}_0, x^{i*}_1, \ldots, x^{i*}_S) \in \mathbb{R}^{S+1}_{++}$ is an interior solution to Problem (PO4') and u^i is strictly increasing (so that the budget constraint (4.43) is satisfied as an equality), the first order optimality conditions imply that

$$-p_n \sum_{r=1}^{S} \pi_r u^i_{x_0}(x^{i*}_0, x^{i*}_r) + \sum_{r=1}^{S} \pi_r u^i_{x_r}(x^{i*}_0, x^{i*}_r)d_{rn} = 0, \qquad \text{for all } n = 1, \ldots, N.$$

By the assumptions on the function u^i, it holds that $u^i_{x_0}(x^{i*}_0, x^{i*}_r) > 0$, for all $r = 1, \ldots, S$, so that the last condition can be rewritten as

$$p_n = \sum_{s=1}^{S} \pi_s \frac{u^i_{x_s}(x^{i*}_0, x^{i*}_s)}{\sum_{r=1}^{S} \pi_r u^i_{x_0}(x^{i*}_0, x^{i*}_r)} d_{sn}, \qquad \text{for all } n = 1, \ldots, N. \qquad (4.45)$$

Similarly to part *(iii)* of Theorem 4.20, equation (4.45) implies that we can define a state price vector $m \in \mathbb{R}^S_{++}$ by letting $m_s := \pi_s u^i_{x_s}(x^{i*}_0, x^{i*}_s)/(\sum_{r=1}^{S} \pi_r u^i_{x_0}(x^{i*}_0, x^{i*}_r))$, for all $s = 1, \ldots, S$. Hence, as in the proof of Proposition 4.22, we can define a risk neutral probability measure $\pi^* = (\pi^*_1, \ldots, \pi^*_S)$ by letting

$$\pi^*_s := \frac{m_s}{\sum_{s=1}^{S} m_s} = \frac{\pi_s u^i_{x_s}(x^{i*}_0, x^{i*}_s)}{\sum_{r=1}^{S} \pi_r u^i_{x_r}(x^{i*}_0, x^{i*}_r)}, \qquad \text{for all } s = 1, \ldots, S.$$

$\square$

In the setting of Proposition 4.31, the implicit risk free rate r_f is given by

$$\frac{1}{r_f} = \sum_{s=1}^{S} m_s = \frac{\sum_{s=1}^{S} \pi_s u^i_{x_s}(x^{i*}_0, x^{i*}_s)}{\sum_{s=1}^{S} \pi_s u^i_{x_0}(x^{i*}_0, x^{i*}_s)}. \qquad (4.46)$$

According to formula (4.44), the risk neutral probability π^*_s associated to the state of the world ω_s is equal to the ratio between the marginal utility of consumption at time $t = 1$ in correspondence of the state of the world ω_s weighted by π_s and the expected value of the marginal utility of consumption. Similarly, in view of equation (4.46), the risk free rate is given by the ratio between the expected marginal utility of consumption at time $t = 0$ and the expected marginal utility of consumption at time $t = 1$. Using the expectation operator $\mathbb{E}[\cdot]$, equation (4.45) can be equivalently rewritten as[6]

$$p_n = \frac{\mathbb{E}\left[u_{\tilde{x}}^i(x_0^{i*}, \tilde{x}^{i*})\tilde{d}_n\right]}{\mathbb{E}\left[u_{x_0}^i(x_0^{i*}, \tilde{x}^{i*})\right]}, \qquad \text{for all } n = 1, \ldots, N, \tag{4.47}$$

where $\tilde{x}^{i*}$ denotes the random variable taking $(x_1^{i*}, \ldots, x_S^{i*})$ as possible values and $\tilde{d}_n$ denotes the random dividend of the n-th security (i.e., the n-th column of the dividend matrix D). The term $u_{\tilde{x}}^i(x_0^{i*}, \tilde{x}^{i*})/\mathbb{E}[u_{x_0}^i(x_0^{i*}, \tilde{x}^{i*})]$ is often called *stochastic discount factor* (or *pricing kernel*), since it represents the random variable which, if taken as a discount factor, relates the random payoff $\tilde{d}_n$ at time $t = 1$ of a security with its price p_n at time $t = 0$. Using the definition of covariance together with equation (4.46), we can equivalently write

$$p_n = \frac{\mathbb{E}[\tilde{d}_n]}{r_f} + \frac{\text{Cov}\left(u_{\tilde{x}}^i(x_0^{i*}, \tilde{x}^{i*}), \tilde{d}_n\right)}{\mathbb{E}\left[u_{x_0}^i(x_0^{i*}, \tilde{x}^{i*})\right]}, \qquad \text{for all } n = 1, \ldots, N. \tag{4.48}$$

The last formula shows that the price at time $t = 0$ of any traded security can be decomposed into two terms: a first term representing the expected discounted dividend (under the statistical probability measure) and a second term represented by the covariance of the dividend with the stochastic discount factor.

Similarly to the risk neutral valuation formula, also the valuation rule (4.47) can be applied to payoffs not traded in the market. Indeed, any stochastic discount factor gives rise to a strictly positive linear pricing functional, which we denote by Q^u, in order to make explicit the dependence on the utility function u. This leads to the *marginal utility pricing rule* (or *stochastic discount factor pricing rule*):

$$Q^u(c) := \frac{\mathbb{E}\left[u_{\tilde{x}}^i(x_0^{i*}, \tilde{x}^{i*})\tilde{c}\right]}{\mathbb{E}\left[u_{x_0}^i(x_0^{i*}, \tilde{x}^{i*})\right]} = \frac{\mathbb{E}[\tilde{c}]}{r_f} + \frac{\text{Cov}\left(u_{\tilde{x}}^i(x_0^{i*}, \tilde{x}^{i*}), \tilde{c}\right)}{\mathbb{E}\left[u_{x_0}^i(x_0^{i*}, \tilde{x}^{i*})\right]}, \tag{4.49}$$

where we denote equivalently by $\tilde{c}$ or by $c = (c_1, \ldots, c_S)$ the random variable having $(c_1, \ldots, c_S)$ as possible values, representing a generic payoff/contingent consumption plan. Clearly, Q^u is a pricing functional, in the sense of Definition 4.19. As in the case of the risk neutral valuation formula (4.35), it can be shown that $Q^u(c)$ is an arbitrage free price. This means that, if the original financial market with N securities is in equilibrium and is extended to include the payoff c at the price $Q^u(c)$, then the extended financial market comprising $N + 1$ securities will be free of arbitrage opportunities.

---

[6]With some abuse of notation, we denote by $u_{\tilde{x}}^i$ the derivative of the function u^i with respect to its second argument, for $i = 1, \ldots, I$.

In particular, if the utility function $u^i : \mathbb{R}^2 \to \mathbb{R}$ is time separable (with a discount factor $0 < \delta \leq 1$), i.e., $u^i(x_0, x_s) = u_0^i(x_0) + \delta u_1^i(x_s)$, formula (4.44) reduces to

$$\pi_s^* = \frac{\pi_s\, u_1^{i'}(x_s^{i*})}{\sum_{r=1}^S \pi_r\, u_1^{i'}(x_r^{i*})}, \qquad \text{for all } s = 1, \ldots, S,$$

where $u_1^{i'}$ denotes the first derivative of the utility function u_1^i. In view of equation (4.46), the risk free rate r_f satisfies

$$\frac{1}{r_f} = \delta \sum_{s=1}^S \pi_s \frac{u_1^{i'}(x_s^{i*})}{u_0^{i'}(x_0^{i*})}, \tag{4.50}$$

where $u_0^{i'}$ denotes the first derivative of the utility function u_0^i. Similarly, equation (4.47) can be rewritten as

$$p_n = \delta\, \mathbb{E}\left[\frac{u_1^{i'}(\tilde{x}^{i*})}{u_0^{i'}(x_0^{i*})} \tilde{d}_n \right], \qquad \text{for all } n = 1, \ldots, N, \tag{4.51}$$

and relation (4.48) can be consequently rewritten as

$$p_n = \frac{\mathbb{E}[\tilde{d}_n]}{r_f} + \delta\, \mathrm{Cov}\left(\frac{u_1^{i'}(\tilde{x}^{i*})}{u_0^{i'}(x_0^{i*})}, \tilde{d}_n \right), \qquad \text{for all } n = 1, \ldots, N.$$

Analogous representations hold true for the valuation rule (4.49).

Formula (4.44) implies that to the optimal consumption plan $(x_0^{i*}, x_1^{i*}, \ldots, x_S^{i*})$ one can associate a likelihood ratio $\ell = (\ell_1, \ldots, \ell_S)$, where

$$\ell_s = \frac{u_{x_s}^i(x_0^{i*}, x_s^{i*})}{\mathbb{E}\left[u_{\tilde{x}}^i(x_0^{i*}, \tilde{x}^{i*}) \right]}, \qquad \text{for } s = 1, \ldots, S.$$

This representation of the likelihood ratio implies that, for $s = 1, \ldots, S$, the risk neutral probability π_s^* is greater than the original probability π_s if and only if the marginal utility of the optimal consumption in correspondence of the state of the world ω_s is larger than the expectation $\mathbb{E}\left[u_{\tilde{x}}^i(x_0^{i*}, \tilde{x}^{i*}) \right]$. In other words, the risk neutral probability introduced in part *(ii)* of Proposition 4.31 embeds risk aversion, in the sense that it assigns a probability greater than the statistical probability to the states of the world characterized by a level of consumption lower than the average and vice-versa.

Proposition 4.31 and the following results have been established in equilibrium considering the individual agents $i = 1, \ldots, I$. However, when markets are (effectively) complete, the result of Proposition 4.31 and formulae (4.47)–(4.51) can also be applied to the representative agent economy. Indeed, if markets are (effectively) complete, then every equilibrium allocation of the economy with I agents is Pareto optimal and, hence, can be characterized in terms of a representative agent economy, where the representative agent's utility function is of the form (4.28)–(4.29). In particular, in the case of utility functions which are separable with respect to time and with a discount factor δ, the valuation formula (4.51) with respect to the representative agent's no-trade equilibrium in correspondence of the aggregate endowment $(e_0, e_1, \ldots, e_S)$ becomes

$$p_n = \delta \, \mathbb{E} \left[\frac{\mathbf{u}_1'(\tilde{e})}{\mathbf{u}_0'(e_0)} \tilde{d}_n \right] = \delta \sum_{s=1}^{S} \pi_s \frac{\mathbf{u}_1'(e_s)}{\mathbf{u}_0'(e_0)} d_{sn}, \qquad \text{for all } n = 1, \ldots, N, \qquad (4.52)$$

where $\tilde{e}$ denotes the random variable taking values $(e_1, \ldots, e_S)$, and the corresponding stochastic discount factor is represented by the random variable taking values

$$\left(\frac{\delta \mathbf{u}_1'(e_1)}{\mathbf{u}_0'(e_0)}, \ldots, \frac{\delta \mathbf{u}_1'(e_S)}{\mathbf{u}_0'(e_0)} \right). \qquad (4.53)$$

As in the case of the stochastic discount factor pricing rule (4.49), also formula (4.52) can be extended to the valuation of payoffs not traded in the original financial market. Note that the representative agent's stochastic discount factor (and, hence, the associated risk neutral probability measure) is greater in correspondence of the states of the world where the aggregate endowment is lower and vice-versa, thus taking into account the agents' risk aversion. Of course, all the considerations made after Proposition 4.31 also apply to a representative agent economy.

The valuation formula (4.52) has interesting applications to asset pricing. In particular, as shown in Exercise 4.36, if the agents' utility functions are assumed to be of the generalized power form (with identical discount factors and cautiousness coefficients) and if the aggregate endowment is lognormally distributed, then one can derive a version of the celebrated Black-Scholes pricing formula for Call options.

The Binomial Model

In this section, we present the classical single-period *binomial model*, which provides a simple example of an economy with two possible states of the world and two securities. The binomial model allows to illustrate in a clear way most of the concepts introduced in the present section and is particularly interesting in view of its practical applications to derivative pricing. At the same time, the binomial model already contains most of the key features of more complex stochastic models widely employed in quantitative finance.

In the binomial model, there are two securities available for trade at the initial date $t = 0$: a risk free security, generically called *bond*, and a risky security, generically called *stock*. Let B_t denote the price of the bond and S_t the price of the stock at time t, for $t \in \{0, 1\}$. Since the bond represents a risk free asset, the value B_1 is constant across the two possible states of the world. More specifically, we suppose that B_0 is conventionally set equal to 1 and $B_1 := r_f$, where $r_f > 0$ denotes the risk free rate. On the contrary, the stock represents a risky asset, so that its price S_1 at time $t = 1$ depends on the realization of the state of the world. More specifically, for some given initial price $S_0 > 0$, we assume that $S_1 = S_0 \tilde{z}$, where $\tilde{z}$ is a random variable taking values $\{u, d\}$, with $u > d > 0$, with probabilities π and $1 - \pi$, respectively (with $\pi > 0$). Summing up, the price processes $(B_t)_{t=0,1}$ and $(S_t)_{t=0,1}$ can be represented as follows:

$$
\begin{array}{cc}
t = 0 & t = 1 \\
\text{---} & \text{-------} \\
B_0 = 1 & B_1 = r_f \\
\end{array}
$$

$$
\begin{array}{ccc}
 & & S_1 = S_0 u \qquad (4.54) \\
 & \pi \nearrow & \\
S_0 & & \\
 & 1 - \pi \searrow & \\
 & & S_1 = S_0 d \\
\end{array}
$$

Equivalently, in terms of the notation previously introduced, we have

$$
p = \begin{pmatrix} 1 \\ S_0 \end{pmatrix} \quad \text{and} \quad D = \begin{pmatrix} r_f & S_0 u \\ r_f & S_0 d \end{pmatrix}. \qquad (4.55)
$$

In the binomial model, the financial market is assumed to satisfy the underlying hypotheses of this section, namely, there are no transaction costs, short sales are allowed, the risk free rate is the same for lending as well as for borrowing and the two securities are perfectly liquid in the market. In this context, a portfolio is simply represented by a vector $z \in \mathbb{R}^2$, where the first component z_1 denotes the number of units of the bond and z_2 the number of units of the stock.

As a preliminary, we show that the market is complete and provide a necessary and sufficient condition for the absence of arbitrage opportunities.

Proposition 4.32 *In the single-period binomial model* (4.55) *the market is complete. Moreover, there are no arbitrage opportunities if and only if $d < r_f < u$.*

Proof Since $u > d$ and $r_f > 0$, it is easy to see that the matrix D is invertible, so that $I(D) = \mathbb{R}^2$, meaning that the market is complete. Due to Theorem 4.20, there are no arbitrage opportunities if and only if there exists a vector $m \in \mathbb{R}^2_{++}$ such that $p = D^{\mathsf{T}} m$. Since the matrix D is invertible it holds that $m = (D^{\mathsf{T}})^{-1} p$, i.e.,

$$
m = \frac{1}{r_f S_0 (u - d)} \begin{pmatrix} -S_0 d & r_f \\ S_0 u & -r_f \end{pmatrix} \begin{pmatrix} 1 \\ S_0 \end{pmatrix} = \begin{pmatrix} \frac{r_f - d}{r_f(u-d)} \\ \frac{u - r_f}{r_f(u-d)} \end{pmatrix}.
$$

Under the standing assumptions that $r_f > 0$ and $u > d$, it is immediate to see that $m \in \mathbb{R}^2_{++}$ if and only if $d < r_f < u$. □

In view of Proposition 4.22, there exists a (unique) risk neutral probability measure π^* if and only if $d < r_f < u$, as shown in the following corollary. Note that, since the market is complete, if a risk neutral probability measure exists, then it is necessarily unique (see Proposition 4.26).

Corollary 4.33 *There exists a risk neutral probability measure $(\pi^*, 1 - \pi^*)$ if and only if $d < r_f < u$, in which case it holds that*

$$\pi^* = \frac{r_f - d}{u - d} \quad and \quad 1 - \pi^* = \frac{u - r_f}{u - d}.$$

Proof As in the proof of Proposition 4.22, it suffices to let $\pi^* := m_1/(m_1 + m_2)$ and rely on the proof of Proposition 4.32. □

The no-arbitrage condition $d < r_f < u$ admits an intuitive interpretation. Indeed, the failure of one of these two inequalities means that one of the two securities dominates the other, in the sense that it has a greater return in each possible state of the world. In this case, it would be possible to create an arbitrage opportunity by investing in the more profitable security and selling short the other. For instance, if $d \geq r_f$, then it would be possible to realize an arbitrage opportunity with the portfolio $z = (-S_0, 1)$. At time $t = 0$, the value of such a portfolio is zero, while at time $t = 1$ the dividend will be either $S_0(u - r_f) > 0$ or $S_0(d - r_f) \geq 0$ in correspondence of the two possible states of the world, thus yielding an arbitrage opportunity. The case $u \leq r_f$ can be treated in a similar way.

Let us now consider the problem of valuing a *derivative contract* in the context of the binomial model. A first and simple approach to the valuation of financial derivatives consists in the application of the risk neutral valuation formula (4.36). In particular, due to Corollary 4.33, if the payoff of the derivative is represented by a vector $c = (c_u, c_d) \in \mathbb{R}^2$, then the price at time $t = 0$ of the derivative, denoted by $Q(c)$, can be computed as

$$Q(c) = \frac{1}{r_f} \mathbb{E}^*[c] = \frac{1}{r_f}\left(\frac{r_f - d}{u - d}c_u + \frac{u - r_f}{u - d}c_d\right). \tag{4.56}$$

In the binomial model, since markets are complete, the valuation formula (4.56) can also be obtained by means of two alternative procedures, namely the value at time $t = 0$ of the portfolio replicating the derivative and the *Delta-hedging* strategy. Let us first consider the replicating portfolio approach. As we have seen in Proposition 4.32, the market is complete and, hence, for any contingent consumption plan $c \in \mathbb{R}^2$ there exists a replicating portfolio $z^c \in \mathbb{R}^2$ such that $c = Dz^c$. More specifically, we have the following proposition.

Proposition 4.34 *Let* $c = (c_u, c_d) \in \mathbb{R}^2$ *represent the payoff of a financial contract. Then there exists a unique portfolio* $z^c \in \mathbb{R}^2$ *such that* $Dz^c = c$, *explicitly given by*

$$z^c = \begin{pmatrix} \frac{1}{r_f} \frac{uc_d - dc_u}{u - d} \\ \frac{1}{S_0} \frac{c_u - c_d}{u - d} \end{pmatrix}.$$

Moreover, the value at time $t = 0$ *of the portfolio* $z^c \in \mathbb{R}^2$ *is given by* (4.56).

Proof Since markets are complete, it holds that

$$z^c = D^{-1} c = \frac{1}{r_f S_0 (u - d)} \begin{pmatrix} -S_0 d & S_0 u \\ r_f & -r_f \end{pmatrix} \begin{pmatrix} c_u \\ c_d \end{pmatrix} = \begin{pmatrix} \frac{1}{r_f} \frac{uc_d - dc_u}{u - d} \\ \frac{1}{S_0} \frac{c_u - c_d}{u - d} \end{pmatrix}.$$

It is then immediate to verify that $p^\top z^c$ coincides with the right-hand side of (4.56).
                                                                                           □

According to the above proposition, the unique arbitrage free price of a payoff $c \in \mathbb{R}^2$ is given by the market value $p^\top z^c$ of the replicating portfolio. As discussed before, any other price for the payoff c would introduce arbitrage opportunities in the extended financial market composed of the two original securities (i.e., the bond and the stock) together with the additional payoff c.

The value of a payoff $c \in \mathbb{R}^2$ can also be obtained in terms of the so-called *Delta hedging* strategy. The Delta hedging technique consists in determining the portfolio, composed of one unit of the contract delivering the payoff c and $-\Delta$ units of the underlying stock, such that the portfolio's dividend at time $t = 1$ is riskless, in the sense that it does not depend on which of the two states of the world will be realized at time $t = 1$. In that sense, the position in the payoff c is *hedged* by the position $-\Delta$ in the underlying security. Let us denote by $V_t(\Delta)$ the value of such a portfolio at time t, for $t \in \{0, 1\}$.

Proposition 4.35 *Let* $c = (c_u, c_d) \in \mathbb{R}^2$ *represent the payoff of a financial contract and let* $V_t(\Delta)$, *for* $t \in \{0, 1\}$, *be defined as above. Then* $V_1(\Delta)$ *does not depend on the state of the world if and only if*

$$\Delta = \frac{c_u - c_d}{S_0(u - d)}. \tag{4.57}$$

Proof By definition, the quantity Δ is such that the value of the portfolio $V_1(\Delta)$ does not depend on the state of the world, i.e.,

$$c_u - \Delta S_0 u = c_d - \Delta S_0 d.$$

This directly implies (4.57), recalling that $u > d$ and $S_0 > 0$.   □

Formula (4.57) also explains the reason of the terminology *Delta hedging*. Indeed, the strategy Δ, which represents the number of units of the underlying asset needed to compensate the riskiness of the random payoff c, is given by the ratio between the difference of the derivative's payoff and the difference of the dividend of the underlying stock in correspondence of the two possible states of the world.

Since the portfolio constructed in Proposition 4.35 is riskless, if there are no arbitrage opportunities (in the market composed by the bond, the stock and the derivative), then its rate of return has to be equal to the risk free rate of return r_f, i.e.,

$$(-Q(c) + \Delta S_0) r_f = -c_u + \Delta S_0 u,$$

where $Q(c)$ denotes the arbitrage free price at time $t = 0$ of the financial contract having payoff c. The above formula implies that

$$Q(c) = \frac{c_u - \Delta S_0 u}{r_f} + \Delta S_0 = \frac{1}{r_f} \frac{u c_d - d c_u}{u - d} + \frac{c_u - c_d}{u - d}.$$

As can be easily seen, this last expression coincides with the right-hand side of (4.56).

The Modigliani-Miller Theorem

By Theorem 4.20, the absence of arbitrage opportunities is equivalent to the existence of a (possibly non-unique) strictly positive linear pricing functional. In particular, the linearity of the pricing functional implies that, in the absence of market frictions, the value of a firm is independent of its capital structure. This result is known as the *Modigliani-Miller theorem* and goes back to Modigliani & Miller [1348, 1349].

In order to state the main result of Modigliani & Miller [1348], let us consider a two-period arbitrage free market and a firm starting its activities at the initial date $t = 0$ and terminating at $t = 1$. We denote by V_1 the revenues generated by the firm at date $t = 1$. We shall consider two capital structures:

(a) an *unleveraged firm*, financing its activities only by issuing equity;
(b) a *leveraged firm*, financing its activities by issuing equity and debt.

We assume that the equity of the firm is composed of shares which are traded in the market and, similarly, that the debt is composed of traded bonds, with a total nominal value equal to $K > 0$. We assume that there are no transaction costs, constraints on trading, taxes or any other market friction.

In case (a), we denote by V_0^{un} the total value of the firm at the initial date 0 and by S_t^{un} the market value of the firm stock at date t, for $t \in \{0, 1\}$. In case (b), we denote by V_0^{lv} the value of the leveraged firm at $t = 0$ and by B_t and S_t^{lv} the value of the firm debt and stock, respectively, at date t, for $t \in \{0, 1\}$. As in the preceding sections of this chapter, we assume that there are S possible states of the world $(\omega_1, \ldots, \omega_S)$.

Let us first consider case (a). In this case, the revenues of the firm at date $t = 1$ are fully distributed among the equity holders, so that the total market value of the firm stock is simply equal to the total firm value, i.e.,

$$S_1^{\text{un}}(\omega_s) = V_1^{\text{un}}(\omega_s), \qquad \text{for all } s = 1, \dots, S.$$

Since the firm stock is traded in the market and the market is assumed to be arbitrage free, Theorem 4.20 implies that there exists a state price vector $m \in \mathbb{R}_{++}^S$ such that

$$V_0^{\text{un}} = S_0^{\text{un}} = \sum_{s=1}^S m_s S_1^{\text{un}}(\omega_s) = \sum_{s=1}^S m_s V_1(\omega_s). \tag{4.58}$$

Let us now consider case (b). In the case of a leveraged firm, with a debt of nominal value K, we assume that equity holders have limited liability. This means that, at date $t = 1$, the amount K has to be repaid to the debt holders if the firm revenues exceed K, while, if the firm revenues do not suffice to repay the liability K, then the debt holders only receive the liquidation value of the firm, due to the limited liability of the equity holders. In other words, it holds that

$$B_1(\omega_s) = \min\{V_1(\omega_s); K\} \qquad \text{and} \qquad S_1(\omega_s) = \max\{V_1(\omega_s) - K; 0\},$$

for all $s = 1, \dots, S$. The value of the firm debt at the initial date $t = 0$ is then given by

$$B_0 = \sum_{s=1}^S m_s B_1(\omega_s) = \sum_{s=1}^S m_s \min\{V_1(\omega_s); K\} = \sum_{s=1}^S m_s \big(K - \max\{K - V_1(\omega_s); 0\}\big) \tag{4.59}$$

and the value of the stock of the leveraged firm at $t = 0$ is given by

$$S_0^{\text{lv}} = \sum_{s=1}^S m_s S_1^{\text{lv}}(\omega_s) = \sum_{s=1}^S m_s \max\{V_1(\omega_s) - K; 0\}. \tag{4.60}$$

In view of (4.60), the market value of the stock of a leveraged firm can be represented as a Call option written on the firm assets, with the exercise price being the nominal value of the debt. In an analogous way, by (4.59), the market value of the debt can be expressed as the difference between its nominal value and the value of a Put option written on the firm assets, with strike price K.

The Modigliani-Miller theorem states that the financial structure of a firm does not affect the total firm value. In view of above relations (4.58), (4.59) and (4.60), we can easily verify this claim. Indeed, it holds that

$$V_0^{\text{lv}} = S_0^{\text{lv}} + B_0 = \sum_{s=1}^{S} m_s \max\{V_1(\omega_s) - K; 0\} + \sum_{s=1}^{S} m_s \min\{V_1(\omega_s); K\}$$

$$= \sum_{s=1}^{S} m_s V_1(\omega_s) = V_0^{\text{un}}.$$

(4.61)

The Modigliani-Miller relation (4.61) does not rely on market completeness, but only on the absence of arbitrage opportunities. Indeed, under the assumption that both the equity and the debt are traded, their market values at the initial date $t = 0$ do not depend on the specific choice of the state price vector $m \in \mathbb{R}_{++}^S$. However, if a firm modifies its financial structure by issuing non-traded securities which are not spanned by the available assets, then the result of Modigliani-Miller does not necessarily hold. As shown in Exercise 4.37, the presence of taxation leads to a modification of the Modigliani-Miller result and it turns out that the value of a firm is increasing with respect to the value of the debt.

As a consequence of relation (4.61), it holds that the cost of capital of a leveraged firm is equal to the cost of capital of an unleveraged firm plus a spread which depends on the debt-equity ratio of the leveraged firm (see Modigliani & Miller [1348, Proposition II]). Indeed, let us define the following quantities:

$$r^{\text{d}} := \mathbb{E}[B_1]/B_0, \qquad r^{\text{lv}} := \mathbb{E}[S_1^{\text{lv}}]/S_0^{\text{lv}} \quad \text{and} \quad r^{\text{un}} := \mathbb{E}[S_1^{\text{un}}]/S_0^{\text{un}} = \mathbb{E}[V_1]/V_0^{\text{un}}.$$

(4.62)

In particular, r^{un} and r^{lv} represent the expected rates of return on the equity for an unleveraged firm and for a leveraged firm, respectively. By relying on the identity

$$S_0^{\text{lv}} = V_0^{\text{lv}} - B_0 = V_0^{\text{un}} - B_0,$$

which is a consequence of (4.61), we can write

$$r^{\text{lv}} = \frac{\mathbb{E}[V_1 - B_1]}{S_0^{\text{lv}}} = \frac{r^{\text{un}} V_0^{\text{un}} - r^{\text{d}} B_0}{S_0^{\text{lv}}}$$

$$= \frac{r^{\text{un}}(S_0^{\text{lv}} + B_0) - r^{\text{d}} B_0}{S_0^{\text{lv}}} = r^{\text{un}} + (r^{\text{un}} - r^{\text{d}}) \frac{B_0}{S_0^{\text{lv}}}.$$

(4.63)

This shows that the cost of capital for a leveraged firm can be decomposed into the sum of the cost of capital for an unleveraged firm and a second component given by the debt-to-equity ratio of the leveraged firm multiplied by the spread $r^{\text{un}} - r^{\text{d}}$. A connection between the Modigliani-Miller result and the CAPM model will be presented in Exercise 5.11 in the following chapter.

4.5   Notes and Further Readings

For more detailed expositions of general equilibrium theory under risk we refer the reader to Mas-Colell et al. [1310], Dothan [581], Magill & Shafer [1288], Dana & Jeanblanc [514], Magill & Quinzii [1286]. Radner [1439, p.932] clearly defines the concept of Radner equilibrium: "agents have common expectations if they associate the same (future) prices to the same events .... I shall say that the plans of the agents are consistent if, for each commodity, each date, and each event at that date, the planned supply of that commodity at that date in that event equals the planned demand, and if a corresponding condition holds for the stock markets. An equilibrium of plans, prices, and price expectations is a set of prices on the current market, a set of common expectations for the future, and a consistent set of individual plans, one for each agent, such that, given the current prices and price-expectations, each individual agent's plan is optimal for him, subject to an appropriate sequence of budget constraints." It is pointed out that "Traders need not agree on the probabilities of future environmental events, and therefore they need not agree on the probability distribution of future prices, but they must agree on which future prices are associated with which events", see Radner [1439, p.940].

On the mutuality principle and risk sharing, the seminal paper is Wilson [1661]. Townsend [1599] performed an empirical analysis of the mutuality principle using micro data from a village in India. He showed that only a small part of income variation is due to aggregate risk and income fluctuations are largely idiosyncratic. Although financial markets are not developed, there is evidence of mutual insurance.

The analysis presented in this chapter concerns economies with a finite number of states of the world. The analysis with infinitely many states requires more refined tools and we refer to the special issue of the *Journal of Mathematical Economics* (1996) on this topic. On the role of options in achieving market completeness in an economy with infinitely many states see Green & Jarrow [822], Brown & Ross [311], Nachman [1365].

It is possible to extend the analysis developed in Sect. 4.2 to assets delivering dividends defined in terms of $L > 1$ goods (real assets), see Magill & Shafer [1288] and Geanakoplos [763]. In this case, we have that $D \in \mathbb{R}^{L \times S \times N}$. Let D_s be the $L \times N$ matrix obtained by specifying D in the state of the world ω_s, for each $s = 1, \ldots, S$. If the economy is *regular* (i.e., for each state of the world ω_s we can extract a row d_s from D_s so that the vectors $(d_1, \ldots, d_S)$ are linearly independent), then the existence of a Radner equilibrium can be proved and the equivalence between Radner equilibrium allocations and Arrow-Debreu equilibrium allocations can be established for an open set of economies of full measure (generic existence result), see Magill & Shafer [1287]. To this end, $N \geq S$. When this equivalence holds true, an equilibrium allocation is Pareto optimal. Also the invariance result with respect to the financial structure is a generic result. If the economy is not regular, then the existence of the Radner equilibrium and its Pareto non-optimality are generic results, see Duffie & Shafer [598], Geanakoplos [763], Magill & Shafer [1288]. A non-existence example with assets paying bundles of goods has been provided in Hart [906]. In proving the existence of the equilibrium, problems arise because the budget correspondences are not always upper semicontinuous. To overcome this problem, constraints on agents' trades in asset markets can be introduced (e.g., no short sales), see Radner [1437].

If *nominal* assets are traded in the market, then the existence of a Radner equilibrium can be established in a complete or incomplete market without short sale constraints under standard regularity conditions, see Cass [375] and Werner [1653]. In Balasko & Cass [114] it is shown that if markets are incomplete and nominal returns are exogenous, then there are multiple equilibria with real indeterminacy of dimension $S - N$ (with N being the number of linearly independent assets). In Geanakoplos & Mas-Colell [764] it is shown that the degree of indeterminacy is $S - 1$ if the current prices of the assets are endogenous to the model, whatever the number of assets. If markets are complete, then the equilibrium allocation is Pareto optimal and the equilibrium is determinate. If markets are incomplete, then generically the equilibrium allocation is not Pareto optimal. An example presented in Hart [906] shows that in incomplete markets with many goods the allocation associated with a Radner equilibrium is not necessarily constrained ex-ante Pareto optimal and equilibria can be Pareto ordered (see Geanakoplos & Polemarchakis [765] and Magill & Shafer [1288] for constrained generic inefficiency results). On these topics see also Magill & Shafer [1288].

For a generic existence result of a Radner equilibrium in a financial market with financial derivatives see Krasa [1130] and Krasa & Werner [1131]. In our analysis, the financial structure is assumed to be exogenous. However, this represents a quite strong assumption in the context of an incomplete market. Market incompleteness means that there is an insurance demand which is not satisfied by the market and that a Pareto improvement can be obtained by introducing enough securities. In this setting, does an incentive emerge to complete the market through financial innovation? The enormous financial innovation observed in the last decades suggests a positive answer. However, this is not always the case. An enlightening example has been provided in Hart [906], showing that the effect of opening new markets is to make everybody worse off. A detailed analysis of this topic lies outside the scope of this book and we refer the reader to Allen & Gale [46] and Duffie & Rahi [597]. The analysis of financial innovation requires in the first place to identify the proposer of the new contract: in the literature it is assumed that the new contract is proposed by a firm issuing new securities or by an agent aiming to gain by trading the new asset. Note that the results provided in the literature are extremely sensitive to the hypotheses of the model. In many cases, financial innovation does not complete the market and does not induce a Pareto improvement. Under some conditions, the equilibrium allocation turns out to be constrained Pareto optimal. Note also that an equilibrium may fail to exist in a general financial market where financial derivatives are traded, see for instance Polemarchakis & Ku [1424]. In Hart [906], the author provides an example of an incomplete market economy with multiple equilibria ranked according to the Pareto order.

On the aggregation property and mutual funds separation in incomplete markets see Milne [1346], Detemple & Gottardi [563], Hens & Pilgrim [937]. Note that, in the special case of exponential utility functions, aggregation in a strong sense can also be established with heterogeneous beliefs and heterogeneous time preferences, as long as all agents agree on the set of events with zero probability, see Huang & Litzenberger [971, Section 5.26].

An exhaustive analysis of the Fundamental Theorem of Asset Pricing is provided in Connor [476], Dothan [581], Dybvig & Ross [611], Duffie [593] and for an overview of the history and the role of this result in the context of mathematical finance we refer to Schachermayer [1501]. In the context of continuous time models, a general formulation of the fundamental theorem of asset pricing has been obtained in Delbaen & Schachermayer [544, 545]. We also mention that, always in a continuous time setting, an asset pricing theory has been developed in Platen & Heath [1420] by relying on a weaker notion of arbitrage (see also Fernholz [683] for a descriptive theory of financial markets based on a weaker notion of arbitrage). See also Fontana [724] for a unifying analysis of several no-arbitrage conditions. The theorem has been extended to an economy with states characterized by null probability in Willard & Dybvig [1660]. An arbitrage opportunity is strictly related to the existence of a couple of consumption plans ordered according to the first order stochastic dominance criterion. In particular, an arbitrage opportunity implies the existence of a couple of consumption plans ordered according to the criterion, but not vice-versa, see Jarrow [1018]. Note that one can also derive a version of the Fundamental Theorem of Asset Pricing by considering the absence of arbitrage opportunities of the second kind. In that case, one obtains the existence of a positive (but not necessarily strictly positive) linear pricing functional or, equivalently, of a solution $m \in \mathbb{R}^S_+$ to the system $p = D^\top m$ (see Theorem 5.2.2 of LeRoy & Werner [1191] and compare also with Exercise 4.17). The capital structure discussed at the end of Sect. 4.4 in the context of the Modigliani-Miller theorem is also at the basis of the seminal model considered in Merton [1331] for the pricing of corporate debt.

4.6   Exercises

Exercise 4.1 Let us consider an economy with two agents a, b, two possible states of the world with strictly positive probabilities $(\pi, 1 - \pi)$ and a single consumption good. Suppose that the endowments of agents a and b satisfy $e_1^a + e_1^b > e_2^a + e_2^b$, so that there is aggregate risk. Show that the prices p_1, p_2 of the contingent goods in correspondence of an interior Pareto optimal allocation satisfy $p_1/p_2 < \pi/(1 - \pi)$.

Exercise 4.2 Consider the setting of Sect. 4.1 in the presence of a single good (i.e., $L = 1$). Show that condition (4.1) characterizing an ex-ante Pareto optimal allocation is equivalent to the Borch condition (4.4).

Exercise 4.3 Let $\{y^i : \mathbb{R}_+ \to \mathbb{R}_+; i = 1, \ldots, I\}$ be a Pareto optimal sharing rule. Show that y^i is linear if and only if

$$t'_{u^i}(y^i(e)) = t'_{u^k}(y^k(e)), \qquad \text{for all } i, k = 1, \ldots, I,$$

where $t'_{u^i}(y^i(e))$ denotes the first derivative of the risk tolerance of agent i computed in correspondence of the Pareto optimal consumption allocation $y^i(e)$, where e denotes an arbitrary realization of the aggregate endowment.

Exercise 4.4 Prove the necessity part of Proposition 4.5.

Exercise 4.5 Consider utility functions of time 1 consumption of the generalized power utility form

$$u^i(x) = \frac{1}{b-1}(\gamma_i + bx)^{\frac{b-1}{b}}, \qquad \text{with } b \notin \{0, 1\}, \text{ for all } i = 1, \ldots, I,$$

where agent i has discount factor δ_i, for $i = 1, \ldots, I$. By relying on condition (4.4), show that the Pareto optimal sharing rule is linear with respect to the aggregate endowment. In an analogous way, prove the same result in the case of exponential utility functions $u^i(x) = -\gamma_i \exp(-x/\gamma_i)$, for all $i = 1, \ldots, I$.

Exercise 4.6 Consider an economy with $L \times S$ future markets open at time $t = 0$, for every contingent good (l, s), for all $l = 1, \ldots, L$ and $s = 1, \ldots, S$, as considered at the beginning of Sect. 4.2. Prove that, in correspondence of an Arrow-Debreu equilibrium allocation, there cannot be incentives to trade at date $t = 1$ after the state of the world has been revealed.

Exercise 4.7 Suppose that markets are complete and let

$$u^i(x_0, x_1, \ldots, x_S) = -\gamma_i \exp(-x_0/\gamma_i) - \gamma_i \delta \sum_{s=1}^{S} \pi_s \exp(-x_s/\gamma_i),$$

for all $i = 1, \ldots, I$. Show that the equilibrium prices $q^* = (q_1^*, \ldots, q_S^*)$ of the S Arrow securities are given by

$$q_s^* = \frac{\delta \pi_s \exp(-e_s/\sum_{i=1}^{I} \gamma_i)}{\exp(-e_0/\sum_{i=1}^{I} \gamma_i)}, \qquad \text{for all } s = 1, \ldots, S.$$

Exercise 4.8 Consider an economy with two traded assets and tree possible states of the world, where the dividend matrix D is given by

$$D = \begin{bmatrix} 1 & 0 \\ 0 & 1 \\ 0 & 1 \end{bmatrix}.$$

Suppose that there are two agents (i.e., $I = 2$), whose utility functions are differentiable, strictly increasing and strictly concave and only depend on consumption at time $t = 1$. The probabilities associated to the three states of the world are given by $(1/4, 1/4, 1/2)$. The endowment of the first agent is given by one unit of the first asset, while that of the second agent is given by one unit of the second asset. Show that the Pareto optimal allocation is given by the consumption plan $(1/4, 1/4, 1/4)$ for the first agent and $(3/4, 3/4, 3/4)$ for the second agent (compare also with LeRoy & Werner [1191], Example 16.5.1).

Exercise 4.9 Consider an economy with two possible states of the world and two agents a and b, with homogeneous beliefs $(\pi, 1 - \pi)$ about the realization of the state of the world and utility functions $u(x) = \sqrt{x}$ defined on consumption at time $t = 1$. Let p_1 and p_2 denote the prices of the two Arrow securities paying one unit of the consumption good in correspondence of each state of the world and let (e_1^i, e_2^i) denote the endowment of agent i, for $i \in \{a, b\}$, in terms of the two contingent consumption goods.

 (i) Determine the equilibrium allocation when the agents' endowments are given by $(e_1^a, e_2^a) = (40, 60)$ and $(e_1^b, e_2^b) = (60, 40)$.
 (ii) Determine the equilibrium allocation when the agents' endowments are given by $(e_1^a, e_2^a) = (40, 60)$ and $(e_1^b, e_2^b) = (50, 50)$.
(iii) How does the equilibrium allocation change if the agents have heterogeneous beliefs such that $\pi^a > \pi^b$?
(iv) Determine the equilibrium allocation in the case of a logarithmic utility function $u(x) = \log(x)$.

Exercise 4.10 Consider an economy with two possible states of the world and two agents a and b. Assume that both agents are characterized by a logarithmic utility function $u(x) = \log(x)$ defined on consumption at time $t = 1$. Let p_1 and p_2 denote the equilibrium prices of the two Arrow securities and suppose that the initial endowments of the agents are given by $(e_1^i, e_2^i) = \alpha^i(e_1, e_2)$, for $i \in \{a, b\}$, for some couple (α^a, α^b) satisfying $\alpha^a + \alpha^b = 1$ and where (e_1, e_2) denotes the economy's aggregate endowment. Show that, if the two agents have homogeneous beliefs $(\pi, 1 - \pi)$ about the possible realization of the state of the world, then they will not trade in equilibrium.

Exercise 4.11 Consider an economy with two possible states of the world, with associated probabilities of occurrence $(1/2, 1/2)$, and a representative agent with expected logarithmic utility function of the form

$$u(x_0, x_1, x_2) = \log(x_0) + \frac{1}{2}\log(x_1) + \frac{1}{2}\log(x_2).$$

In the economy it is possible to trade the two Arrow securities and a risk free asset with constant unitary payoff. Determine the equilibrium prices (q_1^*, q_2^*) of the two Arrow securities and the return r_f of the risk free asset in the representative agent economy when the aggregate endowment is given by $(e_0, e_1, e_2) = (1, 3, 1)$.

Exercise 4.12 Consider the same representative agent economy described in Exercise 4.11, but with the expected utility function

$$u(x_0, x_1, x_2) = x_0^\gamma + \frac{1}{2}x_1^\gamma + \frac{1}{2}x_2^\gamma,$$

with $0 < \gamma < 1$. Determine the equilibrium prices (q_1^*, q_2^*) of the two Arrow securities as well as the return r_f of the risk free asset with constant payoff 1.

Exercise 4.13 Consider the optimal consumption Problem (PO3) for an agent characterized by a time additive state independent utility function of the generalized power form $u(x) = \frac{1}{b-1}(\gamma + bx)^{\frac{b-1}{b}}$, with $b \notin \{0, 1\}$, and discount factor δ. Letting $\bar{e} := e_0 + \sum_{s=1}^{S} p_s e_s$ (i.e., the present value of the endowment, where $p_1, \ldots, p_S$ denote the prices of the S Arrow securities) and x_0^* the optimal consumption at $t = 0$, the quantity $\bar{e} - x_0^*$ represents *intertemporal saving*. Show that intertemporal saving is an affine function of $\bar{e}$ (compare also with Lengwiler [1182, Section 5.4.1]).

Exercise 4.14 Consider an economy with two possible states of the world, with associated probabilities of occurrence $(1/2, 1/2)$, and a representative agent with logarithmic utility function. In the economy there are two traded assets: the first asset, with price $p_1 = 1$, delivers a risk free payoff of 1 in correspondence of both states of the world, while the second asset, with price p_2, delivers the random payoff $(1/2, 2)$. Determine the equilibrium price p_2 of the second asset when the economy's aggregate endowment is given by (e_1, e_2).

Exercise 4.15

(i) Consider an economy with two possible states of the world and two securities (i.e., $S = N = 2$), with

$$D = \begin{bmatrix} 3 & 2 \\ -3 & -2 \end{bmatrix} \quad \text{and} \quad p = \begin{pmatrix} 1 \\ 1 \end{pmatrix}.$$

Show that there exists an arbitrage opportunity of the second kind but there are no arbitrage opportunities of the first kind.

(ii) Consider an economy with two possible states of the world and two securities (i.e., $S = N = 2$), with

$$D = \begin{bmatrix} 3 & 2 \\ 3 & 3 \end{bmatrix} \quad \text{and} \quad p = \begin{pmatrix} 1 \\ 1 \end{pmatrix}.$$

Show that there exists an arbitrage opportunity of the first kind but there are no arbitrage opportunities of the second kind.

(iii) Show that, if there exists a portfolio $\bar{z}$ such that $D\bar{z} > 0$ with $p^\top \bar{z} > 0$, then the existence of an arbitrage opportunity of the second kind implies the existence of an arbitrage opportunity of the first kind.

Exercise 4.16 Consider the same economy described in part *(ii)* of Exercise 4.15. Show that the Law of One Price holds (but, as shown in Exercise 4.15, there are arbitrage opportunities of the first kind).

Exercise 4.17 For a given price-dividend couple (p, D), prove that, if there exists a positive (but not necessarily strictly positive) linear pricing functional Q, then there are no arbitrage opportunities of the second kind.

Exercise 4.18 Show that, if the Law of One Price fails to hold, then every payoff $c \in I(D)$ can be replicated with an arbitrarily small initial wealth.

Exercise 4.19 Let $z, z' \in \mathbb{R}^N$ be two portfolios with riskless payoffs, i.e.,

$$\sum_{n=1}^{N} d_{sn} z_n = c \quad \text{and} \quad \sum_{n=1}^{N} d_{sn} z'_n = c',$$

for some $c, c' > 0$, for all $s =, 1 \dots, S$. Show that, if there are no arbitrage opportunities, then $\frac{c}{p^\top z} = \frac{c'}{p^\top z'}$ (i.e., the two portfolios z and z' have the same rate of return).

Exercise 4.20 Let $c = (c_1, \dots, c_S) \in \mathbb{R}^S$ represent a random payoff and suppose that there exists a portfolio $z^c \in \mathbb{R}^N$ such that $Dz^c = c$. Suppose that it is possible to trade the payoff c for a price $p_{N+1} \neq p^\top z^c$. Show that there exists an arbitrage opportunity in the extended market represented by

$$p' = \begin{pmatrix} p_1 \\ \vdots \\ p_N \\ p_{N+1} \end{pmatrix} \qquad D' = \begin{bmatrix} d_{11} & \dots & d_{1N} & c_1 \\ \vdots & \ddots & \vdots & \vdots \\ d_{S1} & \dots & d_{SN} & c_S \end{bmatrix}.$$

Exercise 4.21 Prove Proposition 4.24.

Exercise 4.22 Suppose that there are no arbitrage opportunities. Prove that Definition 4.9 is well-posed, in the sense that, for any $c \in I(D)$, the market value $V(c)$ is uniquely defined.

Exercise 4.23 Let $c = (c_1, \dots, c_S) \in \mathbb{R}^S$ represent a random payoff which is not attainable in the market, i.e., $c \notin I(D)$. Suppose that it is possible to trade the payoff c for a price $p_{N+1} \notin (q_l(c), q_u(c))$. Show that there exists an arbitrage opportunity in the extended market represented by

$$p' = \begin{pmatrix} p_1 \\ \vdots \\ p_N \\ p_{N+1} \end{pmatrix} \qquad D' = \begin{bmatrix} d_{11} & \dots & d_{1N} & c_1 \\ \vdots & \ddots & \vdots & \vdots \\ d_{S1} & \dots & d_{SN} & c_S \end{bmatrix}.$$

Exercise 4.24 Consider an economy with two possible states of the world and two securities (i.e., $S = N = 2$), as in Exercise 4.15, with

$$D = \begin{bmatrix} 3 & 2 \\ 3 & 3 \end{bmatrix} \quad \text{and} \quad p = \begin{pmatrix} 1 \\ 1 \end{pmatrix}.$$

Show that there are arbitrage opportunities but, nevertheless, the market is complete.

Exercise 4.25 Consider an economy with $S = N = 3$, with the following dividend matrix:

$$D = \begin{bmatrix} 1 & 4 & 3 \\ 6 & 2 & 4 \\ 2 & 3 & 5 \end{bmatrix}.$$

  (i) Show that the market is complete.
 (ii) Given the price vector $p^\top = (2.15, 2.7, 3.35)$, determine the state price vector.
(iii) Determine the risk free rate of return r_f implicit in the couple (p, D) and the portfolio which attains the riskless payoff $(1, 1, 1)$. Verify that the return of such a portfolio coincides with r_f.
 (iv) Determine the portfolio z^c which replicates the payoff $c = (2, 3, 6)$. Verify that $p^\top z^c = Q(c) = m^\top c$.
  (v) Determine the portfolio z^{1_3} which replicates the Arrow security 1_3, with payoff $(0, 0, 1)$, and verify that $p^\top z^{1_3} = m_3$.

Exercise 4.26 Consider an economy with a single traded asset and three possible states of the world (i.e., $N = 1$ and $S = 3$), with dividend $D = (0.5, 1, 2)^\top$ and price $p = 1$. For the payoff $c = (1, 2, 3)^\top$, determine the super replication and sub-replication prices $q_u(c)$ and $q_l(c)$.

Exercise 4.27 Consider an economy with two traded assets and three possible states of the world (i.e., $N = 2$ and $S = 3$), with price-dividend couple

$$D = \begin{bmatrix} 1 & 5 \\ 4 & 2 \\ 3 & 1 \end{bmatrix} \quad \text{and} \quad p = \begin{pmatrix} 2.45 \\ 2.35 \end{pmatrix}.$$

  (i) Characterize the set of possible state prices.
 (ii) Determine the space of attainable payoffs.
(iii) Verify that the arbitrage free price of any attainable payoff does not depend on the specific state price vector chosen.
 (iv) Determine the interval of arbitrage free prices for the payoff $c = (2, 1, 3)^\top$.
  (v) Consider a *Call option* with strike price 2 written on the second security. Does the introduction of such a derivative make the market complete?

Exercise 4.28 Consider an economy with three possible states of the world and three traded securities (i.e., $S = N = 3$), with price-dividend couple

$$
D = \begin{bmatrix} 2 & 3 & 2 \\ 2 & 3 & 5 \\ 5 & 4/3 & 3 \end{bmatrix} \quad \text{and} \quad \begin{pmatrix} 6 \\ 4 \\ k \end{pmatrix}.
$$

   (i) Determine the values of k for which there are no arbitrage opportunities in the market.
  (ii) Determine the range of arbitrage free prices for a *Call option* written on the first security with strike price 3.
 (iii) For a fixed value of k such that arbitrage opportunities are present in the market, construct a portfolio which is an arbitrage opportunity.

Exercise 4.29 Consider an economy with three possible states of the world and three traded securities (i.e., $S = N = 3$), with dividend matrix given by

$$
D = \begin{bmatrix} 3 & 4 & 1 \\ 2 & 2 & 4 \\ 4 & 5 & 2 \end{bmatrix}.
$$

   (i) Verify that the market is complete.
  (ii) Given the price vector $p = (2.85, 3.45, 2.35)^\top$, determine the state price vector.
 (iii) Determine the risk free rate r_f implicit in the price-dividend couple (p, D).
 (iv) Determine the portfolio z^c which replicates the payoff $c = (1, 4, 3)^\top$ and verify that the value at time $t = 0$ of such a portfolio coincides with the arbitrage free price of the payoff c computed in terms of the state price vector m.
  (v) Determine the arbitrage free price of a *Call option* written on the second security with strike price 2. Verify that the value at time $t = 0$ of the replicating portfolio z^{call} coincides with the arbitrage free price of the Call option.

Exercise 4.30 In the setting of Proposition 4.27, show that, if, for any $i = 1, \ldots, I$, the utility function $u^i : \mathbb{R}^2 \to \mathbb{R}$ is strictly increasing in the first argument and non-decreasing in the second argument, then the existence of an optimal portfolio excludes the existence of arbitrage opportunities of the second kind.

Exercise 4.31 In the setting of Proposition 4.27, show that, if, for any $i = 1, \ldots, I$, the utility function $u^i : \mathbb{R}^2 \to \mathbb{R}$ is non-decreasing in the first argument and strictly increasing in the second argument and there exists a portfolio $\hat{z} \in \mathbb{R}^N$ such that $D\hat{z} > 0$, then the existence of an optimal portfolio excludes the existence of arbitrage opportunities.

Exercise 4.32 In the setting of Proposition 4.27, show that:

(i) if the utility function u^i is strictly increasing in its first argument and non-decreasing in the second, then the existence of an optimal portfolio implies the validity of the Law of One Price;

(ii) if the utility function u^i is non-decreasing in its first argument and strictly increasing in the second and there exists a portfolio $\hat{z}$ with $D\hat{z} > 0$, then the existence of an optimal portfolio implies the validity of the Law of One Price.

Exercise 4.33 Consider an economy with two possible states of the world (with equal probabilities of occurrence) and a single traded asset with payoff $(1, 1)$ and price $p \in (0, 1)$. Consider Problem (PO4) with the linear expected utility function $u(x_0, x_1, x_2) = x_0 + x_1/2 + x_2/2$. Show that there are no arbitrage opportunities but, nevertheless, there does not exist an optimal portfolio.

Exercise 4.34 Consider an economy with S possible states of the world, such that $e_1 \leq \ldots \leq e_S$, and such that the price-dividend couple (p, D) is arbitrage free and complete. In correspondence of an equilibrium allocation of an economy populated by risk averse agents, the price q_1^* of the first Arrow security is greater or smaller than $1/r_f$?

Exercise 4.35 Show that in a complete market with no aggregate risk, homogeneous beliefs and a risk free rate equal to one, the price at time $t = 0$ of any security is given by the expectation (with respect to the original probability measure π) of the dividend at time $t = 1$.

Exercise 4.36 (See Huang & Litzenberger [971], Section 6.10) Consider an economy with N traded securities (among which a risk free asset), I agents with utility functions of the generalized power form, as in Proposition 4.15, with $\sum_{i=1}^{I} \gamma_i - 0$ and identical cautiousness coefficient b, and suppose that all the agents are only endowed with units of traded securities. Consider a Call option written on the n-th asset, with strike price k, and denote by $p^{\text{call},n}$ its price at time $t = 0$. Suppose that

$$\left(\log \left(\delta \left(\frac{\tilde{e}}{e_0} \right)^{-\frac{1}{b}} \right), \log \left(\frac{\tilde{d}_n}{p_n} \right) \right) \sim \mathcal{N} \left(\begin{pmatrix} \mu_e \\ \mu_n \end{pmatrix} ; \begin{pmatrix} \sigma_e^2 & \rho\sigma_e\sigma_n \\ \rho\sigma_e\sigma_n & \sigma_n^2 \end{pmatrix} \right),$$

where the random variable $\tilde{e}$ denotes the random aggregate endowment at $t = 1$ and $\tilde{d}_n$ denotes the random dividend of the n-th security, for some correlation coefficient ρ.

(i) Prove that the equilibrium of the economy can be characterized in terms of the no-trade equilibrium of a single representative agent with a utility function of the generalized power form.

(ii) Show that

$$p^{\mathrm{call},n} = \delta \, \mathbb{E}\left[\max\{\tilde{d}_n - k; 0\} \left(\frac{\tilde{e}}{e_0}\right)^{-\frac{1}{b}} \right]. \tag{4.64}$$

(iii) Show that formula (4.64) admits the explicit representation

$$p^{\mathrm{call},n} = p_n \, N(\Delta_1) - \frac{k}{r_f} N(\Delta_2), \tag{4.65}$$

where

$$\Delta_1 := \frac{\log\left(\frac{p_n}{k}\right) + \log(r_f)}{\sigma_n} + \frac{\sigma_n}{2} \quad \text{and} \quad \Delta_2 := \Delta_1 - \sigma_n$$

and $N(x) := \frac{1}{\sqrt{2\pi}} \int_{-\infty}^{x} e^{-z^2/2} dz$.

Exercise 4.37 As at the end of Sect. 4.4, consider an arbitrage free financial market with a firm operating in two dates $t \in \{0, 1\}$. Denote by V_1 the revenues of the firm at date $t = 1$ and suppose that the net firm profits are taxed, in the sense that at date $t = 1$ the following tax has to be payed:

$$T_1(\omega_s) := \tau \, (V_1(\omega_s) - K)^+,$$

for any $s = 1, \ldots, S$, where $\tau > 0$ represents a tax rate and $K \geq 0$ is the nominal value of the firm debt. In particular, if $K = 0$ (*unleveraged firm*), then the tax to be payed at $t = 1$ is simply equal to $\tau V_1(\omega_s)$, for $s = 1, \ldots, S$. We denote by $V_0^{\mathrm{un,tax}}$ the value at date $t = 0$ of an unleveraged firm subject to taxation. On the other hand, if $K > 0$ (*leveraged firm*), then we denote by $V_0^{\mathrm{lv,tax}}$ the value at $t = 0$ of a leveraged firm and by B_t the market value of the debt at date t, for $t \in \{0, 1\}$. Assume that both the firm equity and debt are traded in the financial market. Prove that the value at date $t = 0$ of a leveraged firm subject to taxation is given by

$$V_0^{\mathrm{lv,tax}} = V_0^{\mathrm{un,tax}} + \tau B_0, \tag{4.66}$$

where $V_0^{\mathrm{un,tax}}$ denotes the market value at $t = 0$ of an unleveraged firm subject to taxation. In particular, the value of a leveraged firm subject to taxation is increasing with respect to the market value of the debt.

Chapter 5
Factor Asset Pricing Models: CAPM and APT

> *If stocks are priced rationally, systematic differences in average*
> *returns are due to differences in risk.*
>
> Fama & French (1995)

In Chap. 4, we have discussed the role of financial markets in allocating risk among agents and we have analysed asset prices and returns by means of two different but related approaches (see Sect. 4.4): equilibrium analysis and no-arbitrage valuation. In the present chapter, we will exploit these two approaches to extract information on asset *risk premia*. In particular, we will present two models: the *Capital Asset Pricing Model (CAPM)* and the *Arbitrage Pricing Theory (APT)*. The CAPM is based on equilibrium analysis, while the APT, as the name suggests, on the absence of (asymptotic) arbitrage opportunities. The key feature of both the CAPM and the APT consists in a linear relation between the asset risk premia and the risk premia associated to one or several risk factors. The equilibrium analysis developed in the previous chapter provides us with information about asset prices, returns and risk premia starting from the knowledge of all the ingredients of the economy (in particular, the preferences and the initial endowments of the agents) or, when the aggregation property holds, referring to a representative agent economy. In contrast, under suitable assumptions, the CAPM and the APT provide linear relations for the asset risk premia without requiring a detailed knowledge of all the ingredients of the economy.

This chapter is structured as follows. In Sect. 5.1, we apply the equilibrium analysis developed in the previous chapter to derive the *Consumption Capital Asset Pricing Model (CCAPM)*, which consists of a linear relation between asset risk premia and the risk premium of a market portfolio. In Sect. 5.2, assuming that all the agents' hold portfolios belonging to the portfolio frontier (see Chap. 3), we derive the *Capital Asset Pricing Model (CAPM)*, while in Sect. 5.4 we present the *Arbitrage Pricing Theory (APT)*, with asset returns being generated by a linear multi-factor model. The extensive literature on the empirical analysis of the CAPM and of the APT will be surveyed in Sects. 5.3 and 5.5, respectively. At the end of the chapter, we provide a guide to further readings as well as a series of exercises.

© Springer-Verlag London Ltd. 2017
E. Barucci, C. Fontana, *Financial Markets Theory*,
Springer Finance, DOI 10.1007/978-1-4471-7322-9_5

5.1 The Consumption Capital Asset Pricing Model (CCAPM)

In this section, we derive several fundamental asset pricing relations, first by relying on equilibrium arguments and then on no-arbitrage principles.

Equilibrium and Asset Risk Premia

In an economy with complete markets (or effectively complete markets, see Sect. 4.3), equilibrium allocations are Pareto optimal and, therefore, an equilibrium allocation of the economy can be characterized in terms of the optimum for a representative agent in a no-trade equilibrium. By relying on a representative agent analysis, we now derive general valuation principles which do not rely on specific assumptions on the agents' utility functions or on the distributions of the asset returns.

We assume that the economy comprises I agents, with time additive, state-independent, strictly increasing and risk averse preferences. We assume that the initial endowments of the agents are given in terms of wealth at date $t = 0$ and of units of the available securities. We also assume that markets are effectively complete, so that every equilibrium allocation is Pareto optimal (see Sect. 4.3). In correspondence of the equilibrium, the representative agent consumes the aggregate endowment of the economy. We also assume that a risk free asset (denoted by $n = 0$) is traded in the economy together with N risky assets. For each $i = 1, \ldots, I$, the initial endowment of agent i is represented by the couple (x_0^i, e^i), with $x_0^i > 0$ denoting the initial wealth and $e^i = (e_0^i, e_1^i, \ldots, e_N^i) \in \mathbb{R}^{N+1}$ the endowment in terms of units of the $N + 1$ assets. The aggregate endowment of the economy is then given by $x_0^m := \sum_{i=1}^{I} x_0^i$ at date $t = 0$ and by $x_s^m := \sum_{i=1}^{I} \sum_{n=0}^{N} e_n^i d_{sn}$ at date $t = 1$ in correspondence of the state of the world ω_s, for $s = 1, \ldots, S$, where $d_{s0} = r_f$, for all $s = 1, \ldots, S$, $p_0 = 1$ and where the superscript ^{m} stands for market (i.e., the economy as a whole). We denote by $\tilde{x}^m$ the random variable taking values $(x_1^m, \ldots, x_S^m)$. To simplify the notation, in this chapter we shall refer to equilibrium allocation/prices without denoting them with the superscript *.

In the present setting, letting $\mathbf{u}_0$ and $\mathbf{u}_1$ be the representative agent's utility functions for consumption at time $t = 0$ and consumption at time $t = 1$, respectively, and letting δ be the representative agent's discount factor, in equilibrium it holds that (see equation (4.52))

$$p_n = \delta \, \mathbb{E}\left[\frac{\mathbf{u}_1'(\tilde{x}^m)}{\mathbf{u}_0'(x_0^m)} \tilde{d}_n \right], \qquad \text{for all } n = 0, 1, \ldots, N.$$

Equivalently, assuming that all assets have strictly positive prices (i.e., $p_n > 0$ for all $n = 0, 1, \ldots, N$) and dividing by p_n the above relation, we get

$$\delta \, \mathbb{E}\left[\frac{\mathbf{u}_1'(\tilde{x}^m)}{\mathbf{u}_0'(x_0^m)} \tilde{r}_n \right] = 1, \qquad \text{for all } n = 0, 1, \ldots, N, \tag{5.1}$$

where $\tilde{r}_n$ denotes the random return of the n-th asset. In particular, applying the above formula to the risk free asset, it holds that

$$\delta \, \mathbb{E}\left[\frac{\mathbf{u}_1'(\tilde{x}^m)}{\mathbf{u}_0'(x_0^m)} \right] = \frac{1}{r_f}. \tag{5.2}$$

By combining the above relations, we can write

$$p_n = \frac{\mathbb{E}[\tilde{d}_n]}{r_f} + \delta \, \mathrm{Cov}\left(\frac{\mathbf{u}_1'(\tilde{x}^m)}{\mathbf{u}_0'(x_0^m)}, \tilde{d}_n \right), \qquad \text{for all } n = 1, \ldots, N,$$

$$\delta \, \mathrm{Cov}\left(\frac{\mathbf{u}_1'(\tilde{x}^m)}{\mathbf{u}_0'(x_0^m)}, \tilde{r}_n \right) + \frac{\mathbb{E}[\tilde{r}_n]}{r_f} = 1, \qquad \text{for all } n = 1, \ldots, N.$$

In particular, from the last equation we can immediately derive an expression for the asset risk premia:

$$\mathbb{E}[\tilde{r}_n] - r_f = -r_f \, \mathrm{Cov}\left(\frac{\delta \mathbf{u}_1'(\tilde{x}^m)}{\mathbf{u}_0'(x_0^m)}, \tilde{r}_n \right), \qquad \text{for all } n = 1, \ldots, N. \tag{5.3}$$

Formula (5.3) provides a fundamental relation between the risk premium of an asset and the representative agent's marginal rate of substitution of consumption. Indeed, as we have already seen in the previous chapter, the risk premium of the n-th asset is positive if and only if its random return $\tilde{r}_n$ is negatively correlated with the representative agent's marginal rate of substitution of consumption, for any $n = 1, \ldots, N$. This property is coherent with the diversification and the insurance principles. Indeed, if the return of an asset is negatively correlated with the marginal utility of aggregate consumption, then such an asset will pay more in correspondence of the states of the world where the aggregate consumption is higher and less when the aggregate consumption is lower. For this reason, such an asset will be held by a risk averse agent only in exchange for a positive risk premium. On the contrary, a risk averse agent can accept a negative risk premium for an asset which provides insurance against states of the world where the aggregate consumption is low, i.e., for an asset whose return is positively correlated with the marginal utility of aggregate consumption.

Relation (5.3) also shows that only *systematic risk* is priced in equilibrium. Indeed, if the return of an asset is purely idiosyncratic (i.e., independent from the aggregate consumption), then its risk premium will be null and its expectation will be equal to the risk free rate. This is true since purely idiosyncratic risk can be diversified away and, due to the mutuality principle (see Sect. 4.1), in correspondence of an equilibrium allocation it will not affect any agent. Therefore, purely idiosyncratic risk will not be priced in equilibrium.

Note that, letting $\tilde{m} := \delta \mathbf{u}_1'(\tilde{x}^m)/\mathbf{u}_0'(x_0^m)$ and recalling that $\mathbb{E}[\tilde{m}] = 1/r_f$ (see equation (5.2)), relation (5.3) can be equivalently rewritten as

$$\mathbb{E}[\tilde{r}_n] - r_f = \frac{\text{Cov}(\tilde{m}, \tilde{r}_n)}{\text{Var}(\tilde{m})} \left(-\frac{\text{Var}(\tilde{m})}{\mathbb{E}[\tilde{m}]} \right) = \beta_{nm} \lambda^m, \qquad \text{for all } n = 1, \dots, N,$$

(5.4)

where $\beta_{nm} := \text{Cov}(\tilde{m}, \tilde{r}_n)/\text{Var}(\tilde{m})$ corresponds to the regression coefficient of $\tilde{r}_n$ on $\tilde{m}$ and $\lambda^m := -\text{Var}(\tilde{m})/\mathbb{E}[\tilde{m}]$ can be interpreted as a risk premium on systematic risk, as measured by the representative agent's marginal rate of substitution. Formula (5.4) shows that the risk premium of an asset is proportional to the coefficient appearing in the linear regression of the asset return on the stochastic discount factor $\tilde{m}$.

The above relations between the representative agent's marginal rate of substitution in correspondence of the aggregate consumption and the asset risk premia allow us to derive an upper bound on the *Sharpe ratio* of any traded asset, known as the *Hansen-Jagannathan bound* (see Hansen & Jagannathan [891]). Recall that the Sharpe ratio (with respect to the risk free rate r_f) is defined as the ratio of the return risk premium $\mathbb{E}[\tilde{r}_n] - r_f$ over the standard deviation $\sigma(\tilde{r}_n)$.

Proposition 5.1 *Under the assumptions of the present section, letting* $\tilde{m} := \frac{\delta \mathbf{u}_1'(\tilde{x}^m)}{\mathbf{u}_0'(x_0^m)}$, *the following holds:*

$$\frac{|\mathbb{E}[\tilde{r}_n] - r_f|}{\sigma(\tilde{r}_n)} \leq \frac{\sigma(\tilde{m})}{\mathbb{E}[\tilde{m}]}, \qquad \text{for all } n = 1, \dots, N.$$

(5.5)

Proof From equation (5.3), we can write:

$$\mathbb{E}[\tilde{r}_n] - r_f = -r_f \text{Cov}(\tilde{m}, \tilde{r}_n) = -r_f \rho_{\tilde{m}, \tilde{r}_n} \sigma(\tilde{m}) \sigma(\tilde{r}_n) = -\frac{1}{\mathbb{E}[\tilde{m}]} \rho_{\tilde{m}, \tilde{r}_n} \sigma(\tilde{m}) \sigma(\tilde{r}_n),$$

where $\rho_{\tilde{m}, \tilde{r}_n}$ denotes the correlation coefficient between $\tilde{m}$ and $\tilde{r}_n$ and where the third equality follows from (5.2). Hence, since $|\rho_{\tilde{m}, \tilde{r}_n}| \leq 1$, it holds that:

$$|\mathbb{E}[\tilde{r}_n] - r_f| \leq \frac{1}{\mathbb{E}[\tilde{m}]} \sigma(\tilde{m}) \sigma(\tilde{r}_n),$$

from which (5.5) follows directly, for every $n = 1, \dots, N$. $\qquad\qquad\qquad\qquad\qquad$ □

According to the above result, in equilibrium the Sharpe ratios are bounded from above and from below by the volatility-expectation ratio of the stochastic discount factor. Furthermore, the bound (5.5) shows that all assets lie inside a mean-variance frontier, which is determined by the volatility-expectation ratio of the stochastic discount factor. At the end of the present section we will see that an analogous bound can be derived from no-arbitrage arguments, by replacing the stochastic discount factor with the likelihood ratio of a risk neutral probability measure.

We have so far presented several relations arising from equilibrium arguments on the risk premia of the single traded assets. However, since returns on portfolios composed of the traded assets correspond to linear combinations of the returns on the traded assets, relation (5.3) also applies to returns on arbitrary portfolios of the $N + 1$ available securities. In particular, it can be applied to the *market portfolio*. Denoting by $\tilde{r}^m = (\sum_{i=1}^{I} \sum_{n=0}^{N} e_n^i \tilde{d}_n)/(\sum_{i=1}^{I} \sum_{n=0}^{N} e_n^i p_n)$, the return of the market portfolio, it holds that

$$\mathbb{E}[\tilde{r}^m] - r_f = -r_f \operatorname{Cov}\left(\frac{\delta \mathbf{u}_1'(\tilde{x}^m)}{\mathbf{u}_0'(x_0^m)}, \tilde{r}^m\right). \tag{5.6}$$

This relation implies that, under the assumptions of the present section, the risk premium associated to the market portfolio is positive (i.e., $\mathbb{E}[\tilde{r}^m] > r_f$), as shown in Exercise 5.1. This follows since $r_f > 0$ (see formula (5.2)) and since the marginal utility is a strictly decreasing function.

From relations (5.3) and (5.6) we immediately obtain the following formula, which synthesizes the content of the *Consumption Capital Asset Pricing Model (CCAPM)*:

$$\mathbb{E}[\tilde{r}_n] - r_f = \frac{\operatorname{Cov}\left(\mathbf{u}_1'(\tilde{x}^m), \tilde{r}_n\right)}{\operatorname{Cov}\left(\mathbf{u}_1'(\tilde{x}^m), \tilde{r}^m\right)} \left(\mathbb{E}[\tilde{r}^m] - r_f\right), \qquad \text{for all } n = 1, \ldots, N. \tag{5.7}$$

Relation (5.7) shows that the risk premium of any asset (or portfolio) is proportional to the risk premium of the market portfolio, where the proportionality factor depends on the covariance with the representative agent's marginal utility of aggregate consumption at time $t = 1$.

In the CCAPM equation (5.7), the relation between the risk premium of an asset (or portfolio) and the risk premium associated with the market portfolio is made complex by the appearance of the representative agent's marginal utility of aggregate consumption. A more explicit relation can be derived by considering the portfolio which is minimally correlated with the representative agent's marginal rate of substitution, as shown in the following proposition, the proof of which is given in Exercise 5.2.

Proposition 5.2 *Under the assumptions of the present section, let* $\hat{w} \in \mathbb{R}^N$ *be a portfolio satisfying*

$$\text{Corr}\left(\mathbf{u}_1'(\tilde{x}^m), \tilde{r}^{\hat{w}}\right) = \min_{w \in \mathbb{R}^N} \text{Corr}\left(\mathbf{u}_1'(\tilde{x}^m), \tilde{r}^w\right),$$

where r^w *denotes the return of portfolio w. Then it holds that*

$$\mathbb{E}[\tilde{r}_n] - r_f = \frac{\text{Cov}(\tilde{r}_n, \tilde{r}^{\hat{w}})}{\text{Var}(\tilde{r}^{\hat{w}})}\left(\mathbb{E}[\tilde{r}^{\hat{w}}] - r_f\right), \qquad \text{for all } n = 1, \ldots, N. \tag{5.8}$$

Relation (5.8) shows that the risk premium of every security is proportional to the risk premium of the portfolio having minimal correlation with the stochastic discount factor, with the proportionality factor being given by the coefficient appearing in the linear regression of the random return of the security with respect to the random return of such a portfolio.

The CCAPM relation can be made more explicit by introducing additional assumptions on the preference relations of the agents and/or on the distribution of the asset returns. For instance, if the returns are jointly distributed according to a multivariate normal law, then we can apply Stein's lemma (see Lemma 3.9) to relation (5.3), yielding

$$\mathbb{E}[\tilde{r}_n] - r_f = -r_f\, \mathbb{E}\left[\frac{\delta \mathbf{u}_1''(\tilde{x}^m)}{\mathbf{u}_0'(x_0^m)}\right] \text{Cov}\left(\tilde{x}^m, \tilde{r}_n\right) = -\frac{\mathbb{E}[\mathbf{u}_1''(\tilde{x}^m)]}{\mathbb{E}[\mathbf{u}_1'(\tilde{x}^m)]}\, \text{Cov}(\tilde{x}^m, \tilde{r}_n),$$

for all $n = 1, \ldots, N$, where in the second equality we have made use of formula (5.2). Since this relation holds for all N risky assets, it also holds for arbitrary portfolios, in particular for the market portfolio itself. Hence:

$$\mathbb{E}[\tilde{r}^m] - r_f = -\frac{\mathbb{E}[\mathbf{u}_1''(\tilde{x}^m)]}{\mathbb{E}[\mathbf{u}_1'(\tilde{x}^m)]}\, \text{Cov}(\tilde{x}^m, \tilde{r}^m).$$

Putting together the last two equations, we obtain the relation

$$\mathbb{E}[\tilde{r}_n] - r_f = \frac{\text{Cov}(\tilde{x}^m, \tilde{r}_n)}{\text{Cov}(\tilde{x}^m, \tilde{r}^m)}\left(\mathbb{E}[\tilde{r}^m] - r_f\right) = \frac{\text{Cov}(\tilde{r}^m, \tilde{r}_n)}{\text{Var}(\tilde{r}^m)}\left(\mathbb{E}[\tilde{r}^m] - r_f\right), \tag{5.9}$$

for all $n = 1, \ldots, N$, where the last equality follows from the simple observation that $\tilde{r}^m = \tilde{x}^m / (\sum_{i=1}^I \sum_{n=0}^N e_n^i p_n)$. Formula (5.9) represents the *Capital Asset Pricing Model (CAPM)*, on which we shall focus in Sect. 5.2. Note that this derivation of the CAPM formula (5.9) is based on equilibrium arguments, under the assumption that agents have increasing and strictly concave utility functions and asset returns are jointly distributed according to a multivariate normal distribution. Observe that $-\mathbb{E}[\mathbf{u}_1''(\tilde{x}^m)]/\mathbb{E}[\mathbf{u}_1'(\tilde{x}^m)]$ corresponds to the harmonic mean of the *global absolute risk aversion coefficients* of the agents in the economy (see Constantinides [492]). To this regard, see Exercise 5.3.

The Case of a Power Utility Function

We now specialize the CCAPM relation (5.7) to the case where agents' preferences are represented by power utility functions and asset returns have a general distribution, not necessarily multivariate normal. More specifically, we assume that

$$u^i(x) = \frac{1}{b-1}(\gamma_i + bx)^{1-1/b}, \qquad \text{for all } i = 1, \ldots, I,$$

with $b \notin \{0, 1\}$. Recall that, as shown in Proposition 4.5, in the case of power utility functions with common cautiousness coefficient, Pareto optimal allocations are characterized by a linear sharing rule. Moreover, since the agents' endowments are defined in terms of the available securities and there exists a risk free asset, markets are effectively complete. Hence, Pareto optimality is always attained in correspondence of an equilibrium allocation. In this case, the aggregation property holds and there exists a representative agent with utility functions $\mathbf{u}_0$ and $\mathbf{u}_1$ of the power type (as computed in the proof of Proposition 4.15), with the same cautiousness coefficient b. Equation (5.7) then gives that

$$\mathbb{E}[\tilde{r}_n] - r_f = \frac{\text{Cov}\left((\sum_{i=1}^I \gamma_i + b\tilde{x}^m)^{-1/b}, \tilde{r}_n\right)}{\text{Cov}\left((\sum_{i=1}^I \gamma_i + b\tilde{x}^m)^{-1/b}, \tilde{r}^m\right)}(\mathbb{E}[\tilde{r}^m] - r_f), \qquad \text{for all } n = 1, \ldots, N.$$

$$(5.10)$$

In the special case where $\gamma_i = 0$, for all $i = 1, \ldots, I$, it holds that

$$\mathbb{E}[\tilde{r}_n] - r_f = \frac{\text{Cov}\left((\tilde{x}^m)^{-1/b}, \tilde{r}_n\right)}{\text{Cov}\left((\tilde{x}^m)^{-1/b}, \tilde{r}^m\right)}(\mathbb{E}[\tilde{r}^m] - r_f), \qquad \text{for all } n = 1, \ldots, N,$$

and, similarly, formula (5.1) becomes

$$\delta \mathbb{E}\left[\left(\frac{\tilde{x}^m}{x_0^m}\right)^{-\frac{1}{b}} \tilde{r}_n\right] = 1, \qquad \text{for all } n = 0, 1, \ldots, N. \qquad (5.11)$$

Analogously, equations (5.3) and (5.2) can respectively be written as

$$\mathbb{E}[\tilde{r}_n] - r_f = -r_f \delta \, \text{Cov}\left(\left(\frac{\tilde{x}^m}{x_0^m}\right)^{-\frac{1}{b}}, \tilde{r}_n\right) \quad \text{and} \quad \delta \mathbb{E}\left[\left(\frac{\tilde{x}^m}{x_0^m}\right)^{-\frac{1}{b}}\right] = \frac{1}{r_f}.$$

In particular, from the last relation, which determines the equilibrium risk free rate r_f, we can observe that r_f is high when agents are impatient (i.e., δ is small), when the consumption growth rate $\tilde{x}^m/x_0^m$ is high and when risk aversion $1/b$ is high (or, equivalently, when the elasticity of intertemporal substitution of consumption is low). In this sense, if the elasticity of intertemporal substitution is low, then an increase in the risk free rate has a limited effect on expected saving-consumption growth.

In the case of power utility functions, we can obtain more explicit relations if returns are distributed according to a log-normal law. Indeed, suppose that the aggregate consumption growth rate $\tilde{x}^m/x_0^m$ and the random return $\tilde{r}_n$ are jointly log-normally distributed. Then, the couple $\left(\delta(\tilde{x}^m/x_0^m)^{-1/b}, \tilde{r}_n\right)$ is itself distributed as a bivariate log-normal and, using the elementary properties of the log-normal distribution together with relation (5.11), it holds that[1]

$$\mathbb{E}[\log(\tilde{r}_n)] = -\log(\delta) + \frac{1}{b}\mathbb{E}\big[\log(\tilde{x}^m) - \log(x_0^m)\big]$$

$$-\frac{1}{2}\left(\text{Var}\big(\log(\tilde{r}_n)\big) + \frac{1}{b^2}\text{Var}\big(\log(\tilde{x}^m/x_0^m)\big) - \frac{2}{b}\text{Cov}\big(\log(\tilde{r}_n), \log(\tilde{x}^m/x_0^m)\big)\right),$$

$$\log(r_f) = -\log(\delta) + \frac{1}{b}\mathbb{E}\big[\log(\tilde{x}^m) - \log(x_0^m)\big] - \frac{1}{2b^2}\text{Var}\big(\log(\tilde{x}^m/x_0^m)\big).$$

The last equation for the equilibrium risk free rate r_f is in line with the results of Sect. 3.4 on optimal saving and consumption. Indeed, when the aggregate consumption $\tilde{x}^m$ is more volatile, it is optimal to save more (compare with Proposition 3.26) and, therefore, the equilibrium risk free rate will be lower. It is confirmed that r_f is high when agents are impatient (i.e., δ is small), when the consumption growth rate $\tilde{x}^m/x_0^m$ is high and when risk aversion $1/b$ is high. From the last two relations, we obtain the following expression for the risk premium of the n-th asset:

$$\mathbb{E}[\log(\tilde{r}_n)] - \log(r_f) = -\frac{1}{2}\text{Var}\big(\log(\tilde{r}_n)\big) + \frac{1}{b}\text{Cov}\big(\log(\tilde{r}_n), \log(\tilde{x}^m/x_0^m)\big).$$

The risk premium of the logarithmic return depends positively on the covariance between the logarithmic growth rate of the aggregate consumption and the logarithmic return of the asset and also depends on the coefficient of relative risk aversion. These results are in line with the diversification and insurance principles. A large risk premium is obtained when the coefficient of relative risk aversion is high and/or the covariance is large.

Applying Proposition 5.1 to the present setting of an economy populated by I agents with power utility functions (with identical cautiousness coefficient and with $\gamma_i = 0$ for all $i = 1, \ldots, I$) and under the assumption of a log-normal distribution, we obtain the following bound on the Sharpe ratios:

$$\frac{|\mathbb{E}[\tilde{r}_n] - r_f|}{\sigma(\tilde{r}_n)} \leq \frac{\sigma\left(\delta(\frac{\tilde{x}^m}{x_0^m})^{-1/b}\right)}{\mathbb{E}\left[\delta(\frac{\tilde{x}^m}{x_0^m})^{-1/b}\right]} = \sqrt{e^{\frac{1}{b^2}\sigma^2\left(\log(\tilde{x}^m/x_0^m)\right)} - 1} \approx \frac{1}{b}\sigma\left(\log(\tilde{x}^m/x_0^m)\right).$$

$$(5.12)$$

---

[1] Recall that, if a random variable $\tilde{x}$ is distributed according to $\log(\tilde{x}) \sim \mathcal{N}(\mu, \sigma^2)$, then it holds that $\mathbb{E}[\tilde{x}] = e^{\mu + \sigma^2/2}$ and $\text{Var}(\tilde{x}) = e^{2\mu + \sigma^2}(e^{\sigma^2} - 1)$.

The interpretation of this approximation is that in equilibrium the bound on the Sharpe ratio is higher if the economy is riskier (i.e., the aggregate consumption is more volatile) or if investors are more risk averse.

Note that with $b = -1$ the utility function is quadratic. In this case, formula (5.10) yields the classical CAPM relation, without any additional assumption on the distribution of asset returns:

$$\mathbb{E}[\tilde{r}_n] - r_f = \frac{\text{Cov}(\tilde{x}^m, \tilde{r}_n)}{\text{Cov}(\tilde{x}^m, \tilde{r}^m)} \left(\mathbb{E}[\tilde{r}^m] - r_f \right) = \frac{\text{Cov}(\tilde{r}^m, \tilde{r}_n)}{\text{Var}(\tilde{r}^m)} \left(\mathbb{E}[\tilde{r}^m] - r_f \right), \qquad (5.13)$$

for all $n = 1, \ldots, N$. In particular, observe that relation (5.13) coincides with the CAPM relation (5.9). The CAPM has been previously derived for general utility functions under the assumption of normally distributed returns. On the other hand, equation (5.13) has been derived under the assumption of quadratic utility functions, without any assumption on the specific distribution of the asset returns (to this regard, see also Exercise 5.4). In Sect. 5.2 the CAPM relation will be derived under a different set of assumptions, namely under the key assumption that agents choose to hold portfolios belonging to the mean-variance portfolio frontier, by relying on the results of Sect. 3.2.

Absence of Arbitrage and Asset Risk Premia

So far, we have derived several formulations of the CCAPM relation (as well as of the CAPM relation, under two different sets of assumptions) by relying on equilibrium arguments, together with additional assumptions on the form of the utility functions and/or on the returns' distribution in order to obtain more explicit relations. As we have seen in Sect. 4.4, there is a deep connection between equilibrium analysis and the no-arbitrage paradigm. Hence, it is not a surprise that a CCAPM-type relation can also be derived without explicitly making use of equilibrium arguments, under the assumption that the price-dividend couple (p, D) does not admit arbitrage opportunities. In this framework, CCAPM-type relations can be derived in terms of the *likelihood ratio* defining a risk neutral probability measure (see the discussion following Proposition 4.22).

As in the previous chapter, let us assume that the probability space is finite, meaning that there is a finite number S of elementary states of the world $(\omega_1, \ldots, \omega_S)$, with associated probabilities of occurrence $(\pi_1, \ldots, \pi_S)$. Recall that, in view of Proposition 4.22 and the following discussion, if the price-dividend couple (p, D) does not allow for arbitrage opportunities, then there exists a likelihood ratio ℓ with values $\ell = (\ell_1, \ldots, \ell_S) \in \mathbb{R}_{++}^S$ such that $\pi^* = (\pi_1^*, \ldots, \pi_S^*) = (\ell_1 \pi_1, \ldots, \ell_S \pi_S)$ defines a risk neutral probability measure. Recall also that, when the market is incomplete, there exist infinitely many risk neutral probability measures or, equivalently, infinitely many likelihood ratios (see Proposition 4.26). We can state the following result, which holds with respect to any likelihood ratio $\tilde{\ell}$.

Proposition 5.3 *Let (p, D) be an arbitrage free price-dividend couple and let $\ell = (\ell_1, \ldots, \ell_S)$ be the likelihood ratio associated to some risk neutral probability measure. Then the following hold:*

$$\mathbb{E}[\tilde{r}_n] - r_f = -\operatorname{Cov}(\tilde{r}_n, \tilde{\ell}), \qquad \text{for all } n = 1, \ldots, N, \tag{5.14}$$

and the Sharpe ratio of any traded asset satisfies the following bound:

$$\frac{|\mathbb{E}[\tilde{r}_n] - r_f|}{\sigma(\tilde{r}_n)} \leq \sigma(\tilde{\ell}), \qquad \text{for all } n = 1, \ldots, N. \tag{5.15}$$

Proof Suppose that (p, D) does not admit arbitrage opportunities and let π^* be any risk neutral probability measure, with associated likelihood ratio $\tilde{\ell}$. Then, in view of formula (4.35), it holds that, for any $n = 1, \ldots, N$,

$$0 = \mathbb{E}^*[\tilde{r}_n] - r_f = \mathbb{E}[\tilde{\ell}\,\tilde{r}_n] - r_f = \operatorname{Cov}(\tilde{\ell}, \tilde{r}_n) + \mathbb{E}[\tilde{r}_n]\mathbb{E}[\tilde{\ell}] - r_f,$$

from which (5.14) follows, since $\mathbb{E}[\tilde{\ell}] = 1$. From (5.14), we obtain

$$|\mathbb{E}[\tilde{r}_n] - r_f| = |\rho_{\tilde{\ell}, \tilde{r}_n}|\sigma(\tilde{\ell})\sigma(\tilde{r}_n) \leq \sigma(\tilde{\ell})\sigma(\tilde{r}_n),$$

where $\rho_{\tilde{\ell}, \tilde{r}_n}$ denotes the correlation coefficient between $\tilde{\ell}$ and $\tilde{r}_n$. This implies the bound (5.15). $\qquad \square$

According to the above proposition, in an arbitrage free economy, the risk premium of an asset is equal to the covariance between its return and a likelihood ratio. Of course, this relation can be extended to the risk premium of a portfolio of the available securities. Observe also that the CCAPM relation (5.7) can also be obtained from equation (5.14), recalling that in a representative agent economy a stochastic discount factor (or, equivalently, a likelihood ratio) can be defined in terms of the representative agent's marginal rate of substitution of consumption, as follows from equation (4.53). This observation provides a link between CCAPM relations obtained by equilibrium arguments and CCAPM relations derived from no-arbitrage considerations.

Abstract CAPM-type formulae analogous to relations (5.9) and (5.13) can also be obtained starting from no-arbitrage arguments, as we are now going to show. The return on the market portfolio $\tilde{r}^m$ appearing in relations (5.9) and (5.13) will be replaced by the return on a portfolio replicating a likelihood ratio. Let us first consider the case of a complete market economy (the number of linearly independent traded assets equals the number of possible states of the world). Then, Proposition 5.3 together with the completeness of the market yield the following corollary.

Corollary 5.4 *Let (p, D) be an arbitrage free price-dividend couple. Suppose that the market is complete and let $\ell = (\ell_1, \ldots, \ell_S)$ be the likelihood ratio associated to the (unique) risk neutral probability measure π^*. Then there exists a portfolio*

$z^\ell \in \mathbb{R}^N$ *with associated return* $\tilde{r}^\ell$ *such that the following holds:*

$$\mathbb{E}[\tilde{r}_n] - r_f = \frac{\mathrm{Cov}(\tilde{r}_n, \tilde{r}^\ell)}{\mathrm{Var}(\tilde{r}^\ell)}\left(\mathbb{E}[\tilde{r}^\ell] - r_f\right) = \beta_{n\ell}\left(\mathbb{E}[\tilde{r}^\ell] - r_f\right), \qquad (5.16)$$

where $\beta_{n\ell} := \mathrm{Cov}(\tilde{r}_n, \tilde{r}^\ell)/\mathrm{Var}(\tilde{r}^\ell)$, *for all* $n = 1, \ldots, N$.

Proof If the price-dividend couple (p, D) is arbitrage free and the market is complete, then there exists a unique likelihood ratio $\ell \in \mathbb{R}^S_{++}$, see Proposition 4.26. Due to the completeness of the market, there exists a portfolio z^ℓ such that $Dz^\ell = \ell$. Following the notation introduced in the previous chapter, let us denote by $V(\ell)$ the market value at time $t = 0$ of the portfolio z^ℓ, i.e., $V(\ell) = p^\top z^\ell$, so that

$$\tilde{r}^\ell = Dz^\ell/V(\ell) = \tilde{\ell}/V(\ell).$$

With this notation, relation (5.14) directly implies that

$$\mathbb{E}[\tilde{r}_n] - r_f = -V(\ell)\,\mathrm{Cov}(\tilde{r}_n, \tilde{r}^\ell), \qquad \text{for all } n = 1, \ldots, N.$$

Of course, such a relation also holds for any portfolio composed of the $N + 1$ available assets, due to the linearity of the covariance operator. Hence, in particular, it holds for the portfolio z^ℓ, thus leading to

$$\mathbb{E}[\tilde{r}^\ell] - r_f = -V(\ell)\,\mathrm{Var}(\tilde{r}^\ell).$$

By combining the last two equations we immediately obtain relation (5.16).   □

According to the above corollary, in a complete market economy, the risk premium of an asset is equal to the risk premium of the portfolio replicating the (unique) likelihood ratio multiplied by the $\beta_{n\ell}$ coefficient, which corresponds to the regression coefficient of the return $\tilde{r}_n$ with respect to the random variable $\tilde{r}^\ell$. Note that, despite the similarities, formula (5.16) and formulae (5.9) and (5.13) have been derived under quite different assumptions and, in general, the portfolio z^ℓ will be different from the market portfolio.

The risk premium expression in (5.16) allows us to rewrite the pricing functional $Q(c)$ introduced in Sect. 4.4. Indeed, in a complete arbitrage free market, formula (5.16) can be equivalently rewritten as

$$p_n = \frac{\mathbb{E}[\tilde{d}_n]}{\mathbb{E}[\tilde{r}_n]} = \frac{\mathbb{E}[\tilde{d}_n]}{r_f + \lambda\,\mathrm{Cov}(\tilde{r}_n, \tilde{r}^\ell)}, \qquad \text{for all } n = 1, \ldots, N, \qquad (5.17)$$

where $\lambda := (\mathbb{E}[\tilde{r}^\ell] - r_f)/\mathrm{Var}(\tilde{r}^\ell)$. Hence, due to the market completeness assumption, for any contingent consumption plan $\tilde{c}$ represented by the vector $c \in \mathbb{R}^S$, relation (5.16) implies that

$$Q(c) = \frac{\mathbb{E}[\tilde{c}]}{r_f + \lambda\,\mathrm{Cov}(\tilde{r}^c, \tilde{r}^\ell)}, \qquad (5.18)$$

where $\tilde{r}^c$ denotes the return on the portfolio z^c which replicates the payoff $\tilde{c}$ (due to the market completeness assumption, such a portfolio always exists). In the above formula, the term λ can be interpreted as a correction coefficient for the discount factor which accounts for the risk aversion of the agents, similarly as a risk neutral probability measure differs from the physical probability measure due to the agents' risk aversion. For this reason, the quantity appearing in the denominator of (5.17)–(5.18) is sometimes called *risk adjusted* discount factor. Of course, if all agents are risk neutral, then $\lambda = 0$ and all assets have the same expected return, equal to the risk free rate r_f.

The above results can be extended to incomplete markets. Recall that, in view of Proposition 4.26, in the absence of arbitrage opportunities, the risk neutral probability measure (and, hence, the likelihood ratio) is unique if and only if the market is complete. Hence, if the market is arbitrage free but incomplete then there exist infinitely many likelihood ratios with respect to which relation (5.14) holds true. However, it can be shown that there exists a unique payoff $\hat{\ell} \in I(D)$ such that the relation $Q(c) = \mathbb{E}[\hat{\ell}\,\tilde{c}]$ holds for all attainable contingent consumption plans $\tilde{c}$ (i.e., for all $c \in I(D)$). This can be shown as a consequence of the *Riesz representation theorem*, which in the present context can be stated as follows (compare also with LeRoy & Werner [1191], Theorem 17.7.1 and Section 17.10).[2]

Lemma 5.5 *Let $Q : I(D) \to \mathbb{R}$ be a linear function. Then there exists a unique vector $\hat{\ell} \in I(D)$ such that $Q(c) = \mathbb{E}[\hat{\ell}\,\tilde{c}]$ for every random variable $\tilde{c}$ with values $c \in I(D)$.*

Regardless of whether markets are complete or not, the absence of arbitrage opportunities is equivalent to the existence of a linear pricing functional Q (see Theorem 4.20). Due to the above lemma, there exists a unique element $\hat{\ell} \in I(D)$ such that, for every attainable payoff $c \in I(D)$, the pricing functional $Q(c)$ can be represented as the expectation of the inner product between $\hat{\ell}$ and c. In particular, note that $\hat{\ell}$ is unique regardless of whether markets are complete or not. On the asset span $I(D)$, the vector $\hat{\ell}$ has the same pricing implications of the pricing functional Q, in the sense that $\mathbb{E}[\hat{\ell}c] = Q(c)$, for all $c \in I(D)$, so that the market value of every attainable consumption plan can be equivalently computed by taking the expectation of the inner product of c with the vector $\hat{\ell}$. However, unattainable contingent consumption plans $c \notin I(D)$ cannot in general be priced as $\mathbb{E}[\hat{\ell}\,\tilde{c}]$, since Lemma 5.5 does not ensure in general that the vector $\hat{\ell}$ is strictly positive. As shown in Exercise 5.7, the payoff $\hat{\ell}$ has also interesting properties in relation with mean-variance theory. In particular, the return on the portfolio which replicates $\hat{\ell}$ minimizes the second moment among all possible portfolio returns.

---

[2]Similarly as in the previous chapter, with some abuse of notation we denote equivalently by $\tilde{c}$ and $c \in \mathbb{R}^S$ the random variable $\tilde{c}$ with values $c = (c_1, \ldots, c_S) \in \mathbb{R}^S$. Moreover, we denote by $\mathbb{E}[\hat{\ell}c]$ the expectation of the product between the random variable $\tilde{c}$ and the random variable taking values $\hat{\ell}$.

By relying on Lemma 5.5, we can extend the risk premium relation (5.16) to general incomplete markets as shown in the following proposition.

Proposition 5.6 *Let (p, D) be an arbitrage free price-dividend couple. Then there exists a unique portfolio $z^\ell \in \mathbb{R}^N$ with associated return $\tilde{r}^\ell$ such that relation (5.16) holds true, for all $n = 1, \ldots, N$.*

Proof By Theorem 4.20, the absence of arbitrage opportunities implies the existence of a strictly positive linear pricing functional Q. Hence, in view of Lemma 5.5, there exists a unique vector $\hat{\ell} \in I(D)$ such that

$$p_n = \mathbb{E}[\hat{\ell} \, \tilde{d}_n], \qquad \text{for all } n = 0, 1, \ldots, N.$$

In terms of returns, this can be rewritten as

$$1 = \mathbb{E}[\hat{\ell} \, \tilde{r}_n], \qquad \text{for all } n = 0, 1, \ldots, N.$$

Applied to the risk free asset, this gives $\mathbb{E}[\hat{\ell}] = 1/r_f$. Define $\ell := \hat{\ell}/\mathbb{E}[\hat{\ell}]$. Then, since $\hat{\ell} \in I(D)$, there exists a unique portfolio z^ℓ (under the standing assumption of non-redundant assets) such that $Dz^\ell = \ell$. Thus, denoting by $\tilde{\ell}$ the random variable taking values $\ell = (\ell_1, \ldots, \ell_S)$, we can write, for all $n = 1, \ldots, N$,

$$1 = \frac{1}{r_f} \mathbb{E}[\tilde{\ell} \, \tilde{r}_n] = \frac{1}{r_f} \big(\mathrm{Cov}(\tilde{\ell}, \tilde{r}_n) + \mathbb{E}[\tilde{r}_n]\big) = \frac{1}{r_f} \big(V(\ell) \, \mathrm{Cov}(\tilde{r}^\ell, \tilde{r}_n) + \mathbb{E}[\tilde{r}_n]\big),$$

from which relation (5.16) follows by the same arguments used in the proof of Corollary 5.4. □

5.2 The Capital Asset Pricing Model (CAPM)

In the previous section, we have derived the CAPM relation by equilibrium arguments under the assumption of normally distributed returns or quadratic utility functions (see relations (5.9) and (5.13)). Now, we present an alternative derivation of the CAPM by relying on the assumption that all agents choose to hold portfolios belonging to the mean-variance portfolio frontier (see Sect. 3.2). In a nutshell, if every market participant holds a frontier portfolio, then also the market portfolio, being a convex combination of the individual portfolios, will belong to the portfolio frontier. Hence, Proposition 3.13 (see also Proposition 3.16 when a risk free asset is traded in the market) implies that a linear relation holds true between the expected return of any portfolio and the expected return of the market portfolio. This will lead to the CAPM relation (5.9). Note that the hypotheses of normal distribution or of quadratic utility function imply that agents hold a portfolio belonging to the portfolio frontier (compare with Sect. 2.4).

As in the previous section, we consider an economy populated by I risk averse agents. We assume that N risky assets are available for trade (later we shall also introduce a risk free asset with return r_f), with random returns $(\tilde{r}_1, \ldots, \tilde{r}_N)$ with finite second moments. The N risky assets are in strictly positive supply. As always, we assume a frictionless and perfectly competitive financial market, meaning that there are no transaction costs, taxes or trading constraints and that the economic model is common knowledge among the agents. We denote by w_0^i the wealth of agent i at date $t = 0$ and by w_n^i the proportion of wealth invested by agent i in the n-th asset, for $i = 1, \ldots, I$ and $n = 1, \ldots, N$. We assume that the wealth of the agents is fully invested in the financial assets. The aggregate wealth of the economy (i.e., the total market capitalization) is then given by $w_0^m = \sum_{i=1}^{I} w_0^i = \sum_{i=1}^{I} \sum_{n=1}^{N} e_n^i p_n$ and the proportion of aggregate wealth invested in the n-th asset (i.e., the market capitalization of the n-th asset) is given by $w_n^m = \sum_{i=1}^{I} w_n^i w_0^i / w_0^m = (\sum_{i=1}^{I} e_n^i p_n)/(\sum_{n=1}^{N} \sum_{i=1}^{I} e_n^i p_n)$, where the last equality follows since in equilibrium the aggregate demand of the risky assets equals the aggregate supply. In turn, this implies that $\sum_{n=1}^{N} w_n^m = 1$. We denote by $\tilde{r}^m$ the return on the market portfolio, i.e., $\tilde{r}^m = \sum_{n=1}^{N} w_n^m \tilde{r}_n$. As in Sect. 3.2, we call *portfolio frontier* the set of all portfolios which solve the mean-variance optimization problem (3.14) as the expected return varies. We then have the following proposition, which establishes the CAPM relation.

Proposition 5.7 *Suppose that all the I agents hold portfolios belonging to the mean-variance portfolio frontier. Then, under the assumptions of the present section, the market portfolio w^m belongs to the portfolio frontier. Moreover, if w^m does not coincide with the minimum variance portfolio w^{MVP}, denoting by $\tilde{r}^{zc(m)}$ the return of the frontier portfolio having zero correlation with the market portfolio, the following relation holds true, for all $n = 1, \ldots, N$:*

$$\mathbb{E}[\tilde{r}_n] - \mathbb{E}[\tilde{r}^{zc(m)}] = \frac{\mathrm{Cov}(\tilde{r}_n, \tilde{r}^m)}{\mathrm{Var}(\tilde{r}^m)} \big(\mathbb{E}[\tilde{r}^m] - \mathbb{E}[\tilde{r}^{zc(m)}] \big) =: \beta_{nm} \big(\mathbb{E}[\tilde{r}^m] - \mathbb{E}[\tilde{r}^{zc(m)}] \big).$$

$$(5.19)$$

Proof w^m is a convex combination of the individual portfolios $w^i \in \mathbb{R}^N$, $i = 1, \ldots, I$. Hence, in view of Property 1 of the portfolio frontier (see section "The Case of N Risky Assets"), if w^i belongs to the portfolio frontier, for all $i = 1, \ldots, I$, then also w^m will belong to the portfolio frontier. Relation (5.19) then follows directly from (3.32) by taking $w^p = w^m$.   □

By linearity, relation (5.19) also applies to arbitrary portfolios composed of the N assets. Note that, in the context of the mean-variance portfolio frontier, relation (3.32) is a property which holds with respect to any frontier portfolio w^p. Instead, the CAPM relation (5.19) is based on the assumption that, in equilibrium, the market portfolio belongs to the portfolio frontier.

The result of Proposition 5.7 holds true in the case where the two mutual funds separation property holds, as shown in the following corollary.

Corollary 5.8 *Suppose that the two mutual funds separation property holds. Then, under the assumptions of the present section, the market portfolio w^m belongs to the portfolio frontier. Moreover, if $w^m \neq w^{\mathrm{MVP}}$, then relation (5.19) holds true.*

Proof As shown in section "Mutual Fund Separation and Mean-Variance Portfolio Selection", if the two mutual funds separation property holds, then the two mutual funds belong to the portfolio frontier. Hence, since the portfolio frontier is stable with respect to convex combinations, it follows that the market portfolio also belongs to the portfolio frontier. The result then follows from Proposition 5.7. □

Observe that the sign of the coefficient β_{nm} appearing in (5.19) is determined by the sign of the correlation coefficient between the return of asset n and the return on the market portfolio. In the β-expected return plane, relation (5.19) is represented by a line (*Security Market Line*) which intersects the vertical axis in correspondence of $\mathbb{E}[\tilde{r}^{zc(m)}]$ and with slope $\mathbb{E}[\tilde{r}^m] - E[\tilde{r}^{zc(m)}]$. We have thus established the following important result: in equilibrium, the expected return of a portfolio is a linear affine combination of the expected return of the market portfolio and of the expected return of the portfolio belonging to the portfolio frontier with zero covariance with the market portfolio. The slope $\mathbb{E}[\tilde{r}^m] - \mathbb{E}[\tilde{r}^{zc(m)}]$ of the Security Market Line is positive or negative depending on the efficiency of the market portfolio (see Property 4 in section "The Case of N Risky Assets"). Indeed, if the market portfolio is efficient (i.e., $w^m \in$ EPF, so that $\mathbb{E}[\tilde{r}^m] > \mathbb{E}[\tilde{r}^{zc(m)}]$, always assuming that $w^m \neq w^{\mathrm{MVP}}$), then the Security Market Line has a positive slope, while, if the market portfolio is inefficient (meaning that $\mathbb{E}[\tilde{r}^m] < \mathbb{E}[\tilde{r}^{zc(m)}]$), then the Security Market Line has a negative slope. In the first case, high expected returns correspond to high β coefficients, so that, in order to get a high expected return, one needs to invest in a rather risky portfolio (with risk being measured in terms of the covariance with the return on the market portfolio). As discussed before, this result can be naturally interpreted in terms of the insurance and diversification principles: having the possibility of investing in the market portfolio, the agent invests in a risky asset with a large covariance with the latter (thus increasing the overall variance if an agent is exposed to both the asset and the market portfolio) if and only if the asset expected return is high (see also below for an illustration of this phenomenon). On the contrary, if the Security Market Line has a negative slope, then high expected returns correspond to negative β coefficients.

In particular, the Security Market Line has a positive slope when agents have increasing and strictly concave utility functions and returns are distributed according to a multivariate normal law, as shown in the following corollary.

Corollary 5.9 *Under the assumptions of the present section, suppose that the returns $(\tilde{r}_1, \ldots, \tilde{r}_N)$ are jointly normally distributed with non-identical expectations and that all the I agents have strictly increasing and concave utility functions. Then it holds that*

$$\mathbb{E}[\tilde{r}_n] = \mathbb{E}[\tilde{r}^{zc(m)}] + \beta_{nm}\big(\mathbb{E}[\tilde{r}^m] - \mathbb{E}[\tilde{r}^{zc(m)}]\big), \qquad \text{for all } n = 1, \ldots, N, \qquad (5.20)$$

and $\mathbb{E}[\tilde{r}^m] - \mathbb{E}[\tilde{r}^{zc(m)}] > 0$.

Proof If the N traded assets have jointly normally distributed returns with non-identical expectations, then Proposition 3.21 implies that the two funds separation property holds true with respect to any frontier portfolio w^p and to its zero-correlation portfolio $w^{zc(p)}$. Moreover, since the agents have strictly increasing and concave utility functions, they will choose to hold only efficient portfolios. Indeed, due to Proposition 2.13, the agents' preferences are increasing with respect to the expected return and decreasing with respect to the variance. In turn, this implies that the market portfolio, being a convex combination of the portfolios of the individual agents, also belongs to EPF. The result then follows from Proposition 5.7, while $\mathbb{E}[\tilde{r}^m] - \mathbb{E}[\tilde{r}^{zc(m)}] > 0$ is due to the efficiency of the market portfolio (compare with Property 4 in section "The Case of N Risky Assets"). $\square$

Relation (5.20), together with the inequality $\mathbb{E}[\tilde{r}^m] - \mathbb{E}[\tilde{r}^{zc(m)}] > 0$ (efficiency of the market portfolio), represents the *Zero-β Capital Asset Pricing Model*, see Black [242]. The efficiency of the market portfolio implies that, in order to get a high expected return in equilibrium, one must look for a risky portfolio (i.e., a portfolio characterized by a positive and high covariance with the market portfolio). An asset with a negative covariance will be characterized by a low expected return, lower than the return of the portfolio having zero covariance with the market portfolio.

A conclusion analogous to Corollary 5.9 can be derived under the assumption that all the agents have quadratic utility functions (with possibly different risk aversion coefficients), as shown in the following corollary. Note that we also need to assume that expected returns of portfolios lie in the region where the agents' utility functions are increasing.

Corollary 5.10 *Under the assumptions of the present section, suppose that all the I agents have quadratic utility functions of the form $u^i(x) = x - \frac{b^i}{2}x^2$, for all $i = 1, \ldots, I$. Assume furthermore that $\mathbb{E}[w_0^i \sum_{n=1}^{N} \tilde{r}_n w_n^i] < 1/b^i$ for every $w_0^i \in \mathbb{R}$ and $w^i \in \mathbb{R}^N$, $i = 1, \ldots, I$. Then relation (5.20) holds true and $\mathbb{E}[\tilde{r}^m] - \mathbb{E}[\tilde{r}^{zc(m)}] > 0$.*

Proof As shown in Sect. 2.4, letting $\widetilde{W}^i := w_0^i \sum_{n=1}^{N} \tilde{r}_n w_n^i$, the expected utility of agent i can be written as follows,

$$\mathbb{E}[u^i(\widetilde{W}^i)] = \mathbb{E}[\widetilde{W}^i] - \frac{b^i}{2}\mathbb{E}[(\widetilde{W}^i)^2]$$

$$= \left(1 - \frac{b^i}{2}\mathbb{E}[\widetilde{W}^i]\right)\mathbb{E}[\widetilde{W}^i] - \frac{b^i}{2}\,\text{Var}(\widetilde{W}^i).$$

Since it has been assumed that $\mathbb{E}[\widetilde{W}^i] < 1/b^i$, for all $i = 1, \ldots, I$, the expected utility is strictly increasing with respect to the expected return of a portfolio and strictly decreasing with respect to its variance. As a consequence, every agent will choose to hold a mean-variance efficient portfolio. The result then follows as in the proof of Corollary 5.9. $\square$

It is interesting to compare the result of the above corollary with the CAPM relation (5.13). In both cases, quadratic utility functions have been assumed. Relation (5.13) has been obtained by relying on the existence of a representative agent, while Corollary 5.10 is based on the assumption that the expected returns on optimal portfolios lie in the region where the utility functions are increasing. Note also that market completeness is not required.

Let us now extend the results of the present section to an economy with N risky assets and a risk free asset with return $r_f > 0$. Recall from section "The Case of N Risky Assets and a Risk Free Asset" that, in this case, the shape of the portfolio frontier depends on the relation between the risk free rate r_f and the expected return A/C associated to the minimum variance portfolio composed of the risky assets only. In particular, if $r_f \neq A/C$, then the portfolio frontier is represented by two half-lines and every frontier portfolio can be written as a linear combination of the tangent portfolio w^e and the risk free asset (see Proposition 3.15). On the basis of these observations, we can establish the following simple but important result (compare also with Sharpe [1529], Lintner [1221], Mossin [1357]).

Proposition 5.11 *Suppose that a risk free asset with return $r_f \neq A/C$ is traded and that all the I agents hold portfolios belonging to the mean-variance portfolio frontier. Assume furthermore that the net supply of the risk free asset is zero (and that of the risky assets is strictly positive). Then, under the assumptions of the present section, the market portfolio w^m coincides with the tangent portfolio w^e and, hence, belongs to the portfolio frontier. Moreover, the following relation holds true:*

$$\mathbb{E}[\tilde{r}_n] - r_f = \beta_{nm}(\mathbb{E}[\tilde{r}^m] - r_f), \qquad \text{for all } n = 1, \ldots, N, \qquad (5.21)$$

with $\beta_{nm} = \text{Cov}(\tilde{r}_n, \tilde{r}^m)/\text{Var}(\tilde{r}^m)$.

Proof As explained in section "The Case of N Risky Assets and a Risk Free Asset", if $r_f \neq A/C$, then every portfolio belonging to the portfolio frontier can be represented as a linear affine combination of the tangent portfolio w^e and of the risk free asset. Hence, if all the agents hold portfolios belonging to the portfolio frontier and the aggregate supply of the risk free asset is zero, it follows that the market portfolio coincides with the tangent portfolio w^e. Relation (5.21) then follows directly from (3.41) by taking $w^p = w^m$. □

Relation (5.21) is illustrated in Fig. 5.1. Of course, relation (5.21) immediately extends to arbitrary portfolios composed of the $N+1$ available securities. In the case where $r_f = A/C$, the portfolio frontier including the risk free asset is represented by the two asymptotes to the portfolio frontier composed of the risky assets only. As discussed in section "The Case of N Risky Assets and a Risk Free Asset", in this case the overall net investment in the risky assets is zero and the total wealth is invested in the risk free asset. Hence, the economy is in equilibrium only if the aggregate supply of the risk free asset is strictly positive and the aggregate supply of the risky assets is null. Therefore, under the assumptions of the present section (strictly positive supply of the risky assets), an equilibrium cannot exist and the CAPM does not hold.

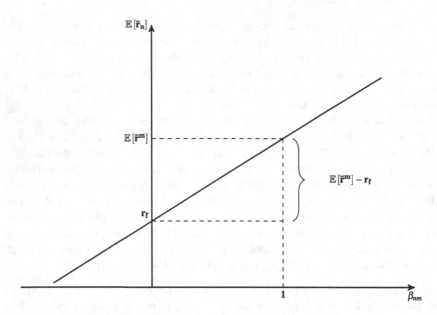

Fig. 5.1 Capital Asset Pricing Model

Proposition 5.11 establishes the CAPM relation in the presence of a risk free asset. However, Proposition 5.11 does not assert the efficiency of the market portfolio, i.e., we do not know a priori whether $\mathbb{E}[\tilde{r}^m] > r_f$. As shown in the next proposition, the efficiency of the market portfolio can be established under rather natural assumptions on the agents' utility functions.

Proposition 5.12 *Under the assumptions of Proposition 5.11, suppose furthermore that all the I agents have strictly increasing and concave utility functions. Then the market portfolio is efficient, i.e., $\mathbb{E}[\tilde{r}^m] > r_f$.*

Proof Under the present assumptions and in view of Proposition 3.7, the optimal portfolio of every agent will have an expected return greater or equal than the risk free rate r_f. Recall also that, due to Proposition 5.11, the market portfolio coincides with the tangent portfolio. Hence, if $r_f > A/C$, then it holds that $\mathbb{E}[\tilde{r}^m] < r_f$ and no investor will choose to invest a strictly positive amount of wealth in the market portfolio, thus leading to a contradiction with the assumption of a strictly positive supply of risky assets. This implies that $r_f < A/C$. In this case, the market portfolio is efficient, since the tangent portfolio belongs to the upper part of the portfolio frontier (see section "The Case of N Risky Assets and a Risk Free Asset").   □

Similarly to the case of formula (5.17), the CAPM relation (5.21) can also be used to establish a valuation rule as follows:

$$p_n = \frac{\mathbb{E}[\tilde{d}_n]}{r_f + \beta_{nm}\big(\mathbb{E}[\tilde{r}^m] - r_f\big)}, \qquad \text{for all } n = 1, \ldots, N. \tag{5.22}$$

The denominator $r_f + \beta_{nm}(E[\tilde{r}^m] - r_f)$ represents a risk adjusted discount factor for the n-th asset. Note that, as long as the market portfolio is efficient (i.e., $\mathbb{E}[\tilde{r}^m] > r_f$), there exists a positive relationship between the β coefficient and the risk adjusted discount factor, or, equivalently, a negative relationship between the price of an asset and its β coefficient.

Summing up, in the Capital Asset Pricing Model, asset risk premia are measured with respect to a single risk factor, represented by the return on the market portfolio. The sensitivity of an asset return with respect to this single factor is captured by the β coefficient, which represents the coefficient in the linear regression of the asset return on the market portfolio return. Indeed, in view of Proposition 3.16, for an arbitrary portfolio w it holds that

$$\tilde{r}^w = r_f + \beta_{wm}(\tilde{r}^m - r_f) + \tilde{\varepsilon}_{wm},$$

where the random variable $\tilde{\varepsilon}_{wm}$ is uncorrelated with the market portfolio return, so that

$$\mathrm{Var}(\tilde{r}^w) = \beta_{wm}^2 \, \mathrm{Var}(\tilde{r}^m) + \mathrm{Var}(\tilde{\varepsilon}_{wm}).$$

According to this representation, the return $\tilde{r}^w$ is decomposed into a constant term, equal to the risk free rate r_f, a risky component related to the return of the market portfolio (*market risk*) and, finally, a residual component $\tilde{\varepsilon}_{wm}$. The latter component is called *idiosyncratic risk*. The market risk is the systematic risk that cannot be diversified away and, therefore, it is compensated through the risk premium $\beta_{wm}(\mathbb{E}[\tilde{r}^m] - r_f)$, unlike the idiosyncratic risk, which can be diversified away (provided that the residual risk components associated to different assets are uncorrelated). For this reason, $|\beta_{wm}|$ is an indicator of the riskiness of a portfolio w. As already mentioned, these facts can be interpreted in terms of the diversification and insurance principles. Indeed, take the point of view of an agent who holds the market portfolio and considers the possibility of investing in the portfolio w. If $\beta_{wm} > 0$, then the return $\tilde{r}^w$ is positively correlated with the market portfolio. Hence, in order to invest a positive amount of wealth in the portfolio w, the investor will require a compensation for the higher risk of his overall position. On the contrary, if $\beta_{wm} < 0$, then the agent can accept a negative risk premium in order to benefit from the negative correlation between $\tilde{r}^w$ and $\tilde{r}^m$ (insurance principle). As shown in the notes at the end of the chapter, the CAPM can be extended to the case where agents exhibit heterogeneous beliefs about the expected returns of the risky assets (see also Hens & Rieger [938, Proposition 3.5]).

5.3   Empirical Tests of the CAPM

The CAPM establishes a linear relation between the risk premium of an asset (or of a portfolio) and its β coefficient with respect to the market portfolio. This is an *ex-ante* relation, since it involves the expectations of random returns, and it cannot be directly tested as such on financial time series. As a matter of fact, no time series for the risk premia and for the β coefficients are directly available. The approaches to test the CAPM fall in two classes: time series approaches and "two pass" approaches. In particular, the latter consist in first estimating the beta coefficients and the expected returns and then testing the validity of the CAPM relation.

The CAPM has been derived in a two-period economy. Therefore, in testing the CAPM we implicitly assume that the CAPM holds period by period (asset returns are assumed to have a stationary probability distribution). The classical assumption is that asset returns are identically and independently distributed (i.i.d.) over time according to a normal multivariate distribution (an assumption consistent with the CAPM).

One of the main problems in testing the CAPM is represented by the fact that the market portfolio is not observable. To overcome this problem, a proxy of the market portfolio is typically employed and usually the proxy is represented by a stock index. Note that the CAPM is a general equilibrium model for a closed economy, so that the market portfolio represents the composition of the wealth of the whole economy, while any stock index is only a proxy of the market portfolio (see Roll [1457] concerning the problems due to the non-observability of the market portfolio). The risk free rate is usually set equal to the interest rate of short term Treasury bills.

Let us briefly illustrate the classical two pass procedure for testing the CAPM. Seminal examples of the two pass procedure are Fama & MacBeth [678], Black et al. [245], Blume & Friend [258]. To implement the first step, we must transform the ex-ante CAPM model into an ex-post model. The ex-ante relation can be transformed into an ex-post relation by making the rational expectations hypothesis: asset returns are generated by a stationary model which is common knowledge among all the agents of the economy. Under this hypothesis, the expected return of an asset should be equal to its historical mean.

At any date $t \in \{0, 1, \ldots\}$, asset returns can be written in the following form

$$\tilde{r}_{n,t} = \mathbb{E}[\tilde{r}_{n,t}] + \beta_{nm,t}(\tilde{r}_t^m - \mathbb{E}[\tilde{r}_t^m]) + \tilde{\epsilon}_{n,t}, \qquad \text{for all } n = 1, \ldots, N, \qquad (5.23)$$

where $\tilde{r}_{n,t}$ denotes the random return of asset n at date t, $\mathbb{E}[\tilde{\epsilon}_t] = 0$, $\mathbb{E}[\tilde{\epsilon}_t \tilde{\epsilon}_t^\top] = \Sigma_t$, for some variance-covariance matrix Σ_t, $\beta_{nm,t} := \mathrm{Cov}(\tilde{r}_{n,t}, \tilde{r}_t^m)/\mathrm{Var}(\tilde{r}_t^m)$ and $\tilde{r}_t^m$ denotes the market portfolio return, with $\mathrm{Cov}(\tilde{r}_t^m, \tilde{\epsilon}_{n,t}) = 0$, for all $n = 1, \ldots, N$ and $t \in \{0, 1, \ldots\}$. As can be easily checked, relation (5.23) is always satisfied, independently of the validity of the CAPM. Suppose now that the returns' distribution is stationary in the sense that $\Sigma_t = \Sigma$ and $\beta_{nm,t} = \beta_{nm}$, for all $t \in \{0, 1, \ldots, \}$ and $n = 1, \ldots, N$. Under the rational expectations hypothesis and imposing the CAPM restrictions in (5.23), the CAPM in *ex-post* form is obtained, for $t \in \{0, 1, \ldots\}$:

$$\tilde{r}_{n,t} - r_f = \alpha_n + \beta_{nm}(\tilde{r}_t^m - r_f) + \tilde{\epsilon}_{n,t}, \qquad \text{for all } n = 1, \ldots, N. \tag{5.24}$$

An equivalent expression is obtained for the zero-β CAPM.

In the first step of the two pass procedure, on the basis of model (5.24), β coefficients are estimated through the ordinary least squares (OLS) estimator (time series regression). Note, however, that in the presence of heteroskedastic and correlated returns OLS estimates are inefficient relative to generalized least square estimators. In the second step, the average returns of the securities are regressed cross-sectionally against the β coefficients estimated in the first step. Letting $\hat{r}_n$ and $\hat{\beta}_n$ be the expected return and the β coefficient estimated for portfolio p in the first step, we can consider the following regression:

$$\hat{r}_n = \gamma_0 + \gamma_1 \hat{\beta}_n + \tilde{v}_n. \tag{5.25}$$

We can test the following implications of the CAPM on relation (5.25):

- validity of the linear relation (5.25) between the risk premium of an asset/portfolio and the β coefficient;
- $\gamma_0 = r_f$;
- the market return is the only risk factor;
- the market portfolio belongs to the efficient part of the portfolio frontier composed of the risky assets together with the risk free asset and its risk premium is equal to γ_1 (and, therefore, $\gamma_1 > 0$).

Roll [1457] points out that the first three implications are a direct consequence of the fact that the market portfolio belongs to the efficient portfolio frontier. So, the only hypothesis to be tested is that the market portfolio belongs to the efficient portfolio frontier. In other words, the above hypotheses cannot be tested independently.

To test the zero-β CAPM, instead of the (observable) risk free rate there is the expected return of the frontier portfolio having zero correlation with the market portfolio, a portfolio with a non-observable return. In this case, the two pass procedure is similar to the one described above for the CAPM, with the hypotheses to be tested on model (5.25) becoming $\gamma_0 \geq 0$ and $\gamma_1 > 0$.

Early Tests

The CAPM could be directly tested on asset returns: unfortunately, the residual (noise) component $\tilde{\epsilon}_t$ for a single asset is very large and, therefore, the estimates of β turn out to be twisted (errors in variables problem). In Miller & Scholes [1343] it was observed that assets with high (low) β are characterized by a risk premium lower (higher) than that predicted by the CAPM. To overcome this drawback (errors in the estimates of the β of single assets), Black et al. [245] and Fama & MacBeth [678] proposed to proceed to an aggregation in the second step of the procedure by building a set of asset portfolios with dispersed β by sorting assets based on estimated betas. On the one hand this aggregation procedure reduces the measurement errors of the β coefficients, on the other hand it reduces the power of regression tests. Moreover, an aggregation process induces selection bias problems (*data snooping biases*). Grouping assets to build portfolios based on some stocks' empirical characteristics (estimated beta, size, price-earnings ratio) creates potentially significant biases in the test statistics, in particular the null hypothesis is quite likely to be rejected even when it is true (see Lo & MacKinlay [1235]). Note that a two pass approach is by its own nature affected by the errors in variables problem.

Early empirical evidence reported in Black et al. [245], Fama & MacBeth [678], Blume & Friend [258] is consistent with the mean-variance efficiency of the market portfolio: the relation between a portfolio risk premium and the β coefficient is linear and there is no other risk factor. The market risk premium is positive but γ_1 is significantly lower than the market risk premium observed empirically (1.08 instead of 1.42 in Black et al. [245]). Moreover, $\gamma_0 - r_f$ turns out to be significantly different from zero (0.519 in Black et al. [245]). However, these findings can be easily interpreted in the context of the zero-β CAPM.

As an illustrative example, let us give some more details on the results of Fama & MacBeth [678]. Their data sample is given by 1926–1968 NYSE stocks and the market portfolio is the NYSE Index. The return frequency is monthly. They start by considering the period 1926–1929 and perform the first pass for each stock estimating the β coefficients. The assets are ranked according to their β and are partitioned in twenty portfolios. For each portfolio the monthly return is computed for the period 1930–1934. The first pass of the procedure is repeated obtaining an estimate of the β coefficients of the portfolios. The second pass is performed for each month in the 1935–1938 period. The procedure is repeated by rolling the estimation and the testing samples.

The use of a market portfolio proxy to test the CAPM can be the origin of a bias. Roll [1457] points out the problem: a negative empirical evidence for the model described above simply means that the market portfolio proxy does not belong to the frontier and nothing can be established about the validity of the CAPM without knowing the relation between the true market portfolio and its proxy. For example, the author argues that the negative results obtained in Black et al. [245] for the CAPM are compatible with the classical CAPM and an ill-specified market portfolio. A test with a proxy of the market portfolio provides implications on the fact that the true market portfolio belongs to the frontier only if its β coefficient is equal to one and the error terms $\tilde{\epsilon}_{n,t}$ are uncorrelated with the market portfolio return. The relevance of this critique (*Roll's critique*) has been analysed in two directions. In Stambaugh [1561] it is shown that the results of an empirical test do not vary in a sensible way with the composition of the proxy of the market portfolio (adding bonds and real estate to the index). In Shanken [1525] and Kandel & Stambaugh [1067] it is shown that a negative result on the CAPM with respect to a proxy of the market portfolio implies the rejection of the CAPM provided that the correlation between the proxy and the true market portfolio is high enough (larger than 0.7). These results led to reconsider the relevance of Roll's critique.

CAPM restrictions can be tested by estimating (5.24) through the time series approach. Under the hypothesis that the CAPM holds, the coefficient α_n should be equal to zero. This fact is a direct consequence of the fact that the market portfolio (or its proxy) belongs to the efficient part of the portfolio frontier composed by the N risky assets together with the risk free asset. We can also test that the CAPM is correctly specified adding other regressors.

CAPM Anomalies

The empirical evidence for the CAPM in the '70s was substantially positive. However, in the '80s a large amount of literature showed that the return of the market portfolio is not the only risk factor. In particular, some characteristics of the stocks turn out to be significant in order to explain asset risk premia (*CAPM anomalies*). Portfolios composed of assets of companies with some characteristics turn out to have a Sharpe ratio higher than that predicted on the basis of the market portfolio proxy. For a survey on this literature see Hawawini & Keim [916] and Schwert [1510]. With no intent to be exhaustive, the list of anomalies include:

Price earnings ratio: in Basu [179], Ball [121, 122] it is shown that the company's price-earnings ratio turns out to be relevant in order to explain the portfolio risk premium: assets with a low (high) price-earnings ratio have an average return higher (lower) than that predicted by the CAPM.

Size: in Banz [140], Reinganum [1443] it is shown that the size of the company, represented by the market capitalization, explains portfolio risk premia better than the β coefficient. Size as a second risk factor, besides the market return, contributes to explain the residual returns variability. The relation is negative: returns of small companies are higher than those predicted by the CAPM. This result suggests the existence of a positive *small cap premium* which has inspired many small cap funds in the last decades. However, note that Dimson & Marsh [578], Schwert [1510] observe that the small cap premium has been small or even negative in the last twenty years.

January effect: Reinganum [1444], Loughran [1246], Keim [1079] showed that a large part of the abnormal returns on small firms occurs during the first month of the year: short selling pressure on small companies reduces the price of small companies in December with a rebound effect in January. Schwert [1510] has shown that this effect is confirmed in a more recent analysis.

Leverage: in Bhandari [221] it is shown that a positive relationship exists between leverage and asset returns. Employing leverage as a third factor together with the market return and size helps explaining the residual returns variability.

Book to market value ratio: in Stattman [1566] it is shown that the ratio of the book value of equity over its market value is positively correlated with the average return of the assets.

Mean reversion and momentum effect: in De Bondt & Thaler [532] it is shown that the past performance of a company turns out to be significant: portfolios composed of assets with a poor performance in the past three/five years exhibit an average return higher than that predicted by the CAPM (*mean reversion effect*). On the other hand, assets with high returns over the past three to twelve months continue to have high returns in the next future (*momentum effect*), see Jegadeesh [1023], Jegadeesh & Titman [1026].

Liquidity: risk premia are related to market liquidity, expected return is an increasing and concave function in the bid-ask spread and other forms of illiquidity, see Amihud & Mendelson [59], Chalmers & Kadlec [393], Brennan & Subrahmanyam [297], Eleswarapu [635], Amihud [57], Easley et al. [615], Bekaert et al. [185]. This phenomenon agrees with the size anomaly.

There is a high degree of interdependence among the above anomalies. These results were not welcomed by the academic community. In the first place, these results have little theoretical foundation and the performance of multi-factor models (APT for instance) are not necessarily better than that of the CAPM. Moreover, these empirical findings are affected by a series of problems (e.g., data snooping, infrequent trading and liquidity bias of small stocks, selection bias, measurement errors in β). The debate on these topics exploded in the '90s after the seminal contribution Fama & French [669]. According to Fama & French [669], the positive relationship between portfolio risk premia and β established in early contributions is not confirmed in the period 1963–1990 (on this point see also Reinganum [1442]). The average slope from the regression of returns on β is found to be 0.15% per month (with t-statistics 0.46, flat relation). For the sample 1981–1990, the market risk premium is even negative. In a cross-sectional regression of asset returns, the β coefficient is less significant than factors like size (-0.15% as a coefficient with t-statistics 2.58), book to market value of equity ratio (0.5% as a coefficient with t-statistics 5.71), leverage, earnings-price ratio. In particular, the size effect is strong and robust but the book to market value ratio is more powerful than size in explaining the cross-sectional variability. High-beta stocks do not exhibit higher returns than low-beta stocks of the same size or with the same book to market equity ratio. Moreover, adding β as a second factor to size, book to market equity, leverage or earnings-price ratio does not help explaining average returns (in the first case the coefficient of the regression is even negative while the coefficients of the other variables are statistically different from zero). The size of the company and the book to market value ratio obscure the leverage and earnings-price ratio effect. This analysis has shown that firms with high book to market value of equity (low market price relative to the book value of assets), earnings to price or cash flow to price ratios (*value stocks*) have higher average returns than stocks with low ratios (*growth stocks*). Value stocks are good candidates to be undervalued by the market and, on the contrary, growth stocks are good candidate to be overvalued by the market.

Three different schools of thought can be identified on the literature on CAPM anomalies: those who criticize the statistical robustness of the anomalies, those who interpret the evidence inside the classical asset pricing theory (i.e., anomalies subsume a risk factor not captured by β), and those who interpret the evidence as a signal of market irrationality (behavioral finance). In our presentation, we follow Boudoukh et al. [273] identifying these three schools respectively as *loyalists, revisionists* and *heretics*.

Loyalists

The results of Fama & French [669] have been discussed in several papers. In Kothari et al. [1126], three different criticisms are put forward. In the first place, due to the low statistical power of the tests for a positive market risk premium, the results of Fama & French [669] provide little support for rejecting the null hypothesis of a market risk premium equal to the one observed historically. This relies on the presence of a strong noise component in the asset returns. The second critique concerns the fact that an estimate of β based on annual returns (instead of monthly returns as in Fama & French [669]) produces a stronger positive relation between average returns and β (the market risk premium is positive and significant, see also Handa et al. [885]). The third point concerns the data set (COMPUSTAT) employed in a large part of the empirical analysis. In this data set new assets are inserted as time goes on: when an asset is inserted in the data set, its time series is integrated by several years, before the year of insertion. This procedure induces a *survivorship bias*: companies characterized at some date t by a high book to market value and subsequently by small returns have a low survival probability and, therefore, are unlikely to be inserted in the data set at a future time $s > t$. On the contrary, companies with high returns are more likely to survive and to be included in the data set. Adding some years to the time series when an asset is inserted in the data set introduces a bias towards the existence of a positive relation between risk premium and the book to market value ratio. Moreover, some authors observe that the COMPUSTAT database does not include many firms experiencing financial distress. Using a different database, some authors have shown that the relation between risk premium and the book to market value ratio is much weaker than that reported in Fama & French [669]. An insignificant effect of the book to market ratio is detected in Breen & Korajczyk [288].

The results obtained in Fama & French [669] and, in particular, the non-existence of a relationship between the β coefficient and expected returns can be traced back to the statistical procedure employed to test the CAPM. A classical resolution relies on the Roll's critique: if the market portfolio proxy does not belong to the portfolio frontier, then other factors besides β turn out to be significant. Roll & Ross [1462] show that there is a region delimited by a parabola inside the mean-variance frontier containing market index proxies with no relation to expected returns. The parabola is not far away from the mean-variance frontier and, therefore, a slight inefficiency of the market proxy with a market portfolio on the frontier may generate the result of Fama & French [669].

The absence of a relation between the β coefficient and expected returns is not confirmed using *generalized least squares* instead of ordinary least squares. In this case, the estimated impact of the β coefficient on expected returns is strong. Using this method, it can be shown that there exists a positive quasi-linear relationship (also under a grossly inefficient market portfolio proxy), with a slope equal to zero occurring only when the mean return of the market proxy is equal to that of the global minimum variance portfolio, see Kandel & Stambaugh [1070] and Lewellen et al. [1213]. The relation becomes linear as the market proxy tends to the efficient frontier and the R^2 coefficient is positively related to the efficiency of the portfolio (on maximum likelihood estimators of the CAPM see Shanken & Zhou [1528]).

In Kim [1091, 1092] it shown that the results obtained in Fama & French [669] are biased because of the two-pass procedure employed to test the CAPM. β estimate errors in the first step result in an underestimation of the price of β risk and in an overestimation of other factors associated with variables observed without error (size, book to market value ratio). Taking into account this phenomenon, the β coefficient has a statistically significant explanatory power for average returns and the size factor becomes much less relevant and in some cases even insignificant (see also Jegadeesh [1025]). On the other hand, book to market value of equity ratio is still significant. Kan & Zhang [1063] show that, due to the error in variables problem, a useless factor (a factor independent of all asset returns) may be priced according to a two pass procedure. On the weak statistical significance of tests showing the capability of explanatory variables (other than the β coefficient) to explain cross-sectionally risk premia see Lewellen et al. [1213]. In Knez & Ready [1107] it is shown that size completely disappears when outliers are removed from the database. Evidence that the book to market effect is driven by extreme observations is also provided. Black [244] and MacKinlay [1269] suggest that the anomalies can be traced back to the statistical methodology employed (e.g., data snooping).

The relevance of the survivorship bias has been discussed in several papers. Using a database not affected by the survivorship bias, Davis [526] shows that the book to market value of equity ratio and the earnings-price ratio are significant for explaining asset returns. In Chan et al. [406], Fama & French [672], Kim [1092] it is shown that the survivorship bias of the COMPUSTAT database does not suffice to explain the relationship between book to market value and risk premium. Moreover, annual and monthly β estimates produce the same inferences about the existence of a positive risk premium. The size effect and the bad performance of the β coefficient in a cross-sectional regression are confirmed in a data set immune to the problems associated with the COMPUSTAT data set, see Breen & Korajczyk [288].

Revisionists and Heretics

The literature surveyed above has shown that some of the anomalies put forward in Fama & French [669] can be explained only partially through biases associated to the statistical procedure employed in testing the CAPM. This evidence opened the door to a debate, with on the one side those in favor of the hypothesis that the anomalies provide evidence of the fact that other factors than the market return are priced by the market and on the other side those supporting the idea of market irrationality.

In Ball [121], Chan et al. [402], Chan & Chen [401], Fama & French [670, 671, 673], Berk [197], Chen & Zhang [425], Cochrane [464], Berk et al. [199], Cohen et al. [472], Vassalou & Xing [1613], Pástor & Veronesi [1408], the connection between book to market value ratio, size, earnings and returns is justified by arguing that these variables (in particular prices) capture some risk components not represented by the β coefficient. These variables capture the financial distress conditions of a company (high book to market value ratio and small size) and are good proxies of future earnings. According to Fama & French [671], the positive relation between book to market value ratio and risk premia can be explained as follows: high book to market value companies (*value companies*) are less profitable than low book to market value companies (*growth companies*) for at least five years before and after portfolios are formed. As a consequence, in equilibrium high book to market value companies are characterized by high expected returns. Moreover, high book to market value companies are riskier than low book to market value companies, thus yielding a distress premium. A similar interpretation has been proposed for the size of a company in Chan & Chen [401], Berk [197]. Campbell & Vuolteenaho [358] and Campbell et al. [350] show that the cash flows of value (growth) stocks are particularly sensitive to permanent (temporary) movements in aggregate stock prices. They conclude that systematic risks of individual stocks with similar accounting characteristics are primarily driven by the systematic risks of their fundamentals.

Gomes et al. [810], Berk et al. [199], Santos & Veronesi [1498], Zhang [1680] theoretically show that the cross-sectional dependency of returns to book to market ratio and to size can be rationalized in terms of a β measurement problem in an equilibrium perspective. Gomes et al. [810] provide a general equilibrium model linking expected returns to firm characteristics such as size and book to market ratio. A one-factor model with the market portfolio as the only factor is consistent with the fact that the above firm characteristics predict future returns since they are correlated with the true (stochastic) β coefficient. Berk et al. [199] provide a similar argument on the basis of a two factor model with time varying risk. Changes in risk are linked to firm-specific variables generating the book to market and the size effect. According to these works, there are no additional risk factors, factor loadings are only due to measurement errors and changes in investment opportunities and apparent risk factors are only related to the market factor.

According to this interpretation, there is a common variation in earnings which is not captured by the β coefficient. On the basis of this observation, Fama & French [670, 673] propose a three factor linear model: a factor associated with the market portfolio, a factor associated with firm size (excess return on a portfolio of small firms over a portfolio of large firms) and a factor associated with the book to market value ratio (excess return on a portfolio of high book to market stocks over a portfolio composed of low book to market stocks). These factors are priced (with positive risk premium) and capture the size, book to market and the other anomalies highlighted in Fama & French [669], thus yielding a large R^2 coefficient in the cross-sectional regression. Extending the relation (5.24), the model tested by the authors is

$$\tilde{r}_{n,t} - r_f = \alpha_n + \gamma_{nm}(\tilde{r}_t^m - r_f) + \gamma_{ns}\text{SMB}_t + \gamma_{nh}\text{HML}_t + \tilde{\epsilon}_{n,t},$$

for all $n = 1, \ldots, N$, where SMB_t is the *small minus big* market capitalization risk factor and HML_t is the *high minus low* value premium risk factor. SMB_t measures the return of a portfolio invested in stocks of companies with relatively small market capitalization, while HML_t measures the return of a portfolio invested in companies with high levels of the book to market value of equity ratio. The abnormal returns resulting from the three factor model are not significantly different from zero for portfolios sorted by size, book to market value, earnings price ratio. The three factor model captures the long term reversals of returns but not the momentum effect, see Fama & French [673] and Brennan et al. [293]. In Carhart [367], a fourth factor associated to the stock performance over the past few months is included in order to capture the momentum effect. Note that the size and book to market premia seem to have diminished over time (see Cochrane [464]). The risk factor interpretation can be rationalized inside the APT (common factors in shocks to expected earnings) or the ICAPM (variation in the investment opportunity set). Brennan et al. [299], Liew & Vassalou [1218], Petkova [1416], Petkova & Zhang [1417], Lettau & Wachter [1201], Bansal et al. [134], Lettau & Ludvigson [1195], Santos & Veronesi [1497] find empirical support for the second interpretation, while Chen [416], including also the momentum effect, finds no support for it (only size may proxy changes of the investment opportunity set). There is little evidence that momentum is a risk factor.

An alternative interpretation of these anomalies is based on the presence of irrational agents in the market, see De Bondt & Thaler [532, 533], Lakonishok et al. [1159], Haugen & Baker [915] and Chap. 9. In this perspective, anomalies are due to irrational traders who extrapolate the strong earnings growth of low book to market value companies too far into the future (high market price) and the poor growth of high book to market value companies (*overreaction*). Low book to market value companies then have low average returns because future earnings growth is weaker than expected and high book to market value companies have high average returns because future earnings growth is stronger than expected.

The two interpretations described above go in opposite directions. The first one (*revisionists*) brings the anomalies back to risk factors and, therefore, in the context of classical asset pricing theory, proposing a multi-factor model for asset returns. The second interpretation (*heretics*) emphasizes agents' irrationality and, therefore, the incapacity of classical asset pricing theory to explain asset risk premia (the stock market is not efficient and assets are incorrectly priced). The empirical analysis in Davis et al. [527], Fama & French [670, 671, 673], Lewellen [1209], Cohen et al. [472], Fama & French [676] favors the first interpretation. MacKinlay [1269] concludes that multi-factor pricing models alone are not enough to solve CAPM anomalies, there is room for other explanations: market imperfections, irrational agents or the inefficiency of the statistical methodology. Moskowitz [1354], analysing the contribution of size, book to market and momentum to the covariance matrix of asset returns, finds that size and market portfolio have a significant explanatory power, while the contribution of the book to market ratio is weak and that of momentum is negligible.

In Haugen & Baker [915], Lakonishok et al. [1159], La Porta et al. [1168], Griffin & Lemmon [826], the empirical evidence favors a market overreaction hypothesis. The underperformance of stocks with low book to market ratios concentrates on earnings announcement dates. In La Porta [1167] a similar conclusion is reached analysing expectations by stock market analysts on earnings growth rates, showing that analysts' forecasts about future earnings growth are too extreme because they excessively extrapolate past earnings movements (i.e., analysts overreact). A conclusion against the multi-factor interpretation is also provided in Daniel & Titman [520], Daniel et al. [523], Daniel & Titman [522]: the authors argue that the three-factor model proposed in Fama & French [670] does not capture risk components (there are no premia for these factors), instead the model appears to explain average returns only because the factor loadings are correlated with firm characteristics (size and book to market value ratio). It is simply the characteristics of the firm and not the covariance of its return with the factor that determine the expected stock return, e.g. firms grouped according to similar characteristics show similar properties. Evidence is provided showing that, after controlling for the size and book-to-market ratio effects, returns are not related to loadings of the above multi-factor model. Baker & Wurgler [110] show that investor sentiment has strong effects on the cross-section of stock prices. This evidence agrees with the irrational interpretation of the anomalies: firms with similar characteristics become incorrectly priced at the same time. Ferson et al. [697], Daniel & Titman [520] and Kan & Zhang [1063] show that attribute-sorted portfolios chosen according to an empirically observed relation to cross-section of stock returns can appear to be risk factors even when this is not the case.

The Conditional CAPM

The above empirical tests concentrate on an *unconditional* CAPM. However, the CAPM may fail unconditionally but nevertheless hold conditionally. The CAPM has been tested allowing for time-varying β, time-varying market variance and time-varying risk premia (*conditional CAPM*; see Ang & Liu [69] for an analytical model). Allowing for time-varying β and expected returns, the conditional mean-variance efficiency of the stock index is rejected in Ferson et al. [694], but a single risk premium model is not rejected if its premium is time-varying and the factor is not restricted to correspond to a market factor. In Bollerslev et al. [264], the CAPM is tested assuming a GARCH model for asset returns and it is shown that the conditional covariance of asset returns with the market portfolio return changes significantly over time with an auto-regressive component and the time variability is significant in order to explain time-varying asset risk premia (time-varying β). Empirical evidence against the conditional CAPM (time-varying expected returns and co-variances) is provided in Bodurtha & Mark [261], Harvey [908, 909], Lewellen & Nagel [1212], Adrian & Franzoni [25], while more positive results are obtained in Brennan et al. [299], Liew & Vassalou [1218], Petkova [1416], Petkova & Zhang [1417], Lettau & Wachter [1201], Vassalou [1612], Avramov & Chordia [93]. Also allowing for time-varying risk premia, the conditional CAPM does not seem to be able to capture size and book to market effects, see He et al. [921] (positive evidence is however reported in Ang & Chen [66]). Allowing for heteroskedasticity in aggregate stock returns and time-varying β, results against the CAPM are obtained (large size effect) in Schwert & Seguin [1511]. These works are unanimous in establishing that conditional co-variances change over time (we will come back to these topics in Sect. 7.3).

The market portfolio misspecification hypothesis has been further investigated in Jagannathan & Wang [1012]. In the papers discussed above, the proxy of the market portfolio is usually given by a stock index. In Jagannathan & Wang [1012], the conditional CAPM allowing for a time-varying β coefficient and expected returns is tested by also considering human capital (labor income growth is a proxy for the return on human capital). Compared with the 1% of the cross-sectional variation in average returns explained by the classical CAPM, the conditional CAPM explains 30% allowing for time-varying β and, when human capital is included, the percentage of explained variation is larger than 50%, with size and book to market value having little explanatory power. The model allows for three betas (for the market, human capital and time variability). Positive results for the conditional CAPM including human capital have been obtained in Jagannathan et al. [1010], Campbell [333] and in Lettau & Ludvigson [1195], Santos & Veronesi [1497] by handling wealth or labor income over total consumption as a conditional variable. These models are able to capture the value effect.

5.4   The Arbitrage Pricing Theory (APT)

The Capital Asset Pricing Model provides a linear relation between the asset risk premia and the risk premium of a single risk factor, represented by the market portfolio. Note also that, in order to derive the CAPM relation, we had to make assumptions on the agents' utility functions and/or on the distribution of the asset returns or, as in Sect. 5.2, on the fact that all the agents hold portfolios belonging to the portfolio frontier. As we have seen in the preceding section, several empirical tests of the CAPM suggest that a multi-factor model is needed to explain asset returns. In the present section, we shall establish a linear relation between the asset risk premia and several risk factors by relying on the absence of arbitrage opportunities and on the existence of a linear model generating asset returns, without explicit assumptions on the agents' utility functions. Such a relation is known as *Arbitrage Pricing Theory (APT)* and goes back to the seminal contributions of Ross (see Ross [1464, 1466]).

The existence of a linear multi-factor model generating asset returns means that the return of every asset can be written as a linear combination of a finite number of random variables (*risk factors*) together with a random component specific for each asset (often called the *idiosyncratic* risk component). As the name suggests, the APT is based on the assumption of absence of arbitrage opportunities. In particular, when there is no idiosyncratic risk component, the APT provides an exact linear relationship between asset risk premia and risk factors. In the presence of an idiosyncratic risk, an analogous relation holds in an approximate form and, under suitable additional assumptions, even in an exact form.

Let us start by discussing the case where there are no idiosyncratic risk components. We assume that N risky assets are traded in the economy, together with a risk free asset with return $r_f > 0$. The returns of the N risky assets are generated by a linear model, with respect to K risk factors $(\tilde{f}_1, \ldots, \tilde{f}_K)$:

$$\tilde{r}_n = \mathbb{E}[\tilde{r}_n] + \sum_{k=1}^{K} b_{nk} \tilde{f}_k, \qquad \text{for all } n = 1, \ldots, N, \tag{5.26}$$

where (without loss of generality) $\mathbb{E}[\tilde{f}_k] = 0$, for all $k = 1, \ldots, K$. The coefficient b_{nk} is the *factor loading* of the risk factor $\tilde{f}_k$ on the n-th asset, for $k = 1, \ldots, K$ and $n = 1, \ldots, N$. Of course, for the linear model (5.26) to be of interest, we assume that $N > K$. Letting $\tilde{f} = (\tilde{f}_1, \ldots, \tilde{f}_K)$, relation (5.26) can be written in vector notation as

$$\tilde{r} = \mathbb{E}[\tilde{r}] + B\tilde{f},$$

where $\mathbb{E}[\tilde{r}]$ is the vector of the expected returns of the N assets and $B \in \mathbb{R}^{N \times K}$ is the matrix of the factor loadings.

Under the assumption of absence of arbitrage opportunities (see Definition 4.17), the following proposition shows that the linear model (5.26) implies that asset risk premia are determined by a linear relationship with respect to the factor loadings matrix B.

Proposition 5.13 *Suppose that asset returns are generated by the linear model (5.26) and that there are no arbitrage opportunities. Then there exists a vector $\lambda \in \mathbb{R}^K$ such that*

$$\mathbb{E}[\tilde{r}] - r_f \mathbf{1} = B\lambda. \tag{5.27}$$

The vector λ satisfies $\lambda_k = \text{Cov}(\tilde{\ell}, \tilde{f}_k)$, for all $k = 1, \ldots, K$, where $\tilde{\ell}$ represents the likelihood ratio of some risk neutral probability measure. Moreover, under the *non-redundancy condition* $\text{rank}(B) = K$, *the quantity* $\text{Cov}(\tilde{\ell}, \tilde{f}_k)$ *does not depend on the specific risk neutral probability measure chosen.*

Proof For any portfolio $w \in \mathbb{R}^N$ (representing the proportions of wealth invested in the N risky assets, thereby normalizing the initial wealth to 1), the associated return is given by $w^\top \tilde{r} = w^\top \mathbb{E}[\tilde{r}] + w^\top B\tilde{f}$. Considering a portfolio w such that $w^\top 1 = 0$ (zero cost at $t = 0$) and $w^\top B = 0$, the absence of arbitrage opportunities implies that $w^\top \mathbb{E}[\tilde{r}] = 0$ (otherwise it would be possible to create a portfolio with zero cost and strictly positive return, contradicting the absence of arbitrage opportunities). Equivalently, this means that $\ker(B^\top) \cap \ker(1^\top) \subseteq \ker(\mathbb{E}[\tilde{r}]^\top)$. Hence, there exists a couple $(\lambda, \lambda_0) \in \mathbb{R}^{K+1}$ such that

$$\mathbb{E}[\tilde{r}] = 1\lambda_0 + B\lambda.$$

If the kernel of $B^\top$ is a subset of the kernel of $1^\top$, then it holds that $1 = B\gamma$, for some $\gamma \in \mathbb{R}^K$. In this case, the last relation implies that

$$\mathbb{E}[\tilde{r}] = B(\lambda_0 \gamma + \lambda) = B(\lambda_0 \gamma - r_f \gamma + \lambda) + r_f 1,$$

thus proving relation (5.27). Suppose now that the kernel of $B^\top$ is not a subset of the kernel of $1^\top$, so that there exists a vector $w \in \mathbb{R}^N$ such that $w^\top B = 0$ but $w^\top 1 \neq 0$. We claim that, due to the absence of arbitrage opportunities, it holds that $\lambda_0 = r_f$. Indeed, suppose on the contrary that $\lambda_0 > r_f$ (the case $\lambda_0 < r_f$ can be treated in an analogous way). Let $\bar{w} := w/(w^\top 1)$ and consider the following strategy: borrow one unit of wealth at the risk free rate r_f and invest that unit of wealth in the N risky assets according to the portfolio $\bar{w}$. This strategy has zero cost at time $t = 0$ and delivers a return of $\bar{w}^\top 1(\lambda_0 - r_f) = \lambda_0 - r_f > 0$ at time $t = 1$, thus contradicting the absence of arbitrage opportunities.

In order to prove the second part of the proposition, recall that, in view of Proposition 4.22, the absence of arbitrage opportunities is equivalent to the existence of a risk neutral probability measure π^* (or, equivalently, of a corresponding likelihood ratio $\tilde{\ell}$). Hence, in view of equation (4.35) together with the linear model (5.26), we have, for all $n = 1, \ldots, N$,

$$r_f = \mathbb{E}^*[\tilde{r}_n] = \mathbb{E}[\tilde{r}_n] + \sum_{k=1}^K b_{nk} \mathbb{E}^*[\tilde{f}_k] = \mathbb{E}[\tilde{r}_n] + \sum_{k=1}^K b_{nk} \text{Cov}(\tilde{\ell}, \tilde{f}_k),$$

where we have used the fact that $\mathbb{E}^*[\tilde{f}_k] = \mathbb{E}[\tilde{\ell}\tilde{f}_k] = \text{Cov}(\tilde{\ell}, \tilde{f}_k) + \mathbb{E}[\tilde{\ell}]\mathbb{E}[\tilde{f}_k]$ together with the assumption that $\mathbb{E}[\tilde{f}_k] = 0$, for all $k = 1, \ldots, K$. Finally, it remains to show that the quantity $\text{Cov}(\tilde{\ell}, \tilde{f}_k)$ does not depend on the specific choice of the likelihood ratio. Indeed, as in Lemma 5.5, let $\hat{\ell}$ be the unique element of $I(D)$ such that $Q(c) = \mathbb{E}[\hat{\ell}\tilde{c}]$, for all contingent consumption plans $\tilde{c}$ taking values $c \in I(D)$. Then, arguing as in the proof of Proposition 5.6, it can be shown that

$$r_f = \frac{\mathbb{E}[\hat{\ell}\tilde{r}_n]}{\mathbb{E}[\hat{\ell}]} = \frac{\mathbb{E}[\hat{\ell}]\mathbb{E}[\tilde{r}_n] + \text{Cov}(\hat{\ell}, \tilde{r}_n)}{\mathbb{E}[\hat{\ell}]} = \mathbb{E}[\tilde{r}_n] + r_f \sum_{k=1}^K b_{nk} \text{Cov}(\hat{\ell}, \tilde{f}_k).$$

The last two equations, together with the non-redundancy condition rank$(B) = K$, imply that $\text{Cov}(\tilde{\ell}, \tilde{f}_k) = r_f \, \text{Cov}(\hat{\ell}, \tilde{f}_k)$, for all $k = 1, \ldots, K$, for any likelihood ratio $\tilde{\ell}$ (and, equivalently, for any risk neutral probability measure). $\qquad\square$

According to the above proposition, the risk premium of the n-th asset satisfies the following linear relation in an exact form:

$$\mathbb{E}[\tilde{r}_n] - r_f = \sum_{k=1}^{K} b_{nk} \, \lambda_k, \qquad \text{for all } n = 1, \ldots, N.$$

The coefficient λ_k admits the interpretation of a risk premium associated to the risk factor $\tilde{f}_k$. The risk premium λ_k is expressed in terms of the covariance between the risk factor $\tilde{f}_k$ and the vector $\hat{\ell}$ appearing in the Riesz representation of a pricing functional (see Lemma 5.5). In particular, this shows that the linear relationship (5.27) does not rely on market completeness. Note also that, as can be seen by examining the proof of Proposition 5.13, if a risk free asset is not traded in the market, then we can still prove the existence of a couple $(\lambda_0, \lambda) \in \mathbb{R}^{K+1}$ such that $\mathbb{E}[\tilde{r}] = \mathbf{1}\lambda_0 + B\lambda$.

As in the case of the CAPM relation (see formula (5.22)), Proposition 5.13 allows to represent the price p_n of the n-th asset as the expectation of the dividend discounted by the risk adjusted discount factor $r_f + \sum_{k=1}^{K} b_{nk} \lambda_k$, i.e.,

$$p_n = \frac{\mathbb{E}[\tilde{d}_n]}{r_f + \sum_{k=1}^{K} b_{nk} \lambda_k}, \qquad \text{for all } n = 1, \ldots, N.$$

Let us now extend the previous analysis by allowing for the presence of idiosyncratic risk components in the linear multi-factor model (5.26). In this case, as will be shown below, a linear relation between the asset risk premia and the factor loadings holds in an approximate form and, under suitable assumptions, even in exact form.

In order to establish an approximate linear relation, we consider a sequence of economies indexed by the number $N \in \mathbb{N}$ of available securities (for the moment, we do not assume that a risk free asset is traded), i.e., we consider a sequence of economies with an increasing number of assets. For each $N \in \mathbb{N}$, the asset returns in the N-th economy are assumed to satisfy the following linear model (the superscript N denotes the N-th economy):

$$\tilde{r}_n^N = \mathbb{E}[\tilde{r}_n^N] + \sum_{k=1}^{K} b_{nk}^N \tilde{f}_k^N + \tilde{\epsilon}_n^N, \qquad \text{for all } n = 1, \ldots, N, \tag{5.28}$$

where, for all $N \in \mathbb{N}$, we assume that $\mathbb{E}[\tilde{\epsilon}_n^N] = \mathbb{E}[\tilde{\epsilon}_m^N \tilde{\epsilon}_n^N] = 0$, for all $m, n = 1, \ldots, N$ with $m \neq n$, $\mathbb{E}[\tilde{f}_k^N] = 0$, for all $k = 1, \ldots, K$, and $\text{Var}(\tilde{\epsilon}_n^N) \leq \bar{\sigma}^2$, for all $n = 1, \ldots, N$, for some constant $\bar{\sigma}^2$. These assumptions imply that the idiosyncratic risk components are uncorrelated among themselves (note, however, that they can be correlated with the risk factors) and have uniformly bounded variances. The linear model (5.28) can be expressed in vector notation as follows:

$$\tilde{r}^N = \mathbb{E}[\tilde{r}^N] + B^N \tilde{f}^N + \tilde{\epsilon}^N,$$

where, for all $N \in \mathbb{N}$, $\mathbb{E}[\tilde{r}^N]$ is the vector of the expected returns, $B^N \in \mathbb{R}^{N \times K}$ is the factor loadings matrix and $\tilde{\epsilon}^N$ is an N-dimensional random vector with zero mean and uncorrelated components. For any given portfolio $w^N \in \mathbb{R}^N$, representing the proportions of wealth invested in the N available assets in the N-th economy, the associated return is given by

$$(w^N)^\top \tilde{r}^N = (w^N)^\top \mathbb{E}[\tilde{r}^N] + (w^N)^\top B^N \tilde{f}^N + (w^N)^\top \tilde{\epsilon}^N.$$

In the present context, we can formulate the following definition, see Huberman [979]. For $N \in \mathbb{N}$, we denote by $\mathbf{1}_N$ the unit vector in $\mathbb{R}^N$.

Definition 5.14 A sequence of portfolios $\{w^N\}_{N \in \mathbb{N}}$ such that $(w^N)^\top \mathbf{1}_N = 0$ is said to be an *asymptotic arbitrage opportunity* if the following hold:

$$\lim_{N \to \infty} \mathbb{E}[(w^N)^\top \tilde{r}^N] = +\infty \quad \text{and} \quad \lim_{N \to \infty} \text{Var}((w^N)^\top \tilde{r}^N) = 0.$$

Intuitively, an asymptotic arbitrage opportunity represents a sequence of investment strategies such that the expected return explodes while the associated risk (as measured by the variance) decreases to zero as the dimension of the economy increases. We have the following proposition (see Ross [1464] and Huberman [979]).

Proposition 5.15 *Suppose that asset returns are generated by the linear model (5.28), for all $N \in \mathbb{N}$, and that there are no asymptotic arbitrage opportunities. Then, for each $N \in \mathbb{N}$, there exists a couple $(\lambda_0^N, \lambda^N) \in \mathbb{R}^{K+1}$ and a positive constant A such that*

$$\sum_{n=1}^{N} \left(\mathbb{E}[\tilde{r}_n^N] - \lambda_0^N - \sum_{k=1}^{K} b_{nk}^N \lambda_k^N \right)^2 \leq A, \quad \text{for all } N \in \mathbb{N}. \tag{5.29}$$

Proof For each $N \in \mathbb{N}$, take the orthogonal projection of the vector $\mathbb{E}[\tilde{r}^N]$ onto the linear space spanned by $\mathbf{1}_N \in \mathbb{R}^N$ and the columns of the matrix B^N, so that

$$\mathbb{E}[\tilde{r}^N] = \lambda_0^N \mathbf{1}_N + B^N \lambda^N + c^N,$$

where $(\lambda_0^N, \lambda^N) \in \mathbb{R}^{K+1}$, for some vector $c^N \in \mathbb{R}^N$ such that $(c^N)^\top \mathbf{1}_N = 0$ and $(c^N)^\top B^N = 0$. Note that

$$\|c^N\|^2 := \sum_{n=1}^{N} (c_n^N)^2 = \sum_{n=1}^{N} \left(\mathbb{E}[\tilde{r}_n^N] - \lambda_0^N - \sum_{k=1}^{K} b_{nk}^N \lambda_k^N \right)^2.$$

Arguing by contradiction, suppose that (5.29) does not hold. Then there exists an increasing subsequence $\{N'\}_{N' \in \mathbb{N}}$ such that

$$\lim_{N' \to \infty} \|c^{N'}\|^2 = +\infty. \tag{5.30}$$

For each $N' \in \mathbb{N}$, define a portfolio $d^{N'} := \alpha_{N'} c^{N'}$, where $\alpha_{N'} := \|c^{N'}\|^{2p}$ for some $p \in (-1, -1/2)$. By the properties of c^N, it holds that $(d^{N'})^\top \mathbf{1}_{N'} = 0$. Noting that $(c^{N'})^\top B^{N'} = 0$, the return of the portfolio $d^{N'}$ is given by

$$(d^{N'})^\top \tilde{r}^{N'} = \alpha_{N'} \|c^{N'}\|^2 + \alpha_{N'} (c^{N'})^\top \tilde{\epsilon}^{N'}.$$

By definition of $\alpha_{N'}$, it holds that $\mathbb{E}[(d^{N'})^\top \tilde{r}^{N'}] = \|c^{N'}\|^{2(1+p)}$, so that, by (5.30), we have that

$$\lim_{N' \to \infty} \mathbb{E}[(d^{N'})^\top \tilde{r}^{N'}] = +\infty$$

and that

$$\mathrm{Var}\big((d^{N'})^\top \tilde{r}^{N'}\big) \leq \bar{\sigma}^2 \alpha_{N'}^2 \|c^{N'}\|^2 = \bar{\sigma}^2 \|c^{N'}\|^{2+4p}$$

and, therefore,

$$\lim_{N' \to \infty} \mathrm{Var}\big((d^{N'})^\top \tilde{r}^{N'}\big) = 0.$$

In view of Definition 5.14, we have thus shown that the failure of (5.29) leads to an asymptotic arbitrage opportunity, thus proving the proposition. □

Proposition 5.15 implies that, in an economy comprising a large number of traded assets, for most of the assets the expected return is approximately linear with respect to the factor loadings matrix, i.e.,

$$\mathbb{E}[\tilde{r}^N] \approx \mathbf{1}\lambda_0^N + B^N \lambda^N.$$

Due to (5.29), the quality of such a linear approximation improves as the dimension of the economy increases. Equivalently, the mean quadratic error of the linear approximation for the expected returns is smaller than A/N and, therefore, it decreases to zero as N increases to infinity. This result hinges on the crucial assumption that the variances of the idiosyncratic risk components ($\tilde{\epsilon}^N$) are uniformly bounded, so that, in the limit, the idiosyncratic risk can be diversified away by choosing a sequence $\{w^N\}_{N \in \mathbb{N}}$ of portfolios such that $\lim_{N \to \infty} \|w^N\|^2 = 0$.

As shown in Huberman [979], if the economies are stationary, in the sense that the vector of expected returns $\mathbb{E}[\tilde{r}]$ as well as the factor loadings matrix B do not depend on N, then in the absence of asymptotic arbitrage opportunities it holds that

$$\sum_{n=1}^{\infty}\left(\mathbb{E}[\tilde{r}_n] - \lambda_0 - \sum_{k=1}^{K} b_{nk}\lambda_k\right)^2 < \infty.$$

Let us now extend the result of Proposition 5.15 to the case where, besides the risky assets, there also exists a risk free asset. More precisely, for all $N \in \mathbb{N}$, we assume that in the N-th economy there is a risk free asset with rate of return $r_f^N > 0$. The proof of the following proposition is analogous to that of Proposition 5.15 (see Exercise 5.16).

Proposition 5.16 *Suppose that asset returns are generated by the linear model* (5.28), *for each* $N \in \mathbb{N}$, *and that there are no asymptotic arbitrage opportunities. Assume furthermore that, for each* $N \in \mathbb{N}$, *a risk free asset with rate of return* $r_f^N > 0$ *is traded in the N-th economy. Then, for each* $N \in \mathbb{N}$, *there exists a vector* $\lambda^N \in \mathbb{R}^K$ *and a positive constant A such that*

$$\sum_{n=1}^{N}\left(\mathbb{E}[\tilde{r}_n^N] - r_f^N - \sum_{k=1}^{K} b_{nk}^N \lambda_k^N\right)^2 \leq A, \qquad for\ all\ N \in \mathbb{N}. \qquad (5.31)$$

In the above approximate relations, the quantity λ^N can be interpreted as the vector of risk premia associated to the risk factors. Moreover, if there exist K portfolios such that, for all $k = 1, \ldots, K$, the return of the k-th portfolio is perfectly correlated (in the limit) with the k-th factor and has zero exposure with all the other risk factors (*mimicking portfolio*), then λ_k^N can be interpreted as the risk premium of such a portfolio, see Ingersoll [999] and Admati & Pfleiderer [17]. Note that, in general, the uniqueness of the coefficients is not guaranteed.

Let us now consider an economy with a fixed number $N \in \mathbb{N}$ of risky assets and a risky free asset (for simplicity of notation, we shall omit the superscript N in the notation of (5.28)). In this economy with N risky assets and a risk free asset, the notion of arbitrage opportunity is defined as in Definition 4.17. Recall that, due to Proposition 4.22, the absence of arbitrage opportunities is equivalent to the existence of a risk neutral probability measure. Hence, denoting by $\tilde{\ell}$ the likelihood ratio of some risk neutral probability measure, then, equation (4.35) together with (5.28) gives

$$r_f = \mathbb{E}[\tilde{\ell}\,\tilde{r}_n] = \mathbb{E}[\tilde{r}_n] + \sum_{k=1}^{K} b_{nk}\mathbb{E}[\tilde{\ell}\tilde{f}_k] + \mathbb{E}[\tilde{\ell}\tilde{\varepsilon}_n], \qquad for\ all\ n = 1, \ldots, N. \quad (5.32)$$

Recall also that, in view of Lemma 5.5, there exists a unique vector $\hat{\ell} \in I(D)$ such that $Q(c) = \mathbb{E}[\hat{\ell}\,\tilde{c}]$ for all random variables $\tilde{c}$ with values $c \in I(D)$ and for any pricing functional $Q(\cdot)$. Hence, we can derive a relation analogous to (5.32) in terms of $\hat{\ell}$ as follows (compare with the proof of Proposition 5.6):

$$r_f = \frac{\mathbb{E}[\hat{\ell}\tilde{r}_n]}{\mathbb{E}[\hat{\ell}]} = \mathbb{E}[\tilde{r}_n] + \sum_{k=1}^{K} b_{nk} \frac{\mathbb{E}[\hat{\ell}\tilde{f}_k]}{\mathbb{E}[\hat{\ell}]} + \frac{\mathbb{E}[\hat{\ell}\tilde{\epsilon}_n]}{\mathbb{E}[\hat{\ell}]}, \qquad \text{for all } n = 1, \dots, N.$$

(5.33)

The above relations directly lead to the following result, which gives a sufficient condition for the validity of the APT relation in exact form (compare also with Proposition 5.13 and see also LeRoy & Werner [1191, Chapter 20]).

Proposition 5.17 *Suppose that asset returns are generated by the linear model (5.28) and that the residual risk $\tilde{\epsilon}_n$ is uncorrelated with the factors $(\tilde{f}_1, \dots, \tilde{f}_K)$, for all $n = 1, \dots, N$. Assume furthermore that there are no arbitrage opportunities and that the likelihood ratio ℓ of some risk neutral probability measure belongs to $\mathrm{span}(1, \tilde{f}_1, \dots, \tilde{f}_K)$. Then the following relation holds:*

$$\mathbb{E}[\tilde{r}_n] = r_f - \sum_{k=1}^{K} b_{nk} \mathbb{E}[\tilde{\ell}\tilde{f}_k] = r_f - \sum_{k=1}^{K} b_{nk} \mathrm{Cov}(\tilde{\ell}, \tilde{f}_k), \qquad \text{for all } n = 1, \dots, N.$$

Proof Note that, since $\tilde{\ell} \in \mathrm{span}(1, \tilde{f}_1, \dots, \tilde{f}_K)$ and

$$\mathbb{E}[\tilde{\epsilon}_n] = \mathbb{E}[\tilde{\epsilon}_n \tilde{f}_k] = \mathrm{Cov}(\tilde{\epsilon}_n, \tilde{f}_k) = 0,$$

for all $k = 1, \dots, K$ and $n = 1, \dots, N$, it holds that $\mathbb{E}[\tilde{\ell}\tilde{\epsilon}_n] = 0$. The claim then follows directly from equation (5.32). $\square$

Note that, in view of Lemma 5.5 and equation (5.33), the APT relation in exact form obtained in the above proposition can also be established under the assumption that the vector $\hat{\ell}$ appearing in the Riesz representation of a pricing functional belongs to $\mathrm{span}(1, \tilde{f}_1, \dots, \tilde{f}_K)$. If $\hat{\ell} \notin \mathrm{span}(1, \tilde{f}_1, \dots, \tilde{f}_K)$ then the APT relation only holds as an approximation, with the error term being given by $\mathbb{E}[\hat{\ell}\tilde{\epsilon}_n]/\mathbb{E}[\hat{\ell}]$, for $n = 1, \dots, N$. In view of equation (5.33), we call *pricing error* the quantity $\|\mathbb{E}[\tilde{r}] - r_f \mathbf{1} + B\,\mathbb{E}[\hat{\ell}\tilde{f}]/\mathbb{E}[\hat{\ell}]\|$. The following proposition provides an upper bound for the pricing error (see LeRoy & Werner [1191, Section 20.4]).

Proposition 5.18 *Suppose that asset returns are generated by the linear model (5.28) and that the residual risk $\tilde{\epsilon}_n$ is uncorrelated with the factors $(\tilde{f}_1, \dots, \tilde{f}_K)$, for all $n = 1, \dots, N$. Assume furthermore that there are no arbitrage opportunities. Then the pricing error satisfies the following bound:*

$$\sqrt{\sum_{n=1}^{N} \left| \mathbb{E}[\tilde{r}_n] - r_f + \sum_{k=1}^{K} b_{nk} \frac{\mathbb{E}[\hat{\ell}\tilde{f}_k]}{\mathbb{E}[\hat{\ell}]} \right|^2} \leq \frac{\overline{\sigma}}{|\mathbb{E}[\hat{\ell}]|} \sqrt{\mathbb{E}[(\hat{\ell} - \widehat{\ell^f})^2]},$$

(5.34)

where $\hat{\ell} \in I(D)$ is the vector appearing in the Riesz representation of a pricing functional Q (see Lemma 5.5) and $\widehat{\ell^f}$ is the orthogonal projection of $\hat{\ell}$ onto the linear space $\mathrm{span}(1, \tilde{f}_1, \dots, \tilde{f}_K)$.

Proof As in Lemma 5.5, let $\hat{\ell}$ be the vector in the Riesz representation of a pricing functional (which exists due to the assumption of absence of arbitrage opportunities, see Theorem 4.20). Denote by $\hat{\ell}^{\tilde{f}}$ the orthogonal projection (with respect to the norm induced by the inner product $(\tilde{x}, \tilde{y}) \mapsto \mathbb{E}[\tilde{x}\tilde{y}]$) of $\hat{\ell}$ onto the linear space $\mathrm{span}(1, \tilde{f}_1, \ldots, \tilde{f}_K)$. As a consequence, $\hat{\ell}$ can be decomposed as $\hat{\ell} = \hat{\ell}^{\tilde{f}} + (\hat{\ell} - \hat{\ell}^{\tilde{f}})$. Moreover, since

$$\hat{\ell} \in \mathrm{span}(r_f, \tilde{r}_1, \ldots, \tilde{r}_N) \subseteq \mathrm{span}(1, \tilde{f}_1, \ldots, \tilde{f}_K) \oplus \mathrm{span}(\tilde{\epsilon}_1, \ldots, \tilde{\epsilon}_N),$$

as follows from equation (5.28), it holds that $\hat{\ell} - \hat{\ell}^{\tilde{f}} \in \mathrm{span}(\tilde{\epsilon}_1, \ldots, \tilde{\epsilon}_N)$. Hence, there exists a vector $\alpha \in \mathbb{R}^N$ such that $\hat{\ell} - \hat{\ell}^{\tilde{f}} = \sum_{n=1}^{N} \alpha_n \tilde{\epsilon}_n$, noting that the elements $(\tilde{\epsilon}_1, \ldots, \tilde{\epsilon}_N)$ are linearly independent and represent a basis for the linear space $\mathrm{span}(\tilde{\epsilon}_1, \ldots, \tilde{\epsilon}_N)$. Since $\mathbb{E}[\tilde{\epsilon}_i \tilde{\epsilon}_j] = 0$ for all $i, j \in \{1, \ldots, N\}$ with $i \neq j$, this implies that, for all $n = 1, \ldots, N$,

$$\mathbb{E}[\hat{\ell}\,\tilde{\epsilon}_n] = \mathbb{E}[(\hat{\ell} - \hat{\ell}^{\tilde{f}} + \hat{\ell}^{\tilde{f}})\tilde{\epsilon}_n] = \mathbb{E}[(\hat{\ell} - \hat{\ell}^{\tilde{f}})\tilde{\epsilon}_n] = \sum_{i=1}^{N} \alpha_i \mathbb{E}[\tilde{\epsilon}_i \tilde{\epsilon}_n] = \alpha_n \, \mathrm{Var}(\tilde{\epsilon}_n)$$

and

$$\mathbb{E}[(\hat{\ell} - \hat{\ell}^{\tilde{f}})^2] = \sum_{n=1}^{N} \alpha_n^2 \, \mathrm{Var}(\tilde{\epsilon}_n).$$

In view of equation (5.33), the pricing error of asset n is given by $\left|\mathbb{E}[\hat{\ell}\,\tilde{\epsilon}_n]/\mathbb{E}[\hat{\ell}]\right|$, so that, using the fact that $\mathrm{Var}(\tilde{\epsilon}_n) \leq \overline{\sigma}$ for all $n = 1, \ldots, N$,

$$\sum_{n=1}^{N} \left|\mathbb{E}[\tilde{r}_n] - r_f + \sum_{k=1}^{K} b_{nk} \frac{\mathbb{E}[\hat{\ell}\,\tilde{f}_k]}{\mathbb{E}[\hat{\ell}]}\right|^2 = \frac{1}{\mathbb{E}[\hat{\ell}]^2} \sum_{n=1}^{N} \mathbb{E}[\hat{\ell}\,\tilde{\epsilon}_n]^2 = \frac{1}{\mathbb{E}[\hat{\ell}]^2} \sum_{n=1}^{N} \alpha_n^2 \, \mathrm{Var}(\tilde{\epsilon}_n)^2$$

$$\leq \frac{\overline{\sigma}^2}{\mathbb{E}[\hat{\ell}]^2} \sum_{n=1}^{N} \alpha_n^2 \, \mathrm{Var}(\tilde{\epsilon}_n) = \frac{\overline{\sigma}^2}{\mathbb{E}[\hat{\ell}]^2} \mathbb{E}[(\hat{\ell} - \hat{\ell}^{\tilde{f}})^2],$$

thus proving the claim. $\square$

The upper bound on the pricing error given in the above proposition is particularly interesting. Indeed, it shows that the pricing error will be small when the vector $\hat{\ell}$ is close to its projection $\hat{\ell}^{\tilde{f}}$ on the factor span. In particular, if it holds that $\hat{\ell} \in \mathrm{span}(1, \tilde{f}_1, \ldots, \tilde{f}_K)$, then $\hat{\ell} = \hat{\ell}^{\tilde{f}}$, so that the pricing error is null, thus confirming the validity of the APT in exact form as obtained in Proposition 5.17. Moreover, note that the upper bound given in (5.34) does not depend on the number N of traded assets, as in Proposition 5.16. In the limit, as the dimension of the economy (i.e., the number of traded assets) increases to infinity, the average pricing error will become arbitrarily small.

Extensions of the Arbitrage Pricing Theory

The Arbitrage Pricing Theory has been extended in several directions (see also Connor [476]). In particular, the assumption that the residual risks are uncorrelated can be significantly relaxed. To this regard, note that the result of Proposition 5.17 does not rely on the absence of correlation among the residual risks, but only on the absence of correlation between the residual risks and the factors $(\tilde{f}_1, \ldots, \tilde{f}_K)$. In Ingersoll [999], the APT in the form of Proposition 5.16 has been generalized to the case where the residual risks $\tilde{\epsilon}_n^N$ are possibly correlated among themselves (but uncorrelated with the risk factors), under the assumption that the elements of the factor loadings matrix are uniformly bounded. In this case, letting Σ^N be the covariance matrix of the residual risks, it can be shown that the absence of asymptotic arbitrage opportunities implies that there exists a positive constant A such that

$$\left(\mathbb{E}[\tilde{r}^N] - \lambda_0^N \mathbf{1}_N - B^N \lambda^N\right)^\top (\Sigma^N)^{-1} \left(\mathbb{E}[\tilde{r}^N] - \lambda_0^N \mathbf{1}_N - B^N \lambda^N\right) \leq A, \qquad \text{for all } N \in \mathbb{N}.$$
(5.35)

In particular, if the matrix Σ^N is diagonal (i.e., the residual risks are uncorrelated), this implies that

$$\lim_{N \to \infty} \frac{1}{N} \sum_{n=1}^{N} \left(\frac{\mathbb{E}[\tilde{r}_n^N] - \lambda_0^N - \sum_{k=1}^{K} b_{nk}^N \lambda_k^N}{\sqrt{\mathrm{Var}(\tilde{\epsilon}_n^N)}}\right)^2 = 0.$$

Note that this result does not rely on the assumption that residual risks have uniformly bounded variances, as considered in Proposition 5.15.

An extension of the APT to the case of correlated residual risks (uncorrelated with the risk factors) has also been proposed in Chamberlain [395] and Chamberlain & Rothschild [397]. In this case, in order to establish a result similar to Proposition 5.15, two hypotheses on the idiosyncratic risk components and on the factors are introduced. Indeed, in the presence of correlated residual risks, the idiosyncratic risk component is not necessarily eliminated via diversification by investing a small proportion of wealth in each asset. We say that a sequence $\{w^N\}_{N \in \mathbb{N}}$ of portfolios is *diversified* if it holds that $(w^N)^\top \mathbf{1}_N = 1$, for all $N \in \mathbb{N}$, and $\lim_{N \to \infty} \|w^N\|^2 = 0$. The first hypothesis requires that the influence of the residual risks $\tilde{\epsilon}_n^N$, for $n = 1, \ldots, N$ and $N \in \mathbb{N}$, can be eliminated in the limit via diversification, in the sense that $\lim_{N \to \infty} \mathbb{E}[((w^N)^\top \tilde{\epsilon}^N)^2] = 0$ for every sequence $\{w^N\}_{N \in \mathbb{N}}$ of diversified portfolios. As shown in Chamberlain [394], this hypothesis is satisfied if the largest eigenvalue of the covariance matrix Σ^N of the residual risks is uniformly bounded from above over all $N \in \mathbb{N}$. The second hypothesis is called the *pervasiveness condition* and amounts to assume that each factor influences a large number of assets in the N-th economy, for each $N \in \mathbb{N}$. This condition requires that the smallest eigenvalue of the matrix $B^N (B^N)^\top$ explodes as N increases to infinity. Under these two hypotheses, asset returns can be described by an *approximate K-factor model* and relations analogous to (5.29) or (5.35) can be established.

Equilibrium versions of the APT have been proposed in Chen & Ingersoll [423], Connor [475], Dybvig [605]. In an economy with a finite number of assets, a linear relation for the asset risk premia holds in exact form if in equilibrium there exists a perfectly diversified portfolio (i.e., not influenced by idiosyncratic risk) representing the optimal choice of an agent (under suitable regularity assumptions on the agent's utility function), as shown in the following proposition (see also Chen & Ingersoll [423]).

Proposition 5.19 *Suppose that the economy comprises a risk free asset with return r_f and N risky assets, whose returns are generated by the linear model (5.28) with (possibly correlated) residual risks $(\tilde{\epsilon}_1, \ldots, \tilde{\epsilon}_N)$ satisfying*

$$\mathbb{E}[\tilde{\epsilon}_n | \tilde{f}_1, \ldots, \tilde{f}_K] = 0, \qquad \text{for all } n = 1, \ldots, N.$$

Then, if there exists a portfolio which is unaffected by the residual risks and represents the optimal choice for an agent with a continuously differentiable, increasing and strictly concave utility function, the APT holds in exact form.

Proof Let us denote by $\widetilde{W}^*$ the agent's optimal wealth. Under the present assumptions, $\widetilde{W}^*$ can be expressed as a function of the risk factors, so that $u'(\widetilde{W}^*) = g(\tilde{f}_1, \ldots, \tilde{f}_K)$, since $\widetilde{W}^*$ is assumed not to be affected by the residual risks $(\tilde{\epsilon}_1, \ldots, \tilde{\epsilon}_N)$. From the first order conditions of the agent's optimal portfolio problem (see equation (3.4)), it holds that

$$\mathbb{E}[u'(\widetilde{W}^*)(\tilde{r}_n - r_f)] = \mathbb{E}[g(\tilde{f}_1, \ldots, \tilde{f}_K)(\tilde{r}_n - r_f)] = 0, \qquad \text{for all } n = 1, \ldots, N.$$

Replacing equation (5.28) into the above expression we get

$$\mathbb{E}\left[g(\tilde{f}_1, \ldots, \tilde{f}_K)\left(\mathbb{E}[\tilde{r}_n] - r_f + \sum_{k=1}^{K} b_{nk}\tilde{f}_k + \tilde{\epsilon}_n\right)\right] = 0,$$

so that, by rearranging the terms, we get, for all $n = 1, \ldots, N$,

$$\mathbb{E}[\tilde{r}_n] = r_f + \sum_{k=1}^{K} b_{nk}\left(-\frac{\mathbb{E}[g(\tilde{f}_1, \ldots, \tilde{f}_K)\tilde{f}_k]}{\mathbb{E}[g(\tilde{f}_1, \ldots, \tilde{f}_K)]}\right) - \frac{\mathbb{E}[g(\tilde{f}_1, \ldots, \tilde{f}_K)\tilde{\epsilon}_n]}{\mathbb{E}[g(\tilde{f}_1, \ldots, \tilde{f}_K)]}.$$

The last term is null, since

$$\mathbb{E}[g(\tilde{f}_1, \ldots, \tilde{f}_K)\tilde{\epsilon}_n] = \mathbb{E}[g(\tilde{f}_1, \ldots, \tilde{f}_K)\mathbb{E}[\tilde{\epsilon}_n | \tilde{f}_1, \ldots, \tilde{f}_K]] = 0,$$

thus proving the claim.   □

In an economy with a finite number of assets, an error bound for each asset has been established in a general equilibrium setting by Dybvig [605] and Grinblatt & Titman [834]. If markets are complete, idiosyncratic noise components are mutually independent and independent of $\tilde{f}_k$, the (representative) agent has an increasing and strictly concave utility function with a non increasing and bounded coefficient of absolute risk aversion, each asset is in positive net supply, then the market portfolio is diversified (in the sense that idiosyncratic risk is diversified away) and a bound to the approximation error holds for each asset, see Dybvig [605]. The bound to the approximation error of an asset depends on the variance of its idiosyncratic risk component, on the bound to the coefficient of absolute risk aversion and its weight on the market portfolio. In an economy with an infinite number of assets, Connor [475] provides an exact APT equilibrium version if the market portfolio is well diversified. In this case, every investor holds a well diversified portfolio (K mutual funds separation, on these results see also Milne [1345]). In this perspective, a problem is represented by Roll's critique: to test an equilibrium version of the APT, the market portfolio should be observable (see Shanken [1524]). In the special case where consumption and idiosyncratic risk components are distributed according to a bivariate normal law, then, if the number of assets grows in such a way that the weight of each asset in the market portfolio tends to zero, the residual also tends to zero, see Constantinides [492].

5.5 Empirical Tests of the APT

The APT cannot be easily tested in its approximate form and, as a consequence, most of the empirical studies concentrate on testing the exact linear relation associated with the APT. A strong critique to testing the APT in exact form was put forward in Shanken [1523] (*Shanken critique*). The returns of two sets of assets generating the same set of portfolio returns may conform to different factor models and even the number of the factors can differ. In particular, the risk factors for a set of returns cannot be identified in a unique way, since they depend on the assets generating them. As a consequence, the implications of the APT in exact form on portfolio risk premia change as reference assets (and factors) change, see also Gilles & LeRoy [781]. The risk premia implications are consistent only if the assets have the same expected return. Transforming assets as in Shanken [1523], factor and idiosyncratic randomness are handled as idiosyncratic noise, making the factor model useless. In a finite economy, an approximation bound is a mathematical tautology with no economic content and, therefore, it is untestable. In Reisman [1445], this critique was further examined, showing that if the APT approximation is to be interpreted as an equality, then expected returns should be a linear function of the β coefficients with respect to virtually any set of reference variables correlated with the true factors. The upper bound to the approximation error increases as the correlation between the true factors and the reference variables decreases (Nawalkha [1372] has shown that under some conditions no loss in pricing accuracy occurs decreasing the correlation with the true factors). The factor model can be manipulated rather arbitrarily by repackaging a given set of securities, so that factors are in practice indeterminate. This is an intrinsic limit of the APT in exact form. These results have been discussed in Grinblatt & Titman [835] and Dybvig & Ross [609], where it is shown that if the APT holds in exact form then the factor model cannot be manipulated as suggested in Shanken [1523]: as the number of assets of the economy tends to infinity, the variance of the returns of some assets becomes unbounded or the inverse of the transformed matrix explodes. Al-Najjar [38] has shown that in a large economy repackaging identifies a unique factor model. For a reply to the above arguments see Shanken [1524]. The Shanken critique suggests considering the approximation error in empirically testing the APT and, therefore, testing models deriving an approximation error bound for each asset. This is accomplished through equilibrium versions of the APT, as discussed at the end of the previous section.

Some methodological considerations on testing the CAPM also apply to APT tests. The APT is a single-period model, therefore one assumes that the factor model holds period by period and that the economy is stationary. The classical hypothesis is that the asset returns vector conditionally on the factors is identically and independently distributed over time as a multivariate normal random variable. As for the CAPM, the (approximate) linear relation of the APT is an ex-ante result, but the factor model is an ex-post model. To test the model one assumes that agents form rational expectations and that the expected return of an asset is equal to the historical average return.

In general, the APT does not provide an exact formula for asset risk premia. In a finite asset economy, the APT provides an approximate linear relation between asset risk premia and factor risk premia: an exact linear relation holds if a portfolio of the factors belongs to the mean-variance frontier or if there exists a well diversified optimal portfolio. Moreover, unless an equilibrium version of the APT is tested, the theory does not provide a bound to the approximation error for each asset, but only a bound to the global approximation error. Empirically, the APT in exact form is usually tested without checking for the validity of the hypothesis yielding an exact pricing relation. Because of the non-uniqueness factor representation result obtained in Shanken [1523], it is difficult to test a violation of the bound on the approximation error.

Imposing the restrictions of the APT in exact form on the multi-factor model (5.28), the following ex-post model is obtained:

$$\tilde{r}_{n,t} = \lambda_0 + \sum_{k=1}^{K} b_{nk}(\lambda_k + \tilde{\delta}_{k,t}) + \tilde{\epsilon}_{n,t}, \qquad \text{for all } n = 1, \dots, N.$$

As for the CAPM, the APT in exact form concerns a linear relation between asset risk premia, factor risk premia and the matrix B. There are two main classes of tests of the APT: cross-sectional regressions and time series regressions. Given a factor structure, in the first case risk premia (λ) are estimated taking as observed the matrix

B, in the second case a joint estimation is performed (i.e., constrained time series estimate of B and λ). The literature on empirical tests of the APT is quite large and we refer to Connor & Korajczyk [481] for a survey. In what follows, we concentrate our attention on cross-sectional regression tests.

A cross-sectional regression test consists of three steps: a) risk factor identification; b) estimation of the matrix B; c) test of a linear relation between risk premia and the coefficients in B. As for the CAPM, a two-pass procedure or a joint estimation procedure is adopted (i.e., betas and relative prices are simultaneously estimated). The methodology to estimate the matrix B and to test the APT depends on the nature of the factors, see Campbell et al. [348]. In some cases, factors are traded portfolio returns (when this is not the case we can look for portfolios mimicking the factors, as explained in the previous section). Given the multi-factor model and the estimated matrix B, risk premia can be estimated by a variety of methods including ordinary least squares and generalized least squares. The two-pass procedure may induce an error in variables problem, see Shanken [1526].

The sequence of steps described above implies that a test of the APT is a joint test of three different hypotheses: a) risk factors are correctly identified; b) the matrix B is correctly estimated; c) the exact linear relation between asset risk premia and coefficients holds.

The risk factor identification problem is a model selection problem. Two main classes of factor models have been considered in the literature, depending on the selection approach: statistical factor models and economic factor models. Let us first consider statistical factor models. There are two main statistical approaches to identify the factors: factor analysis and principal component analysis. In both cases, factors do not have a direct economic interpretation. A factor analysis procedure has been used in Roll & Ross [1461], Chen [422], Lehmann & Modest [1178] and a principal component analysis procedure in Connor & Korajczyk [478] and Jones [1038]. Both methods produce a consistent estimate of B in large samples, while in finite samples there is no clear argument to choose between them. Factor analysis is computationally more expensive than principal component analysis. The number of selected factors is three/four in Roll & Ross [1461], Chan et al. [408], five in Chen [422], Lehmann & Modest [1178] and six in Connor Korajczyk [480]. Factors are chosen by repeating the estimation procedure and varying their number. A test on the number of factors has been proposed in Connor Korajczyk [480]. Typically, no economic interpretation can be given to risk premia signs.

There are two main classes of economic factor models: models with factors referring to firm characteristics and models with factors referring to macroeconomic variables. The first type of factor models was stimulated by the empirical tests of the CAPM showing that factors such as size, book to market value ratio, dividend yield, cash flow to price ratio, turn out to be significant in order to explain asset risk premia, as discussed in Sect. 5.3 (see also Fama & French [670, 673], Chan et al. [408]). Three factors (a market factor, one associated with size and one with book to market value ratio) explain sufficiently well the cross-section of asset risk premia. Building a linear model with macroeconomic factors, one tries to identify those macroeconomic-financial market factors affecting a company's value (i.e., affecting future earnings and the discount factor). In Chen et al. [424] the following five variables have been considered: industrial production growth, unexpected inflation, expected inflation, yield spread between long and short interest rates (maturity premium), yield spread between corporate high and low-grade bonds (default premium). Aggregate consumption growth (as suggested by the CCAPM), market factor (as suggested by the CAPM) and oil price are not significant. Many other macroeconomic factors have been considered in empirical studies. In other studies, an index of the market turns out to be significant. An APT model based on macroeconomic factors has been tested with positive evidence in Burmeister & McElroy [323] and Elton et al. [638].

The APT can be tested in several directions. In a cross-sectional regression setting, given the factors identified through one of the procedures illustrated above we can verify that risk factors are priced (test the statistical significance of the risk premia associated with the factors). In Roll & Ross [1461] it is shown that factors are significant: in the macroeconomic factor model analysed in Chen et al. [424] four factors are significant, only the term spread factor is marginally significant. Another test consists in verifying that other variables are not significant in explaining risk premia. In Roll & Ross [1461] it is shown that the standard deviation of an asset return has no incremental power over the four factors. The result is confirmed in Chen [422]. As far as the size of the company is concerned, results are controversial: in Chen [422] and Chan et al. [402] it is shown that the size anomaly is explained by the macroeconomic APT model proposed in Chen et al. [424]. On the contrary, in Reinganum [1441], Brennan et al. [293] it is shown that the size anomaly is not explained by the APT. The evidence on the other anomalies relative to the CAPM (e.g., book to market value of equity ratio) is mixed, see Brennan et al. [293].

Another strategy to test the APT consists in verifying that the λ coefficients are constant as the set of returns of the economy changes. As reference returns vary, the multi-factor model as well as risk premia change (however if there exists a risk free asset then $\lambda_0 = r_f$ for every set of returns). In Roll & Ross [1461], this hypothesis has not been rejected, while a negative result has been obtained in Brown & Weinstein [316]. Using the time series test approach, a testable implication of the APT in exact form is that $\lambda_0 = r_f$ which is equivalent to the condition established in Grinblatt & Titman [836]. A restriction on the matrix B is derived using results in Huberman & Kandel [982]. Mixed results for exact pricing restrictions were provided in Lehmann & Modest [1178] and Connor & Korajczyk [478], showing that the APT does not explain the size effect. In Connor & Korajczyk [478] the authors compare the APT with the CAPM: the results are mixed with a slight preference for the APT (the same conclusion is reached in Mei [1322]).

In Chen [422] the APT is compared with the CAPM in a cross-sectional regression setting. Expected returns are estimated according to the APT with five factors and to the CAPM: the APT explains returns better than the CAPM. Residuals from the CAPM cross-sectional regression can be explained by factor loadings employed in the APT, while residuals from the APT cannot be explained by the market portfolio β. Data support the APT as a better model for asset returns. A similar conclusion is drawn in Fama & French [673]. The performance of the three classes of factor models has been evaluated in Connor [477]. The ranking is as follows: factor models based on firm's characteristics, statistical factor models, macroeconomic factor models. Firm-specific factors perform well compared with macroeconomic variables and statistical factors, see also Chan et al. [408] and Moskowitz [1354].

Summing up, empirical tests often reject exact linear APT pricing restrictions. Comparing the APT to the CAPM, the APT performs well in explaining cross-sectional differences in asset returns. Moreover, the APT seems to be able to explain some pricing anomalies relative to the CAPM.

5.6 Notes and Further Readings

Dybvig & Ingersoll [607] investigate the relation between the CAPM and the complete markets equilibrium model. The results are striking: in particular, if the CAPM pricing relation holds for all assets, the markets are complete and the market portfolio generates sufficiently high returns in some state of the world, then there exist arbitrage opportunities. As a consequence, if agents' utility functions are always increasing, then in equilibrium the CAPM relation cannot hold for all the assets. If asset returns are unbounded, then the CAPM can hold in equilibrium only if each investor has achieved his level of satiation. Under the complete markets assumption with unbounded returns, if the CAPM holds for all the assets, then in equilibrium agents exhibit quadratic utility functions.

In our analysis, we have focused our attention on asset pricing relations in correspondence of an equilibrium allocation. The existence of the equilibrium can be established by relying on the results presented in Chap. 1. Clearly, an assumption which implies that the agents hold portfolios belonging to the mean-variance efficient portfolio frontier is that the agents' preferences are represented by mean-variance utility functions (see Sect. 2.4). However, this assumption might be problematic in order to establish the equilibrium since quadratic preferences are not always monotone, see Nielsen [1376]. The existence of an equilibrium in a market with a risk free asset is ensured if all the agents agree on the asset expected returns or if their coefficient of absolute risk aversion goes to infinity as the standard deviation of a portfolio goes to infinity. Without a risk free asset, a sufficient condition for the existence of an equilibrium is that the agents agree on the expected returns of the assets and that their risk aversion coefficients are bounded. In particular, this last condition eliminates satiation problems (on this point see Allingham [51] and Nielsen [1375]). Non-monotonicity of preferences can originate negative prices in equilibrium. Imposing bounds on agents' risk aversion, the positivity of the equilibrium prices can be proved as in Nielsen [1377].

The CAPM admits a straightforward generalization to the case where agents exhibit heterogeneous beliefs about the expected returns of the risky assets (nevertheless, all the agents are assumed to have identical beliefs about the covariance matrix V of the traded assets). More precisely, we suppose that, from the point of view of agent i, the expected return of asset n is given by μ_n^i, for $i = 1, \ldots, I$ and $n = 1, \ldots, N$. In order to derive an analogue to the CAPM relation (5.21) in the case of heterogeneous beliefs, we suppose that the agents are characterized by quadratic utility functions of the form

$$u^i(x) = x - \frac{b_i}{2}x^2, \qquad \text{for all } i = 1, \dots, I.$$

Denoting by $\widetilde{W}^{i*}$ the optimal wealth at time $t = 1$ of agent i, for $i = 1, \dots, I$, we suppose that $\mathbb{E}[\widetilde{W}^{i*}] < 1/b_i$, for all $i = 1, \dots, I$ (in order to ensure that the optimal portfolios lie in the region where the utility functions are increasing). Following the notation adopted in section "Closed Form Solutions and Mutual Fund Separation", let us denote by w_n^{i*} the optimal amount of wealth invested at time $t = 0$ by agent i in the n-th asset, for $i = 1, \dots, I$ and $n = 1, \dots, N$. Note that the quantity $\sum_{n=1}^{N} \sum_{i=1}^{I} w_n^{i*}$ represents the *total market capitalization* of the risky assets. Consequently, for each $n = 1, \dots, N$, the *relative market capitalization* of asset n is given by the quantity

$$w_n^m := \frac{\sum_{i=1}^{I} w_n^{i*}}{\sum_{k=1}^{N} \sum_{i=1}^{I} w_k^{i*}},$$

for all $n = 1, \dots, N$. Equivalently, $(w_1^m, \dots, w_N^m)$ denotes the market portfolio in terms of proportions of the total market capitalization invested in each of the risky assets. The return $\tilde{r}^m$ associated to the market portfolio is then given by $\tilde{r}^m = \sum_{n=1}^{N} w_n^m \tilde{r}_n$. Let us also introduce the following notation, for all $n = 1, \dots, N$:

$$\bar{\mu}_n := \sum_{i=1}^{I} \frac{\lambda_i}{\sum_{j=1}^{I} \lambda_j} \mu_n^i,$$

with $\lambda_i := (1 - b_i\mathbb{E}[\widetilde{W}^{i*}])/b_i$, in view of the explicit characterization of the optimal portfolio given in equation (3.9). Recall also that, as explained in section "Closed Form Solutions and Mutual Fund Separation", the quantity λ_i represents the inverse of the coefficient of absolute risk aversion of agent i evaluated in correspondence of his expected optimal wealth. Let us also define

$$\bar{\mu}^M := \sum_{i=1}^{I} \frac{\lambda_i}{\sum_{j=1}^{I} \lambda_j} \mu^{i,M}, \qquad \text{where } \mu^{i,M} := \sum_{n=1}^{N} w_n^m \mu_n^i,$$

We are now in a position to state the following result, the proof of which is given in Exercise 5.17 (see Hens & Rieger [938, Proposition 3.5]).

Proposition 5.20 *Under the above assumptions and using the notation introduced above, the following CAPM relation holds under heterogeneous beliefs:*

$$\bar{\mu}_n - r_f = \beta_{nm}(\bar{\mu}^M - r_f), \qquad \text{for all } n = 1, \dots, N \qquad (5.36)$$

where $\beta_{nm} := \mathrm{Cov}(\tilde{r}^m, \tilde{r}_n)/\mathrm{Var}(\tilde{r}^m)$, *for all* $n = 1, \dots, N$.

Equation (5.36) represents the Security Market Line in an economy where agents have heterogeneous expectations about the asset returns. Note that the expected returns $\bar{\mu}_n$ and $\bar{\mu}^M$ appearing above represent weighted averages of the expected returns as perceived by the individual agents, with the weights depending on the risk aversion of the individual agents. In other words, expected returns as perceived by less risk averse agents play a more important role in the determination of $(\bar{\mu}_1, \ldots, \bar{\mu}_N)$ and of $\bar{\mu}^M$ than the expected returns as perceived by more risk averse agents.

The CAPM has been extended in several directions. In particular, in Brennan [289] and Litzenberger & Ramaswamy [1222] the model has been extended to an economy with dividends and taxes and in Mayers [1313] to an economy with non-tradeable assets and human capital. Furthermore, an analysis of the CAPM with short sales prohibition is given in Sharpe [1530]. An intertemporal version of the CAPM has been introduced in Merton [1330] (for more details, see also Sect. 6.4). Foreign exchange risk has been introduced in the CAPM by Solnik [1555] and in Stulz [1574].

For a survey on the econometric issues and techniques for testing the CAPM, we refer the reader to Shanken [1527] and Campbell et al. [348]. Shanken [1526] provides an econometric view of maximum likelihood methods and of traditional two pass approaches to estimating factor models, also addressing the errors-in-variables problem. The traditional inference procedure, under standard assumptions, overstates the precision of price of risk estimates and in Shanken [1526] an asymptotically valid correction methodology is provided.

The assumption of a normal distribution plays a particularly important role in the literature related to the CAPM. On the one hand, this assumption brings the advantage that finite sample properties of asset pricing models can be derived, on the other hand the normal distribution is only one of the distributions compatible with the CAPM and there is a substantial evidence on the non-normality of asset returns (for instance, heavy tails and heteroskedasticity). An alternative distributional assumption compatible both with the CAPM and with the presence of heavy tails in asset returns is that of a multivariate t-Student distribution. By employing a generalized method of moments estimator, MacKinlay & Richardson [1271] show that under this assumption the bias introduced by the normality hypothesis is small, but it can be relevant when the Sharpe ratio of a portfolio is high and/or the number of degrees of freedom is small. Examples of the time series approach to test the CAPM and the zero-β CAPM are provided in Jobson & Korkie [1033], MacKinlay & Richardson [1271], Gibbons et al. [777], Gibbons [775], Shanken [1527], Ferson [687]. Tests can be performed by means of a maximum likelihood estimator assuming a given distribution for the asset returns (multivariate normality or t-Student) or by means of a generalized method of moments estimator (without distributional assumption).

As reported in Sect. 5.3, the empirical evidence on the CAPM has produced mixed results and several anomalies have been discovered. In recent years, other anomalies have been detected taking the CAPM as the benchmark. Brennan et al. [293], Chordia et al. [439], Datar et al. [525], Liu [1229], Ang et al. [68] have shown a negative and significant relation between returns and turnover rate, trading volume and volatility, taken as proxies of market liquidity. A relation also holds for other measures of market liquidity (e.g., the price impact of trades and fixed-variable cost of trades), see Brennan & Subrahmanyam [297] and Amihud [57]. In particular, according to Brennan & Subrahmanyam [297], there is an increasing relation between adverse selection, illiquidity costs (variable transaction costs) and asset returns. Easley et al. [615] find a positive relationship between the probability of information-based trades and asset returns (adverse selection premium), with the probability of informed trading being lower for high volume stocks, see Easley et al. [618]. In the analysis of Duarte & Young [591], it has been shown that this effect is mainly due to illiquidity and not to adverse selection. Hou & Moskowitz [968] find that a positive risk premium is associated with the delay with which prices respond to information, an effect which is not explained by liquidity and other microstructure effects. Pástor & Stambaugh [1407] find that the stocks whose returns are more exposed to marketwide liquidity fluctuations are characterized by higher expected returns. In Lamont et al. [1163] it is shown that there is a financial constraint factor, in the sense that constrained firms earn lower returns than unconstrained firms. Naranjo et al. [1370] and Brennan et al. [293] show that stock returns are increasing in dividend yields. Ang et al. [67] show that returns are increasing in downside risk (i.e., the risk that asset returns are more correlated with the market when the latter is failing rather than when the market is rising). Harvey & Siddique [910] and Boyer et al. [275] show that conditional skewness helps to explain the cross-sectional variation in expected returns and negative co-skewness with the market return induces a positive risk premium. See Chen et al. [417] for an analysis of the relationship between return skewness and other anomalies. Hou & Robinson [969] show that firms in more concentrated industries earn lower returns and Dittmar [579] finds that kurtosis affects asset returns. In Diether et al. [577], Yu [1668], Sadka & Scherbina [1489], Anderson et al. [61, 62], Zhang [1677], Johnson [1036], Doukas et al. [583] it is shown that analysts' disagreement is related to expected returns. Note that many of these anomalies are not captured by the Fama & French [670] model.

For international evidence on the CAPM anomalies we refer to Hawawini & Keim [916], Fama & French [674], Haugen & Baker [915], Rouwenhorst [1479], Asness et al. [79]. Fama & French [674] propose an international two-factor model with a factor for relative distress in order to capture the international evidence of a value effect (see also Fama & French [677] and Griffin [825]). A positive evidence on the international version of the conditional CAPM is reported in De Santis & Gerard [560].

Considering a non-linear habit formation model, Santos & Veronesi [1498] show that cross-sectional regularities can be explained by the variability of the cash flows of the companies. In Yogo [1665] cross-sectional anomalies are explained by considering a non-separable utility function. Zhang [1680], Chen & Zhang [419], Liu et al. [1228] explain asset price anomalies by assuming costly reversibility of investments through a q-theory model. Bansal et al. [134, 135], Hansen et al. [889] show that value-growth anomalies can be reconciled with an intertemporal model assuming recursive preferences (see Chap. 9) and considering long run risk: the cash flow growth of value (growth, respectively) portfolios has positive (negligible, respectively) covariation with consumption in the long run.

For a survey of theoretical and empirical results on the APT we refer to Connor [476], Huberman [980], Connor & Korajczyk [481]. APT restrictions can be obtained by allowing for private information among the agents on asset returns, see Stambaugh [1562] and Handa & Linn [886]. It can be shown that, under some conditions on the preference relation, there are no asymptotic arbitrage opportunities in the market. The conditions include the existence of an agent with strictly increasing and continuous preferences and of an optimal portfolio (see Jarrow [1019]). Bounds on the pricing kernel in the spirit of the Hansen-Jagannathan bound can be found in Bansal & Lehman [136], Bernardo & Ledoit [204], Cochrane & Saa'-Requejo [468], Snow [1553]. Huberman et al. [984] and Huberman & Kandel [982] have identified necessary and sufficient conditions for the existence of mimicking portfolios: there exists a set of mimicking portfolios if and only if the minimum variance portfolio has positive systematic risk, in the sense that it is affected by the risk factors.

It is of interest to study under which conditions one can obtain a linear relation in an exact form, in the spirit of Propositions 5.13 and 5.17. In Ingersoll [999], assuming an approximate factor model, it is shown that in the limit as the dimension of the economy approaches infinity (i.e., $N \to \infty$), the returns on all well diversified portfolios satisfy a linear relation in exact form. Along the same lines, in Chamberlain [394] it is shown that in the limit as the dimension of the economy approaches infinity a linear pricing relation holds in exact form if and only if there exists a well diversified portfolio which belongs to the mean-variance frontier. A related result has also been obtained in Grinblatt & Titman [836] for an economy with a finite number of assets and a risk free asset. Indeed, assuming that the N traded assets have a non-singular covariance matrix, Grinblatt & Titman [836] consider a set of $K \leq N$ reference portfolios, i.e., a set of K portfolios such that there exists a linear combination of them which does not coincide with the minimum variance portfolio constructed from the N assets. Interpreting the K reference portfolios as risk factors, it can be shown that the linear relation of the APT holds in exact form if and only if there exists a combination of the K reference portfolios which belongs to the mean-variance efficient portfolio frontier. A similar result has also been obtained in Huberman & Kandel [982], discussing the relationship between exact arbitrage pricing and K mutual funds separation. It is worth pointing out that these results provide a bridge between the CAPM, which relies on the fact that the market portfolio belongs to the portfolio frontier, and the APT in exact form, which relies on the fact that a combination of reference portfolios belongs to the portfolio frontier (as a consequence, similar methods aiming at verifying the efficiency of a portfolio can be employed to test the CAPM and the APT, see Shanken [1525]). Reference portfolios playing the role of risk factors are often called *mimicking portfolios*.

The asset pricing models presented in this chapter are all linear models. A non-linear model with the market return playing the role on the only risky factor has been proposed in Bansal & Viswanathan [138] and a multi-factor non-linear model has been introduced in Dittmar [579]. The empirical performance of such non-linear models turns out to be good and outperforms linear models. Moreover, it is shown that both human capital and the market return are relevant risk factors.

5.7 Exercises

Exercise 5.1 Consider the asset pricing relation (5.6). Prove that the covariance between the return on the market portfolio and the stochastic discount factor $\delta u_1'(\tilde{x}^m)/u_0'(x_0^m)$ is negative. Deduce that the risk premium $\mathbb{E}[\tilde{r}^m] - r_f$ of the market portfolio is positive.

Exercise 5.2 Prove Proposition 5.2.

Exercise 5.3 As in Sect. 5.1, let us consider an economy comprising I individuals with increasing and strictly concave utility functions u^i, N risky assets with normally distributed returns and a risk free asset with risk free rate of return r_f. Denoting by $\widetilde{W}^{i*}$ the optimal consumption of agent i, for $i = 1, \ldots, I$, define the *global absolute risk aversion coefficient* as the quantity

$$\theta_i := -\frac{\mathbb{E}[u^{i''}(\widetilde{W}^{i*})]}{\mathbb{E}[u^{i'}(\widetilde{W}^{i*})]}.$$

By relying on equilibrium arguments, express the risk premium of the market portfolio in terms of the global absolute risk aversion coefficients of the agents.

Exercise 5.4 In the setting of Exercise 5.3, remove the assumption that asset returns are normally distributed and assume instead that the I agents have quadratic utility functions of the form

$$u^i(x) = a_i x - \frac{b_i}{2}x^2, \qquad \text{for all } i = 1, \ldots, I.$$

Assume furthermore that the initial endowments of the agents and the asset returns are such that the optimal consumption plans of the agents always remain in the range where the utility functions are increasing (i.e., $\widetilde{W}^{i*} < a_i/b_i$, for all $i = 1, \ldots, I$). By relying on equilibrium arguments, express the risk premium of the market portfolio in terms of the global absolute risk aversion coefficients of the agents.

Exercise 5.5 Consider an economy populated by I agents with exponential utility functions of the form

$$u^i(x) = -\frac{1}{a_i}e^{-a_i x}, \qquad \text{with } a_i > 0 \quad \text{for all } i = 1, \ldots, I.$$

Suppose that there are two traded assets: a risk free asset with rate of return r_f and a risky asset with return distributed according to a normal law with mean μ and variance σ^2. Suppose that the aggregate supply of the risky asset is equal to 1. Determine the risky asset risk premium in correspondence of an equilibrium of the economy.

Exercise 5.6 Consider a representative agent economy as in Sect. 5.1. Show that the risk premium on the market portfolio is increasing with respect to the absolute risk aversion coefficient of the representative agent and, if the representative agent's utility function is DARA, then it is decreasing with respect to the economy's aggregate wealth.

Exercise 5.7 Suppose that there are no arbitrage opportunities and let $Q(\cdot)$ be a pricing functional. Let $\hat{\ell} \in I(D)$ denote the unique vector such that $Q(c) = \mathbb{E}[\hat{\ell}c]$, for all $c \in I(D)$, as in Lemma 5.5. Let $z^\ell \in \mathbb{R}^N$ be the portfolio such that $\hat{\ell} = Dz^\ell$ and denote by $\tilde{r}^\ell$ its return, i.e., $\tilde{r}^\ell = \hat{\ell}/V(\hat{\ell})$. Prove the following properties:

(i) for any arbitrary portfolio $z \in \mathbb{R}^N$ with return $\tilde{r}^z$, it holds that $\mathbb{E}[\tilde{r}^z \tilde{r}^\ell] = \mathbb{E}[(\tilde{r}^\ell)^2]$.

(ii) the portfolio z^ℓ belongs to the mean-variance portfolio frontier.

Exercise 5.8 In an economy with $N + 1$ traded assets, suppose that there are no arbitrage opportunities and let $\tilde{\ell}$ be the likelihood ratio of any risk neutral probability measure. Call the quantity $\tilde{m} := \tilde{\ell}/r_f$ *stochastic discount factor* (see Sect. 4.4). Show that

$$\mathbb{E}[\log(\tilde{r}_n)] \le -\mathbb{E}[\log(\tilde{m})], \qquad \text{for all } n = 1, \dots, N.$$

Exercise 5.9 Consider an economy with a risk free lending rate r_l lower than the corresponding risk free borrowing rate r_b (i.e., $r_b > r_l$), reflecting the presence of transaction costs in the risk free market. Suppose that all the agents choose to hold mean-variance efficient portfolios. By relying on the same arguments adopted in Sect. 5.2 and assuming that the net supply of the risk free asset is zero, show that the following *Zero-β* CAPM relation is obtained:

$$\mathbb{E}[\tilde{r}_n] = \mathbb{E}[\tilde{r}^{zc(m)}] + \beta_{nm}(\mathbb{E}[\tilde{r}^m] - \mathbb{E}[\tilde{r}^{zc(m)}]), \qquad \text{for all } n = 1, \dots, N, \qquad (5.37)$$

with $\mathbb{E}[\tilde{r}^m] - \mathbb{E}[\tilde{r}^{zc(m)}] > 0$ and $r_b \ge \mathbb{E}[\tilde{r}^{zc(m)}] \ge r_l$.

Exercise 5.10 Consider an economy where all the agents choose to hold mean-variance efficient portfolios but it is not possible to borrow at the risk free rate r_f (i.e., only investing in the risk free asset is allowed). By relying on the same arguments adopted in Sect. 5.2 and assuming that the risk free asset is in strictly positive net supply, show that the following *Zero-β* CAPM relation is obtained:

$$\mathbb{E}[\tilde{r}_n] = \mathbb{E}[\tilde{r}^{zc(m)}] + \beta_{nm}(\mathbb{E}[\tilde{r}^m] - \mathbb{E}[\tilde{r}^{zc(m)}]), \qquad \text{for all } n = 1, \dots, N, \qquad (5.38)$$

with $\mathbb{E}[\tilde{r}^m] - \mathbb{E}[\tilde{r}^{zc(m)}] > 0$ and $\mathbb{E}[\tilde{r}^{zc(m)}] \le r_f$.

Exercise 5.11 Consider the setting of the Modigliani-Miller economy described at the end of Sect. 4.4, with an unleveraged firm generating revenues V_1 at date $t = 1$ and, as in (4.62), denote by r^{un} the expected return on the shares of the unleveraged firm. Consider then a leveraged firm generating the same revenues at date $t = 1$ but partly financed by debt with nominal value K. As in (4.62), we denote by r^{lv} and r^d the expected rates of return on the stock and on the debt, respectively, of the leveraged firm. Suppose that the CAPM holds for the stock of the unleveraged firm, the stock of the leveraged firm and the debt (bond) of the leveraged firm, with respect to some market portfolio with expected rate of return $r^m := \mathbb{E}[\tilde{r}^m]$ and a risk free asset with return r_f. Show that the beta coefficient of the stock of the leveraged firm (denoted by β_{lv}) can be expressed as a linear combination of the beta coefficients associated to the stock of the unleveraged firm and to the debt (denoted respectively by β_{un} and β_d).

Exercise 5.12 Consider three risky assets with expected returns, standard deviations and correlations with the market portfolio given by the following vectors:

$$\mu = \begin{bmatrix} 1.07 \\ 1.08 \\ 1.1 \end{bmatrix} \qquad \sigma = \begin{bmatrix} 0.3 \\ 0.2 \\ 0.15 \end{bmatrix} \qquad \rho = \begin{bmatrix} 0.2 \\ 0.4 \\ 0.8 \end{bmatrix}.$$

Suppose that the market portfolio is characterized by expected return $\mu^m = 1.09$ and standard deviation $\sigma^m = 0.1$.

   (i) Does the CAPM relation hold if the risk free rate is equal to $r_f = 1.04$?
  (ii) Compute the risk premium for an asset with $\beta = -0.5$.
 (iii) Show that in equilibrium there cannot exist an asset with expected return $\mu' = 1.2$ and $\beta' = 0.5$.

Exercise 5.13 Consider an economy with three possible states of the world with equal probabilities of occurrence. The economy's aggregate endowment is given by $e = (2, 5, 3)$ and there are three traded assets. The first asset is risk free, has price $p_1 = 1$ at time $t = 0$ and delivers the constant payoff 1 at time $t = 1$. The two remaining assets are risky: the first asset has random return $\tilde{r}_2 = (0, 3, 1)$ and unitary price at time $t = 0$ and the second asset has random payoff $\tilde{d}_3 = (0, 0, 2)$ and price p_3 at time $t = 0$. The aggregate supply of the three assets is given by $(2, 1, 0)$. Determine the equilibrium price of the third asset assuming that the CAPM relation holds.

Exercise 5.14 Consider an economy with two traded assets with returns $\tilde{r}_1$ and $\tilde{r}_2$ together with a market portfolio with return $\tilde{r}^m$. The covariance matrix of the random vector $(\tilde{r}_1, \tilde{r}_2, \tilde{r}^m)$ is given by

$$\begin{pmatrix} 0.16 & 0.02 & 0.064 \\ 0.02 & 0.09 & 0.032 \\ 0.064 & 0.032 & 0.040 \end{pmatrix}.$$

Moreover, it holds that $\mathbb{E}[\tilde{r}^m] = 1.12$ and $r_f = 1.04$. Consider the portfolio w investing $3/4$ and $1/4$ in the first and in the second assets, respectively.

  (i) Compute the β coefficients of the first and of the second asset as well as of the portfolio w with respect to the market portfolio.
 (ii) Write the equation defining the Security Market Line.
(iii) Compute the risk premium of the two assets as well as of the portfolio w.

Exercise 5.15 Consider an economy with three traded assets, whose returns are generated by the following linear model with respect to two risk factors $\tilde{f}_1$ and $\tilde{f}_2$ (with $\mathbb{E}[\tilde{f}_1] = \mathbb{E}[\tilde{f}_2] = 0$):

$$\begin{cases} \tilde{r}_1 = 0.1 + 0.3\tilde{f}_1 - 0.2\tilde{f}_2, \\ \tilde{r}_2 = 0.5 - 0.4\tilde{f}_1 + 0.3\tilde{f}_2, \\ \tilde{r}_3 = 0.2 - 0.2\tilde{f}_1 + 0.4\tilde{f}_2. \end{cases}$$

  (i) Determine the portfolios of the three traded assets which have unitary exposure to one risk factor and zero exposure to the other risk factor.
 (ii) Determine the risk free rate implicit in the economy.
(iii) Consider an additional asset with return $\tilde{r}_4 = \alpha + 0.1\tilde{f}_1 + 0.3\tilde{f}_2$. Determine the constant α such that there is no arbitrage opportunity in the economy extended with this fourth asset.
 (iv) Verify that the APT is satisfied in exact form.
  (v) Verify that an asset with return $\tilde{r}_5 = 0.2 - 0.1\tilde{f}_1 + 0.1\tilde{f}_2$ would generate an arbitrage opportunity.
 (vi) Assuming that the first three assets are available for trading at unitary price, determine the value of a consumption plan equal to $0.4 - 0.2\tilde{f}_1 + 0.4\tilde{f}_2$.

Exercise 5.16 Prove Proposition 5.16.

Exercise 5.17 Prove Proposition 5.20.

Chapter 6
Multi-Period Models: Portfolio Choice, Equilibrium and No-Arbitrage

> *It is perfectly true, as philosophers say, that life must be understood backwards. But they forget the other proposition, that it must be lived forwards... And if one thinks over the proposition it becomes more and more evident that life can never really be understood in time simply because at no particular moment can I find the necessary resting-place to understand it backwards.*
>
> Kierkegaard (Journal of the year 1843)

The analysis developed in the previous chapters refers to a stylized two-period economy, where the uncertainty is related to the possible state of the world realized at $t = 1$ and agents have to take decisions at the initial date $t = 0$. In this chapter, we consider a more general economy in a multi-period setting, where agents can make dynamic decisions and revise their strategies according to the evolution of the uncertainty on the state of the world. In such a dynamic context, a first and basic issue to consider is how to model the evolution of time. To this regard, one can distinguish between discrete time and continuous time models. In this book, despite the powerful mathematical tools that can be employed for the analysis of continuous time models, we shall limit ourselves to a discrete time setting, since this allows us to capture all the main ideas avoiding the mathematical technicalities of continuous time models. More specifically, we shall consider discrete time models on a finite probability space (i.e., the number of elementary states of the world is finite), thus allowing for an easy description of the resolution of uncertainty. The goal of the present chapter is to extend to a multi-period setting the fundamental results on portfolio optimization, equilibrium and no-arbitrage discussed in the previous chapters for a static economy. Chapter 7 contains an extensive presentation of the main empirical findings and puzzles arising in the context of multi period classical asset pricing theory.

We consider a pure exchange economy comprising $T + 1$ dates $t \in \{0, 1, \ldots, T\}$. For simplicity, we assume that there exists a single good, which can be generically interpreted as wealth or as a reference consumption good and is available for trade at every date $t \in \{0, 1, \ldots, T\}$ (at the expense of a heavier notation, the results presented below can be easily generalized to an economy with multiple goods). To represent the uncertainty, we introduce a probability space $(\Omega, \mathscr{F}, \mathbb{P})$, where, as

© Springer-Verlag London Ltd. 2017
E. Barucci, C. Fontana, *Financial Markets Theory*,
Springer Finance, DOI 10.1007/978-1-4471-7322-9_6

usual, $\Omega = \{\omega_1, \ldots, \omega_{|\Omega|}\}$ represents the finite set of all elementary states of the world, $\mathscr{F}$ is the partition generated by all the elements $\omega \in \Omega$ and $\mathbb{P}$ is a probability measure on $(\Omega, \mathscr{F})$ representing agents' beliefs. Let us recall that a partition of Ω is a collection of non-empty pairwise disjoint subsets of Ω (called *events*) such that their union coincides with Ω.

In the previous chapters, we have considered two-period models (i.e., $t \in \{0, 1\}$), the uncertainty being due to the fact that agents at the initial date $t = 0$ do not know which state of the world will be realized at the future date $t = 1$. At the initial date $t = 0$, the agents only know the possible set of states of the world and their beliefs concerning the possible realizations are represented by a probability measure. In a multi-period setting, we need to make information dynamic, in the sense that agents partially learn over time the true state of the world, which will be fully revealed at the terminal date T. In other words, we need to mathematically represent the evolution of information over time. To this effect, we introduce a refining sequence $\mathbb{F} = (\mathscr{F}_t)_{t=0,1,\ldots,T}$ of partitions of Ω satisfying the crucial property that, for every $s, t \in \{0, 1, \ldots, T\}$ with $s < t$, the partition $\mathscr{F}_t$ is finer than $\mathscr{F}_s$, meaning that every event in $\mathscr{F}_s$ can be represented as the union of some events belonging to $\mathscr{F}_t$. Moreover, we assume that $\mathscr{F}_0 = \Omega$, meaning that agents do not have any information about the true state of the world at the initial date $t = 0$, and that $\mathscr{F}_T = \mathscr{F}$, so that the true state of the world is fully revealed at the terminal date $t = T$. The sequence of partitions $\mathbb{F}$ represents the *information flow* of our economy. Indeed, while agents will discover the true state of the world $\omega \in \Omega$ only at the terminal date T, they will accumulate information over time and learn that ω belongs to increasingly smaller subsets of the universe Ω.

For any $t \in \{0, 1, \ldots, T\}$, we denote by A_t an event of the partition $\mathscr{F}_t$ and we say that event A_t is realized if $\omega \in A_t$, i.e., if the true state of the world is an element of A_t. At every date t, the agents know to which event $A_t \in \mathscr{F}_t$ does ω belong but cannot distinguish among different states of the world belonging to the same event A_t. Given a real-valued random variable $\zeta : \Omega \to \mathbb{R}$ and $t \in \{0, 1, \ldots, T\}$, we shall say that ζ is *measurable* with respect to $\mathscr{F}_t$ if the realization $\zeta(\omega)$ is fully known at date t. More specifically, for every date $t \in \{0, 1, \ldots, T\}$, let us introduce the notation $\nu_t := |\mathscr{F}_t|$, so that $\mathscr{F}_t = \{A_t^1, \ldots, A_t^{\nu_t}\}$. We say that a random variable $\zeta : \Omega \to \mathbb{R}$ is $\mathscr{F}_t$-measurable if it can be represented as $\zeta(\omega) = \sum_{s=1}^{\nu_t} \mathbf{1}_{A_t^s}(\omega) \zeta^s$, with $\zeta^s \in \mathbb{R}$, for all $s = 1, \ldots, \nu_t$. The present notions of information flow and measurability can be generalized by assuming that $\mathbb{F}$ is a *filtration*, namely a refining sequence of σ-algebras on Ω.

Due to its refining structure, $\mathbb{F}$ is sometimes referred to as an *event tree*, see Fig. 6.1. An event tree can be represented by a series of knots ordered from the left (starting at the initial date $t = 0$) to the right (up to the final date $t = T$) and connected through the branches of a tree. Every knot of the tree represents an event and, for every $t \in \{0, 1, \ldots, T\}$, the family of all knots of the tree in correspondence of t represents the partition $\mathscr{F}_t$. Conditionally on the realization of some event A_s at date s, an event/knot A_t (for $s < t$) can contain the true state of the world only if A_t can be reached through the branches of the tree departing from the knot A_s (i.e., if

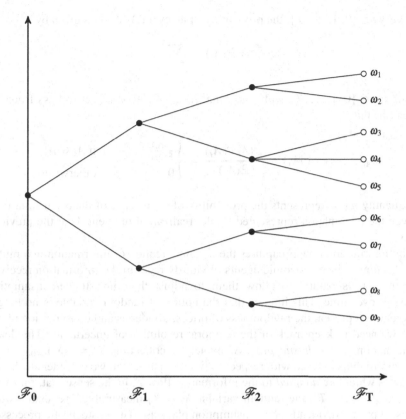

Fig. 6.1 Information flow (event tree)

$A_t \subseteq A_s$). This means that, for every $s, t \in \{0, 1, \dots, T\}$ with $s < t$ and for every $\omega \in \Omega$, if $\omega \in A_s$ then $\omega \notin A_t$ for all the events A_t which are not reachable from A_s.

Since we restrict our attention to information flows represented by refining sequences of partitions (event trees), the knowledge of the event A_t realized at date $t \in \{1, \dots, T\}$ embeds the whole information of the previous events $\{A_s\}_{s=0,\dots,t-1}$ realized before date t. In particular, this implies that the knowledge of the event realized at the current date acts as a sufficient statistic for predicting the realization of future events and past information is not relevant.

As in the previous chapters, we assume that agents exhibit homogeneous beliefs, meaning that all agents agree on a common probability measure $\mathbb{P}$. However, we want to point out that most of the following results can be generalized to the case of heterogeneous beliefs, provided that all agents agree on the set of possible states of the world which can occur with strictly positive probability (i.e., the beliefs of all the agents are equivalent). Moreover, the information flow $\mathbb{F}$ is the same for all the agents, meaning that we do not consider agents having access to some form of privileged information (this situation will be considered in Chap. 8). Every elementary event $\omega \in \Omega$ has a strictly positive probability $\pi_\omega > 0$ of occurrence.

For every $t \in \{0, 1, \ldots, T\}$, the probability of an event $A_t \in \mathscr{F}_t$ is given by

$$\pi_{A_t} := \mathbb{P}(A_t) = \sum_{\omega \in A_t} \pi_\omega.$$

For all $s, t \in \{0, 1, \ldots, T\}$ with $s < t$ and $A_t \in \mathscr{F}_t$, $A_s \in \mathscr{F}_s$, it follows from the Bayes rule that

$$\pi_{A_t|A_s} := \mathbb{P}(A_t|A_s) = \frac{\mathbb{P}(A_s \cap A_t)}{\mathbb{P}(A_s)} = \begin{cases} \frac{\mathbb{P}(A_t)}{\mathbb{P}(A_s)} = \frac{\pi_{A_t}}{\pi_{A_s}} & \text{if } A_t \subseteq A_s; \\ 0 & \text{otherwise.} \end{cases}$$

The quantity $\pi_{A_t|A_s}$ represents the probability of occurrence of the event A_t at time t, given the information represented by the realization of event A_s at the previous date s.

The information flow $\mathbb{F}$ captures the dynamic feature of information in a multi-period setting. Since economic agents obviously react to the information received over time, it is natural to allow them to adjust their investment/consumption strategies over time. This implies that the notion of random variable is no longer sufficient to represent the randomness of prices, dividends and consumption plans, since we need to keep track of the temporal resolution of uncertainty. This leads to the notion of *stochastic process*, namely a collection $X = (X_t)_{t=0,1,\ldots,T}$ of random variables indexed with respect to time. In particular, we consider stochastic processes which are *adapted* to the information flow $\mathbb{F}$, in the sense that, for every date $t \in \{0, 1, \ldots, T\}$, the random variable X_t is $\mathscr{F}_t$-measurable. We shall always represent prices, dividends and consumption plans as adapted stochastic processes. For an adapted stochastic process X, we denote by $X_t(A_t)$ the value of the process X at date t in correspondence of the event A_t, for any $t \in \{0, 1, \ldots, T\}$ and $A_t \in \mathscr{F}_t$.

This chapter is structured as follows. In Sect. 6.1, we analyse by means of dynamic programming techniques the optimal investment-consumption problem for an agent and we present explicit solutions in the case of classical utility functions. In Sect. 6.2, we consider an economy populated by I agents and study the relation between equilibrium and Pareto optimality. Similarly as in Chap. 4, the notion of complete markets will play a crucial role. Section 6.3 extends the fundamental theorem of asset pricing to a multi-period setting, while Sect. 6.4 generalizes the asset pricing relations presented in Chap. 5. In Sect. 6.5 we present some basic concepts related to the existence of speculative bubbles. At the end of the chapter, we provide a guide to further readings as well as a series of exercises.

6.1 Optimal Investment and Consumption Problems

In this section, we present a general approach based on the dynamic programming principle to the solution of the optimal investment-consumption problem of a single agent.

Adopting the setting of Huang & Litzenberger [971, Chapter 7], we assume that the economy comprises N securities paying non-negative (but possibly null) random dividends at the dates $t = 0, 1, \ldots, T$ and available for trade at every date $t = 0, 1, \ldots, T$. In the terminology of Huang & Litzenberger [971], such assets are called *long-lived securities*. Typical examples of long-lived securities are stocks or corporate bonds maturing at the terminal date T. For each $n = 1, \ldots, N$, the n-th security is characterized by its (ex-dividend) price process $s^n = (s^n_t)_{t=0,1,\ldots,T}$ and by its dividend process $d^n = (d^n_t)_{t=0,1,\ldots,T}$ and both processes are assumed to be adapted to the information flow $\mathbb{F}$. As explained in the introduction, this means that, at every date $t = 0, 1, \ldots, T$, the prices and the dividends of the N securities are measurable with respect to the information contained in $\mathscr{F}_t$. This simply means that, at each date $t = 0, 1, \ldots, T$, the prices and the dividends of all N securities are observable. In order to avoid trivial situations, we assume that all securities are in strictly positive supply in the economy and that the dividend process d^n is non-negative and non-null, for all $n = 1, \ldots, N$, in the sense that there exists at least one date $t \in \{0, 1, \ldots, T\}$ such that $d^n_t(A_t) > 0$ in correspondence of some event $A_t \in \mathscr{F}_t$. The price process s^n represents the random evolution of the *ex-dividend* price of the n-th security (i.e., the random variable s^n_t is the price at date t of the n-th security after the delivery of the dividend d^n_t) and is also assumed to be non-negative and non-null, for all $n = 1, \ldots, N$. Moreover, since the economy is supposed to end at the terminal date T, it holds that $s^n_T = 0$, for all $n = 1, \ldots, N$. Let us also define the *cum-dividend* price process $p^n = (p^n_t)_{t=0,1,\ldots,T}$ by $p^n_t := s^n_t + d^n_t$, for all $n = 1, \ldots, N$ and $t = 0, 1, \ldots, T$. We denote by $S = (S_t)_{t=0,1,\ldots,T}$ and $D = (D_t)_{t=0,1,\ldots,T}$ the $\mathbb{R}^N$-valued price and dividend processes of the N securities, respectively, so that $S_t = (s^1_t, \ldots, s^N_t)^\top$ and $D_t = (d^1_t, \ldots, d^N_t)^\top$, for all $t \in \{0, 1, \ldots, T\}$.

Having introduced the general securities available for trade, let us describe how agents can trade, construct portfolios and choose consumption plans. In particular, since the information and the uncertainty evolve dynamically over time, as represented by the information flow $\mathbb{F}$, it is natural to allow agents to rebalance their portfolios over time depending on the resolution of uncertainty and on the available information. For the sake of simplicity, we assume that agents do not receive exogenous income such as labor income. This implies that the agents' wealth is fully determined by their initial endowments, by their consumption plans, by their trading strategies and by the dividend and price processes of the traded securities. As a preliminary, we need to introduce the notion of *predictable* stochastic process: a stochastic process $X = (X_t)_{t=0,1,\ldots,T}$ is said to be predictable if, for all $t = 1, \ldots, T$, the random variable X_t is $\mathscr{F}_{t-1}$-measurable (in particular, note that every predictable stochastic process is adapted). In other words, a stochastic process is predictable if its realization at date t is known one period in advance (at date $t - 1$). We are in a position to formulate the following definition.

Definition 6.1 A couple (θ, c) of stochastic processes is said to be a *trading-consumption strategy* if $\theta = (\theta_t)_{t=0,1,\ldots,T}$ is an $\mathbb{R}^N$-valued predictable stochastic process and $c = (c_t)_{t=0,1,\ldots,T}$ is an $\mathbb{R}_+$-valued adapted stochastic process.

A trading-consumption strategy (θ, c) is said to be *self-financing* if

$$\theta_{t+1}^\top S_t = \theta_t^\top (S_t + D_t) - c_t, \qquad \text{for all } t = 0, 1, \ldots, T - 1, \tag{6.1}$$

$$\theta_T^\top D_T = c_T. \tag{6.2}$$

The processes $\theta = (\theta_t)_{t=0,1,\ldots,T}$ and $c = (c_t)_{t=0,1,\ldots,T}$ appearing in the above definition have the following interpretation. For all $t = 1, \ldots, T$, the random vector $\theta_t = (\theta_t^1, \ldots, \theta_t^N)^\top$ represents the number of units of the N securities held in the portfolio on the time period $[t - 1, t]$ before trading takes place at date t. In particular, the predictability requirement on the process θ ensures that the portfolio constructed at date $t - 1$ and held until date t only depends on the information available at the portfolio formation date $t - 1$. Of course, this represents a minimal informational constraint for a meaningful notion of dynamic trading strategy (excluding clairvoyance of future dividends and asset prices). The vector $\theta_0 = (\theta_0^1, \ldots, \theta_0^N)^\top$ represents the initial holdings in the N securities. Letting $x_0 := \theta_0^\top (S_0 + D_0)$ denote the initial wealth, the self-financing condition can also be formulated as in Definition 6.1 starting from wealth x_0 at date $t = 0$, i.e., requiring that $\theta_1^\top S_0 = x_0 - c$. The consumption process $c = (c_t)_{t=0,1,\ldots,T}$ represents (in units of some reference consumption good or wealth) the quantity consumed by the agent at $t = 0, 1, \ldots, T$. In particular, the process c is adapted, meaning that the quantity consumed at date t is allowed to be contingent on the information released at date t. Note also that condition (6.2) relies on the assumption that the economy ends at the terminal date T and, hence, $S_T = 0$ (so that no trading takes places at the final date T) and the dividends received at T are fully consumed.

To a trading-consumption strategy (θ, c) we can associate a *wealth process* $W(\theta) = (W_t(\theta))_{t=0,1,\ldots,T}$ defined by $W_t(\theta) := \theta_t^\top (S_t + D_t)$, for all $t = 0, 1, \ldots, T$. This process represents the value of the portfolio associated to a strategy θ. In particular, due to equation (6.1), if (θ, c) is a self-financing trading-consumption strategy, then the wealth process $W(\theta)$ satisfies

$$W_t(\theta) = \theta_t^\top (S_t + D_t) = \theta_{t+1}^\top S_t + c_t, \qquad \text{for all } t = 0, 1, \ldots, T - 1. \tag{6.3}$$

The reason for the *self-financing* terminology is made clear by condition (6.3), which requires that the value at date t of the portfolio constructed at the previous date $t - 1$ (on the left-hand side) is equal to the sum of the cost of the new portfolio created at date t and the wealth consumed at date t (on the right-hand side). In particular, there is no injection of wealth at any date $t = 1, \ldots, T$. Condition (6.3) can be equivalently rewritten as follows, for all $t = 0, 1, \ldots, T - 1$:

$$\Delta W_{t+1}(\theta) := W_{t+1}(\theta) - W_t(\theta) = \theta_{t+1}^{\mathsf{T}}(S_{t+1} + D_{t+1} - S_t) - c_t$$
$$= \theta_{t+1}^{\mathsf{T}} \Delta S_{t+1} + \theta_{t+1}^{\mathsf{T}} D_{t+1} - c_t. \tag{6.4}$$

As shown in Exercise 6.1, condition (6.4) easily leads to the following representation of the wealth process $W(\theta)$ associated to a self-financing trading-consumption strategy (θ, c):

$$W_t(\theta) = W_0(\theta) + \sum_{s=1}^{t} \theta_s^{\mathsf{T}} \Delta S_s + \sum_{s=1}^{t} \theta_s^{\mathsf{T}} D_s - \sum_{s=0}^{t-1} c_s, \quad \text{for all } t = 1, \ldots, T. \tag{6.5}$$

Note that the self-financing property of a trading-consumption strategy does not depend on the quantity with respect to which prices and consumption are measured (in other words, the self-financing property is *numéraire invariant*). For instance, if a risk free security is traded in the market, then discounting all the quantities with respect to it will not affect the self-financing property (see Exercise 6.2).

If a trading-consumption strategy (θ, c) is self-financing, then we say that the consumption plan c is *financed* by the trading strategy θ, starting from the initial wealth $x = W_0(\theta)$. We denote by $\mathscr{A}$ the set of all self-financing trading-consumption strategies.

Having described the available securities and introduced the notion of trading-consumption strategy, we are now in a position to formulate the optimal investment-consumption problem of an agent. We assume that the agent is endowed at date $t = 0$ with an initial wealth $x_0 > 0$ (we can also assume that the initial endowment is expressed as a vector $\theta_0 \in \mathbb{R}^N$, with θ_0^n denoting the number of shares of the n-th security, for $n = 1, \ldots, N$, so that the initial wealth is $x_0 = \theta_0^{\mathsf{T}}(S_0 + D_0)$). We introduce the set

$$\mathscr{A}(x_0) := \{(\theta, c) \in \mathscr{A} : W_0(\theta) = x_0\}$$

representing all self-financing trading-consumption strategies starting from initial wealth x_0. Concerning the agent's preferences, we assume that they can be represented by an expected time additive and state independent utility function defined over consumption plans. More specifically, for each $t = 0, 1, \ldots, T$, we assume that the utility associated with consumption at date t is evaluated at $t = 0$ by the discounted expected utility $\delta^t \mathbb{E}[u(c_t)]$, where $u : \mathbb{R}_+ \to \mathbb{R}$ is a strictly increasing, strictly concave and twice differentiable function and δ is a constant discount factor with $0 < \delta \le 1$. Hence, given the initial wealth $x_0 > 0$, the utility function u and the discount factor δ, an agent's optimal investment-consumption problem can be formulated as follows:

$$\max_{(\theta,c)\in\mathscr{A}(x_0)} \left(u(c_0) + \sum_{t=1}^{T} \delta^t \mathbb{E}[u(c_t)] \right). \tag{6.6}$$

In problem (6.6) the maximization is with respect to all self-financing trading-consumption strategies (θ, c) starting from the initial wealth x_0. Whenever it exists (and this will be related to the absence of arbitrage opportunities, see Proposition 6.31), we denote by (θ^*, c^*) the optimal solution to problem (6.6).

Apart from some special cases (see below), the optimal investment-consumption problem (6.6) cannot be solved as a sequence of static single-period optimization problems. Therefore, one cannot apply step-by-step for each date $t = 0, 1, \ldots, T-1$ the static optimization techniques introduced in the previous chapters (even though, as we shall see below, the techniques introduced in the previous chapters for single-period problems will play an important role in the solution of the multi-period problem (6.6)). Nonetheless, if problem (6.6) admits an optimal trading-consumption strategy (θ^*, c^*), then the same strategy should remain optimal at every future date $t = 1, \ldots, T$. In other words, if an optimal solution (θ^*, c^*) to problem (6.6) has been found at the initial date $t = 0$, then (θ^*, c^*) is such that, for every future date $t = 1, \ldots, T$, there does not exist an alternative trading-consumption strategy $(\tilde{\theta}, \tilde{c})$ starting at date t from wealth $W_t(\theta^*)$ which improves the expected utility from date t onwards. This will be shown rigorously in Proposition 6.2 below.

As a preliminary, we need to introduce some notation. For $t \in \{0, 1, \ldots, T-1\}$, we define $\mathscr{C}_t^+(x)$ as the set of all non-negative adapted processes $(c_s)_{s=t,\ldots,T}$ such that there exists an $\mathbb{R}^N$-valued predictable process $(\tilde{\theta}_s)_{s=t+1,\ldots,T}$ (starting at date t) satisfying

$$\tilde{\theta}_{t+1}^{\top} S_t = x - c_t,$$

$$\tilde{\theta}_{s+1}^{\top} S_s = \tilde{\theta}_s^{\top}(S_s + D_s) - c_s, \qquad \text{for all } s = t+1, \ldots, T-1, \tag{6.7}$$

$$\tilde{\theta}_T^{\top} D_T = c_T.$$

In other words, $\mathscr{C}_t^+(x)$ represents the set of all consumption processes financed (from date t onwards) by some trading strategy starting at date t from wealth x.[1] We then have the following result (see also Huang & Litzenberger [971, Section 7.8]).

Proposition 6.2 Let $(\theta^*, c^*) \in \mathscr{A}(x_0)$ be an optimal solution to problem (6.6). Then, for every date $t \in \{0, 1, \ldots, T-1\}$, there does not exist a consumption process

---

[1]The definition of the set $\mathscr{C}_t^+(x)$ can be straightforwardly extended to the case where wealth x is a strictly positive $\mathscr{F}_t$-measurable random variable, as considered in Proposition 6.2.

$(\tilde{c}_s)_{s=t,...,T} \in \mathscr{C}_t^+(W_t(\theta^*))$ *such that, on some event* $\tilde{A}_t \in \mathscr{F}_t$, *it holds that*

$$\sum_{s=t}^{T} \delta^s \sum_{\substack{A_s \in \mathscr{F}_s \\ A_s \subseteq \tilde{A}_t}} \pi_{A_s|\tilde{A}_t} u(\tilde{c}_s(A_s)) > \sum_{s=t}^{T} \delta^s \sum_{\substack{A_s \in \mathscr{F}_s \\ A_s \subseteq \tilde{A}_t}} \pi_{A_s|\tilde{A}_t} u(c_s^*(A_s)), \qquad (6.8)$$

where $\tilde{c}_s(A_s)$ *and* $c_s^*(A_s)$ *denote the values of the processes* $\tilde{c}$ *and* c^*, *respectively, at date* s *in correspondence of the event* A_s, *for* $A_s \in \mathscr{F}_s$ *and* $s \in \{t, \ldots, T\}$.

Proof The proposition can be proved by contradiction. Suppose that there exists a process $(\tilde{c}_s)_{s=t,...,T} \in \mathscr{C}_t^+(W_t(\theta^*))$ such that (6.8) holds in correspondence of some event $\tilde{A}_t \in \mathscr{F}_t$. Since $(\tilde{c}_s)_{s=t,...,T} \in \mathscr{C}_t^+(W_t(\theta^*))$, there exists an $\mathbb{R}^N$-valued predictable process $(\tilde{\theta}_s)_{s=t+1,...,T}$ such that (6.7) holds (with $c = \tilde{c}$). Define then a new trading-consumption strategy (θ', c') as follows, for all events $A_s \in \mathscr{F}_s$ and for all dates $s = 0, 1, \ldots, T$:

$$\theta_s'(A_s) := \begin{cases} \theta_s^*(A_s) & \text{for } s = 0, \ldots, t, \\ \theta_s^*(A_s) + (\tilde{\theta}_s(A_s) - \theta_s^*(A_s))\mathbf{1}_{A_s \subseteq \tilde{A}_t} & \text{for } s = t+1, \ldots, T; \end{cases}$$

$$c_s'(A_s) := \begin{cases} c_s^*(A_s) & \text{for } s = 0, \ldots, t-1, \\ c_s^*(A_s) + (\tilde{c}_s(A_s) - c_s^*(A_s))\mathbf{1}_{A_s \subseteq \tilde{A}_t} & \text{for } s = t, \ldots, T, \end{cases}$$

where $\theta_s^*(A_s)$, $\tilde{\theta}_s(A_s)$, $c_s^*(A_s)$ and $\tilde{c}_s(A_s)$ denote the values of the processes θ^*, $\tilde{\theta}$, c^* and $\tilde{c}$, respectively, at date s in correspondence of $A_s \in \mathscr{F}_s$ and where $\mathbf{1}_{A_s \subseteq \tilde{A}_t}$ takes the value one if A_s is a subset of $\tilde{A}_t$ and zero otherwise. Since $(\theta^*, c^*) \in \mathscr{A}(x_0)$ and $(\tilde{\theta}, \tilde{c})$ satisfies (6.7), it can be easily checked that $(\theta', c') \in \mathscr{A}(x_0)$. Moreover, the definition of c' and inequality (6.8) imply that

$$\sum_{s=t}^{T} \delta^s \sum_{\substack{A_s \in \mathscr{F}_s \\ A_s \subseteq A_t}} \pi_{A_s|A_t} u(c_s'(A_s)) \geq \sum_{s=t}^{T} \delta^s \sum_{\substack{A_s \in \mathscr{F}_s \\ A_s \subseteq A_t}} \pi_{A_s|A_t} u(c_s^*(A_s)),$$

for all events $A_t \in \mathscr{F}_t$, with strict inequality in correspondence of $\tilde{A}_t \in \mathscr{F}_t$ (which is realized with strictly positive probability, since every elementary event $\omega \in \Omega$ has a strictly positive probability of occurrence). Denoting by $\mathbb{E}[\cdot|\mathscr{F}_t]$ the conditional expectation with respect to $\mathscr{F}_t$ and using the tower property of conditional expectation (so that $\mathbb{E}[\cdot] = \mathbb{E}[\mathbb{E}[\cdot|\mathscr{F}_t]]$), it holds that

$$\sum_{s=t}^{T} \delta^s \mathbb{E}[u(c_s')] = \mathbb{E}\left[\sum_{s=t}^{T} \delta^s \mathbb{E}[u(c_s')|\mathscr{F}_t]\right] > \mathbb{E}\left[\sum_{s=t}^{T} \delta^s \mathbb{E}[u(c_s^*)|\mathscr{F}_t]\right] = \sum_{s=t}^{T} \delta^s \mathbb{E}[u(c_s^*)],$$

which implies that

$$\sum_{s=0}^{T} \delta^s \mathbb{E}[u(c_s^*)] = \sum_{s=0}^{t-1} \delta^s \mathbb{E}[u(c_s^*)] + \sum_{s=t}^{T} \delta^s \mathbb{E}[u(c_s^*)]$$

$$< \sum_{s=0}^{t-1} \delta^s \mathbb{E}[u(c_s')] + \sum_{s=t}^{T} \delta^s \mathbb{E}[u(c_s')] = \sum_{s=0}^{T} \delta^s \mathbb{E}[u(c_s')].$$

This contradicts the optimality of (θ^*, c^*), thus proving the claim.   □

According to the above proposition, once an optimal strategy (θ^*, c^*) has been chosen at the initial date $t = 0$, it will never be optimal to revise that strategy at a later date $t = 1, \ldots, T$. The optimal strategy (θ^*, c^*) must remain optimal starting from every knot of the event-tree. In other words, the current optimal decision of an agent already takes into account the future optimal decisions in response to the resolution of the uncertainty. This result is crucial for the solution of the dynamic investment-consumption problem and naturally leads to the *dynamic programming principle*, which provides a backward recursive technique for the solution of stochastic control problems of the type (6.6).

For the investment-consumption problem (6.6), we introduce the *value function* (or, more precisely, *value process*) V as

$$V(x, t) := \sup_{c \in \mathscr{C}_t^+(x)} \left(u(c_t) + \sum_{s=t+1}^{T} \delta^{s-t} \mathbb{E}[u(c_s)|\mathscr{F}_t] \right), \tag{6.9}$$

for $t = 0, 1, \ldots, T - 1$, with $V(x, T) := u(x)$, for all $x \in \mathbb{R}_+$. In (6.9) the supremum is taken over all consumption processes $(c_s)_{s=t,\ldots,T}$ that can be financed by some trading strategy $(\tilde{\theta}_s)_{s=t+1,\ldots,T}$ starting at date t from wealth x, see (6.7). The quantity $V(x, t)$ represents the maximal expected utility that can be reached starting at date t with an endowment of wealth x, given the information $\mathscr{F}_t$ available at t. Definition (6.9) can be extended to the case where x is an $\mathbb{R}_+$-valued $\mathscr{F}_t$-measurable random variable. It is important to remark that, in general, the quantity $V(x, t)$ is an $\mathscr{F}_t$-measurable random variable and cannot be considered as a deterministic function of x and t (see however the special cases considered below in this section). The next proposition states the dynamic programming principle for the investment-consumption problem (6.6) in terms of the value function V.

Proposition 6.3 *For every* $(x, t) \in \mathbb{R}_+ \times \{0, 1, \ldots, T - 1\}$, *the value function V defined in* (6.9) *satisfies the* dynamic programming principle*:*

$$V(x, t) = \sup \left(u(c_t) + \delta \mathbb{E}[V(\theta_{t+1}^\top (S_{t+1} + D_{t+1}), t + 1)|\mathscr{F}_t] \right), \tag{6.10}$$

where the supremum is taken with respect to all $\mathscr{F}_t$*-measurable random variables* $(\theta_{t+1}, c_t) \in \mathbb{R}^N \times \mathbb{R}_+$ *satisfying* $\theta_{t+1}^\top S_t + c_t = x$. *Moreover, if* $(\theta^*, c^*) \in \mathscr{A}(x_0)$ *is* *an optimal solution to problem* (6.6), *then it holds that*

$$V\big(W_t(\theta^*), t\big) = u(c_t^*) + \delta \mathbb{E}\big[V\big(W_{t+1}(\theta^*), t+1\big)\big|\mathscr{F}_t\big], \qquad \text{for all } t = 0, 1, \ldots, T-1. \tag{6.11}$$

Proof Let $(x, t) \in \mathbb{R}_+ \times \{0, 1, \ldots, T-1\}$ and $(c_s)_{s=t,\ldots,T} \in \mathscr{C}_t^+(x)$ (as explained above, this means that $(c_s)_{s=t,\ldots,T}$ is a consumption stream financed by some trading strategy $(\theta_s)_{s=t+1,\ldots,T}$ starting at date t from wealth x). For $t = T-1$, equation (6.10) coincides with the definition (6.9) of the value function V. So, let us suppose that $t \in \{0, 1, \ldots, T-2\}$. Then, it holds that

$$u(c_t) + \sum_{s=t+1}^{T} \delta^{s-t} \mathbb{E}[u(c_s)|\mathscr{F}_t]$$

$$= u(c_t) + \delta \mathbb{E}\bigg[u(c_{t+1}) + \sum_{s=t+2}^{T} \delta^{s-(t+1)} u(c_s)\bigg|\mathscr{F}_t\bigg]$$

$$= u(c_t) + \delta \mathbb{E}\bigg[u(c_{t+1}) + \mathbb{E}\bigg[\sum_{s=t+2}^{T} \delta^{s-(t+1)} u(c_s)\big|\mathscr{F}_{t+1}\bigg]\bigg|\mathscr{F}_t\bigg]$$

$$\leq u(c_t) + \delta \mathbb{E}\bigg[\sup_{c\in\mathscr{C}_{t+1}^+(\theta_{t+1}^\top(S_{t+1}+D_{t+1}))} \bigg(u(c_{t+1}) + \mathbb{E}\bigg[\sum_{s=t+2}^{T} \delta^{s-(t+1)} u(c_s)\big|\mathscr{F}_{t+1}\bigg]\bigg)\bigg|\mathscr{F}_t\bigg]$$

$$= u(c_t) + \delta \mathbb{E}\big[V\big(\theta_{t+1}^\top(S_{t+1} + D_{t+1}), t+1\big)\big|\mathscr{F}_t\big],$$

where the last equality follows from definition (6.9). By taking the supremum over all $\mathscr{F}_t$-measurable couples $(\theta_{t+1}, c_t) \in \mathbb{R}^N \times \mathbb{R}_+$ satisfying $\theta_{t+1}^\top S_t + c_t = x$ in the right-hand side of the above inequality and then the supremum over all $c \in \mathscr{C}_t^+(x)$ in the left-hand side, we obtain

$$V(x, t) \leq \sup_{\substack{(\theta_{t+1}, c_t)\in\mathbb{R}^N\times\mathbb{R}_+ \\ \theta_{t+1}^\top S_t + c_t = x}} \bigg(u(c_t) + \delta \mathbb{E}\big[V\big(\theta_{t+1}^\top(S_{t+1} + D_{t+1}), t+1\big)\big|\mathscr{F}_t\big]\bigg).$$

$$\tag{6.12}$$

To prove the converse inequality, let $(\theta_{t+1}, c_t) \in \mathbb{R}^N \times \mathbb{R}_+$ be a couple of $\mathscr{F}_t$-measurable random variables such that $\theta_{t+1}^\top S_t + c_t = x$. Consider then the value function defined in (6.9) at $t + 1$ in correspondence of wealth $\theta_{t+1}^\top(S_{t+1} + D_{t+1})$. Observe that, by the definition of supremum, for every $\varepsilon > 0$ there exists a consumption stream $(c_s^\varepsilon)_{s=t+1,\ldots,T} \in \mathscr{C}_{t+1}^+(\theta_{t+1}^\top(S_{t+1} + D_{t+1}))$ such that

$$u(c_{t+1}^\varepsilon) + \mathbb{E}\bigg[\sum_{s=t+2}^{T} \delta^{s-(t+1)} u(c_s^\varepsilon)\big|\mathscr{F}_{t+1}\bigg] \geq V\big(\theta_{t+1}^\top(S_{t+1} + D_{t+1})\big) - \varepsilon.$$

This implies that

$$u(c_t) + \delta\mathbb{E}\big[V\big(\theta_{t+1}^\top(S_{t+1} + D_{t+1}), t + 1\big)|\mathscr{F}_t\big]$$

$$\leq u(c_t) + \delta\mathbb{E}\bigg[u(c_{t+1}^\varepsilon) + \mathbb{E}\bigg[\sum_{s=t+2}^{T}\delta^{s-(t+1)}u(c_s^\varepsilon)|\mathscr{F}_{t+1}\bigg]\bigg|\mathscr{F}_t\bigg] + \varepsilon$$

$$= u(c_t) + \sum_{s=t+1}^{T}\delta^{s-t}\mathbb{E}[u(c_s^\varepsilon)|\mathscr{F}_t] + \varepsilon$$

$$\leq V(x, t) + \varepsilon,$$

where the last inequality follows from the fact that $(c_t, c_{t+1}^\varepsilon, \ldots, c_T^\varepsilon) \in \mathscr{C}_t^+(x)$. By the arbitrariness of ε, we get that

$$u(c_t) + \delta\mathbb{E}\big[V\big(\theta_{t+1}^\top(S_{t+1} + D_{t+1}), t + 1\big)|\mathscr{F}_t\big] \leq V(x, t).$$

By taking the supremum in the left-hand side of the last inequality over all elements $(\theta_{t+1}, c_t) \in \mathbb{R}^N \times \mathbb{R}_+$ such that $\theta_{t+1}^\top S_t + c_t = x$, we get the converse inequality to (6.12), thus proving that (6.10) holds as an equality. Finally, in correspondence of an optimal solution $(\theta^*, c^*) \in \mathscr{A}(x_0)$, relation (6.11) follows from Proposition 6.2 together with the dynamic programming principle (6.10). $\square$

The above proposition represents a fundamental result for the solution of dynamic optimal investment-consumption problems of the type (6.6). Indeed, for any $t \in \{0, 1, \ldots, T - 1\}$, the optimal solution to problem (6.6) can be found by first searching for an optimal trading-consumption strategy from time $t + 1$ onwards and then computing the maximum appearing in the right-hand side of (6.10) over the single-period interval $[t, t+1]$. In particular, condition (6.11) shows that, if a trading-consumption strategy $(\theta^*, c^*) \in \mathscr{A}(x_0)$ is an optimal solution to problem (6.6), then, for every $t = 0, 1, \ldots, T - 1$, the couple (θ_{t+1}^*, c_t^*) must be optimal over the time interval $[t, t + 1]$ for the maximization problem (6.10). Equation (6.11) is also known as the *Bellman equation*.

The dynamic programming principle provides a *backward recursive* algorithm for the solution of problem (6.6). At each time step $t \in \{0, 1, \ldots, T - 1\}$, the global maximization problem is reduced to the standard optimization problem (6.10) over $\mathbb{R}^N \times \mathbb{R}_+$ conditionally on the information $\mathscr{F}_t$ available at date t. By means of this approach, the multi-period optimization problem (6.6) is reduced to a sequence of simpler single-period optimization problems, as will be illustrated below. The general strategy in the application of the backward recursive algorithm consists in starting by considering conditions (6.10)–(6.11) at date $T - 1$, with one period remaining until the terminal date T. On this time interval, condition (6.10) reduces to a standard single-period optimization problem, to which the techniques introduced in the previous chapters can be applied. Having computed the optimal consumption and the value function at date $T-1$ conditionally on $\mathscr{F}_{T-1}$, one can then solve (6.10) and compute the optimal consumption for the time interval $[T-2, T-1]$ and then continue backwards in time until the initial date $t = 0$. As a result of this procedure, the optimality of the future decisions is already taken into account at each single step of the optimization problem. This procedure is illustrated in Exercises 6.4–6.7.

The dynamic programming principle leads to the following properties of the value function in correspondence of an optimal solution (θ^*, c^*) to problem (6.6). We denote by $(W_t^*)_{t=0,1,\ldots,T}$ the wealth process associated to the optimal trading-consumption strategy (θ^*, c^*), i.e., $W_t^* := W_t(\theta^*)$, where $W_t(\theta^*)$ is defined as in (6.3). From now on, we always assume that the prices of the N traded securities are strictly positive,[2] i.e., $s_t^n > 0$, for all $n = 1, \ldots, N$ and $t = 0, 1, \ldots, T-1$, and we denote by $r_t^n := p_t^n / s_{t-1}^n$ the random return of asset n in the time interval $[t-1, t]$.

Proposition 6.4 *Let $(\theta^*, c^*) \in \mathscr{A}(x_0)$ be an optimal solution to problem (6.6). If $V(\cdot, t)$ is differentiable with respect to its first argument, for each $t = 1, \ldots, T$, then the following holds, for all $t = 0, 1, \ldots, T-1$:*

$$u'(c_t^*) = \delta \mathbb{E}[V'(W_{t+1}^*, t+1) r_{t+1}^n | \mathscr{F}_t], \qquad \text{for all } n = 1, \ldots, N, \qquad (6.13)$$

and

$$u'(c_t^*) = V'(W_t^*, t), \qquad (6.14)$$

where $V'(\cdot, t)$ denotes the first derivative of $V(\cdot, t)$ with respect to its first argument. In particular, (6.13)–(6.14) together imply that, for all $t = 0, 1, \ldots, T-1$,

$$u'(c_t^*) = \delta \mathbb{E}[u'(c_{t+1}^*) r_{t+1}^n | \mathscr{F}_t], \qquad \text{for all } n = 1, \ldots, N. \qquad (6.15)$$

Proof Due to Proposition 6.3, condition (6.11) holds in correspondence of an optimal solution (θ^*, c^*) to problem (6.6). Hence, using equation (6.3), we get

$$V(W_t^*, t) = u(W_t^* - \theta_{t+1}^{*\mathsf{T}} S_t) + \delta \mathbb{E}[V(\theta_{t+1}^{*\mathsf{T}}(S_{t+1} + D_{t+1}), t+1) | \mathscr{F}_t],$$

for all $t = 0, 1, \ldots, T-1$. Hence, if $V(\cdot, t)$ is differentiable with respect to its first argument, for all $t = 1, \ldots, T$, differentiation with respect to θ_{t+1} leads to the optimality condition

$$0 = -u'(W_t^* - \theta_{t+1}^{*\mathsf{T}} S_t) s_t^n + \delta \mathbb{E}[V'(\theta_{t+1}^{*\mathsf{T}}(S_{t+1} + D_{t+1}), t+1)(s_{t+1}^n + d_{t+1}^n) | \mathscr{F}_t], \qquad (6.16)$$

---

[2]Note that the absence of arbitrage opportunities (see Sect. 6.3) implies that, if $d_t^n(A_t) > 0$ for some $A_t \in \mathscr{F}_t$ and $t \in \{1, \ldots, T\}$, then $s_s^n(A_s) > 0$ for all $A_s \supseteq A_t$ and $s = 0, 1, \ldots, t-1$. In other words, if a security delivers a strictly positive dividend in correspondence of some future event which can be realized with strictly positive probability, then its current price must be strictly positive as well.

for each $n = 1, \ldots, N$. Equation (6.13) then follows by dividing by s_t^n, which is assumed to be strictly positive. Moreover, using in sequence the optimality condition (6.11), the self-financing condition (6.4), equation (6.16) and the self-financing condition (6.3), it holds that

$$
V'(W_t^*, t) = u'(c_t^*)\frac{\partial c_t^*}{\partial W_t^*} + \delta \mathbb{E}\left[V'(W_{t+1}^*, t+1)\frac{\partial W_{t+1}^*}{\partial W_t^*}\bigg|\mathscr{F}_t\right]
$$

$$
= u'(c_t^*)\frac{\partial c_t^*}{\partial W_t^*}
$$

$$
+ \delta \mathbb{E}\left[V'(W_{t+1}^*, t+1)\left(1 - \frac{\partial c_t^*}{\partial W_t^*} + \sum_{n=1}^{N}\frac{\partial \theta_{t+1}^{*n}}{\partial W_t^*}(s_{t+1}^n + d_{t+1}^n - s_t^n)\right)\bigg|\mathscr{F}_t\right]
$$

$$
= u'(c_t^*)\left(\frac{\partial c_t^*}{\partial W_t^*} + \sum_{n=1}^{N}\frac{\partial \theta_{t+1}^{*n}}{\partial W_t^*}s_t^n\right)
$$

$$
+ \delta \mathbb{E}\left[V'(W_{t+1}^*, t+1)\left(1 - \frac{\partial c_t^*}{\partial W_t^*} - \sum_{n=1}^{N}\frac{\partial \theta_{t+1}^{*n}}{\partial W_t^*}s_t^n\right)\bigg|\mathscr{F}_t\right]
$$

$$
= u'(c_t^*),
$$

where in the third equality we have used the $\mathscr{F}_t$-measurability of θ_{t+1}^* (since θ^* is a predictable process). This proves (6.14). □

Proposition 6.4 shows several important properties of a solution to the optimal investment-consumption problem (6.6). In particular, relation (6.14) shows that, in correspondence of an optimum, the marginal utility $u'(c_t^*)$ of consumption at time t is equal to the marginal (indirect) utility of wealth $V'(W_t^*, t)$ at time t. This condition (which is often referred to as the *envelope condition*) intuitively means that, in correspondence of an optimum, one extra unit of consumption at time t or one extra unit of wealth invested in the optimal way from date t onwards should give the same marginal increase of expected utility. Since wealth is only used to finance future consumption, this is equivalent to saying that, in correspondence of an optimal solution, the marginal utility of current consumption should equal the marginal utility of future consumption. Condition (6.15) is often referred to as the *Euler condition* for the investment-consumption problem (6.6). Note also that equation (6.15) can be equivalently rewritten as

$$
s_t^n = \delta \mathbb{E}\left[\frac{u'(c_{t+1}^*)}{u'(c_t^*)}(d_{t+1}^n + s_{t+1}^n)\bigg|\mathscr{F}_t\right], \qquad \text{for all } n = 1, \ldots, N \text{ and } t = 0, 1, \ldots, T-1.
$$

This fundamental relation will be used in Sect. 6.4 to derive several asset pricing relations.

If u is strictly increasing and strictly concave, then it can be shown that $V(\cdot, t)$ is increasing and strictly concave in its first argument (representing wealth), for every $t = 0, 1, \ldots, T$ (see Exercise 6.3). As a consequence, in view of (6.14), there exists for every $t = 0, 1, \ldots, T$ a random function g_t such that $c_t^* = g_t(W_t^*)$, with $g_t(\cdot)$ strictly increasing and $\mathscr{F}_t$-measurable (compare also with Huang & Litzenberger [971, Section 7.10]). However, the function $g_t(\cdot)$ is in general a random function depending on $\mathscr{F}_t$ and, in general, can be explicitly computed only in some special cases, as considered below in this section.

In most of the cases, it is assumed that the available securities include a risk free asset delivering a constant rate of return. In that case, the previous proposition immediately implies the following result, which will be useful for the characterization of optimal strategies.

Corollary 6.5 *Suppose that there are $N + 1$ traded securities $n = 0, 1, \ldots, N$, with security 0 being a risk free asset with constant rate of return $r_f > 0$. Then, under the assumptions of Proposition 6.4, the following condition holds in correspondence of an optimal solution (θ^*, c^*) to problem (6.6), for all $t = 0, 1, \ldots, T - 1$:*

$$\mathbb{E}[V'(W_{t+1}^*, t + 1)(r_{t+1}^n - r_f)|\mathscr{F}_t] = 0, \qquad \text{for all } n = 1, \ldots, N. \tag{6.17}$$

Proof Equation (6.13) implies that, for all $n = 1, \ldots, N$ and $t = 0, 1, \ldots, T - 1$,

$$\delta\mathbb{E}[V'(W_{t+1}^*, t + 1)r_f|\mathscr{F}_t] = u'(c_t^*) = \delta\mathbb{E}[V'(W_{t+1}^*, t + 1)r_{t+1}^n|\mathscr{F}_t],$$

form which (6.17) immediately follows. $\qquad \square$

In view of Proposition 6.4, condition (6.17) can be equivalently rewritten as

$$\mathbb{E}[u'(c_{t+1}^*)(r_{t+1}^n - r_f)|\mathscr{F}_t] = 0, \qquad \text{for all } n = 1, \ldots, N \text{ and } t = 0, 1, \ldots, T - 1.$$

Moreover, equation (6.15) implies that

$$u'(c_t^*) = \delta r_f \mathbb{E}[u'(c_{t+1}^*)|\mathscr{F}_t], \qquad \text{for all } t = 0, 1, \ldots, T - 1.$$

In particular, if $\delta = 1/r_f$, then we get

$$u'(c_t^*) = \mathbb{E}[u'(c_{t+1}^*)|\mathscr{F}_t], \qquad \text{for all } t = 0, 1, \ldots, T - 1,$$

meaning that, in correspondence of an optimum, the marginal utility of the optimal consumption process $(u'(c_t^*))_{t=0,1,\ldots,T}$ is a *martingale* process[3] (if $\delta = 1/r_f$).

---

[3] We recall that an adapted process $X = (X_t)_{t=0,1,\ldots,T}$ is said to be a *martingale* if, for all $s, t \in \{0, 1, \ldots, T\}$ with $s \leq t$, it holds that $\mathbb{E}[X_t|\mathscr{F}_s] = X_s$.

Furthermore, if the utility function is quadratic this reduces to

$$c_t^* = \mathbb{E}[c_{t+1}^* | \mathscr{F}_t], \qquad \text{for all } t = 0, 1, \ldots, T - 1,$$

so that the optimal consumption process c^* follows a random walk of the form

$$c_{t+1}^* = c_t^* + \tilde{\varepsilon}_{t+1}, \qquad \text{for all } t = 0, 1, \ldots, T - 1,$$

where $(\tilde{\varepsilon}_t)_{t=1,\ldots,T}$ is a sequence of random variables satisfying $\mathbb{E}[\tilde{\varepsilon}_t | \mathscr{F}_{t-1}] = 0$, for all $t = 1, \ldots, T$. This result provides a basis for the so-called *life-cycle-permanent income hypothesis*. The random walk hypothesis implies that consumption change is unrelated to predictable and lagged income changes, but it can depend on unanticipated income changes. Consumption growth should be more volatile than income growth if aggregate income exhibits positive serial correlation. There is a large literature on these implications, showing that consumption growth is both excessively smooth relative to unanticipated (labor) income growth and excessively sensitive to lagged-predictable (labor) income growth, see Campbell & Deaton [345], Campbell & Mankiw [349], Carroll [368], Christiano et al. [444], Deaton [530], Flavin [714], Hall [878], Hall & Mishkin [879]. Moreover, consumption is affected by the business cycle. We will return to these topics in the following chapter.

Explicit Solutions to the Optimal Investment-Consumption Problem

We have so far presented a general theory for the solution of optimal investment-consumption problems of the form (6.6). In the remaining part of this section, we derive the optimal investment-consumption plans for two important classes of utility functions, by relying on the dynamic programming principle, following Ingersoll [1000, Chapter 11]. In particular, we shall see that the optimal solution to a multi-period optimal investment-consumption problem does not coincide in general with the sequence of solutions obtained in the static single-period setting with the same utility function. This feature reflects the fact that, when solving problem (6.6), an agent needs to optimally face a trade-off between optimizing single-period returns and hedging against future changes in the set of investment opportunities. In the context of a multi-period binomial model (see Sect. 6.3), Exercise 6.9 derives the explicit solutions to some classical optimal intertemporal consumption problems.

Until the end of this section, we shall assume that the economy comprises $N + 1$ securities, with the 0-th security being a risk free asset delivering a constant rate of return $r_f > 0$, as in Corollary 6.5. Moreover, we restrict our attention to utility functions defined with respect to strictly positive consumption plans.[4]

---

[4]Note that, if we consider utility functions $u : \mathbb{R}_+ \to \mathbb{R}$ satisfying the Inada condition $\lim_{x \searrow 0} u'(x) = +\infty$, then the corresponding optimal consumption plan $(c_t^*)_{t=0,1,\ldots,T}$ will necessarily satisfy $c_t^* > 0$ for all $t = 0, 1, \ldots, T$.

As shown in Lemma 6.19 below, if there are no arbitrage opportunities, then the wealth process $(W_t(\theta))_{t=0,1,\dots,T}$ associated to a trading-consumption strategy $(\theta, c) \in \mathscr{A}(x_0)$ satisfying $c_T > 0$ almost surely necessarily satisfies $W_t(\theta) - c_t > 0$ for all $t = 0, 1, \dots, T-1$. In this case, we can pass to an alternative parametrization of the trading strategies in terms of *proportions of wealth* invested in the $N + 1$ available securities. To this effect, for any $(\theta, c) \in \mathscr{A}(x_0)$ satisfying $W_t(\theta) - c_t > 0$ for all $t = 0, 1, \dots, T - 1$, we define

$$w_{t+1}^n := \frac{\theta_{t+1}^n s_t^n}{W_t(\theta) - c_t}, \tag{6.18}$$

for all $t = 0, 1, \dots, T - 1$ and $n = 0, \dots, N$. The quantity w_{t+1}^n represents the proportion of wealth (which is not consumed at date t) invested in the n-th asset at date t. Note also that, due to equation (6.3), it holds that $\sum_{n=0}^{N} w_{t+1}^n = 1$, for all $t = 0, 1, \dots, T - 1$. Moreover, the self-financing equation (6.3) can be rewritten as follows (for simplicity of notation, if no ambiguity arises about the trading strategy, we write $W_t := W_t(\theta)$):

$$W_{t+1} = \sum_{n=1}^{N} \theta_{t+1}^n (s_{t+1}^n + d_{t+1}^n) + \theta_{t+1}^0 r_f^{t+1}$$

$$= \sum_{n=1}^{N} \theta_{t+1}^n p_{t+1}^n + \left(W_t - c_t - \sum_{n=1}^{N} \theta_{t+1}^n s_t^n\right) r_f$$

$$= (W_t - c_t) \left(\sum_{n=1}^{N} w_{t+1}^n (r_{t+1}^n - r_f) + r_f\right), \tag{6.19}$$

where we have conventionally assumed that $s_0^0 = 1$. In view of equations (6.3) and (6.19), it is clear that the wealth process $(W_t)_{t=0,1,\dots,T}$ can be equivalently characterized by the couple $(\theta, c) \in \mathscr{A}(x_0)$ as well as by the couple (w, c), where $w = (w_t)_{t=1,\dots,T}$ is a predictable stochastic process (defined as in (6.18)) and representing the proportions of wealth (which is not consumed) invested in the $N+1$ assets.

In the remaining part of this section, restricting our attention to utility functions defined with respect to strictly positive consumption streams, we shall use the parametrization of trading strategies introduced in (6.18). In particular, using condition (6.13) (applied to the risk free asset) together with Corollary 6.5 and the $\mathscr{F}_{t-1}$-measurability of the portfolio w_t^*, for all $t = 1, \dots, T$, we get the optimality condition

$$u'(c_t^*) = \delta \mathbb{E}\left[V'(W_{t+1}^*, t+1)r_f | \mathscr{F}_t\right]$$

$$= \delta \sum_{n=1}^{N} w_{t+1}^{*n} \mathbb{E}\left[V'(W_{t+1}^*, t+1)(r_{t+1}^n - r_f) | \mathscr{F}_t\right] + \delta \mathbb{E}\left[V'(W_{t+1}^*, t+1)r_f | \mathscr{F}_t\right]$$

$$= \delta \mathbb{E}\left[V'(W_{t+1}^*, t+1)\left(\sum_{n=1}^{N} w_{t+1}^{*n}(r_{t+1}^n - r_f) + r_f\right)\Big| \mathscr{F}_t\right]. \tag{6.20}$$

Moreover, since $W_t^* - c_t^* > 0$ for all $t = 0, 1, \dots, T - 1$, condition (6.17) holds if and only if

$$(W_t^* - c_t^*)\mathbb{E}\left[V'(W_{t+1}^*, t+1)(r_{t+1}^n - r_f) | \mathscr{F}_t\right] = 0, \quad \text{for all } t = 0, 1, \dots, T - 1, \tag{6.21}$$

for all $n = 1, \dots, N$. As we shall see in the following, condition (6.21) turns out to be useful for characterizing the optimal portfolio process $(w_t^*)_{t=1,\dots,T}$.

Logarithmic Utility Function

In the context of multi-period investment-consumption problems, the case of a logarithmic utility function represents an interesting and special case, not only because an optimal solution is available in explicit form, but also because the optimal solution coincides with a sequence of static single-period optimal investment-consumption plans. In other words, in a dynamic setting, a logarithmic utility functions leads to a *myopic behavior*, in the sense that the portfolio optimization problem at each time step $t \in \{0, 1, \dots, T - 1\}$ is a static optimization problem unaffected by the future investment opportunities, regardless of the probability distribution of the asset returns. Referring to Proposition 6.6 below for a rigorous derivation of this property, the explanation is intuitively clear. Indeed, let us consider for simplicity the problem of maximizing the expected utility $\mathbb{E}[\log(C_T)]$ of terminal consumption at date T. Recalling that $W_T = C_T$, writing $W_T = W_0 \prod_{t=1}^{T} W_t/W_{t-1}$ and using (6.19) we get that

$$\mathbb{E}[\log(C_T)] = \mathbb{E}\left[\log\left(W_0 \prod_{t=1}^{T} \frac{W_t}{W_{t-1}}\right)\right] = \mathbb{E}[\log(W_0)] + \sum_{t=1}^{T} \mathbb{E}\left[\log\left(\frac{W_t}{W_{t-1}}\right)\right]$$

$$= \mathbb{E}[\log(W_0)] + \sum_{t=1}^{T} \mathbb{E}\left[\log\left(\sum_{n=1}^{N} w_t^n(r_t^n - r_f) + r_f\right)\right],$$

from which we immediately see that the maximization of $\mathbb{E}[\log(C_T)]$ reduces to the step-by-step (i.e., for each $t = 0, 1, \dots, T - 1$) maximization of the logarithmic rate of return between $t - 1$ and t which only depends on the distribution of the asset returns at date t and is unaffected by the distribution of future returns. In this sense, the multi-period problem reduces to a sequence of static single-period portfolio optimization problems, which can be solved by relying on the techniques presented in Chap. 3.

The following proposition generalizes the above result to the case of the optimal consumption-investment problem (6.6) (compare also with Ingersoll [1000, Chapter 11] and see also Samuelson [1493], Merton [1327]). The proof is given in Exercise 6.4.

Proposition 6.6 Let $u(x) = \log(x)$ in problem (6.6) and denote by $(W_t^*)_{t=0,1,...,T}$ the corresponding optimal wealth process. Then, the value function satisfies

$$V(W_t^*, t) = f(t)\log(W_t^*) + \psi_t, \qquad \text{for all } t = 0, 1, \ldots, T, \tag{6.22}$$

where the function $f : \{0, 1, \ldots, T\} \to \mathbb{R}$ is explicitly given by $f(t) = \frac{1 - \delta^{T-t+1}}{1-\delta}$ and $(\psi_t)_{t=0,1,...,T}$ is an adapted process defined recursively by $\psi_T = 0$ and

$$\psi_t = \delta f(t+1)\left(\log\left(\frac{f(t)-1}{f(t)}\right) - \frac{\log f(t)}{\delta f(t+1)}\right.$$

$$\left. + \mathbb{E}\left[\log\left(\sum_{n=1}^{N} w_{t+1}^{*n}(r_{t+1}^n - r_f) + r_f\right)\middle|\mathscr{F}_t\right]\right) + \delta\mathbb{E}[\psi_{t+1}|\mathscr{F}_t],$$

for all $t = 0, 1, \ldots, T-1$, where $(w_t^*)_{t=1,...,T}$ denotes the optimal portfolio process. Moreover, the optimal consumption process $(c_t^*)_{t=0,1,...,T}$ satisfies

$$c_t^* = \frac{W_t^*}{f(t)} = \frac{1-\delta}{1-\delta^{T-t+1}}W_t^*, \qquad \text{for all } t = 0, 1, \ldots, T, \tag{6.23}$$

and the optimal portfolio process $(w_t^*)_{t=1,...,T}$ is such that

$$\mathbb{E}\left[\left(\sum_{n=1}^{N} w_{t+1}^{*n}(r_{t+1}^n - r_f) + r_f\right)^{-1}(r_{t+1}^n - r_f)\middle|\mathscr{F}_t\right] = 0, \tag{6.24}$$

for all $n = 1, \ldots, N$ and $t = 0, 1, \ldots, T-1$.

Condition (6.23) shows that, for every date $t = 0, 1, \ldots, T$, the optimal consumption c_t^* is a deterministic (time-varying) fraction of the current optimal wealth W_t^*, regardless of the distribution of the future returns. Moreover, the optimality condition (6.24) characterizing the optimal portfolio process $(w_t^*)_{t=1,...,T}$ corresponds exactly, for every date $t = 0, 1, \ldots, T-1$, to the optimality condition (3.4) obtained in the static single-period portfolio optimization problem, in the presence of N risky assets and a risk free asset. As explained above, this shows that, in the case of a logarithmic utility function, the optimal portfolio problem is unaffected by the future distribution of asset returns and, similarly, the optimal consumption does only depend on the current wealth, regardless of the future investment opportunities (myopic behavior). We want to emphasize that such a myopic behavior is a peculiarity of the logarithmic utility function and is not shared by other typical specifications of utility functions.

By relying on Proposition 6.6 together with equation (6.19) and the optimality condition (6.20), we can prove the following corollary (see Exercise 6.5), which shows an important property of the optimal wealth process in the case of a logarithmic utility function.

Corollary 6.7 *Let $u(x) = \log(x)$ in problem (6.6) and denote by $(W_t^*)_{t=0,1,\ldots,T}$ the corresponding optimal wealth process and by $(c_t^*)_{t=0,1,\ldots,T}$ the optimal consumption process, characterized by equation (6.23). Then, for any self-financing trading-consumption strategy $(\theta, c) \in \mathscr{A}(x_0)$ such that $c_T > 0$, it holds that*

$$\mathbb{E}\left[\frac{W_{t+1}}{W_{t+1}^*}\bigg|\mathscr{F}_t\right] = \frac{W_t - c_t}{W_t^* - c_t^*}, \qquad \text{for all } t = 0, 1, \ldots, T - 1,$$

where $(W_t)_{t=0,1,\ldots,T}$ is the wealth process generated by (θ, c).

In particular, in view of the above corollary, if one considers the simpler problem of maximizing the expected utility from consumption at the terminal date T only (i.e., without intermediate consumption), then an analogous reasoning allows to show that the optimal portfolio process $(W_t^*)_{t=0,1,\ldots,T}$ is such that, for *any* portfolio process $(W_t)_{t=0,1,\ldots,T}$, the ratio $(W_t/W_t^*)_{t=0,1,\ldots,T}$ is a martingale. In view of this important property, the portfolio process $(W_t^*)_{t=0,1,\ldots,T}$ is also called the *numéraire portfolio*.

Power Utility Function

Let us now examine the optimal investment-consumption problem for a power utility function. In this case, as we shall see below, problem (6.6) cannot in general be reduced to a sequence of disconnected single-period problems, unlike the case of a logarithmic utility function considered in Proposition 6.6. The proof of the following proposition is given in Exercise 6.6.

Proposition 6.8 *Let $u(x) = x^\gamma/\gamma$ in problem (6.6), with $\gamma \in (0, 1)$, and denote by $(W_t^*)_{t=0,1,\ldots,T}$ the corresponding optimal wealth process. Then, the value function is explicitly given by*

$$V(W_t^*, t) = a_t^{\gamma-1}\frac{(W_t^*)^\gamma}{\gamma}, \qquad \text{for all } t = 0, 1, \ldots, T, \tag{6.25}$$

where $(a_t)_{t=0,1,\ldots,T}$ is an adapted process defined recursively by $a_T = 1$ and

$$a_t = \left(1 + \left(\delta \mathbb{E} \left[\left(\sum_{n=1}^{N} w_{t+1}^{*n} (r_{t+1}^n - r_f) + r_f \right)^{\gamma} a_{t+1}^{\gamma-1} \Big| \mathscr{F}_t \right] \right)^{\frac{1}{1-\gamma}} \right)^{-1},$$

for all $t = 0, 1, \ldots, T-1$, where $(w_t^*)_{t=1,\ldots,T}$ denotes the optimal portfolio process. Moreover, the optimal consumption process $(c_t^*)_{t=0,1,\ldots,T}$ satisfies

$$c_t^* = a_t W_t^*, \qquad \text{for all } t = 0, 1, \ldots, T, \qquad (6.26)$$

and the optimal portfolio process $(w_t^*)_{t=1,\ldots,T}$ is characterized by

$$\mathbb{E} \left[a_{t+1}^{\gamma-1} \left(\sum_{n=1}^{N} w_{t+1}^{*n} (r_{t+1}^n - r_f) + r_f \right)^{\gamma-1} (r_{t+1}^n - r_f) \Big| \mathscr{F}_t \right] = 0, \qquad (6.27)$$

for all $n = 1, \ldots, N$ and $t = 0, 1, \ldots, T-1$.

Proposition 6.8 shows that, unlike the case of a logarithmic utility function, for a power utility function the optimal consumption c_t^* at date t is not a deterministic fraction of the current wealth W_t^*, for $t = 0, 1, \ldots, T$. Indeed, $(a_t)_{t=0,1,\ldots,T}$ is in general a stochastic process. Analogously, the value function is not a deterministic power function of current optimal wealth, but is state dependent, due to the presence of the $\mathscr{F}_t$-measurable random variable a_t in (6.25). In particular, for every $t = 0, 1, \ldots, T$, the random variable a_t involves the $\mathscr{F}_t$-conditional expectation of a function of the future returns and, hence, the optimal consumption and portfolio processes will depend on the conditional distribution of the future investment opportunities.

The dependence of the optimal portfolio choice on the future investment opportunities is due to the fact that the optimal investment-consumption choice consists in determining the optimal trade-off between optimizing the current portfolio return and hedging against future changes in the investment opportunity set (*intertemporal hedging*, see also Merton [1330]). The fact that the multi-period optimal investment-consumption problem cannot in general be reduced to a sequence of myopic single-period problems is precisely due to the fact that an agent tries to dynamically adjust the portfolio policy in order to hedge against changes in the future investment opportunities. It is important to point out that the hedging problem faced in a multi-period optimal investment-consumption problem cannot be reduced to consumption smoothing. Indeed, even in the absence of intermediate consumption (i.e., considering only the maximization of expected utility of terminal consumption), an investor will choose to dynamically adjust the portfolio through time, depending on the future investment opportunities.

In the case of a power utility function and under additional assumptions, the optimal investment-consumption plan can be shown to be myopic, as in the case of a logarithmic utility function. More specifically, the following corollary (see Exercise 6.7 for a proof) provides a sufficient condition for the optimal portfolio choice to be state independent and such that, for every date $t = 1, \ldots, T$, the portfolio optimality condition coincides with the optimality condition (3.4) arising in the single-period problem.

Corollary 6.9 *Under the assumptions of Proposition 6.8, suppose that, for the optimal portfolio process $(w_t^*)_{t=1,\ldots,T}$,*

$$\mathrm{Cov}\left(a_{t+1}^{\gamma-1}, \left(\sum_{n=1}^{N} w_{t+1}^{*n}(r_{t+1}^n - r_f) + r_f\right)^{\gamma} \middle| \mathscr{F}_t\right) = 0, \tag{6.28}$$

for all $t = 0, 1, \ldots, T-1$, i.e., $a_{t+1}^{\gamma-1}$ and $(\sum_{n=1}^{N} w_{t+1}^{*n}(r_{t+1}^n - r_f) + r_f)^{\gamma}$ are uncorrelated conditionally on $\mathscr{F}_t$. Then the optimal portfolio process $(w_t^*)_{t=1,\ldots,T}$ is characterized by

$$\mathbb{E}\left[\left(\sum_{n=1}^{N} w_{t+1}^{*n}(r_{t+1}^n - r_f) + r_f\right)^{\gamma-1}(r_{t+1}^n - r_f) \middle| \mathscr{F}_t\right] = 0, \tag{6.29}$$

for all $n = 1, \ldots, N$ and $t = 0, 1, \ldots, T-1$.

In particular, the result of Corollary 6.9 does hold if $(\sum_{n=1}^{N} w_{t+1}^{*n}(r_{t+1}^n - r_f) + r_f)^{\gamma}$ and $a_{t+1}^{\gamma-1}$ are independent, for all $t = 0, 1, \ldots, T-1$. In Exercise 6.8 we show that, if the $\mathscr{F}_t$-conditional distribution of the future returns coincides with the unconditional distribution, for every date $t = 0, 1, \ldots, T-1$, then the process $(a_t)_{t=0,1,\ldots,T}$ reduces to a deterministic function of time. In particular, due to equation (6.26), the optimal consumption is then given by a deterministic fraction of the current wealth and the optimal portfolio can be characterized as in (6.29), similarly as in the case of a logarithmic utility function. However, in the presence of correlation, the optimal portfolio choice also takes into account the future investment opportunities and, hence, cannot be reduced to the static optimality condition (3.4).

Power Utility Function with a Markovian State Process

Under the assumption of a power utility function, we now consider the case where an agent is only interested in maximizing the expected utility of consumption at the terminal date T:

$$\max \delta^T \mathbb{E}\left[\frac{C_T^{\gamma}}{\gamma}\right], \tag{6.30}$$

where the maximum is taken with respect to all $\mathbb{R}^{N+1}$-valued predictable self-financing trading strategies $(w_t)_{t=1,\dots,T}$ parametrized in terms of proportions of wealth invested in the $N+1$ available assets (i.e., over all self-financing trading-consumption strategies, in the sense of Definition 6.1, with a null consumption process before date T) satisfying

$$W_{t+1} = W_t \left(\sum_{n=1}^{N} w_{t+1}^n (r_{t+1}^n - r_f) + r_f \right), \qquad \text{for all } t = 0, \dots, T-1, \qquad (6.31)$$

in line with equation (6.19) (recall that $W_T = C_T$).

Similarly as in Brandt [280], we suppose that there exists an $\mathbb{R}^m$-valued (for $m \in \mathbb{N}$) adapted stochastic process $(z_t)_{t=1,\dots,T}$ which can be thought of as a vector of economic state variables. For $t = 1, \dots, T$, let us denote $y_t := (r_t, z_t) \in \mathbb{R}^{N+m}$, where r_t is the return vector at date t of the N risky assets available in the market. We assume that the process $(y_t)_{t=1,\dots,T}$ is a *Markov process*, in the sense that, for all $t = 1, \dots, T-1$, the $\mathscr{F}_t$-conditional distribution of $\{y_{t+1}, \dots, y_T\}$ does only depend on y_t and not on the previous values $\{y_1, \dots, y_{t-1}\}$ of the process. In other words, for any $t = 1, \dots, T$ and for any integrable function f, it holds that

$$\mathbb{E}\left[f(y_{t+1}, \dots, y_T) | \mathscr{F}_t \right] = \mathbb{E}\left[f(y_{t+1}, \dots, y_T) | y_t \right].$$

In this Markovian setting, using equation (6.31), the value function associated to the optimal investment Problem (6.30) can be written as follows, for $x > 0$ and $t = 0, 1, \dots, T-1$:

$$V(x, t) - \sup_{c \in \mathscr{C}_t^+(x)} \delta^{T-t} \mathbb{E}\left[\frac{C_T^\gamma}{\gamma} \middle| \mathscr{F}_t \right]$$

$$= \sup_{(w_s)_{s=t+1,\dots,T}} \delta^{T-t} \mathbb{E}\left[\frac{1}{\gamma} \left(x \prod_{s=t}^{T-1} \left(\sum_{n=1}^{N} w_{s+1}^n (r_{s+1}^n - r_f) + r_f \right) \right)^\gamma \middle| \mathscr{F}_t \right]$$

$$= \sup_{(w_s)_{s=t+1,\dots,T}} \delta^{T-t} \mathbb{E}\left[\frac{1}{\gamma} \left(x \prod_{s=t}^{T-1} \left(\sum_{n=1}^{N} w_{s+1}^n (r_{s+1}^n - r_f) + r_f \right) \right)^\gamma \middle| y_t \right]$$

$$=: \bar{V}(x, t, y_t),$$

where $\bar{V} : \mathbb{R}_+ \times \{0, 1, \dots, T\} \times \mathbb{R}^{N+m} \to \mathbb{R}_+$. In particular, note that $\bar{V}$ is a *deterministic* function, unlike the general value function V introduced in (6.9). Indeed, in this Markovian context, the dependence on $\mathscr{F}_t$ appearing in representation (6.9) can be replaced by a dependence on the vector y_t containing the current realization of the $N+m$ state variables. This is a consequence of the Markov property of the process $(y_t)_{t=1,\dots,T}$.

As shown in Proposition 6.3, the Bellman equation (6.11) holds in correspondence of the optimal portfolio strategy $(w_t^*)_{t=1,\dots,T}$, with associated optimal wealth process $(W_t^*)_{t=0,1,\dots,T}$, where W_0^* denotes a given initial wealth. Hence,

$$\bar{V}(W_t^*, t, y_t) = \delta \mathbb{E}[\bar{V}(W_{t+1}^*, t+1, y_{t+1})|\mathscr{F}_t], \qquad \text{for all } t = 0, 1, \dots, T-1, \tag{6.32}$$

and the characterization (6.17) of the optimal portfolio strategy $(w_t^*)_{t=1,\dots,T}$ can be rewritten as

$$\mathbb{E}[\bar{V}'(W_{t+1}^*, t+1, y_{t+1})(r_{t+1}^n - r_f)|\mathscr{F}_t] = 0, \tag{6.33}$$

for all $n = 1, \dots, N$ and $t = 0, 1, \dots, T-1$, where $\bar{V}'$ denotes the derivative of the function $\bar{V}$ with respect to its first argument. Note that, due to the Markov property of the process $(y_t)_{t=1,\dots,T}$, the conditioning with respect to $\mathscr{F}_t$ appearing in (6.32) and (6.33) can be replaced by a conditioning with respect to y_t. In the present context, the following proposition gives a more explicit representation of the Bellman equation and a characterization of the optimal portfolio process $(w_t^*)_{t=1,\dots,T}$.

Proposition 6.10 *In the context of Problem (6.30), if the process $y = (y_t)_{t=1,\dots,T}$ has the Markov property, then the value function $\bar{V}$ introduced above satisfies*

$$\bar{V}(x, t, y_t) = \sup_{w_{t+1} \in \mathbb{R}^N} \delta \mathbb{E}\left[\frac{\left(x\left(w_{t+1}^{\mathsf{T}}(r_{t+1} - r_f) + r_f\right)\right)^\gamma}{\gamma} \Phi(t+1, y_{t+1}) \Big| y_t \right], \tag{6.34}$$

for all $x > 0$ and $t = 0, 1, \dots, T-1$, where $\Phi : \{1, \dots, T\} \times \mathbb{R}^{N+m} \to \mathbb{R}_+$ is a deterministic function satisfying $\Phi(T, \cdot) = 1$. Moreover, the condition characterizing the optimal portfolio process $(w_t^*)_{t=1,\dots,T}$ is given by

$$\mathbb{E}\left[\left(\sum_{n=1}^N w_{t+1}^{*n}(r_{t+1}^n - r_f) + r_f \right)^{\gamma-1} (r_{t+1}^n - r_f)\Phi(t+1, y_{t+1}) \Big| y_t \right] = 0, \tag{6.35}$$

for all $n = 1, \dots, N$ and $t = 0, 1, \dots, T-1$.

Proof Equation (6.32), together with the definition of the value function $\bar{V}$, the self-financing condition (6.31) and the Markov property of the process $(y_t)_{t=1,\dots,T}$, implies the following representation of the Bellman optimality condition, for all $t = 0, 1, \dots, T-2$:

$\bar{V}(x, t, y_t)$

$$= \sup_{w_{t+1}} \delta \mathbb{E}\left[\sup_{(w_s)_{s=t+2,\dots,T}} \left(\delta^{T-(t+1)} \mathbb{E}\left[\frac{1}{\gamma} \left(x \prod_{s=t}^{T-1} (w_{s+1}^{\mathsf{T}}(r_{s+1} - r_f) + r_f) \right)^\gamma \Big| y_{t+1} \right] \right) \Big| y_t \right]$$

$$= \sup_{w_{t+1}} \delta \mathbb{E} \left[\frac{\left(x \left(w_{t+1}^\top (r_{t+1} - r_f) + r_f \right) \right)^\gamma}{\gamma} \right.$$

$$\left. \times \sup_{(w_s)_{s=t+2,\dots,T}} \left(\delta^{T-(t+1)} \mathbb{E} \left[\left(\prod_{s=t+1}^{T-1} \left(w_{s+1}^\top (r_{s+1} - r_f) + r_f \right) \right)^\gamma \middle| y_{t+1} \right] \right) \middle| y_t \right].$$

Defining the function $\Phi : \{1, \dots, T\} \times \mathbb{R}^{N+m} \to \mathbb{R}_+$ by

$$\Phi(t+1, y_{t+1}) := \sup_{(w_s)_{s=t+2,\dots,T}} \left(\delta^{T-(t+1)} \mathbb{E} \left[\left(\prod_{s=t+1}^{T-1} \left(w_{s+1}^\top (r_{s+1} - r_f) + r_f \right) \right)^\gamma \middle| y_{t+1} \right] \right),$$

for all $t = 0, 1, \dots, T - 2$ and $\Phi(T, \cdot) = 1$, we have thus proved (6.34). Condition (6.35) then easily follows. □

In particular, the above proposition shows that the value function $\bar{V}(x, t, y_t)$ at date t can be represented in terms of the maximization of the expected utility of wealth at the following date $t + 1$ multiplied by the deterministic function Φ which depends on the value y_{t+1} of the Markov process y at date $t + 1$. Moreover, since $\bar{V}(x, t, y_t) = x^\gamma \bar{V}(1, t, y_t)$, for all $t = 0, 1, \dots, T$, it holds that

$$\gamma \bar{V}(1, t, y_t) = \sup_{w_{t+1} \in \mathbb{R}^N} \delta \mathbb{E} \left[\left(w_{t+1}^\top (r_{t+1} - r_f) + r_f \right)^\gamma \Phi(t+1, y_{t+1}) \middle| y_t \right] = \Phi(t, y_t),$$

for all $t = 0, 1, \dots, T$. We have thus shown that

$$\bar{V}(x, t, y_t) = \frac{x^\gamma}{\gamma} \Phi(t, y_t), \qquad \text{for all } t = 0, 1, \dots, T.$$

As a special case, if the processes $(r_t)_{t=1,\dots,T}$ and $(z_t)_{t=1,\dots,T}$ are independent and asset returns at different dates are independent among themselves, then the function Φ can be reduced to a deterministic function of time (i.e., there is no need to keep track of the conditioning information) and the optimality condition (6.35) reduces to the optimality condition characterizing the optimal portfolio in a simple single-period portfolio choice problem (myopic behavior). In the absence of independence, the multi-period optimal policy differs from a sequence of single-period optimal policies, due to the intertemporal hedging effect captured by the interaction with the term $\Phi(t + 1, y_{t+1})$.

6.2   Equilibrium and Pareto Optimality

In this section, we extend to a multi-period setting the equilibrium analysis developed in Chap. 4 in the context of a single-period economy. In particular, we shall focus on the relation between the Pareto optimality of an equilibrium allocation and the completeness of the financial market. Unlike in a single-period economy, the notion of completeness in a multi-period economy will not depend only on the terminal dividends of the available securities but also on their (endogenous) price processes. The presentation of this section is inspired from Huang & Litzenberger [971, Chapter 7].

In a multi-period setting with dates $t \in \{0, 1, \ldots, T\}$, we consider an economy populated by $I \in \mathbb{N}$ agents $i \in \{1, \ldots, I\}$, with homogeneous beliefs and preferences characterized by time additive state independent utility functions. For simplicity, we assume that all the I agents exhibit the same discount factor $\delta \in (0, 1)$. Every agent $i \in \{1, \ldots, I\}$ possesses an *endowment process* $e^i = (e^i_t)_{t=0,1,\ldots,T}$, which can also be represented as $e^i = \{e^i_t(A_t); A_t \in \mathscr{F}_t, t = 0, 1, \ldots, T\}$, with $e^i_t(A_t)$ denoting the endowment of agent i (in units of the reference consumption good) at time t in correspondence of the event $A_t \in \mathscr{F}_t$, with $e^i_0(A_0) := e^i_0$. In this way, the agents' endowment is allowed to be contingent on the events realized over time.[5] We assume that, for all $i = 1, \ldots, I$, the endowment process e^i is non-negative and, to avoid trivial situations, that there exists at least one date $t \in \{0, 1, \ldots, T\}$ and one event $A_t \in \mathscr{F}_t$ such that $e^i_t(A_t) > 0$. Similarly, a consumption process $c^i = (c^i_t)_{t=0,1,\ldots,T}$ can be represented as $c^i = \{c^i_t(A_t); A_t \in \mathscr{F}_t, t = 0, 1, \ldots, T\}$, with $c^i_t(A_t)$ denoting the non-negative consumption of agent i at time t in correspondence of the event $A_t \in \mathscr{F}_t$, with $c^i_0(A_0) := c^i_0$. This represents the fact that the endowment and the consumption processes are adapted to the information flow $\mathbb{F}$.

For every agent $i \in \{1, \ldots, I\}$ and a consumption process c^i, the associated expected utility is given by

$$u^i(c^i_0) + \sum_{t=1}^{T} \delta^t \sum_{A_t \in \mathscr{F}_t} \pi_{A_t} u^i\big(c^i_t(A_t)\big), \qquad (6.36)$$

where $u^i : \mathbb{R}_+ \to \mathbb{R}$ is a strictly increasing and strictly concave differentiable utility function. Observe that the preference functional (6.36) implicitly assumes that the preferences of the agents do not change over time and depend on time only through a geometric discount factor.

---

[5]Note that the present structure of endowment process generalizes the case where the endowment is expressed in terms of traded securities. Indeed, if the economy comprises N traded securities and agent $i \in \{1, \ldots, I\}$ is endowed at the initial date $t = 0$ with $\theta^i_0 \in \mathbb{R}^N$ units of the traded assets, then the corresponding endowment process is given by $e^i = \{e^i_t(A_t); A_t \in \mathscr{F}_t, t = 0, 1, \ldots, T\}$, with $e^i_t(A_t) = \sum_{n=1}^{N} \theta^{i,n}_0 d^n_t(A_t)$, with $d^n_t(A_t)$ denoting the dividend of security n at time t in correspondence of the event $A_t \in \mathscr{F}_t$.

Pareto Optimal Allocations

In the present context, the notion of (ex-ante) *Pareto optimal allocation* can be defined as a straightforward generalization of Definition 4.1 to a multi-period setting. The feasible allocations are given by the collection of non-negative consumption plans $\{c^i; i = 1, \ldots, I\}$ satisfying the feasibility constraint

$$\sum_{i=1}^{I} c_t^i(A_t) \leq \sum_{i=1}^{I} e_t^i(A_t), \qquad \text{for all } A_t \in \mathscr{F}_t \text{ and } t = 0, 1, \ldots, T. \tag{6.37}$$

As discussed in Sect. 4.1, Pareto optimal allocations can be characterized in terms of the no-trade equilibrium in a representative agent economy. Indeed, an allocation $\{c^i; i = 1, \ldots, I\}$ is Pareto optimal if and only if it is a solution to the following optimal consumption problem of a representative agent, for some strictly positive weights $\{a^i; i = 1, \ldots, I\}$:

$$\max_{\{c^i; i=1,\ldots,I\}} \sum_{i=1}^{I} a^i \left(u^i(c_0^i) + \sum_{t=1}^{T} \delta^t \sum_{A_t \in \mathscr{F}_t} \pi_{A_t} u^i(c_t^i(A_t)) \right), \tag{6.38}$$

where the maximization is over all feasible allocations. Due to the strict monotonicity of the utility functions u^i, the feasibility constraint (6.37) can be equivalently expressed as an equality (meaning that the available resources are fully consumed in correspondence of a Pareto optimal allocation). Interior ex-ante Pareto optimal allocations $\{c^{i*}; i = 1, \ldots, I\}$ can then be characterized by the complete allocation of the available resources together with the optimality conditions (compare with (4.5)–(4.6))

$$a^i u^{i'}(c_0^{i*}) = \varphi_0, \tag{6.39}$$

$$a^i \pi_{A_t} \delta^t u^{i'}(c_t^{i*}(A_t)) = \varphi_{A_t}, \qquad \text{for all } A_t \in \mathscr{F}_t \text{ and } t = 1, \ldots, T, \tag{6.40}$$

for all $i = 1, \ldots, I$, where $\{\varphi_0, \varphi_{A_t}; A_t \in \mathscr{F}_t, t = 1, \ldots, T\}$ is a family of strictly positive Lagrange multipliers associated to problem (6.38)–(6.37). The above optimality conditions can also be rewritten as

$$\delta^{t-s} \frac{\pi_{A_t} u^{i'}(c_t^{i*}(A_t))}{\pi_{A_s} u^{i'}(c_s^{i*}(A_s))} = \frac{\varphi_{A_t}}{\varphi_{A_s}}, \quad \text{for all } A_t \in \mathscr{F}_t, A_s \in \mathscr{F}_t \text{ and } s, t = 0, 1, \ldots, T, \tag{6.41}$$

with $\pi_{A_0} := 1$ and $\varphi_{A_0} := \varphi_0$ (recall that we always assume that all the events have a strictly positive probability of occurrence). Observe that the right-hand side of (6.41) does not depend on i. This implies that, as already discussed in Chap. 4, the (probability weighted) marginal rates of substitution between consumption in different dates/states of the world are equal among all the agents in correspondence of a Pareto optimal allocation. In particular, similarly as in Theorem 4.3, this implies the *co-monotonicity* property of Pareto optimal allocations, which means that in correspondence of a Pareto optimal allocation the optimal consumption of every agent only depends on the aggregate endowment and only aggregate risk matters. Note also that, in general, there is a dependence between a Pareto optimal allocation and the set of weights $\{a^i; i = 1, \ldots, I\}$ defining (6.38) (i.e., the aggregation property does not necessarily hold in a strong sense).

Arrow Securities and Static Completeness

In Chap. 4, in the context of a single-period economy, we have seen that market completeness implies the Pareto optimality of equilibrium allocations. As we are going to show, the same result holds true in a general multi-period setting, if we assume that a complete set of *Arrow securities* is available at the initial date $t = 0$. In the present multi-period context, an *Arrow security* for the event $A_t \in \mathscr{F}_t$ is a security which pays a dividend equal to one at date t in correspondence of the event A_t and zero otherwise. Referring to the information flow introduced at the beginning of this chapter, it is clear that the total number of Arrow securities is equal to the total number of knots of the event tree (excluding the knot corresponding to the initial date $t = 0$), i.e., $\sum_{t=1}^{T} \nu_t$, with $\nu_t = |\mathscr{F}_t|$.

Let us first assume that all the $N := \sum_{t=1}^{T} \nu_t$ Arrow securities are available for trade at the initial date $t = 0$. Under this assumption, it is evident that every consumption plan can be attained by holding a suitable quantity of the N Arrow securities. We also assume that markets are only open at the initial date $t = 0$ and we denote by q_{A_t} the price (at date $t = 0$) of the Arrow security associated to the event A_t at date t (with q_0 being the price of one unit of the consumption good at the initial date $t = 0$). The expected utility (6.36) can be maximized by simply determining at the initial date $t = 0$ the quantities to be bought/sold of the N Arrow securities, under the budget constraint that the endowment process e^i suffices to finance such an investment in the N Arrow securities. The optimal consumption-investment problem of agent $i \in \{1, \ldots, I\}$ can then be written as

$$\max_{\{c_0^i, c_t^i(A_t); A_t \in \mathscr{F}_t, t=1,\ldots,T\}} \left(u^i(c_0^i) + \sum_{t=1}^{T} \delta^t \sum_{A_t \in \mathscr{F}_t} \pi_{A_t} u^i\big(c_t^i(A_t)\big) \right), \tag{6.42}$$

under the budget constraint

$$q_0 c_0^i + \sum_{t=1}^{T} \sum_{A_t \in \mathscr{F}_t} q_{A_t} c_t^i(A_t) = q_0 e_0^i + \sum_{t=1}^{T} \sum_{A_t \in \mathscr{F}_t} q_{A_t} e_t^i(A_t). \tag{6.43}$$

Since all the events A_t are assumed to have a strictly positive probability of realization, in equilibrium the prices q_{A_t} are strictly positive. Indeed, it is easy to show that, in correspondence of an equilibrium, the prices q_{A_t} must be strictly positive, otherwise arbitrage opportunities would appear (see also Sect. 6.3).

A feasible allocation $\{c^{i*}; i = 1, \ldots, I\}$ together with a collection of Arrow security prices $\{q_0^*, q_{A_t}^*; A_t \in \mathscr{F}_t, t = 1, \ldots, T\}$ is said to be an *equilibrium* for the economy if, for every $i = 1, \ldots, I$, the consumption plan c^{i*} solves Problem (6.42) for agent i, with the budget constraint (6.43) being satisfied in correspondence of the price system $\{q_0^*, q_{A_t}^*; A_t \in \mathscr{F}_t, t = 1, \ldots, T\}$, and markets clear, i.e.,

$$\sum_{i=1}^{I} c_t^{i*}(A_t) = \sum_{i=1}^{I} e_t^i(A_t), \qquad \text{for all } A_t \in \mathscr{F}_t \text{ and } t = 0, 1, \ldots, T.$$

This definition can be regarded as an extension to the present context of Definition 4.6 of an Arrow-Debreu equilibrium. As shown in the following proposition, it is easy to prove that an equilibrium allocation is Pareto optimal if all N Arrow securities are available (see also Huang & Litzenberger [971, Section 7.3]).

Proposition 6.11 *Suppose that all the $N = \sum_{t=1}^{T} \nu_t$ Arrow securities are available for trade at $t = 0$. Then any feasible strictly positive allocation $\{c^{i*}; i = 1, \ldots, I\}$ corresponding to an equilibrium is Pareto optimal.*

Proof Let $\{c^{i*}; i = 1, \ldots, I\}$ be an equilibrium allocation with associated prices $\{q_0^*, q_{A_t}^*; A_t \in \mathscr{F}_t, t = 1, \ldots, T\}$ for the N Arrow securities. For every $i = 1, \ldots, I$, the consumption plan c^{i*} is a solution to problem (6.42)–(6.43) and, hence, by the optimality condition, it holds that

$$u^{i'}(c_0^{i*}) = \lambda^i q_0^* \tag{6.44}$$

$$\delta^t \pi_{A_t} u^{i'}\left(c^{i*}(A_t)\right) = \lambda^i q_{A_t}^*, \qquad \text{for all } A_t \in \mathscr{F}_t \text{ and } t = 1, \ldots, T, \tag{6.45}$$

where $\lambda^i > 0$ is the Lagrange multiplier for agent i in correspondence of the optimum c^{i*}. Letting $a^i := 1/\lambda^i$, the above optimality conditions reduce to (6.39)–(6.40). Moreover, since markets clear, condition (6.37) is satisfied as an equality, for all $A_t \in \mathscr{F}_t$ and $t = 0, 1, \ldots, T$. This suffices to prove the Pareto optimality of the allocation $\{c^{i*}; i = 1, \ldots, I\}$. $\qquad \square$

It is important to observe that, if all the N Arrow securities are available for trade and markets are only open at the initial date $t = 0$, then the optimal consumption problem (6.42) reduces to a static problem, where the solution is simply given by a suitable investment at the initial date $t = 0$ in the N Arrow securities, which allow to span all possible consumption plans. We express this property by saying that, under the present assumptions, *static completeness* holds. The dynamic feature of the optimal investment-consumption problem considered in Sect. 6.1 has totally disappeared.

Actually, under the assumptions of Proposition 6.11, we can prove a stronger result. Indeed, if all the N Arrow securities are available for trade at the initial date $t = 0$ and markets stay open at the subsequent dates $t = 1, \ldots, T$, then the same equilibrium allocation will be reached if agents form *rational expectations*. In particular, as we are now going to show, there is no need for the markets to remain open at the future dates $t = 1, \ldots, T$ and, if markets are reopened, then there will be no trade in equilibrium.

If markets remain open at the dates $t = 1, \ldots, T$, then the optimal investment-consumption Problem (6.42) becomes a dynamic problem, similarly to problem (6.6). In particular, in order to formulate their decisions, agents have to form expectations about the future prices of the N Arrow securities. In line with the analysis developed in Chap. 4, we assume that agents exhibit rational expectations and, therefore, in equilibrium they will perfectly forecast future prices (perfect foresight or *Radner equilibrium*, see Sect. 4.2). We assume that the N Arrow securities are available for trade at the initial date $t = 0$ with strictly positive prices $\{q_{A_t}^*; A_t \in \mathscr{F}_t, t = 1, \ldots, T\}$. For all $A_s \in \mathscr{F}_s$, $A_t \in \mathscr{F}_t$ and $s, t \in \{1, \ldots, T\}$, define the quantities $q_{A_s|A_t}^*$ by

$$q_{A_s|A_t}^* := \begin{cases} \frac{q_{A_s}^*}{q_{A_t}^*}, & \text{if } s > t \text{ and } A_s \subseteq A_t; \\ 0, & \text{otherwise.} \end{cases} \tag{6.46}$$

We can prove that the quantities introduced in (6.46) correspond to the prices associated with a Radner equilibrium generating the same equilibrium allocation considered in the setting of Proposition 6.11. In the next proposition, $q_{A_s|A_t}^*$ will represent the price at date t in correspondence of the event A_t of the Arrow security paying a unitary amount in correspondence of the event A_s at the future date s.

Proposition 6.12 *Suppose that, for every* $t = 0, 1, \ldots, T - 1$ *and* $A_t \in \mathscr{F}_t$, *all the Arrow securities paying in correspondence of the events* $A_s \subseteq A_t$, *for* $s = t+1, \ldots, T$, *are available for trade. Let* $\{c^{i*}; i = 1, \ldots, I\}$ *be a strictly positive feasible allocation such that* c^{i*} *solves* (6.42)–(6.43), *for every* $i = 1, \ldots, I$. *Then the allocation* $\{c^{i*}; i = 1, \ldots, I\}$ *and the price system* $\{q_0^*, q_{A_s|A_t}^*; A_s \in \mathscr{F}_s, A_t \in \mathscr{F}_t, s, t = 1, \ldots, T\}$ *constitute a Radner equilibrium.*

Proof Let $\{c^{i*}; i = 1, \ldots, I\}$ be a feasible allocation such that c^{i*} is a solution to (6.42)–(6.43), for every $i = 1, \ldots, I$. To prove the proposition, it suffices to show that, for all $i = 1, \ldots, I$ and for every date $t \in \{0, 1, \ldots, T\}$ and event $A_t \in \mathscr{F}_t$, the consumption plan $\{c_s^{i*}(A_s); A_s \subseteq A_t, A_s \in \mathscr{F}_s, s = t, \ldots, T\}$ is a solution to the following problem:

$$\max_{\{c_s(A_s); A_s \in \mathscr{F}_s, s = t, \ldots, T\}} \left(u^i\big(c_t(A_t)\big) + \sum_{s=t+1}^{T} \delta^{s-t} \sum_{A_s \in \mathscr{F}_s} \pi_{A_s|A_t} u^i\big(c_s(A_s)\big) \right),$$

under the budget constraint

$$c_t(A_t) + \sum_{s=t+1}^{T} \sum_{A_s \in \mathscr{F}_s} q^*_{A_s|A_t} c_s(A_s) = c_t^{i*}(A_t) + \sum_{s=t+1}^{T} \sum_{A_s \in \mathscr{F}_s} q^*_{A_s|A_t} c_s^{i*}(A_s).$$

The consumption plan $\{c_s^{i*}(A_s); A_s \subseteq A_t, A_s \in \mathscr{F}_s, s = t, \ldots, T\}$ solves the above problem if (and only if) there exists a strictly positive multiplier $\lambda_{A_t}^i$ such that

$$u^{i'}\left(c_t^{i*}(A_t)\right) = \lambda_{A_t}^i,$$

$$\delta^{s-t} \pi_{A_s|A_t} u^{i'}\left(c_s^{i*}(A_s)\right) = \lambda_{A_t}^i q^*_{A_s|A_t}, \qquad \text{for all } A_s \in \mathscr{F}_s,$$

for all $s = t, \ldots, T$. Due to the optimality conditions (6.44)–(6.45) together with (6.46) (and recalling that $\pi_{A_s|A_t} = \pi_{A_s}/\pi_{A_t}$, for all $A_s \in \mathscr{F}_s$ with $A_s \subseteq A_t$), it can be easily checked that the above conditions are satisfied by setting

$$\lambda_{A_t}^i := \lambda^i \frac{q^*_{A_t}}{\delta^t \pi_{A_t}}.$$

Since $A_t \in \mathscr{F}_t$ and $t \in \{0, 1, \ldots, T\}$ are arbitrary, the proposition is proved. $\qquad\square$

Proposition 6.12 shows that, if the complete set of Arrow securities is available for trade at the initial date $t = 0$, then there is no trade if markets remain open at the subsequent dates $t = 1, \ldots, T$. Indeed, if an equilibrium allocation is reached at $t = 0$, then the same allocation will be an equilibrium when agents are allowed to trade at the subsequent dates $t = 1, \ldots, T$, with corresponding prices given by (6.46). Actually, agents will not trade at the dates $t = 1, \ldots, T$. The quantity $q^*_{A_s|A_t}$ represents the price at date t in event A_t of the Arrow security paying a unitary dividend in correspondence of the event A_s and $q^*_{A_s|A_t} = 0$ for all events A_s which are incompatible with the event A_t (see also Exercise 6.10).

Long-Lived Securities and Dynamic Completeness

We have so far considered economies where the whole set of Arrow securities is available for trade at the initial date $t = 0$. In this case, the market is (statically) complete and equilibrium allocations are Pareto optimal allocations (see Proposition 6.11). In particular, Pareto optimality is reached by trading in the optimal way at the initial date $t = 0$, without any trading activity taking place at the subsequent dates $t = 1, \ldots, T$, even if markets remain open (see Proposition 6.12). We now consider the case where the complete set of Arrow securities is not necessarily available at $t = 0$. In this case, unlike the situation considered in Proposition 6.12, the possibility of trading at the following dates $t = 1, \ldots, T$

can become valuable, since it allows agents to reallocate wealth depending on the resolution of uncertainty. As we shall see, under suitable conditions and if markets are *dynamically complete*, Pareto optimality can be reached in equilibrium with a number of traded securities smaller than $\sum_{t=1}^{T} \nu_t$.

We adopt the setting of Sect. 6.1. The economy is assumed to comprise N securities paying non-negative (but possibly null) random dividends at the dates $t = 0, 1, \ldots, T$ and available for trade at every date $t = 0, 1, \ldots, T$ (*long-lived securities*, following the terminology of Huang & Litzenberger [971]). For each $n = 1, \ldots, N$, the n-th security is characterized by its (ex-dividend) price process $s^n = (s_t^n)_{t=0,1,\ldots,T}$ and by its dividend process $d^n = (d_t^n)_{t=0,1,\ldots,T}$, both assumed to be adapted to the information flow $\mathbb{F}$. The cum-dividend price process is then given by $p^n = s^n + d^n$, for all $n = 1, \ldots, N$. Similarly as at the beginning of Sect. 6.1, we shall use the notation $S_t = (s_t^1, \ldots, s_t^N)^\top$, $D_t = (d_t^1, \ldots, d_t^N)^\top$ and $P_t = (p_t^1, \ldots, p_t^N)^\top$.

Since the N securities are available for trade at all dates $t = 0, 1, \ldots, T$, it is natural to allow agents to dynamically reallocate their wealth by means of *self-financing trading-consumption strategies*, in the sense of Definition 6.1. Each agent $i = 1, \ldots, I$ is assumed to be endowed with a given quantity $\theta_0^i = (\theta_0^{i,1}, \ldots, \theta_0^{i,N})^\top$ of the N traded securities at date $t = 0$ and, without loss of generality, we assume that the total supply of each security is normalized to 1 (i.e., $\sum_{i=1}^{I} \theta_0^{i,n} = 1$, for all $n = 1, \ldots, N$). If there is not a complete set of Arrow securities available at $t = 0$ and, hence, the investment-consumption problem (6.6) cannot be reduced to a static optimization problem at the initial date $t = 0$ (as considered in the first part of this section), then problem (6.6) is a truly dynamic problem, which can be solved by relying on the dynamic programming approach developed in Sect. 6.1.

In the context of an economy with N securities traded at the dates $t = 0, 1, \ldots, T - 1$, the notion of Radner equilibrium can be formulated as follows, extending in a natural way Definition 4.7 (recall that, since the economy is assumed to end at date T, it always holds that $S_T = 0$).

Definition 6.13 Let $D = (D_t)_{t=0,1,\ldots,T}$ be a given dividend process, i.e., a non-negative non-null adapted process. A tuple $\{S^*, D; (\theta^{i*}, c^{i*}), i = 1, \ldots, I\}$, with $S^* = (S_t^*)_{t=0,1,\ldots,T}$ being a non-negative non-null adapted process and (θ^{i*}, c^{i*}) a self-financing trading-consumption strategy, for all $i = 1, \ldots, I$, constitutes a *Radner equilibrium* if

(i) for every $i = 1, \ldots, I$, the strategy (θ^{i*}, c^{i*}) solves problem (6.6) for agent i in correspondence of the price-dividend couple (S^*, D);

(ii) $\sum_{i=1}^{I} \theta_t^{i*,n} \leq 1$, for all $n = 1, \ldots, N$ and $t = 0, 1, \ldots, T$.

It is easy to show that in equilibrium, if the inequality appearing in part (ii) of Definition 6.13 holds as an equality, then the aggregate consumption equals the aggregate dividends delivered by the N available securities. In other words, it holds that $\sum_{i=1}^{I} c_t^{i*} = \sum_{n=1}^{N} d_t^n$, for all $t = 0, 1, \ldots, T$ (see Exercise 6.11).

As we are going to show, Pareto optimality can be reached in equilibrium if the market comprising N long-lived securities is complete in the sense of the following definition (see also Kreps [1135]), which is formulated with respect to a given price-dividend couple of processes (S, D) (recall that, following the notation introduced in Sect. 6.1, the set $\mathscr{C}_0^+(x)$ contains all consumption streams which can be financed starting from a given initial wealth x at date $t = 0$).

Definition 6.14 A consumption stream $c = (c_t)_{t=0,1,\ldots,T}$ is said to be *attainable* (or *replicable*) if there exists an initial wealth $x \in \mathbb{R}_+$ such that $c \in \mathscr{C}_0^+(x)$. We say that the market is *dynamically complete* if every non-negative adapted process (i.e., every consumption stream) is attainable.

According to Definition 6.14, a consumption stream is attainable if there exists some initial wealth from which one can construct a trading strategy financing the consumption stream. Clearly, the initial wealth x can be considered as the (market) value of the consumption stream $c \in \mathscr{C}_0^+(x)$ at the initial date $t = 0$. Note that, if all the $\sum_{t=1}^{T} v_t$ Arrow securities are available at $t = 0$, then the market is trivially complete in the sense of Definition 6.14, since every consumption stream can be attained by means of a suitable static portfolio of the Arrow securities. However, as we are now going to show, dynamic completeness can be more realistically reached by allowing agents to trade on a much smaller number of long-lived securities.

As a preliminary, following Dothan [581], let us define the so-called *bifurcation index* $v_t(A_t) := |\{A_{t+1} \in \mathscr{F}_{t+1} : A_{t+1} \subseteq A_t\}|$, for $t = 0, 1, \ldots, T-1$. In other words, the quantity $v_t(A_t)$ represents the number of distinct events that can occur at the next date $t + 1$ given the realization of the event A_t at date t. Referring to the event tree described at the beginning of this chapter, the bifurcation index simply counts the number of branches leaving from each node of the tree, so that $v_t(A_t)$ is the number of knots at $t + 1$ which can be reached from the knot corresponding to the event A_t at date t. Let us also define $v^* := \max_{t=0,1,\ldots,T-1} \max_{A_t \in \mathscr{F}_t} v_t(A_t)$, representing the maximum number of branches leaving from any node of the tree. For every date $t = 0, 1, \ldots, T - 1$ and every event $A_t \in \mathscr{F}_t$, we collect in the following matrix the cum-dividend prices which can be realized at the subsequent date $t + 1$:

$$
\mathfrak{P}_{t+1}(A_t) = \begin{pmatrix}
p_{t+1}^1(A_{t+1}^1) & p_{t+1}^2(A_{t+1}^1) & \cdots & p_{t+1}^N(A_{t+1}^1) \\
p_{t+1}^1(A_{t+1}^2) & p_{t+1}^2(A_{t+1}^2) & \cdots & p_{t+1}^N(A_{t+1}^2) \\
\vdots & \vdots & \ddots & \vdots \\
p_{t+1}^1(A_{t+1}^{v_t(A_t)}) & p_{t+1}^2(A_{t+1}^{v_t(A_t)}) & \cdots & p_{t+1}^N(A_{t+1}^{v_t(A_t)})
\end{pmatrix}.
\tag{6.47}
$$

For every $n = 1, \ldots, N$, the n-th column of the above matrix represents the possible prices of the n-th security in correspondence of all the events which might become true at date $t + 1$, given that event A_t is realized at the preceding date t. Similarly, for every $s = 1, \ldots, v_t(A_t)$, the s-th row coincides with the vector $P_{t+1}(A_{t+1}^s)$ of the prices of the N securities at date $t + 1$ in correspondence of the event $A_{t+1}^s \subseteq A_t$.

The following proposition (the proof of which is given in Exercise 6.12) provides a characterization of dynamic completeness. For any $t = 1, \ldots, T$ and $A_t \in \mathscr{F}_t$, we say that the Arrow security paying a unitary dividend in correspondence of event A_t

at date t (and zero otherwise) is attainable if there exists an attainable consumption stream which coincides with the dividend process of the Arrow security.

Proposition 6.15 *Given the price-dividend couple of processes* (S, D), *the following are equivalent:*

(i) the market is dynamically complete;

(ii) for every $t = 0, 1, \ldots, T - 1$ *and* $A_t \in \mathscr{F}_t$, *it holds that* $\mathrm{rank}(\mathfrak{P}_{t+1}(A_t)) = \nu_t(A_t)$;

(iii) all the $\sum_{t=1}^{T} \nu_t$ *Arrow securities are attainable.*

Part *(ii)* of this proposition makes clear that market completeness is strongly related to the temporal resolution of uncertainty. Indeed, if the uncertainty is gradually resolved over time (i.e., the number of distinct branches leaving each node of the tree is small), then it suffices to trade in a small number of long-lived securities in order to achieve market completeness. On the contrary, if the uncertainty is resolved in a more abrupt way (i.e., there exist nodes from which many distinct branches of the tree originate), then many securities are needed to reach market completeness. This also reflects the fact that in a dynamic setting agents learn over time that the true state of the world belongs to a refining sequence of events and, hence, adjust their strategies according to a learning process. However, unlike in the case of the single-period economy considered in Chap. 4, dynamic completeness does not only depend on the asset dividends but also on their (endogenous) price processes.

Part *(ii)* of Proposition 6.15 implies that a necessary condition for the market to be dynamically complete is that $N \geq \nu^*$, since $N \geq \nu_t(A_t)$ has to hold for every $t = 0, 1, \ldots, T - 1$ and $A_t \in \mathscr{F}_t$. In other words, a necessary condition for dynamic completeness is that, in correspondence of every node of the event tree, the number of traded securities is greater or equal than the number of distinct branches leaving from that node. In view of part *(iii)* of Proposition 6.15, if markets are dynamically complete, then every Arrow security can be replicated by trading dynamically in the available long-lived securities. As a consequence (if there are no arbitrage opportunities, see Sect. 6.3), the market value at the initial date $t = 0$ of an Arrow security paying a unitary dividend in correspondence of event A_t at date t (and zero otherwise) corresponds to the initial wealth needed to finance a consumption plan c such that $c_s(A_s) = \mathbf{1}_{A_s = A_t}$ if $s = t$ and zero otherwise.

We are now in a position to prove that, if the market is dynamically complete, then an equilibrium allocation is Pareto optimal. In view of the equivalence *(i)* $\Leftrightarrow$ *(iii)* in Proposition 6.15, it is easy to see that, if the market is dynamically complete, then the equilibrium allocation obtained in the presence of N long-lived securities (Radner equilibrium, see Definition 6.13) coincides with the equilibrium allocation obtained by allowing agents to trade at the initial date all the $\sum_{t=1}^{T} \nu_t$ Arrow securities at the prices implicit in the price-dividend couple (S^*, D) corresponding to the Radner equilibrium. This simply follows from the fact that, if the market is dynamically complete, then the complete set of Arrow securities can be attained by dynamic trading in the N long-lived securities. We have thus established the following result.

Proposition 6.16 *Let $D = (D_t)_{t=0,1,\ldots,T}$ be a given dividend process, i.e., a non-negative non-null adapted process. Let the tuple $\{S^*, D; (\theta^{i*}, c^{i*}), i = 1, \ldots, I\}$ correspond to a Radner equilibrium, in the sense of Definition 6.13, and suppose that the market is dynamically complete in correspondence of the price-dividend couple (S^*, D). Then the allocation $\{c^{i*}; i = 1, \ldots, I\}$ is Pareto optimal.*

If markets are incomplete, then only a subset of the Arrow securities can be replicated by dynamic trading in the N long-lived securities. In this case, an allocation corresponding to a Radner equilibrium is not necessarily Pareto optimal. Note also that the constrained Pareto optimality result obtained in Diamond [571] under market incompleteness in a two-period economy (see Definition 4.11 in Sect. 4.2) does not extend to a multi-period setting. Indeed, equilibrium allocations in an incomplete multi-period market are typically Pareto inefficient, in the sense that feasible reallocations of wealth can lead to Pareto improvements of the equilibrium allocation (see also Magill & Quinzii [1286, Chapter 25]).

The condition $N \geq \nu^*$ is also sufficient for the market to be dynamically complete in a generic sense (see Kreps [1135] for full details). Indeed, fixing the information flow, the state space and agents' preferences, an Arrow-Debreu equilibrium allocation $\{c^{i*}; i = 1, \ldots, I\}$ (i.e., an equilibrium allocation in the economy where all the $\sum_{t=1}^{T} \nu_t$ Arrow securities are traded) can be implemented as a Radner equilibrium (see Definition 6.13) with $N \geq \nu^*$ long-lived securities for almost every dividend process of those N securities (i.e., for every dividend process belonging to a set such that the closure of its complement has Lebesgue measure zero). In correspondence of such a Radner equilibrium, the market will be dynamically complete and prices determined according to the agents' optimality conditions.

Representative Agent Analysis

Having discussed the notions of equilibrium and market completeness, let us now generalize to a multi-period setting the representative agent analysis introduced in a two-period economy in Sect. 4.1, referring the reader to Constantinides [488], Milne [1344], Huang & Litzenberger [971], Duffie [593] for a more detailed presentation.

As above, we restrict ourselves to an economy populated by I agents characterized by homogeneous beliefs, the same discount factor δ and time additive and state independent preferences represented by a family of strictly increasing and strictly concave utility functions $\{u^i; i = 1, \ldots, I\}$. Letting $e^i = \{e_t^i(A_t); A_t \in \mathscr{F}_t, t = 0, 1, \ldots, T\}$ be the endowment process of agent i, for $i = 1, \ldots, I$, the *aggregate endowment* process can be represented as $e = \{e_t(A_t); A_t \in \mathscr{F}_t, t = 0, 1, \ldots, T\}$, with $e_t(A_t) := \sum_{i=1}^{I} e_t^i(A_t)$. Under the present assumptions, the result of Proposition 1.6 can be extended to a multi-period risky setting. More specifically, the same arguments used in the proof of Proposition 4.4 allow to show that the representative agent's utility function can be represented in the form (6.36):

$$\mathbf{u}(c_0) + \sum_{t=1}^{T} \delta^t \sum_{A_t \in \mathscr{F}_t} \pi_{A_t} \overline{\mathbf{u}(c_t(A_t))}, \tag{6.48}$$

where the function $\mathbf{u} : \mathbb{R}_+ \to \mathbb{R}$ is given by

$$\mathbf{u}(x) = \max_{\substack{(x^1, \ldots, x^I) \in \mathbb{R}_+^I \\ \sum_{i=1}^{I} x^i \leq x}} \left(\sum_{i=1}^{I} a^i u^i(x^i) \right), \tag{6.49}$$

where $\{a^i; i = 1, \ldots, I\}$ is a collection of strictly positive weights characterizing the representative agent's utility function. In particular, observe that the utility function (6.48) preserves the time additivity and state independence of the individual preference structures, with the same discount factor δ. It can also be shown that the function $\mathbf{u}$ is strictly increasing and concave (and the inequality constraint appearing in (6.49) can be equivalently replaced by an equality constraint).

As explained in Sect. 1.3, given the weights $\{a^i; i = 1, \ldots, I\}$ and the aggregate endowment process e, a feasible allocation $\{c^i; i = 1, \ldots, I\}$ which defines (in the sense explained before Proposition 1.6) the representative agent's utility function (6.48) is Pareto optimal. Given a complete market where all the $\sum_{t=1}^{T} \nu_t$ Arrow securities are traded (as considered in the first part of Sect. 6.2) and an aggregate endowment process e, a feasible allocation $\{c^{i*}; i = 1, \ldots, I\}$ corresponding to an equilibrium of the economy defines the utility function of a representative agent, where the weight a^i is given by the reciprocal of the Lagrange multiplier λ^i of the optimal investment-consumption problem of agent i at the optimum, for every $i = 1, \ldots, I$ (see Exercise 6.13). In this case, the representative agent will simply consume the aggregate endowment process (*no-trade equilibrium*).

Relaxing the assumption that the whole set of Arrow securities is available for trade at the initial date $t = 0$, the same result can be obtained under the assumption that there are N long-lived securities which make the market dynamically complete (in the sense of Definition 6.14). Indeed, as shown in Proposition 6.15, in this case the whole set of Arrow securities can be replicated by means of long-lived securities and an allocation corresponding to a Radner equilibrium coincides with an equilibrium allocation (see Proposition 6.16 and the preceding discussion). In this case, in correspondence of the no-trade equilibrium, the representative agent will simply hold the aggregate supply of long-lived securities and consume the aggregate dividends generated by the N securities.

The optimality conditions corresponding to a no-trade equilibrium of the representative agent economy (given by the first order conditions of the optimization Problem (6.48) in correspondence of the aggregate endowment process e) allow to characterize the prices of the whole set of Arrow securities in terms of the marginal rate of substitution of the representative agent's utility function defined in (6.48)–(6.49). Indeed, normalizing $q_0 = 1$, it holds that

$$q^*_{A_t} = \delta^t \frac{\pi_{A_t} \mathbf{u}'\left(e(A_t)\right)}{\mathbf{u}'(e_0)}, \qquad (6.50)$$

and

$$q^*_{A_s|A_t} = \delta^{s-t} \frac{\pi_{A_s} \mathbf{u}'\left(e(A_s)\right)}{\pi_{A_t} \mathbf{u}'\left(e(A_t)\right)}, \qquad (6.51)$$

for all $A_s \in \mathscr{F}_s$, $A_t \in \mathscr{F}_t$ and $s,t \in \{1,\dots,T\}$ with $s > t$ and $A_s \subseteq A_t$. It is important to remark that, in general, the economy with I agents and the representative agent economy only share the same equilibrium prices of the Arrow securities, meaning that the representative agent economy is only locally representative of the original economy. Indeed, as pointed out in Sect. 4.3, the representative agent's utility function depends on the weights $\{a^i; i = 1,\dots,I\}$ and, therefore, on the distribution of the aggregate endowment among the I agents.

As in a two-period economy (see Sect. 4.3 and compare also with Rubinstein [1485] and Milne [1344]), the aggregation property holds in a strong sense if agents exhibit homogeneous beliefs, the same discount factor δ, time separable and state independent preferences with utility functions belonging to the HARA class (i.e., generalized power, exponential or logarithmic utility functions with the same cautiousness coefficient, compare with Propositions 4.5 and 4.15). In this case, the representative agent's utility function $\mathbf{u}$ will also belong to the HARA class (with the same cautiousness coefficient of the individual utility functions) and there exists a representative agent in a strong sense, meaning that the economy with I agents and the single agent economy are observationally equivalent and equilibrium prices do not depend on the initial resources distribution, since the preferences of the representative agent will not depend on the latter. Moreover, if a sufficiently rich set of claims on the aggregate endowment process can be traded, then markets can be *effectively complete* (compare with the discussion at the end of Sect. 4.3) and an equilibrium allocation can be Pareto optimal. In this sense, the validity of formulae (6.50)–(6.51) can be extended to incomplete markets.

6.3   The Fundamental Theorem of Asset Pricing

In this section, we generalize to a multi-period setting the fundamental results presented in Sect. 4.4 in the context of a two-period economy. As we shall see, most of the results can be extended to a dynamic economy in a rather straightforward way and the interpretation of the major findings remains unchanged. In particular, the equivalence between the absence of arbitrage opportunities, the existence of a strictly positive linear pricing functional (and, equivalently, of a risk neutral probability measure) and the existence of an equilibrium for an economy continues to hold in a dynamic setting. In line with the structure of Sect. 4.4, we first characterize the absence of arbitrage opportunities via the existence of a risk neutral probability measure and then we study the relation between the absence of arbitrage opportunities and the existence of an equilibrium.

In this section, we assume that the economy comprises $N + 1$ long-lived securities, characterized by their price and dividend processes $S = (S_t)_{t=0,1,\dots,T}$ and $D = (D_t)_{t=0,1,\dots,T}$, respectively. As before, the processes S and D are supposed to be adapted to the information flow $\mathbb{F}$ and the cum-dividend price process is defined as $P := S + D$. Security 0 is supposed to be a risk free asset paying a constant rate $r_f > 0$, so that $p_t^0 = s_t^0 = r_f^t$, for all $t = 0, 1, \dots, T - 1$, and $p_T^0 = d_T^0 = r_f^T$ (without loss of generality, the price of the risk free security at the initial date $t = 0$ is normalized to $p_0^0 = 1$).[6] The remaining N securities represent risky assets paying non-negative random dividends at each date $t = 0, 1, \dots, T$. In this environment, we assume that the agents' trading activity can be described by means of self-financing trading-consumption strategies, in the sense of Definition 6.1.

No-Arbitrage and Risk Neutral Probability Measures

The notion of arbitrage opportunity introduced in Definition 4.17 can be formulated as follows in a multi-period setting.

Definition 6.17 For a given couple (S, D) of price-dividend processes, a non-negative consumption process $c = (c_t)_{t=0,1,\dots,T}$ such that $c \in \mathscr{C}_0^+(0)$ is said to be

(i) an *arbitrage opportunity of the first kind* if there exist a date $t \in \{1, \dots, T\}$ and an event $A_t \in \mathscr{F}_t$ such that $c_t(A_t) > 0$;

(ii) an *arbitrage opportunity of the second kind* if $c_0 > 0$.

---

[6]We want to point out that the results presented in this section can be extended to the case where the assumption of a constant risk free rate r_f is relaxed by assuming a deterministic time-varying rate or a stochastic rate given by a predictable stochastic process.

A non-negative consumption process $c = (c_t)_{t=0,1,...,T}$ is simply said to be an *arbitrage opportunity* if it is an arbitrage opportunity of the first kind and/or an arbitrage opportunity of the second kind.

In other words, an arbitrage opportunity is a non-negative consumption stream c which is strictly positive in correspondence of some event (at the initial date or at some future date, with strictly positive probability) and which can be financed from zero initial wealth, i.e., $c \in \mathscr{C}_0^+(0)$. More specifically, an arbitrage opportunity of the first kind is a consumption stream which does not require any initial wealth at the initial date $t = 0$ and yields a non-negative and non-null stream of payoffs at the dates $t \in \{1, ..., T\}$. An arbitrage opportunity of the second kind is a consumption stream which can be financed by a strictly negative amount of wealth at $t = 0$ (in the sense that $\theta_1^\top S_0 = -c_0 < 0$) and leads to a non-negative (but possibly null) stream of payoffs at dates $t = 1, ..., T$. Clearly, there are no arbitrage opportunities if and only if $\mathscr{C}_0^+(0) = \{0\}$, with 0 denoting the process which is identically null at all dates $t = 0, 1, ..., T$ (recall that the set $\mathscr{C}_0^+$ consists of non-negative adapted processes).

As in the case of a two-period economy (see Proposition 4.18), the absence of arbitrage opportunities implies the validity of the *Law of One Price*, meaning that identical consumption streams are necessarily financed by identical initial wealths. The proof of the next proposition is given in Exercise 6.15.

Proposition 6.18 *Let* $c = (c_t)_{t=0,1,...,T}$ *and* $\tilde{c} = (\tilde{c}_t)_{t=0,1,...,T}$ *be two consumption processes financed by the strategies* $\theta = (\theta_t)_{t=0,1,...,T}$ *and* $\tilde{\theta} = (\tilde{\theta}_t)_{t=0,1,...,T}$, *respectively, and such that* $\mathbb{P}(c_t = \tilde{c}_t) = 1$ *for all* $t = 0, 1, ..., T$. *Then the following hold:*

(i) *if there are no arbitrage opportunities of the second kind, then* $W_0(\theta) = W_0(\tilde{\theta})$;
(ii) *if there are no arbitrage opportunities of the first kind, then* $W_t(\theta) = W_t(\tilde{\theta})$ *with probability one, for all* $t = 1, ..., T$.

For any consumption process $c = (c_t)_{t=0,1,...,T} \in \mathscr{C}_0^+(x)$, the initial wealth x represents the cost at $t = 0$ of financing the consumption stream. More generally, for a self-financing trading-consumption strategy $(\theta, c) \in \mathscr{A}(x_0)$, the quantity $W_t(\theta)$ represents the cost (or market value) at time t of the future consumption stream $(c_s)_{s=t,...,T}$, for $t \in \{0, 1, ..., T\}$, i.e., the amount of wealth at time t that allows to finance the future consumption stream. In this sense, the Law of One Price (Proposition 6.18) implies that if two consumption processes can be financed by trading in the available securities and deliver identical payment streams, then they are financed by the same wealth. The absence of arbitrage opportunities also implies that the wealth associated to a self-financing trading-consumption strategy must always be greater (or equal) than the current consumption, as shown in the following lemma (see Exercises 6.16 and 6.17 for the proofs).

Lemma 6.19 *Suppose that the price-dividend couple* (S, D) *does not admit arbitrage opportunities and let* $(\theta, c) \in \mathscr{A}(x_0)$. *Then the following hold:*

(i) $\mathbb{P}(W_t(\theta) \geq c_t) = 1$, *for all* $t = 0, 1, ..., T$;

(ii) for all $t = 0, 1, \ldots, T$ and $A_t \in \mathscr{F}_t$, if there exist $s \in \{t + 1, \ldots, T\}$ and $A_s \subseteq A_t$ such that $c_s(A_s) > 0$, then it holds that $W_t(\theta)(A_t) > c_t(A_t)$, with $W_t(\theta)(A_t)$ denoting the value of the process $W(\theta)$ in correspondence of event A_t and date t.

Observe that the strategy $\bar{\theta} := \theta_{t+1}\mathbf{1}_{A_t}$ considered in the proof of Lemma 6.19 (see Exercise 6.16) realizes an arbitrage opportunity (in the sense of Definition 4.17) in the single trading period $[t, t + 1]$ with respect to the couple (S_t, P_{t+1}). Indeed, it holds that $\bar{\theta}^\top S_t \leq 0$, with strict inequality holding in correspondence of the event A_t, and $\bar{\theta}^\top (S_{t+1} + D_{t+1}) = W_{t+1}(\theta)\mathbf{1}_{A_t} \geq c_{t+1}\mathbf{1}_{A_t} \geq 0$. This remark will turn out to be useful in the proof of Lemma 6.20 below. Indeed, the absence of arbitrage opportunities in a multi-period economy can be characterized in terms of the absence of arbitrage opportunities in each single trading interval $[t, t + 1]$, for $t = 0, 1, \ldots, T - 1$. The proof of the following lemma is given in Exercise 6.18.

Lemma 6.20 *Let (S, D) be a couple of price-dividend processes. The following are equivalent:*

(i) the couple (S, D) admits an arbitrage opportunity;
(ii) there exists $t \in \{0, 1, \ldots, T - 1\}$ such that the couple (S_t, P_{t+1}) admits an arbitrage opportunity in the sense of Definition 4.17 in the period $[t, t + 1]$.

According to the above lemma and recalling the event tree structure described at the beginning of the present chapter, there are no arbitrage opportunities in the multi-period economy if and only if there is no arbitrage opportunity (in the sense of Definition 4.17) in every elementary single-period sub-tree. This observation turns out to be useful for characterizing the absence of arbitrage in a multi-period setting, as shown in Theorem 6.23 below. As a preliminary, let us define the notion of *risk neutral probability measure*, extending Definition 4.21 to the present setting.

Definition 6.21 A probability measure $\mathbb{P}^*$ on $(\Omega, \mathscr{F})$ such that $\mathbb{P}^*(\omega) > 0$ for all $\omega \in \Omega$ is a *risk neutral probability measure* for a couple (S, D) of price-dividend processes if, for all $t \in \{0, 1, \ldots, T - 1\}$, it holds that

$$\frac{1}{r_f}\mathbb{E}^* \left[s_{t+1}^n + d_{t+1}^n | \mathscr{F}_t\right] = s_t^n, \qquad \text{for all } n = 0, 1, \ldots, N, \tag{6.52}$$

where $\mathbb{E}^*[\cdot | \mathscr{F}_t]$ denotes the $\mathscr{F}_t$-conditional expectation under the probability measure $\mathbb{P}^*$.

Note that, by definition, a risk neutral probability measure $\mathbb{P}^*$ is *equivalent* to the original probability measure $\mathbb{P}$, in the sense that $\mathbb{P}$ and $\mathbb{P}^*$ share the same null sets.

Let us first derive some useful properties of a risk neutral probability measure. The proof of the following proposition is based on Definition 6.21 together with some simple computations and is given in detail in Exercise 6.19.

Proposition 6.22 *Let* $\mathbb{P}^*$ *be a risk neutral probability measure for* (S, D). *Then, for all* $t \in \{0, 1, \dots, T - 1\}$, *it holds that*

$$s_t^n = \sum_{s=t+1}^{T} \mathbb{E}^* \left[\frac{d_s^n}{r_f^{s-t}} \middle| \mathscr{F}_t \right], \qquad \text{for all } n = 0, 1, \dots, N. \qquad (6.53)$$

As a consequence, the process

$$\left(\frac{s_t^n}{r_f^t} + \sum_{s=0}^{t} \frac{d_s^n}{r_f^s} \right)_{t=0,1,\dots,T}$$

is a martingale under the probability measure $\mathbb{P}^*$, *for all* $n = 0, 1, \dots, N$. *Moreover, for any trading-consumption strategy* $(\theta, c) \in \mathscr{A}(x_0)$ *and for all* $s, t \in \{0, 1, \dots, T\}$ *with* $s < t$, *it holds that*

$$\mathbb{E}^* \left[\frac{W_t(\theta)}{r_f^t} + \sum_{r=s}^{t-1} \frac{c_r}{r_f^r} \middle| \mathscr{F}_s \right] = \frac{W_s(\theta)}{r_f^s}. \qquad (6.54)$$

The above proposition shows that, under a risk neutral probability measure $\mathbb{P}^*$, the (ex-dividend) price of a traded security is simply given by the conditional expectation of the future dividend stream, discounted with respect to the risk free rate r_f, thus explaining the *risk neutral* terminology. Equivalently, recalling that the return r_{t+1}^n of the *n*-the security is defined as $r_{t+1}^n = p_{t+1}^n / s_t^n$ (under the implicit assumption that all prices are strictly positive at every date $t = 0, 1, \dots, T - 1$), it holds that

$$\mathbb{E}^* [r_{t+1}^n | \mathscr{F}_t] = r_f, \qquad \text{for all } t = 0, 1, \dots, T - 1 \text{ and } n = 1, \dots, N, \qquad (6.55)$$

meaning that the expected return of each security under a risk neutral probability measure is equal to the risk free rate.

Proposition 6.22 shows an important *martingale property* of self-financing trading-consumption strategies. In fact, equation (6.54) means that, for every trading-consumption strategy $(\theta, c) \in \mathscr{A}(x_0)$, the process

$$\left(\frac{W_t(\theta)}{r_f^t} + \sum_{r=0}^{t-1} \frac{c_r}{r_f^r} \right)_{t=0,1,\dots,T}$$

is a martingale under a risk neutral probability measure $\mathbb{P}^*$. This property has interesting consequences for the analysis of optimal investment-consumption problems of the type (6.6) (see Corollary 6.28 below and compare also with Exercise 6.23). In particular, the martingale property (6.54) holds for any self-financing trading-consumption strategy $(\theta, 0)$.

The following result represents the counterpart to Proposition 4.22 for the case of a multi-period economy and shows that the absence of arbitrage opportunities can be characterized in terms of the existence of a risk neutral probability measure.

Theorem 6.23 *Let (S, D) be a couple of price-dividend processes. The following are equivalent:*

 (i) *there are no arbitrage opportunities;*
 (ii) *there exists a risk neutral probability measure $\mathbb{P}^*$, in the sense of Definition 6.21.*

Proof $(i) \Rightarrow (ii)$: by Lemma 6.20, the absence of arbitrage opportunities in the sense of Definition 6.17 implies that each single-period sub-tree is arbitrage free in the sense of Definition 4.17. More specifically, for every $t = 0, 1, \ldots, T-1$ and $A_t \in \mathscr{F}_t$, the couple $(S_t(A_t), \mathfrak{P}_{t+1}(A_t))$ is arbitrage free in the sense of Definition 4.17, with $\mathfrak{P}_{t+1}(A_t)$ being defined as in (6.47) (here extended to the case of $N + 1$ securities). Hence, in view of Proposition 4.22, for every $t = 0, 1, \ldots, T - 1$ and $A_t \in \mathscr{F}_t$, there exists a vector $q_{t+1}(A_t) = (q(A_{t+1}^1|A_t), \ldots, q(A_{t+1}^{\nu_t(A_t)}|A_t))^\top \in \mathbb{R}_{++}^{\nu_t(A_t)}$ such that

$$\frac{1}{r_f}\mathfrak{P}_{t+1}(A_t)^\top q_{t+1}(A_t) = S_t(A_t), \tag{6.56}$$

where $S_t(A_t) = (s_t^0(A_t), s_t^1(A_t), \ldots, s_t^N(A_t))^\top$ are the prices of the $N + 1$ securities at date t in correspondence of event A_t. Since $q_{t+1}(A_t) \in \mathbb{R}_{++}^{\nu_t(A_t)}$ and it holds that $\sum_{s=1}^{\nu_t(A_t)} q(A_{t+1}^s|A_t) = 1$, the elements of $q_{t+1}(A_t)$ can be taken as the quantities characterizing a probability measure (conditional on the event A_t) over the events in $\mathscr{F}_{t+1}$ that are subsets of A_t.

We now construct a risk neutral probability measure $\mathbb{P}^*$ by aggregating all the $q_{t+1}(A_t)$, recalling that the information flow $\mathbb{F} = (\mathscr{F}_t)_{t=0,1,\ldots,T}$ is a refining sequence of partitions of Ω and that the final partition $\mathscr{F}_T$ consists of all the elementary events. For any fixed $\omega \in \Omega$ and for every date $t = 1, \ldots, T$, let $A_t(\omega)$ denote the event of $\mathscr{F}_t$ which contains ω. For $t = 0, 1, \ldots, T - 1$, let us also define $q(A_{t+1}(\omega)|A_t(\omega))$ as the element of the vector $q_{t+1}(A_t(\omega))$ corresponding to the event $A_{t+1}(\omega)$ containing ω. Define then

$$\pi_\omega^* := \prod_{t=0}^{T-1} q\big(A_{t+1}(\omega)|A_t(\omega)\big).$$

The probability measure $\mathbb{P}^*$ defined by $(\pi_1^*, \ldots, \pi_{|\Omega|}^*)$ is a risk neutral probability measure, in the sense of Definition 6.21. Indeed, it holds that $\pi_\omega^* > 0$ for all $\omega \in \Omega$.

Moreover,

$$\sum_{\omega \in \Omega} \pi_\omega^* = \sum_{\omega \in \Omega} \prod_{t=0}^{T-1} q\big(A_{t+1}(\omega)|A_t(\omega)\big)$$

$$= \sum_{A_1 \in \mathscr{F}_1} q(A_1|A_0) \sum_{A_2 \in \mathscr{F}_2} q(A_2|A_1) \cdots \sum_{\omega \in \mathscr{F}_T} q(\omega|A_{T-1}) = 1,$$

with $A_0 = \Omega$. According to the measure $\mathbb{P}^*$, the probability of an event $A_t \in \mathscr{F}_t$, for $t = 1, \ldots, T$, is given by

$$\mathbb{P}^*(A_t) = \sum_{\omega \in A_t} \pi_\omega^* = \sum_{\omega \in A_t} \prod_{s=0}^{t-1} q\big(A_{s+1}(\omega)|A_s(\omega)\big),$$

while the vector $q_{t+1}(A_t)$ represents the conditional probabilities associated with $\mathbb{P}^*$. In fact:

$$\mathbb{P}^*(A_{t+1}|A_t) = \frac{\mathbb{P}^*(A_{t+1} \cap A_t)}{\mathbb{P}^*(A_t)} = \frac{\pi_{A_{t+1} \cap A_t}^*}{\pi_{A_t}^*} = \frac{\pi_{A_{t+1}}^*}{\pi_{A_t}^*} = q(A_{t+1}|A_t),$$

for any $A_{t+1} \in \mathscr{F}_{t+1}$ such that $A_{t+1} \subseteq A_t$. Together with (6.56), this implies the validity of condition (6.52), thus showing that $\mathbb{P}^*$ is a risk neutral probability measure.

(*ii*) $\Rightarrow$ (*i*): suppose that there exists a risk neutral probability measure $\mathbb{P}^*$ and let $c \in \mathscr{C}_0^+(0)$. Then, due to equation (6.54), it holds that

$$0 = c_0 + \sum_{t=1}^{T} \frac{1}{r_f^t} \mathbb{E}^*[c_t],$$

so that the process c is identically null. This shows that $\mathscr{C}_0^+(0) = \{0\}$, meaning that there are no arbitrage opportunities. $\square$

The proof of Theorem 6.23 is constructive, in the sense that it explicitly constructs a risk neutral probability measure $\mathbb{P}^*$ (in the sense of Definition 6.21) by decomposing the multi-period economy into a family of elementary single-period models and by aggregating all the risk neutral probability measures obtained for each single-period sub-tree. However, Theorem 6.23 can also be proved in a more abstract way by relying on a separating hyperplane argument, analogously to Theorem 4.20. Note that the risk neutral probability measure $\mathbb{P}^*$ is not unique in general (see Theorem 6.26 below).

As in Sect. 4.4, the absence of arbitrage opportunities can be equivalently characterized in terms of (the properties of) pricing functionals. Let us first introduce the notion of *pricing functional* in a dynamic setting. Let $c = (c_t)_{t=0,1,\ldots,T} \in \mathscr{C}_0^+(x)$ be a consumption process, for some initial wealth $x \in \mathbb{R}_+$. As discussed after

Proposition 6.18, the quantity x can be regarded as the initial cost (or market value) at $t = 0$ for the consumption stream $(c_t)_{t=0,1,\ldots,T}$. Moreover, due to the Law of One Price, such initial cost is uniquely defined if there are no arbitrage opportunities. Therefore, if $c \in \mathscr{C}_0^+(x)$, we can write $V(c_0, c_1, \ldots, c_T) := x$, where V is the function which gives the market value of an attainable consumption plan. Note also that, if (S, D) is a couple of price-dividend processes not admitting arbitrage opportunities, then relation (6.54) implies that V can be represented as

$$V(c_0, c_1, \ldots, c_T) = x = \sum_{s=0}^{T} \frac{1}{r_f^s} \mathbb{E}^*[c_s],$$

where $\mathbb{P}^*$ is a risk neutral probability measure. This representation shows that the initial wealth needed to finance an attainable consumption plan can be expressed as the risk neutral expectation of future discounted consumption.

The following definition is a natural extension of Definition 4.19 to the present setting. We let $\mathscr{K}$ denote the set of all real-valued stochastic processes adapted to the information flow $\mathbb{F}$.

Definition 6.24 A mapping $Q : \mathscr{K} \to \mathbb{R}$ is said to be a *pricing functional* for (S, D) if, for every $x \in \mathbb{R}_+$ and for every process $c = (c_t)_{t=0,1,\ldots,T} \in \mathscr{C}_0^+(x)$, it holds that

$$Q(c) := Q(c_0, c_1, \ldots, c_T) = V(c_0, c_1, \ldots, c_T) = x.$$

A pricing functional $Q : \mathscr{K} \to \mathbb{R}$ is said to be *linear* if

$$Q(\alpha c + \beta c') = \alpha Q(c) + \beta Q(c'),$$

for all $c, c' \in \mathscr{K}$ and $\alpha, \beta \in \mathbb{R}$. Moreover, the functional $Q : \mathscr{K} \to \mathbb{R}$ is said to be *strictly positive* if, for every non-negative and non-null process $c \in \mathscr{K}$, it holds that $Q(c) > 0$. Similarly as in Sect. 4.4, a pricing functional extends the notion of market value to consumption plans that are not necessarily financed by some trading strategy. Note that the pricing functional is uniquely defined for every attainable consumption plan.

We are now in a position to prove that the absence of arbitrage opportunities is equivalent to the existence of a strictly positive linear pricing functional, as shown in the following simple corollary.

Corollary 6.25 *Let (S, D) be a couple of price-dividend processes. The following are equivalent:*

(i) there are no arbitrage opportunities;
(ii) there exists a risk neutral probability measure $\mathbb{P}^*$;
(iii) there exists a strictly positive linear pricing functional $Q : \mathscr{K} \to \mathbb{R}$.

Proof (i) $\Leftrightarrow$ (ii): this equivalence is the content of Theorem 6.23.

(ii) $\Rightarrow$ (iii): if $\mathbb{P}^*$ is a risk neutral probability measure for (S, D), then a strictly positive linear pricing functional $Q : \mathscr{K} \to \mathbb{R}$ for (S, D) can be defined as

$$Q(c) = Q(c_0, c_1, \ldots, c_T) := \sum_{t=0}^{T} \frac{1}{r_f^t} \mathbb{E}^*[c_t].$$

The linearity of Q is evident by construction, while the strict positivity follows from the fact that the probability measure $\mathbb{P}^*$ is equivalent to $\mathbb{P}$.

(iii) $\Rightarrow$ (i): suppose that there exists a strictly positive linear pricing functional $Q : \mathscr{K} \to \mathbb{R}$ and let $c \in \mathscr{C}_0^+(0)$ be an arbitrage opportunity. Then, in view of Definition 6.24, it holds that

$$Q(c_0, c_1, \ldots, c_T) = 0.$$

Since the pricing functional Q is strictly positive, this implies that the process c is identically zero and, hence, cannot be an arbitrage opportunity. $\qquad\square$

The proof of the above corollary shows that, similarly to the case of a two-period economy, there is a one-to-one relation between strictly positive linear pricing functionals and risk neutral probability measures. Recalling the event-tree structure described at the beginning of the present chapter, a consumption process $c = (c_t)_{t=0,1,\ldots,T}$ can be represented as $c = \{c_0, c_t(A_t); A_t \in \mathscr{F}_t, t = 1, \ldots, T\}$, with $c_t(A_t)$ denoting the level of consumption at date t in correspondence of event A_t. By the Riesz representation theorem, if $Q : \mathscr{K} \to \mathbb{R}$ is a linear pricing functional, then it can be represented in terms of a collection $\{m_0, m_t(A_t); A_t \in \mathscr{F}_t, t = 1, \ldots, T\}$, so that, for any $c \in \mathscr{K}$,

$$Q(c) = m_0 c_0 + \sum_{t=1}^{T} \sum_{A_t \in \mathscr{F}_t} m_t(A_t) c_t(A_t). \tag{6.57}$$

Similarly, if $\mathbb{P}^*$ is a risk neutral probability measure, the proof of Theorem 6.23 shows that it can be represented by the collection $\{\pi_{A_t}^*; A_t \in \mathscr{F}_t, t = 1, \ldots, T\}$, so that

$$\sum_{t=0}^{T} \frac{1}{r_f^t} \mathbb{E}^*[c_t] = c_0 + \sum_{t=1}^{T} \frac{1}{r_f^t} \sum_{A_t \in \mathscr{F}_t} \pi_{A_t}^* c_t(A_t). \tag{6.58}$$

As discussed after Theorem 4.20, the quantities $\{m_0, m_t(A_t); A_t \in \mathscr{F}_t, t = 1, \ldots, T\}$ appearing in (6.57) admit the interpretation of *state prices* associated to one unit of consumption in correspondence of the different dates/events (*Arrow securities*). In particular, consider the consumption plan $\mathbf{1} = (0, \ldots, 0, 1)$ which is identically null before the final date T and always equal to one at T. Of course, the consumption plan $\mathbf{1}$ can be attained by simply holding a quantity $1/r_f^T$ of the risk free security.

Hence, in view of Definition 6.24, it holds that

$$\sum_{\omega \in \Omega} m_T(\omega) = Q(\mathbf{1}) = V(\mathbf{1}) = \frac{1}{r_f^T}.$$

The same argument can be applied for $t = 0, 1, \ldots, T - 1$, thus showing that

$$\sum_{A_t \in \mathscr{F}_t} m_t(A_t) = \frac{1}{r_f^t}, \qquad \text{for all } t = 0, 1, \ldots, T - 1.$$

These observations imply that, if Q is a strictly positive linear pricing functional, then the quantities

$$\pi^*_{A_t} = \frac{m_t(A_t)}{\sum_{A \in \mathscr{F}_t} m_t(A)}, \qquad \text{for } A_t \in \mathscr{F}_t \text{ and } t = 0, 1, \ldots, T,$$

define a risk neutral probability measure. Conversely, similarly as in the proof of Corollary 6.25, if $\mathbb{P}^*$ is a risk neutral probability measure then the quantities

$$m_t(A_t) = \frac{\pi^*_{A_t}}{r_f^t}, \qquad \text{for } A_t \in \mathscr{F}_t \text{ and } t = 0, 1, \ldots, T,$$

define a strictly positive linear pricing functional. This last relation shows how state prices can be generated from a risk neutral probability measure $\mathbb{P}^*$.

The following theorem generalizes Theorem 4.26 and characterizes dynamic market completeness in terms of risk neutral probability measures (this result is typically referred to as the *second Fundamental Theorem of Asset Pricing*).

Theorem 6.26 *Let (S, D) be a couple of price-dividend processes not admitting arbitrage opportunities. The following are equivalent:*

 (i) the market is dynamically complete, in the sense of Definition 6.14;
 (ii) there exists a unique risk neutral probability measure $\mathbb{P}^*$.

Proof In view of Proposition 6.15, the market is dynamically complete if and only if, for all $A_t \in \mathscr{F}_t$ and $t = 0, 1, \ldots, T - 1$, it holds that $\text{rank}(\mathfrak{P}_{t+1}(A_t)) = \nu_t(A_t)$. Hence, the equivalence $(i) \Leftrightarrow (ii)$ follows by noting that equation (6.56) in the proof of Theorem 6.23 has a unique solution for all $A_t \in \mathscr{F}_t$ and $t = 0, 1, \ldots, T - 1$ if only if condition (ii) of Proposition 6.15 is satisfied. $\square$

In view of the above result, if the market is free of arbitrage opportunities and dynamically complete then there exists a unique risk neutral probability measure or, equivalently, a unique strictly positive linear pricing functional. Furthermore, the market is dynamically complete if and only if there exists a unique set of strictly positive prices for the Arrow securities (state prices), which fully characterize the strictly positive linear pricing functional. Note that, in a dynamically complete market, the whole set of prices of the Arrow securities is fully determined by the couple (S, D). Of course, this is no longer true in the case of incomplete markets. Note, however, that there is no general implication between the absence of arbitrage opportunities and market completeness, in the sense that one can construct examples of dynamically complete markets admitting arbitrage opportunities.

The discussion made in Sect. 4.4 after Proposition 4.22 can be extended to the present multi-period context, with the same interpretation. In particular, if the market is dynamically complete, in the sense of Definition 6.14, then every non-negative adapted process can be attained, meaning that $\mathscr{K}^+ = \bigcup_{x \in \mathbb{R}_+} \mathscr{C}_0^+(x)$, where $\mathscr{K}^+$ denotes the set of non-negative adapted stochastic processes. In other words, given an appropriate initial wealth, every non-negative consumption plan can be financed by trading in the $N + 1$ available securities. Moreover, it can be easily verified that, if a market is dynamically complete in the sense of Definition 6.14, then every real-valued process $c = (c_t)_{t=0,1,\ldots,T} \in \mathscr{K}$ can be financed by a trading strategy. This simply follows by considering the decomposition $c = c^+ - c^-$ of c into its positive and negative parts and by the linearity of the self-financing condition. Hence, in a dynamically complete market it holds that $\mathscr{K} = \bigcup_{x \in \mathbb{R}} \mathscr{C}_0(x)$, where we denote by $\mathscr{C}_0(x)$ the set of real-valued adapted processes $c = (c_t)_{t=0,1,\ldots,T}$ such that there exists a trading strategy $\theta = (\theta_t)_{t=1,\ldots,T}$ satisfying the self-financing condition (6.7) with respect to x. In a dynamically complete market, the arbitrage free price (market value) of a process $c \in \mathscr{K}$ is uniquely given by

$$Q(c) = V(c) = c_0 + \sum_{t=1}^{T} \frac{1}{r_f^t} \mathbb{E}^*[c_t].$$

By following the same arguments given in Exercise 4.20, it can be shown that, if it was possible to attain a consumption plan c starting from an initial wealth $x \neq Q(c)$, then the market would admit arbitrage opportunities.

Under the assumption of a complete market, the above discussion leads us to the *risk neutral valuation* paradigm. Indeed, a *contingent claim* X can be represented by an adapted process $X = (x_t)_{t=1,\ldots,T} \in \mathscr{K}$, with x_t denoting the random payoff of the contingent claim at date t. We say that a contingent claim is *European* if the process X is identically null before the terminal date T, i.e., the claim only delivers a non-null payoff at its expiration date (maturity). The previous arguments imply the following simple but important result.

Proposition 6.27 *Suppose that the couple (S, D) of price-dividend processes does not admit arbitrage opportunities and is such that the market is dynamically complete. Let $X = (x_t)_{t=1,\ldots,T}$ be a contingent claim. Then, the unique arbitrage free price $q_0(X)$ of X at date $t = 0$ is given by*

$$q_0(X) = Q(X) = \sum_{t=1}^{T} \frac{1}{r_f^t} \mathbb{E}^*[x_t]. \tag{6.59}$$

In particular, in the case of a European contingent claim X, it holds that

$$q_0(X) = Q(X) = \frac{1}{r_f^T} \mathbb{E}^*[x_T].$$

Proof Since the market is dynamically complete, there exists a trading strategy $\theta = (\theta_t)_{t=0,1,...,T}$ such that the couple $(\theta, (0, x_1, \ldots, x_T))$ is self-financing, starting from some initial wealth $x_0 \in \mathbb{R}$ (i.e., $(\theta, (0, x_1, \ldots, x_T)) \in \mathscr{A}(X_0)$). Hence, in view of equation (6.54) together with Definition 6.24, it holds that

$$Q(X) = V(0, x_1, \ldots, x_T) = W_0(\theta) = \sum_{t=1}^{T} \frac{1}{r_f^t} \mathbb{E}^*[x_t] = x_0.$$

□

As observed above (compare also with Exercise 4.20), $q_0(X) = Q(X)$ is the unique price such that, if the contingent claim X is introduced in the market with price $q_0(X)$, then no arbitrage opportunity arises.

If the market is incomplete then the situation is in general more complicated and there exist elements of $\mathscr{H}$ that cannot be attained. In this case, one has to distinguish between the elements of $\mathscr{H}$ that can be attained and the unattainable consumption streams/contingent claims. In the first case, i.e., for the elements belonging to $\bigcup_{x\in\mathbb{R}} \mathscr{C}_0(x)$, the arbitrage free price is still uniquely defined as in Proposition 6.27 above, regardless of the choice of the risk neutral probability measure (see Exercise 6.21). In the case of unattainable consumption plans/contingent claims, i.e., for the elements of $\mathscr{H} \setminus (\bigcup_{x\in\mathbb{R}} \mathscr{C}_0(x))$, there exist infinitely many arbitrage free prices. Similarly as in Definition 4.23, one can prove the existence of an interval of arbitrage free prices by introducing the *super-replication* price and the *sub-replication* price. In particular, analogously to Proposition 4.25, this interval of arbitrage free prices can be shown to be generated by the family of discounted risk neutral expectations of the consumption stream, with respect to all risk neutral probability measures. Any candidate price not belonging to such an interval would inevitably lead to arbitrage opportunities (compare with Exercise 4.23).

The existence of a risk neutral probability measure allows to give an alternative representation of the budget constraint in the optimal investment-consumption problem (6.6), as shown in the following corollary, under the assumption of complete markets (see Exercise 6.23 for a related result in the case of possibly incomplete markets). Recall that, in the original formulation of problem (6.6), an agent maximizes the expected utility of consumption over all self-financing trading-consumption strategies (θ, c) starting from some given initial wealth $x > 0$.

Corollary 6.28 *Let (S, D) be a couple of price-dividend processes not admitting arbitrage opportunities. Suppose that the market is dynamically complete and let $\mathbb{P}^*$ be the (unique) risk neutral probability measure. Then, for any $x > 0$, the maximization in problem (6.6) can be taken over all non-negative adapted processes*

$c = (c_t)_{t=0,1,...,T}$ *satisfying*

$$c_0 + \sum_{t=1}^{T} \frac{1}{r_f^t} \mathbb{E}^*[c_t] = x. \tag{6.60}$$

Proof Observe that the maximization in problem (6.6) is over all $c \in \mathscr{C}_0^+(x)$. Since the market is complete and $\mathbb{P}^*$ is the risk neutral probability measure, the claim follows since

$$\mathscr{C}_0^+(x) = \left\{ c \in \mathscr{K}^+ : c_0 + \sum_{t=1}^{T} \frac{1}{r_f^t} \mathbb{E}^*[c_t] = x \right\}.$$

Indeed, if $c \in \mathscr{C}_0^+(x) \subseteq \mathscr{K}^+$, Proposition 6.22 implies that $c_0 + \sum_{t=1}^{T} \frac{1}{r_f^t} \mathbb{E}^*[c_t] = x$. On the contrary, if $c \in \mathscr{K}^+$ and the market is complete, then there exists a trading strategy $(\theta_t)_{t=0,1,...,T}$ which finances the consumption process c, so that, in view of Proposition 6.22,

$$W_0(\theta) = c_0 + \sum_{t=1}^{T} \frac{1}{r_f^t} \mathbb{E}^*[c_t] = x,$$

for some $x > 0$, thus showing that $c \in \mathscr{C}_0^+(x)$.   □

Corollary 6.28 provides an important insight to address optimal investment-consumption problems of the type (6.6) and leads to the so-called *martingale approach* (see, e.g., Cox & Huang [504]). In a nutshell, in the context of complete markets, the martingale approach corresponds to a two-step approach to the solution of problem (6.6): in the first step the preference functional in (6.6) is simply maximized over all non-negative adapted processes $c \in \mathscr{K}^+$ satisfying the budget constraint in the form (6.60), with respect to the risk neutral probability measure $\mathbb{P}^*$. In a second step, by relying on the completeness of the market, the optimal trading strategy is characterized as the trading strategy which finances the optimal consumption process.

No-Arbitrage and Equilibrium

After having characterized the absence of arbitrage opportunities in terms of risk neutral probability measures or, equivalently, strictly positive linear pricing functionals, we now address the relation between the absence of arbitrage and the existence of an equilibrium. As we shall see, the following results can be regarded as rather straightforward extensions to the present setting of the results obtained in Sect. 4.4 in the case of an economy with two dates.

We consider an economy comprising $N + 1$ traded securities (long-lived securities) characterized by the couple (S, D) of price-dividend processes, with security 0 being a risk free asset with rate of return $r_f > 0$. As in Sect. 6.2, we suppose that the economy is populated by I agents, with state independent and time additive preferences described by the utility functions $\{u^i; i = 1, \ldots, I\}$. Each agent $i = 1, \ldots, I$ is endowed with $\bar{\theta}_0^i \in \mathbb{R}^{N+1}$ shares of the $N + 1$ traded securities (as in Sect. 6.2, without loss of generality, we assume that $\sum_{i=1}^I \bar{\theta}_0^{i,n} = 1$, for all $n = 0, 1, \ldots, N$). A first and basic result shows that the absence of arbitrage opportunities is a necessary condition for the existence of a solution to the optimal investment-consumption problem (6.6) and, hence, for the existence of an equilibrium of the economy. This generalizes to a multi-period economy the result of Proposition 4.27 (see also Corollary 4.28).

Proposition 6.29 *For any $i = 1, \ldots, I$, suppose that $u^i : \mathbb{R}_+ \to \mathbb{R}$ is strictly increasing and that there exists an optimal solution $(\theta^{i*}, c^{i*}) \in \mathscr{A}(x)$ to problem (6.6), for some $x \in \mathbb{R}_+$, in correspondence of the couple (S, D) of price-dividend processes. Then there are no arbitrage opportunities. As a consequence, if the couple (S, D) corresponds to a Radner equilibrium, in the sense of Definition 6.13, then there are no arbitrage opportunities.*

Proof Arguing by contradiction, let $(\theta^{i*}, c^{i*}) \in \mathscr{A}(x)$ be a solution to problem (6.6) for agent i and let $\tilde{c} \in \mathscr{C}_0^+(0)$ be an arbitrage opportunity, financed by some strategy $\tilde{\theta} = (\tilde{\theta}_t)_{t=0,1,\ldots,T}$. Consider then the strategy $(\hat{\theta}^i, \hat{c}^i)$ defined by $\hat{\theta}^i := \theta^{i*} + \tilde{\theta}$ and $\hat{c}^i := c^{i*} + \tilde{c}$. Since $(\tilde{\theta}, \tilde{c}) \in \mathscr{A}(0)$, it follows that $(\hat{\theta}^i, \hat{c}^i) \in \mathscr{A}(x)$ and

$$W_0(\hat{\theta}^i) = W_0(\theta^{i*}) + W_0(\tilde{\theta}) = W_0(\theta^{i*}) = x$$

implies that $\hat{c}^i \in \mathscr{C}_0^+(x)$. Moreover, in view of Definition 6.17, there exist a date $t \in \{0, 1, \ldots, T\}$ and an event $A_t \in \mathscr{F}_t$ such that $\hat{c}_t^i = c_t^{i*} + \tilde{c}_t > c_t^{i*}$ on A_t. Since u^i is strictly increasing, this contradicts the optimality of (θ^{i*}, c^{i*}). In view of Definition 6.13, the last claim of the proposition is obvious.   □

Of course, since an arbitrage opportunity corresponds to the possibility of financing a non-null consumption stream out of nothing, the above result is hardly surprising.

We now study the converse and more interesting implication, namely whether the absence of arbitrage opportunities implies the existence of an equilibrium of the economy. As a first step, allowing for risk neutral agents, the following version of Proposition 4.29 can be established by extending to the present multi-period setting the proof of Proposition 4.29 (see Exercise 6.25).

Proposition 6.30 *Let (S, D) be a couple of price-dividend processes not admitting arbitrage opportunities. Then there exists an economy populated by I risk neutral agents endowed with $\bar{\theta}_0^i \in \mathbb{R}^{N+1}$ shares of the $N + 1$ securities, for all $i = 1, \ldots, I$, with beliefs given by risk neutral probabilities and a risk free discount factor $\delta := 1/r_f$, such that the couple (S, D) corresponds to a Radner equilibrium of this economy.*

As explained in the case of an economy with two dates, the above proposition also gives an explanation for the *risk neutral* terminology (compare also with the discussion following Proposition 4.29).

Let us now move to the more interesting case of an economy populated by risk averse agents. As a preliminary, similarly as in the case of a two-period economy (compare with Proposition 4.30 and see Exercise 6.27 for more details), under suitable assumptions on the utility function u^i, it can be shown that the absence of arbitrage opportunities is not only necessary (as shown in Proposition 6.29 above) but also sufficient in order to ensure the existence of a solution to the optimal investment-consumption problem (6.6).

Proposition 6.31 *For any $i = 1, \ldots, I$, suppose that $u^i : \mathbb{R}_+ \to \mathbb{R}$ is continuous and strictly increasing and concave. Then, there exists an optimal solution to the investment-consumption problem (6.6) in correspondence of the couple (S, D) of price-dividend processes if and only if (S, D) does not admit arbitrage opportunities.*

The last proposition implies that, given an arbitrage free couple (S, D) of price-dividend processes, there exists a single-agent economy such that (S, D) is associated to a (no-trade) equilibrium of such an economy. This can be proved as a direct extension of Proposition 4.31, by endowing the single agent at date $t = 0$ with the optimal consumption plan $\{c^{i*}(A_t); A_t \in \mathscr{F}_t, t = 0, 1, \ldots, T\}$, recalling the notation introduced at the beginning of Sect. 6.2. The interesting implication of this observation is given by the fact that it allows to construct a risk neutral probability measure in terms of the marginal rates of substitution at the optimum, as shown in the following proposition.

Proposition 6.32 *For any $i = 1, \ldots, I$, let $u^i : \mathbb{R}_+ \to \mathbb{R}$ be strictly increasing and concave and denote by $c^{i*} = (c^{i*}_t)_{t=0,1,\ldots,T}$ the corresponding consumption process solving problem (6.6) with respect to the couple (S, D) of price-dividend processes. Then a risk neutral probability measure $\mathbb{P}^*$ can be defined by letting*

$$\pi^*_\omega := (\delta r_f)^T \frac{u^{i'}\left(c^{i*}_T(\omega)\right)}{u^{i'}\left(c^{i*}_0\right)} \pi_\omega, \qquad \text{for all } \omega \in \Omega. \tag{6.61}$$

Furthermore, the conditional probabilities associated to $\mathbb{P}^*$ are given by

$$\pi^*_{A_t|A_s} := \mathbb{P}^*(A_t|A_s) = (\delta r_f)^{t-s} \frac{u^{i'}\left(c^{i*}_t(A_t)\right)}{u^{i'}\left(c^{i*}_s(A_s)\right)} \pi_{A_t|A_s}, \tag{6.62}$$

for all $s, t \in \{0, 1, \ldots, T\}$ with $t > s$ and $A_s \in \mathscr{F}_s$, $A_t \in \mathscr{F}_t$ such that $A_t \subseteq A_s$.

Proof If c^{i*} is an optimal consumption process associated to the utility function u^i, then Corollary 6.5 (in the case where a risk free security is traded) together with (6.14) implies that

$$u^{i'}(c^{i*}_t) = \delta r_f \mathbb{E}[u^{i'}(c^{i*}_{t+1})|\mathscr{F}_t], \qquad \text{for all } t = 0, 1, \ldots, T-1. \tag{6.63}$$

By iterating and using the tower property of the conditional expectation, this implies that $u^{i'}(c_0^{i*}) = (\delta r_f)^T \mathbb{E}[u^{i'}(c_T^{i*})]$, so that

$$\sum_{\omega \in \Omega} \pi_\omega^* = \sum_{\omega \in \Omega} (\delta r_f)^T \frac{u^{i'}(c_T^{i*}(\omega))}{u^{i'}(c_0^{i*})} \pi_\omega = \frac{(\delta r_f)^T \mathbb{E}[u^{i'}(c_T^{i*})]}{u^{i'}(c_0^{i*})} = 1.$$

Since u^i is assumed to be strictly increasing, this shows that $\{\pi_\omega^*; \omega \in \Omega\}$ describes a probability measure on $(\Omega, \mathscr{F})$. Moreover, in view of (6.15) and using repeatedly equation (6.63), it holds that, for any $n = 0, 1, \ldots, N$ and $t = 0, 1, \ldots, T-1$,

$$u^{i'}(c_t^{i*})s_t^n = \delta \mathbb{E}[u^{i'}(c_{t+1}^{i*})p_{t+1}^n | \mathscr{F}_t] = (\delta r_f)^{T-t} \mathbb{E}\left[u^{i'}(c_T^{i*})\frac{p_{t+1}^n}{r_f} \Big| \mathscr{F}_t\right],$$

so that, using again the fact that u^i is strictly increasing and equation (6.63),

$$s_t^n = \frac{(\delta r_f)^T \mathbb{E}\left[u^{i'}(c_T^{i*})\frac{p_{t+1}^n}{r_f} \big| \mathscr{F}_t\right]}{(\delta r_f)^t u^{i'}(c_t^{i*})} = \frac{(\delta r_f)^T \mathbb{E}\left[u^{i'}(c_T^{i*})\frac{p_{t+1}^n}{r_f} \big| \mathscr{F}_t\right]}{(\delta r_f)^T \mathbb{E}\left[u^{i'}(c_T^{i*}) | \mathscr{F}_t\right]}$$

$$= \frac{\mathbb{E}\left[\frac{(\delta r_f)^T u^{i'}(c_T^{i*})}{u^{i'}(c_0^{i*})} \frac{p_{t+1}^n}{r_f} \big| \mathscr{F}_t\right]}{\mathbb{E}\left[\frac{(\delta r_f)^T u^{i'}(c_T^{i*})}{u^{i'}(c_0^{i*})} \big| \mathscr{F}_t\right]} = \mathbb{E}^*\left[\frac{p_{t+1}^n}{r_f} \Big| \mathscr{F}_t\right],$$

where the last equality follows from the conditional version of the Bayes rule.[7] This shows that $\mathbb{P}^*$ is a risk neutral probability measure, in the sense of Definition 6.21. The conditional probabilities given in (6.62) can be obtained as shown in Exercise 6.20.   □

In particular, representation (6.62) (by taking $s = 0$ and $A_s = \Omega$) yields

$$\pi_{A_t}^* = (\delta r_f)^t \frac{u^{i'}(c_t^{i*}(A_t))}{u^{i'}(c_0^{i*})} \pi_{A_t} =: \ell_t(A_t)\pi_{A_t}, \quad \text{for all } A_t \in \mathscr{F}_t \text{ and } t = 0, 1, \ldots, T,$$

(6.64)

where $\{\ell_t(A_t); A_t \in \mathscr{F}_t, t = 0, 1, \ldots, T\}$ is defined by

$$\ell_t(A_t) := (\delta r_f)^t \frac{u^{i'}(c_t^{i*}(A_t))}{u^{i'}(c_0^{i*})}, \quad \text{for } A_t \in \mathscr{F}_t \text{ and } t = 0, 1, \ldots, T,$$

---

[7] Recall that the conditional version of the Bayes rule can be stated as follows. Let $\mathbb{P}$ and $\mathbb{P}^*$ be two probability measures on the space $(\Omega, \mathscr{F})$ such that $\mathbb{P} \sim \mathbb{P}^*$ (i.e., $\mathbb{P}$ and $\mathbb{P}^*$ are equivalent). Then, for any bounded random variable X, it holds that $\mathbb{E}^*[X|\mathscr{F}_t] = \mathbb{E}[X\frac{d\mathbb{P}^*}{d\mathbb{P}}|\mathscr{F}_t]/\mathbb{E}[\frac{d\mathbb{P}^*}{d\mathbb{P}}|\mathscr{F}_t]$, for each $t = 0, 1, \ldots, T$.

and represents the *likelihood ratio* between the risk neutral probability measure $\mathbb{P}^*$ and the original probability measure $\mathbb{P}$. The process $(M_t)_{t=0,1,\ldots,T}$ defined by

$$M_t := \delta^t \frac{u^{i'}(c_t^{i*})}{u^{i'}(c_0^{i*})}, \qquad \text{for } t = 0, 1, \ldots, T, \tag{6.65}$$

represents the *stochastic discount factor* (or *pricing kernel*). Similarly as in the basic case of a two-period economy (see Sect. 4.4), this shows that the likelihood ratio and the stochastic discount factor are determined by the marginal utility of an agent in correspondence of the optimal consumption stream. In Sect. 6.4, we shall derive several asset pricing relations by relying on expression (6.64).

Note also that, since $\sum_{A_t \in \mathscr{F}_t} \pi_{A_t}^* = 1$ for all $t = 0, 1, \ldots, T$, relation (6.64) implies that

$$\frac{1}{(\delta r_f)^t} = \frac{\mathbb{E}[u^{i'}(c_t^{i*})]}{u^{i'}(c_0^{i*})}, \qquad \text{for all } t = 0, 1, \ldots, T,$$

which, in turn implies that

$$\pi_{A_t}^* = \frac{u^{i'}\left(c_t^{i*}(A_t)\right)}{\mathbb{E}[u^{i'}(c_t^{i*})]} \pi_{A_t}, \qquad \text{for all } A_t \in \mathscr{F}_t \text{ and } t = 0, 1, \ldots, T.$$

Furthermore, if the subjective discount factor δ coincides with $1/r_f$, relation (6.62) can be expressed as

$$\pi_{A_t|A_s}^* = \frac{u^{i'}\left(c_t^{i*}(A_t)\right)}{u^{i'}\left(c_s^{i*}(A_s)\right)} \pi_{A_t|A_s},$$

for all $s, t \in \{0, 1, \ldots, T\}$ with $t > s$ and $A_s \in \mathscr{F}_s$, $A_t \in \mathscr{F}_t$ such that $A_t \subseteq A_s$.

Proposition 6.32 as well as the above relations concern the optimal investment-consumption problem of an individual agent. Under suitable assumptions, we can extend these results to the case of an economy populated by I risk averse agents by relying on the existence of a representative agent, as discussed in Sect. 6.2. More specifically, if the I agents exhibit state independent time homogeneous preferences characterized by the utility functions $\{u^i; i = 1, \ldots, I\}$ and by the same discount factor δ and if the market is complete, then the equilibrium prices can be supported by the no-trade equilibrium of a single agent (*representative agent*) economy, with the representative agent's utility function $\mathbf{u}$ being constructed as in (6.48). Under these assumptions, Proposition 6.32 and, hence, relation (6.64) can be established with respect to the representative agent's utility function $\mathbf{u}$, so that

$$\pi_{A_t}^* = (\delta r_f)^t \frac{\mathbf{u}'\left(e_t(A_t)\right)}{\mathbf{u}'\left(e_0\right)} \pi_{A_t} =: \ell_t(A_t)\pi_{A_t}, \quad \text{for all } A_t \in \mathscr{F}_t \text{ and } t = 0, 1, \ldots, T,$$

$$\tag{6.66}$$

where $e = \{e_t(A_t); A_t \in \mathscr{F}_t, t = 0, 1, \ldots, T\}$ denotes the aggregate endowment process. Hence, the representative agent's marginal rate of substitution in correspondence of the aggregate endowment generates a stochastic discount factor. Under additional hypotheses on the agents' utility functions (see the discussion at the end of Sect. 6.2), the aggregation property can be shown to hold in a strong sense, meaning that the representative agent's utility function only depends on the aggregate endowment and not on the specific distribution among the I agents. In this case, the stochastic discount factor (6.66) constructed from the representative agent's utility function **u** will also be independent on how the aggregate endowment is distributed among the agents (see also Sect. 4.4 for a related discussion).

The Multi-Period Binomial Model

We close this section by briefly extending to a multi-period context the simple binomial model introduced at the end of Sect. 4.4. Following the seminal paper Cox et al. [507], we assume that the market comprises two securities: a risk free security (*bond*), whose price evolves deterministically as $s_t^0 = r_f^t$, for all $t = 0, 1, \ldots, T$, with r_f being the risk free interest rate, and a risky security (*stock*) which does not pay dividends and such that its price process $(s_t^1)_{t=0,1,\ldots,T}$ evolves according to

$$s_t^1 = s_{t-1}^1 \varepsilon_t, \qquad \text{for all } t = 1, \ldots, T,$$

with $s_0^1 > 0$ and where $(\varepsilon_t)_{t=1,\ldots,T}$ is a sequence of random variables taking values in $\{d, u\}$, for some $0 < d < u$, and such that

$$\mathbb{P}(\varepsilon_t = u) = p_t \qquad \text{and} \qquad \mathbb{P}(\varepsilon_t = d) = 1 - p_t,$$

with $p_t \in (0, 1)$, for all $t = 1, \ldots, T$. We assume that the information flow of the economy corresponds to the information generated by the stochastic evolution of the stock price. Note that the risk free asset does not generate any information, since its price process is deterministic.

The following proposition summarizes the properties of the multi-period binomial model.

Proposition 6.33 *In the context of the multi-period binomial model described above, the market is dynamically complete. Moreover, there are no arbitrage opportunities if and only if $d < r_f < u$. In this case, the risk neutral probability measure $\mathbb{P}^*$ is characterized by*

$$\mathbb{P}^*(\varepsilon_t = u | \mathscr{F}_t) = \pi_u^* := \frac{r_f - d}{u - d} \qquad \text{and} \qquad \mathbb{P}^*(\varepsilon_t = d | \mathscr{F}_t) = \pi_d^* := \frac{u - r_f}{u - d}, \tag{6.67}$$

for all $t = 1, \ldots, T$.

Proof Similarly as in the proof of Proposition 4.32, since $r_f > 0$ and $u > d$, it holds that $\operatorname{rank}(\mathfrak{P}_{t+1}(A_t)) = 2 = v_t(A_t)$ for every $t = 0, 1, \ldots, T - 1$. Therefore, in view of Proposition 6.15, the market is dynamically complete. By Theorem 6.23, there are no arbitrage opportunities if and only if there exists a risk neutral probability measure $\mathbb{P}^*$. In the present case of the multi-period binomial model, condition (6.52) reads

$$\frac{1}{r_f}\left(u\mathbb{P}^*(\varepsilon_{t+1} = u|\mathscr{F}_t) + d\mathbb{P}^*(\varepsilon_{t+1} = d|\mathscr{F}_t)\right) = 1,$$

for all $t = 0, 1, \ldots, T - 1$. It is easy to see that the above equation admits a unique strictly positive solution $(\mathbb{P}^*(\varepsilon_{t+1} = u|\mathscr{F}_t), \mathbb{P}^*(\varepsilon_{t+1} = d|\mathscr{F}_t))$ such that

$$\mathbb{P}^*(\varepsilon_{t+1} = u|\mathscr{F}_t) + \mathbb{P}^*(\varepsilon_{t+1} = d|\mathscr{F}_t) = 1$$

if and only $d < r_f < u$. In that case, the probability measure $\mathbb{P}^*$ is characterized by (6.67), for all $t = 1, \ldots, T$.   □

According to the above proposition, if there are no arbitrage opportunities, then the risk neutral probabilities of an upward movement and of a downward movement in each period $[t, t+1]$ are given by $(r_f - d)/(u - d)$ and $(u - r_f)/(u - d)$, respectively. In particular, observe that, under the risk neutral probability measure $\mathbb{P}^*$, the sequence $(\varepsilon_t)_{t=1,\ldots,T}$ is composed of independent and identically distributed random variables, regardless of their distribution under the original probability measure $\mathbb{P}$. Note also that the condition $d < r_f < u$ characterizing the absence of arbitrage opportunities in the multi-period binomial model coincides with the condition obtained in the case of the single-period binomial model (see Proposition 4.32). Indeed, in view of Lemma 6.20, there is an arbitrage opportunity in the multi-period model if and only if there is an arbitrage opportunity in some elementary single-period sub-tree of the model.

Let us conclude by considering the problem of valuing a European contingent claim paying the random payoff $X = f(s_T^1)$ at the maturity T, for some deterministic function $f : \mathbb{R}_+ \to \mathbb{R}_+$. For instance, in the case of an *European Call* option, the payoff $f(s_T^1)$ is given by $\max\{s_T^1 - K; 0\}$, for some fixed strike price $K > 0$ (see also Exercises 6.24 and 6.26 for explicit examples of binomial models). As explained above, since the market is dynamically complete, any contingent claim can be attained by trading in the two available securities.

Corollary 6.34 *In the context of the multi-period binomial model described above, the arbitrage free price $q_0(f(s_T^1))$ of an European contingent claim paying the* *random payoff* $f(s_T^1)$ *at the expiration date* T *can be explicitly computed as*

$$q_0\big(f(s_T^1)\big) = \frac{1}{r_f^T}\mathbb{E}^*\big[f(s_T^1)\big] = \frac{1}{r_f^T}\sum_{j=0}^{T}\binom{T}{j}(\pi_u^*)^j(\pi_d^*)^{T-j}f\big(s_0^1 u^j d^{T-j}\big),$$

where $\binom{T}{j} = \frac{T!}{j!(T-j)!}$ *is the binomial coefficient.*

Proof The claim directly follows from Proposition 6.27 by noting that the random variables $(\varepsilon_t)_{t=1,\dots,T}$ are independent and identically distributed as Bernoulli random variables under the risk neutral probability measure $\mathbb{P}^*$. Moreover, the number of realizations that have exactly j upwards movements is given by the binomial coefficient $T!/(j!(T-j)!)$.   □

6.4 Asset Pricing Relations

In the present section, we extend to a multi-period economy the asset pricing relations presented in Chap. 5 in the context of a single-period economy. As outlined at the beginning of the present chapter, one of the fundamental features of a multi-period economy is represented by the dynamic evolution of information. Hence, in this section we shall derive *conditional* asset pricing relations, meaning that, at each date $t \in \{0, 1, \dots, T-1\}$, the relation will be expressed in terms of expectations and covariances conditioned on the current information $\mathscr{F}_t$ available at date t. Of course, by the tower property of conditional expectations, conditional asset pricing relations also lead to unconditional asset pricing relations. However, we want to point out that the validity of a conditional linear factor pricing model does not necessarily imply the validity of an unconditional linear factor pricing model (see Cochrane [465, Chapter 8] for a discussion on this point).

By relying on the equilibrium analysis developed in the preceding section, we aim at explaining the conditional expected excess returns of the traded securities in terms of the aggregate consumption process. Note that, unlike in a two-period economy, in a multi-period economy the aggregate wealth does not coincide with the aggregate consumption (except, of course, at the terminal date $t = T$). Indeed, as can be deduced from relation (6.3), the aggregate wealth at date $t = 0, 1, \dots, T-1$, is equal to the aggregate consumption plus the aggregate ex-dividend prices of the existing securities, so that the aggregate consumption only represents a fraction of the aggregate wealth. Hence, unlike in Sect. 5.1, we can no longer freely interchange aggregate wealth and aggregate consumption.

In this section, we assume that the economy is populated by $I \in \mathbb{N}$ agents with state independent and time homogeneous preferences, homogeneous beliefs and with identical discount factor $\delta \in (0, 1)$. We also assume that the market is (effectively) complete, so that equilibrium prices can be supported by the no-trade equilibrium of a single agent (representative agent) economy, where the representative agent's utility function $\mathbf{u}(\cdot)$ can be constructed as in (6.48). As explained at the beginning of Sect. 6.3, we suppose that there are $N + 1$ traded securities, with security 0 being a risk free security paying the constant rate r_f. Without loss of generality, the total supply of each security is normalized to one. As above, we denote by $e = (e_t)_{t=0,1,\ldots,T}$ the aggregate endowment process, which is entirely determined by the dividend processes of the traded securities, i.e., $e_t = \sum_{n=0}^{N} d_t^n$, for all $t = 0, 1, \ldots, T$. As explained in Sect. 6.2, the aggregate consumption process coincides with the aggregate dividends paid by the traded securities. In correspondence of a no-trade equilibrium, the representative agent simply consumes at each date $t = 0, 1, \ldots, T$ the aggregate dividends paid at date t by the $N + 1$ available securities.

As shown in the previous section (see Proposition 6.32 and equation (6.66)), under the present assumptions the following fundamental asset pricing relation holds, for all $t = 0, 1, \ldots, T - 1$, with respect to the representative agent's marginal rate of substitution[8] between consumption at date $t + 1$ and consumption at date t:

$$s_t^n = \mathbb{E}\left[\delta \frac{\mathbf{u}'(e_{t+1})}{\mathbf{u}'(e_t)}(d_{t+1}^n + s_{t+1}^n)\middle|\mathscr{F}_t\right], \qquad \text{for all } n = 0, 1, \ldots, N. \qquad (6.68)$$

In turn, by applying the tower property of conditional expectation and recalling that $s_T^n = 0$, relation (6.68) implies that, for all $t = 0, 1, \ldots, T - 1$,

$$s_t^n = \sum_{s=1}^{T-t} \mathbb{E}\left[\delta^s \frac{\mathbf{u}'(e_{t+s})}{\mathbf{u}'(e_t)} d_{t+s}^n \middle|\mathscr{F}_t\right], \qquad \text{for all } n = 0, 1, \ldots, N. \qquad (6.69)$$

In terms of returns, equation (6.68) can be rewritten as follows:

$$\begin{aligned}
1 &= \mathbb{E}\left[\delta \frac{\mathbf{u}'(e_{t+1})}{\mathbf{u}'(e_t)} r_{t+1}^n \middle|\mathscr{F}_t\right] \\
&= \mathbb{E}\left[\delta \frac{\mathbf{u}'(e_{t+1})}{\mathbf{u}'(e_t)}\middle|\mathscr{F}_t\right]\mathbb{E}[r_{t+1}^n|\mathscr{F}_t] + \mathrm{Cov}\left(\delta \frac{\mathbf{u}'(e_{t+1})}{\mathbf{u}'(e_t)}, r_{t+1}^n \middle|\mathscr{F}_t\right),
\end{aligned} \qquad (6.70)$$

for all $n = 0, 1, \ldots, N$ and $t = 0, 1, \ldots, T - 1$. Moreover, since the representative agent's utility function $\mathbf{u}$ is assumed to be strictly increasing (so that $\mathbf{u}'(\cdot) > 0$),

---

[8]In view of the optimality conditions of Proposition 6.4, the asset pricing relations presented in this section can also be obtained in an incomplete market with respect to an individual agent's utility function evaluated in correspondence of the individual agent's optimal consumption plan.

relation (6.70) can be further rewritten as

$$\mathbb{E}[r_{t+1}^n|\mathcal{F}_t] = \frac{1 - \mathrm{Cov}\left(\delta \frac{\mathbf{u}'(e_{t+1})}{\mathbf{u}'(e_t)}, r_{t+1}^n \middle| \mathcal{F}_t\right)}{\mathbb{E}\left[\delta \frac{\mathbf{u}'(e_{t+1})}{\mathbf{u}'(e_t)} \middle| \mathcal{F}_t\right]},$$

for all $n = 0, 1, \ldots, N$ and $t = 0, 1, \ldots, T-1$. In particular, for the risk free security $n = 0$, this gives

$$\frac{1}{r_f} = \frac{\delta \mathbb{E}[\mathbf{u}'(e_{t+1})|\mathcal{F}_t]}{\mathbf{u}'(e_t)}, \qquad \text{for all } t = 0, 1, \ldots, T - 1, \qquad (6.71)$$

where we have used the fact that the aggregate endowment process e is adapted to the information flow $\mathbb{F}$, so that e_t is $\mathcal{F}_t$-measurable. In particular, note that the right-hand side of relation (6.71) determines the equilibrium risk free interest rate for the time interval $[t, t+1]$, which equals the expected rate of intertemporal substitution of consumption of the representative agent. In particular, the equilibrium risk free rate will be constant if the representative agent is risk neutral or if there is no variation over time in the aggregate endowment process. Furthermore, relation (6.71) implies that the equilibrium risk free rate follows a predictable stochastic process, since the risk free rate for the interval $[t, t + 1]$ is measurable with respect to $\mathcal{F}_t$, for all $t = 0, 1, \ldots, T - 1$. See also Exercise 6.28 for a related representation of the equilibrium prices of traded securities.

The following simple proposition gives the general asset pricing relation explaining conditional expected excess returns in terms of the conditional covariance with the representative agent's intertemporal rate of substitution. This represents the result at the basis of the *Consumption Capital Asset Pricing Model* (CCAPM).

Proposition 6.35 *Under the assumptions of the present section, the following holds, for all $n = 1, \ldots, N$ and $t = 0, 1, \ldots, T - 1$:*

$$\mathbb{E}[r_{t+1}^n|\mathcal{F}_t] - r_f = -r_f \, \mathrm{Cov}\left(\delta \frac{\mathbf{u}'(e_{t+1})}{\mathbf{u}'(e_t)}, r_{t+1}^n \middle| \mathcal{F}_t\right) \qquad (6.72)$$

$$= -\frac{\mathrm{Cov}(\mathbf{u}'(e_{t+1}), r_{t+1}^n|\mathcal{F}_t)}{\mathbb{E}[\mathbf{u}'(e_{t+1})|\mathcal{F}_t]}. \qquad (6.73)$$

Proof Relation (6.72) immediately follows by (6.70) and (6.71). To prove relation (6.73), it suffices to notice that, in view of equation (6.71), it holds that $\mathbf{u}'(e_t)/(\delta r_f) = \mathbb{E}[\mathbf{u}'(e_{t+1})|\mathcal{F}_t]$, for all $t = 0, 1, \ldots, T - 1$. $\square$

The interpretation of the above result is analogous to that of relation (5.3) obtained in a two-period economy. Indeed, relation (6.72) shows that the conditional expected excess return of a traded security is determined by the conditional covariance of its return with the representative agent's intertemporal rate of substitution in correspondence of the aggregate endowment. The risk premium of a security will be positive if and only if the return is negatively correlated with the representative agent's intertemporal rate of substitution. As a consequence, all traded securities will exhibit a null risk premium (meaning that $\mathbb{E}[r_t^n] = r_f$, for all $n = 1, \ldots, N$ and $t = 1, \ldots, T$) if the representative agent is risk neutral or if there is no variation over time in the aggregate endowment process or if the aggregate endowment follows a predictable process (so that e_{t+1} is $\mathscr{F}_t$-measurable, for every $t \in \mathbb{N}$, and the conditional covariance appearing in (6.72)–(6.73) becomes null). As explained in Sect. 5.1, relation (6.72) can be interpreted in terms of the diversification and insurance principles and shows that only systematic risk will be priced in equilibrium. However, unlike in Sect. 5.1, in general we cannot replace aggregate endowment/consumption with aggregate wealth for $t = 0, 1, \ldots, T - 1$. Similarly, we cannot guarantee the existence of a portfolio (*market portfolio*) whose return is perfectly correlated with the aggregate endowment/consumption.

If the aggregate endowment and the asset returns are jointly distributed according to a multivariate normal distribution, at every date $t = 1, \ldots, T$, then we can make relation (6.73) more explicit by relying on Stein's lemma. In this case, the conditional risk premium takes the form

$$\mathbb{E}[r_{t+1}^n|\mathscr{F}_t] - r_f = -\frac{\mathbb{E}[\mathbf{u}''(e_{t+1})|\mathscr{F}_t]}{\mathbb{E}[\mathbf{u}'(e_{t+1})|\mathscr{F}_t]} \operatorname{Cov}\left(e_{t+1}, r_{t+1}^n|\mathscr{F}_t\right),$$

for all $n = 1, \ldots, N$ and $t = 0, 1, \ldots, T - 1$, thus showing that the conditional risk premium of a security is proportional to the conditional covariance of its return with the aggregate endowment.

As shown in Proposition 6.32 (applied with respect to the representative agent's utility function $\mathbf{u}$), the quantity

$$m_t := \frac{\delta \mathbf{u}'(e_t)}{\mathbf{u}'(e_{t-1})}, \qquad \text{for } t = 1, \ldots, T,$$

can be interpreted as the *one-period stochastic discount factor*, in the sense that, in view of (6.68), it holds that

$$\mathbb{E}\left[m_{t+1}r_{t+1}^n|\mathscr{F}_t\right] = 1, \qquad \text{for all } n = 0, \ldots, N \text{ and } t = 0, 1, \ldots, T - 1.$$

In view of (6.65), the sequence $(m_t)_{t=1,\ldots,T}$ of one-period stochastic discount factors generates the stochastic discount factor $(M_t)_{t=0,1,\ldots,T}$ via $M_t := \prod_{s=1}^t m_s$, with $M_0 = 1$. Letting $z_{t+1}^n := r_{t+1}^n - r_f$ be the excess return of security n, for $n = 1, \ldots, N$ at date $t + 1$, relation (6.72) can be equivalently rewritten in terms of m_{t+1} as

$$\frac{\mathbb{E}[z_{t+1}^n|\mathscr{F}_t]}{\sqrt{\operatorname{Var}(z_{t+1}^n|\mathscr{F}_t)}} = -\frac{\operatorname{Cov}\left(m_{t+1}, z_{t+1}^n|\mathscr{F}_t\right)}{\sqrt{\operatorname{Var}(z_{t+1}^n|\mathscr{F}_t)}\sqrt{\operatorname{Var}(m_{t+1}|\mathscr{F}_t)}} \frac{\sqrt{\operatorname{Var}(m_{t+1}|\mathscr{F}_t)}}{\mathbb{E}[m_{t+1}|\mathscr{F}_t]},$$

for all $n = 1, \ldots, N$ and $t = 0, 1, \ldots, T - 1$, where we have used the fact that $\mathbb{E}[m_{t+1}|\mathscr{F}_t] = 1/r_f$, due to (6.71). The last relation has the same interpretation of (5.4) and immediately leads to the following result, which extends to the present multi-period setting the Hansen-Jagannathan bounds obtained in Proposition 5.1 for a two-period economy.

Proposition 6.36 *Under the assumptions of the present section, letting* $m_t := \frac{\delta u'(e_t)}{u'(e_{t-1})}$, *for all* $t = 1, \ldots, T - 1$, *the following holds:*

$$\frac{\left|\mathbb{E}[r_{t+1}^n|\mathscr{F}_t] - r_f\right|}{\sqrt{\mathrm{Var}(r_{t+1}^n|\mathscr{F}_t)}} \leq \frac{\sqrt{\mathrm{Var}(m_{t+1}|\mathscr{F}_t)}}{\mathbb{E}[m_{t+1}|\mathscr{F}_t]}, \quad \textit{for all } n = 1, \ldots, N \textit{ and } t = 0, 1, \ldots, T-1.$$

As explained in Sect. 5.1, Proposition 6.36 shows that in equilibrium the conditional Sharpe ratio of each traded security at every date t is bounded from above by the conditional volatility-expectation ratio of the one-period stochastic discount factor m_{t+1}.

Similarly as in Sect. 5.1, since the asset pricing relations presented so far are linear with respect to the asset returns, they also hold for any portfolio composed of the $N + 1$ traded securities. More specifically, if $(w_t)_{t=1,\ldots,T}$ is an $\mathbb{R}^{N+1}$-valued predictable stochastic process, with w_t representing the proportions of wealth invested in the $N + 1$ traded assets at date t (compare with (6.18)), and if $r_t^w := \sum_{n=0}^{N} w_t^n r_t^n$ denotes the return of the portfolio w_t on the interval $[t-1, t]$, then relation (6.73) implies that

$$\mathbb{E}[r_{t+1}^w|\mathscr{F}_t] - r_f = -\frac{\mathrm{Cov}(u'(e_{t+1}), r_{t+1}^w|\mathscr{F}_t)}{\mathbb{E}[u'(e_{t+1})|\mathscr{F}_t]}. \tag{6.74}$$

In the case of a multi-period economy, there does not exist an analogous to the market portfolio appearing in relation (5.6). Nevertheless, we can still find a special portfolio with respect to which an analogue of relation (5.9) can be established, as shown in the following proposition.

Proposition 6.37 *Under the assumptions of the present section, let* $(\hat{w}_t)_{t=1,\ldots,T}$ *be an* $\mathbb{R}^{N+1}$-valued predictable stochastic process such that, for all $t = 0, 1, \ldots, T-1$, the quantity $\hat{w}_{t+1}$ satisfies $\mathbf{1}^\top \hat{w}_{t+1} = 1$ and*

$$\mathrm{Corr}\left(r_{t+1}^\top \hat{w}_{t+1}, u'(e_{t+1})\big|\mathscr{F}_t\right) = \min_{\substack{w \in \mathbb{R}^{N+1} \\ w^\top \mathbf{1} = 1}} \mathrm{Corr}\left(r_{t+1}^\top w, u'(e_{t+1})\big|\mathscr{F}_t\right).$$

Then, for all $n = 1, \ldots, N$ *and* $t = 0, 1, \ldots, T - 1$, *it holds that*

$$\mathbb{E}[r_{t+1}^n|\mathscr{F}_t] - r_f = \beta_{ne,t}\left(\mathbb{E}[r_{t+1}^{\hat{w}}|\mathscr{F}_t] - r_f\right), \tag{6.75}$$

with $\beta_{ne,t} := \mathrm{Cov}(r_{t+1}^{\hat{w}}, r_{t+1}^n|\mathscr{F}_t)/\mathrm{Var}(r_{t+1}^{\hat{w}}|\mathscr{F}_t)$ *and* $r_{t+1}^{\hat{w}} := r_{t+1}^\top \hat{w}_{t+1}$.

Proof As shown in Exercise 6.29, the portfolio process $(\hat{w}_t)_{t=1,\dots,T}$ satisfies

$$\mathrm{Cov}\left(\mathbf{u}'(e_{t+1}), r^n_{t+1}|\mathscr{F}_t\right) = \frac{\mathrm{Cov}\left(r^{\hat{w}}_{t+1}, r^n_{t+1}|\mathscr{F}_t\right)}{\mathrm{Var}\left(r^{\hat{w}}_{t+1}|\mathscr{F}_t\right)}\,\mathrm{Cov}\left(\mathbf{u}'(e_{t+1}), r^{\hat{w}}_{t+1}|\mathscr{F}_t\right),$$

(6.76)

for all $n = 1,\dots,N$ and $t = 0, 1, \dots, T-1$. Equations (6.73) and (6.74) (applied to the portfolio $\hat{w}$) imply that

$$\mathbb{E}[r^n_{t+1}|\mathscr{F}_t] - r_f = \frac{\mathrm{Cov}(\mathbf{u}'(e_{t+1}), r^n_{t+1}|\mathscr{F}_t)}{\mathrm{Cov}(\mathbf{u}'(e_{t+1}), r^{\hat{w}}_{t+1}|\mathscr{F}_t)}\left(\mathbb{E}[r^{\hat{w}}_{t+1}|\mathscr{F}_t] - r_f\right),$$

so that, in view of equation (6.76),

$$\mathbb{E}[r^n_{t+1}|\mathscr{F}_t] - r_f = \frac{\mathrm{Cov}(r^{\hat{w}}_{t+1}, r^n_{t+1}|\mathscr{F}_t)}{\mathrm{Var}(r^{\hat{w}}_{t+1}|\mathscr{F}_t)}\left(\mathbb{E}[r^{\hat{w}}_{t+1}|\mathscr{F}_t] - r_f\right),$$

thus proving relation (6.75).   □

The asset pricing relation (6.75) has the typical β-form, with respect to a risk factor represented by the return of the portfolio having minimal conditional correlation with the representative agent's marginal utility of aggregate consumption. This makes more explicit the general CCAPM relations (6.72)–(6.73) (see also Rubinstein [1487] and Breeden & Litzenberger [287]).

We have so far presented general asset pricing relations, without making specific assumptions on the form of the utility function **u** or on the distribution of the asset returns or of the aggregate endowment. In the remaining part of this section, we derive more explicit asset pricing relations under specific assumptions on the utility functions and/or on the distributions. In particular, we shall see that, under suitable assumptions, the risk premium of an asset is determined by the covariance of the asset return with the growth rate of the aggregate consumption process.

Quadratic Utility Function

Let us suppose that the representative agent's utility function has a quadratic form, i.e., $\mathbf{u}(x) = ax - \frac{b}{2}x^2$, with $a, b > 0$. Since the function **u** is increasing only on the domain $[0, a/b]$ (compare with the discussion in Sect. 2.2), we assume that the aggregate endowment process is such that $e_t \in [0, a/b]$, for all $t = 0, 1, \dots, T$. In this case, the asset pricing relations obtained in Proposition 6.35 take the form

$$\mathbb{E}[r_{t+1}^n|\mathcal{F}_t] - r_f = \frac{b}{a - b\mathbb{E}[e_{t+1}|\mathcal{F}_t]} \operatorname{Cov}\left(e_{t+1}, r_{t+1}^n|\mathcal{F}_t\right)$$

$$= \frac{b}{a - b\mathbb{E}[e_{t+1}|\mathcal{F}_t]} \operatorname{Cov}\left(e_{t+1}, z_{t+1}^n|\mathcal{F}_t\right),$$

for all $n = 1, \ldots, N$ and $t = 0, 1, \ldots, T - 1$. This shows that, in the case of a quadratic utility function, the conditional expected risk premium is determined by the conditional covariance of the excess return with the aggregate endowment. Similarly, one can establish a version of Proposition 6.37 with respect to the portfolio process $(\hat{w}_t)_{t=1,\ldots,T}$ having maximal correlation with the aggregate endowment, i.e., such that the quantity $\hat{w}_{t+1}$ satisfies, for all $t = 0, 1, \ldots, T - 1$,

$$\operatorname{Corr}\left(r_{t+1}^\top \hat{w}_{t+1}, e_{t+1}|\mathcal{F}_t\right) = \max_{\substack{w \in \mathbb{R}^{N+1} \\ w^\top 1 = 1}} \operatorname{Corr}\left(r_{t+1}^\top w, e_{t+1}|\mathcal{F}_t\right).$$

Power Utility

Assume now that the representative agent's utility function is of the power form, i.e., $\mathbf{u}(x) = x^\gamma/\gamma$, with $\gamma \in (0, 1)$. In this case, the asset pricing relations obtained in Proposition 6.35 take the form

$$\mathbb{E}[r_{t+1}^n|\mathcal{F}_t] - r_f = -\frac{\operatorname{Cov}\left(\left(\frac{e_{t+1}}{e_t}\right)^{\gamma-1}, r_{t+1}^n\big|\mathcal{F}_t\right)}{\mathbb{E}\left[\left(\frac{e_{t+1}}{e_t}\right)^{\gamma-1}\big|\mathcal{F}_t\right]} = -\frac{\operatorname{Cov}\left(\left(\frac{e_{t+1}}{e_t}\right)^{\gamma-1}, z_{t+1}^n\big|\mathcal{F}_t\right)}{\mathbb{E}\left[\left(\frac{e_{t+1}}{e_t}\right)^{\gamma-1}\big|\mathcal{F}_t\right]},$$

for all $n = 1, \ldots, N$ and $t = 0, 1, \ldots, T - 1$. This shows that, in the case of a power utility function, the conditional risk premium is positive if and only if the excess return is negatively correlated with the growth rate of the aggregate consumption process. Moreover, the equilibrium risk free rate r_f is determined by the conditional expected growth rate of aggregate consumption. Indeed, due to equation (6.71), it holds that

$$\frac{1}{r_f} = \delta \mathbb{E}\left[\left(\frac{e_{t+1}}{e_t}\right)^{\gamma-1}\big|\mathcal{F}_t\right]. \tag{6.77}$$

Log-Normal Distribution

Under the assumption that the joint conditional distribution of the asset returns and the representative agent's intertemporal rate of substitution is log-normal and homoskedastic, a simple formula can been derived for the asset risk premia. More specifically, we have the following proposition, the proof of which is given in Exercise 6.30.

Proposition 6.38 *Under the assumptions of the present section, let* $m_t := \frac{\delta u'(e_t)}{u'(e_{t-1})}$, *for all* $t = 1, \ldots, T$, *and suppose that* $(\log m_t, \log r_t^n)$ *conditionally on* $\mathscr{F}_{t-1}$ *follows a bivariate normal distribution with mean* $\mu^n \in \mathbb{R}^2$ *and covariance* $\Sigma^n \in \mathbb{R}^{2 \times 2}$, *for all* $n = 1, \ldots, N$ *and* $t = 1, \ldots, T$. *Then the following holds:*

$$
\mathbb{E}[\log r_{t+1}^n | \mathscr{F}_t] - \log r_f = -\frac{\Sigma_{22}^n}{2} - \Sigma_{12}^n
$$

$$
= -\frac{\operatorname{Var}\left(\log r_{t+1}^n | \mathscr{F}_t\right)}{2} - \operatorname{Cov}\left(\log m_{t+1}, \log r_{t+1}^n | \mathscr{F}_t\right),
$$

(6.78)

for all $n = 1, \ldots, N$ *and* $t = 0, 1, \ldots, T - 1$.

A similar result has been obtained in Hansen & Singleton [896] (see also Singleton [1549]) by assuming that the representative agent's utility function is of the power form $\mathbf{u}(x) = x^\gamma / \gamma$ and that the aggregate consumption growth rate and the return of each asset are jointly distributed according to a log-normal distribution. Indeed, if $(\log(e_t/e_{t-1}), \log r_t^n)$ conditionally on $\mathscr{F}_{t-1}$ has a bivariate normal distribution with mean $\mu^n \in \mathbb{R}^2$ and covariance $\Sigma^n \in \mathbb{R}^{2 \times 2}$, for all $n = 1, \ldots, N$ and $t = 1, \ldots, T$, then it holds that

$$
\mathbb{E}[\log r_{t+1}^n | \mathscr{F}_t] - \log r_f = -\frac{\Sigma_{22}^n}{2} - (\gamma - 1)\Sigma_{12}^n
$$

$$
= -\frac{\operatorname{Var}\left(\log r_{t+1}^n | \mathscr{F}_t\right)}{2}
$$

$$
- (\gamma - 1)\operatorname{Cov}\left(\log(e_{t+1}/e_t), \log r_{t+1}^n | \mathscr{F}_t\right),
$$

for all $n = 1, \ldots, N$ and $t = 0, 1, \ldots, T - 1$. This relation follows by the same arguments used in Exercise 6.30 to prove (6.78). The equilibrium risk free rate r_f is determined by

$$
\log r_f = -\log \delta - (\gamma - 1)\mathbb{E}\left[\log(e_{t+1}/e_t) | \mathscr{F}_t\right] - \frac{(\gamma - 1)^2}{2}\operatorname{Var}\left(\log(e_{t+1}/e_t) | \mathscr{F}_t\right).
$$

In the case of a power utility function and a log-normal distribution, the above formulae relate the conditional expected return to the conditional expectation and variance of the aggregate consumption growth rate. In particular, the logarithmic risk free return is linear in the expected logarithmic consumption growth rate, with a slope equal to the coefficient of relative risk aversion of the representative agent $(1 - \gamma)$. In this sense, a large consumption growth rate leads to a high risk free rate or to a low risk aversion coefficient (high elasticity of intertemporal substitution) in order to induce agents to transfer money intertemporally smoothing the consumption stream. The variance term in the equilibrium risk free rate is due to a demand for precautionary saving. All the returns are decreasing in the representative agent's discount factor δ. The conditional risk premium of an asset is linear in the conditional variance of the logarithmic return and in the conditional covariance of the logarithmic return with the logarithmic consumption growth rate with a slope equal to the coefficient of relative risk aversion of the representative agent.

By relying on the Hansen-Jagannathan bound established in Proposition 6.36 and on the properties of the log-normal distribution, one can also obtain an upper bound for the conditional Sharpe ratio of a traded security in terms of the conditional variance of the aggregate consumption growth rate, similarly as in the case of a two-period economy (see formula (5.12) in Sect. 5.1).

Multi-Factor Models

In view of Propositions 6.35 and 6.37, the CCAPM can be regarded as a one-factor asset pricing model, where the risk premia of the traded securities are described in terms of their beta coefficients with respect to the representative agent's marginal rate of intertemporal substitution of consumption (or, in view of Proposition 6.37, with respect to the portfolio minimally correlated with the representative agent's rate of intertemporal substitution).

A multi-factor model has been proposed through the conditional *Intertemporal Capital Asset Pricing Model* (ICAPM) (originally proposed in continuous time by Merton [1330]; we base our presentation on Constantinides [492]). Suppose that there exists an adapted stochastic factor process $(y_t)_{t=0,1,\dots,T}$ such that y_t is $\mathbb{R}^K$-valued, for all $t = 0, 1, \dots, T$, for some $K \in \mathbb{N}$, and such that the aggregate endowment e_t at date t can be written as an $\mathscr{F}_{t-1}$-measurable function $f_{t-1}(\cdot)$: $\mathbb{R}^K \to \mathbb{R}_+$ of y_t, i.e., $e_t = f_{t-1}(y_t)$, for all $t = 1, \dots, T$. In other words, y_t can be thought of as a vector of economic variables which determine the state of the economy. Furthermore, assume that the $(K + N + 1)$-dimensional vector (y_t, r_t) is distributed according to a multivariate normal distribution. Under the present assumptions, relation (6.73) together with an application of Stein's lemma implies that

$$\mathbb{E}[r_{t+1}^n|\mathscr{F}_t] - r_f = -\sum_{k=1}^{K} \frac{\mathbb{E}[\mathbf{u}''(e_t)f_{k,t}(y_{t+1})|\mathscr{F}_t]}{\mathbb{E}[\mathbf{u}'(e_{t+1})|\mathscr{F}_t]} \operatorname{Cov}(y_{t+1}^k, r_{t+1}^n|\mathscr{F}_t),$$

for all $n = 1, \ldots, N$ and $t = 0, 1, \ldots, T-1$, where $f_{k,t}(\cdot)$ denotes the first derivative of the function $f_t(\cdot)$ with respect to its k-th argument. If the utility function is logarithmic, then the conditional ICAPM reduces to the conditional CAPM (see Constantinides [492, Section V] for more details).

A multi-factor asset pricing model can also be derived under the assumption that the conditional one-period stochastic discount factor can be expressed as a linear function of K factors (see also Campbell [335], Ferson & Jagannathan [696], Ferson [688]). More specifically, suppose that there exists a K-dimensional adapted stochastic process $(y_t)_{t=0,1,\ldots,T}$, for some $K \in \mathbb{N}$, such that

$$m_t = a_t - \sum_{k=1}^{K} b_t^k y_t^k, \qquad \text{for all } t = 1, \ldots, T,$$

for some predictable real-valued and $\mathbb{R}^K$-valued stochastic processes $(a_t)_{t=1,\ldots,T}$ and $(b_t)_{t=1,\ldots,T}$, respectively. Then formula (6.72) implies that

$$\mathbb{E}[r_{t+1}^n | \mathscr{F}_t] - r_f = -r_f \operatorname{Cov}\left(m_{t+1}, r_{t+1}^n | \mathscr{F}_t\right) = r_f \sum_{k=1}^{K} b_{t+1}^k \operatorname{Cov}(y_{t+1}^k, r_{t+1}^n | \mathscr{F}_t)$$

$$= r_f \sum_{k=1}^{K} b_{t+1}^k \operatorname{Var}(y_{t+1}^k | \mathscr{F}_t) \frac{\operatorname{Cov}(y_{t+1}^k, r_{t+1}^n | \mathscr{F}_t)}{\operatorname{Var}(y_{t+1}^k | \mathscr{F}_t)} =: r_f \sum_{k=1}^{K} \lambda_t^k \beta_t^{nk},$$

for all $n = 1, \ldots, N$ and $t = 0, 1, \ldots, T-1$, with

$$\lambda_t^k := b_{t+1}^k \operatorname{Var}(y_{t+1}^k | \mathscr{F}_t) \qquad \text{and} \qquad \beta_t^{nk} := \frac{\operatorname{Cov}\left(y_{t+1}^k, r_{t+1}^n | \mathscr{F}_t\right)}{\operatorname{Var}(y_{t+1}^k | \mathscr{F}_t)},$$

where we have used the predictability of the processes a and $(b^1, \ldots, b^K)$.

An intertemporal discrete time competitive equilibrium version of the Arbitrage Pricing Theory (APT) has been obtained in Connor & Korajczyk [479] and Bossaerts & Green [270] assuming a factor model for asset dividends. In these models, the beta coefficients and the intertemporal risk premia vary over time. Similarly as above, a multi-factor representation of risk premia can be obtained assuming a linear factor model for dividends and normality of the relevant random variables by expanding (6.72) be means of Stein's lemma (see Constantinides [492]). A non-linear no-arbitrage multi-factor model has been proposed in Bansal & Viswanathan [138].

The Term Structure of Interest Rates

The equilibrium analysis and the asset pricing relations presented so far provide a natural framework for studying the *term structure of interest rates*. Indeed, unlike in the case of a simple two-period economy, in a multi-period economy the risk free interest rate is allowed to vary through time (as long as it follows a predictable stochastic process), depending on the equilibrium price of intertemporal consumption. In particular, as we have already seen in relation (6.71), the equilibrium risk free rate is related to the conditional expected growth of the aggregate consumption. The study of the term structure of interest rates takes explicitly into account how discount factors behave in correspondence of different time horizons.

For $t \in \{0, 1, \dots, T-1\}$ and $\tau \in \{1, \dots, T-t\}$, let $B(t, t+\tau)$ denote the price at date t of a *zero-coupon risk free bond* maturing at the future date $t + \tau$. This means that $B(t, t+\tau)$ is the price at date t of a security delivering a riskless unitary dividend at (and only at) the future date $t + \tau$. Adopting the representative agent's valuation, the equilibrium pricing relation (6.69) implies that

$$B(t, t + \tau) = \mathbb{E}\left[\delta^\tau \frac{u'(e_{t+\tau})}{u'(e_t)} \middle| \mathscr{F}_t \right] \tag{6.79}$$

and the yield implicit in the bond price $B(t, t + \tau)$, i.e., the rate $i(t, t + \tau)$ which satisfies $1/B(t, t + \tau) = (1 + i(t, t + \tau))^\tau$, can then be expressed as

$$i(t, t + \tau) = \frac{1}{\delta}\left(\mathbb{E}\left[\frac{u'(e_{t+\tau})}{u'(e_t)} \middle| \mathscr{F}_t \right]\right)^{-1/\tau} - 1. \tag{6.80}$$

For every date $t \in \{0, 1, \dots, T-1\}$, the collection $\{i(t, t+\tau); \tau = 1, \dots, T-t\}$ represents the (spot) *term structure of interest rates* at date t. The above relations show that the term structure of interest rates encodes the conditional expectations about the growth of the aggregate endowment over time, as measured through the representative agent's marginal utility.

By no arbitrage arguments, the interest rates $i(t, t + \tau)$ defined in (6.80) allow to recover the *forward interest rates*. For any $t \in \{0, 1, \dots, T-1\}$ and $\tau_1, \tau_2 \in \{0, 1, \dots, T-t\}$ with $\tau_1 < \tau_2$, the forward rate $f(t; t + \tau_1, t + \tau_2)$ is defined as the interest rate set at date t which satisfies the relation

$$\left(1 + i(t, t + \tau_2)\right)^{\tau_2} = \left(1 + i(t, t + \tau_1)\right)^{\tau_1}\left(1 + f(t; t + \tau_1, t + \tau_2)\right)^{\tau_2 - \tau_1}. \tag{6.81}$$

In other words, $f(t; t+\tau_1, t+\tau_2)$ is the interest rate set at date t for lending/borrowing over the future time interval $[t + \tau_1, t + \tau_2]$ which is coherent with the absence of arbitrage opportunities if all zero-coupon risk free bonds are available in the market (see also Exercise 6.33). As a matter of fact, the same amount of wealth is obtained by investing over the horizon $[t, t + \tau_2]$ (yielding the amount of wealth $(1 + i(t, t + \tau_2))^{\tau_2}$, corresponding to the left-hand side of (6.81)) or by investing over the horizon $[t, t+\tau_1]$ and then reinvesting at date $t+\tau_1$ the wealth over the horizon $[t+\tau_1, t+\tau_2]$ (thus yielding the amount of wealth $(1 + i(t, t + \tau_1))^{\tau_1} (1 + f(t; t + \tau_1, t + \tau_2))^{\tau_2-\tau_1}$, corresponding to the right-hand side of (6.81)).

Consider a fixed date $t \in \{0, 1, \ldots, T-2\}$ and suppose that at date t it is possible to trade a zero-coupon risk free bond with maturity $t+1$ having price $B(t, t+1)$ and a zero-coupon bond with maturity $t+2$ having price $B(t, t+2)$. Suppose furthermore that at the subsequent date $t+1$ it will be possible to trade a zero-coupon bond with maturity $t+2$ for the price $B(t+1, t+2)$ (which is measurable with respect to the information at date $t+1$). Let $\mathbb{P}^*$ be a risk neutral probability measure. Then, as shown in Exercise 6.31, a no-arbitrage argument implies that (see also Pascucci & Runggaldier [1404, Proposition 4.5])

$$B(t, t+2) = B(t, t+1)\mathbb{E}^*[B(t+1, t+2)|\mathscr{F}_t], \qquad (6.82)$$

with $\mathbb{E}^*[\cdot|\mathscr{F}_t]$ denoting the $\mathscr{F}_t$-conditional expectation under $\mathbb{P}^*$. However, this relation does not hold in general under the physical/statistical probability measure $\mathbb{P}$. We say that the *expectation hypothesis* holds if relation (6.82) holds under the physical/statistical probability measure $\mathbb{P}$, i.e., if

$$B(t, t+\tau) = B(t, t+1)\mathbb{E}[B(t+1, t+2)\ldots B(t+\tau-1, t+\tau)|\mathscr{F}_t].$$

As follows from the next proposition, the expectation hypothesis is satisfied only in rather special cases under the physical probability measure $\mathbb{P}$.

Proposition 6.39 *Under the assumptions of the present section, the following hold:*

(i) *if the representative agent is risk neutral, then $i(t, t+\tau) = 1/\delta - 1$ for every $t = 0, 1, \ldots, T-1$ and $\tau \in \{1, \ldots, T-t\}$;*

(ii) *if the representative agent is risk averse and the aggregate endowment process is constant over time, then $i(t, t+\tau) = 1/\delta - 1$ for all $t = 0, 1, \ldots, T-1$ and every $\tau \in \{1, \ldots, T-t\}$.*

Moreover, for all $t \in \{0, 1, \ldots, T-2\}$, it holds that

$$B(t, t+2) = B(t, t+1)\mathbb{E}[B(t+1, t+2)|\mathscr{F}_t] + \operatorname{Cov}\left(\delta\frac{u'(e_{t+1})}{u'(e_t)}, \delta\frac{u'(e_{t+2})}{u'(e_{t+1})}\Big|\mathscr{F}_t\right).$$
$$(6.83)$$

Proof The first two claims follow immediately from relation (6.80), while equation (6.83) can be obtained by using the tower property of the conditional expectation together with relation (6.79):

$$
\begin{aligned}
B(t, t+2) &= \mathbb{E}\left[\delta^2 \frac{\mathbf{u}'(e_{t+2})}{\mathbf{u}'(e_t)}\Big|\mathscr{F}_t\right] = \mathbb{E}\left[\delta\frac{\mathbf{u}'(e_{t+1})}{\mathbf{u}'(e_t)}\delta\frac{\mathbf{u}'(e_{t+2})}{\mathbf{u}'(e_{t+1})}\Big|\mathscr{F}_t\right] \\
&= \mathbb{E}\left[\delta\frac{\mathbf{u}'(e_{t+1})}{\mathbf{u}'(e_t)}\Big|\mathscr{F}_t\right]\mathbb{E}\left[\delta\frac{\mathbf{u}'(e_{t+2})}{\mathbf{u}'(e_{t+1})}\Big|\mathscr{F}_t\right] \\
&\quad + \mathrm{Cov}\left(\delta\frac{\mathbf{u}'(e_{t+1})}{\mathbf{u}'(e_t)}, \delta\frac{\mathbf{u}'(e_{t+2})}{\mathbf{u}'(e_{t+1})}\Big|\mathscr{F}_t\right) \\
&= B(t, t+1)\mathbb{E}[B(t+1, t+2)|\mathscr{F}_t] + \mathrm{Cov}\left(\delta\frac{\mathbf{u}'(e_{t+1})}{\mathbf{u}'(e_t)}, \delta\frac{\mathbf{u}'(e_{t+2})}{\mathbf{u}'(e_{t+1})}\Big|\mathscr{F}_t\right).
\end{aligned}
$$

□

The above proposition shows that in the special case where the representative agent is risk neutral or when there is no variation in the aggregate consumption process, then the term structure is flat. In the general case of a risk averse representative agent, formula (6.83) shows that the price $B(t, t+2)$ differs from the price obtained under the expectation hypothesis due to the presence of an additional covariance term which corresponds to $B(t, t+2) - B(t, t+1)\mathbb{E}[B(t+1, t+2)|\mathscr{F}_t]$ and is called *expected term premium*. In turn, this implies that the expectation hypothesis will be satisfied if, given the information available at the previous date, the one-period stochastic discount factor (the intertemporal rate of substitution of the representative agent) is not serially correlated. In other words, the expectation hypothesis holds if the current growth rate of aggregate consumption has no predictive power on the future growth rate of aggregate consumption. Under suitable assumptions (see Exercise 6.32), we can also obtain an approximate linear relation between the equilibrium yield $i(t, t + \tau)$ and the conditional expected logarithmic growth rate of aggregate consumption.

A Markov Chain Model in Infinite Horizon

We close this section by briefly presenting some asset pricing relations in the context of an economy with an infinite time horizon (i.e., $T = \infty$) where the aggregate endowment process follows a Markov chain (we refer to Ljungqvist & Sargent [1231, Chapter 13] for full details). This model has been used as the basis for several important studies, notably in the analysis of the equity premium puzzle in Mehra & Prescott [1319].

Suppose that the aggregate endowment process $(e_t)_{t\in\mathbb{N}}$ evolves as a Markov chain taking values in the K-dimensional state space $\{e(1), \ldots, e(K)\}$, for some $K \in \mathbb{N}$,

so that

$$\mathbb{P}\big(e_{t+1} = e(j)|e_t = e(i)\big) = \pi_{ij}, \qquad \text{for all } i,j \in \{1,\dots,K\},$$

with $\sum_{j=1}^{K} \pi_{ij} = 1$, for all $i = 1,\dots,K$. We denote by $\Pi \in \mathbb{R}^{K \times K}$ the associated transition matrix composed of the elements $\{\pi_{ij}; i,j = 1,\dots,K\}$. As above in the present section, we adopt the representative agent's perspective and assume that his utility function is given by $\mathbf{u}$. We also assume that the market only contains a single traded security which, at every date $t \in \mathbb{N}$, pays the aggregate endowment as dividend. We denote by $s(k)_t$ the price of such a security at date t if the aggregate endowment at date t is in state k, for $k \in \{1,\dots,K\}$. The pricing relation (6.69) (here extended to an infinite time horizon) yields the following equilibrium price for the aggregate consumption stream in correspondence of the event $\{e_t = e(k)\}$:

$$s(k)_t = \sum_{s=1}^{\infty} \mathbb{E}\left[\delta^s \frac{\mathbf{u}'(e_{t+s})}{\mathbf{u}'(e_t)} e_{t+s}\bigg|\mathscr{F}_t\right] = \frac{1}{\mathbf{u}'(e(k))} \sum_{s=1}^{\infty} \delta^s \mathbb{E}\left[\mathbf{u}'(e_{t+s})e_{t+s}|e_t = e(k)\right],$$

where the second equality follows from the Markov property of the aggregate endowment process $(e_t)_{t \in \mathbb{N}}$. In turn, the right-hand side of the above relation implies that the equilibrium price of the aggregate consumption stream does not depend on time t but only on the current state of the aggregate endowment, so that we can simply write $s(k)$ for the price (at any date $t \in \mathbb{N}$) of the security in correspondence of state k of the aggregate endowment. Together with the asset pricing relation (6.68), this implies that

$$s(k) = \delta \sum_{j=1}^{K} \frac{\mathbf{u}'(e(j))}{\mathbf{u}'(e(k))}\big(e(j) + s(j)\big)\pi_{kj},$$

for all $k = 1,\dots,K$. Let us define $\varrho(k) := s(k)\mathbf{u}'(e(k))$, for all $k = 1,\dots,K$, and $\omega(k) := \delta \sum_{j=1}^{K} \mathbf{u}'(e(j))e(j)\pi_{kj}$, so that

$$\varrho(k) = \omega(k) + \delta \sum_{j=1}^{K} \pi_{kj}\varrho(j), \qquad \text{for all } k = 1,\dots,K.$$

In vector notation, this last equation can be rewritten as $\varrho = \omega + \delta\Pi\varrho$, which admits the unique solution

$$\varrho = (I - \delta\Pi)^{-1}\omega,$$

where $I \in \mathbb{R}^{K \times K}$ is the identity matrix (see Ljungqvist & Sargent [1231, Example 13.7.2]).

Let us now assume that the representative agent's utility function is of the power type, i.e., $\mathbf{u}(x) = x^\gamma/\gamma$ and that the growth rate of the aggregate endowment (rather than the aggregate endowment itself, as considered above) follows a finite state Markov chain (under the standing assumption that the aggregate endowment is always strictly positive). More precisely, assume that the quantity $x_t := e_t/e_{t-1}$, for $t \in \mathbb{N}$, satisfies

$$\mathbb{P}(x_{t+1} = \lambda(j)|x_t = \lambda(k)) = \pi_{kj}, \qquad \text{for all } j,k \in \{1,\dots,K\},$$

for all $t \in \mathbb{N}$, where $\{\lambda(1),\dots,\lambda(K)\}$ represents the K-dimensional state space of the Markov chain $(x_t)_{t\in\mathbb{N}}$. As above, in view of equation (6.69), the equilibrium price s_t at date t of the traded security which delivers the aggregate endowment as dividend satisfies

$$s_t = \frac{1}{e_t^{\gamma-1}} \sum_{s=1}^{\infty} \delta^s \mathbb{E}[e_{t+s}^\gamma|\mathscr{F}_t] = e_t \sum_{s=1}^{\infty} \delta^s \mathbb{E}[x_{t+s}^\gamma|\mathscr{F}_t] = e_t \sum_{s=1}^{\infty} \delta^s \mathbb{E}[x_{t+s}^\gamma|x_t],$$

where the last equality uses the Markov property of the process $(x_t)_{t\in\mathbb{N}}$. Defining the process $(\xi_t)_{t\in\mathbb{N}}$ by $\xi_t := s_t/e_t$, for all $t \in \mathbb{N}$, the above equation shows that ξ_t does only depend on the state of x_t and not on all information contained in $\mathscr{F}_t$ at date t. Hence, we can simply denote by $\xi(k)$ the value of ξ_t if $x_t = \lambda(k)$, for some $k \in \{1,\dots,K\}$, for every $t \in \mathbb{N}$. In other words, $\xi(k)$ represents the price-dividend ratio of the security in correspondence of state k of the aggregate endowment growth. By relying on this observation and using equation (6.68) (dividing by e_t and rearranging terms), this gives

$$\xi(k) = \delta \sum_{j=1}^{K} \lambda(j)^\gamma \big(1 + \xi(j)\big)\pi_{kj}, \qquad \text{for all } k \in \{1,\dots,K\}. \tag{6.84}$$

This last equation has been used in Mehra & Prescott [1319] to compute equilibrium prices. In this setting, at every date $t \in \mathbb{N}$ and in state k, the price of a zero-coupon bond which matures at the subsequent date $t + 1$ is determined by

$$B(t,t+1)(k) = \delta \sum_{j=1}^{K} \lambda(j)^{\gamma-1}\pi_{kj}, \qquad \text{for all } k = 1,\dots,K.$$

6.5   Infinite Horizon Economies and Rational Bubbles

As we have seen in the previous section, in an economy with a finite time horizon $T < \infty$, the equilibrium prices of the traded securities are determined by the conditional expectation of the future discounted dividends, weighted by the representative agent's marginal rate of substitution in correspondence of the aggregate endowment. In this section, we remove the assumption of a finite time horizon and allow for infinitely lived economies. In this context, we will see that asset prices can exhibit a *bubble* component, which represents a deviation from the conditional expectation of future discounted dividends. This is an interesting possibility, taking into account the numerous historical episodes where asset prices appeared to be incompatible with plausible values of the underlying dividends/fundamentals (think for instance of the Dutch tulip bubble at the beginning of the 16th century).

For simplicity, we consider an economy with two traded securities (the analysis can be easily extended to economies with multiple securities): a risk free asset paying the constant rate r_f and a risky asset described by the couple of non-negative price-dividend processes (s, d). We assume that markets are open at every date $t \in \mathbb{N}$ (infinite time horizon). As in the previous section, we adopt the representative agent's paradigm and suppose that the representative agent's utility function and discount factor are given by $\mathbf{u}$ and $\delta \in (0, 1)$, respectively. In the present context, the asset pricing relation (6.68) implies that

$$\mathbf{u}'(e_t)s_t = \delta \mathbb{E}\left[\mathbf{u}'(e_{t+1})(s_{t+1} + d_{t+1})|\mathscr{F}_t\right], \qquad \text{for all } t \in \mathbb{N}, \tag{6.85}$$

where $(e_t)_{t\in\mathbb{N}}$ denotes the aggregate endowment/consumption process. Given the processes $(e_t)_{t\in\mathbb{N}}$ and $(d_t)_{t\in\mathbb{N}}$, equation (6.85) does not fully determine the equilibrium price process $(s_t)_{t\in\mathbb{N}}$ because it is defined backward in time and involves a conditional expectation (see also Pesaran [1414] on the solution of models under rational expectations). Indeed, as shown in the following proposition, there does not exist in general a unique solution $(s_t)_{t\in\mathbb{N}}$ to equation (6.85) in an infinitely lived economy.

Proposition 6.40 *Suppose that* $\sum_{s=1}^{\infty} \delta^s \mathbb{E}[\mathbf{u}'(e_s)d_s] < \infty$. *Then, under the assumptions of the present section, an* $\mathbb{F}$-*adapted non-negative stochastic process* $(s_t)_{t\in\mathbb{N}}$ *satisfies equation (6.85) if and only if* $s_t = s_t^* + \beta_t$, *for all* $t \in \mathbb{N}$, *where*

$$s_t^* := \frac{1}{\mathbf{u}'(e_t)} \sum_{s=1}^{\infty} \delta^s \mathbb{E}[\mathbf{u}'(e_{t+s})d_{t+s}|\mathscr{F}_t], \qquad \text{for all } t \in \mathbb{N}, \tag{6.86}$$

and $(\beta_t)_{t\in\mathbb{N}}$ *is an* $\mathbb{F}$-*adapted stochastic process satisfying*

$$\mathbf{u}'(e_t)\beta_t = \delta \mathbb{E}[\mathbf{u}'(e_{t+1})\beta_{t+1}|\mathscr{F}_t], \qquad \text{for all } t \in \mathbb{N}.$$

Proof Let $(s_t)_{t\in\mathbb{N}}$ be a non-negative $\mathbb{F}$-adapted process satisfying (6.85). Then, for any $t \in \mathbb{N}$, the tower property of the conditional expectation implies that

$$\mathbf{u}'(e_t)s_t = \delta \mathbb{E}\left[\mathbf{u}'(e_{t+1})(s_{t+1} + d_{t+1})|\mathscr{F}_t\right]$$
$$= \delta \mathbb{E}\left[\mathbf{u}'(e_{t+1})d_{t+1} + \delta \mathbb{E}\left[\mathbf{u}'(e_{t+2})(s_{t+2} + d_{t+2})|\mathscr{F}_{t+1}\right]|\mathscr{F}_t\right]$$
$$= \delta \mathbb{E}[\mathbf{u}'(e_{t+1})d_{t+1}|\mathscr{F}_t] + \delta^2 \mathbb{E}[\mathbf{u}'(e_{t+2})d_{t+2}|\mathscr{F}_t] + \delta^2 \mathbb{E}[\mathbf{u}'(e_{t+2})s_{t+2}|\mathscr{F}_t].$$

By iterating the same argument, we get that, for any $t, T \in \mathbb{N}$,

$$\mathbf{u}'(e_t)s_t = \sum_{s=1}^{T} \delta^s \mathbb{E}[\mathbf{u}'(e_{t+s})d_{t+s}|\mathscr{F}_t] + \delta^T \mathbb{E}[\mathbf{u}'(e_{t+T})s_{t+T}|\mathscr{F}_t].$$

Taking the limit for $T \to \infty$ in the last equation, we get

$$\mathbf{u}'(e_t)s_t = \sum_{s=1}^{\infty} \delta^s \mathbb{E}[\mathbf{u}'(e_{t+s})d_{t+s}|\mathscr{F}_t] + \lim_{T \to \infty} \delta^T \mathbb{E}[\mathbf{u}'(e_{t+T})s_{t+T}|\mathscr{F}_t]$$

$$=: \mathbf{u}'(e_t)s_t^* + \mathbf{u}'(e_t)\beta_t,$$

(6.87)

where we define $\beta_t := \lim_{T \to \infty} \delta^T \mathbb{E}[\mathbf{u}'(e_{t+T})s_{t+T}|\mathscr{F}_t]/\mathbf{u}'(e_t)$, for all $t \in \mathbb{N}$. Note that s_t^* is well defined due to the assumption that $\sum_{s=1}^{\infty} \delta^s \mathbb{E}[\mathbf{u}'(e_s)d_s] < \infty$. It remains to show that $\mathbf{u}'(e_t)\beta_t = \delta \mathbb{E}[\mathbf{u}'(e_{t+1})\beta_{t+1}|\mathscr{F}_t]$, for all $t \in \mathbb{N}$. This easily follows by replacing $s_t = s_t^* + \beta_t$ into equation (6.85), so that

$$\sum_{s=1}^{\infty} \delta^s \mathbb{E}[\mathbf{u}'(e_{t+s})d_{t+s}|\mathscr{F}_t] + \mathbf{u}'(e_t)\beta_t$$

$$= \delta \mathbb{E}\left[\sum_{s=1}^{\infty} \delta^s \mathbb{E}[\mathbf{u}'(e_{t+1+s})d_{t+1+s}|\mathscr{F}_{t+1}]\Big|\mathscr{F}_t\right] + \delta \mathbb{E}[\mathbf{u}'(e_{t+1})d_{t+1}|\mathscr{F}_t]$$

$$+ \delta \mathbb{E}[\mathbf{u}'(e_{t+1})\beta_{t+1}|\mathscr{F}_t]$$

$$= \sum_{s=1}^{\infty} \delta^s \mathbb{E}[\mathbf{u}'(e_{t+s})d_{t+s}|\mathscr{F}_t] + \delta \mathbb{E}[\mathbf{u}'(e_{t+1})\beta_{t+1}|\mathscr{F}_t],$$

where the first equality follows from the hypothesis that $s_t^* + \beta_t$ satisfies (6.85) and where in the last equality we have used the tower property of the conditional expectation. Conversely, suppose that $(\beta_t)_{t \in \mathbb{N}}$ is an $\mathbb{F}$-adapted stochastic process such that $\mathbf{u}'(e_t)\beta_t = \delta \mathbb{E}[\mathbf{u}'(e_{t+1})\beta_{t+1}|\mathscr{F}_t]$, for all $t \in \mathbb{N}$. Then it is immediate to verify that the process $(s_t)_{t \in \mathbb{N}}$ defined by $s_t := s_t^* + \beta_t$, for all $t \in \mathbb{N}$, satisfies equation (6.85). □

According to the above proposition, any price process $(s_t)_{t \in \mathbb{N}}$ which solves equation (6.85) can be decomposed into two components: the *fundamental value* $(s_t^*)_{t \in \mathbb{N}}$, which is completely determined by the conditional expectation of the future discounted dividends, and the *bubble component* $(\beta_t)_{t \in \mathbb{N}}$, which is an $\mathbb{F}$-adapted process such that $(\delta^t \mathbf{u}'(e_t)\beta_t)_{t \in \mathbb{N}}$ is a martingale. Given a dividend stream, the fundamental value of a security can be computed by relying on equation (6.86) (see also Exercises 6.36–6.37 for some explicit computations of the fundamental value of a security).

As we have seen above, provided that the martingale property of the process $(\delta^t \mathbf{u}'(e_t)\beta_t)_{t \in \mathbb{N}}$ is satisfied, then any adapted process $(\beta_t)_{t \in \mathbb{N}}$ can represent a bubble. In particular, we can have a deterministic bubble if $\beta_t = \beta(t)$, for some deterministic function $\beta : \mathbb{N} \to \mathbb{R}_+$. For instance, a simple deterministic bubble can be defined as $\beta_t = \alpha \beta_{t-1}$, for all $t \in \mathbb{N}$, thus implying that $\beta(t) = \alpha^t \beta_0$. This represents a bubble which will continue to expand forever and never bursts. More interesting bubbles can be defined by letting $(\beta_t)_{t \in \mathbb{N}}$ be a stochastic process. For instance, in Blanchard & Watson [248] a speculative bubble has been introduced by letting

$$
\mathbf{u}'(e_{t+1})\beta_{t+1} = \begin{cases} \frac{1}{\delta \pi} \mathbf{u}'(e_t)\beta_t + \varepsilon_{t+1}, & \text{with probability } \pi, \\ \varepsilon_{t+1}, & \text{with probability} 1 - \pi, \end{cases}
$$

with $\pi \in (0,1)$ and where $(\varepsilon_t)_{t \in \mathbb{N}}$ is a sequence of random variables such that $\mathbb{E}[\varepsilon_{t+1}|\mathscr{F}_t] = 0$, for all $t \in \mathbb{N}$. A stochastic process $(\beta_t)_{t \in \mathbb{N}}$ of this form does not depend on the fundamental value of a security and represents a *bursting bubble*, which can grow to higher and higher values and then suddenly collapse (with probability $1 - \pi$). When the bubble collapses, its expected value is zero and a new bubble can start again. Another type of bubble process is represented by *intrinsic bubbles*, which depend on the dividend process itself (see Exercise 6.39 and Froot & Obstfeld [745]).

In general, a process $(\beta_t)_{t \in \mathbb{N}}$ can be thought of as the bubble component because it explodes in expectation as time goes to infinity. Indeed, by the martingale property of the process $(\delta^t \mathbf{u}'(e_t)\beta_t)_{t \in \mathbb{N}}$ and recalling that $\delta \in (0,1)$,

$$
\lim_{T \to \infty} \mathbb{E}[\mathbf{u}'(e_{t+T})\beta_{t+T}|\mathscr{F}_t] = \lim_{T \to \infty} \delta^{-T} \mathbf{u}'(e_t)\beta_t = \begin{cases} -\infty & \text{if } \beta_t < 0; \\ +\infty & \text{if } \beta_t > 0. \end{cases}
$$

This exploding behavior illustrates why the process $(\beta_t)_{t \in \mathbb{N}}$ represents the bubble component of an asset price. It is important to remark that, as we have seen, asset prices exhibiting a non-null bubble component are *rational* in the sense that they are coherent with no-arbitrage and equilibrium, being solutions to the fundamental equation (6.85).

Observe that, in view of equation (6.87), a price process $(s_t)_{t \in \mathbb{N}}$ satisfying equation (6.85) coincides with the fundamental value if and only if the following *transversality condition* holds:

$$
\lim_{T \to \infty} \delta^T \mathbb{E}[\mathbf{u}'(e_{t+T})s_{t+T}|\mathscr{F}_t] = 0, \qquad \text{for all } t \in \mathbb{N}. \tag{6.88}
$$

Note also that, if the time horizon T is finite, then the standing assumption that prices are null at the terminal date (i.e., $s_T = 0$) implies that the transversality condition is always satisfied and, hence, the bubble component is null. This shows that rational bubbles can only exist in infinite horizon economies.

In equilibrium, the bubble component of any security in strictly positive supply is always non-negative. Indeed, since $\mathbf{u}'(e_t)\beta_t = \lim_{T\to\infty} \delta^T \mathbb{E}[\mathbf{u}'(e_{t+T})s_{t+T}|\mathscr{F}_t]$, if $\beta_t < 0$, then there exists some future date T at which $\mathbb{P}(s_T < 0) > 0$. However, a negative price for an asset delivering a non-negative stream of dividends is not consistent with market equilibrium, because the demand for such an asset would be infinite at date T in correspondence of the event $\{s_T < 0\}$. This shows that $\beta_t \geq 0$ for all $t \in \mathbb{N}$. Note, however, that this observation does not automatically imply that the price of an asset characterized by a speculative bubble is always larger than the corresponding fundamental value. This holds if the dividend stream and the discount factor are not affected by the presence of a non-null bubble component. Instead, if the presence of a bubble has an impact on the future dividend stream and on the discount factor, then there are situations where the price s_t can be smaller than the fundamental value s_t^* (see, e.g., Weil [1649]). For instance, in an overlapping generations model, a positive bubble can induce an increase of the discount factor and, consequently, a decrease of the fundamental value.

It is interesting to note that, under the present assumptions, a rational bubble cannot restart once it bursts (see also Diba & Grossman [573, 575]). In other words, a non-null rational bubble must be present from the initial date $t = 0$ of the economy and cannot appear at some later date, meaning that $\beta_t = 0$ implies that $\beta_{t+s} = 0$ for all $s \in \mathbb{N}$ (see Exercise 6.38 for a proof). Moreover, as shown in Exercise 6.38, an asset having a price process which is uniformly bounded from above (i.e., when it holds that $0 \leq s_t \leq K$, for all $t \in \mathbb{N}$ for some constant $K > 0$; think for instance of a European Put option written on a stock) cannot exhibit a bubble.

It is important to emphasize that the absence of arbitrage opportunities and the existence of an equilibrium of the economy do not suffice to exclude the existence of bubble solutions to the fundamental equation (6.85). However, suitable theoretical considerations restrict the behavior of bubble solutions. In particular, as we have seen above, bubbles are restricted to be non-negative, must be present from the beginning and cannot restart and can only exist if the price process is unbounded from above. Under appropriate assumptions, we can exclude the existence of non-trivial bubble components. For instance, if we assume that the market (in an infinite time horizon) is complete and that the economy is populated by infinitely lived agents, then the absence of arbitrage opportunities implies that there cannot exist a bubble solution. Indeed, assuming that the presence of a bubble does not affect the fundamental value of the asset, suppose on the contrary that there exists a date $t \in \mathbb{N}$ such that $\mathbb{P}(\beta_t > 0) = \mathbb{P}(s_t > s_t^*) > 0$. Then, at date t on the event $\{s_t > s_t^*\}$, an agent could realize an arbitrage opportunity by selling short the security (thus receiving the amount s_t) and investing in the portfolio of the Arrow securities replicating the future dividend stream $(d_s)_{s=t,t+1,\dots}$ at the cost of $s_t^* < s_t$, thus realizing an arbitrage profit at date t. Note, however, that this arbitrage argument relies on the assumption of infinitely lived agents. If agents are finitely lived, then the absence of arbitrage opportunities does not suffice to exclude the presence of speculative bubbles.

Since speculative bubbles are non-negative, the presence of a bubble component implies that the asset price s_t will diverge as time goes to infinity (indeed, the transversality condition (6.88) cannot be satisfied by the price process $(s_t)_{t \in \mathbb{N}}$ if the bubble component is non-null). As shown in Brock [305], in a complete market economy with an infinitely lived representative agent, speculative bubbles cannot exist since the transversality condition (6.88) necessarily holds in correspondence of the representative agent's optimum. In the seminal paper Tirole [1594], it has been shown that speculative bubbles cannot exist in correspondence of a dynamic rational expectations equilibrium in an economy populated by a finite number of infinitely lived risk neutral agents for an asset delivering an exogenously specified dividend stream. This non-existence result is robust with respect to the introduction of short-sale restrictions and heterogeneous information among the agents. However, asset prices can exhibit a bubble if agents have a myopic behavior (in the sense that, at each date $t \in \mathbb{N}$, they base their portfolio choice only on the comparison between the price at date t and the distribution of the price at the next date $t + 1$).

The introduction of short-sales and borrowing constraints has important consequences for the (non-)existence of speculative bubbles. For instance, Montrucchio & Privileggi [1351] have shown that, in a general economy (without restrictions on the stochastic process describing the dividends) populated by infinitely lived homogeneous agents with a zero short-sales constraint and a uniformly bounded relative risk aversion coefficient, speculative bubbles cannot occur regardless of the specific form of the agents' utility functions. Always assuming a zero short-sales constraint, the non-existence of speculative bubbles for unbounded utility functions (including logarithmic and power utilities) has been established in Kamihigashi [1062], with conditions depending only on the asymptotic behavior of the marginal utility at zero and infinity. In Santos & Woodford [1496], the authors have established that typical examples of speculative bubbles found in the literature represent rather exceptional cases. More specifically, they establish that in an economy populated by a finite number of infinitely lived agents with potentially incomplete markets, incomplete participation (thus allowing for overlapping generation models) and (price/time dependent) borrowing constraints, speculative bubbles cannot exist for an asset in positive net supply, under the assumption that the present value of the aggregate endowment is finite. However, bubbles can exist for an asset having infinite maturity (or having finite maturity but in zero net supply; on this point see also Magill & Quinzii [1285] and Huang & Werner [972, 973]).

Concerning the possible existence of speculative bubbles, Kocherlakota [1111] has shown that, in the context of a deterministic economy with a finite number of infinitely lived agents, the presence of constraints on debt accumulation makes bubbles possible in equilibrium. Such constraints prevent agents from engaging in *Ponzi schemes*[9] and ensure the existence of an equilibrium. At the same time, the same constraints prevent agents from transforming the presence of a bubble into an arbitrage opportunity and, hence, in some cases bubbles can exist in equilibrium. In

---

[9]In an economy with an infinite time horizon, a *Ponzi scheme* represents a strategy where the debt is rolled over indefinitely and never repaid (see, e.g., Lengwiler [1182, Section 6.2.2]), taking advantage of the infinite horizon of the economy.

particular, if agents face a wealth constraint, then an asset can exhibit a bubble if and only if it is in zero net supply. Conditions are also obtained for the existence of a bubble in the presence of exogenous short-sales constraints.

In Tirole [1595], the author considers a deterministic overlapping generations economy populated by finitely lived agents and provides necessary and sufficient conditions for the existence of a speculative bubble in this setting. It is shown that a bubble can exist if the growth rate of the economy is larger than the long run equilibrium bubble-free interest rate (dynamically inefficient economy). However, the equilibrium is asymptotically bubble-free (see Tirole [1595, Proposition 1]). On the other hand, if the interest rate exceeds the growth rate of the economy, then no bubble can exist in equilibrium. A similar result was obtained in Weil [1647] in the context of a stochastic economy. In an overlapping generation economy with agents living more than two periods, we can observe a non-divergent bubble similar to that proposed in Blanchard & Watson [248] (see Leach [1171]).

More generally, it is difficult to prove the non-existence of speculative bubbles in the case of assets for which the fundamentals are not clearly defined. In an asymmetric information environment with common priors, a condition for a bubble to exist is that the bubble is not common knowledge among agents (see Allen et al. [49] and Conlon [474]).

Summing up, one can affirm that there are many theoretical arguments against the presence of speculative bubbles in asset prices. We also want to point out that the hypothesis of common priors among the agents is crucial for the above results to hold. In Chap. 8 we shall discuss the existence of speculative bubbles in an economy with asymmetrically informed agents.

6.6   Notes and Further Readings

In Sect. 6.1, we have analysed the optimal investment-consumption problem by means of the dynamic programming approach and we refer the reader to Pascucci & Runggaldier [1404, Section 2.3] for an illustration of the dynamic programming approach in the context of several examples of discrete time models. The continuous time versions of the investment problems considered in Sect. 6.1 have been first considered in the seminal papers Merton [1327, 1328]. As we have discussed after Corollary 6.28, an alternative approach to the solution of optimal investment-consumption problems is represented by the *martingale approach*, which is especially simple in the case of a complete markets. According to the martingale approach, the solution of the optimal investment-consumption problem is decomposed into two successive steps: first, the optimization problem is solved over an abstract space of random variables and, as a second step, the optimal strategy is determined as the hedging strategy for the optimal random variable obtained in the first step. This approach was first developed for complete markets in Cox & Huang [504, 505], Pliska [1421] and then extended to an incomplete market setting by He & Pearson [919]. A detailed presentation of the martingale approach in the context of general continuous time models based on Itô-processes can be found in Karatzas & Shreve [1073] and further generalizations to the semimartingale setting have been developed in Kramkov & Schachermayer [1128, 1129]. In the context of discrete time models, the martingale approach is also discussed in Pascucci & Runggaldier [1404, Chapter 2].

On the basis of the theoretical implications of the solutions to optimal portfolio problems, during the last three decades the financial literature has paid significant attention to the comparison of the theoretical results with the empirical findings. In particular, there have been empirical studies on household portfolio choices (see Guiso et al. [857], Campbell & Viceira [356], Campbell [337]) reporting evidence of an *asset allocation puzzle*. Canner et al. [360], analysing stocks, bonds and cash allocations recommended by four financial advisors in the U.S., find that young people (i.e., investors with a long horizon) as well as less risk averse investors are advised to invest in risky assets (high stocks-bonds ratio) more than old or strongly risk averse investors. In Jagannathan & Kocherlakota [1009], the opposite phenomenon has been reported for investment in bonds (horizon effect). Moreover, it is shown that such asset allocation recommendations are inconsistent with the two mutual funds separation theorem and with the optimal trading-consumption strategy obtained under the assumption of identically and independently distributed asset returns and of a power or logarithmic utility function.

The empirical analysis on portfolio holdings provides mixed results on the relationship between age and investment in stocks: in many cases an increasing or a hump-shaped pattern in age is observed (see Ameriks & Zeldes [55], Bertaut & Starr-McCluer [210], Guiso et al. [860], Poterba & Samwick [1429]). In particular, as reported in Ameriks & Zeldes [55], the fraction of investment in risky assets reaches a peak in the interval 50–59 years. Moreover, even though there is a hump-shaped pattern in the ownership of risky assets with respect to age, the risky asset holding is almost constant with respect to age conditionally on shares ownership. It is also observed that rich people save more and hold more risky assets with respect to what the theory would predict (see Carroll [371], Dynan et al. [614]). It is empirically observed that the ownership of risky assets and the participation rate in the financial market tend to be increasing with respect to wealth, contradicting the optimal portfolios suggested in the seminal papers Samuelson [1493], Merton [1327]. Moreover, Heaton & Lucas [929] report that households with high and variable business income invest less wealth in stocks than other similarly wealthy households, thus showing that entrepreneurial income risk has a significant impact on portfolio choices.

In the empirical literature analysing households' portfolios, limited diversification and a high degree of heterogeneity have been reported. Moreover, there exists a *participation puzzle*, meaning that a rather low participation rate in the stock market is detected. Mankiw & Zeldes [1298] have shown that at the time of their study only one quarter of the U.S. families owned stocks. Since that time, the fraction of population trading in the market has increased, but still nowadays a large part of the population does not own stocks (about one half of the population in the U.S., according to Bertaut & Starr-McCluer [210], Ameriks & Zeldes [55], Haliassos &

Bertaut [876]). In other countries, the fraction of population active in the financial market is even lower. Moreover, this figure does not increase significantly by including in the analysis also investors holding stocks through pension funds. Typically, active participation to the stock market occurs rather late in the life span of an agent. Grinblatt et al. [832] show that stock market participation is positively related to education and cognitive ability. It is difficult to explain the participation puzzle in the context of the models presented in this chapter, based on classical asset pricing theory.

Further puzzles have emerged when testing the implications of the theory in the context of international asset pricing models (i.e., models with several financial markets corresponding to different countries). In particular, a *home bias puzzle* has been detected both in stock holdings as well as in consumption: on the one hand, the proportion of foreign assets held by investors is too limited (see French & Poterba [740]) and, on the other hand, domestic investors do not share risk with foreign investors, with the consequence that consumption growth rates do not move together across different countries as much as international risk sharing would suggest (see Lewis [1216]). Huberman [981] reports a geographical bias, showing that shareholders of a Regional Bell Operating Company tend to live in the area which it serves. In a related direction, Benartzi [190] shows that agents tend to invest significantly in their employer's stock and this behavior is not due to the presence of inside information. Coval & Moskowitz [501, 502] show that mutual funds and stockholders invest locally, thus providing further evidence on the existence of a home bias. Note that, as shown in DeMarzo et al. [553], a general equilibrium model where agents care about their consumption relative to the per capita consumption in their community can explain this portfolio bias, since agents bias their portfolios towards the (undiversified) portfolios held by the other members of the community (see also Sect. 9.2 for a discussion of similar models). Moskowitz & Vissing-Jorgensen [1356] show that the investment in private equity is extremely concentrated: even though private equity returns are generally not higher than the market return on publicly traded stocks, there is a substantial investment in a single privately held firm.

From a theoretical point of view, several papers have studied the impact on optimal portfolio choices of specific features of the investment opportunity set. For instance, optimal portfolio problems in the presence of a stochastic risk free interest rate have been considered in Brennan & Xia [300, 301], Wachter [1634], Campbell & Viceira [355]. In a continuous time model with a constant risk free rate, Kim & Omberg [1099] analyse the optimal consumption-investment problem in the presence of a single risky asset whose expected return follows a mean reverting process. Under the assumption of a HARA utility function defined with respect to consumption at the terminal date, a closed form solution to the optimal portfolio problem is obtained. Moreover, it is shown that, if risk premia innovations are negatively correlated with asset price movements and the risk premium is positive, then the hedging demand is positive if the agent is more risk averse than in the case of a logarithmic utility function. This implies that, in the presence of mean reversion, we expect an agent with a long horizon to hold a portfolio larger than an agent with a short horizon. The optimal portfolio is increasing in the investment horizon because mean reversion in stock returns reduces the variance of cumulative returns over long horizons. Other studies analysing the implications of mean reversion on consumption and optimal portfolios include Barberis [146], Campbell & Viceira [354], Campbell et al. [340, 342], Wachter [1636], Brennan [290], Brennan et al. [296], Brandt [279], Lynch & Balduzzi [1259], Jurek & Viceira [1055]. Analysing the optimal investment-consumption problem in the presence of a stochastic volatility, Liu [1226] and Chacko & Viceira [384] obtain results similar to those obtained under mean reversion: if volatility innovations are negatively correlated with asset price movements, the risk premium is positive and the agent is more risk averse than in the logarithmic case, then the hedging demand is positive and the risky asset demand is increasing with respect to the investment horizon.

When the investment horizon is long, long term bonds with a stochastic interest rate present similar characteristics to a non-risky investment. As a consequence, the investment in long term bonds should be increasing with respect to an investor's risk aversion. A similar effect is obtained also in the presence of stochastic inflation. Assuming a coefficient of relative risk aversion greater than one, these effects contribute to explain the asset allocation puzzle (see Campbell & Viceira [355], Campbell et al. [340], Wachter [1634], Brennan & Xia [300, 301]). It is optimal for conservative long term investors to buy (inflation indexed) long term bonds and agents should invest significantly in risky assets when they are young, reducing progressively their exposure when their age increases (see Campbell et al. [339], Bajeux-Besnainou et al. [182, 181]).

The possible predictability of asset returns also has important implications on portfolio choices (see Campbell & Viceira [354], Balduzzi & Lynch [118], Han [883], Wachter & Warusawitharana [1638], Handa & Tiwari [887], De Miguel et al. [557]). In particular, Lynch [1258] has analysed portfolio choices based on characteristics like size and the book-to-market ratio, when returns can be predicted via the dividend yield. Compared with the investors' allocation in their last period, the presence of predictability leads the investor early in life to tilt his portfolio away from high book-to-market stocks and from small stocks. Portfolio returns exhibit both the size and the book-to-market effect. Moreover, it is shown that the utility costs due to ignoring the predictability of returns are relevant.

If asset returns are distributed independently over time, the utility function is of the logarithmic, exponential or power type and the investors are infinitely lived, then it can be shown that consumption and portfolio weights are linear with respect to wealth (see Hakansson [875]). A *portfolio turnpike* result for a finite horizon economy with an agent maximizing expected utility of final wealth is proved in Huberman & Ross [985]: the optimal policy converges to a time-independent one as the investment horizon grows to infinity if the coefficient of relative risk aversion converges as wealth increases. We also want to mention that investment-consumption problems have been also extended by including housing services (see Campbell & Cocco [341], Cocco [455], Yao & Zhang [1664]), illiquid assets (see Ang et al. [70], Longstaff [1244], Tepla [1585]), luxury goods (see Carroll [371], Wachter & Yogo [1639], Ait-Sahalia et al. [32]), health risk and portfolio-saving decisions of the elders (see Palumbo [1399], Edwards [629], French [737], French & Jones [738], De Nardi et al. [558], Yogo [1667]), annuities (see Milevsky & Young [1340], Chai et al. [386], Horneff et al. [966]) and durable goods (see Grossman & Laroque [844]). A different strand of literature has considered the presence of non-diversifiable labor income and we refer to Sect. 9.5 for a discussion on this point. Note that non-diversifiable labor income cannot be spanned by the securities available for trading in the market and, therefore, leads to market incompleteness.

In the context of a pure exchange economy, classical asset pricing theory has been mainly developed in the seminal contributions of LeRoy [1183], Rubinstein [1487], Lucas [1252], Harrison & Kreps [904]. Considering also a production economy, equilibrium models have beed developed in Brock [305], Cox et al. [506], Cochrane [460, 462], Rouwenhorst [1477], Jermann [1032]. In this case, instead of explaining asset returns through marginal rates of substitution over intertemporal consumption, asset returns are explained through marginal rates of transformation and, therefore, through the firm's investment demand and investment returns. In particular, these models allow to relate asset returns to economic fluctuations and can explain the variations in expected returns both cross-sectionally and over time. Empirically, models based on a production economy are not rejected and have a performance comparable to that of the CAPM and of the APT in the version proposed in Chen et al. [424] and outperform the CCAPM.

In a dynamically complete market, the existence of an equilibrium is obtained under standard regularity conditions. In an incomplete market setting, the existence of an equilibrium has been studied in Duffie & Shafer [598, 599] (see also Magill & Shafer [1288]). Duffie [592] has considered the existence of an equilibrium in an economy comprising nominal assets. In the presence of an infinite horizon, establishing the existence of an equilibrium is more challenging, as explained in Duffie [593]. In that setting, Ponzi schemes have to be avoided (see also Hernandez & Santos [939], Araujo et al. [75]). To this end, borrowing or short sale constraints are introduced in order to ensure a suitable transversality condition. See also Huang & Werner [973] on the implementation of an Arrow-Debreu equilibrium in an infinite horizon economy.

In an economy with more than one good and real dividends, a generic result has been obtained in Magill & Shafer [1287]. More specifically, they consider equilibrium allocations corresponding to contingent markets (i.e., markets open at the initial date for every good in correspondence of every time-event couple) and a sequential system of spot and real asset markets, where each real asset delivers a vector of goods (as considered in Sect. 4.2). They provide a necessary and sufficient condition on the asset structure for the two equilibrium allocations to coincide in a generic sense (i.e., on an open set of full measure). In particular, this condition requires that, for every $t = 0, 1, \ldots, T-1$ and $A_t \in \mathscr{F}_t$, the number of traded assets equals $\nu_t(A_t)$, using the notation introduced in Sect. 6.2.

As we have seen in Sect. 6.2, the notion of market completeness plays an important role in the equilibrium analysis of multi-period economies. Guesnerie & Jaffray [853] have shown that it is possible to complete the market through short-lived securities. To this effect, it suffices to require that, for all events $A_t \in \mathscr{F}_t$, for $t = 0, 1, \ldots, T - 1$, and $A_{t+1} \subseteq A_t$ with $A_{t+1} \in \mathscr{F}_{t+1}$, there exists a security which is traded at date t in correspondence of the event A_t and pays a unitary dividend at the subsequent date $t + 1$ in correspondence of the event A_{t+1} (and zero otherwise). As in a two-period economy, contingent claims may help to complete the market also in a multi-period setting. However, unlike in a two-period economy (see Sect. 4.3), in a multi-period discrete time setting European Call options are not always useful in order to complete the market (see Bajeux-Besnainou & Rochet [183], but note that this result does not hold in continuous time). However, exotic path-dependent options can allow to dynamically complete the market. The result on dynamic market completeness in the presence of a sufficient number of traded securities (see Proposition 6.15) has been extended in Duffie & Huang [596] to a continuous time economy where the price processes are described by diffusion processes. Loosely speaking, in this case the infinite dimensional state space is matched by the possibility of infinitely frequent trading.

In the case of incomplete markets, an arbitrage free price is not univocally determined. In this case, several different approaches have been proposed in the literature in order to determine suitable arbitrage free prices (see Musiela & Rutkowski [1364] for an overview). Recall that, if markets are incomplete, then there exist contingent claims which cannot be perfectly replicated by trading according to some self-financing strategy. Among the different approaches proposed in the context of incomplete markets, several valuation techniques are based on the idea of minimizing the risk of the unhedgeable part of a contingent claim, according to some utility function or to a mean-variance criterion. Alternatively, one can insists on the self-financing requirement, minimizing the quadratic cost of the replication strategy (*risk minimization*, see Föllmer & Schweizer [722], Föllmer & Sondermann [723] and the surveys by Pham [1418] and Schweizer [1508]). Another important approach to determine an arbitrage free price in the case of incomplete markets is represented by the *utility indifference pricing*, see Henderson & Hobson [936] for an overview.

Similarly as in the case of Theorem 4.20 established in the context of a two-period economy, one can prove a version of Theorem 6.23 by only assuming the absence of arbitrage opportunities of the second kind. In that case, the same arguments used in the proof of Theorem 6.23 allow to show the existence of a probability measure $\mathbb{P}^*$ satisfying the martingale condition (6.52) but not necessarily equivalent to $\mathbb{P}$, in the sense that it may happen that some events have null probability under the measure $\mathbb{P}^*$. To this effect, see also LeRoy & Werner [1191, Theorem 24.2.2].

We also mention that the CCAPM has been extended to a continuous time setting in Breeden [285], Duffie & Zame [600] by assuming that asset returns follow diffusion processes and a continuous time version of the CAPM has been developed in Chamberlain [396].

We want to point out that, as argued in Shleifer & Summers [1540], the no-arbitrage analysis has some intrinsic limits. For instance, it is typically assumed that the agents' investment horizon is equal to the horizon of the economy. However, if the agents' horizon is shorter than that of the economy, then a buy-and-hold strategy may not work and an agent may not be able to exploit an arbitrage opportunity. In this case, an agent will not only consider the underlying dividend stream paid by a security, but also its future price when closing a position. In this setting, an arbitrage opportunity may persist in the market. Dow & Gorton [584] show that a limited horizon may induce an agent with private information not to trade on an arbitrage opportunity. The informed trader trades if there is a chain of informed agents trading in the future spanning the event date, i.e., at the end of his horizon there is an informed trader who makes the price move in the direction of the information. This effect induces an agent not to trade until the event date is close.

In Sect. 6.5, we have considered economies with an infinite horizon where asset prices may exhibit a non-null bubble component. Assuming a finite number of agents in the economy, a bubble can originate by a payoff stream that occurs at infinity (lack of countable additivity in asset valuation; see LeRoy & Gilles [1186], Gilles & LeRoy [781, 783, 784]). In Timmermann [1591], it is shown that there is no bubble when there is a feedback of prices on dividends. The presence of such a feedback effect is motivated by studies such as Lintner [1220], Marsh & Merton [1306], showing that managers typically define dividends as proportions of the fundamental value of the company which is reflected by the market price of the company. A model with feedback effects admits multiple rational expectations equilibria and no speculative bubble. The existence of a bubble solution in continuous time models has been addressed in Loewenstein & Willard [1242]. For positive net supply assets, there are no bubbles under weak conditions. Bubbles may exist for zero net supply assets. Heston et al. [940] study the existence of a bubble in a market containing derivative assets. Intrinsic bubbles (fundamental dependent bubbles) have been characterized in continuous time in Ikeda & Shibata [997]. Bewley [219] suggests that the value of fiat money can be interpreted as a bubble. On the interpretation of the dot-com bubble in terms of rational bubbles see Ofek & Richardson [1382], LeRoy [1185], Pástor & Veronesi [1409], Griffin et al. [828]. On the existence of speculative bubbles under endogenous borrowing constraints see also Kocherlakota [1114], Hellwig & Lorenzoni [930], Werner [1654]. On the existence of bubbles with heterogeneous opinions and short-sale constraints see also Miller [1342], Varian [1611], Harris & Raviv [901], Kandel & Pearson [1066], Harrison & Kreps [903], Chen et al. [418], Hong & Stein [962], Scheinkman & Xiong [1502], Abreu & Brunnermeier [10], Kyle & Wang [1150]. In mathematical finance, financial bubbles are often modeled by strict local martingales (i.e., local martingales which are not true martingales) and we refer to Protter [1435] for an overview of this approach.

6.7 Exercises

Exercise 6.1 Let (θ, c) be a self-financing trading-consumption strategy and let $W(\theta)$ be the corresponding wealth process. Show that condition (6.5) holds.

Exercise 6.2 Suppose that the economy contains a risk free security paying a constant and deterministic risk free rate $r_f > 0$. Define the *discounted quantities* $\bar{S}_t := S_t/r_f^t$ and $\bar{D}_t := D_t/r_f^t$, for all $t = 0, 1, \ldots, T$. For any trading-consumption strategy (θ, c), define the discounted portfolio value $\bar{W}_t(\theta) := W_t(\theta)/r_f^t$ and the discounted amount of consumption $\bar{c}_t := c_t/r_f^t$, for all $t = 0, 1, \ldots, T$. Show that the self-financing condition (6.4) is stable with respect to discounting, in the sense that, for any strategy $(\theta, c) \in \mathscr{A}(x_0)$:

$$\Delta \bar{W}_{t+1}(\theta) = \theta_{t+1}^\top \Delta \bar{S}_{t+1} + \theta_{t+1}^\top \bar{D}_{t+1} - \bar{c}_t.$$

In particular, it holds that

$$\bar{W}_T(\theta) = \theta_0^\top (\bar{S}_0 + \bar{D}_0) + \sum_{t=1}^{T} \theta_t^\top (\Delta \bar{S}_t + \bar{D}_t) - \sum_{t=0}^{T-1} \bar{c}_t$$

$$= x + \sum_{t=1}^{T} \theta_t^\top (\Delta \bar{S}_t + \bar{D}_t) - \sum_{t=0}^{T-1} \bar{c}_t.$$

Exercise 6.3 Let the value function be defined as in (6.9) with respect to a utility function $u : \mathbb{R}_+ \to \mathbb{R}$ strictly increasing and concave. Prove that the value function is strictly increasing and concave in its first argument (wealth).

Exercise 6.4 Prove Proposition 6.6.

Exercise 6.5 Prove Corollary 6.7.

Exercise 6.6 Prove Proposition 6.8.

Exercise 6.7 Prove Corollary 6.9.

Exercise 6.8 In the context of Proposition 6.8, suppose that, for all $t = 1, \ldots, T-1$, the $\mathscr{F}_t$-conditional distribution of the asset returns r_{t+1}^n, $n = 1, \ldots, N$, coincides with the unconditional distribution of r_1^n, $n = 1, \ldots, N$. Prove that the optimal portfolio process $(w_t^*)_{t=1,\ldots,T}$ can be characterized as in equation (6.29), does not depend on time and that the process $(a_t)_{t=0,1,\ldots,T}$ appearing in Proposition 6.8 reduces to a deterministic function of time.

Exercise 6.9 Let us consider a multi-period economy $t = 0, 1, \ldots, T$ with two assets: a risk free asset, whose price is given by $B_t = r_f^t$, for all $t = 0, 1, \ldots, T$, with $r_f > 0$ being the risk free rate of return, and a risky asset with price process $(S_t)_{t=0,1,\ldots,T}$. We assume that, at any date $t = 1, \ldots, T$, given the price S_{t-1} at the previous date $t - 1$, the risky asset can take the two possible values $S_{t-1}u$ and $S_{t-1}d$

with probabilities p and $1 - p$, respectively (this represents a special case of the multi-period binomial model presented at the end of Sect. 6.3). Consider an optimal investment-consumption problem of the type (6.6) with consumption only at the terminal date T, i.e., the agent is interested in maximizing $\mathbb{E}[u(W_T)]$ over all self-financing trading strategies $(\theta_t)_{t=1,...,T}$. Prove the following claims:

(i) if $u(x) = \sqrt{x}$, then the value function introduced in (6.9) is given by

$$V(x, t) = c^{T-t}\sqrt{x}, \qquad \text{for all } t = 0, 1, \ldots, T \text{ and } x > 0,$$

for a constant $c > 0$, and the optimal number of units of the risky asset is given by

$$\theta_t^* = K\frac{W_{t-1}}{S_{t-1}}, \qquad \text{for all } t = 1, \ldots, T,$$

for a constant $K \in \mathbb{R}$ which can be explicitly computed.

(ii) If $u(x) = \log(x)$, then the value function introduced in (6.9) is given by

$$V(x, t) = c(T - t) + \log(x), \qquad \text{for all } t = 0, 1, \ldots, T \text{ and } x > 0,$$

for a constant $c > 0$, and the optimal number of units of the risky asset is given by

$$\theta_t^* = K\frac{W_{t-1}}{S_{t-1}}, \qquad \text{for all } t = 1, \ldots, T,$$

for a constant $K \in \mathbb{R}$ which can be explicitly computed.

(iii) If $u(x) = x^\gamma/\gamma$, with $\gamma \in (0, 1)$, then the value function introduced in (6.9) is given by

$$V(x, t) = c^{T-t}\frac{x^\gamma}{\gamma}, \qquad \text{for all } t = 0, 1, \ldots, T \text{ and } x > 0,$$

for a constant $c > 0$, and the optimal number of units of the risky asset is given by

$$\theta_t^* = K\frac{W_{t-1}}{S_{t-1}}, \qquad \text{for all } t = 1, \ldots, T,$$

for a constant $K \in \mathbb{R}$ which can be explicitly computed.

Exercise 6.10 In the context of Proposition 6.12, let $t \in \{0, 1, \ldots, T - 1\}$ and $A_t \in \mathscr{F}_t$ and suppose that, for some $s \in \{t+1, \ldots, T\}$, there exists an event $A_s \in \mathscr{F}_s$ such that $A_s \subseteq A_t$ and $q_{A_s|A_t}^* \neq q_{A_s}^*/q_{A_t}^*$. Prove that such a price system admits arbitrage opportunities and, hence, cannot correspond to an equilibrium.

Exercise 6.11 Let $\{S^*, D; (\theta^{1*}, c^{1*}), \ldots, (\theta^{I*}, c^{I*})\}$ be a Radner equilibrium, in the sense of Definition 6.13. Suppose that the total supply of the N securities is normalized to $\sum_{i=1}^{I} \theta_0^{i,n} = 1$, for all $n = 1, \ldots, N$, and that all the agents have strictly increasing utility functions $u^i : \mathbb{R}_+ \to \mathbb{R}$. Show that in equilibrium it holds that $\sum_{i=1}^{I} c_t^{i*} = \sum_{n=1}^{N} d_t^n$, for all $t = 0, 1, \ldots, T$.

Exercise 6.12 Prove Proposition 6.15.

Exercise 6.13 As in the first part of Sect. 6.2, consider an economy with I agents (with homogeneous beliefs, the same discount factor δ, time additive and state independent preferences represented by strictly increasing and strictly concave differentiable utility functions u^i, $i = 1, \ldots, I$) and where all the $\sum_{t=1}^{T} \nu_t$ Arrow securities are traded. Let $e = \{e_t(A_t); A_t \in \mathscr{F}_t, t = 0, 1, \ldots, T\}$ be the aggregate endowment process and $\{c^{i*}; i = 1, \ldots, I\}$ a feasible allocation corresponding to an (Arrow-Debreu) equilibrium, in correspondence of the prices $\{q_0, q_{A_t}; A_t \in \mathscr{F}_t, t = 1, \ldots, T\}$ for the Arrow securities. Show that the allocation $\{c^{i*}; i = 1, \ldots, I\}$ defines the utility function of a representative agent and that the prices $\{q_0, q_{A_t}; A_t \in \mathscr{F}_t, t = 1, \ldots, T\}$ are supported by a no-trade equilibrium in the representative agent economy.

Exercise 6.14 Consider a representative agent economy where the representative agent's utility function is given by $\mathbf{u}(x) = \log(x)$ and the aggregate endowment process is $e = \{e_0, e(A_t); A_t \in \mathscr{F}_t, t = 1, \ldots, T\}$. Show that, in correspondence of a no-trade equilibrium in the representative agent economy, the value at the initial date $t = 0$ of the aggregate endowment process e is given by $e_0(1 - \delta^{T+1})/(1 - \delta)$ (normalizing $q_0 = 1$, as at the end of Sect. 6.2).

Exercise 6.15 Prove Proposition 6.18.

Exercise 6.16 Prove Lemma 6.19.

Exercise 6.17 Suppose that the price-dividend couple (S, D) does not admit arbitrage opportunities and let $(\theta, c) \in \mathscr{A}(x_0)$. Use Proposition 6.22 to give an alternative and simple proof of Lemma 6.19.

Exercise 6.18 Prove Lemma 6.20.

Exercise 6.19 Prove Proposition 6.22.

Exercise 6.20 In the setting of Proposition 6.32, prove the representation (6.62) for the conditional probabilities associated to the risk neutral probability measure $\mathbb{P}^*$ defined by (6.61) (compare also with Huang & Litzenberger [971, Section 8.5]).

Exercise 6.21 Suppose that the market is incomplete, so that there exist elements of $\mathscr{K}^+$ that are not contained in $\bigcup_{x \in \mathbb{R}_+} \mathscr{C}_0^+(x)$. Show that the arbitrage free price of an attainable consumption process $c \in \bigcup_{x \in \mathbb{R}_+} \mathscr{C}_0^+(x)$ is uniquely defined, regardless of the specific risk neutral probability measure $\mathbb{P}^*$ chosen.

Exercise 6.22 For any $n = 1, \ldots, T$, define the *discounted gain* process $g^n = (g^n_t)_{t=1,\ldots,T}$ by

$$g^n_{t+1} = \frac{1}{r^t_f}\left(\frac{1}{r_f}(d^n_{t+1} + s^n_{t+1}) - s^n_t\right), \qquad \text{for all } t = 0, 1, \ldots, T-1.$$

Suppose that (S, D) is a couple of price-dividend processes not admitting arbitrage opportunities and let $\mathbb{P}^*$ be a risk neutral probability measure. Let $(\theta_t)_{t=1,\ldots,T}$ be a predictable $\mathbb{R}^N$-valued stochastic process. Show that, for every $t = 1, \ldots, T$, it holds that

$$\mathbb{E}^*\left[\sum_{n=1}^{N} \theta^n_t g^n_t \,\Big|\, \mathscr{F}_{t-1}\right] = 0$$

and, as a consequence,

$$\sum_{t=1}^{T}\sum_{n=1}^{N} \mathbb{E}^*[\theta^n_t g^n_t] = 0$$

i.e., the cumulative expected discounted gains of any trading strategy are null under a risk neutral probability measure.

Exercise 6.23 Let (S, D) be a couple of price-dividend processes not admitting arbitrage opportunities and let $\mathbb{P}^*$ be a risk neutral probability measure (which exists by Theorem 6.23). As in Exercise 6.2, define the *discounted quantities* $\bar{S}_t := S_t/r^t_f$ and $\bar{D}_t := D_t/r^t_f$, for all $t = 0, 1, \ldots, T$. Similarly, for any trading-consumption strategy (θ, c), define the discounted portfolio value $\bar{W}_t(\theta) := W_t(\theta)/r^t_f$ and the discounted consumption $\bar{c}_t := c_t/r^t_f$, for all $t = 0, 1, \ldots, T$. Prove that, for every $x \in \mathbb{R}_+$, a non-negative adapted process $c = (c_t)_{t=0,1,\ldots,T}$ belongs to $\mathscr{C}^+_0(x)$ if and only if the following two requirements are satisfied:

(i) $\bar{c}_T = x + \sum_{t=1}^{T} \theta^\top_t (\Delta\bar{S}_t + \bar{D}_t) - \sum_{t=0}^{T-1} \bar{c}_t$, for some predictable process $(\theta_t)_{t=0,1,\ldots,T}$;

(ii) $\sum_{t=0}^{T} \mathbb{E}^*[\bar{c}_t] = x$.

Exercise 6.24 Let us consider a model with dates $t = 0, 1, 2$ with four states of nature $\{\omega_1, \omega_2, \omega_3, \omega_4\}$ and the following information flow

$$\mathscr{F}_0 = \{\omega_1, \omega_2, \omega_3, \omega_4\},$$

$$\mathscr{F}_1 = \{\mathscr{F}^1_1 = \{\omega_1, \omega_2\}, \mathscr{F}^2_1 = \{\omega_3, \omega_4\}\},$$

$$\mathscr{F}_2 = \{\{\omega_1\}, \{\omega_2\}, \{\omega_3\}, \{\omega_4\}\}.$$

Two assets are traded in the market, with dividends at $t = 2$ given by

$$D = \begin{bmatrix} 4 & 2 \\ 2 & 3 \\ 2 & 9 \\ 4 & 3 \end{bmatrix}$$

the prices of the two assets are $(3, 2)$ in $\mathscr{F}_1^1$, $(2, 3)$ in $\mathscr{F}_1^2$, and $(1.1, 0.9)$ in $\mathscr{F}_0$.

(i) Is the market dynamically complete?
(ii) Are there arbitrage opportunities in the market?
(iii) If it exists, compute a risk neutral probability measure.
(iv) Determine the no-arbitrage price at the initial date $t = 0$ of a European Call option written on the first asset with strike price $K = 3$ and maturity $T = 2$.

Exercise 6.25 Suppose that (S, D) is a couple of price-dividend processes not admitting arbitrage opportunities. Generalize to the multi-period case the proof of Proposition 4.29 and prove Proposition 6.30.

Exercise 6.26 Consider a binomial model with three dates as introduced at the end of Sect. 6.3 (with $t \in \{0, 1, 2\}$) and a contingent claim with payoff $f(s_2)$, where s_2 denotes the price of the risky asset at the terminal date $T = 2$. By using a backward induction procedure, compute the arbitrage free price of the claim as well as the associated hedging strategy. Verify that the arbitrage free price computed at the initial date $t = 0$ coincides with the discounted risk neutral expectation (see Corollary 6.34).

Exercise 6.27 Let (S, D) be a couple of price-dividend processes not admitting arbitrage opportunities and let $u : \mathbb{R}_+ \to \mathbb{R}$ be a continuous, strictly increasing and concave utility function. Prove that, for all $x > 0$, the optimal investment-consumption problem (6.6) admits a solution in correspondence of (S, D).

Exercise 6.28 Under the assumptions of Sect. 6.4, prove that the following asset pricing relation holds true, for all $n = 0, 1, \ldots, N$ and $t = 0, 1, \ldots, T - 1$:

$$s_t^n = \sum_{s=1}^{T-t} \frac{\mathbb{E}[d_{t+s}^n | \mathscr{F}_t]}{r_f^s} + \sum_{s=1}^{T-t} \mathrm{Cov}\left(\delta^s \frac{\mathbf{u}'(e_{t+s})}{\mathbf{u}'(e_t)}, d_{t+s}^n \Big| \mathscr{F}_t \right),$$

Exercise 6.29 In the context of Proposition 6.37, let $(\hat{w}_t)_{t=1,\ldots,T}$ be an $\mathbb{R}^{N+1}$-valued predictable process satisfying $\mathbf{1}^\top \hat{w}_t = 1$ for all $t = 1, \ldots, T$ and such that, for every $t = 0, 1, \ldots, T - 1$,

$$\mathrm{Corr}\left(r_{t+1}^\top \hat{w}_{t+1}, \mathbf{u}'(e_{t+1}) | \mathscr{F}_t \right) = \min_{w \in \mathbb{R}^{N+1} : w^\top \mathbf{1} = 1} \mathrm{Corr}\left(r_{t+1}^\top w, \mathbf{u}'(e_{t+1}) | \mathscr{F}_t \right).$$

Show that relation (6.76) holds true.

Exercise 6.30 Prove Proposition 6.38

Exercise 6.31 In the setting of Sect. 6.4, consider a fixed $t \in \{0, 1, \ldots, T - 2\}$ and suppose at date t it is possible to trade a zero-coupon bond with maturity $t + 1$ having price $B(t, t + 1)$ and a zero-coupon bond with maturity $t + 2$ having price $B(t, t + 2)$. Suppose furthermore that at the future date $t + 1$ it will be possible to trade a zero-coupon bond with maturity $t + 2$ for the price $B(t + 1, t + 2)$. Let $\mathbb{P}^*$ be a risk neutral probability measure for the economy. Prove that

$$B(t, t + 2) = B(t, t + 1)\mathbb{E}^*[B(t + 1, t + 2)|\mathscr{F}_t].$$

Exercise 6.32 In the context of Sect. 6.4, let $g(t, \tau) := (e_{t+\tau}/e_t)^{1/\tau}$ and suppose that the representative agent has a power utility function of the type

$$\mathbf{u}(x) = \frac{x^{1-\gamma}}{1 - \gamma}.$$

Show that the following relation holds in an approximate form:

$$\log\left(1 + i(t, t + \tau)\right) = -\log \delta + \gamma\, \mathbb{E}[\log g(t, \tau)|\mathscr{F}_t],$$

for all $t \in \{0, 1, \ldots, T - 1\}$ and $\tau \in \{0, 1, \ldots, T - t\}$.

Exercise 6.33 Under the assumptions of Sect. 6.4, suppose that $T = 2$ and consider a representative agent with logarithmic utility function $\sum_{t=0}^{2} \delta^t \log(c_t)$. Suppose furthermore that the aggregate endowment process $(e_t)_{t=0,1,2}$ starts from a deterministic value e_0 at the initial date $t = 0$ and, at date $t = 1$, takes the two possible values $e_1(u)$ and $e_1(d)$ with probabilities p and $1 - p$, respectively, and then evolves deterministically on the period $[1, 2]$, taking values $e_2(u)$ and $e_2(d)$, respectively (the price of aggregate endowment is supposed to be always normalized to one). In other words, the uncertainty of the economy is represented by the uncertainty about two possible states of the world realized at the intermediate date $t = 1$. Show that

$$1 + i(0, 1) = \frac{1}{\delta \mathbb{E}\left[\frac{1}{x(0,1)}\right]}, \quad 1 + i(0, 2) = \frac{1}{\delta\sqrt{\mathbb{E}\left[\frac{1}{x(0,2)}\right]}}, \quad 1 + f(0; 1, 2) = \frac{1}{\delta}\frac{\mathbb{E}\left[\frac{1}{x(0,1)}\right]}{\mathbb{E}\left[\frac{1}{x(0,2)}\right]},$$

where $x(0, 1) := e_1/e_0$ and $x(0, 2) := e_2/e_0$ denote the growth rates of the aggregate endowment on the periods $[0, 1]$ and $[0, 2]$, respectively. Show also that on the period $[1, 2]$ it holds that

$$
\begin{aligned}
1 + i(1, 2; u) &= \frac{x(1, 2; u)}{\delta} \qquad \text{if } e_1 = e_1(u), \\
\\
1 + i(1, 2; d) &= \frac{x(1, 2; d)}{\delta} \qquad \text{if } e_1 = e_1(d),
\end{aligned}
\tag{6.89}
$$

with $i(1, 2; u)$ and $x(1, 2; u)$ ($i(1, 2; d)$ and $x(1, 2; d)$, resp.) denoting respectively the interest rate and the aggregate endowment growth rate on the period $[1, 2]$ if state u (state d, resp.) is realized.

Exercise 6.34 Under the assumptions of Sect. 6.4, suppose that the economy is infinitely-lived (i.e., $T = \infty$) and let $B(t, t + \tau)$ denote the price at date t of a zero-coupon risk free bond maturing at date $t + \tau$, for $t \in \mathbb{N}$ and $\tau \in \mathbb{N}$. Suppose in addition that there exists a representative agent characterized by the utility function $\sum_{t=0}^{\infty} \delta^t \log(c_t)$ and that the aggregate endowment process $(e_t)_{t \in \mathbb{N}}$ evolves according to

$$e_{t+1} = \rho e_t u_{t+1}, \qquad \text{for all } t \in \mathbb{N},$$

where $(u_t)_{t \in \mathbb{N}}$ is a sequence of independent and identically distributed positive random variables and $\rho > 0$. Show that, for each $t \in \mathbb{N}$, it holds that

$$B(t, t + 1) = \frac{\delta}{\rho} \mathbb{E}\left[\frac{1}{u_1}\right] \quad \text{and} \quad B(t, t + 2) = \frac{\delta^2}{\rho^2} \left(\mathbb{E}\left[\frac{1}{u_1}\right]\right)^2.$$

Exercise 6.35 Under the assumptions of Sect. 6.4, suppose that the economy is infinitely-lived (i.e., $T = \infty$), with a representative agent characterized by the utility function $\sum_{t=0}^{\infty} \delta^t c_t^{1-\alpha}/(1 - \alpha)$ and that the growth rate process $(x_t)_{t \in \mathbb{N}}$ of the aggregate endowment (i.e., $x_t = e_t/e_{t-1}$, for $t \in \mathbb{N}$) evolves as a two-state Markov chain with values h and l (with $h > l$) and transition probabilities

$$\Pi = \begin{bmatrix} \pi_{ll} & \pi_{lh} \\ \pi_{hl} & \pi_{hh} \end{bmatrix},$$

where $\pi_{lh} = \mathbb{P}(x_{t+1} = h | x_t = l)$, for all $t \in \mathbb{N}$, and similarly for the other quantities. Compute

(i) the risk free rates $r_f(l)$ and $r_f(h)$ corresponding to the two possible states of aggregate endowment growth rate;
(ii) the price of the security delivering the aggregate endowment as dividend;
(iii) the expected return of the above security.

Exercise 6.36 Under the assumptions of Sect. 6.5, suppose that the representative agent is risk neutral and consider the following securities in an infinite time horizon:

(i) a security paying a constant dividend stream $d_t = \bar{d}$ at all dates $t = 0, 1, 2, \ldots$ (in an infinite horizon). Show that the fundamental value of this security is given by $s_t^* = \frac{\delta}{1-\delta}\bar{d}$, for all $t \in \mathbb{N}$;
(ii) a security paying a constant dividend stream $d_t = \bar{d}(1)$. At some date t_0, a new dividend policy is unexpectedly being announced to start at the future date $\bar{t} > t_0$. Such a new policy consists in the distribution of the constant dividend

$d_t = \bar{d}(2)$. Show that the fundamental value of this security is given by

$$
s_t^* = \begin{cases}
\frac{\delta}{1-\delta}\bar{d}(1), & \text{if } t < t_0; \\
\frac{\delta}{1-\delta}\left(\bar{d}(1) + \delta^{\bar{t}-t}(\bar{d}(2) - \bar{d}(1))\right), & \text{if } t_0 \le t < \bar{t}; \\
\frac{\delta}{1-\delta}\bar{d}(2), & \text{if } t \ge \bar{t}.
\end{cases}
$$

Exercise 6.37 Under the assumptions of Sect. 6.5, suppose that the representative agent is risk neutral.

(i) Consider a security paying the following dividend stream:

$$
d_t = \bar{d} + \varrho(d_{t-1} - \bar{d}) + \varepsilon_t, \qquad \text{for all } t \in \mathbb{N},
$$

with $\mathbb{E}[\varepsilon_t | \mathscr{F}_{t-1}] = 0$, for all $t \in \mathbb{N}$. Show that the fundamental value of this security is given by

$$
s_t^* = \left(\frac{\delta}{1-\delta} - \frac{\delta\varrho}{1-\delta\varrho}\right)\bar{d} + \frac{\delta\varrho}{1-\delta\varrho}d_t, \qquad \text{for all } t \in \mathbb{N}.
$$

(ii) Consider a security paying the following dividend stream:

$$
d_t = \beta d_{t-1}, \qquad \text{for all } t \in \mathbb{N},
$$

for some $\beta > 1$ such that $\beta\delta < 1$. Show that the fundamental value of this security is given by

$$
s_t^* = \frac{\delta\beta}{1-\delta\beta}d_t.
$$

(iii) Consider a security paying the following dividend stream:

$$
\log d_t = \mu + \log d_{t-1} + \varepsilon_t, \qquad \text{for all } t \in \mathbb{N},
$$

for some $\mu \in \mathbb{R}$ and where $(\varepsilon_t)_{t\in\mathbb{N}}$ is a sequence of independent and identically distributed normal random variables with zero mean and variance σ^2. Show that, if $\log\delta + \mu + \frac{\sigma^2}{2} < 0$, the fundamental value of this security is given by

$$
s_t^* = \frac{\delta e^{\mu+\frac{\sigma^2}{2}}}{1 - \delta e^{\mu+\frac{\sigma^2}{2}}}, \qquad \text{for all } t \in \mathbb{N}.
$$

Exercise 6.38 Under the assumptions of Proposition 6.40, prove the following claims:

(i) a rational bubble cannot restart once it bursts, i.e., $\beta_t = 0$ implies that $\beta_{t+s} = 0$ for all $s \in \mathbb{N}$;

(ii) suppose that a price process is uniformly bounded from above, i.e., there exists a constant $K > 0$ such that $\mathbb{P}(s_t \in [0, K]) = 1$ holds for all $t \in \mathbb{N}$. Prove that $s_t^* = s_t$ holds for all $t \in \mathbb{N}$, i.e., there is no bubble component.

Exercise 6.39 Under the assumptions of Proposition 6.40, suppose that the dividend process $(d_t)_{t \in \mathbb{N}}$ is defined by

$$\log d_{t+1} = \mu + \log d_t + \varepsilon_{t+1},$$

with $\mu > 0$ and where $(\varepsilon_t)_{t \in \mathbb{N}}$ is a sequence of independent and identically distributed normal random variables with zero mean and variance σ^2. Show that there exists an (intrinsic) bubble of the form $\mathbf{u}'(e_t)\beta_t = c d_t^\lambda$ where c is a constant and λ is the solution of the equation $\lambda^2 \sigma^2 / 2 + \lambda \mu + \log \delta = 0$ (see Froot & Obstfeld [745]).

Chapter 7
Multi-Period Models: Empirical Tests

> *I believe there is no other proposition in economics which has more solid empirical evidence supporting it than the efficient market hypothesis. That hypothesis has been tested and, with very few exceptions, found consistent with the data in a wide variety of markets.*
>
> Jensen (1978)

Differently from single-period asset pricing models, which can be empirically tested only by assuming that they hold period by period, the implications of multi-period models can be directly tested on financial time series. In the present chapter, we aim at discussing some of the most important empirical findings on the properties of asset prices and returns. We will refer to the asset pricing theory described in Chap. 6 (in particular Sect. 6.4) as *classical asset pricing theory*.

Any formulation of the classical asset pricing theory necessarily consists of two essential ingredients: an equilibrium model and the information available to market participants, represented by some information flow $\mathbb{F}$. Moreover, any specification of a model requires some hypothesis on the agents' preferences and on the technology of the economy, in particular concerning the agents' utility functions, the agents' endowment and the available investment opportunities. Of course, different implications can be obtained from different specifications. Furthermore, the results obtained depend on the choice of the reference probability measure. Indeed, one can either work under the historical/statistical probability measure $\mathbb{P}$ (i.e., the probability measure describing the real economic model) or under a risk neutral probability measure $\mathbb{P}^*$. Note that, as pointed out above, heterogeneous beliefs (i.e., "private" probability measures) are allowed provided that they agree on null sets.

Under a risk neutral probability measure $\mathbb{P}^*$, the expected return of every security is equal to the risk free rate and, hence, all risk premia are null and the discounted asset price follows a martingale in case of a constant risk free rate (see Proposition 6.22). As a consequence, the discounted wealth of a self-financing portfolio is a martingale (see equation (6.54)). Under the assumption of a constant risk free rate, asset returns are not serially correlated and future returns cannot be predicted on the basis of the information currently available in the market. Indeed,

© Springer-Verlag London Ltd. 2017
E. Barucci, C. Fontana, *Financial Markets Theory*,
Springer Finance, DOI 10.1007/978-1-4471-7322-9_7

in view of equation (6.55), it holds that

$$\text{Cov}^*(r_{t+1}, r_t) = \mathbb{E}^*[r_{t+1}r_t] - \mathbb{E}^*[r_{t+1}]\mathbb{E}^*[r_t]$$
$$= \mathbb{E}^*\big[\mathbb{E}^*[r_{t+1}|\mathscr{F}_t]r_t\big] - \mathbb{E}^*[r_{t+1}]\mathbb{E}^*[r_t] = 0, \tag{7.1}$$

since $\mathbb{E}^*[r_{t+1}|\mathscr{F}_t] = r_f$, for all t, with $\text{Cov}^*(\cdot,\cdot)$ denoting the covariance under a risk neutral probability measure $\mathbb{P}^*$. Note that (7.1) relies on the assumption of a constant risk free rate r_f. If the risk free rate is time varying (recall that the risk free rate can also be a stochastic process, as long as it is predictable, see Sect. 6.4), then future expected returns are not constant and might be predictable on the basis of the current information.

Under the historical probability measure $\mathbb{P}$, property (7.1) does not hold in general. In the special case where agents are risk neutral, in equilibrium the historical probability measure $\mathbb{P}$ is also a risk neutral probability measure, with the equilibrium risk free rate being equal to the inverse of the agents' discount factor, so that the risk free rate can vary over time only if the agents' discount factor does. However, this case represents a rather special situation. In the more interesting case where agents are risk averse, the equilibrium risk free rate and the asset risk premia are typically time varying and depend on the agents' preferences and on the structure of the economy (in particular, on the endowment and on the available information). In a representative agent economy, the equilibrium risk free rate and the asset risk premia are determined by (6.71)–(6.72). It is important to remark that, under the historical probability measure, future returns could be predictable on the basis of the information available in the market and can be serially correlated. The fact that the equilibrium risk free rate and the asset risk premia vary over time is due to changes in the agents' preferences and/or to changes in the structure of the economy (e.g., set of available investment opportunities).

According to the above observations, returns should be unpredictable in a risk neutral economy or, under suitable assumptions, in an economy with risk averse agents and constant preferences and investment opportunities. In view of the above discussion, it is important to point out that a test providing evidence in favor of predictability of asset returns is not necessarily to be interpreted as an evidence against the validity of the classical asset pricing theory. Such an evidence would only mean that agents are not risk neutral and that the set of investment opportunities is changing over time.

The literature on the empirical tests of classical asset pricing theory can be classified in three main groups, depending on the type of implications investigated:

1. restrictions on the asset price-dividend-return time series;
2. restrictions on the asset risk premia, risk free rate and consumption process;
3. information in financial markets.

The first group of contributions aims at testing the implications of no-arbitrage and equilibrium conditions on financial time series (prices, dividends, returns), while the contributions of the second group mainly test the structural restrictions established in Sect. 6.4 on asset risk premia, consumption and risk free rate. The third group of contributions concerns the role of information on asset price restrictions obtained in equilibrium or under the hypothesis of absence of arbitrage opportunities. Of course, a statistical test providing negative evidence of classical asset pricing theory can be explained in two opposite ways: a) the specification of the theory is not confirmed in the real world; b) the set of information and/or the model on which the theory is being tested is not correctly specified. In the remaining part of this chapter we shall discuss empirical studies that belong to the first and to the second of the above three groups, referring to the next chapter for an analysis of the role of information in financial markets.

This chapter is structured as follows. In Sect. 7.1, we survey the empirical literature testing the existence of bubbles and the excess volatility of asset prices. Section 7.2 reviews the empirical evidence on return predictability, also discussing its relation with market efficiency. In Sect. 7.3, we provide a brief overview of the empirical evidence on the validity of equilibrium asset pricing models and discuss the equity premium puzzle as well as several related asset pricing puzzles. At the end of the chapter, we provide a guide to further readings as well as a series of exercises. In this chapter, we will mention several concepts related to the statistical analysis of time series. We refer the reader to the textbooks Campbell et al. [348], Hamilton [881], Tsay [1601] for a thorough presentation of the estimation methods and of the econometric analysis of time series.

7.1 Tests on the Price-Dividend Process: Bubbles and Excess Volatility

The first group of contributions mentioned above can be further classified into three subgroups: studies on the presence of bubbles (testing whether prices are in agreement with the fundamental solution (6.86) of the no arbitrage equation (6.85)); studies on the volatility of financial time series; studies on return predictability. In many cases, the analysis concentrates on a specific asset pricing model, typically represented by a risk neutral economy with constant investment opportunities and discount factor. Again, we want to stress that empirical evidence against the implications of a model (for instance, evidence of return predictability) does not necessarily provide evidence against classical asset pricing theory as whole, but may only indicate that risk neutrality is violated and/or that the structure of the economy is changing over time.

Before starting our survey of the empirical literature, we want to mention that the analysis of financial time series started even before the modern theory of financial markets. Indeed, the conjecture that the no-arbitrage condition implies unpredictability of price changes can be traced back to Bachelier (see Bachelier [96]). Bachelier suggested that French government bonds were characterized by an evolution which is well described by a random walk stochastic process (see Campbell et al. [348] for a detailed presentation of the random walk model). Bachelier's contribution remained unrecognized for a long time, since the academic and the financial community were skeptical about an asset pricing theory involving random processes. For a long time, financial practitioners (well represented by Graham & Dodd [818]) identified the fundamental value of an asset as the discounted flow of future dividends (*fundamentalist approach*). A stochastic approach to the analysis of financial markets was also outside the economic theory paradigm: at that time, neoclassical economics was looking for an equilibrium foundation of market prices. According to that theory, prices are only related to the technology and the preferences of the economy, without an explicit use of probabilistic tools. The seminal contribution which marks the reconciliation between the fundamentalist approach and Bachelier's intuition is Samuelson [1491], where, assuming risk neutral agents and using no-arbitrage arguments together with the law of iterated expectations, it was shown that prices follow a martingale process.

Bubble Solutions vs. Fundamental Solution

The literature on the presence of speculative bubbles (in the sense of Sect. 6.5) in financial time series is quite large (see West [1656], Flood & Hodrick [719], Gurkaynak [866] for a survey). First of all, the condition established in Tirole [1595] for the existence of a bubble in a growing economy (dynamic inefficiency) has not been confirmed empirically in developed countries (see Abel et al. [8]). Recalling that a rational bubble is a non-negative process, a consequence of the presence of a speculative bubble in financial time series is that prices grow at a rapid rate. Consequently, if the dividend grows more slowly that the long run real interest rate, then the dividend-price ratio should decrease. The empirical evidence in this direction is weak.

If dividends are generated by a linear non-stationary stochastic process, then the linearity of the no-arbitrage equation implies that, in the absence of a speculative bubble, the stock price and the dividend processes are cointegrated (see Hamilton [881, Chapter 19] for a description of cointegration). This feature of a financial time series can be tested empirically. In particular, a stationarity test based on the non-cointegration property of a bubble solution immune to the problems pointed out in Hamilton & Whiteman [882] has been proposed in Diba & Grossman [574] in a constant discount rate model, allowing for possible unobservable variables. The empirical evidence obtained is against the presence of a bubble and this result is confirmed by a large part of the subsequent literature. On the other hand, results in favor of the presence of an intrinsic bubble in stock prices time series are obtained in Froot & Obstfeld [745]. Note that stationarity tests are not able to detect periodically collapsing bubbles of the type proposed in Blanchard & Watson [248] (see Evans [656]).

The presence of a bubble has also been associated with the excess volatility phenomenon, meaning that the asset price volatility is too high and is not compatible with the no-arbitrage equation, as we are going to explain more precisely below. In Blanchard & Watson [248] and Tirole [1595], depending on the correlation of innovations of the bubble (i.e., the differences $\beta_t - \mathbb{E}[\beta_t | \mathscr{F}_{t-1}]$) with those of the fundamental solution, it is claimed that a violation of volatility bounds as observed for example in Shiller [1533] (see below) can be attributed to the presence of a bubble. Moreover, if the innovations of the bubble component are positively correlated with the innovations of the dividend process, then the presence of a bubble leads to an increase of the variance. However, in Ikeda & Shibata [997], considering an intrinsic bubble, it is shown that the volatility of a price process exhibiting a speculative bubble can be both larger or smaller than the volatility corresponding to the fundamental solution. Similarly, Drees & Eckwert [590] show that, in the absence of correlation in asset returns, the volatility of a price process exhibiting an intrinsic bubble can be either smaller or greater than the volatility of the fundamental solution depending on agent's risk attitude. Flood & Hodrick [718] and Flood et al. [720] remark that many excess volatility tests effectively embed bubbles into the null hypothesis, so that a failure of variance bounds test cannot be attributed to the presence of a speculative bubble. A test on the presence of speculative bubbles in a constant discount rate model based on the comparison of two sets of estimates of the parameters needed to calculate the expected discounted value of a given stock's dividend stream has been proposed in West [1655]. The author finds evidence in favor of the rejection of the hypothesis of absence of a speculative bubble and claims that a bubble can be the origin of the excess volatility phenomenon (see also Blanchard & Watson [248]).

A test on the presence of a bubble may be ambiguous, since it may be difficult to distinguish empirically between bubbles and other phenomena such as irrationality, noise, structural changes, non-stationarities and (rational) contribution to the asset prices of factors observed by the market participants but not by the econometrician (in this direction, see Hamilton & Whiteman [882], Flood & Hodrick [719], West [1656]). Moreover, many tests are affected by the fact that the null hypothesis is composite and the statistical power of the test is rather low. According to Hamilton & Whiteman [882], the presence of a bubble is empirically untestable and can be interpreted within the classical asset pricing framework through the presence of unobservable factors.

Excess Volatility

The literature on asset price *excess volatility* is based on the observation that the empirical volatility of asset prices is larger than what predicted on the basis of the no-arbitrage equation (6.85). The literature on this topic is quite large and originated from LeRoy & Porter [1189], Shiller [1533]. We refer the reader to LeRoy [1184], Cochrane [458], Shiller [1538], West [1656], Gilles & LeRoy [782] for good surveys on the topic.

Let us start by explaining the excess volatility phenomenon, adopting the setting of Sect. 6.5. Besides the assumptions of Sect. 6.5, assume furthermore that the representative agent is risk neutral and is characterized by a discount factor $\delta = 1/r_f \in (0, 1)$. In this case, equation (6.86) implies that the fundamental value of a security paying the dividend stream $(d_t)_{t\in\mathbb{N}}$ can be expressed as

$$s_t^* = \sum_{s=1}^{\infty} \frac{1}{r_f^s} \mathbb{E}[d_{t+s}|\mathscr{F}_t], \qquad \text{for all } t \in \mathbb{N}.$$

We denote by $(s_t^e)_{t\in\mathbb{N}}$ the *ex-post rational* (or *perfect foresight*) price process of the security, corresponding to the price of the security in the hypothetical case where the dividend stream is fully known in advance:

$$s_t^e := \sum_{s=1}^{\infty} \frac{1}{r_f^s} d_{t+s}, \qquad \text{for all } t \in \mathbb{N}. \tag{7.2}$$

By comparing the representations of s_t^* and s_t^e above, we obtain the decomposition

$$s_t^e = s_t^* + u_t, \qquad \text{for all } t \in \mathbb{N}, \tag{7.3}$$

where the process $(u_t)_{t\in\mathbb{N}}$ represents the deviation of the present value of the future dividends from its conditional expectation based on the information $\mathscr{F}_t$ available at time t and is defined by $u_t = \sum_{s=1}^{\infty} \frac{1}{r_f^s} \varepsilon_{t+s}$, where the process $(\varepsilon_t)_{t\in\mathbb{N}}$ represents the unexpected component of the variation in the stock price and is given by

$$\varepsilon_t := s_t^* + d_t - r_f s_{t-1}^*. \tag{7.4}$$

Note that $\mathbb{E}[\varepsilon_t|\mathscr{F}_{t-1}] = 0$, for all $t \in \mathbb{N}$. Moreover, since $\frac{1}{r_f}\mathbb{E}[s_t^* + d_t|\mathscr{F}_{t-1}] = s_{t-1}^*$, for all $t \in \mathbb{N}$, the process $(\varepsilon_t)_{t\in\mathbb{N}}$ is serially uncorrelated. The decomposition (7.3) of the perfect foresight price s_t^e follows by noting that, for any $t \in \mathbb{N}$,

$$\sum_{s=1}^{\infty} \frac{1}{r_f^s} \varepsilon_{t+s} = s_t^e + \sum_{s=1}^{\infty} \left(\frac{1}{r_f^s} s_{t+s}^* - \frac{1}{r_f^{s-1}} s_{t+s-1}^* \right) = s_t^e - s_t^*,$$

as long as $\lim_{T\to\infty} s_{t+T}^* / r_f^T = 0$, for all $t \in \mathbb{N}$.

Note that, in view of the definition of the process $(\varepsilon_t)_{t\in\mathbb{N}}$, it holds that

$$\mathbb{E}[u_t|\mathscr{F}_t] = \sum_{s=1}^{\infty} \frac{1}{r_f^s} \mathbb{E}[\varepsilon_{t+s}|\mathscr{F}_t] = 0,$$

meaning that the fundamental value s_t^* represents an unbiased estimate of the perfect foresight price s_t^e given the information available at time t. In particular, this implies that $\operatorname{Cov}(u_t, s_t^* | \mathscr{F}_{t-1}) = 0$, for all $t \in \mathbb{N}$, and, hence,

$$\operatorname{Var}\left(s_t^e | \mathscr{F}_{t-j}\right) = \operatorname{Var}\left(s_t^* | \mathscr{F}_{t-j}\right) + \operatorname{Var}\left(u_t | \mathscr{F}_{t-j}\right) \geq \operatorname{Var}\left(s_t^* | \mathscr{F}_{t-j}\right), \tag{7.5}$$

for all $t \in \mathbb{N}$ and $j \in \{1, \ldots, t-1\}$. Furthermore, provided that unconditional variances are well-defined, it holds that

$$\operatorname{Var}\left(s_t^e\right) = \operatorname{Var}\left(s_t^*\right) + \operatorname{Var}\left(u_t\right) \geq \operatorname{Var}\left(s_t^*\right), \qquad \text{for all } t \in \mathbb{N}. \tag{7.6}$$

The intuition behind the above inequalities is the following: the fundamental value is an unbiased forecast of the perfect foresight price and, therefore, its variance is smaller than the variance of the perfect foresight price (being a conditional expectation of the latter). Inequality (7.6) is always valid regardless of the information available to market participants and of the specific dividend process considered, under the standing assumption that the second moments exist. These features make inequality (7.6) empirically testable. However, in order to evaluate the statistical significance of a violation of inequality (7.6), one needs to specify a model for the asset dividends.

As shown in (7.6), the difference between the variance of the perfect foresight price and the variance of the fundamental value is given by $\operatorname{Var}(u_t)$. Clearly, the process $(u_t)_{t \in \mathbb{N}}$ cannot be observed in real markets. However, if the distribution of the unexpected component ε of the variation in the stock price is stationary, so that $\operatorname{Var}(\varepsilon_{t+s}) = \operatorname{Var}(\varepsilon_t)$, for all $t, s \in \mathbb{N}$, then it holds that

$$\operatorname{Var}(u_t) = \sum_{s=1}^{\infty} \frac{1}{r_f^{2s}} \operatorname{Var}(\varepsilon_{t+s}) = \frac{1}{r_f^2 - 1} \operatorname{Var}(\varepsilon_t), \qquad \text{for all } t \in \mathbb{N}. \tag{7.7}$$

It has been remarked in LeRoy & Porter [1189] that the variance of asset prices increases with the information available to market participants, while the variance of unexpected returns goes in the opposite direction. Indeed, let us consider two information flows $\mathbb{F}$ and $\mathbb{G}$ with the property that $\mathscr{F}_t \subseteq \mathscr{G}_t$, for all $t \in \mathbb{N}$ (intuitively, the information represented by $\mathscr{G}_t$ is finer than the information represented by $\mathscr{F}_t$) and let us denote by $s_t^{\mathbb{F}}$ and $s_t^{\mathbb{G}}$ the conditional expectations (with respect to $\mathscr{F}_t$ and $\mathscr{G}_t$, respectively) of the perfect foresight price, at date t. Then, as shown in Exercise 7.1, it holds that

$$\operatorname{Var}\left(s_t^{\mathbb{F}}\right) \leq \operatorname{Var}\left(s_t^{\mathbb{G}}\right) \tag{7.8}$$

and, under the assumption that the distributions of the unexpected components $\varepsilon^{\mathbb{F}}$ and $\varepsilon^{\mathbb{G}}$ are stationary, so that $\operatorname{Var}(\varepsilon_{t+s}^{\mathbb{F}}) = \operatorname{Var}(\varepsilon_t^{\mathbb{F}})$ and $\operatorname{Var}(\varepsilon_{t+s}^{\mathbb{G}}) = \operatorname{Var}(\varepsilon_t^{\mathbb{G}})$, for all $t, s \in \mathbb{N}$, it holds that

$$\operatorname{Var}\left(\varepsilon_t^{\mathbb{F}}\right) \geq \operatorname{Var}\left(\varepsilon_t^{\mathbb{G}}\right). \tag{7.9}$$

Since the rational ex-post value s_t^e given in (7.2) cannot be observed in real markets, one needs to estimate s_t^e on the basis of market data. In the seminal paper Shiller [1533], the infinite sum appearing in the rational ex-post value (7.2) is estimated as (compare with Marsh & Merton [1306])

$$\hat{s}_t^e = \sum_{s=1}^{T-t-1} \frac{1}{r_f^s} d_{t+s} + \frac{1}{r_f^{T-t}} \bar{s}_T^e, \qquad \text{for } t = 0, 1, \ldots, T-1, \tag{7.10}$$

where T represents a large enough time horizon of the dividend time series and $\bar{s}_T^e$ an arbitrary terminal value (for instance, $\bar{s}_T^e$ can be taken as the average of the historically observed prices, see Grossman & Shiller [848]). Note that this truncation of the infinite sum (7.2) does not automatically exclude the presence of a non-null bubble component. The sample variance of $\hat{s}_t^e$ is then taken as an estimator of the variance $\mathrm{Var}(s_t^e)$ of the ex-post rational price. The idea of Shiller [1533] is that, if the time series is sufficiently long, then a reasonable estimate of the variance of s_t^e can be obtained on the basis of (7.10).

In Shiller [1533], inequality (7.6) has been tested on the Standard and Poor index time series over the period 1871–1979. The author verified in the first place that the time series is characterized by a long run exponential growth rate. Removing this trend, inequality (7.6) turns out to be empirically violated: the volatility of the stock index is dramatically higher than the volatility of the rational ex-post price (*excess volatility* phenomenon). As no specific model for dividends was assumed (in other words, the test procedure is model-independent), the statistical significance of this violation of inequality (7.6) cannot be statistically assessed. Nonetheless, the remarkable difference between the two variances motivated Shiller [1533] to claim that the empirical evidence is in favor of a strong violation of the implications of classical asset pricing theory. This conclusion is reinforced by the comparison between the time series of the stock price index and that of the rational ex-post price: the latter appears to be substantially smoother and more stable than the first one. Always in Shiller [1533], alternative tests were conducted by comparing the variances of (de-trended) price innovations and the variance of the dividend process, with empirical results always in contradiction with the implications of classical asset pricing theory: stock prices appear to be excessively volatile.

In LeRoy & Porter [1189], excess volatility was also detected. The authors assumed that dividends and stock prices (suitably taking into account a trend component) are generated by a stationary bivariate linear process. Inequality (7.6) was empirically tested together with the absence of correlation between prices and return forecasting errors. In particular, inequality (7.6) is shown to hold empirically in the opposite direction and volatility bounds are empirically violated. The null hypothesis of the absence of correlation test is rejected but the statistical test leads to very wide confidence intervals and does not always lead to statistically significant results.

These results, in particular those reported in the seminal paper Shiller [1533], stimulated a large debate on the excess volatility phenomenon. First of all, the results were discussed concerning the statistical methodology employed. In particular, two statistical issues deserve attention: the hypothesis of stationarity of the price-dividend time series and the methodology used to estimate the volatility of the time series. In Flavin [715], it is observed that the variances of s_t^* and s_t^e are both estimated with a downward bias in small samples due to the presence of serial correlation and, furthermore, this bias is more pronounced for the time series of s_t^e. The bias derives from the estimation of the population mean through the sample mean and, in the case of the time series of the rational ex-post price, from the fact that s_t^e is computed as a moving average. Moreover, the methodology employed by Shiller [1533] to compute the rational ex-post price s_t^e introduces a further downward bias (see Gilles & LeRoy [782]). This remark is important since such a downward bias can be large enough to provide a potential explanation for the apparent violation of the volatility bounds, with the effect that variance bound tests tend to be biased in favor of a rejection. Moreover, as we have already pointed out, without specifying the model generating the dividends, it is difficult to quantify the bias and, therefore, to take it properly into account. An additional problem is represented by the presence of *nuisance parameters*, due to the fact that inequality (7.6) holds independently of the agents' information. The sample distribution of the test statistic is affected by parameters which are unrestricted under the null hypothesis of validity of the variance bounds. The agents' information turns out to be relevant in order to establish a critical value of the volatility bound test. As a consequence, unless the agents' information is precisely defined, it is not possible to evaluate the statistical relevance of a violation to the volatility bounds (see, e.g., LeRoy & Parke [1188] and LeRoy & Steigerwald [1190]).

Kleidon [1104] deeply criticized the results reported in Shiller [1533]. His criticism concerns two main issues: (i) the distinction between the relative smoothness of the time series of s_t^e and s_t^* and the variance inequality (7.6); (ii) the stationarity of the dividend process (and note that both criticisms are not concerned with the sample size). First of all, Kleidon [1104] points out that the procedure adopted in Shiller [1533] to estimate the rational ex-post price s_t^e (see equation (7.10)) relies on ex-post information which is available only after the date when prices are set (uncertainty in future dividends) and, therefore, the estimate of the conditional variance of the ex-post rational price can be biased. If dividends follow an autoregressive linear process (either stationary or non-stationary, see Hamilton [881] for a detailed presentation) and conditional variances are calculated properly, then inequality (7.5) is satisfied and the pattern of the rational ex-post price $\hat{s}_t^e$ calculated according to the Shiller procedure can be smoother than that of the price s_t^* computed according to the fundamental no-arbitrage solution. In particular, this smoothness effect is obtained in the presence of a sufficiently strong autoregressive component. Moreover, Kleidon [1104] shows that, if dividends follow a geometric random walk (so that log-returns are non-stationary), then the time series of the rational ex-post price is always smoother than that of the fundamental no-arbitrage price and inequality (7.5) between the conditional variances is satisfied (even if the unconditional variances cannot be defined). This observation leads to the conclusion that there is no direct relation between the relative smoothness of the two time series and the variance inequalities (7.5)–(7.6). Moreover, a variance inequality has to be regarded as a cross-sectional evaluation and, therefore, one simple plot of the two time series and a comparison of their smoothness is rather uninformative. In other words, ex-post one does only observe one of the many ex-ante possible realizations of the economy and, therefore, one cannot look at different values of s_t^e, each corresponding to a different realization, to test whether inequalities (7.5)–(7.6) are satisfied. This observation becomes particularly critical when the dividend time series is non-stationary. Kleidon [1104] relies on this argument to explain the results in Shiller [1533], arguing that a geometric random walk for dividends is a good model for the time series at hand. In this case, unconditional variances are not well-defined and the use of sample variances as estimators is invalid, so that the test procedure adopted by Shiller [1533] is uninformative. Moreover, Kleidon [1104] shows by means of a Monte Carlo analysis that Shiller's procedure for the computation of the unconditional variance on a small sample assuming a geometric random walk process (with a normally distributed noise term) for dividends leads to an empirical violation of inequality (7.6). Such a violation occurs in more than 70% of the cases and in most of the cases the size of the violation is similar to that detected by Shiller, while a conditional variance test for the Standard and Poor time series is not violated. This different behavior should be attributed to the non-stationarity of the de-trended time series investigated by Shiller. According to Kleidon [1104], an unconditional variance test is uninformative, since the unconditional variance is not well-defined, while the conditional variance inequality (7.5) is satisfied.

Shiller [1537] replied to the above arguments by observing that the assumptions adopted by Kleidon [1104] produce unrealistic price-dividend ratios and by showing (through Monte Carlo simulations, as in Kleidon [1104]) that violations with a size equal to that detected in Shiller [1533] happen with small probability (smaller than 1%) in the case of a log-normal process of the dividends with realistic values for the price-dividend ratio and in the case of other non-stationary dividend processes.

Summing up, we can affirm that an empirical violation of inequality (7.6) according to the test procedure proposed by Shiller is in favor of a rejection of classical asset pricing theory (due to the excess volatility phenomenon) if the dividend process, once the trend component is removed, is assumed to be stationary. However, the same conclusion cannot be drawn if de-trended dividends (or log-dividends) are non-stationary (*random walk process*), because in that case the unconditional variance is not well-defined. In the latter case, the empirical variances estimated on small samples often do not satisfy inequality (7.6).

The first generation of excess volatility tests suffered for small sample biases and non-stationarity problems. To overcome these deficiencies, a second generation of volatility tests has been proposed in the literature, see in particular Mankiw et al. [1294], West [1657], Campbell & Shiller [351], Mankiw et al. [1295]. In Mankiw et al. [1294], a "naive forecast" of the future discounted dividend stream is considered and a test is proposed which does not suffer for small sample biases and does not depend on the assumption of stationarity of the dividend time series. Suppose that, at each date $t \in \mathbb{N}$, one uses a subset of the available information to make a "naive forecast" (which may not be a rational forecast) of future dividends and denote by $F_t(\cdot)$ the corresponding forecast operator, so that $F_t(d_{t+s})$ represents the naive forecast at date t of the dividend paid at the future date $t + s$. Define then

$$s_t^f := \sum_{s=1}^{\infty} \frac{1}{r_f^s} F_t(d_{t+s}), \qquad \text{for all } t \in \mathbb{N}.$$

Similarly as in the case of inequality (7.5), the following inequalities hold by definition, as shown in Exercise 7.2 (see also Mankiw et al. [1294]):

$$\mathbb{E}[(s_t^e - s_t^f)^2 | \mathscr{F}_t] \geq \mathbb{E}[(s_t^e - s_t^*)^2 | \mathscr{F}_t] \quad \text{and} \quad \mathbb{E}[(s_t^e - s_t^f)^2 | \mathscr{F}_t] \geq \mathbb{E}[(s_t^* - s_t^f)^2 | \mathscr{F}_t].$$
(7.11)

Moreover, by the tower property of the conditional expectation, the above inequalities also imply that

$$\mathbb{E}[(s_t^e - s_t^f)^2 | \mathscr{H}_t] \geq \mathbb{E}[(s_t^e - s_t^*)^2 | \mathscr{H}_t] \quad \text{and} \quad \mathbb{E}[(s_t^e - s_t^f)^2 | \mathscr{H}_t] \geq \mathbb{E}[(s_t^* - s_t^f)^2 | \mathscr{H}_t],$$

for any subset $\mathscr{H}_t \subset \mathscr{F}_t$ of the information available at date t, for every $t \in \mathbb{N}$ (in particular, one can take $\mathscr{H}_t = \mathscr{F}_{t-\delta}$, for some $\delta \in \mathbb{N}$). In particular, as long as the above conditional expectations are taken with respect to the information available a finite amount of time before date t, the non-stationary of the dividend time series does not pose problems concerning the existence of the conditional expectations. The above inequalities mean that the market price is a better forecast (in the mean-square sense) of the perfect foresight price (which is constructed as in Shiller [1536]) than the naive forecast. In their empirical application, Mankiw et al. [1294] suppose that the naive forecast is completely myopic, in the sense that future dividends are naively forecasted to be always equal to the current dividend, and show that the above inequalities are violated, but do not discuss the statistical significance of such a violation. In Mankiw et al. [1295], a Monte Carlo sampling distribution of the test is proposed and it is shown that, assuming a constant discount factor smaller than 6%, the violation of the inequality turns out to be statistically significant, while assuming higher discount factors leads to violations that are not necessarily statistically significant.

In Blanchard & Watson [248] and West [1657], a test based on the information available to market participants is proposed. For every date $t \in \mathbb{N}$, consider two information flows $\mathbb{F} = (\mathscr{F}_t)_{t \in \mathbb{N}}$ and $\mathbb{G} = (\mathscr{G}_t)_{t \in \mathbb{N}}$ such that the first one contains less information than the second one (i.e., $\mathscr{F}_t \subset \mathscr{G}_t$, for all $t \in \mathbb{N}$) and let $s_t^{\mathbb{F}}$ and $s_t^{\mathbb{G}}$ be the rational forecasts of s_t^e based on the two information sets $\mathscr{F}_t$ and $\mathscr{G}_t$, respectively, for all $t \in \mathbb{N}$. Assuming the stationarity of the forecasting errors with respect to both information flows $\mathbb{F}$ and $\mathbb{G}$, the following inequality holds (see Exercise 7.3):

$$\mathbb{E}[(s_t^{\mathbb{F}} + d_t - \mathbb{E}[s_t^{\mathbb{F}} + d_t | \mathscr{F}_{t-1}])^2] \geq \mathbb{E}[(s_t^{\mathbb{G}} + d_t - \mathbb{E}[s_t^{\mathbb{G}} + d_t | \mathscr{G}_{t-1}])^2]. \qquad (7.12)$$

A similar inequality can be established about the return variance (both conditional and unconditional), see LeRoy & Porter [1189]. In West [1657], an empirical test of inequality (7.12) has been proposed. In particular, such a test does not require the knowledge of the rational ex-post (perfect foresight) price s_t^e and does not depend on the stationarity of the dividend process. The two information sets $\mathscr{G}_t$ and $\mathscr{F}_t$ are respectively represented by the information available in the market at date t and by the observation of the time series of the dividends. Inequality (7.12) is empirically violated, with the violation being highly statistically significant. Moreover, Monte Carlo simulations show that the test does not suffer for small sample biases. However, Ackert et al. [13] remark that earlier studies have only considered ordinary cash dividends, neglecting the effects of share repurchases and takeover distributions. Taking into account these phenomena, the variance bounds are empirically satisfied and no sign of excess volatility is detected (see Ackert et al. [13]).

Summing up, the second generation of volatility tests provided a general evidence of excess volatility, with excess volatility being detected also when the dividend time series is allowed to be non-stationary. However, the statistical significance of the results is not always strong.

A Critical Assessment

On the basis of the results reported above, a large debate started in the literature on the interpretation of the excess volatility phenomenon. In this regard, one can distinguish three main lines of thought: (i) studies pointing out statistical problems; (ii) studies allowing for non-constant investment opportunities and (iii) studies supportive of approaches going beyond classical asset pricing theory. In particular, the contributions from the second group propose an explanation of the excess volatility phenomenon within the realm of the classical asset pricing theory, while the contributions from the third group suggest that the excess volatility originates from some form of market irrationality which is not taken into account by classical asset pricing theory (behavioral finance).

Remaining in the context of classical asset pricing theory, non-constant investment opportunities (non-constant expected returns) can be introduced by assuming a risk averse representative agent. Indeed, in view of equation (6.71), a constant risk free rate is obtained in the case of a risk neutral representative agent or when the aggregate endowment process is constant over time. Risk aversion has been introduced in LeRoy & LaCivita [1187], Grossman & Shiller [848], West [1657]. As shown in Sects. 6.4–6.5, classical asset pricing theory establishes that the equilibrium price of an asset at date t is equal to the discounted expected value of its future dividends (together with a possible bubble component), with the (stochastic) discount factor at date t for consumption at the future date $t + s$ being given by the representative agent's marginal rate of intertemporal substitution between dates $t + s$ and t. In particular, this implies that the discount factor is stochastic if the representative agent is not risk neutral (in which case the discount factor would be simply given by $1/r_f$, see equation (6.71)).

In the presence of a risk averse representative agent, equation (6.69) suggests that the variability of the ex-post rational price (i.e., perfectly knowing the whole future dividend stream) could happen to be larger than that of an ex-post rational price corresponding to a constant discount factor. In other words, the risk aversion of the representative agent can lead to more volatile asset prices. The argument can intuitively be explained as follows. In a pure exchange representative agent economy, in equilibrium the representative agent consumes the aggregate endowment of the economy (no-trade equilibrium). Of course, in equilibrium the price system should be compatible with this choice. In general, a risk averse agent would like to smooth consumption over time. However, this cannot be realized at the aggregate level. Hence, equilibrium prices need to be such that the representative agent optimally chooses not to smooth consumption and simply consumes the aggregate endowment at each date. In order to induce the representative agent not to buy stocks in good periods and sell stocks in bad periods, stock prices must be procyclical and, therefore, highly volatile (see, e.g., LeRoy [1184]). Assuming a power utility function, in Grossman & Shiller [848] it has been observed that the variability of rational ex-post prices increases with the agent's risk aversion, with asset prices being supposed to be coherent with (6.69). Assuming a coefficient of relative risk aversion equal to four, a time series of the rational ex-post price compatible with the empirical time series can be reproduced. These results have been also confirmed in Cochrane & Hansen [467], where it is shown that the time variability of expected returns allows to explain excess volatility if agents are strongly risk averse or if their preferences are characterized by a habit formation process. However, as the consumption time series are relatively smooth over time, the risk aversion coefficient should be very high in order to reconcile classical asset pricing theory with the empirical data: this observations is consistent with the *equity premium puzzle* (see below). In addition, Mehar & Sah [1321] show that fluctuations in the subjective discount factor and in the attitude towards risk may also explain the excess volatility phenomenon.

A methodology that allows to investigate the presence of excess volatility with time varying expected returns was developed in Campbell & Shiller [351, 352] (see also Cochrane [465, Chapter 20]). The methodology is based on a simple log-linear approximation. For all dates $t \in \mathbb{N}$, let $R_{t+1} := \log((s_{t+1} + d_{t+1})/s_t)$ be the logarithmic return of the asset. Then the following log-linear approximation of the logarithmic return (centered on the logarithm of the average value of the dividend-price ratio) can be obtained (see Exercise 7.4):

$$R_{t+1} \approx k + \rho \log s_{t+1} + (1 - \rho) \log d_{t+1} - \log s_t, \tag{7.13}$$

where

$$\rho := 1/\big(1 + \exp(\log d - \log s)\big) \quad \text{and} \quad k := -\log(\rho) - (1 - \rho)\log(1/\rho - 1).$$

In particular, when the dividend-price ratio is constant, then $\rho = s_t/(s_t + d_t)$, for all $t \in \mathbb{N}$, with $\log(s/d)$ representing the average value of the logarithm of the price-dividend ratio. The precision of this approximation is rather satisfactory and, moreover, it holds as an exact relation if the dividend-price ratio does not vary over time.

Iterating equation (7.13) and assuming that a transversality condition of the type $\lim_{\tau \to \infty} \rho^\tau (s_{t+\tau} + d_{t+\tau}) = 0$ holds, the following approximation for the logarithm of the price-dividend ratio is then obtained:

$$\log d_t - \log s_t \sim -\frac{k}{1-\rho} + \sum_{s=0}^{\infty} \rho^s (R_{t+1+s} - \Delta \log d_{t+1+s}), \qquad (7.14)$$

with $\Delta \log d_{t+1+s} := \log d_{t+1+s} - \log d_{t+s}$, for all $t, s \in \mathbb{N}$. Approximation (7.14) holds ex-post but, by taking the $\mathscr{F}_t$-conditional expectation on the right-hand side of (7.14), we can obtain an analogous ex-ante relation between the logarithm of the price-dividend ratio and the forecast at date t of future changes in dividends and returns, so that the logarithm of the price-dividend ratio turns out to be approximately equal to the conditional expectation of future discounted (logarithmic) returns and of future dividend growth rates. As a consequence, the dividend-price ratio will be large if dividends grow slowly or if future returns are forecasted to be large. Note, however, that expected dividend growth and expected returns are positively correlated, thus yielding an offsetting effect on the logarithm of the dividendŰ-price ratio (see Lettau & Ludvigson [1196]).

Exploiting the linearity of the above approximation, a vector autoregression (see Hamilton [881, Chapter 11]) has been estimated in Campbell & Shiller [351] for the logarithm of the dividend-price ratio, the first difference of the logarithm of dividends and the logarithm of earnings relative to price. The restrictions on the regression coefficients implied by the approximate relation (7.14) are empirically rejected, thus showing that the dividend-price ratio time series do not agree with (several versions of) this model. The test procedure employed in Campbell & Shiller [351] does not suffer from possible non-stationarity problems. Moreover, Monte Carlo simulations show that the test works well in small samples. Excess volatility is detected assuming a constant or a variable expected rate of return (a constant plus the expected real return on commercial paper).

Some authors have pointed out that allowing for time varying expected returns does not suffice to explain the excess volatility phenomenon. Indeed, as shown in Shiller [1533], Mankiw et al. [1295], Campbell & Shiller [351], variance inequalities of asset prices are empirically violated even under the assumption of non-constant expected returns (estimated as the risk free rate plus a time varying risk premium component), thus proving that time varying discount factors only explain part of the excess volatility phenomenon. To reproduce the asset price volatility observed historically, discount rates should have a much higher standard deviation than what is observed in real time series (see, e.g., Shiller [1533], West [1657], Poterba & Summers [1430]).

According to the fundamental solution (see equation (6.86)), changes in asset prices are due to changes in expected dividends and in the discount rates. This basic observation motivated several studies on the relationship between ex-post changes in economic fundamentals and ex-post changes in stock price-returns. In Roll [1459], the time series of the orange juice future contract has been analysed. Due to the specific characteristics of this commodity, its fundamental is well proxied by the weather conditions in Florida. As a consequence, the price of the future contract should reflect the predictable part of weather patterns and, therefore, future prices should change in response to predictable changes in weather patterns. Instead, the author finds that prices react to unanticipated weather changes, but these unexpected changes only explain a small portion of the daily price variation and, actually, around 90% of the daily price variability cannot be explained by economic fundamentals. Similar results have been obtained for stock returns in Roll [1460], taking into account economy, industry and firm specific factors. Considering both CAPM and multi-factor models and including an industry factor, less than 35% of the daily/monthly variations of individual stock returns can be explained in terms of changes of economic fundamentals. Similar results have been obtained in Campbell & Shiller [351], Cutler et al. [510].

In Campbell [331], Campbell & Ammer [338], Campbell [334], Cochrane [463], by relying on the log-linear approximation (7.13), it has been shown that the variance of aggregate unexpected returns can be explained in terms of the variability of expected returns-risk premia (expected return news) rather than of the variability of future cash flows (dividend news). The opposite holds for individual firms (see Vuolteenaho [1633]) and for the book to market ratio (see Cohen et al. [472]). Only a small fraction of the variation in book-to-market ratios is explained by variations in expected returns. Campbell & Shiller [351] show that the dividend-price ratio predicts future expected returns rather than dividend changes. These results confirm that expected returns are time varying and that their variability over time is important in order to explain price/return movements.

Summing up, we observe that a large part of the excess volatility tests has dealt with models with a constant discount factor (risk neutral agents), showing that this assumption does not allow to explain the magnitude of the variability observed in financial time series. This empirical evidence can be partly explained within the context of classical asset pricing theory by introducing risk aversion and time varying expected returns. However, in order to explain the volatility of financial markets, the recent literature has shown that either agents exhibit a very strong risk aversion or discount rates are much more variable than what empirically observed. As suggested by Fama [663] and Cochrane [458, 461], returns are predictable and we do not have a good model to explain their dynamics: this is the origin of excess volatility results.

An alternative interpretation of excess volatility has been proposed in Shiller [1535]. According to his interpretation, the excess volatility phenomenon is due to the inefficiency of the market: in particular, there are some forms of irrationality (noise traders, feedback trading, irrational expectations) in the market that make market prices deviate from rational prices (we will return to this topic in Chap. 9).

As a matter of fact, regardless of the interpretation of the results reported above, excess volatility test rejections are strictly related to the predictability of future returns (see, e.g., Campbell & Shiller [352]). Indeed, both volatility tests and absence of correlation tests are based on the fact that u_t and s_t^* are uncorrelated (see equation (7.3)). If this is not the case, then future returns can be (at least partially) predicted through the available information. A volatility bound violation may be due to the fact that s_t^* and future returns are negatively correlated. With some caution, this result can be interpreted as an evidence of negative serial correlation of returns (see LeRoy [1184]). Indeed, referring to the decomposition (7.3), the presence of excess volatility can be interpreted in terms of negative correlation between a weighted average of past returns (the quantity s_t^*) and a weighted average of future returns (the quantity u_t).

7.2 Tests on the Price-Dividend Process: Return Predictability

In the '80s, the literature on excess volatility animated the debate on the *efficient market hypothesis*. Afterwards, the debate moved to the more general theme of asset return predictability (we refer to LeRoy [1184], Fama [663], Kaul [1076] for good surveys on the topic). As we have seen above, return predictability is intrinsically related to the excess volatility phenomenon.

According to the classical asset pricing theory as presented in Sect. 6.4, returns are unpredictable under a risk neutral probability measure or, under the historical probability measure, in the presence of risk neutral agents (with a constant discount factor). Indeed, in view of Proposition 6.35, under one of these two assumptions, the conditional expected return of any asset (or portfolio of assets) is simply equal to the risk free rate, so that the information available at date t (represented by $\mathscr{F}_t$) does not yield any useful information for predicting future returns. However, as we have already pointed out, in the absence of one of these hypotheses, classical asset pricing theory does not necessarily imply that returns are unpredictable. It is important to emphasize this point since return predictability has often been erroneously interpreted as an evidence against market efficiency/classical asset pricing theory.

As before, we assume that a risk free asset is available in the market, yielding the constant risk free rate $r_f > 0$. We also assume that agents are risk neutral (a similar analysis can be performed with respect to a risk neutral probability measure). Denoting by r_t the return of an asset over the time period $[t-1, t]$, we let $z_t := r_t - r_f$ denote the excess return of the asset over the same time period, for $t \in \mathbb{N}$. Risk neutrality together with Proposition 6.35 implies that

$$\mathbb{E}[z_{t+1}|\mathscr{F}_t] = 0, \qquad \text{for all } t \in \mathbb{N}. \tag{7.15}$$

In other words, the excess return of any asset represents a *fair game*: at every date t, the $\mathscr{F}_t$-conditional expectation of the asset excess return is simply equal to zero.

Equation (7.15) implies an ex-ante restriction on the asset prices time series, which translates into an ex-post relation of the following form:

$$r_{t+1} = r_f + v_{t+1}, \tag{7.16}$$

where $(v_t)_{t\in\mathbb{N}}$ is a sequence of random variables such that $\mathbb{E}[v_{t+1}|\mathscr{F}_t] = 0$, for all $t \in \mathbb{N}$. Relation (7.16) is related to a *random walk* process. Following Campbell et al. [348, Chapter 2], three different types of random walk processes can be identified depending on the features of the sequence $(v_t)_{t\in\mathbb{N}}$: (1) a random walk with independent and identically distributed innovations; (2) a random walk with independent innovations; (3) a random walk with serially uncorrelated innovations. Clearly, a random walk of the first type is also a random walk of the second type and, similarly, a random walk of the second type is also a random walk of the third type. Note also that a random walk of the second type allows for heteroskedastic return time series (i.e., time varying volatility, see Campbell et al. [348] and Tsay [1601]) and a random walk of the third type allows for conditional heteroskedasticity in asset returns. Time varying volatility in financial time series is a widely documented phenomenon (see Schwert [1509]). In particular, conditional heteroskedasticity is typically evident as volatility is a highly persistent phenomenon with clustering patterns.[1]

The fundamental conditions (7.15)–(7.16) only require the return process $(r_t)_{t\in\mathbb{N}}$ to follow a random walk of the weakest form, with serially uncorrelated innovations. In particular, (conditional) heteroskedasticity is perfectly compatible with the hypothesis of efficient markets in the form (7.15)–(7.16).

Condition (7.15) captures the notion of returns unpredictability and has important implications on the properties of asset returns time series. Indeed, (7.15) shows that the information contained in $\mathscr{F}_t$ does not allow to make predictions about future returns and, as a consequence, returns cannot be serially correlated (recall that, by definition, the random variable r_t is $\mathscr{F}_t$-measurable). Moreover, as shown in the next proposition (see Exercise 7.6 for a proof), there does not exist a trading strategy which yields a non-null excess expected return with respect to any information set contained in the market information $\mathscr{F}_t$.

Proposition 7.1 *Under the assumptions of Sect. 6.4, suppose furthermore that agents are risk neutral. Let $\mathbb{K} = (\mathscr{K}_t)_{t\in\mathbb{N}}$ be an information flow contained in the market information flow $\mathbb{F} = (\mathscr{F}_t)_{t\in\mathbb{N}}$ (i.e., $\mathscr{K}_t \subseteq \mathscr{F}_t$, for all $t \in \mathbb{N}$) and such that s_t^n and d_t^n are $\mathscr{K}_t$-measurable, for all $n = 1,\ldots,N$ and $t \in \mathbb{N}$. Then the following hold:*

---

[1]This phenomenon is particularly relevant in high frequency data and can be described by relying on ARCH and GARCH models (see Akgiray [36], Pagan & Schwert [1392], Pagan [1391], Campbell et al. [348] and Tsay [1601] for a textbook presentation of ARCH and GARCH models in finance).

(i) returns are not serially correlated conditionally on the information $\mathbb{K} = (\mathscr{K}_t)_{t \in \mathbb{N}}$, i.e., $\mathrm{Cov}(r_{t+s}, r_t | \mathscr{K}_k) = 0$ for all $k, t, s \in \mathbb{N}$;

(ii) for any $t \in \mathbb{N}$, the value of any $\mathscr{K}_t$-measurable portfolio strategy $\theta_{t+1} \in \mathbb{R}^{N+1}$ is given by its market value $\sum_{n=0}^{N} \theta_{t+1}^n s_t^n$ and the $\mathscr{K}_t$-conditional expected excess return of such a strategy is null.

The above proposition provides a theoretical foundation for the *efficient market hypothesis*: under the assumption of risk neutral agents (or, under a risk neutral probability measure), it is not possible to "beat the market" (excess return greater than zero) by trading on the basis of any information set contained in the available market information, since the (conditional) expected return of any portfolio is equal to the risk free rate. Of course, since the risk free rate r_f is assumed to be deterministic, asset returns are not serially correlated if and only if the asset risk premium is not correlated with past returns.

In view of the above observations (see in particular part *(i)* of Proposition 7.1), a straightforward test of classical asset pricing theory consists in verifying the absence of serial correlation in asset returns. However, Summers [1578] has shown that this type of test has a rather low power against the alternative hypothesis of a persistent autoregressive component in stock prices, so that the inability of such a test to reject market efficiency does not provide a strong evidence for its validity.

More powerful tests have been proposed exploiting the linearity of the variance with respect to the return time horizon. For instance, as suggested in Cochrane [457], if the log-price is described by a first-difference stationary linear process, where the error terms are independent and identically distributed with constant variance σ_ϵ^2, then the variance of the k-period return divided by k should be equal to the constant term σ_ϵ^2 (which also represents the variance of single period returns). Testing this property leads to the so-called *variance ratio tests*, see Lo & MacKinlay [1233, 1234] and Campbell et al. [348]. Variance ratio tests are strictly related to serial correlation tests: indeed, the variance of the k-period return divided by the variance of the single period return is equal to one plus a positively weighted sum of return autocorrelations (see Cochrane [457, Appendix A]).

Depending on the information set considered (corresponding to the information flow $\mathbb{K} = (\mathscr{K}_t)_{t \in \mathbb{N}}$ considered in Proposition 7.1), the notion of *market efficiency* takes three possible forms: *weak* market efficiency ($\mathscr{K}_t$ only contains the information of past and current market prices), *semi-strong* market efficiency ($\mathscr{K}_t$ contains all public information available at date t) and *strong* market efficiency ($\mathscr{K}_t$ contains all possible information, including private information). In this section, we shall limit our attention to the restrictions imposed by classical asset pricing theory on the price-dividend processes and, hence, on the weak and semi-strong forms of market efficiency, referring to Chap. 8 for a discussion of the role of private information in financial markets. In particular, Proposition 7.1 implies that future excess returns cannot be predicted on the basis of current and past returns or on the basis of any information set $\mathscr{K}_t$ contained in the set $\mathscr{F}_t$ of available market information (including public information such as macroeconomic factors and accounting variables but excluding private information). According to the above classification of market efficiency, this corresponds to the weak and semi-strong forms of market efficiency.

As reported in the seminal paper Fama [661], the first empirical results provided positive evidence in favor of the efficient market hypothesis. In particular, no serial correlation was detected in the return time series. Fama [660] showed that first order serial correlation of daily/monthly returns is positive, but in general only a weak significance of serial correlation was detected. In the same years, negative empirical evidence was only reported for the strong form of market efficiency. The analysis of trading rules typically adopted by practitioners also provided support for the efficient market hypothesis. For instance, Fama & Blume [665] analysed a filter trading strategy, consisting in buying an asset when its price rises by a fixed percentage and selling it when its price drops by the same percentage. Analysing this type of trading strategy, it was shown that a filter strategy with a reference percentage comprised between 0.5 and 1.5 was able to produce excess returns and outperformed simple buy-and-hold strategies by trading on the basis of very short term price swings. However, the presence of small transaction costs suffices to eliminate the excess profits generated by such a strategy.

In the '80s, the debate on market efficiency blazed up and many recurrent anomalies were reported in the literature. Let us first discuss the so-called *seasonality effects*, consisting in some calendar anomalies depending on the frequency of the considered time series. In monthly time series, a *January effect* was observed: stock returns are higher in January than in other months, especially in the case of small capitalization stocks, see Rozeff & Kinney [1481], Roll [1458], Keim [1079, 1080], Reinganum [1444]. The January effect was shown to be robust with respect to different time horizons and different markets. A possible explanation of the January effect comes from the turn-of-the-year tax related trading (tax loss selling), see Constantinides [490], Poterba & Weisbenner [1431], Grinblatt & Moskowitz [833]. On the other hand, on daily time series, a *weekend effect* was observed: returns are on average negative from the closing of the trading activities on Friday to the opening of trading activities on Monday, see French [739], Keim & Stambaugh [1082], Jaffe & Westerfield [1008]. Moreover, returns are typically higher on the day preceding a holiday, on the last day of a month and before the end of a year (see Lakonishok & Smidt [1160]). Some of these phenomena are due to a *window dressing* behavior by institutional investors: often, before reporting dates, the composition of a portfolio is changed in order to improve the appearance of the portfolio managed by the investor.

The results obtained in the '70s providing positive evidence in support of return unpredictability mostly considered short term returns. Afterwards, in the '80s, predictability patterns were reported in many studies based on returns over longer time horizons. One of the first papers casting doubts about the efficient market hypothesis in the finance community was De Bondt & Thaler [532], where the authors showed that portfolios composed by extreme "winner" stocks over the past three to five years (i.e., portfolios composed by stocks which exhibited in the past three to five years a return higher than the market return) have a relatively poor performance and, moreover, are outperformed by portfolios composed by extreme "loser" stocks. The excess return difference between these two portfolios in three years is around 25%: the average return on "winner" portfolios is about 5% less than the market return, whereas the average return on "loser" portfolios is about 20% more than the market return. These results agree with CAPM anomalies associated with size, earnings-price ratio and book-to-market ratio (see Sect. 5.3), i.e., firms with high ratios tend to be losers in the past (value stocks, see De Bondt & Thaler [533]). The authors attributed these results to *market irrationality* and, more specifically, to an *overreaction* phenomenon: market participants tend to overreact to unexpected and dramatic events and the Bayes rule is typically not applied in practice since agents tend to overweight recent observations (see also Kahneman & Tversky [1061] and Chap. 9). These results suggest that asset prices cannot be well represented by a random walk process: there is a mean reversion effect and stock prices may contain autoregressive components with a small decay rate.

In Summers [1578], it was shown that serial correlation tests based on short term returns have a very low statistical power to discriminate the hypothesis that prices depart from their fundamental value by exhibiting a temporary component with slow decay. The model alternative to market efficiency proposed in Summers [1578] for asset prices is formulated as follows:

$$s_t = s_t^* + u_t,$$

$$u_t = \alpha u_{t-1} + v_t,$$

for all $t \in \mathbb{N}$, where s_t^* denotes the asset fundamental value (see expression (6.86)) and where $0 \leq \alpha \leq 1$ (so that deviations from the fundamental value persist but do not grow forever) and $(v_t)_{t\in\mathbb{N}}$ is a sequence of independent random variables with zero mean and constant variance. In this model, $\alpha = 0$ is the null hypothesis, representing market efficiency (note also that this model well represents Shiller's suggestion according to which an asset price is given by the sum of the fundamental value and of an autoregressive component describing investors' irrationality). When α is close to one (representing persistent price departures), it is unlikely that serial correlation tests of monthly returns will reject the null hypothesis, without being able to really discriminate the existence of the inefficient component u_t (using daily returns does not alter the conclusion). Furthermore, since it is difficult to detect mean reversion phenomena in short horizon returns, results showing the absence of serial correlation in returns should be interpreted with caution.

The debate on the presence of serial correlation in asset returns was animated by the results obtained in Poterba & Summers [1430], Fama & French [666], Lo & MacKinlay [1233]. In Poterba & Summers [1430], Fama & French [666], it has been shown that returns computed on a horizon longer than one year are negatively serially correlated, while in Lo & MacKinlay [1233] positive serial correlation was detected for weekly and monthly returns. The methodology employed in Poterba & Summers [1430], Lo & MacKinlay [1233] was based on a variance ratio test, while in Fama & French [666] the test directly concerned serial correlation in the asset return time series.

Elaborating on the model of Summers [1578], the following model has been proposed in Fama & French [666] for the evolution of log-prices:

$$\log s_t = q_t + z_t,$$

$$q_t = \mu + q_{t-1} + \eta_t,$$

$$z_t = \psi z_{t-1} + \epsilon_t,$$

for all $t \in \mathbb{N}$, where $(\eta_t)_{t\in\mathbb{N}}$ and $(\epsilon_t)_{t\in\mathbb{N}}$ are two *white noise* sequences (i.e., sequences composed of independent random variables with zero mean). According to the above model, the log-price is decomposed into the sum of two components: a random walk component $(q_t)_{t\in\mathbb{N}}$, with the parameter μ representing a drift parameter, and an auto-regressive component $(z_t)_{t\in\mathbb{N}}$ with parameter $\psi < 1$ (to ensure stationarity). The stationary component $(z_t)_{t\in\mathbb{N}}$ represents the deviations of the log-price from the fundamental value, deviations which tend to disappear as time goes on (mean reversion effect). The mean reversion of the stationary component $(z_t)_{t\in\mathbb{N}}$ induces a negative autocorrelation in log-returns. Note, however, that positive serial correlation in log-returns cannot be generated by the model proposed in Fama & French [666]. More precisely, if the asset log-price time series is generated according to the above model, then the serial correlation of log-returns computed for a small horizon is almost null and instead approaches the value -0.5 for large time horizons. The authors tested this model by estimating the serial correlation of log-returns of some asset portfolios (based on industry and size variables) in correspondence of different time horizons in the period 1926–1985. While serial correlation of one year returns is found to be negligible, negative serial correlation is observed for horizons longer than one year. More specifically, plotting the values of the estimated serial correlation coefficient in correspondence of increasing return horizons reveals a U-shaped pattern, with the highest degree of correlation being obtained in correspondence of a three to five years horizon. This pattern is consistent with the hypothesis that log-prices have a slowly decaying stationary component, while the random walk component dominates in long horizon returns. In the case of three to five year returns, between 30% and 45% of the variance of log-returns is generated by the stationary mean reverting component $(z_t)_{t\in\mathbb{N}}$. Moreover, return predictability turns out to be more significant for small capitalization firms than for large capitalization firms. However, the negative autocorrelation is shown to be significantly weaker in the post-Second World War period and does not show the U-shaped pattern in the overall 1926–1985 period. Since the sample size is relatively small in the case of long horizon returns, the authors establish the significance of their results by means of Monte Carlo simulations, mimicking the properties of the observed log-returns time series. Fama & French [666] explain these empirical results through time varying equilibrium expected returns generated by changing investment opportunities and not through market irrationality. Similar results have been obtained in Poterba & Summers [1430] by relying on a variance ratio test, with the main difference being that the highest degree of mean reversion is observed for a six to eight years horizon. Long horizon negative serial correlation is also detected in Cutler et al. [510]. In particular, these studies show that the presence of serial correlation in asset returns is not observed if one does only look at short term returns as previously considered in many tests of the efficient market hypothesis.

The presence of serial correlation for short horizon returns turned out to be a controversial topic. Indeed, French & Roll [741], Lo & MacKinlay [1233] show that weekly and monthly (up to one year) asset returns do not exhibit (or exhibit to a small extent) negative serial correlation, while significant negative correlation was reported in Conrad & Kaul [483], Lehman [1177], Jegadeesh [1023]. In particular, Lo & MacKinlay [1233] show that the random walk hypothesis is strongly rejected on the basis of weekly returns on the period 1962–1985, using aggregate return indexes as well as size-related portfolios (with more pronounced effects in the case of small capitalization firms), and report significant positive serial correlation for weekly and monthly returns (*momentum effect*, see also Lo & MacKinlay [1237], Poterba & Summers [1430], Cutler et al. [510], Jegadeesh & Titman [1026], Chan et al. [405]). In Lo & MacKinlay [1233], the first order serial correlation coefficient of an equally weighted index is estimated to be approximately 30%, while, somehow surprisingly, the serial correlation of individual securities is typically found to be negative. In the subsequent paper Lo & MacKinlay [1237], using weekly data, it has been reported that the serial correlation of a small size companies portfolio (the smallest quintile) is equal to 42%, while portfolios composed by large companies exhibit a weaker serial correlation. The absence of serial correlation in asset returns can be explained by the presence of significant noise components in the single assets, while negative correlation may be due to a lack of liquidity and to short term price pressure. It is also interesting to remark that, as shown in Lo & MacKinlay [1237], a strong positive *cross-covariance effect* between returns of individual stocks (capturing the situation where a high return on a given stock today implies that the return on another stock will probably be high tomorrow) can generate positive serial correlation in the returns of an index or of a portfolio.

Conrad et al. [487] propose a solution to the short-horizon serial correlation puzzle by assuming that security returns are generated by three independent components: a positively autocorrelated common component (representing time varying expected returns), a negatively autocorrelated idiosyncratic component (due to microstructure effects such as transaction costs) and a white noise component. Using weekly returns on an index, the authors show that the first two components allow to explain about 24% of the variance of returns. At the level of individual securities, market microstructure effects dominate, thus producing negative serial correlation, while at the portfolio level returns are mostly driven by the common component (since bid-ask errors are diversified away, being cross-sectionally uncorrelated), thus leading to positive serial correlation. In this sense, the model of Conrad et al. [487] is able to reproduce the regularities observed empirically in the return time series. An alternative explanation for the presence of serial correlation in asset returns is provided by *non-synchronous trading*: assets are exchanged in different time periods and with a different frequency and, therefore, they incorporate new information at different times. This fact can induce positive serial correlation in stock index returns (in particular, equally weighted indexes). Mech [1316] shows that transaction costs and bid-ask spreads may lead to serial correlation in portfolio returns even if the returns of individual assets are serially uncorrelated. However, Lo & MacKinlay [1233, 1236], Cutler et al. [510], Lo & MacKinlay [1237] show that non-synchronous trading only accounts for a small part of the observed positive serial correlation in portfolio returns. On the other hand, Boudoukh et al. [273] and Ahn et al. [28] suggest that the effect of non-synchronous trading on small stocks has been underestimated. An example of an equilibrium model generating negative serial correlation in long horizon returns is provided in Exercise 7.7.

Momentum and Contrarian Strategies

Another way to empirically test the predictability of future asset returns consists in evaluating the performance of suitable trading strategies. Indeed, in view of part *(ii)* of Proposition 7.1 (under the assumption of risk neutrality), there exists no trading strategy based on the available information yielding a non-null excess return on average. Quoting Jensen (see Jensen [1031]): "a market is efficient with respect to information set Θ, if it is impossible to make economic profits by trading on the basis of Θ. By economic profits we mean the risk-adjusted rate of return, net of all costs". Since this result relies on the absence of serial correlation in asset returns, the existence of serial correlation in asset returns might lead to trading strategies yielding non-null excess returns.

In this perspective, market efficiency has been evaluated by assessing the excess returns of trading strategies obtained from *technical analysis* methods (i.e., trading strategies based on the analysis of past and recent movements in the returns time series). In Brock et al. [308], it has been shown that technical trading rules based on two moving averages and resistance-support levels generate returns that are not consistent with the hypothesis of a random walk, of a first-order autoregressive model or of a GARCH model (see Tsay [1601]), supporting the claim that technical analysis has predictive power on future returns. Along the same lines, Lo et al. [1238] provided evidence that classical technical analysis may have a predictive power.

In De Bondt & Thaler [532], the performance of *contrarian strategies* has been evaluated. Such strategies consists in forming portfolios with long positions on the assets that have performed badly in a previous period (loser assets) and short positions on the assets with a good performance in the previous period (winner assets). It has been empirically observed that contrarian strategies produce excess returns in the long run because of the agents' overreaction (the profitability of such strategies has been first observed by Graham & Dodd [818]). On a short horizon, the existence of excess returns generated by contrarian strategies has been reported in Jegadeesh [1023] and Lehman [1177].

Somehow in an opposite direction, it has also been observed that strategies buying (selling, resp.) winner assets (loser assets, resp.) in the recent past generate excess returns, exploiting short horizon trends in asset returns time series. Trading strategies exploiting this phenomenon are called *momentum strategies*. The capability of momentum strategies to generate excess returns over a medium time horizon has been documented in Jegadeesh & Titman [1026] for the U.S. market: assets with high returns over the past three-twelve months generate significant positive returns. However, after twelve months, the performance tends to reverse (see Jegadeesh & Titman [1028]). Excess returns are not due to changes in the riskiness of the assets or to a delayed reaction of the asset price to common risk factors. The momentum effect is stronger and persistent in small firms, growth firms (rather than value firms), low trading volume firms, high volume markets, firms with low analysts' coverage, as documented in Rouwenhorst [1478], Moskowitz & Grinblatt [1355], Hong et al. [958], Lee & Swaminathan [1175], Chan et al. [405]. In Conrad & Kaul [486] it is shown that contrarian and momentum strategies are equally likely to be successful: contrarian strategies earn profits in the long run (although the profits are significant only in the 1926–1947 period), while momentum strategies are usually profitable over a medium horizon (three-twelve months), confirming the results reported above.

A Critical Assessment

The empirical results showing the presence of serially correlated asset returns, the existence of excess returns generated by contrarian and momentum strategies stimulated an intense debate on the serial correlation of asset returns. As in the excess volatility debate, one can distinguish three main lines of thought on the profitability of contrarian/momentum strategies: the existence of statistical pitfalls in the test procedures, the existence of risk factors not taken into account by the models being tested (coherently with the classical asset pricing theory) and, finally, interpretations going beyond the classical asset pricing theory (behavioral finance).

Starting from the first of the above three lines of thought, the evidence of negative serial correlation in asset returns has been criticized from a statistical point of view. In particular, the sample size of long horizon returns can be very small when the return horizon is large compared to the length of the time series, thus reducing the statistical power of the test. Moreover, when working with long horizon returns, overlapping observations typically occur. Due to these issues, variance ratio tests as well as serial correlation tests have a rather low statistical power and are typically biased towards the rejection of the classical random walk hypothesis, see Richardson & Smith [1449], Richardson & Stock [1450], Kim et al. [1093]. In these studies, taking into account the statistical issues discussed above, the results previously reported in Fama & French [666] and Poterba & Summers [1430] have not been confirmed and represent spurious deviations from the null hypothesis. Moreover, in Richardson [1448] it has been shown that the U-shaped autocorrelation structure reported in Fama & French [666] can also be generated when the log-price process follows a random walk.

In Kim et al. [1093] and Jegadeesh [1024] it has been shown that mean reversion is peculiar to a limited time period: indeed, after the Second World War, no mean reversion is observed but rather persistence in long horizon returns (*mean aversion*). In Jegadeesh [1024] it has been shown that an equally weighted index of stocks exhibits mean reversion, but there is little evidence of mean reversion in the case of a value-weighted index. Moreover, the mean reversion phenomenon is typically concentrated in the month of January (*January effect*). Motivated by Fama & French [666], Lamoureux & Zhou [1165] adopt a Bayesian methodology and focuses on testing return predictability per se (as opposed to a specific null hypothesis that implies the absence of serial correlation in asset returns), providing evidence in favor of the random walk hypothesis. However, the evidence on the profitability of contrarian/momentum strategies cannot be disregarded from a statistical point of view. Indeed, even when taking into account most of the statistical issues discussed above, Balvers et al. [127] report the existence of mean reversion in post-war data in the case of national equity indexes of well-developed countries. Strong mean reversion effects have been also detected in Daniel [515].

On a different ground, the existence of excess returns generated by short horizon contrarian strategies (i.e., trading strategies exploiting weekly/monthly mean reversion) may be due to order imbalance, lack of liquidity and price pressure, as documented in Lehman [1177], Jegadeesh [1023]. Conrad & Kaul [485] have shown that the long run excess returns generated by a contrarian strategy (computed as cumulative short term returns) are typically overestimated due to the presence of bid-ask spreads, price discreteness and non-synchronous trading. Taking into account these effects and computing portfolio returns with a holding period of up to three years, the excess returns generated by a contrarian strategy can be even negative. Moreover, as reported above, contrarian profits are in general associated with the existence of a January effect, see also Zarowin [1674], De Bondt & Thaler [533] (however, a different interpretation is proposed in Loughran & Ritter [1247]).

Remaining within the context of classical asset pricing theory, the anomalies observed in asset returns are typically explained in terms of time varying expected returns. For instance, serially correlated mean reverting expected returns have been considered in Fama & French [666], Ball & Kothari [124], Chan [399], Fama [662], Kothari & Shanken [1124], Berk et al. [199]. In Cecchetti et al. [378], the authors propose an intertemporal general equilibrium model producing negative serial correlation in long horizon returns. Agents are characterized by constant relative risk aversion and the growth rate of the endowment follows a Markov switching process (see Exercise 7.7 for more details). An intertemporal equilibrium model producing similar results has been also proposed in Kandel & Stambaugh [1069]. In Conrad & Kaul [483], the following model allowing for time varying expected returns has been tested:

$$r_t = \mathbb{E}[r_t | \mathscr{F}_{t-1}] + \epsilon_t,$$

$$\mathbb{E}[r_t | \mathscr{F}_{t-1}] = \mu + \psi \mathbb{E}[r_{t-1} | \mathscr{F}_{t-2}] + u_{t-1},$$

for all $t \in \mathbb{N}$, where $\psi \leq 1$ and $(\epsilon_t)_{t \in \mathbb{N}}$ and $(u_t)_{t \in \mathbb{N}}$ are two white noise sequences. The expected return is evaluated by means of a Kalman filtering technique and the data lead to a rejection of the hypothesis of constant expected returns. Moreover, the empirical analysis shows that this model fits well the time series of weekly returns. The variation over time of expected returns explains up to 26% of the variance of the returns of a portfolio of stocks of small firms (this percentage decreases when considering a portfolio composed by stocks of large firms). Similar results were obtained in Conrad & Kaul [484] in the case of monthly returns.

Leaving the realm of classical asset pricing theory, De Bondt & Thaler [532, 533] suggest that the existence of long run mean reversion phenomena is due to the agents' overreaction to recent observations: loser assets are excessively underestimated by non-rational agents due to a poor recent performance while winner assets are excessively overestimated due to a very good recent performance (*overreaction*). Lehman [1177] supports this behavioral interpretation. This interpretation of the excess returns generated by contrarian strategies is similar to the "irrational" explanation of the CAPM anomalies presented in Sect. 5.3. An alternative interpretation has been proposed in Ball & Kothari [124] and Chan [399]: considering the CAPM as the reference model, the authors observe that negative serial correlation in returns can be induced by time varying expected returns generated in turn by time varying expected returns on the market portfolio. Hence, expected returns can change over time, the β coefficients are correlated with the market portfolio expected return and change over time. In the period comprised between the date of portfolio formation and the date of performance evaluation, the expected returns and the β coefficients of winner assets typically decrease while the expected returns and the β coefficients of loser assets increase. This implies that winner assets will be less risky while loser assets will be riskier with respect to the date of portfolio formation. One of the main causes of changes in riskiness is asset leverage. Indeed, since leverage is decreasing with respect to past asset returns, a negative series of abnormal returns will increase leverage and, consequently, the β coefficient (typically, the opposite occurs for winner assets). It has been observed that the β coefficient of loser assets typically exceeds the β coefficient of winner assets by 0.62 following the date of portfolio formation. By allowing for time varying expected returns in an equilibrium model (e.g., the CAPM model extended with time varying β coefficients), it can be shown that contrarian strategies do not produce significant excess returns. Similarly, in Zarowin [1673, 1674] no sign of overreaction is detected and it is shown that the profits generated by contrarian strategies are mainly due to a size effect, which is not captured by the β coefficient. Indeed, while loser assets typically outperform winner assets, loser assets are also typically smaller than winner assets. When loser assets are matched with winner assets of equal size, then there is little statistical evidence of a different performance. Fama & French [673] and Brennan et al. [293] show that the multi-factor model proposed in Fama & French [670] captures contrarian profits but not momentum profits. Korajczyk & Sadka [1120] show that momentum profits persist even when transaction costs are taken into account.

Chopra et al. [431] address the robustness of the profitability of contrarian strategies with respect to the above criticisms. By allowing for changing β coefficients of winner and loser assets, taking into account the existence of a size effect and considering January and non-January strategies, they still find evidence of non-null excess returns generated by contrarian strategies (5–10% in excess return per year in the years following the portfolio formation). This phenomenon is more pronounced for small firms rather than for large firms and, as already mentioned above, there is a strong January seasonality pattern. Similarly to De Bondt & Thaler [532], they conclude that the excess profits of contrarian strategies are mainly due to the presence of non-rational agents overreacting to news together with the fact that small firms are more frequently owned by individuals. Lehman [1177] reached a similar conclusion, analysing contrarian strategies over a weekly time horizon.

Summing up, we can remark that the explanations reported above for the excess profits generated by contrarian/momentum strategies resemble the explanations proposed for the CAPM anomalies, namely (besides the issues related to the statistical methodologies) the incorrect specification of the asset pricing model (see, e.g., Fama & French [673]) and the presence of irrational agents in the market (see Lakonishok et al. [1159]). Hence, some of the arguments presented in Sect. 5.3 about the CAPM anomalies can also be applied to the present context. Assuming a behavioral finance perspective, the presence of negative serial correlation (mean reversion) and excess profits generated by contrarian strategies are often referred to as overreaction of asset prices to news. On the other hand, positive serial correlation and the momentum effect are typically referred to as *underreaction* (or delayed reaction of asset prices to news). Both phenomena, at different time horizons, are simultaneously present in financial markets.

Adopting a behavioral finance point of view (see Lakonishok et al. [1159], La Porta et al. [1168]), the excess profits generated by contrarian strategies can be traced back to the presence of irrational agents in the market. Such agents are irrational in the sense that they form expectations about future prices by extrapolating recent earnings growth rates, giving an excessive weight to recent news about earnings growth rates. In this regard, La Porta [1167] and Dechow & Sloan [536] show that the analysts' forecasts of future earnings growth rates are significantly and systematically biased in the direction of overreaction (in other words, analysts' forecasts are typically extreme). Exploiting the errors in the analysts' forecasts to build a contrarian strategy, excess returns are obtained. Since analysts' forecasts can be regarded as a relatively good proxy of the agents' forecasts, this evidence of overreaction can explain a part of the excess returns generated by contrarian strategies (and there is no evidence of low return stocks being riskier than high return stocks). Long term overreaction in analysts' recommendations has also been detected in La Porta [1167], Dechow & Sloan [536], De Bondt & Thaler [534], while short horizon underreaction has been reported in Abarbanell & Bernard [1], Chan et al. [407], Michaely & Womack [1339]. The presence of heterogeneous types of agents (feedback and noise traders) in the market has also been used to explain the negative/positive serial correlation in asset returns (see Cutler et al. [510] and Chap. 9).

A severe criticism of behavioral finance came from Fama [664]. The author pointed out that underreaction is equally frequent as overreaction and, moreover, long term anomalies are sensitive to the factor model being tested as well as to the statistical methodology being employed. Typically, the presence of asset pricing anomalies is not robust to changes in the model or in the statistical methodology and, hence, if the model and the testing procedures are well-chosen, no anomaly should be detected. Almost all the empirical studies on market anomalies test the hypothesis of market efficiency without clearly specifying an alternative hypothesis (i.e., without satisfying the Khun methodological approach). In other words, market efficiency can only be replaced by a better specific model of price formation, while in most of the empirical studies the alternative hypothesis is just vaguely meant to represent market inefficiency. Moreover, many of the models proposed by the behavioral finance literature manage to explain a specific anomaly but predict other facts that are not always empirically confirmed in real financial markets. The evaluation of Fama [664] of the empirical literature is in favor of the efficient market hypothesis, potentially allowing for time varying expected returns. The author suggests that "the expected value of abnormal returns is zero, but chance generates apparent anomalies that split randomly between overreaction and underreaction" (see Fama [664]).

Lo & MacKinlay [1237] investigate whether the profitability of contrarian strategies does provide evidence of overreaction. The authors give a negative answer if there are sufficiently many traded securities in the market: the returns generated by a contrarian strategy can be attributed to a lead-lag relation among assets (covariance across stocks) and not to overreaction (or mean reversion) phenomena. They attribute over one half of the expected profits generated by a contrarian strategy to such cross correlation effects and not to the negative serial correlation in the individual stocks. Hence, the profitability of contrarian strategies does not necessarily imply overreaction. The authors report a positive correlation (about 28%) between weekly returns of small size companies and lagged returns of large size companies with a lead-lag relation (so that the returns of large capitalization stocks almost always lead those of smaller stocks, but not vice versa). This effect also generates positive serial correlation in weekly return indexes together with weak asset returns serial correlation, as documented in Lo & MacKinlay [1233]. The phenomenon can also be explained in terms of a different diffusion of the information concerning common risk factors: a quick diffusion of information for large size companies and a slow diffusion of information for small size companies (see McQueen et al. [1315]). In particular, there is evidence of a slow response by some small stocks to good, but not to bad, common news. In Jegadeesh & Titman [1027] it is shown that stock prices react with a delay to common factors while overreact to firm-specific factors: this difference generates a size related lead-lag relation in stock returns (see also Hou [967] on the existence of lead-lag effects). However, in contrast with the above observations, contrarian profits are mainly due to overreaction to firm-specific information and not to delayed reaction to common factors. Lead-lag response to common factors is also a source of profits of momentum strategies (see Lewellen [1210]).

The existence of a lead-lag relation between the returns on stocks of large size firms and of small size firms suggests to build a trading strategy which suitably buys and sells portfolios of large and small companies. In Knez & Ready [1106], it has been shown that a strategy of this type (switching between portfolios of small and large firms on the basis of the returns observed in the previous week) generates excess annual returns of around 15%. However, effective spreads cancel these gains and, once spreads are appropriately taken into account, the strategy will be outperformed by simple buy-and-hold strategies. In Jegadeesh & Titman [1027] it is shown that a contrarian strategy applied to size-sorted portfolios does not generate abnormal returns.

A risk-based explanation of the profitability of momentum strategies has been proposed in Conrad & Kaul [486] and Berk et al. [199], in the sense that the profits generated by momentum strategies are not abnormal returns but rather the compensation for bearing systematic risks changing in predictable ways. More specifically, Conrad & Kaul [486] observe that a momentum strategy gains from any cross-sectional dispersion in the unconditional mean returns, while Chordia & Shivakumar [436] and Berk et al. [199] explain the profits generated by momentum strategies through cross-sectional variations in time varying (conditional) expected returns and β coefficients driven by macroeconomic variables related to the business cycle (rather than stock-specific factors). Moreover, Conrad et al. [487] report momentum effects in common factors affecting stock returns. As shown in Moskowitz & Grinblatt [1355], the presence of a momentum effect related to a specific industry can explain the profits of momentum strategies, in line with the cross-sectional correlation explanation of momentum profitability. A risk-adjusted analysis of momentum strategies based on industry factors can explain a large part of momentum profits (around 50%, see Ahn et al. [29]). In addition, Ang et al. [67] and Harvey & Siddique [910] have shown that returns on momentum strategies are related to their sensitivity (high exposure) to downside risk.

In Fama & French [673], Brennan et al. [293], Grundy & Martin [850], it has been observed that the profitability of momentum strategies cannot be explained by mean reversion, size and book-to-market effects. Grundy & Martin [850] provide evidence against risk-based explanations, observing that momentum profitability reflects momentum in the stock-specific components of returns which are not associated with risk factors. Jegadeesh & Titman [1029] show that the explanation of momentum profits through cross-sectional differences in expected returns provided in Conrad & Kaul [486] is affected by small sample biases: taking this issue into account, then no evidence in support of the cross-sectional interpretation is found. Chan et al. [407], Jegadeesh & Titman [1026, 1028], Hvidkjaer [995], Hong et al. [958], Grundy & Martin [850], Moskowitz [1354], Hvidkjaer [995], Lee & Swaminathan [1175] report positive evidence that momentum strategy profits are due to initial market underreaction/overreaction followed by a delayed reaction to firm-specific news (in particular to earnings news). Jegadeesh & Titman [1028], Lee & Swaminathan [1175] find reversals in momentum portfolio returns after one year, a result that contrasts with a risk-based explanation and is in favor of the delayed reaction explanation. Underreaction to firm-specific information (slow diffusion of information) as the origin of the momentum effect is supported by the observed negative relation between the momentum effect and the number of analysts following the asset (see Hong et al. [958]). In Lewellen [1210], no sign of underreaction to news or of positive serial correlation in single asset returns is detected and the main source of momentum profits is identified as the negative auto- and cross-sectional correlation (lead-lag effect).

Predictability and Event Studies

The limited statistical power of tests based on univariate returns led many researchers to study the capability of specific firm-related and macroeconomic variables to predict future returns. Recall that, if the asset price follows a random walk, then no subset of the available information set $\mathscr{F}_t$ allows to predict future returns (however, we want to stress that return predictability is compatible with classical asset pricing theory if expected returns are allowed to be time varying). As far as monetary and macroeconomic variables are concerned, the variables that turn out to have some predictive power are the (expected) inflation, monetary growth, short term interest rates, term spread, default spread, changes in industrial production, output, aggregate consumption-wealth ratio, aggregate wealth, labor income-consumption ratio, market liquidity, bid-ask spread, equity issues proportion of new securities issues and volatility (see Balvers et al. [126], Fama & Schwert [679], Chen [421], Keim & Stambaugh [1083], Campbell [330], Fama & French [667, 668], French et al. [742], Cutler et al. [510], Marshall [1308], Patelis [1410], Pesaran & Timmermann [1415], Lettau & Ludvigson [1194], Santos & Veronesi [1497], Jones [1039], Baker & Wurgler [109], Flannery & Protopapadakis [713], Bollerslev et al. [265]).

In Rozeff [1480], Shiller [1535], Campbell & Shiller [352], Kothari & Shanken [1125], it has been documented that the aggregate dividend-price ratio (dividend yield), the earnings-price ratio and the book-to-market value, respectively, have a predictive power on future market returns (in particular, Shiller [1535] shows that stock prices appear to overreact to dividends). These results have been confirmed in many papers (see Fama & French [667], Campbell & Shiller [351], Fama & French [668], Cutler et al. [510], Hodrick [951], Flood et al. [720], Campbell & Shiller [353], Pontiff & Schall [1427], Campbell [334], Lewellen [1209, 1211]). A theoretical foundation of the predictability of future returns through the dividend yield can be found in the dividend growth model (7.14). In general, the price is low when expected returns are high and, therefore, high dividend yields typically predict high returns. The predictability of future returns through these variables increases in the return horizon: in correspondence of a four year period, the dividend-price ratio explains about 25% of the return variance (see Fama & French [667]). Note that it is easy to reconcile this evidence with the presence of negative serial correlation in asset returns: indeed, the predictability (with a positive coefficient) of future returns through the current dividend yield provides an evidence of mean reversion in the return time series. Weak predictability in the '90s and a poor out of sample performance (due to parameter instability) have been detected in Cochrane [463], Goyal & Welch [817], Schwert [1510]. In Lamont [1162] it is shown that also the aggregate dividend-earnings ratio predicts future returns, in the sense that high dividends forecast high returns and high earnings forecast low returns. Furthermore, accounting variables that are useful to explain asset returns cross-sectionally are also useful at the aggregate level to predict market returns.

As we have already pointed out in the case of other market anomalies, the empirical evidence on return predictability through the dividend yield and other fundamental variables can be interpreted within or outside the classical asset pricing theory. In the first case (see, e.g., Rozeff [1480], Fama & French [667, 668], Fama [663], Cochrane [464], Pontiff & Schall [1427]), dividends and earnings are regarded as fundamental variables. For instance, according to Rozeff [1480], changes in dividend yield proxy for variations in the risk premium of stocks. An equilibrium monetary asset pricing model relating inflation to returns has been provided in Marshall [1308], while an intertemporal equilibrium model relating asset returns to macroeconomic fluctuations (output) is provided in Balvers et al. [126]. On the other side, return predictability can be interpreted as an evidence of the existence of noise traders and market irrationality (see, e.g., Shiller [1535], Campbell & Shiller [351], Cutler et al. [510]). According to this interpretation, low dividend yields are associated with overvalued stocks and, therefore, with lower future returns on stocks.

The existence of underreaction/delayed reaction has also been studied in correspondence of specific events (*event studies*), see Daniel et al. [516], Fama [664], Hirshleifer [945], Kothari [1123], MacKinlay [1270]. In particular, in correspondence of some public events, a post-event drift has been observed, in the sense that the average return on the date of the event has the same sign of the subsequent average long run abnormal performance, up to a three to five years period. Such events typically include earning announcements/surprises (see Ball & Brown [123], Bernard & Thomas [202], Jegadeesh & Titman [1026], Chan et al. [407]), dividend initiations and omissions (see Michaely et al. [1336]), seasoned issues of common stocks and initial public offerings (IPO) (see Ibbotson & Ritter [996]) and stock market repurchases (see Ikenberry et al. [998]). In particular, a post-earnings drift is observed: stocks with surprisingly good news tend to outperform those with bad news. Analysts' earnings forecasts and stock recommendations underreact to earnings news, see Abarbanell & Bernard [1], Womack [1662], Michaely & Womack [1339], Chan et al. [407]. As we have already pointed out, stock underreaction can be related to analysts' underreaction, but this only provides a partial explanation (and the analysts' behavior may also be unrelated to the stock price overreaction).

7.3   Tests on Intertemporal Equilibrium Models

The intertemporal general equilibrium models presented in Chap. 6 have been exten-
sively empirically tested in many different forms. The literature on the empirical
tests of multi-period asset pricing theory is extremely rich and we refer to Singleton
[1549], Ferson [687], Ferson & Jagannathan [696], Kocherlakota [1112], Campbell
[334, 335, 336] for good surveys. In broad terms, we can distinguish three main
classes of empirical tests:

1. tests verifying the relationship between asset returns and the intertemporal rate
   of substitution of consumption of the representative agent of the economy
   (stochastic discount factor);
2. tests on intertemporal risk premium factor models (CCAPM, ICAPM);
3. calibration studies: the ingredients of an equilibrium model (notably the parameters describing the preference structure and the technology) are chosen in order
   to match some empirical properties of asset returns.

Before starting our survey of the empirical literature we want to point out that a
large part of the empirical literature on multi-period asset pricing models is based
on the assumption of a time additive utility function of the power form $u(x) = x^{1-\alpha}/(1 - \alpha)$. However, this type of utility function is affected by a significant
constraint: the parameter α represents both the coefficient of relative risk aversion
as well as the reciprocal of the elasticity of the intertemporal rate of substitution (see
Hall [880] and Exercise 7.8). The identification through a single parameter of the
willingness of an agent to diversify wealth over different states of the world and to
substitute wealth intertemporally is not justified on an empirical basis (for instance,
in Barsky et al. [169] it has been shown that these two features are essentially
unrelated across individuals). In view of empirically testing the implications of
classical asset pricing theory, this represents a severe limitation. Considering a non-
power time additive utility function, the identity between the coefficient of relative
risk aversion and the reciprocal of the elasticity of intertemporal substitution holds
in an approximate form (see Exercise 7.8). The time additivity of the utility function
implies that agents tend to dislike variations in their consumption stream and
transfer wealth from periods of high consumption to periods of low consumption
in order to smooth consumption over time, similarly as risk averse agents try to
diversify consumption across different states of the world. In the remaining part of
this chapter, we shall restrict our attention to time additive utility functions, referring
to Chap. 9 for a discussion of alternative preference functionals.

Tests on the Euler Conditions

Considering a representative agent economy, a possible way to test the validity of
classical asset pricing theory consists in verifying empirically the validity of the

optimality conditions (6.68)–(6.71) (*Euler conditions*). These optimality conditions can be seen as (conditional) orthogonality conditions between the representative agent's one-period stochastic discount factor and the asset excess returns. However, such relations depend non-linearly on the state variables (consumption/endowment) and on the unknown preference structure. To deal with this issue, a generalized instrumental variable estimation methodology has been developed in Hansen & Singleton [895]. Assuming a log-normal stationary joint distribution of consumption and monthly returns and a power utility function, the Euler condition (6.78) has been empirically rejected. Furthermore, excluding the case of value-weighted portfolios, the empirical evidence reported in Hansen & Singleton [896] is against the validity of the Euler conditions. The estimated coefficient of relative risk aversion belongs to the interval $(0, 2)$.

A semi-nonparametric test performed in Gallant & Tauchen [753] rejects the joint hypothesis that consumption and returns follow a multivariate normal distribution and that the utility function is time additive. The empirical evidence is in favor of the presence of durable goods among the arguments of the utility function (see Sect. 9.2), in the sense that there exist assets yielding an utility flow which extends beyond the date of acquisition of the asset, thus contradicting the time separability of the utility function. Evidence against a time separable utility function has been also provided in Grossman et al. [845]: the authors suggest to consider a utility function representing risk aversion and elasticity of intertemporal substitution through different parameters (compare with Exercise 7.8). Note also that errors in the measurement of consumption do not provide a sufficient explanation for the negative results of these tests on classical asset pricing theory (see Singleton [1549]).

A non-parametric methodology to test asset pricing models has been developed in Hansen & Jagannathan [891] on the basis of the Hansen-Jagannathan bound established in Proposition (6.36) (see also Cochrane & Hansen [467], Cecchetti et al. [380], Gallant et al. [750], Balduzzi & Kallal [117], Hansen & Jagannathan [892], Ferson & Siegel [698], Lettau & Uhlig [1199], Bekaert & Liu [188] and Hansen et al. [890] for an extension to an economy with transaction costs and market imperfections). In particular, note that a test of the Hansen-Jagannathan bound does not require precise assumptions on the utility functions and, hence, on the intertemporal rate of substitution of the representative agent. According to Proposition 6.36, the excess return of every asset (normalized by its standard deviation) should be bounded from above by the standard deviation-mean ratio of the one-period stochastic discount factor. However, it has been pointed out in Ferson & Siegel [699] that this type of test (in particular in an unconditional form) is typically biased towards the rejection of the validity of the bound. The results reported in Balduzzi & Kallal [117], Hansen & Jagannathan [891], Cecchetti et al. [380], Cochrane & Hansen [467] show that the stochastic discount factor is insufficiently volatile compared to stock returns for the bound to be satisfied, thus providing empirical evidence against the validity of an intertemporal equilibrium model with a time additive utility function. In particular, it has been shown that an intertemporal equilibrium model with a time separable power utility function only agrees with the data for a very high coefficient of relative risk aversion. Intertemporal complementarity in preferences (habit formation, see Chap. 9) may help to satisfy the bound.

Tests on the CCAPM and on the ICAPM

A second approach (which can however be related to the previous one) to test the implications of classical asset pricing theory consists in empirically testing the validity of the CCAPM. An empirical test of the CCAPM with positive results has been implemented in Breeden et al. [286]. After adjusting reported consumption data for aggregation and measurement errors, the authors have tested cross-sectionally that expected returns are linear in the β coefficients with respect to the return of a portfolio which is maximally correlated with the consumption process (compare with Proposition 6.37). The risk premium of such a portfolio is found to be positive and the relation between the expected returns of the assets and of the portfolio is approximately linear, especially if one does not include the 1929–1939 period. Consumption aggregation and measurement errors decrease the variance of measured consumption growth and lead to an underestimation of the covariance with asset returns, thus inducing a bias towards the rejection of the CCAPM. The results of Breeden et al. [286] have been confirmed in Wheatley [1659] as well as in Cecchetti & Mark [382], assuming a very high risk aversion coefficient. These results are mainly based on seasonally adjusted data, which could in principle lead to biases and erroneous inferences. However, even using unadjusted data, Ferson & Harvey [692] still obtain a rejection of the model. Again, an implausibly high coefficient of risk aversion is required to fit the data, as already observed in the case of tests based on the Euler conditions. Bansal et al. [134] propose an intertemporal equilibrium model explaining cross-sectional asset returns through the exposure of asset dividends to aggregate consumption. The model explains more than 50% of the cross-sectional variation in risk premia across different assets (momentum, size and book-to-market sorted portfolios). The logarithm of the aggregate consumption is modeled through an ARIMA(1,1,1) process (see Hamilton [881]) with a small predictable component in the growth rates.

In this family of tests, the measurement of the aggregate consumption has a crucial importance. In Jagannathan & Wang [1013], consumption betas of stocks have been computed using yearly consumption growth rates (based on the fourth quarter of the year). With this methodology, the CCAPM is shown to explain the cross-section of stock returns similarly as the three-factor Fama & French [670] model. Positive evidence on the CCAPM in explaining the cross-section of expected returns has been also obtained in Parker & Julliard [1401], considering the covariance with respect to consumption growth cumulated over many quarters following the return period.

A conditional version of the CCAPM explains the cross-section of average returns on size and book-to-market sorted portfolios when the consumption-wealth ratio, idiosyncratic consumption risk or the consumption/labor income ratio are used as conditioning variables, see Lettau & Ludvigson [1195], Jacobs & Wang [1006] and Santos & Veronesi [1497], respectively. These variables summarize the investors' expectations about asset returns. Note that the use of the conditioning technique improves the fit of the model because some stocks are more highly correlated with consumption growth in bad times than in good times. Clearly, this effect cannot be captured by unconditional models, since the latter assume constant risk premia.

The CCAPM has been tested against the CAPM, i.e., the hypothesis that asset expected returns are proportional to the consumption β has been tested against the hypothesis that they are proportional to the β coefficient with respect to a market portfolio. The empirical results are mostly in favor of the CAPM. In particular, Mankiw & Shapiro [1297], Attanasio [80], Campbell & Cochrane [344] have shown that the CAPM has a better explanatory power than the CCAPM when considering cross-sectional average returns. In a similar direction, Chen et al. [424] have shown that the multi-factor APT performs better than the CCAPM in explaining returns cross-sectionally. The main reasons for the empirical failure of the CCAPM are represented by the lack of variability in the intertemporal rate of substitution of consumption (which corresponds to the one-period stochastic discount factor) and the lack of covariance between consumption growth and asset returns. This limitation has been partly addressed by considering conditional versions of the CCAPM. Note also that these tests of the CCAPM are often affected by a joint hypothesis problem, since the rejection of the CCAPM can also be attributed to a failure of one of the many auxiliary assumptions needed for its specification.

In the previous section, we have reported many empirical results showing that asset returns have a predictable component. As we have already mentioned, in an efficient market the predictability of returns can be generated by time varying investment opportunities and, hence, this feature should be captured by an intertemporal general equilibrium asset pricing model. Balvers et al. [126] have proposed a general equilibrium model where the returns are related to macroeconomic fluctuations. Asset returns depend on consumption which is in turn linked to aggregate output. As a consequence, asset returns can be predicted if the output itself is predictable. Consistently with conventional macroeconomic models, the output is typically serially correlated and, hence, predictable. An intertemporal asset pricing model with fluctuating mean and variance of consumption growth generating mean reversion in asset returns has also been considered in Kandel & Stambaugh [1068] and Cecchetti et al. [378]. Both consumption growth moments and risk premia exhibit business cycle effects. Ferson & Harvey [691] employ a macroeconomic multi-factor model (ICAPM) of the type described in Chen et al. [424] to capture the time varying nature of returns and show that the stock market is the most important factor in order to capture predictable variations of stock portfolio returns. Moreover, the variation over time in the premium for the β-risk is more important than changes in the β coefficient itself. Ferson & Korajczyk [695] have shown that a macroeconomic or a statistical multi-factor model captures the predictability of long horizon returns, allowing for time varying beta and risk premia. It is shown that variations in risk premia are the primary source of predictability and five factors suffice to explain about 80% of the variability of asset returns. An analogous task is accomplished in Evans [658] with a two-factor model (returns on the stock market and on corporate bonds), time varying β and risk premia. Allowing for time varying first and second moments, Attanasio [80] has shown that the traditional CAPM accounts for predictability of returns through the dividend yield (mixed results have been instead obtained in Kirby [1103]). Huberman & Kandel [983] provide positive evidence on the capability of an intertemporal equilibrium model to capture return predictability through time varying returns. Furthermore, Ferson [685] shows that conditional β changes are associated with interest rates, providing an explanation of return predictability through interest rates.

Avramov [92] reported substantial deviations from the capability of the three-factor model of Fama & French [670] to explain asset return predictability as well as from the model based on firm characteristics proposed by Daniel & Titman [520]. Ferson & Harvey [693] have shown that predetermined variables used to predict the time series of stock and bond returns also provide cross-sectional explanatory power for stock returns on top of the three factors used in Fama & French [670]. The conditional three-factor model is not able to capture common dynamics in returns (e.g. those captured by the variables employed in Fama & French [667]).

The Equity Premium and Other Asset Pricing Puzzles

As we have seen in the preceding sections, the empirical literature has reported several anomalies which apparently conflict with classical asset pricing theory. In particular, let us emphasize the following striking empirical facts (see Campbell [334]):

- the average asset return is rather high (in the postwar period, the average real annual stock return has been 7.6%);
- the average risk free rate is rather small (in the postwar period, the three-month Treasury bills average rate has been 0.8% per year);
- the time series of aggregate consumption exhibits a rather low variability (the standard deviation of the consumption growth rate in the United States is around 1%);
- the volatility of the interest rate is rather small (the standard deviation of the real return on U.S. Treasury bills is 1.8%);
- the correlation between consumption growth and real asset returns is rather weak (0.22 on the basis of U.S. quarterly data).

In particular, it is not easy to explain within the context of classical asset pricing theory a large equity premium (*equity premium puzzle*, see Mehra & Prescott [1319]), a low risk free rate (*risk free rate puzzle*, see Weil [1648]) and a low variability of consumption growth (*consumption smoothing puzzle*, see Hall [880], Campbell & Deaton [345]). We refer to Kocherlakota [1112], Mehra & Prescott [1320], Campbell [336], De Long & Magin [546] for an evaluation of these puzzles. As shown also in cross-sectional tests, at the origin of these puzzles lies the fundamental observation that the time series of consumption growth is too smooth if compared with income growth and asset returns. Note that, in view of relation (6.71), it holds that

$$\mathbb{E}[\mathbf{u}'(e_t)|\mathscr{F}_{t-1}] = \frac{1}{\delta r_f}\mathbf{u}'(e_{t-1}), \qquad \text{for all } t \in \mathbb{N},$$

so that the quantity $\mathbf{u}'(e_t) - \mathbf{u}'(e_{t-1})/(\delta r_f)$ is orthogonal to all the information available at date $t - 1$, for every $t \in \mathbb{N}$. In particular, assuming a quadratic utility function of the form $\mathbf{u}(x) = ax - bx^2/2$ and letting $\delta = 1/r_f$, we obtain the linear model

$$e_t = e_{t-1} + \varepsilon_t, \qquad \text{for all } t \in \mathbb{N},$$

where $(\varepsilon_t)_{t\in\mathbb{N}}$ is a sequence of uncorrelated random variables with zero mean.

As we have already remarked, many specifications of classical asset pricing theory have considered utility functions of the power form, for which the coefficient of relative risk aversion coincides with the reciprocal of the elasticity of intertemporal substitution of consumption (see Exercise 7.8). Assuming a log-normal distribution, the results presented after Proposition 6.38 imply that the equity premium is increasing in the coefficient of relative risk aversion (assuming that the covariance term is positive), while the relation between the coefficient of relative risk aversion and the equilibrium risk free interest rate depends on both the average consumption growth rate and the variance of the consumption growth rate. This shows that, in the case of a power utility function, both the equity premium puzzle and the risk free rate puzzle are related through the coefficient of relative risk aversion. A large risk aversion coefficient, as required to match the observed values of the equity premium, would imply a strong preference for consumption smoothing, meaning that the agent dislikes fluctuations in consumption and is unwilling to substitute consumption over time. On the other hand, since the variance of the consumption growth rate is typically small, a large value of the coefficient of relative risk aversion would imply unrealistically large values for the risk free rate. This shows the difficulty of solving at the same time the equity premium and the risk free rate puzzles in this setting. Note also that, as shown in Exercise 7.8, the inverse relation between the coefficient of relative risk aversion and the elasticity of intertemporal substitution holds in an approximate form for any time additive utility function. As a consequence, the difficulty of solving the two puzzles together does not only depend on the choice of a power utility function.

The estimate of the equity premium reported above, of course depending on the estimation methodology as well as on the underlying time series, has been criticized from a methodological point of view. In Mehra & Prescott [1319], the risk premium has been estimated as the average historical return, so that the precision of the estimate clearly depends on the length of the time series. However, it has to be noted that long time series are also more sensitive to structural breaks (for instance, the 1929 crisis). Indeed, Cogley & Sargent [470] have shown that the large equity premium estimate can be attributed to the effects of the great depression: even assuming that agents update their beliefs according to the Bayes rule, the great depression induced pessimism that persisted over time inducing agents to ask for a significant risk premium. In this direction, incorporating structural breaks in the time series, Pástor & Stambaugh [1406] have estimated a risk premium comprised between 4% and 6% percent in the last two centuries, with a sharp decline in the last half century. In Fornari [726], using a conditional variance model, the equity premium has been estimated around 5–6%, with a countercyclical behavior. Moreover, computing the equity premium by an arithmetic average leads to an overestimation when asset returns are mean reverting and noisy, as shown in Siegel [1542]. Evidence of an overestimation of stock returns was also reported in Siegel [1543] when computing the equity premium as the historical average over a long time series, together with an underestimation of fixed income returns. In more recent years, the equity premium seems to have declined (see for instance Mehra & Prescott [1320] for more recent estimates).

To take into account these effects, model-based estimates of the risk premium have been proposed. In general, these methods still report a positive equity premium, but typically smaller than the estimate produced on the basis of average historical returns (see, e.g., Blanchard [247]). A decline of the equity premium in recent years has been established in Fama & French [675] and Jagannathan et al. [1011] by relying on a dividend-earnings growth model (Gordon-type model) to estimate the equity premium through fundamentals (dividends and earnings). The results reported show that in the period 1951–2000 the equity premium estimates on the basis of dividend and earnings growth models are 2.55% and 4.32%, respectively, far below the equity premium estimate computed on the basis of average returns (note also that in the late 20th century unusually high expected returns have occurred). According to Lettau et al. [1197], this reduction of the equity premium can be rationalized through a reduction in macroeconomic risk. Campbell & Shiller [353], using the price-dividend ratio, predicted a conditional equity premium smaller than its sample average. On the other hand, even considering model-based estimates of the equity premium, Constantinides [494] reported large values for the (unconditional) equity premium. Note, however, that the equity premium puzzle can be partly explained by recognizing that agents face uninsurable and idiosyncratic income shocks, together with borrowing constraints. Moreover, a survivorship bias similar to that detected in Brown et al. [315] for asset returns can be at the origin of the high equity premium in the U.S. economy. Indeed, the fact that we only observe returns of surviving firms induces an upward bias in the estimation of the equity premium. In Goetzmann & Jorion [797], the authors show that the U.S. market had the highest uninterrupted real rate of appreciation of all countries (around 5% per year, while for other countries the median real appreciation rate has been around 1.5% per a year), concluding that the large equity premium observed in the U.S. market could be the exception rather than the rule. However, some analyses have shown that the equity premium puzzle is pervasive in most financial markets (see Campbell [334], Mehra & Prescott [1320], Campbell [336]).

Mehra & Prescott [1319] reported that the average annual real rate of return on short term bills was 0.8% and the average annual real rate of return on stocks was 6.98% in the U.S. market during the period 1889–1978. In order to produce these estimates, the authors assumed a representative agent economy with a time additive power utility function of the form $\sum_{t=0}^{\infty} \delta^t c_t^{1-\alpha}/(1-\alpha)$, supposing that the aggregate consumption growth rate follows a two-state Markov chain (see the last part of Sect. 6.4 for a detailed presentation of the model). In Mehra & Prescott [1319], the Markov chain was calibrated to market data by matching the average consumption growth rate, together with the sample standard deviation and first-order serial correlation. In correspondence of an equilibrium, the aggregate consumption coincides with the aggregate dividends and the parameters α and δ describing the preference structure of the representative agent were calibrated in order to match the equity premium and the risk free rate observed empirically. With a discount factor smaller than one, the authors verified that it is not possible to reproduce the observed equity premium with a parameter $\alpha \leq 10$ (the largest equity premium which could be obtained by the model was computed to be 0.35). In order to match the empirically observed equity premium, the coefficient of relative risk aversion has to take unrealistically large values. The covariance of asset returns with consumption is driven by the variance of the consumption growth rate, which is typically rather low and, therefore, a large coefficient of relative risk aversion is needed to reproduce the average equity return. The smoothness of the consumption growth rate plays a fundamental role in understanding the equity premium puzzle (the lack of variability of consumption is partly due to the presence of adjustment costs in consumption that artificially reduce the variability of consumption, see Gabaix & Laibson [748]). These results are partly related to those in Hall [880], where it is shown that the consumption growth rate in the postwar years had a small variance while the risk free rate was low and equity returns and risk free rate variations over time were large (in general, large real interest rates and large consumption growth rates are typically not observed simultaneously). These results support the conclusion that the elasticity of intertemporal substitution is unlikely to be larger than 0.1, a value confirmed by experimental results in Barsky et al. [169], which is larger than the reciprocal of the coefficient of relative risk aversion needed to match the historical equity premium. Summing up, these observations suggest that the relation existing between the coefficient of relative risk aversion and the elasticity of intertemporal substitution (see Exercise 7.8) represents a severe limitation of classical asset pricing theory.

A significant drawback of a large coefficient of relative risk aversion, needed to match the observed equity premium, is represented by the very large value that it implies for the equilibrium risk free rate (*risk free rate puzzle*), in view of relation (6.77). Therefore, we can summarize as follows the risk free rate puzzle: given a large coefficient of relative risk aversion (and, hence, a small elasticity of intertemporal substitution), an agent prefers to smooth consumption over time, but this contrasts with the saving behavior of typical investors, which generates an average pro capita consumption growth rate of around 2% per year. This phenomenon can be explained by a negative time preference (i.e., $\delta > 1$) or by a very large risk free interest rate. However, this is in contradiction with the historical observations. A low risk free rate and a low elasticity of intertemporal substitution can only be explained by a consumption growth rate much lower than the 2% observed historically (see, e.g., Deaton [530], Kocherlakota [1112], Weil [1648]).

The large values of the coefficient of relative risk aversion obtained above have been confirmed by many calibration experiments: for instance, Cecchetti & Mark [382], Kocherlakota [1109] have estimated $\alpha = 13.7$, Weil [1648] have estimated $\alpha \geq 20$ and Kandel & Stambaugh [1069] have estimated $\alpha = 29$ (see also Kandel & Stambaugh [1068]). The seasonality adjustments of consumption do not suffice to explain the puzzle: indeed, as shown in Ferson & Harvey [692], seasonally unadjusted consumption data still require large values of risk aversion in order to solve the puzzle. These results agree with the test proposed in Hansen & Jagannathan [891], where a large risk aversion is needed to satisfy the volatility bound established in Proposition 6.36 on the intertemporal rate of substitution. Hence, the large equity premium can only be explained by assuming enough volatility in the stochastic discount factor. Note that Blume & Friend [258] have estimated through a panel data analysis a small risk aversion coefficient (around 3), while the experimental analysis in Barsky et al. [169] has produced larger estimates but still not large enough to match the historical risk premium. However, these estimates have been questioned by Wheatley [1659], Kocherlakota [1109], Kandel & Stambaugh [1068], suggesting that a larger risk aversion coefficient may indeed be plausible.

One of the crucial assumptions in the analysis of Mehra & Prescott [1319] is that aggregate consumption coincides with aggregate dividends. In Cecchetti et al. [379], this assumption has been relaxed and non-traded labor income has been introduced, so that the aggregate consumption is given by the sum of aggregate dividends and labor income. A bivariate log-normal process with a two-state Markov drift is estimated for the consumption and dividend growth processes. In this setting, the equity premium and the risk free rate empirically observed can be reproduced through plausible parameters. However, the model does not allow to simultaneously match the first and the second moments of returns. Considering equity as a levered claim on aggregate consumption (without labor income) and calibrating the degree of leverage (treated as a free parameter) to match the volatility of stock returns, Benninga & Protopapadakis [192], Kandel & Stambaugh [1068, 1069] match the first and the second moments of the risk free rate and of equity returns. Bonomo & Garcia [266] estimate a heteroskedastic joint bivariate Markov process for consumption and dividends and try to reproduce the first and the second moments of real and excess equity returns together with negative serial correlation of excess returns and predictive power of the dividend-price ratio. While the model captures the main features of real returns data, the main failure comes from excess returns: the premium is still small and the dividend-price ratio does not appear to have significant predictive power with respect to excess returns.

The equity premium puzzle has also been addressed by introducing a small probability of a future crash in the market. In Rietz [1451], introducing an event crash in correspondence of which the output falls by 50% of its value with a probability of 0.4%, the U.S. equity premium can be matched with a risk aversion coefficient equal to $\alpha = 5$. Moreover, the risk aversion needed to explain the equity premium puzzle decreases as the probability of a crash event increases. As shown in Salyer [1490], the agents' intertemporal rate of substitution is consistent with the bound determined in Hansen & Jagannathan [891] (see Proposition 6.36), as calibrated to real data, however the volatility of excess returns as predicted by the model is significantly smaller than what observed in the data. Barro [166] further analyses the possibility that a disaster may explain the equity premium. The author estimates that "disaster events" yielding an output fall between 15% and 64% occur with a probability comprised in the interval [1.5%,2%] and allowing for a probability of a disaster (calibrated as a drop of more than 29%) of 1.7% allows to match the equity premium with coefficient of relative risk aversion equal to 4 (see also Barro & Jin [167]). A rare disaster (eventually with time varying intensity) also contributes to explain other puzzles such as the presence of excess volatility and return predictability, see Gabaix [747] and Wachter [1637]. We refer to Exercise 7.9 for a presentation of the model considered in Gabaix [747]. However, Julliard & Ghosh [1054] and Gourio [816] express a skeptical view on the possibility that the probability of a disaster can explain the observed values of the equity risk premium. In Labadie [1152], the effects of stochastic inflation on the equity premium are investigated: assuming a cash in advance constraint, inflation risk increases the equity premium predicted by the model, but the latter is still less than half of the equity premium observed in historical data.

7.4 Notes and Further Readings

The excess volatility results detected in Shiller [1533] focused on the (non-) stationarity of the dividend process. Marsh & Merton [1306] consider the case of a stochastic dividend process depending on the asset price process and conclude that the Shiller's approach cannot be used to test the validity of classical asset pricing theory. This type of dividend process is motivated by the empirical analysis of Lintner [1220], showing that managers typically set dividends having a target pay-out ratio and choose dividend policies that smooth the dividend changes required to meet their goal. As a consequence, dividends represent weighted averages of past earnings. Assuming a constant expected (real) rate of return and that the stock price reflects investor beliefs (which are in turn given by the expectation of future discounted dividends), the de-trended dividend process can then be written as a moving average of current and past de-trended stock prices (see also Marsh & Merton [1307] and, for a microfoundation of this feedback effect, Timmermann [1591]). According to this policy, managers typically deviate from the long run growth path in response to changes in permanent earnings. The authors claim that such a dividend policy well fits the observed dividend time series. Assuming this type of dividend policy, inequality (7.6), computed according to the procedure proposed by Shiller, holds in the opposite direction. The violation is due to the fact that the de-trended dividend time series is non-stationary and the Shiller's procedure suffers from a joint hypothesis problem concerning the stationarity of the dividend process. However, other dividend smoothing policies in the spirit of Lintner [1220] do not generate a violation of the volatility inequality (7.6) (see Shiller [1536] for more details). In this context, see also Exercise 7.5.

The (non)-stationarity of the dividend process has been deeply investigated in the literature. Shiller [1534, 1536, 1537], Campbell & Shiller [351] report evidence in favor of a trend-stationary dividend time series, while Kleidon [1104] and Marsh & Merton [1306] find a negative evidence. If the time series has a trend, then a geometric random walk test (without a trend component) is biased towards accepting the null hypothesis of a unit root (see Hamilton [881, Chapter 17] for a study of univariate processes with a unit root). Including a trend component, the test still has a low power against trend stationary alternatives (see DeJong & Whiteman [541]). Using a Bayesian approach, DeJong & Whiteman [541] conclude that the dividend and the U.S. price time series are more likely to be trend-stationary than integrated. A possible way to avoid non-stationarity problems is to perform excess volatility tests on the price-dividend ratio time series, which is more likely to be stationary than the dividend time series. Both absence of correlation and volatility bound tests can be performed on the price-dividend ratio time series. In this perspective, see Cochrane [461] for a result not indicating striking rejections of the volatility bounds and, on the contrary, Campbell & Shiller [351, 352], LeRoy & Parke [1188] for results in favor of excess volatility.

The presence of excess volatility can also be tested by running a linear regression of s_t^e on s_t^* and it can be shown that this type of test is analogous to an orthogonality test on price volatility. If asset prices are consistent with classical asset pricing theory, then the intercept of the regression must be equal to zero and the regression coefficient must be equal to one. However, this hypothesis is not verified empirically (see Scott [1513]). The test is well defined in the case of stationary dividend-price time series, while, if this is not the case, then a similar test can be implemented for the price-dividend ratio. Even in this case, it has been shown that the intercept is positive and the regression coefficient is not significantly different from zero.

Positive evidence on the capability of fundamentals to explain asset return variations has been provided in Fama [662]. The author identifies three sources of variations in returns: shocks to expected cash flows, shocks to expected returns and time varying expected returns. Taking into account these three components, around 58% of the variation in ex-post annual stock returns can be explained. Moreover, as shown in Kothari & Shanken [1124], variables that proxy for market expectations of future dividends explain over 70% of variations in returns over time. Evans [659], allowing for non-stationary aggregate dividends and discount rates, establishes that changing forecasts of future dividend growth account for 90% of the predictable variations in the dividend-price ratio.

Positive evidence on time varying discount rates as a source of excess volatility is reported in Cochrane [461]. Bansal & Lundblad [137] show that cash flow growth rates can be modeled as a stationary ARIMA(1,0,1) process (i.e., they contain a small predictable long run component). By assuming this process together with a time varying systematic risk factor, about 70% of the volatility of asset prices can be explained.

The relevance of microstructure and mispricing effects (in particular liquidity and bid-ask spreads) on the measurement of the long run performance of contrarian strategies has been also pointed out in Ball et al. [125]. Moreover, Ball et al. [125] observe that the profitability of a trading strategy depends on the month of portfolio formation: contrarian strategies formed in December have positive excess returns, while those formed in June have negative excess returns. Again, there is evidence of a January effect. On the other hand, momentum strategies earn negative profits in January, while positive profits are associated with the remaining months. There are seasonality patterns in the profits generated by momentum and contrarian strategies as well as in the predictability of future returns through past returns, with the seasonality effect being more pronounced in the months of December and January (therefore, it can be at least partly attributed to tax loss selling), when effective capital gain tax rates are expected to decrease, see Grinblatt & Keloharju [830], Grinblatt & Moskowitz [833], Hvidkjaer [995]. Momentum strategies produce their greatest profits towards the end of the year, due to the increasing selling pressure on loser assets and their greatest losses towards the beginning of the year. The opposite occurs for contrarian strategies. On monthly seasonality effects see also Heston & Sadka [941].

Return seasonality is robust across many different markets, see Hawawini & Keim [916], Fama [663], Schwert [1510], Malkiel [1290] and also Chap. 10 for intraday patterns in stock returns. Winner-loser reversals have been observed in national stock market indices (see Richards [1447]). Buckley & Tonks [322] evaluate a trading strategy based on excess volatility consisting in buying-selling the market portfolio when it is far away from its mean, showing that such a strategy can yield excess returns. Moreover, the existence of profitable trading strategies based on technical analysis has been confirmed by more recent studies (see Park & Irwin [1400], Po-Hsuan & Chung-Ming [1423], Han et al. [884]). The profitability of momentum strategies has been confirmed in Chan et al. [407] and an analogous phenomenon has been observed in European and emerging markets (see Rouwenhorst [1478], Chan et al. [405], Bhojraj & Swaminathan [229]). Moreover, the profitability of momentum strategies is confirmed by the investment strategies frequently adopted by mutual fund managers, whose performance is evaluated over a short horizon. Note also that tax loss selling practices may induce a momentum effect around the month of January, when agents sell badly performing assets to offset capital gains (see Grinblatt & Moskowitz [833]).

In Chan et al. [407] and Barber et al. [141] it has been documented that investment strategies based on analysts' recommendations typically earn abnormal gross returns. This empirical result can be interpreted as an evidence against market efficiency, either due to the existence of mispricings or due to analysts' private information. Moreover, the performance of mutual funds performance is highly persistent and this can also be interpreted as an evidence against market efficiency (see Brown & Goetzmann [314]).

The predictability of future returns through variables such as dividend yields, earnings and interest rates has been criticized from a statistical point of view (see, e.g., Kaul [1076]). Two main issues have been pointed out, especially in the case of long horizon returns, see Kirby [1102], Hodrick [951], Richardson & Stock [1450], Richardson [1448], Goetzmann & Jorion [795], Nelson & Kim [1373]. First, overlapping observations induce serial correlation in errors and, therefore, the validity of the statistical procedure is negatively affected. Second, there is a small sample size problem. The regression on endogenous lagged variables is not unbiased in finite samples and problems mainly arise when the predictor variable is persistent and its innovations are highly correlated with returns. As a result, these biases may induce spurious predictability. Taking into account these potential biases, weak predictability through the dividend yield and earnings yield has been detected in Goetzmann & Jorion [795], Ang et al. [65], Campbell et al. [348], while Nelson & Kim [1373], Hodrick [951], Lewellen [1211], Campbell & Yogo [359], Campbell & Thompson [357], Cochrane [466], Lettau & Van Nieuwerburgh [1200] still find some evidence of return predictability through the dividend yield, the book-to-market and the earnings-price ratio. Moreover, as in the case of serial correlation, predictability seems to be mostly restricted to the pre-Second World War period (see Nelson & Kim [1373]).

In addition, all the empirical results reporting predictability of future returns through variables such as past returns, dividends and earnings suffer from a strong *survivorship bias*. Indeed, structural relations between macroeconomic/monetary or firm-specific variables and asset returns are only observed ex-post, conditionally on the survival of the asset on which the returns are observed. In other words, especially in the long run, we only observe the returns of the assets which did not default during the period, thus inducing a bias in the empirical analysis. This bias can introduce spurious mean reversion and predictability in asset returns. Indeed, among the firms with a poor performance in the recent past, we can only observe the firms with a relatively good recent performance which have survived so far. In Goetzmann & Jorion [796] it is shown that the survivorship bias generates predictability of future returns through the dividend-price ratio. Taking this fact into account, only marginal predictability of returns through the dividend yield is observed. This problem is similar to that identified by Kothari et al. [1126] in the case of the CAPM (see also Brown et al. [315] on the survivorship bias).

Several studies have investigated the origins of the lead-lag effect: according to Boudoukh et al. [273], the lead-lag effect is to be attributed to the presence of non-synchronous trading and transaction costs and not to the existence of delayed reaction of small sized stocks to information or to time varying expected returns. However, McQueen et al. [1315] and Lo & MacKinlay [1237] provide evidence against transaction costs, non-synchronous trading, time varying expected returns as a source of lead-lag effects and in favor of delayed reaction of small firms to information. Moreover, Badrinath et al. [104] show that past returns on stocks held by institutional (informed) traders are positively correlated with contemporaneous returns on stocks held by non-institutional (uninformed) traders. Somehow analogously, Brennan et al. [295] show that stocks followed by many analysts tend to lead stocks followed by fewer analysts, while Chordia & Swaminathan [440] show that there is a lead-lag relationship between returns of high trading volume stocks and returns of low trading volume stocks. In general terms, this empirical evidence supports the interpretation of delayed diffusion of information as a source of lead-lag effects.

Balduzzi & Kallal [117], Balduzzi & Robotti [119] show that multi-beta factor models can be tested by means of the Hansen-Jagannathan bound (see Proposition 6.36). Balduzzi & Kallal [117] find evidence against the three-factor model proposed by Fama & French [670], while Balduzzi & Robotti [119] find that several macroeconomic factors are indeed priced by the market with a significant risk premium (in particular: the return on a stock market proxy, the consumption growth, the slope of the term structure, the interest rate and the default premium) and the sign of the risk premium is generally consistent with the intuition of the ICAPM. In general, the ICAPM has a better performance than the CAPM, the CCAPM and the three-factor model of Fama & French [670].

In Gibbons & Ferson [776], Ferson et al. [690], Ferson [686, 687], Bekaert & Hodrick [186], the variation over time of asset returns has been described by relying on latent variables models, where expected returns and β coefficients are conditional on a set of latent variables. Expected returns vary over time but conditional β are fixed parameters. In Gibbons & Ferson [776] it is shown that a model based on a single latent variable is not rejected empirically, while subsequent papers suggest the presence of more than one latent variable.

Besides the equity premium puzzle, it has also been observed that classical asset pricing theory encounters difficulties in explaining the risk premia observed in the term structure of interest rates (see Sect. 6.4). Indeed, as shown in Backus et al. [103], an intertemporal general equilibrium asset pricing model with time additive utility is not capable of reproducing the sign, the magnitude and the variability of the risk premia observed in forward interest rates over expected future spot interest rates. A very large risk aversion coefficient (greater than 8) would be needed to generate average risk premia as large as those observed in the Treasury bill markets. In particular, classical asset pricing theory does not easily explain the positive relation existing between maturity and risk premium. This phenomenon is typically referred to as the *term premium puzzle*.

Finally, we refer to Fama [664], Barber & Lyon [142], Kothari & Warner [1127], Kothari [1123] for a discussion of the statistical problems associated to the measurement of abnormal long horizon returns associated with event studies. In particular, there is often a bias towards the rejection of the market efficiency hypothesis.

7.5   Exercises

Exercise 7.1 Prove inequalities (7.8)–(7.9).

Exercise 7.2 Prove the two inequalities in (7.11).

Exercise 7.3 As in equation (7.4), define $\varepsilon_t^{\mathrm{F}} := s_t^{\mathrm{F}} + d_t - r_f s_{t-1}^{\mathrm{F}}$, for all $t \in \mathbb{N}$, with an analogous definition for the information flow $\mathbb{G}$. Suppose that $r_f > 1$ and that

$$\mathrm{Var}(\varepsilon_{t+s}^{\mathrm{F}}) = \mathrm{Var}(\varepsilon_t^{\mathrm{F}}) \quad \text{and} \quad \mathrm{Var}(\varepsilon_{t+s}^{\mathrm{G}}) = \mathrm{Var}(\varepsilon_t^{\mathrm{G}}), \qquad \text{for all } t, s \in \mathbb{N}.$$

Under this assumption, show that inequality (7.12) holds.

Exercise 7.4 Derive the log-linear approximation (7.13).

Exercise 7.5 Consider an infinitely lived economy (i.e., $T = \infty$) and suppose that the dividend process of a risky security satisfies the following dynamics:

$$d_{t+1} = (1 + g)d_t + \sum_{k=0}^{N} \gamma_k \big(E_{t-k+1} - (1 + g)E_{t-k} \big), \qquad \text{for all } t \in \mathbb{N}, \qquad (7.17)$$

for some $N \in \mathbb{N}$, with $d_0 = 0$ and where $(E_t)_{t\in\mathbb{N}}$ denotes the earnings process and $\{\gamma_k\}_{k=0,1\ldots,N}$ is a family of non-negative weight factors. This model represents the situation where the dividend process is set according to a growth rate equal to g, but managers deviate from this long run growth path in response to changes in earnings that deviate from their long run growth path. Define by $\bar{d}_t := d_t/(1 + g)^t$ the de-trended dividend and, similarly, $\bar{E}_t := E_t/(1 + g)^t$, for all $t \in \mathbb{N}$. Suppose that the earnings process $(E_t)_{t\in\mathbb{N}}$ is related to the firm value process $(V_t)_{t\in\mathbb{N}}$ by $E_t = r_f V_t$, for all $t \in \mathbb{N}$, where r_f is the constant risk free rate and suppose also that stocks are priced rationally, i.e., $s_t = V_t$, for all $t \in \mathbb{N}$. Show that, under the present assumption, the rational ex-post price as defined in (7.10) (with the terminal value $\bar{s}_T^e$ being defined as $\bar{s}_T^e := \sum_{t=0}^{T-1} \bar{s}_t/T$, with $(\bar{s}_t)_{t\in\mathbb{N}}$ denoting the de-trended observed price process) admits the representation

$$\hat{s}_t^e = \sum_{k=-N}^{T-1} w_{tk} \bar{s}_k, \qquad \text{for all } 0 \le t \le T, \qquad (7.18)$$

for a suitable family $\{w_{tk}\}$ of weight factors.

Exercise 7.6 Prove Proposition 7.1.

Exercise 7.7 In this exercise, we introduce and solve the general equilibrium asset pricing model proposed by Cecchetti et al. [378] with the purpose of showing that negative serial correlation in long horizon stock returns is consistent with an equilibrium model of asset pricing.

Consider an infinitely lived economy (i.e., $T = \infty$) admitting a representative agent with power utility function of the form $u(x) = x^{1+\gamma}/(1 + \gamma)$ and discount factor δ and where a single risky security is traded, with price process $(s_t)_{t \in \mathbb{N}}$ and paying the aggregate endowment as dividend. Assume that the aggregate endowment process $(e_t)_{t \in \mathbb{N}}$ is modeled as follows:

$$e_t = e_{t-1} e^{\alpha_0 + \alpha_1 X_{t-1} + \varepsilon_t}, \qquad \text{for all } t \in \mathbb{N},$$

where $(\varepsilon_t)_{t \in \mathbb{N}}$ is a sequence of independent random variables identically distributed according to a normal law with mean zero and variance σ^2 and where $(X_t)_{t \in \mathbb{N}}$ is a Markov process with state space $\{0, 1\}$ characterized by the transition probabilities

$$\mathbb{P}(X_t = 1 | X_{t-1} = 0) = 1 - q \quad \text{and} \quad \mathbb{P}(X_t = 0 | X_{t-1} = 1) = 1 - p.$$

This means that the logarithm of the aggregate endowment process follows a random walk with a stochastic drift, where the process $(X_t)_{t \in \mathbb{N}}$ is meant to represent the high/low growth state of the economy. The Markov chain $(X_t)_{t \in \mathbb{N}}$ is assumed to be independent of the sequence $(\varepsilon_t)_{t \in \mathbb{N}}$.

Show that the equilibrium price s_t is given by

$$s_t = \varrho(X_t) e_t, \qquad \text{for all } t \in \mathbb{N},$$

where the function $\varrho : \{0, 1\} \to \mathbb{R}$ is defined by

$$\varrho(0) := \tilde{\delta}\big(1 - \tilde{\delta}\tilde{\alpha}_1(p + q - 1)\big)/\Delta,$$
$$\varrho(1) := \tilde{\delta}\tilde{\alpha}_1\big(1 - \tilde{\delta}(p + q - 1)\big)/\Delta,$$

with

$$\tilde{\delta} := \delta e^{\alpha_0(1+\gamma)+(1+\gamma)^2\sigma^2/2}, \qquad \tilde{\alpha}_1 := e^{\alpha_1(1+\gamma)}$$

and

$$\Delta := 1 - \tilde{\delta}(p\tilde{\alpha}_1 + q) + \tilde{\delta}^2\tilde{\alpha}_1(p + q - 1).$$

Exercise 7.8 Consider a multi-period economy with dates $\{0, 1, \ldots, T\}$, for some $T \in \mathbb{N}$, and a time additive utility function of the form:

$$U(c_0, c_1, \ldots, c_N) = \sum_{t=0}^{T} \delta^t u_t(c_t),$$

with $\delta \in (0, 1)$ and $u_t : \mathbb{R}_+ \to \mathbb{R}$, for all $t = 0, 1, \ldots, T$. Let us define by $r^r_{u_t}$ the coefficient of relative risk aversion of the utility function u_t (see Sect. 2.2), for all $t = 0, 1, \ldots, T$, and define the elasticity of intertemporal substitution of U as

$$\eta_U(s, t) := \frac{d \log (c_t/c_s)}{d \log \left(\frac{\partial U}{\partial c_s} / \frac{\partial U}{\partial c_t} \right)}, \qquad \text{for all } s, t \in \{0, 1, \ldots, T\}.$$

Show that the following relation holds for every $s, t \in \{0, 1, \ldots, T\}$:

$$\eta_U(s, t) = \frac{1 - \frac{c_t}{c_s} \frac{\delta^{t-s} u'_t(c_t)}{u'_s(c_s)}}{r^r_{u_t}(c_t) - \frac{c_t}{c_s} \frac{\delta^{t-s} u'_t(c_t)}{u'_s(c_s)} r^r_{u_s}(c_s)}. \tag{7.19}$$

Moreover, if the utility functions $(u_t)_{t=0,1,\ldots,T}$ are CRRA with the same coefficient of relative risk aversion for all dates $t = 0, 1, \ldots, T$, then it holds that

$$\eta_U(s, t) = \frac{1}{r^r_{u_t}(c_t)}, \qquad \text{for all } s, t = 0, 1, \ldots, T.$$

Exercise 7.9 Consider a representative agent economy, where the representative agent's utility function is given by $\sum_{t=0}^{\infty} e^{-\delta t} (c_t^{1-\gamma} - 1)/(1-\gamma)$. Suppose that, at each period $t + 1$, a disaster event can occur with probability p_t, with p_t denoting the $\mathscr{F}_t$-conditional probability of occurrence, for all $t \in \mathbb{N}$. Depending on the occurrence of a disaster, the aggregate endowment/consumption growth satisfies

$$\frac{e_{t+1}}{e_t} = \begin{cases} e^g, & \text{if there is no disaster,} \\ e^g B_{t+1}, & \text{otherwise,} \end{cases}$$

where B_{t+1} is a random variable taking values in $(0, 1)$, for all $t \in \mathbb{N}$, and g represents the logarithmic rate of growth of the aggregate endowment. Define $R := \delta + \gamma g$. Suppose also that the occurrence of the disaster affects the dividend process $(d_t)_{t\in\mathbb{N}}$ of a risky security (whose cum-dividend price process is denoted by $(p_t)_{t\in\mathbb{N}}$) in the following way:

$$\frac{d_{t+1}}{d_t} = \begin{cases} e^{gd}(1 + \varepsilon_{t+1}), & \text{if there is no disaster,} \\ e^{gd}(1 + \varepsilon_{t+1}) F_{t+1}, & \text{otherwise,} \end{cases}$$

where g_d is the logarithmic rate of growth of the dividend process, $(\varepsilon_t)_{t\in\mathbb{N}}$ is a sequence of random variables (independent from the disaster event) with zero mean and F_t is a random variable taking values in $[0, 1]$, for all $t \in \mathbb{N}$. We also assume that the random variable ε_{t+1} and the random variable representing the occurrence of a disaster are $\mathscr{F}_t$-conditionally independent, for all $t \in \mathbb{N}$. Define also the *asset resilience* H_t by

$$H_t := p_t \mathbb{E}[B_{t+1}^{-\gamma} F_{t+1} - 1 | \{\text{there is a disaster at } t + 1\} \vee \mathscr{F}_t]$$

and suppose that the resilience process $(H_t)_{t\in\mathbb{N}}$ satisfies the following dynamics:

$$H_{t+1} = H^* + \frac{1 + H^*}{1 + H_t} e^{-\phi}(H_t - H^*) + \varepsilon_{t+1}, \qquad \text{for all } t \in \mathbb{N},$$

where $H^* > -1$ represents a long term equilibrium resilience level and $(\varepsilon_t)_{t\in\mathbb{N}}$ is a sequence of random variables (independent from all the other random variables introduced above) with zero mean.

Show that the equilibrium cum-dividend price p_t of the risky security is given by

$$p_t = \frac{d_t}{1 - e^{-r}}\left(1 + \frac{e^{-R+g_d}(H_t - H^*)}{1 - e^{-r-\phi}}\right), \qquad \text{for all } t \in \mathbb{N}, \tag{7.20}$$

where $r := R - g_d - \log(1 + H^*)$.

Chapter 8
Information and Financial Markets

> *We must look at the price system as a [...] mechanism for communicating information if we want to understand its real function.*
>
> Hayek (1945)

The preceding chapters have been devoted to the analysis of economies populated by agents with homogeneous beliefs. We have always assumed that the structure of the probability space is *common knowledge* among all the agents, in the sense that every agent knows the laws of the random variables describing asset prices, dividends, returns and endowments. Moreover, agents are assumed to be price takers and rational, but do not recognize any informational value of asset prices concerning the uncertainty of the economy. This assumption is certainly plausible when agents have homogeneous beliefs and information, but it becomes quite restrictive in an economy where agents are endowed with private information.

In this chapter, we consider economies populated by agents with different information, in the sense that the random variables describing asset prices, dividends and returns have different probability laws from the point of view of different agents. In this sense, we say that agents have *private information*. More specifically, we distinguish between *heterogeneous information*, when all the agents of the economy observe a private signal, specific to each agent, and *asymmetric information*, when the economy is populated by informed and uninformed agents, with only the informed agents having access to some additional information about the economy. In both cases, we assume that agents are rational and, hence, are able to identify the informational content of asset prices. When the information is heterogeneous, asset prices will *aggregate* the agents' private information, possibly making information fully homogeneous among the agents, while, when the information is asymmetric, asset prices will *transmit* information from the informed to the uninformed agents. It has to be noted that, when the information is heterogeneous or asymmetric, agents will trade for two main motivations: for risk sharing, similarly as in the analysis developed in the preceding chapters, and for informational reasons (*speculation*). Due to this second motivation for trading, especially when the information is asymmetric, there will be an *adverse selection* problem: less informed agents will be afraid to trade with more informed agents and this may preclude them for trading. The central question of the present chapter can be phrased in the following terms:

© Springer-Verlag London Ltd. 2017   397
E. Barucci, C. Fontana, *Financial Markets Theory*,
Springer Finance, DOI 10.1007/978-1-4471-7322-9_8

in a perfectly competitive market, when agents have private (heterogeneous or asymmetric) information, can asset prices aggregate/transmit the agents' private information?

The above question can be easily addressed in the context of a simple two-period economy (i.e., $t \in \{0, 1\}$), where agents take their investment-consumption decisions at the initial date $t = 0$ and consume at $t = 1$. In order to take decisions at $t = 0$, agents need to evaluate the probability distribution of future consumption at $t = 1$ and, therefore, they need to evaluate the probability distribution of the random variables describing the dividends delivered at $t = 1$. Suppose that the information is heterogeneous because each agent can observe at the initial date $t = 0$ a private signal with a non-trivial informational content. By observing the private signal, each agent will revise his beliefs concerning the probability distribution of the random variables associated with the investment opportunities. Since each agent is assumed to be an expected utility maximizer, his optimal choice at $t = 0$ (in particular his asset demand) will depend on the probability distribution updated on the basis of the information conveyed by the observation of the private signal. This implies that, by trading in the market, an agent will somehow reveal his beliefs. Investment decisions and, therefore, the resulting market prices will carry a non-trivial informational content, resulting from the aggregation of the agents' private information. By observing asset prices, agents can infer the information of other agents and, as a consequence, update their beliefs. In this chapter, we aim at understanding to what extent asset prices make public, aggregate and transmit private information in a perfectly competitive economy. In particular, we are interested in understanding when this process of price observation and revision of beliefs will lead to homogeneous beliefs among the agents, with market prices fully revealing the agents' private information. The relevance of this aggregation/transmission of information phenomenon depends on the organization of the market. In this chapter, we consider perfectly competitive markets, while in Chap. 10 we will extend the analysis to alternative market settings.

If the answer to the above question is positive, then markets are *information-ally efficient*. It goes without saying that a positive answer would be a strong argument in favor of a free market: indeed, not only does the market allocate resources in an efficient way (allocative efficiency, first welfare theorem), but it also aggregates/transmits private information, thereby eliminating any advantage associated with private information. In this chapter, we want to analyse the aggregation/transmission of information function of prices and evaluate under which conditions markets are informationally efficient. Of course, it suffices to consider the insider trading legislation adopted in most countries and the amount of resources spent in market research by investors to understand that the message from the efficient markets theory is not fully satisfactory in reality (see also Sect. 10.5).

The fact that asset prices aggregate/transmit private information implies that the classical concept of Arrow-Debreu equilibrium does no longer suffice to describe a stable condition of the economy. To this end, the appropriate notion of equilibrium is represented by the *Green-Lucas equilibrium*. The limitations of the Arrow-Debreu equilibrium in an economy where agents have private information can be understood by considering the following example, taken from Huang & Litzenberger [971, Section 9.2].

Consider a two-period economy (i.e., $t \in \{0, 1\}$) with five possible states of the world $\Omega = \{\omega_1, \ldots, \omega_5\}$. There exists a single consumption good which is traded at $t = 0$ and consumed at $t = 1$. We assume that markets are complete, in the sense that at $t = 0$ there are five markets open for the five Arrow securities paying one unit of the single consumption good in correspondence of each state of the world ω_s, $s = 1, \ldots, 5$. Suppose that the economy is populated by two agents a and b. Both agents are endowed (ex-ante) with the same beliefs, in the sense that they agree on the probabilities of occurrence of each of the five states of the world. However, at the initial date $t = 0$, both agents observe a private signal, represented by a binary random variable $\tilde{y}^i$, for $i = a, b$, taking two possible values $\{L, H\}$. We assume that

$$\tilde{y}^a(\omega) = \begin{cases} L, & \text{if } \omega \in \{\omega_1, \omega_2, \omega_3\}, \\ H, & \text{if } \omega \in \{\omega_4, \omega_5\}; \end{cases} \quad \text{and} \quad \tilde{y}^b(\omega) = \begin{cases} L, & \text{if } \omega \in \{\omega_1, \omega_2\}, \\ H, & \text{if } \omega \in \{\omega_3, \omega_4, \omega_5\}. \end{cases}$$

The signal induces heterogeneous information: if agent a observes $\{\tilde{y}^a = L\}$ then he will know that the state of the world belongs to the subset $\{\omega_1, \omega_2, \omega_3\}$. On the other hand, if agent b observes the realization $\{\tilde{y}^b = L\}$, he will know that the state of the world belongs to $\{\omega_1, \omega_2\}$. After the observation of the private signals, the beliefs of agent a will be represented by the partition $\{\{\omega_1, \omega_2, \omega_3\}, \{\omega_4, \omega_5\}\}$, whereas those of agent b will be described by $\{\{\omega_1, \omega_2\}, \{\omega_3, \omega_4, \omega_5\}\}$. Furthermore, we assume that the structure of these two partitions is common knowledge among the two agents. In the context of the present example, the two agents exhibit *heterogeneous beliefs*. Indeed, after the observation of the signal, the two agents believe in different probability distributions, here represented by different partitions of Ω. In particular, after the observation of their private signals, they can disagree about the possibility of occurrence of the state of the world ω_3.

Let us assume that the two agents a and b have strictly increasing preferences. In this case, it can be easily shown that, if one adopts the classical notion of equilibrium introduced in Chap. 4, then there does not exist an equilibrium of the economy. Indeed, suppose that the realized state of the world is ω_1. In this case, the signal $\tilde{y}^a$ received by agent a takes the value L and the signal $\tilde{y}^b$ received by agent b takes the value L as well. After the observation of the private signal, agent a learns that the true state of the world belongs to $\{\omega_1, \omega_2, \omega_3\}$, while agent b learns that the true state of the world belongs to the smaller subset $\{\omega_1, \omega_2\}$. This difference in the agents' beliefs implies that there cannot exist an equilibrium price p_3 of the Arrow security corresponding to the state of the world ω_3. Indeed, suppose that such a security has a strictly positive price. In that case, agent b would like to short sell an unlimited amount of the security, since he knows that the state of the world cannot be ω_3, and use the proceeds of this short sale to finance consumption in the two states of the world $\{\omega_1, \omega_2\}$. On the other hand, agent a, who assigns a strictly positive probability of occurrence to the state of the world ω_3, would be interested in investing in the Arrow security corresponding to ω_3, but certainly his demand would not balance the supply of agent b. Hence, markets cannot clear and an Arrow-Debreu equilibrium cannot exist if markets are complete. A similar argument can be applied if the Arrow security corresponding to ω_3 has a non-positive price.

By examining the above example, we can easily see why the Arrow-Debreu equilibrium does not represent the appropriate notion of equilibrium in the presence of heterogeneous beliefs. In the classical definition of equilibrium, agents consider prices as given and only use prices to define their budget constraints. In other words, agents do not assign any informational value to prices. However, in the context of the above example, since the structure of the partitions is assumed to be of common knowledge among the two agents, it is rather unrealistic to assume that agent a will not revise his beliefs after observing a huge short sale from agent b, for any strictly positive price p_3 of the Arrow security paying one unit of the consumption good in correspondence of ω_3. Somehow, the trading activity of agent b reveals his private information to the market. In this case, a price equal to zero for the Arrow security would be of equilibrium if, after observing the private signal $\{\tilde{y}^a = L\}$ *and* the price $p_3 = 0$, agent a would infer that the realized state of the world cannot be ω_3. Note also that, after the revision of beliefs, the ex-ante difference in the agents' private information would be completely eliminated ex-post and both agents would believe that the state of the world belongs to $\{\omega_1, \omega_2\}$.

The above discussion leads us to the introduction of the *Green-Lucas equilibrium*, which is a rational expectations equilibrium. In a Green-Lucas equilibrium, all agents determine their optimal demands by fully exploiting the information contained in asset prices. In the context of the above example, a Green-Lucas equilibrium with $p_3 = 0$ can be constructed. If $\{\tilde{y}^a = \tilde{y}^b = L\}$ and $p_3 = 0$, agent b will find $p_3 = 0$ coherent with his private information and agent a, knowing the partition of agent b, will learn from observing $p_3 = 0$ that agent b knows for sure that the realized state of the world cannot correspond to ω_3. Therefore, by updating his beliefs via the Bayes rule, agent a will infer that the state of the world cannot be ω_3. As explained above, the (ex-ante) difference in the agents' private information will be fully equalized (ex-post) by equilibrium prices.

Radner [1439, p. 941] points out that "an agent who has a good understanding of the market is in a position to use market prices to make inferences about the (non-price) information received by other agents... These inferences are derived, explicitly or implicitly, from an individual's "model" of the relationship between the non-price information received by market participants and the market prices... An equilibrium of this system, in which the individual models are identical with the true model, is called a *rational expectations equilibrium*... The concept of equilibrium is more subtle, of course, than the ordinary concept of the equilibrium of supply and demand. In a rational expectation equilibrium, not only are prices determined so as to equate supply and demand, but individual economic agents correctly perceive the true relationship between the non-price information received by the market participants and the resulting equilibrium market prices. This contrasts with the ordinary concept of equilibrium in which the agents respond to prices but do not attempt to infer other agents' non-price information from actual market prices".

In this chapter, we shall distinguish between *heterogeneous beliefs*, i.e., when agents exhibit different (posterior) probability distributions conditioned on the observation of private signals related to the economic environment, and *heterogeneous opinions*, i.e., when agents exhibit different probability distributions not related to the observation of private signals informative about the economic environment. Typically, information about the economic environment is meant to be information about the distribution of asset payoffs/dividends. The difference between heterogeneous beliefs and heterogeneous opinions may seem subtle but has quite distinct implications. In particular, an agent will not change his behavior or revise his probability distribution by knowing that another agent exhibits a different probability distribution if the latter is not motivated by private information related to the economic conditions (heterogeneous opinions). On the contrary, an agent will revise his behavior if he knows that the other agent's probability distribution is generated by private information about the economy (heterogeneous beliefs). In general, heterogeneous opinions can be treated within the paradigm described in the preceding chapters by introducing heterogeneous agents. In this chapter we shall be mostly concerned with heterogeneous beliefs. We will briefly discuss the case of heterogeneous opinions in Sect. 8.5.

The question of whether market prices fully reflect the agents' information represents the central question of efficient markets theory and goes back to seminal contributions in the financial markets literature. A microeconomic approach to this issue is somehow more recent, going back to the contributions of the rational expectations theory (see Grossman [842]). Moreover, this question has important implications from an empirical perspective, as discussed in Chap. 7. From this point of view, the analysis of market efficiency can be reduced to the following problem: given an equilibrium market model, do asset prices reflect the information available in the market?

Any answer to this question requires on the one hand a definition of an equilibrium model and, on the other hand, a definition of the notion of available information. Hence, testing market efficiency corresponds to jointly testing the validity of an equilibrium model and the fact that a given information set is embedded in market prices. This is intimately related to the fundamental no-arbitrage equation presented in Sect. 6.5 and discussed from an empirical perspective in Chap. 7. In particular, the fundamental solution (6.86) to the no-arbitrage equation (6.85) shows that equilibrium prices are determined by three key ingredients: the dividend process $(d_t)_{t \in \mathbb{N}}$, the information flow $\mathbb{F} = (\mathscr{F}_t)_{t \in \mathbb{N}}$ and the stochastic discount factor. In particular, under the assumption of risk neutrality, future excess returns cannot be predicted on the basis of the current information $\mathscr{F}_t$ and expected excess returns are null (see Proposition 7.1). If an information set $\mathscr{F}_t$ does not allow to predict future excess returns, then asset prices are informationally *efficient* in the sense that they correctly reflect the available information. If this is the case, then expected returns are equal to the risk free rate r_f and any trading strategy will not perform better on average than the market portfolio or simple buy-and-hold strategies in any of the traded security. Somehow more generally, following Jensen [1031], we can define (informational) *market efficiency* with respect to some information flow $\mathbb{F}$ as the impossibility of making profits by trading on the basis of the information contained in $\mathbb{F}$.

Following a classification widely accepted in the literature, we can identify three different forms of market efficiency, depending on the specification of the information flow (see Sect. 7.2 and Fama [661]):

- *Efficiency in weak form*, when the information set includes all past realizations of asset prices and returns;
- *Efficiency in semi-strong form*: when the information set includes all publicly available information, including all past realizations of asset prices and returns;
- *Efficiency in strong form*: when the information set includes all publicly available information and all private information.

According to this classification, three different information flows are associated to three different levels of market efficiency. Clearly, these three notions of market efficiency are increasingly stronger, in the sense that strong efficiency implies semi-strong efficiency and, in turn, semi-strong efficiency implies weak efficiency. This simply follows by the inclusions among the different information sets considered in the above classification.

The typical way of empirically testing market efficiency consists in building trading strategies on the basis of the information contained in some information set. If such a strategy can be shown to perform better than the market, then this is an indication that the considered information is not fully incorporated in market prices and, hence, some form of market efficiency is rejected. In Chap. 7, discussing the empirical results on the implications of the fundamental no-arbitrage equation, we have considered market efficiency in the weak and in the semi-strong form. As we have seen, there is empirical evidence showing that asset returns exhibit predictable components. However, as we have already mentioned, these empirical findings do not necessarily have to be interpreted as an evidence against market efficiency (or against agents' rationality). Indeed, as a matter of fact, returns can be predictable because of time-varying investment opportunities or of risk averse agents. The empirical evidence reported in the literature on market efficiency in the strong form is mostly negative: trading strategies constructed on the basis of private information are shown to yield positive excess returns with respect to the market. This phenomenon is intimately related to insider trading (see Sect. 10.5).

This chapter is structured as follows. In Sect. 8.1, we study the value of information first from the point of view of an individual decision maker and then for the economy as a whole, introducing the notion of Green-Lucas equilibrium and the role of information transmission/aggregation of market prices. Sections 8.2 and 8.3 respectively deal with the possibility and the impossibility of informationally efficient markets, also discussing the acquisition of information by market participants. In Sect. 8.4, we provide a brief overview on the role of information in dynamic multi-period market models. Section 8.5 is concerned with the equilibrium and asset pricing implications of economies where the agents exhibit different opinions. In Sect. 8.6, we survey the empirical evidence on information-based models. At the end of the chapter, we provide a guide to further readings as well as a series of exercises.

8.1 The Role of Information in Financial Markets

In the most simple context, the role of information in financial markets can be explained by analogy to the example of the farmer given at the beginning of Chap. 4, by introducing an intermediate date between the sowing and the harvesting dates. At this intermediate date, the farmer observes a private signal carrying some information about the future weather conditions. The present simple example consists of three dates (i.e., $t \in \{-1, 0, 1\}$): at $t = -1$ the farmers sow, at the intermediate date $t = 0$ they receive private signals and at $t = 1$ they harvest. Note that, even though all the farmers start at $t = -1$ with the same prior beliefs about the future weather conditions, if they receive different private signals at $t = 0$ then they will form *heterogeneous beliefs* about the future weather conditions.

Generalizing this example, let us consider an economy populated by $I \in \mathbb{N}$ agents ($i = 1, \ldots, I$). At the starting date $t = -1$, all agents share homogeneous beliefs about the future state of the world, described by a common probability space $(\Omega, \mathscr{F}, \mathbb{P})$. For simplicity, we assume that the probability space is finite, i.e., the set Ω consists of $S \in \mathbb{N}$ elementary states of the world $\{\omega_1, \ldots, \omega_S\}$ and we denote by $\pi_s := \mathbb{P}(\omega_s) > 0$ the probability of occurrence of state ω_s, for all $s = 1, \ldots, S$. The structure of the probability space $(\Omega, \mathscr{F}, \mathbb{P})$ is assumed to be *common knowledge* among the I agents. We assume that each agent is characterized by the couple (u^i, e^i), where $u^i : \mathbb{R}_+ \to \mathbb{R}$ is a strictly increasing and strictly concave utility function and the vector $e^i = (e^i_1, \ldots, e^i_S) \in \mathbb{R}^S_+$ denotes the endowment of agent i (expressed in terms of wealth or some reference consumption good) in the S possible states of the world, for all $i = 1, \ldots, I$. Similarly as in Chap. 4, a consumption plan is represented by a vector $x = (x_1, \ldots, x_S) \in \mathbb{R}^S_+$, with x_s representing consumption in the state of the world ω_s. We assume that consumption is non-negative and only takes place at the final date $t = 1$. All the elements of the economy introduced so far are assumed to be common knowledge among the agents. We want to point out that the present setting can be generalized to the case of state dependent utility functions.

At the intermediate date $t = 0$, each agent i observes a private signal represented by the random variable $\tilde{y}^i$. The signals are private in the sense that each agent i only observes his own signal $\tilde{y}^i$, without having access to the signals of the other agents. For simplicity, we assume that each random variable $\tilde{y}^i$ can take $J \in \mathbb{N}$ possible values $\{y^i_1, \ldots, y^i_J\}$, for all $i = 1, \ldots, I$, and we define $\pi(y^i_j) := \mathbb{P}(\tilde{y}^i = y^i_j)$. We denote by $\tilde{y}$ the I-dimensional random vector $(\tilde{y}^1, \ldots, \tilde{y}^I)$ and let $\pi(y) := \mathbb{P}(\tilde{y} = y)$, for all $y \in \mathscr{Y}$, with $\mathscr{Y}$ denoting the set of all possible realizations of the random vector $\tilde{y}$. We assume that the probability distribution $\{\pi(y); y \in \mathscr{Y}\}$ of the random vector $\tilde{y}$ is common knowledge among the I agents and is known at the initial date $t = -1$. Note that we allow for $y^i_j \neq y^{i'}_j$, for $i, i' \in \{1, \ldots, I\}$, and it may happen that agent i observes the realization y^i_j while agent i' observes the realization $y^{i'}_{j'}$, for some $j, j' \in \{1, \ldots, J\}$ with $j \neq j'$.

As soon as an agent observes his private signal at the intermediate date $t = 0$, he will update his beliefs according to the Bayes rule. Let $y = (y^1, \ldots, y^I) \in \mathscr{Y}$ denote a realization of the random vector $\tilde{y}$ of the private signals. For each $i = 1, \ldots, I$, the updated beliefs of agent i are represented by the conditional distribution

$$\left(v^i_1(y^i), \ldots, v^i_S(y^i) \right), \tag{8.1}$$

with $v^i_s(y^i) := \mathbb{P}(\omega_s | \tilde{y}^i = y^i)$ representing the probability of the state of the world ω_s conditionally on the observation of signal $y^i \in \{y^i_1, \ldots, y^i_J\}$ by agent i, for all $i = 1, \ldots, I$. At date $t = 0$, the agents have *heterogeneous beliefs* since they observe different signals, leading to different conditional probability distributions for the unknown state of the world. It is important to remark that the private signals are assumed not to directly affect the agents' endowments nor their utility functions. However, the signals convey useful information on the unknown state of the world.

Summing up, in the present setting we can distinguish three different dates:

- $t = -1$ (*ex-ante*): nature selects a state of the world;
- $t = 0$ (*interim*): agents observe private signals and update their beliefs;
- $t = 1$ (*ex-post*): the state of the world is revealed and agents consume.

In the remaining part of this section, we shall first analyse the value of receiving a private signal from the individual point of view and then the effect of information from a social point of view, introducing the notion of Green-Lucas equilibrium. The presentation of this section is inspired from Laffont [1154, Chapters 4 and 9].

The Value of Information

In order to analyse the impact of additional information on the optimal choice of an individual, let us consider an optimal choice problem of the following form, for a strictly increasing and strictly concave utility function $u : \mathbb{R}_+ \to \mathbb{R}$ (since we now consider an individual optimization problem, we drop the superscript i):

$$\max_{(x_1, \ldots, x_S) \in B} \sum_{s=1}^{S} \pi_s u(x_s), \tag{8.2}$$

where $B \subset \mathbb{R}^S_+$ represents the set of feasible consumption plans, with respect to a budget constraint defined in terms of a given endowment $(e_1, \ldots, e_S)$. We assume that an optimal solution $(x^*_1(\pi), \ldots, x^*_S(\pi)) \in B$ exists, where we emphasize the dependence on the probability distribution $\pi = (\pi_1, \ldots, \pi_S)$ representing the prior beliefs. We denote by $U^*(\pi)$ the optimal value of the expected utility associated to problem (8.2) in correspondence of the prior beliefs $\pi = (\pi_1, \dots, \pi_S)$, i.e., $U^*(\pi) := \sum_{s=1}^{S} \pi_s u(x_s^*(\pi))$.

Consider the case of an agent who can observe a signal $\tilde{y}$ carrying some information about the unknown state of the world. We assume that $\tilde{y}$ takes values in the set $\{y_1, \dots, y_J\}$, for some $J \in \mathbb{N}$, with corresponding probabilities $\pi^{\tilde{y}} = (\pi(y_1), \dots, \pi(y_J))$. As explained above, after the observation of the signal, the agent will update his beliefs via the Bayes formula, leading to the conditional distribution $(v_1(y_j), \dots, v_S(y_j))$, where y_j denotes the realization of the random variable $\tilde{y}$, for some $j = 1 \dots, J$. The agent takes a decision after the observation of the signal but before the revelation of the state of the world. Hence, given the observation of the signal $\{\tilde{y} = y_j\}$, for some $j = 1, \dots, J$, the optimal choice problem (8.2) can be transformed into

$$\max_{(x_1, \dots, x_S) \in B} \sum_{s=1}^{S} v_s(y_j) u(x_s). \tag{8.3}$$

As in the case of problem (8.2), the optimal consumption plan solving problem (8.3) will depend on the conditional probability distribution $(v_1(y_j), \dots, v_S(y_j))$. We denote the optimal solution to (8.3) by $(x_1^*(v(y_j)), \dots, x_S^*(v(y_j)))$, assuming that an optimal solution to (8.3) exists. We denote by $U^*(v(y_j))$ the optimal value of the expected utility conditionally on the realization $\{\tilde{y} = y_j\}$, i.e.,

$$U^*(v(y_j)) := \sum_{s=1}^{S} v_s(y_j) u(x_s^*(v(y_j))),$$

for $y_j \in \{y_1, \dots, y_J\}$. We are now in a position to establish the following proposition, which compares the maximal expected utilities associated to problems (8.2) and (8.3).

Proposition 8.1 *Let $U^*(\pi)$ denote the maximal expected utility associated to problem (8.2) and, for each possible realization $y_j \in \{y_1, \dots, y_J\}$ of the random variable $\tilde{y}$, let $U^*(v(y_j))$ denote the maximal expected utility associated to problem (8.3) conditionally on the event $\{\tilde{y} = y_j\}$, for $j = 1, \dots, J$. Then it holds that*

$$U^*(\pi) \leq \sum_{j=1}^{J} \pi(y_j) U^*(v(y_j)). \tag{8.4}$$

Proof Let $j \in \{1, \dots, J\}$. By definition, since the set B does not depend on the probability distribution, it holds that

$$\sum_{s=1}^{S} v_s(y_j) u(x_s^*(\pi)) \leq \sum_{s=1}^{S} v_s(y_j) u(x_s^*(v(y_j))),$$

for every $j = 1, \ldots, J$. In turn, multiplying both sides by $\pi(y_j)$, summing over all $j = 1, \ldots, J$ and noting that $v_s(y_j)\pi(y_j) = \mathbb{P}(\omega_s, \tilde{y} = y_j)$, for all $s = 1, \ldots, S$ and $j = 1, \ldots, J$, we have that

$$U^*(\pi) = \sum_{s=1}^{S} \pi_s u\big(x_s^*(\pi)\big) = \sum_{j=1}^{J} \pi(y_j) \sum_{s=1}^{S} v_s(y_j) u\big(x_s^*(\pi)\big)$$

$$\leq \sum_{j=1}^{J} \pi(y_j) \sum_{s=1}^{S} v_s(y_j) u\big(x_s^*(v(y_j))\big) = \sum_{j=1}^{J} \pi(y_j) U^*\big(v(y_j)\big),$$

thus proving the claim.   □

The result of Proposition 8.1 is in line with economic intuition. Indeed, it shows that the optimal expected utility of an uninformed agent (corresponding to problem (8.2)) is always smaller than the expected optimal utility achieved by an informed agent (corresponding to problem (8.3)). In other words, at least from an individual point of view, additional information is always beneficial and has a non-negative value. In the worst possible case, the additional information conveyed by the signal $\tilde{y}$ has no value and the informed agent can at least do as well as the uninformed agent. Note that this result does not require any assumption on the utility function (in particular, it does not rely on risk aversion) nor on the feasibility set B (apart from the fact that it is not state dependent) and is a consequence of the fact that the optimal expected utility $U^*(\pi)$ is convex with respect to the vector of probabilities $\pi = (\pi_1, \ldots, \pi_S)$, see Exercise 8.1. The introduction of additional information also affects optimal saving decisions (see Sect. 3.4). Indeed, as shown in Exercise 8.2, if the additional information leads to an earlier resolution of uncertainty, then it can induce a reduction in the level of saving by a prudent agent.

The observation of the signal $\tilde{y}$ induces an information structure represented by the couple $(\pi^{\tilde{y}}, v)$, where we denote by v the $(S \times J)$-dimensional matrix with elements $v_s(y_j)$, for all $s = 1, \ldots, S$ and $j = 1, \ldots, J$. Each column of the matrix v represents the conditional probability distribution of the S states of the world given one of the J possible realizations of the signal. Note that, by the Bayes rule, it holds that $\pi = v\pi^{\tilde{y}}$, where we recall that $\pi = (\pi_1, \ldots, \pi_S)$ denotes the prior probability distribution of the S states of the world. Considering the two extreme cases, the information structure $(\pi^{\tilde{y}}, v)$ can be *fully informative* if the observation of the signal fully reveals the state of the world or *completely uninformative* if the conditional probabilities coincide with the prior probabilities. In the latter case, the result of Proposition 8.1 holds with an equality and the signal does not convey any useful information (see also Gollier [800]). In view of Proposition 8.1, it is evident that a fully informative information structure is always preferred to a completely uninformative one.

Proposition 8.1 shows that the value of information is always non-negative (we call this effect the *Blackwell effect*). Let us now assume that an agent can choose between the observation of two possible signals $\tilde{y}_a$ and $\tilde{y}_b$. Both signals convey some information on the unknown state of the world, but they do not generate the same information structure. In this context, an agent is interested in determining which one of the two signals is more informative. For simplicity, we assume that $\tilde{y}_a$ and $\tilde{y}_b$ take values in the same set $\{y_1, \ldots, y_J\}$ but have different probability distributions $\pi^a = (\pi_1^a, \ldots, \pi_J^a)$ and $\pi^b = (\pi_1^b, \ldots, \pi_J^b)$ and lead to different conditional distributions $v^a(y) = (v_1^a(y), \ldots, v_S^a(y))$ and $v^b(y) = (v_1^b(y), \ldots, v_S^b(y))$, for each $y \in \{y_1, \ldots, y_J\}$, where, for $\ell = a, b$, we denote by v^ℓ the $(S \times J)$-dimensional matrix where the j-th column $v^\ell(y_j)$ is the probability distribution of the S states of the world conditionally on the observation of $\{\tilde{y}^\ell = y_j\}$, for some $j = 1, \ldots, J$.

In this context, we say that the information structure represented by (π^a, v^a) is *finer* (or more precise) than the information structure (π^b, v^b) if, for any utility function $u : \mathbb{R}_+ \to \mathbb{R}$ and for any feasible set of choices $B \subset \mathbb{R}_+^S$, it holds that

$$\sum_{j=1}^J \pi^a(y_j) U^*\left(v^a(y_j)\right) \geq \sum_{j=1}^J \pi^b(y_j) U^*\left(v^b(y_j)\right), \tag{8.5}$$

where $U^*(v^\ell(y_j))$ is defined as in (8.3) with respect to the vector of conditional probabilities $(v_1^\ell(y_j), \ldots, v_S^\ell(y_j))$, for $\ell = a, b$ and $j = 1, \ldots, J$. In other words, (π^a, v^a) is finer than (π^b, v^b) if the expected maximal utility associated to (π^a, v^a) is greater than the expected maximal utility associated to (π^b, v^b).

The following proposition, due to Blackwell [246], provides a necessary and sufficient condition for an information structure to be finer than another (see Exercise 8.3 for a proof of the sufficiency part).

Proposition 8.2 *Let* (π^a, v^a) *and* (π^b, v^b) *be two information structures as described above. Then* (π^a, v^a) *is finer than* (π^b, v^b) *if and only if there exists a* $(J \times J)$-*dimensional matrix* K, *satisfying* $K_{ij} \geq 0$, *for all* $i, j = 1, \ldots, J$, *and* $\sum_{i=1}^J K_{ij} = 1$, *for all* $j = 1, \ldots, J$, *such that*

$$v^b = v^a K \qquad and \qquad \pi^a = K \pi^b. \tag{8.6}$$

Condition (8.6) can be equivalently formulated as the property that the conditional probability distribution associated to the information structure (π^b, v^b) is a *mean-preserving spread*[1] of the conditional distribution associated to (π^a, v^a).

---

[1]Given two random variables $\tilde{x}_1$ and $\tilde{x}_2$, we say that the law of $\tilde{x}_2$ is a mean-preserving spread of the law of $\tilde{x}_1$ if it holds that $\tilde{x}_2$ is equal in law to $\tilde{x}_1 + \tilde{z}$, where $\tilde{z}$ is a random variable satisfying $\mathbb{E}[\tilde{z} \,|\, \tilde{x}_1] = 0$ (compare with part *(iii)* of Proposition 2.10).

Pareto Optimality and Green-Lucas Equilibrium

In the previous subsection, we have analysed the value of information from the point of view of an individual decision maker. Let us now consider the role of information in an economy populated by I agents, as introduced at the beginning of this section. In comparison with the two-period economy considered in Chap. 4, the present setting features an intermediate date in correspondence of which agents observe the private signals represented by the random vector $\tilde{y}$. Hence, besides the notions of *ex-ante* and *ex-post Pareto optimality* (see Definitions 4.1 and 4.2), we can formulate a third notion of Pareto optimality.

Definition 8.3 A feasible allocation $\{x^i \in \mathbb{R}_+^S; i = 1, \dots, I\}$ is *interim Pareto optimal* if there is no feasible allocation $\{x^{i'} \in \mathbb{R}_+^S; i = 1, \dots, I\}$ such that

$$\sum_{s=1}^S v_s^i(y_j^i) u^i(x_s^{i'}) \geq \sum_{s=1}^S v_s^i(y_j^i) u^i(x_s^i), \qquad \text{for all } i = 1, \dots, I \text{ and } j = 1, \dots, J,$$

with at least one strict inequality, for some $i \in \{1, \dots, I\}$ and $j \in \{1, \dots, J\}$.

The notions of ex-ante, interim and ex-post Pareto optimality differ with respect to the date at which the optimality of an allocation is evaluated, respectively at date $t = -1$, before any information is released (*ex-ante*), at date $t = 0$, when agents observe the signals but do not known the state of the world (*interim*), and at date $t = 1$, when the state of the world is revealed (*ex-post*). If all prior and conditional probabilities are strictly positive, it is easy to see that ex-ante optimality implies interim optimality and, in turn, interim optimality implies ex-post optimality (see Exercise 8.4 and compare also with Holmstrom & Myerson [956]).

Having introduced the notions of ex-ante, interim and ex-post Pareto optimality, it is interesting to analyse the interplay between information and allocative efficiency. In particular, an important result due to Milgrom & Stokey [1341] shows that, if an allocation is ex-ante Pareto optimal, then the observation of private signals cannot create any incentive to trade. This result is sometimes called the *no-trade theorem*. To analyse this property, we adopt the setting introduced at the beginning of the present section, with an economy $\{(u^i, e^i); i = 1, \dots, I\}$ and S possible states of the world. Let the random vector $\tilde{y} = (\tilde{y}^1, \dots, \tilde{y}^I)$ represent the private signals observed by the agents. Following Milgrom & Stokey [1341], we say that $\{z^i : \{1, \dots, S\} \times \{y_1^i, \dots, y_J^i\} \to \mathbb{R}; i = 1, \dots, I\}$ is a set of *feasible signal-contingent trades* if

$$e_s^i + z^i(s, y_j^i) \geq 0, \qquad \text{for all } s = 1, \dots, S, \ i = 1, \dots, I \text{ and } j = 1, \dots, J,$$

and

$$\sum_{i=1}^{I} z^i(s, y^i) \le 0, \quad \text{for all } s = 1, \dots, S \text{ and for every realization } (y^1, \dots, y^I) \in \mathcal{Y}.$$

If $\{z^i : \{1, \dots, S\} \times \{y_1^i, \dots, y_J^i\} \to \mathbb{R}; i = 1, \dots, I\}$ does not depend on the signal (i.e., each $z^i(\cdot)$ is a function only of the state of the world and not of the signal and, thus, reduces to a function of one variable) and is feasible, then it is simply said to be a set of *feasible trades*. We are now in a position to establish the following proposition.

Proposition 8.4 *Let* $\{z^i : \{1, \dots, S\} \times \{y_1^i, \dots, y_J^i\} \to \mathbb{R}; i = 1, \dots, I\}$ *be a set of feasible signal-contingent trades. Then there always exists a set of feasible trades (i.e., not depending on the signal) which is ex-ante (weakly) preferred by every agent.*

Assume furthermore that there exists at date $t = -1$ *a complete set of markets for the S Arrow securities and that the allocation* $\{e^i; i = 1, \dots, I\}$ *is an equilibrium allocation at* $t = -1$. *Then the following hold:*

(i) *the allocation* $\{e^i; i = 1, \dots, I\}$ *cannot be ex-ante Pareto improved by any set of feasible signal-contingent trades;*
(ii) *suppose that the markets for the S Arrow securities are open also at the interim date* $t = 0$. *Then every agent chooses not to trade at date* $t = 0$.

Proof Let $\{z^i : \{1, \dots, S\} \times \{y_1^i, \dots, y_J^i\} \to \mathbb{R}; i = 1, \dots, I\}$ be a set of feasible signal-contingent trades and define

$$\bar{z}^i(s) := \sum_{j=1}^{J} \mathbb{P}(\tilde{y}^i = y_j^i | \omega_s) z^i(s, y_j^i), \quad \text{for all } s = 1, \dots, S \text{ and } i = 1, \dots, I.$$

The fact that $\{z^i : \{1, \dots, S\} \times \{y_1^i, \dots, y_J^i\} \to \mathbb{R}; i = 1, \dots, I\}$ is feasible implies that $\{\bar{z}^i : \{1, \dots, S\} \to \mathbb{R}; i = 1, \dots, I\}$ is also feasible, i.e., it is a set of feasible trades. Moreover, since the utility function u^i is assumed to be concave, the conditional version of Jensen's inequality implies that

$$\sum_{s=1}^{S} \pi_s u^i \left(e_s^i + \bar{z}^i(s)\right) \ge \sum_{s=1}^{S} \sum_{j=1}^{J} \mathbb{P}(\tilde{y}^i = y_j^i, \omega_s) u^i \left(e_s^i + z^i(s, y_j^i)\right),$$

for all $i = 1, \dots, I$, thus proving the first part of the proposition.

Furthermore, if markets are complete and the initial allocation $\{e^i; i = 1, \dots, I\}$ is of equilibrium, then $\{e^i; i = 1, \dots, I\}$ is also ex-ante Pareto optimal (see Sect. 4.2) and claim *(i)* of the proposition follows.

In order to prove part *(ii)*, arguing by contradiction, suppose that at date $t = 0$ the S markets reopen and each agent i observes a signal $y_{k_i}^i$, for some $k_i \in \{1, \dots, J\}$, for $i = 1, \ldots, I$, such that $\{z_{k_i}^i : \{1, \ldots, S\} \to \mathbb{R}; i = 1, \ldots, I\}$ is a set of feasible trades, meaning that $e_s^i + z_{k_i}^i(s) \geq 0$ and $\sum_{i=1}^I z_{k_i}^i(s) \leq 0$, for all $s = 1, \ldots, S$ and $i = 1, \ldots, I$, and such that

$$\sum_{s=1}^S v_s^i(y_{k_i}^i) u^i(e_s^i + z_{k_i}^i(s)) \geq \sum_{s=1}^S v_s^i(y_{k_i}^i) u^i(e_s^i), \qquad \text{for all } i = 1, \ldots, I, \qquad (8.7)$$

with strict inequality holding for some $i \in \{1, \ldots, I\}$. Define then the set of feasible signal-contingent trades $\{\hat{z}^i : \{1, \ldots, S\} \times \{y_1^i, \ldots, y_J^i\} \to \mathbb{R}; i = 1, \ldots, I\}$ by

$$\hat{z}^i(s, y_j^i) := \begin{cases} z_{k_i}^i(s) & \text{if } j = k_i, \\ 0 & \text{otherwise}, \end{cases}$$

for all $s = 1, \ldots, S$ and $i = 1, \ldots, I$. For every $i = 1, \ldots, I$, it then holds that

$$\sum_{s=1}^S \sum_{j=1}^J \mathbb{P}(\tilde{y}^i = y_j^i, \omega_s) u^i(e_s^i + \hat{z}^i(s, y_j^i))$$

$$= \sum_{j=1}^J \pi(y_j^i) \sum_{s=1}^S v_s^i(y_j^i) u^i(e_s^i + \hat{z}^i(s, y_j^i))$$

$$= \sum_{\substack{j=1 \\ j \neq k_i}}^J \pi(y_j^i) \sum_{s=1}^S v_s^i(y_j^i) u^i(e_s^i) + \pi(y_{k_i}^i) \sum_{s=1}^S v_s^i(y_{k_i}^i) u^i(e_s^i + z_{k_i}^i(s))$$

$$\geq \sum_{\substack{j=1 \\ j \neq k_i}}^J \pi(y_j^i) \sum_{s=1}^S v_s^i(y_j^i) u^i(e_s^i) + \pi(y_{k_i}^i) \sum_{s=1}^S v_s^i(y_{k_i}^i) u^i(e_s^i) = \sum_{s=1}^S \pi_s u^i(e_s^i),$$

with strict inequality holding for some $i \in \{1, \ldots, I\}$ (under the standing assumption that $\pi(y_j^i) > 0$, for all $i = 1, \ldots, I$ and $j = 1, \ldots, J$). However, this contradicts part (i) of the proposition and, hence, inequality (8.7) must hold as an equality, for all $i = 1, \ldots, I$. Moreover, since the utility functions u^i are assumed to be strictly concave, for all $i = 1, \ldots, I$, this implies that $z_{k_i}^i(s) = 0$, for all $s = 1, \ldots, S$ and $i = 1, \ldots, I$. Indeed, if this was not the case, then the strict concavity of the utility functions would imply that the set of trades $\{z_{k_i}^i/2 : \{1, \ldots, S\} \to \mathbb{R}; i = 1, \ldots, I\}$ is feasible and yields a strict ex-ante Pareto improvement of the initial allocation, contradicting part (i) of the proposition. $\square$

Even though the assumptions of Proposition 8.4 are rather strong, the result is important. Indeed, it shows that, if a Pareto optimal allocation has been reached ex-ante, then the arrival of new information does not provide by itself a sufficient incentive for trading, even if markets reopen. In other words, letting the markets be open after the arrival of the additional information represented by the signal $\tilde{y}$ does not change the equilibrium allocation (provided that markets are complete). This result is in line with the fact that ex-ante Pareto optimality implies interim Pareto optimality and, in turn, interim Pareto optimality implies ex-post Pareto optimality (see Exercise 8.4).

It is important to remark that the result of Proposition 8.4 strongly relies on the implicit assumptions that agents form rational expectations, have homogeneous prior beliefs and that the feasibility of a set of trades is common knowledge in the economy. In particular, the fact that the feasibility of a set of trades is common knowledge is crucial for obtaining the no-trade result of part *(ii)* of Proposition 8.4. Indeed, since the initial allocation is assumed to be Pareto optimal, any further trading activity at $t = 0$ cannot be motivated by insurance or risk-sharing purposes. As a consequence, any trading activity at $t = 0$ must be driven purely by speculation (i.e., agents trade hoping to find advantageous bets with other agents), after the revision of beliefs on the basis of the private signals received. However, if the feasibility of a trade is common knowledge, then, as stated by Milgrom & Stokey [1341], "the willingness of the other traders to accept their parts of the bet is evidence to at least one trader that his own part is unfavorable". In other words, at least one agent will suffer a loss from trading and this suffices to prevent agents from participating into the trading. This precludes trading motivated by differences in private information (to this regard, compare also with the result of Exercise 8.5).

Arrow-Debreu and Green-Lucas Equilibria

In view of the no-trade result of Proposition 8.4, we now focus on the case of an economy where markets are only open at date $t = 0$, after the observation of the private signals by the agents. Recall that, after the arrival of this additional information, agents exhibit *heterogeneous beliefs*. We consider an economy where N securities are traded at $t = 0$. Each security n pays a random dividend $\tilde{d}_n$ at $t = 1$, taking the possible values $\{d_{1n}, \ldots, d_{Sn}\}$. We assume that the total supply of each traded security is normalized to zero. In this setting, for all $i = 1, \ldots, I$, the optimal choice problem of agent i can be represented as follows:

$$\max_{(z_1^i, \ldots, z_N^i) \in \mathbb{R}^N} \sum_{s=1}^{S} v_s^i(y^i) u^i \left(e_s^i + \sum_{n=1}^{N} z_n^i d_{sn} \right), \tag{8.8}$$

where, as above, $v_s^i(y^i)$ denotes the conditional probability $\mathbb{P}(\omega_s | \tilde{y}^i = y^i)$, for every $s = 1, \ldots, S$ and $y^i \in \{y_1^i, \ldots, y_J^i\}$. In problem (8.8), the budget constraint is given by

$$\sum_{n=1}^{N} z_n^i p_n \leq 0, \tag{8.9}$$

where $(p_1, \ldots, p_N)$ is the price vector of the N traded securities. Note that problem (8.8) can be also generalized to the case of state dependent utility functions u^i. Given a realization $y = (y^1, \ldots, y^I) \in \mathscr{Y}$ of the signal random vector $\tilde{y}$, we denote by $z^{i*}(y^i) \in \mathbb{R}^N$ the optimal solution to (8.8)–(8.9) given the observation of y^i by agent i, for each $i = 1, \ldots, I$.

In the present setting, the notion of Arrow-Debreu equilibrium can be specialized as follows (compare with Definition 4.6).

Definition 8.5 Let $y = (y^1, \ldots, y^I) \in \mathscr{Y}$ be a realization of the signal random vector $\tilde{y} = (\tilde{y}^1, \ldots, \tilde{y}^I)$. A price-trade vector couple $(p^*(y), z^{1*}(y^1), \ldots, z^{I*}(y^I))$, with $p^*(y) \in \mathbb{R}^N$ and $z^{i*}(y^i) \in \mathbb{R}^N$, for all $i = 1, \ldots, I$, constitutes an *Arrow-Debreu equilibrium* contingent on $\{\tilde{y} = y\}$ if

(i) for every $i = 1, \ldots, I$, given the price vector $p^*(y)$, the trade vector $z^{i*}(y^i)$ is a solution of the optimum problem (8.8)–(8.9) of agent i;

(ii) $\sum_{i=1}^I z_n^{i*}(y^i) \leq 0$, for all $n = 1, \ldots, N$.

According to the above definition, the Arrow-Debreu equilibrium at date $t = 0$ of the economy depends on the realization of the random signal $\tilde{y}$. Indeed, while the utility functions and the endowments of the agents do not depend on the realization of the signals, the beliefs of the agents do depend on the signals received and, hence, also the equilibrium price vector will depend on the realization of the signals.

We always work under the implicit but crucial assumption that all the relevant ingredients of the economy (i.e., the initial allocation, the utility functions of all the agents, the prior distributions of the endowment and of the private signals, the feasibility of a trade vector) are *common knowledge* among the agents. Unlike in the analysis developed in Chap. 4, in the present setting the notion of equilibrium is contingent on the realization of the random signal and the equilibrium price vector will depend on the signals received by the agents. We represent this property by introducing a function $\phi : \mathscr{Y} \to \mathbb{R}^N$ such that $p^*(y) = \phi(y)$, for every possible realization $y \in \mathscr{Y}$ of the signal random vector $\tilde{y}$. We call ϕ the *equilibrium price functional*.

Since signals are private, each agent cannot observe the realizations of the signals received by the other agents. However, the fact that equilibrium prices depend on the information received by every agent is common knowledge. In particular, since agents have a perfect knowledge of the structure of the economy, they also know the structure of the equilibrium price functional ϕ. In turn, the knowledge of the equilibrium price functional implies that each agent can try to infer the signals received by the other agents on the basis of the observation of the equilibrium prices, which depend on the signals received by all the agents, as explained above.

In other words, Arrow-Debreu equilibrium prices *transmit information*. In the classical setting considered in Chap. 4, prices were only used to define the agents' budget constraints. In the present setting, besides defining the budget constraints, prices also carry additional information: every agent, when confronted with an equilibrium price vector p^* will try to infer the signals received by the other agents by determining the set of realizations $y \in \mathscr{Y}$ that are compatible with the price vector p^*. In this way, each agent will learn that the realization of the signal random vector $\tilde{y}$ belongs to the set $\{y \in \mathscr{Y} : \phi(y) = p^*\}$. In particular, if the equilibrium price functional is injective, then every agent can perfectly infer the signals received by every other agent by simply looking at equilibrium prices. If ϕ is not injective, then an agent will determine the set $\phi^{-1}(p^*) := \{y \in \mathscr{Y} : \phi(y) = p^*\} \subseteq \mathscr{Y}$ which is compatible with the prices observed. Of course, this represents an additional information and, being assumed to be rational, agents will revise their beliefs taking into account the information conveyed by market prices. This revision of the agents' beliefs leads to a new conditional distribution

$$\left(v_1^i(y^i, \phi^{-1}(p^*)), \ldots, v_S^i(y^i, \phi^{-1}(p^*)) \right), \qquad \text{for all } i = 1, \ldots, I \text{ and } y^i \in \{y_1^i, \ldots, y_J^i\}, \tag{8.10}$$

where, for every realization $y^i \in \{y_1^i, \ldots, y_J^i\}$ of the signal $\tilde{y}^i$ and for every price vector p, we denote by $v_S^i(y^i, \phi^{-1}(p))$ the probability of ω_s, from the point of view of agent i, conditionally on the information contained in the signal y^i and in the price vector p. In general, the conditional distribution (8.10) is different from the conditional distribution (8.1), since it incorporates an additional source of information. In turn, this implies that the optimal solution to problem (8.8)–(8.9) will no longer be optimal, since the optimal behavior of the agents will now be determined as the solution to

$$\max_{(z_1^i, \ldots, z_N^i) \in \mathbb{R}^N} \sum_{s=1}^{S} v_s^i\left(y^i, \phi^{-1}(p^*)\right) u^i \left(e_s^i + \sum_{n=1}^{N} z_n^i d_{sn} \right), \tag{8.11}$$

for all $i = 1, \ldots, I$. Therefore, the price vector $p^* = \phi(y)$ obtained from the Arrow-Debreu equilibrium will no longer be of equilibrium. This observation shows that the notion of Arrow-Debreu equilibrium contingent on a realization of the signal random vector $\tilde{y}$ (see Definition 8.5) is not the appropriate notion of equilibrium, since the associated price vector leads agents to revise their beliefs and, therefore, their optimal choices. In the present context, an appropriate notion of equilibrium should represent a stable point in this process of price observation - revision of beliefs and agents should not want to revise their behavior once they observe the market clearing prices.

As outlined in the introduction to this chapter, an appropriate notion of equilibrium is represented by the *Green-Lucas equilibrium* (see Lucas [1251], Green [820]), which captures the idea that at equilibrium there should be no incentive to revise beliefs and trades. For a price vector $p \in \mathbb{R}^N$, we denote by $z_n^{i*}(y_j^i, p)$ the optimal solution to problem (8.11) for agent i given the observation of the private signal y_j^i and of the price vector p.

Definition 8.6 Let $\phi^* : \mathscr{Y} \to \mathbb{R}^N$ and $z^* = \{z^{i*} : \{y_1^i, \ldots, y_J^i\} \times \mathbb{R}^N \to \mathbb{R}^N;$ $i = 1, \ldots, I\}$. The couple (ϕ^*, z^*) constitutes a *Green-Lucas equilibrium* if the following two conditions hold, for every realization $y = (y^1, \ldots, y^I) \in \mathscr{Y}$ of the signal random vector $\tilde{y}$:

(i) the equilibrium price vector p^* is given by $p^* = \phi^*(y)$;
(ii) $\sum_{i=1}^I z_n^{i*}(y^i, p^*) \leq 0$, for all $n = 1, \ldots, N$.

Definition 8.6 captures the fact that equilibrium prices transmit information. Agents are rational and are assumed to know how private signals affect equilibrium prices and, therefore, optimally take this information into account.

A Green-Lucas equilibrium is said to be *fully revealing* if the equilibrium price functional ϕ^* is injective, so that the observation of the equilibrium prices completely determines the realization of the signals received by all the agents. Otherwise, we say that the equilibrium is only *partially revealing*, since the observation of the equilibrium prices only allows agents to determine a set of possible signals which are compatible with the observed prices. Of course, in view of studying market efficiency from an informational point of view, as discussed in the introduction to this chapter, it is of particular interest to determine under which conditions a Green-Lucas equilibrium is fully revealing. Indeed, if the equilibrium is fully revealing, then all available (private) information will be conveyed by market prices.

Given an economy as described in the present section, in order to determine a Green-Lucas equilibrium, one can proceed along the following successive steps (see also Brunnermeier [319, Section 1.2.1]):

- define the agents' prior beliefs, the prior distributions of all the random variables appearing in the model and specify a conjectured price functional ϕ;
- taking the conjectured price functional ϕ as given, derive the posterior beliefs of each agent, conditionally on the observation of the price vector and of the private signal, and compute the optimal trade vector for each agent, as a function of the price functional, of the signals received and of the endowment;
- impose the market clearing condition to derive the actual relation between the conjectured price functional and the signals;
- impose the rational expectations condition (i.e., the conjectured price functional corresponds to the true price functional; in other words, the price conjecture must be self-fulfilling) in order to fully determine the equilibrium price functional.

Under suitable conditions, a closed form Green-Lucas equilibrium exists. Classical conditions to find a closed form equilibrium with a linear price functional are multivariate normality together with exponential utility functions (compare also with the models presented in Chap. 10). In the following sections, we shall consider several instances where the above steps can be solved and a Green-Lucas equilibrium can be explicitly constructed.

8.2   On the Possibility of Informationally Efficient Markets

As we have seen at the end of the previous section, equilibrium prices convey information about the private signals received by the agents. Under suitable conditions, a Green-Lucas equilibrium can be *fully revealing*. If this is the case, then equilibrium prices aggregate and transmit all private information and markets are informationally efficient (in a strong form). Sufficient conditions for Green-Lucas equilibrium prices to fully reveal the private information of the agents have been established in a series of papers, see Grossman [839, 840, 841].

To analyse the aggregation-transmission of information by Green-Lucas equilibrium prices, it is useful to introduce an *artificial economy*, where every agent is assumed to observe all private signals received by all agents. In this artificial economy, each agent solves the optimal choice problem (8.8), with the probability distribution of the state of the world being conditioned on the realization of the whole signal random vector $\tilde{y} = (\tilde{y}^1, \ldots, \tilde{y}^I)$. As in Sect. 8.1, we assume that the signal random vector takes values in a set $\mathscr{Y}$ and denote by $y = (y^1, \ldots, y^I) \in \mathscr{Y}$ a generic realization of $\tilde{y}$. Similarly as in (8.1), for $y \in \mathscr{Y}$, we define the conditional probability distribution $v(y) = (v_1(y), \ldots, v_S(y))$, with

$$v_s(y) := \mathbb{P}(\omega_s | \tilde{y} = y) = \mathbb{P}(\omega_s | \tilde{y}^1 = y^1, \ldots, \tilde{y}^I = y^I), \qquad \text{for all } s = 1, \ldots, S, \tag{8.12}$$

representing the probability of the state of the world ω_s conditionally on the realization $\{\tilde{y} = y\}$. Observe that in the artificial economy there are no heterogenous beliefs, since every agent observes the full set of private signals.

Similarly as in Grossman [841], we assume that at $t = 0$ there exist S open markets for the S Arrow securities corresponding to the S states of the world (hence, markets are complete). As in Sect. 8.1, the economy is populated by I agents with strictly increasing and strictly concave utility functions $u^i : \mathbb{R}^S_+ \to \mathbb{R}$ satisfying the Inada conditions (i.e., $\lim_{x \searrow 0} u^{i\prime}(x) = +\infty$ and $\lim_{x \to +\infty} u^{i\prime}(x) = 0$, for all $i = 1, \ldots, I$). We also assume that the initial endowment of every agent is strictly positive. For every $i = 1, \ldots, I$, we denote by $z^{i*}(y, p)$ the optimal solution to problem (8.8) in correspondence of the vector of prices $p \in \mathbb{R}^S$ and of the probability distribution (8.12) conditional on the realization $\{\tilde{y} = y\}$. We adopt the notation $z^{i*}(y, p)$ in order to emphasize the dependence on the vector of prices p. Note that the Inada conditions ensure that the optimal consumption is strictly positive in correspondence of each state of the world.

For this artificial economy, we can adopt the notion of Arrow-Debreu equilibrium contingent on the realization of the signal random vector (see Definition 8.5). Under the present assumptions, it can be shown that, for every realization $y \in \mathscr{Y}$ of the signal random vector $\tilde{y}$, the artificial economy admits an Arrow-Debreu equilibrium contingent on $\{\tilde{y} = y\}$. In particular, the vector $p^{a*}(y)$ of Arrow-Debreu equilibrium prices of the artificial economy (where the superscript ^{a} stands for artificial) depends on the realization y of $\tilde{y}$ and, similarly as in the discussion following Definition 8.5, we can regard $p^{a*}(\cdot) : \mathscr{Y} \to \mathbb{R}^S$ as an equilibrium price functional.

The following lemma provides a sufficient condition for the equilibrium price functional p^{a*} to be injective (compare also with Laffont [1154, Chapter 9]).

Lemma 8.7 *Let* $p^{a*} : \mathscr{Y} \to \mathbb{R}^S$ *denote the price functional associated to an Arrow-Debreu equilibrium of the artificial economy introduced above and suppose that* $v_s(y) > 0$, *for all* $s = 1, \ldots, S$ *and* $y \in \mathscr{Y}$. *Let* $y, y' \in \mathscr{Y}$ *be two realizations of* $\tilde{y}$ *such that* $v(y) \neq v(y')$. *If*

$$z^{j*}\big(y, p^{a*}(y)\big) \neq z^{j*}\big(y', p^{a*}(y')\big), \qquad \text{for some } j \in \{1, \ldots, I\},$$

then it holds that $p^{a*}(y) \neq p^{a*}(y')$.

Proof Arguing by contradiction, suppose that $p^{a*}(y) = p^{a*}(y') =: \bar{p} \in \mathbb{R}^S$, meaning that the two realizations y and y' generate the same equilibrium prices and, consequently, the same budget constraint. Then, it holds that

$$\sum_{s=1}^{S} v_s(y)u^i\big(e_s^i + z_s^{i*}(y, \bar{p})\big) \geq \sum_{s=1}^{S} v_s(y)u^i\big(e_s^i + z_s^{i*}(y', \bar{p})\big),$$

for all $i = 1, \ldots, I$, with strict inequality holding for $i = j$, due to the strict concavity of the utility function u^j. For brevity of notation, let us denote $z_s^{i*} := z_s^{i*}(y, \bar{p})$ and $z_s^{i*'} := z_s^{i*}(y', \bar{p})$, for all $s = 1, \ldots, S$ and $i = 1, \ldots, I$. The concavity of u^i, for all $i = 1, \ldots, I$, implies that

$$\sum_{s=1}^{S} v_s(y)\big(z_s^{i*} - z_s^{i*'}\big)u^{i'}\big(e_s^i + z_s^{i*'}\big) \geq 0, \qquad \text{for all } i = 1, \ldots, I, \tag{8.13}$$

with $u^{i'}$ denoting the first derivative of the utility function u^i and where a strict inequality holds for $i = j$. The optimality conditions for problem (8.8) in correspondence of y', recalling that the optimal consumption is strictly positive (as a consequence of the Inada conditions), imply that

$$v_s(y')u^{i'}\big(e_s^i + z_s^{i*'}\big) = \lambda^i \bar{p}_s, \qquad \text{for all } s = 1, \ldots, S \text{ and } i = 1, \ldots, I, \tag{8.14}$$

where λ^i denotes the Lagrange multiplier associated to the optimal choice problem of agent i, for $i = 1, \ldots, I$. By combining conditions (8.13) and (8.14), it follows that

$$\sum_{s=1}^{S} v_s(y)\big(z_s^{i*} - z_s^{i*'}\big)\frac{\lambda^i \bar{p}_s}{v_s(y')} \geq 0, \qquad \text{for all } i = 1, \ldots, I,$$

with strict inequality holding for $i = j$. In turn, dividing by λ^i and summing over all agents, this implies that

$$\sum_{s=1}^{S} \frac{\bar{p}_s v_s(y)}{v_s(y')} \sum_{i=1}^{I} (z_s^{i*} - z_s^{i*'}) > 0,$$

which is in contradiction with the market clearing condition $\sum_{i=1}^{I} z_s^{i*} = \sum_{i=1}^{I} z_s^{i*'}$, for $s = 1, \ldots, S$. $\qquad\square$

Lemma 8.7 shows that, in the artificial economy where each agent observes the full set of private signals, if different posterior distributions lead to different optimal demands then the equilibrium prices must be different. In turn, this result allows us to prove the following proposition, which shows that the Arrow-Debreu equilibrium of the artificial economy corresponds to a Green-Lucas equilibrium of the original economy.

Proposition 8.8 *Let $p^{a*} : \mathcal{Y} \to \mathbb{R}^S$ denote the price functional associated to an Arrow-Debreu equilibrium of the artificial economy introduced above and suppose that $v_s(y) > 0$, for all $s = 1, \ldots, S$ and $y \in \mathcal{Y}$. Then p^{a*} is a Green-Lucas equilibrium price functional of the original economy in which each agent i observes his private signal and the equilibrium prices.*

Proof Define the candidate Green-Lucas equilibrium price functional $\phi^* : \mathcal{Y} \to \mathbb{R}^S$ by $\phi^*(y) := p^{a*}(y)$, for every $y \in \mathcal{Y}$. We want to prove that the definition

$$z^{i*}\left(y^i, \phi^*(y)\right) := z^{i*}\left(y, p^{a*}(y)\right), \qquad \text{for all } i = 1, \ldots, I, \tag{8.15}$$

for every realization $y = (y^1, \ldots, y^I) \in \mathcal{Y}$, is well posed and, together with the functional ϕ^*, defines a Green-Lucas equilibrium of the original economy, in the sense of Definition 8.6. In (8.15), the right hand side represents the optimal demand of agent i in the artificial economy, while the left hand side represents the candidate optimal demand of agent i in the original economy, associated to the candidate equilibrium price functional ϕ^*.

Suppose first that the functional $p^{a*} : \mathcal{Y} \to \mathbb{R}^S$ is injective. In this case, the observation of $p^{a*}(y)$ corresponds exactly to the observation of the realization $\{\tilde{y} = y\}$ of the whole set of signals. Therefore, the conditional distribution $v(y)$ considered in the artificial economy (see (8.12)) coincides with the conditional distribution introduced in (8.10). Indeed, for all $i = 1, \ldots, I$, it holds that

$$v_s^i\left(y^i, \phi^{*,-1}(p^{a*}(y))\right) = \mathbb{P}(\omega_s | \tilde{y}^i = y^i, \tilde{y} = y) = \mathbb{P}(\omega_s | \tilde{y} = y) = v_s(y), \tag{8.16}$$

for all $i = 1, \ldots, I$, $s = 1, \ldots, S$ and $y \in \mathcal{Y}$. In particular, this shows that the information available to each agent in the original economy coincides with the information available in the artificial economy, where every agent observes the full set of private signals. In view of the formulation of problem (8.11), the claim then follows.

Suppose now that p^{a*} is not injective. Clearly, there is nothing to prove if, for every $i = 1, \ldots, I, s = 1, \ldots, S$ and $y \in \mathscr{Y}$, relation (8.16) holds. Therefore, let us assume that for some $i \in \{1, \ldots, I\}$ and $j \in \{1, \ldots, J\}$ and $\bar{p} \in \mathbb{R}^S$, the set

$$\mathscr{Y}_j^i := \{y \in \mathscr{Y} : y^i = y_j^i \text{ and } p^{a*}(y) = \bar{p}\}$$

contains at least two elements y and y' for which $v(y) \neq v(y')$. In other words, the set $\mathscr{Y}_j^i$ contains all the realizations of the signal random vector $\tilde{y}$ which generate different conditional distributions but are indistinguishable from the point of view of agent i, given the observation of his private signal y_j^i and the vector of prices $\bar{p}$. In view of Lemma 8.7, since $p^{a*}(y) = p^{a*}(y') = \bar{p}$, it necessarily holds that $z^{i*}(y, \bar{p}) = z^{i*}(y', \bar{p})$. In other words, the optimal demand of agent i in the artificial economy is constant across all the elements of the set $\mathscr{Y}_j^i$ and, hence, the definition $z^{i*}(y^i, \phi^*(y)) := z^{i*}(y, p^{a*}(y))$ is well posed. Since market clearing holds in the artificial economy and, hence, in the original economy as well, we have constructed a Green-Lucas equilibrium of the original economy in the sense of Definition 8.6.

$\square$

Together with Lemma 8.7, the proof of the above proposition shows that, if two realizations y and y' of the signal random vector $\tilde{y}$ generate different conditional distributions $v(y)$ and $v(y')$ but generate the same Arrow-Debreu equilibrium prices in the artificial economy, then the optimal demand of every agent in the artificial economy will be the same in correspondence of y and y'. In turn, this implies that the optimal demand at the equilibrium of the original economy, where each agent observes his private signal and the market prices, coincides with the optimal demand at the equilibrium in the artificial economy. In particular, a Green-Lucas equilibrium in the original economy can be constructed by considering the Arrow-Debreu equilibrium in the artificial economy.

We have reached an important result. Indeed, under the present assumptions, in correspondence of equilibrium prices all information is embedded in market prices and, hence, the market is efficient in a strong form from an informational point of view. Indeed, while the equilibrium price functional ϕ^* is not necessarily injective (and, hence, agents are not always able to infer the realization of the whole random vector $\tilde{y}$ given the observation of equilibrium prices), the optimal demands of the agents are constant across all realizations of $\tilde{y}$ that lead to the same equilibrium prices. Hence, under the assumption of complete markets, in a rational expectation equilibrium the price system fully symmetrizes all the differences in information among the agents. Moreover, if prices fully symmetrize the informational differences, then agents will only trade for hedging purposes and there is no *speculation*, in the sense that there is no trading activity purely motivated by differences in information/beliefs. An explicit example of a fully revealing Green-Lucas equilibrium will be provided at the end of this section.

The relation between Green-Lucas equilibria of the original economy with private information and Arrow-Debreu equilibria of the artificial economy can be analysed in a more general setting where the private signals are represented by random variables admitting probability density functions. To this end, let us introduce the notion of *sufficient statistic*. Let $\tilde{z}$, $\tilde{x}$ and $\tilde{y}$ be three random variables and denote respectively by z, x and y their generic realizations. We say that $\tilde{z}$ is a sufficient statistic for the conditional density $f(y, z|x)$ if and only if there exist functions $g_1(\cdot)$ and $g_2(\cdot)$ such that, for every (z, x, y), it holds that

$$f(y, z|x) = g_1(y, z)g_2(z, x).$$

As shown in Exercise 8.6, $\tilde{z}$ is a sufficient statistic for the conditional density $f(y, z|x)$ if and only if the conditional distribution of $\tilde{x}$ given $\{\tilde{y} = y, \tilde{z} = z\}$ does not depend on y (see also Huang & Litzenberger [971, Section 9.8]).

The notion of sufficient statistic can be applied in the context of a Green-Lucas equilibrium. More specifically, assuming that the market comprises N traded securities paying dividends $(\tilde{d}_1, \ldots, \tilde{d}_N)$ and given a vector of equilibrium prices p^*, we say that p^* is a *sufficient statistic* if the probability distribution of the dividends $(\tilde{d}_1, \ldots, \tilde{d}_N)$ conditioned on the observation of the private signal $\tilde{y}^i$ and of the prices p^* does not depend on the realization of the private signal $\tilde{y}^i$, for each $i = 1, \ldots, I$. In this sense, if p^* is a sufficient statistic, then the information conveyed by the private signals becomes redundant once agents observe market prices. This property implies that equilibrium prices embed all useful information and, hence, the market is informationally efficient.

In Grossman [840], an economy with $N + 1$ traded assets and I agents is considered. The $N + 1$ assets are available for trading at $t = 0$ and include N risky assets (with dividends represented by the random variables $(\tilde{d}_1, \ldots, \tilde{d}_N)$) and one riskless asset yielding the riskless return r_f. As above, each agent $i = 1, \ldots, I$ observes at $t = 0$ (before the opening of the market) a private signal represented by the random variable $\tilde{y}^i$. In Grossman [840], it is assumed that the random vector $(\tilde{d}_1, \ldots, \tilde{d}_N, \tilde{y}^1, \ldots, \tilde{y}^I)$ is distributed according to a multivariate normal law and, under this distributional assumption, the following results are obtained:

- there exists a Green-Lucas equilibrium price functional which is a sufficient statistic if and only if there exists an equilibrium price functional of the artificial economy (where every agent observes all private signals) which is a sufficient statistic (see Exercise 8.8 for a proof of this property). Note that this result does not require the assumption of a multivariate normal distribution;
- the vector of expected dividends conditioned on the observation of $\tilde{y}$ (i.e., the vector composed by $m_n(y) := \mathbb{E}[\tilde{d}_n|\tilde{y} = y]$, for $n = 1, \ldots, N$) is a sufficient statistic;
- the equilibrium price functional of the artificial economy, whenever it exists, depends on the realization of the signals only through the vector $(m_1(y), \ldots, m_N(y))$.

- as a consequence of the multivariate normality assumption, agents act as if there were only two assets: the riskless asset and a mutual fund composed by the N risky assets (two mutual funds separation; see Chap. 3). Moreover, if there exists an equilibrium price functional for the artificial economy, then it is an invertible function of $(m_1(y), \ldots, m_N(y))$ and a sufficient statistic. Under these conditions and if the demand of the risky mutual fund is a decreasing function of its price (i.e., the mutual fund is not a *Giffen good*), then there exists a Green-Lucas equilibrium which is a sufficient statistic (see Grossman [840, Theorem 1]).

A sufficient condition for the risky mutual fund not to be a Giffen good is represented by a non-increasing coefficient of absolute risk aversion of the agents' utility functions. Together with the joint normality assumption, this condition suffices to ensure that the rational expectations equilibrium price is a sufficient statistic. In particular, note that the results of Grossman [840] do not require markets to be complete. Moreover, generalizing a previous result in Grossman [839], Grossman [840] shows that, if the rational expectations equilibrium price is a sufficient statistic, then the equilibrium allocation cannot be Pareto improved by a central planner having access to all available information. In this sense, Grossman [840] argues that the appropriate notion of informationally efficient market is the one in which the price vector is a sufficient statistic.

Of course, not every Green-Lucas equilibrium price functional, in the sense of Definition 8.6, is fully revealing nor corresponds to informationally efficient markets. Sometimes, only *partially revealing* equilibria exist. For an example of an economy which does not admit fully revealing equilibria, admits fully revealing equilibria or partially revealing equilibria depending on the preference parameters, we refer to Exercise 8.7 and to Radner [1438]. A simple model where a fully revealing Green-Lucas equilibrium does not exist is reported in Exercise 8.12.

The Blackwell and Hirshleifer Effects

In the first part of Sect. 8.1, we have analysed the value of information from the point of view of an individual decision maker. From this perspective, we have shown that information always has a non-negative value (see Proposition 8.1). Let us now analyse the value of information from a social point of view. In a nutshell, we aim at understanding whether the introduction of additional information can increase or decrease the social welfare in the economy.

As argued above, a fully revealing Green-Lucas equilibrium implies market efficiency in a strong form from the informational point of view. Moreover, from the allocative point of view, a fully revealing Green-Lucas equilibrium under market completeness is also ex-post Pareto efficient, as a direct consequence of the First Welfare theorem (see also Grossman [841]). However, interim Pareto optimality and ex-ante Pareto optimality do not necessarily hold (see Laffont [1153] for an example of a Green-Lucas equilibrium which is interim Pareto inefficient). Moreover, partially revealing equilibria are typically ex-post Pareto inefficient (see Laffont [1153]).

The study of the social impact of information in financial markets requires the understanding of two competing effects: the *Blackwell effect* and the *Hirshleifer effect*. As we have already pointed out, better information always leads to better individual choices and, hence, to a higher personal welfare. This is indeed the content of Propositions 8.1 and 8.2 (*Blackwell effect*). On the contrary, the *Hirshleifer effect* is due to a lack of optimal risk sharing: Hirshleifer [948] remarked that the revelation of information can eliminate profitable risk sharing opportunities. In order to explain the Hirshleifer effect, let us consider the simple situation (as illustrated in Sect. 4.1) where there is no aggregate risk and agents do not have private information. In a complete market economy, the agents are willing to trade in order to share the individual risks, thus improving their expected utilities. In particular, in the absence of aggregate risk, they can fully eliminate the randomness of their future consumption plans. However, if the true state of the world is revealed before the opening of the markets, then optimal risk sharing becomes impossible, since the agent whose endowment is larger in the realized state of the world would not participate to the trade. This elementary example shows that an equilibrium where the information is revealed cannot be Pareto efficient, since it implies a sub-optimal risk sharing to the extent that it induces agents not to trade at all. In other words, realized risks cannot be insured and, hence, the revelation of information in the economy may eliminate profitable insurance opportunities. Due to the Hirshleifer effect, information can be detrimental from the social point of view and the Pareto efficiency of the equilibrium allocation can in some cases be improved if the equilibrium is only partially revealing. An example of a simple economy where the Hirshleifer effect is particularly evident is given in Exercise 8.9.

The above discussion can be rephrased in the setting introduced at the beginning of the present section. Let us consider an economy with I agents who can all observe, before the opening of the markets, the same signal $\tilde{y}$, here represented by a random variable taking values in the set $\{y_1, \ldots, y_J\}$, for some $J \in \mathbb{N}$, with corresponding probabilities $(\pi(y_1), \ldots, \pi(y_J))$. We assume that the complete set of S Arrow securities is available for trading (complete market economy). Conditionally on a realization $\{\tilde{y} = y_j\}$ of the common signal, let us denote by $\{x^{i*}(y_j) \in \mathbb{R}^S_+; i = 1, \ldots, I\}$ an Arrow-Debreu equilibrium allocation contingent on the realization $\{\tilde{y} = y_j\}$ (see Definition 8.5), where $x^{i*}_s(y_j)$ denotes the optimal consumption of agent i in the state of the world ω_s conditionally on the observation of $\{\tilde{y} = y_j\}$. In particular, note that this equilibrium allocation is allowed to be contingent on the realization of the random signal. Let us also denote by $\{x^{i*,0} \in \mathbb{R}^S_+; i = 1, \ldots, I\}$ the "uninformative" equilibrium allocation reached before the agents are allowed to observe the common signal. We can now establish the following proposition, which can be proved by the same arguments used in the first part of the proof of Proposition 8.4 (see also Gollier [800, Proposition 97]).

Proposition 8.9 *Under the above assumptions, for every $j = 1, \ldots, J$, the equilibrium allocation $\{x^{i*}(y_j) \in \mathbb{R}^S_+ ; i = 1, \ldots, I\}$ contingent on the signal realization $\{\tilde{y} = y_j\}$ observed by the agents cannot ex-ante Pareto dominate the allocation $\{x^{i*,0} \in \mathbb{R}^S_+ ; i = 1, \ldots, I\}$ obtained when the agents cannot observe $\tilde{y}$.*

The above proposition shows that when more precise (common) information becomes available in the economy, then it cannot be beneficial to all market participants. There exists at least one agent whose expected utility decreases after the arrival of new information.

In general, the Blackwell and the Hirshleifer effects go in opposite directions. While information increases the individual expected utility, it may lead to suboptimal risk-sharing and, therefore, it may reduce the social welfare. Indeed, as shown in Gottardi & Rahi [813], the effect on social welfare of a change in the information available prior to trading can be either beneficial or adverse. Schlee [1504] shows that the introduction of information is Pareto inferior if one of the following conditions is satisfied: there exists a risk neutral agent, there is no aggregate risk or agents have linear absolute risk tolerance with common slope (see also Gollier [800, Proposition 98] and Exercise 8.10). Eckwert & Zilcha [626] show that, in an economy with production, information has a positive value if the information refers to non-tradable risks, while information related to tradable risks may reduce the welfare if the consumers are strongly risk averse.

Due to the Hirshleifer effect, in an economy with asymmetric information (informed and uninformed agents), an incomplete financial structure may be optimal, as shown in Marin & Rahi [1305]. In that paper, uninformed agents face an adverse selection problem due to the presence of informed agents. In particular, uninformed agents are afraid to trade with more informed agents. In this context, the revelation of information induces two different effects: the Hirshleifer effect and a reduction of the adverse selection phenomenon. As discussed above, the Hirshleifer effect leads to negative consequences for the welfare of the economy, while the second effect has positive consequences for the uninformed agents. In the model considered in Marin & Rahi [1305], markets are initially incomplete and the equilibrium is only partially revealing. At a second stage, a new security is introduced in order to complete the market, leading to a fully revealing equilibrium. When passing from the partially revealing to the fully revealing equilibrium, the Hirshleifer effect and the adverse selection reduction come into play and the overall impact on the social welfare depends on the relative strength of these two effects. If the Hirshleifer effect outweighs the reduction of adverse selection, then it is better to have incomplete markets and a partially revealing equilibrium, because less information is revealed and this leads to better risk sharing possibilities. Note also that the two effects also affect the trading volume. In order to add noise to the information transmitted by prices, it may be optimal to introduce a purely speculative security unrelated to endowments and preferences (see Marin & Rahi [1304]). Related results have been obtained in Dow & Rahi [586], assuming that the asset is not perfectly correlated with the initial risk exposure represented by the agents' endowment. In this context, it is confirmed that greater revelation of information that agents wish to insure against reduces their hedging opportunities (negative Hirshleifer effect). On the contrary, early revelation of information that is uncorrelated with hedging needs allows agents to hedge risks more efficiently.

A Fully Revealing Equilibrium

The existence of a fully revealing equilibrium has been established in the context of a tractable model in Grossman [839]. Let us consider a two-period economy ($t \in \{0, 1\}$), with a single consumption good which is consumed at $t = 1$. There are two traded assets: a riskless asset paying the constant rate of return r_f and a risky asset delivering a random dividend $\tilde{d}$ at $t = 1$. The price of the risky asset at $t = 0$ is denoted by p and the total supply of the riskless asset is normalized to zero. Similarly as above, we assume that the economy is populated by I agents $i = 1, \ldots, I$. The initial endowment of agent i is expressed in terms of e^i units of the risky asset and we suppose that $\sum_{i=1}^{I} e^i = 1$.

Letting w^i be the demand of the risky asset by agent i (in terms of units of the asset), his wealth at date $t = 1$ is described by the random variable

$$\widetilde{W}^i = (e^i - w^i)pr_f + w^i\tilde{d}, \qquad \text{for all } i = 1, \ldots, I.$$

We assume that each agent i is characterized by a negative exponential utility function of the form $u^i(x) = -\mathrm{e}^{-a^i x}$, with $a^i > 0$, for all $i = 1, \ldots, I$.

Ex-ante (at $t = -1$), all the agents believe that $\tilde{d}$ is distributed as a normal random variable with mean $\bar{d}$ and variance σ^2. At the interim date (at $t = 0$), each agent i observes a private signal $\tilde{y}^i$ of the form

$$\tilde{y}^i = \tilde{d} + \tilde{\varepsilon}^i,$$

where $\tilde{\varepsilon}^i$ is a random variable such that the random vector $(\tilde{d}, \tilde{\varepsilon}^1, \ldots, \tilde{\varepsilon}^I)$ is distributed according to a normal multivariate law with uncorrelated (and, hence, independent) components. As above, we denote by $\tilde{y}$ the random vector $(\tilde{y}^1, \ldots, \tilde{y}^I)$, with generic realization $y = (y^1, \ldots, y^I) \in \mathbb{R}^I$. We assume that $\mathbb{E}[\tilde{\varepsilon}^i] = 0$ and we denote by $\sigma_{\tilde{\varepsilon}^i}^2$ the variance of $\tilde{\varepsilon}^i$, for all $i = 1, \ldots, I$. The I agents have private information and, after the observation of their signals, revise their beliefs via the Bayes rule. More specifically, by the properties of the multivariate normal distribution, for every $i = 1, \ldots, I$, the conditional distribution of $\tilde{d}$ given $\tilde{y}^i$ is normal and characterized by

$$\mathbb{E}[\tilde{d}|y^i] = \bar{d} + \beta^i(y^i - \bar{d}) \qquad \text{and} \qquad \mathrm{Var}(\tilde{d}|y^i) = \sigma^2(1 - \beta^i),$$

where we denote $\mathbb{E}[\tilde{d}|y^i] := \mathbb{E}[\tilde{d}|\tilde{y}^i = y^i]$, similarly for the conditional variance $\text{Var}(\tilde{d}|y^i)$, and where $\beta^i := \sigma^2/(\sigma^2 + \sigma^2_{\tilde{\varepsilon}i})$, for all $i = 1, \ldots, I$. Note that the conditional variance does not depend on the realization of the signal. After agents observe their private signals, markets open and trading occurs. As a consequence of formula (3.12), the optimal demand of agent i given the observation of the realization $\{\tilde{y}^i = y^i\}$ of his private signal is given by

$$ w^{i*} = \frac{\mathbb{E}[\tilde{d}|y^i] - pr_f}{a^i \, \text{Var}(\tilde{d}|y^i)}. \tag{8.17} $$

Observe that, besides the properties discussed in section "Closed Form Solutions and Mutual Fund Separation", the optimal demand w^{i*} is an increasing function of the precision of the information (if $\mathbb{E}[\tilde{d}|y^i] \geq pr_f$), here measured by the quantity $1/\text{Var}(\tilde{d}|y^i)$.

The market clearing condition $\sum_{i=1}^{I} w^{i*} = 1$ then implies that, for every realization $y = (y^1, \ldots, y^I) \in \mathbb{R}^I$, the equilibrium price p^* of the risky asset is given by

$$ p^*(y) = \left(\frac{\sum_{i=1}^{I} \frac{\mathbb{E}[\tilde{d}|y^i]}{a^i \, \text{Var}(\tilde{d}|y^i)}}{\sum_{i=1}^{I} \frac{1}{a^i \, \text{Var}(\tilde{d}|y^i)}} - \left(\sum_{i=1}^{I} \frac{1}{a^i \, \text{Var}(\tilde{d}|y^i)} \right)^{-1} \right) \frac{1}{r_f}. \tag{8.18} $$

The price p^* computed in equation (8.18) represents the Arrow-Debreu equilibrium price contingent on $\{\tilde{y} = y\}$ (see Definition 8.5). Moreover, $p^*(y)$ has several nice and intuitive features: it is related to a weighted average of the conditional expectations of the dividend by the I agents, where the weight associated to each agent is given by the reciprocal of the coefficient of absolute risk aversion multiplied by the agent's estimate of the variance of the dividend. Hence, the more an agent is risk averse and the smaller is the precision of his private information, the smaller is his weight in the determination of $p^*(y)$. The rationale for this property is that an agent exhibiting a large coefficient of absolute risk aversion and with a large conditional variance will also exhibit a small demand of the risky asset and, hence, will not influence significantly equilibrium prices.

The Arrow-Debreu equilibrium price $p^*(y)$ is a function of the realization of the signals $\tilde{y} = (\tilde{y}^1, \ldots, \tilde{y}^I)$. Hence, as discussed above, agents can learn about the other agents' private signals by observing equilibrium prices. Motivated by this observation, we now aim at deriving the Green-Lucas equilibrium of this economy in the presence of heterogeneous information.

Let us assume that $\sigma^2_{\tilde{\varepsilon}i} = \sigma^2_\varepsilon$, for all $i = 1, \ldots, I$, for some $\sigma^2_\varepsilon > 0$. Under this additional assumption, there exists a fully revealing Green-Lucas equilibrium, due to the normality assumption together with the fact that agents exhibit constant absolute risk aversion. Let $\bar{\tilde{y}} := (\sum_{i=1}^{I} \tilde{y}^i)/I$ denote the average of the signals received by the agents. By exploiting the assumption of multivariate normality, it can be easily shown that $\bar{\tilde{y}}$ is a sufficient statistic for the distribution of $(\tilde{y}^i, \bar{\tilde{y}})$ conditioned on $\tilde{d}$, for all $i = 1, \ldots, I$ (see Grossman [839, Lemma 1]). This implies that

$$\mathbb{E}[\tilde{d}|\tilde{y}^i = y^i, \tilde{\bar{y}} = \bar{y}] =: \mathbb{E}[\tilde{d}|y^i, \bar{y}] = \mathbb{E}[\tilde{d}|\bar{y}] = \bar{d} + \bar{\beta}(\bar{y} - \bar{d}),$$
$$\mathrm{Var}(\tilde{d}|\tilde{y}^i = y^i, \tilde{\bar{y}} = \bar{y}) =: \mathrm{Var}(\tilde{d}|y^i, \bar{y}) = \mathrm{Var}(\tilde{d}|\bar{y}) = \sigma^2(1 - \bar{\beta}),$$

(8.19)

for every realization y^i and $\bar{y}$ of the random variables $\tilde{y}^i$ and $\tilde{\bar{y}}$ and where we define $\bar{\beta} := \sigma^2/(\sigma^2 + \sigma_\epsilon^2/I)$. Moreover, one can also show that $\tilde{\bar{y}}$ is a sufficient statistic for the joint distribution of $(\tilde{y}^1, \ldots, \tilde{y}^I, \tilde{\bar{y}})$ conditioned on $\tilde{d}$ and it holds that

$$\mathbb{E}[\tilde{d}|y^1, \ldots, y^I, \bar{y}] = \mathbb{E}[\tilde{d}|\bar{y}] \qquad \text{and} \qquad \mathrm{Var}(\tilde{d}|y^1, \ldots, y^I, \bar{y}) = \mathrm{Var}(\tilde{d}|\bar{y}).$$

We are now in a position to establish the following result, which corresponds to Grossman [839, Theorem 1].

Proposition 8.10 *In the context of the model described above (see Grossman [839]), there exists a fully revealing Green-Lucas equilibrium, characterized by the equilibrium price functional $\phi^* : \mathbb{R}^I \to \mathbb{R}$ defined by*

$$\phi^*(y^1, \ldots, y^I) = \frac{\bar{d} + \bar{\beta}(\bar{y} - \bar{d})}{r_f} - \frac{\sigma^2(1 - \bar{\beta})}{r_f} \left(\sum_{i=1}^I \frac{1}{a^i} \right)^{-1}.$$

(8.20)

Proof In view of Proposition 8.8, to show that (8.20) defines a Green-Lucas equilibrium price functional of the economy, we first show that it corresponds to the equilibrium price functional in an artificial economy where every agent observes all the private signals $\tilde{y} = (\tilde{y}^1, \ldots, \tilde{y}^I)$. As a preliminary, recall that $\tilde{\bar{y}}$ is a sufficient statistic for the joint distribution of $(\tilde{y}^1, \ldots, \tilde{y}^I, \tilde{\bar{y}})$ conditioned on $\tilde{d}$. Hence, the equilibrium of the artificial economy where every agent observes the full set of private signals coincides with the equilibrium of an artificial economy where every agent only observes the sufficient statistic $\tilde{\bar{y}}$. In this artificial economy, the optimal demand of agent i at equilibrium is given similarly as in (8.17) by

$$w^{i*} = \frac{\mathbb{E}[\tilde{d}|\bar{y}] - p^{a*}(\bar{y})r_f}{a^i \mathrm{Var}(\tilde{d}|\bar{y})}, \qquad \text{for every } i = 1, \ldots, I,$$

where $p^{a*}(\bar{y})$ denotes the equilibrium price corresponding to the artificial economy contingent on the realization $\{\tilde{\bar{y}} = \bar{y}\}$. Imposing the market clearing condition, this leads to the price functional

$$p^{a*}(y_1, \ldots, y_I) := p^{a*}(\bar{y}) = \frac{\mathbb{E}[\tilde{d}|\bar{y}]}{r_f} - \mathrm{Var}(\tilde{d}|\bar{y}) \left(\sum_{i=1}^I \frac{1}{a^i} \right)^{-1} \frac{1}{r_f},$$

which corresponds to the right hand side of (8.20), making use of the conditional expectation and variance given in (8.19). It order to show that the Green-Lucas equilibrium price functional ϕ^* of the original economy corresponds to p^{a*}, it suffices to compute the optimal demand of every agent in the original economy when the price of the risky asset is equal to $p^{a*}(\bar{y})$. Indeed, noting that the map $\bar{y} \mapsto p^{a*}(\bar{y})$ is invertible, it holds that

$$\frac{\mathbb{E}[\tilde{d}|y^i, p^{a*}(\bar{y})] - p^{a*}(\bar{y})r_f}{a^i \operatorname{Var}(\tilde{d}|y^i, p^{a*}(\bar{y}))} = \frac{\mathbb{E}[\tilde{d}|y^i, \bar{y}] - p^{a*}(\bar{y})r_f}{a^i \operatorname{Var}(\tilde{d}|y^i, \bar{y})}$$

$$= \frac{\mathbb{E}[\tilde{d}|\bar{y}] - p^{a*}(\bar{y})r_f}{a^i \operatorname{Var}(\tilde{d}|\bar{y})} = w^{i*},$$

where we have used the fact that $\tilde{\bar{y}}$ is a sufficient statistic for the joint distribution of $(\tilde{y}^i, \tilde{\bar{y}})$ conditioned on $\tilde{d}$, for every $i = 1, \ldots, I$. We have thus shown that the optimal demand in the original economy of each agent i, when faced with the price functional $p^{a*} =: \phi^*$, corresponds exactly to w^{i*}. Since the price functional clears the market, this proves the proposition. $\square$

The equilibrium price functional obtained in Proposition 8.10 has several intuitive properties. For instance, we see from (8.20) that ϕ^* is increasing with respect to the average signal $\bar{y}$ observed by the agents and is decreasing with respect to the coefficients of risk aversion of the agents. Moreover, the equilibrium price is a sufficient statistic and, therefore, the private information becomes redundant once agents are able to observe the equilibrium price. This shows that, as we have already remarked after Proposition 8.8, the equilibrium price functional eliminates the heterogeneity in the agents' information and implies that the market is informationally efficient.

However, as remarked in Admati [16], the fully revealing equilibrium constructed in Proposition 8.10 presents some problems. First, we can observe that the optimal demands of the agents are independent of their initial wealth (as a consequence of the assumption of exponential utility functions), of the private information (indeed, as a consequence of the fact that the equilibrium is fully revealing, the optimal demand only depends on the sufficient statistic) and, most surprisingly, of the price of the risky asset in equilibrium (compare also with Exercise 8.11). In particular, the independence of the optimal demand on the private information leads to a paradox that will be discussed later in this chapter: since prices perfectly aggregate and transmit information, there is no incentive for the agents to collect costly private information. But then it is difficult to understand how equilibrium prices can aggregate private information if the optimal demands of the agents are independent from their private information. If equilibrium prices are a sufficient statistic for the private information, agents do not have any incentive to collect private information, thus leading to the paradox that there is no private information to aggregate. The property that in equilibrium the optimal demand is independent of the price of the risky asset is due to the fact that, in addition to the classical substitution and income effects, in the present context a price change also has an *information effect*. The substitution and information effects go in opposite directions: on the one hand, if agents' expectations are constant, a price increase induces the agents to invest more in the riskless asset; on the other hand, a price increase is also interpreted as a signal that the risky asset is more valuable and the consequent revision of beliefs offsets the substitution effect. DeMarzo & Skiadas [555] show that the equilibrium of the Grossman [839] model is unique, but minor changes in the normality assumption lead to indeterminacy and partially revealing equilibria.

8.3   On the Impossibility of Informationally Efficient Markets

In the previous section, we have presented conditions under which Green-Lucas equilibrium prices transmit and aggregate private information and make agents' beliefs homogeneous, thereby providing a microfoundation to the efficient markets theory. However, by means of the example discussed at the end of the previous section, we have also seen that a fully revealing equilibrium is problematic, in the sense that, if equilibrium prices are a sufficient statistic, then the agents' optimal demands no longer depend on the private information and agents have no incentive to collect costly private information.

This paradox does not occur if equilibrium prices are only *partially revealing*. Indeed, if equilibrium prices are not a sufficient statistic for the agents' private information, then the optimal demand will depend not only on market prices but also on the private information and, hence, agents have an incentive to collect private information, since it can improve the optimal expected utility. In this case, market efficiency does not hold in the strong sense. Typical settings where a rational expectations equilibrium is only partially revealing are represented by the presence of random supply of the assets, costly observation of private information and imperfect competition. In this section, we focus our attention on the first two cases, referring to Chap. 10 for a discussion of the role of imperfect competition.

Random Supply and Heterogeneous Information

Following Hellwig [931], let us consider a generalized version of the model introduced in Grossman [839] and discussed in Sect. 8.2. As before, we assume that the economy is populated by $I \in \mathbb{N}$ agents $i = 1, \ldots, I$, with preferences characterized by the utility functions $u^i(x) = -\mathrm{e}^{-a^i x}$, with $a^i > 0$, for all $i = 1, \ldots, I$. In the economy, there are two traded assets: a riskless asset paying the constant rate of return $r_f > 0$ and a risky asset delivering the random dividend $\tilde{d}$ at date $t = 1$. The total supply of the riskless asset is normalized to zero, while the total supply of the risky asset is assumed to be random and given by the realization of the random variable $\tilde{u}$. The total supply of the risky asset can be random due to the presence of random trading by agents trading for liquidity reasons (*noise* or *liquidity traders*, see also Chap. 10). Each agent $i = 1, \ldots, I$ observes a private signal of the form

$$\tilde{y}^i = \tilde{d} + \tilde{\varepsilon}^i.$$

We assume that the random vector $(\tilde{u}, \tilde{d}, \tilde{\varepsilon}^1, \ldots, \tilde{\varepsilon}^I)$ is distributed according to a multivariate normal law, with mean vector $(\bar{u}, \bar{d}, 0, \ldots, 0)$ and a diagonal covariance matrix with the vector $(\sigma_u^2, \sigma_d^2, \sigma_{\varepsilon^1}^2, \ldots, \sigma_{\varepsilon^I}^2)$ along the diagonal. In particular, this implies that the random variables $(\tilde{u}, \tilde{d}, \tilde{\varepsilon}^1, \ldots, \tilde{\varepsilon}^I)$ are mutually independent.

Ex-ante (i.e., at date $t = -1$), all agents share the same beliefs and the structure of the economy is assumed to be common knowledge. In particular, every agent knows that the total market supply of the risky asset is a realization of the random variable $\tilde{u}$ but, of course, cannot observe the exact realization of $\tilde{u}$. In this context, we can establish the following proposition, which derives a partially revealing equilibrium price functional.

Proposition 8.11 *In the context of the model described above (see Hellwig [931]), there exists a partially revealing Green-Lucas equilibrium, characterized by the equilibrium price functional $\phi^* : \mathbb{R}^{I+1} \to \mathbb{R}$ defined by*

$$\phi^*(u, y^1, \ldots, y^I) = \gamma_0 + \sum_{i=1}^{I} \gamma_i y^i - \gamma u, \tag{8.21}$$

where the coefficients $\gamma_0, \gamma_i \in \mathbb{R}$, $i = 1, \ldots, I$, and $\gamma \neq 0$ can be explicitly determined as the solutions of a system of non-linear equations (see Hellwig [931]).

Proof In order to construct the Green-Lucas equilibrium price functional, we follow the steps outlined at the end of Sect. 8.1. First, we conjecture a linear price functional of the form (8.21). Given this conjectured price functional, we then compute the agents' optimal demands, given the observation of the price and of the private signals. Finally, we aggregate the individual optimal demands and impose market clearing in order to characterize the coefficients appearing in (8.21).

As a preliminary, note that the random variable $\tilde{\phi} := \phi^*(\tilde{u}, \tilde{y}^1, \ldots, \tilde{y}^I)$ given as in the right hand side of (8.21) is normally distributed, being a linear combination of normally distributed random variables. Moreover, the random vector $(\tilde{d}, \tilde{y}^1, \ldots, \tilde{y}^I, \tilde{\phi})$ is multivariate normal. By the properties of the multivariate normal law, the conditional distribution of $\tilde{d}$ given the observation of $\{\tilde{\phi} = \phi\}$ and $\{\tilde{y}^i = y^i\}$, for some $\phi \in \mathbb{R}$ and $y^i \in \mathbb{R}$, $i \in \{1, \ldots, I\}$, is again normal, with conditional mean and variance given by

$$\mathbb{E}[\tilde{d}|\phi, y^i] = \alpha_{0i} + \alpha_{1i}y^i + \alpha_{2i}\phi \qquad \text{and} \qquad \text{Var}(\tilde{d}|\phi, y^i) = \beta_i,$$

where the coefficients α_{0i}, α_{1i}, α_{2i} and β_i can be explicitly computed, for $i = 1, \ldots, I$, and depend on the parameters of the model and on the coefficients appearing in the right hand side of (8.21). Under the conjecture that the equilibrium price is given by ϕ, the optimal demand of each agent $i = 1, \ldots, I$ is then given by, similarly as in (8.17),

$$w^{i*} = \frac{\mathbb{E}[\tilde{d}|\phi, y^i] - \phi r_f}{a^i \operatorname{Var}(\tilde{d}|\phi, y^i)} = \frac{\alpha_{0i} + \alpha_{1i} y^i + (\alpha_{2i} - r_f)\phi}{a^i \beta_i}.$$

Market clearing implies that, for every realization $\{\tilde{u} = u\}$ of the random aggregate supply of the risky asset, it holds that

$$u = \sum_{i=1}^{I} w^{i*} = \sum_{i=1}^{I} \frac{\alpha_{0i} + \alpha_{1i} y^i + (\alpha_{2i} - r_f)\phi}{a^i \beta_i}.$$

Solving for ϕ then gives

$$\phi = \left(\sum_{i=1}^{I} \frac{r_f - \alpha_{2i}}{a^i \beta_i} \right)^{-1} \left(\sum_{i=1}^{I} \frac{\alpha_{0i} + \alpha_{1i} y^i}{a^i \beta_i} - u \right).$$

In turn, the last relation allows us to identify the coefficients appearing in (8.21) as the solutions to

$$\gamma = \left(\sum_{i=1}^{I} \frac{r_f - \alpha_{2i}}{a^i \beta_i} \right)^{-1},$$

$$\gamma_i = \frac{\gamma}{a^i} \frac{\alpha_{1i}}{\beta_i}, \qquad\qquad (8.22)$$

$$\gamma_0 = \gamma \sum_{i=1}^{I} \frac{\alpha_{0i}}{a^i \beta_i},$$

for all $i = 1, \ldots, I$. Recalling that the coefficients α_{0i}, α_{1i}, α_{2i} and β_i, for every $i = 1, \ldots, I$, in turn depend on γ_0, γ_i and γ, system (8.22) represents a system of non-linear equations for γ_0, γ_i and γ, which can be explicitly solved as shown in Hellwig [931, Section 3]. It remains to show that $\gamma \neq 0$. Arguing by contradiction, suppose that $\gamma = 0$. In that case, (8.22) would imply that $\gamma_0 = \gamma_i = 0$, for all $i = 1, \ldots, I$. In turn, this implies that the equilibrium price functional given as in the right hand side of (8.21) becomes identically equal to zero and, hence, $\alpha_{2i} = 0$, for all $i = 1, \ldots, I$. But then the first relation in (8.22) would imply that $\gamma \neq 0$, thus yielding a contradiction. □

In particular, the fact that $\gamma \neq 0$ in (8.21) implies that the equilibrium price functional ϕ^* is only partially revealing: equilibrium prices are determined not only by the private signals, via the term $\sum_{i=1}^{I} \gamma_i y^i$, but also by the random aggregate supply. Relation (8.22) also shows that the coefficient γ_i is inversely proportional to agent i's risk aversion and to the variance of the private signal observed by agent *i*. Hence, in equilibrium, very risk averse agents and agents with very imprecise information will affect the equilibrium price in a limited way.

In the context of Proposition 8.11, it is interesting to study the limit behavior of the Green-Lucas equilibrium when $\sigma_u^2 \to +\infty$ (i.e., the noise due to the random supply dominates): in this case (see Hellwig [931, Proposition 4.2]), variations in prices reflect mostly variations in the supply of the risky asset rather than variations in the agents' private signals and, therefore, agents cannot extract any useful information from prices. Hence, if $\sigma_u^2 \to +\infty$, then the equilibrium price functional converges to the equilibrium price functional obtained in an economy where agents condition their expected utilities only on their private signals. On the other hand, if $\sigma_u^2 \to 0$, then the equilibrium price functional obtained in Proposition 8.11 converges to that obtained in Proposition 8.10, where the aggregate supply is known ex-ante and the equilibrium price is a sufficient statistic for the agents' private signals (see Hellwig [931, Proposition 4.3]). In general, the presence of the noise component represented by the random asset supply prevents the equilibrium price from being a sufficient statistic. In Hellwig [931, Proposition 5.2] it is also shown that, as the number of agents increases to infinity, the private information effect vanishes and the equilibrium price only depends on the realization of $\tilde{d}$ (positively) and $\tilde{u}$ (negatively). In particular, $\tilde{d}$ represents the common element in the agents' private signals (in the limit, the noise terms $\tilde{\varepsilon}_i$ vanish by the law of large numbers). Note that, even in this limit case, the equilibrium price functional is only partially revealing because agents cannot distinguish whether a high price is due to a high realization of $\tilde{d}$ or to a low realization of $\tilde{u}$.

A model similar to that considered in Hellwig [931] has been analysed in Diamond & Verrecchia [569], under the assumption that the (random) aggregate supply is given by the sum of the random endowments of the I agents (in terms of units of the risky asset), so that $\tilde{u} = \sum_{i=1}^{I} \tilde{e}^i$, where $\tilde{e}^i$ denotes the random endowment of agent i and is assumed to be normally distributed with zero mean and variance σ_e^2, for all $i = 1, \ldots, I$. The random endowments are assumed to be mutually independent and independent of all the other random variables appearing in the model. In addition to observing the private signal and the market price, as considered above, every agent also observes his own endowment $\tilde{e}^i$. Under the additional assumptions that $a^i = a > 0$ and $\sigma_{\varepsilon^i}^2 = 1$, for all $i = 1, \ldots, I$, the same arguments given in the proof of Proposition 8.11 allow to show that the Green-Lucas equilibrium price functional ϕ^* is linear with respect to the aggregate supply $\tilde{u}$ and the average private signal $\tilde{\bar{y}} := \sum_{i=1}^{I} \tilde{y}^i / I$ observed by the agents. More precisely, as shown in Exercise 8.13, it holds that:

$$\phi^*(u, \bar{y}) = \alpha \bar{d} + \frac{\eta}{I} \sum_{i=1}^{I} y^i - \gamma u = \alpha \bar{d} + \eta \bar{y} - \gamma u, \tag{8.23}$$

with

$$\alpha = \frac{\sigma_e^2 + 1}{1 + I\sigma^2 + \sigma_e^2(1 + \sigma^2)}, \qquad \eta = \frac{\sigma^2(\sigma_e^2 + I)}{1 + I\sigma^2 + \sigma_e^2(1 + \sigma^2)} \qquad \text{and} \qquad \gamma = \frac{\eta}{I},$$

where we recall that $\tilde{d}$ is assumed to be normally distributed with mean $\bar{d}$ and variance σ^2.

The equilibrium price functional (8.23) is increasing with respect to the average signal received by the agents and decreasing with respect to the aggregate supply. In the presence of a random endowment and differently from the result of Proposition 8.10, the optimal demand of each agent depends on the private signal because the presence of the random supply does not make the private signal irrelevant. Hence, there exists an incentive to collect private information, thus solving the paradox generated by fully revealing equilibria discussed at the end of Sect. 8.2.

Random Supply and Asymmetric Information

A general model with asymmetric information has been proposed in Vives [1630, Chapter 4], encompassing several variations of the models originally proposed in Diamond & Verrecchia [569] and Hellwig [931]. As above, we consider a two-period economy ($t \in \{0, 1\}$) with two traded securities: a riskless asset paying a constant rate of return $r_f = 1$ and a risky asset delivering at $t = 1$ the random dividend $\tilde{d}$. We assume that the economy is populated by a continuum of risk averse agents i indexed in the interval $[0, 1]$ endowed with the Lebesgue measure. Similarly to the models considered above, the preferences of each agent $i \in [0, 1]$ are characterized by a negative exponential utility function of the form $u^i(x) = -\mathrm{e}^{-a^i x}$, with $a^i > 0$. The initial wealth of the agents is supposed to be non-random and normalized to zero (this is without loss of generality, due to the assumption of CARA utility functions). Under this hypothesis, denoting by $w^i \in \mathbb{R}$ the demand of the risky asset by agent i, the corresponding wealth at $t = 1$ is given by the random variable $\widetilde{W}^i = w^i(\tilde{d} - p)$, for $i = 1, \ldots, I$, where p denotes the price at date $t = 0$ of the asset.

In the economy, there are *informed* and *uninformed agents*. A priori (i.e., at date $t = -1$), all agents believe that the random dividend $\tilde{d}$ is normally distributed with mean $\bar{d}$ and variance σ^2. At the interim date $t = 0$, each informed trader i observes a private signal $\tilde{y}^i$ of the form $\tilde{y}^i = \tilde{d} + \tilde{\varepsilon}^i$. On the contrary, uninformed traders do not have access to any private information and can only infer information from market prices. The proportion of informed traders is denoted by $\mu \in [0, 1]$, with the proportion of uninformed traders being consequently $1 - \mu$. We assume that the two classes of agents have homogeneous risk aversion, in the sense that $a^i = a^{\mathrm{inf}} > 0$ for all $i \in [0, \mu]$ and $a^i = a^{\mathrm{un}} > 0$ for all $i \in (\mu, 1]$, where the superscripts ^{inf} and ^{un} stand for informed and uninformed, respectively. The random variables $\tilde{d}$ and $\tilde{\varepsilon}^i$, are assumed to be normally distributed and mutually independent, with $\mathbb{E}[\tilde{\varepsilon}^i] = 0$ and $\text{Var}(\tilde{\varepsilon}^i) = \sigma_\varepsilon^2$ for all $i \in [0, \mu]$ (i.e., all informed traders observe signals of the same precision and private signals are uncorrelated across agents). Besides informed and uninformed traders, there are also *noise traders* and we assume that their aggregate demand is represented by the random variable $\tilde{u}$, which is normally distributed, uncorrelated with all the random variables introduced above, with zero mean and variance σ_u^2. Noise traders represent investors trading for exogenous liquidity reasons. As above, we assume that the structure of the economy as well as all the distributional assumptions are common knowledge among all the agents.

In deriving a rational expectations equilibrium for this economy, we focus our attention on *symmetric equilibria*, meaning that traders of the same class exhibit optimal demands of the same form. Let us denote by $w^{\text{inf}}(y^i, p)$ the demand of each informed trader $i \in [0, \mu]$ (which depends on both the private signal y^i and the price information p) and by $w^{\text{un}}(p)$ the demand of each uninformed trader $i \in (\mu, 1]$ (depending on the price information only). In this context, a symmetric rational expectations equilibrium is characterized by two conditions: both informed and uninformed agents maximize their expected utilities conditionally on their information sets and markets clear. In particular, the market clearing condition is given by

$$\int_0^\mu w^{\text{inf}}(y^i, p)\mathrm{d}i + \int_\mu^1 w^{\text{un}}(p)\mathrm{d}i + u = 0, \tag{8.24}$$

where $u \in \mathbb{R}$ is a realization of the random variable $\tilde{u}$ representing the aggregate demand of noise traders. In this context, we can establish the following proposition (compare with Vives [1630, Proposition 4.1] and see Exercise 8.14 for a proof), where we say that a rational expectations equilibrium is *linear* if the associated demand functions are given by linear functions.

Proposition 8.12 *In the context of the model described above (see Vives [1630]), suppose that $\int_0^\mu \tilde{\varepsilon}^i \mathrm{d}i = 0$ almost surely, so that $\frac{1}{\mu} \int_0^\mu \tilde{y}^i \mathrm{d}i = \bar{d}$ almost surely. Then there exists a unique partially revealing rational expectations symmetric linear equilibrium characterized by the price functional $\phi^* : \mathbb{R}^2 \to \mathbb{R}$ given by*

$$\phi^*(d, u) = \frac{\frac{\mu}{a^{\text{inf}}\sigma_\varepsilon^2}(d - \bar{d}) + u}{\mu\left(\frac{1}{a^{\text{inf}}\sigma_\varepsilon^2} + b^{\text{inf}}\right) + (1 - \mu)b^{\text{un}}} + \bar{d}, \tag{8.25}$$

where

$$b^{\text{inf}} = \frac{\frac{1}{\sigma^2}}{a^{\text{inf}} + \frac{\mu\left(\frac{\mu}{a^{\text{inf}}} + \frac{1-\mu}{a^{\text{un}}}\right)}{\sigma_\varepsilon^2 \sigma_u^2}} \qquad \text{and} \qquad b^{\text{un}} = \frac{a^{\text{inf}}}{a^{\text{un}}}b^{\text{inf}}.$$

Moreover, for $p := \phi^*(d, u)$, *the optimal demands* $w^{\mathrm{inf},*}$ *and* $w^{\mathrm{un},*}$ *of the informed agents and of the uninformed agents are respectively given by*

$$w^{\mathrm{inf},*}(y^i, p) = \frac{1}{a^{\mathrm{inf}}\sigma_\varepsilon^2}(y^i - p) - b^{\mathrm{inf}}(p - \bar{d}), \qquad \text{for all } i \in [0, \mu], \qquad (8.26)$$

and

$$w^{\mathrm{un},*}(p) = -b^{\mathrm{un}}(p - \bar{d}), \qquad \text{for all } i \in (\mu, 1], \qquad (8.27)$$

The rational expectations equilibrium characterized in Proposition 8.12 is not fully revealing: similarly as in the models considered in the first part of the present section, traders observing a high price cannot determine whether such a price is motivated by a high dividend or by a large demand by the noise traders. In other words, the presence of noise traders does not allow the equilibrium price to fully reveal the private information of the informed agents. As shown in Exercise 8.15, the equilibrium price obtained in Proposition 8.12 can be represented as a weighted average of the conditional expectations of the future dividend of the two classes of agents plus a noise term due to the presence of noise traders. If $a^{\mathrm{un}} \to 0$, then the equilibrium price functional converges to the conditional expectation $\mathbb{E}[\tilde{d}|p]$.

The result of Proposition 8.12 leads to an interesting interpretation of the behavior of the two classes of agents. Since $b^{\mathrm{un}} > 0$, formula (8.27) shows that uninformed agents trade against the market (i.e., they sell the asset when its price is above the a priori expected dividend). Uninformed agents face an adverse selection problem, because the equilibrium is not fully revealing and they do not know whether trading is motivated by informed traders or by noise traders. The behavior of informed traders is determined by two distinct components, as shown by (8.26): informed agents tend to buy (sell, resp.) the risky asset if its price is below (above, resp.) their private signal and tend to sell (buy, resp.) the asset if its price is above (below, resp.) the a priori expected dividend. In particular, the first component reflects a speculative behavior motivated by private information. For the informed traders, the sensitivity to the private signal is measured by the quantity $1/(a^{\mathrm{inf}}\sigma_\varepsilon^2)$, which is inversely proportional to the coefficient of absolute risk aversion and directly proportional to the precision of the private signal. Moreover, the quantity $\mu(1/(a^{\mathrm{inf}}\sigma_\varepsilon^2) + b^{\mathrm{inf}}) + (1 - \mu)b^{\mathrm{un}}$ provides a measure of *market depth*, i.e., how much the market can absorb a change in the noise traders' demand without a significant price change. This measure of market depth is a weighted sum of the parameters describing the traders' responsiveness to the market price, as can be deduced from Proposition 8.12.

Costly Private Information and the Grossman-Stiglitz Paradox

The impossibility of market efficiency when the acquisition of private information is costly has been shown in Grossman & Stiglitz [849]. This paper is concerned with a phenomenon already discussed in Sect. 8.2: if acquiring private information is costly and market prices transmit private information, then it can be convenient for some agents to adopt a *free riding behavior* and not buy private information, inferring instead information from market prices, since market prices reflect (at least partially) the information acquired by the other agents.

The model considered in Grossman & Stiglitz [849] is a two-period model ($t \in \{0, 1\}$), with an economy populated by a continuum of agents indexed in the interval $[0, 1]$ and two traded assets: a riskless security with return $r_f > 1$ and a risky asset delivering at date $t = 1$ the random dividend $\tilde{d}$. More specifically, the random variable $\tilde{d}$ is assumed to be of the form

$$\tilde{d} = \tilde{d}^o + \tilde{\varepsilon},$$

where the pair of random variables $(\tilde{d}^o, \tilde{\varepsilon})$ is assumed to have a bivariate normal distribution, with

$$\mathbb{E}[\tilde{\varepsilon}] = 0, \qquad \text{Cov}(\tilde{d}^o, \tilde{\varepsilon}) = 0 \quad \text{and} \quad \text{Var}(\tilde{d}|\tilde{d}^o = d^o) = \text{Var}(\tilde{\varepsilon}) =: \sigma_\varepsilon^2,$$

for every $d^o \in \mathbb{R}$. Furthermore, the per capita supply of the risky asset is random and given by the realization of the random variable $\tilde{u}$, which is independent of $(\tilde{d}^o, \tilde{\varepsilon})$ and normally distributed with mean $\bar{u}$ and variance σ_u^2. Let also $\sigma^2 := \sigma^2(\tilde{d}) = \sigma^2(\tilde{d}^o) + \sigma_\varepsilon^2$.

Letting w_0^i and w^i denote the initial wealth[2] and the demand of the risky asset, respectively, of agent i, the random wealth at date $t = 1$ is given by

$$\widetilde{W}_1^i = (w_0^i - w^i p)r_f + w^i \tilde{d}, \qquad \text{for all } i \in [0, 1],$$

where p denotes the price at date $t = 0$ of the risky asset. Similarly as in the models presented above, we assume that the preferences of each agent i are characterized by a negative exponential utility function of the form $u^i(x) = -e^{-ax}$, with $a > 0$, for all $i \in [0, 1]$. In particular, the coefficient of absolute risk aversion is assumed to be the same for all the agents.

At date $t = 0$, agents can choose to observe the realization of the random variable $\tilde{d}^o$ by paying a fixed cost $c > 0$, while the residual component $\tilde{d} - \tilde{d}^o = \tilde{\varepsilon}$ remains unobservable. As usual, we suppose that each agent maximizes his expected

---

[2]For the sake of simplicity and without loss of generality, we assume that the initial wealth of each agent is non-random. In the original paper Grossman & Stiglitz [849], the initial wealth is specified in terms of the two traded securities and, hence, depends on the equilibrium price, which is in turn random.

utility, exploiting all the available information and deciding whether to buy or not the information related to the observation of $\tilde{d}^o$. In other words, each agent has to take two decisions: whether to buy or not the additional information and how much to invest in the risky asset. We say that an agent is *informed* if he has chosen to observe the component $\tilde{d}^o$ and *uninformed* otherwise (in the following, we shall use the superscript ^{inf} when referring to informed agents and the superscript ^{un} when referring to uninformed agents). Agents have to decide ex-ante whether to buy the information or not. Apart from the information set used to take the investment decisions, the two types of agents are identical and the structure of the economy as well as all the distributional assumptions are common knowledge among the agents. We denote by $\lambda \in [0, 1]$ the proportion of informed agents in the economy. In a first step, we shall consider λ as an exogenous variable and then, in a second step, determine the endogenous value of λ corresponding to an overall equilibrium of the economy.

Considering λ as a fixed exogenous variable, a Green-Lucas equilibrium of the economy, similarly as in Definition 8.6 (and compare also with Proposition 8.11), is characterized by an equilibrium price functional $\phi_\lambda^*(d^o, u) : \mathbb{R}^2 \to \mathbb{R}$ in correspondence of which the market clears. The equilibrium price functional ϕ^* depends on the realization of the random variables $\tilde{d}^o$ and $\tilde{u}$ (representing the information available to the informed agents and the random per capita supply, respectively). The subscript λ makes explicit the dependence of the equilibrium price functional on the proportion of informed agents. In the following proposition, which corresponds to Grossman & Stiglitz [849, Theorem 1], we derive the optimal demands of the informed and uninformed agents and the equilibrium price functional, for λ fixed.

Proposition 8.13 *In the context of the model described above (see Grossman & Stiglitz [849]), let the proportion $\lambda \in [0, 1]$ of informed agents be fixed. Then the equilibrium price functional $\phi_\lambda^* : \mathbb{R}^2 \to \mathbb{R}$ is given by*

$$\phi_\lambda^*(d^o, u) = \alpha_1(\lambda) + \alpha_2(\lambda)\eta_\lambda(d^o, u), \qquad \text{for all } d^o \in \mathbb{R} \text{ and } u \in \mathbb{R}, \qquad (8.28)$$

where

$$\begin{cases} \eta_\lambda(d^o, u) := d^o - \frac{a\sigma_\varepsilon^2}{\lambda}(u - \bar{u}), & \text{if } \lambda > 0, \\ \eta_0(d^o, u) := -u, & \text{if } \lambda = 0, \end{cases}$$

and $\alpha_1(\lambda) \in \mathbb{R}$ and $\alpha_2(\lambda) > 0$. Moreover, for every realization d^o and u, the optimal demand of each informed agent in correspondence of the price $p = \phi_\lambda^*(d^o, u)$ of the risky asset is given by

$$w^{\text{inf},*}(d^o, p) = \frac{d^o - pr_f}{a\sigma_\varepsilon^2}, \qquad (8.29)$$

and the optimal demand of each uninformed agent is given by

$$w^{\text{un},*}(p) = \frac{\mathbb{E}[\tilde{d}|\phi_\lambda^*(\tilde{d}^o,\tilde{u}) = p] - pr_f}{a \operatorname{Var}(\tilde{d}|\phi_\lambda^*(\tilde{d}^o,\tilde{u}) = p)}. \tag{8.30}$$

Proof Similarly as in the case of the models considered earlier in this section, we conjecture that the equilibrium price functional is a linear function of the random variables $\tilde{d}^o$ and $\tilde{u}$ as in the right-hand side of (8.28).

Let us start by computing the optimal demand of an informed agent. By assumption, each informed agent observes the realization of the random variable $\tilde{d}^o$. Since $\tilde{u}$ and $\tilde{\varepsilon}$ are independent, an informed agent observing the realization of $\tilde{d}^o$ does not receive any useful further information from market prices. Hence, it holds that

$$\mathbb{E}[\tilde{d}|\tilde{d}^o = d^o, \phi_\lambda^*(\tilde{d}^o,\tilde{u}) = p] = \mathbb{E}[\tilde{d}|\tilde{d}^o = d^o] = d^o,$$

where the last equality follows since $\operatorname{Cov}(\tilde{d}^o,\tilde{\varepsilon}) = 0$. Similarly, it holds that

$$\operatorname{Var}(\tilde{d}|\tilde{d}^o = d^o, \phi_\lambda^*(\tilde{d}^o,\tilde{u}) = p) = \operatorname{Var}(\tilde{d}|\tilde{d}^o = d^o) = \sigma_\varepsilon^2.$$

As a consequence, by the assumption of an exponential utility function, the optimal demand of each informed agent is given by (8.29), for every realization $d^o \in \mathbb{R}$. Considering now the case of an uninformed agent, (8.30) simply follows by computing the optimal demand of an agent maximizing an exponential utility function (together with the assumption of a normal distribution, compare with formula (3.11)), conditionally on the information conveyed by the observation of the market price.

In order to explicitly characterize the equilibrium price functional, consider first the case $\lambda = 0$ (i.e., there are no informed agents). In this case, the market clearing condition reduces to the condition $w^{\text{un},*}(p) = u$, for every realization $u \in \mathbb{R}$. Let us conjecture that the equilibrium price functional is more specifically given by

$$\phi_0^*(d^o, u) = \frac{\mathbb{E}[\tilde{d}^o]}{r_f} - \frac{a\sigma^2}{r_f}u, \tag{8.31}$$

for every $(d^o, u) \in \mathbb{R}^2$. Note that $\phi_0^*(\tilde{d}^o, \tilde{u})$ is independent from $\tilde{d}$, since it does not depend explicitly on $\tilde{d}^o$ (because there are no informed agents) and $\tilde{u}$ is assumed to be independent from $\tilde{d}$. Hence:

$$\mathbb{E}[\tilde{d}|\phi_0^*(\tilde{d}^o,\tilde{u}) = p] = \mathbb{E}[\tilde{d}] = \mathbb{E}[\tilde{d}^o]$$

and $\mathrm{Var}(\tilde{d}|\phi_0^*(\tilde{d}^o,\tilde{u}) = p) = \sigma^2$. This implies that the optimal demand of every uninformed agent in correspondence of the price $p = \phi_0^*(d^o,u)$ is given by

$$w^{\mathrm{un},*}(p) = \frac{\mathbb{E}[\tilde{d}^o] - p r_f}{a\sigma^2}.$$

By construction, the equilibrium price functional defined in (8.31) satisfies the market clearing condition $w^{\mathrm{un},*}(\phi_0^*(d^o,u)) = u$ and the linear representation (8.28) of the equilibrium price functional is also satisfied. Moreover, the optimal demand is clearly of the form (8.30).

Let us now consider the more interesting case where $\lambda \in (0,1]$. Define

$$\alpha_1(\lambda) := \frac{\frac{(1-\lambda)\mathbb{E}[\tilde{d}|\eta_\lambda(d^o,u)]}{a\,\mathrm{Var}(\tilde{d}|\eta_\lambda(d^o,u))} - \mathbb{E}[\tilde{u}]}{r_f\left(\frac{\lambda}{a\sigma_\varepsilon^2} + \frac{1-\lambda}{a\,\mathrm{Var}(\tilde{d}|\eta_\lambda(d^o,u))}\right)} \quad \text{and} \quad \alpha_2(\lambda) := \frac{\frac{\lambda}{a\sigma_\varepsilon^2}}{r_f\left(\frac{\lambda}{a\sigma_\varepsilon^2} + \frac{1-\lambda}{a\,\mathrm{Var}(\tilde{d}|\eta_\lambda(d^o,u))}\right)},$$

where we denote by $\mathbb{E}[\tilde{d}|\eta_\lambda(d^o,u)]$ the conditional expectation of $\tilde{d}$ given the observation of $\{\eta_\lambda(\tilde{d}^o,\tilde{u}) = \eta_\lambda(d^o,u)\}$ (and similarly for the conditional variance), and let us conjecture that the equilibrium price functional is given by

$$\phi_\lambda^*(d^o,u) = \alpha_1(\lambda) + \alpha_2(\lambda)\eta_\lambda(d^o,u), \tag{8.32}$$

for every realization $(d^o,u) \in \mathbb{R}^2$ of the random variables $\tilde{d}^o$ and $\tilde{u}$, where $\eta_\lambda(d^o,u)$ is defined as in the statement of the proposition. By the properties of the normal multivariate distribution, it holds that

$$\mathbb{E}[\tilde{d}|\eta_\lambda(d^o,u)] = \mathbb{E}[\tilde{d}^o] + \frac{\sigma^2(\tilde{d}^o)}{\mathrm{Var}(\eta_\lambda(\tilde{d}^o,\tilde{u}))}(\eta_\lambda(d^o,u) - \mathbb{E}[\tilde{d}^o]),$$

$$\mathrm{Var}(\tilde{d}|\eta_\lambda(d^o,u)) = \sigma^2(\tilde{d}^o)\left(1 - \frac{\sigma^2(\tilde{d}^o)}{\mathrm{Var}(\eta_\lambda(\tilde{d}^o,\tilde{u}))}\right) + \sigma_\varepsilon^2, \tag{8.33}$$

$$\mathrm{Var}(\eta_\lambda(\tilde{d}^o,\tilde{u})) = \sigma^2(\tilde{d}^o) + \left(\frac{a\sigma_\varepsilon^2}{\lambda}\right)^2 \sigma_u^2.$$

Moreover, since the map $(d^o,u) \mapsto \phi_\lambda^*(d^o,u)$ is linear and $\alpha_2(\lambda) \neq 0$, it holds that

$$\mathbb{E}[\tilde{d}|\phi_\lambda^*(d^o,u)] = \mathbb{E}[\tilde{d}|\eta_\lambda(d^o,u)] \quad \text{and} \quad \mathrm{Var}(\tilde{d}|\phi_\lambda^*(d^o,u)) = \mathrm{Var}(\tilde{d}|\eta_\lambda(d^o,u)).$$

By using (8.29), it can be checked that the market clearing condition

$$u = \lambda w^{\text{inf},*}\big(d^o, \phi^*(d^o, u)\big) + (1 - \lambda)w^{\text{un},*}\big(\phi^*(d^o, u)\big)$$

$$= \lambda \frac{d^o - \phi^*(d^o, u)r_f}{a\sigma_\varepsilon^2} + (1 - \lambda)\frac{\mathbb{E}[\tilde{d}|\eta_\lambda(d^o, u)] - \phi^*(d^o, u)r_f}{a\,\text{Var}(\tilde{d}|\eta_\lambda(d^o, u))}$$

holds for every $(d^o, u) \in \mathbb{R}^2$, thus showing that definition (8.32) gives indeed the equilibrium price functional, satisfying the linear representation (8.28).

$\square$

The equilibrium price functional obtained in Proposition 8.13 is only partially revealing. Indeed, uninformed agents cannot learn the realization of $\tilde{d}^o$ by observing the equilibrium price because the equilibrium price is also affected by the random aggregate supply of the risky asset. Hence, changes in the equilibrium price can be due to changes in the informed agents' information as well as to changes in the supply of the risky asset.

The equilibrium price $\phi_\lambda^*(\tilde{d}^o, \tilde{u})$ is informationally equivalent to the random variable $\eta_\lambda(\tilde{d}^o, \tilde{u})$, which is in turn a mean-preserving spread of $\tilde{d}$ (this simply follows by the fact that $\mathbb{E}[\eta_\lambda(\tilde{d}^o, \tilde{u})|\tilde{d}^o = d^o] = d^o$, since the random variables $\tilde{d}^o$ and $\tilde{u}$ are assumed to be independent). In particular, the noise produced by the random supply $\tilde{u}$ prevents uninformed traders from perfectly learning the realization of $\tilde{d}^o$ from the observation of $\eta_\lambda(\tilde{d}^o, \tilde{u})$. More precisely, the quality of the information conveyed by $\eta_\lambda(\tilde{d}^o, \tilde{u})$ to uninformed agents is measured by

$$\text{Var}\big(\eta_\lambda(\tilde{d}^o, \tilde{u})|\tilde{d}^o = d^o\big) = \frac{a^2\sigma_\varepsilon^4}{\lambda^2}\sigma_u^2.$$

We observe that, if the variance σ_u^2 of the supply is large, then the equilibrium price is scarcely informative about the component $\tilde{d}^o$ of the random dividend of the risky asset. Furthermore, the informativeness of the equilibrium price is decreasing with respect to the coefficient of absolute risk aversion and, in line with intuition, is increasing with respect to the proportion of informed traders in the economy. Note also that, if $\lambda = 0$ (i.e., there are no informed agents), then the equilibrium price does not convey any information on $\tilde{d}^o$.

So far, the proportion $\lambda \in [0, 1]$ of informed agents in the economy has been considered as an exogenous fixed variable. Indeed, the equilibrium price functional obtained in Proposition 8.13 is explicitly characterized for a given $\lambda \in [0, 1]$. However, the proportion of informed agents should be considered as an endogenous variable in the model: indeed, agents can a priori choose whether to buy the information (paying the cost $c > 0$) or not. Hence, we can define an *overall equilibrium* as a couple $(\lambda^*, \phi_{\lambda^*}^*(\tilde{d}^o, \tilde{u}))$, where λ^* is such that

- if $\lambda^* = 0$, then, in correspondence of the price $\phi_0^*(\tilde{d}^o, \tilde{u})$, the optimal expected utility of an informed agent is less or equal than the optimal expected utility of an uninformed agent;

- if $\lambda^* \in (0, 1)$, then, in correspondence of the price $\phi_\lambda^*(\tilde{d}^o, \tilde{u})$, the optimal expected utility of an informed agent is equal to the optimal expected utility of an uninformed agent;
- if $\lambda^* = 1$, then, in correspondence of the price $\phi_\lambda^*(\tilde{d}^o, \tilde{u})$, the optimal expected utility of an informed agent is greater or equal than the optimal expected utility of an uninformed agent.

If an agent $i \in [0, 1]$ chooses to buy the information, then his optimal wealth at $t = 1$ in correspondence of the equilibrium price $\phi_\lambda^*(\tilde{d}^o, \tilde{u})$ will be given by

$$\widetilde{W}_{1,\lambda}^{\text{inf},i,*} := \left(w_0^i - c - w^{\text{inf},*}(\tilde{d}^o, \phi_\lambda^*(\tilde{d}^o, \tilde{u}))\phi_\lambda^*(\tilde{d}^o, \tilde{u})\right)r_f + w^{\text{inf},*}(\tilde{d}^o, \phi_\lambda^*(\tilde{d}^o, \tilde{u}))\tilde{d},$$

while, if agent $i \in \{1, \ldots, I\}$ chooses not to buy the information,

$$\widetilde{W}_{1,\lambda}^{\text{un},i,*} := \left(w_0^i - w^{\text{un},*}(\phi_\lambda^*(\tilde{d}^o, \tilde{u}))\phi_\lambda^*(\tilde{d}^o, \tilde{u})\right)r_f + w^{\text{un},*}(\phi_\lambda^*(\tilde{d}^o, \tilde{u}))\tilde{d},$$

where $w^{\text{inf},*}(\tilde{d}^o, \phi_\lambda^*(\tilde{d}^o, \tilde{u}))$ and $w^{\text{un},*}(\phi_\lambda^*(\tilde{d}^o, \tilde{u}))$ denote the optimal demands of informed and uninformed agents, respectively, as considered in Proposition 8.13. Observe that, due to the assumption of an exponential utility function, the optimal demand of the risky asset does not depend on the initial wealth and, hence, does not depend on the cost c. Since w_0^i is supposed to be non-random and since we consider exponential utility functions, it is easy to see that the ratio $\mathbb{E}[u(\widetilde{W}_{1,\lambda}^{\text{inf},i,*})]/\mathbb{E}[u(\widetilde{W}_{1,\lambda}^{\text{un},i,*})]$ does not depend on i and, hence, we can define the function $\gamma : [0, 1] \to \mathbb{R}_+$ by

$$\gamma(\lambda) := \frac{\mathbb{E}[u(\widetilde{W}_{1,\lambda}^{\text{inf},i,*})]}{\mathbb{E}[u(\widetilde{W}_{1,\lambda}^{\text{un},i,*})]},$$

for $\lambda \in [0, 1]$. We can now establish the following proposition, which characterizes the overall equilibrium $(\lambda^*, \phi_{\lambda^*}^*(\tilde{d}^o, \tilde{u}))$ where the proportion of informed traders is endogenous (compare with Grossman & Stiglitz [849, Theorem 3]).

Proposition 8.14 *In the context of the model described above (see Grossman & Stiglitz [849]), the function $\lambda \mapsto \gamma(\lambda)$ is strictly increasing. For $\lambda = 0$ it is given by*

$$\gamma(0) = e^{acr_f}\sqrt{\frac{1}{1+n}}$$

and, for $\lambda > 0$, it admits the representation

$$\gamma(\lambda) = e^{acr_f}\sqrt{\frac{\text{Var}(\tilde{d}|\tilde{d}^o = d^o)}{\text{Var}(\tilde{d}|\eta_\lambda(\tilde{d}^o, \tilde{u}) = \eta_\lambda(d^o, u))}} = e^{acr_f}\sqrt{\frac{1+m}{1+m+nm}}, \qquad (8.34)$$

where

$$m := \left(\frac{a\sigma_\varepsilon^2}{\lambda} \right)^2 \frac{\sigma_u^2}{\sigma^2(\tilde{d}^o)} \qquad and \qquad n := \frac{\sigma^2(\tilde{d}^o)}{\sigma_\varepsilon^2}.$$

Moreover, the following hold:

  (i) *if, for some* $\lambda^* \in [0,1]$, *it holds that* $\gamma(\lambda^*) = 1$, *then the couple* $(\lambda^*, \phi_{\lambda*}^*(\tilde{d}^o, \tilde{u}))$ *is an overall equilibrium,*
 (ii) *if* $\gamma(1) < 1$, *then the couple* $(1, \phi_1^*(\tilde{d}^o, \tilde{u}))$ *is an overall equilibrium,*
(iii) *if* $\gamma(0) > 1$, *then the couple* $(0, \phi_0^*(\tilde{d}^o, \tilde{u}))$ *is an overall equilibrium,*

where $\phi_\lambda^*(\tilde{d}^o, \tilde{u})$ *is defined as in Proposition 8.13, for every* $\lambda \in [0,1]$. *Moreover, for every equilibrium price functional* ϕ_λ^* *which is a monotone function of* η_λ *(as defined in Proposition 8.13), there exists a unique overall equilibrium* $(\lambda^*, \phi_{\lambda*}^*)$.

Proof The fact that the function $\lambda \mapsto \gamma(\lambda)$ is strictly increasing and can be represented as in (8.34) is proved in Exercise 8.16. The second part of the proposition easily follows from the fact that $\lambda \mapsto \gamma(\lambda)$ is strictly increasing together with the assumption of negative exponential utility functions and the definition of overall equilibrium. $\square$

The quantity n introduced in Proposition 8.14 can be regarded as a measure of the quality of the information accessible by informed agents, while the quantity $1/m$ represents a measure of the quality of the information conveyed by the equilibrium price. Indeed, if $\lambda^* \in (0,1)$ (so that $\gamma(\lambda^*) = 1$ in correspondence of the overall equilibrium), then the correlation between the equilibrium price and the signal $\tilde{d}^o$ (and, therefore, the degree of informativeness of equilibrium prices) can be expressed as

$$\varrho := \left| \text{Corr}(\tilde{d}^o, \phi_{\lambda*}^*(\tilde{d}^o, \tilde{u})) \right| = \frac{1}{\sqrt{1+m}} = \sqrt{1 - \frac{e^{2acr_f} - 1}{n}}. \tag{8.35}$$

The following proposition presents several comparative statics for the equilibrium characterized in Proposition 8.14 (see Grossman & Stiglitz [849, Section II.H]).

Proposition 8.15 *In the context of the model described above (see Grossman & Stiglitz [849]), the overall equilibrium* $(\lambda^*, \phi_{\lambda*}^*(\tilde{d}^o, \tilde{u}))$ *satisfies the following properties:*

  (i) *if* $\lambda^* \in (0,1)$, *then* ϱ *is increasing with respect to n and decreasing with respect to c and a;*
 (ii) *if* $\lambda^* \in (0,1)$, *then* ϱ *is constant with respect to changes in* σ_u^2 *or with respect to changes in* σ^2 *if n is constant;*
(iii) *if* $\lambda^* \in (0,1)$, *then* λ^* *is increasing with respect to* σ_u^2, *decreasing with respect to c and increasing with respect to* σ^2 *if n is constant;*

(iv) if $\lambda^* \in (0, 1)$ and σ^2 is constant, then there exists a value $\bar{n}$ such that λ^* is decreasing with respect to n for $n > \bar{n}$ and increasing with respect to n for $n < \bar{n}$;

(v) there exists $\bar{c}$ such that $\lambda^* = 1$ for every $c \leq \bar{c}$;

(vi) if $\sigma_\varepsilon^2 \to 0$ and σ^2 is constant (so that $\sigma^2(\tilde{d}^o)$ increases to σ^2), then the equilibrium price functional becomes fully revealing and λ^* converges to zero;

(vii) if $\sigma_u^2 \to 0$, then λ^* converges to zero and the informational content of the equilibrium price is unchanged;

(viii) if $\sigma_u^2 = 0$ and $e^{a\bar{c}r_f} < \sqrt{1+n}$, then an overall equilibrium does not exist;

(ix) if $\sigma_\varepsilon^2 = 0$ (i.e., the signal $\tilde{d}^o$ is fully informative about the random variable $\tilde{d}$), then an overall equilibrium does not exist;

(x) the mean and the variance of the trading volume are zero for c sufficiently large or small and the mean and the variance of the trading volume converge to zero for $n \to +\infty$.

Proof Parts (i)–(ii) are direct consequences of (8.35), while part (iii) easily follows from the explicit expression of $\gamma(\lambda)$ given in (8.34) together with the fact that the map $\lambda \mapsto \gamma(\lambda)$ is strictly increasing. Part (iv) is proved in Exercise 8.17. Part (v) simply follows from the observation that there exists a value $\bar{c}$ satisfying

$$e^{a\bar{c}r_f} \sqrt{\frac{\mathrm{Var}(\tilde{d}|\tilde{d}^o = d^o)}{\mathrm{Var}(\tilde{d}|\eta_1(\tilde{d}^o, \tilde{u}) = \eta_1(d^o, u))}} = 1.$$

Concerning part (vi), the fact that the equilibrium price functional becomes fully revealing can be deduced from (8.35), since if $\sigma_\varepsilon^2 \to 0$ and $\sigma^2(\tilde{d}^o) \to \sigma^2$, then the quantity n introduced in Proposition 8.14 explodes, so that $\varrho \to 1$ (fully informative price system). Moreover, representation (8.34) implies that, as $\sigma_\varepsilon^2 \to 0$, the quantities m and nm converge to zero for any fixed $\lambda > 0$, thus implying that $\gamma(\lambda) \to e^{a\bar{c}r_f} > 1$, for any fixed $\lambda > 0$. Therefore, for the equilibrium to be maintained, λ must also converge to zero. Parts (vii), (viii) and (ix) are proved in Exercise 8.18. Finally, we refer the reader to Grossman & Stiglitz [849, Theorem 6] for the proof of part (x).   □

Proposition 8.15 shows several interesting properties of the overall equilibrium characterized in Proposition 8.14. In particular, the informativeness of the equilibrium price (as measured by the correlation coefficient ϱ introduced in (8.35)) is high if the quality of information is high (i.e., if n is large), if the coefficient of absolute risk aversion is small and if the cost of purchasing the information is small (see part (i) of Proposition 8.15). Moreover, (8.35) shows that any change in the parameters of the model that does not affect a, c or n leaves unchanged the informational content of the equilibrium price.

If the variance σ_u^2 of the random supply increases, then, for any fixed λ, the informational content of the equilibrium price decreases. However, in view of Proposition 8.14, an increase in σ_u^2 also increases the proportion of informed traders, since it increases the benefits of acquiring information. In equilibrium, as shown in part *(ii)* of Proposition 8.15, these two effects offset exactly and, therefore, the informational content of the equilibrium price is not affected by changes in σ_u^2.

As shown in part *(iii)* of Proposition 8.15, an increase in the variance σ_u^2 of the random supply, an increase in the variance σ^2 of the dividend (provided that n is constant) or a decrease in the cost of information will increase the proportion of informed agents. Moreover, part *(iv)* shows that the proportion of informed agents is increasing with respect to the quality of the information (as measured by n) if the latter is sufficiently poor, otherwise it will be decreasing.

This last result leads to a paradox (*Grossman-Stiglitz paradox*): as the economy converges to the ideal conditions of absence of noise ($\sigma_u^2 = 0$) or of perfect information ($\sigma_\varepsilon^2 = 0$), the proportion of informed agents converges to zero (see parts *(vi)–(vii)* of Proposition 8.15), since the equilibrium price will perfectly transmit all the private information to the uninformed agents. As explained in Grossman & Stiglitz [849], "there is a fundamental conflict between the efficiency with which markets spread information and the incentives to acquire information". This is to say that there exists a positive information externality in the activity of information acquisition by informed agents. As the economy tends to ideal conditions, the equilibrium price becomes fully revealing and provides a very accurate estimate of the dividend. In turn, this leads agents to behave as *free riders*: since the cost of acquiring the information is strictly positive, agents prefer not to buy the information, learning it from the price observation. But if there is no incentive to buy information, then the only possible equilibrium is the one with no informed agents. On the other hand, if everyone is uninformed and c is small enough, then it is clearly optimal for some agent to acquire information, hence the paradox of the inexistence of an overall equilibrium (compare also with the discussion in Grossman & Stiglitz [849, Section IV]). This is the content of parts *(viii)–(ix)* of Proposition 8.15, showing that an overall equilibrium with costly information acquisition does not exist if markets are informationally efficient. The behavior of the trading volume (part *(x)* of Proposition 8.15) confirms this result. Note that the key feature of the Grossman & Stiglitz [849] model is that the proportion of informed agents is considered as an endogenous variable (see also Sect. 8.7 for a further discussion of the features of the model of Grossman & Stiglitz [849]).

An economy with costly private information has been analysed in Verrecchia [1618] by relying on the model proposed in Hellwig [931]. Similarly as above, the dividend $\tilde{d}$ of the risky asset is assumed to be distributed as a normal random variable with mean $\bar{d}$ and variance σ^2. The preferences of each agent i are characterized by a constant coefficient of absolute risk aversion $a^i > 0$ (CARA preferences). The private information is represented by the observation of a signal of the form $\tilde{y}^i = \tilde{d} + \tilde{\varepsilon}^i$, where $\tilde{\varepsilon}^i$ is a normal random variable with zero mean and variance $\sigma^2(\tilde{\varepsilon}^i) = 1/s_i$, with $s_i > 0$, for all $i = 1, \ldots, I$. The supply of the risky asset is distributed as a normal random variable with zero mean and all the random variables introduced in the model are assumed to be mutually independent. The quantity s_i can be regarded as a measure of the precision of the information of agent i, for each $i = 1, \ldots, I$. The distinguishing feature of the model of Verrecchia [1618] is that the quality of the information is endogenously determined and agents choose the precision of their signal before trading. More specifically, the cost of observing a signal with precision s is determined by a continuous cost function $s \mapsto c(s)$ satisfying $c(s) > 0$, $c'(s) > 0$ and $c''(s) \geq 0$, for every $s \in (0, \infty)$ (i.e., c is a strictly increasing convex function). Similarly as in the models considered earlier in this section, the fact that the aggregate supply is random yields an equilibrium price functional which is not fully revealing. As above, CARA preferences together with the assumption of multivariate normality represent crucial assumptions in order to derive closed-form solutions. It is shown that there exists a Green-Lucas equilibrium with an endogenous level of precision of information. In equilibrium, the level of precision of the information acquired by an agent is a non-decreasing function of his risk tolerance and a non-increasing function of the informational content of the equilibrium price (which is in turn decreasing with respect to the variance of the supply and the cost of acquiring information).

Diamond [568] considers a model similar to Diamond & Verrecchia [569] and Hellwig [931], with infinitely many traders and two traded assets and where the traders have the possibility of acquiring information about a firm's return. In equilibrium (see Diamond [568, Lemma 3]), it is shown that all the agents can decide not to buy information (in the case of a high cost and a low precision of the signal) or all the agents buy information (in the case of a small cost and a very precise information). In all the intermediate cases, only a fraction of the agents will acquire information. In a model similar to Diamond & Verrecchia [569] and Grossman & Stiglitz [849], two equilibria with strategic substitution/complementarity effects may emerge when the aggregate endowment and the individual endowments are positively correlated (see also Ganguli & Yang [754]). These two equilibria have opposing properties concerning the informational content of equilibrium prices. In the context of a model with a convex cost of precision of information and where the valuation of each agent has both a private and a common component, Vives [1631] shows that there exists a fully revealing equilibrium and an incentive to acquire information, provided that the common value component does not dominate. Furthermore, the introduction of derivative assets provides an incentive to acquire costly information and is shown to positively affect the price efficiency (see Cao [361]).

Markets for Information

In all the models presented so far, it has been always assumed that information is a good with features and a cost defined ex-ante. In Admati & Pfleiderer [18], these hypotheses have been relaxed by assuming the existence of a market for information, with a monopolistic agency selling information about the random dividend of the risky asset (information is represented by a noisy signal of the dividend). The agency can decide the features of the information sold to the agents, by determining the level of noise added to the signal, by deciding the price of the information, by selling "personalized" signals to different agents and by deciding how many agents can become informed. Similarly as in the models considered above, there is a noise component (random aggregate supply) in the economy that makes equilibrium prices only partially revealing. As above, agents are assumed to have exponential utility functions and the random variables appearing in the model are jointly normally distributed. Note that, as discussed in the case of the Grossman & Stiglitz [849] model, although prices are partially revealing, agents can behave as free riders by choosing not to buy any information, learning it from the observation of market prices. This phenomenon is particularly severe when prices reveal information precisely (i.e, the noise due to the random supply is small) and when the information is accurate.

If the agency sells the same information to every agent and the information is precise, then the agency will decide to add a noise component to the information sold. In the case where very precise information is sold, then it is used very aggressively by informed agents. The agency can contrast this phenomenon by deciding to sell imprecise information. When the equilibrium price is a relatively precise signal of the information sold (so that the externality effect in the acquisition of information is important), the agency does sell information only to a fraction of the agents and the information sold will be made imprecise. On the contrary, if the noise component of the economy is large (so that equilibrium prices will transmit less information) and the signal is by itself imprecise, then the agency will sell information to a large portion of the market without adding further noise.

As a further step, Admati & Pfleiderer [18] analyse the case where the agency can sell personalized information (i.e., signals with an idiosyncratic noise component) to different agents. Selling personalized information in a large market increases the informational content at the aggregate level and, therefore, encourages a free riding behavior. On the other hand, personalization can make information more attractive to each agent. If the agency chooses to sell identical information to every agent, then informed traders do not learn any additional information from prices. On the contrary, when agents can buy personalized signals, all agents will infer information from the equilibrium price. The agency, which is assumed to maximize profits, aims at maximizing the amount of information sold to the informed agents and, at the same time, at minimizing the amount of information embedded in the price, in order to reduce the free riding behavior. To compensate these two effects, it is optimal for the agency to add noise to the signals in such a way that its realization does not affect the equilibrium price. When the economy is large, selling personalized information is generally better than selling the same photocopied signal and the agency will choose to sell identically distributed signals to the entire market. In particular, in order to reduce the possibility of a free riding behavior by uninformed agents, it is better to add personalized noise components and not to restrict the number of agents who can acquire information. Moreover, if the variance of the noise component in the economy is sufficiently large or if the agency's own information is very imprecise, then it is optimal for the agency not to add any noise component and to sell the information to the entire market.

8.4 Information in Dynamic Market Models

In this chapter, we have so far discussed the role of heterogeneous/asymmetric information in the context of economies with a single trading period. In this section, we give an overview of the possible extensions to economies where agents are allowed to trade over several periods and receive private information over time. At each trading date, (informed) agents receive a private signal and can also infer information from the observation of market prices. Note that, at each trading date, agents can infer information not only from private signals and market prices at the current date, but also from past signals and past prices. This is also related to technical analysis, in which the past price information is exploited to construct trading strategies.

The main difficulty of multi-period models with heterogeneous/asymmetric information consists in the definition and the computation of the rational expectations equilibrium. Indeed, in a rational expectations equilibrium agents will *forecast the forecasts of the others*: market participants try to infer information about the beliefs of the other market participants, who in turn try to infer information about the beliefs of the other market participants, with the latter doing the same thing and so on (*higher-order beliefs*). When the private signals received over time are serially correlated, this phenomenon leads to an *infinite regression problem*, as shown in the seminal paper Townsend [1598], since agents try to infer information over time from endogenous variables and, hence, the dimensionality of the state variables increases over time. Models of this type are complex and in some cases can only be solved numerically (see, e.g., Hussman [994], Singleton [1548], Zhou [1681], Makarov & Rytchkov [1289]). Typical ways to avoid the infinite regression problem are represented by the assumption that agents are *hierarchically informed* (i.e., the information of a group of agents includes the information of the other agents).

The no-trade result of Proposition 8.4 has been extended to a multi-period economy. In particular, considering a finite number of risk neutral agents and an asset paying a dividend at each trading date, it has been shown in Tirole [1594] that, if agents exhibit common prior distributions and the initial allocation is ex-ante Pareto optimal, then there are no speculative bubbles in a rational expectations equilibrium (regardless of whether short sales are allowed or not). As a consequence, there is no speculative behavior in correspondence of a rational expectations equilibrium, in the sense that no agent is willing to pay for an asset more than what he would pay if he was obliged to hold the asset forever (this notion of speculative behavior is due to Harrison & Kreps [903]). As stated in Tirole [1594], "speculation relies on inconsistent plans and is ruled out by rational expectations". This result is intimately connected to the impossibility of speculative bubbles in a dynamic rational expectations equilibrium (compare with Sect. 6.5). Keeping the assumption of a finite number of risk neutral agents with common priors (and considering an asset with no fundamental value, a feature assumed to be of common knowledge in the economy), Bhattacharya & Lipman [226] show the possibility of speculative bubbles if each agent's initial endowment is private information. In this context, assuming that the initial allocation is interim Pareto optimal (i.e., conditionally on the private information of each agent), a bubble can exist in equilibrium.

In a finite horizon economy with risk averse agents and a single risky asset, Allen et al. [49] distinguish between *expected* and *strong* bubbles: a strong bubble occurs if there exists a state of the world in correspondence of which every agent knows that the asset price is greater than the expected value of the future dividends, while an expected bubble occurs if there exists a state of the world in correspondence of which the asset price is greater than every agent's marginal valuation of the asset. Allen et al. [49] show that necessary conditions for a strong bubble to occur are that: *(i)* each agent has private information in the date and state in which the bubble occurs; *(ii)* each agent is short sale constrained at some date in the future with positive probability; *(iii)* agents' trades are not common knowledge. In particular, as shown by an example in Allen et al. [49], the fact that agents' beliefs are not common knowledge plays a key role: even if every agent realizes that the asset is overpriced (and it is not common knowledge that everybody believes that its price will fall) but assigns a positive probability to the event of reselling the asset at a greater price to some other agent at some future date/state, then a bubble may arise and the no-trade result does not hold (provided that the short sale constraint is binding). Every agent knows that the bubble will eventually burst, but there is uncertainty on when this will happen.

In the models surveyed so far, all informed agents are assumed to receive their private information simultaneously. Hirshleifer et al. [947] consider an economy where some agents can observe a private signal about the fundamental value of a risky asset earlier than other agents, showing that the timing at which investors receive information may be even more important than the accuracy of the information itself. The equilibrium where some agents receive information before other agents can be fundamentally different from the equilibrium where all the agents receive information simultaneously. Referring to Exercise 8.19 for a detailed presentation of the model and for the derivation of the equilibrium price functional, Hirshleifer et al. [947] consider an economy with two successive trading periods, populated by a continuum of risk averse agents, where *early informed* agents can observe a random component of the terminal value of a risky asset at date $t = 1$, while *late informed* agents can observe this component only at the successive date $t = 2$. The economy is also populated by liquidity traders (noise traders) and by risk neutral competitive market makers (who do not have any private information and set prices in a competitive way ensuring market clearing). As shown in Exercise 8.19, the equilibrium price functional is linear, the optimal demand of the early informed traders at the first trading date ($t = 1$) is determined by two components (see equation (8.48)): a first component exploiting the expected price appreciation between $t = 1$ and $t = 2$ and a second component to lock in at the current price the expected optimal demand at the successive date $t = 2$. The early informed agents reverse a fraction of their previous position at date $t = 2$, when the trades of the late informed agents cause the price to more fully reflect the available information. For the late informed agents, it is optimal not to trade during the first trading period. As shown in part *(ii)* of Exercise 8.19, the price changes at $t = 1$ and $t = 2$ are both positively correlated with the private signal. Moreover, as shown in Hirshleifer et al. [947, Proposition 2], the late informed agents appear to "follow the leader", in the sense that their trades are positively correlated with those of the early informed agents. The model also allows for an explanation of the "herding" behavior among investors.

As discussed at the beginning of Sect. 8.3, the introduction of a random supply (noise) in the economy is one of the most typical ways to prevent the problems associated with a fully revealing rational expectations equilibrium. In the presence of noise, the agents' private information does not become redundant after observing market prices. In Hellwig [932], considering a multi-period economy where one riskless asset and one risky asset are traded and agents formulate their optimal demands according to a mean-variance criterion, it is shown that, if all agents condition their expectations on past market prices and the time span between two successive trading dates converges to zero, then the equilibrium price converges towards the fully revealing rational expectations equilibrium price but the returns of being informed are still bounded away from zero. In this sense, the market will approximate full informational efficiency arbitrarily closely. The model of Hellwig [931] has been extended to a continuous time setting in Naik [1366], assuming a continuum of infinitely lived agents with CARA preferences (with possibly heterogeneous risk aversion). If the asset supply is random and agents observe noisy private signals of different precision on the growth rate of the dividend process (which includes a random unobservable component), then the stationary rational expectations equilibrium price reflects the information of the common part of the private signals of a large set of agents. However, the agents' private information does not become redundant and the equilibrium is only partially revealing. In Naik [1367], the above model is analysed under the assumption of the presence of a monopolistic seller of information endowed with perfect information. It is shown that the optimal strategy for the information seller consists in selling information only to a fraction of the market, without adding noise components to the information sold.

The transmission of information in a continuous time economy with informed and uninformed traders is analysed in Wang [1643] (see also Biais et al. [233] for an analysis of the effects of asymmetric information in an overlapping generations model with multiple assets, in line with Admati [15]). In the model of Wang [1643], informed agents possess information on the future dividend growth rate. The information flow is hierarchical (thus avoiding the infinite regression problem), in the sense that informed agents know all the information known by uninformed agents. Since the growth rate of the dividend process determines the rate of appreciation of the stock price, changes in prices provide a signal on the future dividend growth rate and uninformed agents rationally extract this information from the observation of market prices. The aggregate supply of the risky asset is assumed to be random and, hence, equilibrium prices do not fully reflect the agents' private information. Wang [1643] shows that the presence of uninformed agents increases the asset risk premium, can increase price volatility and induces negative serial correlation in returns (due to the mean reversion in the underlying state variables affecting the expected excess returns). In particular, the risk premium is increasing in the number of uninformed agents because of the adverse selection problem (i.e., uninformed agents demand an additional premium for the risk of trading against better informed agents) and because the price is less informative and there is more uncertainty in the stock's future cash flows. The effect of the presence of uninformed agents on price volatility depends on two effects: on the one hand, uninformed agents are characterized by a less volatile expected cash flow, thus reducing price volatility; on the other hand, future cash flows are more uncertain, so that investors demand a greater risk premium for noise trading and prices become more sensitive to supply shock, thus increasing price volatility. If noise trading is significant, then the price volatility is increasing with respect to the number of uninformed traders.

It is interesting to remark that the presence of informed/uninformed agents can also lead to an explanation of the equity premium puzzle (see Sect. 7.3). Indeed, Zhou [1682] shows that a high risk premium together with a low risk free rate and a plausible level of risk aversion can be obtained in a model with asymmetric information under the assumption that uninformed agents cannot buy the market portfolio. In this model, the adverse selection effect plays a crucial role.

Information, Trading Volume and Price Volatility

In the context of multi-period economies, the presence of private/asymmetric information has also interesting implications on the trading volume and on the volatility of the equilibrium price process. In the context of a multi-period economy with private information and random aggregate supply, He & Wang [920] show that, in correspondence of the (partially revealing) equilibrium, trading volume is significantly correlated with the inflow of information (both private and public) in the economy. In the model considered in He & Wang [920], the information flow is not hierarchical and the economy is populated by infinitely many investors. At each trading date, investors receive both private and public information concerning the fundamental value of a stock. When the information is private and dispersed, agents trade in the market even after the arrival of information and, therefore, trading volume can be serially correlated and may lag behind the information inflow, being also related to the private information previously received. When the information is public, agents trade only simultaneously with the arrival of information. While the information inflow generates trading volume and price changes, trading generated by existing information is not accompanied by price changes. Note that movements in asset prices may not necessarily be caused by external news, but only by the revelation of information by the trading process itself, if the equilibrium aggregates the private information in an imperfect way (see Romer [1463] and Grundy & Kim [852]).

The relation between trading volume and asymmetric information has been also explored by Watanabe [1646] in the context of an overlapping generations model with multiple assets, heterogeneous information and random supply. The model generates multiple equilibria that can exhibit strong or weak correlations between asset returns. Trading volume has a hump-shaped relation with respect to the average level of information accuracy and is positively correlated with absolute price changes. Less informed agents trade in the same direction of price changes and behave like trend followers, while better informed agents trade by adopting contrarian strategies.

Campbell et al. [346] propose a model with asymmetric information to analyse the relationship between trading volume and the serial correlation of asset returns (see Gallant et al. [751] and Sect. 8.6 for some empirical evidence). A detailed presentation of this model together with the derivation of the equilibrium price is provided in Exercise 8.20. In the economy, there are two traded assets (a riskless asset and a risky asset) and two classes of agents: risk averse "market makers" (utility maximizers) and "liquidity" traders (with changing risk aversion), who trade for exogenous reasons. In the model there is no private information, in the sense that all the agents are assumed to observe at each trading date a common noisy signal about the future dividend shock. Price changes are determined by three sources: innovations in the current dividend, innovations in the information about future dividends and innovations in the time-varying risk aversion of the liquidity traders. By construction, in equilibrium the trading volume is positive only due to changes in the risk aversion and the serial correlation of asset returns is negatively affected by the trading volume. The rationale of this result can be explained as follows: if there is a price change and the trading volume is large, then the price change is due to preference shocks (uninformative trades), while, if the trading volume is small, then the price change is due to the inflow of new public information. As a consequence, price changes accompanied by high volume will tend to be reversed (thereby generating negative serial correlation in asset returns).

The relationship between trading volume and asymmetric information has been further analysed in Wang [1644], considering an economy with uninformed and informed agents and showing that a different nature of investors' heterogeneity implies different behaviors for the trading volume. While all the agents are assumed to have CARA preferences with common risk aversion parameter and discount factor, the two classes of agents differ in their information sets as well as in their investment opportunities. Informed agents have access to private information about the stock's future dividends and trade both for informational reasons (speculation) and liquidity reasons. Moreover, informed agents have access to a private investment opportunity, which is not accessible to uninformed agents. Uninformed agents, besides the realized dividends and market prices, observe a noisy public signal of the private information received by informed agents (note that this hierarchical information flow avoids the infinite regression problem). Informed agents trade when they receive private information (informational trading) as well as when their private investment opportunity changes (non-informational trading). Uninformed agents only trade for non-informational reasons and rationally extract information from the publicly available information. The aggregate supply of the risky asset is constant and the structure of the economy is supposed to be common knowledge. In this setting, the no-trade result does not hold, since uninformed agents are willing to trade with informed agents since the latter also trade for non-informational reasons. However, similarly as in other models with asymmetric information, uninformed agents face an adverse selection problem, because they do not know whether informed agents are trading for informational reasons or not. Due to this adverse selection problem, trading volume is decreasing with respect to the asymmetry of information between informed and uninformed agents. However, the correlation between trading volume and the inflow of public information, excess returns and absolute price-dividend changes is increasing with respect to the information asymmetry. Since agents are risk averse, trading is always accompanied by price changes, the trading volume is positively correlated with absolute price changes and public information on the stock's future dividends generates abnormal trading. The implications of the current return and trading volume on future returns depends on two effects: a high return accompanied by high trading volume implies high future returns if the trading activity of informed agents is mostly due to informational reasons, while it implies low future returns if the trading activity of informed agents is mostly due to non-informational reasons (similarly as in Campbell et al. [346]). A momentum effect associated with informative trading is due to the fact that prices do not fully reveal private information. Similar implications are obtained and tested with positive evidence in Llorente et al. [1232].

Note that, in almost all the models presented above, the trading volume is related to price changes as well as to the quality of information. In models with symmetric-private information, trading volume is increasing with respect to the investors' heterogeneity, while in models with asymmetric information (i.e., models where there are informed and uninformed traders) high volume results when the information asymmetry is limited (i.e., when the adverse selection effect is limited). In particular, as considered in Bhattacharya & Spiegel [227], if the adverse selection problem is particularly severe (more precisely, when the variance of the signal which is observable to informed agents and unobservable to uninformed agents exceeds a critical value) and there is a single monopolistic informed investor, then uninformed agents refrain from trading, thus leading to a *market breakdown* (see Bhattacharya & Spiegel [227, Proposition 3] and compare also with Chap. 10). This result has been extended in Spiegel & Subrahmanyam [1559], showing that, if uninformed agents do not know the variance of the informational variable, then the equilibrium price functional is non-linear and a market breakdown occurs if the upper bound of the support of the variance exceeds the variance of the liquidity component of the demand. In this setting, public disclosure by firms can bound the variance uncertainty and, therefore, mitigate the market breakdown problem.

8.5 Difference of Opinions

As explained in the introduction to the present chapter, agents are said to have *heterogeneous opinions* if they exhibit different probability distributions which do not result from conditioning on information related to the fundamentals of the economy. The analysis of economies populated by agents with heterogeneous opinions is qualitatively different from the case of heterogeneous beliefs discussed so far. Indeed, unlike heterogeneous beliefs, heterogeneous opinions are not related to the observation of informative private signals. The most important consequence of this difference is that, in the case of heterogeneous opinions, agents do not assign any informational value to market prices (i.e., there is no private information to aggregate and to infer from the observation of market prices). In this section, we give an overview of some of the main results in the case of economies with heterogeneous opinions.

No-Trade Results under Heterogeneous Opinions

The seminal no-trade result of Tirole [1594] that trading motivated by speculation cannot occur in a rational expectations equilibrium does no longer hold if agents have heterogeneous prior distributions (*opinions*). Indeed, in this case agents can *agree to disagree*, since the fact that they exhibit different probability distributions is unrelated to the fundamentals of the economy and does not embed private information related to future dividends/payoffs. In comparison with the case of heterogeneous beliefs, differences in opinions are not due to private information and this prevents the no-trade result, since the willingness to trade is not interpreted by other agents as the evidence of superior private information. Therefore, trade may occur in equilibrium and the difference in opinions simply means that the agents interpret in different ways the available information.

The fact that differences in opinions can lead to trade has been verified in Varian [1611], Shalen [1522], Biais & Bossaerts [232]: if agents have different opinions (i.e., different prior distributions about the asset's fundamental value or about the relationship between the realization of a signal and the value of the asset), then the no-trade result does not hold and agents are willing to trade. Moreover, trading volume and volatility are positively related to the dispersion of opinions and trading volume and price changes are correlated. Morris [1352] deals with the problem of identifying the different varieties of heterogeneous prior distributions (opinions) that can lead to trade, pointing out that differences in opinions leading to different interpretations of new information may represent the most important source of trading in response to new information. Considering an economy with a finite number of risk averse agents each observing a private signal and exhibiting heterogeneous prior distributions, Morris [1352] shows that the no-trade result may also hold in the presence of heterogeneous opinions, under suitable conditions:

differences in opinions of observing one's own signal will not lead to trade; if agents are allowed to make trades contingent on some event before the arrival of information, then differences in opinions about that event will not lead to trade; differences in opinions where agents undervalue their own signal will not lead to trade (assuming that trades are incentive compatible); if trade is to be common knowledge, then differences in beliefs about events which are not publicly revealed will not lead to trade.

In Harris & Raviv [901] it is shown that the no-trade result does not hold when agents receive public information but interpret information according to different models (there is no private information but only different opinions). In this case, a public announcement induces trading volume also in an economy without private information and difference of opinions among agents leads to a positive correlation between absolute price changes and trading volume as well as positive serial correlation in the trading volume. Moreover, trading volume is positively related to the heterogeneity of opinions (see also Kondor [1118], Banerjee & Kremer [130], Cao & Ou-Yang [362]). In Kandel & Pearson [1066], it is shown that trading volume can be positive in the presence of a public announcement even in the absence of a price change, as observed empirically.

Asset Pricing Under Heterogeneous Opinions

In Miller [1342], Harrison & Kreps [903], Morris [1353] it is shown that, in the presence of heterogeneous opinions, short sale constraints can generate overpricing. A simple example from Miller [1342] provides a clear illustration of this phenomenon (we follow the presentation of Scheinkman & Xiong [1503, Section 2]). Let us consider a two-period economy (i.e., $t \in \{0, 1\}$), with a riskless asset paying the constant rate of return $r_f = 1$ and one risky asset. The random dividend $\tilde{d}$ of the risky asset is given by

$$\tilde{d} = \bar{d} + \tilde{\varepsilon}, \tag{8.36}$$

where $\tilde{\varepsilon}$ is a normal random variable with zero mean and variance σ_ε^2. Consumption occurs only at date $t = 1$. The economy is populated by a continuum of agents and each agent exhibits a personal opinion about the distribution of $\tilde{d}$. More specifically, instead of the true distribution (8.36), each agent believes that $\tilde{d}$ is distributed according to a normal law with mean $\bar{d}_i$ and variance σ_ε^2. In other words, all the agents agree on the variance of the random dividend, but have different opinions on the expected dividend. Since each agent is assumed to be characterized by his view on the expected dividend, we index agents by $\bar{d}_i$ and assume that $\bar{d}_i$ is uniformly distributed on the interval $[\bar{d} - k, \bar{d} + k]$, where k can be regarded as a measure of the heterogeneity of opinions in the economy. Note that the average opinion in the market is $\bar{d}$, corresponding to the true expected dividend. The preference structure of each agent $\bar{d}_i$ is characterized by a negative exponential utility function, with a common risk aversion parameter a, and agent $\bar{d}_i$ is endowed with initial wealth w_0^i. The aggregate supply of the risky asset is supposed to be constant and given by u. In this context, we can establish the following result.

Proposition 8.16 *In the context of the model described above, if agents are not allowed to short sell the risky asset, then its equilibrium price p^* at date $t = 0$ is given by*

$$p^* = \begin{cases} \bar{d} - a\sigma_\varepsilon^2 u, & \text{if } k < a\sigma_\varepsilon^2 u; \\ \bar{d} + k - 2\sqrt{ka\sigma_\varepsilon^2 u}, & \text{otherwise.} \end{cases} \tag{8.37}$$

Proof Let us denote by w^i the demand of the risky asset by agent $\bar{d}_i$ in terms of units of the asset. The corresponding random wealth at $t = 1$, in correspondence of a price p for the risky asset, is then given by $\widetilde{W}_1^i = w_0^i + w^i(\tilde{d} - p)$. Given the short sale constraint and since each agent has a negative exponential utility function, the optimal demand w^{i*} is given by

$$w^{i*} = \max\left(\frac{\bar{d}_i - p}{a\sigma_\varepsilon^2}; 0\right), \qquad \text{for all } \bar{d}_i \in [\bar{d} - k, \bar{d} + k].$$

Given the above optimal demand, the market clearing condition requires that

$$\frac{1}{2k} \int_{\max(p, \bar{d} - k)}^{\bar{d} + k} \frac{\bar{d}_i - p}{a\sigma_\varepsilon^2} d\bar{d}_i = u.$$

The latter condition directly implies that the equilibrium price p^* is given by (8.37).

$\square$

As can be seen by examining the proof of Proposition 8.16, the agents believing that the expected dividend is less than the current market price do not take part to the trade, due to the prohibition of short sales. In the absence of short sale constraints, the equilibrium price p^* would simply be given by $\bar{d} - a\sigma_\varepsilon^2 u$. It is easy to see that $\bar{d} - a\sigma_\varepsilon^2 u \leq \bar{d} + k - 2\sqrt{ka\sigma_\varepsilon^2 u}$. Formula (8.37) shows that, when the difference in opinions is limited (i.e., when $k < a\sigma_\varepsilon^2 u$), then the short sales constraint is not binding for any investor and the equilibrium price coincides with the one obtained under homogeneous expectations given by the average opinion $\bar{d}$. On the contrary, when heterogeneity becomes more relevant (i.e., when $k \geq a\sigma_\varepsilon^2 u$), then the equilibrium price only reflects the opinions of the more optimistic investors with the equilibrium price being higher than the one obtained under homogeneous opinions. In this sense, difference of opinions together with short sale constraints may lead to an overvaluation of the asset. In particular, observe that an increase in the dispersion of opinions (as measured by k) will lead to an increase in the equilibrium price p^*.

This model has been extended in Chen et al. [418] by allowing for the presence of a group of fully rational arbitrageurs who can take either long or short positions in the risky asset. Similarly as in the above setting, Chen et al. [418] consider a two-period economy where a riskless asset with constant rate of return $r_f = 1$ is traded together with a single risky asset with random dividend $\tilde{d} = \bar{d} + \tilde{\varepsilon}$, where $\tilde{\varepsilon}$ is a normally distributed random variable with zero mean and unitary variance. The total supply of the risky asset is supposed to be constant and given by u. There are two groups of agents. The first group is composed of risk averse agents (whose preferences are represented by negative exponential utility functions with a common risk aversion parameter a) with different opinions about the expected dividend of the risky asset: similarly as above, each agent believes that the expected dividend is given by $\bar{d}_i$ and we assume that $\bar{d}_i$ is uniformly distributed on the interval $[\bar{d} - k, \bar{d} + k]$. For all the agents belonging to this first group, short sales are prohibited. The second group of agents is composed of fully rational arbitrageurs (correctly believing that the expected dividend is equal to $\bar{d}$) who can take either long or short positions in the risky asset. The agents of this second group have negative exponential utility functions and their aggregate risk aversion is represented by the parameter a_{arb}. Under these hypotheses, we can establish the following version of Proposition 8.16 (the proof follows the same steps of the proof of Proposition 8.16 and is given in Exercise 8.21).

Proposition 8.17 *In the context of the model described above (see Chen et al. [418]), if the agents of the first group are not allowed to short sell the risky asset and the agents of the second group are not constrained, then the equilibrium price p^* at date $t = 0$ of the risky asset is given by*

$$
p^* = \begin{cases} \bar{d} - \left(\frac{1}{a} + \frac{1}{a_{\text{arb}}}\right)^{-1} u, & \text{if } k < \left(\frac{1}{a} + \frac{1}{a_{\text{arb}}}\right)^{-1} u; \\ \bar{d} + k + 2ak \left(\frac{1}{a_{\text{arb}}} - \sqrt{\frac{1}{a_{\text{arb}}^2} + \frac{1}{aa_{\text{arb}}} + \frac{u}{ak}}\right), & \text{otherwise.} \end{cases}
$$

(8.38)

Chen et al. [418] also analyse the *breadth of ownership*, defined as the fraction of agents of the first group who hold long positions in the stock, and show that as the divergence of opinions increases (i.e., k increases) then both the breadth of ownership and the expected return $\bar{d} - p^*$ decrease. Observe that the result of Proposition 8.16 can be recovered from Proposition 8.17 (in the special case $\sigma_\varepsilon^2 = 1$) by letting $a_{\text{arb}} \to +\infty$, representing the fact that arbitrageurs are not trading in the market.

The interplay of heterogeneous opinions and short sale constraints has been also analysed in the context of a multi-period model in Harrison & Kreps [903]. They show that the combined presence of these two features may generate speculative behavior and may lead to an equilibrium price which is greater than the fundamental value of the asset (in other words, there exists a bubble, see Sect. 6.5). According to Harrison & Kreps [903], agents are said to exhibit *speculative behavior* if "the right to resell a stock makes them willing to pay more for it than what they would pay if obliged to hold it forever". In the model proposed in Harrison & Kreps [903] (compare also with the presentation in Scheinkman & Xiong [1503, Section 3]), the economy is characterized by an infinite trading horizon, with trading dates $t \in \mathbb{N}$. Similarly as in Chap. 6, the information flow is represented by $\mathbb{F} = (\mathscr{F}_t)_{t \in \mathbb{N}}$. In the economy, there is one unit of a single risky asset, delivering a non-negative random dividend d_t at each date $t \in \mathbb{N}$. The dividend process $(d_t)_{t \in \mathbb{N}}$ is assumed to be adapted to $\mathbb{F}$. In this context, a *price process* for the risky asset is represented by a non-negative stochastic process $(p_t)_{t \in \mathbb{N}}$ adapted to $\mathbb{F}$. The economy is populated by $A \in \mathbb{N}$ homogeneous classes of risk neutral agents. Each agent is assumed to believe in a subjective probability measure (on a common measurable space) and agents belonging to the same class share the same probability measure, but agents belonging to different classes may exhibit different probability measures (opinions). For each $a \in \{1, \dots, A\}$, we denote by $\mathbb{P}^a$ the probability measure (opinion) of the agents belonging to class a, with $\mathbb{E}^a[\cdot]$ denoting the associated expectation. Each class consists of an infinite number of agents. All the agents have a common discount factor δ.

In the above setting, Harrison & Kreps [903] introduce the crucial hypothesis that agents are prohibited from short selling the risky asset. Hence, together with the assumption of risk neutrality, the analysis of the trading strategies at date t reduces to the analysis of strategies consisting in selling one unit of the risky asset at a future date $T \geq t$ (we also allow for the possibility of $T = +\infty$, i.e., the asset is held forever). Following Harrison & Kreps [903], a price process $(p_t)_{t \in \mathbb{N}}$ is said to be *consistent* if

$$p_t = \max_{a \in \{1, \dots, A\}} \sup_{T \geq t} \mathbb{E}^a \left[\sum_{k=t+1}^{T} \delta^{k-t} d_k + \delta^{T-t} p_T \Big| \mathscr{F}_t \right], \qquad \text{for all } t \in \mathbb{N}. \quad (8.39)$$

As explained in Harrison & Kreps [903, Section IV], condition (8.39) represents a natural condition for a price process $(p_t^*)_{t \in \mathbb{N}}$ to correspond to an equilibrium of the economy. In fact, the quantity

$$\sup_{T \geq t} \mathbb{E}^a \left[\sum_{k=t+1}^{T} \delta^{k-t} d_k + \delta^{T-t} p_T \Big| \mathscr{F}_t \right]$$

represents the maximum present value that an agent belonging to class a can realize by holding the risky asset at date t and selling it at some optimal future date T, given the information available at date t. By the assumption of risk neutrality, this represents the utility associated to owning the asset at date t for the agents belonging to class a. Hence, the right-hand side of (8.39) is the maximum value of the risky asset from the point of view of any agent in the economy. Indeed, if such an amount was strictly greater than a candidate equilibrium price p_t^*, then competition among the (infinite number of) agents belonging to the class assigning the largest value to the asset would make the price increase. A symmetric argument shows that the maximum value of the risky asset from the point of view of any agent in the economy cannot be smaller than p_t^*, thus showing that an equilibrium price process must be consistent in the sense of (8.39) (see also Harrison & Kreps [903, Section 4]).

The following proposition gives a characterization of consistent price processes and should be compared with the fundamental no-arbitrage equation (6.85) (under the assumption of risk neutrality). The proof of the following proposition is given in Exercise 8.22.

Proposition 8.18 *In the context of the model described above (see Harrison & Kreps [903]), a price process $(p_t)_{t \in \mathbb{N}}$ is consistent if and only if it satisfies the following condition:*

$$p_t = \max_{a \in \{1,\dots,A\}} \delta \mathbb{E}^a \left[d_{t+1} + p_{t+1} | \mathscr{F}_t \right], \qquad \text{for all } t \in \mathbb{N}. \tag{8.40}$$

Note also that, for any consistent price process $(p_t)_{t \in \mathbb{N}}$, since $T = +\infty$ is always a feasible selling strategy, condition (8.39) implies that

$$p_t \geq \max_{a \in \{1,\dots,A\}} \mathbb{E}^a \left[\sum_{k=t+1}^{\infty} \delta^{k-t} d_k \middle| \mathscr{F}_t \right] \qquad \text{for all } t \in \mathbb{N}. \tag{8.41}$$

In (8.41), the right-hand side represents the value of holding the asset forever. Speculative behavior arises if the inequality is strict: in this case the option to resell the risky asset at a future date has a strictly positive value and this option will become viable when there are agents who are relatively more optimistic than the current owner about the future dividends' growth. This leads to a bubble with a positive price wedge with respect to the fundamental no-arbitrage solution. Summing up, the fundamental insight of Harrison & Kreps [903] is represented by the observation that, in the presence of heterogeneous opinions and short sale prohibition, agents can agree to disagree and prices may exceed fundamental values.

Related results have been also obtained in Scheinkman & Xiong [1502], considering the combined effect of short sale constraints and heterogeneous opinions in the formation of speculative bubbles. In the context of a continuous time model, Scheinkman & Xiong [1502] assume that *overconfidence* (i.e., the belief of an agent that his own information is more accurate than what in fact it is) generates disagreement among agents regarding asset fundamentals. More specifically, agents disagree about the interpretation of publicly available signals, with different groups of agents overestimating the informativeness of different signals. As information flows into the market, investors' forecasts change and, in turn, these changes generate trading volume. Furthermore, an increase in the degree of overconfidence induces an increase in the value of the bubble component (see Scheinkman & Xiong [1502, Lemma 3]). Related results have been also obtained in Hong et al. [960] in the context of a discrete time multi-period model. In this context, see also Dumas et al. [601] for portfolio choice analysis and Xiong & Hongjun [1671] for models on the term structure and asset float.

Hong & Stein [962] analyse the impact of differences of opinions and short sale constraints on the possibility of market crashes. Their model considers two short sales constrained investors A and B, each of whom can observe a private signal about the stock's terminal payoff, together with a group of rational arbitrageurs (not short sales constrained). The model is able to generate market crashes and the rationale of this result is that the presence of short sale constraints prevents pessimistic agents' beliefs to be fully reflected in market prices. Indeed, suppose that at some date $t = 1$ agent B receives a more pessimistic signal than agent A and decides to stay out of the market. In this case, at date $t = 1$, trading will only occur between agent A and the arbitrageurs. The arbitrageurs (who cannot observe the private signals of the two agents) will only infer that the signal of agent B is more pessimistic than that of agent A, but cannot determine to what extent. Hence, market prices at $t = 1$ do not fully reflect the pessimistic view of agent B. Suppose now that at a later date $t = 2$ agent A receives a bad signal and wants to bail out the market. If the signal received at the previous date $t = 1$ by agent B was very bad, then agent B won't offer buying support to the sell orders of agent A. In turn, this event will be interpreted as a bad news by the arbitrageurs, beyond the direct bad news already represented by the desire to sell of agent A. This behavior can generate a cascade effect. In this model, large trading volume (due to the difference of opinions) forecasts a negative skewness in asset returns.

In the context of a continuous time model with heterogeneous opinions, the role of equilibrium prices in aggregating different opinions is analysed in Detemple & Murthy [564]. They consider a model where agents have heterogeneous opinions about the unobservable expected growth rate of the aggregate production (while past realizations of the production growth rate are publicly observable). Agents have different prior distributions, update their beliefs rationally and disagree on the interpretation of economic news. As a consequence, there is persistent disagreement among the agents (i.e., posterior distributions differ). The equilibrium risk free rate and asset prices are weighted averages of those obtained in equilibrium for the corresponding homogeneous economies (i.e., the economies where all the agents share the same opinions of the respective agents of the original heterogeneous economy), with the weights being determined by the wealth distribution among the agents. In a similar continuous time setting for a pure exchange economy, incomplete (symmetric) information and heterogeneous opinions about a process which is unrelated to asset fundamentals can generate a low risk free rate and excess volatility in equilibrium (see Basak [175]). In this setting, the perceived risk premium of a security is not purely explained by the covariance with the aggregate consumption (compare with Proposition 6.35), but also by the covariance with the exogenous uncertainty (see also Abel [3] and Zapatero [1672] for related results).

8.6   Empirical Analysis

It is difficult to empirically test the informational efficiency of financial markets. Such a difficulty is also due to the *joint hypothesis problem*, since any test of market efficiency is a joint test on market efficiency as well as on the specific equilibrium model on which market efficiency is being tested. In very general terms, the goal of an empirical analysis of market efficiency consists in testing whether and to what extent market prices transmit private information and aggregate the agents' beliefs.

Information Aggregation

The empirical analysis of the role of market prices in aggregating private information goes back to Figlewski [702], where it was investigated whether horse quotes reflected the opinions of professional bidders. The empirical analysis reported in Figlewski [702] provided positive evidence.

In Lang et al. [1166], a test has been developed based on the analysis of trading volume and, in particular, of its correlation with market prices and some variables representing agents' expectations/beliefs. The test allows to distinguish among four of the equilibrium models presented above: competitive equilibrium, competitive equilibrium with random supply, Green-Lucas equilibrium and Green-Lucas equilibrium with random supply. The test concerns the market behavior when earning announcements are released by firms. The empirical evidence seems to be consistent with a Green-Lucas equilibrium model with random supply: asset prices reveal only part of the private information. Agents are rational, update their beliefs by exploiting the private information and identify the informational content of market prices. However, as explained earlier in this chapter, the presence of a noise component prevents a perfect transmission/aggregation of private information by market prices. Similar results on the capacity of financial markets to aggregate dispersed information have been obtained in Huberman & Schwert [986] in the case of information about inflation in a bond market.

In this regard, some interesting results come from experimental economics (see, e.g., Camerer [327], Forsythe & Lundholm [728], Sunder [1581], Smith et al. [1552], Plott & Sunder [1422], McKelvey & Page [1314], Bossaerts et al. [271]). Noisy (i.e., in the presence of a noise component such as random supply) rational expectations equilibria have been shown to be an adequate representation of the market behavior if some of the agents are informed and some are not. However, the same equilibrium concept does not perform well if every agent in the economy observes a signal with a private noise component: in this case, the aggregation of the information of the agents is not always reached in experimental markets (and, even when the aggregation is reached, it may be reached after many iterations). Furthermore, in the presence of a monopolistic position by the agent possessing private information, market crashes and bubbles are typically observed before the convergence is reached. In this context, it is difficult to observe a market behavior compatible with a rational expectations equilibrium, so that experimental results confirm only in part the theoretical implications of rational expectations equilibrium theory (see however Barner et al. [165] for results in favor of fully revealing equilibria).

Volatility, Volume and Returns

In Chap. 7, we have reported some empirical evidence pointing out that classical asset pricing theory fails to explain the volatility of asset prices. The empirical literature surveyed in that chapter is mostly concerned with low frequency data. High frequency data show that asset prices/returns are typically much more volatile during trading hours than during non-trading hours (see French & Roll [741]). Indeed, the variance of stock returns from the opening to the close of the market is greater than the variance from the close to the opening (see also the related discussion in Sect. 10.6). Three explanations of this phenomenon have been proposed: more public information arrives during normal business hours; private information affects prices only during trading hours; there exist pricing errors. Ross [1470] has proposed a model relating asset price volatility to the rate of information flow into the market. Examining returns during trading holidays, it turns out that a large part of the variance difference is due to private information. Ex-post public information about fundamentals allows to explain only a small part of the variance and this finding has been confirmed by many other studies (see, e.g., Mitchell & Mulherin [1347], Barclay et al. [160], Barclay & Warner [161], Berry & Howe [208], Roll [1459, 1460], Cutler et al. [510]). An interpretation in favor of public information as a major source of volatility has been proposed in Ederington & Lee [628], Jones et al. [1040], Stoll & Whaley [1571]: after controlling for these announcement effects, volatility is basically flat. Madhavan et al. [1278] show that both public information and trading frictions are important factors in explaining intraday price volatility.

Empirical studies have also shown the existence of a *high volume puzzle*: it is difficult to explain the huge amount of trading observed in financial markets by relying on a rational expectations equilibrium framework. The empirical literature has also reported a strong positive correlation between trading volume and contemporaneous as well as delayed return volatility/price changes (price changes lead to volume movements) and positive serial correlation in trading volume (see, e.g., Karpoff [1075], Gallant et al. [751, 752]). In Jones et al. [1041], Easley et al. [617], Chan & Fong [404], it is shown that the number of transactions rather than trading volume is related to volatility. In general terms, these findings suggest that volume and volatility are driven by the same factors. Note that the above regularities represent empirical confirmations of some of the implications of the models presented earlier in this chapter in the presence of random supply and heterogeneous beliefs/opinions (see, e.g., Kim & Verrecchia [1095], Blume et al.

[257], Harris & Raviv [901], Foster & Viswanathan [730], Shalen [1522], Wang [1644], He & Wang [920], Bernardo & Judd [203]). Many factors can contribute to the trading volume observed in the market: differences in endowments, liquidity trading, differences/shifts in preferences, heterogeneous information, heterogeneous valuations/beliefs/opinions and the arrival of new information. While differences in the endowments do not suffice to explain the large trading volume observed in real markets, the other factors have been empirically investigated.

According to Kim & Verrecchia [1095], Kandel & Pearson [1066], Lang et al. [1166], Harris & Raviv [901], Foster & Viswanathan [730], Hong & Stein [962], He & Wang [920], the arrival of public information in the market generates trading. Trading volume is positively related to price changes and the strength of the relation is increasing with respect to the differential quality of private information, the degree of disagreement (opinions/valuations/beliefs) and the market noise. To this effect, some empirical evidence is reported in Bamber et al. [128], based on data on analysts' revisions of forecasts of annual earnings after the announcement of quarterly earnings. The relation between news and volume has been tested empirically with positive evidence in Berry & Howe [208], Kandel & Pearson [1066], Bessembinder et al. [215] and with mixed results in Mitchell & Mulherin [1347].

Trading volume is related to returns. Brennan & Subrahmanyam [297], Amihud [57], Easley et al. [615], Kelly & Ljungqvist [1084] report a positive relation between adverse selection proxies and risk premia. Campbell et al. [346] show empirically (and theoretically) that return serial correlation (for both stock indexes and individual large stocks) tends to decline with past volume (that is mainly due to liquidity trading). This regularity has been empirically confirmed in Conrad et al. [482] and Avramov et al. [94]: price reversals are typically observed for widely traded stocks, while returns of less traded securities tend to be positively auto-correlated. The implications derived in Wang [1644] on the relationship between return dynamics and the effects on trading volume of informational/liquidity trading have been confirmed empirically in Llorente et al. [1232]: informational trading induces positively autocorrelated returns and liquidity trading induces negatively autocorrelated returns. This relation determines whether returns accompanied by trading volume exhibit negative or positive serial correlation (compare with the discussion in Sect. 8.4). In Chae [385] it is shown that trading volume before public announcements is decreasing with respect to adverse selection proxies (see also Tetlock [1586] on public announcements and trading volume).

The relation between trading volume and the profitability of momentum strategies has also been investigated: in Lee & Swaminathan [1175] and Chan et al. [405] it is shown that assets with high trading volume in the past earn lower returns in the future and exhibit faster price reversals than assets with low trading volume (this result is aligned with the results reported in Campbell et al. [346]). Chordia & Swaminathan [440] show that volume is a significant determinant of the lead-lag patterns observed in stock returns: daily/weekly returns on high volume portfolios predict returns on low volume portfolios and returns on low volume portfolios respond more slowly to information in market returns. This effect is due to differentiated diffusion of market information and shows that the differential speed of adjustment to information is a significant source of the cross-autocorrelation patterns observed in short horizon returns.

Difference of Opinions

As pointed out in Hong & Stein [963], a nice feature of models with heterogeneous opinions is represented by the fact that they allow to capture the joint behavior of returns and trading volume. In particular, as mentioned above, the large trading volume observed in financial markets is difficult to reconcile with classical asset pricing theory, even allowing for heterogeneous/asymmetric beliefs and noise/liquidity traders. Indeed, while these models are able to relate volatility and trading volume, they do not match other empirical regularities such as the fact that overvalued assets (i.e., high market-to-book ratio assets) tend to have higher volume than undervalued assets and that stocks with large volume tend to have low future returns. In particular, it is difficult to explain the size of trading volume observed in real financial markets and the excess volatility phenomenon by relying on the hypothesis of heterogeneous beliefs. The presence of liquidity traders does not suffice by itself to explain these phenomena. On the other hand, models with difference of opinions are able to provide a theoretical explanation of these empirical facts, being able to generate a large trading volume even when asset prices do not move relative to the fundamentals.

As we have seen in Sect. 8.5, models with difference of opinions often assume the existence of short sale constraints. In Figlewski [703], using observed short interest rates as a proxy for the level of short sale constraints, some empirical evidence is reported showing that more heavily shorted stocks underperform less shorted stocks. Boehme et al. [262] find empirical evidence of significant overvaluation for stocks characterized by a large dispersion of investors' opinions and short sale constraints, in line with the results of Propositions 8.16 and 8.17. Similarly, Diether et al. [577] show that stocks with a larger dispersion in analysts' earnings forecasts (that can be regarded as a proxy for differences in opinions) earn lower future returns than otherwise similar stocks (in this direction, see also Yu [1668], Sadka & Scherbina [1489], Chordia et al. [432]). This evidence is consistent with the hypothesis that prices reflect the optimistic view whenever investors with the lowest valuations do not trade (see however Doukas et al. [583] for negative evidence). Diether et al. [576] provide positive evidence to the short sales constraint overvaluation hypothesis by showing that a strong short selling activity predicts negative future returns. By means of an experimental analysis, Palfrey & Wang [1397] report evidence of speculative overpricing both in complete and incomplete markets, where the information flow is represented by a gradually revealed sequence of imperfect public signals about the state of the world.

In recent years, a growing literature is providing positive empirical evidence of the main findings of models with differences of opinions. For instance, Chen et al. [417, 418] confirm the main implications of Hong & Stein [962]: negative skewness in returns/negative returns are more likely to occur in periods with high trading volume (which is in turn a proxy for the degree of disagreement). In general, "glamour stocks" (i.e., stocks which are overvalued according to typical market-based ratios) tend to have high volume and to be characterized by low future returns, see Brennan et al. [293], Chordia et al. [439] and Chap. 5. Kandel & Pearson [1066] empirically document the relationship between volume, disagreement and return volatility in correspondence of public announcements.

Banerjee [129] provides a comparison of the hypotheses of heterogeneous beliefs and heterogeneous opinions. When investors have rational expectations and infer private information from the observation of market prices (as in the case of heterogeneous beliefs), the degree of dispersion of beliefs is positively related to expected returns, return volatility and correlation between volume and absolute returns, but negatively related to return autocorrelation. On the contrary, when investors do not condition on market prices (as in the case of heterogeneous opinions) these relationships are reversed. In both cases, assets exhibiting a higher disagreement have a higher expected volume as well as a higher variance of volume. The empirical evidence reported by Banerjee [129] is in favor of the hypothesis that agents condition on market prices (and, therefore, infer private information from market prices) on the basis of quarterly and longer horizons, but reject both models at the monthly horizon.

From a behavioral perspective, a crucial point in order to understand the empirical findings reported above concerns the identification of a suitable proxy of the *investor sentiment* (i.e., beliefs about future cash flows that are not related by fundamentals), see Baker & Wurgler [111]. In line with the theoretical predictions of several difference of opinions models, stocks or markets characterized by a high trading volume exhibit lower subsequent returns. Moreover, when the investor sentiment is high, future returns are low on stocks in which there are limits to arbitrage and on stocks that are difficult to value (see Baker & Stein [108], Datar et al. [525], Brennan et al. [293], Chordia et al. [439], Lee & Swaminathan [1175] and also Baker & Wurgler [110], Stambaugh et al. [1563], Baker et al. [112] for an analysis with a refined market sentiment indicator). We also want to mention that the empirical evidence reported above on the relation between heterogeneity of opinions and trading volume/asset returns has been also identified as an evidence of information uncertainty (*information risk*, see Zhang [1677] and Johnson [1036]).

8.7   Notes and Further Readings

In Sect. 8.1, we have considered the role of information in financial markets. A more detailed analysis of the value of information from the point of view of an individual agent can be found in Gollier [800, Part VIII], relating information and risk aversion.

The existence of a Green-Lucas equilibrium and its fully revealing property have been addressed in Radner [1438] in the context of an economy with a single good, incomplete markets, real assets, a finite number of elementary events and finite set of possible signals for the agents. Under these assumptions, it is shown that a fully revealing Green-Lucas equilibrium exists for a generic set of economies. The assumption on the finiteness of the states of the world and of the signals observed by the agents has been relaxed in Allen [42, 43], where it is shown that a fully revealing Green-Lucas equilibrium generically exists if the dimension of the signal space is less than the number of relative prices (equal to $N - 1$ if the market comprises N securities). In this context, the dimension of the signal space can be interpreted as the number of sources of information for each agent. This result can also be extended to models with infinitely many states and/or with signals having continuous distributions. In Jordan & Radner [1047] it is shown that, if the number of relative prices is equal to the dimension of the signal space, then there exists an open set of economies which do not admit an equilibrium and open sets of economies admitting fully revealing equilibria. In particular, this result implies that a fully revealing Green-Lucas equilibrium may not exist in an economy with a risky asset, a risk free asset and a private signal taking values in $\mathbb{R}$. If the dimension of the signal space is larger than the number of relative prices, then the existence of a Green-Lucas equilibrium is ensured for a dense set of economies, but full revelation does not hold generically. As shown in Jordan [1045], a Green-Lucas equilibrium is generically partially revealing and approximately fully revealing. Moreover, Jordan [1046] shows that the equilibrium is fully revealing in three possible cases: (i) when all investors are risk neutral; (ii) when all agents have identical constant relative risk aversion; (iii) when all agents have constant absolute risk aversion.

Examples of the non-existence of fully revealing equilibria in incomplete markets have been given in Kreps [1134], Radner [1438], Jordan & Radner [1047]. In particular, a generic non-existence result has been established in Jordan & Radner [1047]. The analysis has been extended in Pietra & Siconolfi [1419] to economies with multiple numéraire assets and physical commodities, with finitely many states of private information. In this context, it is shown that all rational expectations equilibria are fully revealing for a generic dense subset of economies. Classes of economies characterized by (robust) non-fully revealing equilibria have been identified in Ausubel [90]. DeMarzo & Skiadas [556, 555] identify a class of economies with private information (*quasi-complete economies*) which always admit a unique fully revealing equilibrium and, under some conditions, partially revealing equilibria. An economy is quasi-complete if a risk neutral probability measure exists and, given the private information and the information revealed by prices, equilibrium allocations are interim Pareto optimal (see Definition 8.3). Under asymmetric information, assuming that all agents exhibit linear risk tolerance (i.e., HARA utility functions) with a common slope and that their endowments are tradable, demand aggregation holds even under partially revealing equilibria and the economy is quasi-complete (see also Madrigal & Smith [1283]). In this case, when agents exhibit quadratic utility functions, a conditional version of the CAPM holds.

Assuming that nominal assets are traded, it has been shown that partially revealing rational expectations equilibria exist in an incomplete markets economy with a finite dimensional signal and state space (see Polemarchakis & Siconolfi

[1425]). In Rahi [1440], it has been shown that any information flow revealed by prices (from fully revealing to fully non-revealing equilibria) that is consistent with the absence of arbitrage opportunities can be obtained in a rational expectations equilibrium (n this context, see also Citanna & Villanacci [453]).

As we have mentioned in Sect. 8.1, Proposition 8.4 can be regarded as a no-trade result. Considering an economy with a single asset, a no-trade result has been established in Kreps [1134] and in Tirole [1594]. In particular, in Tirole [1594, Proposition 1] it is shown that in a purely speculative market (in an economy where the agents' endowments are uncorrelated with the return of the asset, relative to the information available to each agent) at equilibrium no trade occurs if agents are risk averse. At equilibrium, risk neutral agents may trade, but they do not expect any gain from trading. An early version of the no-trade theorem has been established in Rubinstein [1486]. We also want to mention that, in the presence of both informed and uninformed agents, the no-trade result may not hold at equilibrium even in the absence of noise (see, e.g., Dow & Gorton [585], under the assumption that uninformed agents cannot buy the market portfolio).

Orosel [1386] provides an analysis of the Hirshleifer effect in an overlapping generations model, where at each period all agents receive information about the future dividends of the assets. The author shows that there exists a unique stationary equilibrium and, in the presence of market incompleteness, informationally efficient prices do not lead to constrained Pareto optimal allocations. In this context, more information may be socially harmful. On the efficiency of rational expectations equilibria see also Green [821], Hankansson et al. [888], Schlee [1504], Eckwert & Zilcha [625, 626], Gottardi & Rahi [813].

In Sect. 8.3, we have presented several models admitting a partially revealing rational expectations equilibrium. In this context, it is important to remark that the results of Grossman & Stiglitz [849] turn out to be very sensitive to the hypotheses of the model. For instance, several of the comparative statics results given in Proposition 8.15 do not hold in the case of CRRA preferences with log-normally distributed returns (see Bernardo & Judd [203]). Moreover, strategic complementarity in the acquisition of information can solve the Grossman-Stiglitz paradox. Manzano & Vives [1301] and Chamley [398] show the robustness of the substitution effect in the process of information acquisition. Models where a costly information acquisition coexists with fully revealing equilibrium prices have been proposed in Krebs [1133] and Muendler [1359]. Furthermore, in a model with risk neutral agents and in the presence of no-borrowing constraints (i.e., agents cannot spend more than their original endowments), Barlevy & Veronesi [163] show that the paradox is solved and prices do not become more informative as more agents acquire information, under different distributional assumptions. In the same model, uninformed agents may generate a market crash (see Barlevy & Veronesi [164]): if they observe a low price generated for example by pure liquidity trading, then they can infer that informed agents have observed a bad signal and revise consequently their expectations, thereby generating a selling panic among uninformed traders. Garcia & Strobl [756] show that complementarities in the acquisition of information arise if each agent's satisfaction with his own consumption depends on how much other agents are consuming (relative wealth concerns, see also Sect. 9.2). We also want to mention that an extension of the model proposed by Grossman & Stiglitz [849] to markets with multiple assets has been considered in Veldkamp [1615]: if information is costly, rational investors only buy information related to a subset of the available assets. This behavior induces strategic complementarities in the acquisition of information and common movements in asset prices.

The model of Hellwig [931] presented at the beginning of Sect. 8.3 has been extended in Admati [15] to a setting with multiple risky assets and infinitely many agents. Kim & Verrecchia [1095] analyse a model with costly private information similar to that proposed in Verrecchia [1618] (briefly discussed at the end of Sect. 8.3), with the goal of understanding how anticipating a public announcement through information gathering affects the market reaction to the announcement. There are two trading periods: in the first one agents with common priors and private information with different precision trade in the market, in the second there is a public announcement about the dividend of the stock. Supply in the market is noisy. The anticipated impact of the public announcement affects the acquisition of private information. The main result of the analysis is that trading volume in the second period is equal to the absolute price change from period one to period two multiplied by an aggregate measure of the agents' individual idiosyncrasy. This measure is mainly due to the differential quality of private information across agents, adjusted for the investors' preferences for risk, and can be interpreted as a proxy of information heterogeneity. Volume sensitivity to price changes is increasing in the agents' information heterogeneity. The model generates a positive relationship between volume and absolute price changes (volatility). Information asymmetry first increases with the precision of the public announcement and then decreases, so that volume is sensitive to absolute price changes when the public announcement precision is neither too high nor too low. The variance of price changes and expected trading volume are decreasing in the quality of prior information and increasing in market noise and in the precision of the public announcement. Similar results are obtained in Lang et al. [1166], Foster & Viswanathan [730] by allowing agents to acquire private information after a public announcement. Both volume and volatility depend on the public announcement and volume is positively serially correlated. On the relationship between public announcement, private information and trading volume see also Kim & Verrecchia [1096, 1097].

The analysis on partially revealing equilibria developed in Sect. 8.3 refers to a single trading period. Since markets are open for a single period, it is difficult to detect agents' private information through prices if the aggregate supply is random. However, if agents trade over several periods, then they can learn more information over time. In Brown & Jennings [310] and Grundy & Martin [851] it is shown that, if agents are allowed to trade over several periods, then they will eventually identify the information through market prices. This result provides a theoretical foundation for the use of trading strategies based on technical analysis. In other words, prices can be regarded as a noisy signal of private information: by observing prices over several periods the underlying information can be identified more precisely and, consequently, the agents' trading strategies will depend on past prices. A similar conclusion is drawn in Blume et al. [257] and Schmeidler [1506] by allowing agents to observe market volume. In particular, in Blume et al. [257] agents observe private signals of different precision and the precision of each agent's private signal is not common knowledge. Due to this assumption, trading volume has a predictive power and agents can learn useful information by observing market volume (technical analysis based on price and volume). In equilibrium, it is shown that absolute price changes and volume are strongly correlated and, in the limit, prices converge to the full information value but trading volume does not vanish. See also Brunnermeier [319] on the role of technical analysis in models with partial revelation of information.

In Sect. 8.3, we have considered the possibility of acquiring private information from an equilibrium point of view. However, the possibility of acquiring costly information also affects portfolio choices. Van Nieuwerburgh & Veldkamp [1608] show that agents tend to acquire information on a limited set of assets and hold an undiversified portfolio. Moreover, the information acquisition process can explain the *home bias* (i.e., the fact that individual and institutional investors in most countries typically hold modest amounts of foreign stocks). Investors overweight the assets belonging to the domestic market in their portfolio and exhibit a scarce international diversification, thus holding non-optimal portfolios (see Van Nieuwerburgh & Veldkamp [1607]). Peress [1412] shows that the possibility of acquiring costly information in conjunction with DARA preferences allows to explain why wealthier people invest a large proportion of their wealth in risky assets (see also the correction note Peress [1413]). The presence of asymmetric information has been proposed as a possible explanation of the home bias puzzle, by considering that foreign investors are less informed than domestic investors about the domestic market and, as a consequence, domestic investors hold a large part of their portfolio in domestic assets. A noisy rational expectations model to explain this puzzle through asymmetric information (in the sense that domestic investors have privileged information) has been proposed in Brennan & Cao [291] and Zhou [1681] (see also Kang & Stulz [1072]). According to this model, there should be a positive correlation between contemporaneous local equity returns and the net purchase of equity made by foreign investors (trend-follower pattern). Foreigners are more likely to buy when the price is increasing. This trend-following pattern of foreign investors has been confirmed empirically.

At the end of Sect. 8.3, we have briefly discussed the model of Admati & Pfleiderer [18], where there exists a market for information. In this direction, Garcia & Sangiorgi [755] study information sales in financial markets with strategic risk averse traders. The optimal selling strategy turns out to be similar to the one obtained in Admati & Pfleiderer [18]: either sell a very imprecise information to as many agents as possible, or sell a very precise information to a limited number of agents. As risk sharing considerations prevail over the negative effects of competition, the first strategy dominates the second one. Another approach to the modeling of markets for information is represented by assuming that the agency acts as a *mutual fund* by building a portfolio exploiting its own information and then selling shares of the fund in the market. In this case, the information is being sold indirectly. This case has been studied in Admati & Pfleiderer [22] by considering a monopolistic mutual fund (see Garcia & Vanden [757] for an analysis under imperfect competition). If information is being sold indirectly in an economy with homogeneous agents, then there is no incentive to introduce a noise component. Moreover, an indirect sale of information is more profitable than a direct sale if the agency can establish a per share price and a fixed participation fee. However, when agents are heterogeneous, an indirect sale can be either more or less profitable than a direct one (see Biais & Germain [235] on the optimal contract for an indirect sale of information by a mutual fund). Under some conditions, it is optimal for a broker to sell information in return for a brokerage commission (see Brennan & Chordia [292]). See also Fishman & Hagerty [711] for an analysis of the incentives for an informed trader to sell private information.

In the present chapter, we have not discussed the role of information disclosure by firms in the market (see Verrecchia [1620] for a survey on this topic and see also Chap. 10). In Diamond [568], it is shown that public announcements by firms can induce a Pareto improvement (more efficient risk sharing), lesser costs of acquiring information and of issuing capital (see also Diamond & Verrecchia [570]). In general, disclosure leads to liquid and more efficient markets, more efficient investment decisions and reduces the cost of capital for firms. As a consequence, firms should voluntarily disclose their private information to the market. In Fishman & Hagerty [708] it is shown that, if firms need to compete in order to capture the attention of traders, then an overdisclosure of information can occur. In a related context, Boot & Thakor [268] show that voluntary disclosure is beneficial for the economy and the firm, but when the disclosed information complements the information which is already present in the market, then agents buy more private information.

Despite the positive effects for the economy and for the companies, full voluntary disclosure is quite rare. Admati & Pfleiderer [24] build a model on regulation of firm disclosure based on an externality that leads individual firms not to internalize the full social value of the information released. The externality is due to the presence of correlation in the firms' values and, as a consequence, the information disclosed by one firm is used by investors to evaluate other firms. The Nash equilibrium of a voluntary disclosure game is often inefficient: both underinvestment and overinvestment in disclosure are possible, depending on the degree of correlation among the firms. In this context, there is room for a regulation encouraging firm's disclosure. Under some conditions, high type (low type) firms withhold good (bad) news and disclose bad (good) news (see Teoh & Hwang [1584]). Other works studying the relations between corporate disclosure, corporate behavior and capital markets are Leuz & Verrecchia [1202], Healy & Palepu [922], Verrecchia [1619, 1621], Lambert et al. [1161].

In Sect. 8.4, we have briefly mentioned the phenomenon of higher-order beliefs, which play a crucial role in the analysis of multi-period market models. In particular, considering a multi-period economy with overlapping generations of risk averse agents who can observe a private signal at each date, Allen et al. [50] show that the law of iterated expectations does not hold for the average belief. Bacchetta & van

Wincoop [95] show that higher-order beliefs induce a wedge (called higher-order wedge) between the equilibrium price of an asset and the equilibrium price of an asset if higher-order expectations were replaced by first-order expectations (i.e., in the case of homogeneous information). Furthermore, it is shown that the wedge at each date t depends on the average expectational error at time t about the vector of average private signals that remain informative about future dividends at $t + 1$ (see Bacchetta & van Wincoop [95, Proposition 1]). Moinas & Pouget [1350] provide an experimental analysis showing that higher-order beliefs may generate a bubble: the probability to enter into a bubble is increasing with respect to the number of steps of iterated reasoning needed to reach an equilibrium. More recently, Banerjee et al. [131] have shown that, in a dynamic setting with homogeneous beliefs, higher-order difference of opinions is a necessary condition for the existence of a drift in the price process. This result stands in contrast to that of Allen et al. [50], since Banerjee et al. [131] show that in a rational expectations equilibrium heterogeneous beliefs do not lead to the existence of a drift.

In Sect. 8.5 we have introduced the possibility that agents exhibit heterogeneous opinions. In that regard, Simsek [1547] analyse a model with difference of opinions and borrowing constraints. Optimistic agents are assumed to have limited wealth and need to borrow in order to take positions in line with their beliefs, borrowing from more pessimistic agents using loans collateralized by the asset itself. However, due to the fact that the collateral is provided by the asset itself, pessimistic agents are reluctant to lend to the optimistic agents, thus inducing an endogenous constraint on the ability of optimistic agents to influence market prices. This creates an asymmetric disciplining effect, due to the asymmetry in the shape of the payoffs of collateralized loans. In this setting, the level of opinion disagreement has ambiguous effects on the equilibrium price.

It is important to mention that investors' heterogeneity can arise from several possible features of the economy. In a rather general setting, in Jackson & Peck [1004] (compare also with Jackson [1003]) it is shown that a *sunspot equilibrium* can exist in an overlapping generations economy where agents receive private signals that are totally unrelated to fundamentals (i.e., variables not informative about future dividends, like information about the psychology of the market).[3] On the basis of this information, agents trade with the aim of realizing a gain which is not motivated by asset fundamentals. This shows the existence of speculative trading generated by differences in information, even in the case of risk neutrality and of an asset with no fundamental value. In equilibrium, it is shown that the asset price can be larger than the fundamental value and bubbles can exist.

Among other possible sources of investors' heterogeneity, an interesting aspect is represented by *tax heterogeneity*. Indeed, agents can exhibit different evaluations if they are subject to different taxation schemes. In particular, in Michaely et al.

---

[3]Sunspots represent publicly observable extrinsic events that do not affect the fundamentals of the economy, i.e., do not affect the agents' endowments, preferences, budget constraints and information.

[1338] and Michaely & Vila [1337] it is shown that tax heterogeneity may generate trading volume around ex-dividend dates. Another possible source of heterogeneity is represented by the *timing of information*. In Berk & Uhlig [200], it is shown that agents may choose to release information early (at a sufficiently small cost), resulting in incompleteness of the market. Therefore, markets are unlikely to be complete if the timing of information is endogenous and an equilibrium may not even exist (see Berk [198]). As shown in Caplin & Lehay [364], the diffusion of information is limited if agents change their behavior infrequently. In this setting, the evolution of information is discontinuous and a small additional piece of information can generate a market crash.

8.8 Exercises

Exercise 8.1 Let $\pi = (\pi_1, \dots, \pi_S)$ be a vector of probabilities, i.e., $\pi \in \Delta_S$, where Δ_S is the simplex

$$\Delta_S := \left\{ \pi \in \mathbb{R}_+^S : \sum_{s=1}^S \pi_s = 1 \right\}.$$

Consider the optimal choice problem (8.2) and suppose that there exists an optimal solution $(x_1^*(\pi), \dots, x_S^*(\pi)) \in \mathbb{R}_S^+$, for each $\pi \in \Delta_S$, and define

$$U^*(\pi) := \sum_{s=1}^S \pi_s u(x_s^*(\pi)).$$

Show that the map $\pi \mapsto U^*(\pi)$ from Δ_S onto $\mathbb{R}$ is convex in the vector of probabilities.

Exercise 8.2 This exercise is taken from Eeckhoudt et al. [631, Section 8.2.2]. Consider the optimal saving problem of an agent over three dates $t \in \{0, 1, 2\}$, assuming that the agent is endowed with initial wealth $w_0 > 0$, has a constant discount factor normalized to one and is characterized by a strictly increasing and strictly concave utility function u such that $u''' > 0$ (i.e., the agent is *prudent*). At the final date $t = 2$, the agent receives the random income $\tilde{x}$. Consider the following optimal saving problem, where α_0 and α_1 denote saving at dates $t = 0$ and $t = 1$, respectively:

$$\max_{(\alpha_0, \alpha_1) \in \mathbb{R}^2} \left(u(w_0 - \alpha_0) + \mathbb{E}[u(\alpha_0 - \alpha_1) + u(\tilde{x} + \alpha_1)] \right). \tag{8.42}$$

(i) Consider first the case where there is no early resolution of uncertainty, i.e., the random variable $\tilde{x}$ is only observed at the final date $t = 2$. Show that in this case the optimal saving decision satisfies $\alpha_1^* = 2\alpha_0^* - w_0$ (i.e., it is optimal to smooth consumption over the first two dates).

(ii) Consider then the case where the realization of the random variable $\tilde{x}$ is observed at the intermediate date $t = 1$. Show that in this case the optimal saving α_0^{i*} (where the superscript ^{i} stands for *informed*) from date $t = 0$ to $t = 1$ satisfies the first order condition

$$ u'(w_0 - \alpha_0^{i*}) = \mathbb{E}\left[u'\left(\frac{\alpha_0^{i*} + \tilde{x}}{2} \right) \right]. \tag{8.43} $$

(iii) Deduce that $\alpha_0^* \geq \alpha_0^{i*}$, i.e., an early resolution of uncertainty reduces the optimal level of saving before the resolution of uncertainty.

Exercise 8.3 Let (π^a, v^a) and (π^b, v^b) be two information structures as described before Proposition 8.2. Prove that the information structure (π^a, v^a) is finer than the information structure (π^b, v^b) if there exists a $(J \times J)$-dimensional matrix K, satisfying $K_{ij} \geq 0$, for all $i, j = 1, \ldots, J$, and $\sum_{i=1}^{J} K_{ij} = 1$, for all $j = 1, \ldots, J$, such that condition (8.6) holds.

Exercise 8.4 In the setting of Sect. 8.1, prove that ex-ante Pareto optimality implies interim Pareto optimality and, in turn, interim Pareto optimality implies ex-post Pareto optimality.

Exercise 8.5 (A No-Trade Result) This exercise is inspired by Scheinkman & Xiong [1503, Proposition 2]. Consider a two-period (i.e., $t \in \{0, 1\}$) economy populated by I risk neutral agents $i = 1, \ldots, I$ and where a single risky asset is traded, delivering at date $t = 1$ the random dividend $\tilde{d}$. Denoting by z^i the demand of agent i (assumed to be bounded), the corresponding profits at $t = 1$ are given by $(\tilde{d} - p)z^i$, with p denoting the price at date $t = 0$ of the asset. All the agents are assumed to share the same prior probability distribution, but observe different private signals $\tilde{y}^i$. At equilibrium, every trader (by the assumption of risk neutrality) maximizes his expected profit, conditioning on the information represented by his private signal and the observation of the market price, as considered in Sect. 8.1. Show that, in correspondence of a Green-Lucas equilibrium, the expected gains from trading of each agent are null. Therefore, in correspondence of a Green-Lucas equilibrium, there always exists another Green-Lucas equilibrium associated with null optimal demands.

Exercise 8.6 Let $\tilde{z}$, $\tilde{x}$ and $\tilde{y}$ be three random variables. Show that $\tilde{z}$ is a sufficient statistic for the conditional density $f(y, z|x)$ if and only if the conditional distribution of $\tilde{x}$ given $\{\tilde{y} = y, \tilde{z} = z\}$ does not depend on y.

Exercise 8.7 This exercise is inspired by Laffont [1154, Section 9.2] (see also Radner [1438] for a general analysis). Consider an exchange economy with two goods, two agents and two possible states of the world. The preference relation of agent i, for $i = 1, 2$, is represented by the following state dependent utility function

$$u^i(x_1^i, x_2^i, \omega) = \alpha_\omega^i \log\big(x_1^i(\omega)\big) + (1 - \alpha_\omega^i) \log\big(x_2^i(\omega)\big), \qquad \text{for } \omega \in \{\omega_1, \omega_2\},$$

with $x_n^i(\omega)$ denoting the quantity of good $n = 1, 2$ consumed by agent i in correspondence of the state of the world ω. The initial endowments of the agents are given by

$$e_1^i(\omega_1) = e_1^i(\omega_2) =: e_1^i > 0,$$
$$e_2^i(\omega_1) = e_2^i(\omega_2) =: e_2^i > 0, \qquad \text{for } i = 1, 2,$$

i.e., the endowments do not depend on ω. Suppose that $\mathbb{P}(\omega_1) = \pi$, with $\pi \in (0, 1)$. At date $t = 0$, the first agent receives a private signal $\tilde{y}^1$ which perfectly reveals the state of the world, while the second agent receives an uninformative signal. At date $t = 0$ (non-contingent) markets for the two consumption goods are open. Note that, in the present setting, markets are incomplete.

(i) Suppose that the equilibrium prices of the two goods at date $t = 0$ are normalized to $(p^*, 1 - p^*)$. Show that the price p^* corresponding to a competitive (Arrow-Debreu) equilibrium at date $t = 0$ is given by

$$p^*(\omega_1) = \frac{\alpha_{\omega_1}^1 e_2^1 + \bar{\alpha}^2 e_2^2}{\alpha_{\omega_1}^1 e_2^1 + (1 - \alpha_{\omega_1}^1) e_1^1 + \bar{\alpha}^2 e_2^2 + (1 - \bar{\alpha}^2) e_1^2}, \qquad \text{on } \{\omega = \omega_1\},$$

$$p^*(\omega_2) = \frac{\alpha_{\omega_2}^1 e_2^1 + \bar{\alpha}^2 e_2^2}{\alpha_{\omega_2}^1 e_2^1 + (1 - \alpha_{\omega_2}^1) e_1^1 + \bar{\alpha}^2 e_2^2 + (1 - \bar{\alpha}^2) e_1^2}, \qquad \text{on } \{\omega = \omega_2\},$$

$$(8.44)$$

where $\bar{\alpha}^2 := \pi \alpha_{\omega_1}^2 + (1 - \pi) \alpha_{\omega_2}^2$.

(ii) Deduce that, if $p^*(\omega_1) = p^*(\omega_2)$, then (8.44) corresponds to an Arrow-Debreu equilibrium of the economy, where prices do not reveal any information. On the contrary, if $p^*(\omega_1) \neq p^*(\omega_2)$, then there cannot exist an Arrow-Debreu equilibrium and the only candidate for an equilibrium of the economy is a Green-Lucas equilibrium.

(iii) Show that, if the uninformed agent extracts information from the observation of market prices, then the equilibrium prices associated to a Green-Lucas equilibrium are characterized by

$$\tilde{p}^*(\omega_1) = \frac{\alpha_{\omega_1}^1 e_2^1 + \alpha_{\omega_1}^2 e_2^2}{\alpha_{\omega_1}^1 e_2^1 + (1 - \alpha_{\omega_1}^1) e_1^1 + \alpha_{\omega_1}^2 e_2^2 + (1 - \alpha_{\omega_1}^2) e_1^2}, \qquad \text{on } \{\omega = \omega_1\},$$

$$\tilde{p}^*(\omega_2) = \frac{\alpha_{\omega_2}^1 e_2^1 + \alpha_{\omega_2}^2 e_2^2}{\alpha_{\omega_2}^1 e_2^1 + (1 - \alpha_{\omega_2}^1) e_1^1 + \alpha_{\omega_2}^2 e_2^2 + (1 - \alpha_{\omega_2}^2) e_1^2}, \qquad \text{on } \{\omega = \omega_2\}.$$

$$(8.45)$$

(iv) Deduce that, if $\tilde{p}^*(\omega_1) \neq \tilde{p}^*(\omega_2)$, then the price functional obtained in part *(iii)* corresponds indeed to a Green-Lucas equilibrium price functional. However, if $p^*(\omega_1) \neq p^*(\omega_2)$ and $\tilde{p}^*(\omega_1) = \tilde{p}^*(\omega_2)$, then the economy admits no Green-Lucas equilibrium.

Exercise 8.8 In the context of the model considered by Grossman [840] (as presented in Sect. 8.2), prove that there exists a Green-Lucas equilibrium price functional $\phi^* : \mathscr{Y} \to \mathbb{R}^N$ which is a sufficient statistic if and only if there exists an equilibrium price functional $p^{a*} : \mathscr{Y} \to \mathbb{R}^N$ of the artificial economy (where every agent observes the full set of private signals) which is a sufficient statistic.

Exercise 8.9 This exercise is taken from Mas-Colell et al. [1310, Example 19.H.1]. Consider a one-period exchange economy with two agents $i \in \{1, 2\}$, two possible states of the world $\{\omega_1, \omega_2\}$ and two consumption goods. Suppose that ω_1 and ω_2 have the same probability of occurrence, i.e., $\mathbb{P}(\omega_1) = \mathbb{P}(\omega_2) = 1/2$. The endowments of the agents, expressed in terms of units of the two goods, are given by $e^1 = (1, 0)$ and $e^2 = (0, 1)$, in correspondence of both states of the world. In aggregate terms, there is one unit of each of the two goods in both states of the world. The preferences of the two agents are characterized by the state dependent utility function

$$u^i\big(\omega, x_1^i(\omega), x_2^i(\omega)\big) = \beta(\omega)\sqrt{x_1^i(\omega)} + (1 - \beta(\omega))\sqrt{x_2^i(\omega)}, \qquad \text{for } \omega \in \{\omega_1, \omega_2\},$$

where $\beta(\omega_1) = 1$ and $\beta(\omega_2) = 0$, with $x_n^i(\omega)$ denoting the demand by agent i of good n in correspondence of state ω. This specification means that agents do not receive any utility from the second good in correspondence of the state of the world ω_1.

 (i) In this economy, there are no security markets, but only a spot market (i.e., agents trade immediately before the realization of the state of nature). Suppose first that there is no private information. Show that in equilibrium each agent consumes $1/2$ of each good and has an expected utility equal to $\sqrt{1/2}$.
 (ii) Suppose now that the state of the world is perfectly revealed to the two agents before the opening of the market. Show that in equilibrium each agent only consumes his initial endowment, getting an utility of 1 in one state of the world and 0 in the other state of the world.
(iii) Deduce that the revelation of the state of the world before the opening of the market decreases the social welfare (*Hirshleifer effect*, see Sect. 8.2).

Exercise 8.10 This exercise is inspired by Schlee [1504, Example 1]. Consider an economy with I agents characterized by power utility functions $u^i(x) = \frac{1}{\alpha}x^\alpha$, with common risk aversion parameter $0 \neq \alpha < 1$, and endowment $(e_1^i, \dots, e_S^i) \in \mathbb{R}_+^S$, for all $i = 1, \dots, I$. Assume that at $t = 0$ there are S markets for the S Arrow securities paying in correspondence of the S states of the world, with prices $(p_1^*, \dots, p_S^*)$. Show that in correspondence of an equilibrium of this economy the optimal expected utility of each agent is concave with respect to the vector of probabilities of the S states of the world. Deduce that additional information will decrease the ex-ante expected utility of each agent as long as the individual endowments are not proportional to the aggregate endowment, which we denote by $(e_1, \ldots, e_S)$.

Exercise 8.11 In the setting of Proposition 8.10, show that at equilibrium the optimal demand of every agent does not depend on the price of the risky asset.

Exercise 8.12 This exercise in inspired by Kreps [1134, Section 4]. Consider an economy with two agents: an informed agent and an uninformed agent. The preferences of both agents are characterized by negative exponential utility functions with a common risk aversion parameter a. The possible states of the world are partitioned into two events Ω_1 and Ω_2, with $\mathbb{P}(\Omega_1) = \mathbb{P}(\Omega_2) = 1/2$. In the economy, there is a single asset, which delivers at date $t = 1$ a random dividend $\tilde{d}$. Moreover, agents can trade in $t = 0$ a future contract delivering one unit of the asset in $t = 1$. The price of the future contract at $t = 0$ is denoted by q and the supply of the future contract is null. Conditionally on Ω_1, the dividend $\tilde{d}$ at $t = 1$ is normally distributed with mean μ_1 and variance σ^2 while, conditionally on Ω_2, it is normally distributed with mean μ_2 and variance σ^2, where $\mu_1 \neq \mu_2$. The first agent is assumed to be informed, meaning that he observes a signal $\tilde{y}$ taking the value 1 if $\omega \in \Omega_1$ and the value 2 if $\omega \in \Omega_2$. The second agent receives no information (uninformed agent). The uninformed agent receives an exogenous income at date $t = 1$, defined in terms of units of the asset. More specifically, conditionally on Ω_i, the uninformed agent receives $k_i \geq 0$ units of the asset, for $i = 1, 2$. In this economy, an equilibrium price functional for the forward contract is represented by a functional $q^* = \phi^* : \{1, 2\} \to \mathbb{R}$, which maps the realization of the signal $\tilde{y}$ into a price at $t = 0$ for the forward contract. The equilibrium price functional is fully revealing if $\phi^*(1) \neq \phi^*(2)$.

Show that, if $a = 1$, $\mu_1 = 4$, $\mu_2 = 5$, $k_1 = 2$, $k_2 = 4$ and $\sigma^2 = 1$, then there does not exist a fully revealing Green-Lucas equilibrium.

Exercise 8.13 Consider an economy as described at the beginning of Sect. 8.3, under the additional assumptions that $a^i = a > 0$ and $\sigma_{\varepsilon^i}^2 = 1$, for all $i = 1, \ldots, I$. Assume furthermore that each agent has a random initial endowment $\tilde{e}^i$ (in terms of units of the risky asset), which is normally distributed with zero mean and variance σ_e^2, for all $i = 1, \ldots, I$. Show that this economy admits the partially revealing Green-Lucas equilibrium price functional given in (8.23).

Exercise 8.14 (See Vives [1630], Proposition 4.1) Prove Proposition 8.12.

Exercise 8.15 In the setting of Proposition 8.12, show that the equilibrium price is a weighted average of the agents' conditional expectations of the future dividend plus a noise term. Moreover, show that, if uninformed traders become risk neutral (i.e., $a^{\mathrm{un}} \to 0$), then the equilibrium price functional $\phi^*(d, u)$ converges to $\mathbb{E}[\tilde{d}|p]$.

Exercise 8.16 In the context of Proposition 8.14, prove that the function $\lambda \mapsto \gamma(\lambda)$ is strictly increasing and admits the representations given in (8.34).

Exercise 8.17 Prove part *(iv)* of Proposition 8.15.

Exercise 8.18 Prove parts *(vii)*, *(viii)* and *(ix)* of Proposition 8.15.

Exercise 8.19 (The Equilibrium in the Hirshleifer et al. [947] Model) Consider an economy with two trading dates $t \in \{1, 2\}$ and two traded assets: a riskless asset with constant rate of return $r_f = 1$ and a risky asset whose liquidation value after date $t = 2$ is given by the random variable

$$\tilde{f} = \bar{f} + \tilde{\theta} + \tilde{\varepsilon},$$

where $\mathbb{E}[\tilde{\theta}] = \mathbb{E}[\tilde{\varepsilon}] = 0$ and the couple $(\tilde{\theta}, \tilde{\varepsilon})$ follows as a bivariate normal distribution with independent components, with $\mathrm{Var}(\tilde{\theta}) =: \sigma_\theta^2$ and $\mathrm{Var}(\tilde{\varepsilon}) =: \sigma_\varepsilon^2$.

The economy is populated by a continuum of agents indexed in the interval $[0, N]$. All agents are assumed to have negative exponential utility functions with a common risk aversion parameter a. In this economy, there are *early informed* and *late informed* agents. More specifically, a mass $[0, M]$ of the total population $[0, N]$ of agents is early informed, meaning that at date $t = 1$ they can observe the realization of the random variable $\tilde{\theta}$. On the other hand, late informed agents can only observe the realization of $\tilde{\theta}$ at the successive date $t = 2$. The error term $\tilde{\varepsilon}$ remains unobservable to both classes of agents. Furthermore, every agent is assumed to have an initial endowment of e_0 units of the riskless asset.

Besides early and late informed agents, there are also *liquidity traders*, whose net trades (demand shocks) at dates $t = 1$ and $t = 2$ are represented by the random variables $\tilde{z}_1$ and $\tilde{z}_2$, respectively. The random variables $\tilde{z}_1$ and $\tilde{z}_2$ are normally distributed, with zero mean and variance $\mathrm{Var}(\tilde{z}_1) = \mathrm{Var}(\tilde{z}_2) =: \sigma_z^2$ and are mutually independent as well as independent of $(\tilde{\theta}, \tilde{\varepsilon})$.

Finally, besides early and late informed agents and liquidity traders, there is a group of competitive risk neutral *market makers*, who possess no information on the fundamental value of the risky asset. Market makers are willing to absorb the net demand of the other traders at competitive prices.

If we denote by $\mathscr{G}_1$ and $\mathscr{G}_2$ the information sets of the early informed agents at date $t = 1$ and $t = 2$, respectively, then the information set of the late informed agents at date $t = 2$ coincides with $\mathscr{G}_2$, while the information set of the late informed agents at $t = 1$ coincides with the information set of the risk neutral market makers (which contains only the observation of the aggregate demand or, equivalently, the equilibrium price functional). As usual, all the agents are assumed to rationally extract information from the observation of equilibrium prices. Letting p_1 and p_2 denote the price of the risky asset at dates $t = 1$ and $t = 2$, respectively, we denote by $x_1(\theta, p_1)$ and $x_2(\theta, p_2)$ the demand of an early informed agent at $t = 1$ and $t = 2$, respectively, and by $y_1(p_1)$ and $y_2(\theta, p_2)$ the demand of a late informed agent at $t = 1$ and $t = 2$, respectively. Taking into account the effect of the liquidity traders, the net aggregate demands are then given by

$$D_1(p_1) := M x_1(\theta, p_1) + (N - M) y_1(p_1) + \tilde{z}_1$$
$$D_2(p_2) := M x_2(\theta, p_2) + (N - M) y_2(\theta, p_2) + \tilde{z}_1 + \tilde{z}_2,$$

at $t = 1$ and $t = 2$, respectively. Since market makers are risk neutral and competitive, they set prices which are equal to the expectation of the terminal value of the risky asset conditionally on their information set (which only consists of the aggregate demand function of all other traders), so that

$$p_1 = \mathbb{E}[\tilde{f}|D_1(\cdot)] \quad \text{and} \quad p_2 = \mathbb{E}[\tilde{f}|D_1(\cdot), D_2(\cdot)]. \tag{8.46}$$

(i) Let us conjecture that the equilibrium price functional is of the form

$$p_2 = \phi_2^*(\bar{f}, \theta, z_1, z_2) = \bar{f} + \alpha\theta + \beta z_1 + \gamma z_2$$
$$p_1 = \phi_1^*(\bar{f}, \theta, z_1) = \bar{f} + \eta\theta + \phi z_1,$$

where $\alpha, \beta, \gamma, \eta, \phi$ are suitable coefficients to be determined. Show that in correspondence of the equilibrium prices the optimal demands of the early and late informed agents are given by

$$x_2(\theta, p_2) = y_2(\theta, p_2) = \frac{\bar{f} + \theta - p_2}{a\sigma_\varepsilon^2} \tag{8.47}$$

$$x_1(\theta, p_1) = \frac{\mathbb{E}[\phi_2^*(\bar{f}, \tilde{\theta}, \tilde{z}_1, \tilde{z}_2)|\mathscr{G}_1] - p_1}{a} \left(\frac{1}{\mathrm{Var}(\phi_2^*(\bar{f}, \tilde{\theta}, \tilde{z}_1, \tilde{z}_2)|\mathscr{G}_1)} + \frac{1}{\sigma_\varepsilon^2} \right)$$
$$+ \frac{\bar{f} + \theta - \mathbb{E}[\phi_2^*(\bar{f}, \tilde{\theta}, \tilde{z}_1, \tilde{z}_2)|\mathscr{G}_1]}{a\sigma_\varepsilon^2} \tag{8.48}$$

$$y_1(p_1) = 0. \tag{8.49}$$

(ii) As shown in Hirshleifer et al. [947, Lemma 1], the conjectured price functional is indeed the equilibrium price functional and the coefficients $\alpha, \beta, \gamma, \eta, \phi$ can be explicitly determined. Moreover, it holds that $\eta \in (0, 1)$ and $\alpha > \eta$. In this case, show that

$$\mathrm{Cov}\left(\phi_1^*(\bar{f}, \tilde{\theta}, \tilde{z}_1) - \bar{f}, \tilde{\theta}\right) > 0 \quad \text{and} \quad \mathrm{Cov}\left(\phi_2^*(\bar{f}, \tilde{\theta}, \tilde{z}_1, \tilde{z}_2) - \phi_1^*(\bar{f}, \tilde{\theta}, \tilde{z}_1), \tilde{\theta}\right) > 0. \tag{8.50}$$

Exercise 8.20 Consider the model proposed in Campbell et al. [346, Section III] and discussed in Sect. 8.4. In this exercise, following Campbell et al. [346, Appendix A], we compute the equilibrium price of the risky asset.

The economy is supposed to have an infinite time horizon (i.e., the trading dates are $t \in \mathbb{N}$) and there are two traded assets: a riskless asset paying the constant rate of return $r_f > 1$ and a risky asset paying a random dividend d_t at each trading date $t \in \mathbb{N}$. For each $t \in \mathbb{N}$, the random variable d_t is supposed to be of the form $d_t = \bar{d} + \tilde{d}_t$, where $\bar{d}$ represents the average dividend and $\mathbb{E}[\tilde{d}_t] = 0$, for all $t \in \mathbb{N}$.

The sequence of random variables $\{\tilde{d}_t\}_{t\in\mathbb{N}}$ is assumed to follow the process

$$\tilde{d}_t = \alpha_d \tilde{d}_{t-1} + \tilde{u}_t, \qquad \text{for all } t \in \mathbb{N},$$

where $\alpha_d \in [0, 1]$ and $(\tilde{u}_t)_{t\in\mathbb{N}}$ is a sequence of i.i.d. normally distributed random variables with zero mean and constant variance σ_u^2. The per capita supply of the risky asset is fixed and normalized to one.

There are two classes A and B of agents. The preferences of both types of agents are represented by negative exponential utility functions, where the risk aversion parameter of class A agents is a constant a, while the risk aversion parameter of class B agents is time-varying and denoted by b_t, for $t \in \mathbb{N}$. Let λ denote the proportion of type A agents in the economy.

At each trading date $t \in \mathbb{N}$, every agent observes a signal $\tilde{y}_t$, which is supposed to be of the form

$$\tilde{y}_t = \tilde{u}_{t+1} - \tilde{\varepsilon}_{t+1},$$

where $(\tilde{y}_t, \tilde{\varepsilon}_{t+1})$ are jointly normally distributed, for all $t \in \mathbb{N}$, with

$$\mathbb{E}[\tilde{\varepsilon}_t] = \mathbb{E}[\tilde{y}_t] = 0, \quad \text{Var}(\tilde{\varepsilon}_t) = \sigma_\varepsilon^2, \quad \text{Var}(\tilde{y}_t) = \sigma_y^2 \quad \text{and} \quad \mathbb{E}[\tilde{u}_{t+1}|\tilde{y}_t] = \tilde{y}_t.$$

For each $t \in \mathbb{N}$, define the following variable z_t that can be interpreted as the risk aversion of the marginal investor:

$$z_t := \frac{ab_t}{(1-\lambda)a + \lambda b_t},$$

and assume that $z_t = \bar{z} + \tilde{z}_t$, for all $t \in \mathbb{N}$, where

$$\tilde{z}_t = \alpha_z \tilde{z}_{t-1} + \tilde{\eta}_t, \qquad \text{for all } t \in \mathbb{N},$$

where $\alpha_z \in [0, 1]$ and $(\tilde{\eta}_t)_{t\in\mathbb{N}}$ is a sequence of i.i.d. normally distributed random variables with $\mathbb{E}[\tilde{\eta}_t] = 0$ and $\text{Var}(\eta_t) = \sigma_\eta^2$, for all $t \in \mathbb{N}$, independent of all other random variables introduced so far.

Prove that the equilibrium price process $(p_t^*)_{t\in\mathbb{N}}$ of the risky asset is given by

$$p_t^* = f_t - d_t + (\phi_0 + \phi_z z_t), \tag{8.51}$$

where

$$\phi_0 = \frac{(1-\alpha_z)\phi_z \bar{z}}{r_f - 1} < 0 \quad \text{and} \quad \phi_z = -\frac{r_f - \alpha_z}{2\sigma_\eta^2}\left(1 - \sqrt{1 - (\sigma_\eta^2/\sigma^{2,*})}\right),$$

and, for all $t \in \mathbb{N}, f_t$ denotes the cum-dividend fundamental value of the risky asset in the hypothetical case of risk neutral agents and is given by

$$f_t = \sum_{s=0}^{\infty} \mathbb{E}\left[\frac{\tilde{d}_{t+s}}{r_f^s}\Big|\tilde{d}_t, \tilde{y}_t\right] = \frac{r_f\bar{d}}{r_f - 1} + \frac{r_f}{r_f - \alpha_d}\tilde{d}_t + \frac{1}{r_f - \alpha_d}\tilde{y}_t,$$

under the assumption that $\sigma_\eta^2 \le \sigma^{2,*} := (r_f - \alpha_z)^2/(4\sigma_f^2)$, where

$$\sigma_f^2 := \frac{r_f^2}{(r_f - \alpha_d)^2}\sigma_\varepsilon^2 + \frac{1}{(r_f - \alpha_d)^2}\sigma_\eta^2$$

denotes the innovation variance of f_t (see Campbell et al. [346, Theorem 1]).

Exercise 8.21 Prove Proposition 8.17.

Exercise 8.22 Prove Proposition 8.18.

Chapter 9
Uncertainty, Rationality and Heterogeneity

*If you consider men as infinitely egoistic and infinitely
farsighted. The first hypothesis can be accepted in first
approximation, but the second would maybe need a few reserves.*
Poincaré to L. Walras (letter of October 1st, 1901)

*Investment based on genuine long-term expectation is so
difficult as to be scarcely practicable. He who attempts it must
surely lead much more laborious days and run greater risks
than he who tries to guess better than the crowd how the crowd
will behave; and given equal intelligence, he may make more
disastrous mistakes.*
Keynes (1936)

The results presented in the previous chapters rely on a set of classical assumptions concerning the agents' preferences, the economy and the modeling of randomness. More specifically, we have assumed the validity of several fundamental hypotheses, which can be summarized as follows:

- classical probability theory;
- expected utility theory;
- substantial rationality;
- rational expectations;
- homogeneous agents.

In this chapter, as well as in the following one, we will critically analyse the role played by these assumptions and present several alternative approaches which go beyond the classical paradigm. In our analysis, we adopt the perspective illustrated by the two quotations by Solow and Kuhn at the beginning of Chap. 1. We will consider the relation between hypotheses and results: each hypothesis will be evaluated both for its own realism as well as for its implications in generating theoretical results or in explaining empirical facts. Alternative behavioral hypotheses will be evaluated not only on the basis of their capability to explain the asset pricing anomalies illustrated in the previous chapters but also in terms of their capability to provide an alternative paradigm compatible with what classical asset pricing theory is already able to explain. Indeed, one of the main problems of going beyond the classical framework of full rationality-expected utility theory is

© Springer-Verlag London Ltd. 2017

E. Barucci, C. Fontana, *Financial Markets Theory*,
Springer Finance, DOI 10.1007/978-1-4471-7322-9_9

represented by the fact that, typically, each alternative theory is only able to explain a specific puzzle arising in the context of classical asset pricing theory but cannot reproduce other facts, thus making new puzzles arise. Moreover, the robustness of the results of each alternative theory should be carefully evaluated with respect to different behavioral assumptions.

In this chapter, we shall focus our attention on the assumptions made on the modeling of randomness, on the agents' behavior and on the existence of market imperfections. For the time being, we continue to make the standing assumption of *perfect competition*, which will be relaxed in the following chapter.

This chapter is structured as follows. In Sect. 9.1, we discuss the fundamental notions of probability, risk, uncertainty and information ambiguity, pointing out their implications for the modeling of preferences. In Sect. 9.2, we present several alternative preference functionals which generalize the classical time additive expected utility representation, including behavioral models, recursive utilities, habit formation preferences and alternative forms of discounting. In Sect. 9.3, we relax the perfect rationality hypothesis and discuss models with noise traders, overconfident traders, feedback traders and the role of market selection in the survival of irrational agents. In Sect. 9.4, we provide an overview of the concepts of bounded rationality, learning and distorted beliefs. Section 9.5 contains a discussion of the impact of several market imperfections. At the end of the chapter, we provide a guide to further readings as well as a series of exercises.

9.1   Probability, Risk and Uncertainty

In the analysis developed in the previous chapters, we have always assumed that randomness is described through a probability space and that this probability space is common knowledge among the agents. More specifically, we adopted the classical axiomatic approach to probability theory and we assumed the existence of a probability space $(\Omega, \Theta, \mathbb{P})$, where the set Ω is the set of all elementary states of the world, Θ is a σ-algebra on Ω and $\mathbb{P}$ represents an objective probability measure (see the introduction to Chap. 2). In this section, we will discuss the relevance of this hypothesis and we present several alternative paradigms for the modeling of randomness, also on the basis of experimental evidence.

Even though it is a very natural assumption, representing agents' beliefs-opinions by means of a probability space $(\Omega, \Theta, \mathbb{P})$ does not come without consequences. In particular, while the couple (Ω, Θ) seems to be a necessary tool to describe the set of possible realizations of randomness, the assumption of the existence of an *objective* probability measure $\mathbb{P}$ is more delicate. In particular, recall that, according to the standard axiomatic approach to probability theory, the measure $\mathbb{P}$ is assumed to be countably additive. Many philosophers, economists and mathematicians (including Ramsey, Keynes, Knight, De Finetti, Savage amongst others) have pointed out the limits of the classical axiomatic probability theory and proposed alternative approaches.

The objectivist approach to probability theory has a long tradition, dating back to the origins of probability theory with the study of gambling games. In this context, the likelihood of an event can be defined as the ratio of favorable outcomes over all possible outcomes of a random experiment, with all elementary events being assumed to have an equal probability of realization. This approach led to the *frequentist approach* to probability theory, where the probability of occurrence of an event is given by the limit of its relative frequency of occurrence. As in the classical axiomatic probability theory, this interpretation refers to an objective probability measure.

The frequentist approach to probability is particularly well suited to random phenomena for which experiments can be repeated and reproduced in order to assess the probability of an event, like gambling games. However, this is clearly not the case in economic problems. In this direction, the *subjectivist approach* seems more promising. According to this approach, the likelihood of an event is given by the confidence that an agent places on its realization, thus providing an estimate of the probability of occurrence without resorting to repeated experiments. In particular, according to the subjectivist interpretation of probability, dating back to contributions of De Finetti, Ramsey and Savage (we refer to Gilboa & Marinacci [779] for a survey of their approaches), the probability of an event $E \in \Theta$ from the point of view of an agent is given by the price that he considers fair for the contract delivering one unit of a reference good whenever the event E occurs (see De Finetti [538]). According to this interpretation, the probability is subjective since it depends on the agent's assessment of the likelihood of the event. The probability of an event can therefore be associated with the certainty equivalent of a specific gamble and, in the case of an elementary event, it corresponds to the price of an Arrow security. De Finetti also required that the probability evaluation be coherent, meaning that no combination of bets should allow for a surely positive gain. Note that the classical and the subjectivist approaches to probability share some features: indeed, in both approaches, it holds that $\mathbb{P}(\Omega) = 1$, $\mathbb{P}(\emptyset) = 0$ and the probability of two incompatible events is equal to the sum of the probabilities of the two events. However, according to De Finetti's approach, the probability measure is only finitely additive and is not assumed to be countably additive. This represents a significant departure from the classical axiomatic approach to probability theory.

Expected utility theory is based on the existence of a probability measure known ex-ante by the agents. As discussed in Chap. 2, this reflects the *riskiness* of the economic environment, where agents do not know which state of the world will be realized at a future date but have a complete knowledge of the underlying probability distribution. Savage has extended expected utility theory to a setting with *unknown probabilities*, i.e., a setting where agents do not know ex-ante the probabilities of realization of the events (see Savage [1499]). According to Savage, even though their beliefs cannot be represented in terms of a probability measure, agents still have preferences over gambles and can formulate choices among different gambles. Denoting by $\mathscr{R}$ the preference relation over gambles, if $\mathscr{R}$ satisfies suitable axioms, then it is possible to infer a subjective probability distribution and a utility function such that $\mathscr{R}$ can be represented by the expectation of this utility function with respect to the subjective probability distribution (compare with Sect. 2.1). In this setting, it is important to remark that the subjective probabilities are derived from the agent's preferences and, hence, the preference relation $\mathscr{R}$ embeds both an agent's preferences with respect to wealth under certainty as well as his beliefs. We refer to Kreps [1136, Chapter 9] for a detailed analysis of the subjectivist approach to decision theory.

According to Savage's subjectivist approach, even though an agent does not know ex-ante the probability distribution, the decision making problem can be analysed ex-post in the context of expected utility theory, with the agent behaving as if the probability distribution was known and where the utility function and the probability distribution are implied by the preference relation $\mathscr{R}$. However, this type of behavior is not confirmed empirically. In particular, a deep criticism was expressed by Knight, who distinguished between *uncertainty* and *risk* (or uncertainty in a strong sense and in a weak sense, respectively), see Knight [1108]. As explained in the introduction to Chap. 2, uncertainty arises when agents are unable to assign a probability to some event, typically because of missing information (*information ambiguity*), and do not behave (ex-post) according to the expected utility paradigm, with respect to some objective or subjective probability measure.

The Ellsberg paradox illustrated in Ellsberg [636] provides an enlightening example of a violation of the expected utility theory in the presence of uncertainty-information ambiguity. Consider an agent who has to draw a ball from an urn containing nine colored balls: three balls are red and each of the remaining six balls is either yellow or black and the agent does not know how many of the six balls are yellow. This example cannot be described in the classical setting considered in the previous chapters. Indeed, while the agent knows the probability of extracting a red ball $(1/3)$, he does not know the probability of extracting a yellow (or a black) ball. The probability of such an event is comprised between 0 and $2/3$. This situation provides an example of information ambiguity, since the agent does not know how many balls in the urn are yellow and/or black.

Continuing the description of the Ellsberg paradox, let us consider four gambles A, B, C and D, whose payoffs depend on the colour of the first ball drawn from the urn (we write $\{R\}$, $\{Y\}$ and $\{B\}$ to denote respectively the event where a red, a yellow or a black ball is drawn):

- gamble A: payoff 1 if $\{R\}$, payoff 0 if $\{Y\}$ or $\{B\}$;
- gamble B: payoff 0 if $\{R\}$, payoff 1 if $\{B\}$ and payoff 0 if $\{Y\}$;
- gamble C: payoff 1 if $\{R\}$, payoff 0 if $\{B\}$ and payoff 1 if $\{Y\}$;
- gamble D: payoff 0 if $\{R\}$ and payoff 1 if $\{Y\}$ or $\{B\}$.

Observe that the payoffs of gambles A and D are not affected by the information ambiguity. On the contrary, the payoffs of gambles B and C are affected by information ambiguity. When confronted with the above four gambles, most of the agents will typically choose according to the following order: A $\mathscr{P}$ B and D $\mathscr{P}$ C. However, this preference ordering is not compatible with the expected utility theory à la Savage. As a matter of fact, if an agent's preferences can be represented by some expected utility function, then A $\mathscr{P}$ B implies that the agent considers the probability of the event $\{B\}$ less than $1/3$ and the probability of the event $\{Y\}$ bigger than $1/3$. Instead, D $\mathscr{P}$ C implies that the agent estimates the probability of the event $\{B\}$ to be bigger than $1/3$ and the probability of the event $\{Y\}$ to be smaller than $1/3$. This experimental result shows that the expected utility theory, even in its subjectivist version à la Savage, is unable to describe the agents' behavior in a context of information ambiguity-uncertainty (see Camerer & Weber [329] for a survey on the experimental evidence on decision making under information ambiguity). Typically, it is shown that agents are averse towards gambles characterized by payoffs with uncertain probabilities, meaning that agents dislike information ambiguity (see Epstein [640] and Ghirardato & Marinacci [774]).

This empirical evidence motivated the development of new decision theories aiming at explaining the agents' behavior under information ambiguity (see again Camerer & Weber [329] for a survey). In this section, we briefly present two approaches, both based on an axiomatic description of agents' preferences: the *maxmin expected utility theory* and the *Choquet expected utility theory*. By relaxing the independence axiom (see Assumption 2.3), these two approaches incorporate the distinction between risk and uncertainty and, in the special case of no information ambiguity, they reduce to the classical expected utility theory as presented in Chap. 2. In a nutshell, the maxmin expected utility theory is based on the idea that agents formulate their decisions with respect to a set of probability measures compatible with the available information, while the Choquet expected utility theory relies on non-additive probabilities. On the axiomatization of decision theory under information ambiguity see also Chateauneuf et al. [413], Epstein & Zhang [648], Maccheroni et al. [1262, 1263], Seo [1516], Klibanoff et al. [1105].

In the seminal paper Gilboa & Schmeidler [780], the *maxmin expected utility* (or expected utility in the worst-case scenario) has been proposed to represent preferences under information ambiguity. In order to illustrate the concept, let us consider an economy with a finite number $S \in \mathbb{N}$ of elementary states of the world. Under information ambiguity, there does not exist a single probability measure commonly known to all the agents, but rather a set of probability measures which are compatible with the available information. For instance, in the case of the Ellsberg paradox, there are seven probability measures compatible with the available information, each one corresponding to a specific number of yellow (or black) balls in the urn. We denote the set of these probability measures by $\Pi = \{\pi^1, \pi^2, \dots, \pi^K\}$, for some $K \in \mathbb{N}$, where $\pi^k \in \mathbb{R}_+^S$ and $\sum_{s=1}^{S} \pi_s^k = 1$, for all $k = 1, \dots, K$. Similarly as in (2.1), we represent a gamble as a measurable mapping $\tilde{x} : \Omega \to \mathbb{R}_+$, with $x(\omega_s)$ denoting the realization of the gamble in correspondence of ω_s, for $s = 1, \dots, S$. Following Schmeidler [1505], the preference relation $\mathscr{R}$ is said to exhibit *uncertainty aversion* if, for any three gambles $\tilde{x}_1, \tilde{x}_2$ and $\tilde{x}_3$,

$$\tilde{x}_1 \mathscr{R} \tilde{x}_3 \text{ and } \tilde{x}_2 \mathscr{R} \tilde{x}_3 \Rightarrow \alpha \tilde{x}_1 + (1 - \alpha)\tilde{x}_2 \mathscr{R} \tilde{x}_3, \text{ for all } \alpha \in [0, 1].$$

Assuming that the preference relation $\mathscr{R}$ satisfies suitable axioms and that the agent is *uncertainty averse* (see Gilboa & Schmeidler [780, Axiom A.5]), there exists a utility function $u : \mathbb{R}_+ \to \mathbb{R}$ such that, for every two gambles $\tilde{x}_1$ and $\tilde{x}_2$, it holds that

$$\tilde{x}_1 \mathscr{R} \tilde{x}_2 \iff \min_{\pi \in \Pi} \mathbb{E}^\pi [u(\tilde{x}_1)] \geq \min_{\pi \in \Pi} \mathbb{E}^\pi [u(\tilde{x}_2)], \tag{9.1}$$

where, for each $\pi \in \Pi$, we denote by $\mathbb{E}^\pi [\cdot]$ the expectation with respect to the probability distribution π (see Gilboa & Schmeidler [780, Theorem 1]). Note that, in Gilboa & Schmeidler [780, Theorem 1], the probability space is general (i.e., it is not necessarily finite) and the existence of a non-empty, closed and convex set Π of (finitely additive) probability distributions such that (9.1) holds is actually part of the result. Moreover, similarly to the result of Theorem 2.1, the utility function u is unique up to positive linear transformations. Relation (9.1) captures the fact that the knowledge of the probability structure is incomplete and agents are averse to uncertainty-information ambiguity. This corresponds to evaluating a gamble with respect to the probability distribution associated with the lowest possible level of expected utility (worst-case scenario), among all probability distributions compatible with the available information. As a consequence, if $\mathscr{X}$ denotes a given set of gambles, an agent will optimally choose a gamble $\tilde{x}^* \in \mathscr{X}$ by solving the problem

$$\max_{\tilde{x} \in \mathscr{X}} \min_{\pi \in \Pi} \mathbb{E}^\pi [u(\tilde{x})],$$

where u is the utility function appearing in the representation (9.1) of the preference relation $\mathscr{R}$. In particular, it can be shown that the maxmin expected utility is consistent with the Ellsberg paradox.

An axiomatization of subjective expected utility theory different from the maxmin expected utility has been formulated in Gilboa [778] and Schmeidler [1505] by relying on non-additive probabilities (*capacities*). The expected utility with respect to a non-additive probability can be computed by using the Choquet integral, making the expected utility criterion applicable to cases where classical expected utility theory is not applicable. By weakening the independence axiom (and adopting the notion of *comonotonic independence*, see Schmeidler [1505]), it is shown that preferences can be represented through the expected utility where the expectation is computed with respect to a monotonic non-additive subjective probability μ (*Choquet expected utility*). The fact that the probability is monotonic and non-additive means that, for any two events $A, B \in \Theta$ such that $A \subseteq B$, it holds that $\mu(A) \leq \mu(B)$, together with the basic properties $\mu(\Omega) = 1$ and $\mu(\emptyset) = 0$. Note that, for a non-additive probability μ (also known as capacity), it holds that

$$\mu(A) + \mu(B) \neq \mu(A \cup B) + \mu(A \cap B), \tag{9.2}$$

for two events $A, B \in \Theta$. In particular, a non-additive probability μ is said to be convex if the left-hand side of (9.2) is less or equal than the right-hand side. In particular, μ is convex if and only if an agent is *uncertainty averse* (see Schmeidler [1505]). The inequality reduces to an equality when the agent faces a standard risky situation, as considered in Chap. 2. Note that μ does not represent an objective measure, but only allows for a representation of an agent's preferences in the form of non-additive expected utility. Similarly as above, the utility function is unique up to linear transformations. As in the case of the maximin expected utility, the approach based on the Choquet expected utility is consistent with the Ellsberg paradox. Note also that the Choquet expected utility theory has a partial overlap with the maxmin expected utility theory. Indeed, when the non-additive probability is convex, the Choquet expected utility can be described in terms of a suitable maxmin expected utility (see Gilboa & Schmeidler [780]).

Information Ambiguity: Portfolio Choices, Equilibrium and Asset Pricing Implications

The literature on the implications of information ambiguity on the modeling of financial markets is quite large. In the remaining part of this section, we limit ourselves to a brief overview of some important results.

An interesting effect of information ambiguity on the optimal demand of a risky asset has been shown in Dow & Werlang [588]. The authors consider a two-period economy (i.e., $t \in \{0, 1\}$) and an agent characterized by a utility function u such that $u' > 0$ and $u'' \leq 0$ having the possibility of investing in a risky asset delivering the random dividend $\tilde{d}$ at date $t = 1$. The risk free rate is normalized to one. In the classical setting, i.e., when the probability measure in the expected utility representation of the agent's preferences is additive (no information ambiguity), then the agent invests a positive amount of wealth in the risky asset if and only if the expected dividend is higher than the current price (see Proposition 3.1) and the optimal demand decreases with respect to the agent's risk aversion (see Proposition 3.2). On the contrary, if the agent's preferences imply a non-additive probability measure and the agent is uncertainty averse, then these results do not hold anymore. More precisely, we have the following proposition, which corresponds to Dow & Werlang [588, Theorem 4.2].

Proposition 9.1 *Consider a risk averse or risk neutral agent (with a constant initial wealth $w_0 > 0$) who is uncertainty averse and has the possibility of investing in a risky asset delivering the random dividend $\tilde{d}$ for a price $p > 0$. Denote by w^* the agent's optimal demand of the risky asset. Then the following hold:*

(i) $w^* > 0$ if $p < \mathbb{E}[\tilde{d}]$ and only if $p \leq \mathbb{E}[\tilde{d}]$;
(ii) $w^* < 0$ if $p > -\mathbb{E}[-\tilde{d}]$ and only if $p \geq -\mathbb{E}[-\tilde{d}]$,

where $\mathbb{E}[\cdot]$ denotes the expectation with respect to a non-additive probability.

Proof We only provide an outline of the proof, referring to Dow & Werlang [588] for full details. In the present context (as in Schmeidler [1505]), the agent's preference relation can be represented in expected utility terms with respect to a non-additive probability. Moreover, as discussed in Dow & Werlang [588, Appendix], the expectation defined with respect to a non-additive probability is such that Jensen's inequality holds and the expectation of an affine transformation of a random variable is given by the affine transformation of the expectation of the random variable. The claim then follows by arguments similar to those used in the proof of Proposition 3.1. $\square$

As a consequence of the above proposition, there exists an interval of prices, given by $(\mathbb{E}[\tilde{d}], -\mathbb{E}[-\tilde{d}])$, in correspondence of which the optimal demand of a risk averse or risk neutral agent is null. Note that, in the axiomatic approach proposed by Schmeidler [1505], it holds that $\mathbb{E}[\tilde{d}] < -\mathbb{E}[-\tilde{d}]$. This result has been extended to the case of multiple beliefs characterized by a set of normal distributions, see Easley & O'Hara [621]: in this case, ambiguity averse agents do not trade in the market for a set of prices. Microstructure models showing that illiquidity in the market may arise from ambiguity aversion have been analysed in Easley & O'Hara [622, 623] and Routledge & Zin [1475], while the implications of ambiguity on portfolio and equilibrium analysis in a mean-variance framework have been analysed in Maccheroni et al. [1264], Garlappi et al. [759], Boyle et al. [276].

The above argument has been further investigated in Mukerji & Tallon [1361], where the authors show that uncertainty aversion may lead to market incompleteness (in a related context, see also Rinaldi [1454]). Mukerji & Tallon [1361] adopt the Choquet expected utility framework of Schmeidler [1505], where the agents' subjective beliefs are represented by a convex non-additive probability, thus introducing uncertainty aversion in the economy. In a complete market economy, in the presence of a severe information ambiguity, the agents can decide not to trade financial assets which are affected by idiosyncratic risk (in the sense that their payoffs do not only depend on realized endowments). Therefore, uncertainty aversion reduces the risk sharing opportunities offered by the financial market, with the consequence that agents cannot fully diversify the idiosyncratic risk. In turn, this leads to an equilibrium allocation that does not correspond to the allocation obtained in a complete market economy. Note that this no-trade result holds under information ambiguity as long as the initial allocation is Pareto optimal. This result can also explain the limited use of indexed debt (i.e., loan contracts indexed to the inflation rate) in real financial markets: if agents are uncertainty averse, then under some conditions there is no trade in indexed bonds in equilibrium (see Mukerji & Tallon [1362]).

Pareto optimal allocations and sharing rules under uncertainty aversion have been investigated in Chateauneuf et al. [412]. When agents believe in the same convex set of probability distributions, then the set of Pareto optimal allocations is independent of that set and coincides with the set of Pareto optimal allocations obtained in a classical economy where agents have a standard expected utility function and homogeneous beliefs. Pareto optimal allocations depend only on aggregate risk and are comonotonic (compare with Theorem 4.3). Moreover, in the absence of aggregate risk, Chateauneuf et al. [412] provide sufficient conditions in order to obtain an allocation with full insurance. In Epstein [641], considering a simple two-period economy with heterogeneous agents and ambiguity, it is shown that agents' consumption levels are only positively correlated with aggregate shocks, but are not necessarily perfectly correlated, unlike in the classical setting considered in the previous chapters (see Epstein & Miao [643] for a continuous time extension of this result). Dana [513], Rigotti & Shannon [1452], Rigotti et al. [1453] provide general characterizations of equilibrium and Pareto optimality, also analysing the determinacy of equilibria.

In Epstein & Wang [646, 647] the maxmin expected utility approach of Gilboa & Schmeidler [780] has been applied to a multi-period economy in an infinite time horizon by relying on the recursive utility framework (see Sect. 9.2 below). In a dynamic setting with information ambiguity, in correspondence of an event $A_t \in \mathscr{F}_t$ an agent does not believe in a unique conditional probability distribution for the events belonging to $\mathscr{F}_{t+1}$, but rather in a set of conditional probability distributions forming a probability kernel correspondence. In particular, this reflects both the presence of uncertainty and the agents' aversion to uncertainty. In this discrete time setting, the utility function is represented by a utility process solving a recursive relation (the existence of a utility process is established in Epstein & Wang [646, Theorem 1]). Referring to the next section for a presentation of the recursive utility framework, we just mention here that, in the setting considered in Epstein & Wang [646], the utility is computed at each date $t \in \mathbb{N}$ by taking into account the certainty equivalent of the future utility, which is computed by taking the minimum among all certainty equivalents associated to the probability distributions belonging to the kernel. In a pure exchange economy with a representative agent, the existence of an equilibrium is established and asset prices are characterized by means of an Euler inequality. Equilibria may be indeterminate and non-unique. Necessary and sufficient conditions for equilibrium indeterminacy can be found in Epstein & Wang [646]. More precisely, there may exist uncountably many equilibria, with the consequence that equilibrium asset prices are not uniquely defined but rather only belong to an interval. Equilibrium prices may be discontinuous with respect to the previous realizations, thus inducing high volatility. Moreover, in Epstein & Wang [647] the presence of uncertainty aversion is linked to the possibility of abrupt changes in asset prices and may be incompatible with the existence of a risk neutral representation of asset prices.

A continuous time utility functional incorporating information ambiguity with multiple probability distributions has been proposed in Chen & Epstein [426]. Under information ambiguity, it is shown that the equity premium is given by the sum of a risk premium and an *ambiguity premium*. Epstein & Schneider [645] show that excess returns are higher when future information quality is more uncertain (see also Kogan & Wang [1117] for a two-factor CAPM extension). It is important to remark that the presence of an ambiguity premium represents an interesting perspective to address the equity premium and interest rate puzzles, the home bias puzzle and the participation puzzle, see Epstein & Miao [643], Easley & O'Hara [621], Cao et al. [363], Maenhout [1284], Trojani & Vanini [1600], Leippold et al. [1179], Uppal & Wang [1605], Ju & Miao [1053], Barillas et al. [162], Gollier [802]. In general, the demand of the risky asset under information ambiguity is lower than in the classical setting (see also Epstein & Schneider [644] for an analysis of portfolio choices under information ambiguity and learning).

Asset pricing results under information ambiguity have been obtained by relying on robust-risk sensitive control theory in Anderson et al. [63], Hansen & Sargent [893]. More precisely, in Hansen & Sargent [893] the authors refer to *model uncertainty* rather than ambiguity and consider a continuous time perturbed Markov process in an infinite horizon. Each perturbation corresponds to an absolutely continuous measure and represents a specification error on the model. The optimization criterion is formulated by taking the infimum over all such measures (worst case scenario), in line with the maxmin approach of Gilboa & Schmeidler [780] (on this point see also Maccheroni et al. [1262]). Agents facing model uncertainty are averse to specification errors and make decisions that are robust to this type of uncertainty. Risk sensitivity and preference for robustness contribute (positively) to generate a high risk premium together with a positive *model misspecification premium* and induce a precautionary saving motive, see Hansen et al. [894] and Maenhout [1284]. Portfolio choices with ambiguity averse investors and (partly) predictable returns have been analysed in Ait-Sahalia & Brandt [31] (see also Chen et al. [415] and Liu [1223]), showing that the stock demand is decreasing with respect to the ambiguity aversion, similarly as in the classical case of risk aversion (in this context, see also Epstein & Schneider [644]).

9.2   On Expected Utility Theory

In the previous chapters, we have always assumed that an agent's preference relation can be represented by the expectation of an utility function. In particular, as shown in Sect. 2.1, if an agent's preference relation satisfies a set of axioms, then there exists a utility function and a probability distribution (objective or subjective, as in the Savage framework) such that the corresponding expected utility induces the same ordering of the preference relation. In this section, we will discuss the relevance of the set of axioms introduced in Sect. 2.1 and present several alternative approaches that have been proposed in the literature in order to overcome the limitations of classical expected utility theory. Moreover, we will also discuss the structure of the utility functional (6.6) in a multi-period economy and the assumptions of time additivity and of constant preferences over time.

As before, we denote by $\mathscr{R}$ the preference relation of an agent. The rationality axiom introduced in Assumption 2.1 implies that $\mathscr{R}$ is complete, reflexive and transitive. The completeness of $\mathscr{R}$ means that, for every couple of gambles, an agent is always able to express a preference between the two gambles (either a strict preference or an indifference relation), while the transitivity property implies that the agent is able to rank different gambles according to his preference criterion. The continuity axiom (Assumption 2.2) mainly plays a technical role: the main inconvenient of this hypothesis is that some preference relations do not satisfy it (e.g., lexicographic preferences). The *independence axiom* (Assumption 2.3) represents probably the most crucial and controversial assumption. Indeed, the independence axiom requires that the preference relation between two arbitrary gambles is not affected if each of the two gambles is mixed (with the same weight) with a third arbitrary gamble. In other words, an agent whose preferences satisfy the independence axiom evaluates two gambles by only taking into account what is different among the two gambles. As can be seen from the proof of Theorem 2.1, the independence axiom implies that the functional representing the agent's preferences is linear with respect to the probabilities.

The validity of the independence axiom has been deeply discussed in the decision theory literature, the driving force of the debate being experimental (laboratory) evidence showing that agents do not behave according to the independence axiom. In this context, the most famous example violating the independence axiom is the Allais paradox presented in Sect. 2.5. Starting from the choices over two couples of gambles and applying repeatedly the independence axiom, we have shown that agents' choices are inconsistent with a representation of preferences by means of the expectation of a utility function. A second set of experiments contradicting the expected utility paradigm has been illustrated in Kahneman & Tversky [1060]. For instance, Problems 3 and 4 in Kahneman & Tversky [1060] consider the case of an agent who has to choose between a lottery offering a 25% chance of winning 3.000 and a lottery offering a 20% chance of winning 4.000. It is empirically observed that 65% of the agents will choose the second lottery. On the contrary, if agents have to choose between a 100% chance of winning 3.000 and an 80% chance of winning 4.000, then 80% of the agents will choose the first lottery. In the context of this example, according to classical expected utility theory, agents should not choose differently in the two situations, since the probabilities associated by the two lotteries to the two scenarios have been only multiplied by the same factor. This distortion has been called *certainty effect* in Kahneman & Tversky [1060], i.e., agents prefer certain outcomes over random outcomes and overweight outcomes considered certain, with the consequences that preferences are in general non-linear with respect to the probabilities. There are by now many experimental paradoxes of this type: see Machina [1267, 1268], Camerer [328], Hens & Rieger [938], Starmer [1564] for some surveys of the most significant ones. However, it has to be noted that most of these experiments (including the Allais paradox) are obtained by considering gambles with extreme probabilities (probabilities close to zero or one). On the contrary, the application of expected utility theory seems to be more appropriate when considering gambles without extreme probabilities.

In the decision theory literature, the independence axiom has been widely discussed and several weaker formulations have been proposed, mainly by introducing non-linearities (with respect to the probabilities) in the expected utility through different preference axiomatizations. Quiggin [1436] has introduced the concept of *rank dependent expected utility*, i.e., a utility functional defined as the expectation of a utility function with respect to a transformation of the probability distribution based on the rank of the outcome. This allows to relax the validity of the independence axiom, preserving at the same time the completeness, the reflexivity and the transitivity of the preference relation. In Machina [1266], the independence axiom has been replaced by a smoothness condition of the utility functional (i.e., the utility functional is assumed to be Fréchet differentiable on the space of probability distribution functions) and it is shown that most of the basic tools and results of classical expected utility theory can be derived under this weaker assumption, notably first and second order stochastic dominance, risk aversion and the Arrow-Pratt risk aversion comparison (compare with Chap. 2). Moreover, the utility functional introduced in Machina [1266] is consistent with Allais-type paradoxes and other experimental evidence (in a related context, see also Chew et al. [428]).

In a seminal paper on behavioral finance, Kahneman & Tversky [1060] have proposed a utility functional that is additive but non-linear with respect to probabilities, such that small probabilities are overweighted and, conversely, large probabilities are underweighted (*prospect theory*). The utility functional is defined over changes (gains/losses) with respect to a reference point representing the status quo and is steeper for losses than for gains, concave for gains and convex for losses. This type of utility functional is consistent with the certainty effect and captures several behavioral aspects of decision making, like *loss aversion* and different risk attitudes in the case of losses or gains. Prospect theory preferences will be discussed in more detail below in this section.

In Chap. 6, we have assumed that an agent's preferences can be represented by a time additive expected utility function, with the implicit assumption that the utility at each date t only depends on consumption at date t and not on consumption at different dates. Moreover, we have typically assumed that the utility functions associated to consumption at different dates coincide, thus implying that the agent's preferences do not change over time. These assumptions, however, have a weak behavioral justification (see Machina [1268]). In a multi-period setting, besides the inherent limitations of classical expected utility theory discussed above, a major problem is represented by the fact that, assuming a utility functional of the form (6.6), the function u determines both the degree of risk aversion and the degree of intertemporal substitution of consumption. For instance, if the function u is nearly linear, then the risk aversion is very low and, at the same time, the rate of intertemporal substitution is almost insensitive to consumption levels (large elasticity of intertemporal substitution of consumption), indicating a large propensity to substitute consumption intertemporally. In the special case of a power utility function, the relative risk aversion coefficient is exactly the inverse of the elasticity of the intertemporal rate of substitution (see also Sect. 7.3 and Exercise 7.8). Therefore, a utility functional of the form (6.6) can only capture high risk aversion together with low intertemporal substitution or low risk aversion together with high intertemporal substitution. It turns out that there is no behavioral motivation for a connection of this type between risk aversion and elasticity of intertemporal substitution of consumption. Indeed, risk aversion represents the willingness to diversify over different states of the world, while intertemporal substitution represents the willingness to diversify over different periods in time. Experimental results in Barsky et al. [169] have shown that there is no relation between risk aversion and the intertemporal rate of substitution.

In the literature, two main approaches have been proposed for relaxing this relation between risk aversion and intertemporal substitution. A first approach is based on *recursive preferences*, where the utility at every date is also affected by the expected utility from future consumption. The second approach is represented by *habit formation* utilities, where the utility also depends on past consumption or other variables. These two approaches will be discussed in more detail below.

Prospect Theory, Loss Aversion and Disappointment Aversion

As put forward in Kahneman & Tversky [1060], the *loss aversion* phenomenon refers to the tendency of individuals to be more sensitive to reductions/losses in their levels of consumption or wealth than to increases/gains (for experimental evidence on loss aversion see Bateman et al. [180], Abdellaoui et al. [2] and Levy & Levy [1206]). In the context of *prospect theory*, this asymmetry has been taken into account by considering utility functions defined over changes (gains and losses) with respect to a reference point representing the status quo. The fact that utilities are defined with respect to a reference point also captures the so-called *framing effect*, meaning that individuals tend to compare alternatives to a given frame or reference point. The reference point can be defined with respect to consumption, wealth, capital gains, etc. (on reference-dependent preferences see also Koszegi & Rabin [1121], where the reference point is determined endogenously by rational expectations). The utility function is assumed to be steeper for losses than for gains and, typically, it is concave in the region of gains and convex in the region of losses. This last feature captures a risk averse behavior with respect to gains and a risk seeking behavior with respect to losses.

According to prospect theory, the utility functional $U(\cdot)$ representing preferences over gambles $\tilde{x} = \{x_1, \ldots, x_S; \pi_1, \ldots, \pi_S\}$, with x_s being a gain/loss with respect to some reference point, for $s = 1, \ldots, S$, is provided by

$$U(\tilde{x}) = \sum_{s=1}^{S} u(x_s) w(\pi_s), \qquad (9.3)$$

where $u(\cdot)$ is the *value function* and $w(\cdot)$ the *weighting function*. The value function $u(\cdot) : \mathbb{R} \to \mathbb{R}$ is assumed to be continuous and increasing, strictly concave for gains and strictly convex for losses and, at zero, it is steeper for losses than for gains. The weighting function $w(\cdot) : [0, 1] \to [0, 1]$ is continuous and increasing, with $w(0) = 0$ and $w(1) = 1$, satisfies $w(\pi) > \pi$ for small values of $\pi > 0$ and $w(\pi) < \pi$ for large values of $\pi < 1$ and $\sum_{s=1}^{S} w(\pi_s) = 1$. The value function plays an analogous role to the classical utility function in the expected utility representation (2.2), while the weighting function w transforms probabilities into subjective probabilities. We call the functional U appearing in (9.3) the *prospect utility functional*.

A drawback of prospect theory preference functionals of the form (9.3) involves potential violations of first order stochastic dominance (see Sect. 2.3). To see this, as explained in Kahneman & Tversky [1060], it suffices to consider a probability space with two possible states of the world and two gambles $\tilde{x}_1$ and $\tilde{x}_2$ given by

$$\tilde{x}_1 = \{x, y; \pi, 1 - \pi\} \qquad \text{and} \qquad \tilde{x}_2 = \{x, y; \pi', 1 - \pi'\},$$

with $0 < y < x$ and $1 > \pi > \pi' > 0$. Clearly, $\tilde{x}_1$ dominates $\tilde{x}_2$ according to the first order stochastic dominance criterion (compare with Proposition 2.8). For a given value function $u(\cdot)$ and a weighting function $w(\cdot)$, $\tilde{x}_1$ is preferred to $\tilde{x}_2$ if and only if

$$w(\pi)u(x) + w(1 - \pi)u(y) \geq w(\pi')u(x) + w(1 - \pi')u(y),$$

i.e., if and only if

$$\frac{w(\pi) - w(\pi')}{w(1 - \pi') - w(1 - \pi)} \geq \frac{u(y)}{u(x)}.$$

Since $u(\cdot)$ is assumed to be continuous, the right-hand side of the last inequality converges to 1 as $y \to x$. Hence, $w(\pi) - w(\pi')$ approaches $w(1 - \pi') - w(1 - \pi)$ as y converges to x. Since $\pi + (1 - \pi) = \pi' + (1 - \pi') = 1$, this means that the above inequality can be satisfied only if the weighting function $w(\cdot)$ is essentially linear. In this way, for suitable non-linear weighting functions, one can derive explicit examples violating first order stochastic dominance.

A suitable notion of stochastic dominance has been formulated in the context of prospect theory by Levy & Wiener [1208] introducing the *prospect stochastic dominance* criterion. Similarly as in Sect. 2.3, let us consider gambles $\tilde{x}$ defined on a general (not necessarily discrete) probability space taking values in the bounded interval $[-1, 1]$ (recall that the realizations of $\tilde{x}$ represent changes with respect to a status quo).

Definition 9.2 Let $\tilde{x}_1$ and $\tilde{x}_2$ be two gambles taking values in $[-1, 1]$ and let $w(\cdot)$ be a weighting function. We say that $\tilde{x}_1$ dominates $\tilde{x}_2$ according to the *prospect stochastic dominance* criterion, i.e., $\tilde{x}_1 \succeq_{\mathrm{PSD}} \tilde{x}_2$, if $U(\tilde{x}_1) \geq U(\tilde{x}_2)$ holds for every continuous and increasing value function $u(\cdot)$ such that $u''(x) \leq 0$ for $x \geq 0$ and $u''(x) \geq 0$ for $x \leq 0$, where the preference functional $U(\cdot)$ is defined as in (9.3).

Similarly to the results presented in Sect. 2.3, the prospect stochastic dominance criterion admits a characterization in terms of the distribution functions of two random variables, as shown in the following proposition. The distribution functions can be defined with respect to either the objective probability measure (i.e., the probability measure represented by $(\pi_1, \dots, \pi_S)$) or the subjective probability (i.e., the probability transformed by the weighting function $w(\cdot)$). We denote by $F_i(\cdot)$ the cumulative distribution function of the random variable $\tilde{x}_i$, for $i = 1, 2$.

Proposition 9.3 *For any two random variables $\tilde{x}_1$ and $\tilde{x}_2$ taking values in $[-1, 1]$, the following are equivalent:*

(i) $\tilde{x}_1 \succeq_{PSD} \tilde{x}_2$;
(ii) $\int_y^x (F_1(z) - F_2(z))dz \leq 0$, for all $-1 \leq y \leq 0 \leq x \leq 1$.

Moreover, if $\tilde{x}_1 \succeq_{FSD} \tilde{x}_2$, then it also holds that $\tilde{x}_1 \succeq_{PSD} \tilde{x}_2$.

Proof The proof of the equivalence between properties *(i)–(ii)* is given in Exercise 9.1 and follows arguments similar to those used in the proof of Proposition 2.10. In view of Proposition 2.8, $\tilde{x}_1 \succeq_{FSD} \tilde{x}_2$ holds if and only if $F_1(x) \leq F_2(x)$, for every $x \in [-1, 1]$. The last claim of the proposition then follows directly from the implication *(ii)*⇒*(i)* above. □

In Proposition 9.3, the weighting function (which in turn determines the distribution functions F_1 and F_2 of the two random variables $\tilde{x}_1$ and $\tilde{x}_2$) is given. However, Levy & Wiener [1208, Theorem 8] shows that prospect stochastic dominance is stable with respect to all monotonic transformations of the distribution function that are concave for gains and convex for losses.

Portfolio choices with loss averse preferences have been analysed in several papers (see, e.g., Jarrow & Zhao [1022] for an analysis of optimal portfolios considering the lower partial moment as a measure of downside risk). In particular, despite the fact that prospect theory is in contrast with the foundations of mean-variance theory (see Sect. 2.4), it has been shown in Levy & Lèvy [1207] that, when diversification is allowed, the set of mean-variance efficient and *prospect theory efficient* portfolios (i.e., portfolios that are not dominated by any other portfolio according to the prospect theory stochastic dominance criterion) almost coincide. More precisely, we have the following proposition, which corresponds to Levy & Levy [1207, Theorem 2].

Proposition 9.4 *Let us consider a financial market with normally distributed returns and such that there does not exist a couple of assets with perfectly correlated returns. Then, in the absence of portfolio restrictions and for any prospect utility functional that does not violate first order stochastic dominance, the set of prospect theory efficient portfolios is contained in the set of mean-variance efficient portfolios.*

Proof To prove the claim, it suffices to prove that any mean-variance inefficient portfolio is also prospect theory inefficient. Let $\tilde{r}$ denote the return of an arbitrary mean-variance inefficient portfolio. This means that there exists another portfolio such that its return $\tilde{r}'$ satisfies $\sigma^2(\tilde{r}) = \sigma^2(\tilde{r}')$ and $\mathbb{E}[\tilde{r}'] > \mathbb{E}[\tilde{r}]$. Due to the assumption of normal distribution, this can be seen to imply that $\tilde{r}' \succeq_{FSD} \tilde{r}$ under the objective probability distribution. By assumption, $\tilde{r}' \succeq_{FSD} \tilde{r}$ also holds with respect to the subjective probability distribution. The result then follows by Proposition 9.3. □

In the above proposition, the prospect stochastic dominance criterion can refer to subjective probabilities (probabilities suitably transformed by a weighting function).

The result can be refined in the case where the objective probabilities are employed (see Levy & Levy [1207, Theorem 1]), with a more precise description of the set of mean-variance efficient portfolios that do not belong to the set of prospect theory efficient portfolios.

In order to solve the inconsistency between (classical) stochastic dominance and prospect theory, Tversky & Kahneman [1603] introduce *cumulative prospect theory*. Consider a gamble $\tilde{x} = \{x_1, \ldots, x_S; \pi_1, \ldots, \pi_S\}$, with outcomes $x_1 < x_2 < \ldots < x_S$. According to cumulative prospect theory, the preference functional (9.3) is replaced by the following functional

$$U(\tilde{x}) = \sum_{s=1}^{S} u(x_s) \left(w\left(\sum_{k=1}^{s} \pi_k \right) - w\left(\sum_{k=1}^{s-1} \pi_k \right) \right), \tag{9.4}$$

where $u(\cdot)$ still plays the role of the value function and $w(\cdot)$ of the weighting function. Note that $\sum_{k=1}^{s} \pi_k$ represents the cumulative probability distribution of the outcome x_s, for $s = 1, \ldots, S$. According to cumulative prospect theory, only probabilities associated to extreme outcomes are overweighted, while probabilities associated to outcomes in the middle of the distribution are underweighted. As an example, the following specification has been proposed in Tversky & Kahneman [1603]:

$$u(x) = x^{\alpha} \quad \text{for } x \geq 0 \qquad \text{and} \qquad u(x) = -\lambda(-x)^{\beta} \quad \text{for } x \leq 0,$$

with $\alpha, \beta \in (0, 1)$ and $\lambda > 1$ and $w(\pi) = \pi^{\gamma}(\pi^{\gamma} + (1 - \pi)^{\gamma})^{-1/\gamma}$. As shown in Exercise 9.2, cumulative prospect theory is compatible with first order stochastic dominance. Moreover, unlike prospect theory in the form (9.3), cumulative prospect theory can be easily extended to gambles on general (not necessarily discrete) probability spaces.

Adopting the cumulative prospect theory of Tversky & Kahneman [1603], Benartzi & Thaler [191] have shown that a utility functional taking into account loss aversion allows for a solution to the equity premium and the risk free rate puzzles. More specifically, Benartzi & Thaler [191] assume a short evaluation period together with loss aversion. The combined effect of loss aversion and a short evaluation period has been named *myopic loss aversion*. It is shown that a one-year evaluation period allows to explain the equity premium puzzle (see also Thaler et al. [1588], Gneezy & Potters [792], Gneezy et al. [791] for experimental evidence on myopic loss aversion and on the interplay between risk aversion and the length of the evaluation period). In this context, the agents' behavior shows a framing effect, in the sense that agents are sensitive to changes in financial wealth with respect to some status quo, as discussed above.

In Barberis et al. [150] and Barberis & Huang [148], several asset pricing anomalies are addressed by assuming loss aversion with respect to changes in financial wealth, with a loss after prior gains being less painful than a loss after previous losses. More specifically, Barberis et al. [150] assume a multi-period economy in an infinite time horizon, with a single risky asset (in unit supply) paying a dividend stream $(d_t)_{t \in \mathbb{N}}$ satisfying

$$\log d_{t+1} = \log d_t + g_d + \sigma_d \varepsilon_{t+1}, \qquad \text{for all } t \in \mathbb{N}, \tag{9.5}$$

where $(\varepsilon_t)_{t \in \mathbb{N}}$ is a sequence of i.i.d. standard normal random variables, g_d represents an average logarithmic growth rate and σ_d a volatility coefficient. Moreover, the market also contains a risk free asset (in zero net supply) paying the constant return r_f (a priori allowed to be time-varying in Barberis et al. [150]). The aggregate per capita consumption process $(\bar{c}_t)_{t \in \mathbb{N}}$ is modeled as a separate process (i.e., it does not coincide with the dividend process) satisfying the following dynamics:

$$\log \bar{c}_{t+1} = \log \bar{c}_t + g_c + \sigma_c \eta_{t+1}, \qquad \text{for all } t \in \mathbb{N}, \tag{9.6}$$

where $(\eta_t)_{t \in \mathbb{N}}$ is a sequence of i.i.d. standard normal random variables such that $\mathrm{Cov}(\eta_t, \varepsilon_t) = \rho$, for all $t \in \mathbb{N}$. The difference between the dividend and the aggregate consumption process can arise from the fact that agents have other sources of income besides dividends. Indeed, it is assumed that each agent in the economy receives an income stream $(y_t)_{t \in \mathbb{N}}$ such that $\bar{c}_t = d_t + y_t$, for all $t \in \mathbb{N}$, with $((d_t, y_t))_{t \in \mathbb{N}}$ being a joint Markov process.

Investors' preferences are described by the following functional, defined with respect to self-financing trading-consumption strategies $(w, c) = ((w_t, c_t))_{t \in \mathbb{N}}$, where w_t denotes the amount of wealth invested in the risky asset in the period $[t-1, t]$:

$$U(c, w) := \mathbb{E}\left[\sum_{t=0}^{\infty} \left(\delta^t u(c_t) + b_t \delta^{t+1} v(X_{t+1}, w_{t+1}, z_t) \right) \right], \tag{9.7}$$

where $u(x) = x^{1-\gamma}/(1-\gamma)$, with $\gamma > 0$ being the coefficient of relative risk aversion, and where $\delta \in (0, 1)$ is the discount factor. The processes appearing in (9.7) have the following interpretation:

- $(c_t)_{t \in \mathbb{N}}$ denotes the consumption process;
- $(w_t)_{t \in \mathbb{N}}$ represents the trading strategy parameterized in terms of the amount of wealth invested in the risky asset between date $t-1$ and date t (recall also that $(w_t)_{t \in \mathbb{N}}$ is a predictable stochastic process);
- $(X_t)_{t \in \mathbb{N}}$ is a process describing the gains or losses realized between date $t-1$ and date t;
- $(z_t)_{t \in \mathbb{N}}$ measures the gains or losses realized prior to date t as a fraction of w_t, for each $t \in \mathbb{N}$;
- $(b_t)_{t \in \mathbb{N}}$ is a process of scaling factors specified as $b_t = b_0 \bar{c}_t^{-\gamma}$, with $b_0 \geq 0$.

Letting $b_0 = 0$ allows to recover the classical expected utility functional in an infinite horizon economy (compare with Sect. 6.5). The term $v(X_{t+1}, w_{t+1}, z_t)$ appearing in the preference functional (9.7) represents the utility derived from gains or losses. In particular, the presence of w_{t+1} and z_t implies that this utility does not only depend on the gains/losses realized at the current date (captured by X_{t+1}), but also on the previous performance, which affects how current gains/losses are perceived by the investor.

Let us explain in detail how the processes $(X_t)_{t\in\mathbb{N}}$ and $(z_t)_{t\in\mathbb{N}}$ are defined. Denoting by $(r_t)_{t\in\mathbb{N}}$ the return process of the risky asset, the process $(X_t)_{t\in\mathbb{N}}$ describing the gains/losses realized at each trading date is defined by

$$X_{t+1} := w_{t+1}(r_{t+1} - r_f), \qquad \text{for all } t \in \mathbb{N}.$$

This corresponds to assuming that the reference point (status quo) at date $t+1$ is the amount of wealth invested at the preceding date t in the risky asset (measured by the quantity w_{t+1}) multiplied by the risk free rate r_f. The process $(z_t)_{t\in\mathbb{N}}$ captures the fact that the pain of a loss also depends on the investment performance prior to the loss, with z_t representing a benchmark level. In this sense, $z_t < 1$ or $z_t > 1$ means that the agent has realized gains or losses, respectively, prior to date t, while $z_t = 1$ means that the agent has realized neither prior gains nor losses. The benchmark process $(z_t)_{t\in\mathbb{N}}$ is modeled via

$$z_{t+1} = \beta\left(z_t \frac{\bar{r}}{r_{t+1}}\right) + (1-\beta), \qquad \text{for all } t \in \mathbb{N},$$

where $\beta \in [0,1]$ and $\bar{r}$ is a parameter which will be endogenously determined by requiring that, in equilibrium, the median value of z_t is equal to one.

The function $v(\cdot)$ is then defined as follows, if $z_t \le 1$:

$$v(X_{t+1}, w_{t+1}, z_t) := \begin{cases} w_{t+1}(r_{t+1} - r_f), & \text{for } r_{t+1} \ge z_t r_f, \\ w_{t+1}(z_t - 1)r_f + \lambda w_{t+1}(r_{t+1} - z_t r_f), & \text{for } r_{t+1} < z_t r_f, \end{cases}$$
(9.8)

where $\lambda > 1$. On the other hand, if $z_t > 1$, the function $v(\cdot)$ is defined as

$$v(X_{t+1}, w_{t+1}, z_t) := \begin{cases} w_{t+1}(r_{t+1} - r_f), & \text{for } r_{t+1} \ge r_f, \\ \lambda(z_t)w_{t+1}(r_{t+1} - r_f), & \text{for } r_{t+1} < r_f, \end{cases}$$
(9.9)

where $\lambda < \lambda(z_t) := \lambda + k(z_t - 1)$, for $k > 0$ (this represents the fact that the larger z_t is the more painful subsequent losses will be). Referring to the original paper Barberis et al. [150] for a more detailed explanation of the above specification, we just mention that the specification (9.8)–(9.9) captures three main behavioral aspects:

- If $z_t = 1$, then the investor is more sensitive to reductions in financial wealth than to increases (*loss aversion*). This can be seen from (9.8), noting that for $z_t = 1$ losses are amplified by the factor $\lambda > 1$.

- If $z_t < 1$ (meaning that the investor has accumulated prior gains), then small losses are not heavily penalized, but once the loss exceeds a certain amount (i.e., if $r_{t+1} < z_t r_f$) then it is penalized at a more severe rate. This is reflected in (9.8).
- If $z_t > 1$ (meaning that the investor has previously realized losses), then losses are more heavily penalized, capturing the idea that losses following previous losses are more painful to the investor.

In the context of the above model, Barberis et al. [150] build a one-factor Markov equilibrium, in the sense that the Markov state variable z_t determines the distribution of future stock returns and the risk free rate is constant. Specifically, Barberis et al. [150] assume that the equilibrium price-dividend ratio of the risky asset is given by

$$\frac{p_t^*}{d_t} = f(z_t), \qquad \text{for all } t \in \mathbb{N},$$

for a suitable function $f(\cdot)$. The following proposition, the proof of which is given in Exercise 9.3, characterizes the equilibrium of the economy.

Proposition 9.5 *In the context of the above model (see Barberis et al. [150]), suppose that*

$$\log \delta - \gamma g_c + g_d + \frac{\gamma^2 \sigma_c^2 - 2\gamma\rho\sigma_c\sigma_d + \sigma_d^2}{2} < 0.$$

Then, in equilibrium the risk free rate r_f satisfies

$$r_f = \frac{1}{\delta} e^{\gamma g_c - \gamma^2 \frac{\sigma_c^2}{2}} \tag{9.10}$$

and the function $f(\cdot)$ determining the equilibrium price of the risky asset satisfies, for all $t \in \mathbb{N}$,

$$1 = \delta e^{g_d - \gamma g_c + \frac{\gamma^2 \sigma_c^2 (1-\rho^2)}{2}} \mathbb{E}\left[\frac{1 + f(z_{t+1})}{f(z_t)} e^{(\sigma_d - \gamma\rho\sigma_c)\varepsilon_{t+1}} \middle| \mathscr{F}_t \right]$$
$$+ b_0 \delta \mathbb{E}\left[\hat{v}\left(\frac{1 + f(z_{t+1})}{f(z_t)} e^{g_d + \sigma_d \varepsilon_{t+1}}, z_t \right) \middle| \mathscr{F}_t \right], \tag{9.11}$$

where, for $z_t \leq 1$,

$$\hat{v}(x, z_t) = \begin{cases} x - r_f, & \text{for } x \geq z_t r_f, \\ (z_t - 1)r_f + \lambda(x - z_t r_f), & \text{for } x < z_t r_f; \end{cases}$$

and, for $z_t > 1$,

$$\hat{v}(x, z_t) = \begin{cases} x - r_f, & \text{for } x \geq r_f, \\ \lambda(z_t)(x - r_f), & \text{for } x < r_f. \end{cases}$$

The above model permits to rationalize several asset pricing anomalies discussed in Sects. 7.2 and 7.3. In particular, it is consistent with a low volatility of consumption growth together with a high volatility and a large risk premium for stock returns. Moreover, a low risk free rate can be easily generated by the model. The model predicts a low correlation between consumption growth and stock returns and generates long horizon predictability in stock returns. The crucial feature is that, depending on the previous investment performance, agents exhibit a changing risk aversion (low risk aversion after a market run up and high risk aversion after a market fall), so that expected returns change over time and a risk premium for downside risk is generated. Note that this model differs from habit formation models (see below), since the changes in risk aversion are not due to changes in the level of consumption. Barberis & Huang [149] analyse asset pricing implications of cumulative prospect theory, focusing on the effects of the specification of the weighting function. They show that the CAPM can hold when securities are normally distributed. However, a positively skewed security can be overpriced and earn low average returns.

Prospect theory also allows us to rationalize the so-called *disposition effect*, i.e., the tendency to hold loser assets too long and to sell winner assets too soon (see Shefrin & Statman [1531], Ferris et al. [684], Odean [1380], Gomes [805], Grinblatt & Keloharju [829], Barberis & Xiong [156, 155]). The rationale of this behavior is that, in the case of a winner asset, the investor is in the concave region and, therefore, he is risk averse. On the contrary, in the case of a loser asset, the investor is in the convex region and, therefore, he is risk lover. Intraday empirical evidence supporting loss aversion is provided in Coval & Shumway [503].

The disposition effect may generate underreaction to information and a momentum effect in stock returns, as shown in Grinblatt & Han [827]. More specifically, Grinblatt & Han [827] consider a multi-period equilibrium model with a single risky asset (with supply normalized to one). The fundamental value $(f_t)_{t \in \mathbb{N}}$ of the risky asset is assumed to follow the random walk

$$f_{t+1} = f_t + \varepsilon_{t+1}, \qquad \text{for all } t \in \mathbb{N}, \tag{9.12}$$

where $(\varepsilon_t)_{t \in \mathbb{N}}$ is a sequence of i.i.d. random variables with zero mean. At each date $t \in \mathbb{N}$, public news about the fundamental value f_t arrive just prior to trading. The economy is populated by two classes of agents: rational traders and *prospect theory traders*. It is assumed that prospect theory traders constitute a proportion $\mu \in (0, 1)$ of the overall population. The optimal demand of the risky asset by the two classes of agents is assumed to be of the following form:

$$
\begin{aligned}
\theta_{t+1}^{\text{PT}} &= 1 + b_t \big((f_t - p_t) + \lambda (x_t - p_t) \big), \\
\theta_{t+1}^{\text{RA}} &= 1 + b_t (f_t - p_t),
\end{aligned}
\tag{9.13}
$$

where θ_{t+1}^{PT} and θ_{t+1}^{RA} denote the optimal demand of a prospect theory agent and of a rational agent, respectively, for $t \in \mathbb{N}$. In (9.13), $(p_t)_{t \in \mathbb{N}}$ denotes the market price of the risky asset, $(b_t)_{t \in \mathbb{N}}$ is allowed to be an adapted stochastic process and depends on the choice of the utility function as well as on the structure of the economy. For the purposes of the present discussion, we do not need to specify $(b_t)_{t \in \mathbb{N}}$ further. The process $(x_t)_{t \in \mathbb{N}}$, supposed to be a predictable stochastic process, represents a reference price of the risky asset with respect to which prospect theory investors measure their gains and losses. Finally, the coefficient $\lambda > 0$ represents the relative importance of the capital gain component in the demand of prospect theory investors. The perturbation term appearing in the demand function of prospect theory investors is inversely related to the unrealized profit on the risky asset. As we are going to show, this specification allows us to obtain an explicit description of the deviation of the risky asset market price from the equilibrium price that would prevail if all the investors were rational.

The reference price process $(x_t)_{t \in \mathbb{N}}$ is assumed to satisfy the following relation:

$$x_{t+1} = v_t p_t + (1 - v_t) x_t, \qquad \text{for all } t \in \mathbb{N}, \tag{9.14}$$

capturing the fact that the reference price gets updated as shares are exchanged between investors at each trading date, so that new reference price is a weighted average of the old reference price and the current market price. The process $(v_t)_{t \in \mathbb{N}}$, taking values in the interval $[0, 1]$, represents the weighting factors and captures the mental accounting of prospect theory traders.

Proposition 9.6 *In the context of the above model (see Grinblatt & Han [827]), the following hold:*

(i) the equilibrium market price $(p_t^*)_{t \in \mathbb{N}}$ of the risky asset is given by

$$p_t^* = w f_t + (1 - w) x_t, \qquad \text{for all } t \in \mathbb{N},$$

where $w := 1/(1 + \mu \lambda)$;
(ii) at each date $t \in \mathbb{N}$, the expected return of the risky asset satisfies

$$\mathbb{E}\left[\frac{p_{t+1} - p_t}{p_t} \Big| \mathscr{F}_t \right] = (1 - w) v_t \frac{p_t - x_t}{p_t}. \tag{9.15}$$

Proof Part *(i)* simply follows by aggregating the optimal demands given in (9.13) and equating the aggregate demand with the market supply (which is equal to one at all dates $t \in \mathbb{N}$). In order to prove part *(ii)*, note that

$$\mathbb{E}[p_{t+1} - p_t | \mathscr{F}_t] = w \mathbb{E}[f_{t+1} - f_t | \mathscr{F}_t] + (1 - w) \mathbb{E}[x_{t+1} - x_t | \mathscr{F}_t]$$
$$= w \mathbb{E}[\varepsilon_{t+1} | \mathscr{F}_t] + (1 - w) v_t \mathbb{E}[p_t - x_t | \mathscr{F}_t]$$
$$= (1 - w) v_t (p_t - x_t),$$

where we have used part *(i)* of the proposition together with relations (9.12) and (9.14).   □

Part *(i)* of Proposition 9.6 shows that the risky assets underreact to information about the fundamental value. The degree of underreaction is measured by the term w, which is in turn determined by the proportion of prospect theory traders in the economy and by the parameter λ characterizing the optimal demand of prospect theory traders. The implication is that in equilibrium stocks that have performed well in the past tend to be undervalued and, on the contrary, stocks that have badly performed tend to be overvalued. The smaller the number of prospect theory traders the closer the market price will be to the fundamental value. The right-hand side of (9.15) is related to the (percentage) unrealized capital gain. Relation (9.15) implies that any variable which proxies for the unrealized capital gain (for instance, past returns) has a predictive power on future returns. In this sense, this model is able to explain the profitability of momentum strategies.

In the decision theory literature, *disappointment* or *anticipation* effects have also been considered. Disappointment occurs when an agent assigns more importance to the bad outcomes than to the good outcomes of a lottery, while anticipation refers to the fact that current decisions are affected by suspense and anxiety about the future. Agents may experience disappointment by comparing an outcome with its past expectation. On the other hand, agents may anticipate future utility and, therefore, the current utility will be affected by the expectation of future utility. An axiomatic framework for *disappointment aversion* has been first proposed in Gul [863], including classical expected utility as a special case and rationalizing the Allais paradox. More specifically, preferences are characterized in Gul [863] in terms of a utility function u and a parameter $\beta > -1$ (the case of classical expected utility corresponding to $\beta = 0$). The case $\beta \geq 0$ captures disappointment aversion and it is shown that risk aversion implies disappointment aversion. The effects of disappointment will also be discussed below in the context of recursive preferences.

The asset pricing results obtained in Ang et al. [64], Routledge & Zin [1476], Bonomo et al. [267], Liu & Miao [1225], Fielding & Stracca [700], assuming a utility function capturing disappointment aversion (good outcomes are underweighted with respect to bad outcomes), confirm that disappointment may help to explain the equity premium and the risk free rate puzzles. Furthermore, disappointment aversion can also explain the low participation rate in the market with a counter-cyclical risk aversion. There is evidence that disappointment is priced in the market: Ang et al. [67] show that returns are increasing in downside risk, i.e., the risk that asset returns are more correlated with the market when the latter is falling rather than when it is rising. The impacts of anticipation effects on asset prices have been investigated in Caplin & Lehay [365], where the authors show that *anxiety* (the agent is anxious for future wealth-consumption) contributes to explain both the risk free rate and the equity premium puzzle, due to the time inconsistency in investors' preferences caused by anticipation.

Recursive Preferences

In Chap. 6, we have considered time additive utility functions. As explained in Sect. 7.3, one of the drawbacks of this specification consists in the fact that the relative risk aversion coefficient determines the elasticity of intertemporal substitution of consumption. Time additive preferences have been generalized by introducing *recursive* utilities (also known as generalized expected utility), see Epstein & Zin [649] and Weil [1648]. In particular, one of the main advantages of this preference structure consists in the possibility of separating risk aversion from the elasticity of intertemporal substitution of consumption.

In order to illustrate the theory of recursive preferences, we adopt the same setting introduced in Sect. 6.1 for multi-period optimal consumption-investment problems. We consider an economy with $T \in \mathbb{N}$ trading dates $t \in \{0, 1, \ldots, T\}$ and N traded securities. Trading-consumption strategies are defined as in Definition 6.1 and denoted by (θ, c), with $(\theta_t)_{t=0,1,\ldots,T}$ being an $\mathbb{R}^N$-valued predictable stochastic process representing the trading strategy and $(c_t)_{t=0,1,\ldots,T}$ a non-negative adapted stochastic process representing the consumption stream. As in Sect. 6.1, we define by $\mathscr{C}_t(x)$ the set of all consumption processes $(c_s)_{s=t,\ldots,T}$ which can be financed by some trading strategy starting at date t with wealth x. We shall always assume that trading is done in a self-financing way and, for simplicity, we do not consider the presence of an exogenous income stream.

Extending the classical time additive expected utility function, we assume that preferences are defined with respect to self-financing trading-consumption strategies (c, θ) and are characterized by a time-dependent preference functional $U_t(c, \theta)$ satisfying the following recursive relation:

$$U_t(c, \theta) = v\Big(c_t, \mu_t\big(U_{t+1}(c, \theta)\big)\Big), \qquad \text{for all } t = 0, 1, \ldots, T-1, \qquad (9.16)$$

with $U_T(c, \theta) = u(c_T)$, for some utility function $u(\cdot)$. In (9.16), the term c_t represents the current consumption at date t, while $\mu_t(U_{t+1}(c, \theta))$ represents the certainty equivalent, as measured at date t, of the future utility at the following date $t + 1$. More precisely, the functional μ_t is specified as

$$\mu_t\big(U_{t+1}(c, \theta)\big) = \tilde{u}^{-1}\Big(\mathbb{E}\big[\tilde{u}\big(U_{t+1}(c, \theta)\big)|\mathscr{F}_t\big]\Big), \qquad \text{for all } t = 0, 1, \ldots, T-1, \tag{9.17}$$

for some increasing and concave function $\tilde{u}(\cdot)$. The function $v : \mathbb{R}_+ \times \mathbb{R} \to \mathbb{R}$ appearing in (9.16) is also assumed to be increasing and concave. The role of the function v consists in aggregating the utility derived from current consumption and the future utility. With this specification, risk aversion and intertemporal substitution of consumption can be easily separated. Indeed, risk aversion is encoded in the functional μ_t, while the propensity of substituting consumption intertemporally is determined by the function v. Note that (9.16) implies that the utility associated to a consumption stream $(c_t)_{t=0,1,\dots,T}$ can be computed as the solution to a backward stochastic difference equation. As shown in Exercise 9.4, the time additive expected utility of the form (6.6) represents a special case of (9.16).

In Epstein & Zin [649, 651], the following specification of recursive preferences has been proposed (in the original formulation, an infinite horizon economy was considered):

$$v(c,y) = \left((1-\delta)c^{1-\varrho} + \delta y^{1-\varrho}\right)^{\frac{1}{1-\varrho}} \quad \text{and} \quad \tilde{u}(x) = \frac{x^{1-\alpha}}{1-\alpha}, \tag{9.18}$$

for $\varrho \in (0,1)$, $\delta \in (0,1)$ and $\alpha \in (0,1)$. With this specification, the recursive relation (9.16) can be rewritten in the following form:

$$U_t(c,\theta) = \left((1-\delta)c_t^{1-\varrho} + \delta \left(\mathbb{E}[U_{t+1}(c,\theta)^{1-\alpha}|\mathscr{F}_t]\right)^{\frac{1-\varrho}{1-\alpha}}\right)^{\frac{1}{1-\varrho}}, \tag{9.19}$$

for all $t = 0, 1, \dots, T-1$. We also assume that $u(c_T) = c_T^{1-\alpha}/(1-\alpha)$. The parameter α represents the coefficient of relative risk aversion, while the parameter $1/\varrho$ represents the elasticity of intertemporal substitution of consumption (hence, the preference functional (9.19) exhibits constant elasticity of intertemporal substitution of consumption and constant relative risk aversion). As shown in Exercise 9.4, this specification allows to recover the classical time additive expected utility representation (6.6) with a power utility function if $\alpha = \varrho$ (i.e., relative risk aversion coincides with the inverse of the elasticity of intertemporal substitution of consumption). Moreover, taking the limit for $\alpha \to 1$, a logarithmic utility specification is obtained (in a time additive form if $\alpha = \varrho = 1$). Finally, the parameter δ plays the role of a (subjective) discount factor.

An important feature of the recursive utility functional (9.19) is that it allows to model different preferences with respect to the resolution of uncertainty. Indeed, if $\alpha = \varrho$ then the agent is indifferent between early and late resolution of uncertainty, while, if $\alpha > \varrho$, then early resolution of uncertainty is preferred (as it seems intuitively plausible) and, if $\alpha < \varrho$, then late resolution of uncertainty is preferred (see Epstein & Zin [649]).

In order to analyse optimal investment-consumption problems under recursive preferences, proceeding similarly as in Sect. 6.1, let us introduce the *value function* V defined by

$$V(x,t) := \sup_{c \in \mathscr{C}_t(x)} U_t(c,\theta) = \sup_{c \in \mathscr{C}_t(x)} v\Big(c_t, \mu_t\big(U_{t+1}(c,\theta)\big)\Big), \tag{9.20}$$

for all $t = 0, 1, \dots, T-1$ and for an arbitrary initial wealth $x \in \mathbb{R}_+$, with $V(x,T) = u(x)$. Note that, if the function v is assumed to be homogeneous of degree one, as in (9.18), then the value function V is proportional to wealth (see Epstein & Zin [649, 651]). The value function satisfies a suitable version of the Bellman optimality principle, as shown in the following proposition, which can be proved by relying on the same arguments given in the proof of Propositions 6.3 and 6.4.

Proposition 9.7 *Consider the recursive preference functional of the general form (9.16) and the corresponding value function V defined in (9.20). Then, for every* $(x, t) \in \mathbb{R}_+ \times \{0, 1, \ldots, T-1\}$, *the value function V satisfies the following recursive relation:*

$$V(x, t) = \sup v\Big(c_t, \mu_t\big(V(W_{t+1}(\theta), t+1)\big)\Big), \tag{9.21}$$

where the supremum is taken with respect to all $\mathscr{F}_t$-*measurable random variables* $(\theta_{t+1}, c_t) \in \mathbb{R}^N \times \mathbb{R}_+$ *satisfying* $\theta_{t+1}^\top S_t + c_t = x$ *and where* $W_{t+1}(\theta) = \theta_{t+1}^\top (S_{t+1} + D_{t+1})$. *Moreover, in correspondence of an optimal solution* (θ^*, c^*), *it holds that*

$$V(W_t^*, t) = v\Big(c_t^*, \mu_t\big(V(W_{t+1}^*, t+1)\big)\Big), \qquad \text{for all } t = 0, 1, \ldots, T-1, \tag{9.22}$$

with $W_t^* := W_t(\theta^*)$, *and the following envelope condition holds:*

$$\frac{\partial}{\partial c_t} v\Big(c_t^*, \mu_t\big(V(W_{t+1}^*, t+1)\big)\Big) = V'(W_t^*, t), \qquad \text{for all } t = 0, 1, \ldots, T-1,$$

with $V'(\cdot, t)$ *denoting the first derivative of the value function with respect to its first argument, for* $t = 0, 1, \ldots, T$.

Formula (9.21) corresponds to the equation of the dynamic programming principle in the context of general recursive preferences. According to (9.22), in correspondence of the optimal trading-consumption strategy (θ^*, c^*), the optimal utility at each date t corresponds to the utility derived from current consumption at date t and from the certainty equivalent of the future date's optimal utility, aggregated by the function v.

Proposition 9.7 implicitly characterizes the optimal trading-consumption strategy (θ^*, c^*) maximizing the recursive preference functional (9.16). In particular, in the case of recursive preferences of the form (9.19), explicit Euler conditions can be formulated, as shown in the following proposition, the proof of which is given in Exercise 9.5.

Proposition 9.8 *Consider the recursive preference functional defined in (9.19), with parameters* $\varrho \in (0, 1)$, $\delta \in (0, 1)$ *and* $\alpha \in (0, 1)$ *and with* $U_T(c, \theta) = u(c) := c(1 - \delta)^{1/(1-\varrho)}$. *Denote by* $(W_t^*)_{t=0,1,\ldots,T}$ *the optimal wealth process associated to the trading-consumption strategy* (θ^*, c^*) *maximizing the recursive preference functional* $U_0(c, \theta)$ *defined in (9.19). Then the following Euler condition holds:*

$$1 = \delta \mathbb{E}\left[\left(\frac{V(W_{t+1}^*, t+1)}{\mu_t\big(V(W_{t+1}^*, t+1)\big)}\right)^{\varrho-\alpha} \left(\frac{c_{t+1}^*}{c_t^*}\right)^{-\varrho} r_{t+1}^n \,\Big|\, \mathscr{F}_t\right], \tag{9.23}$$

for all $t = 0, 1, \ldots, T - 1$ and $n = 1, \ldots, N$. In particular, assuming that there exists a risk free asset with constant rate of return $r_f > 0$, it holds that

$$\mathbb{E}\left[\left(\frac{V(W_{t+1}^*, t+1)}{\mu_t(V(W_{t+1}^*, t+1))}\right)^{\varrho-\alpha} \left(\frac{c_{t+1}^*}{c_t^*}\right)^{-\varrho} (r_{t+1}^n - r_f)\bigg| \mathscr{F}_t\right] = 0, \tag{9.24}$$

for all $t = 0, 1, \ldots, T - 1$ and $n = 1, \ldots, N$. Moreover, the optimal wealth process $(W_t^*)_{t=0,1,\ldots,T}$ satisfies

$$W_t^* = \frac{V(W_t^*, t)^{1-\varrho}}{1 - \delta}(c_t^*)^\varrho, \qquad \text{for all } t = 0, 1, \ldots, T, \tag{9.25}$$

and the return process $(r_t^*)_{t=1,\ldots,T}$ on the optimal wealth process satisfies

$$r_{t+1}^* := \frac{W_{t+1}^*}{W_t^* - c_t^*} = \frac{1}{\delta}\left(\frac{V(W_{t+1}^*, t+1)}{\mu_t(V(W_{t+1}^*, t+1))}\right)^{1-\varrho} \left(\frac{c_{t+1}^*}{c_t^*}\right)^\varrho, \tag{9.26}$$

for all $t = 0, 1, \ldots, T - 1$. Therefore, assuming the existence of a risk free asset with constant rate of return $r_f > 0$, the Euler condition (9.24) can be rewritten as

$$\mathbb{E}\left[\left(\frac{c_{t+1}^*}{c_t^*}\right)^{-\varrho\gamma} (r_{t+1}^*)^{\gamma-1}(r_{t+1}^n - r_f)\bigg| \mathscr{F}_t\right] = 0, \tag{9.27}$$

for all $t = 0, 1, \ldots, T$ and $n = 1, \ldots, N$, with $\gamma := (1 - \alpha)/(1 - \varrho)$.

It is interesting to compare Proposition 9.8 with the results obtained in the context of optimal investment-consumption problems for classical time additive expected utility functions, as considered in Sect. 6.1, in the case of a power utility function. Starting with the Euler conditions (9.23)–(9.24), note first that, in the special case where $\alpha = \varrho$, then the first term inside the conditional expectation vanishes and the conditions thus reduce to the classical Euler conditions obtained for a time additive power utility function (compare with Exercise 6.6). This is a consequence of the fact that (9.19) reduces to a classical time additive power utility function if $\alpha = \varrho$, as shown in Exercise 9.4.

While Proposition 9.8 has been established for $\alpha \in (0, 1)$ and $\varrho \in (0, 1)$, it can be shown that if $\alpha = 1$ and $\varrho \neq 1$, the Euler condition (9.27) reduces to

$$\mathbb{E}\left[(r_{t+1}^*)^{\gamma-1}(r_{t+1}^n - r_f)\big| \mathscr{F}_t\right] = 0, \tag{9.28}$$

for all $t = 0, 1, \ldots, T$ and $n = 1, \ldots, N$ (see Epstein & Zin [651] and Attanasio & Weber [86]). When $\alpha = 1$, Giovannini & Weil [785] have shown that the portfolio choice is myopic (Samuelson's result) for any value of ϱ. On the other hand, if $\varrho = 1$ then c_t^*/W_t^* is constant over time for any value of α (see Giovannini & Weil [785] and compare also with relation (9.25)).

Relation (9.25) gives a precise link between the optimal wealth process and the optimal consumption process. In particular, due to the specification (9.18), it can be shown that the value function is linear with respect to x (see Epstein & Zin [651]). As a consequence, relation (9.25) implies that the optimal consumption c_t^* is proportional to the current optimal wealth W_t^*. This result is analogous to the case of a classical time additive power utility function and has to be compared with relation (6.26) in Proposition 6.8.

In the general case where $\alpha \neq \varrho$, the Euler conditions (9.23)–(9.24) show that the stochastic discount factor $(M_t)_{t=0,1,\ldots,T}$ associated to the recursive preferences (9.19) is given by

$$M_t = \delta^t \left(\frac{c_t^*}{c_0^*} \right)^{-\varrho} \prod_{s=0}^{t-1} \left(\frac{V(W_{s+1}^*, s+1)}{\mu_s\big(V(W_{s+1}^*, s+1)\big)} \right)^{\varrho-\alpha}, \qquad \text{for all } t = 1, \ldots, T,$$
$$(9.29)$$

with $M_0 = 1$ (see also Exercise 9.5). Recall that the parameter ϱ represents the reciprocal of the elasticity of intertemporal substitution of consumption. In comparison with the classical structure (6.64) of the stochastic discount factor obtained with a time additive utility function, (9.29) shows that the stochastic discount factor under recursive preferences depends not only on the optimal consumption process but also on the certainty equivalent of future's utility. Moreover, as a consequence of (9.26)–(9.27), the stochastic discount factor admits the following alternative representation:

$$M_t = \delta^{\gamma t} \left(\frac{c_t^*}{c_0^*} \right)^{-\varrho\gamma} \prod_{s=0}^{t} (r_s^*)^{\gamma-1}, \qquad \text{for all } t = 1, \ldots, T, \qquad (9.30)$$

with $M_0 = 1$. This shows that the stochastic discount factor depends on the return process $(r_t^*)_{t=1,\ldots,T}$ associated to the optimal portfolio in the maximization of the recursive preference functional (9.19). Relation (9.30) also shows that the stochastic discount factor associated to the recursive preference functional (9.19) is a geometric average of the stochastic discount factor associated to a standard time additive power utility function and the stochastic discount factor associated to the logarithmic case (compare with relation (9.28) above). The weight between these two components is determined by the parameter γ.

As a consequence of the Euler condition (9.27), asset risk premia depend on the covariance between asset excess returns and a geometric average of the growth rate of the optimal consumption process and of the return on the optimal portfolio. In particular, when $\alpha = \varrho$ (so that $\gamma = 1$), asset risk premia only depend on the covariance between the excess returns and the growth rate of the optimal consumption process, thus recovering the classical CCAPM result with a time additive power utility function (see Sect. 6.4). On the other hand, if $\alpha = 1$ (so that $\gamma = 0$), then asset risk premia are determined similarly as in the CAPM, with the reciprocal of the return on the optimal portfolio playing the role of the single risk factor. For other values of γ, both the consumption growth rate and the optimal return are necessary for determining asset risk premia. More precisely, relation (9.27) implies that

$$\mathbb{E}[r_{t+1}^n|\mathscr{F}_t] - r_f = -r_f \operatorname{Cov}\left(\left(\frac{c_{t+1}^*}{c_t^*}\right)^{-\varrho\gamma}(r_{t+1}^*)^{\gamma-1}, r_{t+1}^n \middle| \mathscr{F}_t\right), \qquad (9.31)$$

for all $t = 0, 1, \ldots, T-1$ and $n = 1, \ldots, N$. Moreover, by approximating the geometric average with the arithmetic average, we obtain

$$\mathbb{E}[r_{t+1}^n|\mathscr{F}_t] - r_f \approx r_f(1-\gamma) \operatorname{Cov}(r_{t+1}^*, r_{t+1}^n|\mathscr{F}_t) + r_f\varrho\gamma \operatorname{Cov}\left(\frac{c_{t+1}^*}{c_t^*}, r_{t+1}^n \middle| \mathscr{F}_t\right). \qquad (9.32)$$

Once more, if $\gamma = 1$, then we recover the classical asset pricing relation of the CCAPM, while, if $\gamma = 0$, relation (9.32) reduces to the CAPM relation. Moreover, under suitable assumptions on the joint conditional distribution of the optimal consumption growth rate and of the returns, relation (9.32) can be shown to hold in an exact form, as shown in the following proposition, the proof of which is given in Exercise 9.6 (compare also with Munk [1363, Theorem 9.4]).

Proposition 9.9 *Under the same assumptions of Proposition 9.8, let us denote by $(r_t^*)_{t=1,\ldots,T}$ and by $(c_t^*)_{t=0,1,\ldots,T}$ the return process on the optimal portfolio and the optimal consumption process, respectively, associated to the trading-consumption strategy (θ^*, c^*) maximizing the recursive preference functional (9.19). Suppose that, for every $t = 0, 1, \ldots, T-1$, conditionally on $\mathscr{F}_t$, it holds that*

$$\log\left(\frac{c_{t+1}^*}{c_t^*}\right) \sim \mathscr{N}(\mu_c, \sigma_c^2) \qquad \text{and} \qquad \log r_{t+1}^* \sim \mathscr{N}(\mu_r, \sigma_r^2),$$

with $\operatorname{Cov}(\log(c_{t+1}^*/c_t^*), \log r_{t+1}^*|\mathscr{F}_t) = \sigma_{cr}$, for all $t = 0, 1, \ldots, T-1$. Then, the equilibrium risk free return r_f is given by

$$\log r_f = -\log \delta + \varrho\mu_c + \frac{\sigma_r^2}{2}(\gamma-1) - \frac{\gamma\varrho^2\sigma_c^2}{2}. \qquad (9.33)$$

Suppose furthermore that $(\log(c_{t+1}^*/c_t^*), \log r_{t+1}^1, \ldots, \log r_{t+1}^N)$ is jointly normally distributed conditionally on $\mathscr{F}_t$, for all $t = 0, 1, \ldots, T-1$. Then the asset risk premia in equilibrium satisfy

$$\mathbb{E}[\log r_{t+1}^n|\mathscr{F}_t] - \log r_f = -\frac{\operatorname{Var}(\log r_{t+1}^n|\mathscr{F}_t)}{2} - (\gamma-1)\operatorname{Cov}\left(\log r_{t+1}^*, \log r_{t+1}^n|\mathscr{F}_t\right)$$

$$+ \varrho\gamma \operatorname{Cov}\left(\log\left(\frac{c_{t+1}^*}{c_t^*}\right), \log(r_{t+1}^n) \middle| \mathscr{F}_t\right), \qquad (9.34)$$

for all $t = 0, 1, \ldots, T-1$ and $n = 1, \ldots, N$.

In particular, if $\alpha = \varrho$ (so that $\gamma = 1$), then the result of the above proposition is in line with Proposition 6.38 (compare the asset pricing relations (6.78) and (9.34)). In the case of general distributions for the return and consumption processes, the asset pricing relations stated in Proposition 9.9 hold in an approximate form (see Campbell & Viceira [354, Section III]).

As we have already remarked, an important feature of the recursive preference functionals (9.16)–(9.19) consists in the separation between risk aversion and the elasticity of intertemporal substitution of consumption. This feature seems to be promising in order to provide an explanation for the equity premium puzzle and the risk free rate puzzle (see Sect. 7.3). In particular, relation (9.33) shows that the equilibrium risk free rate is determined not only by the degree of risk aversion (encoded in the parameter γ), but also by the average growth rate of the consumption process and by the elasticity of intertemporal substitution (measured by the parameter ϱ). As a consequence, a high degree of risk aversion is no longer incompatible with a low risk free rate. This allows us to rationalize the risk free rate puzzle. Moreover, a high degree of risk aversion is no longer the only way of explaining the equity premium puzzle, since asset premia depend on both the covariance with consumption growth and the covariance with the return on a market portfolio. Unfortunately, as we are going to discuss below, recursive preferences have been empirically shown to provide only a partial resolution to the equity premium puzzle.

Assuming i.i.d. aggregate consumption growth rates, Weil [1648], Kocherlakota [1110], Epstein & Melino [642] show that a representative agent economy with a utility functional of the type (9.19) is observationally equivalent to a classical economy with a time additive expected utility function. The parameters α and ϱ cannot be identified separately. On the other hand, assuming a Markovian process for the evolution of the aggregate consumption growth rates, Wang [1645] obtains a non-equivalence result.

It has been empirically observed that the equity premium puzzle cannot be easily explained by simply assuming recursive preferences, see Weil [1648], Kandel & Stambaugh [1069], Kocherlakota [1110, 1112]. Unlike the risk free rate puzzle, the equity premium puzzle does not depend on the relation between risk aversion and the elasticity of intertemporal substitution of consumption imposed by the classical time additive expected utility framework. The equity premium puzzle originates from the fact that the consumption process is too smooth and, therefore, the degree of risk aversion needs to be dramatically high in order to reproduce in equilibrium the risk premia observed historically. Instead, recursive preferences may explain the risk free rate puzzle, as shown in Kocherlakota [1112]. Indeed, as we have already mentioned, the essential ingredient of the risk free rate puzzle is the connection between risk aversion and the intertemporal substitution of consumption. Considering recursive preferences of the form (9.19), both risk aversion and the elasticity of intertemporal substitution can be simultaneously high. As a consequence, in order to generate a low risk free rate in equilibrium, it suffices to choose an appropriate value of the elasticity of intertemporal substitution of consumption. However, the case of the equity premium puzzle is different. In order to fit both the risk free rate and the risk premium, Weil [1648] need to choose a relative risk aversion coefficient around 45 and an elasticity of intertemporal substitution equal to 0.10 (higher than 1/45). As shown in Kandel & Stambaugh [1069], a low elasticity of intertemporal substitution generates predictable asset returns and a high volatility of returns. Volatility is determined primarily by the elasticity of intertemporal substitution. Kandel & Stambaugh [1069] manage to explain the risk free rate puzzle as well as the equity premium puzzle, but find it difficult to reproduce first and second moments of returns.

A more positive evidence on the possibility of explaining the above phenomena by relying on recursive preferences has been provided in Attanasio & Vissing-Jørgensen [85] and Guvenen [867, 868] considering limited stock market participation and heterogeneous agents. In Bansal & Yaron [139], it is shown that a representative agent economy with recursive preferences of the type (9.19) can simultaneously explain the equity premium puzzle, the risk free rate puzzle and the volatility of equity returns. In their model, the consumption and the dividend growth rates contain a small long run predictable component and a fluctuating economic uncertainty (consumption volatility). As argued in Bansal & Yaron [139], in order to reproduce the above phenomena, it is important to consider both recursive preferences and suitable dynamics for the consumption and dividend growth rates (see however Constantinides & Ghosh [497] for some negative evidence concerning the estimation of the Bansal & Yaron [139] model). In a related context, positive evidence for models with recursive preferences has been provided in Ai [30] and Lettau et al. [1197], showing that a plausible level of risk aversion is consistent with the observed risk premia. Moreover, long run risk and recursive preferences allow also to explain risk premia anomalies cross-sectionally (see Bansal et al. [134, 135], Hansen et al. [889], Malloy et al. [1291]).

The implication of the Euler conditions stated in Proposition 9.8 have been tested in Epstein & Zin [651] on consumption and monthly returns time series by applying a methodology similar to that employed in Hansen & Singleton [895]. It is shown that the elasticity of intertemporal substitution is typically small (thus confirming the result of Hall [880]), risk preferences do not differ statistically from the logarithmic case and investors prefer a late resolution of uncertainty. The empirical evidence reported in Epstein & Zin [651] is in favor of recursive preferences and leads to a rejection of the hypothesis of a time additive utility function. However, Epstein & Zin [651] report that the performance of the model and of the tests is sensitive to the choice of the instrumental variables. In Jorion & Giovannini [1048], it has been empirically shown that a recursive utility function does not improve the performance of a time additive utility function.

Empirical tests of intertemporal asset pricing models require consumption time series. Consumption data are typically measured with errors and are aggregated. This may have serious consequences on asset pricing tests, with a bias towards rejection. Campbell [332], assuming that asset returns and news about future returns are jointly log-normally distributed and employing a log-linear approximation of the budget constraint (see Exercise 9.7), replaces consumption with wealth in the asset pricing formulae obtained above. The approximation is accurate when the variation of the consumption-wealth ratio is small. Similarly as in Proposition 9.9, the logarithmic risk premium of an asset is related to the variance of the logarithmic return of the asset and to its covariances with the logarithmic return of invested wealth (market portfolio) and with news about future returns on invested wealth (changing investment opportunities). Under suitable conditions (notably in the case $\alpha = 1$ or in the presence of constant investment opportunities), the last term can be ignored, thus obtaining an approximate asset pricing relation where the market risk is the single risk factor. Otherwise, a multi-factor asset pricing model is obtained, with the market portfolio (and not consumption) being one of the risk factors. This result can also be derived in the presence of heteroskedasticity.

In Epstein & Zin [650], a recursive utility characterized by *first order risk aversion* has been proposed. This type of utility function, introduced in Segal & Spivak [1514], is such that the risk premium for a small gamble is proportional to its standard deviation rather than to its variance (compare with Sect. 2.2). Since the aggregate consumption process is smooth, its standard deviation is typically larger than its variance. This type of utility functional seems to be promising in order to explain the equity premium and the risk free rate puzzles. Indeed, unlike the case of a classical utility function, a high equity premium is compatible with a low risk free rate. By calibrating the model, the authors show that a low risk free rate is compatible with a risk premium larger than the one associated with a standard utility function but is still significantly lower than the historical value, thus providing only a partial resolution of the equity premium puzzle. In Bekaert et al. [187], an equilibrium model with a utility function characterized by first order risk aversion has been adopted to model time varying risk premia and to reproduce the observed predictability of asset returns. Increasing first order risk aversion substantially increases the variance of the risk premia, but the effect on return predictability does not suffice to fit the observed time series.

Portfolio choices with recursive preferences have been analysed by Campbell & Viceira [354], Campbell et al. [340], Chacko & Viceira [387], Gomes & Michaelides [808], Cocco et al. [456] in various settings, also taking into account the presence of mean reversion and stochastic volatility. In particular, as shown in Exercise 9.7, in Campbell & Viceira [354] an approximate formula for the optimal demand of the risky asset has been obtained by taking a log-linear approximation of the Euler condition and of the self-financing condition, assuming recursive preferences of the form (9.19). The approximate formula derived in Campbell & Viceira [354] is composed of two terms: the first term is related to the risk premium of the asset, while the second term reflects the intertemporal hedging demand (see Exercise 9.7 for more details).

Habit Formation and Durable Goods

Besides recursive preferences, another generalization of the classical time additive expected utility theory consists in relaxing the separability of the preference functional with respect to time. In other words, we can consider utility functionals where the current utility at date t depends not only on consumption c_t at date t, but also on a process z_t related to consumption at the previous dates. More specifically, let us consider a preference functional of the following form, defined with respect to a consumption process $c = (c_t)_{t=0,1,\dots,T}$:

$$\sum_{t=0}^{T} \delta^t \mathbb{E}[u(c_t, z_t)]. \tag{9.35}$$

The process $(z_t)_{t=0,1,\dots,T}$ can be specified as a function of the previous consumption or can also play the role of an exogenous factor influencing the preferences (this is for instance the case of the "*Keeping up with the Joneses*" preferences discussed below). For the sake of the present discussion, let us suppose that $(z_t)_{t=0,1,\dots,T}$ is given by a weighted average of past consumption, i.e.,

$$z_t = b^t z_0 + a \sum_{s=1}^{t} b^s c_{t-s}, \qquad \text{for all } t = 1, \dots, T, \tag{9.36}$$

with $a > 0$, $0 < b < 1$ and z_0 a given initial value. Equivalently, for a given consumption process $(c_t)_{t=0,1,\dots,T}$, the process $(z_t)_{t=0,1,\dots,T}$ can be characterized as the solution to the recursive relation

$$z_t = b z_{t-1} + a b c_{t-1}, \qquad \text{for all } t = 1, \dots, T,$$

with initial value z_0. In a continuous time formulation, the specification (9.36) can be naturally generalized to $z_t = e^{-bt} z_0 + a \int_0^t e^{-b(t-s)} c_s ds$, for some $a, b > 0$.

The utility function u appearing in (9.35) now depends on two arguments. Depending on the behavior of u with respect to the second argument, two different situation arise (in the following, we denote by $\partial_z u(c, z)$ the first partial derivative of the function u with respect to its second argument):

- If $\partial_z u(c, z) < 0$, then the preference functional (9.35) represents *habit formation* (or *habit persistence*): an agent compares his current consumption with the level of past consumption. If the agent has experienced in the past a high level of consumption, then he continues to desire a high level of consumption and the agent will be disappointed if he has to reduce the current level of consumption. In other words, we can think of the process $(z_t)_{t=0,1,\dots,T}$ as representing a *standard of living*. The discounting of past consumption in (9.36) captures the fact that the effect of past consumption is declining over time and that the habit mostly depends on recent rather than remote consumption.

- If $\partial_z u(c, z) > 0$, then the preference functional (9.35) represents the presence of *durable goods* and we can think of the process $(z_t)_{t=0,1,\ldots,T}$ as modeling durable goods that an agent accumulates over time (a typical example being housing).

According to these two interpretations, we can have complementarity or substitutability effects between consumption at different points in time. Indeed, if there exists a complementarity (substitutability, resp.) effect between consumption at different dates, then the second cross derivative of the function u will be negative (positive, resp.). When goods are durable, consumptions at nearby dates are almost perfect substitutes (see also below).

As suggested in Constantinides [493], a preference functional of the form (9.35) with a habit formation process $(z_t)_{t=0,1,\ldots,T}$ allows to generate a wedge between the coefficient of relative risk aversion and the reciprocal of the elasticity of intertemporal substitution of consumption, thus making possible the resolution of several asset pricing puzzles, notably the risk free rate puzzle and the equity premium puzzle (see Constantinides [493]). The utility function proposed in Constantinides [493] is of the following form, for all $t = 0, 1, \ldots, T$:

$$
u(c_t, z_t) = \begin{cases} \frac{(c_t - z_t)^{1-\alpha}}{1-\alpha}, & \text{if } c_t \geq z_t, \\ -\infty, & \text{if } c_t < z_t, \end{cases} \tag{9.37}
$$

with $(z_t)_{t=0,1,\ldots,T}$ representing the habit formation process, given by a weighted average of past consumption, and $\alpha > 0$. By adopting this specification, in a continuous time formulation, Constantinides [493] shows that the risk premium observed historically can be fitted with $\alpha < 3$ (see however Chapman [410] and Otrok et al. [1388] for a critical analysis of this result). The specification (9.37) is sometimes called the *difference model*. A similar result was obtained in Abel [4] assuming

$$
u(c_t, z_t) = \frac{(c_t / z_t^\gamma)^{1-\alpha}}{1-\alpha}, \tag{9.38}
$$

with $\gamma \in [0, 1]$. Specification (9.38) is sometimes called the *ratio model*. Note that the difference model generates a time-varying coefficient of relative risk aversion, while the ratio model generates a constant coefficient of relative risk aversion. Indeed, in the difference model it holds that

$$
r_u^r(c_t, z_t) = -c_t \frac{\partial_{cc} u(c_t, z_t)}{\partial_c u(c_t, z_t)} = \frac{\alpha c_t}{c_t - z_t},
$$

while in the ratio model it holds that

$$
r_u^r(c_t, z_t) = -c_t \frac{\partial_{cc} u(c_t, z_t)}{\partial_c u(c_t, z_t)} = \alpha,
$$

for all $t = 0, 1, \ldots, T$, where we denote by ∂_c and ∂_{cc} the first and the second partial derivatives, respectively, of the function u with respect to its first argument.

For the sake of the present discussion, let us consider a preference functional of the form (9.35), defined with respect to a habit formation process satisfying (9.36), with a general utility function u. Consider a multi-period economy with $N + 1$ traded assets, where asset 0 is a risk free asset paying the constant rate of return r_f and the remaining N assets are risky and deliver random dividends. Similarly as in Sect. 6.1, we consider self-financing trading-consumption strategies (θ, c). As shown in Exercise 9.8, the Euler condition associated to the optimal investment-consumption problem under habit formation preferences can be written in the following form (compare also with Proposition 6.4):

$$\mathbb{E}\left[\frac{\partial_c u(c^*_{t+1}, z^*_{t+1}) + a \sum_{s=1}^{T-t-1}(\delta b)^s \mathbb{E}[\partial_z u(c^*_{t+1+s}, z^*_{t+1+s})|\mathscr{F}_{t+1}]}{\partial_c u(c^*_t, z^*_t) + a \sum_{s=1}^{T-t}(\delta b)^s \mathbb{E}[\partial_z u(c^*_{t+s}, z^*_{t+s})|\mathscr{F}_t]}(r^n_{t+1} - r_f)\Big|\mathscr{F}_t\right] = 0,$$
(9.39)

for all $t = 0, 1, \ldots, T - 1$ and $n = 1, \ldots, N$, where we denote by $(z^*_t)_{t=0,1,\ldots,T}$ the habit formation process associated to the optimal consumption stream $(c^*_t)_{t=0,1,\ldots,T}$. The above Euler condition is rather complicated, mainly due to the endogeneity of the habit formation process $(z^*_t)_{t=0,1,\ldots,T}$.

A simpler Euler condition can be obtained by assuming that the habit formation process $(z_t)_{t=0,1,\ldots,T}$ is simply given by the level of consumption at the previous date, i.e., $z_t = c_{t-1}$, for all $t = 1, \ldots, T$. Following Kocherlakota [1112], suppose that the utility function $u(c, z)$ appearing in (9.35) is given by

$$u(c_t, z_t) = \frac{(c_t - \lambda c_{t-1})^{1-\alpha}}{1-\alpha},$$
(9.40)

for $\lambda > 0$. The economic intuition is that an agent having consumed c_{t-1} at date $t - 1$ has become familiar with that level of consumption and, therefore, he compares his current consumption at date t with the previous level of consumption. In other words, lagged consumption endogenously determines a subsistence level. Note that the instantaneous utility at date t is a decreasing function of consumption at the previous date $t - 1$ (provided that $c_t - \lambda c_{t-1} > 0$). See also Munk [1363, Example 6.3] for a simple two-period example of habit formation preferences of the type (9.40). As shown in Exercise 9.8, in the case of the utility function (9.40), the Euler condition (9.39) reduces to

$$\mathbb{E}\left[\frac{(c^*_{t+1} - \lambda c^*_t)^{-\alpha} - \delta\lambda\mathbb{E}[(c^*_{t+2} - \lambda c^*_{t+1})^{-\alpha}|\mathscr{F}_{t+1}]}{(c^*_t - \lambda c^*_{t-1})^{-\alpha} - \delta\lambda\mathbb{E}[(c^*_{t+1} - \lambda c^*_t)^{-\alpha}|\mathscr{F}_t]}(r^n_{t+1} - r_f)\Big|\mathscr{F}_t\right] = 0,$$
(9.41)

for all $t = 0, 1, \ldots, T - 1$ and $n = 1, \ldots, N$. Observe that both Euler conditions (9.39) and (9.41) depend on the investor's ability to predict future consumption growth. This also implies that the stochastic discount factor depends on the individual information.

In Sect. 7.3, we have observed that the aggregate consumption time series is typically smooth, an observation which does not agree with several asset pricing results obtained with a time additive utility function. Sundaresan [1580] has shown that a utility function with habit persistence generates a consumption process which is less sensitive to wealth shocks than the one obtained with a time additive utility function (excess smoothness of consumption). This result has been confirmed in Detemple & Zapatero [567] for a generic utility function $u(c, z)$ decreasing with respect to the second argument (habit formation), under suitable conditions. Similar results are also obtained assuming that the instantaneous utility depends negatively on the current value of the habit process and on the conditional expectation of future habit, see Antonelli et al. [73].

In Detemple & Zapatero [567] and Sundaresan [1580], two-factor and multi-factor versions of the CCAPM are derived. In general, the presence of habit formation leads to an increase of the risk premium in comparison to the classical case. Moreover, the elasticity of intertemporal substitution becomes more volatile. However, a resolution of the risk premium puzzle comes at the expense of a rather poor performance in fitting the variance of asset returns and, in particular, of the short term interest rate, see Heaton [924], Heaton & Lucas [927], Cochrane [463]. Habit formation preferences help to resolve the risk free rate puzzle, since the introduction of habit persistence allows to disentangle risk aversion from the elasticity of intertemporal substitution, see Daniel & Marshall [519] and Kocherlakota [1112].

Consumption and portfolio choices under habit formation preferences have been analysed in Gomes & Michaelides [808], Polkovnichenko [1426], Díaz et al. [572]. In Alessie & Lusardi [39], a closed-form solution is derived, with optimal current consumption also depending on past consumption. The optimal consumption problem in the presence of durable goods has been investigated in several papers, see Dunn & Singleton [602], Eichenbaum & Hansen [633], Cuoco & Liu [508], Grossman & Laroque [844], Hindy & Huang [943], Campbell & Cocco [341], Cocco [455], Yao & Zhang [1664], Yogo [1665], Villaverde & Krueger [1624], Bertola et al. [211], Flavin & Nakagawa [716]. In some of these models, there exists a single durable good (typically housing), while in other models there are both durable and non-durable goods. In the first case, the utility function can be a function only of z_t.

Similarly as in the habit formation case, durability introduces a wedge between risk aversion and the elasticity of intertemporal substitution. However, durable goods and habit formation have different (and sometimes contrary) asset pricing implications. In Hindy & Huang [943], the risk premium in the presence of durable goods turns out to be smaller than in the classical case. As a matter of fact, when the levels of consumption at nearby dates are substitutes, agents tend to be less risk averse. In Grossman & Laroque [844], it is shown that, in the presence of an illiquid durable good, the optimal consumption is not a smooth function of wealth. Indeed, it is optimal to wait until a large shock in wealth before adjusting the consumption.

As a consequence, the CCAPM clearly fails but the CAPM still holds (see also Hindy & Huang [943], Cuoco & Liu [508]). Yogo [1665] show that a model with non-separable utility in non-durable and durable consumption can explain both the cross-sectional variation in expected stock returns and the time variation in the equity premium.

The empirical research on the presence of a habit persistence, durable goods and intertemporal complementarity/substitutability of consumption is quite large. Testing Euler conditions on monthly time series, Dunn & Singleton [602], Eichenbaum & Hansen [633], Eichenbaum et al. [634], Gallant & Tauchen [753] find evidence of the presence of durable good effects (substitutability over short periods) rather than of habit persistence. On the other hand, limited evidence of a habit formation effect has been detected in Dynan [613]. The analysis in Ferson & Constantinides [689] and Braun et al. [282] on monthly, quarterly and annual data shows that habit persistence effects dominate over durability effects. The empirical evidence reported in Ferson & Harvey [692] leads to a rejection of time separable preferences, even when the model incorporates seasonality and allows for seasonal heteroskedasticity. A form of seasonal habit persistence, rather than goods durability, is empirically detected. The habit persistence model performs better than the classical model with a time separable utility function and is not rejected by Euler condition tests. However, the equity premium puzzle cannot be explained. The mean-variance bound on the intertemporal rate of substitution is more likely to be satisfied by a habit persistence preference functional showing intertemporal complementarity than by a utility with durable goods characterized by intertemporal substitution, see Hansen & Jagannathan [891], Gallant et al. [750], Cecchetti et al. [380], Ferson & Harvey [692]. Consumption durability helps to satisfy Euler conditions tests, while habit persistence helps to match the first two moments of asset returns. Consumption durability lowers the volatility of the marginal rate of substitution. Chen & Ludvigson [420] find evidence of a (non-linear) internal habit rather than external habit (keeping up with the Joneses preferences, see below). A model with habit persistence can also explain a cross section of returns on size and book-market sorted portfolio better than the CAPM and several of the classical multi-factor models discussed in Sect. 5.3.

An empirical analysis of a continuous time representative agent economy with non-separable preferences has been proposed in Heaton [923]. It is shown that the empirical performance of an intertemporal consumption model depends on the frequency of consumption data as well as on their aggregation. Monthly observations of seasonally adjusted consumption and quarterly observations of seasonally unadjusted consumption are at odds with the hypothesis that aggregate consumption follows a martingale process. However, the martingale hypothesis and its implications are typically stated under the assumption of a representative agent economy with time separable preferences. In contrast, non-separable preferences can account for different dynamics of consumption. Using observations of consumption on non-durable goods, there is strong evidence in favor of a model where consumption is substitutable over time (consumption is durable). No evidence is found for habit persistence alone. Using seasonally unadjusted data, there is also evidence for habit persistence at seasonal frequencies.

A similar analysis has been carried out in Heaton [924], examining the asset pricing implications of a representative agent economy with non-separable preferences. Evidence is found in favor of a preference structure where consumption at nearby dates is substitutable (*local substitution*) and where habit over consumption develops slowly. Moreover, there is evidence of long run habit persistence. The two effects (local substitutability and long run habit persistence) have an important interaction, in the sense that habit persistence substantially improves the performance of the model only if local substitutability is also present. These results suggest that there is consumption substitution over short periods and consumption complementarity over longer periods. A preference structure of this type is consistent with the Hansen & Jagannathan [891] bounds, generates a low volatility of the risk free rate as well as positively autocorrelated returns but it is rejected using a simulated methods of moments approach and cannot reproduce the historical equity premium as well as the volatility of stock returns. A utility function capturing local substitution (durability) over short periods and long run complementarity (habit formation) has been also proposed in Hindy et al. [944] and Detemple & Giannikos [562]. Habit and durability also provide an explanation of the excess sensitivity of consumption to past income changes.

Keeping Up with the Joneses and External Habit

So far, we have considered preference functionals exhibiting an *internal* habit formation, in the sense that the habit process $(z_t)_{t=0,1,...,T}$ is defined as in (9.36) as a function of past consumption and, hence, is endogenously determined. In Abel [4, 5] and Gali [749], preference functionals with an *external* habit process have been proposed. Typically, the external habit process $(z_t)_{t=0,1,...,T}$ is assumed to represent the aggregate consumption per capita (as in the case of the *keeping up with the Joneses* preferences described below) or simply a fully exogenous process (see Clark et al. [454] for a review).

The following preference functional has been proposed in Kocherlakota [1112] (originally in an infinite horizon economy):

$$\sum_{t=0}^{T} \delta^t \mathbb{E}\left[\frac{c_t^{1-\alpha}}{1-\alpha} \bar{c}_t^{\gamma} \bar{c}_{t-1}^{\lambda} \right], \tag{9.42}$$

for $\alpha \geq 0$ and $\gamma, \lambda < 1$, where $c = (c_t)_{t=0,1,...,T}$ denotes as usual the consumption process of an agent and $\bar{c} = (\bar{c}_t)_{t=0,1,...,T}$ represents the aggregate per capita consumption in the economy. The parameters γ and λ model the agent's evaluation of his own consumption at each date t with respect to the per capita consumption at the same date and the one-period lagged per capita consumption. In other words, an agent with preferences (9.42) is concerned about how well he performs with respect to the average agent in the economy. This utility function captures an externality effect in consumption: if an agent is jealous, then γ and λ will be negative (negative externality, the agent is unhappy when other individuals perform better than him). On the other hand, if an individual is altruistic, then γ and λ will be positive (positive externality). When solving the optimal consumption problem, an individual takes the per capita aggregate consumption as an exogenous datum, while in equilibrium the agent's consumption will coincide with the per capita consumption in a homogeneous economy. In this case, as shown in Kocherlakota [1112], the Euler condition for the optimal consumption-investment problem can be written as

$$\mathbb{E}\left[\left(\frac{\bar{c}_{t+1}}{\bar{c}_t}\right)^{\gamma-\alpha}\left(\frac{\bar{c}_t}{\bar{c}_{t-1}}\right)^{\lambda}(r_{t+1}^n - r_f)\middle|\mathscr{F}_t\right] = 0,$$

for all $n = 1, \ldots, N$ and $t = 0, 1, \ldots, T - 1$. This condition can be derived along the lines of Exercise 9.5 (compare also with Proposition 6.4).

In Abel [4], the author introduces a preference functional that nests three classes of preferences: classical time separable utility, habit formation preferences and external habit preferences. More specifically, Abel [4] considers the following preference functional:

$$\sum_{t=0}^{T} \delta^t \mathbb{E}[u(c_t, z_t)], \tag{9.43}$$

where $c = (c_t)_{t=0,1,\ldots,T}$ denotes the consumption process and where the habit process $(z_t)_{t=0,1,\ldots,T}$ depends both on an external and on an internal habit:

$$z_t = \left(c_{t-1}^{\beta} \bar{c}_{t-1}^{1-\beta}\right)^{\gamma}, \qquad \text{for all } t = 1, \ldots, T, \tag{9.44}$$

where $\gamma \geq 0$ and $\beta \geq 0$. Clearly, if $\gamma = 0$, then (9.43) reduces to the classical time additive utility function. If $\gamma > 0$ and $\beta = 0$, then the habit process (9.44) is purely external, since it only depends on the per capita aggregate consumption, as considered in (9.42). Finally, if $\gamma > 0$ and $\beta = 1$, then (9.43) reduces to the preference functional (9.35) with an endogenous habit process. Abel [4] considers a utility function of the form

$$u(c_t, z_t) = \frac{(c_t/z_t)^{1-\alpha}}{1-\alpha}, \qquad \text{for all } t = 0, 1, \ldots, T,$$

with $\alpha > 0$, corresponding to the ratio model discussed above. In the context of the Abel [4] model, the Euler condition and the equilibrium price-dividend ratio of a stock are derived in Exercise 9.9.

Considering a consumption process with i.i.d. growth rates calibrated to the U.S. economy, Abel [4] shows that both the risk premium and the risk free rate observed empirically can be reproduced with $\alpha = 6$ and $\gamma \neq 0$ (non-separable preferences). Similarly to the case of habit persistence preferences, the main empirical drawback is that the model generates a return variance (in particular of the risk free rate) which is much larger than that observed historically. In Abel [5], it is shown that this drawback can be avoided by introducing a leverage factor in the wealth of the economy. Setting $\lambda = 0$ in the preferences (9.42), Gali [749] shows that there exists an inverse relation between γ and the risk premium, so that sufficiently negative values of γ allow to solve the equity premium puzzle. However, large negative values of γ also induce a large risk free rate.

Campbell & Cochrane [343] consider a utility function with an *external habit*, where each agent compares his consumption with past aggregate consumption. More specifically, the preference functional is defined by the utility function

$$u(c_t, z_t) = \frac{(c_t s_t)^{1-\alpha} - 1}{1 - \alpha}, \qquad \text{where } s_t := \frac{c_t - z_t}{c_t}, \qquad (9.45)$$

for all $t = 0, 1, \ldots, T$ and with $\alpha \in (0, 1)$. The process $(s_t)_{t=0,1,\ldots,T}$ represents the surplus consumption ratio and $(z_t)_{t=0,1,\ldots,T}$ the external habit process. It is assumed that the logarithm of the surplus process satisfies the following dynamics:

$$\log s_{t+1} = \bar{s} + \varphi(\log s_t - \bar{s}) + \lambda(s_t)(\log c_{t+1} - \log c_t - g), \qquad (9.46)$$

for all $t = 0, 1, \ldots, T - 1$, where φ, $\bar{s}$ and g are given parameters. According to (9.46), the logarithm of the surplus process fluctuates around the value $\bar{s}$, thus capturing business cycle fluctuations. The function $\lambda(\cdot)$ is called the sensitivity function and will be specified further below. The aggregate endowment process (expressed as the average endowment across all the agents of the economy) satisfies the following dynamics:

$$\Delta \log c_{t+1} := \log c_{t+1} - \log c_t = g + v_{t+1}, \qquad (9.47)$$

where $(v_t)_{t=1,\ldots,T}$ is a sequence of i.i.d. normal random variables with zero mean and variance σ^2. Note that, due to the specification (9.46)–(9.47), the surplus process is perfectly correlated with the consumption growth, in the sense that

$$\mathrm{Cov}(\Delta \log s_{t+1}, \Delta \log c_{t+1}) = 1, \qquad \text{for all } t = 0, 1, \ldots, T - 1.$$

In equilibrium, identical individuals choose the same level of consumption. We can then establish the following proposition, the proof of which is given in Exercise 9.10.

Proposition 9.10 *Consider the Campbell & Cochrane [343] model introduced above, where the surplus process and the consumption growth process satisfy the* *dynamics (9.46) and (9.47), respectively. In equilibrium, the intertemporal marginal rate of substitution process $(m_t)_{t=1,...,T}$ is given by*

$$m_{t+1} = \delta e^{-\alpha(g+(1+\lambda(s_t))v_{t+1}+(\varphi-1)(\log s_t-\bar{s}))}, \qquad (9.48)$$

and it holds that

$$\frac{\mathbb{E}[m_{t+1}|\mathscr{F}_t]}{\sqrt{\mathrm{Var}(m_{t+1}|\mathscr{F}_t)}} = \sqrt{e^{\alpha^2(1+\lambda(s_t))^2\sigma^2}-1} \approx \alpha(1+\lambda(s_t))\sigma, \qquad (9.49)$$

for all $t = 0, 1, \ldots, T-1$. Moreover, the equilibrium risk free rate r_f is given by

$$r_f = \frac{1}{\delta}e^{\alpha(g+(\varphi-1)(\log s_t-\bar{s}))-\alpha^2(1+\lambda(s_t))^2\frac{\sigma^2}{2}}, \qquad \text{for all } t = 1, \ldots, T. \qquad (9.50)$$

Note that m_t corresponds to the one-period stochastic discount factor introduced in Sect. 6.4. In particular, relation (9.49) represents the Hansen & Jagannathan [891] bound for the maximal Sharpe ratio among all traded assets (compare with Proposition 5.1). In the expression for the equilibrium risk free rate (which is a priori time dependent), the term $\log s_t - \bar{s}$ reflects intertemporal substitution, while the last term appearing in the exponential in (9.50) represents a precautionary saving term. As pointed out in Campbell & Cochrane [343], the approximation obtained in (9.49) is helpful for the specification of the model. To generate time-varying Sharpe ratios, the term $\lambda(s_t)$ must vary with s_t and it seems desirable that the function $\lambda(\cdot)$ is decreasing. In Campbell & Cochrane [343], the function $\lambda(\cdot)$ is chosen such that the equilibrium risk free rate is constant, the habit process is predetermined at the steady state $\bar{s}$ and habit moves non-negatively with consumption. This leads to the specification

$$\lambda(s_t) = \begin{cases} e^{-\bar{s}}\sqrt{1-2(\log s_t-\bar{s})}-1, & \text{if } \log s_t \leq s_{\max}, \\ 0, & \text{otherwise}, \end{cases}$$

where $s_{\max}$ is given by $s_{\max} := \bar{s} + (1-e^{2\bar{s}})/2$ and $\bar{s} = \log\sigma + \frac{1}{2}\log(\alpha/(1-\varphi))$. With this specification of $\lambda(\cdot)$, formula (9.50) reduces to

$$r_f = \frac{1}{\delta}e^{\alpha g+\frac{\alpha}{2}(\varphi-1)}.$$

The Campbell & Cochrane [343] model allows to reproduce a wide variety of asset pricing phenomena. In particular, it is possible to reproduce the historically observed risk free rate, the average excess stock return and its variance. In particular, the model fits the equity premium without generating a high volatility in the risk free rate, unlike most of the habit formation models, and generates high volatility. Moreover, the model explains business cycle movements of asset prices and risk premia (high risk premia during recessions and small risk premia during booms, i.e., counter-cyclical variations in the Sharpe ratios) as well as the presence of serial correlation in asset returns and predictability via the price-dividend ratio. The rationale is that during a recession the consumption of an agent decreases towards the level of the habit, thereby increasing risk aversion and the risk premium. This type of utility function generates high and time-varying risk aversion. According to this model, there is an inverse relation between consumption surplus and expected returns. The model also generates a low correlation between asset returns and consumption growth, as observed empirically. As pointed out in Campbell & Cochrane [343], the main rationale for these implications is the presence of a slowly time-varying, counter-cyclical risk premium. On asset pricing models under this type of preferences see also Cochrane [463], Tallarini & Zhang [1582], Wachter [1635], Li [1217], Yogo [1666]. As argued in Campbell & Cochrane [344], this type of model may explain the poor performance of the CCAPM relative to the CAPM. The model also performs well in explaining the cross-sectional variation in conditional expected returns, showing that asset characteristics can play a role (growth premium rather than value premium), see Santos & Veronesi [1498], Wachter [1635], Lettau & Wachter [1201].

Time Preferences and Hyperbolic Discounting

In most of the analysis developed so far, we have assumed that multi-period preferences are modeled by assuming a constant discount factor that grows exponentially with respect to time. However, experimental studies have shown that this way of representing preferences over time is typically violated (see Frederick et al. [736] for a review of time discounting and time preferences). In order to overcome the shortcomings of this specification of the discount factor, alternative forms of discounting have been proposed. In particular, *hyperbolic discounting* has been introduced in order to capture the empirical fact that discount rates should be declining with respect to time (see, e.g., Angeletos et al. [71]). Loewenstein & Prelec [1248] enumerate a series of anomalous findings on time preferences and introduce an hyperbolic discount function $\delta : \mathbb{R}_+ \to \mathbb{R}_+$ of the following form:

$$\delta(t) = (1 + \alpha t)^{-\beta/\alpha}, \qquad \text{for all } t \geq 0,$$

with $\alpha, \beta > 0$, thus yielding the preference functional

$$\sum_{t=0}^{T} \delta(t)\mathbb{E}[u(c_t)],$$

where $c = (c_t)_{t=0,1,\dots,T}$ denotes as usual a consumption stream. This alternative specification of time preferences is consistent with several empirical facts and with intertemporal choice policies which are incompatible with the classical assumption

of exponential discounting. In Laibson [1156], a *quasi-hyperbolic* discounting
function has been proposed, leading to the preference functional:

$$\mathbb{E}\left[u(c_0) + \beta \sum_{t=1}^{T} \delta^t u(c_t)\right].$$

When $\beta < 1$, quasi-hyperbolic discounting mimics the qualitative properties of
hyperbolic discounting, while preserving the analytical tractability of exponential
discounting.

Non-exponential discounting introduces time inconsistency in optimal choice
problems: a decision taken at date $t = 0$ about something that is going to happen
at a future date $t > 0$ may be evaluated in a different way at a later date (e.g., at
date $t = 1$). This phenomenon, which is not observed in the classical setting of
exponential discounting, induces personal strategic conflicts, since a future option
may be valued in different ways as time goes. Gul & Pesendorfer [864, 865] provide
an axiomatization of this attitude in the context of a two-period model where an ex-
ante inferior choice may tempt the decision-maker in the second period (*temptation*
and *self-control* preferences). These axioms yield a representation that identifies the
agent's commitment ranking, the temptation ranking and the cost of self-control.
By analysing household data, Huang et al. [974] have found an empirical support
for temptation and self-control preferences. A life-cycle model including temptation
and self-control preferences has been studied in Laibson et al. [1157]. Building on
Gul & Pesendorfer [864], Krusell et al. [1139] analyse a general equilibrium model
where a small subset of the agents exhibits a negative short run discount factor
(short run "urge to save"), at the same time avoiding time inconsistency. Agents
are fully rational, but are subject to temptation and self-control. It is shown that, in
equilibrium, the model can generate a low risk free rate together with a high risk
premium. However, DeJong & Ripoll [542] show that the self-control preferences
introduced by Gul & Pesendorfer [864] do not allow for a resolution of the asset
pricing puzzles.

9.3   Beyond Substantial Rationality

In the preceding sections of this chapter, we have focused our attention on how
agents behave in the presence of risk and uncertainty and we have introduced
alternative preference functionals that depart from the classical time additive
expected utility paradigm. However, we have always assumed that the agents'
behavior is characterized by *substantial rationality*, i.e., agents pursue their goals
in the most appropriate way under the constraints imposed by the environment, as
already discussed in Sect. 1.1. In a perfectly competitive economy, the assumption
of substantial rationality implies that the decision problem of each agent can be
represented by the maximization of a preference functional. In this section (as

well as in the following one), we will depart from the substantial rationality assumption, entering the domain of *non (substantial) rationality*. The notion of non-rationality (as well as bounded rationality and near-rationality, as discussed in the following section) means that agents do not maximize an objective functional representing their preferences and/or that agents do not handle risk/uncertainty in an optimal way (i.e., by updating their beliefs via Bayes' rule). Alternative behavioral assumptions are mainly motivated by two arguments: first, the assumption of substantial rationality (together with the rational expectations hypothesis) seems rather implausible to psychologists and economists; second, the hypothesis of perfect rationality leads to very complex decision problems and, in some cases, no explicit solution to an optimal choice problem exists. Moreover, as discussed in the previous sections, there is a consistent experimental evidence reporting empirical regularities that are incompatible with classical asset pricing theory. Based on this observation, Simon [1545] proposed a *satisfying* approach: agents possess limited computing skills and the inherent complexity of typical decision problems leads agents to adopt what is improperly defined as a *non-fully rational behavior* (e.g., adopting rules of thumb or mental accounting rules).

We will address the impact of non-rationality in a setting with heterogeneous agents. Note that heterogeneity can produce interesting results going beyond classical asset pricing theory only if it is considered in conjunction with non-rationality. Indeed, under suitable assumptions, if agents are rational and markets are complete, then even under heterogeneity an aggregation result holds, so that the prices of the economy with heterogeneous agents are observationally equivalent to an economy with a single representative agent (compare with Sects. 1.3 and 4.3). In what follows, heterogeneity will typically not refer to differences in the agents' resources, preferences, or information, but rather in their goals, in their motives to trade and in the way they pursue their goals.

Before presenting alternative asset pricing models based on a non-rational behavior, we want to point out that classical asset pricing theory based on full rationality (and rational expectations) will always play an important role, both in a normative and in a descriptive sense. Indeed, classical asset pricing theory represents the main reference point, because it describes what happens in the economy when agents fully exploit the available information and act in an optimal way. Alternative asset pricing theories will always be compared to classical asset pricing theory: recalling Kuhn's point of view (quoted at the beginning of Chap. 1), alternative behavioral theories will be evaluated not only for their capability to explain anomalies appearing within classical asset pricing theory but also for their capability to provide alternative paradigms which can explain what classical asset pricing theory is already able to explain.

Apart from the normative interpretation, Friedman [743] proposed two arguments in favor of full rationality and rational expectations. The first argument is that a theory should be evaluated on the validity of its implications and not necessarily on the validity of the assumptions (this line of thought is well exemplified by the two statements by Solow and Kuhn quoted at the beginning of Chap. 1). The second argument is that, in a competitive market, non-rational agents will tend to perform poorly when compared to rational agents, with the consequence that in the long run non-rational agents will be wiped out of the market by rational agents. In this section, we will carefully discuss both arguments. We do not aim at providing an exhaustive survey on the topic, limiting ourselves to highlighting some of the most important results. More precisely, we shall consider economies characterized by the presence of the following types of non-rational agents: a) noise traders; b) overconfident agents; c) feedback traders. At the end of the present section, we will also discuss the possibility that non-rational agents survive in the long run and the role of rational agents in stabilizing the market. This section is also related to the analysis developed in Sect. 8.5 on the difference of opinions.

Noise Traders

De Long et al. [549, 550] and Shleifer & Summers [1540] analyse an economy with *noise traders*. The underlying idea is that noise traders are characterized by incorrect beliefs on the fundamental value of an asset and underestimate or overestimate the future expected price of the asset. From the point of view of rational traders, the presence of noise traders creates an additional risk in the economy and prices can diverge significantly from fundamental values. In this way, several anomalies appearing in the context of classical asset pricing theory can be addressed.

   In De Long et al. [549], the authors consider an overlapping generations model with agents living for two periods, with no consumption occurring in the first period and where agents are endowed with exogenous resources to invest in the market. Agents have only to choose a portfolio in the first period (when agents are "young"), while in the second period (when agents are "old") all the wealth generated by the chosen portfolio will be consumed. There are two available assets. The first asset is a safe asset yielding the dividend r_f in each period and is available in perfectly elastic supply, with a price fixed at one. The second asset is interpreted as the aggregate equity available in the financial market and, since it is not in elastic supply, it is a risky asset. The risky asset yields the constant dividend r_f, equal to the dividend of the safe asset, in each period, but is available in unitary fixed supply. We denote by p_t the price of the risky asset at date t, for $t \in \mathbb{N}$. Note that the safe asset and the risky asset pay identical dividends.

   If all agents were rational (so that the prices of both assets would be equal to the net present value of the corresponding future dividends), then the safe and the risky assets would be perfect substitutes and would be traded for a price equal to one at every date. However, in the presence of noise traders, the equality $p_t = 1$ does not necessarily hold. De Long et al. [549] assume that the economy is populated by two types of agents: there are sophisticated investors with rational expectations and there are *noise traders*. It is assumed that noise traders represent a proportion $\mu \in [0, 1]$ of the economy. All agents of a given type are identical and every agent is allowed to take unlimited short positions in the assets when choosing a portfolio in the first period. At each date $t \in \mathbb{N}$, every young agent chooses his portfolio by maximizing an expected utility function given his own beliefs at date t and the ex-ante distribution of the price p_{t+1} of the risky asset at the subsequent date $t + 1$. All the agents know the economic model and are characterized by an exponential utility function of the form $u(x) = -e^{-2\gamma x}$, with 2γ being the coefficient of absolute risk aversion. The utility function of every agent is defined with respect to wealth when the agent is old. The non-rational behavior of noise traders is modeled by assuming that they misperceive the future expected price of the risky asset by a random variable ϱ_t. More specifically, at date t a noise trader believes that the expected future price of the risky asset is given by

$$\mathbb{E}[p_{t+1}|\mathscr{F}_t] + \varrho_t,$$

where $(\mathscr{F}_t)_{t\in\mathbb{N}}$ represents the information flow of the economy. It is assumed that the sequence $(\varrho_t)_{t\in\mathbb{N}}$ is composed of i.i.d. random variables with common distribution $\mathscr{N}(\bar{\varrho}, \sigma_\varrho^2)$. The mean value $\bar{\varrho}$ represents the average "bullishness" of noise traders, while σ_ϱ^2 represents the variance of noise traders' misperceptions. On the contrary, sophisticated investors correctly perceive the conditional distribution of the future price.

In the context of the economy described above, we can characterize the equilibrium price process $(p_t^*)_{t\in\mathbb{N}}$ of the risky asset. Moreover, we can compute the expected relative returns of noise and rational traders in equilibrium. This is the content of the following proposition, the proof of which is given in Exercise 9.11.

Proposition 9.11 *In the context of the above model (see De Long et al. [549]), consider a steady-state equilibrium, where the unconditional distribution of the equilibrium price p_t^* does not depend on the date $t \in \mathbb{N}$. Then in a steady state equilibrium the following hold:*

(i) the price process $(p_t^*)_{t\in\mathbb{N}}$ of the risky asset satisfies

$$p_t^* = 1 + \frac{\mu(\varrho_t - \bar{\varrho})}{1 + r_f} + \frac{\mu\bar{\varrho}}{r_f} - \frac{2\gamma}{r_f}\frac{\mu^2\sigma_\varrho^2}{(1 + r_f)^2}, \qquad \text{for all } t \in \mathbb{N}; \qquad (9.51)$$

(ii) the expectation of the difference ΔR between noise traders' and rational investors' total returns (given equal initial wealth) is given by

$$\mathbb{E}[\Delta R_{t+1}] = \bar{\varrho} - \frac{(1 + r_f)^2\bar{\varrho}^2 + (1 + r_f)^2\sigma_\varrho^2}{2\gamma\mu\sigma_\varrho^2}, \qquad (9.52)$$

for all $t \in \mathbb{N}$.

As shown in Exercise 9.11, in comparison to the optimal demand of a rational trader, the optimal demand of the risky asset by a noise trader contains an additional term. This additional term is due to the noise trader's misperception of the future expected price of the risky asset: when noise traders overestimate expected returns, they demand more of the risky asset than rational traders do. The variance appearing in the optimal demand of rational and noise traders is purely due to the presence of noise traders, since there is no aggregate risk in the economy.

The equilibrium price process $(p_t^*)_{t\in\mathbb{N}}$ characterized in part *(i)* of Proposition 9.11 admits an interesting interpretation. Note first that, if the distribution of ϱ_t converges to a point mass at zero, then the equilibrium price is constant and always equal to one (the fundamental value), as previously mentioned. If this is not the case, then the presence of noise traders affects the equilibrium price in three distinct ways, each of them corresponding to one of the three terms appearing on the right-hand side of (9.51) (and observe that the relevance of these three terms increases as the proportion μ of noise traders increases):

- the term $\mu(\varrho_t - \bar{\varrho})/(1+r_f)$ captures the variability in the equilibrium price due to the variations in the misperception of noise traders. Even though the risky asset does not have any fundamental risk (since it delivers a constant dividend), its price varies depending on the noise traders' misperceptions of its future price. In particular, if the young generation of noise traders at date t is characterized by a misperception greater than its average ($\varrho_t > \bar{\varrho}$), then this leads to an increase of the equilibrium price;
- the term $\mu\bar{\varrho}/r_f$ captures deviations of the equilibrium price from the fundamental value of the asset due to the fact that the average misperception of noise traders is different from zero ($\bar{\varrho} \neq 0$). If noise traders are bullish on average ($\bar{\varrho} > 0$), then this price pressure effect makes the equilibrium price of the risky asset increase;
- the last term $2\gamma\mu^2\sigma_\varrho^2/(r_f(1+r_f)^2)$ plays a crucial role: because of the presence of noise traders in the market, the future price of the risky asset is random and, therefore, its return is risky. This fact induces rational and noise traders to require a positive risk premium in order to invest in the risky asset and this risk premium drives the price down and the return up, with the risk premium being proportional to the conditional variance of the future price (see Exercise 9.11). Note that this effect plays a role despite the fact that both rational and noise traders always hold portfolios without any fundamental risk. The last term in (9.51) means that *noise traders create their own space in the market* and their trading activity is the only source of risk in the economy.

It has to be remarked that the results described above depend strongly on the assumptions of the model, as pointed out in Loewenstein & Willard [1249]. In particular, the assumptions that agents can take unlimited short positions, that the risk free asset is in perfectly elastic supply and the two-period overlapping generations structure are crucial for the main results of De Long et al. [549] to hold. If rational agents live for two periods, then they care about tomorrow's price and require a risk premium in order to compensate the risk generated by noise traders. On the contrary, if rational agents are long-lived (in particular, if they have an infinite time horizon), then they will not care too much about tomorrow price and they will adopt strategies (buy and hold strategies) aimed at exploiting the mispricing opportunities created by noise traders. However, in a mean-variance setting, Bhushan et al. [230] have shown that "trading myopia" is neither a necessary nor a sufficient condition for noisy prices. They show that a unique noisy equilibrium (in the sense of noise traders affecting prices) exists only if agents are rational regarding volatility and irrational regarding expected returns and noise traders are allowed to take unbounded positions. If these two crucial hypotheses (which are satisfied in De Long et al. [549]) are not verified, then multiple noise equilibria can coexist with a classical rational equilibrium. Furthermore, noisy equilibria exist only on a subset of the parameter space, while the rational equilibrium exists for all parameter values.

Let us now comment on part *(ii)* of Proposition 9.11. As already mentioned in the first part of the present section, the presence of non-rational traders in the economy has often been denied on the basis of an evolutionary argument (see Friedman [743]): non-rational agents tend to have a poor investment performance, in comparison to rational agents, and, therefore, only rational agents will survive and dominate the economy in the long run. In other words, non-rational agents should earn lower returns on average and, therefore, economic selection will make rational agents prevail in the market. In the model of De Long et al. [549], this conjecture is not necessarily true. Indeed, in correspondence of a steady-state equilibrium, the expected difference in the returns on the optimal portfolios of noise and rational traders is given by formula (9.52) in Proposition 9.11. This formula shows that a necessary condition for the expected return of noise traders to be greater or equal than that of rational agents is $\bar{\varrho} > 0$. The economic intuition behind this fact is rather simple: if $\bar{\varrho} > 0$ then noise traders will take larger positions in the risky asset and, therefore, they will earn a larger share of the rewards to risk bearing. On the other hand, if $\bar{\varrho} < 0$ then the expected return of noise traders is lower than that of rational agents. The first term in the numerator on the right-hand side of (9.52) captures the "price pressure" effect, i.e., the fact that the demand of bullish noise traders drives up the equilibrium price of the risky asset, thus decreasing the return to risk bearing and, therefore, the expected return difference $\mathbb{E}[\Delta R_{t+1}]$. The second term in the numerator captures the fact that noise traders have the worst possible market timing: they buy the risky asset when all the other noise traders are buying it (high price and low return). The more variable ϱ_t is, the stronger the bad timing effect is. This effect goes against the investment performance of noise traders and reduces their expected return, thus supporting the evolutionary conjecture of Friedman [743]. The denominator on the right-hand side of (9.52) captures the fact that noise traders create their own space in the market and, as the variability of the misperceptions of noise traders increases, the risk premium increases. In order to exploit the mispricing created by the misperceptions of noise traders, rational investors must bear a greater risk. Since rational investors are risk averse, they reduce the extent to which they bet against noise traders as a response to this increased risk. Summing up, the first term as well as the denominator on the right-hand side of (9.52) lead to an increase of the expected return of noise traders, while the numerator tends to a decrease. A priori neither of these two effects clearly dominates on the other. The expected return difference $\mathbb{E}[\Delta R_{t+1}]$ is quadratic in $\bar{\varrho}$, so that noise traders will have an expected return lower than that of rational agents if on average they are bearish or if they are excessively bullish (for a large value of

$\bar{\varrho}$). On the contrary, for intermediate values of $\bar{\varrho}$, the expected return of noise traders can be greater than that of rational agents. However, the optimal expected utility of noise traders is always less or equal than that of rational agents (see De Long et al. [549, Section II.B]).

So far, we have considered a fixed proportion μ of noise traders. De Long et al. [549] extend the basic model by introducing population dynamics. More specifically, they assume that new generations of investors enter into the economy and decide on which investment strategy to follow (rational or non-rational) depending on the past performance of those strategies, without accurately assessing the ex-ante risks involved in each strategy. In this context, the impact of noise traders on market prices does not necessarily diminish over time. In De Long et al. [549, Section III.A], the authors introduce the following population dynamics for the noise traders, based on imitation among agents:

$$\mu_{t+1} = \max\left\{0, \min\{1, \mu_t + \xi_t(r_t^n - r_t^r)\}\right\}, \qquad \text{for all } t \in \mathbb{N}, \qquad (9.53)$$

where r_t^n and r_t^r denote respectively the wealth returns of noise and rational traders at date t and where ξ_t represents the rate at which new investors entering into the economy choose to be noise traders (per unit difference in realized returns), with $\lim_{t\to\infty} \xi_t = 0$. According to the population dynamics (9.53), the proportion of noise traders increases if they overperform rational investors. The dynamics can be interpreted as the result of an imitation process, in the sense that the most profitable behavior (between being rational or being non-rational) tends to be adopted by new agents entering the market (success breeds imitation), with some inertia represented by the parameter ξ_t.

The model introduced above with the population dynamics (9.53) can be easily solved if ξ_t is close to zero. In this case, returns can be computed under the hypothesis that the proportion of noise traders in the economy remains approximately constant over two consecutive dates. Formula (9.51) under the population dynamics (9.53), with μ changed to μ_t, characterizes the equilibrium price of the risky asset in the limit as ξ_t converges to zero. Relation (9.52) (in the conditional expectation form, as given in Exercise 9.11), with μ changed to μ_t and in the limit as ξ_t converges to zero, has an analogous interpretation and can be stated as follows:

$$\mathbb{E}[\Delta R_{t+1}|\mathscr{F}_t] = \varrho_t - \frac{(1+r_f)^2 \varrho_t^2}{2\gamma\mu_t\sigma_\varrho^2}. \qquad (9.54)$$

Over time, the proportion μ_t of noise traders will grow or shrink according to whether the right-hand side of (9.54) is greater or lower than zero. Equivalently, if μ_t is below the critical value

$$\mu^* := \frac{(\bar{\varrho}^2 + \sigma_\varrho^2)^2(1 + r_f)^2}{2\bar{\varrho}\gamma\sigma_\varrho^2},$$

then μ_t tends to shrink. On the contrary, if $\mu_t > \mu^*$, then noise traders induce so much risk to make rational investors reluctant to speculate against them, with the consequence that the proportion of noise traders grows and rational traders disappear in the long run.

The model is further extended in De Long et al. [549, Section III.B] considering, together with the population dynamics (9.53), also the possibility that the risky asset delivers at each date $t \in \mathbb{N}$ a random dividend equal to $r_f + \tilde{\varepsilon}_t$, where $(\tilde{\varepsilon}_t)_{t \in \mathbb{N}}$ is a sequence of i.i.d. normal random variables with zero mean and constant variance, independent of the noise traders' misperceptions $(\varrho_t)_{t \in \mathbb{N}}$. In particular, as shown in Exercise 9.12, in correspondence of a steady-state equilibrium and in the limit as ξ_t converges to zero, the difference between the expected return of noise and rational traders can be greater than zero and the "create space" effect is more pronounced, as rational investors are less willing to trade in the risky asset, due to the increased riskiness. In the long run, either noise traders dominate the economy or μ_t converges (in expectation) to an equilibrium where both types of agents coexist. On the other hand, if the imitation process is based on the comparison of the utility obtained at the previous date, then in the long run noise traders disappear from the economy.

The model of De Long et al. [549] provides several insights that allow us to address some of the anomalies encountered in the context of classical asset pricing theory. First, noise traders induce an excess variability of the price process $(p_t^*)_{t \in \mathbb{N}}$ with respect to what would be implied by a fully rational equilibrium. Moreover, the volatility of market prices is not due to variations in the asset fundamentals, but only to the presence of noise traders. Hence, the introduction of noise traders in the economy helps to explain the excess volatility phenomenon discussed in Sect. 7.2. Furthermore, as considered in De Long et al. [549, Section IV.A], if the noise traders' misperceptions $(\varrho_t)_{t \in \mathbb{N}}$ follow a first-order autoregressive process instead of being a sequence of i.i.d. random variables, then the risky asset price will exhibit mean reversion. In this setting, a contrarian strategy can earn excess profits. Moreover, if misperceptions are serially correlated, then noise traders can cause larger deviations of asset prices from the fundamental values. Finally, the model can also shed some light on the equity premium puzzle (see Sect. 7.3), since the presence of noise trader risk drives down the price of the risky asset and, therefore, the risky asset yields a higher return than the risk free asset.

In De Long et al. [550], in the context of a slightly different model, the dynamics of the population is analysed. The economy is assumed to be populated by noise and rational traders, but noise traders do not affect market prices. Noise traders misperceive the distribution of the asset return (in particular, they misperceive its mean and variance). It is shown that, if noise traders only misperceive the variance, then they earn expected returns higher than those of rational traders (see De Long et al. [550, Proposition 1]). Moreover, if the noise traders' utility function exhibits a larger risk aversion than a logarithmic utility function and they misperceive the variance in a limited way, then noise traders survive in the market, in the sense that the share of the aggregate wealth held by noise traders does not converge almost surely to zero as time passes. Under some conditions, noise traders even dominate the market, meaning that after a sufficient time the probability that noise traders hold a share of the aggregate wealth larger than that of rational traders is greater than $1/2$ (see De Long et al. [550, Proposition 2]). On the other hand, if noise traders strongly underestimate the variance, then they are more likely to default.

In De Long et al. [547], the welfare consequences of the presence of noise traders are evaluated in the context of an overlapping generations model. It is shown that the presence of noise traders typically reduces the stock of capital and consumption in comparison to an economy populated only by rational agents and that the consumption process is more volatile. The driving force behind these results is the risk generated by the presence of noise traders. In many cases, the wealth of rational agents is negatively affected by the presence of noise traders and the benefits generated by profitable trading opportunities for skillful rational traders are not compensated by the costs of bearing the additional risk generated by noise trading. In particular, the welfare costs of noise trading can be significant if the magnitude of noise in aggregate stock prices is large.

Overconfidence

As we have explained above, models allowing for the presence of noise traders allow to explain mean reversion phenomena in asset prices. However, models with noise traders typically fail to explain the *momentum effect* (see Sect. 7.2). In the literature, models with *overconfident agents* have also been considered and, in particular, it has been shown that these models allow to generate both overreaction and underreaction of asset prices, thus addressing the momentum effect.

Overconfidence can account for many different behavioral patterns: agents tend to overestimate the precision of their information, agents tend to overestimate their ability to perform well and overestimate their own role when they have obtained a positive performance in the past, agents consider themselves above the average of the population as well as several other related phenomena. In the financial literature, it is typically assumed that agents are *overconfident about information*, i.e., agents overestimate the precision of the information available to them. More specifically, agents tend to overestimate the precision of their private information, while they are not overconfident concerning the publicly available information (see Odean [1379] and Daniel et al. [516]). Moreover, agents can also be overconfident on their knowledge of the private information of the other agents, as considered for instance in Benos [193]. Overconfidence dynamically changes over time, as it tends to increase after a positive performance (see Daniel et al. [516] and Gervais & Odean [773]). In these models, the posterior beliefs of overconfident agents are too precise and overweight (underweight, respectively) private information (public information, respectively) in comparison to a rational Bayesian updating. As a consequence, overconfidence leads to suboptimal portfolio choices and, since posterior beliefs are excessively precise, overconfident agents tend to exhibit a larger risk exposure in comparison to rational agents.

A model based on overconfidence and variations in confidence arising from *biased self-attribution* of investment outcomes has been proposed in Daniel et al. [516]. Agents are overconfident about their private information and, therefore, they overreact to private information and underreact to public information. This overreaction-underreaction phenomenon can generate positive serial correlation of returns in the short run together with negative serial correlation in the long run (agents overreact to their private information and then gradually revise their beliefs according to the public information) and excess volatility. Biased self-attribution of investment outcomes means that agents are biased in updating their confidence on their own ability: they tend to attribute their success to their own ability and, on the contrary, bad events to external forces. When agents receive a public signal that confirms the validity of their actions, then the agents' confidence rises. On the contrary, the agents' confidence is not affected by a public signal that does not confirm their actions. This process is consistent with a momentum effect in the short run together with long run reversals in asset prices.

Let us present in more detail the model of Daniel et al. [516], first considering the case of a static confidence (meaning that overconfidence does not change over time). The economy is populated by two classes of agents: uninformed risk averse agents and informed risk neutral agents. Apart from the different attitude towards risk, informed agents differ from uninformed agents since they can observe a private signal. Each individual is endowed with a basket of security shares and units of the risk free asset paying one unit of wealth at the terminal date. The model consists of dates $t \in \{0, 1, 2, 3\}$. At the initial date $t = 0$, agents start with their endowments, identical prior beliefs and trade only for risk sharing purposes. At date $t = 1$, the informed agents observe a common noisy private signal about the fundamental value of the security and they trade with uninformed agents. At date $t = 2$, everyone observes a noisy public signal and further trade occurs. Finally, at the terminal date $t = 3$, all the uncertainty is resolved, the security pays its liquidating dividend and every agent consumes. All the random variables appearing in the model are assumed to be mutually independent and normally distributed.

The dividend delivered by the risky security at the terminal date $t = 3$ is represented by the random variable $\tilde{d}$ which is supposed to be a normal random variable with mean $\bar{d}$ and variance σ_d^2 (without loss of generality, we shall assume from now on that $\bar{d} = 0$). The private signal observed by the informed agents at date $t = 1$ is a realization of

$$\tilde{s}_1 = \tilde{d} + \tilde{\varepsilon},$$

where $\tilde{\varepsilon} \sim \mathcal{N}(0, \sigma_\varepsilon^2)$. Uninformed agents correctly estimate the variance of the noise component $\tilde{\varepsilon}$ (even though they do not observe the signal), whereas informed agents underestimate it believing that the variance of $\tilde{\varepsilon}$ is given by σ_c^2, with $\sigma_c^2 < \sigma_\varepsilon^2$. The different beliefs about the variance of the noise of the signal are common knowledge. The public signal released at date $t = 2$ is a realization of

$$\tilde{s}_2 = \tilde{d} + \tilde{\eta},$$

where $\eta \sim \mathcal{N}(0, \sigma_\eta^2)$. Concerning the variance of the noise component of the public signal, every agent believes in the correct variance σ_η^2. In the model, it is essential that some noisy public information is released after the private signal.

The following proposition, the proof of which is given in Exercise 9.13, collects some fundamental properties of the Daniel et al. [516] model described above.

Proposition 9.12 *In the context of the above model (see Daniel et al. [516]), the equilibrium price of the risky security is given by*

$$p_1^* = \frac{\sigma_d^2}{\sigma_d^2 + \sigma_c^2}(\tilde{d} + \tilde{\varepsilon}),$$

$$p_2^* = \frac{\sigma_d^2(\sigma_c^2 + \sigma_\eta^2)}{D}\tilde{d} + \frac{\sigma_d^2\sigma_\eta^2}{D}\tilde{\varepsilon} + \frac{\sigma_d^2\sigma_c^2}{D}\tilde{\eta},$$

(9.55)

and $p_3^* = \tilde{d}$, where $D := \sigma_d^2(\sigma_c^2 + \sigma_\eta^2) + \sigma_c^2\sigma_\eta^2$. As a consequence, it holds that

$$\mathrm{Cov}(p_3^* - p_2^*, p_2^* - p_1^*) > 0,$$

$$\mathrm{Cov}(p_2^* - p_1^*, p_1^* - p_0^*) < 0,$$

$$\mathrm{Cov}(p_3^* - p_1^*, p_1^* - p_0^*) < 0,$$

$$\mathrm{Cov}(p_3^* - p_2^*, p_1^* - p_0^*) < 0.$$

(9.56)

As shown in Exercise 9.13, due to the assumption of risk neutral informed agents, the equilibrium price p_1^* is given by the expectation of the dividend $\tilde{d}$ conditionally on the information received by the informed agents at date $t = 1$, whereas the equilibrium price p_2^* is given by the expectation of the dividend $\tilde{d}$ conditionally on the information received by the informed agents at date $t = 1$ and the public signal released at $t = 2$. Moreover, formula (9.55) shows the impact of overconfidence on the dynamics of the equilibrium price. The overconfidence of informed agents in the precision of their private signal makes the equilibrium price at date $t = 1$ overreact to the information conveyed by the observation of $\tilde{d} + \tilde{\varepsilon}$. This overreaction is partially corrected at the following date $t = 2$ when the public signal $\tilde{\eta}$ is released and fully corrected at the terminal date $t = 3$, when the fundamental value is revealed. The effect of the correction to the initial overreaction is captured by the covariances given in (9.56). Indeed, the overconfidence of informed investors induces a positive correlation between the price change in response to the arrival of public information and later price changes, while the price change resulting from the arrival of private information is on average partially reversed in the long run. This latter property can be seen by noting that the price changes in the first two periods are negatively correlated, while the price changes in the last two periods are positively correlated (see Proposition 9.12). Arguing as in Daniel et al. [516], if the econometrician is equally likely to pick either pair of consecutive dates, then the overall autocorrelation is negative. Moreover, in the presence of overconfidence, the volatility induced by private information is higher than the volatility that would be observed in a fully rational economy, whereas the volatility due to the arrival of public information can be higher or lower than the corresponding volatility in a fully rational economy (see Daniel et al. [516, Proposition 3]).

In the setting described above, overconfidence accounts for long run reversals in asset prices (negative long lag serial correlation) but not for short run momentum effects (positive short lag serial correlation). In order to overcome this limitation, Daniel et al. [516] propose an extended version of the model where the degree of overconfidence is allowed to change in response to previous actions and outcomes (*biased self-attribution*). In this version of the model, the public signal released at date $t = 2$ is assumed to be of the form

$$\tilde{s}_2 = \begin{cases} +1, & \text{with probability } p; \\ -1, & \text{with probability } 1 - p. \end{cases}$$

It is allowed (but not necessarily required) that informed investors are initially overconfident, meaning that $\sigma_c^2 \leq \sigma_\varepsilon^2$. From the perspective of an informed trader, the public signal observed at date $t = 2$ confirms his beliefs if

$$\text{sign}(d + \varepsilon) = \text{sign}(s_2),$$

with $d + \varepsilon$ and s_2 denoting the realizations of the random variables $\tilde{d} + \tilde{\varepsilon}$ and $\tilde{s}_2$, respectively. In this case, the public signal makes the confidence of an informed investor increase, leading to an estimation of the variance of the noise component equal to $\sigma_c^2 - k$, for some $k \in (0, \sigma_c^2)$. On the contrary, if

$$\text{sign}(d + \varepsilon) \neq \text{sign}(s_2),$$

then the public signal does not confirm the beliefs of the informed agent and his confidence remains constant at the level σ_c^2. In other words, the confidence of the informed trader on the precision of his private information depends on whether the public signal released at date $t = 2$ confirms or not his beliefs.

Assuming an outcome dependent overconfidence, we can establish the following version of Proposition 9.12 (the proof is given in Exercise 9.14).

Proposition 9.13 *In the context of the above model with outcome dependent overconfidence (see Daniel et al. [516, Section III]), the equilibrium price of the risky security is given by*

$$p_1^* = \frac{\sigma_d^2}{\sigma_d^2 + \sigma_c^2}(\tilde{d} + \tilde{\varepsilon}),$$

$$p_2^* = \begin{cases} p_1^*, & \text{if } \tilde{s}_2 = +1, \\ \frac{\sigma_d^2}{\sigma_d^2 + \sigma_c^2 - k}(\tilde{d} + \tilde{\varepsilon}), & \text{if } \tilde{s}_2 = -1, \end{cases} \tag{9.57}$$

and $p_3^* = \tilde{d}$. As a consequence, it holds that

$$\mathrm{Cov}(p_2^* - p_1^*, p_1^* - p_0^*) > 0,$$
$$\mathrm{Cov}(p_3^* - p_1^*, p_1^* - p_0^*) < 0, \qquad\qquad (9.58)$$
$$\mathrm{Cov}(p_3^* - p_2^*, p_2^* - p_1^*) < 0.$$

The fact that $\mathrm{Cov}(p_2^* - p_1^*, p_1^* - p_0^*) > 0$ shows that the overreaction phase, and not only the correction phase, can contribute positively to short term momentum, yielding positively serially correlated returns in the short run. In particular, positive autocorrelation is due to continuing (delayed) overreaction and this allows to explain the profitability of momentum strategies. The empirical analysis in Jegadeesh & Titman [1028] provides support to this interpretation (however, difficulties arise in producing a mean reversion effect in the long run, see Fama [664]). The Daniel et al. [516] model also allows for an explanation of post-event continuation (e.g., post-earnings announcement drift, see Chap. 7).

A model similar to the one presented above with overconfident informed investors allows to address several cross-sectional anomalies discovered when testing the CAPM, as shown in Daniel et al. [517]. In the latter paper, the authors consider a model with static overconfidence in a single period. Risk averse investors use their information incorrectly when choosing their investment strategies and, as a consequence, in equilibrium securities are mispriced. In turn, this implies that variables that proxy for the mispricing are informative about future returns. According to this perspective, it is shown that overconfidence allows to rationalize the ability of the dividend yield, size, earnings/price and book to market value ratio to predict cross-sectional differences in returns. Moreover, these variables can dominate the β coefficient in explaining the cross section of expected returns. The rationale is that overconfidence leads to a misvaluation which is reflected in market prices and, hence, in the above variables. Misvaluation due to overreaction to favorable news leads to a high price, low ratios and low future returns, whereas underreaction implies the opposite relationship. It is important to remark that this perspective does not provide a risk based explanation of CAPM anomalies, but rather it represents a departure from classical asset pricing theory. Note also that rational investors exploit the pricing errors induced by overconfident agents, but they cannot fully eliminate the mispricing because of risk aversion.

A variant of the above model has been considered in Daniel & Titman [521]: in this model agents are overconfident about intangible information (e.g., management decisions' information) and not about tangible information (e.g., firm performance information). Therefore, agents tend to overreact to intangible private information and to underreact to tangible public information. The implications of these models are confirmed empirically and the book to market value ratio is shown to be a good proxy of the intangible return and, therefore, it allows to forecast future returns.

Odean [1379] analyses some models with private information, assuming that agents are rational in all respects except how they process information. Odean

[1379] considers both a perfectly competitive market with noisy supply and disseminated information (as considered for instance in Diamond & Verrecchia [569] and Hellwig [931]) as well as a market with strategic insider trading and concentrated information (as considered for instance in Kyle [1147]). Coherently with the Daniel et al. [516] model presented above, agents overweight their private signals and underweight market price signals. In a perfectly competitive market, trading volume and volatility are increasing with respect to the agents' overconfidence. Furthermore, overconfidence worsens the informational quality of market prices and the expected utility of overconfident agents is lower than that of rational agents (and, typically, overconfident agents exhibit less diversified portfolios). If overconfident agents underestimate (overestimate, resp.) new information, then price changes exhibit positive (negative, resp.) serial correlation and underreaction (overreaction, resp.). The degree of under/overreaction depends on the fraction of traders who under/over weight information. In a strategic market model with an insider, trading volume, market depth and volatility increase with respect to the insider's overconfidence (see also Benos [193] and Deaves [531] for experimental evidence on the market effects of overconfidence). In particular, the effect on the trading volume is the most robust effect of overconfidence. Overconfident insiders improve price quality, but overconfident price takers worsen it. Insider expected profits decrease in his overconfidence. However, in the context of a similar model, Kyle & Wang [1150] show that an overconfident trader may get higher expected profits-utility than a rational trader and, as a consequence, overconfident agents may persist and survive in the long run in a market with a market maker.

An analogous result is obtained in Hirshleifer & Luo [946] in the context of a perfectly competitive market: overconfident traders taking more risks perform better than rational traders at exploiting the mispricing induced by liquidity and noise traders. Assuming an imitation based population dynamics of the form (9.53), overconfident agents are shown to represent a substantial proportion of the population in correspondence of the long run steady-state equilibrium (similar results are also obtained in Fischer & Verrecchia [705]). In Hirshleifer & Luo [946, Proposition 1], it is shown that the higher the volatility of the underlying security payoff is, the higher is the proportion of overconfident traders in equilibrium and, on the contrary, the greater the confidence of the overconfident traders is, the lower is the proportion of them surviving in equilibrium. Moreover, the more volatile is liquidity/noise trading, the higher the proportion of overconfident traders is. In a multi-period economy, overconfidence can generate disagreement among agents (see Scheinkman & Xiong [1502]) and disagreement can lead to a bubble (asset prices are above the corresponding fundamentals, see Sect. 6.5), trading volume, excess volatility and return predictability (see also Dumas et al. [601]).

In several models, it has been shown that overconfidence leads to a significant trading volume and to suboptimal portfolio choices by overconfident investors and, therefore, to lower levels of expected utility (see Biais et al. [238] for experimental evidence). These implications have been empirically tested with positive evidence, see Chuang & Lee [447], Glaser & Weber [786], Grinblatt & Keloharju [831]. In Barber & Odean [143, 145] it is shown that households who trade frequently earn an annualized geometric return smaller than that of similar households trading less frequently. In this context, overconfidence can explain the high trading levels and the resulting poor performance of individual investors. Psychological research has shown that women are typically less overconfident than men: supporting this fact, Barber & Odean [144] and Graham et al. [819] show that men trade more frequently and that their average return is smaller than the average return obtained by women. In Odean [1381], it is shown that securities purchased by small traders underperform securities sold by the same agents, meaning that small investors trade too much and in the wrong direction. Note also that the tendency of typical traders to hold losers and to sell winners can be explained in terms of overconfidence. With biased self-attribution, the level of investor overconfidence and, hence, trading volume varies with past returns. Statman et al. [1565] test the trading volume predictions of formal overconfidence models and find that the share turnover is positively related to lagged returns.

We have pointed out that overconfident agents may survive in the long run, since under suitable assumptions they can earn a higher utility than rational agents. Gervais & Odean [773] analyse patterns in trading volume, volatility and expected prices as well as the profits resulting from endogenous overconfidence. In this setting, agents may learn to become overconfident by overestimating the degree to which they are responsible for their own success (*self-serving attribution bias*). There is an endogenous tendency to overconfidence: overconfidence does not make traders wealthy, but the process of becoming wealthy can make traders overconfident (in other words, traders take too much credit for their own success). Wang [1641] examines the survival of non-rational investors in an evolutionary game: underconfidence or pessimism cannot survive, but moderate overconfidence or optimism can survive and even dominate the economy, especially when the fundamental risk is significant.

Feedback Traders

Price dynamics in the presence of *feedback traders* have been analysed in Cutler et al. [509] and in De Long et al. [548]. The main feature of economies with feedback traders is that such agents buy or sell an asset on the basis of its past returns, rather than on the basis of the expectation of future fundamentals. More specifically, positive (negative, respectively) feedback traders buy (sell, respectively) an asset after a run up of its price. Positive feedback trading could result from many different phenomena, including stop loss orders, portfolio insurance strategies, technical analysis rules, risk aversion decreasing in wealth, positive wealth elasticity of the demand of risky assets, contagion among the agents in the market, asset price extrapolation, trend chasing. On the other hand, negative feedback trading can be the outcome of profit taking strategies or strategies targeting a constant share of wealth in different assets.

In Cutler et al. [509] it has been shown that the introduction of feedback traders allows to generate positive serial correlation in asset returns over short horizons as well as negative serial correlation over longer horizons. The authors consider a market for a future contract, with a well defined fundamental value which is equal to the terminal value of the contract and without dividend payments. It is assumed that the asset is in zero net supply. The economy is populated by three types of agents: *rational traders*, *fundamental traders* and *feedback traders*. The model is linear and it is assumed that the proportion of the three different types of agents in the economy is the same and does not change over time.

Rational traders invest in the asset on the basis of rational forecasts of its future returns, holding a higher proportion of the asset when expected returns are high. Denoting by θ^r_{t+1} the quantity of the asset held by the rational agents in the period $[t, t+1]$, it is assumed that

$$\theta^r_{t+1} = \gamma\big(\mathbb{E}[r_{t+1}|\mathscr{F}_t] - \rho\big), \tag{9.59}$$

for some constant $\gamma > 0$, with r_{t+1} and ρ denoting respectively the return of the asset on the period $[t, t+1]$ and the required return for holding the asset, respectively. In particular, if the expected return equals the required rate of return, then rational agents do not trade. The rational agents' demand is linear in the expected excess return of the asset.

Fundamental traders trade on the basis of the expected returns with respect to some perceived fundamentals: when prices are high relative to perceived fundamentals, then demand is low. This approach is represented by the investment rule

$$\theta^f_{t+1} = \beta\big(p_t - \alpha(L)f_t\big), \tag{9.60}$$

where θ^f_{t+1} denotes the quantity of the asset held by fundamentalist traders in the period $[t, t+1]$, $\beta < 0$ and where $\alpha(L)$ represents a delay polynomial operator defining the weights associated to the previous fundamental values, with $\alpha(1) = 1$. The process $(f_t)_{t\in\mathbb{N}}$ represents the fundamental value of the asset and is assumed to follow a random walk, i.e.,

$$f_{t+1} = f_t + \tilde{\varepsilon}_{t+1}, \qquad \text{for all } t \in \mathbb{N},$$

where $(\tilde{\varepsilon}_t)_{t\in\mathbb{N}}$ is a sequence of i.i.d. random variables with zero mean and constant variance. In (9.60), we allow $\alpha(L)$ to be different from one, thus allowing for the possibility that perceived fundamentals reflect true fundamentals with a time lag.

Finally, there are *feedback traders*, who base their demand on the past returns of the asset. More specifically, denoting by θ^{fb}_{t+1} the quantity of the asset held by feedback traders in the period $[t, t+1]$, it is assumed that

$$\theta^{fb}_{t+1} = \delta(L)(r_t - \varrho), \tag{9.61}$$

where $\delta(L)$ is a delay polynomial operator, where the sign of the weights defining it characterizes the feedback trader as a positive or a negative feedback trader.

Imposing the equilibrium condition

$$\theta_t^r + \theta_t^f + \theta_t^{fb} = 0, \qquad \text{for all } t \in \mathbb{N},$$

assuming a constant required return ϱ equal to zero and making use of (9.59)–(9.61), we obtain the following rational expectations difference equation for the equilibrium price process $(p_t^*)_{t \in \mathbb{N}}$ of the asset:

$$\mathbb{E}[p_{t+1}^*|\mathscr{F}_t] - p_t^* = -\frac{\beta}{\gamma}\left(p_t^* - \alpha(L)f_t\right) - \frac{1}{\gamma}\delta(L)(p_t^* - p_{t-1}^*), \qquad \text{for all } t \in \mathbb{N}.$$

As explained in Cutler et al. [509], solving this equation under the rational expectations hypothesis (and considering the fundamental solution, compare with Sect. 6.5), the equilibrium price of the asset can be expressed as a function of past prices, expected future fundamentals and past fundamentals. Furthermore, the fundamental innovations $(\bar{\varepsilon}_t)_{t \in \mathbb{N}}$ are fully reflected in equilibrium prices.

The model illustrated above can generate positive serial correlation in asset returns in three different ways:

- first, if fundamental traders learn about the true fundamental value of the asset with a delay (i.e., if $\alpha(L) \neq 1$), then, in the absence of noise traders ($\delta(L) = 0$), asset returns will be positively serially correlated for as many periods as it takes for the information on the asset fundamentals to be incorporated in the demand of fundamental traders;
- second, negative feedback traders (i.e., $\delta(L) < 0$) are a source of positive serial correlation. Indeed, assuming that there are no fundamental traders ($\beta = 0$), if the asset price increases (positive return), then negative feedback traders reduce their demand of the asset. This implies that in equilibrium rational agents have to hold a larger share of the asset and this can happen only if the expected return is higher than in the past. Therefore, on average successive returns will be positively serially correlated;
- finally, positive serial correlation can be due to the presence of feedback traders, whose demand depends on the asset returns on several previous periods. Hence, if the asset returns in one period affect the demand of feedback traders for many subsequent periods, feedback traders will persistently exhibit long or short positions. In turn, this implies that required returns for rational investors will be above or below the average for several periods, respectively, and this induces a positive serial correlation in asset returns. Note that this effect can occur also in the absence of fundamental traders.

Observe also that, in the first two cases, the initial reaction of the asset price to the news is incomplete. Positive feedback traders with long memory induce price overreaction to fundamental news and in the long run returns will be negatively serially correlated. A model of this type with fundamental and feedback traders provides a rationale for the presence of serial correlation in asset returns at different horizons by acting on two key factors: delay in the incorporation of fundamental news in the demand of fundamental traders and feedback trading effects (see also De Long et al. [548] for a related model).

A similar setting has been analysed in Hong & Stein [961] by considering two classes of agents with limited rationality: "newswatchers" and "momentum traders". Traders of the first type make forecasts based on private information without taking into account market prices (i.e., newswatchers fail to extract other newswatchers' information from market prices), while momentum traders use only past prices in order to form their strategies. Neither type of agent is fully rational and information gradually diffuses among agents. The first class induces an underreaction phenomenon, in the sense that firm-specific information is incorporated into prices only gradually. On the other hand, momentum traders induce an overreaction effect, since their trading strategy conditioned on past prices is similar to that of positive feedback traders. Momentum traders try to exploit the underreaction generated by newswatchers but are only able to partially eliminate the mispricing and in doing so they create an excessive momentum in prices that will ultimately lead to overreaction. In line with the implications of the model presented above, price changes are positively correlated over short horizons and negatively correlated over longer horizons. This model has been tested with positive evidence in Hong et al. [958]. Sentana & Wadhwani [1515] find that when volatility is low returns are positively autocorrelated and, on the contrary, when volatility is high returns exhibit negative autocorrelation. This suggests that positive feedback traders have a greater influence on the price when the volatility is high, while negative feedback traders dominate when the volatility is low.

Heterogeneous Agents, Market Selection and Speculation

Concerning the assumption of rationality, two conjectures have been made in the literature: i) rational agents having more sophisticated information perform better than less informed agents; ii) the speculation activity of rational agents helps to stabilize the market. The first conjecture can also be extended to an economy populated by both rational and non-rational agents, since it can be conjectured that non-rational agents earn less profits than rational agents and, hence, they are more likely to default. As a consequence, in the long run only rational and well informed agents should survive. These conjectures have been formulated by many authors, notably in Friedman [743] (*Friedman's conjecture*). The first conjecture is based on an evolutionary argument, in the sense that the market selects the best agents (the better informed agents and/or the rational agents). Of course, the validity of these two conjectures depends on the characteristics and on the population of the economy. If every agent is rational, then agents having access to better information should perform better than less informed agents (compare with Sect. 8.1). Furthermore, rational agents exploit mispricings and, therefore, their trading activity stabilizes market prices around the corresponding fundamental values. However, in an economy populated by heterogeneous agents, the two conjectures are not necessarily valid. For instance, we have already seen in this section that noise traders are not necessarily eliminated from the market in the long run. The evolution of the population of an economy has been analysed by considering the wealth dynamics of different types of agents or the evolution of the agents' trading behavior via methods developed in the evolutionary game literature. For instance, as considered in (9.53), agents tend to imitate the trading strategy which has better performed in the past.

In Figlewski [701], the validity of the first conjecture has been discussed and related to the efficiency of the market (in a strong form). The author considers an economy with two types of agents, which can differ in price expectations, risk aversion, predictive ability and wealth. Each of the two groups of agents is composed of identical individuals. The two types of agents receive different information and their prediction error (with respect to the fundamental value of the asset) is normally distributed, with zero mean and a variance which depends on the type of agent. It is assumed that every agent determines his demand by maximizing a mean-variance utility function. Figlewski [701] shows that in the short run the market tends towards efficiency, but perfect efficiency is not reached, not even in the long run. Indeed, in the long run the economy exhibits deviations from the distribution of wealth that would prevail in a fully efficient market. In the long run, better informed agents hold more wealth than less informed agents (see Figlewski [701, Table 1]). A model exhibiting convergence towards market efficiency has been proposed in Luo [1254].

As already mentioned, non-rational (noise or overconfident) agents are not necessarily eliminated from the market but can survive in the long run and may earn higher returns than rational agents. In some cases, non-rational agents can even dominate the market. The results obtained in Biais & Shadur [241] are along these lines, thus providing a counter-argument to Friedman's conjecture. Biais & Shadur [241] consider a market with rational agents and non-rational overoptimistic sellers (or pessimistic buyers), with the dynamics of the population depending on the realized utility. It is shown that, in the long run, non-rational agents are concentrated on one side of the market generating under or overpricing. Moreover, the survival of non-rational traders reduces the opportunities for profitable risk sharing and induces a welfare loss for the economy. This model also offers an alternative interpretation of bubbles observed in speculative markets suggesting that, even though bubbles can be generated by investors' irrationality, they fail to be corrected by the forces of market selection. In a similar direction, it has been shown in Blume & Easley [255] and Shefrin & Statman [1532] that non-rational agents (i.e., agents who do not update their beliefs according to Bayes' rule) may survive in the market.

On the other hand, Blume & Easley [256] have provided a formal proof of the market selection hypothesis, see also Sandroni [1495]. The authors consider a general equilibrium model in infinite horizon and show that, if markets are complete, assuming that the aggregate endowment is bounded from above and away from zero and controlling for the discount factor, then only the traders with correct beliefs can survive in the long run. In other words, the market selects those agents who update their beliefs according to Bayes' rule and for whom the true distribution belongs to the support of the prior distributions. Moreover, agents' preferences (risk aversion) do not affect the long run fate of traders, only the discount factor matters: a low discount rate leads to a high saving rate and, hence, a large consumption in the future. Therefore, irrational agents with a discount rate lower than that of rational agents may survive in the long run: due to incorrect beliefs, irrational agents take bad investment decisions but nevertheless they save more, thus consume more in the future and in the long run they may even survive. However, this result can fail when markets are incomplete, when consumption grows too quickly, or when discount factors and beliefs are correlated. In particular, the boundedness of the aggregate endowment represents a strong assumption. Indeed, assuming an economy with an unbounded growth rate, the market selection hypothesis does not hold in general. Kogan et al. [1116] have shown that the market selection hypothesis does not hold under the assumption that the aggregate endowment evolves as a geometric Brownian motion and that irrational traders have incorrect opinions on the drift of the endowment process. In this case, the expected aggregate endowment grows as an exponentially linear function and agents disagree on the coefficient appearing in the exponential. Kogan et al. [1116, Proposition 4] shows that moderately optimistic irrational agents (who overestimate the rate of growth of the aggregate endowment) may dominate the market in the long run and rational traders may disappear. In a related context, Yan [1663] has shown that irrational agents with a sufficiently low discount rate (or a sufficiently high elasticity of intertemporal substitution) save more and are more likely to survive in the future. For an analysis in a mean reverting environment we refer to Barucci & Casna [170]. The survivorship of irrational agents has been confirmed in a bounded rationality framework in Berrada [207].

In the absence of full rationality, agents typically adopt a "rule of thumb" when forming forecasts. This behavior can be justified by an evolutionary argument when agents have to choose how to form their forecasts. In the context of a simple asset pricing problem, Brock & Hommes [306] have analysed an economy with heterogeneous agents: fundamental traders and agents who forecast the future price by looking at past deviations of the price from the fundamental value (such agents can represent trend followers, contrarians or technical analysts). Similarly as in the imitation-based population dynamics (9.53), agents switch to the more successful forecasting rule on the basis of its past performance. In Brock & Hommes [306] it is shown that this behavior yields a highly irregular evolution of the economy, with chaotic asset price fluctuations when the switching intensity from one forecasting rule to another is high. This result finds confirmation in LeBaron et al. [1172], where the agents' evolution is modeled via a genetic algorithm. A model of this type allows to generate serially correlated returns and patterns of volatility and volume which are consistent with those empirically observed (see Brock & LeBaron [309], Gaunersdorfer [762], Boswijk et al. [272], Chiarella & He [429]). In Routledge [1474] a process of adaptive (or evolutionary) learning is analysed in the context of the model proposed in Grossman & Stiglitz [849] (see Sect. 8.3). Agents must choose whether or not they wish to become informed on the dividend of a risky asset and the proportion of informed agents in the economy evolves by means of experimentation and imitation, in the sense that successful behavior is imitated. Therefore, if being informed allows to reach a higher level of utility, then more agents will buy information in the next period. Unlike in Grossman & Stiglitz [849], both the choice of whether to buy information and the inference are determined by adaptive learning rather than optimization, thus capturing non-rationality. Routledge [1474] shows that the equilibrium defined in Grossman & Stiglitz [849] is asymptotically stable with respect to the learning process.

The second conjecture reported above (namely, that the speculation activity of rational traders helps to stabilize the market) also does not hold in general. Indeed, as shown in Hart & Kreps [907], even speculation with storage in a commodity market may lead to a destabilization of the market. Stein [1567] shows that the introduction of a new group of speculators may destabilize prices and cause welfare losses to the economy.

The stabilizing effect of rational trading (speculation) has been analysed in a large literature (see, e.g., De Long et al. [548, 549]). The analysis developed in De Long et al. [549] suggests that, in the presence of a mean reverting process for the noise traders' misperception, the optimal investment strategy is not necessarily the strategy of investing on the basis of the expected return of the asset (rational speculation). Indeed, the optimal strategy for sophisticated investors is a market timing strategy, with an increased exposure to stocks after their prices have fallen and a decreased exposure after they have risen. In other words, the optimal trading strategy for sophisticated investors seems to be in line with the Keynes' quotation reported at the beginning of this chapter: buy when noise traders want to sell and sell when noise traders want to buy (a contrarian strategy, see De Long et al. [549, Section IV]). Note also the similarity of this strategy to the trading behavior of feedback traders. In this setting, in an economy with non-rational agents, rational agents (sophisticated investors) must consider the beliefs and the behavior of non-rational agents. In this setting, rational speculation is always stabilizing, but the average returns earned by rational speculators is not necessarily as high as those earned by noise traders, thus showing that "destabilizing speculation" (i.e., the trading activity of noise traders) is not necessarily unprofitable.

De Long et al. [548] show that speculation can destabilize the market in the presence of positive feedback traders. Moreover, feedback traders and rational agents together induce an increase in asset price volatility. The rationale of this effect can be traced back to Keynes' quotation: suppose that rational agents receive positive news about the asset and buy it, making the price climb up, then positive feedback traders react to this price increase by buying the asset and, therefore, feedback traders make the price increase even further. Rational agents anticipate this behavior of feedback traders and, therefore, they act to increase the price well above the fundamental value (by buying more of the asset today) in order to sell the asset at a higher price at a later date. In other words, the fact that feedback traders will buy the asset at a later date induces rational traders to make the price of the asset overreact to the information arrived in the market. The combined effect of these two classes of traders generates positive serial correlation of returns. Note that in this setting rational agents buy an asset when its price is likely to run up and this may also occur when the price is already above the corresponding fundamental value. In this way, speculation can destabilize the market. Similar results have been obtained in Hirshleifer et al. [947] in the case where agents acquire the same information sequentially.

Summing up, the analysis developed in the present section shows that rational agents have two motives to trade in the market: buy an asset in a buy and hold perspective in order to receive its future dividends and buy an asset in order to sell it at a later date at a more favorable price. If there are rational agents with the same time horizon of the economy and liquidity traders (price inelastic demand), then the first motive for trade is more relevant and, therefore, speculation stabilizes the market. On the other hand, in a market where non-rational agents are present the second motive plays an important role, especially in a short time horizon, and speculation may destabilize the market.

9.4   Bounded Rationality and Distorted Beliefs

In the analysis developed in the previous chapters, one of the main hypotheses is that of *rational expectations*. The rational expectations hypothesis consists of two main assumptions: i) agents perfectly know the economic model; ii) agents are able to efficiently elaborate all the available information and rationally update their beliefs according to Bayes' rule.

Clearly, the rational expectations hypothesis represents a strong assumption on the behavior of the agents. In this regard, Radner assessed as follows the role of the rational expectations and perfect foresight hypotheses (see Radner [1439]): "the perfect foresight approach is contrary to the spirit of much of competitive market theory in that it postulates that individual traders must be able to forecast, in some sense, the equilibrium prices that will prevail in the future under alternative states of the environment. [...] This approach still seems to require of the traders a capacity for imagination and computation far beyond what is realistic. An equilibrium of plans and price expectations might be appropriate as a conceptualization of the ideal goal of indicative planning, or of long run steady state toward which the economy might tend in a stationary stochastic environment".

In this section, we consider two different ways of relaxing the rational expectations (perfect foresight) hypothesis (see Thaler [1587] for a survey on the topic):

- *bounded rationality and learning*, meaning that agents are assumed to believe in a simplified model (different from the true underlying model) and update their beliefs via some learning mechanism;
- *distorted beliefs*, meaning that agents' beliefs are modeled by taking into account psychological biases;

In the notes at the end of the chapter, we shall also briefly address models with incomplete information, where the economy is affected by some unobservable factors which have to be estimated by the agents.

Bounded Rationality and Learning

In a nutshell, *bounded rationality* is based on two main ideas: i) agents do not have a complete knowledge of the true underlying economic model but rather believe in some simplified model; ii) agents use the simplified model when formulating their expectations and update their beliefs via some recursive learning mechanism (see Evans & Honkapohja [657] for an overview of this strand of literature).

In order to illustrate an economy where agents exhibit bounded rationality, let us consider the fundamental equation (6.85). For simplicity, let us assume the existence of a risk neutral representative agent and suppose furthermore that the dividend process $(d_t)_{t\in\mathbb{N}}$ is predictable, meaning that the dividend at date d_{t+1} is known at date t, for each $t \in \mathbb{N}$. In this case, equation (6.85) can be rewritten as

$$s_t = \frac{1}{r_f} \left(\mathbb{E}[s_{t+1}|\mathscr{F}_t] + d_{t+1} \right), \qquad \text{for all } t \in \mathbb{N}, \qquad (9.62)$$

where $1/r_f$ is the discount factor of the representative agent, $(s_t)_{t\in\mathbb{N}}$ is the price process of the security paying the dividend stream $(d_t)_{t\in\mathbb{N}}$ and the filtration $(\mathscr{F}_t)_{t\in\mathbb{N}}$ represents the information flow of the economy. Of course, $\mathbb{E}[s_{t+1}|\mathscr{F}_t]$ is the expectation of the price prevailing at date $t + 1$ on the basis of all available information up to date t. Let us introduce the further assumption that the dividend process $(d_t)_{t\in\mathbb{N}}$ follows a first order autoregressive process of the form

$$d_{t+1} = \alpha + \psi d_t + \tilde{\varepsilon}_t, \qquad \text{for all } t \in \mathbb{N}, \qquad (9.63)$$

where $(\tilde{\varepsilon}_t)_{t\in\mathbb{N}}$ is a sequence of i.i.d. random variables with zero mean and where $\alpha \in \mathbb{R}$ and $\psi \in (-1, 1)$ are some parameters which are assumed to be unknown to the agents.

As shown in Exercise 9.15, under the above assumptions, the rational expectations price process $(s_t)_{t\in\mathbb{N}}$ of the security paying the dividend stream $(d_t)_{t\in\mathbb{N}}$ is given by

$$s_t = \frac{\alpha}{(r_f - 1)(r_f - \psi)} + \frac{1}{r_f - \psi} d_{t+1} + \beta_t, \qquad \text{for all } t \in \mathbb{N}, \qquad (9.64)$$

where $(\beta_t)_{t\in\mathbb{N}}$ represents the bubble component (see Sect. 6.5). In particular, the fundamental (bubble free) solution under rational expectations is obtained by taking $\beta = 0$. Formula (9.64) shows that the fundamental value of a security under rational expectations is linear with respect to the dividend released at date

$t + 1$. According to the rational expectations hypothesis, the representative agent formulates his expectation of the future price of a security by exploiting all the available information.

On the other hand, in a bounded rationality setting, the agent formulates his expectations on the basis of a simplified model which can be taken of the following form:

$$s_t = \gamma_{t-1} + \lambda_{t-1} d_{t+1} + \beta_t, \qquad \text{for all } t \in \mathbb{N}, \tag{9.65}$$

where $(\gamma_t)_{t \in \mathbb{N}}$ and $(\lambda_t)_{t \in \mathbb{N}}$ are two sequences of parameters which are estimated on the basis of past prices and dividends according to some *learning rule* (for instance, ordinary least squares). The simplified model (9.65) is misspecified and coincides with the rational expectations solution only when it is correctly specified, i.e., when γ_t and λ_t coincide with the true values of the coefficients given in (9.64). The asymptotic behavior of γ_t and λ_t as a result of a learning procedure can be described by a system of differential equations having as equilibrium the rational expectations solution given in (9.64). In this sense, the bounded rationality hypothesis has interesting implications. Indeed, even in the most simple asset pricing model (as considered in Timmermann [1590]), the rational expectations solution admits an evolutionary justification by considering the limit behavior of the agents' learning process. In other words, if agents do not know the true underlying economic model, but they use the public information to make their forecasts and they update their beliefs by means of a learning algorithm, then, under suitable conditions, the economy will converge to the rational expectations equilibrium price.

In Evans [654, 655] and Adam & Marcet [14] it is shown that the fundamental solution is stable, without restrictions on the parameters of the model. Depending on the model specification, it is also possible that the learning process converges to a bubble solution, as considered for instance in Branch & Evans [278] by assuming a learning procedure based on ordinary least squares. The convergence of the bounded rationality equilibrium price to the rational expectations price can also be established without introducing the assumption that the dividend process is predictable, as considered above. In this case, an agent must estimate both the expected dividend and the expected price, typically by relying on an autoregressive (misspecified) model for the dividend and on a model of the type described above for the asset price (see Timmermann [1590, 1593] and Sogner & Mitlohner [1554]). On the other hand, the convergence of the learning procedure is not ensured in the presence of a feedback effect of the price on the dividend in the case of multiple rational expectations equilibria, see Timmermann [1592].

As explained above, an important characteristic of a bounded rationality economy is that, under suitable conditions, the learning procedure converges to the rational expectations equilibrium in the long run. However, in a finite time horizon, a bounded rationality economy may significantly depart from a rational expectations economy. In turn, this fact opens the door to the introduction of bounded rationality as a potential explanation for many of the asset pricing anomalies encountered when testing the implications of classical asset pricing theory (see Chap. 7).

In particular, Buckley & Tonks [321] suggest that bounded rationality may explain the excess volatility phenomenon. The authors remark that most of the criticisms on excess volatility results have focused on the statistical properties of the tests, without paying sufficient attention to the fact that most of the volatility tests assume a strong form of the rational expectations hypothesis, namely that the agents know and employ the true underlying economic model when forming expectations (for instance, the detrending procedure adopted in the seminal work of Shiller [1533] is based on the full observation sample). Hence, the empirical evidence reported in the literature on excess volatility can also point at violations of the rational expectations hypothesis. Starting from this observation, Buckley & Tonks [321] suggest to test excess volatility by assuming a weak form of rationality (*bounded rationality*), assuming that agents form expectations by relying on a model estimated from historical data using unbiased estimation techniques, where at each date t the expectations are formed on the basis of the data available only up to date t.

Let us present in more detail the volatility test proposed by Buckley & Tonks [321]. Following Shiller [1533], the perfect foresight rational price (or ex-post rational price) of a security paying the dividend stream $(d_t)_{t \in \mathbb{N}}$ is given by $s_t^e := \sum_{s=1}^{\infty} d_{t+s}/r_f^s$, for $t \in \mathbb{N}$, as in formula (7.2). Buckley & Tonks [321] consider the following weak form rational expectation of the perfect foresight price, given the information set $\mathscr{F}_t$ available at date t:

$$\mathbb{E}_{\hat{\theta}_t}[s_t^e | \mathscr{F}_t] = \sum_{s=1}^{\infty} \frac{1}{r_f^s} \mathbb{E}_{\hat{\theta}_t}[d_{t+s} | \mathscr{F}_t], \qquad \text{for all } t \in \mathbb{N}, \tag{9.66}$$

where θ is a vector of parameters describing the model generating the dividends and $\hat{\theta}_t$ denotes an estimation of θ given the information $\mathscr{F}_t$ available up to date t. Observe the double role of the information in relation (9.66): on the one hand, the available information appears directly as the conditioning information $\mathscr{F}_t$, representing the current realization of the explanatory variables of the model. On the other hand, the information is used for computing the estimates $\hat{\theta}_t$ of the parameters θ, which are then used for computing the conditional expectation appearing in (9.66). In order to derive the volatility test of Buckley & Tonks [321], we start by expressing the perfect foresight price s_t^e in the following form:

$$s_t^e = \mathbb{E}_{\theta}[s_t^e | \mathscr{F}_t] + \tilde{v}_t, \tag{9.67}$$

for all $t \in \mathbb{N}$, where $\mathbb{E}_{\theta}[\cdot | \mathscr{F}_t]$ denotes the $\mathscr{F}_t$-conditional expectation assuming that the true underlying model is known and where $\tilde{v}_t$ is a random variable uncorrelated with $\mathbb{E}_{\theta}[s_t^e | \mathscr{F}_t]$ and with mean zero. On the other hand, under weak form rational expectations, in an efficient market the actual price s_t is given by

$$s_t = \mathbb{E}_{\hat{\theta}_t}[s_t^e | \mathscr{F}_t]. \tag{9.68}$$

Assuming that agents form their forecasts by relying on unbiased estimation techniques, we may write

$$\mathbb{E}_{\hat{\theta}_t}[s_t^e | \mathscr{F}_t] = \mathbb{E}_\theta[s_t^e | \mathscr{F}_t] + \tilde{u}_t, \tag{9.69}$$

where $\tilde{u}_t$ represents the forecast error due to the use of an estimated model. From (9.67) and (9.69) we get

$$s_t^e = \mathbb{E}_{\hat{\theta}_t}[s_t^e | \mathscr{F}_t] - \tilde{u}_t + \tilde{v}_t,$$

for all $t \in \mathbb{N}$. Assuming that the random variables $\tilde{u}_t$ and $\tilde{v}_t$ are uncorrelated, for every $t \in \mathbb{N}$, the last relation yields the following inequality:

$$\mathbb{E}\left[\left(s_t^e - \mathbb{E}_{\hat{\theta}_t}[s_t^e | \mathscr{F}_t]\right)^2\right] \geq \mathbb{E}\left[\left(s_t - \mathbb{E}_{\hat{\theta}_t}[s_t^e | \mathscr{F}_t]\right)^2\right], \tag{9.70}$$

for all $t \in \mathbb{N}$. This inequality captures the fact that the actual price s_t, being an expected value (see (9.68)), should be closer than the realization s_t^e to the predicted value of s_t^e. In particular, inequality (9.70) should be compared to the analogous inequality (7.6) adopted in the context of the Shiller [1533] volatility test.

Buckley & Tonks [321] apply this methodology to UK data, providing empirical evidence that the volatility bound (9.70) is not violated (however, the inequality is rejected on U.S. data, see Shiller [1538]). Part of the excess volatility can be attributed to revisions of the estimated parameters $\hat{\theta}_t$ of the dividend model by agents with bounded rationality, since the true parameters θ are not known to the agents. In a related context, Timmermann [1592] shows that, before reaching convergence to the rational expectations equilibrium, asset prices exhibit a high volatility, thus suggesting that the existence of a learning process may explain excess volatility.

In Timmermann [1590] it is shown that learning is a natural candidate in order to explain the predictability of asset returns and the excess volatility of asset prices. The author considers a geometric random walk for the evolution of dividends:

$$\log d_{t+1} = \mu + \log d_t + \tilde{\varepsilon}_t, \qquad \text{for all } t \in \mathbb{N}, \tag{9.71}$$

where $(\tilde{\varepsilon})_{t \in \mathbb{N}}$ is a sequence of i.i.d. random variables with mean zero and constant variance σ_ε^2 and where $\mu + \sigma_\varepsilon^2/2 < \log r_f$. Similarly as above, by adopting the rational expectations equation (6.85) (under the assumption of a risk neutral representative agent), it follows that the fundamental (bubble free) rational expectations solution is given by

$$s_t^* = \frac{g}{r_f - g} d_t, \qquad \text{for all } t \in \mathbb{N},$$

where $g := \exp(\mu + \sigma_\varepsilon^2/2)$, thus showing that the fundamental rational expectations price is proportional to the dividend. However, in the presence of bounded rationality, the agents do not know the true values of the parameters μ and σ_ε^2. Agents need to estimate the values of those parameters on the basis of past dividend observations. In the long run, agents learn the true values of the parameters μ and σ_ε^2. However, in the transient period the learning process induces excess price volatility and correlation between returns and the lagged dividend yield. The rationale behind this result is simple, noting that agents need to estimate the dividend growth rate. Indeed, suppose that an agent has underestimated the dividend growth rate at some date t (i.e., the estimated growth rate is lower than the true unknown growth rate). In this case, the agent's forecasts of future dividends will be smaller than their "true" values and the stock price, computed as the present value of expected future dividends, will be lower than the price computed on the basis of the true growth rate of the dividends. With significant probability, the dividend realized at the following date $t + 1$ will be higher than the forecasted value and this will induce the agent to revise upwards the estimated dividend growth rate, thus increasing the estimated price of the asset. This learning effect produces a positive correlation between dividend yields and future returns, whereas returns display only weak serial correlation.

The impact of learning on volatility is more difficult to assess. As shown above, in correspondence of the rational expectations fundamental solution, the asset price is proportional to the dividend, with a proportionality factor that does not depend on time. This implies that dividend shocks will be reflected in proportional shocks in the stock price. Under bounded rationality, learning implies an additional effect on stock prices since the estimated dividend growth rate is also influenced by the dividend shock. A priori, it is not clear how the two effects relate, i.e., whether they offset each other or whether they act in the same direction. In Timmermann [1590], by relying on a simulation analysis, it is shown that learning leads to an increased volatility, in the presence of a sufficiently small sample. These results have been confirmed assuming different dynamical models for the dividends (see Timmermann [1593], Guidolin & Timmermann [855], Branch & Evans [278, 277], Hong & Stein [964]). Similar results have been also obtained in Lewellen & Shanken [1214] and Ozguz [1390] with Bayesian learning in a constant or time-varying expected returns setting: they show that learning contributes to explain CAPM anomalies and the predictability of the market return through the aggregate dividend yield. See however Carceles-Poveda & Giannitsarou [366] for a skeptical view on the capability of bounded rationality to explain excess volatility and return predictability.

In Barsky & De Long [168], bounded rationality is invoked in order to explain the overreaction of prices to dividend shocks, i.e., the fact that a long run 1% increase in the level of dividends is associated with an approximately 1.5% increase in equity values (in other words, the price elasticity of the market index with respect to dividends is greater than one). As explained above, if the logarithmic dividends follow a random walk, then the dividend yield is constant and this overreaction phenomenon cannot occur. For this reason, Barsky & De Long [168] assume that

the long run dividend growth rate is uncertain and time-varying and propose the
following dynamics:

$$\Delta \log d_t = \tilde{\varepsilon}_t + \sum_{k=1}^{t-1}(1-\theta)\tilde{\varepsilon}_{t-k} + g_0 =: \tilde{\varepsilon}_t + g_t, \qquad \text{for all } t \in \mathbb{N}, \qquad (9.72)$$

where g_0 is the permanent dividend growth rate as of date $t = 0$ and $(g_t)_{t \in \mathbb{N}}$ is
defined recursively by

$$g_{t+1} = g_t + (1-\theta)\tilde{\varepsilon}_t, \qquad \text{for all } t \in \mathbb{N}.$$

The random variables $(\tilde{\varepsilon}_t)_{t \in \mathbb{N}}$ represent shocks to the dividend growth rates that not
only affect $\Delta \log d_t$ (through the first term on the right-hand side of (9.72)) but also
have an attenuated but permanent effect on future dividend growth rates (through
the second term on the right-hand side of (9.72)). By relying on (9.72), the expected
dividend growth rate, conditionally on the information available at date t, is given
by

$$\mathbb{E}[\Delta \log d_{t+j}|\mathscr{F}_t] = g_t = \sum_{k=1}^{t-1}(1-\theta)\tilde{\varepsilon}_{t-k} + g_0,$$

for all $t, j \in \mathbb{N}$. As can be easily verified, the permanent dividend growth rate can
therefore be written as a geometric average of past dividend shocks:

$$g_t = (1-\theta)\sum_{k=0}^{t-1}\theta^k \Delta \log d_{t-1-k} + \theta^t g_0.$$

This formula shows that agents extrapolate past dividend growth into the future and
compute prices by taking the expectation of future dividends estimated according to
the above growth rate. This model allows to generate overreaction effects in financial
time series (long swings in stock prices after a shock in the dividends).

In Barucci et al. [173], assuming an autoregressive stationary process for the dividends, the authors analyse a bounded rationality economy when agents update their
price cum dividend expectation according to an adaptive scheme. They show that the
learning procedure introduces serial correlation in returns and correlation between
lagged dividend yields and returns. In particular, serial correlation is positive when
returns are computed over a small time window and negative when the window
is long. The bounded rationality learning mechanism generates positive/negative
serial correlation in returns, as documented in the asset pricing empirical literature.
Overreaction and delayed overreaction of prices to dividend news induce the above
serial correlation effects. The memory of the learning mechanism plays a crucial
role: a longer memory induces a smaller degree of dependence when the horizon of
the return is long (the mean reversion effect is weaker), while over a short horizon

the serial correlation of returns and their correlation with the dividend yield is always significant and positive. The interesting point is that long run dependencies in financial time series can be explained by short memory in the learning mechanism of the agents. When the dividend at each time can be zero or constant with uncertain probability, Bayesian learning induces mean reversion in asset returns (see Cassano [377]).

Veronesi [1616] analyses an economy where stock dividends are generated by a Gaussian diffusion process, with the dividend growth rate (the drift of the diffusion) following a non observable two-state Markov process, with the two states representing respectively a high and a low growth rate. The crucial point is that agents cannot observe the dividend growth rate but rather have to estimate it on the basis of realized dividends. It is shown that the equilibrium price of the asset is an increasing and convex function of the investor's posterior probability of the high state of the unobservable Markov process. This effect is due to the fact that, when the posterior probability is around $1/2$, the agent requires a positive risk premium for the high uncertainty which is absent when the model becomes close to full observability (i.e., a posterior probability close to zero or one). As a consequence, prices show overreaction to (bad) news in good times and underreaction to (good) news in bad times, an asymmetric behavior widely observed empirically. Moreover, this non-linearity increases with the investors' degree of risk aversion. The model also generates persistence in volatility changes, excess volatility (with a maximum when there is a high degree of uncertainty and in recessions) and time-varying expected returns. In a similar model, assuming a very small probability for the economy to enter into a very long recession state (a small probability of switching to a low dividend growth rate regime), Veronesi [1617] shows that, if the deep recession does not occur, then the model may generate ex-post a high equity premium, as suggested in Rietz [1451].

Finally, it is worth to mention that bounded rationality may also help to explain the equity premium puzzle, see Brandt et al. [281], Guidolin [854], Cogley & Sargent [470], Weitzman [1651] and also Liu et al. [1227] for a model with rare disasters. Assuming heterogeneous agents endowed with diverse correlated beliefs and bounded rationality (agents do not know the true probability distribution of the dividend process), it is possible to address the equity premium, the risk free rate puzzle and excess volatility, see Kurz & Motolese [1145]. In the latter work, the authors argue that most of the observed volatility in financial markets is generated by the agents' beliefs and the asset pricing puzzles are all driven by the structure of market expectations. In this context, there is a premium for heterogeneous beliefs.

Distorted Beliefs

A model based on bounded rationality and generating underreaction or overreaction phenomena (as depicted in Bernard & Thomas [202]) and, therefore, positive and negative serial correlation in asset returns has been proposed in Barberis et al. [153].

The model is based on two well grounded behavioral patterns: *representativeness heuristic* and *conservatism*. Representativeness heuristic means that agents view events as typical or representative of some specific class and ignore the true probability law, whereas conservatism consists in the slow updating in face of new evidence.

In the Barberis et al. [153] model, the earnings process $(x_t)_{t \in \mathbb{N}}$ is supposed to follow a random walk of the form

$$x_{t+1} = x_t + \tilde{y}_t, \qquad \text{for all } t \in \mathbb{N},$$

where $\tilde{y}_t$ represents the shock to earnings at date t and is supposed to take one of two possible values $\{-y, +y\}$. All the earnings are paid out as dividends. The agents do not realize that earnings follow a random walk. Instead, they believe that the value of $\tilde{y}_t$ is determined by one of two models, depending on the state of the economy. According to the first model, it holds that

$$\mathbb{P}(\tilde{y}_{t+1} = +y \mid \tilde{y}_t = +y) = \pi_L \quad \text{and} \quad \mathbb{P}(\tilde{y}_{t+1} = +y \mid \tilde{y}_t = -y) = 1 - \pi_L.$$

According to the second model, it holds that

$$\mathbb{P}(\tilde{y}_{t+1} = +y \mid \tilde{y}_t = +y) = \pi_H \quad \text{and} \quad \mathbb{P}(\tilde{y}_{t+1} = +y \mid \tilde{y}_t = -y) = 1 - \pi_H.$$

The difference between the two models is represented by the fact that

$$0 < \pi_L < 1/2 < \pi_H < 1.$$

This hypothesis captures the idea that, according to the first model, a positive shock is likely to be reversed (mean reversion), while, according to the second model, a positive shock is likely to be followed by another positive shock (trend). The agents are convinced that they know the probabilities π_L and π_H. Moreover, they are convinced that the process $(\eta_t)_{t \in \mathbb{N}}$ governing the transition from the first model to the second model is also a Markov process (so that the state of the economy tomorrow only depends on the state of the economy today), taking values in $\{1, 2\}$ and with transition probabilities

$$\mathbb{P}(\eta_{t+1} = 2 \mid \eta_t = 1) = \lambda_1 \quad \text{and} \quad \mathbb{P}(\eta_{t+1} = 1 \mid \eta_t = 2) = \lambda_2.$$

If $\eta_t = 1$, then the earnings shocks $(\tilde{y})_{t \in \mathbb{N}}$ evolve according to the first model (mean reversion), while, if $\eta_t = 2$, then they evolve according to the second model (trend). Barberis et al. [153] focus on small values for the parameters λ_1 and λ_2, meaning that transitions from one regime to the other occur rarely. In particular, it is assumed that $\lambda_1 + \lambda_2 < 1$ and $\lambda_1 < \lambda_2$. Since $\mathbb{P}(\eta_t = 1) = \lambda_2/(\lambda_1 + \lambda_2)$, agents think that the mean reverting model is more likely than the trend model.

In order to value the security paying the earnings $(x_t)_{t \in \mathbb{N}}$ as dividends, agents have to forecast future earnings. According to the above description, the evolution of earnings is driven by one of two possible Markov models and the investor tries to understand, on the basis of the earnings' observations, in which of the two regimes is the state of the economy. At each date $t \in \mathbb{N}$, the conditional probability of the economy being in state 1 (where the earnings are governed by the mean reverting model) is given by

$$q_t = \mathbb{P}(\eta_t = 1 | y_t, y_{t-1}, q_{t-1}),$$

where q_t is computed by relying on the Bayes rule (see Barberis et al. [153, Section 4.2] for the explicit expression of q_t). Assuming the existence of a risk neutral representative agent with discount factor δ, the security price is given by the conditional expectation of discounted future earnings. If the representative agent was able to realize that the earning process is a random walk, then the asset price would be equal to x_t/δ, at every date $t \in \mathbb{N}$. However, the representative agent does not realize that earnings are simply driven by a random walk process, but rather believes in some combination of the two models described above, neither of which is a random walk. In this context, the following proposition holds (we refer to the Appendix of Barberis et al. [153] for the proof).

Proposition 9.14 *In the context of the above model (see Barberis et al. [153, Section 4.2]), the equilibrium price process* $(s_t)_{t \in \mathbb{N}}$ *of the security paying the earnings* $(x_t)_{t \in \mathbb{N}}$ *is given by*

$$s_t = \frac{x_t}{\delta} + y_t(p_1 - p_2 q_t), \qquad \text{for all } t \in \mathbb{N}, \tag{9.73}$$

where p_1 *and* p_2 *are two constants depending on* π_L, π_H, λ_1 *and* λ_2. *Moreover, if*

$$k_* p_2 < p_1 < k^* p_2 \qquad \text{and} \qquad p_2 \geq 0,$$

where k_* *and* k^* *are two constants depending on* π_L, π_H, λ_1 *and* λ_2, *then the price process* $(s_t)_{t \in \mathbb{N}}$ *given in (9.73) exhibits both underreaction and overreaction to earnings.*

By means of numerical simulations, Barberis et al. [153] show that, according to the above proposition, underreaction and overreaction are generated for a wide range of the values of the model's parameters. Moreover, the earnings to price ratio has predictive power.

Uncertainty on the dividend growth rate under incomplete information (see also below in this section) and different learning mechanisms have been analysed in Brandt et al. [281]. The authors study the effects on the risk premium and volatility of Bayesian and suboptimal learning rules, optimism, pessimism, conservatism and limited memory. Bayesian learning leads to realistic variations in the conditional equity risk premium, return volatility and Sharpe ratio, while significantly different results are obtained under alternative learning rules. Moreover, it is shown that, when agents are conscious of their limits and take them into account, then the economy with rational (Bayesian) learning and that with suboptimal learning are almost indistinguishable. When investors have to estimate an unknown parameter which is relevant for the valuation, conservatism (prior beliefs receive excessive weights) and representativeness heuristic (agents overweight recent observations) have been also analysed in Brav & Heaton [284], showing that overreaction and underreaction arise in both cases.

In Cecchetti et al. [381], distorted beliefs without learning have been considered in the context of a model which is classical in every respect, apart from the fact that the beliefs on the endowment growth are distorted. Agents observe the true state of the economy, but they do not know the true transition probabilities (as estimated from U.S. data by means of maximum likelihood techniques) between the high growth regime and the low growth regime as well as the true growth rates. Assuming pessimism about the persistence of the expansion state and optimism about the persistence of the contraction state (in other words, beliefs are distorted towards a lower growth rate for the economy), the risk free rate and the risk premium observed historically are matched with a time separable utility function and a rather low coefficient of risk aversion (below 10). If beliefs about transition probabilities fluctuate randomly about their subjective (distorted) mean values, then the model also generates excess volatility and predictability in asset returns. Similar results are obtained in Abel [7] assuming *pessimism* and *doubt* on the distribution of the growth rates of aggregate consumption. Pessimism corresponds to a leftward translation of the objective distribution of the logarithmic growth rate (the subjective distribution is dominated by the objective distribution according to first order stochastic dominance, see Sect. 2.3). Doubt means that the subjective distribution is a mean preserving spread of the objective distribution. As shown in Cabrales & Hoshi [326], a model with optimist and pessimist agents may generate time-varying conditional variance of asset returns and explain the relationship between volume and volatility. On the relevance of pessimism as an explanation of the equity premium puzzle see also Cogley & Sargent [470] and Guidolin [854] and Jouini & Napp [1051, 1052] and Bhamra & Uppal [220] in the case of heterogeneous beliefs.

9.5 Incomplete and Imperfect Markets

In most of the analysis developed in the previous chapters, we have assumed that financial markets were *frictionless*, in the sense that there are no transaction costs, taxes, trading constraints or liquidity constraints. In the present section, we discuss the relevance of these hypotheses and present a broad overview of some of the main results obtained in the financial economics literature by relaxing those assumptions. In particular, we have seen that many crucial results rely on the assumption of *market completeness* and we start this section by analysing the role of the market completeness hypothesis in financial economics.

Market Incompleteness

As we have seen in Sect. 4.2, the assumption of market completeness has important implications from the point of view of equilibrium theory, especially for the Pareto optimality of equilibrium allocations. Indeed, market completeness implies that agents are unrestricted in their possibilities to share risks and, therefore, every equilibrium allocation will be Pareto optimal and all the idiosyncratic risk will be diversified away. In correspondence of an equilibrium, the optimal consumption of every agent depends only on aggregate risk and the marginal rates of substitution are identical across all agents. Moreover, the analysis of an economy with heterogeneous agents is simplified if markets are complete. Indeed, under suitable conditions (see Chap. 4), the original economy with multiple agents is observationally equivalent to a representative agent economy.

In an incomplete market setting, things are not so simple and, in correspondence of an equilibrium, the variability of individual consumption may exceed that of aggregate consumption and asset prices may differ from those obtained in a representative agent economy. As a consequence, *market incompleteness* and heterogeneous agents can offer an interesting perspective in order to reconcile classical asset pricing theory with the asset pricing anomalies described above. For instance, Mehra & Prescott [1319] suggest that market incompleteness may help to explain the equity premium puzzle.

In this context, let us briefly present and discuss the model proposed in Weil [1650], which illustrates the asset pricing implications of market incompleteness due to the presence of uninsurable labor income risk. Consider a two-period economy (i.e., $t \in \{0, 1\}$) populated by a continuum of identical agents $i \in I$. Each agent $i \in I$ is supposed to maximize over all feasible consumption plans a time additive expected utility of the form

$$u_0(c_0^i) + \mathbb{E}[u_1(c_1^i)],$$

where the utility functions u_0 and u_1 are the same for all the agents and are assumed to be strictly increasing, strictly concave and twice differentiable. Every agent receives a known endowment e_0 at date $t = 0$ (identical across all the agents) and a random endowment $\tilde{e}_1^i$ at date $t = 1$. It is assumed that the probability distribution of $\tilde{e}_1^i$ is the same for all the agents (so that the endowments of all the agents are ex-ante identical), but the realizations of $\tilde{e}_1^i$ are idiosyncratic to each agent (so that, the endowments are ex-post different). Since the number of agents is infinite, the law of large numbers implies that $\bar{e}_1 := \mathbb{E}[\tilde{e}_1^i]$, meaning that the realized per capita aggregate endowment is identical to the ex-ante expected endowment of every agent. Therefore, there is no aggregate risk in the endowment at $t = 1$. It is crucial to assume that the realization of $\tilde{e}_1^i$ is a private information of agent i, for each $i \in I$. As a consequence, the agents are restricted in sharing the endowment risk, since they cannot insure against unobservable realizations of the individual random endowment of the other agents. According to Weil [1650], this feature should model the unavailability of insurance against fluctuations in labor income (labor risk).

Besides the endowments e_0 and $\tilde{e}_1^i$, $i \in I$, agents are also endowed with one unit of a security paying a dividend d_0 at the initial date $t = 0$ and the non-negative random dividend $\tilde{d}_1$ at date $t = 1$. We denote by p the price of the security at $t = 0$. In the model, the dividend risk represents the aggregate risk and it is assumed that the dividend risk and the idiosyncratic risks (i.e., the endowment risks) are independent.

In this setting, we can write as follows the budget constraint of each agent $i \in I$:

$$c_0^i = d_0 - (w^i - 1)p + e_0, \tag{9.74}$$

$$c_1^i = w^i \tilde{d}_1 + \tilde{e}_1^i, \tag{9.75}$$

where w^i represents the demand of the security by agent i. Due to the budget constraints (9.74)–(9.75), the optimal consumption problem faced by agent $i \in I$ can be formulated as follows:

$$\max_{w^i \in \mathbb{R}} \left(u_0(d_0 - (w^i - 1)p + e_0) + \mathbb{E}[u_1(w^i \tilde{d}_1 + \tilde{e}_1^i)] \right).$$

The first order optimality condition leads to

$$p\, u_0'(d_0 - (w^{i*} - 1)p + e_0) = \mathbb{E}[u_1'(w^{i*}\tilde{d}_1 + \tilde{e}_1^i)\tilde{d}_1],$$

for every $i \in I$. Since all the agents are ex-ante identical (because they have the same preferences, the same endowments at $t = 0$ and random endowments at $t = 1$ with identical distributions) and in equilibrium the market must clear, it holds that $w^{i*} = 1$, for all $i \in I$. Hence, the equilibrium price p^* of the security is given by

$$p^* = \frac{\mathbb{E}[\tilde{d}_1 u_1'(\tilde{d}_1 + \tilde{e}_1^i)]}{u_0'(d_0 + e_0)}. \tag{9.76}$$

The equilibrium expected rate of return r^* is given by

$$r^* = \frac{\mathbb{E}[\tilde{d}_1]}{p^*} = \frac{\mathbb{E}[\tilde{d}_1] u_0'(d_0 + e_0)}{\mathbb{E}[\tilde{d}_1 u_1'(\tilde{d}_1 + \tilde{e}_1^i)]} \tag{9.77}$$

and the equilibrium risk free rate r_f is given by

$$r_f = \frac{u_0'(d_0 + e_0)}{\mathbb{E}[u_1'(\tilde{d}_1 + \tilde{e}_1^i)]}. \tag{9.78}$$

In correspondence of this equilibrium allocation, each agent does not trade and simply consumes his endowment and the dividends delivered by the unit of the security in his endowment. This equilibrium is Pareto dominated by the "risk pooling" allocation where each agent consumes $d_0 + e_0$ at date $t = 0$ and $\tilde{d}_1 + \bar{e}_1$ at date $t = 1$. Indeed, due to the concavity of the utility function u_1, Jensen's inequality implies that

$$
\begin{aligned}
\mathbb{E}[u_1(\tilde{d}_1 + \tilde{e}_1^i)] &= \mathbb{E}\big[\mathbb{E}[u_1(\tilde{d}_1 + \tilde{e}_1^i)|\tilde{d}_1]\big] \\
&\leq \mathbb{E}\big[u_1(\tilde{d}_1 + \mathbb{E}[\tilde{e}_1^i|\tilde{d}_1])\big] \\
&= \mathbb{E}[u_1(\tilde{d}_1 + \bar{e}_1)],
\end{aligned}
$$

where in the last equality we have used the independence of the random variables $\tilde{d}_1$ and $\tilde{e}_1$.

Let us now consider a fictitious representative agent economy, assuming that each agent has received the average endowment $\bar{e}$ at date $t = 1$, neglecting the fact that individuals have their idiosyncratic and uninsurable labor risks. In other words, this representative agent economy corresponds to the allocation where each agent is supposed to receive the Pareto optimal consumption plan $(d_0 + e_0, \tilde{d}_1 + \bar{e}_1)$ described above. In this case, by following the above line of reasoning, the equilibrium price $\hat{p}^*$ of the security in the representative agent economy would be given by

$$
\hat{p}^* = \frac{\mathbb{E}[\tilde{d}_1 u_1'(\tilde{d}_1 + \bar{e}_1)]}{u_0'(d_0 + e_0)}
$$

and the equilibrium expected return $\hat{r}^*$ and risk free rate $\hat{r}_f$ by

$$
\hat{r}^* = \frac{\mathbb{E}[\tilde{d}_1]}{p^*} = \frac{\mathbb{E}[\tilde{d}_1] u_0'(d_0 + e_0)}{\mathbb{E}[\tilde{d}_1 u_1'(\tilde{d}_1 + \bar{e}_1)]}, \tag{9.79}
$$

$$
\hat{r}_f = \frac{u_0'(d_0 + e_0)}{\mathbb{E}[u_1'(\tilde{d}_1 + \bar{e}_1)]}. \tag{9.80}
$$

The following proposition compares the pricing implications of the competitive equilibrium allocation with that of the fictitious representative agent equilibrium (see Exercise 9.16 for a proof).

Proposition 9.15 *In the context of the above model (see Weil [1650]), it holds that*

$$
\hat{r}_f \geq r_f \qquad and \qquad \hat{r}^* \geq r^*
$$

if and only if $u_1''' > 0$.

This proposition provides an example of the implications of market completeness, since in the setting considered above the presence of uninsurable labor risk makes the market incomplete. When $u''' > 0$ (the utility function u exhibits *prudence*, see Sect. 3.3), there is a *precautionary saving* motive: the presence of uninsurable labor risk increases the demand for both the risk free and the risky asset. In equilibrium, since the risk free asset is in zero net supply and the risky asset is in fixed unit supply (per capita), the increased demand must be compensated by a higher price of the two securities or, equivalently, by a lower return. On the other hand, the fictitious representative agent neglects this precautionary saving motive, thus generating an overestimation of the expected rates of return of the two securities. This result allows to explain the risk free rate puzzle (see also Kocherlakota [1112]). Concerning the equity risk premium, as shown in Weil [1650, Proposition3], if the utility function u_1 exhibits decreasing absolute risk aversion and exhibits decreasing absolute prudence, then the equity premium will be underestimated in the fictitious representative agent economy in comparison to the original competitive economy with uninsurable labor income. See also Mankiw [1293] and Gollier & Schlesinger [804] for related models.

In the financial economics literature, even in a multi-period setting allowing for dynamic trading, incomplete markets can be generated by an exogenous labor income process affecting the wealth dynamics. In the presence of labor income, the classical results obtained under the assumption of market completeness may change substantially. As a first effect, the wealth process associated to a self-financing strategy does not follow anymore the dynamics (6.4), since on the right hand side an additional random variable y_t appears, representing the labor income received at date t. In this setting, as illustrated in the simple context of the two-period model of Weil [1650], a precautionary saving motive emerges for a prudent agent: the agent saves more in order to face the uncertainty of future labor income, reducing current consumption.

The presence of a precautionary saving motive has been invoked in the macroeconomics literature in order to reconcile the properties of observed consumption time series with a general equilibrium model of the type described in Chap. 6. As a matter of fact, several anomalies appear when one tries to interpret consumption data in the setting of the classical life cycle permanent income hypothesis (in this regard, see Deaton [530] and Chap. 6). However, the relevance of a precautionary motive as a possible explanation of consumption data is unclear. Indeed, some authors emphasize the relevance of precautionary saving quantitatively and in explaining some puzzles such as the excess sensitivity of consumption to anticipated and lagged income changes, the excess smoothness of consumption with respect to unanticipated income changes and the high growth of consumption with a low risk free rate (this is related to the risk free rate puzzle, see Sect. 7.3), see Zeldes [1676], Caballero [325], Banks et al. [132], Skinner [1550], Carroll [369], Carroll & Samwick [374], Carroll [368], Dynan et al. [614], Gourinchas & Parker [815]. Carroll [368] and Carroll & Samwick [373] show theoretically and empirically that agents with a greater predictable income uncertainty exhibit lower current consumption, hold more wealth and the wealth held in order to face future income fluctuations represents a relevant part of the total wealth. This component helps to explain a high consumption growth rate in the presence of a low risk free rate.

Impatient and prudent agents with a borrowing constraint adopt a *"buffer stock"* behavior (see Deaton [529], Carroll [369]). According to this approach, impatient agents have a wealth target: below that level prudence and precautionary saving dominate, so that the agent saves, while above that level the agent consumes and does not save. The buffer stock model is shown to provide an adequate description of the consumers' behavior up to the age of fifty, while afterwards the standard life cycle hypothesis gives an appropriate representation of the agent's behavior (see Gourinchas & Parker [815]). On the other hand, Aiyagari [33] and Guiso et al. [859] show that the contribution of uninsured idiosyncratic risk to aggregate saving is rather modest and Dynan [612] estimates a small coefficient of relative prudence and, consequently, a limited precautionary motive.

Portfolio choices in the presence of uninsurable labor income have been analysed in several papers. In particular, uninsurable labor income has three different effects on portfolio choices: a *mean effect*, a *variance effect*, and a *correlation effect*. As a first observation, note that an exogenous riskless labor income is equivalent to a position in the risk free asset. As a consequence, the presence of a riskless labor income increases the demand of risky assets for agents with DARA utility functions (mean effect, see also Jagannathan & Kocherlakota [1009]). This happens in particular when the agent is young, thus generating a horizon effect. On the other hand, the presence of a labor income that is perfectly correlated to stock returns should induce the agent to invest more in the risk free asset. Assuming the presence of a non-perfect correlation between labor income shocks and stock market returns, the hedging demand of risky assets has a sign which is the opposite to that of the correlation (correlation effect). Moreover, ex-post labor supply flexibility induces an agent to take more risks ex-ante, thus providing an explanation to the asset allocation puzzle (young agents have more labor supply flexibility, such as the retirement date choice, and, therefore, they invest in risky assets more than old agents, see Bodie et al. [260], Dybvig & Liu [608], Farhi & Panageas [680], Barucci & Marazzina [172]). As we have already remarked in a two-period setting in Sect. 3.3, when facing a saving-portfolio problem, a prudent agent with decreasing absolute prudence and decreasing absolute risk aversion who is exposed to an uninsurable risk in the future (background risk) will reduce his demand of the risky asset (variance effect), see also Elmendorf & Kimball [637]. In Elmendorf & Kimball [637] it is also shown that in this case a reduction in labor income risk (which is independent of asset returns and does not make the agent worse off) leads the agent to invest more in risky assets. This result is confirmed in a multi-period setting, with an increase in labor income risk inducing an agent to save more and to reduce his positions in the risky assets (see Koo [1119], Viceira [1622], Campbell & Viceira [356], Henderson [935], Benzoni et al. [195], Gomes et al. [806], Lynch & Tan [1260]). Summing up, the presence of non-financial income has a (positive) mean effect and a (negative) variance effect on the risky asset demand. Some empirical evidence on these effects has been reported in Guiso et al. [860], Haliassos & Bertaut [876], Carroll & Samwick [373], Vissing-Jorgensen [1625], while small effects of labor income risk on portfolio choices have been detected in Letendre & Smith [1192].

Multi-period portfolio choices in the presence of an idiosyncratic labor income component which is uncorrelated with stock returns show that young (employed) agents hold more risky assets than old (retired) agents, see Viceira [1622]. Young agents borrow to invest in risky assets. The stock demand of a young agent increases if the human capital decreases over the life span and is increasing with respect to the expected retirement date. Positive correlation between non-financial wealth and asset returns induces a negative hedging demand and a young agent may hold less risky assets than a retired agent (positive evidence on the correlation effect has been detected in Heaton & Lucas [929], while weak evidence is reported in Vissing-Jorgensen [1625]). Heaton & Lucas [929] observe that households with high and variable proprietary business income, which is highly correlated with stock returns, hold less risky assets than otherwise similar wealthy people, but they still hold a significant amount of risky assets (thus contradicting the variance effect).

If markets are complete, then full mutual risk insurance implies that consumption should not vary across agents in response to purely idiosyncratic shocks, since the consumption plan of every agent depends only on the aggregate risk. In particular, considering a power utility function, optimal consumption growth rates should not respond to idiosyncratic risks, they should be highly correlated and more correlated than income growth rates. In the literature, several empirical tests of the implications of optimal risk sharing on consumption have been proposed. In particular, Cochrane [459] has regressed consumption growth rates on a set of idiosyncratic variables, showing that full insurance is rejected for long illness and involuntary job loss, but not for unemployment and involuntary move. Negative results for full insurance have been also obtained in Attanasio & Davis [83], Banks et al. [132], Blundell et al. [259] by analysing the effects of changes in the relative hourly wage across birth cohorts and education groups on the distribution of household consumption. Perfect risk sharing among agents of different countries is also rejected when analysing consumption growth rates of different countries (*risk sharing home bias*), see Lewis [1216, 1215]. However, considering utility functions with a high risk aversion near the subsistence level and a decreasing risk aversion in its neighborhood, positive evidence for risk sharing is obtained in Ogaki & Zhang [1383]. Assuming homothetic preferences, Mace [1265] tests the full insurance hypothesis by regressing individual consumption on aggregate consumption and some idiosyncratic risk variables. Any variable other than the change in the aggregate consumption is predicted to be insignificant in explaining changes in the agents' consumption. The empirical results reported in Mace [1265] agree with the full insurance hypothesis in the case of an exponential utility function but not for a power utility function.

The market completeness hypothesis has also important implications for the validity of the aggregation property (the existence of a representative agent) in financial markets. Several empirical tests of the aggregation property have been proposed in the literature. Considering a power utility function and a joint log-normal distribution for the real interest rate and the consumption growth rate, the relations derived from (6.78) can be tested for the representative agent (with respect to aggregate consumption) as well as for cohort data. The empirical results reported in Attanasio & Weber [87] show the existence of aggregation imperfections, namely the estimates of the elasticity of intertemporal substitution of consumption obtained from aggregate data are substantially lower than those obtained from average cohort data. Moreover, Euler conditions are rejected on aggregate data, but they are not rejected on average cohort data. Similar results are also obtained in Attanasio & Weber [88] for non-durable expenditure in U.S. Non-linearities and demographic factors are at the origin of these differences. Aggregation may also explain the excess sensitivity of consumption to income shocks, while the effect disappears when including demographic variables (see Attanasio & Browning [82]).

Jacobs [1005] has argued that market incompleteness is not a relevant phenomenon, in the sense that many well documented asset pricing puzzles do not result from aggregation problems. Indeed, the Euler conditions for a time separable power utility function can be tested for individual consumers (households), even in incomplete markets. If markets are complete, then the conclusions of tests based on an individual consumer should agree with those of tests based on a representative agent. The empirical evidence reported in Jacobs [1005] indicates that the restrictions implied by the Euler conditions are rejected. Similar results have been also obtained in Telmer [1583] assuming a non-diversifiable idiosyncratic component in the agents' endowments. A limited set of securities suffices to approach almost perfect risk sharing and this result is even reinforced if idiosyncratic risk is correlated with the market return. Market incompleteness introduces very little variability in the intertemporal marginal rate of substitution (the test proposed in Hansen & Jagannathan [891] is violated). Levine & Zame [1203] propose a general equilibrium model with a single good showing little relevance of market incompleteness on welfare, prices and consumption when agents are patient and have a long horizon, a risk free bond is traded and shocks are transitory.

While there is empirical evidence of market incompleteness, its effects on asset prices in a multi-period economy are limited because agents can trade repeatedly over time (see Brav et al. [283], Kocherlakota [1112], Huggett [992], Lucas [1250], Heaton & Lucas [926, 927], Cochrane [463], Lettau [1193], Marcet & Singleton [1302], Krusell & Smith [1140]). It is difficult to explain asset pricing anomalies (notably, the equity premium puzzle) by introducing uninsurable endowment components. Typically, market incompleteness induces a risk free rate lower than the risk free rate that would prevail under market completeness (due to the presence of a precautionary saving motive, as discussed at the beginning of the present section) but the reduction of the risk free rate depends on the set of available assets as well as on the presence of other market frictions.

Assuming a power utility function, Krueger & Lustig [1138] show that the absence of insurance markets for idiosyncratic labor risk has no impact on the premium for aggregate risk if the distribution of idiosyncratic risk is independent of aggregate shocks and the aggregate consumption growth rate is independent over time. A common feature of many models is that the time series of the ratio of each consumer's labor income over the aggregate labor income is assumed to be a stationary Markov process with low persistence. If this is the case, then agents can effectively self-insure in the asset market against temporary income changes. Constantinides & Duffie [496] have relaxed the above hypothesis and shown that the joint hypothesis of incomplete insurance and agents' heterogeneity enriches the pricing implications of a representative agent model. They assume that individual income processes are heteroskedastic with a high degree of persistence (e.g., idiosyncratic income shocks follow a random walk) and, moreover, that the variance of idiosyncratic risk increases when the market declines (countercyclical conditional variance). In an incomplete market, if labor income shocks are permanent, then agents cannot easily self-insure against them. Under this assumption, agents' heterogeneity and incomplete insurance help to address several asset pricing puzzles without the need of introducing other market frictions. Assuming a time additive utility function with constant relative risk aversion and given a set of processes of arbitrage free prices, dividends and aggregate income satisfying some restrictions, Constantinides & Duffie [496] show that there exists an equilibrium supporting the price processes by a judicious choice of the agents' labor income process. Therefore, the Euler conditions hold.

The result of Constantinides & Duffie [496] has been extended by relaxing the distributional restrictions in Krebs [1132] and providing an explanation of some asset pricing anomalies. The pricing density is affected by the aggregate consumption growth and by the cross-sectional variance of consumption growth. If the variance of idiosyncratic shocks is negatively correlated with equity returns and aggregate shocks, then the economy with incomplete markets generates a larger equity premium and a lower risk free rate than the complete market economy (compare also with the result Proposition 9.15 above). The key features to obtain this result are the strong heterogeneity and the high persistence in shocks to consumers' income with high volatility during recessions. The model also addresses the countercyclical behavior of the equity premium. The equity premium is positively affected by the covariance of security returns with the cross-sectional variance of individual agents' consumption growth. Moments of the cross-sectional distribution of the household consumption growth rate affect equity returns. Jacobs & Wang [1006] show that agents are not fully insured and that the cross-sectional variance of consumption growth is also a priced factor. The resulting two factor consumption-based asset pricing model significantly outperforms the CAPM.

Some empirical evidence in favor of persistence in idiosyncratic shocks and higher volatility during recessions has been provided in Storesletten et al. [1572, 1573]. Introducing life cycle effects (young agents face more idiosyncratic risks than old agents), the authors can explain the U.S. equity premium and Sharpe ratio with a plausible level of risk aversion. However, estimating the processes of idiosyncratic shocks on U.S. data, Heaton & Lucas [927] find a low degree of persistence and obtain a risk free rate similar to that obtained under the hypothesis of complete markets (in this regard, see also Kocherlakota [1112], Cochrane [463], Lettau [1193]). Cogley [469] shows that the cross-sectional dispersion of logarithmic consumption growth is only weakly correlated with stock returns and the correlation is not strong enough to generate an equity premium comparable to that observed in the U.S. financial market. Krusell & Smith [1140] show that with a plausible degree of heterogeneity (i.e., with an equilibrium distribution of income and wealth matched to the U.S. data) the model cannot address the equity premium puzzle.

Market Imperfections

Market imperfections can refer to many different aspects of financial markets: borrowing constraints, short sale constraints, transaction costs, solvency constraints, bid-ask spreads as well as other types of market frictions.

The fundamental theorem of asset pricing has been extended to markets with transaction costs (bid-ask spread) in several papers, see Jouini & Kallal [1049], Naik [1369], Schachermayer [1500], Kabanov & Safarian [1056] in discrete time models. In a nutshell, it holds that a market with bid-ask spreads does not allow for arbitrage opportunities if and only if there exists at least one equivalent probability measure that transforms a process taking values between the bid and the ask prices into a martingale. Similarly as in the context of Sect. 4.4, such a probability measure can be interpreted as a stochastic discount factor or as the intertemporal marginal rate of substitution of a maximizing agent. In turn, the latter allows to define arbitrage bounds on a the price of a contingent claim. For a version of the fundamental theorem of asset pricing under short sale constraints see Jouini & Kallal [1050], under solvency constraints see Hindy [942], under credit constraints see Loewenstein & Willard [1243]. Holding costs allow the price of an asset to deviate from its fundamental value without yielding riskless arbitrage opportunities, see Tuckman & Vila [1602]. See Basak & Croitoru [177] and Detemple & Murthy [565] for general equilibrium models under portfolio constraints.

Constantinides [491] has shown that proportional transaction costs (a transaction fee which is proportional to the prices) induce agents to reduce the frequency and the volume of trading. In particular, there is a no-trade region which is increasing with respect to the transaction costs. However, the effect on an agent's utility is limited, the (positive) liquidity premium and, therefore, the effect on asset prices induced by transaction costs are small. The result has been confirmed in Vayanos [1614] assuming an overlapping generations model and it is also shown that a stock price may increase in transaction costs. Empirical evidence supporting these results (namely that transaction costs induce a small positive risk premium and reduce trading volume) has been provided in Barclay et al. [159]. A strong discount effect on asset prices (or, equivalently, a large risk premium) has been derived in Lo et al. [1239] assuming fixed transaction costs. When asset returns are predictable, ignoring transaction costs induces a substantial utility cost, see Balduzzi & Lynch [118] and Garleanu & Pedersen [761]: in this setting, there is a state dependent no-trade region, which is greater than in the case without predictability and the classical horizon effect is obtained (agents with a long horizon hold more stocks, see Lynch & Balduzzi [1259]). The horizon effect can be also obtained in a model with transaction costs and a finite horizon.

The bid-ask spread can be regarded as a proxy for the liquidity of an asset and is positively related to its expected return, as documented in Amihud & Mendelson [59], Chalmers & Kadlec [393], Brennan & Subrahmanyam [297], Eleswarapu [635], Amihud [57], Easley et al. [615], Bekaert et al. [185]. The rationale is simple: an agent anticipates the cost associated with trading in the future an illiquid asset and, for this reason, requires a higher risk premium as a compensation. Equilibrium models studying the relation between liquidity and risk premia have been considered in Huang [975] and Acharya & Pedersen [11]. Transaction costs are shown to have a relevant effect on asset returns if agents are constrained from borrowing against future income (a similar effect is obtained in Longstaff [1244]). Market liquidity as a predictor of future returns (in the sense that an increase of liquidity predicts lower future returns) is explained in Baker & Stein [108] by relying on a model with irrational traders and short sale constraints. Cho et al. [430] show that, allowing agents to choose assets to trade with trading costs, then lead-lag effects of the type described in Lo & MacKinlay [1237] arise between portfolios sorted by size.

Asset prices with transaction costs in an incomplete market have been analysed in Aiyagari & Gertler [34]. In this model, agents are exposed to idiosyncratic shocks to their income (with no aggregate shocks) and this risk component cannot be hedged in financial markets, similarly as in the seminal model of Weil [1650] presented at the beginning of this section. Two types of assets are traded in the market: stocks and short term government bonds, with stock trading incurring in transaction costs. Market incompleteness generates a low risk free rate and the presence of transaction costs increases the equity premium. This type of model produces a smooth aggregate consumption and a highly variable individual consumption. However, very large transaction costs are needed in order to match the historically observed values, as confirmed in Jang et al. [1014].

Transaction costs in consumption have been analysed in Cuoco & Liu [508], Grossman & Laroque [844], Liu & Loewenstein [1224], showing that transaction costs have an effect on consumption smoothing. Fixed adjustment costs in consumption can account for much of the discrepancy between the observed low variance of aggregate consumption growth and predictions of the CCAPM. Adjustment consumption costs make agents respond to changes in asset prices with a lag and, therefore, consumption growth covaries weakly with equity returns. Adjustment costs help to address the equity premium puzzle (see Lynch [1257] and Marshall & Parekh [1309]) and yield a strong upward bias in the estimation of the coefficient of relative risk aversion by means of an Euler equation. Taking into account this effect, Euler equation tests do not reject the classical CCAPM.

Borrowing, credit and liquidity constraints are relevant for the agents' decisions: many empirical studies have shown that these constraints are binding for some groups of agents (for instance, low income and young agents in the market for consumer loans, see Attanasio et al. [84]). Similarly, Jappelli [1015] estimates that 20% of the families are rationed in the credit market and also Hayashi [917] and Gross & Souleles [838] report evidence supporting the fact that liquidity constraints are binding. In Zeldes [1675] and Jappelli et al. [1017], by relying on tests on the Euler conditions, it is shown that liquidity-borrowing constraints are binding. On the other hand, constraints are shown to be non-relevant in the tests proposed in Runkle [1488] and Meghir & Weber [1318] (except for young households).

The *buffer stock* model with precautionary saving motives, impatience and restrictions on borrowing allows to explain both excess sensitivity, life cycle and business cycle patterns as well as excess smoothness of consumption (see Carroll

[368], Deaton [529], Carroll [369], Ludvigson & Michaelides [1253], Gross & Souleles [838], Browning & Crossley [317], Deaton [530], Carroll [370]). When borrowing constraints are binding (i.e., wealth is below a given target), agents do not save and consume the wealth generated by the assets. In this sense, assets act like a buffer stock protecting the level of consumption against low income. In this case, the marginal propensity to consume is higher than in the standard model. Gourinchas & Parker [815] show that agents behave like buffer stock consumers early in their life, while afterwards they behave according to the life-cycle hypothesis. Huggett [993] show that idiosyncratic income shocks and liquidity constraints induce positive precautionary saving, independently of the third derivative of u (compare with Proposition 9.15 above). A generalization of the model with a hyperbolic discount factor (see the last part of Sect. 9.2) has been proposed in Harris & Laibson [897].

Guiso et al. [860] show that in a setting with labor income risk, the presence of liquidity constraints reduces the investment in risky assets and increases the saving rate: liquidity constrained agents hold less risky assets than agents who are not. The rationale of this phenomenon is that liquidity constraints reduce the effect of labor income and agents invest more in the risk free asset in order to prevent future income shocks (see also Jappelli & Pagano [1016] and Koo [1119]). Portfolio choices in an incomplete market (due to labor income) under borrowing and short sale constraints have been analysed in Heaton & Lucas [928], Cocco et al. [456], Haliassos & Michaelides [877]. Unless returns and labor income shocks are (significantly) positively correlated, agents hold only stocks. If agents are not constrained, then they short sell the risk free asset. The empirical evidence on a significantly positive correlation is weak. With low positive correlation, labor income is a close substitute for the risk free asset and, therefore, it induces the classical horizon effect (a young agent invests a large fraction of wealth in risky assets and afterwards reduces his stock holdings when he becomes older, see Cocco et al. [456]). Including return predictability (mean reversion), it is shown that there are substantial effects on the optimal consumption but only limited effects on portfolio choices (in particular, a result similar to the one in Heaton & Lucas [928] is obtained), see Campbell et al. [342]. Michaelides [1335] analysed the relevance of the hedging demand in the presence of returns predictability, incomplete markets and borrowing constraints.

Asset prices with short sale and borrowing constraints in an incomplete market have been analysed in Lucas [1250]. The model takes into account the presence of non-diversifiable labor income idiosyncratic and serially uncorrelated shocks. The incompleteness of the market is shown to reduce the risk free rate, but does not suffice to solve the equity premium puzzle. The model reproduces both the observed risk free rate and the equity premium only when asset markets are closed or the borrowing constraint becomes severe. With limited access to capital markets, asset prices will be similar to those obtained in a representative agent economy. Individual consumption is highly correlated with aggregate consumption and the volatility of pricing kernels is increased. Similar results are also obtained in Marcet & Singleton [1302], Telmer [1583], Krusell & Smith [1140], Detemple & Serrat [566], while more positive results have been obtained in Alvarez & Jermann [54] with endogenous solvency constraints. Moreover, as shown in Zhang [1679], strong borrowing constraints allow to match the mean and the variance of discount bond returns but not higher moments (like the skewness and the kurtosis). Assuming heterogeneous preferences (namely, different degrees of risk aversion among the agents) with a time additive utility function, heterogeneous borrowing constraints, incomplete markets and a stochastic investment opportunity set, a low risk free rate and a high risk premium are obtained in equilibrium with a tightening of the borrowing constraint, while volatility increases with the cross-sectional dispersion of risk aversion (see Kogan et al. [1115] and Gomes & Michaelides [809]). Capital market restrictions help to explain the lack of international consumption risk sharing (see Lewis [1215]).

Heaton & Lucas [925, 927] analyse an incomplete market model assuming systematic and idiosyncratic labor income risk. There are borrowing constraints, short sale constraints and transaction costs. If there are (large) transaction costs both in the stock and in the bond market and there is a binding borrowing constraint, then the risk free rate is low and a large equity premium can be reproduced (compare also with Huggett [992]). Constantinides et al. [495] show that the presence of borrowing constrains in an overlapping generations economy (where young agents face borrowing constraints) generates a risk premium higher than that of an unconstrained economy. Once more, market incompleteness alone does not help to solve the puzzles. In general, the main drawback of these models is represented by the high volatility of the risk free rate. Luttmer [1255] and He & Modest [918] show that a combination of proportional transaction costs, short-sale constraints, borrowing constraints and solvency constraints helps to explain consumption data in the context of a standard intertemporal equilibrium asset pricing model: the presence of these market frictions generates low comovements in consumption and returns and reduces the required volatility in the intertemporal marginal rate of substitution needed to satisfy the bound of Hansen & Jagannathan [891]. A power utility function with a low risk aversion cannot be rejected by the data. Bansal & Coleman [133] propose a monetary model with transaction costs which can explain the equity premium, the risk free rate and the term premium puzzle.

If transaction costs and frictions affect asset prices, then they should primarily affect high frequency returns and have a weak effect on long horizon returns. This claim is confirmed empirically in Daniel & Marshall [519], where it is shown that asset pricing models with keeping up with the Joneses or habit formation preferences are rejected on quarterly data but perform well at longer horizons. Correlation between consumption growth and market return increases with respect to the time horizon and both the equity premium and the risk free rate are reproduced at longer horizons. However, in many models transaction costs must be very high in order to be able to reproduce the observed equity premium and the risk free rate. Under some conditions, Luttmer [1256] estimates a fixed costs of at least 3% of monthly per capita consumption for a logarithmic utility function. The estimated fixed cost decreases if risk aversion is increased, if preferences exhibit habit persistence or if short sale constraints or bid-ask spreads are introduced. In this perspective, the decline of the U.S. risk premium in the last three decades can be attributed to a reduction of market imperfections.

Market Participation

So far, we have always assumed that all the agents of the economy participate to the market. However, this hypothesis is not confirmed in reality, as already mentioned in Sect. 6.6.

There is evidence of a high degree of heterogeneity in portfolio choices, agents change their portfolio infrequently, their entry in the financial market does not occur early in the life span and there is a hump shaped (or increasing) age profile in the share of stocks in household portfolios. Portfolio diversification is mostly observed in wealthy people. These results are puzzling, since the classical theory (also considering precautionary saving) predicts significant positions in stocks, early entrance in the market and a large participation in the market. The low participation rate also contrasts with the observed high equity premium, i.e., given a high risk premium agents should invest a large fraction of their wealth in stocks. Fixed (information) participation costs, preference structures different from expected utility, borrowing constraints and uninsurable labor income may contribute to explain the heterogeneity in portfolio choices, the low participation rate and the late entry in the stock market (see Haliassos & Bertaut [876], Haliassos & Michaelides [877], Gomes & Michaelides [807, 808], Vissing-Jorgensen [1625], Bertaut [209], Constantinides et al. [495], Storesletten et al. [1572], Davis et al. [528], Guiso & Sodini [862]). Abel [6] analyses a model with a fixed cost to participate in the stock market and borrowing constraints for young agents: in equilibrium, only high income consumers trade in financial markets, while young agents are inclined to remain outside of the market. More recent studies have shown that several non-financial factors affect the decision to invest in stocks: cognitive abilities, trust in the stock market, awareness of financial assets, financial literacy, familiarity (geographical or professional proximity) and social interaction (see Christelis et al. [443], Guiso et al. [861], Guiso & Jappelli [858], Grinblatt et al. [832], Massa & Simonov [1311], Hong et al. [957], Huberman [981], van Rooij et al. [1609], Brown et al. [313]).

The observations reported above suggest to explicitly consider *limited participation* as an important factor in the analysis. The main source of the equity premium puzzle is that consumption growth covaries too little with asset returns (or the intertemporal rate of substitution is not variable enough). Mankiw & Zeldes [1298], Brav et al. [283], Attanasio et al. [81] have shown that stockholders' consumption is more volatile and more correlated with the stock market than with the economy as a whole and this fact reduces the risk aversion level needed to match the data (approximately one third of the level of risk aversion based on aggregate consumption), thus providing a partial resolution of the equity premium puzzle. Brav et al. [283] show that the unexplained equity premium decreases as limited participation and incomplete insurance are taken into account. Considering only asset holders with total assets exceeding a certain threshold, the coefficient of relative risk aversion needed to explain the equity premium decreases with the threshold. Euler conditions on asset holders consumption are not rejected with a coefficient of risk aversion between two and four. In Attanasio et al. [81] it is shown that the CCAPM is not rejected when it is tested with respect to stockholders' consumption. A plausible degree of risk aversion to replicate the observed equity premium is obtained in Ait-Sahalia et al. [32] considering luxury and basic goods with non-homotetic preferences (a certain amount of basic goods is always required) and consumption only of wealthy investors. Vissing-Jorgensen [1626] shows that the elasticity of intertemporal substitution estimated considering only asset holders is larger than the value obtained considering also non-asset holders. The value for asset holders is around 0.4, while for non-asset holders it is insignificantly different from zero. These results help to explain the equity risk premium: as a matter of fact, considering also non-stockholders, the coefficient of relative risk aversion is five times the value obtained by considering only stockholders (see also Paiella [1396], Attanasio & Vissing-Jørgensen [85], Guvenen [867, 868]).

Merton [1332] has shown that limited participation in the market allows explaining some of the CAPM anomalies. Limited participation and agents' heterogeneity together with market incompleteness are analysed in Balduzzi & Yao [120]: on an empirical basis, it is shown that these two features reduce the coefficient of risk aversion needed to explain the historical equity premium. Moreover, the model satisfies Euler condition tests with plausible levels of risk aversion. Similar conclusions are reached in a similar setting (limited participation and incomplete markets) in Danthine & Donaldson [524] introducing wage negotiation between workers and firms. Basak & Cuoco [178] have proposed an asset pricing model confirming the above results. In this model, a risky asset and a risk free asset are traded in the market and some agents with a logarithmic utility function cannot invest in the risky asset due to the presence of frictions-information costs. In equilibrium, the historical risk premium and the risk free rate can be reproduced with a coefficient of relative risk aversion of the representative agent equal to 1.3 and the risk free rate is smaller than in the standard model. Furthermore, the volatility of consumption of unrestricted agents is higher than that of per capita consumption.

As suggested by the above analysis, stock market participation can be included as an endogenous variable in asset pricing models. In Allen & Gale [45] and Pagano [1393], the participation of the agents in the financial market is determined endogenously assuming a fixed entry cost. It is shown that the volatility and the size of the financial market are strictly related. By increasing market depth, each additional trader in the market generates a positive externality for the other traders by reducing asset volatility. This effect leads other traders to participate to the market, lowers risk, increases the asset price and, therefore, induces the firm to issue more stocks. In this context, multiple equilibria arise: low volatility-small markets and high volatility-large markets. An equilibrium with a large market Pareto dominates an equilibrium with a small market. When the fixed cost is small, there is full participation. Orosel [1387] analyses an overlapping generations model with fixed costs of stock market participation and dividends following a Markov process. In equilibrium, it is shown that market participation is positively correlated with past dividends and current prices (positive feedback trading). Endogenous participation in the market contributes to increase the asset price volatility and generates a mean reversion effect. Hong et al. [957] propose a model of stock market participation with social interaction and multiple equilibria and show that social households are more likely to invest in stocks.

Incomplete risk sharing has been modeled by considering a participation constraint such that agents are never willing to revert permanently to autarchy (i.e., only consuming their private endowment, see Kehoe & Levine [1078], Zhang [1678], Alvarez & Jermann [53, 54]). An intertemporal asset pricing equilibrium model including this endogenous participation constraint has been analysed. An agent exits the financial market, defaults on his debt and reverts to autarchy if the expected utility derived from consuming the future private endowment without trading in the financial market is larger than the expected utility derived from participating in the market. Imposing the participation constraint that autarchy is never preferred, interest rates are shown to be smaller than in an standard economy without constraints. Under some conditions, incomplete risk sharing is obtained in equilibrium. Under plausible conditions on the endowment process (relative shocks are correlated with aggregate shocks) and on risk aversion, the model satisfies the volatility constraint on the intertemporal rate of substitution established in Hansen & Jagannathan [891] and generates a large equity premium (see Alvarez & Jermann [53, 54]).

9.6 Notes and Further Readings

Considering the role of uncertainty aversion, in Dow & Werlang [589] a high price volatility is obtained under information ambiguity if the preferences of the agents are represented by an expected utility with a non-additive probability measure. In a related direction, Kelsey & Milne [1086] show that the equilibrium version of the APT obtained in Connor [475] can be extended to an economy with uncertainty averse agents whose preferences are represented by an expected utility with a non-additive probability measure. We refer to Chateauneuf [411] for a simplified presentation of the framework of Schmeidler [1505]. In a purely subjective setting, Alon & Schmeidler [52] introduces some axioms for the derivation of the maxmin expected utility rule. For a survey see Etner et al. [653] and Guidolin & Rinaldi [856].

The impact of non-classical preference functionals on portfolio choices and asset pricing has been studied in several papers. For instance, Chew et al. [428] show that higher aversion with a rank dependent utility function induces a smaller risky asset demand in the presence of a single asset. In Dekel [543], considering a utility function à la Machina, it is shown that risk aversion does not imply risk diversification. Kelsey & Milne [1086] shows that the APT results of Connor [475] can be extended to preferences satisfying the smoothness condition imposed in Machina [1266] or represented by rank dependent utility functions.

Considering loss aversion, a model for portfolio choice and stock trading volume has been proposed in Gomes [805], showing that the optimal demand of risky assets is discontinuous and non-monotonic. As wealth reaches a threshold level, investors follow a generalized portfolio insurance strategy (see Gomes [805], Berkelaar et al. [201]). Loss-averse investors will not hold stocks unless the equity premium is sufficiently high. The model generates positive correlation between trading volume and stock return volatility. We refer to Bowman et al. [274] for an analysis of consumption with loss averse preferences. Note also that, as we have already shown in the context of the Barberis et al. [150] model, loss aversion can also generate a horizon effect on portfolio choices when returns are predictable, in the sense that a long horizon induces agents to hold more stocks (see Ait-Sahalia & Brandt [31]).

A non-parametric version of recursive utility has been proposed in Epstein & Melino [642]. Assuming positive serial correlation in consumption growth, Epstein & Melino [642] show that their model can explain the equity premium puzzle and the risk free rate puzzle but is not able to match second order moments of returns. Increasing the elasticity of intertemporal substitution, the risk free rate and the equity volatility decrease. Confirming Kandel & Stambaugh [1069], it is shown that it is not possible to match simultaneously the risk free rate and the volatility of equity returns. The empirical evidence reported above shows that recursive preferences do not allow to simultaneously explain three empirically observed phenomena: a high risk premium, a low risk free rate and a high volatility of asset returns.

In Campbell [333], the asset pricing methodology of Campbell [332] has been implemented by including human capital as a factor. The author proposes a multi-factor model including the market return and variables that help to predict future stock returns and future labor income growth (which proxies for human capital). It is shown that the aggregate stock market represents the main risk factor and the model is able to explain cross-sectional returns and to capture mean reversion effects. Positive evidence on this model is also reported in Hodrick & Zhang [952].

In an analysis of the role of habit formation and its relation to behavioral patterns, Antonelli et al. [74] propose a preference functional which takes into account disappointment and anticipation, by assuming that the habit process is affected by past utility. When the utility functional is increasing (decreasing, respectively) in the habit, then the anticipation (disappointment, respectively) effect is captured. In the first case, when savoring in the past a high level of expected utility, the agent gets a high level of utility from the current consumption rate. In the second case, a high level of expected utility in the past induces the agent to require a higher consumption rate today. Asset pricing results show that an anticipation effect generates a risk premium smaller than the one obtained with an additive expected utility, whereas the disappointment effect leads to a higher risk premium. Therefore, disappointment yields an interesting perspective to address the equity premium puzzle. Mankiw et al. [1296] and Eichenbaum et al. [634] have introduced leisure in the utility function. Empirical tests have shown that leisure does not help to improve the fit of the intertemporal equilibrium model.

Recursive preferences have been formulated in continuous time in Duffie & Epstein [594, 595]. In Bergman [196], an ICAPM is obtained with a recursive utility function characterized by a discount factor which is a function of consumption (*Uzawa utility*). Epstein & Zin [652] extend the recursive utility approach by relaxing the independence axiom. However, such a generalization does not seem to enhance the performance of an asset pricing model.

A different formulation of the keeping up with the Joneses phenomenon has been proposed in Bakshi & Chen [113], assuming that the wealth level of the agent is compared to a social wealth index. A volatility bound test of the type described in Hansen & Jagannathan [891] and a generalized method of moments test provide positive evidence for this model (see also Smith [1551]). DeMarzo et al. [554] show that this type of preferences may induce a bubble in the market. As shown in DeMarzo et al. [553], if agents care about consumption relative to per capita consumption in their community, then they tend to hold undiversified portfolios similar to those of the other members of the community. Linking the keeping up with the Joneses phenomenon to market participation, Hong et al. [957] show that social investors (investors with social activities) find the market more attractive when more of their peers participate.

An axiomatization of intrinsic and internal habit has been proposed in Rozen [1484]. In Jermann [1032] and Boldrin et al. [263], habit formation has been included in a real business cycle model (endogenous endowment process). Assuming capital adjustment costs or limited inter-sectoral factor mobility, the model can explain the historical equity premium, the average risk free rate and the asset Sharpe ratio. However, the model incurs in some problems when trying to reproduce the second moments of asset returns, in particular of the risk free rate. Note that some form of input market inflexibility is necessary in order to obtain the above results, otherwise consumption will be smoother than in a pure exchange economy (see Boldrin et al. [263] and Lettau & Uhlig [1198]). Standard business cycle properties are replicated by the models. Carroll et al. [372] show that habit formation may reproduce the phenomenon that an increase in growth leads to an increase in saving, as observed empirically.

External habit preferences also allow to reproduce some stylized facts on return predictability: Menzly et al. [1325] show that time-varying risk preferences induce a positive relation between dividend yields and expected returns. Countercyclical variations in the Sharpe ratio of stock returns and in the risk aversion of the economy are obtained in Chan & Kogan [409] by considering a general equilibrium model with external habit, heterogeneous preferences and constant risk aversion. The agents' risk aversion is constant over time but varies across the population. External habit preferences imply that risk aversion and, hence, portfolio choices are time-varying. However, according to Brunnermeier & Nagel [320], this is not consistent with the empirical data, which typically exhibit inertia in portfolio decisions with respect to changes in wealth.

For a survey on psychology, decision making and finance see Camerer [328], Barberis & Thaler [154], Daniel et al. [518], Hirshleifer [945]. In an imperfectly competitive market, Palomino [1398] shows that noise traders may earn higher expected returns and higher expected utility than rational traders. Barberis et al. [147] show that a CCAPM with rational traders and agents extrapolating future returns from past returns provides a behavioral explanation of return predictability. In Barucci & Landi [171], the convergence of an agent's learning process is analysed in the model proposed in Cutler et al. [509] (a model with rational, fundamentalist and feedback traders). Agents are able to learn the fundamental solution without restrictions on the parameters of the model.

Concerning the role of heterogeneity, Gollier [801] investigates the effect of wealth inequality on asset prices in the case of identical agents differing with respect to their initial endowments. In the context of a two-period model, the author shows that wealth inequality affects asset prices in comparison to an egalitarian economy if and only if the absolute risk tolerance is non-linear (non-hyperbolic utility functions). Wealth inequality increases the equity premium when absolute risk tolerance is concave and, introducing a precautionary saving component, wealth inequality also decreases the risk free rate if the inverse of absolute prudence is concave.

Considering irrational noise traders with erroneous beliefs, a model of the type of De Long et al. [549] has been proposed in order to explain the difference between the closed-end fund price and the net value of the assets held in the fund. In Lee et al. [1174] the authors report evidence that the small investor sentiment affects the risk of common stocks, i.e., it is priced by the market (see however Elton et al. [639] for a different interpretation).

The presence of noise traders as introduced in Shiller [1535] (price inelastic random demand) has been analysed in Campbell & Kyle [347] in the context of a continuous time model. In this model, there are two classes of agents: infinitely lived rational agents maximizing an exponential utility function with a constant coefficient of absolute risk aversion and noise traders. The demand of noise traders is described by an Ornstein-Uhlenbeck stochastic process and changes in the levels of (de-trended) dividends and prices are represented by normal random variables with constant variance. It is shown that the presence of noise traders affects the behavior of rational agents by introducing an additional risk component. The analysis of Campbell & Kyle [347] allows the identification of three components in the asset price volatility: a first component associated with the information on future dividends contained in the dividend time series, a second component associated with the information on future dividends not contained in the dividend time series and a third component associated with noise trading. Campbell & Kyle [347] calibrate their model on the Standard and Poor's time series and show that the observed stock price movements can be explained by assuming a low discount rate (lower than 4%) and a high risk aversion (or, in the case of a discount factor larger than 5%, by assuming the presence of a relevant noise traders' component highly correlated with fundamental values). In particular, it is argued that the presence of noise traders can explain several asset pricing anomalies, including excess volatility, overreaction and mean reversion.

The presence of noise traders has been empirically confirmed in Kelly [1085] by looking at the rate of participation to the financial market. The author tests with positive evidence that the rate of noise traders' participation is a negative predictor of future stock returns. In Lee et al. [1176] (using the Investors' Intelligence of New Rochelle as an investors' sentiment index), it is shown that investors' sentiment is a systematic risk that is priced in the market. Excess returns are contemporaneously positively correlated with shifts in sentiment and, moreover, the magnitude of bullish (bearish) changes in sentiment leads to downward (upward) revisions in volatility and higher (lower) future excess returns.

The effect of the speculation activity of *convergence traders* has been analysed in Xiong [1670] and Kyle & Xiong [1151] in an economy populated by noise and fundamental traders. Convergence traders bet that the price difference between two assets with similar (but not identical) characteristics will narrow in the future. In particular, convergence traders exploit short term profitable investment opportunities generated by noise traders. Under some conditions, it is shown that convergence traders destabilize the market, i.e., they unwind their positions by buying (selling) when prices are high (low) after a capital loss. This type of trading strategy acts as an amplification and contagion mechanism. In Abreu & Brunnermeier [9] it is shown that rational traders "time the market", rather than correcting the mispricing right away (*delayed arbitrage*) and delay trading when they are uncertain about the time their peers will be informed and will exploit the mispricing.

Infrequent feedback trading has been analysed in Balduzzi et al. [115]. The authors consider an economy where bonds and stocks are traded and where there are two classes of agents: "*speculators*" and "*feedback traders*". Speculators maximize their utility from consumption and have rational expectations, while feedback traders submit infrequent and discrete market orders, mechanically responding to price changes. It is shown that positive feedback traders increase the volatility of stock returns and the response of prices to news on dividends, while negative feedback traders induce the opposite effect. In both cases, feedback trading generates predictability and heteroskedasticity in returns, even though dividend growth rates are assumed to be i.i.d. In a related setting, Balduzzi et al. [116] consider an economy with agents trading continuously and maximizing an expected utility and agents buying and selling assets only when the price reaches certain triggering thresholds. The second type of behavior derives from the adoption of optimal strategies with transaction costs or portfolio insurance strategies and generates resistance-support levels, jumps in asset prices and volatility when the price barrier is reached.

Barberis & Shleifer [151] have analysed an economy populated by fundamentalist (rational) agents and a class of agents adopting a *style investing* strategy, meaning that they classify assets in some groups (according to size, industry or other characteristics) and allocate wealth among the different groups (styles) rather than among the individual securities, switching between different groups according to their past performance. In other words, investors get into styles that have performed well in the past relative to other styles, financing the reallocation of wealth by withdrawing money from styles that have performed poorly. It is shown that style investing leads to comovements between stocks belonging to the same style (see also Barberis et al. [152]). Furthermore, asset returns are positively serially correlated at short horizons and mean reverting in the long run. Prices deviate from fundamental values but there is a long run pressure towards fundamental values. The volatility turns out to be higher in comparison to an economy populated only by fundamentalist agents.

Assuming that agents have confidence on information that depends on its reliability, Bloomfield et al. [250] show by means of experiments that agents tend to moderate their confidence on their prior expectation of reliability and, as a consequence, overreact to unreliable information (overestimating its reliability) and underreact to reliable information (underestimating its reliability). Evidence of underreaction to cash flow news by individuals, and not by institutions, is provided in Cohen et al. [471].

In an intertemporal setting, the optimal investment-consumption problem has been analysed in Chap. 6 under the assumption that agents perfectly know the economic model and fully observe the realization of the random variables describing asset returns and the state of the economy. However, these two assumptions can be significantly relaxed, allowing in particular for the presence of non-observable or hidden economic factors (*incomplete information*).

Assuming incomplete information or partial observability of the economic variables, the solution to the optimal investment-consumption problem requires to estimate the unobserved random variables. Moreover, several implications of the classical models change in the presence of unobservable factors. For instance, assuming parameter uncertainty, the results obtained on return predictability can admit different interpretations, as shown for instance in Kandel & Stambaugh [1071] in the context of a two-period economy. In this setting, there is an *estimation risk* as well as a *learning risk*. Estimation risk refers to the fact that an agent does not know with certainty the parameter values, while learning risk refers to the fact that an agent will learn more in the future, so that there is a risk that he will discover bad news (low mean returns). Avramov [91] introduces *model uncertainty*, namely uncertainty about which economic variables should appear in the return forecasting model, and shows that this component significantly affects portfolio choices. Barberis [146] examines how the evidence of return predictability affects the optimal portfolio choice for investors with long horizons. The author shows that, even after incorporating parameter uncertainty, return predictability makes investors allocate more wealth to stocks, the longer their horizon. Moreover, there is evidence of estimation risk.

In a continuous time setting, Gennotte [767] (see also Dothan & Feldman [582] and Detemple [561]) shows that a *certainty equivalence result* is valid: in the presence of unobservable parameters the optimal investment-consumption problem can be solved as in the classical full information setting by replacing the unknown parameters with their estimates. In Brennan [290] a continuous time optimal portfolio problem is analysed assuming that the investor knows that the investment opportunity set is constant but he is uncertain about the mean return of the risky asset (the volatility instead is known), learning about the mean rate of return from the observation of historical returns. This learning process affects the portfolio problem, inducing a learning risk and a hedging demand because the investor attempts to protect himself against learning bad news. This hedging component is null when the utility function is logarithmic: in this case the investor holds the same position in the risky asset as in the full information economy where the mean return of the risky asset is equal to the investor's current assessment.

Otherwise, the investor takes a larger or a smaller position in the risky asset in comparison to the full information case, depending on whether he is more or less risk tolerant than a logarithmic investor. This effect increases with the investor's time horizon and decreases with the volatility of the market return. The learning effect joined by possibly predictable returns makes the relationship between investment horizon and risky allocation quite complex (see Xia [1669]): in particular, it is no longer true that young agents should hold more risky assets than old agents.

Incomplete information and partial observation help to address some asset pricing puzzles. Brennan & Xia [302] analyse a representative agent economy with dividends described by a stochastic log-normal process with a mean reverting drift that is not directly observable but must be estimated from the realized growth rates of dividends and aggregate consumption. The learning procedure increases stock price volatility and allows to match the interest rate level and the equity premium observed in U.S. capital markets with a coefficient of relative risk aversion of 15. Moreover, learning can help to resolve the apparent discrepancy between high volatility of stock prices and low volatility of dividends and consumption. Learning by rational agents also provides an explanation to value and size anomalies of asset prices, see Pástor [1405] and Brennan & Xia [303].

The APT in a setting characterized by incomplete information on the parameters of the model generating asset returns has been analysed in Handa & Linn [886]. Under some conditions, a linear relation for expected returns still holds. However, beta coefficients and prices differ from those computed under complete information. Agents attribute more systemic risk to an asset with low information than to an asset with high information, leading to relatively lower betas for high information assets in comparison to low information assets. In an equilibrium model, Coles et al. [473] show that for low information securities the beta coefficients and expected returns tend to be higher than without estimation risk, while the opposite is verified in the case of high information assets.

In Kahn [1059], it is shown that assuming imperfect information and moral hazard leads to incomplete insurance. To illustrate the phenomenon, consider a risk averse agent who supplies unobservable labor in order to produce some output. If labor is observable, then complete risk sharing is obtained, while, if it is not observable, then there is imperfect insurance (market incompleteness). In equilibrium agents bear idiosyncratic risk in consumption and the variance of a representative agent's consumption is larger than the variance of per capita consumption. This fact contributes to a partial resolution of the equity premium puzzle. Kocherlakota [1113] analyses a model with moral hazard under the hypothesis of complete markets and that all the trades are public information. In this setting, it is shown that the equity premium is lower than the one obtained in a representative agent economy.

The asset pricing implications of market imperfections have been studied in Heaton & Lucas [926], surveying the related literature. The main message of the authors is that, despite the different forms of market frictions and the differences in the models considered, market frictions allow for the resolution of some of the anomalies encountered in the empirical tests of classical asset pricing theory.

In a continuous time setting, the mathematical theory of transaction costs has been the subject of a considerable amount of research in mathematical finance in the last years (see Muhle-Karbe & Guasoni [1360] for a survey on portfolio optimization in the presence of transaction costs). Longstaff [1244] analyse the intertemporal optimal consumption problem in continuous time under the constraint that trading strategies have bounded variation and show that an investor facing this type of constraint behaves very differently from an unconstrained investor. It is shown that the optimal trading strategy consists in trading as much as possible, whenever possible, and the investor acts as if facing borrowing and short selling constraints. Mean-variance portfolio strategies under solvency constraints are analysed in Nguyen & Portait [1374], while portfolio choices with illiquid assets have been studied in Garleanu [760] and Longstaff [1245].

Liquidity constraints, margin requirements, and, in general, financial constraints may weaken classical no-arbitrage arguments used to exclude asset pricing anomalies and, therefore, may allow for an explanation of return predictability, overreaction, excess volatility, inefficient allocations and also a financial crisis with divergence of prices from their fundamentals, see Aiyagari & Gertler [35], Chowdhry & Nanda [442], Shleifer & Vishny [1541], Liu & Longstaff [1230], Gromb & Vayanos [837], Kupiec [1143], Hsieh & Miller [970], Kupiec & Sharpe [1144], Chabakauri [383]. When short sale constraints are binding, assets exhibit high valuations and low subsequent returns (see Jones & Lamont [1042]). As shown in Gromb & Vayanos [837] and Liu & Longstaff [1230], arbitrageurs may even underinvest in an arbitrage opportunity if margin requirements are present. In a related direction, pointing out the limits of arbitrage, Shleifer & Vishny [1541] show that if arbitrageurs invest wealth of outsider investors, who direct their funds according to past performance of arbitrageurs, then their activity may not be effective in exploiting arbitrage opportunities in the market and, therefore, in reducing the gap between asset fundamentals and prices. The reason is that performance-based funds inflow induces the arbitrageur not to take risky investments (arbitrage opportunities). The arbitrage argument is particularly ineffective in extreme circumstances.

A form of market friction is represented by the presence of taxation. No-arbitrage analysis has been extended to an economy with taxes in Dybvig & Ross [610] and Ross [1469]. On the optimal consumption-portfolio problem with capital gains taxation see Constantinides [489]. A general equilibrium model with taxes has been proposed in Basak & Croitoru [176]. Michaely et al. [1338], Michaely & Vila [1337] show that taxation generates trading volume around ex-dividend days, while transaction costs reduce trading volume. Moreover, trading volume is positively related to the degree of tax heterogeneity. Supporting empirical evidence is also provided.

Considering the model proposed in Hong & Stein [962] with differences of opinions as well as short sale constraints, Chen et al. [418] show that the number of short sale constrained agents can be regarded as a proxy of *pessimism* in the market and, therefore, predicts low future returns. Moreover, this proxy and, therefore, short sale constraints themselves are strictly related to the momentum effect. Lamont et al.

[1163] shows the relevance of financing frictions-constraints of a firm in the cross-sectional analysis of asset returns (see however Gomes et al. [811] for a different conclusion).

9.7 Exercises

Exercise 9.1 By relying on arguments similar to those used in the proof of Proposition 2.10, complete the proof of Proposition 9.3.

Exercise 9.2 Consider a preference functional of the cumulative prospect theory form (9.4). Let $\tilde{x}_1$ and $\tilde{x}_2$ be two gambles on the same finite probability space:

$$\tilde{x}_1 = \{x_1, \ldots, x_S; \pi_1, \ldots, \pi_S\} \qquad \text{and} \qquad \tilde{x}_1 = \{x_1, \ldots, x_S; \pi'_1, \ldots, \pi'_S\},$$

with $x_1 < \ldots < x_S$ and $\sum_{s=1}^{S} \pi_s = \sum_{s=1}^{S} \pi'_s = 1$. Suppose that $\tilde{x}_1$ dominates $\tilde{x}_2$ in the sense of first order stochastic dominance (see Sect. 2.3). Show that, for any choice of the value and of the weighting functions $u(\cdot)$ and $w(\cdot)$, the preference functional (9.4) satisfies $U(\tilde{x}_1) \geq U(\tilde{x}_2)$ (compare also with Hens & Rieger [938, Proposition 2.39]).

Exercise 9.3 Consider the Barberis et al. [150] model described by equations (9.5)–(9.9). In this exercise, following the Appendix of Barberis et al. [150], we provide the proof of Proposition 9.5. Denote by $(w, c) := ((w_t, c_t))_{t \in \mathbb{N}}$ a self-financing trading-consumption strategy, where the investment strategy w_{t+1} is parameterized in terms of the amount of wealth invested in the single risky asset between date t and date $t + 1$. In the context of the present exercise, we assume that a trading-consumption strategy (w, c) is admissible if w_t is bounded, for all $t \in \mathbb{N}$. Similarly as in relations (6.4) and (6.19), it is easy to see that a trading-consumption strategy (θ, c) satisfies the following self-financing condition:

$$W_{t+1} = (W_t - c_t) r_f + w_{t+1}(r_{t+1} - r_f) + y_{t+1}, \qquad \text{for all } t \in \mathbb{N},$$

where $(W_t)_{t \in \mathbb{N}}$ denotes the self-financing wealth process associated with the strategy (w, c) and $(y_t)_{t \in \mathbb{N}}$ is the exogenous income stream.

(i) Let $(w^*, c^*) = ((w_t^*, c_t^*))_{t \in \mathbb{N}}$ be a trading-consumption strategy satisfying the above self-financing condition. Let $(w, c) = ((w_t, c_t))_{t \in \mathbb{N}}$ be an alternative self-financing strategy and define the strategy

$$(w^* + \alpha w, c^* + \alpha c) = ((w_t^* + \alpha w_t, c_t^* + \alpha c_t))_{t \in \mathbb{N}},$$

for $\alpha \in \mathbb{R}_+$. Show that, according to the preference functional (9.7), the difference in the expected utilities associated to $(c^* + \alpha c, w^* + \alpha w)$ and (c^*, w^*)

satisfies

$$\mathbb{E}\left[\sum_{t=0}^{\infty}\left(\delta^t u(c_t^* + \alpha c_t) + b_t \delta^{t+1} v(X_{t+1}, w_{t+1}^* + \alpha w_{t+1}, z_t)\right)\right]$$

$$-\mathbb{E}\left[\sum_{t=0}^{\infty}\left(\delta^t u(c_t^*) + b_t \delta^{t+1} v(X_{t+1}, w_{t+1}^*, z_t)\right)\right]$$

$$\leq \Delta(\alpha c, \alpha w) := \mathbb{E}\left[\sum_{t=0}^{\infty}\left(\delta^t u'(c_t^*)\alpha c_t + b_t \delta^{t+1}\alpha w_{t+1}\hat{v}(r_{t+1}, z_t)\right)\right],$$

where the function $\hat{v}$ is defined as in the statement of Proposition 9.5.

(ii) Suppose that, for all $t \in \mathbb{N}$, it holds that

$$u'(c_t^*) = r_f \delta \mathbb{E}[u'(c_{t+1}^*)|\mathscr{F}_t], \tag{9.81}$$

$$u'(c_t^*) = \delta \mathbb{E}[u'(c_{t+1}^*)r_{t+1}|\mathscr{F}_t] + \delta b_t \mathbb{E}[\hat{v}(r_{t+1}, z_t)|\mathscr{F}_t], \tag{9.82}$$

with $u(x) = x^{1-\gamma}/(1 - \gamma)$ and where the function $\hat{v}(\cdot)$ is defined as in Proposition 9.5 and the process $(b_t)_{t\in\mathbb{N}}$ is specified by $b_t = b_0 \bar{c}_t^{-\gamma}$, for all $t \in \mathbb{N}$. Show that, for all $t \in \mathbb{N}$, it holds that

$$\mathbb{E}\left[u'(c_t^*)\alpha c_t + \alpha w_{t+1}\delta b_t \hat{v}(r_{t+1}, z_t)|\mathscr{F}_t\right] = u'(c_t^*)\alpha W_t - \delta \mathbb{E}\left[u'(c_{t+1}^*)\alpha W_{t+1}\Big|\mathscr{F}_t\right], \tag{9.83}$$

where $(W_t)_{t\in\mathbb{N}}$ is the wealth process associated to the self-financing strategy (w, c).

(iii) Deduce that

$$\Delta(\alpha c, \alpha w) = u'(c_0^*)\alpha W_0 - \lim_{T\to+\infty}\delta^T \mathbb{E}[u'(c_T^*)\alpha W_T].$$

As explained in Barberis et al. [150, Appendix], the right-hand side of the last equality vanishes if $W_0 = 0$ and the following condition holds:

$$\log\delta - \gamma g_c + g_d + \frac{\gamma^2\sigma_c^2 - 2\gamma\rho\sigma_c\sigma_d + \sigma_d^2}{2} < 0.$$

This shows that the Euler conditions (9.81)–(9.82) are necessary and sufficient for the optimality of the strategy $((\theta_t^*, c_t^*))_{t\in\mathbb{N}}$.

(iv) Show that the risk free return r_f given in (9.10) satisfies the Euler condition (9.81) when the consumption process $(c_t^*)_{t\in\mathbb{N}}$ is given by the aggregate per capita consumption process $(\bar{c}_t)_{t\in\mathbb{N}}$.

(v) Suppose that the equilibrium price-dividend ratio of the risky asset satisfies

$$\frac{p_t^*}{d_t} = f(z_t), \qquad \text{for all } t \in \mathbb{N},$$

for a suitable function $f(\cdot)$. Show that

$$\mathbb{E}\left[r_{t+1} \left(\frac{\bar{c}_{t+1}}{\bar{c}_t} \right)^{-\gamma} \Big| \mathscr{F}_t \right] = e^{g_d - \gamma g_c + \frac{\gamma^2 \sigma_c^2 (1-\rho^2)}{2}} \mathbb{E}\left[\frac{1 + f(z_{t+1})}{f(z_t)} e^{(\sigma_d - \gamma \rho \sigma_c)\varepsilon_{t+1}} \Big| \mathscr{F}_t \right].$$

Deduce that condition (9.11) of Proposition 9.5 implies that the Euler optimality condition (9.82) holds.

(vi) Deduce that the strategy of consuming $c_t^* = \bar{c}_t = d_t + y_t$, for all $t \in \mathbb{N}$, and holding the market supply of the risky asset satisfies the Euler conditions (9.81)–(9.82) and the claim of Proposition 9.5 follows.

Exercise 9.4

(i) Consider the general form of recursive preferences as defined in (9.16). Suppose that the functions v and $\tilde{u}$ are of the following form:

$$v(c, y) = u(c) + \delta y \qquad \text{and} \qquad \tilde{u}(x) = x.$$

Show that in this case the preference functional (9.16) reduces to the classical time additive utility.

(ii) Consider the recursive utility functional of the Epstein & Zin [651] form given in (9.19). Show that, if $\alpha = \varrho$, then the recursive preference functional reduces to the classical time additive expected utility with a power utility function.

Exercise 9.5 In this exercise, we prove Proposition 9.8, considering a recursive preference functional of the form (9.19), with $U_T(c, \theta) = u(c) = (1-\delta)^{1/(1-\rho)} c$.

(i) By following the arguments given in the proof of Proposition 6.4, show that the Euler conditions (9.23)–(9.24) hold.

(ii) Denoting by $(W_t^*)_{t=0,1,\dots,T}$ and $(c_t^*)_{t=0,1,\dots,T}$ the optimal wealth and consumption processes, respectively, in the maximization of the recursive preference functional $U_0(c, \theta)$ defined in (9.19), define the adapted stochastic process $(\xi_t)_{t=0,1,\dots,T}$ as the solution to the recursive relation

$$\xi_{t+1} := \delta \left(\frac{V(W_{t+1}^*, t+1)}{\mu_t(V(W_{t+1}^*, t+1))} \right)^{\varrho - \alpha} \left(\frac{c_{t+1}^*}{c_t^*} \right)^{-\varrho} \xi_t \qquad \text{and} \qquad \xi_0 = 1.$$

Consider the backward stochastic difference equation

$$X_t = c_t^* + \mathbb{E}\left[\frac{\xi_{t+1}}{\xi_t} X_{t+1} \Big| \mathscr{F}_t \right], \qquad \text{for } t = 0, 1, \dots, T-1, \text{ with } X_T = \frac{V(W_T^*, T)}{u'(c_T^*)},$$

the solution of which is an adapted stochastic process $(X_t)_{t=0,1,\ldots,T}$. Show that this backward stochastic difference equation admits a unique solution which is given by $X_t = W_t^*$, for all $t = 0, 1, \ldots, T$. Note also that the process $(\xi_t)_{t=0,1,\ldots,T}$ corresponds to the stochastic discount factor introduced in (9.29).

(iii) Deduce that the optimal wealth process $(W_t^*)_{t=0,1,\ldots,T}$ satisfies relation (9.25) and that the return process $(r_t^*)_{t=1,\ldots,T}$ associated to the optimal wealth process can be represented as in (9.26).

(iv) Deduce the validity of the Euler condition (9.27).

Exercise 9.6 In this exercise, we prove Proposition 9.9.

(i) As a first step, show that relation (9.23), suitably rewritten in terms of the optimal return $(r_t^*)_{t=1,\ldots,T}$ process, implies that

$$\mu_r = -\log \delta + \varrho \mu_c - \frac{\gamma \sigma_r^2}{2} - \frac{\gamma \varrho^2 \sigma_c^2}{2} + \gamma \varrho \sigma_{cr}.$$

(ii) Using part (i) of the exercise together with relation (9.23) with respect to the risk free asset with return r_f, deduce that the equilibrium risk free rate r_f is given by (9.33).

(iii) Suppose furthermore that $(\log(c_{t+1}^*/c_t^*), \log r_{t+1}^1, \ldots, r_{t+1}^N)$ are jointly normally distributed conditionally on $\mathscr{F}_t$, for all $t = 0, 1, \ldots, T - 1$. Deduce that asset risk premia satisfy relation (9.34).

Exercise 9.7 (An approximate formula for the optimal demand of the risky asset with recursive preferences, see Campbell & Viceira [354]). Consider the recursive preference functional defined in (9.19), with parameters $\alpha \in (0, 1)$, $\varrho \in (0, 1)$ and $\delta \in (0, 1)$, in an economy with a single risky asset with return process $(r_t)_{t=1,\ldots,T}$ and a risk free asset delivering the constant return r_f. Denote by $(\theta_t)_{t=0,1,\ldots,T}$ a trading strategy, measured in terms of the number of units of the risky asset held in the portfolio, and let the couple (θ^*, c^*) be the trading-consumption self-financing strategy associated with the maximization of the recursive preference functional (9.19). Letting $(W_t^*)_{t=0,1,\ldots,T}$ be the associated wealth process, the corresponding optimal return process $(r_t^*)_{t=1,\ldots,T}$ satisfies

$$W_{t+1}^* = (W_t^* - c_t^*) r_{t+1}^*, \qquad \text{for all } t = 0, 1, \ldots, T - 1.$$

As mentioned after Proposition 9.9, relation (9.34) holds as an approximation for the risk premium of the risky asset, i.e.,

$$\mathbb{E}[\log r_{t+1}|\mathscr{F}_t] - \log r_f + \frac{\mathrm{Var}(\log r_{t+1}|\mathscr{F}_t)}{2} \approx (1 - \gamma) \, \mathrm{Cov}(\log r_{t+1}^*, \log r_{t+1}|\mathscr{F}_t)$$

$$+ \varrho \gamma \, \mathrm{Cov}(\Delta \log c_{t+1}^*, \log r_{t+1}),$$

$$\tag{9.84}$$

for all $t = 0, 1, \ldots, T - 1$, where $\Delta \log c_{t+1}^* := \log c_{t+1}^* - \log c_t^*$ denotes the logarithmic growth rate of the optimal consumption process.

(i) Starting from the relation $W_{t+1}^* = (W_t^* - c_t^*) r_{t+1}^*$, prove the following log-linear approximation of the self-financing condition:

$$\Delta \log W_{t+1}^* \approx \log r_{t+1}^* + \left(1 - \frac{1}{\lambda} \right) (\log c_t^* - \log W_t^*) + K,$$

for all $t = 0, 1, \ldots, T - 1$, where

$$K := \log \lambda + (1 - \lambda) \frac{\log(1 - \lambda)}{\lambda} \qquad \text{and} \qquad \lambda := 1 - e^{\mathbb{E}[\log c_t^* - \log W_t^*]}.$$

(ii) Suppose that the return process $(r_t^*)_{t=1,\ldots,T}$ associated to the strategy (θ^*, c^*) satisfies the following approximate relation:

$$\log r_{t+1}^* \approx \theta_{t+1}^* (\log r_{t+1} - \log r_f) + \log r_f + \frac{\theta_{t+1}^*}{2}(1 - \theta_{t+1}^*) \operatorname{Var}(\log r_{t+1} | \mathscr{F}_t), \tag{9.85}$$

for all $t = 0, 1, \ldots, T - 1$. By relying on part (i) of the exercise, show that the asset pricing relation (9.84) can be rewritten in the following form:

$$\mathbb{E}[\log r_{t+1} | \mathscr{F}_t] - \log r_f + \frac{\operatorname{Var}(\log r_{t+1} | \mathscr{F}_t)}{2} \approx (1 - \gamma) \theta_{t+1}^* \operatorname{Var}(\log r_{t+1} | \mathscr{F}_t)$$

$$+ \varrho \gamma \Big(\operatorname{Cov}(\log r_{t+1}, \log c_{t+1}^* - \log W_{t+1}^* | \mathscr{F}_t) + \theta_{t+1}^* \operatorname{Var}(\log r_{t+1} | \mathscr{F}_t) \Big). \tag{9.86}$$

(iii) Deduce that the optimal demand θ_{t+1}^* of the risky asset satisfies the following approximate relation:

$$\theta_{t+1}^* \approx \frac{1}{\alpha} \frac{\mathbb{E}[\log r_{t+1} | \mathscr{F}_t] - \log r_f + \frac{\operatorname{Var}(\log r_{t+1} | \mathscr{F}_t)}{2}}{\operatorname{Var}(\log r_{t+1} | \mathscr{F}_t)}$$

$$- \frac{\alpha - 1}{\alpha} \frac{\varrho}{\varrho - 1} \frac{\operatorname{Cov}(\log r_{t+1}, \log c_{t+1}^* - \log W_{t+1}^* | \mathscr{F}_t)}{\operatorname{Var}(\log r_{t+1} | \mathscr{F}_t)}.$$

This approximate relation shows that the optimal demand of the risky asset can be decomposed into two terms: a first term which depends exclusively on the risk premium of the asset and a second term which reflects the intertemporal hedging demand (see Campbell & Viceira [354, Section III] for more details).

Exercise 9.8 Consider the habit formation preferences introduced in (9.35)–(9.36), with respect to a general utility function u.

(i) Derive the Euler condition (9.39) (compare also with Munk [1363, Theorem 6.9]).

(ii) Consider the habit formation preferences introduced in (9.40). Show that in this case the general Euler condition (9.39) reduces to (9.41).

Exercise 9.9 Consider the habit preferences introduced in (9.43)–(9.44). In this exercise, following the original paper Abel [4], we derive the Euler conditions and the equilibrium price-dividend ratio of a risky asset.

(i) For a consumption process $c = (c_t)_{t=0,1,...,T}$, let us denote

$$U(c) = \sum_{t=0}^{T} \delta^t u(c_t, z_t).$$

As a first step, show that, for all $t = 0, 1, \ldots, T - 1$, it holds that

$$\frac{\partial}{\partial c_t} U(c) = \delta^t \left(1 - \delta \gamma \beta \left(\frac{c_{t+1}}{c_t}\right)^{1-\alpha} \left(\frac{z_t}{z_{t+1}}\right)^{1-\alpha}\right) \left(\frac{c_t}{z_t}\right)^{1-\alpha} \frac{1}{c_t}. \tag{9.87}$$

(ii) Let the process $(d_t)_{t=0,1,...,T}$ represent the total amount of consumption good per capita delivered by the capital stock (dividend). Assume that, in equilibrium, all dividend is consumed as soon as it is delivered. If we furthermore assume that all the agents of the economy are identical, this implies that $c_t^* = \bar{c}_t = d_t$, for all $t = 0, 1, \ldots, T$, so that

$$\frac{c_{t+1}^*}{c_t^*} = \frac{\bar{c}_{t+1}}{\bar{c}_t} = \frac{d_{t+1}}{d_t} =: x_{t+1},$$

where the process $(x_t)_{t=1,...,T}$ represents the growth rate of the dividend. Rewrite (9.87) as follows:

$$\frac{\partial}{\partial c_t} U(c^*) = \delta^t \left(1 - \delta \gamma \beta x_{t+1}^{1-\alpha} x_t^{-\gamma(1-\alpha)}\right) z_t^{\alpha-1} (c_t^*)^{-\alpha} \tag{9.88}$$

and deduce that the following Euler condition holds:

$$\delta \mathbb{E}\left[\frac{1 - \delta \gamma \beta x_{t+2}^{1-\alpha} x_{t+1}^{-\gamma(1-\alpha)}}{1 - \mathbb{E}[\delta \gamma \beta x_{t+1}^{1-\alpha} x_t^{-\gamma(1-\alpha)} | \mathscr{F}_t]} x_t^{\gamma(\alpha-1)} x_{t+1}^{-\alpha} r_{t+1} \middle| \mathscr{F}_t\right] = 1,$$

for all $t = 0, 1, \ldots, T - 1$ and $n = 1, \ldots, N$, where $(r_t)_{t=1,...,T}$ is the return process of the stock.

(iii) Let the process $(p_t^*)_{t=0,1,\dots,T}$ denote the equilibrium price process of the stock. Show that the equilibrium price dividend ratio process $(p_t^*/d_t)_{t=0,1,\dots,T}$ satisfies

$$\frac{p_t^*}{d_t} = \delta \mathbb{E}\left[\frac{1 - \delta\gamma\beta x_{t+2}^{1-\alpha} x_{t+1}^{-\gamma(1-\alpha)}}{1 - \mathbb{E}[\delta\gamma\beta x_{t+1}^{1-\alpha} x_t^{-\gamma(1-\alpha)}|\mathscr{F}_t]} x_t^{\gamma(\alpha-1)} x_{t+1}^{-\alpha}\left(1 + \frac{p_{t+1}^*}{d_{t+1}}\right) x_{t+1} \,\middle|\, \mathscr{F}_t\right],$$

for all $t = 0, 1, \dots, T-1$.

Exercise 9.10 Consider the Campbell & Cochrane [343] model with external habit preferences, where the surplus process and the consumption growth process satisfy the dynamics (9.46)–(9.47), respectively. Prove Proposition 9.10.

Exercise 9.11 Consider the De Long et al. [549] model described in Sect. 9.3 with a proportion $\mu \in (0,1)$ of noise traders. Prove Proposition 9.11.

Exercise 9.12 Consider the De Long et al. [549] model introduced in Sect. 9.3 and extended as in De Long et al. [549, Section III.B] to the case where the risky asset yields at each date $t \in \mathbb{N}$ a dividend d_t equal to

$$d_t = r_f + \tilde{\varepsilon}_t, \qquad \text{for all } t \in \mathbb{N},$$

where $(\tilde{\varepsilon}_t)_{t\in\mathbb{N}}$ is a sequence of i.i.d. normal random variables with zero mean and constant variance σ_ε^2, independent of the noise traders' misperceptions $(\varrho_t)_{t\in\mathbb{N}}$. By following the arguments given in Exercise 9.11, show that:

(i) the optimal demand of the risky asset by rational and noise traders is respectively given by

$$\theta_{t+1}^{r,*} = \frac{r_f + \mathbb{E}[p_{t+1}|\mathscr{F}_t] - (1+r_f)p_t}{2\gamma(\mathrm{Var}(p_{t+1}|\mathscr{F}_t) + \sigma_\varepsilon^2)},$$

$$\theta_{t+1}^{n,*} = \frac{r_f + \mathbb{E}[p_{t+1}|\mathscr{F}_t] - (1+r_f)p_t}{2\gamma(\mathrm{Var}(p_{t+1}|\mathscr{F}_t) + \sigma_\varepsilon^2)} + \frac{\varrho_t}{2\gamma(\mathrm{Var}(p_{t+1}|\mathscr{F}_t) + \sigma_\varepsilon^2)};$$

(ii) the steady-state equilibrium price process $(p_t^*)_{t\in\mathbb{N}}$ of the risky asset satisfies

$$p_t^* = 1 + \frac{\mu_t\bar{\varrho}}{r_f} - \frac{2\gamma}{r_f}\left(\sigma_\varepsilon^2 + \frac{\mu^2\sigma_\varrho^2}{(1+r_f)^2}\right) + \frac{\mu_t(\varrho_t - \bar{\varrho})}{1+r_f};$$

(iii) in correspondence of a steady-state equilibrium, the difference $\mathbb{E}[\Delta R_{t+1}(\mu)]$ in the expected optimal return between the noise and the rational traders is

given by

$$\mathbb{E}[\Delta R_{t+1}(\mu)] = \bar{\varrho} - \frac{\bar{\varrho}^2 + \sigma_\varrho^2}{2\gamma \left(\frac{\sigma_\varrho^2 \mu}{(1+r_f)^2} + \frac{\sigma_\varepsilon^2}{\mu} \right)}$$

if $\mu > 0$ and $\mathbb{E}[\Delta R_{t+1}(\mu)] = \bar{\varrho}$ if $\mu = 0$, for all $t \in \mathbb{N}$.

Exercise 9.13 Consider the model proposed in Daniel et al. [516] and described in Sect. 9.3, with uninformed agents and overconfident informed agents. Prove Proposition 9.12.

Exercise 9.14 Consider the model proposed in Daniel et al. [516, Section III] and described in Sect. 9.3, with uninformed agents and overconfident agents, assuming that the overconfidence of informed traders depends on the outcome of the public signal $\tilde{s}_2$ released at date $t = 2$. Prove Proposition 9.13.

Exercise 9.15 Consider a risk neutral representative agent economy under the assumption of rational expectations, with a risky security paying the dividend stream $(d_t)_{t \in \mathbb{N}}$, which is supposed to follow a first order autoregressive process of the form (9.63). Assume furthermore that the dividend process $(d_t)_{t \in \mathbb{N}}$ is predictable, meaning that the random variable d_{t+1} is $\mathscr{F}_t$-measurable, for every $t \in \mathbb{N}$. Starting from equation (9.62), show that the rational expectations solution for the price process $(s_t)_{t \in \mathbb{N}}$ of the risky security is given by (9.64).

Exercise 9.16 Consider the Weil [1650] model introduced in Sect. 9.5. By applying Jensen's inequality, prove Proposition 9.15.

Chapter 10
Financial Markets Microstructure

> *Noise trading is essential to the existence of liquid markets [...]*
> *Noise makes financial markets possible, but also makes them*
> *imperfect [...] Noise creates the opportunity to trade profitably,*
> *but at the same time makes it difficult to trade profitably.*
>
> Black (1986)

In the previous chapters, we have considered financial markets as abstract places where trading occurs. In reality, financial markets are not idealized entities and are characterized by a well defined institutional setting. In particular, taking the institutional setting into account is fundamental in order to understand the functioning of real financial markets. To this effect, it is important to analyse the different roles of market makers, dealers and brokers. These types of investors trade according to their own objectives and they also coordinate and supervise the market by making sure that the demand meets the supply.

In this chapter, we focus our attention on financial markets microstructure by considering the effects of different institutional settings on the functioning of financial markets. To this purpose, we relax the Walrasian-perfectly competitive market hypothesis and introduce market models with *imperfect competition*. We shall be concerned with the following features of financial markets:

a) *Order-driven* and *quote-driven* (or price-driven) markets. In the case of an order-driven market, buying and selling orders arrive in the market and prices are then defined through some automatic mechanism matching the orders or through the action of market makers who supervise the market and set prices with the commitment to satisfy the net market order at that price. In the case of a quote-driven market, some agents (dealers) provide quotes for bid prices (corresponding to sell orders) and ask prices (corresponding to buy orders) and, under some conditions, are committed to satisfy market orders at the quoted prices.

b) *Consolidated* and *fragmented* markets. This feature reflects the fact that, in some cases, trades occur in a single market (consolidated market), while, in other cases, it is possible to trade the same asset in more than one market (fragmented markets).

© Springer-Verlag London Ltd. 2017

E. Barucci, C. Fontana, *Financial Markets Theory*,
Springer Finance, DOI 10.1007/978-1-4471-7322-9_10

c) *Broker* and *dealer* markets. This distinction concerns the fact that brokers act only for their own customers while dealers also trade on their own account.

d) Possible *competition among market makers (dealers)*, meaning that several market makers can manage the same security simultaneously.

Moreover, one can also consider discrete time or continuous time markets, meaning that market orders can be assumed to arrive either in correspondence of a finite set of predetermined dates or at any time during the trading period. If an auction market functions in discrete time, then orders (in whatever form they are specified) arrive in the market and are matched among themselves or satisfied thanks to the intervention of market makers. On the other hand, if an auction market functions in continuous time, so that orders can arrive at any point in time, it is typically assumed that market makers determine the trading price and clear the markets in correspondence of a set of predetermined dates (*batch auction market*). We refer to Foucault et al. [735, Chapter 1] for a complete overview of different market structures and trading mechanisms.

One of the most interesting perspectives to analyse the market is represented by welfare analysis. Even though this is not always an easy task, welfare analysis can be useful to evaluate some of the features of financial markets, for instance:

a) *information revelation*, namely the possibility that market prices aggregate and transmit information in the context of non-perfectly competitive financial markets;

b) *price discovery*, meaning that in some markets agents do not observe trading prices before sending orders to the market;

c) *execution risk*, consisting in the risk due to the possible uncertainty of whether and when an order will be executed;

d) *market liquidity*, where liquidity is generically understood as the cost that an agent must support when buying and selling the same asset at the same time. Typically, the bid-ask spread is taken as a measure of market liquidity, but liquidity is not always totally reflected in the bid-ask spread and is also related to the trading volume;

e) *market depth*, meaning the capacity of the market to absorb large market orders without producing large price changes;

f) *transparency*, consisting in the capability of agents to recover information concerning trades in the market (prices of executed trades, bid and ask prices, trading volume, identity of the agents).

This chapter is structured as follows. In Sect. 10.1, by relying on an equilibrium analysis, we discuss the interplay between private information and imperfect competition. In Sect. 10.2, we present the seminal Kyle (1985) model and several extensions of it. Section 10.3 deals with the modeling of quote-driven markets. In Sect. 10.4, we provide a brief overview on the dynamic modeling of market microstructure, also discussing the existence of intraday and intraweek regularities in asset prices. Section 10.5 is devoted to a discussion of insider trading behavior, while Sect. 10.6 contains a survey on the effects of the different institutional features of financial markets. At the end of the chapter, we provide a guide to further readings as well as a series of exercises.

In this chapter, if $\tilde{x}$ and $\tilde{y}$ are two random variables, we shall denote (with some abuse of notation) by $\mathbb{E}[\tilde{x}|\tilde{y}]$ the conditional expectation of $\tilde{x}$ given $\tilde{y}$, viewed as a random variable, and by $\mathbb{E}[\tilde{x}|y] := \mathbb{E}[\tilde{x}|\tilde{y} = y]$ the conditional expectation of $\tilde{x}$ given the realization y of the random variable $\tilde{y}$.

10.1 The Role of Information Under Imperfect Competition

In Chap. 8, we have analysed the role of information in the context of perfectly competitive markets, where agents are assumed to be price takers. In particular, we have focused our attention on whether the prices associated to a Green-Lucas equilibrium of the economy aggregate and transmit the agents' private information. In this section, we are going to address the same problem in the context of an imperfectly competitive market, focusing on the model introduced in Kyle [1148]. In this model, an imperfectly competitive market with private information is analysed, assuming that agents can submit to a central auctioneer a demand function which depends on the asset price and on the possible observation of a private signal. In an imperfectly competitive market, asset prices are affected by the agents' demand. Agents are aware of this effect and, therefore, they do not behave as price takers but act strategically, exploiting the dependence of market prices on their demand. This argument induces a rational agent to maximize his expected utility considering the informational content of prices as well as the effect of his demand on market prices.

The model proposed in Kyle [1148], which has some similarities to the model of Grossman [839] discussed in Chap. 8, provides a partial equilibrium analysis in a single-period setting (i.e., $t \in \{0, 1\}$). It is assumed that there is a single risky asset, with liquidation value (dividend) $\tilde{d}$ at date $t = 1$, traded at date $t = 0$ at some market clearing price p. The economy is populated by three homogeneous classes of agents: *noise traders*, *informed speculators* and *uninformed speculators*. Noise traders trade an aggregate quantity of the asset represented by the random variable $\tilde{z}$, capturing the fact that the aggregate demand of noise traders is exogenous and does not result from a maximizing behavior. There are N informed speculators $n = 1, \ldots, N$ and each informed speculator receives a private signal $\tilde{y}_n$, supposed to be of the form

$$\tilde{y}_n = \tilde{d} + \tilde{\varepsilon}_n, \qquad \text{for every } n = 1, \ldots, N,$$

where the random variables $\tilde{d}, \tilde{z}, \tilde{\varepsilon}_1, \ldots, \tilde{\varepsilon}_N$ are mutually independent and normally distributed with variances

$$\text{Var}(\tilde{d}) = \tau_d^{-1}, \qquad \text{Var}(\tilde{z}) = \sigma_z^2 \quad \text{and} \quad \text{Var}(\tilde{\varepsilon}_n) = \tau_\varepsilon^{-1},$$

for all $n = 1, \ldots, N$, with τ_d and τ_ε representing the precision. The hypothesis that the random variable $\tilde{z}$ is independent from the other random variables means that noise trading has no informational content, while the assumption that the random variables $\tilde{\varepsilon}_1, \ldots, \tilde{\varepsilon}_N$ are independent simply means that different informed speculators have access to independent private information. After observing a realization y_n of his private signal $\tilde{y}_n$ and conditionally on the observation of the market price p, the demand of the n-th informed speculator is given by $X_n(p, y_n)$, for $n = 1, \ldots, N$.

Besides noise traders and informed speculators, there are M uninformed speculators ($m = 1, \ldots, M$), who do not receive private signals. The demand of uninformed speculators conditionally on a market clearing price p is given by the quantities $Y_m(p)$, for $m = 1, \ldots, M$. Similarly as in Chap. 8, uninformed agents have only access to the public information represented by the observation of the market price. All speculators have deterministic initial endowment, normalized at zero for simplicity.

Every speculator is characterized by an exponential utility function, with a coefficient of absolute risk aversion given by α_{inf} or α_{un}, for informed or uninformed speculators, respectively. Hence, given the demand $X_n(p, y_n)$ and $Y_m(p)$ of an informed speculator and of an uninformed speculator, respectively, the realized utilities at date $t = 1$ in correspondence of a market clearing price p and of the observation of the signal y_n are given by

$$u_n^{\text{inf}}\big((d - p)X_n(p, y_n)\big) := -e^{-\alpha_{\text{inf}}(d-p)X_n(p,y_n)}, \quad \text{for all } n = 1, \ldots, N,$$

and

$$u_m^{\text{un}}\big((d - p)Y_m(p)\big) := -e^{-\alpha_{\text{un}}(d-p)Y_m(p)}, \quad \text{for all } m = 1, \ldots, M,$$

where d denotes a generic realization of the random variable $\tilde{d}$. In the setting considered in Kyle [1148], the *demand schedules* $X_n(\cdot, y_n)$ and $Y_m(\cdot)$ (i.e., the demands of informed and uninformed speculators, respectively, viewed as functions of the price and, in the case of informed agents, given the observation of the private signal) are allowed to be arbitrary convex upper-semicontinuous correspondences, taking values in $[-\infty, +\infty]$. In particular, such a general specification includes the order trades typically taking place in organized exchanges, such as market orders, limit orders and stop orders (together with all the linear combinations thereof).

In this setting, under the assumption of zero aggregate supply of the risky asset, the market clearing condition in correspondence of a price p requires that

$$\sum_{n=1}^{N} X_n(p, \tilde{y}_n) + \sum_{m=1}^{M} Y_m(p) + \tilde{z} = 0. \tag{10.1}$$

Similarly as in Sect. 8.1, a *perfectly competitive Green-Lucas rational expectations equilibrium* consists of an equilibrium price functional $\phi^* : \mathbb{R}^{N+1} \to \mathbb{R}$ and of a family of demand schedules $X_1(\cdot, \cdot), \ldots, X_N(\cdot, \cdot), Y_1, (\cdot), \ldots, Y_M(\cdot)$ such that the market clearing condition (10.1) holds almost surely in correspondence of the equilibrium price $\tilde{p}^* = \phi^*(\tilde{y}_1, \ldots, \tilde{y}_N, \tilde{z})$ and

$$\mathbb{E}\left[u_n^{\mathrm{inf}}\left((\tilde{d} - \tilde{p}^*)X_n(\tilde{p}^*, \tilde{y}_n)\right) \big| \tilde{p}^* = \phi^*(y_1, \ldots, y_n, z), \tilde{y}_n = y_n \right]$$

$$\geq \mathbb{E}\left[u_n^{\mathrm{inf}}\left((\tilde{d} - \tilde{p}^*)X_n'(\tilde{p}^*, \tilde{y}_n)\right) \big| \tilde{p}^* = \phi^*(y_1, \ldots, y_n, z), \tilde{y}_n = y_n \right],$$

$$\mathbb{E}\left[u_m^{\mathrm{un}}\left((\tilde{d} - \tilde{p}^*)Y_m(\tilde{p}^*)\right) \big| \tilde{p}^* = \phi^*(y_1, \ldots, y_n, z) \right] \tag{10.2}$$

$$\geq \mathbb{E}\left[u_m^{\mathrm{un}}\left((\tilde{d} - \tilde{p}^*)Y_m'(\tilde{p}^*)\right) \big| \tilde{p}^* = \phi^*(y_1, \ldots, y_n, z) \right],$$

for all $n = 1, \ldots, N$ and $m = 1, \ldots, M$, for every realization $(y_1, \ldots, y_N, z)$ of the random variables $(\tilde{y}_1, \ldots, \tilde{y}_N, \tilde{z})$ and where $X_1'(\cdot, \cdot), \ldots, X_N'(\cdot, \cdot), Y_1'(\cdot),$ $\ldots, Y_M'(\cdot)$ is any alternative family of demand schedules. Condition (10.2) means that both informed and uninformed speculators maximize their expected utilities taking market prices as given and conditioning on the observation of the market price. Note that both informed and uninformed agents use their knowledge of the equilibrium price functional ϕ^* to extract all the available information from the market price. Moreover, informed agents also condition on the realization of their own private signal. This notion of equilibrium is in line with the perfect rationality hypothesis, according to which every agent fully exploits all the available information, under the perfect competition hypothesis. Observe also the similarity between the present notion of equilibrium and the results presented in Sect. 8.3.

In the setting considered in Kyle [1148], relaxing the perfect competition assumption, each trader submits to a central auctioneer his demand schedule. The auctioneer aggregates the demand schedules submitted by all the traders in the economy, sets a price which satisfies the market clearing condition (10.1) and allocates quantities in order to satisfy the traders' demands (see Kyle [1148, Section 3] for a description of the way the auctioneer operates in the case of an infinite demand by some traders or in the case of a multiplicity of clearing prices). In this context, *imperfect competition* is present since every trader realizes that, by submitting to the central auctioneer a demand schedule, he can influence the resulting equilibrium price. Speculators are assumed to be rational and, therefore, they are aware of the fact that their behavior can influence the market price, since by changing his demand, an agent changes both the quantity he trades and the market clearing price at which he trades that quantity. In order to emphasize the dependence of the equilibrium price on the demand schedules submitted by the informed and uninformed speculators we let $\tilde{p}^* = \psi^*(X, Y, z)$, where X and Y are the vectors of demand schedules defined by $X := (X_1(\cdot, \cdot), \ldots, X_N(\cdot, \cdot))$ and $Y := (Y_1(\cdot), \ldots, Y_M(\cdot))$.

The functional ψ^* admits a conjectural as well as an institutional interpretation. The conjectural interpretation follows the reasoning presented in Chap. 8 when discussing the Green-Lucas equilibrium. Speculators are aware of the fact that their behavior affects market prices and they conjecture a functional relating their demand to the market price. Having conjectured a price functional, speculators then take their decisions according to such a conjecture and act strategically (knowing also the other agents' conjectures and that other agents do the same). In equilibrium, the agents' conjectures are confirmed. According to the institutional interpretation, the functional ψ^* derives from a specific price formation rule determined by the market institutional setting. Agents know it and exploit it when defining their demand schedules.

The above discussion leads us to introduce the notion of *rational expectations equilibrium under imperfect competition*, defined as a couple (X, Y) of vectors of demand schedules $X = (X_1(\cdot, \cdot), \ldots, X_N(\cdot, \cdot))$ and $Y = (Y_1(\cdot), \ldots, Y_M(\cdot))$ such that the market clearing condition (10.1) is satisfied in correspondence of the equilibrium price $\tilde{p}^* = \psi^*(X, Y, z)$ and

$$
\begin{aligned}
&\mathbb{E}\left[u_n^{\text{inf}}\left((\tilde{d} - \tilde{p}^*)X_n(\tilde{p}^*, \tilde{y}_n)\right) \middle| \tilde{p}^* = \psi^*(X, Y), \tilde{y}_n = y_n\right] \\
&\geq \mathbb{E}\left[u_n^{\text{inf}}\left((\tilde{d} - \tilde{p}^*)X_n^{(n)}(\tilde{p}^*, \tilde{y}_n)\right) \middle| \tilde{p}^* = \psi^*(X^{(n)}, Y), \tilde{y}_n = y_n\right], \\
&\mathbb{E}\left[u_m^{\text{un}}\left((\tilde{d} - \tilde{p}^*)Y_m(\tilde{p}^*)\right) \middle| \tilde{p}^* = \psi^*(X, Y)\right] \\
&\geq \mathbb{E}\left[u_m^{\text{un}}\left((\tilde{d} - \tilde{p}^*)Y_m^{(m)}(\tilde{p}^*)\right) \middle| \tilde{p}^* = \psi^*(X, Y^{(m)})\right],
\end{aligned}
\tag{10.3}
$$

for all $n = 1, \ldots, N$ and $m = 1, \ldots, M$, for every realization $(y_1, \ldots, y_N, z)$ of the random variables $(\tilde{y}_1, \ldots, \tilde{y}_N, \tilde{z})$. In (10.3), $X^{(n)}$ denotes an arbitrary demand vector $X^{(n)} = (X_1^{(n)}, \ldots, X_N^{(n)})$ such that $X_j^{(n)} \neq X_j$ only for $j = n$. Similarly, $Y^{(m)}$ denotes an arbitrary demand vector $Y^{(m)} = (Y_1^{(m)}, \ldots, Y_M^{(m)})$ such that $Y_j^{(m)} \neq Y_j$ only for $j = m$. Condition (10.3) defines a *Nash equilibrium* in trading strategies: given his private information, each (informed) speculator chooses a demand schedule in order to maximize his expected utility, taking into account the effect of his demand on the market price and taking as given the demand functions of the other agents.

In Kyle [1148], the analysis is focused on symmetric linear equilibria, namely couples (X, Y) of demand schedules defining a rational expectations equilibrium under imperfect competition, so that conditions (10.1) and (10.3) are satisfied, and such that

$$
\begin{aligned}
X_n(p, y_n) &= \mu_{\text{inf}} + \beta y_n - \gamma_{\text{inf}} p, \\
Y_m(p) &= \mu_{\text{un}} - \gamma_{\text{un}} p,
\end{aligned}
\tag{10.4}
$$

for some constants $\mu_{\text{inf}}, \beta, \gamma_{\text{inf}}, \mu_{\text{un}}$ and γ_{un}, for all $n = 1, \ldots, N$ and $m = 1, \ldots, M$, for all realizations p and y_n of the market clearing price and of the random signal, respectively. As shown in Exercise 10.1, the market clearing condition (10.1) implies that, in correspondence of a symmetric linear equilibrium, the equilibrium price $\tilde{p}^*$ must satisfy

$$
\tilde{p}^* = \lambda \left(N\beta\tilde{d} + \beta \sum_{n=1}^{N} \tilde{\varepsilon}_n + \tilde{z} + N\mu_{\text{inf}} + M\mu_{\text{un}} \right),
\tag{10.5}
$$

where $\lambda := 1/(N\gamma_{inf} + M\gamma_{un})$ is a measure of market depth (i.e., the order flow necessary to induce the equilibrium price to rise or fall by one unit).

In particular, the linear structure (10.4) allows for a clear analysis of the informativeness of the equilibrium price (see Kyle [1148, Section 4]). To this effect, let us define $\tau_F := 1/\text{Var}(\tilde{d}|\tilde{y}_1,\dots,\tilde{y}_N)$ and note that, due to the assumption of multivariate normality, the quantity $\text{Var}(\tilde{d}|\tilde{y}_1,\dots,\tilde{y}_N)$ is constant and does not depend on the realization $(y_1,\dots,y_N)$ of the conditioning random variables $(\tilde{y}_1,\dots,\tilde{y}_N)$. The quantity τ_F can be regarded as a measure of the precision of the forecast of the liquidation value $\tilde{d}$ given all private information in the economy. The distributional assumptions on $(\tilde{z},\tilde{d},\tilde{y}_1,\dots,\tilde{y}_N)$ imply that

$$\tau_F = \tau_d + N\tau_\varepsilon.$$

Recalling that τ_d measures the precision of the prior forecast of $\tilde{d}$ (i.e., the reciprocal of the unconditional variance of $\tilde{d}$), this shows that the full information precision τ_F is equal to the prior precision τ_d plus τ_ε units of precision for each private signal. Of course, informed and uninformed speculators do not observe all the realizations $(y_1,\dots,y_N)$ of the private signals but only the market price and, in the case of informed agents, their own private signals. We can define as follows the precisions corresponding to an informed and to an uninformed agent, respectively:

$$\tau_{inf} := \frac{1}{\text{Var}(\tilde{d}|\tilde{p}^*,\tilde{y}_n)} \quad \text{and} \quad \tau_{un} := \frac{1}{\text{Var}(\tilde{d}|\tilde{p}^*)},$$

for all $n = 1,\dots,N$. Again, due to the assumption of multivariate normality, the quantities τ_{inf} and τ_{un} are constant and do not depend on the realizations. Moreover, τ_{inf} is the same for every informed agent (i.e., it does not depend on n). The following proposition, corresponding to Kyle [1148, Theorem 4.1, Corollaries 4.1 and 4.2], characterizes the informativeness of the equilibrium price (see Exercise 10.1 for the proof).

Proposition 10.1 *In the context of the above model (see Kyle [1148]), it holds that*

$$\tau_{inf} = \tau_d + \tau_\varepsilon + \varphi_{inf}(N-1)\tau_\varepsilon \quad \text{and} \quad \tau_{un} = \tau_d + \varphi_{un} N\tau_\varepsilon, \qquad (10.6)$$

where φ_{inf} and φ_{un} are two constants belonging to the interval $[0,1]$. Moreover, in correspondence of a symmetric linear equilibrium (X,Y) of the form (10.4), it holds that

$$\varphi_{inf} = \frac{(N-1)\beta^2}{(N-1)\beta^2 + \sigma_z^2\tau_\varepsilon} \quad \text{and} \quad \varphi_{un} = \frac{N\beta^2}{N\beta^2 + \sigma_z^2\tau_\varepsilon}. \qquad (10.7)$$

The expected dividend for informed and uninformed speculators is respectively given by

$$\mathbb{E}[\tilde{d}|\tilde{p}^*, \tilde{y}_n] = \frac{(1 - \varphi_{\inf})\tau_\varepsilon}{\tau_{\inf}}\tilde{y}_n + \frac{\varphi_{\inf}\tau_\varepsilon}{\beta\tau_{\inf}}\left(\frac{\tilde{p}^*}{\lambda} - N\mu_{\inf} - M\mu_{\mathrm{un}}\right),$$
$$\mathbb{E}[\tilde{d}|\tilde{p}^*] = \frac{\varphi_{\mathrm{un}}\tau_\varepsilon}{\beta\tau_{\mathrm{un}}}\left(\frac{\tilde{p}^*}{\lambda} - N\mu_{\inf} - M\mu_{\mathrm{un}}\right),$$

(10.8)

where $\lambda := 1/(N\gamma_{\inf} + M\gamma_{\mathrm{un}})$ is the slope of the aggregate excess demand schedule and the constants $\mu_{\inf}$, β, $\gamma_{\inf}$, μ_{un} and γ_{un} are as in (10.4). Furthermore, if $\mu_{\inf} = \mu_{\mathrm{un}} = 0$, then $\mathbb{E}[\tilde{d}|\tilde{p}^*] = \tilde{p}^*$ holds if and only if $\varphi_{\mathrm{un}}\tau_\varepsilon = \beta\lambda\tau_{\mathrm{un}}$. Finally, in correspondence of a symmetric linear equilibrium, the informational efficiency parameters $\varphi_{\inf}$ and φ_{un} satisfy $0 \le \varphi_{\inf} \le \varphi_{\mathrm{un}} \le 1$, with

$$\varphi_{\inf} = \frac{(N-1)\varphi_{\mathrm{un}}}{N - \varphi_{\mathrm{un}}}, \qquad \varphi_{\mathrm{un}} = \frac{N\varphi_{\inf}}{N - (1 - \varphi_{\inf})},$$
$$\varphi_{\mathrm{un}} - \varphi_{\inf} = \frac{1}{N}\varphi_{\mathrm{un}}(1 - \varphi_{\inf})$$

(10.9)

and $\tau_{\inf} - \tau_{\mathrm{un}} = (1 - \varphi_{\mathrm{un}})(1 - \varphi_{\inf})\tau_\varepsilon$.

In the above proposition, the parameters $\varphi_{\inf}$ and φ_{un} measure the informational efficiency with which the equilibrium price aggregates the private information of informed speculators. More precisely, φ_{un} represents the fraction of the precision of the N informed speculators revealed by the market price to uninformed speculators. Analogously, $\varphi_{\inf}$ represents the fraction of the precision of the signals of the other $N - 1$ informed agents transmitted by the equilibrium price to each informed agent. Note also that the informational efficiency parameters $\varphi_{\inf}$ and φ_{un} depend only on β and not on the parameters appearing in (10.4). This can be easily understood by recalling that β represents the sensitivity of the informed speculators' demand with respect to their private information, as shown in (10.4). If $\varphi_{\inf} = 0$ or $\varphi_{\mathrm{un}} = 0$, then the equilibrium price does not reveal any information. On the contrary, if $\varphi_{\inf} = 1$ or $\varphi_{\mathrm{un}} = 1$, then the price is fully revealing.

In Kyle [1148, Theorem 7.1] it is shown that the value of $\varphi_{\inf}$ associated to an equilibrium under imperfect competition is less than the value obtained in the corresponding perfectly competitive equilibrium (i.e., when agents do not perceive any influence of their actions on market prices). The intuition for this result is rather simple: an informed speculator facing an upward-sloping residual supply curve will restrict the quantity he trades, thus reducing his demand elasticity with respect to his private information (i.e., β decreases). In turn, this makes the price less informative under imperfect competition in comparison to the case of perfect competition.

The following proposition concerns the existence and the uniqueness of a symmetric linear equilibrium of the economy (see Kyle [1148, Theorem 5.1]).

Proposition 10.2 *In the context of the above model (see Kyle [1148]), suppose that* $\sigma_z^2 > 0$ *and* $\tau_\varepsilon > 0$. *If* $N \geq 2$ *and* $M \geq 1$, *or if* $N \geq 3$ *and* $M = 0$, *or if* $M \geq 3$ *and* $N = 0$, *then there exists a unique symmetric linear equilibrium. If* $N = 1$ *and* $M \geq 2$, *then a unique symmetric linear equilibrium exists if* M *is sufficiently large and does not exist if* α_{un} *is sufficiently large. If* $N + M \leq 2$, *then a symmetric linear equilibrium does not exist.*

Proof We shall only give an outline of the proof, referring to the original paper Kyle [1148] for full details. Suppose first that there exists a symmetric linear equilibrium of the form (10.4). As shown in Exercise 10.1, the market clearing condition (10.1) implies that the equilibrium price $\tilde{p}^*$ must be of the form (10.5). Hence, a first necessary condition for the existence of an equilibrium is that the quantity $\lambda = 1/(N\gamma_{inf} + M\gamma_{un})$ must be well defined, i.e., $N\gamma_{inf} + M\gamma_{un} \neq 0$, because otherwise condition (10.1) would imply an infinite equilibrium price, in which case the demand schedules $X_n(\cdot, \cdot)$ and $Y_m(\cdot)$ would not be optimal and, therefore, could not correspond to an equilibrium. Furthermore, again the market clearing condition (10.1) allows to define a residual supply curve for each speculator, of the form

$$\tilde{p}^* = \tilde{p}_{inf,n} + \lambda_{inf} \tilde{x}_{inf,n} \qquad \text{and} \qquad \tilde{p}^* = \tilde{p}_{un,m} + \lambda_{un} \tilde{x}_{un,m},$$

where $\tilde{x}_{inf,n}$ and $\tilde{x}_{un,m}$ denote the demand of the n-th informed agent and of the m-th uninformed agent, respectively, for each $n = 1,\dots,N$ and $m = 1,\dots,M$, and where

$$\lambda_{inf} = \frac{1}{(N-1)\gamma_{inf} + M\gamma_{un}} \qquad \text{and} \qquad \lambda_{un} = \frac{1}{N\gamma_{inf} + (M-1)\gamma_{un}},$$

and

$$\tilde{p}_{inf,n} = \lambda_{inf} \left(\beta \sum_{\substack{k=1 \\ k \neq n}}^{N} \tilde{y}_k + \tilde{z} + (N-1)\mu_{inf} + M\mu_{un} \right),$$

$$\tilde{p}_{un,m} = \lambda_{un} \left(\beta \sum_{n=1}^{N} \tilde{y}_n + \tilde{z} + N\mu_{inf} + (M-1)\mu_{un} \right).$$

For an equilibrium to be well defined, it is also necessary that the quantities λ_{inf} and λ_{un} introduced above are well defined (i.e., the residual supply curves must have finite slopes). Considering then the optimization problem faced by an informed speculator, the residual supply curve introduced above, together with the linear demand schedule (10.4), implies that

$$(1 + \lambda_{inf}\gamma_{inf})\tilde{p}^* = \tilde{p}_{inf,n} + \lambda_{inf}\beta\tilde{y}_n + \lambda_{inf}\mu_{inf}.$$

As shown in Kyle [1148, Section 5], in this case the demand function $X_n(\cdot, \cdot)$ of an informed speculator can be written in the following form:

$$X_n(p^*, y_n) = \frac{\mathbb{E}[\tilde{d}|p^*, y_n] - p^*}{\lambda_{\text{inf}} + \alpha_{\text{inf}}/\tau_{\text{inf}}}, \qquad \text{for every } n = 1, \ldots, N,$$

for all realizations p^* and y_n of the equilibrium price and of the random signal, respectively. The second order condition of the associated optimization problem is given by $2\lambda_{\text{inf}} + \alpha_{\text{inf}}/\tau_{\text{inf}} > 0$. In particular, this implies that, in line with economic intuition, informed speculators increase their demand when they receive bullish signals and, on the contrary, reduce their demand when they receive bearish information. Similarly, considering the optimization problem of an uninformed speculator, it holds that

$$Y_m(p^*) = \frac{\mathbb{E}[\tilde{d}|p^*] - p^*}{\lambda_{\text{un}} + \alpha_{\text{un}}/\tau_{\text{un}}}, \qquad \text{for every } m = 1, \ldots, M,$$

for every realization p^* of the equilibrium price, with the second order condition $2\lambda_{\text{un}} + \alpha_{\text{un}}/\tau_{\text{un}} > 0$. We have thus derived a set of necessary conditions for a symmetric linear equilibrium to exist. The rest of the proof then proceeds by verifying that this set of conditions is actually sufficient for the existence of a symmetric linear equilibrium, making use of the result of Proposition 10.1. We refer to Kyle [1148, Appendix B] for the detailed calculations and for the explicit expressions of the coefficients appearing in (10.4) in correspondence of the equilibrium.   □

According to Proposition 10.2, a unique symmetric linear equilibrium exists provided that there are enough speculators to generate a sufficiently competitive trading environment. Moreover, as pointed out in Kyle [1148, Section 5], the linearity of the demand schedules is part of the result, in the sense that for each speculator, the equilibrium linear strategy dominates all non-linear alternatives. Clearly, this result strongly depends on the assumption of exponential utility functions together with a multivariate normal distribution.

The incidence of the private information on the equilibrium price can be measured by the quantity

$$\zeta = \beta\lambda\frac{\tau_{\text{inf}}}{\tau_\varepsilon}.$$

The quantity ζ measures the variation in the equilibrium price $\tilde{p}^*$ associated with a unitary change in the valuation of the asset by the informed agent n as a result of a higher realization of his private signal $\tilde{y}_n$, for every $n = 1, \ldots, N$. In correspondence of an imperfectly competitive equilibrium, Kyle [1148, Theorem 7.2] shows that

$$0 \le \varphi_{\text{inf}} \le \zeta \le \frac{1}{2} \qquad \text{and} \qquad \varphi_{\text{inf}} < \varphi_{\text{un}}.$$

Under imperfect competition, the equilibrium price does never reveal more than half of the private precision of the informed agents. This result contrasts with the price revelation obtained in a perfectly competitive setting (see Chap. 8), where the equilibrium price reveals almost all the private information when there is very little noise trading. Moreover, when the variance of the aggregate demand of noise traders vanishes (i.e., $\sigma_z^2 \to 0$), or when the risk aversion parameter α_{inf} converges to zero, or when the precision τ_ε increases to infinity, or, finally, when the number of informed agents converges to infinity, then the quantity ζ converges to the upper bound $1/2$, while φ_{inf} remains strictly below $1/2$, so that perfect revelation is never reached. On the contrary, under the same conditions but in the presence of perfect competition, the quantity φ_{inf} converges to one, meaning that the equilibrium price becomes fully informative. At first sight, the fact that the equilibrium price does not become fully revealing as noise trading disappears from the economy looks like a paradox. However, in the presence of imperfect competition, when the noise traders trade a small amount, then the market becomes so illiquid that informed traders are induced to trade a proportionally small amount as well and this prevents the equilibrium price from fully revealing their private information.

The quantities φ_{inf} and φ_{un} are increasing with respect to the number M of uninformed speculators and decreasing with respect to the risk aversion coefficient α_{un} of the uninformed speculators (see Kyle [1148, Theorem 7.3]). The rationale for this result is that an increase in M or a decrease in α_{un} flattens the residual supply curve faced by informed agents. As a consequence, informed agents will exhibit a more aggressive demand schedule and, therefore, prices will be more affected by their private information. If $M \to \infty$, then $\mathbb{E}[\tilde{d}|\tilde{p}^*] = \tilde{p}^*$, meaning that the equilibrium price becomes unbiased even in the presence of imperfect competition (see Kyle [1148, Theorem 7.4]). On the other hand, when M is finite, then the equilibrium price is always a biased forecast of the dividend, even when uninformed speculators are risk neutral. Moreover, as shown in Kyle [1148, Theorem 7.5], the following conditions are equivalent (under the assumption that $N \geq 2$ and $N + M \geq 3$):

 (i) $\tau_\varepsilon \to \infty$, i.e., the precision of the informed speculators converges to infinity;
 (ii) $\beta \to \infty$, i.e., informed trading is infinitely sensitive to private information;
(iii) informed speculators dominate trading, i.e., $\gamma_{\text{inf}}\lambda \to 1/N$ and $\gamma_{\text{un}}\lambda \to 0$;
 (iv) $\tau_{\text{un}} \to \infty$, i.e., the equilibrium price is infinitely accurate;
 (v) $\mathbb{E}[(\tilde{d} - \tilde{p}^*)^2] \to 0$, i.e., the equilibrium price converges in mean square to the dividend $\tilde{d}$.

In the presence of any of these conditions, speculative profits vanish. Furthermore, both under perfect and imperfect competition, an increase in the number N of informed speculators leads to an increase of the informational efficiency parameter φ_{inf} (see Kyle [1148, Theorem 8.2]). Under some conditions (see Kyle [1148, Theorem 9.2]), if $N \to \infty$ then the imperfectly competitive equilibrium converges to the perfectly competitive equilibrium as far as information transmission is concerned. Note that, in the model, agents trade for both speculative and hedging reasons.

In Kyle [1148, Section 10], the model is extended by considering the decision to become informed as an endogenous decision, in the spirit of the model of Grossman & Stiglitz [849] presented in Sect. 8.3. More specifically, Kyle [1148] assumes that each informed speculator has ex-ante the possibility of paying a positive cost in order to become informed (if he does not pay this cost, then he remains uninformed). Costs are allowed to be different across speculators. In Kyle [1148, Theorem 10.2] it is shown that, both in the perfectly competitive and in the imperfectly competitive market, there exists a unique symmetric linear equilibrium with endogenous acquisition of private information, under the assumption of free entry of uninformed speculators. However, the equilibrium can be such that no agent chooses to become informed. In particular, it is shown that, in the presence of imperfect competition, the Grossman-Stiglitz paradox does not appear, since agents keep prices inefficient enough to create profitable trading opportunities which are sufficient to encourage traders to purchase costly private information.

The above discussion shows that the Grossman-Stiglitz paradox is strongly dependent on the assumption of perfect competition. In Jackson [1002], in an imperfectly competitive market (agents are not price takers) and assuming that the asset dividend is distributed as an exponential random variable and that agents are risk neutral, it is shown that there exist fully revealing equilibria with costly information acquisition. Since the price is fully revealing, buying information does not give an advantage to an agent and, actually, agents who acquire information are worse off than those who do not acquire information. However, they would be even worse off if they did not buy information, given that the other agents expect them to do so. Agents are willing to invest in costly information even if they know that the information will be completely revealed to the other agents by equilibrium prices. Complementarities emerge in the acquisition of information because there is a negative externality effect associated with the decision of not buying information. This induces agents to invest in costly information.

If the adverse selection effect is severe, meaning that the informational disadvantage of uninformed agents is severe (i.e., the variance of the random variable which is observed by informed agents exceeds a critical value) and the agent with private information acts as a monopolist, then uninformed agents can refuse to trade because the adverse selection effect outweighs the hedging motive to trade. In this case, as shown in Bhattacharya & Spiegel [227], a *market breakdown* can occur.

10.2 Order-Driven Markets

In an order-driven market, there is a market maker who handles a book (specialist book). In a *batch auction market*, orders arrive to the market maker, the market maker looks at the order book (order imbalance) and then sets a price at which the market is cleared, matching the aggregate demand. In this section, we first present the model proposed in the seminal work Kyle [1147] and then we discuss several extensions of this model.

The Kyle (1985) Model

In Kyle [1147], a simple two-period economy is considered, where a single risky asset is traded. The liquidation value (dividend) of the asset is represented by the random variable $\tilde{d}$, observed at date $t = 1$. The economy is populated by *noise traders*, an *insider trader* and a *market maker*. Noise traders trade an aggregate quantity represented by the random variable $\tilde{z}$, assumed to be independent of $\tilde{d}$. Moreover, it is assumed that

$$\tilde{d} \sim \mathcal{N}(\bar{d}, \sigma_d^2) \qquad \text{and} \qquad \tilde{z} \sim \mathcal{N}(0, \sigma_z^2).$$

The trading mechanism is assumed to be decomposed into two successive steps:

1. the insider trader learns the realization of the dividend, i.e., he observes the realization of the random variable $\tilde{d}$. On the basis of this information, the insider trader determines his optimal demand by choosing to trade a quantity $\tilde{x} := X(\tilde{d})$ of the asset, for some measurable function $X : \mathbb{R} \to \mathbb{R}$. Note that the insider trader cannot observe the realization of the random variable $\tilde{z}$;
2. the market maker determines the price $\tilde{p}$ at which the market clears, observing the aggregate market demand $\tilde{x} + \tilde{z}$. The market maker is assumed to be able to only observe the aggregate demand and does not observe the two components $\tilde{x}$ and $\tilde{z}$ separately. The price set by the market maker can therefore be represented as $\tilde{p} = P(\tilde{x} + \tilde{z})$, for some measurable function $P : \mathbb{R} \to \mathbb{R}$ of the aggregate market demand.

The insider trader is assumed to be risk neutral and determines his demand by maximizing the conditional expectation of his profit given the observation of the dividend. The profit of the insider trader is given by

$$\tilde{\pi} := (\tilde{d} - \tilde{p})\tilde{x} = (\tilde{d} - \tilde{p})X(\tilde{d}).$$

Recalling that the price $\tilde{p}$ is set by the market maker according to the measurable function P, let us write $\tilde{\pi} = \tilde{\pi}(X, P)$ in order to emphasize the dependence of the profit of the insider trader on the functions X and P introduced above. In a similar way, let us also write $\tilde{p} = \tilde{p}(X, P)$ for the price set by the market maker. In this context, an equilibrium of the economy is defined as a couple (X, P) of measurable functions such that

(i) the insider trader maximizes his expected profit, given the observation d of the dividend $\tilde{d}$, i.e., the insider trader demand function X satisfies

$$\mathbb{E}[\tilde{\pi}(X, P)|d] \geq \mathbb{E}[\tilde{\pi}(X', P)|d], \tag{10.10}$$

for any alternative measurable function $X' : \mathbb{R} \to \mathbb{R}$ and for every realization d;

(ii) the market is efficient, in the sense that the price $\tilde{p}$ set by the market maker
satisfies

$$\tilde{p}(X, P) = \mathbb{E}[\tilde{d}|x + z], \tag{10.11}$$

for every realization $x + z$ of the random variable $\tilde{x} + \tilde{z}$.

The price setting rule (10.11) implies that the expected profit of the market maker
are equal to zero. This can be justified by considering the presence of several market
makers competing among themselves, thus making their profits converge to zero and
leading to the market efficiency condition (10.11).

The insider trader exploits his private information and takes into account the fact
that the quantity he chooses to trade will influence the market clearing price set by
the market maker. In other words, the demand function X also depends on the price
setting rule P. The insider trader takes the price setting rule used by the market
maker as given and is not allowed to condition the quantity he decides to trade on
the market price (i.e., the insider trader can only submit *market orders* and not *limit
orders*, see also below in this section).

In this setting, the equilibrium admits an explicit characterization, as shown in the
following proposition (compare with Kyle [1147, Theorem 1]), the proof of which
is given in Exercise 10.2.

Proposition 10.3 *In the model considered above (see Kyle [1147]), there exists a
unique linear equilibrium* (X, P), *where* $X : \mathbb{R} \to \mathbb{R}$ *and* $P : \mathbb{R} \to \mathbb{R}$ *are two
measurable functions such that*

$$X(\tilde{d}) = \beta(\tilde{d} - \bar{d}) \qquad and \qquad P(\tilde{x} + \tilde{z}) = \bar{d} + \lambda(\tilde{x} + \tilde{z}), \tag{10.12}$$

where the constants β *and* λ *are given by*

$$\beta = \sqrt{\frac{\sigma_z^2}{\sigma_d^2}} \qquad and \qquad \lambda = \frac{1}{2}\sqrt{\frac{\sigma_d^2}{\sigma_z^2}}.$$

Proposition 10.3 yields several interesting insights. First, the fact that the optimal
demand of the insider trader is linear is not an assumption but rather a derived result,
in the sense that linear strategies are optimal even when non-linear strategies would
be a priori allowed. The equilibrium couple (X, P) is completely determined by
the exogenous parameters $\bar{d}$, σ_d^2 and σ_z^2. In particular, observe that the larger the
variance of the demand of the noise traders, the higher the coefficient β representing
the sensitivity of the demand of the insider trader to his private information. The
rationale of this result is simple: if the demand of the noise traders exhibits a
large variance, then the insider trader can actively trade in the market knowing

that his private information will be more easily hidden to the eyes of the market maker and, therefore, less reflected in the market price. The quantity $1/\lambda$ can be regarded as a measure of *market depth* (i.e., the order flow necessary to induce the equilibrium price to rise or fall by a unitary amount). This quantity is increasing with respect to the variance of the demand of the noise traders and decreasing with respect to the variance of the dividend, which represents the private information of the insider trader. The sensitivity of the demand of the insider trader, measured by the coefficient β, is proportional to the market depth. Moreover, as shown in Exercise 10.3, the expected profits of the insider trader (conditionally on the observation of the private information $\tilde{d} = d$) are equal to $(d - \bar{d})^2/(4\lambda)$ and, therefore, are proportional to the market depth. Ex-ante (i.e., before the observation of the private information), the expected profits of the insider trader are equal to $\sigma_d \sigma_z/2$.

In this model, the insider trader perfectly knows the dividend before trading, while from the point of view of noise traders the dividend is a random variable with variance equal to σ_d^2. As shown in Exercise 10.3, in equilibrium it holds that $\text{Var}(\tilde{d}|p) = \sigma_d^2/2$. Intuitively, this means that half of the insider trader's private information is incorporated in the equilibrium price. Therefore, the equilibrium is not fully revealing. Note also that, unlike in the previous section, the conditional variance of the dividend given the observation of the equilibrium price does not depend on the variance of the noise traders' demand. Since an increase in noise trading brings forth more informed trading, noise trading does not destabilize prices. This result is due to the fact that the insider trader is risk neutral: as the variance of the noise traders' demand increases, the insider will behave more aggressively, without changing the informational content of the equilibrium price. On average (by definition of market efficiency), the profits of the market maker are zero. The market maker loses money when trading with the insider trader but gains an offsetting amount of money when trading with noise traders. Therefore, the profits of the insider trader are also the noise traders' costs.

In Kyle [1147, Section 3], the above two-period model is extended to a dynamic setting in discrete time, assuming that a series of trading rounds take place sequentially, as will be discussed in more detail in Sect. 10.4. The model is structured in such a way that, at each trading round, the equilibrium price reflects the information contained in the present and past order flows and the insider, at each trading round, maximizes his expected profits, taking into account his effect on prices in the current trading round as well as in the future trading rounds. The informed agent trades in such a way that his private information becomes incorporated in the equilibrium price gradually. In the limit, as the time step between two consecutive trading rounds goes to zero, a model of continuous trading emerges and the price follows a Brownian motion.

Extensions

Risk Averse Informed Traders

In Subrahmanyam [1575], the model of Kyle [1147] has been extended by removing the assumption that the insider trader is risk neutral and allowing for the presence of $N \geq 1$ risk averse insider traders. It is assumed that each insider trader is characterized by a negative exponential utility function with coefficient of absolute risk aversion a (the same for all insider traders). Similarly as in Kyle [1147], the economy also comprises a certain number of noise traders, whose aggregate demand is described by the random variable $\tilde{z}$, and a risk neutral market maker who absorbs the net demand of the other traders and sets the market price in such a way that he earns zero expected profits, conditionally on observing the aggregate demand. The liquidation value of the asset at date $t = 1$ is represented by the random variable $\tilde{d}$, decomposed as

$$\tilde{d} = \bar{d} + \tilde{\varepsilon},$$

where $\bar{d}$ is a constant known to every agent in the economy and $\tilde{\varepsilon}$ is a random variable. Informed agents observe a noisy signal $\tilde{y}$ given by a realization of the random variable $\tilde{y} = \tilde{\varepsilon} + \tilde{\eta}$. The three random variables $\tilde{z}$, $\tilde{\varepsilon}$ and $\tilde{\eta}$ are mutually independent and normally distributed, with zero mean and variances $\text{Var}(\tilde{z}) = \sigma_z^2$, $\text{Var}(\tilde{\eta}) = \sigma_\eta^2$ and $\text{Var}(\tilde{\varepsilon}) = 1$ (without loss of generality).
    Similarly as above, we denote by $\tilde{x} = X(\tilde{\varepsilon} + \tilde{\eta})$ the demand of an informed agent, for some measurable function $X : \mathbb{R} \to \mathbb{R}$ representing the demand of an informed agent as a function of his private signal, and we denote by $\tilde{p} = P(N\tilde{x} + \tilde{z})$ the price set by the market maker, for some measurable function $P : \mathbb{R} \to \mathbb{R}$ representing the price as a function of the total market demand. Similarly as in Kyle [1147], the attention is focused on proving the existence and the uniqueness of a linear equilibrium, represented by a couple (X, P), where each of the N informed traders maximizes his expected utility, conditionally on the observation of the private signal, and the market efficiency condition (10.11) holds (i.e., the risk neutral market maker obtains zero expected profits). This is the content of the following proposition, the proof of which is given in Exercise 10.5.

Proposition 10.4 *In the context of the above model (see Subrahmanyam [1575]), there exists a unique linear equilibrium (X, P), where $X : \mathbb{R} \to \mathbb{R}$ and $P : \mathbb{R} \to \mathbb{R}$ are two measurable functions such that*

$$X(\tilde{\varepsilon} + \tilde{\eta}) = \beta(\tilde{\varepsilon} + \tilde{\eta}) \qquad and \qquad P(N\tilde{x} + \tilde{z}) = \bar{d} + \lambda(N\tilde{x} + \tilde{z}), \qquad (10.13)$$

where the constant β is given by

$$\beta = \frac{1}{(1 + \sigma_\eta^2)\lambda(N + 1) + a(\sigma_\eta^2 + \lambda^2 \sigma_z^2(1 + \sigma_\eta^2))} \qquad (10.14)$$

and λ satisfies the quintic equation

$$\lambda \left(N^2(1 + \sigma_\eta^2) + \frac{1}{\beta^2}\sigma_z^2 \right) = \frac{N}{\beta}.$$

In Subrahmanyam [1575, Proposition 1] it is shown that the above quintic equation for the coefficient λ admits a unique solution. In equilibrium, the value of λ is decreasing with respect to the variance of the aggregate demand of noise traders. Moreover, in equilibrium λ is either a unimodal function (i.e., first increasing and then decreasing) of the number N of informed traders or is a monotonically decreasing function of N. Similarly as in the model of Kyle [1147] presented above, the quantity $1/\lambda$ represents a measure of market depth. Hence, these results imply that market depth is increasing with respect to the variance of the noise traders' demand and, in some cases, also with respect to the number of informed agents in the economy. Similarly, with respect to the parameters a and σ_η^2, the equilibrium value of λ is either a unimodal function or a monotonically decreasing function. As shown in Exercise 10.6, when the informed traders are risk neutral (similarly as in the model of Kyle [1147]), the equilibrium value of λ is decreasing in the number of informed traders and in the variance of the noise traders' demand and is increasing in the precision of information, as represented by the quantity $1/(1 + \sigma_\eta^2)$.

Unlike in the Kyle [1147] model, one of the main conclusions of Subrahmanyam [1575] is that the equilibrium price reveals less information if the variance of the noise traders' demand is large (and this result is confirmed if the market maker is risk averse, see Subrahmanyam [1575, Proposition 2]). Moreover, if informed traders are risk averse, increasing the level of informational asymmetry between the informed and the uninformed traders can yield an improvement of the terms of trade for the uninformed agents. As shown in Subrahmanyam [1575, Proposition 2], if informed traders are risk averse, price efficiency is decreasing with respect to the risk aversion coefficient of informed traders. The rationale for this result is that more risk averse informed agents trade less aggressively than less risk averse informed agents, with the consequence that less private information will be reflected in the equilibrium price.

Risk Averse Hedgers

The model of Kyle [1147] has been generalized in Spiegel & Subrahmanyam [1558] considering an economy comprising three classes of agents: K risk neutral insider traders, N risk averse uninformed traders and a competitive risk neutral market maker. In particular, in this model the presence of noise traders is replaced by the presence of risk averse uninformed expected utility maximizers and it is shown that this leads to a significant modification of several results obtained in the original model of Kyle [1147].

In the model considered in Spiegel & Subrahmanyam [1558], there is a single risky asset, which is traded at the initial date $t = 0$ and liquidated at $t = 1$. The liquidation value (dividend) of the asset is represented by the random variable $\tilde{d}$:

$$\tilde{d} = \bar{d} + \tilde{\varepsilon},$$

where the constant term $\bar{d}$ is known to every agent at date $t = 0$. Each informed trader $k \in \{1, \ldots, K\}$ observes a private signal $\tilde{y}_k$ of the form

$$\tilde{y}_k = \tilde{\varepsilon} + \tilde{\eta}_k, \qquad \text{for every } k = 1, \ldots, K,$$

where $(\tilde{\eta}_k)_{k=1,\ldots,K}$ is a family of mutually independent random variables with normal distribution with zero mean and identical variance σ_η^2. The random variable $\tilde{\varepsilon}$ is also normally distributed with zero mean and variance σ_ε^2 and is assumed to be independent of $(\tilde{\eta}_k)_{k=1,\ldots,K}$. Every risk averse hedger $n \in \{1, \ldots, N\}$ has a random endowment $\tilde{e}_n$ of the asset and is characterized by a negative exponential utility function with risk aversion parameter a. The family $(\tilde{e}_n)_{n=1,\ldots,N}$ is assumed to be composed of i.i.d. random variables with normal distribution with zero mean and identical variance σ_e^2. Furthermore, it is assumed that the family $(\tilde{e}_n)_{n=1,\ldots,N}$ is independent of $(\tilde{\varepsilon}, \tilde{\eta}_1, \ldots, \tilde{\eta}_K)$.

The price formation mechanism is similar to that considered in the models presented above in this section. Namely, informed and uninformed traders submit their demands to the risk neutral market maker, who then sets a price which equals the expected liquidation value of the security, given the observation of the aggregate demand. The analysis is focused on proving the existence and the uniqueness of a linear equilibrium of the economy. In this context, a linear equilibrium can be represented by a triplet (X, W, P) of measurable functions, where X represents the demand of each informed trader, given as a function of the private signal, W represents the demand of each uninformed trader, given as a function of his endowment, and P represents the price set by the market maker, given as a function of the total market flow. The following proposition, whose proof is given in Exercise 10.7, provides an explicit characterization of the linear equilibrium of the economy (see Spiegel & Subrahmanyam [1558, Proposition 1 and Appendix A]).

Proposition 10.5 *In the context of the above model (see Spiegel & Subrahmanyam [1558]), if*

$$a^2 N \sigma_e^2 (\sigma_\varepsilon^2 + 2\sigma_\eta^2)^2 > 4K(\sigma_\varepsilon^2 + \sigma_\eta^2), \tag{10.15}$$

then there exists a unique linear equilibrium (X, W, P), where $X : \mathbb{R} \to \mathbb{R}$, $W : \mathbb{R} \to \mathbb{R}$ and $P : \mathbb{R} \to \mathbb{R}$ are measurable functions explicitly given by

$$X(\tilde{\varepsilon} + \tilde{\eta}_k) = \beta(\tilde{\varepsilon} + \tilde{\eta}_k), \qquad W(\tilde{e}_n) = \gamma \tilde{e}_n, \tag{10.16}$$

for all $k = 1, \ldots, K$ and $n = 1, \ldots, N$, and

$$P\left(\sum_{k=1}^{K} X(\bar{\varepsilon} + \tilde{\eta}_k) + \sum_{n=1}^{N} W(\tilde{e}_n)\right) = \bar{d} + \lambda\left(\sum_{k=1}^{K} X(\bar{\varepsilon} + \tilde{\eta}_k) + \sum_{n=1}^{N} W(\tilde{e}_n)\right),$$
(10.17)

where the constants β, γ and λ are explicitly given by

$$\lambda = \frac{a\sigma_\varepsilon^2 \sqrt{K(\sigma_\varepsilon^2 + \sigma_\eta^2)}\left((\sigma_\varepsilon^2 + 2\sigma_\eta^2)^2 + K\sigma_\varepsilon^2\sigma_\eta^2 + (K(N-1)/N)\sigma_\varepsilon^2(\sigma_\varepsilon^2 + \sigma_\eta^2)\right)}{\left((1+K)\sigma_\varepsilon^2 + 2\sigma_\eta^2\right)^2 \left(a\sqrt{N\sigma_e^2}(\sigma_\varepsilon^2 + 2\sigma_\eta^2) - 2\sqrt{K(\sigma_\varepsilon^2 + \sigma_\eta^2)}\right)},$$

$$\beta = \frac{\left((1+K)\sigma_\varepsilon^2 + 2\sigma_\eta^2\right)\left(a\sqrt{N\sigma_e^2}(\sigma_\varepsilon^2 + 2\sigma_\eta^2) - 2\sqrt{K(\sigma_\varepsilon^2 + \sigma_\eta^2)}\right)}{a\sqrt{K(\sigma_\varepsilon^2 + \sigma_\eta^2)}\left((\sigma_\varepsilon^2 + 2\sigma_\eta^2)^2 + K\sigma_\varepsilon^2\sigma_\eta^2 + (K(N-1)/N)\sigma_\varepsilon^2(\sigma_\varepsilon^2 + \sigma_\eta^2)\right)},$$

$$\gamma = \frac{(1+K)\sigma_\varepsilon^2 + 2\sigma_\eta^2}{\sqrt{N\sigma_e^2}} \frac{a\sqrt{N\sigma_e^2}(\sigma_\varepsilon^2 + 2\sigma_\eta^2) - 2\sqrt{K(\sigma_\varepsilon^2 + \sigma_\eta^2)}}{a\left((\sigma_\varepsilon^2 + 2\sigma_\eta^2)^2 + K\sigma_\varepsilon^2\sigma_\eta^2 + (K(N-1)/N)\sigma_\varepsilon^2(\sigma_\varepsilon^2 + \sigma_\eta^2)\right)}.$$

If condition (10.15) is not satisfied, then a linear equilibrium does not exist.

Condition (10.15) means that a linear equilibrium exists if and only if the risk aversion of the uninformed traders, or the variance of their endowment, or their number, is sufficiently high and the number of informed traders is sufficiently low. When condition (10.15) fails to hold, a market breakdown occurs, since the demand of the informed traders overwhelms that of uninformed traders, with the consequence that the market maker incurs in systematic losses, thus preventing the existence of an equilibrium. Note that this is in contrast with the model of Kyle [1147] presented above, where a linear equilibrium always exists.

It can be shown that the equilibrium value of λ given in Proposition 10.5 is decreasing with respect to a and σ_e^2, meaning that market depth is increasing in the coefficient of absolute risk aversion of uninformed agents and in the variance of their endowments (see Spiegel & Subrahmanyam [1558, Proposition 2]). The behavior of λ with respect to N, K, σ_ε^2 and σ_η^2 is not necessarily monotone and depends on the parameters of the model. Note that this implies the counterintuitive result that having more uninformed traders is not necessarily beneficial from the point of view of market liquidity. However, the market becomes infinitely liquid (i.e., $\lambda \to 0$) if $\sigma_\eta^2 \to \infty$ or $N \to \infty$ and infinitely illiquid (i.e., $\lambda \to \infty$) if $\sigma_\varepsilon^2 \to \infty$ (see Spiegel & Subrahmanyam [1558, Corollary to Proposition 1]). Moreover, there exists a finite value K^* such that as K converges to K^* the market becomes infinitely illiquid and, for $K > K^*$, a linear equilibrium does not exist. The interpretation of this last result is that, if there are too many insider traders, then the adverse selection effect is too severe and a (linear) equilibrium cannot exist.

By relying on the distributional assumptions of the model and on the properties of the bivariate normal distribution, it can be easily shown that in correspondence of the linear equilibrium it holds that

$$\text{Var}\left(\tilde{d} \Bigg| \sum_{k=1}^{K} X(\tilde{\varepsilon} + \tilde{\eta}_k) + \sum_{n=1}^{N} W(\tilde{e}_n)\right) = \frac{\sigma_\varepsilon^2(K\beta^2\sigma_\eta^2 + N\gamma^2\sigma_e^2)}{K\beta^2(K\sigma_\varepsilon^2 + \sigma_\eta^2) + N\gamma^2\sigma_e^2}$$

and, by replacing the equilibrium value of the coefficient β as given in Proposition 10.5, we get

$$\text{Var}\left(\tilde{d} \Bigg| \sum_{k=1}^{K} X(\tilde{\varepsilon} + \tilde{\eta}_k) + \sum_{n=1}^{N} W(\tilde{e}_n)\right) = \frac{\sigma_\varepsilon^2(\sigma_\varepsilon^2 + 2\sigma_\eta^2)}{(1 + K)\sigma_\varepsilon^2 + 2\sigma_\eta^2},$$

see Spiegel & Subrahmanyam [1558, Section 1.2]. In particular, this shows that in equilibrium the posterior variance of the liquidation value of the asset (and, hence, the informational efficiency of the equilibrium price) does not depend on any of the parameters a, σ_e^2 and N. This result depends on the assumption of risk neutrality of the informed traders. The expected profits of informed traders are increasing with respect to a and σ_e^2, while the welfare of uninformed agents is decreasing with respect to K and σ_ε^2 (see Spiegel & Subrahmanyam [1558, Propositions 4 and 5]).

Informed Traders Submitting Limit Orders

In the model of Kyle [1147], it is assumed that insider traders can only submit *market orders*, i.e., orders which do not depend on the price. In Vives [1628], the author develops a model with heterogeneous risk averse informed agents who are also allowed to submit orders depending on the market price (*limit orders*).

The model developed in Vives [1628] is similar to the noisy rational expectations equilibrium models considered in Admati [15], Diamond & Verrecchia [569], Hellwig [931], Grossman & Stiglitz [849] (see Sect. 8.3). There is a single risky asset, with liquidation value $\tilde{d}$, and a risk free asset with unitary return. The economy is populated by three types of agents:

- a continuum of informed risk averse agents, indexed in the interval $[0, 1]$, characterized by a negative exponential utility function with common risk aversion parameter a;
- noise traders, who trade an aggregate quantity $\tilde{z}$ of the risky asset;
- a risk neutral competitive marker maker, similarly as in the models considered above in the present section.

Each informed agent i observes a private signal $\tilde{y}_i$ of the form

$$\tilde{y}_i = \tilde{d} + \tilde{\varepsilon}_i, \qquad \text{for every } i \in [0, 1].$$

The initial wealth of informed agents is normalized to zero. Similarly as in the model of Kyle [1148], each informed agent $i \in [0, 1]$ submits to the market a demand schedule $X_i(\cdot, \tilde{y}_i)$, taking into account the equilibrium functional relationship of prices with respect to the random variables describing the economy. We can think of the demand schedule $X_i(\cdot, \tilde{y}_i)$ as a form of generalized limit order contingent on private information. Similarly as in Sect. 10.1, if the market clearing price of the risky asset is p, then trader i demands $X_i(p, y_i)$ units of the risky asset, conditionally on the observation of a realization y_i of his private signal $\tilde{y}_i$.

The market maker is assumed to be competitive and risk neutral. Hence, observing the aggregate demand schedule

$$L(\cdot) := \int_0^1 X_i(\cdot, \tilde{y}_i) \, di + \tilde{z},$$

the market maker sets an efficient price $\tilde{p}$ for the risky asset characterized by the condition $\tilde{p} = \mathbb{E}[\tilde{d} | L(\cdot)]$. It is assumed that all the random variables appearing in the model are normally distributed:

$$\tilde{d} \sim \mathcal{N}(\bar{d}, \sigma_d^2), \qquad \tilde{\varepsilon}_i \sim \mathcal{N}(0, \sigma_\varepsilon^2) \quad \text{and} \quad \tilde{z} \sim \mathcal{N}(0, \sigma_z^2).$$

Moreover, the random variables $\tilde{d}$ and $\tilde{z}$ are assumed to be independent and the family of random variables $(\tilde{\varepsilon}_i)_{i \in [0,1]}$ is assumed to be composed of independent random variables, independent of $\tilde{d}$ and $\tilde{z}$. We define as follows the precision of the random variables appearing in the model:

$$\tau_d := 1/\sigma_d^2, \qquad \tau_\varepsilon := 1/\sigma_\varepsilon^2 \quad \text{and} \quad \tau_z := 1/\sigma_z^2.$$

It is furthermore assumed that

$$\int_0^1 \tilde{\varepsilon}_i \, di = 0, \qquad \text{almost surely.}$$

All the distributional assumptions are common knowledge among all the agents in the economy.

Similarly as in the models considered above, in Vives [1628] the attention is focused on proving the existence and the uniqueness of a linear equilibrium of the economy. This is the content of the following proposition, corresponding to Vives [1628, Proposition 1.1] and proved in Exercise 10.8.

Proposition 10.6 *In the context of the above model (see Vives [1628]), there exists a unique symmetric linear equilibrium, where the demand schedules $X_i(\tilde{p}, \tilde{y}_i)$ of the informed agents are given by*

$$X_i(\tilde{p}, \tilde{y}_i) = \frac{\tau_\varepsilon}{a}(\tilde{y}_i - \tilde{p}), \qquad \text{for every } i \in [0, 1], \tag{10.18}$$

and the equilibrium price is given by

$$\tilde{p} = \left(1 - \lambda\frac{\tau_\varepsilon}{a}\right)\bar{d} + \lambda\left(\frac{\tau_\varepsilon}{a}\tilde{d} + \tilde{z}\right), \tag{10.19}$$

where

$$\lambda = \frac{\tau_\varepsilon\tau_z/a}{\tau_d + \tau_\varepsilon^2\tau_z/a^2}.$$

In particular, the above proposition shows that, in equilibrium, informed traders trade more intensively if the precision of their private signal is higher or their risk aversion is lower. Moreover, the trading intensity is independent of the amount of noise trading. The depth of the market, as measured by the quantity $1/\lambda$, is increasing in noise trading and monotonic (increasing or decreasing) in risk aversion and in the precision of the information of informed traders. Expected trading volume is increasing in noise trading (see Vives [1628, Corollary 2.1]).

Monopolistic Market Maker

In the model of Kyle [1147] presented at the beginning of this section, the market maker sets the equilibrium price in such a way that his expected profits are null. This price setting rule can be motivated by the presence of perfect competition among several market makers. However, in some markets, the market maker operates as a *monopolist* and, therefore, he does not set the market price according to the zero expected profits condition. On the contrary, a monopolist market maker maximizes his expected profits.

In Glosten [787] it is shown that, in a model with asymmetric information among traders, the presence of a monopolistic market maker allows to reach a better risk sharing in comparison with a competitive market maker when the adverse selection effect is significant. Recall that, in the presence of traders with heterogeneous and private information, the market maker faces an adverse selection problem, due to the presence of informed agents in the market. The presence of competition among market makers entails two different welfare effects. On the one hand, competition should offer better prices to liquidity traders. On the other hand, competition can prevent the market makers from effectively screening the different types of traders, thereby inducing an adverse selection effect which negatively affects the welfare. If the adverse selection effect is extreme, then a market breakdown can occur (i.e., an equilibrium fails to exist), since the competitive market maker expects to lose money. On the other hand, in such a situation, the presence of a monopolistic market maker does not necessarily lead to a market breakdown, since the monopolist can decide to keep the market open and learn some of the information of the informed agents by trading, thus reducing the adverse selection problem and making his subsequent trades more profitable, offsetting the previous losses. In other words, the monopolist has the possibility of averaging his profits across different trades, thus implying a better liquidity even in the presence of extensive trading based on private information. When the adverse selection effect is significant, the welfare of informed and liquidity traders is higher in a monopolistic market maker economy than in a perfectly competitive market makers economy.

The model of Glosten [787] can be described as follows. In the market, there are two traded securities: a risk free security with constant rate of return (normalized to one for simplicity) and a risky security with liquidation value $\tilde{d}$, supposed to be normally distributed with mean $\bar{d}$ and variance $1/\tau_d$. There is an informed trader characterized by a negative exponential utility function with risk aversion parameter a and endowed at date $t = 0$ with w_0 units of the risk free asset and with $\tilde{w}$ units of the risky asset. It is assumed that $\tilde{w}$ is a random variable, independent of $\tilde{d}$, with a normal distribution with zero mean and variance $1/\tau_w$. The informed trader observes a private signal $\tilde{y}$ of the form $\tilde{y} = \tilde{d} + \tilde{\varepsilon}$, where $\tilde{\varepsilon}$ is a normally distributed random variable, independent of $\tilde{d}$ and $\tilde{w}$, with zero mean and variance $1/\tau_\varepsilon$. The market maker is assumed to be risk neutral and cannot observe neither the realization of the private signal $\tilde{y}$ of the informed trader nor the realization of the random endowment $\tilde{w}$.

Conditionally on the observation of a realization y of the private signal $\tilde{y}$ and on a realization w of the random endowment $\tilde{w}$, the informed trader defines his demand schedule $X := X_P(y, w)$ by maximizing the expected utility of his wealth at date $t = 1$, given the price schedule $P : x \mapsto P(x)$ set by the market maker (with the interpretation that a trade of x will lead to a transfer of money of $xP(x)$ from the trader to the market maker). The notation $X_P(y, w)$ emphasizes that the demand schedule of the informed trader depends on the pricing rule P as well as on the signal and on the endowment. Analogously to Kyle [1147], a competitive market maker sets the market price according to the zero expected profits rule, i.e., he sets a market price defined by the condition $P_c(X) = \mathbb{E}[\tilde{d}|X]$. On the contrary, a monopolistic market maker defines the price schedule $P_m(X)$ by maximizing his expected profits, given the demand schedule X of the informed trader.

In Glosten [787, Proposition 1], it is shown that an equilibrium with a competitive market maker exists if and only if

$$\gamma := \frac{a^2 \tau_d}{a^2 \tau_d + \tau_\varepsilon \tau_w (\tau_d + \tau_\varepsilon)} > \frac{1}{2}.$$

If this condition is not satisfied, then an equilibrium cannot exist and the market breaks down. The quantity γ can be regarded as a measure of adverse selection. If there is no private information (i.e., $\tau_\varepsilon = 0$), then $\gamma = 1$, while if τ_ε is large enough (meaning that the adverse selection effect is significant) then the above inequality is not satisfied and a competitive equilibrium does not exist. As shown in Glosten [787, Section 2], there exists a value $\hat{\gamma}$ such that a monopolistic market maker is preferred by traders when $\gamma < \hat{\gamma}$ (strong adverse selection), while the competitive market makers setting is preferred when $\gamma > \hat{\gamma}$ (weak adverse selection).

In a dynamic setting, Leach & Madhavan [1170] show that a monopolistic market maker can also help the diffusion of information by experimenting with prices, i.e., by setting prices different from the conditional expectation of the liquidation value in order to learn the agents' information by observing the resulting market order flow. This activity is costly, but a monopolistic market maker can recover the costs in an intertemporal setting. However, this is not possible in the presence of competing market makers.

Markets with a Limit Order Book

In several financial markets, a *limit order book* has been adopted in order to increase the liquidity of the market. As mentioned above, one can distinguish between two main types of orders: *market orders* (orders executed immediately at the best price available in the market) and *limit orders* (orders which will be executed only at some specified price). A market order typically leads to an immediate execution of the order, but there is no certainty about the price at which the order will be executed. For this reason, market orders are well suited for liquidity trades. On the contrary, limit orders are characterized by an uncertain execution time (and, in some cases, may also remain unexecuted) but will be executed at a known price and, hence, are well suited for patient traders. With market orders, the cost of the immediate execution of the order is represented by the fact that the price is determined by the market, while with limit orders the cost of setting the execution price of the order is represented by the fact that the execution of the order can be delayed. Limit orders provide liquidity to future trades but are exposed to adverse informational price changes, in the sense that a limit order might turn out to be mispriced ex-post. The limit order book registers all the limit orders previously posted in the market (i.e., bids and offers specifying quantities and prices).

An analysis of a limit order book market system has been proposed in Glosten [788], assuming that (risk neutral) liquidity suppliers behave competitively (i.e., they earn zero expected profits), facing informed traders in the market. The limit order book is modeled as a publicly visible screen providing bids and offers together with the corresponding prices and quantities. Transactions against the book pick off the limit orders at their limit prices. The source of bids and offers is a large pool of risk neutral liquidity suppliers. It is shown that the system works well in the presence of strong adverse selection, providing as much liquidity as possible compared to other market settings, and it allows to average profits and gains across trades as in Glosten [787]. Limit order traders gain from small trades and liquidity-driven price changes but lose from information-driven price changes. There is a positive small-trade bid-ask spread and the limit order book is immune to competition from third market dealers.

Biais et al. [240] provide an analysis with strategic liquidity suppliers (imperfect competition) who earn positive expected profits. The trading volume is shown to be lower than in the case of perfect competition. Under some conditions, in the presence of imperfect competition and no asymmetric information, risk averse agents prefer a hybrid market (dealer market together with a limit order book) to a dealer market or to a limit order book (see Viswanathan & Wang [1627]): small trades are directed towards the limit order book, while large orders are handled by dealers. A limit order book is preferred by risk neutral traders and the dealer market is preferred when there are many market makers. Biais et al. [234] show that a limit order market provides better risk sharing and narrower spreads than a dealer market (see also Parlour & Seppi [1403] on the efficiency of a limit order book setting).

10.3 Quote-Driven Markets

In many financial markets, there exist market dealers who quote bid and ask prices for a set of assets. A bid price corresponds to the price that the dealer accepts to pay in order to execute a sell order in the market, while an ask price corresponds to the price that the dealer accepts to receive in order to execute a buy order in the market. The dealer is committed to satisfy the orders by applying the quoted prices, under some conditions. In many cases, the commitment of the dealer only holds for orders smaller than a certain amount and the dealer is allowed to revise the bid and ask prices in the case of large orders. Typically, in a dealer market orders can arrive at any point in time (hence, a dealer market usually operates in continuous time). The main features of a dealer market are immediacy and quick price discovery.

By quoting bid-ask prices and accepting to execute orders, the dealer provides a service to the market. This service exposes the dealer to at least two different types of costs: first, the dealer is forced to accept a non-optimal position in the traded assets since he has to keep an inventory of the assets; second, the dealer faces an adverse selection problem due to possibility of trading with better informed investors. Because of the presence of these costs, the dealer asks for a remuneration for his service and this remuneration is reflected in the existence of a *bid-ask spread*.

The literature on dealer markets and, in particular, on the determination of the bid-ask spread can be divided into two main groups of contributions, depending on the relative relevance of the two types of costs described above (see O'Hara [1384] and De Jong & Rindi [540] for a detailed account on these topics). A first series of contributions concentrates on the first cost, emphasizing the fact that the dealer is forced to hold a non-optimal portfolio and assuming that the dealer is risk averse. The dealer manages an inventory of the asset, quotes bid and ask prices and faces a stochastic order flow (often described by a Poisson process). The dealer considers the risk of the inventory position with respect to some target portfolio, revising the quoted bid and ask prices in order to manage the risk. Optimal inventory management and bid-ask strategies have been determined in Amihud & Mendelson [58], Ho & Stoll [949], O'Hara & Oldfield [1385], Biais [231], showing that the bid-ask spread is increasing in the dealer's risk aversion (but is independent of the inventory).

Let us consider the portfolio management problem of a dealer in the context of a simple two-period model first proposed by Stoll [1570], following the presentation in O'Hara [1384, Section 2.2] and De Jong & Rindi [540, Section 5.1]. In this model, the dealer is risk averse and demands a compensation when acting as a liquidity provider in the market. In Stoll [1570], a dealer simply represents a market participant who is willing to alter his own portfolio away from optimality in order to match the demands of other agents.

Consider an economy with N risky assets, where each asset $n \in \{1,\dots,N\}$ is traded at some price p_n at the initial date $t = 0$ and delivers the random dividend $\tilde{d}_n$ at the following date $t = 1$, where $\tilde{d}_n \sim \mathcal{N}(\bar{d}_n, \sigma_n^2)$. Consider a dealer endowed with initial wealth w_0 and characterized by a negative exponential utility function with risk aversion parameter a. At the initial date $t = 0$, the dealer chooses a portfolio of the N risky assets by maximizing the expected utility function of the random wealth realized at date $t = 1$. We denote by $\theta^* = (\theta_1^*,\dots,\theta_N^*)^\top \in \mathbb{R}^N$ the optimal portfolio of the dealer, with θ_n^* representing the number of units of the n-th asset held in the portfolio, for $n = 1,\dots,N$, and we denote by $\widetilde{W}^*$ the corresponding optimal wealth realized at date $t = 1$ when no trade occurs in the market between date $t = 0$ and date $t = 1$, i.e.,

$$\widetilde{W}^* = \sum_{n=1}^N \theta_n^*(\tilde{d}_n - p_n) = \sum_{n=1}^N \theta_n^* \tilde{d}_n - w_0.$$

Recall that, as explained in Sect. 3.2, under the assumptions of an exponential utility function and of a normal distribution, the optimal portfolio belongs to the mean-variance efficient portfolio frontier.

Let us suppose that, at $t = 0$ and after the dealer has formed his optimal portfolio, a trade of x_n units of the n-th asset arrives to the dealer, for some $n \in \{1,\dots,N\}$. The dealer has to execute this order and, of course, after the execution of the order, the initial optimal portfolio of the dealer will no longer be optimal. To compensate for this risk, the dealer sets the bid and the ask prices for the n-th asset in such a way that the modified portfolio, after the execution of the order, has the same mean-variance profile of his original optimal portfolio θ^*. Let us denote by p_n^{ask} and p_n^{bid} the ask and bid prices, respectively, set by the dealer for the n-th asset at date $t = 0$. The dealer's wealth at date $t = 1$, after the execution of the order x_n, is given by

$$\widetilde{W}' = \sum_{k=1}^N \theta_k^*(\tilde{d}_k - p_k) - x_n(\tilde{d}_n - \bar{p}_n) = \widetilde{W}^* - x_n(\tilde{d}_n - \bar{p}_n),$$

where

$$\bar{p}_n = p_n^{\text{ask}} \mathbf{1}_{\{x_n>0\}} + p_n^{\text{bid}} \mathbf{1}_{\{x_n<0\}},$$

meaning that $\bar{p}_n$ is the ask price if x_n is a buy order and the bid price if x_n is a sell order. As a compensation for the commitment of executing the order, the dealer has the possibility of setting the bid and ask prices. In the model of Stoll [1570], the dealer chooses p_n^{ask} and p_n^{bid} in such a way that he is indifferent (in the mean-variance sense) between the original optimal portfolio and the modified portfolio after the execution of the order, i.e.,

$$\mathbb{E}[\widetilde{W}^*] - \frac{a}{2}\operatorname{Var}(\widetilde{W}^*) = \mathbb{E}[\widetilde{W}'] - \frac{a}{2}\operatorname{Var}(\widetilde{W}'). \tag{10.20}$$

In the present setting, the mean-variance problem (10.20) can be explicitly solved and the optimal bid and ask prices can be determined, as shown in the following proposition, whose proof is given in Exercise 10.9.

Proposition 10.7 *In the context of the above model (see Stoll [1570]), for each $n = 1, \ldots, N$, the optimal bid price p_n^{bid} and the optimal ask price p_n^{ask}, determined as the solution to the mean-variance problem (10.20), are given by*

$$p_n^{\text{bid}} = \bar{d}_n - \frac{a}{2}\sigma_n^2|x_n| - a\operatorname{Cov}(\widetilde{W}^*, \tilde{d}_n),$$

$$p_n^{\text{ask}} = \bar{d}_n + \frac{a}{2}\sigma_n^2|x_n| - a\operatorname{Cov}(\widetilde{W}^*, \tilde{d}_n),$$

where $\widetilde{W}^*$ denotes the original optimal wealth of the dealer, and the associated bid-ask spread is given by

$$p_n^{\text{ask}} - p_n^{\text{bid}} = a\sigma_n^2|x_n|.$$

According to the above proposition, the bid-ask spread set by the dealer is increasing in the dealer's risk aversion, in the variance of the dividend of the traded asset as well as in the size of the order to be executed.

The adverse selection problem faced by the dealer in the presence of informed traders has been analysed in several papers and, in particular, we refer to the contributions of Bagehot [105], Copeland & Galai [498], Glosten & Milgrom [790]. In Copeland & Galai [498], a two-period dealer economy with liquidity and informed traders has been analysed. In the market, there exists a risky asset, whose current price at date $t = 0$ is given by p_0 and whose underlying value is given by $\tilde{p}$, where $\tilde{p}$ is a random variable with density function $f(\cdot)$. The price p_0 represents the "true" price of the asset as perceived by the dealer at date $t = 0$. It is assumed that in the economy there are no taxes nor restrictions on short selling and that the density function $f(\cdot)$ is common knowledge among market participants. The dealer makes a commitment to buy a fixed quantity of the asset at the bid price p^{bid} and to sell a fixed quantity of the asset at the ask price p^{ask}. In the market, there are informed traders and liquidity traders. Informed traders know ex-ante the fundamental value $\tilde{p}$ and trade on the basis of this information, while liquidity traders trade for exogenous reasons and do not know the fundamental value. Both types of traders are assumed to be risk neutral and decide whether to post a buy order, a sell order, or do nothing, depending on the bid and ask prices quoted by the dealer. The dealer does not know

the fundamental value and also cannot distinguish between informed and liquidity
traders. The dealer is risk neutral and, hence, is only interested in maximizing his
expected profits (similarly as in the models presented before in this chapter, this can
be justified by the presence of competition among several dealers). In the model, it
is assumed that the probability that a trade is motivated by superior information
is exogenously given and equal to $\pi_{\mathrm{inf}} > 0$, with $\pi_{\mathrm{liq}} := 1 - \pi_{\mathrm{inf}}$ being the
probability that a trade is due to liquidity trading. The probability π_{inf} is known
to the dealer. Furthermore, it is assumed that a liquidity trader submits a buy order
with probability $\pi_{\mathrm{B,liq}}$ and a sell order with probability $\pi_{\mathrm{S,liq}}$.

The dealer's problem consists in setting the bid and the ask prices of the asset. If
the dealer sets a very large bid-ask spread, then he can reduce the potential losses
due to trading with better informed traders, but he risks to lose the profits generated
by trading with liquidity traders. On the other hand, if the dealer sets a very tight
bid-ask spread, then he can profit from trading with liquidity traders, but he is more
exposed to the risk of trading with informed traders. The optimal bid-ask spread
must be chosen as a compromise between these two situations. Due to the structure
of the model, the dealer's expected profit from trading with liquidity traders is given

$$\pi_{\mathrm{B,liq}}(p^{\mathrm{ask}} - p_0) + \pi_{\mathrm{S,liq}}(p_0 - p^{\mathrm{bid}}).$$

On the other hand, the expected loss of the dealer when trading with the better
informed traders are given by

$$\int_{p^{\mathrm{ask}}}^{\infty} (p - p^{\mathrm{ask}})f(p)\mathrm{d}p + \int_{0}^{p^{\mathrm{bid}}} (p^{\mathrm{bid}} - p)f(p)\mathrm{d}p.$$

The integrals with respect to the density function $f(\cdot)$ represent the fact that, unlike
an informed trader, the dealer does not know the true underlying value of the asset.
The (risk neutral) dealer sets the bid and ask prices p^{bid} and p^{ask} by maximizing the
following functional:

$$\pi_{\mathrm{liq}} \left(\pi_{\mathrm{B,liq}}(p^{\mathrm{ask}} - p_0) + \pi_{\mathrm{S,liq}}(p_0 - p^{\mathrm{bid}}) \right)$$

$$- \pi_{\mathrm{inf}} \left(\int_{p^{\mathrm{ask}}}^{\infty} (p - p^{\mathrm{ask}})f(p)\mathrm{d}p + \int_{0}^{p^{\mathrm{bid}}} (p^{\mathrm{bid}} - p)f(p)\mathrm{d}p \right).$$

In the model, there exists a bid-ask spread as long as there is a positive probability
that some of the trades are motivated by superior information, even if the dealer is
risk neutral. In other words, risk aversion is not necessary in order to explain the
emergence of a bid-ask spread. The bid-ask spread decreases when we shift from
a monopolistic market dealer to an economy where several dealers compete among
themselves (see also below in this section for a discussion of this aspect). Moreover,
if the proportion of informed traders increases, the difference between the bid-ask
spreads in the case of monopoly and in the case of competition will decrease. If

the elasticity of demand for liquidity trading decreases then the ask price will also decrease (see Copeland & Galai [498]).

In the presence of superiorly informed agents, a dealer faces an adverse selection problem, as shown in Copeland & Galai [498]. However, this adverse selection problem also has an intertemporal dimension which is not reflected in the above model. The intertemporal dimension is related to the fact that, by trading with better informed agents over time, the dealer can eventually learn their private information. A dynamic version of the above model which allows to take into account this aspect has been developed in the seminal paper Glosten & Milgrom [790]. In this work, the authors consider a discrete time economy with a finite number of trading dates and a pure dealership market, meaning that all orders are market orders (limit orders are not allowed). In the economy, there is a single risky asset and market orders are constrained to be for one unit of the asset (so that the dealer is not allowed to quote different bid/ask prices for different quantities). The market mechanism considered in Glosten & Milgrom [790] can be described by the following iterative sequence of steps:

1. the dealer quotes ask and bid prices for buy and sell orders of one unit of the asset;
2. a trader arrives at the dealer and, on the basis of the quoted bid/ask prices and his information, decides whether to submit a market order to the dealer;
3. if the trader decides to post a market order, then the order is executed;
4. after the execution of the order and before the arrival of the next trader, the dealer can revise his quotes for the bid and ask prices.

According to this mechanism, after executing a buy or a sell order, the dealer updates his beliefs and is allowed to revise his quotes for the bid and ask prices of the asset.

The risky asset has a fundamental value (per share) represented by the non-negative random variable $\tilde{v}$ with finite variance. In the economy, there are informed traders (insider traders) and pure liquidity traders. It is assumed that informed traders know ex-ante the realization of the random variable $\tilde{v}$, while this is unknown to liquidity traders as well as to the dealer. It is also assumed that there exists a finite time T_0 at which the fundamental value $\tilde{v}$ is revealed to every market participant. In this sense, T_0 can represent a future date at which the information becomes homogeneous and the informational advantage of informed traders disappears. Every market participant (informed traders, liquidity traders and the dealer) is risk neutral. In the market, exactly one investor arrives at the dealer at each date $t \in \{1, \ldots, T_0\}$ and the dealer is not able to distinguish between informed investors and liquidity traders (i.e., the market is anonymous). Since only one trader arrives at the dealer at each date, we can identify investors by their time of arrival. Different traders are characterized by different time preferences and these are identified by a parameter δ_t, which describes the preferences of the investor arriving at date t between current consumption and future consumption derived from the ownership of the asset. The utility of agent t can described by

$$\delta_t x \tilde{v} + c,$$

where x denotes the number of shares of the asset traded by investor t and c denotes current consumption (for the dealer, it is conventionally assumed that $\delta = 1$). According to this preference structure, a high value of δ_t represents a desire to invest for the future, while a low value of δ_t indicates a preference for immediate consumption. Since the market is anonymous, the dealer cannot discriminate between different investors. Accordingly, from the perspective of the dealer, the collection $(\delta_t)_{t=1,...,T_0}$ can be viewed as a stochastic process and it is assumed to be independent of the fundamental value $\tilde{v}$. Note that, since every market participant is risk neutral, it is necessary that different agents exhibit different values of the time preference parameter δ, otherwise the no-trade result of Milgrom & Stokey [1341] (see Sect. 8.1) would imply that the bid-ask spread is set large enough to preclude any trade.

In this model, it is important to distinguish the different information flows corresponding to the different types of traders. Similarly as in Chap. 6, we denote by $\mathbb{F} = (\mathscr{F}_t)_{t=1,...,T_0}$ the information flow of liquidity traders, consisting of all the information generated by the observation of past transaction prices, the current bid and ask prices as well as any publicly available information. We then denote by $\mathbb{G} = (\mathscr{G}_t)_{t=1,...,T_0}$ the information flow of informed traders, which consists of all the information available to liquidity traders together with the knowledge of the fundamental value v of the asset (the realization of the random variable $\tilde{v}$). In the model, it is assumed that the knowledge of the process $(\delta_t)_{t=1,...,T_0}$ does not bring any useful information on the other random variables. In particular, it is assumed that

$$\mathbb{E}[\tilde{v}|\mathscr{G}_t \vee \sigma(\delta_t)] = \mathbb{E}[\tilde{v}|\mathscr{G}_t] \quad \text{and} \quad \mathbb{E}[\tilde{v}|\mathscr{F}_t \vee \sigma(\delta_t)] = \mathbb{E}[\tilde{v}|\mathscr{F}_t], \quad (10.21)$$

for all $t = 1,...,T_0$, with $\sigma(\delta_t)$ denoting the σ-algebra generated by the random variable δ_t. This means that the knowledge of the time preference parameter δ_t does not permit to make better forecasts about the fundamental value of the asset.

For each $t = 1,...,T_0$, let p_t^{ask} and p_t^{bid} denote the ask and bid prices, respectively, quoted by the dealer at date t. At each date t, an investor arrives at the dealer, observes the ask and bid prices quoted by the dealer and decides whether to trade and, in the case he decides to trade, whether to submit a buy or a sell order. The investor $t \in \{1,...,T_0\}$ decides according to the following rule:

$$\text{buy if } Z_t > p_t^{\text{ask}},$$
$$\text{sell if } Z_t < p_t^{\text{bid}},$$
$$(10.22)$$

where the process $(Z_t)_{t=1,...,T_0}$ is defined by

$$Z_t := \delta_t(1 - U_t)\mathbb{E}[\tilde{v}|\mathscr{G}_t] + \delta_t U_t \mathbb{E}[\tilde{v}|\mathscr{F}_t], \quad (10.23)$$

where U_t is an indicator variable which is equal to one or zero depending on whether the investor arriving at date t is a liquidity trader or an informed trader, respectively.

Arguing as in Glosten & Milgrom [790], by including the specification of who is informed in the sample space, we can generically represent the information flow of a trader by $\mathbb{K} = (\mathcal{K}_t)_{t=1,\dots,T_0}$. In this way, we can rewrite (10.23) as

$$Z_t = \delta_t \mathbb{E}[\tilde{v}|\mathcal{K}_t], \qquad \text{for every } t = 1, \dots, T_0.$$

Given the structure of the investors' preferences, the dealer determines the bid and the ask prices by adopting a zero expected profits condition. We denote by $\mathbb{D} = (\mathscr{D}_t)_{t=1,\dots,T_0}$ the information flow of the dealer. Given the investors' behavior described in (10.22)–(10.23) and the information $\mathscr{D}_t$ available to the dealer at date t, the dealer's expected profit at date t is given by

$$\mathbb{E}\left[(p_t^{\text{ask}} - \tilde{v})\mathbf{1}_{\{Z_t > p_t^{\text{ask}}\}} + (\tilde{v} - p_t^{\text{bid}})\mathbf{1}_{\{Z_t < p_t^{\text{bid}}\}}\,\big|\,\mathscr{D}_t\right],$$

which can be equivalently rewritten as

$$\begin{aligned}
&\left(p_t^{\text{ask}} - \mathbb{E}\big[\tilde{v}|\mathscr{D}_t, Z_t > p_t^{\text{ask}}\big]\right)\mathbb{P}(Z_t > p_t^{\text{ask}}|\mathscr{D}_t) \\
&\quad - \left(p_t^{\text{bid}} - \mathbb{E}\big[\tilde{v}|\mathscr{D}_t, Z_t < p_t^{\text{bid}}\big]\right)\mathbb{P}(Z_t < p_t^{\text{bid}}|\mathscr{D}_t).
\end{aligned} \tag{10.24}$$

The zero expected profits condition implies that (10.24) has to be equal to zero at all dates $t = 1, \dots, T_0$. Hence, the zero expected profits equilibrium (if it exists) at each date $t \in \{1, \dots, T_0\}$ can be described by a couple of random variables $(p_t^{\text{ask}}, p_t^{\text{bid}})$ satisfying

$$p_t^{\text{ask}} = \mathbb{E}\big[\tilde{v}|\mathscr{D}_t, Z_t > p_t^{\text{ask}}\big] \quad \text{and} \quad p_t^{\text{bid}} = \mathbb{E}\big[\tilde{v}|\mathscr{D}_t, Z_t < p_t^{\text{bid}}\big]. \tag{10.25}$$

In general, showing the existence of two processes $(p_t^{\text{ask}})_{t=1,\dots,T_0}$ and $(p_t^{\text{bid}})_{t=1,\dots,T_0}$ satisfying condition (10.25) at all $t = 1, \dots, T_0$ is not easy. In the special case where the dealer's information flow $\mathbb{D}$ contains the information flow $\mathbb{G}$ of the informed traders (i.e., if $\mathscr{G}_t \subseteq \mathscr{D}_t$, for all $t = 1, \dots, T_0$), then it holds that $p_t^{\text{ask}} = p_t^{\text{bid}} = \mathbb{E}[\tilde{v}|\mathscr{D}_t]$, for all $t = 1, \dots, T_0$. On the other hand, if the dealer's information flow $\mathbb{D}$ coincides with the information flow $\mathbb{F}$ of the liquidity trader (i.e., $\mathscr{D}_t = \mathscr{F}_t$, for all $t = 1, \dots, T_0$), then it holds that

$$p_t^{\text{ask}} = \inf\big\{a : \mathbb{E}[\tilde{v}|\mathscr{F}_t, Z_t > a] \leq a\big\} \quad \text{and} \quad p_t^{\text{bid}} = \sup\big\{b : \mathbb{E}[\tilde{v}|\mathscr{F}_t, Z_t < b] > b\big\},$$

for all $t = 1, \dots, T_0$. As pointed out in Glosten & Milgrom [790], according to this notion of equilibrium, the dealer does not regret ex-post any trade that he is committed to execute. The ask price corresponds to the revised expectation of the fundamental value of the asset if the dealer is executing a buy order, while the bid price corresponds to the revised expectation of the fundamental value of the asset if the dealer is executing a sell order. In other words, equilibrium prices are set equal to the dealer's conditional expectations of the fundamental value of the asset given the type of order submitted. In this sense, the type of order is interpreted by the dealer as an informative signal and the bid and ask prices set by the dealer will incorporate the information that each trade has revealed.

A first and fundamental property to be checked is that the bid price is less or equal than the ask price and, moreover, that the conditional expectation of the fundamental value of the asset (on the basis of all publicly available information) lies between the bid price and the ask price. This is confirmed by the next proposition, which corresponds to Glosten & Milgrom [790, Proposition 1] and the proof of which is given in Exercise 10.10.

Proposition 10.8 *In the context of the above model (see Glosten & Milgrom [790]), suppose that there exist two processes* $(p_t^{\text{ask}})_{t=1,\ldots,T_0}$ *and* $(p_t^{\text{bid}})_{t=1,\ldots,T_0}$ *satisfying the equilibrium condition* (10.25). *Then it holds that*

$$p_t^{\text{bid}} \leq \mathbb{E}[\tilde{v} | \mathscr{D}_t \wedge \mathscr{F}_t] \leq p_t^{\text{ask}}, \qquad \text{for all } t = 1, \ldots, T_0,$$

where $\mathscr{D}_t \wedge \mathscr{F}_t$ *denotes the smallest* σ*-algebra generated by the sets belonging to both* $\mathscr{D}_t$ *and* $\mathscr{F}_t$.

At each date $t \in \{1, \ldots, T_0\}$, if a trade is being executed and it is a buy order, then the transaction price at t will correspond to the ask price p_t^{ask}. On the contrary, if at date t a trade is executed and it is a sell order, then the transaction price at date t will correspond to the bid price p_t^{bid}. Hence, in view of rule (10.22), if a transaction happens at t, then the corresponding transaction price p_t can be written as

$$p_t = p_t^{\text{ask}} \mathbf{1}_{\{Z_t > p_t^{\text{ask}}\}} + p_t^{\text{bid}} \mathbf{1}_{\{Z_t < p_t^{\text{bid}}\}}.$$

In equilibrium, the ask price and the bid price are determined according to (10.25), so that the last relation can be rewritten as

$$p_t = \mathbb{E}[\tilde{v} | \mathscr{D}_t, Z_t > p_t^{\text{ask}}] \mathbf{1}_{\{Z_t > p_t^{\text{ask}}\}} + \mathbb{E}[\tilde{v} | \mathscr{D}_t, Z_t < p_t^{\text{bid}}] \mathbf{1}_{\{Z_t < p_t^{\text{bid}}\}} = \mathbb{E}[\tilde{v} | \mathscr{D}_t],$$
$$(10.26)$$

conditionally on the fact that a trade takes place at date t (a buy or a sell order). Suppose that N trades take place between the initial date $t = 0$ and date T_0 and denote by $(T_n)_{n=1,\ldots,N} \subseteq \{1, \ldots, T_0\}$ the collection of dates at which a transaction has taken place. Then, relation (10.26) can be equivalently rewritten as

$$p_{T_n} = \mathbb{E}[\tilde{v} | \mathscr{D}_{T_n}], \qquad \text{for all } n = 1, \ldots, N, \qquad (10.27)$$

where the process $(p_{T_n})_{n=1,\ldots,N}$ collects all the prices at which transactions have occurred. As shown in the following simple proposition (corresponding to Glosten & Milgrom [790, Proposition 2]), transaction prices form a martingale with respect to the information flow of the dealer.

Proposition 10.9 *In the context of the above model (see Glosten & Milgrom [790]), suppose that there exist two processes* $(p_t^{\text{ask}})_{t=1,\ldots,T_0}$ *and* $(p_t^{\text{bid}})_{t=1,\ldots,T_0}$ *satisfying* *the equilibrium condition* (10.25). *Suppose that N trades take place at dates* $(T_n)_{n=1,\ldots,N}$. *Then the transaction price process* $(p_{T_n})_{n=1,\ldots,N}$ *is a martingale with respect to the information flow* $(\mathscr{D}_{T_n})_{n=1,\ldots,N}$, *i.e., it holds that*

$$p_{T_n} = \mathbb{E}[p_{T_{n+1}}|\mathscr{D}_{T_n}], \qquad \text{for all } n = 1,\ldots,N-1.$$

Proof The claim simply follows from property (10.27) together with the tower property of the conditional expectation:

$$p_{T_n} = \mathbb{E}[\tilde{v}|\mathscr{D}_{T_n}] = \mathbb{E}\big[\mathbb{E}[\tilde{v}|\mathscr{D}_{T_{n+1}}]|\mathscr{D}_{T_n}\big] = \mathbb{E}[p_{T_{n+1}}|\mathscr{D}_{T_n}],$$

for every $n = 1,\ldots,N-1$. $\qquad\square$

A first implication of the above proposition is that transaction prices are informationally efficient in the semi-strong sense (see Chap. 8), in the sense that they are efficient with respect to the information flow of the dealer, which is assumed to contain all publicly available information. The fact that the dealer is risk neutral (or, analogously, that there is perfect competition among several dealers), implies that there are no profit opportunities arising from the information known to the dealer, due to the martingale property of transaction prices. A second implication of the above proposition is that the first differences of transaction prices are serially uncorrelated, as a direct consequence of the fact that martingale increments are uncorrelated.

After having presented the general framework of the model of Glosten & Milgrom [790], let us illustrate some of its main implications in a very simple setting, following O'Hara [1384, Section 3.3] and De Jong & Rindi [540, Section 4.1]. We suppose that the random variable $\tilde{v}$ representing the fundamental value of the asset can take only two possible values $\overline{v}$ and $\underline{v}$, with $\overline{v} > \underline{v}$, with probabilities π and $1-\pi$, respectively. As above, the probability π is known to every market participant, but only the insider traders observe the realization of $\tilde{v}$. The dealer faces an informed trader with probability α, while with probability $1-\alpha$ he faces a liquidity trader. The liquidity trader randomly decides to submit buy or sell orders, i.e., the probability that a liquidity trader decides to buy one unit of the asset is given by $1/2$ and the probability that he decides to sell one unit of the asset is also given by $1/2$. On the other hand, an informed agent trades by maximizing his expected profit, so that he will submit a (one unit) buy order if he knows that the fundamental value is equal to $\overline{v}$ and, on the contrary, a (one unit) sell order if the fundamental value is $\underline{v}$.

Similarly as above, the dealer sets the bid and the ask prices for the asset, according to the zero expected profits rule. More specifically, the bid and the ask prices are set according to the rule (10.25), i.e., bid and ask prices represent the expectations of the fundamental value of the asset (from the point of view of the dealer), conditionally on the information revealed by the trade itself. Hence:

$$p^{\text{ask}} = \mathbb{E}[\tilde{v}|\text{buy order}] \qquad \text{and} \qquad p^{\text{bid}} = \mathbb{E}[\tilde{v}|\text{sell order}],$$

where $\mathbb{E}[\tilde{v}|\text{buy order}]$ and $\mathbb{E}[\tilde{v}|\text{sell order}]$ denote the conditional expectations of the fundamental value of the asset, given the information that a buy or a sell order has been submitted, respectively. In the present simple setting, as shown in Exercise 10.11, the equilibrium bid and ask prices can be easily computed in closed form, by making use of Bayes' rule:

$$p^{\text{ask}} = \overline{v}\frac{\pi(1+\alpha)}{1-\alpha+2\pi\alpha} + \underline{v}\frac{(1-\pi)(1-\alpha)}{1-\alpha+2\pi\alpha},$$

$$p^{\text{bid}} = \overline{v}\frac{\pi(1-\alpha)}{1+\alpha-2\pi\alpha} + \underline{v}\frac{\pi(1+\alpha)}{1+\alpha-2\pi\alpha}. \tag{10.28}$$

Moreover, if we furthermore assume that $\pi = 1/2$, we then get

$$p^{\text{ask}} = \overline{v}\frac{1+\alpha}{2} + \underline{v}\frac{1-\alpha}{2} = \frac{\overline{v}+\underline{v}}{2} + \frac{\alpha}{2}(\overline{v}-\underline{v}),$$

$$p^{\text{bid}} = \overline{v}\frac{1-\alpha}{2} + \underline{v}\frac{1+\alpha}{2} = \frac{\overline{v}+\underline{v}}{2} - \frac{\alpha}{2}(\overline{v}-\underline{v})$$

and the corresponding bid-ask spread is given by

$$p^{\text{ask}} - p^{\text{bid}} = \alpha(\overline{v}-\underline{v}).$$

In this simple setting, we can observe that the bid-ask spread is increasing in the proportion of informed traders in the economy and it is due to the presence of adverse selection costs. Moreover, we can regard the difference $\overline{v}-\underline{v}$ as a measure of the volatility of the fundamental value of the asset. In this sense, the above relation shows that the bid-ask spread is increasing in volatility. The existence of a positive bid-ask spread amounts to a compensation of the dealer for the risk of trading with better informed investors. Of course, even in this simple setting, the transaction prices exhibit the martingale property established in Proposition 10.9, thus implying that transaction prices are semi-strong efficient from an informational point of view (with respect to the information flow of the dealer).

The model of Glosten & Milgrom [790] yields several additional interesting insights. For instance, markets exhibiting a large trading volume will have smaller spreads and vice versa. Glosten & Milgrom [790, Proposition 4] show that, under suitable technical conditions, the dealer's conditional expectation of the fundamental value of the asset converges to the insider trader's conditional expectation of the fundamental value as the number of trading dates increases. Intuitively, this means that the dealer learns over time by updating his beliefs on the basis of the information revealed by the trades themselves. Glosten & Milgrom [790, Proposition 5] shows that, at any given date $t \in \{1, \ldots, T_0\}$, the bid-ask spread increases when, all other things being equal, the insider traders' information becomes more precise, the ratio of informed to uninformed traders arrival rates at t increases and the elasticity of uninformed supply and demand at date t increases. This is due to the fact that the adverse selection problem is worse the greater the fraction of informed traders and the better their information. These results have been confirmed in Madhavan [1272]: if the adverse selection problem is severe, then the dealer may set prices in such a way that no trade occurs (market breakdown).

The model of Glosten & Milgrom [790] considers standardized trades where the traded quantity is fixed to one unit of the asset. By making the trade size endogenous, several interesting implications can be obtained. A model in the spirit of Glosten & Milgrom [790] but with possibly different trade sizes has been developed in Easley & O'Hara [619], motivated by the fact that large trades are usually carried out at less favorable prices than small trades. As a consequence, the size of an order also reveals some information to the dealer. If investors are allowed to choose the size of their trades, then the size of the trade can affect the bid-ask spread, since better informed agents prefer to trade larger quantities in comparison to uninformed agents. In Easley & O'Hara [619], it is shown that two types of equilibria may arise, depending on the values of the parameters of the model (see Exercise 10.12): either informed traders submit both small and large orders, so that they will not be separated from liquidity traders (*pooling equilibrium*) or informed traders only submit large orders, so that they can be separated from liquidity traders (*separating equilibrium*). In particular, in correspondence of a separating equilibrium, the adverse selection problem does not appear in correspondence of small market orders, since the latter are only submitted by liquidity traders. Moreover, the ask price associated to a large order in a separating equilibrium is greater than the ask price associated to a large order in a pooling equilibrium. Again, the rationale of this phenomenon is the adverse selection effect. More details on this model are given in Exercise 10.12.

Not surprisingly, competition among market dealers affects the bid-ask spread. Inventory-based models predict that the bid-ask spread is a decreasing function of the number of dealers in the market (see Amihud & Mendelson [58], Grossman & Miller [846], Ho & Stoll [950], Biais [231]). For instance, in the context of a model similar to that proposed by Stoll [1570], the bid-ask spread is inversely proportional to the number of market dealers, as shown in Exercise 10.13. Moreover, considering a financial market where multiple assets are traded, the presence of substitutability among the traded securities in case of a monopolistic dealer leads to a decrease of the bid-ask spread, as considered in Hagerty [873]. In the presence of asymmetric information, the effects of the competition among dealers on the bid-ask spread is more ambiguous. In Dennert [559], assuming risk neutral insider investors, it is shown that an increase in the number of risk neutral market makers leads to a higher risk exposure for each individual market maker and, therefore, to larger bid-ask spreads and transaction costs. The rationale for this effect is that the total amount of uninformed liquidity trading is limited and remains constant and, therefore, the adverse selection problem for each individual dealer becomes more pronounced as the number of dealers increases. Moreover, in a competitive dealers market, the informational advantage of each dealer is smaller than in a monopolistic dealer market (this effect is similar to that described in Glosten [787]). Under some conditions, liquidity traders prefer a monopolistic market maker. A similar result is obtained in an order-driven market by allowing agents to submit orders to more than one risk neutral market maker (see Bernhardt & Hughson [206]). In Madhavan [1272], tight bid-ask spreads are obtained in a setting with adverse selection and competition among several dealers.

In Dutta & Madhavan [604], a game theoretic model has been developed in order to analyse the dynamic pricing strategies of competing market makers in a dealer market. In particular, the authors aims at studying whether competing market makers can sustain bid-ask spreads that are above the competitive levels, even if they do not explicitly cooperate to fix prices. It is shown that, under some conditions, the market makers can act in a non-cooperative way in setting prices and, at the same time, they may still set bid-ask spreads that are higher than the corresponding bid-ask spreads obtained in a competitive environment. This is a form of *implicit collusion*, which results from the dealers adopting Nash strategies. In some cases, the dealers' strategies under implicit collusion coincide with the strategies obtained under explicit collusion. However, implicit collusion is not feasible in every market. Indeed, if the dealers are sufficiently impatient or if there are no entry barriers, then the only possible equilibrium of the economy is the competitive equilibrium (in this regard, see also Grossman et al. [847]). The implicit costs represented by the costs of establishing a reputation and forming relationships to gain access to the order flow (*order preferencing arrangements*) are sufficient to reduce price competition and to ensure implicit collusion. When dealers receive preferenced order flow, in equilibrium the bid-ask spread increases. Furthermore, collusion is more frequent in an actively traded market and when dealers are of a similar size.

Dealers provide immediacy to the market. Assuming that traders experience liquidity needs, Grossman & Miller [846] develop an equilibrium model for the number of dealers in the market. The number of market makers will adjust until in equilibrium the dealers' costs of maintaining a continuous presence are equal to the expected returns of trading. As the number of market makers increases, the market liquidity increases (see also Biais [231] for a related equilibrium analysis of the number of dealers in a market). Wahal [1640] shows that the number of dealers is increasing in the trading intensity.

According to the above analysis, three different types of costs are at the origin of the bid-ask spread: order processing costs, inventory costs and adverse selection costs. A cross-sectional analysis shows that volume, risk, price and firm size explain most of the variability of the bid-ask spread or of the price impact of a trade. While the relevance of order processing costs is widely recognized, the relevance of the other two cost components is more debated. The relevance of adverse selection and inventory effects has been pointed out in George et al. [768], Glosten & Harris [789], Hasbrouck [911], Huang & Stoll [976, 978], Easley et al. [618], Lin et al. [1219], Hasbrouck & Sofianos [913], Madhavan & Smidt [1279, 1280], Hendershott & Seasholes [934], showing that the adverse selection effect increases with trade size (see Bloomfield [249] for some experimental evidence). Madhavan & Sofianos [1281] and Manaster & Mann [1292] suggest that dealers manage their inventory by timing the size and the direction of their trades rather than by adjusting their quotes. Note also that quote adjustments due to adverse selection and inventory management go in the same direction (e.g., a buy order leads to an increase in the price), but quote adjustments due to inventory management tend to be reversed over time (since the dealer wants to reconstruct the optimal inventory position), while quote adjustments for adverse information are not. Therefore, inventory effects induce negative serial correlation in orders, price changes and returns at high frequencies (on the other hand, under asymmetric information the quote revision is persistent). Confirming the results of Glosten & Milgrom [790] and Easley & O'Hara [619], large trades have a large price impact (i.e., the price is shown to be an increasing and concave function of the trade size) and widen the spread (see Hasbrouck [911, 912]). The price impact is increasing in the degree of adverse selection (see Koski & Michaely [1122]).

The positive relation detected empirically between bid-ask spreads and expected returns (see Amihud & Mendelson [59], Chalmers & Kadlec [393], Datar et al. [525], Brennan & Subrahmanyam [297], Eleswarapu [635], Amihud [57], Easley et al. [615], Brennan et al. [294], Bekaert et al. [185]) can be explained in terms of the adverse selection effect, leading to a higher probability of trades based on superior information (and, therefore, higher expected returns) and widening the bid-ask spread. A theoretical model accounting for this phenomenon has been developed in Easley & O'Hara [620], while evidence in favor of the Kyle [1147] microfoundation has been documented in Chordia et al. [433]. Easley et al. [618] show that the probability of informed trading is lower for high volume stocks. Moreover, there is a positive premium for informational trading. A decrease of the bid-ask spread and of its adverse selection component is observed in Greene & Smart [823] when the noise traders' component is more significant in the market. Although a secular relation between the reduction of transaction costs and trading activity has been identified (see Chordia et al. [435]), no significant relation has been observed between volume/trading activity and the bid-ask spread of stocks (see Jones [1039] and Johnson [1037]).

10.4 Multi-Period Models of Market Microstructure

The seminal two-period model developed in Kyle [1147, Section 2] and discussed at the beginning of Sect. 10.2 has been extended to a dynamic setting in Kyle [1147, Section 3]. The dynamic version of the model preserves the fundamental features already discussed in Sect. 10.2 and, in particular, at each date the equilibrium price reflects the information contained in the past and present order flow and the insider trader maximizes his expected profits, taking into account the effect of his actions on the prices both at the current date as well as at the future dates.

More specifically, Kyle [1147] considers a multi-period economy where N auctions take place sequentially over time, at dates $t_1 < \ldots < t_N$, where $t_N = 1$ and with t_n denoting the date at which the n-th auction takes place. Let $\Delta t_n := t_n - t_{n-1}$, for all $n = 1, \ldots, N$, with $t_0 := 0$, denote the time step between two successive auction dates. As in the two-period version of the model presented in Sect. 10.2, we assume that a single risky asset is traded, with liquidation value $\tilde{d}$, where

$\tilde{d} \sim \mathcal{N}(\bar{d}, \sigma_d^2)$. Let the process $(\tilde{z}_t)_{t\in[0,1]}$ denote a scaled Brownian motion, so that the increment $\Delta \tilde{z}_n := \tilde{z}_{t_n} - \tilde{z}_{t_{n-1}}$ is distributed as a normal random variable with mean zero and variance $\sigma_z^2 \Delta t_n$, for every $n = 1, \dots, N$, for some scaling parameter $\sigma_z^2 > 0$. For every $n = 1, \dots, N$, the random variable $\Delta \tilde{z}_n$ represents the quantity of the asset traded by noise (liquidity) traders at the n-th auction. Note that the independence of Brownian increments implies that the quantities traded by noise traders in correspondence of different auctions are independent. It is also assumed that the process $(\tilde{z}_t)_{t\in[0,1]}$ is independent of the random variable $\tilde{d}$. Similarly as in the version of the model considered in Sect. 10.2, the insider trader knows the realization of the liquidation value $\tilde{d}$ of the asset. We denote by $\tilde{x}_n$ the aggregate position of the insider trader after the n-th auction, for all $n = 1, \dots, N$, so that $\Delta \tilde{x}_n := \tilde{x}_n - \tilde{x}_{n-1}$ represents the quantity traded by the insider at the n-th auction. For each $n = 1, \dots, N$, we denote by $\tilde{p}_n$ the market clearing price at the n-th auction viewed as a random variable and by p_n its generic realization.

In correspondence of each auction, the trading mechanism follows the two steps described at the beginning of Sect. 10.2, suitably modified in order to include the relevant information generated by the previous auctions. Namely, in correspondence of each auction date t_n, $n = 1, \dots, N$:

1. the insider, knowing the realization of the liquidation value of the asset and the transaction prices $p_1, \dots, p_{n-1}$ realized at the previous auctions, determines his optimal position $\tilde{x}_n = X_n(p_1, \dots, p_{n-1}, \tilde{d})$, where $X_n : \mathbb{R}^n \to \mathbb{R}$ is a measurable function, for each $n = 1, \dots, N$;
2. the market maker determines the price $\tilde{p}_n$ at which the market clears, observing the current order flow $\Delta \tilde{x}_n + \Delta \tilde{z}_n$ at date t_n and observing also the past values of the order flow. Accordingly, the market clearing price is determined by a rule of the form $\tilde{p}_n = P_n(\tilde{x}_1 + \tilde{z}_1, \dots, \tilde{x}_n + \tilde{z}_n)$, where $P_n : \mathbb{R}^n \to \mathbb{R}$ is a measurable function, for each $n = 1, \dots, N$.

Note that the information flow of the insider trader includes the quantities he has traded in the past. Similarly, the information flow of the market maker includes the prices he has set in the past. For brevity of notation, let us define the family of functions $X := (X_1, \dots, X_N)$ and $P := (P_1, \dots, P_N)$. Following Kyle [1147], the family X can be referred to as the insider trader's *trading strategy* and P as the market maker's *pricing rule*.

Similarly as in the static version of the model presented in Sect. 10.2, it is assumed that the market maker is risk neutral and sets the market price according to the zero expected profits condition. The insider trader is also risk neutral and, at each auction date t_n, determines his demand of the asset by maximizing his expected profit. At each date t_n the profits obtained by the insider trader from the auctions taking places at the dates $\{t_n, \dots, t_N\}$ are given by

$$\tilde{\pi}_n := \sum_{k=n}^{N} (\tilde{d} - \tilde{p}_k)\tilde{x}_k = \sum_{k=n}^{N} (\tilde{d} - \tilde{p}_k) X_k(p_1, \dots, p_{k-1}, \tilde{d}),$$

with $p_1, \ldots, p_{k-1}$ denoting the realizations of $\tilde{p}_1, \ldots, \tilde{p}_{k-1}$. Recalling that the price process $(\tilde{p}_n)_{n=1,\ldots,N}$ is determined dynamically by the market maker according to the pricing rule P, we write $\tilde{\pi}_n = \tilde{\pi}_n(X, P)$ in order to emphasize the dependence of the profits of the insider trader on the family of functions X and P. Analogously, following the notation of Kyle [1147], we also write $\tilde{p}_n = \tilde{p}_n(X, P)$ and $\tilde{x}_n = \tilde{x}_n(X, P)$.

In this context, similarly as in Sect. 10.2, an *equilibrium* of the economy (*sequential auction equilibrium*) is defined as a couple (X, P) of families of measurable functions such that, at every date t_n, $n = 1, \ldots, N$:

(i) the insider trader maximizes his expected profits, given the observation of the realization d of the liquidation value $\tilde{d}$ and the realized past transaction prices, i.e., the insider trader demand function X satisfies

$$\mathbb{E}[\tilde{\pi}_n(X, P)|p_1, \ldots, p_{n-1}, d] \geq \mathbb{E}[\tilde{\pi}_n(X', P)|p_1, \ldots, p_{n-1}, d], \qquad (10.29)$$

for any alternative family $X' = (X'_1, \ldots, X'_N)$ of measurable functions such that $X'_k = X_k$, for all $k = 1, \ldots, n-1$, and for all possible realizations $p_1, \ldots, p_{n-1}, d$ of the random variables $\tilde{p}_1, \ldots, \tilde{p}_{n-1}, \tilde{d}$;

(ii) the market is efficient, in the sense that the price $\tilde{p}_n$ set by the market maker satisfies

$$\tilde{p}_n = \tilde{p}_n(X, P) = \mathbb{E}[\tilde{d}|x_1 + z_1, \ldots, x_n + z_n], \qquad (10.30)$$

for all realizations $x_1 + z_1, \ldots, x_n + z_n$ of $\tilde{x}_1 + \tilde{z}_1, \ldots, \tilde{x}_n + \tilde{z}_n$.

Apart from the different information structures appearing in the above definition of equilibrium, the interpretation is completely analogous to the notion of equilibrium discussed in Sect. 10.2. In the present dynamic setting, it is important to remark that the market efficiency condition (10.30) implies that the price process $(\tilde{p}_n)_{n=1,\ldots,N}$ is a martingale with respect to the information flow generated by the total order flow of the market. This property immediately follows from the tower property of the conditional expectation together with the fact that, for each $n = 1, \ldots, N$, the price $\tilde{p}_n$ is given in (10.30) by the conditional expectation of the terminal random variable $\tilde{d}$.

In the present dynamic setting, an equilibrium (X, P) is said to be a *recursive linear equilibrium* if each component of the family of functions (X, P) is linear and if there exist constants $\lambda_1, \ldots, \lambda_N$ such that

$$\tilde{p}_n = \tilde{p}_{n-1} + \lambda_n(\Delta\tilde{x}_n + \Delta\tilde{z}_n), \qquad \text{for every } n = 1, \ldots, N.$$

Similarly to Proposition 10.3, it can be shown that there exists a unique recursive linear equilibrium, which furthermore admits an explicit characterization (see Kyle [1147, Theorem 2]). In particular, for every $n = 1, \ldots, N$, it holds that

$$\Delta \tilde{x}_n = \beta_n (\tilde{d} - \tilde{p}_{n-1}) \Delta t_n \quad \text{and} \quad \Delta \tilde{p}_n = \lambda_n (\Delta \tilde{x}_n + \Delta \tilde{z}_n),$$

where the constants $\beta_1, \ldots, \beta_N$ and $\lambda_1, \ldots, \lambda_N$ are given as the unique solution to a system of difference equations. The parameters $\beta_1, \ldots, \beta_N$ measure the sensitivity of the demand of the insider trader with respect to his private information, while the parameters $\lambda_1, \ldots, \lambda_N$ measure the depth of the market, with a small value of λ corresponding to a deep market. In particular, note the similarity with the structure of the linear equilibrium described in Proposition 10.3. In equilibrium, it holds that $\lambda_n = \beta_n \Sigma_n / \sigma_z^2$, where Σ_n is the variance of the liquidation value of the asset conditionally on the observation of the market flow up to date t_n. The quantity Σ_n measures how much of the insider trader's information is not yet incorporated into the market price (as estimated by the market maker). Referring to the original paper Kyle [1147] for full details, we only mention that the proof of this result proceeds by backward induction.

Many of the properties of the equilibrium discussed in Sect. 10.2 continue to hold in the present dynamic setting. The parameter Σ_n, which measures the informativeness of the equilibrium price, is decreasing with respect to n, reflecting the fact that, as time goes on, more of the private information of the insider trader is reflected in the market price. Note, however, that $\Sigma_N > 0$, meaning that the insider trader's information is only partially incorporated into the equilibrium price at the terminal date. The parameters λ_n, $n = 1, \ldots, N$, are inversely proportional to the parameter σ_z^2, representing the variance of the noise traders' demand. This means that increasing the amount of liquidity trading increases the depth of the market, increases proportionately the profits of the insider trader and leaves the informativeness of the equilibrium price unchanged, similarly as in the simple two-period model discussed in Sect. 10.2.

Kyle [1147, Section 5] analyses the limiting behavior of the equilibrium when the time step between two successive auction dates goes to zero. As Δt_n converges to zero, the stochastic processes of the price and of the order flow converge to a system of stochastic differential equations driven by Brownian motion. In the limit, the parameter λ does not depend on time and equals σ_d / σ_z, meaning that, in the continuous time limit, the market depth is constant in equilibrium (see Kyle [1147, Theorem 3]). On the other hand, the conditional variance of the liquidation value of the asset decreases linearly with time and, moreover, by the end of trading, the insider trader's private information will be perfectly reflected into the equilibrium price. The equilibrium price process is a martingale with constant instantaneous variance.

The properties of the equilibrium characterized in Kyle [1147] reflect the fact that an insider trader acts in a way that does not lead to an immediate revelation of his private information, but rather acts in a strategic way and his private information is gradually incorporated into the equilibrium price. Such a result provides a potential explanation for the fact that market prices typically need some time in order to incorporate new information arriving into the market (underreaction). Considering endogenous liquidity traders in a market with long-lived asymmetric information, the dissemination of information turns out to be slower, as shown in Mendelson & Tunca [1324]. Moreover, as considered in Chau & Vayanos [414], if new private information arrives at each trading date, then the insider trader acts aggressively and his private information is quickly revealed to the market. Multiple insider traders have been considered in Holden & Subrahmanyan [953], assuming that each individual insider knows the liquidation value of the asset before starting to trade. In this setting, the presence of competition among insider traders induces them to trade aggressively and leads to a rapid incorporation of their private information into the equilibrium prices. Moreover, the trade of an insider in the presence of competition is larger than that of a monopolistic insider, the market quickly becomes very deep and the conditional variance of the liquidation value of the asset quickly converges to zero. Increasing the number of informed traders, the intensity of these effects is even more pronounced. In the limit case where the time step between two successive trading dates converges to zero, the price becomes immediately fully revealing and the market very deep. These results also hold in markets with risk averse insiders (see Holden & Subrahmanyan [954]). However, competition among insiders can be much less intense if agents are endowed with heterogeneous (and correlated) private information, as considered in Foster & Viswanathan [732]. When the correlation among heterogeneous signals observed by different insiders is low, then less information is revealed in the presence of multiple insiders than in the case of a unique insider (see Back et al. [101]).

It is important to remark that, in the model of Kyle [1147], insider traders never *manipulate* the market. Indeed, the trading strategies of the insiders are linear and insiders always trade following the direction of their private information. The model has been extended in Hong & Rady [959] by considering the case where the market maker is better informed than the insider trader on liquidity trades, in the sense that the market maker knows the variance of liquidity trades while the insider trader knows it imperfectly. In this context, insider traders optimally take into account the uncertainty on the variance and gradually learn about market liquidity from past prices and trading volume. One of the main implications of this model is that past prices and trading volume affect future trades via a learning effect. Wang [1642] allows for heterogeneous prior beliefs (i.e., agents have different distributional assumptions concerning an informed trader's private signal) and asymmetric information in the above model. In equilibrium, it turns out that volume and absolute price changes-volatility are positively correlated and trading volume is positively serially correlated.

In the above setting, no public disclosure of insider trades has been considered. In many financial markets public disclosure of insider trades is mandatory, in the sense that insiders associated with a firm (i.e., officers, directors and major shareholders) must report to some market authority the trades they make in the stock of the firm (see also Sect. 10.5). The rationale for this rule is that public disclosure of insider trades should be beneficial to uninformed agents and, therefore, should facilitate price discovery. Huddart et al. [989] confirm this effect in the context of the Kyle [1147] dynamic model discussed above, with a single insider trader and a market maker who observes insider trades at the end of each round of trading. In comparison with the original model of Kyle [1147], in this setting the insider trades by adding a random noise component to his demand, except in the last trading round. Intuitively, the insider trader tries to dissimulate his private information in order to limit its dissemination in the market. Public disclosure of insider trades accelerates the price discovery process and lowers the trading costs with respect to the non-disclosure case and, moreover, the insider trader's expected profits are lower in comparison with the classical setting of Kyle [1147]. These effects become stronger as the number of trading rounds increases. However, these results are not necessarily confirmed in the presence of several insider traders. In Huddart et al. [988], it is shown that public disclosure of insider trades may inhibit the price discovery process by dampening competition among insiders. Assuming repeated (short-lived) private information arrivals, public disclosure of insider trades induces a tacit coordination by allowing insiders to monitor each others' trades. Insider traders coordinate among themselves and trade less aggressively in order to exploit their private information. In the case of delayed disclosure, Cheng et al. [427] have considered the possibility of price manipulation.

Under the assumption of a perfectly competitive market, we have seen that informed traders trade more aggressively than uninformed agents and trading volume is increasing in the precision of the private signals. This result is only partially confirmed when agents trade strategically. Indeed, in Kyle [1147], Admati & Pfleiderer [20], Foster & Viswanathan [729], it is shown that informed agents act strategically, revealing to the market their private information in a gradual way and trying to dissimulate their trades in order to better exploit their informational advantage. However, the empirical literature does not provide a clear conclusion on this aspect. On the one hand, Hasbrouck [912], Easley et al. [616], Seppi [1518] show that trade size has an effect on price adjustments (in the sense that large trades are more informative), while in Barclay & Warner [161], Chan & Fong [404], Chakravarty [390] it is shown that medium size trades are the more informative (*stealth trading*) and in Jones et al. [1041], Easley et al. [617], Chan & Fong [404] it is shown that transaction numbers rather than volume determine volatility. In a related direction, Meulbroek [1333] and Cornell & Sirri [499] show that insider traders use medium size trades in order to better conceal their information. Block trades can also be motivated by liquidity (uninformative) reasons: taking this observation into account, Seppi [1518] shows that the informational content is increasing in trade size. Whether an insider trader acts competitively or strategically has been investigated empirically. According to the Kyle [1147] model, in a strategic perspective the trade size should be inversely related to market illiquidity (as measured by the parameter λ introduced above) and positively related to trading volume. These results have been confirmed empirically in Brennan & Subrahmanyam [298].

Intraday and Intraweek Regularities

Market microstructure models have also been considered as a possible explanation of some well-known intraday regularities appearing in asset prices time series. Indeed, in some financial markets, it has been observed that volume, volatility and bid-ask spreads follow a U-shaped pattern during a trading day. They are at the highest point at the opening of the trading day, decrease rapidly to lower levels during the day and then rise again towards the end of the day. These empirical regularities have been reported for instance in Goodhart & O'Hara [812], Foster & Viswanathan [731], Brock & Kleidon [307], Chan et al. [403], Werner & Kleidon [1652], Chordia et al. [434] and are difficult to explain in the context of classical asset pricing theory. While this U-shaped pattern is by itself a puzzle, one of the main issues is represented by the fact that trading volume is high when trading costs are large. From a theoretical point of view, it is difficult to explain such a relation, while, on the other hand, the relation between price volatility and spread can be supported by theoretical models based on information asymmetry.

A model allowing to explain the intraday patterns described above has been proposed in Admati & Pfleiderer [20]. The main feature of the model is that noise traders can decide the timing of their trades during a time span (day, week, month). In Admati & Pfleiderer [20], the authors consider a multi-period economy, with trading dates $t = 1, \ldots, T$, where a single risky asset is traded. The value of the asset at the terminal date T is exogenously given and equal to

$$\tilde{d} = \bar{d} + \sum_{t=1}^{T} \tilde{\varepsilon}_t,$$

where the family $(\tilde{\varepsilon}_t)_{t=1,\ldots,T}$ is composed of independent random variables with zero mean. While at the terminal trading date T every market participant has full information on the liquidation value $\tilde{d}$ of the asset, at previous dates the information on $\tilde{d}$ is partially revealed through public and private sources. More specifically, it is assumed that, at each date $t = 1, \ldots, T$, the random variable $\tilde{\varepsilon}_t$ becomes publicly known, but there also exist informed agents who observe a signal $\tilde{y}_t$ of the form

$$\tilde{y}_t = \tilde{\varepsilon}_{t+1} + \tilde{\eta}_t, \qquad \text{for each } t = 1, \ldots, T-1,$$

with $\text{Var}(\eta_t) = \sigma_t^{2,\eta}$. This represents the fact that informed agents receive a signal about the information that will become publicly known at the successive trading date (or, according to an alternative interpretation, informed agents are able to process the public information in a faster or more efficient way). Note that in this model the private information becomes useless at the subsequent trading date, since it will become part of the publicly available information. In this sense, the private information is short-lived and informed traders, therefore, have no incentive to restrict their trading activity in order to have a larger informational advantage in the future, unlike in most of the models mentioned above.

In the economy, all the traders are risk neutral, similarly as in Kyle [1147]. At each date $t = 1, \ldots, T$, it is assumed that there are N_t informed traders (in Admati & Pfleiderer [20] the model is also extended by making the decision of becoming informed endogenous). Differently from the models considered above, the economy comprises two types of liquidity traders: in each period, there are *non-discretionary liquidity traders*, who must trade a given number of shares in that period, and *discretionary liquidity traders*, who have liquidity demands that do not need to be satisfied immediately but only before some future date. At each trading date $t = 1, \ldots, T$, non-discretionary liquidity traders have a total demand represented by the random variable $\tilde{z}_t$, while discretionary liquidity traders determine their demand by minimizing their expected trading costs. It is assumed that, in each period, there exists a constant number M of discretionary liquidity traders.

Similarly as in Kyle [1147], at each trading date the market price is set by a risk neutral market maker who sets the market price according to the zero expected profits condition (forced by competition). The market price set by the market maker is given by the expectation of the liquidation value of the asset, conditionally on the publicly available information and the observation of the market flow up to the current date. Similarly as in the models considered above, under the assumption that all the random variables appearing in the model are normally distributed, it can be shown that the equilibrium of the economy is characterized by a linear pricing rule of the form

$$p_t = \bar{d} + \sum_{s=1}^{t} \tilde{\varepsilon}_t + \lambda_t \tilde{\omega}_t, \qquad \text{for every } t = 1, \ldots, T,$$

where $\tilde{\omega}_t$ denotes the aggregate market demand (order flow) at date t (see Admati & Pfleiderer [20, Section 1.2]). Moreover, in equilibrium the demand $\tilde{x}_t$ of an informed trader at date t is given by

$$\tilde{x}_t = \beta_t \tilde{y}_t, \qquad \text{for every } t = 1, \ldots, T,$$

where

$$\beta_t = \sqrt{\frac{\Psi_t}{N_t(\mathrm{Var}(\tilde{\varepsilon}_{t+1}) + \sigma_t^{2,\eta})}} \quad \text{and } \lambda_t = \frac{\mathrm{Var}(\tilde{\varepsilon}_{t+1})}{N_t + 1} \sqrt{\frac{N_t}{\Psi_t(\mathrm{Var}(\tilde{\varepsilon}_{t+1}) + \sigma_t^{2,\eta})}},$$

where Ψ_t is the variance of the aggregate liquidity trades at date t (i.e., the variance of the total demand by non-discretionary liquidity traders and discretionary liquidity traders), see Admati & Pfleiderer [20, Lemma 1].

The coefficient λ_t is decreasing with respect to N_t and Ψ_t. Similarly as in the model of Kyle [1147], the reciprocal of the parameter λ_t measures the market depth at date t. This implies that, as the number of informed agents increases, the competition among them becomes more intense, since all the informed agents are assumed to observe the same private signal, and this leads to an increase of the depth of the market. A similar effect is observed as the (endogenous, due to the presence of discretionary liquidity traders) variance of noise traders' aggregate demand increases. As far as the insider trader's trading intensity is concerned (as measured by the coefficient β_t), it increases with respect to the liquidity traders' aggregate trading volume and to the precision of the private signal, while it is decreasing with respect to the number of insider traders.

In equilibrium, each discretionary liquidity trader follows a trading policy aiming at minimizing the expected trading costs, subject to the constraint of meeting the liquidity need at some future date. It can be shown that this induces discretionary liquidity traders to trade in the period where the market is more deep (i.e., when the parameter λ_t takes the smallest value). Therefore, there always exist equilibria where discretionary liquidity trading is concentrated in the same period (see Admati & Pfleiderer [20, Proposition 1]). As shown above, the parameter λ_t is small when the variance Ψ_t is large. In turn, the large variance of liquidity trading allows insider traders to more easily conceal their trades in the aggregate liquidity trades and this phenomenon is reflected in the expression for the coefficient β_t given above. In other words, the informed traders trade more aggressively in the period when all the discretionary liquidity traders trade together. Similarly as in Kyle [1147], the variance of price changes and the amount of information incorporated into the equilibrium price do not depend on the variance of the aggregate liquidity trades (the rationale for this effect is the same as in the Kyle [1147] model), as long as the number of informed traders is kept fixed and the precision of information is constant over time.

Admati & Pfleiderer [20] also consider an extended version of their model where the decision to become informed is endogenous, so that the number of informed traders becomes an endogenous variable and is determined as part of the equilibrium. In this case, the concentration of trades becomes even more intense: the fact that at some date there is more liquidity trading attracts more informed trading, which can be more easily concealed, and this makes even more attractive for liquidity traders to trade in the same period. In the presence of endogenous information acquisition, the clustering of trades increases the informativeness of the equilibrium price and the variance of price changes over time. Traders collect more information when there is more liquidity trading. The model of Admati & Pfleiderer [20] can explain the relation between large trading volume and large price changes observed empirically. These results are in part confirmed when information is heterogeneous. When the private information is long-lived, liquidity trades (volume) and volatility of prices are still related (see Back & Pedersen [102]).

The model of Admati & Pfleiderer [20] presented above does not explain the fact that large trading volume is often associated with large bid-ask spreads. Indeed, one would expect the opposite relation to hold, with a large trading volume being associated to a tight bid-ask spread. Information-based models suggest that bid-ask spread and volatility decline monotonically over the day, as a consequence of the presence of a learning process reducing the information asymmetry, with the largest volatility being at the opening of the day. This intraday volatility pattern has been confirmed for many financial markets (see Gerety & Mulherin [772] and Stoll & Whaley [1571]): the variance of open-to-open returns is larger than the variance of close-to-close returns. The result can be attributed to the procedure for opening stocks or to private information released by prices. Moreover, prices tend to reverse around the opening of the trading day. Madhavan et al. [1278] argue that, while information costs decrease over the day (as a consequence of the fact that agents learn from the trading process itself), inventory costs increase and generate the U-shaped pattern of the bid-ask spread.

The intraday pattern of trading volume and bid-ask spreads has been explained by introducing market closure inside the dynamic optimal investment-consumption problem without assuming private-asymmetric information (see Brock & Kleidon [307] and Gerety & Mulherin [771]). Market closure induces agents to trade with more intensity and less price elasticity for hedging reasons towards the beginning and the end of the trading period. As information is diffused without trading opportunities, portfolios at the opening differ in general from optimal positions and typically involve large trades. Similarly, in preparation for an overnight non-trading period, agents will trade more. Heterogeneity across agents regarding preferences and risk tolerance leads to high volume in the opening and before the market closure. Under some conditions, also the volatility increases and a monopolist dealer will perform price discrimination, resulting in a wide bid-ask spread associated with large trading volume.

Hong & Wang [965] allow for trades motivated by both informational and hedging reasons with market closures. Market closures preclude not only trades but also learning, so that information asymmetry increases during market closures. There is a time-varying hedging effect and a time-varying information asymmetry effect. Under some conditions, these two effects generate U-shaped patterns in volume, mean and volatility of returns. Moreover, open-to-open returns are more volatile than close-to-close returns and returns over trading periods are more volatile than returns over non-trading periods.

Intraweek regularities have been also observed in asset prices (see Harris [898] and Hawawini & Keim [916]). In particular, returns earned during the weekend are typically negative (see French [739]). A model in the spirit of Admati & Pfleiderer [20] allowing to describe this type of regularities has been proposed in Admati & Pfleiderer [21]. The key ingredient is that in equilibrium buy and sell liquidity trades and, therefore, buying and selling trading volume are concentrated in different periods. When buy liquidity traders trade in the market, prices are high and vice versa, so that patterns in order imbalance emerge. All discretionary liquidity buyers (sellers) trade in the same period, but liquidity buyers and liquidity sellers are concentrated in different periods. Moreover, systematic patterns emerge due to the presence of privately informed traders. By assuming that information remains private for more than one trading period and that liquidity traders act strategically (in the sense that they can postpone their trades), intraday regularities can be generated as shown in Foster & Viswanathan [729]. Confirming the empirical observations (see Foster & Viswanathan [731]), if the insider has more information on Monday, then the variance of price changes and trading costs (bid-ask spreads and the market maker's sensitivity to order flow) on Monday are highest and volume is lower than on Tuesday (liquidity traders do not trade on Monday).

10.5 Market Abuse: Insider Trading and Market Manipulation

Market abuse refers to a *misuse* of the market by its participants. This definition is of course quite vague and for a more precise definition one has to refer to specific financial market regulations. There are two main classes of market abuse: *insider trading* and *market manipulation*. Insider trading refers to the fact that an agent (insider trader) exploits his privileged information when trading in the market, thereby earning positive excess returns. Market manipulation consists of many different actions, some of them identified by the financial markets regulation. Of course, insider trading and market manipulation are two strictly related phenomena and it is difficult to establish a precise boundary between the two. However, one can identify three main differences. First, insider trading is always based on privileged information, while market manipulation is not necessarily based on privileged information. Second, market manipulation affects market prices, whereas insider traders typically do not want to affect market prices. Third, trades based on inside information are typically followed by other trades based on information, while trades originated by market manipulation are more likely to be reversed.

Insider Trading

The notion of insider trading depends on the regulation adopted in a specific financial market. Typically, a class of market participants is identified by the regulator as potential insider traders (e.g., CEOs and agents having a privileged relation with a firm). In Bhattacharya & Daouk [223], it is shown that among 103 countries having a stock market, 87 have adopted an insider trading regulation. Insider trading has effects on the market and on the functioning of the firms. In the literature, benefits and costs of insider trading have been identified. The following arguments have been put forward in favor of a light insider trading regulation, pointing out some of the benefits generated by the presence of insider traders in the market:

- there is more diffusion of information in the market, market prices are more efficient, price discovery is quicker and, therefore, agents have better risk sharing opportunities;
- because of reduced risk, asset prices are higher and the cost of capital is lower;
- the opportunity to engage in profitable insider trading activities provides a way to compensate the managers of a firm;

- the presence of insider traders in the market leads to lower information acquisition costs for the outsiders;
- greater informational efficiency in the market should encourage efficient investment decisions by the firms.

On the other hand, the following arguments have been put forward in favor of a more pervasive insider trading regulation, pointing out the possible drawbacks of the presence of insider traders in the market:

- insider trading generates an adverse selection problem and, therefore, low liquidity, large bid-ask spreads, high costs of capital and, under some conditions, a market breakdown can even occur;
- insider trading generates a cost to the shareholders since it exploits information which is not publicly available to outsiders, resulting in a lower efficiency of the firms;
- managers may have an incentive to mostly consider their insider trading profits.

Summing up, on the one hand insider trading increases the diffusion of information in the market, on the other hand it induces an adverse selection effect which may discourage agents from trading in the market.

In previous sections of this chapter, we have discussed the effects of the presence of insider traders in the market by considering market indicators such as efficiency, liquidity and volatility. In the following, we concentrate on the welfare implications of insider trading. In this direction, Bhattacharya & Nicodano [222] show that the presence of insiders can have beneficial effects on the outsiders' welfare when the equilibrium trades of the insiders are small compared to the liquidity based trades of the outsiders and when the adverse selection effect is not too strong. In their model, Bhattacharya & Nicodano [222] assume inflexible ex-ante aggregate investment choices by the agents. In this setting, the improvement of risk sharing with more efficient prices (due to insider trading) compensates the losses due to the adverse selection effect. On the contrary, when the adverse selection is strong and investment choices are flexible (i.e., outsiders reduce their investment because of the presence of insider traders), a prohibition of insider trading may lead to a Pareto improvement (see also Ausubel [89]). Dow & Rahi [586] show that the welfare effects of insider trading depend on how the revelation of information changes the risk sharing opportunities in the market. Fishman & Hagerty [709] show that insider trading may make a (non-perfectly competitive) security market less efficient if outsiders gather less information. In a model with endogenously determined and price-sensitive investment, Leland [1181] shows that, if insider trading is permitted then stock prices, expected real investments and firm's profits are higher and the market is less liquid, more volatile and more efficient (in the sense that prices better reflect the information). Insider trading decreases both the expected return and the risk of outsiders' investments. The total welfare may increase as well as decrease in the presence of insider trading, while outside investors and liquidity traders suffer a welfare loss. On the contrary, insiders and owners of the company benefit from insider trading. In the model of Leland [1181], the key factor in determining the welfare consequences of insider trading is represented by the sensitivity of investment of the company to the current price. In the presence of a high sensitivity, insider trading turns out to be beneficial.

The empirical analysis of insider trading phenomena concerns three main aspects: market effects (and, in particular, diffusion of privileged information), insider traders' profits and the informational content of insider trades.

The capability of insider trading to quickly disseminate private information has been tested with positive evidence in Damodaran & Liu [512], Cornell & Sirri [499], Meulbroek [1333], Aktas et al. [37]. It is shown that illegal insider trading is associated with quick price discovery: the stock market detects insider trades and incorporates a large fraction (about 50%) of their information in prices before the information is made public. Trading days where insider trading activities have occurred are characterized by a high volume, but the proportion of that volume due to insiders is small, thus implying that there are many market participants who just follow the insider traders. Moreover, Bhattacharya et al. [225] have shown that in some emerging markets there is little market response to public market announcements, since the new information was already incorporated in market prices, due to the presence of unrestricted insider trading. On the other hand, Chakravarty & McConnell [392] find that insider trading does not affect market prices in a significant way.

Analysing legal transactions by corporate insiders, Jaffe [1007], Seyhun [1520, 1521], Lakonishok & Lee [1158], Jeng et al. [1030] show that insider trading is profitable (i.e., there is evidence of abnormal returns) and that insider traders are actually informed. However, negative abnormal returns have been reported in Eckbo & Smith [624]. Insiders' purchases (sales, respectively) are a good (bad, respectively) news and predict positive (negative, respectively) future returns. Furthermore, insider purchases are more informative and earn more excess returns than insider sales. Seyhun [1520] and Rozeff & Zaman [1482] show that, taking into account transaction costs, outsiders do not earn excess returns when they try to imitate insiders. Confirming asset overreaction, insider buying tends to increase as a stock changes from a growth stock to a value stock (see Rozeff & Zaman [1483]).

The insider trading regulation can take different forms. In what follows, we shall concentrate on the following types of regulation:

- public disclosure of trades by insiders and trade restrictions;
- company-level regulation of trades by insiders;
- disclosure of information by firms;
- legal prosecution.

In most financial markets of developed economies, public disclosure of insider trades is mandatory. The effects of public disclosure on insider trading have been considered in Huddart et al. [989], in the context of the Kyle [1147] model, as well as in Huddart et al. [988], as we have already mentioned in Sect. 10.4. Furthermore, in some financial markets the financial regulation prohibits short sales by insiders and the opportunity of profiting from short term offsetting transactions.

Concerning the second element in the above list, company-level regulation of insider trading is widespread: typically, there is a *blackout period* during which the company prohibits trading by its insiders (e.g., around quarterly earnings announcements, mergers, takeover bids). Corporate policies are designed to minimize the costs for the company, given the constraints imposed by the financial regulation, improve the liquidity of the market and maximize the value of the firm. During a blackout period, it is shown in Bettis et al. [218] that insider trading is lower than in normal days and this fact induces a greater market liquidity. Note also that, a single company and the society as a whole may disagree on the opportunity of imposing insider trading restrictions, since they exhibit different cost-incentives (see Khanna et al. [1088]).

Public disclosure of inside information by firms precludes profits by insider trading. A manager may decide not to disclose some information in order to exploit it by trading in the market. An equilibrium model with a manager whose compensation is increasing in the stock price is presented in Narayanan [1371]. It is shown that, depending on the pay-performance sensitivity of the manager, full disclosure, nondisclosure or partial disclosure can be observed in equilibrium. Partial disclosure implies that the manager does not disclose bad news and only discloses good news. If the manager's pay-performance sensitivity is low, then the manager will not disclose bad and good news, while full disclosure occurs when the manager's sensitivity is high. Leuz & Verrecchia [1202] show that disclosure by firms reduces the bid-ask spread and increases volume.

Legal prosecution is made against trades motivated by non-public information. While legal prosecution is adopted in almost all developed countries, it is not easy to prosecute an agent for insider trading and in some countries legal prosecutions are indeed rare events. The enforcement of insider trading regulations is costly, in the sense that there is a social cost in enforcing insider trading regulations (costly investigation). DeMarzo et al. [552] model the optimal enforcement of insider trading regulations by maximizing the expected utility of uninformed traders. It turns out that a non-random threshold policy is optimal, i.e., the regulator starts an insider trading investigation if and only if the market volume exceeds some threshold and the optimal penalty is a high fixed penalty for trading above a critical value and nothing below. Therefore, from a social welfare point of view, it is optimal to allow for limited insider trading. Shin [1539] reaches a similar conclusion by minimizing the liquidity traders' expected trading costs when insider trading regulation improves the precision of the information of market professionals.

Estimating undetected insider trading before acquisition by means of the abnormal volume before the announcement, Bris [304] shows that insider trading enforcement increases both the incidence and the profitability of insider trading. However, there is a negative relationship between the insider trading profits and the toughness of the law. During the '80s, the sanctions to insider trading behavior have been increased in the U.S. market. However, according to Seyhun [1521], the increase in the sanctions did not induce a decline in insider trading activity. The cost of equity is not affected by the introduction of insider trading laws, but decreases (and the price informativeness increases) after the first prosecution (see Bhattacharya & Daouk [223, 224], Fernandes & Ferreira [682]).

Market Manipulation

Following Allen & Gale [44] and Kyle & Viswanathan [1149], we can identify three different forms of market manipulation: *trade based manipulation, action based manipulation,* and *information based manipulation.* Trade based manipulation occurs when some market participant buys and/or sells stocks over time by trading against his information or preferences, with the purpose of inducing a price movement that will allow for profitable trading opportunities in the future. Trade based manipulation can be based on privileged information or not, by simply exploiting market power (squeezes, corners). Note also that an agent can manipulate the market by exploiting the false opinions of other agents who believe that he is informed. Action based manipulation consists in actions that change the actual or the perceived value of an asset, while information based manipulation consists in releasing false information or spreading false rumors in the market.

Trade based manipulation based on market power has been analysed in Jarrow [1020], assuming the presence in the market of a large trader (i.e., a trader whose trades change market prices, with no private information). In this context, a market manipulation strategy is a strategy which generates positive wealth with no risk. Sufficient conditions under which the large trader cannot manipulate the market are established. Under general conditions, if the price process depends on the past sequence of the large trader's holdings (as opposed to only his current holdings), then market manipulation is possible.

The possibility of trade based manipulation based on private information has been analysed in several papers. Allen & Gale [44] show that a uninformed trader may manipulate the market provided that other agents in the market assign a positive probability to the manipulator being an informed trader. Allen & Gorton [47] show that manipulation can occur in a market microstructure model similar to the model of Kyle [1147] by introducing asymmetric noise traders (i.e., liquidity sales are more likely than liquidity purchases). Information dissimulation by an insider trader in the Kyle [1147] model may become manipulation (meaning that the informed trader trades against his information) if the market maker does not perfectly know that an informed agent exists in the market and does not know the nature of his information (see Chakraborty & Yilmaz [388, 389]). When agents become informed sequentially, they trade aggressively in the initial period and then reverse partially their trades. Brunnermeier [318] shows that the insider may manipulate the market if he is allowed to trade twice before and after the information is made public. John & Narayanan [1034] and Fishman & Hagerty [710] have shown that a trade disclosure rule may induce the insider to manipulate the market by trading against his information. This result is due to the inability of the market to observe the motive for trading of the insider (information based or liquidity based).

Mandatory disclosure leads to profitable trading opportunities for the (informed or uninformed) insider. Manipulation occurs when there are many liquidity traders or when the informational advantage is small. In this case, market efficiency is reduced by mandatory disclosure and mandatory disclosure can make insiders better off and outsiders worse off. The presence of competition among insiders reduces the likelihood of manipulation.

Models of action based manipulation have been developed in Vila [1623], Bagnoli & Lipman [107], Gerard, & Nanda [770]: agents trade and make profits in correspondence of a takeover bid announcement or of seasoned equity offerings by depressing the market before the offer. Models of information based manipulation have been proposed in Vila [1623] and Benabou & Laroque [189]. In this case, an agent manipulates the market by releasing false information. If an agent has a reputation of being informed about a stock and the information is noisy (i.e., it is difficult to verify), then he can successfully manipulate the market by releasing distorted information.

10.6 Market Design

A financial market is not an abstract entity: in this chapter as well as in Chap. 8, we have seen that the functioning of a market and, in particular, its efficiency (from a welfare and from an informational point of view) depend on its institutional features. In this section, we address market design issues, focusing on the following topics: transaction taxes, concentration-fragmentation, order/quote-driven market, continuous/discrete time market, market transparency, and circuit breakers. We refer to Madhavan [1275] for a survey on market microstructure and market design.

Transaction Taxes

A highly integrated financial system may go under pressure because of temporary imbalance between supply and demand. Such an order imbalance can be due to the fact that traders are highly reactive to the arrival of new information in the market (aggressive speculation), to noise trading by uninformed agents, or to the adoption of portfolio insurance strategies by institutional traders. We have already mentioned in Sect. 9.3 that speculation (a trading strategy aiming to revert asset prices towards the corresponding fundamentals) does not necessarily stabilize the market. In order to limit the impact of speculation and of noise trading and to reduce market volatility, some authors have suggested the introduction of *transaction taxes*, see Tobin [1597], Summers & Summers [1579], Stiglitz [1569]. Supporters of transaction taxes follow Keynes [1087]: "It is usually agreed that casinos should in public interest be inaccessible and expensive. And perhaps the same is true of stock exchanges". In several cases, taxes are paid on stocks held for a short period against short term speculation and noise trading. There are other motivations for the introduction of a transaction tax: (short term) speculation may come from privileged information (insider trading) and the revelation of information may also induce a cost for the society (restricted risk sharing opportunities due to the Hirshleifer effect, see Chap. 8). Stiglitz [1569] points out that financial analysts invest too many resources in financial research, thereby inducing an inefficient outcome: in this case, the introduction of a transaction tax may help to reduce this inefficiency. A tax on short term trading induces agents to extend their trading horizon and reduces the number of myopic traders. However, it has to be noted that transaction taxes increase the required rates of return and, therefore, the cost of capital. In this sense, the introduction of transaction taxes can lead to a deterioration of the liquidity and of the efficiency of the market. Moreover, it is not easy to assess the distributional effects and costs of a transaction tax. We refer to Schwert & Seguin [1512] for a survey on the benefits and the costs generated by the introduction of transaction taxes.

Subrahmanyam [1577] has addressed the effect of transaction taxes on financial markets, showing that, in the presence of several informed traders, a transaction tax leads to a decrease of market liquidity and of the profits of informed traders (informed traders trade less aggressively). However, the opposite conclusion is obtained in a market with a monopolist informed trader. A positive effect of a transaction tax is that agents invest less resources to get early acquisition of information. A transaction tax induces agents to acquire long term information rather than short term information. Recall that, as we have discussed in Chap. 8, the revelation of information by market prices leads to two competing effects: on the one hand, the revelation of information concerning the agents' risk exposure leads to a welfare loss because agents have limited risk sharing opportunities (Hirshleifer effect), on the other hand, the revelation of information on extra risk factors can generate a welfare benefit, allowing for more efficient hedging possibilities. Under some conditions, Dow & Rahi [587] show that a speculator may be better off in the presence of a transaction tax. Moreover, a tax may lead to a Pareto improvement. Indeed, a transaction tax always reduces the informativeness of market prices: in this perspective, the introduction of a transaction tax may also induce a positive welfare effect if the Hirshleifer effect is very strong.

The empirical evidence is mostly against the introduction of transaction taxes (see, e.g., Schwert & Seguin [1512]). Considering the Swedish stock market, Umlauf [1604] shows that transaction taxes do not reduce volatility, while Jones & Seguin [1044], Aliber et al. [40] show that a reduction in transaction costs has led to an increase in volume and a decrease in volatility in the NYSE and exchange rate markets. Song & Zhang [1557] have developed a theoretical model showing that a transaction tax may produce either an increase in market volatility or a decrease. Hsieh & Miller [970] and Chowdhry & Nanda [442] have shown that the introduction of margin requirements does not produce effects on volatility and, under some conditions, margin requirements may destabilize the market.

Concentration and Fragmentation

The concentration/fragmentation of exchanges in an economy with asymmetric-private information and liquidity traders is a complex issue. In general, market depth and liquidity are positively related to trading volume and to the agents' rate of participation in the market. However, this does not suffice to induce all the agents to trade in a single market. In support of market concentration, there are microstructure models (see in particular Admati & Pfleiderer [18] and Kyle [1147]) showing that liquidity traders tend to concentrate their trades in order to reduce transaction costs (high volume is associated with high liquidity and low transaction costs) and that the insider trader's intensity of trade is positively related to the liquidity of the market and, therefore, to the trading volume generated by liquidity traders. Liquidity traders create volume, which is further amplified by the demand of informed traders, as already discussed above. On the other hand, there are many examples of assets simultaneously traded in more than one market (cross listing) and of financial markets organized through more than one trading system (with different features), depending on the order size and/or on the agents.

The analysis of market concentration/fragmentation depends on the presence of informed traders in the market. Let us first consider the case of agents endowed with homogeneous information. In Pagano [1394], considering an imperfectly competitive market with hedgers having a stochastic initial asset endowment and without private information, it is shown that exchanges tend to concentrate in a single market (in the absence of differential transaction costs). Volume and a greater number of agents in the market induce high liquidity and low transaction costs and, therefore, all the agents trade in the same market. If transaction costs differ, then in equilibrium an asset can be traded in two markets. In this case, there is a separation phenomenon: traders will cluster together according to the size of their transactions. In the more expensive market there are more traders and greater diversity in the initial endowment. One of the two markets can be organized through a direct search of the trading partner. In general, concentration of trades is Pareto superior to two-market fragmentation. In Mendelson [1323], it is shown that fragmentation reduces the expected trading volume, increases price variance and decreases expected gains from trade, while it improves the quality of price signals. The existence of a two-market equilibrium has been shown in Chowdhry & Nanda [441] in the context of an economy populated by agents with short-living information and by liquidity traders, assuming that some of them are discretionary liquidity traders who are allowed to place their orders in both markets (looking for the market with lower transaction costs). In equilibrium, it is shown that discretionary liquidity traders place their orders in the market which has more non-discretionary liquidity traders, thereby creating a dominating market. In turn, the mechanism described in Admati & Pfleiderer [20] works and the dominating market will also attract more informed traders.

When two or more locations compete for listing decisions by firms, then two different effects emerge: liquidity traders prefer disclosure requirements with low trading costs, whereas corporate insiders (those who actually make the listing decision) prefer low disclosure requirements. Huddart et al. [987] show that locations engage in a race for the top with an increase in disclosure requirements and low trading costs: again, the preferences of liquidity traders drive the decisions of insiders, since the latter prefer to trade in a deep market. Under some conditions, liquidity traders flow to low disclosure exchanges. However, there are other driving forces behind the decision to list in more than one market (for instance, the desire to increase the set of potential shareholders, see Karolyi [1074]). By analysing the listings in the NYSE, Doidge et al. [580] show that companies want to increase the level of protection of shareholders.

In many financial markets, a separation is observed between a market for small and medium size orders (*downstairs market*) and a market for block trades (*upstairs market*). In Seppi [1517], a market is analysed where a large institution trades either to exploit private information or for liquidity reasons. The institution can execute a large order or can split it in a sequence of small orders. It is shown that the institution trading for liquidity reasons always trades a large order, while the institution may prefer to trade a sequence of small orders when trading for informational reasons. By trading a block, a trader signals that there is no informational content in his trade. However, in some cases a block order can also contain information (see Seppi [1518]). Order separation mitigates the adverse selection problem. A separating equilibrium is observed in many financial markets: for instance, in the NYSE there is a downstairs market for medium-small orders and an upstairs market for block trading. The downstairs market works through a continuous intraday market, while the upstairs market works through a search-brokerage mechanism where prices are determined through negotiation. The downstairs market is more efficient, while the upstairs market is deeper. Keim & Madhavan [1081] show that as a block order is sent upstairs, information leakage occurs before the trade is being executed and the temporary price impact or liquidity effect is a concave function of the order size (upstairs intermediation costs). Madhavan & Cheng [1276] confirm the result of Seppi [1517]: in the NYSE, a large part of block trades are executed in the downstairs market and only traders who can signal in a credible way that they are trading for liquidity reasons trade in the upstairs market. Brokers in the upstairs market having some information about the agent and the motivation of a trade (liquidity or information) help this separation between liquidity and informed block trading. As the upstairs market is not anonymous, an agent can develop a reputation of being an informationless trader. The price impact of a large block trade in the upstairs market is smaller than that of trades executed downstairs.

An alternative explanation for the existence of an upstairs and a downstairs market has been provided in Grossman [843]. The motivation is that agents do not participate continuously in all the markets and intermediaries are repositories of information about unexpressed demands of non-participating customers. They may know what states of nature are likely to induce agents to trade and have information about unexpressed order flow. This knowledge increases the effective liquidity of the upstairs market, making it well suited for block trades. On the other hand, customers bear search costs. Under some conditions, the downstairs and the upstairs

markets coexist in equilibrium. The empirical evidence provided in Bessembinder
& Venkataraman [217] supports the analysis of Grossman [843] and Seppi [1518].

Trading Systems

A comparative analysis of different trading systems requires a welfare analysis,
focusing in particular on the effects on uninformed-liquidity trading. As we have
already discussed in the previous sections of this chapter, some indirect measures of
welfare are represented by market liquidity, price impact, bid-ask spread, volatility
and informational efficiency.

Madhavan [1272] compares the performance of a continuous quote-driven
market (i.e., dealers set prices before order submission) with that of a continuous
order-driven market (i.e., traders submit orders before prices are set). In this
context, continuity means that an order is executed immediately upon submission.
In the model, dealers are assumed to be risk neutral and investors trade for both
informational and hedging reasons. When the adverse selection is strong, in the
sense that the private information is very precise, then an equilibrium does not
exist (market breakdown). It is shown that a dealer market is more robust than
a continuous auction market with respect to adverse selection. Indeed, for a high
degree of adverse selection, the continuous order-driven market does not admit
an equilibrium, while an equilibrium exists in a dealer market. In the dealer
market, transaction prices follow a martingale process and prices are efficient in
a semi-strong form. Moreover, prices are less variable than those corresponding
to a continuous time market. However, these properties are not satisfied in the
continuous auction market. Furthermore, allowing for the free entry of market
makers in the continuous auction market, the two price systems turn out to coincide.
The bid-ask spread is increasing in the degree of adverse selection, while it is
decreasing in the risk aversion of the agents. However, it has to be noted that a
discrete time system does not ensure immediate order execution and provides less
information to the investors. The different robustness of discrete and continuous
auction systems with respect to adverse selection is due to an argument similar to
that proposed in Glosten [787]. A discrete call auction is less risky for uninformed
traders, since the negative effects associated with trading with an insider are weaker
than in a continuous auction market with pooled orders.

A continuous auction market is an expensive trading mechanism and is typically
less deep than a discrete time call auction market. The disadvantages of a discrete
time call auction are that information diffusion is slow with high volatility and that
there is a loss of immediacy and, therefore, risk management opportunities can be
negatively affected. In this sense, there is a trade-off in organizing an auction market
in continuous time or in discrete time. In Schnitzlen [1507], the performance of a
continuous and of a discrete call auction have been evaluated through laboratory

experiments. Confirming theoretical results, a discrete call auction is shown to be more robust than a continuous auction with respect to adverse selection. Liquidity is also greater and noise traders incur in smaller transaction costs than in a continuous market. Amihud et al. [60], Amihud & Mendelson [59] have empirically shown that shifting from a daily call auction to a repeated continuous trading session improves the liquidity and the efficiency of the market.

Market Transparency and Disclosure

Financial markets can exhibit different degrees of transparency, depending on the extent to which market participants (i.e., market makers, brokers, dealers and traders) know the order flows and the quotes of the other agents. Even though market transparency represents a highly debated topic, in recent years there has been a general trend towards a greater transparency in most financial markets. Transparency helps uninformed traders to learn private information and, therefore, helps to reduce adverse selection problems. However, in a transparent market informed investors trade less aggressively and, therefore, this leads to a reduction of the information disseminated in the market by insider traders. Moreover, in a transparent market agents have a smaller incentive to buy information and the Hirshleifer effect may produce negative welfare effects. In this sense, a greater transparency can reduce the risk sharing opportunities. This effect is particularly strong for dealers who have less time to manage their inventory position. In a transparent market, transaction costs for a dealer who has to execute a large uninformative trade or for an informed trader can be high. In general, a dealer market is more (pre-trade) quote-transparent than an auction market and a continuous auction market is more order-transparent than a dealer market. Market transparency has many dimensions and we can distinguish between *order (imbalance) transparency, trader anonymity, pre- and post-trade quote disclosure*.

Order transparency has been investigated in Pagano & Roell [1395], where the authors propose the following ranking in terms of order transparency: (i) transparent auction (a batch auction where agents know all the individual orders); (ii) continuous time auction (where agents observe all the individual past orders); (iii) dealer market (where each dealer does not observe the orders of other dealers). Two specifications of a dealer market are given, under the assumption that dealers are risk neutral: immediate last trade publication and no trade publication. The latter specification represents the less transparent setting, while the former has a transparency similar to that of a continuous time auction. Note that a batch auction cannot be ranked with respect to the above market specifications: indeed, in a batch auction market the dealer sets the price observing the total order flow and not the individual orders. In a model with an insider trader, Pagano & Roell [1395] show that the expected transaction costs for uninformed traders are lower in a transparent auction than in a dealer market. The costs in the batch and in the continuous auction cannot be compared with the others. These results are obtained by assuming that the insider trader adopts the same strategy in all three types of market. Under suitable conditions, this ranking in transaction costs also holds when the insider trader is allowed to modify his trading strategy depending on the market. In general, transparency is associated with lower transaction costs.

Anonymity of trading is strictly related to market transparency. In Forster & George [727] it is shown that, when market makers have some information about the trading motives of the investors, execution costs for liquidity traders and gains for informed traders are lower if there is enough competition among informed agents. If brokers have information about the order flow motivation (informed/ uninformed), then the bid-ask spread decreases (see Benveniste et al. [194]). Specialist markets are less anonymous than dealer markets (as the NASDAQ), in particular when orders are channelled through brokers as in the NYSE (see Benveniste et al. [194] and Garfinkel & Nimalendran [758]).

Madhavan [1274] analyses information disclosure about the order flow and the order imbalance. In particular, the author analyses the effect of publicizing the order imbalance due to price inelastic demands (liquidity trades) prior to trading. This demand can be interpreted as the order submitted by a large uninformed trader and transparency can be interpreted as *sunshine trading*: the large trader announces his trade declaring that the trade is not based on an informational motive. In the model, it is assumed that agents are endowed with symmetric private information. Disclosure of an order imbalance due to liquidity trades (or uninformative trades) should have a stabilizing effect. It is shown that in a transparent market prices are more informative, volatility is reduced and market liquidity is increased, provided that the market is sufficiently large and liquid. However, if this is not the case, then market transparency may increase the volatility generated by transitory order imbalances and may lead to less liquidity and higher transaction costs. Moreover, if the agents' information is very precise, in a transparent market there is no equilibrium, even though an equilibrium would exist in an opaque market. The rationale of this result is that market transparency may lead to a market failure: agents endowed with very precise information prefer not to trade in a transparent market. Madhavan et al. [1277] have empirically investigated the pre-trade transparency about investors' latent demands present in the limit order book by relying on data from the Toronto stock exchange, reporting that greater transparency (i.e., public disclosure of the limit order book) is associated with larger execution costs and volatility. A welfare analysis of sunshine trading has been carried out in Admati & Pfleiderer [23]. In the model, some liquidity traders can pre-announce the size of their orders, declaring also their nature (uninformative). It is shown that such agents have lower trading costs, but the costs to informed and liquidity traders who cannot pre-announce their trades increase. Sunshine trading increases the informational content of prices, while the effect on volatility is not clear a priori.

In Biais [231], in the context of a model without asymmetric-private information, it is shown that information about quotes, trades and inventories of other intermediaries may affect the bid-ask spread. In the model, the number of (risk averse) dealers-market makers is endogenous. In a centralized (auction) market, there are market makers and limit order traders, who can monitor the positions-quotes of their competitors. In a fragmented market, there are dealers who cannot monitor the positions of other intermediaries. The expected bid-ask spread in a centralized market is equal to the expected bid-ask spread in a decentralized market where dealers do not know positions. However, the spread is more volatile in centralized than in fragmented markets. Note that fragmentation (associated with low transparency) provides the dealer with a monopolistic surplus (in the sense of wide bid-ask spreads), which is decreasing in the number of traders. Assuming private-asymmetric information, risk averse dealers may prefer a non-transparent inter-dealer market. In Lyons [1261], it is observed that the Hirshleifer effect works against dealers in a transparent market: greater order flow transparency accelerates the diffusion of information by prices, leaving dealers less time to manage their inventory position. In this context, market transparency can limit the risk sharing possibilities. The optimal level of transparency for dealers is at an intermediate level, while both high and low levels of transparency restrict risk sharing. In this sense, dealers prefer imperfect market transparency (i.e., the market order flow can be observed up to a noise component).

Madhavan [1273] has analysed the post-trade disclosure of prices by dealers in the context of an economy populated by liquidity and informed traders. Large liquidity and informed traders prefer a market with no price disclosure because in that case they can more easily conceal their trades, thus reducing the execution costs. Non-transparency allows dealers to set wide spreads since there is less price competition. Moreover, non-transparency generates an informational advantage and, therefore, large gains for the dealer. Because of these effects, dealers, informed and large traders will not disclose their trades-prices unless disclosure is mandatory. Porter & Weaver [1428] show that dealers in the NASDAQ delay large trades disclosure and the delay is associated with the informational content of the trade. In this setting, transparency reduces volatility and increases price efficiency. In a model with asymmetric information, the welfare effects of post-trade price disclosure for a risk averse investor are ambiguous (see Naik et al. [1368]). Disclosure of trades reduces the adverse selection effects but increases the price revision risk for the dealer. The liquidity effects of price disclosure in a dealer market have been empirically investigated in Gemmill [766] on the basis of London Stock Exchange data. It is shown that transparency (as measured by the delay in the publication of trades) does not affect execution costs, liquidity, price adjustment and the diffusion of information.

In Bloomfield & O'Hara [252, 253] and Flood et al. [717], market transparency has been evaluated by means of laboratory experiments. In Bloomfield & O'Hara [252], it is shown that the disclosure of trades and quotes increases the informational efficiency of transaction prices, but also increases the bid-ask spread by reducing competition among dealers. As a result, market makers benefit from trade disclosure at the expense of liquidity and informed traders who cannot time their trades. Trade disclosure can have important effects on the market behavior, whereas quote disclosure does not. Flood et al. [717] investigate quote disclosure (pre-trade transparency) in a quote-driven market where trade information is never revealed. Quote transparency reduces opening spreads and increases volume, but price discovery is much faster in a opaque market and prices are also more efficient. This is due to the low competition among dealers in a transparent market, leading to a trade-off between liquidity and price efficiency. Market transparency affects competition between different trading systems. Note that liquidity traders prefer transparent markets, but informed traders prefer opaque markets. In Bloomfield & O'Hara [253], it is shown that when there are two markets, the less transparent one tends to become the dominant location. Less transparent dealers are more aggressive in the early rounds of trading, attract order flow and informative trades and use the resulting informational advantage to quote narrower spreads and attract liquidity trades. Bloomfield & O'Hara [253] support these findings with laboratory experiments. It is also shown that most dealers prefer low transparency. However, the existence of a more transparent market is ensured by the fact that the informational advantage of non-transparent dealers decreases with the number of dealers.

Circuit Breakers, Trading Halts and Price Continuity

After the 1987 market breakdown, the Presidential Task Force on Market Mechanism was nominated to investigate the crash. Among other policy measures, the committee recommended the adoption of *circuit breakers* in order to close the market when it does not work "properly": "circuit breaker mechanisms (such as price limits and coordinated *trading halts*) should be formulated and implemented to protect the market system" (Presidential Task Force on Market Mechanism [1434]). The possibility of adopting circuit breakers is a highly debated issue from a theoretical and an empirical point of view. Indeed, the effects of circuit breakers are typically ambiguous. A trading halt is called by specialists when a unusual order imbalance that cannot be matched occurs or by market officials when news with a strong price impact are announced or expected. During a market closure, the specialist engages in price exploration, receives commitments to trade and then reopens the market through a call auction matching the commitments to trade. In some financial markets, the stock exchange authority can call for a trading halt and then reopen the market when it is less under pressure. Bhattacharya & Spiegel [228] show that suspension occurs at the NYSE when a firm announces impending news or when there is a severe order imbalance.

A circuit breaker is needed because of a malfunctioning of the market or in the presence of strong adverse selection/poor information transmission by prices. In Chap. 8 as well as in the present chapter, we have seen that uninformed traders can refuse to trade when they recognize that there is a strong adverse selection effect in the market. According to this interpretation, a circuit breaker is designed to remove a (technical) market malfunctioning and/or to improve the diffusion of information and the price efficiency in the market by weakening adverse selection effects and execution risk. As a consequence, a circuit breaker should reduce volatility and volume and prevent a chaotic market behavior. To this end, the characteristics of the market opening system are crucial. Note also that a circuit breaker should not affect the long run tendency of the market.

Theoretical arguments are mostly against the adoption of circuit breakers. In a non-fully revealing rational expectations model with private-asymmetric information and noise supply, if agents have the possibility of observing intertemporally trading volume and price, then agents are able to detect private information and prices reveal more information (see Blume et al. [257], Brown & Jennings [310], Grundy & Martin [851]). In this setting, circuit breakers would have a negative effect. The effects of circuit breakers have also been discussed in studies explaining intraday patterns of asset prices, showing that high volatility, volume and spreads typically occur before a natural market closure and also that volatility is high at the opening of the market (see Brock & Kleidon [307], Gerety & Mulherin [771, 772], Stoll & Whaley [1571] and see also Sect. 10.4). When expecting a market closure, similar phenomena should be observed. In Subrahmanyam [1576], it is shown that agents sub-optimally make trades in advance when they expect a trading halt and, therefore, a circuit breaker may actually increase the trading volume, liquidity, price variability and the probability of the price crossing the circuit breaker bound and decrease price efficiency. Moreover, in a two-market setting, if a circuit breaker is adopted in the liquid dominant market, both price variability and liquidity and volume migrate to the satellite market.

In support of circuit breakers, Greenwald & Stein [824] show that when there is a large (liquidity or information based) trade and uncertainty on the number of traders aiming to compete for it, traders face a large transactional risk in an auction (i.e., agents do not know the transaction price and fear cascade dynamics). The uncertainty on the number of traders may be due to adverse selection in the market or to price uncertainty/volatility. In general, there is a high transactional risk when prices are uninformative. This transactional risk restrains agents from trading and in equilibrium the price will drop substantially. In these situations, circuit breakers will help a trader by reducing the uncertainty about the traders' behavior, transactional risk and adverse selection. Stein [1567] proposes a formal model showing that, during periods of price uncertainty, continuous trading can decrease the informational content of prices, while informative halts can improve market quality. However, note that a market closure preventing trades always induces hedging-liquidity costs, thus exacerbating liquidity problems. Instead of closing the market, it may be a good solution to switch to a trading system that is more robust with respect to adverse selection. In this sense, Madhavan [1272] suggests to switch to a periodic auction market. Edelen & Gervais [627] show that trading halts serve as a discipline device for market makers. Indeed, in some cases market makers may have an incentive to quote a privately optimal pricing schedule (e.g., not executing orders immediately) and this type of behavior has a negative externality effect for the market as a whole in terms of loss of order flow in the future when the market enjoys a reputation for execution quality. A trading halt called by the specialist or by the exchange commission can act as a coordination device preventing this type of behavior and increasing exchange profits.

The empirical evidence on the adoption of circuit breakers is ambiguous. While some positive effects of circuit breakers have been detected in Lauterbach & Ben-Zion [1169] (low opening order imbalance), in Lee et al. [1173], Edelen & Gervais [627], Bhattacharya & Spiegel [228] it is observed that NYSE firm-specific trading halts increase both post-halt volume and volatility. A possible explanation of these effects is that the reopening mechanism is inefficient and limits the diffusion of information. Corwin & Lipson [500] confirm the results of Lee et al. [1173] and show that NYSE trading halts reduce order book depth around the event, increase the spread but, on the other hand, allow traders to effectively reposition their trades. The experimental results of Ackert et al. [12] confirm that an imminent trading halt increases the pre-halt trading volume.

Price continuity requirements are designed to stabilize the market. According to these requirements, upper limits are placed on absolute price changes during trading hours. The upper limit depends on past price evolution and/or on the volume or liquidity of the market. Dealers/market makers set prices in order to satisfy these requirements. The effect of a bound on the absolute price change given by a multiple of past order flow has been analysed in Dutta & Madhavan [603]. It is shown that, under stringent continuity requirements, informed traders will trade more aggressively and, as a consequence, the price efficiency is unaffected or even increased if more agents buy private information. This type of rule creates a trend in asset prices. There is a loss for dealers/market makers and a gain for liquidity traders. Moreover, price limits might increase market stability (see Chowdhry & Nanda [442] as well as Sect. 9.5). Kim [1094] empirically observes that volatility does not usually decrease when price limits are made more restrictive.

10.7   Notes and Further Readings

We refer the reader to Madhavan [1275], O'Hara [1384], Biais et al. [236], De Jong & Rindi [540], Foucault et al. [735] for detailed surveys on financial markets microstructure .

In the Kyle [1147] model discussed in Sect. 10.2, if the linearity condition or the normality assumption are not imposed, then multiple equilibria arise. However, regardless of the normality assumption, the uniqueness of the equilibrium holds if the insider trader can observe noise trades (see Rochet & Vila [1455]). The model has been extended to a multi-security setting in Caballé & Krishnam [324]. Considering a generalized version of the model of Kyle [1147], Vives [1629] shows that the presence of the market maker contributes to the convergence towards the fundamental value: the rate of convergence is $1/\sqrt{n}$, where n is the number of rounds in the adjustment process. On the other hand, in the absence of a market marker, the rate of convergence is much slower $(1/\sqrt{n^{1/3}})$. A continuous time extension of the Kyle [1147] model has been developed in Back [97] and Back

& Baruch [99]. In the context of equilibrium models under imperfect competition, it is also worth to mention that non-fully revealing equilibrium prices have been obtained in Laffont & Maskin [1155]. In Biais & Hillion [237], in the presence of imperfect competition, it is shown that adverse selection and market breakdown can be avoided by introducing derivative securities.

It is interesting to consider the possibility that information can be bought and sold in the financial markets. In this direction, Admati & Pfleiderer [19] investigate the direct and indirect sale of information in the context of the Kyle (1985) model: an insider trader can sell information to other agents or exploit it by trading the asset. The private information consists in the observation of a signal about the fundamental value of the asset. In the market, there are N potential buyers and noise traders. The insider trader sells his information without manipulating its content. Agents who do not buy the information quit the market because they know that the price being determined according to the market efficiency condition (10.11) will be unfair. Therefore, the number of agents in the market, excluding noise traders, is equal to the number of those who decide to buy information. Agents have a constant coefficient of absolute risk aversion. If the information owner is risk neutral, then he does not want to sell his information to any other trader. On the other hand, if the information owner is risk averse, then he can have an incentive to sell his private information. Competition with other insider traders reduces the profits associated with speculation but creates risk sharing opportunities (with a positive effect for the insider's welfare). There is a trade-off between competition among insider traders and risk sharing possibilities. For plausible values of the parameters of the model, the profit of informed agents is a unimodal function of the number of agents acquiring information and, therefore, there is a unique optimal number of insider traders. Typically, if the risk aversion of the information owner is low, then he does not sell the information, preferring to fully exploit it by trading in the market. Furthermore, if potential buyers are weakly risk averse then the information owner sells the information and commits not to trade. For intermediate degrees of risk aversion, the information owner trades in the market and at the same time sells the information to other agents. Less precise information is sold to more traders and leads to lower profits. The indirect sale of information through a mutual fund is also analysed. By setting appropriate fees (a fixed fee and a per share fee), the information owner can control the effects of competition among informed traders and can increase his profits.

The strategy of a market maker endowed with private information has been analysed in Gould & Verrecchia [814]. The market maker sets a price and then a trader chooses the quantity he wants to trade. Both the market maker and the trader in the market can be privately informed. If the market maker is privately informed, then he will introduce a noise component in the price with respect to the one that maximizes his expected profits. In Madrigal & Scheinkman [1282], the model has been extended to an economy populated by noise traders and risk averse insider traders endowed with private and heterogeneous information. In this setting, the (bid and ask) prices quoted by the market maker aggregate information with some noise. The risk neutral market maker sets prices to maximize expected profits conditionally on the observation of market orders. Prices always reveal information, but never fully. The equilibrium price is a discontinuous function of market orders, with the consequence that a slight change in the parameters of the model can lead to a crash of the market.

Limit order book models and an analysis of the trade-off between market and limit orders have been developed in Foucault [733], Foucault et al. [734], Hasbrouck & Schwartz [914], Parlour [1402], Goettler et al. [794], Chakravarty & C. Holden [391], Rosu [1471], Back & Baruch [100]. For empirical studies on the functioning of limit order books we refer to Harris & Hasbrouck [900], Biais et al. [239], Harris & Venkatesh [902], Hollifield et al. [955, 955] (see also Bloomfield et al. [254] for an experimental analysis). Comparing the NYSE with the NASDAQ, Chung et al. [450, 451] confirm that limit orders decrease execution costs with respect to a pure quote-driven market. Insufficient depth in the limit order book relative to theoretical predictions is reported in Sandas [1494]. In some markets, a specialist or a dealer faces competition from limit order traders and competition contributes to lower the bid-ask spread and to increase price efficiency (see Seppi [1519], Brown & Zhang [312], Baruch [174], Hendershott & Mendelson [933]). In Wahal [1640], the author provides empirical evidence in favor of a reduction of the bid-ask spread as competition among market makers increases. A model on dealers' competition with payments for order flow and preferencing arrangements has been developed in Kandel & Marx [1065].

In Chan [400], the author suggests that financial markets microstructure can generate cross-autocorrelation among stock returns, explaining for example the coexistence of weak serial correlation of asset returns and positive serial correlation of index returns and the lead-lag effect. Let us consider a monopolist specialist market with many assets, where the value of a stock is given by the sum of a market-wide component and a stock-specific component. A market maker observes a noisy signal about the value of his stock without instantaneously observing signals about the value of other stocks. Each signal contains market-wide information and uncorrelated noise. In a multi-period setting, if the market maker revises his quotes by observing previous price changes, then stock returns will be serially uncorrelated and positively cross-autocorrelated. It is shown that the lead-lag effect between large and small firms can be explained by this argument. Foster & Viswanathan [730] show that if the relevant random variables have an elliptically contoured distribution, then volume and price variance are correlated and volume is serially correlated. Chordia & Subrahmanyan [438] analyse and provide empirical evidence on the relation between order imbalance and returns, showing that the price pressure due to autocorrelated imbalances generates a positive relation between lagged imbalances and returns.

In Chap. 8, we have remarked that in a model with heterogeneous beliefs or opinions, trading volume is positively related to the degree of heterogeneity. However, this is not necessarily true in a dealer market model with transaction costs when liquidity trading is elastic with respect to transaction costs (see George et al. [769]). Trading volume can increase or decrease with respect to the informational asymmetry and the dispersion of beliefs: volume increases with respect to the informational asymmetry (or diversity in beliefs) if and only if liquidity trading decreases in transaction costs at an increasing rate.

Brokers and, in general, financial intermediaries can be allowed or forbidden to trade for their own account. The practice of trading both for customers and for their own account is known as *dual capacity*. The effects of dual capacity have been analysed in Fishman & Longstaff [712] and Roell [1456]. On the one hand, the dual capacity can introduce a conflict of interest between brokers and their customers. On the other hand, the dual capacity can reduce the costs of trading. The effects of the dual capacity on market depth, liquidity and welfare are ambiguous in a model with heterogeneous agents (see Locke et al. [1241]). In a model with asymmetric-private information, the observation by the broker of the customers' order flow can have an informational content.

Insider trading may either increase or decrease the value of a firm, see Manne [1299] and Manove [1300] for results in favor of the two different conclusions. Depending on the trade-off between diffusion of information and adverse selection and its consequences on firm investment, positive or negative welfare consequences of insider trading have been obtained in Bernhardt et al. [205]. Repullo [1446] shows that the positive effects of insider trading detected in Leland [1181] are not robust to the introduction of a noise component in the insider's information (see also Medrano & Vives [1317]). As pointed out in Sect. 10.5, allowing for insider trading can represent a way to compensate managers, since they can use their private information to earn profits by trading in the market. However, by allowing for this possibility, firm managers can operate in a detrimental way for other shareholders and this effect would in turn discourage corporate investment, thus decreasing the efficiency of the firm. Moreover, the possibility of insider trading affects the choice among investment projects, since insider traders tend to choose riskier investment projects in order to take advantage of the greater volatility (see Bebchuk & Fershtman [184]). In Fischer [704], it is shown that insider trading does not affect the firm if there are no other agency-related or asymmetric information problems. Otherwise, prohibiting or imposing delayed registration of insider trades may be valuable.

According to asymmetric information models of corporate finance, managers should issue common stocks or convertible debts when stock prices are overvalued (i.e., managers have a negative private information). This hypothesis is confirmed by insider trades analysis, showing that insider sales increase and purchases decrease prior to issues of convertible debt and equity and firms with abnormal insider selling underperform in the long run (see Kahle [1057]). Noe [1378] shows that managers exploit market reaction to management earnings forecasts: insider sales (purchases) increase after a positive (negative) price reaction. Insiders trade after voluntary disclosures (when the information asymmetry with outsiders is low) and not before. These trades earn abnormal returns and are not correlated with management forecast errors. In a signalling model, insider trades complement dividends: dividends convey positive information (positive price response) when they are joined by insider buying and negative information when they are joined by insider selling (see John & Lang [1035] for a theoretical model and empirical evidence). Bagnoli

& Khanna [106] show that insider trades may preclude signaling through observable actions (e.g., dividend announcements). A relation between information asymmetry and insider trades is also detected in Huddart & Ke [990], Ke et al. [1077], Huddart et al. [991]: insiders possess and trade on the basis of specific and significant information about the firm (e.g., forthcoming accounting disclosures). The evidence is consistent with the signaling role of insider trades.

A particular type of manipulation can arise by simultaneously trading in derivatives and in the underlying assets. Kumar & Seppi [1142] show that a uninformed trader may manipulate the market through a sequence of trades in the spot and in the future markets against informed agents (see also Kyle [1146] on future markets manipulation). Assuming the presence of a large trader in a model similar to that proposed by Jarrow [1020] with traders having symmetric information and complete markets, Jarrow [1021] shows that allowing for trading in a derivative security allows for manipulation even in a model which otherwise would not admit manipulation without the derivative security. Manipulation is avoided by excluding corners through quantity limits and by facilitating information diffusion concerning trades in two markets. It is generally shown that manipulation increases volatility, liquidity and returns (we refer to Merrick et al. [1326], Allen et al. [48], Aggarwal & Wu [27] for an empirical analysis of manipulation by trading on derivatives).

The shareholders' monitoring activity is affected by market liquidity, since shareholders can trade in the stock market as an informed agent after having monitored the company, see Maug [1312]. If monitoring is costly, a liquid market motivates a large shareholder to monitor the firm and trade on private information. A liquid market mitigates free riding by small shareholders on large shareholders. In some cases, it is optimal for the large shareholder to not exert pressure on the management of the firm and to exploit (negative) information on the firm to trade in the stock market (see Kahn & Winton [1058] and Faure-Grimaud & Gromb [681]).

Concerning the trading mechanisms of real financial markets, the minimum tick size has been reduced over the years in many markets. On the one hand, a reduced tick size should reduce the bid-ask spread through dealers' competition and, therefore, benefit liquidity traders. On the other hand, a small tick size could be detrimental to liquidity providers by decreasing their profits (see Grossman & Miller [846] and Chordia & Subrahmanyan [437]). The tick size reduction from eighths to sixteenths in U.S. financial markets has been analysed in several papers, showing that spreads and depths declined, liquidity traders trading small quantities are better off but traders placing large orders are damaged (see Jones & Lipson [1043], Goldstein & Kavajecz [798], Van Ness et al. [1606]). A large tick size on the NYSE generates the payment for order flow practice and precludes competition (see Chordia & Subrahmanyan [437]).

Christie & Schultz [446], Christie et al. [445], Barclay [157], Bessembinder [212], Kandel & Marx [1064], Barclay et al. [159] have remarked the absence of odd-eighth quotes for 70 of the 100 most capitalized NASDAQ stocks (the phenomenon is not observed in the NYSE) and that spreads are quite large (at least 0.25$). There is a positive relation between spread and price rounding (see Harris [899], Barclay [157], Bessembinder [212], Kandel & Marx [1064], Simaan et al. [1544], Bessembinder [214]) and internalization-preferencing of the order flow (see Huang & Stoll [977], Godek [793], Chung et al. [449]). Some authors suggest that dealers tacitly collude to set wide spreads. Bloomfield & O'Hara [251] show through experiments that order preferencing can increase bid-ask spreads and also reduce informational efficiency. Information ambiguity also affects the bid-ask spread: Routledge & Zin [1475] show that information ambiguity (or model uncertainty, see Sect. 9.1) increases the bid-ask spread set by an uncertainty averse monopolist market maker and reduces the market liquidity.

In Huang & Stoll [977] and Bessembinder & Kaufman [216], it is shown that execution costs on the NASDAQ (before the 1997 reform) were higher than on the NYSE. The difference can be attributed to higher order processing costs (see Affleck-Graves et al. [26]), avoidance of odd-eighth quotes and lack of competition in the NASDAQ market (collusion) (see Christie et al. [445] and Barclay [157]) or order preferencing arrangements (see Huang & Stoll [977]). The reform of the NASDAQ market and the reduction of the tick size have decreased the (still positive) difference between execution costs on the NASDAQ and on the NYSE (see Bessembinder [213], Barclay et al. [158], Weston [1658], Bessembinder [214]). Based on an analysis of dual listed stocks traded both in Paris (automated market) and in London (dealer market), De Jong et al. [539] show that effective spreads are lower in Paris than in London. In Kim et al. [1098], analysing how NYSE and NASDAQ prices incorporate new information (analyst recommendations), it is shown that a call auction market incorporates the new information before the dealer market does. However, information asymmetry can create a loss of liquidity in a call auction market (i.e., time is needed to find a price) which is instead not observed in a dealer market. Lamoureux & Schnitzlen [1164] compare through an experimental investigation the performance of a dealer market when (informed and liquidity) traders can decide to opt for bilateral search. It is shown that this opportunity induces dealers to trade more aggressively and that their profits decrease substantially, while on the contrary price discovery increases.

In Barclay et al. [158] and Chung & Van Ness [452], the 1997 NASDAQ reform has been analysed. The reform was designed to offer investors more competitive quotes through the mandatory display of limit orders when they are better than quotes posted by market makers and the dissemination of superior prices placed in proprietary trading systems. In this way, the public can compete directly through limit orders with dealers. This reform has reduced fragmentation, increased market transparency and introduced some auction features in the market. It is shown that spreads have tightened by approximately 30%. Chung & Van Ness [452] show that the decline is particularly large during midday and this can be attributed to the intraday variation in competition among limit order traders. According to Weston [1658], the reform has also increased competition and reduced the dealers' profits.

10.8 Exercises

Exercise 10.1 Consider the model proposed by Kyle [1148] and described in Sect. 10.1. By making use of the multivariate normality assumption on the random variables $(\tilde{d}, \tilde{y}_1, \ldots, \tilde{y}_N, \tilde{z})$, prove Proposition 10.1 (compare also with Kyle [1148, Appendix A]).

Exercise 10.2 Consider the model proposed by Kyle [1147] and described in Sect. 10.2.

(i) Suppose that the functions X and P are linear:

$$X(\tilde{d}) = \alpha + \beta \tilde{d} \qquad \text{and} \qquad P(\tilde{x} + \tilde{z}) = \mu + \lambda(\tilde{x} + \tilde{z}),$$

for some constants α, β, μ and λ to be determined. Show that the profit maximization constraint (10.10) implies that

$$\beta = \frac{1}{2\lambda} \qquad \text{and} \qquad \alpha = -\mu\beta.$$

(ii) By relying on the result obtained in the first step, prove that the market efficiency condition (10.11) implies that

$$\lambda = \frac{\beta \sigma_d^2}{\beta^2 \sigma_d^2 + \sigma_z^2} \qquad \text{and} \qquad \mu - \bar{d} = -\lambda(\alpha + \beta \bar{d}).$$

Deduce that Proposition 10.3 holds.

Exercise 10.3 In the context of the model introduced in Kyle [1147] and discussed in Sect. 10.2, let the couple (X, P) represent the linear equilibrium, as stated in Proposition 10.3. Prove the following claims:

(i) Letting $\tilde{p}$ denote the equilibrium price, it holds that

$$\mathrm{Var}(\tilde{d}|\tilde{p}) = \frac{\sigma_d^2}{2};$$

(ii) The optimal profits of the insider trader (i.e., the profits associated to his optimal demand $X(d)$), conditionally on the observation of the private signal $\tilde{d} = d$, are given by

$$\frac{(d - \bar{d})^2}{4\lambda}$$

Ex-ante (i.e., before the observation of the realization of $\tilde{d}$), the expected profits of the insider trader are equal to $\sigma_d^2/(4\lambda)$.

Exercise 10.4 Suppose that all the assumptions of the model of Kyle [1147] as presented in Sect. 10.2 are satisfied, apart from the fact that informed agents do not observe perfectly the liquidation value $\tilde{d}$ of the asset, but rather an imprecise signal $\tilde{y}$ of the form

$$\tilde{y} = \tilde{d} + \tilde{\varepsilon},$$

where $\tilde{\varepsilon}$ is a normally distributed random variable independent of all the other random variables appearing in the model, with mean zero and variance $\mathrm{Var}(\tilde{\varepsilon}) = \sigma_\varepsilon^2$. Show that, under this assumption, the economy admits a unique linear equilibrium (X, P), where $X : \mathbb{R} \to \mathbb{R}$ and $P : \mathbb{R} \to \mathbb{R}$ are two measurable functions such that

$$X(\tilde{y}) = \alpha + \beta(\tilde{y} - \bar{d}) \qquad \text{and} \qquad P(\tilde{x} + \tilde{z}) = \mu + \lambda(\tilde{x} + \tilde{z}),$$

where the constants β and λ are explicitly given by

$$\beta = \sqrt{\frac{\sigma_z^2}{\sigma_d^2 + \sigma_\varepsilon^2}} \qquad \text{and} \qquad \lambda = \frac{1}{2}\sqrt{\frac{\sigma_d^4}{(\sigma_d^2 + \sigma_\varepsilon^2)\sigma_z^2}}$$

and where the constants α and μ are given by

$$\mu = \bar{d}\,\frac{\frac{\beta^2\sigma_\varepsilon^2 + \sigma_z^2}{\beta^2(\sigma_d^2 + \sigma_\varepsilon^2) + \sigma_z^2}}{1 - \frac{\beta\sigma_d^2}{\beta^2(\sigma_d^2 + \sigma_\varepsilon^2) + \sigma_z^2}\frac{1}{2\lambda}} \qquad \text{and} \qquad \alpha = -\frac{\mu}{2\lambda}.$$

In particular, recalling that the market depth is measured by the quantity $1/\lambda$, this implies that market depth is increasing with respect to the variance of the error term appearing in the informed agent's observation (or, equivalently, decreasing with respect to the precision of the informed agent's signal). Similarly, the intensity of trading of the informed agent, as measured by the coefficient β, is increasing with respect to the precision of his signal, as measured by the quantity $1/\sigma_\varepsilon^2$.

Exercise 10.5 Consider the model developed in Subrahmanyam [1575] and presented in Sect. 10.2. In this exercise, we are going to prove Proposition 10.4.

(i) Start from the conjecture that the risk neutral market maker adopts a linear pricing rule, i.e., letting $N\tilde{x} + \tilde{z}$ denoting the total market demand, the market maker sets a price $\tilde{p} = P(N\tilde{x} + \tilde{z})$ given by

$$P(N\tilde{x} + \tilde{z}) = \bar{d} + \lambda(N\tilde{x} + \tilde{z}),$$

for some constant λ to be determined. Consider the expected utility maximization problem of an informed trader, conditionally on the observation of the private signal $\tilde{y} = \tilde{\varepsilon} + \tilde{\eta}$. Assume that each informed trader conjectures that the other informed traders have an asset demand of the form $\tilde{x} = \beta(\tilde{\varepsilon} + \tilde{\eta})$, for some constant β to be determined. Denoting by $\tilde{x}$ the optimal demand of an arbitrary informed trader, show that

$$
\tilde{x}^* = \frac{\tilde{\varepsilon} + \tilde{\eta}}{(1 + \sigma_\eta^2)(2\lambda + a\lambda^2\sigma_z^2) + a\sigma_\eta^2} - \frac{(N-1)\beta\lambda(1 + \sigma_\eta^2)(\tilde{\varepsilon} + \tilde{\eta})}{(1 + \sigma_\eta^2)(2\lambda + a\lambda^2\sigma_z^2) + a\sigma_\eta^2}.
$$
(10.31)

(ii) By relying on (10.31), show that, for a given value of the constant λ, the parameter β corresponding to a linear equilibrium of the economy is given by

$$
\beta = \frac{1}{(1 + \sigma_\eta^2)\lambda(N+1) + a(\sigma_\eta^2 + \lambda^2\sigma_z^2(1 + \sigma_\eta^2))}.
$$

(iii) By relying on the equation satisfied in equilibrium by the coefficient β defining the demand of the informed traders, prove that the coefficient λ is given by the solution to the quintic equation

$$
\lambda\left(N^2(1 + \sigma_\eta^2) + \frac{1}{\beta^2}\sigma_z^2\right) = \frac{N}{\beta},
$$

thus completing the proof of Proposition 10.4.

Exercise 10.6 Consider the model proposed in Subrahmanyam [1575] and presented in Sect. 10.2. In the setting of Proposition 10.4, suppose that all the N informed traders are risk neutral, in the sense that $a \to 0$. Show that, in this case, the equilibrium value of λ is given by

$$
\lambda = \frac{1}{N+1}\sqrt{\frac{N}{1 + \sigma_\eta^2}}.
$$

Exercise 10.7 Consider the model developed in Spiegel & Subrahmanyam [1558] and presented in Sect. 10.2. In this exercise, following Spiegel & Subrahmanyam [1558, Appendix A], we are going to prove Proposition 10.5.

(i) Assume that each informed trader $k \in \{1, \ldots, K\}$ conjectures that

- every uninformed agent $n \in \{1, \ldots, N\}$ demands a quantity $\tilde{w}_n = \gamma\tilde{e}_n$ of the security, for some constant γ to be determined;
- every other informed trader $l \in \{1, \ldots, K\} \setminus \{k\}$ demands a quantity $\tilde{x}_l = \beta(\tilde{\varepsilon} + \tilde{\eta}_l)$ of the security, for some constant β to be determined;

- the market maker sets a price $\tilde{p}$ according to the rule

$$\tilde{p} = \bar{d} + \lambda \left(\sum_{k=1}^{K} \tilde{x}_k + \sum_{n=1}^{N} \tilde{w}_n \right),$$

for some constant λ to be determined.

Show that the optimal quantity $\tilde{x}_k^*$ of the informed agent $k \in \{1, \ldots, K\}$, conditionally on the observation of the private signal $\tilde{\varepsilon} + \tilde{\eta}_k$, is given by

$$\tilde{x}_k^* = \sigma_\varepsilon^2 \frac{1 - \lambda(K-1)\beta}{2\lambda(\sigma_\varepsilon^2 + \sigma_\eta^2)} (\tilde{\varepsilon} + \tilde{\eta}_k).$$

Deduce that, in equilibrium, the constant β must satisfy

$$\beta = \frac{\sigma_\varepsilon^2}{\lambda((1+K)\sigma_\varepsilon^2 + 2\sigma_\eta^2)}. \tag{10.32}$$

(ii) Consider then the expected utility maximization problem of each uninformed trader. Suppose that each uninformed trader $n \in \{1, \ldots, N\}$ conjectures that

- every informed trader $k \in \{1, \ldots, K\}$ demands a quantity $\tilde{x}_k = \beta(\tilde{\varepsilon} + \tilde{\eta}_k)$ of the security;
- each other uninformed agent $m \in \{1, \ldots, N\} \setminus \{n\}$ demands a quantity $\tilde{w}_m = \gamma \tilde{e}_m$ of the security, for some constant γ to be determined;
- the market maker sets a price $\tilde{p}$ according to the rule

$$\tilde{p} = \bar{d} + \lambda \left(\sum_{k=1}^{K} \tilde{x}_k + \sum_{n=1}^{N} \tilde{w}_n \right),$$

for some constant λ to be determined.

Show that the optimal quantity $\tilde{w}_n^*$ demanded by the risk averse uninformed agent $n \in \{1, \ldots, N\}$, conditionally on his endowment $\tilde{e}_n$, is given by

$$\tilde{w}_n^* = -\frac{a\sigma_\varepsilon^2(1 - K\lambda\beta)\tilde{e}_n}{2\lambda + a\left(\sigma_\varepsilon^2(1-K\lambda\beta)^2 + K\sigma_\eta^2\lambda^2\beta^2 + \lambda^2(N-1)\gamma^2\sigma_e^2\right)}. \tag{10.33}$$

Deduce that, in equilibrium, the constant γ must satisfy the equation

$$a\lambda^2\gamma^3(N-1)\sigma_e^2 + \gamma\left(aK\sigma_\eta^2\lambda^2\beta^2 + a\sigma_\varepsilon^2(1-K\lambda\beta)^2 + 2\lambda\right) + a\sigma_\varepsilon^2(1-K\lambda\beta) = 0. \tag{10.34}$$

(iii) By relying on the zero expected profits condition for the risk neutral market maker, together with the linear price setting rule, deduce that in equilibrium the parameters γ and λ satisfy the equation

$$K\sigma_\varepsilon^4(\sigma_\varepsilon^2 + \sigma_\eta^2) = N\gamma^2\lambda^2\sigma_e^2\big((1+K)\sigma_\varepsilon^2 + 2\sigma_\eta^2\big)^2 \qquad (10.35)$$

and, as a consequence, γ is given by

$$\gamma = -\frac{\sqrt{K\sigma_\varepsilon^4(\sigma_\varepsilon^2 + \sigma_\eta^2)}}{\sqrt{N\sigma_e^2}\lambda\big((1+K)\sigma_\varepsilon^2 + 2\sigma_\eta^2\big)}.$$

(iv) On the basis of the previous steps, deduce the result of Proposition 10.5.

Exercise 10.8 Consider the model proposed in Vives [1628] and presented in Sect. 10.2. In this exercise, we are going to prove the existence of a unique linear symmetric equilibrium of the economy, characterized as in Proposition 10.6.

(i) As usual, start from the conjecture that there exists a linear equilibrium. A first consequence of the linearity of the equilibrium, together with the distributional assumptions of the model, is that the equilibrium price will be normally distributed. Consider then the expected utility maximization problem of an arbitrary informed trader $i \in [0, 1]$. Show that, conditionally on the observation of the realization p of the equilibrium price $\tilde{p}$ and of the realization y_i of the private signal $\tilde{y}_i$, the demand schedule of the i-th informed trader satisfies

$$X_i(p, y_i) = \frac{\mathbb{E}[\tilde{d}|p, y_i] - p}{a\operatorname{Var}(\tilde{d}|p, y_i)}.$$

(ii) For any price p, let $L(p) := \int_0^1 X_i(p, y_i)di + \tilde{z}$ be the aggregate demand in correspondence of the realization y_i of the private signal $\tilde{y}_i$. Suppose that

$$X_i(p, y_i) = \alpha y_i + bp + c, \qquad \text{for every } i \in [0, 1],$$

for some constants α, b and c, for every realization y_i. Show that the equilibrium price set by the risk neutral market maker via the rule $\tilde{p} = \mathbb{E}[\tilde{d}|L(\cdot)]$ is given by

$$\tilde{p} = (1 - \lambda\alpha)\bar{d} + \lambda(\alpha\tilde{d} + \tilde{z}),$$

where $\lambda = \alpha\tau_z/(\alpha^2\tau_z + \tau_d)$.

(iii) By making use of the results established in the two previous steps of the exercise, deduce that the optimal demand schedules of the informed traders are given as in (10.18), so that Proposition 10.6 holds.

Exercise 10.9 Let us consider the two-period model of a market dealer economy originally proposed in Stoll [1570] and presented in Sect. 10.3. Prove Proposition 10.7.

Exercise 10.10 Consider the model proposed in Glosten & Milgrom [790] and presented in Sect. 10.3. In this exercise, by relying mainly on the tower property of the conditional expectation and following the original arguments of Glosten & Milgrom [790], we are going to prove Proposition 10.8. Suppose that there exist two processes $(p_t^{\mathrm{ask}})_{t=1,\ldots,T_0}$ and $(p_t^{\mathrm{bid}})_{t=1,\ldots,T_0}$ satisfying the equilibrium condition (10.25). We shall prove the inequality stated in Proposition 10.8 for the ask price, the proof of the inequality for the bid price being completely analogous. Let $t \in \{1, \ldots, T_0\}$ and, for simplicity of notation, define the event C_t by

$$C_t := \{Z_t > p_t^{\mathrm{ask}}\} = \{\mathbb{E}[\tilde{v}|\mathscr{K}_t] > p_t^{\mathrm{ask}}/\delta_t\},$$

where the second equality simply follows by the definition of Z_t, using the information flow $\mathbb{K} = (\mathscr{K}_t)_{t=1,\ldots,T_0}$ introduced after (10.23). Note that

$$p_t^{\mathrm{ask}} = \mathbb{E}[\tilde{v}|\mathscr{D}_t, C_t].$$

For brevity of notation, in the remaining part of the exercise, we shall denote by $\mathbb{E}_t[\cdot]$ the conditional expectation $\mathbb{E}[\cdot|\mathscr{D}_t \wedge \mathscr{F}_t]$, for $t = 1, \ldots, T_0$.

  (i) By relying on the tower property of the conditional expectation, prove that

$$p_t^{\mathrm{ask}} = \mathbb{E}_t[\tilde{v}|C_t].$$

 (ii) By relying on condition (10.21), show that

$$\mathbb{E}_t[\tilde{v}|C_t] = \mathbb{E}_t\Big[\mathbb{E}\big[\mathbb{E}[\tilde{v}|\mathscr{K}_t]|\mathscr{D}_t \wedge \mathscr{F}_t \vee \sigma(\delta_t), C_t\big]\big|C_t\Big].$$

(iii) Prove that, for any random variable X with a finite expectation, it holds that

$$\mathbb{E}[X|X > a] \geq \mathbb{E}[X],$$

for any $a \in \mathbb{R}$. Moreover, argue that the same relationship also holds for a conditional expectation of the form $\mathbb{E}[X|\mathscr{A}, X > a]$, where $\mathscr{A}$ is an arbitrary σ-algebra.

(iv) By relying on the results established in steps *(ii)–(iii)* of the exercise, show that

$$\mathbb{E}_t[\tilde{v}|C_t] \geq \mathbb{E}_t\Big[\mathbb{E}_t\big[\mathbb{E}[\tilde{v}|\mathscr{K}_t]|\sigma(\delta_t)\big]\big|C_t\Big]$$

and deduce that

$$p_t^{\mathrm{ask}} \geq \mathbb{E}_t[\tilde{v}],$$

thus completing the proof of the inequality for the ask price stated in
Proposition 10.8.

Exercise 10.11 Consider the simple version of the model of Glosten & Milgrom
[790] described in Sect. 10.3, where the fundamental value $\tilde{v}$ of the asset is a random
variable taking the two possible values $\overline{v}$ and $\underline{v}$ with probabilities π and $1 - \pi$,
respectively, and where the probability that an agent is informed is α. Suppose
furthermore that an uninformed agent is equally likely to submit a buy or a sell
order. By making use of Bayes' rule, show that the equilibrium bid and ask prices
set by the dealer by applying the zero expected profits condition are explicitly given
as in (10.28).

Exercise 10.12 Consider the model of Easley & O'Hara [619], where traders are
allowed to post market orders of different sizes, as presented in De Jong & Rindi
[540, Section 4.2]. More specifically, consider the same setting of the simplified
Glosten & Milgrom [790] model presented in Sect. 10.3, characterized by the
following assumptions:

- the fundamental value of the asset is represented by the random variable $\tilde{v}$ which
  can take the two possible values 1 and 0 with equal probabilities;
- in the economy, there are informed and liquidity traders, with informed traders
  knowing the exact realization of $\tilde{v}$;
- there is an exogenous probability α that an order is being submitted by an
  informed trader and a probability $1 - \alpha$ that an order is being submitted by a
  liquidity trader;
- liquidity traders randomly submit buy and sell orders with equal probability;
- informed traders submit buy or sell orders by maximizing their expected profit
  (i.e., knowing the realization of $\tilde{v}$, they submit a buy order if $\tilde{v} = 1$ and, on the
  contrary, a sell order if $\tilde{v} = 0$).

Assume furthermore that buy orders can be of two different sizes B_1 and B_2, with
$0 < B_1 < B_2$, and sell orders can be of sizes S_1 and S_2, with $0 < S_1 < S_2$. We
let $\beta/2$ and $(1 - \beta)/2$ be the probabilities that an uninformed trader makes large
(i.e., B_2 or S_2) orders and small (i.e., B_1 or S_1) orders, respectively. For simplicity,
suppose that

$$B_1 = S_1 = 1 \quad \text{and} \quad B_2 = S_2 = 2.$$

In this model, the dealer is risk neutral and has the possibility of fixing bid and ask
prices for the asset depending on the size of the trade, i.e., the dealer sets $p^{ask}(1)$,
$p^{ask}(2)$, $p^{bid}(1)$ and $p^{bid}(2)$, where $p^{ask}(i)$ and $p^{bid}(i)$ represent respectively the ask
and bid prices for an order of size i, with $i \in \{1, 2\}$.

(i) Suppose that informed traders submit large orders with probability $\mu \in$
    $(0, 1)$ and small orders with probability $1 - \mu$. By arguing similarly as in
    Exercise 10.11, show that the equilibrium ask prices $p^{ask}(1)$ and $p^{ask}(2)$ are

given by

$$p^{\text{ask}}(1) = \frac{\alpha(1-\mu) + \frac{(1-\alpha)(1-\beta)}{2}}{\alpha(1-\mu) + (1-\alpha)(1-\beta)} \quad \text{and} \quad p^{\text{ask}}(2) = \frac{\alpha\mu + \frac{(1-\alpha)\beta}{2}}{\alpha\mu + (1-\alpha)\beta}.$$

(*ii*) Suppose furthermore that the following condition holds:

$$2(1 - p^{\text{ask}}(2)) = 1 - p^{\text{ask}}(1),$$

meaning that the profit realized by the informed agent when submitting a large buy order coincides with the profit realized by the informed agent when submitting a small buy order (of course, this is a necessary condition for the insider trader to post orders of both sizes). Show that the probability μ is then endogenously determined as

$$\mu = \beta + \frac{\beta(1-\beta)}{\alpha(1+\beta)}$$

and, as a consequence, the equilibrium ask prices $p^{\text{ask}}(1)$ and $p^{\text{ask}}(2)$ are given by

$$p^{\text{ask}}(1) = \frac{1 - \beta + \alpha(1+\beta)}{2} \quad \text{and} \quad p^{\text{ask}}(2) = \frac{3 - \beta + \alpha(1+\beta)}{4}.$$

The ask prices $p^{\text{ask}}(1)$ and $p^{\text{ask}}(2)$ then corresponds to a *pooling equilibrium* of the economy, where informed traders have incentive to submit both small and large market orders. In this case, the equilibrium bid prices can be computed in an analogous way. In the remaining part of the exercise, we are going to characterize the equilibrium ask prices corresponding to a *separating equilibrium* of the economy, where informed traders only submit large market orders.

(*iii*) Suppose that

$$2(1 - p^{\text{ask}}(2)) \geq 1 - p^{\text{ask}}(1),$$

so that informed traders have incentive to submit only large orders (i.e., in the above notation, it holds that $\mu = 1$). Show that in this case the equilibrium ask prices $p^{\text{ask}}(1)$ and $p^{\text{ask}}(2)$ are given by

$$p^{\text{ask}}(1) = \frac{1}{2} \quad \text{and} \quad p^{\text{ask}}(2) = \frac{(1-\alpha)\beta + \alpha}{2(1-\alpha)\beta + 2\alpha}$$

and deduce that a necessary condition for a separating equilibrium to exist is $\beta \geq \alpha/(1-\alpha)$.

Exercise 10.13 In this exercise, taken from De Jong & Rindi [540, Section 5.2.1] and related to the models discussed in Sect. 10.3, we present a simple model of an economy where several risk averse dealers operate. Consider an economy where a single risky asset with liquidation value $\tilde{d}$ is traded. It is assumed that $\tilde{d} \sim \mathcal{N}(\bar{d}, \sigma_d^2)$. Suppose that the economy is populated by the two following types of agents:

- M risk averse dealers submitting limit orders to the market, with each dealer $m \in \{1, \ldots, M\}$ being endowed with E_m units of the asset. The utility function of each dealer is assumed to be negative exponential, with a common risk aversion parameter a.
- N liquidity traders, submitting market orders. The aggregate demand of liquidity traders is represented by the random variable $\tilde{z}$, supposed to be normally distributed with mean $\bar{z}$ and variance σ_z^2.

(i) For each $m \in \{1, \ldots, M\}$, let $X_m : \mathbb{R} \to \mathbb{R}$ represent the demand schedule of the m-th dealer, representing the number of units of the asset demanded as a function of the current market price. Show that

$$X_m(\tilde{p}) = \frac{\bar{d} - \tilde{p}}{a\sigma_d^2} - E_m, \qquad \text{for every } m = 1, \ldots, M.$$

(ii) By imposing market clearing, deduce that the equilibrium price p satisfies

$$\tilde{p} = \left(\frac{\tilde{z}}{M} - \overline{E} \right) a\sigma_d^2 + \bar{d},$$

where $\overline{E} := \sum_{m=1}^{M} E_m / M$, with the consequence that the bid and ask prices associated to an aggregate quantity z traded by liquidity traders are given by

$$p^{\text{ask}} = \bar{d} - \overline{E} a\sigma_d^2 + \frac{a\sigma_d^2}{M} |z|, \qquad \text{if } z > 0,$$

$$p^{\text{bid}} = \bar{d} - \overline{E} a\sigma_d^2 - \frac{a\sigma_d^2}{M} |z|, \qquad \text{if } z < 0,$$

with a bid-ask spread given by

$$p^{\text{ask}} - p^{\text{bid}} = \frac{2a\sigma_d^2}{M} |z|.$$

Observe that the term $a\sigma_d^2 / M$ represents the price impact of a trade: the greater the dealers' risk aversion and the volatility of the asset, the greater the price impact. Increasing the number M of dealers leads to a reduction of the price impact.

(iii) By using the results established in steps *(i)–(ii)* of the exercise, show that the quantity traded by each dealer is given by

$$x_m^* = \overline{E} - E_m - \frac{z}{M},$$

where z is the realization of the random variable $\tilde{z}$.

Chapter 11
Solutions of Selected Exercises

In this appendix, we provide the detailed solutions to a selection of the exercises proposed at the end of the chapters. The solutions to the exercises that are not solved in this appendix can be found in the solutions manual.

11.1 Exercises of Chap. 2

Solution of Exercise 2.1 Let the expected utility function $U(\tilde{x}) = \sum_{s=1}^{S} \pi_s u(x_s)$ represent the preference relation $(\mathcal{M}, \mathcal{R})$ and let b, c be strictly positive numbers. Define $\tilde{U}(\tilde{x}) := b + cU(\tilde{x})$ and $\tilde{u}(x_s) := b + cu(x_s)$, for every $s = 1, \ldots, S$. Then, since $\sum_{s=1}^{S} \pi_s = 1$, we have that

$$\tilde{U}(\tilde{x}) = b + cU(\tilde{x}) = b + c\sum_{s=1}^{S} \pi_s u(x_s) = \sum_{s=1}^{S} \pi_s \tilde{u}(x_s),$$

thus showing that the function $\tilde{U}$ has the expected utility form. Moreover, since $b, c > 0$, we have that, for every $\tilde{x}_1, \tilde{x}_2 \in \mathcal{M}$,

$$\tilde{x}_1 \mathcal{R} \tilde{x}_2 \iff U(\tilde{x}_1) \geq U(\tilde{x}_2) \iff \tilde{U}(\tilde{x}_1) \geq \tilde{U}(\tilde{x}_2).$$

Solution of Exercise 2.3 For $k = 0$, the random variable $\tilde{x}$ coincides with its expectation, so that $\tilde{x} \equiv \mathbb{E}[\tilde{x}]$. In this case, for any utility function $u : \mathbb{R} \to \mathbb{R}$, it holds that $u(\mathbb{E}[\tilde{x}]) = u(\tilde{x}) = \mathbb{E}[u(\tilde{x})]$. Definition 2.2 directly implies that $\rho_u(\tilde{x}, 0) = 0$. The second claim can be proved by differentiating with respect to k condition (2.4) which defines the risk premium:

$$-u'\big(\mathbb{E}[\tilde{x}] - \rho_u(\tilde{x}, k)\big)\frac{\partial \rho_u(\tilde{x}, k)}{\partial k} = \mathbb{E}[u'(\tilde{x})\tilde{\epsilon}],$$

© Springer-Verlag London Ltd. 2017
E. Barucci, C. Fontana, *Financial Markets Theory*,
Springer Finance, DOI 10.1007/978-1-4471-7322-9_11

so that

$$\frac{\partial \rho_u(\tilde{x}, k)}{\partial k} = -\frac{\mathbb{E}\left[u'\left(\mathbb{E}[\tilde{x}] + k\tilde{\epsilon}\right)\tilde{\epsilon}\right]}{u'\left(\mathbb{E}[\tilde{x}] - \rho_u(\tilde{x}, k)\right)}.$$

Since $\mathbb{E}[\tilde{\epsilon}] = 0$ and $\rho_u(\tilde{x}, 0) = 0$, we get

$$\left.\frac{\partial \rho_u(\tilde{x}, k)}{\partial k}\right|_{k=0} = -\frac{u'\left(\mathbb{E}[\tilde{x}]\right)\mathbb{E}[\tilde{\epsilon}]}{u'\left(\mathbb{E}[\tilde{x}]\right)} = 0.$$

Solution of Exercise 2.7   Let $\rho_{u^a}(\tilde{x})$ and $\rho_{u^b}(\tilde{x})$ denote the risk premia of agents a and b, respectively, in correspondence of a random variable $\tilde{x}$. Let us consider the random variable $\tilde{x}$ which takes values $x_0 \pm \epsilon$ with equal probabilities. Then, by definition of risk premium:

$$x_0 - \rho_{u^a}(\tilde{x}) = u^a\left(\mathbb{E}[\tilde{x}] - \rho_{u^a}(\tilde{x})\right) = \mathbb{E}\left[u^a(\tilde{x})\right] = \frac{1}{2}(x_0 - \epsilon) + \frac{1}{2}(x_0 + a\epsilon)$$

thus showing that $\rho_{u^a}(\tilde{x}) = \epsilon(1 - a)/2$. Analogously, in the case of agent b, we have that $\rho_{u^b}(\tilde{x}) = \epsilon(1 - b)/2$. Agent b is more risk averse than agent a only if $\rho_{u^b}(\tilde{x}) \geq \rho_{u^a}(\tilde{x})$, i.e., only if $b \leq a$.

Note that in the context of the present exercise the expected utility of a generic random variable $\tilde{x}$ can be equivalently rewritten as follows, for $k = a, b$:

$$\mathbb{E}[u^k(\tilde{x})] = \mathbb{E}\left[\tilde{x}\mathbf{1}_{\{\tilde{x} \leq x_0\}} + x_0\mathbf{1}_{\{\tilde{x} > x_0\}}\right] + k\,\mathbb{E}\left[(\tilde{x} - x_0)^+\right],$$

where, for any event E, the symbol $\mathbf{1}_E$ denotes the function (indicator function) that takes value one if event E occurs and zero otherwise and $(\cdot)^+$ denotes the positive part function. Due to the above expression, if $a \geq b$, then agent a will always accept a gamble if agent b does, when both agents start from the same initial wealth. Hence, agent b is more risk averse than agent a.

Solution of Exercise 2.8

(i): The claim follows directly by differentiating the absolute risk aversion coefficient $r_u^a(x)$:

$$\frac{\mathrm{d}}{\mathrm{d}x}r_u^a(x) = -\frac{u'''(x)u'(x) - (u''(x))^2}{(u'(x))^2}.$$

(ii): An increasing concave function u satisfies $p_u^a(x) > r_u^a(x)$, for every $x \in \mathbb{R}$, if and only if the following condition holds:

$$\frac{u'''(x)}{u''(x)} < \frac{u''(x)}{u'(x)}, \qquad \text{for every } x \in \mathbb{R}.$$

Equivalently, the above condition holds if and only if $u'''(x)u'(x) > (u''(x))^2$, for every $x \in \mathbb{R}$. The claim then follows from the proof of part a).

(iii): The claim follows by noting that $\frac{d}{dx}r_u^a(x) = r_u^a(x)(r_u^a(x) - p_u^a(x))$.

Solution of Exercise 2.9   It suffices to note that $r_u^a(x) = -\frac{d}{dx}\log\left(u'(x)\right)$, for every $x \in \mathbb{R}$.

Solution of Exercise 2.10

(i): The claim can be readily verified by differentiating $r_u^a(x) = 1/(a + bx)$ and $r_u^r(x) = x/(a + bx)$ with respect to x.

(ii): Let us consider the three different cases: $b = 0$, $b = 1$ and $b \notin \{0, 1\}$.

   If $b = 0$, then $-u''(x)/u'(x) = 1/a$ and, as a consequence (compare with Exercise 2.9),

$$u'(x) = ke^{-x/a} \quad \text{and} \quad u(x) = -ake^{-x/a} + h, \qquad \text{for every } x \in \mathbb{R},$$

for suitable constants $k > 0$ and $h \in \mathbb{R}$. Hence, up to an increasing linear transformation, we have $u(x) = -a\exp(-x/a)$, where the utility function is defined for every $x \in \mathbb{R}$.

   If $b = 1$, then $-u''(x)/u'(x) = 1/(a + x)$ and, as a consequence,

$$u'(x) = k(a + x)^{-1} \quad \text{and} \quad u(x) = k\log(x + a) + h \qquad \text{for every } x > -a,$$

for suitable constants $k > 0$ and $h \in \mathbb{R}$. Hence, up to an increasing linear transformation, we have $u(x) = \log(x+a)$, where the utility function is defined for $x > -a$.

   In the remaining cases (i.e., if $b \neq 0$ and $b \neq 1$), an analogous reasoning leads to

$$u(x) = \frac{k}{b}\frac{1}{1 - \frac{1}{b}}(a + bx)^{1-1/b} + h, \qquad \text{for every } x \in \mathbb{R},$$

for suitable constants $k > 0$ and $h \in \mathbb{R}$. Hence, up to an increasing linear transformation, we have $u(x) = \frac{1}{b-1}(a + bx)^{\frac{b-1}{b}}$. Let us now determine the domain of the utility function u. If $b > 0$, which corresponds to assuming that u is of the DARA type, we must have $x > \underline{x} := -a/b$. If $b < 0$, which corresponds to assuming that u is of the IARA type, then the function u is defined for every $x \in \mathbb{R}$ (however, in order to exclude negative risk aversion, we should restrict the domain to $x < \bar{x} := -a/b$).

Solution of Exercise 2.12

(i): It holds that $r_u^r(x) = b$, for some $b \in \mathbb{R}$, if and only if $r_u^a(x) = b/x$. As shown in Sect. 2.2, this implies that the coefficient of relative risk aversion is constant if and only if the utility function u belongs to the HARA class and has the form $u(x) = \alpha x^{1-b} + \beta$, with $b \neq 1$, for some $\alpha > 0$ and $\beta \in \mathbb{R}$.

(ii): As for part *(i)*, we have $r_u^r(x) = 1$ if and only if $r_u^a(x) = 1/x$ or, equivalently, if the utility function u belongs to the HARA class and has the form $u(x) = \alpha \log(x) + \beta$ for some $\alpha > 0$ and $\beta \in \mathbb{R}$.

Solution of Exercise 2.13   The first part follows directly from the definitions. To show that the converse implication does not necessarily hold, it suffices to consider two agents with exponential utility functions: $u^a(x) = -\mathrm{e}^{-ax}$ and $u^b(x) = -\mathrm{e}^{-bx}$ with $a > b$. In this case, agent a is more risk averse than agent b. Take $x_1, x_2 \in \mathbb{R}$ such that $x_1 - x_2 > \frac{\log a - \log b}{a-b}$. Then, it is easy to check that:

$$\inf_{z \in \mathbb{R}} \frac{u^{a''}(z)}{u^{b''}(z)} \le \frac{u^{a''}(x_1)}{u^{b''}(x_1)} = \frac{a^2}{b^2} \mathrm{e}^{-(a-b)x_1} < \frac{a}{b} \mathrm{e}^{-(a-b)x_2} = \frac{u^{a'}(x_2)}{u^{b'}(x_2)} \le \sup_{z \in \mathbb{R}} \frac{u^{a'}(z)}{u^{b'}(z)}.$$

This shows that agent a is not strongly more risk averse than agent b.

Solution of Exercise 2.14   We shall always assume that we can interchange expectation and differentiation. By differentiating the function V with respect to its first argument we get

$$\frac{\partial V}{\partial \mu}(\mu, \sigma^2) = \mathbb{E}[u'(\mu + \sigma \tilde{z})] = \mathbb{E}[u'(\tilde{x})] \ge 0,$$

since the function u is assumed to be increasing. This shows that V is increasing with respect to the first argument. Let us now differentiate V with respect to σ^2:

$$\frac{\partial V}{\partial \sigma^2}(\mu, \sigma^2) = \frac{1}{2\sigma} \mathbb{E}[u'(\mu + \sigma \tilde{z})\tilde{z}].$$

Let us denote by $\varphi : \mathbb{R} \to \mathbb{R}$ the density function of the standard Normal random variable $\tilde{z}$. Then

$$\mathbb{E}[u'(\mu + \sigma \tilde{z})\tilde{z}] = \int_{-\infty}^{+\infty} u'(\mu + \sigma z) z \varphi(z) \mathrm{d}z$$

$$= \int_0^{+\infty} u'(\mu + \sigma z) z \varphi(z) \mathrm{d}z + \int_{-\infty}^0 u'(\mu + \sigma z) z \varphi(z) \mathrm{d}z$$

$$= \int_0^{+\infty} u'(\mu + \sigma z) z \varphi(z) \mathrm{d}z - \int_0^{+\infty} u'(\mu - \sigma z) z \varphi(z) \mathrm{d}z,$$

where we have used the symmetry of the density function φ. Using the fact that the function $x \mapsto u'(x)$ is decreasing (since the function u is assumed to be concave), we obtain

$$\mathbb{E}[u'(\mu + \sigma \tilde{z})\tilde{z}] = \int_0^{+\infty} \left(u'(\mu + \sigma z) - u'(\mu - \sigma z) \right) z \varphi(z) \mathrm{d}z \le 0,$$

thus showing that $\partial V(\mu, \sigma^2)/\partial \sigma^2 \le 0$.

Solution of Exercise 2.15 It suffices to compute the ratio (2.14) for the specific cases of an exponential and a quadratic utility function, respectively, using the Laplace transform of a normal random variable,

$$-\frac{\mathbb{E}\left[e^{-a\tilde{x}\left(\frac{\tilde{x}-\mu}{\sigma}\right)}\right]}{\mathbb{E}[e^{-a\tilde{x}}]} = -\frac{-\frac{1}{\sigma}\frac{d}{da}\mathbb{E}[e^{-a\tilde{x}}] - \frac{\mu}{\sigma}\mathbb{E}[e^{-a\tilde{x}}]}{\mathbb{E}[e^{-a\tilde{x}}]} = a\sigma;$$

$$-\frac{\mathbb{E}\left[(1 - b\tilde{x})\left(\frac{\tilde{x}-\mu}{\sigma}\right)\right]}{\mathbb{E}[1 - b\tilde{x}]} = \frac{b\,\mathbb{E}\left[\tilde{x}\left(\frac{\tilde{x}-\mu}{\sigma}\right)\right]}{\mathbb{E}[1 - b\tilde{x}]} = \frac{\frac{b}{\sigma}(\sigma^2 + \mu^2) - \frac{b}{\sigma}\mu^2}{1 - b\mu} = \frac{b\sigma}{1 - b\mu}.$$

Solution of Exercise 2.16 Let F_i be the distribution function of the random variable $\tilde{x}_i$, for $i = 1, 2$. Due to Proposition 2.10, in order to show that $\tilde{x}_1 \succeq_{SSD} \tilde{x}_2$, we need to prove that

$$G(y) = \int_{-\infty}^{y} \left(F_1(z) - F_2(z)\right)dz \leq 0, \qquad \text{for every } y \in \mathbb{R}.$$

As a preliminary, note that, if $\sigma^2(\tilde{x}_1) \leq \sigma^2(\tilde{x}_2)$, then $F_1(x) \leq F_2(x)$ for every $x \leq \mu$ and $F_2(x) \leq F_1(x)$ for every $x \geq \mu$, with $F_1(\mu) = F_2(\mu)$. Note also that the function $y \mapsto G(y)$ is negative and decreasing for $y \leq \mu$, reaches its minimum for $y = \mu$ and is increasing afterwards. Hence, to prove our claim it suffices to show that $\lim_{y\to+\infty} G(y) = 0$. Using integration by parts and the fact that $\mathbb{E}[\tilde{x}_1] = \mathbb{E}[\tilde{x}_2] = \mu$, we get

$$\lim_{y \to +\infty} G(y) = \int_{-\infty}^{+\infty} \left(F_1(z) - F_2(z)\right)dz$$

$$= z\left(F_1(z) - F_2(z)\right)\Big|_{-\infty}^{+\infty} - \int_{-\infty}^{+\infty} z\left(\varphi_1(z) - \varphi_2(z)\right)dz = \mathbb{E}[\tilde{x}_2] - \mathbb{E}[\tilde{x}_1] = 0,$$

where $\varphi_i : \mathbb{R} \to \mathbb{R}$ denotes the density function of the random variable $\tilde{x}_i$, for $i = 1, 2$, and where $z\left(F_1(z) - F_2(z)\right)\Big|_{-\infty}^{+\infty} = 0$ follows from the fact that $F_1(-\infty) - F_2(-\infty) = F_1(\infty) - F_2(\infty) = 0$ together with the fact that the tails of a normal distribution decay exponentially fast. The converse implication can be shown by applying Definition 2.9 to the concave function $x \mapsto -(x - \mu)^2$.

Solution of Exercise 2.17

(i): Due to Proposition 2.8, we have $\tilde{x}_1 \succeq_{FSD} \tilde{x}_2$ if and only if $F_1(x) \leq F_2(x)$ for every $x \in [0, 1]$. In the present case, this condition holds if and only if $\pi \in [0.3, 1]$ and $x \leq 1$.

(ii): Due to Proposition 2.10, we have $\tilde{x}_1 \succeq_{SSD} \tilde{x}_2$ if and only if $\mathbb{E}[\tilde{x}_1] = \mathbb{E}[\tilde{x}_2]$ and $\int_0^y \left(F_1(z) - F_2(z)\right)dz \leq 0$ for all $y \in [0, 1]$. In the present context, the first condition amounts to $x(1 - \pi) = 0.7$. For the second one, note that:

$$F_1(y) - F_2(y) = \begin{cases} 0.3 - \pi, & \text{for } y \in [0, x); \\ 0.3 - 1, & \text{for } y \in [x, 1); \\ 0, & \text{for } y = 1. \end{cases}$$

Hence, the condition $\int_0^y \big(F_1(z) - F_2(z)\big)dz \leq 0$, for all $y \in [0, 1]$, holds if and only if $\pi \geq 0.3$.

(iii): Due to Proposition 2.12, we have $\tilde{x}_1 \succeq_{\text{SSD}}^M \tilde{x}_2$ if and only if $\mathbb{E}[\tilde{x}_1] \geq \mathbb{E}[\tilde{x}_2]$ and $\int_0^y \big(F_1(z) - F_2(z)\big)dz \leq 0$ for all $y \in [0, 1]$. In the present context, analogously as in part b), these conditions hold if and only if $x(1 - \pi) \leq 0.7$ and $0.3 \leq \pi$.

(iv): We have $\tilde{x}_1 \succeq_{\text{MV}} \tilde{x}_2$ if and only if $\mathbb{E}[\tilde{x}_2] = (1 - \pi)x \leq 0.7 = \mathbb{E}[\tilde{x}_1]$ and $\sigma^2(\tilde{x}_2) = x^2 \pi(1 - \pi) \geq 0.21 = \sigma^2(\tilde{x}_1)$.

Solution of Exercise 2.18 It is easy to show that $\int_{-1}^y \big(F_1(z) - F_2(z)\big)dz \leq 0$, for all $y \in [-1, 1]$, if and only if $\gamma \geq \pi$. The claim then follows by Proposition 2.10.

Solution of Exercise 2.19 The first implication follows directly from Proposition 2.8. For a counterexample, it suffices to take any two non-negative random variables $\tilde{x}_1$ and $\tilde{x}_2$ with $\mathbb{E}[\tilde{x}_1] \geq \mathbb{E}[\tilde{x}_2]$ and $\mathbb{E}[\tilde{x}_1^2] \leq \mathbb{E}[\tilde{x}_2^2]$. Since the function $u :$ $\mathbb{R}_+ \to \mathbb{R}_+$ defined by $u(x) := x^2$ is non-decreasing, we have $\mathbb{E}[u(\tilde{x}_1)] \leq \mathbb{E}[u(\tilde{x}_2)]$, thus showing that $\tilde{x}_1 \succeq_{\text{FSD}} \tilde{x}_2$ does not hold.

Solution of Exercise 2.20 The answer to both questions is negative. Indeed, it suffices to consider two random variables $\tilde{x}_1$ and $\tilde{x}_2$ such that $\tilde{x}_1 \succeq_{\text{SSD}}^M \tilde{x}_2$ but $\sigma^2(\tilde{x}_1) > \sigma^2(\tilde{x}_2)$. For instance, let $\tilde{x}_1$ be an arbitrary random variable and $\tilde{\xi}$ a non-negative random variable such that $\text{Cov}(\tilde{x}_1, \tilde{\xi}) > \sigma^2(\tilde{\xi})/2$ and let $\tilde{x}_2 := \tilde{x}_1 - \tilde{\xi}$. Then, due to Proposition 2.12, we have $\tilde{x}_1 \succeq_{\text{SSD}}^M \tilde{x}_2$ but

$$\sigma^2(\tilde{x}_2) = \sigma^2(\tilde{x}_1) - 2\,\text{Cov}(\tilde{x}_1, \tilde{\xi}) + \sigma^2(\tilde{\xi}) < \sigma^2(\tilde{x}_1).$$

Solution of Exercise 2.21 Let us consider the first order conditions for the solution of the minimization problem (2.20):

$$\frac{\partial L(w_1^*, \ldots, w_N^*, \lambda)}{\partial w_n} = 2\, w_n^*\, \sigma^2(\tilde{x}_n) - \lambda = 0, \quad \text{for every } n = 1, \ldots, N;$$

$$\frac{\partial L(w_1^*, \ldots, w_N^*, \lambda)}{\partial \lambda} = \sum_{n=1}^N w_n^* - 1 = 0.$$

Since $\sigma^2(\tilde{x}_n) > 0$, for every $n = 1, \ldots, N$, the above first order conditions are both necessary and sufficient for the optimal solution to the problem. Hence:

$$w_n^* = \frac{\lambda}{2\,\sigma^2(\tilde{x}_n)}, \quad \text{for every } n = 1, \ldots, N.$$

Together with the condition $\sum_{n=1}^{N} w_n^* = 1$, this leads to:

$$\lambda = \frac{2}{\sum_{n=1}^{N} 1/\sigma^2(\tilde{x}_n)},$$

which immediately implies that the optimal solution w_n^* is given by expression (2.21), for every $n = 1, \ldots, N$.

Solution of Exercise 2.22 Let define the random variables

$$\tilde{x}_1^n = \{0, n; \ 1 - 1/n^2, 1/n^2\}, \text{ for all } n \in \mathbb{N},$$

and let $\tilde{x}_2$ be the trivial random variable which always takes value zero. It is easy to show that $\mathbb{E}[\tilde{x}_1^n] = 1/n$ and $\sigma^2(\tilde{x}_1^n) = 1 - 1/n^2$. For $n \to \infty$, it holds that $\mathbb{E}[\tilde{x}_1^n] \to 0$, while $\sigma^2(\tilde{x}_1^n) \to 1$. Hence, since the function V is assumed to be continuous, increasing with respect to the mean and decreasing with respect to the variance, we have that, for sufficiently large n,

$$\mathbb{E}[u(\tilde{x}_1^n)] = V\big(\mathbb{E}[\tilde{x}_1^n], \sigma^2(\tilde{x}_1^n)\big) < V(0, 0) = \mathbb{E}[u(\tilde{x}_2)].$$

However, it is clear that $\mathbb{P}(\tilde{x}_1^n \geq \tilde{x}_2) = 1$ and $\mathbb{P}(\tilde{x}_1^n > \tilde{x}_2) > 0$, since the random variable $\tilde{x}_1^n$ satisfies $\tilde{x}_1^n \geq \tilde{x}_2$ and $\mathbb{P}(\tilde{x}_1^n = n > 0) = 1/n^2 > 0$.

Solution of Exercise 2.23 The claim easily follows by considering the random variables $\tilde{x}_1^n$ and $\tilde{x}_2$ introduced in the previous exercise. Indeed, if Assumption 2.3 was satisfied, then it would imply that the sure payoff of zero is preferred to the sure payoff of n, thus yielding a clear contradiction with the strictly increasing feature of the mean-variance utility function V.

11.2 Exercises of Chap. 3

Solution of Exercise 3.1 By the first order optimality condition, for an agent to invest all his initial wealth w_0 (or more) in the risky asset, it must be that

$$\mathbb{E}[u'(w_0 \tilde{r})(\tilde{r} - r_f)] \geq 0.$$

Since the utility function u is supposed to be twice differentiable, we can apply a Taylor approximation of the marginal utility function u' up to the first order:

$$u'(w_0 \tilde{r}) \approx u'(w_0 r_f) + u''(w_0 r_f) w_0 (\tilde{r} - r_f).$$

By combining the two above expressions, the claim follows directly.

Solution of Exercise 3.3 We already know that, by the concavity of the utility function u, the optimal wealth $\widehat{W}^*$ solving problem (3.2) is unique. Arguing by contradiction, suppose that there exist two distinct portfolios $w^1, w^2 \in \mathbb{R}^N$ such that

$$\widetilde{W}^* = w_0 \, r_f + \sum_{n=1}^{N} w_n^1 (\tilde{r}_n - r_f) = w_0 \, r_f + \sum_{n=1}^{N} w_n^2 (\tilde{r}_n - r_f).$$

If $\sum_{n=1}^{N} w_n^1 \neq \sum_{n=1}^{N} w_n^2$, then the last relation implies that

$$r_f = \frac{\sum_{n=1}^{N} (w_n^1 - w_n^2) \tilde{r}_n}{\sum_{n=1}^{N} (w_n^1 - w_n^2)},$$

thus contradicting the assumption that there are no redundant assets in the economy. If $\sum_{n=1}^{N} w_n^1 = \sum_{n=1}^{N} w_n^2$, then the existence of two distinct portfolios w^1, w^2 allows to obtain an analogous contradiction.

Solution of Exercise 3.4   Let $w^* \in \mathbb{R}^N$ denote the optimal portfolio, with $\sum_{k=1}^{N} w_k^* = 1$, for some given initial wealth w_0. Then, for $n \in \{1, \dots, N\}$, consider the following auxiliary maximization problem:

$$\max_{\eta \in \mathbb{R}} \mathbb{E}\left[u\left(\sum_{\substack{k=1 \\ k \neq n}}^{N} w_k^* \, \tilde{r}_k + \eta \tilde{r}_n + (w_n^* - \eta) \sum_{\substack{k=1 \\ k \neq n}}^{N} \lambda_k \tilde{r}_k \right) \right]. \tag{11.1}$$

Due to the optimality of $w^* \in \mathbb{R}^N$, problem (11.1) is solved by $\eta^* = w_n^*$. Whether $w_n^* > 0$ holds depends on the sign of the derivative of the function to be optimized in problem (11.1) with respect to η, evaluated at $\eta = 0$. Such a derivative can be computed as

$$\mathbb{E}\left[u'\left(\sum_{\substack{k=1 \\ k \neq n}}^{N} (w_k^* + w_n^* \lambda_k) \tilde{r}_k \right) \left(\tilde{r}_n - \sum_{\substack{k=1 \\ k \neq n}}^{N} \lambda_k \tilde{r}_k \right) \right]$$

$$= \mathbb{E}\left[u'\left(\sum_{\substack{k=1 \\ k \neq n}}^{N} (w_k^* + w_n^* \lambda_k) \tilde{r}_k \right) \tilde{\epsilon}_n \right] = \mathbb{E}\left[u'\left(\sum_{\substack{k=1 \\ k \neq n}}^{N} (w_k^* + w_n^* \lambda_k) \tilde{r}_k \right) \right] \mathbb{E}[\tilde{\epsilon}_n],$$

where the last equality follows by iterated conditioning and (3.55), since condition (3.55) implies that

$$\mathbb{E}[f(\tilde{r}_1, \dots, \tilde{r}_{n-1}, \tilde{r}_{n+1}, \dots, \tilde{r}_N) \tilde{\epsilon}_n] = \mathbb{E}[f(\tilde{r}_1, \dots, \tilde{r}_{n-1}, \tilde{r}_{n+1}, \dots, \tilde{r}_N)] \mathbb{E}[\tilde{\epsilon}_n]$$

for any integrable function $f : \mathbb{R}^{N-1} \to \mathbb{R}$. Since the utility function u is assumed to be strictly increasing, we see that the sign of the derivative is determined by the expected value $\mathbb{E}[\tilde{\epsilon}_n]$, thus proving the claim.

Solution of Exercise 3.5 The claim follows from Exercise 3.4, by writing $\tilde{r}_n = r_f + \tilde{\epsilon}_n$ and noting that, under the present assumptions, the random variable $\tilde{\epsilon}_n$ satisfies condition (3.55), for all $n = 1, \dots, N$.

Solution of Exercise 3.7 The reasoning is analogous to that used in the proof of Proposition 3.3. Indeed, since $w^*(w_0) > 0$ for all $w_0 \in \mathbb{R}_+$, it holds that $\widetilde{W}^* \geq w_0\, r_f$ on the event $\{\tilde{r} \geq r_f\}$, while $\widetilde{W}^* \leq w_0\, r_f$ on the event $\{\tilde{r} \leq r_f\}$. Hence, since the mapping $x \mapsto r'(x)$ is decreasing, we have that

$$\widetilde{W}^* \frac{u''(\widetilde{W}^*)}{u'(\widetilde{W}^*)} = -r'(\widetilde{W}^*) \geq -r'(w_0\, r_f) = w_0\, r_f \frac{u''(w_0\, r_f)}{u'(w_0\, r_f)}, \qquad \text{on } \{\tilde{r} \geq r_f\},$$

with the converse inequality holding on the event $\{\tilde{r} \leq r_f\}$. Hence:

$$\mathbb{E}\big[\widetilde{W}^* u''(\widetilde{W}^*)(\tilde{r} - r_f)\big] \geq -r'(w_0\, r_f)\mathbb{E}\big[u'(\widetilde{W}^*)(\tilde{r} - r_f)\big] = 0,$$

where the last equality follows from the optimality condition (3.4).

Solution of Exercise 3.10 Note that, as a consequence of Theorem 3.12, the two portfolios w^{*1} and w^{*2} admit the representation

$$w^{*i} = \frac{B(V^{-1}\mathbf{1}) - A(V^{-1}e)}{D} + \frac{C(V^{-1}e) - A(V^{-1}\mathbf{1})}{D}\, \mathbb{E}[\tilde{r}_{w^{*i}}], \qquad \text{for } i = 1, 2.$$

As a consequence, it holds that

$$\mathrm{Cov}(\tilde{r}_{w^{*1}}, \tilde{r}_{w^{*2}}) = w^{*1\mathsf{T}} V w^{*2}$$

$$= \left(\frac{B(V^{-1}\mathbf{1}) - A(V^{-1}e)}{D} + \frac{C(V^{-1}e) - A(V^{-1}\mathbf{1})}{D}\, \mathbb{E}[\tilde{r}_{w^{*1}}] \right)^{\mathsf{T}} V$$

$$\times \left(\frac{B(V^{-1}\mathbf{1}) - A(V^{-1}e)}{D} + \frac{C(V^{-1}e) - A(V^{-1}\mathbf{1})}{D}\, \mathbb{E}[\tilde{r}_{w^{*2}}] \right)$$

$$= \left(\frac{B(V^{-1}\mathbf{1}) - A(V^{-1}e)}{D} + \frac{C(V^{-1}e) - A(V^{-1}\mathbf{1})}{D}\, \mathbb{E}[\tilde{r}_{w^{*1}}] \right)^{\mathsf{T}}$$

$$\times \left(\frac{B\mathbf{1} - Ae}{D} + \frac{Ce - A\mathbf{1}}{D}\, \mathbb{E}[\tilde{r}_{w^{*2}}] \right).$$

Developing the product, using the relation $D = BC - A^2$ and simplifying, relation (3.24) follows.

Solution of Exercise 3.11   Since the minimum variance portfolio w^{MVP} belongs to the portfolio frontier, it suffices to apply formula (3.24) for $w^{*1} = w^{\text{MVP}}$ and $w^{*2} = w^*$. Indeed, recalling that $\mathbb{E}[\tilde{r}_{w^{\text{MVP}}}] = A/C$, the first term on the right-hand side of formula (3.24) becomes null, thus proving the claim.

Solution of Exercise 3.12   As can be easily verified from formula (3.24), if there exists a portfolio w^{zc} such that $\text{Cov}(\tilde{r}_{w^*}, \tilde{r}_{w^{\text{zc}}}) = 0$, then its expected return must be necessarily given by the expression on the right-hand side of (3.30). Since, due to Theorem 3.12 there exists a unique frontier portfolio for any given expected return $\mu \in \mathbb{R}$, the zero correlation portfolio exists and is given by the unique frontier portfolio with expected return given by the expression on the right-hand side of (3.30).

Solution of Exercise 3.13   Given any frontier portfolio (other than the minimum variance portfolio w^{MVP}), the expected return of its zero correlation portfolio is given by expression (3.30). Hence, due to (3.25), the variance of the return of the zero correlation portfolio w^{zc} with respect to a frontier portfolio w^* is given by

$$\sigma^2(\tilde{r}_{w^{\text{zc}}}) = \frac{C}{D}\left(\mathbb{E}[\tilde{r}_{w^{\text{zc}}}] - \frac{A}{C}\right)^2 + \frac{1}{C} = \frac{C}{D}\left(\frac{D/C^2}{\mathbb{E}[\tilde{r}_{w^*}] - A/C}\right)^2 + \frac{1}{C}.$$

Hence, the frontier portfolio w^* satisfies $\sigma^2(\tilde{r}_{w^*}) = \sigma^2(\tilde{r}_{w^{\text{zc}}})$ if and only if the following equality holds, due to (3.25):

$$\frac{C}{D}\left(\mathbb{E}[\tilde{r}_{w^*}] - \frac{A}{C}\right)^2 + \frac{1}{C} = \frac{C}{D}\left(\frac{D/C^2}{\mathbb{E}[\tilde{r}_{w^*}] - A/C}\right)^2 + \frac{1}{C}.$$

The solution to the above equation is given by $\mathbb{E}[\tilde{r}_{w^*}] = A/C + \sqrt{D}/C$. Then, due to Theorem 3.12, the unique frontier portfolio w^* satisfying $\mathbb{E}[\tilde{r}_{w^*}] = A/C + \sqrt{D}/C$ is given by

$$w^* = g + h\left(\frac{A}{C} + \frac{\sqrt{D}}{C}\right),$$

where g and h are defined as in Theorem 3.12.

Solution of Exercise 3.14   Recall that the minimum variance portfolio w^{MVP} is represented in the variance-expected return plane by the point $(1/C, A/C)$, while, due to expression (3.25), the frontier portfolio w^* is represented by the point $\left(C/D(\mathbb{E}[\tilde{r}_{w^*}]-A/C)^2+1/C, \mathbb{E}[\tilde{r}_{w^*}]\right)$. In the variance-expected return plane (σ^2, μ), the line connecting the two points satisfies

$$\mu = \frac{A}{C} + \frac{\mathbb{E}[\tilde{r}_{w^*}] - A/C}{(\mathbb{E}[\tilde{r}_{w^*}] - A/C)^2\, C/D}\left(\sigma^2 - \frac{1}{C}\right).$$

Hence, the intersection of the above line with the vertical axis is found by letting $\sigma^2 = 0$, yielding

$$\mu = \frac{A}{C} - \frac{1}{C}\frac{\mathbb{E}[\tilde{r}_{w*}] - A/C}{(\mathbb{E}[\tilde{r}_{w*}] - A/C)^2 \, C/D} = \frac{A}{C} - \frac{D/C^2}{\mathbb{E}[\tilde{r}_{w*}] - A/C} = \mathbb{E}[\tilde{r}_{w^{zc}}].$$

In the standard deviation - expected return plane (σ, μ), by taking the total differential of (3.25), we obtain that the slope of the tangent to the portfolio frontier at the point $\left(\sigma^2(\tilde{r}_{w*}), \mathbb{E}[\tilde{r}_{w*}]\right)$ is given by

$$\frac{d\mu}{d\sigma} = \frac{\sigma(\tilde{r}_{w*})D}{C\mathbb{E}[\tilde{r}_{w*}] - A}.$$

Hence, the intercept of the tangent with the vertical axis is given by

$$\mathbb{E}[\tilde{r}_{w*}] - \frac{d\mu}{d\sigma}\sigma(\tilde{r}_{w*}) = \frac{A}{C} - \frac{D/C^2}{\mathbb{E}[\tilde{r}_{w*}] - A/C} = \mathbb{E}[\tilde{r}_{w^{zc}}],$$

where the first equality follows from (3.25) and the second from (3.30).

Solution of Exercise 3.15 The portfolio w^q is given as the solution to the following problem:

$$\min_{w \in \mathbb{R}^N} \frac{1}{2} w^{\top} V w, \qquad \text{subject to } w^{\top} V w^p = 0 \text{ and } w^{\top} \mathbf{1} = 1.$$

The above problem can be solved by optimizing the following Lagrangian:

$$L(w, \lambda, \mu) = \frac{1}{2} w^{\top} V w - \lambda w^{\top} V w^p - \mu(w^{\top} \mathbf{1} - 1).$$

The first order conditions give

$$\frac{\partial L(w^*, \lambda, \mu)}{\partial w^*} = V w^* - \lambda V w^p - \mu \mathbf{1} = 0,$$

so that

$$w^* = \lambda w^p + \mu V^{-1} \mathbf{1},$$

where the optimal values for the Lagrange multipliers are given by

$$\lambda = \frac{1}{1 - C\sigma^2(\tilde{r}_{wp})} \qquad \text{and} \qquad \mu = -\frac{\sigma^2(\tilde{r}_{wp})}{1 - C\sigma^2(\tilde{r}_{wp})},$$

using the notation introduced in Theorem 3.12. Hence, the optimal portfolio w^q which solves the above problem can be represented as

$$
w^q = \frac{1}{1 - C\sigma^2(\tilde{r}_{w^p})} w^p - \frac{\sigma^2(\tilde{r}_{w^p})}{1 - C\sigma^2(\tilde{r}_{w^p})} V^{-1} \mathbf{1}
$$

$$
= \frac{1}{1 - C\sigma^2(\tilde{r}_{w^p})} w^p - \frac{C\sigma^2(\tilde{r}_{w^p})}{1 - C\sigma^2(\tilde{r}_{w^p})} w^{MVP},
$$

recalling that the minimum variance portfolio w^{MVP} is given by $w^{MVP} = V^{-1}\mathbf{1}/C$. The expected return $\mathbb{E}[\tilde{r}_{w^q}]$ can be easily computed as

$$
\mathbb{E}[\tilde{r}_{w^q}] = \frac{1}{1 - C\sigma^2(\tilde{r}_{w^p})} \mathbb{E}[\tilde{r}_{w^p}] - \frac{C\sigma^2(\tilde{r}_{w^p})}{1 - C\sigma^2(\tilde{r}_{w^p})} \frac{A}{C} = \frac{\mathbb{E}[\tilde{r}_{w^p}] - A\sigma^2(\tilde{r}_{w^p})}{1 - C\sigma^2(\tilde{r}_{w^p})}.
$$

As can be easily checked (compare Exercise 3.14), the value $\mathbb{E}[\tilde{r}_{w^q}]$ coincides with the intersection with the vertical axis of the line connecting the two points $(\sigma^2(\tilde{r}_{w^p}), \mathbb{E}[\tilde{r}_{w^p}])$ and $(\sigma^2(\tilde{r}_{w^{MVP}}), \mathbb{E}[\tilde{r}_{w^{MVP}}])$ in the expected return - variance plane.

Solution of Exercise 3.17   The claim follows directly from formula (3.27) by noting that, if $\mathbb{E}[\tilde{r}_{w^q}] = \mathbb{E}[\tilde{r}_{w^p}]$, then it holds that

$$
\mathrm{Cov}(\tilde{r}_{w^q}, \tilde{r}_{w^p}) = \mathrm{Cov}(\tilde{r}_{w^p}, \tilde{r}_{w^p}) = \sigma^2(\tilde{r}_{w^p}).
$$

Solution of Exercise 3.18   Let us consider the frontier portfolio $w^* \in PF^*$ with expected return $\mu := \mathbb{E}[\tilde{r}_{w^*}]$, given by (see Proposition 3.14)

$$
w^* = V^{-1}(e - r_f \mathbf{1}) \frac{\mu - r_f}{K},
$$

where $K = B - 2Ar_f + Cr_f^2 > 0$. Consider an arbitrary linear combination (with weights α and $1 - \alpha$) of the tangent portfolio w^e and the risk free asset, respectively, i.e., consider the portfolio $w(\alpha) := \alpha w^e + (1 - \alpha)w^0$, with $w^0 := 0 \in \mathbb{R}^N$. Due to Proposition 3.15, such a portfolio is characterized by

$$
w(\alpha) = \alpha w^e = \alpha V^{-1} \frac{e - r_f \mathbf{1}}{\mathbf{1}^\top V^{-1}(e - r_f \mathbf{1})}. \tag{11.2}
$$

Due to equation (3.39), the expected return associated to the portfolio $w(\alpha)$ is given by

$$
\mathbb{E}[\tilde{r}_{w(\alpha)}] = (1 - \alpha)r_f + \alpha \frac{Ar_f - B}{Cr_f - A}.
$$

Hence, by choosing $\alpha = (\mu - r_f)(A - Cr_f)/K$, we have that $\mathbb{E}[\tilde{r}_{w(\alpha)}] = \mathbb{E}[\tilde{r}_{w*}]$. As can be easily verified, if we replace this value of α into equation (11.2), we recover the portfolio w^*, thus proving the claim.

Solution of Exercise 3.19 Fix an arbitrary value $\alpha > \mu$ and consider the following auxiliary problem:

$$\max \frac{w^\top e - \mu}{\sqrt{w^\top V w}} \qquad \text{over all } w \in \Delta_N \text{ such that } w^\top e = \alpha. \qquad (11.3)$$

Clearly, the solution to this auxiliary optimization problem corresponds to the solution to Problem (3.14) with respect to the fixed expected return α, with optimal solution

$$w^*(\alpha) = g + h\alpha,$$

where g and h are defined as in Theorem 3.12. In view of Property 2 of the portfolio frontier (see formula (3.25)), the optimal value of problem (11.3) is then given by

$$\frac{\alpha - \mu}{\sqrt{\frac{C\alpha^2 - 2A\alpha + B}{D}}},$$

Optimizing over $\alpha > \mu$ the last expression gives the optimal value

$$\alpha^* = \frac{\mu A - B}{\mu C - A},$$

so that the optimal solution to the original problem (3.56) is given by

$$w^* = w^*(\alpha^*) = g + h\frac{\mu A - B}{\mu C - A}$$

$$= \frac{1}{D}(B\mu C - AB - \mu A^2 + AB)\frac{V^{-1}1}{\mu C - A} + \frac{1}{D}(\mu AC - BC - \mu AC + A^2)\frac{V^{-1}e}{\mu C - A}$$

$$= \frac{V^{-1}(1\mu - e)}{\mu C - A} = \frac{V^{-1}(e - 1\mu)}{1^\top V^{-1}(e - 1\mu)},$$

thus proving the claim.

Solution of Exercise 3.20 By Proposition 3.15, the tangent portfolio w^e is given by

$$w^e = V^{-1}\frac{e - r_f 1}{1^\top V^{-1}(e - r_f 1)}.$$

If the risky asset returns are uncorrelated then the matrix V^{-1} is diagonal, with diagonal elements $(V_{n,n}^{-1})_{n=1}^{N}$. Hence, it holds that, for all $n = 1, \dots, N$,

$$w_n^e = \frac{V_{n,n}^{-1}(e_n - r_f)}{\sum_{i=1}^{N} V_{i,i}^{-1}(e_i - r_f)}.$$

Since $e_n > r_f$ for all $n = 1, \dots, N$, this directly implies that the tangent portfolio is diversified, i.e., $w_n^e \in (0, 1)$ for all $n = 1, \dots, N$.

Solution of Exercise 3.24   If the portfolio w^* satisfies the one fund separation property, then the random variable $\tilde{r}_{w^*}$ dominates (in the sense of second order stochastic dominance) the random return $\tilde{r}_w$ of any portfolio $w \in \Delta_N$. Hence, due to Proposition 2.10, it holds that $\mathbb{E}[\tilde{r}_{w^*}] = \mathbb{E}[\tilde{r}_w]$ for all $w \in \Delta_N$. In particular, this implies that all portfolios have the same expected return. Moreover, as discussed after Proposition 2.10, we also have $\sigma^2(\tilde{r}_{w^*}) \leq \sigma^2(\tilde{r}_w)$ for any portfolio $w \in \Delta_N$, thus showing that w^* is the minimum variance portfolio.

Solution of Exercise 3.25   Define the random variables

$$\tilde{r}^* := \frac{1}{N} \sum_{n=1}^{N} (\tilde{r}_n - r_f) \quad \text{and} \quad \tilde{\epsilon}_n := \tilde{r}_n - \tilde{r}^* - r_f, \qquad \text{for all } n = 1, \dots, N.$$

Let $a_i = 1/N$ for all $i = 1, \dots, N$. Then, it is easy to check that $\sum_{i=1}^{N} a_i \tilde{\epsilon}_i = 0$ and $\sum_{i=1}^{N} a_i = 1$. Moreover, since the returns $(\tilde{r}_1, \dots, \tilde{r}_N)$ are i.i.d., the conditional expectation $\mathbb{E}[\tilde{\epsilon}_n | \tilde{r}^*]$ does not depend on n and, hence,

$$\mathbb{E}[\tilde{\epsilon}_1 | \tilde{r}^*] = \frac{1}{N} \sum_{n=1}^{N} \mathbb{E}[\tilde{\epsilon}_n | \tilde{r}^*] = \mathbb{E}\left[\frac{1}{N} \sum_{n=1}^{N} \tilde{r}_n - \tilde{r}^* \middle| \tilde{r}^* \right] - r_f = 0.$$

Since the random vector $\tilde{\mathbf{r}}$ can be equivalently written as $\tilde{\mathbf{r}} = r_f \mathbf{1} + b\tilde{r}^* + \tilde{\epsilon}$, where $\tilde{\epsilon} = (\tilde{\epsilon}_1, \dots, \tilde{\epsilon}_N)$, Proposition 3.18 then implies that the two fund monetary separation property holds.

Solution of Exercise 3.26   Note first that, for any portfolio $w \in \Delta_N$, we can write $\tilde{r}_w = \tilde{r}_{w\text{MVP}} + \tilde{\epsilon}_w$, for some random variable $\tilde{\epsilon}_w$ with $\mathbb{E}[\tilde{\epsilon}_w] = 0$. Moreover, due to equation (3.28), it holds that $\text{Cov}(\tilde{r}_w, \tilde{r}_{w\text{MVP}}) = \sigma^2(\tilde{r}_{w\text{MVP}})$. In turn, this implies that $\text{Cov}(\tilde{\epsilon}_w, \tilde{r}_{w\text{MVP}}) = 0$. Hence, due to the normal distribution of $\tilde{r}_w$, $\tilde{r}_{w\text{MVP}}$ and $\tilde{\epsilon}_w$ and since uncorrelated normal random variables are independent, we have that

$$\mathbb{E}[\tilde{\epsilon}_w | \tilde{r}_{w\text{MVP}}] = \mathbb{E}[\tilde{\epsilon}_w] = 0.$$

Then claim then follows by the same arguments used in the proof of Proposition 3.20.

Solution of Exercise 3.28 The first order condition (3.46) for the optimal consumption problem gives

$$x_1^* = \left(\frac{p_1}{p_2} \frac{1 - \pi}{\pi} \right)^{\frac{1}{\gamma - 1}} x_2^*.$$

If $p_1/p_2 > \pi/(1 - \pi)$, this directly implies that $x_1^* < x_2^*$.

Solution of Exercise 3.29 From the proof of Proposition 3.24, let us consider the first order condition evaluated at $w = 0$, using the basic properties of the covariance:

$$\mathbb{E}\big[u'(w_0 - \tilde{x})(\tilde{x} - (1 + \lambda)\mathbb{E}[\tilde{x}])\big] = \mathrm{Cov}\,(\tilde{x}, u'(w_0 - \tilde{x})) - \lambda\,\mathbb{E}[u'(w_0 - \tilde{x})]\mathbb{E}[\tilde{x}].$$

Hence, the first order condition will be negative for all values $\lambda > \lambda^*$. In correspondence of such values, the optimal insurance demand will be $w^*=0$.

Solution of Exercise 3.30 The insurance problem can be formulated as follows:

$$\max_{w \in \mathbb{R}_+}\left(-\frac{1}{a}\mathbb{E}\Big[\exp\Big(- a\big(w_0 - (1 - w)\tilde{x} - w(1 + \lambda)\mathbb{E}[\tilde{x}]\big)\Big)\Big]\right),$$

or, equivalently, using the moment generating function of the normal distribution,

$$\max_{w \in \mathbb{R}_+}\left(w_0 - (1 - w)\mu - w(1 + \lambda)\mu - \frac{a}{2}(1 - w)^2\sigma^2\right)$$

and, hence, the optimal value w^* is given by

$$w^* = \left(1 - \frac{\mu\lambda}{a\sigma^2}\right) \vee 0.$$

Hence, under the assumption that $\mu > 0$, it immediately follows that $w^* < 1$ if (and only if) $\lambda > 0$.

Solution of Exercise 3.31 Consider an alternative insurance contract with the same price p and with a different indemnity function $\bar{I}(\cdot)$. Since $\mathbb{E}[I^*(\tilde{x})] = p/(1 + \lambda) = \mathbb{E}[\bar{I}(\tilde{x})]$, the expected value of $\bar{I}(\tilde{x})$ must be equal to the expected value of $I^*(\tilde{x}) = \max\{0; \tilde{x} - K\}$. Hence, if the indemnity is increased for the loss x_j, for some $j \in \{1, \ldots, N\}$, then it must be decreased for the losses x_i, with $i \neq j$. So, consider an increase $\epsilon_j > 0$ of the indemnity in correspondence of the loss level x_j, for some $j \in \{1, \ldots, N\}$, and a decrease $\epsilon_i \geq 0$ of the indemnities in correspondence to the loss levels x_i, for all $i \in \{1, \ldots, N\}$ with $i \neq j$, so that $\sum_{i \neq j} \epsilon_i \pi_i = \epsilon_j \pi_j$, in order to ensure that $\mathbb{E}[I^*(\tilde{x})] = \mathbb{E}[\bar{I}(\tilde{x})]$. Note that, since the indemnity cannot be reduced to negative values, we must decrease the indemnity only for the loss levels $x_i > K$.

Summing up, the indemnity $\bar{I}(\tilde{x})$ can be described as follows:

$$\bar{I}(x_j) = \max\{0; x_j - K\} + \epsilon_j,$$

$$\bar{I}(x_i) = x_i - K - \epsilon_i, \qquad\qquad \text{for all } i \neq j \text{ such that } x_i \geq K + \epsilon_i.$$

Let denote by $w_1^*(\tilde{x})$ the net wealth at time $t = 1$ associated to the insurance contract with indemnity $I^*(\tilde{x})$ and price p and, analogously, denote by $\bar{w}_1(\tilde{x})$ the net wealth at time $t = 1$ associated to the insurance contract with indemnity $\bar{I}(\tilde{x})$ and price p. Note that, due to the definition of the new indemnity $\bar{I}(\tilde{x})$, it holds that

$$\bar{w}_1(x_j) = w_1^*(x_j) + \epsilon_j,$$

$$\bar{w}_1(x_i) = w_1^*(x_i) - \epsilon_i \mathbf{1}_{\{x_i \geq K + \epsilon_i\}}, \qquad \text{for all } i \neq j. \tag{11.4}$$

Note that, if $i \neq j$ and $x_i \geq K + \epsilon_i$, we have that

$$w_1^*(x_i) = w_0 - p - x_i + I^*(x_i) = w_0 - p - K$$

$$\leq w_0 - p - \min\{x_j; K\} = w_0 - p - x_j + I^*(x_j) = w_1^*(x_j). \tag{11.5}$$

From equations (11.4)–(11.5), we can observe that the indemnity scheme $\bar{I}(\tilde{x})$ decreases the wealth at time $t = 1$ for small levels of wealth, while it increases the wealth at time $t = 1$ for large levels of wealth. This implies that the original indemnity scheme $I^*(\tilde{x})$ dominates $\bar{I}(\tilde{x})$ according to the second order stochastic dominance criterion, thus establishing the result.

Solution of Exercise 3.32 Since the background risk $\tilde{y}$ is assumed to be independent of the return $\tilde{r}$ of the risky asset, the portfolio optimization problem (3.50) can be rewritten as follows, using the tower property of (conditional) expectations:

$$\max_{w \in \mathbb{R}} \mathbb{E}\big[u\big(w_0\, r_f + \tilde{y} + w(\tilde{r} - r_f)\big)\big]$$

$$= \max_{w \in \mathbb{R}} \mathbb{E}\Big[\mathbb{E}\big[u\big(w_0\, r_f + \tilde{y} + w(\tilde{r} - r_f)\big)|\tilde{r}\big]\Big]$$

$$= \max_{w \in \mathbb{R}} \mathbb{E}\big[v\big(w_0\, r_f + w(\tilde{r} - r_f)\big)\big]$$

where the utility function $v(\cdot)$ is defined by $v(x) := \mathbb{E}[u(x + \tilde{y})]$, for all $x \in \mathbb{R}$. At this stage, in view of Proposition 3.2, in order to show that $w^{**} < w^*$, it suffices to show that the function v corresponds to a greater risk aversion in comparison with the utility function u. In other words, we must check whether

$$r_v^a(x) = -\frac{v''(x)}{v'(x)} = -\frac{\mathbb{E}[u''(x + \tilde{y})]}{\mathbb{E}[u'(x + \tilde{y})]} \geq -\frac{u''(x)}{u'(x)} = r_u^a(x). \tag{11.6}$$

It holds that

$$-\mathbb{E}[u''(x+\tilde{y})] = \mathbb{E}[r_u^a(x+\tilde{y})u'(x+\tilde{y})]$$

$$= \mathrm{Cov}\left(r_u^a(x+\tilde{y}), u'(x+\tilde{y})\right) + \mathbb{E}[r_u^a(x+\tilde{y})]\mathbb{E}[u'(x+\tilde{y})]$$

$$\geq \mathbb{E}[r_u^a(x+\tilde{y})]\mathbb{E}[u'(x+\tilde{y})]$$

$$\geq r_u^a\left(\mathbb{E}[x+\tilde{y}]\right)\mathbb{E}[u'(x+\tilde{y})] = r_u^a(x)\mathbb{E}[u'(x+\tilde{y})],$$

where the first two inequalities use the assumption that u is concave and that the mapping $x \mapsto r_u^a(x)$ is decreasing and convex, respectively, together with Jensen's inequality (for the last inequality). In particular, the fact that the covariance between $r_u^a(x+\tilde{y})$ and $u'(x+\tilde{y})$ is positive follows from the monotonicity of the functions r_u^a and u' (see, e.g., Thorisson [1589]). This implies that condition (11.6) is satisfied. Finally, note that, if the mapping $x \mapsto r_u^a(x)$ is decreasing, then it holds that $dr_u^a(x)/dx \leq 0$, i.e.,

$$\frac{dr_u^a(x)}{dx} = -\frac{u'''(x)u'(x) - \left(u''(x)\right)^2}{\left(u'(x)\right)^2} \leq 0,$$

which can be easily seen to imply that $p_u^a(x) \geq r_u^a(x)$, for all $x \in \mathbb{R}$.

Solution of Exercise 3.33 :Since the utility function u is assumed to be strictly concave, the denominator of (3.53) is always strictly negative. Hence, in order to study the behavior of s^* with respect to changes in the risk free rate r_f, it suffices to study the sign of the numerator of (3.53), which can be rewritten as follows:

$$-u''(w_1 + s^* r_f)s^* r_f - u'(w_1 + s^* r_f)$$

$$= u'(w_1 + s^* r_f)\left(-\frac{u''(w_1 + s^* r_f)}{u'(w_1 + s^* r_f)}(w_1 + s^* r_f) - 1\right) + u''(w_1 + s^* r_f)w_1$$

$$= u'(w_1 + s^* r_f)\left(r_u^r(w_1 + s^* r_f) - 1\right) + u''(w_1 + s^* r_f)w_1,$$

where r_u^r denotes the coefficient of relative risk aversion associated to the utility function u. By the strict concavity of u, it is clear that the above quantity is negative if $r_u^r(w_1 + s^* r_f) < 1$.

Solution of Exercise 3.34   Recall that the coefficient of relative risk aversion is given by $r_u^r(x) = -xu''(x)/u'(x)$. By differentiating, we get

$$\frac{dr_u^r(x)}{dx} = -\frac{u''(x)}{u'(x)}\left(1 + x\frac{u'''(x)}{u''(x)} - x\frac{u''(x)}{u'(x)}\right) = r_u^a(x)\left(1 - xp_u(x) + r_u^r(x)\right).$$

If the mapping $x \mapsto r_u^r(x)$ is decreasing, then the above expression is negative. Hence, if $r_u^r(x) > 1$ for all $x \in \mathbb{R}$, this implies that $x p_u(x) > 2$, for all $x \in \mathbb{R}$, thus proving the claim.

11.3 Exercises of Chap. 4

Solution of Exercise 4.1 From the analysis of Sect. 4.1, we know that an ex-ante Pareto optimal allocation $\{x_1^{a*}, x_2^{a*}, x_1^{b*}, x_2^{b*}\}$ satisfies

$$\frac{\pi\, u^{a'}(x_1^{a*})}{(1-\pi)u^{a'}(x_2^{a*})} = \frac{\pi\, u^{b'}(x_1^{b*})}{(1-\pi)u^{b'}(x_2^{b*})},$$

so that $u^{a'}(x_1^{a*})/u^{a'}(x_2^{a*}) = u^{b'}(x_1^{b*})/u^{b'}(x_2^{b*}) =: k$. It holds that $k < 1$. Indeed, suppose on the contrary that $k \geq 1$. In that case, one would have $u^{a'}(x_1^{a*}) \geq u^{a'}(x_2^{a*})$ and $u^{b'}(x_1^{b*}) \geq u^{b'}(x_2^{b*})$, which in turns implies that $x_1^{a*} \leq x_2^{a*}$ and $x_1^{b*} \leq x_2^{b*}$, due the concavity of the utility functions. But then:

$$e_1^a + e_1^b = x_1^{a*} + x_1^{b*} \leq x_2^{a*} + x_2^{b*} = e_2^a + e_2^b,$$

thus yielding a contradiction with the hypothesis of the exercise. Hence, if (p_1, p_2) are the prices of the contingent goods in correspondence of a Pareto optimal allocation it holds that

$$\frac{p_1}{p_2} = \frac{\pi\, u^{a'}(x_1^{a*})}{(1-\pi)u^{a'}(x_2^{a*})} = \frac{\pi}{1-\pi}k < \frac{\pi}{1-\pi},$$

thus proving the claim.

Solution of Exercise 4.2 If condition (4.4) holds, then

$$u^{i'}(x_s^{i*}) = \frac{\lambda_j}{\lambda_i} u^{j'}(x_s^{j*}), \qquad \text{for all } i, j = 1, \ldots, I \text{ and } s = 1, \ldots, S.$$

Hence, for every $i, j = 1, \ldots, I$ and $s, r = 1, \ldots, S$, it holds that

$$\frac{\pi_s u^{i'}(x_s^{i*})}{\pi_r u^{i'}(x_r^{i*})} = \frac{\pi_s}{\pi_r} \frac{\lambda_j}{\lambda_i} \frac{\lambda_i}{\lambda_j} \frac{u^{j'}(x_s^{j*})}{u^{j'}(x_r^{j*})} = \frac{\pi_s u^{j'}(x_s^{j*})}{\pi_r u^{j'}(x_r^{j*})},$$

thus proving that condition (4.1) holds. Conversely, if condition (4.1) holds, then condition (4.4) can be obtained by taking

$$\lambda_i := \frac{1}{u^{i'}(x_1^{i*})}, \qquad \text{for all } i = 1, \ldots, I,$$

so that

$$\frac{\pi_s}{\pi_1}\lambda_i u^{i'}(x_s^{i*}) = \frac{\pi_s}{\pi_1}\lambda_j u^{j'}(x_s^{j*}), \qquad \text{for all } i,j = 1,\ldots,I \text{ and } s = 1,\ldots,S,$$

thus proving that the Borch condition (4.4) holds.

Solution of Exercise 4.3 The sharing rule is linear if and only if $\frac{d^2 y^i(e)}{de^2} = 0$ for all $i = 1,\ldots,I$. Hence, in view of Condition (4.11), this means that the sharing rule is linear if and only if

$$t_{u^i}'(y^i(e))\frac{dy^i(e)}{de}\sum_{j=1}^{I}t_{u^j}(y^j(e)) = t_{u^i}(y^i(e))\frac{d}{de}\left(\sum_{j=1}^{I}t_{u^j}(y^j(e))\right),$$

for all $i = 1,\ldots,I$. Substituting condition (4.11) in the above expression and simplifying, we get

$$t_{u^i}'(y^i(e)) = \frac{d}{de}\left(\sum_{j=1}^{I}t_{u^j}(y^j(e))\right),$$

for all $i = 1,\ldots,I$. Since the right hand side of the last expression does not depend on i, the claim is proved.

Solution of Exercise 4.4 We follow Huang & Litzenberger [971, Section 5.14]. Suppose that the sharing rule $\{y^i : \mathbb{R}_+ \rightarrow \mathbb{R}_+; i = 1,\ldots,I\}$ is linear in correspondence of every possible Pareto optimal allocation. Note first that, since Pareto optimal allocations are parameterized by the weights $a = (a^1,\ldots,a^I)$, the sharing rule also depends on the weights a, so that we get the representation

$$y^i(e) = \alpha_i(a) + \beta_i(a)e, \qquad \text{for all } i = 1,\ldots,I,$$

for some constants $\alpha_i(a)$ and $\beta_i(a)$ depending on the weights a, for every possible realization e of the aggregate endowment. In the present context, condition (4.6) implies that

$$a^i u^{i'}(y^i(e)) = a^j u^{j'}(y^j(e)), \tag{11.7}$$

for all $i,j = 1,\ldots,I$. By differentiating (11.7) with respect to e we obtain

$$a^i u^{i''}(y^i(e))\frac{dy^i(e)}{de} = a^j u^{j''}(y^j(e))\frac{dy^j(e)}{de},$$

for all $i,j = 1,\ldots,I$. Moreover, by differentiating (11.7) with respect to a_i and using the linear form of the sharing rule, we get

$$u^{i'}(y^i(e)) + a^i u^{i''}(y^i(e))(\alpha_{ii}(a) + \beta_{ii}(a)e) = a^j u^{j''}(y^j(e))(\alpha_{ji}(a) + \beta_{ji}(a)e),$$

for all $i,j = 1, \ldots, I$, where $\alpha_{ii}(a) := \frac{d\alpha_i(a)}{da^i}$ and $\alpha_{ji}(a) := \frac{d\alpha_j(a)}{da^i}$ and similarly for β_{ii} and β_{ji}. Combining the last two expressions and making use of the linear form of the sharing rule, we get

$$-\frac{u^{i'}\left(y^i(e)\right)}{u^{i''}\left(y^i(e)\right)} = A_i + B_i\, y^i(e), \tag{11.8}$$

for all $i = 1, \ldots, I$, where A_i and B_i are constants. Now, as follows from Exercise 4.3, the linearity of the sharing rule implies that, for every agent $i = 1, \ldots, I$, the first derivative of the risk tolerance of the utility function u^i computed in correspondence of the Pareto optimal allocation $y^i(e)$ does not depend on i. This implies that $B_i = B_j$ for all $i,j = 1, \ldots, I$. Since we assumed that the sharing rule is linear for all possible Pareto optimal allocations and since the latter are parameterized by the weights $a = (a^1, \ldots, a^I)$, equation (11.8) must hold for all possible weights a, thus showing that condition (4.15) has to be satisfied.

Solution of Exercise 4.5 In the case of generalized power utility functions, condition (4.4) leads to

$$\left(\frac{\lambda_i\delta_i}{\lambda_j\delta_j}\right)^{-b}(\gamma_i + bx_s^{i*}) = \gamma_j + bx_s^{j*}, \qquad \text{for all } i,j = 1, \ldots, I \text{ and } s = 1, \ldots, S.$$

Summing over j, we obtain

$$x_s^{i*} = \frac{be_s + \sum_{j=1}^I \gamma_j}{b(\lambda_i\delta_i)^{-b}\sum_{j=1}^I (\lambda_j\delta_j)^b} - \frac{\gamma_i}{b}, \qquad \text{for all } i = 1, \ldots, I \text{ and } s = 1, \ldots, S,$$

thus proving the linearity of the Pareto optimal sharing rule.

In the case of exponential utility functions, condition (4.4) leads to

$$\lambda_i\delta_i \exp(-x_s^{i*}/\gamma_i) = \lambda_j\delta_j \exp(-x_s^{j*}/\gamma_j), \qquad \text{for all } i,j = 1, \ldots, I \text{ and } s = 1, \ldots, S.$$

Taking logarithms and summing over j, we obtain, for all $i = 1, \ldots, I$ and $s = 1, \ldots, S$,

$$x_s^{i*} = \gamma_i \log(\lambda_i\delta_i) + \frac{\gamma_i}{\sum_{j=1}^I \gamma_j}\left(e_s - \sum_{j=1}^I \gamma_j \log(\lambda_j\delta_j)\right).$$

Solution of Exercise 4.6 Let the price-allocation couple $(q^*; x^{1*}, \ldots, x^{I*})$ correspond to an Arrow-Debreu equilibrium. Suppose that at time $t = 1$ the state of the world ω_s is realized, for some $s = 1 \ldots, S$, and suppose that (spot) markets are open at time $t = 1$, so that agents have the possibility to trade at $t = 1$ the L consumption goods at prices $q_s^* \in \mathbb{R}_{++}^L$. By definition of Arrow-Debreu equilibrium, for every

$i = 1, \ldots, I$, the allocation x^{i*} solves Problem (PO1) for agent i, subject to the feasibility constraint (4.16). In particular, this implies that $x_s^{i*} \in \mathbb{R}_{++}^L$ maximizes the function $x \mapsto U^i(x_1^{i*}, \ldots, x_{s-1}^{i*}, x, x_{s+1}^{i*}, \ldots, x_S^{i*})$ subject to the constraint

$$\sum_{l=1}^{L} q_{ls}^*(x - e_{ls}^i) \leq \sum_{l=1}^{L} q_{ls}(x_{ls}^{i*} - e_{ls}^i).$$

This means that agent i has no incentive to trade at time $t = 1$ in correspondence of the equilibrium prices q^*. Since $i = 1, \ldots, I$ and $s = 1, \ldots, S$ are arbitrary, the claim is proved.

Solution of Exercise 4.7 We proceed as in the proof of Proposition 4.15. The first order conditions of (4.29) yield, for all $s = 0, 1, \ldots, S$ and $i = 1, \ldots, I$,

$$a^i \exp(-x_s^{i*}/\gamma_i) = \theta_s \qquad \text{and} \qquad \sum_{i=1}^{I} x_s^{i*} = e_s.$$

Hence, solving for $\exp(-x_s^{i*})$,

$$\exp(-x_s^{i*}) = \theta_s^{\gamma_i}(a^i)^{-\gamma_i},$$

which, by multiplying over all i, gives that, for all $s = 0, 1, \ldots, S$,

$$\exp(-e_s) = \theta_s^{\sum_{i=1}^{I} \gamma_i} \prod_{i=1}^{I}(a^i)^{-\gamma_i},$$

and, solving for θ_s,

$$\theta_s = \exp\left(-\frac{e_s}{\sum_{i=1}^{I} \gamma_i}\right) \prod_{i=1}^{I}(a^i)^{\frac{\gamma_i}{\sum_{i=1}^{I} \gamma_i}}.$$

Hence, the utility function **u** of the representative agent is defined by

$$\mathbf{u}_0(e_0) = -\exp\left(-\frac{e_0}{\sum_{i=1}^{I} \gamma_i}\right) \prod_{i=1}^{I}(a^i)^{\frac{\gamma_i}{\sum_{i=1}^{I} \gamma_i}} \sum_{i=1}^{I} \gamma_i,$$

and similarly for $\mathbf{u}_1(e_s)$, for all $s = 1, \ldots, S$. Hence, by condition (4.32), the equilibrium price vector q^* of the S Arrow securities is given by

$$q_s^* = \frac{\delta \pi_s \exp(-e_s/\sum_{i=1}^{I} \gamma_i)}{\exp(-e_0/\sum_{i=1}^{I} \gamma_i)}, \qquad \text{for all } s = 1, \ldots, S,$$

thus proving the claim. Note that, as in the case of Proposition 4.15, the equilibrium price vector $q^* \in \mathbb{R}^S$ does not depend on the resource allocation but only on the aggregate endowment of the economy.

Solution of Exercise 4.10 For $i = a, b$, let us denote by (x_1^{i*}, x_2^{i*}) the optimal consumption plan of agent i. If the two agents have homogeneous beliefs, then the optimality conditions give

$$\frac{\pi \, x_2^{i*}}{(1 - \pi) x_1^{i*}} = \frac{p_1}{p_2}, \qquad \text{for } i = a, b,$$

thus implying that $x_2^{i*} = \frac{p_1}{p_2} \frac{1 - \pi}{\pi} x_1^{i*}$, for $i = a, b$. For each agent $i = a, b$, the budget constraint then implies that

$$\alpha^i (p_1 e_1 + p_2 e_2) = p_1 x_1^{i*} + p_2 x_2^{i*} = \frac{p_1}{\pi} x_1^{i*}.$$

In turn, the last condition implies that $x_1^{a*}/x_1^{b*} = \alpha^a/\alpha^b$ and an analogous reasoning allows to show that $x_2^{a*}/x_2^{b*} = \alpha^a/\alpha^b$. Since $e_s^a/e_s^b = \alpha^a/\alpha^b$, for $s = 1, 2$, and the feasibility constraint implies that $x_s^{a*} + x_s^{b*} = e_s^a + e_s^b$, for $s = 1, 2$, it easily follows that $x_j^{i*} = e_j^i$, for $i = a, b$ and $j = 1, 2$, thus showing that the optimal choice of the two agents is simply given by consuming their initial endowments.

Solution of Exercise 4.13 Under the present assumptions, the optimal consumption Problem (PO3) is given by

$$\max_{(x_0, x_1, \ldots, x_S) \in \mathbb{R}_+^{S+1}} \frac{1}{b-1} \left((\gamma + bx_0)^{\frac{b-1}{b}} + \delta \sum_{s=1}^{S} \pi_s (\gamma + bx_s)^{\frac{b-1}{b}} \right),$$

subject to the budget constraint (note that the budget constraint can be expressed as an equality since the utility function is strictly increasing)

$$x_0 + \sum_{s=1}^{S} p_s x_s = e_0 + \sum_{s=1}^{S} p_s e_s = \bar{e}.$$

Denoting by $(x_0^*, x_1^*, \ldots, x_S^*)$ the optimal consumption plan for Problem (PO3), the optimality conditions give

$$(\gamma + bx_0^*)^{-1/b} = \lambda \qquad \text{and} \qquad \delta \pi_s (\gamma + bx_s^*)^{-1/b} = \lambda p_s,$$

for all $s = 1, \ldots, S$, so that

$$\delta \pi_s \left(\frac{\gamma + bx_s^*}{\gamma + bx_0^*} \right)^{-1/b} = p_s, \qquad \text{for all } s = 1, \ldots, S.$$

Solving for $\gamma + bx_s^*$ and using the budget constraint, we get

$$\gamma + bx_s^* = \left(\frac{\delta \pi_s}{p_s}\right)^b \left(\gamma + b\left(\bar{e} - \sum_{r=1}^{S} p_r x_r^*\right)\right),$$

so that, multiplying by p_s and summing over all s, we obtain

$$\gamma \sum_{s=1}^{S} p_s + b \sum_{s=1}^{S} p_s x_s^* = \gamma \delta^b \sum_{s=1}^{S} \pi_s^b p_s^{1-b} + b\bar{e}\delta^b \sum_{s=1}^{S} \pi_s^b p_s^{1-b} - b\delta^b \sum_{s=1}^{S} p_s x_s^* \sum_{s=1}^{S} \pi_s^b p_s^{1-b}.$$

In turn, the last equation implies that

$$\bar{e} - x_0^* = \sum_{s=1}^{S} p_s x_s^* = \frac{\gamma}{b} \frac{\delta^b \sum_{s=1}^{S} \pi_s^b p_s^{1-b} - \sum_{s=1}^{S} p_s}{1 + \delta^b \sum_{s=1}^{S} \pi_s^b p_s^{1-b}} + \frac{\delta^b \sum_{s=1}^{S} \pi_s^b p_s^{1-b}}{1 + \delta^b \sum_{s=1}^{S} \pi_s^b p_s^{1-b}} \bar{e},$$

thus proving that the saving $\bar{e} - x_0^*$ is given by an affine function of $\bar{e}$.

Solution of Exercise 4.17   Let $z \in \mathbb{R}^N$ be an arbitrage opportunity of the second kind. By Definition 4.17, we have that $Dz \geq 0$. Since the pricing functional Q is positive, we have that $Q(Dz) \geq 0$ and, since $Dz \in I(D)$, it also holds that

$$0 \leq Q(Dz) = V(Dz) = p^\top z,$$

thus showing that z cannot be an arbitrage opportunity of the second kind.

Solution of Exercise 4.18   Due to the equivalence between statements *(i)-(ii)* of Proposition 4.18, the failure of the Law of One Price implies the existence of a portfolio $\hat{z} \in \mathbb{R}^N$ such that $D\hat{z} = 0$ and $p^\top \hat{z} \neq 0$. This implies that the price of the payoff c can be arbitrarily modified by adding multiples of the zero-payoff portfolio $\hat{z}$ to the portfolio z such that $Dz = c$.

Solution of Exercise 4.19   Note first that, due to part *(iii)* of Proposition 4.18, the absence of arbitrage opportunities implies that $p^\top z > 0$ and $p^\top z' > 0$, so that $\frac{c}{p^\top z}$ and $\frac{c'}{p^\top z'}$ are well defined. Arguing by contradiction, suppose that $\frac{c}{p^\top z} > \frac{c'}{p^\top z'}$. Define then the portfolio $\bar{z} := z - \frac{p^\top z}{p^\top z'} z'$. Then, the portfolio $\bar{z}$ satisfies $p^\top \bar{z} = 0$ and $D\bar{z} = c - \frac{p^\top z}{p^\top z'} c' > 0$, thus showing that $\bar{z}$ is a riskless arbitrage opportunity (see Proposition 4.18). This contradicts the assumption that there are no arbitrage opportunities. The case $\frac{c}{p^\top z} < \frac{c'}{p^\top z'}$ can be treated in an analogous way.

Solution of Exercise 4.20   Suppose that $p_{N+1} > p^\top z^c$. Then, consider the portfolio $z := (z^c, -1) \in \mathbb{R}^{N+1}$. At time $t = 0$, the value of the portfolio z is given by $p'^\top z = p^\top z^c - p_{N+1} < 0$, while, at time $t = 1$, the associated payoff is given by $D'z = Dz^c - c = 0$. Hence, the portfolio z is an arbitrage opportunity (of the second kind) in the extended market represented by (p', D'). An analogous argument can be applied in the case $p_{N+1} < p^\top z^c$, thus showing that $p_{N+1} = p^\top z^c$ must necessarily hold in the absence of arbitrage opportunities.

Solution of Exercise 4.21

(i): Let $z^u \in \mathbb{R}^N$ be a portfolio such that $Dz^u \geq c$ and, analogously, let $z^l \in \mathbb{R}^N$ be a portfolio such that $Dz^l \leq c$. Then, if there are no arbitrage opportunities, we must have

$$p^\top z^u \geq p^\top z^l. \tag{11.9}$$

   Indeed, the inequality $p^\top z^u < p^\top z^l$ would easily imply that the portfolio $\bar{z} := z^u - z^l$ is an arbitrage opportunity, since $D\bar{z} \geq 0$ and $p^\top \bar{z} < 0$. By taking the minimum and the maximum on the left- and right-hand sides of (11.9), the claim follows. In particular, note that it is enough to exclude the existence of arbitrage opportunities of the second kind in order to establish the claim.

(ii): If $c \in I(D)$, there exists a portfolio z^c such that $Dz^c = c$. The claim then follows from the previous step, noting that $q_u(c) \leq p^\top z^c \leq q_l(c)$ and recalling that $V(c) = p^\top z^c$.

(iii): By the first step, it suffices to prove that $c \notin I(D)$ implies that $q_u(c) \neq q_l(c)$. Arguing by contradiction, suppose that $q_u(c) = q_l(c)$, so that there exist two portfolios $z^u, z^l \in \mathbb{R}^N$ such that

$$Dz^u \geq c \qquad \text{and} \qquad Dz^l \leq c,$$

   with $p^\top z^l = p^\top z^u$. Moreover, since $c \notin I(D)$, at least one of the above two inequalities must be a strict inequality. Then, the portfolio $\bar{z} := z^u - z^l$ satisfies $p^\top \bar{z} = 0$ and $D\bar{z} > 0$, i.e., $\bar{z}$ is an arbitrage opportunity. Since the market is assumed to be free of arbitrage opportunities, this shows that $c \notin I(D)$ implies that $q_u(c) > q_l(c)$.

Solution of Exercise 4.22 This property is a consequence of the Law of One Price. Suppose that there exist two portfolios $z^c, z^{c'} \in \mathbb{R}^N$ such that $Dz^c = Dz^{c'}$, while $p^\top z^c > p^\top z^{c'}$. In this case, it would be possible to create wealth at time $t = 0$ out of nothing, by simply observing that the portfolio $\bar{z} := z^c - z^{c'}$ satisfies $D\bar{z} = 0$ while $p^\top \bar{z} > 0$. In this case, at the initial time $t = 0$, an agent could invest in the portfolio $z^{c'}$ and sell short the portfolio z^c. Such an investment strategy yields a strictly positive amount of wealth at time $t = 0$ and does not affect an agent's consumption plan at $t = 1$, i.e., the portfolio $\bar{z}$ is an arbitrage opportunity. Clearly, any non-satiated agent would invest in an unlimited way into this strategy and markets could not clear, thus contradicting the existence of an equilibrium. Of course, a similar reasoning holds in the case $p^\top z^c < p^\top z^{c'}$, thus showing that we must have $p^\top z^c = p^\top z^{c'}$.

Solution of Exercise 4.23 Suppose that $p_{N+1} \geq q_u(c)$ (the case $p_{N+1} \leq q_l(c)$ can be treated in an analogous way). In view of Definition 4.23, there exists a portfolio

$z \in \mathbb{R}^N$ such that $Dz \geq c$ and $p^\top z = q_u(c)$. In particular, since $c \notin I(D)$, the inequality $Dz \geq c$ cannot be an equality, so that $Dz > c$. Consider then the portfolio $z' := (z, -1) \in \mathbb{R}^{N+1}$. Then, the value at time $t = 0$ of such a portfolio is given by $p^\top z - p_{N+1} \leq 0$, while the payoff at time $t = 1$ is given by $Dz - c > 0$. We have thus shown that the portfolio z' is an arbitrage opportunity in the extended market represented by (p', D').

Solution of Exercise 4.31   Arguing by contradiction, suppose that $\bar{z} \in \mathbb{R}^N$ is an arbitrage opportunity. If $\bar{z}$ is an arbitrage opportunity of the first kind (or if $p^\top \hat{z} \leq 0$, so that the portfolio $\hat{z}$ is an arbitrage opportunity of the first kind), the claim follows exactly as in the proof of Proposition 4.27, since u^i is strictly increasing in its second argument. Suppose that $p^\top \hat{z} > 0$ and that $\bar{z}$ is an arbitrage opportunity of the second kind, so that $p^\top \bar{z} < 0$. Consider the portfolio $\bar{z}^i := z^{i*} + \bar{z} - \frac{p^\top \bar{z}}{p^\top \hat{z}}\hat{z}$. Such a portfolio satisfies

$$p^\top \bar{z}^i = p^\top z^{i*} \qquad \text{and} \qquad D\bar{z}' = Dz^{i*} + D\bar{z} - \frac{p^\top \bar{z}}{p^\top \hat{z}}D\hat{z} > Dz^{i*}.$$

Since u^i is strictly increasing in its second argument, this contradicts the optimality of z^{i*}.

Solution of Exercise 4.32

(i): This statement follows from Exercise 4.30 together with the same argument used in the proof of part (i) of Proposition 4.18.

(ii): This statement follows from Exercise 4.31 together with Proposition 4.18.

Solution of Exercise 4.33   In view of Definition 4.17, it is straightforward to check that there are no arbitrage opportunities for any price $p \in (0, 1)$. Since the utility function u is strictly increasing, the budget constraint (4.41) is satisfied as an equality in correspondence of an optimal portfolio, so that the portfolio optimization problem can be rewritten as

$$\max_{z \in \mathbb{R}}\left(e_0 + \frac{e_1}{2} + \frac{e_2}{2} + z(1 - p)\right).$$

Clearly, since $p \in (0, 1)$ and consumption is not restricted to be non-negative, this problem does not admit an optimal solution.

Solution of Exercise 4.34   We can apply the representative agent analysis, denoting by $\mathbf{u}_0$ and $\mathbf{u}_1$ the representative agent's utility functions for consumption at time $t = 0$ and at time $t = 1$, respectively. The price q_1^* of the first Arrow security is equal to the corresponding state price m_1, which in turn is given by, due to equation (4.52),

$$m_1 = \delta\frac{\mathbf{u}_1'(e_1)}{\mathbf{u}_0'(e_0)} = \frac{1}{r_f}\frac{\mathbf{u}_1'(e_1)}{\sum_{s=1}^{S}\pi_s \mathbf{u}_1'(e_s)},$$

where the second equality follows from equation (4.50). Since the aggregate endowment in correspondence of the different states of the world satisfies $e_1 \leq \ldots \leq e_S$, it follows that $m_1 \geq 1/r_f$, due to the concavity of $\mathbf{u}_1$.

Solution of Exercise 4.35   If markets are complete and there is no aggregate risk, then any equilibrium allocation is Pareto efficient and characterized by an equal optimal consumption in correspondence of any state of the world at time $t = 1$ (i.e., the risk is perfectly diversified). The claim then follows directly from equation (4.48), since the covariance with respect to a constant random variable is zero.

Solution of Exercise 4.36

(*i*):   From the analysis of Sect. 4.3, since the agents' endowments are expressed in terms of units of the traded securities, we know that the equilibrium allocation of the original economy with I agents is Pareto optimal and characterized by a linear sharing rule (see Proposition 4.5). As a consequence, the equilibrium allocation can be obtained as the no-trade equilibrium of a single representative agent. Moreover, as shown in the proof of Proposition 4.15, the utility function of the representative agent is of the generalized power form.

(*ii*):   In view of equation (4.52), in the representative agent economy, the stochastic discount factor valuation rule, applied to the payoff of the Call option (recall that, as explained at the end of Sect. 4.4, the stochastic discount factor valuation rule (4.52) can be also applied to payoffs not traded in the financial market, compare also with (4.49)), together with the explicit form of the representative agent's utility function obtained in the proof of Proposition 4.15, gives that (recall that $\sum_{i=1}^{I} \gamma_i = 0$)

$$p^{\text{call},n} = \delta\,\mathbb{E}\left[\max\{\tilde{d}_n - k; 0\}\frac{\mathbf{u}_1'(\tilde{e})}{\mathbf{u}_0'(e_0)} \right] = \delta\,\mathbb{E}\left[\max\{\tilde{d}_n - k; 0\}\left(\frac{\tilde{e}}{e_0}\right)^{-\frac{1}{b}} \right].$$

(*iii*):   Thanks to the distributional assumptions, the expected value (4.64) can be explicitly calculated. Indeed,

$$p^{\text{call},n} = p_n \int_{-\infty}^{\infty} \int_{\log\left(\frac{k}{p_n}\right)}^{\infty} \left(e^y - \frac{k}{p_n} \right) e^x f(x, y)\,dy\,dx$$

$$= p_n \int_{-\infty}^{\infty} \int_{\log\left(\frac{k}{p_n}\right)}^{\infty} e^{x+y} f(x, y)\,dy\,dx - k \int_{-\infty}^{\infty} \int_{\log\left(\frac{k}{p_n}\right)}^{\infty} e^x f(x, y)\,dy\,dx,$$

$$\tag{11.10}$$

where f denotes the joint density of the random variables $(\log(\delta(\frac{\tilde{e}}{e_0})^{-\frac{1}{b}})$, $\log(\frac{\tilde{d}_n}{p_n}))$. Recall that, from the basic properties of the bivariate normal distribution, the conditional distribution of $\log(\delta\tilde{e}/e_0)^{-1/b}$ given that $\log(\tilde{d}_n/p_n) = y$, with density denoted by $f(x|y)$, is normal with mean $\mu_e + \rho(\sigma_e/\sigma_n)(y - \mu_n)$ and variance $(1 - \rho^2)\sigma_e^2$. This allows us to compute the second term of (11.10) as follows:

$$
\begin{aligned}
\int_{-\infty}^{\infty} \int_{\log\left(\frac{k}{p_n}\right)}^{\infty} e^x f(x, y) dy dx &= \int_{\log\left(\frac{k}{p_n}\right)}^{\infty} f(y) \int_{-\infty}^{\infty} e^x f(x|y) dx \, dy \\
&= \int_{\log\left(\frac{k}{p_n}\right)}^{\infty} e^{\mu_e + \rho(\sigma_e/\sigma_n)(y - \mu_n) + \frac{(1-\rho^2)\sigma_e^2}{2}} f(y) dy \\
&= \frac{e^{\mu_e + \sigma_e^2/2}}{\sqrt{2\pi}\sigma_n} \int_{\log\left(\frac{k}{p_n}\right)}^{\infty} e^{\rho(\sigma_e/\sigma_n)(y - \mu_n) - \frac{\rho^2\sigma_e^2}{2} - \frac{(y - \mu_n)^2}{2\sigma_n^2}} dy \\
&= \frac{e^{\mu_e + \sigma_e^2/2}}{\sqrt{2\pi}\sigma_n} \int_{\log\left(\frac{k}{p_n}\right)}^{\infty} e^{-\frac{(y - \mu_n - \rho\sigma_e\sigma_n)^2}{2\sigma_n^2}} dy \\
&= \frac{e^{\mu_e + \sigma_e^2/2}}{\sqrt{2\pi}\sigma_n} \int_{\frac{\log\left(\frac{k}{p_n}\right) - \mu_n}{\sigma_n} - \rho\sigma_e}^{\infty} e^{-\frac{y^2}{2}} dy \\
&= e^{\mu_e + \sigma_e^2/2} N\left(\frac{\log\left(\frac{p_n}{k}\right) + \mu_n}{\sigma_n} + \rho\sigma_e \right).
\end{aligned}
$$
$$ (11.11) $$

Similarly, for the first term of (11.10) we have

$$
\begin{aligned}
\int_{-\infty}^{\infty} \int_{\log\left(\frac{k}{p_n}\right)}^{\infty} e^{x+y} f(x, y) dy \, dx &= \int_{\log\left(\frac{k}{p_n}\right)}^{\infty} e^y f(y) \int_{-\infty}^{\infty} e^x f(x|y) dx dy \\
&= \int_{\log\left(\frac{k}{p_n}\right)}^{\infty} e^{y + \mu_e + \rho(\sigma_e/\sigma_n)(y - \mu_n) + \frac{(1-\rho^2)\sigma_e^2}{2}} f(y) dy \\
&= \frac{e^{\mu_e + \frac{\sigma_e^2}{2}}}{\sqrt{2\pi}\sigma_n} \int_{\log\left(\frac{k}{p_n}\right)}^{\infty} e^{y + \rho(\sigma_e/\sigma_n)(y - \mu_n) - \frac{\rho^2\sigma_e^2}{2} - \frac{(y - \mu_n)^2}{2\sigma_n^2}} dy \\
&= \frac{e^{\mu_e + \mu_n + \frac{\sigma_e^2 + 2\rho\sigma_e\sigma_n + \sigma_n^2}{2}}}{\sqrt{2\pi}\sigma_n} \int_{\log\left(\frac{k}{p_n}\right)}^{\infty} e^{-\frac{(y - \mu_n - \rho\sigma_e\sigma_n - \sigma_n^2)^2}{2\sigma_n^2}} dy \\
&= e^{\mu_e + \mu_n + \frac{\sigma_e^2 + 2\rho\sigma_e\sigma_n + \sigma_n^2}{2}} N\left(\frac{\log\left(\frac{p_n}{k}\right) + \mu_n}{\sigma_n} + \rho\sigma_e + \sigma_n \right).
\end{aligned}
$$
$$ (11.12) $$

Note that

$$ e^{\mu_e + \frac{\sigma_e^2}{2}} = \delta\, \mathbb{E}\left[\left(\frac{\tilde{e}}{e_0} \right)^{-\frac{1}{b}} \right] = \frac{1}{r_f}, \qquad (11.13) $$

since the above expression represents the arbitrage free price of a risk free asset which pays one unit of wealth in correspondence of every possible state of the world. Similarly, applying formula (4.52) to the underlying security of the Call option we get

$$e^{\mu_e + \mu_n + \frac{\sigma_e^2 + 2\rho\sigma_e\sigma_n + \sigma_n^2}{2}} = \delta \, \mathbb{E}\left[\frac{\tilde{d}_n}{p_n}\left(\frac{\tilde{e}}{e_0}\right)^{-\frac{1}{b}}\right] = 1. \tag{11.14}$$

Hence, putting together (11.10), (11.11), (11.12), (11.13) and (11.14), we get

$$p^{\text{call},n} = p_n N\left(\frac{\log(\frac{p_n}{k}) + \mu_n}{\sigma_n} + \rho\sigma_e + \sigma_n\right) - \frac{k}{r_f}N\left(\frac{\log\left(\frac{p_n}{k}\right) + \mu_n}{\sigma_n} + \rho\sigma_e\right)$$

$$= p_n N\left(\frac{\log(\frac{p_n}{k}) + \log(r_f)}{\sigma_n} + \frac{\sigma_n}{2}\right) - \frac{k}{r_f}N\left(\frac{\log(\frac{p_n}{k}) + \log(r_f)}{\sigma_n} - \frac{\sigma_n}{2}\right),$$

where the last equality follows from (11.13)–(11.14), which imply that $\mu_n + \rho\sigma_e\sigma_n + \sigma_n^2/2 = \log(r_f)$. This yields the explicit formula (4.65).

Solution of Exercise 4.37   Consider first the case of an unleveraged firm subject to taxation. In this case, at date $t = 1$, a tax equal to $\tau V^1(\omega_s)$ has to be payed and, similarly to (4.58), the firm value at the initial date $t = 0$ is then given by

$$V_0^{\text{un,tax}} = (1 - \tau) \sum_{s=1}^{S} m_s V_1(\omega_s),$$

for any choice of the state price vector $m \in \mathbb{R}^S_{++}$. On the other hand, the value at $t = 0$ of a leveraged firm with debt of nominal value K and subject to taxation is given by

$$V_0^{\text{lv,tax}} = (1 - \tau) \sum_{s=1}^{S} m_s \max\{V_1(\omega_s) - K; 0\} + \sum_{s=1}^{S} m_s \min\{V_1(\omega_s); K\}$$

$$= (1 - \tau) \sum_{s=1}^{S} m_s V_1(\omega_s) + \tau \sum_{s=1}^{S} \min\{V_1(\omega_s); K\}$$

$$= V_0^{\text{un,tax}} + \tau B_0,$$

where the last equality makes use of relation (4.59), thus proving the claim.

11.4   Exercises of Chap. 5

Solution of Exercise 5.1   Observe first that

$$\text{Cov}\left(\frac{\delta \mathbf{u}_1'(\tilde{x}^m)}{\mathbf{u}_0'(x_0^m)}, \tilde{r}^m\right) = \frac{\delta}{\mathbf{u}_0'(x_0^m)}\,\text{Cov}\left(\mathbf{u}_1'(\tilde{x}^m), \tilde{r}^m\right) = \frac{\delta}{\mathbf{u}_0'(x_0^m)}\,\text{Cov}\left(\mathbf{u}_1'(w_0^m \tilde{r}^m), \tilde{r}^m\right),$$

where $w_0^m := \sum_{i=1}^{I}\sum_{n=0}^{N} e_n^i\, p_n$ denotes the value at the initial date $t = 0$ of the market portfolio. Since the representative agent's utility function is strictly concave and $w_0^m > 0$, the map $x \mapsto \mathbf{u}_1'(w_0^m x)$ is strictly decreasing. It then follows that, using the definition of covariance,

$$\begin{aligned}
\text{Cov}\left(\mathbf{u}_1'(w_0^m \tilde{r}^m), \tilde{r}^m\right) &= \mathbb{E}\left[\left(\mathbf{u}_1'(w_0^m \tilde{r}^m) - \mathbb{E}[\mathbf{u}_1'(w_0^m \tilde{r}^m)]\right)(\tilde{r}^m - \mathbb{E}[\tilde{r}^m])\right] \\
&= \mathbb{E}\left[\left(\mathbf{u}_1'(w_0^m \tilde{r}^m) - \mathbf{u}_1'(w_0^m \mathbb{E}[\tilde{r}^m]) + \mathbf{u}_1'(w_0^m \mathbb{E}[\tilde{r}^m]) - \mathbb{E}[\mathbf{u}_1'(w_0^m \tilde{r}^m)]\right)\right. \\
&\qquad \left. (\tilde{r}^m - \mathbb{E}[\tilde{r}^m])\right] \\
&= \mathbb{E}\left[\left(\mathbf{u}_1'(w_0^m \tilde{r}^m) - \mathbf{u}_1'(w_0^m \mathbb{E}[\tilde{r}^m])\right)(\tilde{r}^m - \mathbb{E}[\tilde{r}^m])\right] \\
&= \mathbb{E}\left[\left(\mathbf{u}_1'(w_0^m \tilde{r}^m) - \mathbf{u}_1'(w_0^m \mathbb{E}[\tilde{r}^m])\right)(\tilde{r}^m - \mathbb{E}[\tilde{r}^m])\, \mathbf{1}_{\{\tilde{r}^m > \mathbb{E}[\tilde{r}^m]\}}\right] \\
&\quad + \mathbb{E}\left[\left(\mathbf{u}_1'(w_0^m \tilde{r}^m) - \mathbf{u}_1'(w_0^m \mathbb{E}[\tilde{r}^m])\right)(\tilde{r}^m - \mathbb{E}[\tilde{r}^m])\, \mathbf{1}_{\{\tilde{r}^m \le \mathbb{E}[\tilde{r}^m]\}}\right] \\
&\le 0,
\end{aligned}$$

from which the claim then follows directly.

Solution of Exercise 5.2   Let us denote by $w \in \mathbb{R}^N$ a portfolio, parameterized in terms of proportions of initial wealth invested in the N risky assets (with $1 - w^\top \mathbf{1}$ being the proportion of wealth invested in the riskless asset). The corresponding random return is given by

$$\tilde{r}^w = r_f + w^\top(\tilde{\mathbf{r}} - r_f),$$

where $\tilde{\mathbf{r}} = (\tilde{r}_1, \ldots, \tilde{r}_n)^\top$. Hence:

$$\text{Corr}\left(\mathbf{u}_1'(\tilde{x}^m), \tilde{r}^w\right) = \text{Corr}\left(\mathbf{u}_1'(\tilde{x}^m), w^\top \tilde{\mathbf{r}}\right).$$

Consider then the minimization problem

$$\min_{w \in \mathbb{R}^N}\text{Corr}\left(\mathbf{u}_1'(\tilde{x}^m), w^\top \tilde{\mathbf{r}}\right) = \min_{w \in \mathbb{R}^N}\frac{\mathbb{E}\left[\left(\mathbf{u}_1'(\tilde{x}^m) - \mathbb{E}[\mathbf{u}_1'(\tilde{x}^m)]\right)w^\top(\tilde{\mathbf{r}} - \mathbb{E}[\tilde{\mathbf{r}}])\right]}{\sqrt{\text{Var}(\mathbf{u}_1'(\tilde{x}^m))}\sqrt{\text{Var}(w^\top \tilde{\mathbf{r}})}}.$$

The first order condition for the above problem yields

$$\text{Cov}\left(\mathbf{u}_1'(\tilde{x}^m), \tilde{r}_n\right)\text{Var}\left((\hat{w})^\top \tilde{\mathbf{r}}\right) - \text{Cov}\left(\mathbf{u}_1'(\tilde{x}^m), (\hat{w})^\top \tilde{\mathbf{r}}\right)\text{Cov}\left((\hat{w})^\top \tilde{\mathbf{r}}, \tilde{r}_n\right) = 0,$$

for all $n = 1, \ldots, N$. Equivalently, we have that

$$\frac{\mathrm{Cov}\left(\mathbf{u}_1'(\tilde{x}^m), \tilde{r}_n\right)}{\mathrm{Cov}\left(\mathbf{u}_1'(\tilde{x}^m), \tilde{r}^{\hat{w}}\right)} = \frac{\mathrm{Cov}\left(\tilde{r}^{\hat{w}}, \tilde{r}_n\right)}{\mathrm{Var}\left(\tilde{r}^{\hat{w}}\right)}, \qquad \text{for all } n = 1, \ldots, N. \qquad (11.15)$$

Relation (5.3) applied to the portfolio $\hat{w}$ gives

$$\mathbb{E}[\tilde{r}^{\hat{w}}] - r_f = -r_f \frac{\delta \, \mathrm{Cov}\left(\mathbf{u}_1'(\tilde{x}^m), \tilde{r}^{\hat{w}}\right)}{\mathbf{u}_0'(x_0^m)}$$

from which (5.8) follows by relying on (5.3) and making use of (11.15).

Solution of Exercise 5.3   For simplicity of notation, we shall omit the superscript * for denoting the optimal solution. In correspondence of an equilibrium allocation, the optimal consumption $\widetilde{W}^i$ of every agent i has to satisfy the first order optimality condition (3.4), so that

$$\mathbb{E}[u^{i'}(\widetilde{W}^i)](\mathbb{E}[\tilde{r}_n] - r_f) = -\mathrm{Cov}\left(u^{i'}(\widetilde{W}^i), \tilde{r}_n\right), \qquad \text{for all } n = 1, \ldots, N.$$

Due to the assumption of normally distributed returns, the couple $(\widetilde{W}^i, \tilde{r}_n)$ is distributed according to a bivariate normal law, for all $n = 1, \ldots, N$. Hence, we can apply Stein's lemma (see Lemma 3.9):

$$\mathbb{E}[u^{i'}(\widetilde{W}^i)](\mathbb{E}[\tilde{r}_n] - r_f) = -\mathbb{E}[u^{i''}(\widetilde{W}^i)]w_0^i \, \mathrm{Cov}\left(\tilde{r}^{w^i}, \tilde{r}_n\right), \qquad \text{for all } n = 1, \ldots, N,$$

where $\tilde{r}^{w^i}$ denotes the return on the optimal portfolio of the i-th agent and w_0^i denotes the total wealth allocated by agent i at date $t = 0$ in the $N + 1$ available assets (recall also that $\tilde{r}^{w^i} = \widetilde{W}^i / w_0^i$). Dividing by $\mathbb{E}[u^{i''}(\widetilde{W}^i)]$, summing over all $i = 1, \ldots, I$ and rearranging terms, we get

$$\mathbb{E}[\tilde{r}_n] - r_f = w_0^m \left(\sum_{i=1}^{I} (\theta^i)^{-1}\right)^{-1} \mathrm{Cov}(\tilde{r}^m, \tilde{r}_n), \qquad \text{for all } n = 1, \ldots, N, \qquad (11.16)$$

where $w_0^m := \sum_{i=1}^{I} w_0^i$ is the economy's aggregate wealth and $\tilde{r}^m = \sum_{i=1}^{I} \widetilde{W}^i / w_0^m$. Note that the quantity $(\sum_{i=1}^{I} (\theta^i)^{-1})^{-1}$ represents the harmonic average of the individual global absolute risk aversion coefficients. Moreover, relation (11.16) also holds for arbitrary portfolios of the traded assets and, hence, in particular for the market portfolio itself, thus leading to

$$\mathbb{E}[\tilde{r}^m] - r_f = w_0^m \left(\sum_{i=1}^{I} (\theta^i)^{-1}\right)^{-1} \mathrm{Var}(\tilde{r}^m). \qquad (11.17)$$

We have thus shown that the risk premium on the market portfolio can be expressed in terms of the harmonic mean of the individual global absolute risk aversion coefficients. Note also that the quantity $w_0^m(\sum_{i=1}^I (\theta^i)^{-1})^{-1}$ can be thought of as an *aggregate relative risk aversion coefficient* of the economy. Moreover, by combining equations (11.16) and (11.17) we immediately recover the CAPM relation (5.9).

Solution of Exercise 5.4 Assuming a quadratic utility function, the global absolute risk aversion coefficient of agent i is given by

$$\theta_i = \left(\frac{a_i}{b_i} - \mathbb{E}[\widetilde{W}^i]\right)^{-1}, \qquad \text{for all } i = 1, \ldots, I.$$

Then, the same arguments employed in the solution of Exercise 5.3 allow to show that

$$\mathbb{E}[\tilde{r}^m] - r_f = w_0^m \left(\sum_{i=1}^I \left(\frac{a^i}{b^i} - \mathbb{E}[\widetilde{W}^m]\right)\right)^{-1} \text{Var}(\tilde{r}^m).$$

Solution of Exercise 5.6 Recall that, in the representative agent economy, the equilibrium allocation is characterized in terms of the representative agent's optimal consumption problem. In particular, the market portfolio represents the optimal portfolio for the representative agent in a single agent economy. Hence, we can apply the results of Sect. 2.2 on the optimal portfolio problem of a single agent. In particular, the result follows directly from Propositions 2.5 and 2.6.

Solution of Exercise 5.7

(i): As in the proof of Proposition 5.6, note that $\mathbb{E}[\hat{\ell}\,\tilde{r}_n] = 1$, for all $n = 1, \ldots, N$. Hence, letting $V(\hat{\ell}) = p^\top z^\ell$ be the value at time $t = 0$ of the portfolio z^ℓ, it holds that, for an arbitrary portfolio $z \in \mathbb{R}^N$ such that $z^\top \mathbf{1} = 1$,

$$1 = \sum_{n=1}^N z_n = \sum_{n=1}^N z_n \mathbb{E}[\hat{\ell}\,\tilde{r}_n] = \mathbb{E}[\hat{\ell}\,\tilde{r}^z] = V(\hat{\ell})\mathbb{E}[\tilde{r}^\ell\,\tilde{r}^z].$$

In particular, taking $z = z^\ell$, this implies that $\mathbb{E}[\tilde{r}^z\,\tilde{r}^\ell] = \mathbb{E}[(\tilde{r}^\ell)^2]$.

(ii): We first prove that the return of the portfolio z^ℓ minimizes the second moment among all possible returns. Indeed, in view of part (i), for any portfolio z, it holds that

$$\mathbb{E}[(\tilde{r}^z)^2] = \mathbb{E}[(\tilde{r}^z - \tilde{r}^\ell + \tilde{r}^\ell)^2] = \mathbb{E}[(\tilde{r}^z - \tilde{r}^\ell)^2] + \mathbb{E}[(\tilde{r}^\ell)^2] + 2\mathbb{E}[\tilde{r}^\ell(\tilde{r}^z - \tilde{r}^\ell)]$$

$$= \mathbb{E}[(\tilde{r}^z - \tilde{r}^\ell)^2] + \mathbb{E}[(\tilde{r}^\ell)^2] \geq \mathbb{E}[(\tilde{r}^\ell)^2],$$

where the third equality follows from the first part of the exercise. The claim then follows by noting that $\mathrm{Var}(\tilde{r}^z) = \mathbb{E}[(\tilde{r}^z)^2] - (\mathbb{E}[\tilde{r}^z])^2$, so that the portfolio z^ℓ minimizes the variance among all portfolios having expected return equal to $\mathbb{E}[\tilde{r}^\ell]$.

Solution of Exercise 5.8 Defining the stochastic discount factor as $\tilde{m} = \tilde{\ell}/r_f$, equation (4.35) implies that

$$1 = \mathbb{E}[\tilde{m}\,\tilde{r}_n], \qquad \text{for all } 1, \ldots, N.$$

Recall that $\tilde{m} > 0$, due to the absence of arbitrage opportunities (see Proposition 4.22). Hence, if $\tilde{r}_n$ can take non-positive values with strictly positive probability, then there is nothing to prove, since $\mathbb{E}[\log(\tilde{r}_n)] = -\infty$. So, suppose that $\tilde{r}_n > 0$, for all $n = 1, \ldots, N$. Then, applying Jensen's inequality, we obtain

$$0 = \log\big(\mathbb{E}[\tilde{m}\,\tilde{r}_n]\big) \geq \mathbb{E}\big[\log(\tilde{m}\,\tilde{r}_n)\big],$$

so that

$$\mathbb{E}\big[\log(\tilde{r}_n)\big] \leq -\mathbb{E}\big[\log(\tilde{m})\big],$$

with the equality holding only in the case where $\tilde{m}\tilde{r}_n$ is constant.

Solution of Exercise 5.9 We restrict our attention to the case where $A/C > r_b > r_\ell$, using the notation introduced in section "The Case of N Risky Assets and a Risk Free Asset". As shown in Fig. 3.5, in this case there are two tangent portfolios, denoted by $w^{e,b}$ and $w^{e,\ell}$, corresponding to the two risk free rates r_b and r_ℓ, respectively. As made clear by Fig. 3.5, the portfolio frontier consists of four regions:

(A) positive investment in the risk free asset with return r_ℓ and short sale of the tangent portfolio $w^{e,\ell}$;
(B) positive investment in both the risk free asset with return r_ℓ and in the tangent portfolio $w^{e,\ell}$;
(C) all the wealth invested in the risky assets;
(D) borrowing at the rate r_b and positive investment in the tangent portfolio $w^{e,b}$.

If agents choose to hold efficient portfolios, then they will only choose portfolios belonging to the regions B, C and D. In any of those regions, the risky component of a frontier portfolio will be represented by a linear combination of the two portfolios $w^{e,b}$ and $w^{e,\ell}$ (due to the properties of the portfolio frontier discussed in section "The Case of N Risky Assets"). Hence, if the aggregate supply of the risk free asset is zero, then the market portfolio will be only composed by the risky assets and given as a convex linear combination of the two tangents portfolios $w^{e,b}$ and $w^{e,\ell}$. Hence, the market portfolio belongs to the portfolio frontier and relation (5.37) follows by the same arguments used in the proof of Proposition 5.7. The efficiency of the market portfolio (i.e., the fact that $\mathbb{E}[\tilde{r}^m] > \mathbb{E}[\tilde{r}^{zc(m)}]$) follows from the fact that the market portfolio is given by a convex linear combination of the two tangent portfolios $w^{e,b}$ and $w^{e,\ell}$ and the latter are efficient since $A/C > r_b > r_\ell$. Finally, it remains to show that

$$r_b \geq \mathbb{E}[\tilde{r}^{zc(m)}] \geq r_\ell.$$

To this effect, recall that, due to equation (3.30), the portfolio $w^{zc(m)}$ satisfies

$$\mathbb{E}[\tilde{r}^{zc(m)}] = \frac{A}{C} - \frac{D/C^2}{\mathbb{E}[\tilde{r}^m] - A/C}.$$

Moreover, it also holds that $\mathbb{E}[\tilde{r}^m] \geq \mathbb{E}[\tilde{r}^{w^{e,\ell}}]$, so that, applying equation (3.39) (replacing r_f with r_ℓ), we get

$$\mathbb{E}[\tilde{r}^{zc(m)}] \geq \frac{A}{C} - \frac{D/C^2}{\mathbb{E}[\tilde{r}^{w^{e,\ell}}] - A/C} = \frac{A}{C} - \frac{D/C^2}{(Ar_\ell - B)/(Cr_\ell - A) - A/C} = r_\ell.$$

Noting that $\mathbb{E}[\tilde{r}^m] \leq \mathbb{E}[\tilde{r}^{w^{e,b}}]$, an analogous computation shows that $\mathbb{E}[\tilde{r}^{zc(m)}] \leq r_b$.

Solution of Exercise 5.10 We restrict our attention to the case where $A/C > r_f$, using the notation introduced in section "The Case of N Risky Assets and a Risk Free Asset". As shown in Fig. 3.4, denoting by w^e the tangent portfolio, borrowing would occur only at the right of the point $(\sigma(\tilde{r}_{w^e}), \mathbb{E}[\tilde{r}_{w^e}])$, while, on the left of that point, an agent would invest a positive amount of wealth both in the risk free asset and in the tangent portfolio. Since borrowing is not allowed and all the agents choose to hold efficient portfolios, this implies that the market portfolio is a convex linear combination of the risk free asset and the tangent portfolio w^e. In particular, the market portfolio is efficient (i.e., $\mathbb{E}[\tilde{r}^m] > \mathbb{E}[\tilde{r}^{zc(m)}]$) and relation (5.38) can be obtained similarly as in the proof of Proposition 5.7. It remains to show that $\mathbb{E}[\tilde{r}^{zc(m)}] \leq r_f$. To this effect, recall that, due to equation (3.30), the portfolio $w^{zc(m)}$ satisfies

$$\mathbb{E}[\tilde{r}^{zc(m)}] = \frac{A}{C} - \frac{D/C^2}{\mathbb{E}[\tilde{r}^m] - A/C}.$$

Moreover, it also holds that $\mathbb{E}[\tilde{r}^m] \leq \mathbb{E}[\tilde{r}_{w^e}]$, so that, applying equation (3.39), we get

$$\mathbb{E}[\tilde{r}^{zc(m)}] \leq \frac{A}{C} - \frac{D/C^2}{\mathbb{E}[\tilde{r}_{w^e}] - A/C} = \frac{A}{C} - \frac{D/C^2}{(Ar_f - B)/(Cr_f - A) - A/C} = r_f.$$

Solution of Exercise 5.11 Under the present assumptions, it holds that

$$r^d = r_f + \beta_d(r^m - r_f),$$
$$r^{un} = r_f + \beta_{un}(r^m - r_f),$$
$$r^{lv} = r_f + \beta_{lv}(r^m - r_f).$$

On the other hand, relation (4.63) gives that

$$r^{\text{lv}} = r^{\text{un}} + (r^{\text{un}} - r^{\text{d}})\frac{B_0}{S_0^{\text{lv}}}.$$

By rewriting the last identity using the CAPM relations for r^{un} and r^{d}, we get that

$$r^{\text{lv}} = r_f + \beta_{\text{un}}(r^m - r_f) + (\beta_{\text{un}} - \beta_{\text{dm}})(r^m - r_f)\frac{B_0}{S_0^{\text{lv}}},$$

thus implying that

$$\beta_{\text{lv}} = \left(1 + \frac{B_0}{S_0^{\text{lv}}}\right)\beta_{\text{un}} + \frac{B_0}{S_0^{\text{lv}}}\beta_{\text{d}}.$$

Solution of Exercise 5.16 Proposition 5.16 can be proved by relying on arguments analogous to those used in the proof of Proposition 5.15. Suppose that, for all $N \in \mathbb{N}$, in the N-th economy it is possible to trade a risk free asset with rate of return $r_f^N > 0$. For each economy $N \in \mathbb{N}$, consider then the expected excess returns $\mathbb{E}[\tilde{r}_n^N] - r_f^N$ of the N risky assets with respect to the risk free rate r_f^N. Take the orthogonal projection of the expected excess returns onto the linear space spanned by the columns of the matrix B^N, so that

$$\mathbb{E}[\tilde{r}^N] - r_f^N = B^N \lambda^N + c^N,$$

for some $\lambda^N \in \mathbb{R}^K$ and $c^N \in \mathbb{R}^N$ with $(c^N)^\top B^N = 0$. At this point, one can follow exactly the proof of Proposition 5.15 and the result is obtained by replacing λ_0^N with r_f^N.

Solution of Exercise 5.17 By equation (3.9), note that, for all $i = 1, \ldots, I$,

$$Vw^i = \lambda_i(\mu^i - r_f\mathbf{1}),$$

with $\mu^i = (\mu_1^i, \ldots, \mu_N^i)^\top \in \mathbb{R}^N$ denoting the vector of expected returns with respect to the beliefs of agent i. Hence, using the notation introduced before Proposition 5.20, it holds that, for all $n = 1, \ldots, N$,

$$\text{Cov}(\tilde{r}^m, \tilde{r}_n) = \sum_{k=1}^{N} w_k^m V_{nk} = \sum_{k=1}^{N}\sum_{i=1}^{I} \frac{w_k^i}{\sum_{l=1}^{N}\sum_{j=1}^{I} w_l^j} V_{nk} = \frac{1}{\sum_{l=1}^{N}\sum_{j=1}^{I} w_l^j} \sum_{i=1}^{I}\sum_{k=1}^{N} w_k^i V_{nk}$$

$$= \frac{1}{\sum_{l=1}^{N}\sum_{j=1}^{I} w_l^j} \sum_{i=1}^{I} \lambda_i(\mu_n^i - r_f),$$

and, moreover,

$$\text{Var}(\tilde{r}^m) = \sum_{n=1}^{N} w_n^m \text{Cov}(\tilde{r}^m, \tilde{r}_n) = \frac{1}{\sum_{l=1}^{N} \sum_{j=1}^{I} w_l^j} \sum_{n=1}^{N} w_n^m \sum_{i=1}^{I} \lambda_i (\mu_n^i - r_f)$$

$$= \frac{1}{\sum_{l=1}^{N} \sum_{j=1}^{I} w_l^j} \sum_{i=1}^{I} \lambda_i \sum_{n=1}^{N} w_n^m (\mu_n^i - r_f) = \frac{1}{\sum_{l=1}^{N} \sum_{j=1}^{I} w_l^j} \sum_{i=1}^{I} \lambda_i (\mu^{i,M} - r_f).$$

Dividing the last two expressions by $\sum_{j=1}^{I} \lambda_j$, we get

$$\text{Cov}(\tilde{r}^m, \tilde{r}_n) = \frac{1}{\sum_{l=1}^{N} \sum_{j=1}^{I} w_l^j} \sum_{i=1}^{I} \frac{\lambda_i}{\sum_{j=1}^{I} \lambda_j} (\mu_n^i - r_f) = \frac{1}{\sum_{l=1}^{N} \sum_{j=1}^{I} w_l^j} (\bar{\mu}_n - r_f)$$

and

$$\text{Var}(\tilde{r}^m) = \frac{1}{\sum_{l=1}^{N} \sum_{j=1}^{I} w_l^j} \sum_{i=1}^{I} \frac{\lambda_i}{\sum_{j=1}^{I} \lambda_j} (\mu^{i,M} - r_f) = \frac{1}{\sum_{l=1}^{N} \sum_{j=1}^{I} w_l^j} (\bar{\mu}^M - r_f).$$

Relation (5.36) follows directly by combining the last two equations.

11.5 Exercises of Chap. 6

Solution of Exercise 6.2 By definition, it holds that $\bar{W}_t(\theta) = \theta_t^\top (\bar{S}_t + \bar{D}_t)$, for all $t = 0, 1, \dots, T$. Moreover, it is immediate to see that the self-financing conditions (6.1)–(6.2) hold as well with respect to discounted quantities (it suffices to divide both sides of the equality by r_f^t). By combining these two observations, it follows that

$$\Delta \bar{W}_{t+1}(\theta) := \bar{W}_{t+1}(\theta) - \bar{W}_t(\theta) = \theta_{t+1}^\top (\bar{S}_{t+1} + \bar{D}_{t+1} - \bar{S}_t) - \bar{c}_t$$
$$= \theta_{t+1}^\top \Delta \bar{S}_{t+1} + \theta_{t+1}^\top \bar{D}_{t+1} - \bar{c}_t,$$

thus proving the claim.

Solution of Exercise 6.3 Let $x' > x > 0$. For every $\varepsilon > 0$, there exists a consumption plan $(c_s^\varepsilon)_{s=t,\dots,T} \in \mathscr{C}_t^+(x)$ such that

$$V(x, t) \le u(c_t^\varepsilon) + \sum_{s=t+1}^{T} \delta^{s-t} \mathbb{E}[u(c_s^\varepsilon) | \mathscr{F}_t] + \varepsilon.$$

Then, since the function u is strictly increasing, it holds that

$$V(x,t) < u(c_t^\varepsilon + x' - x) + \sum_{s=t+1}^{T} \delta^{s-t}\mathbb{E}[u(c_s^\varepsilon)|\mathscr{F}_t] + \varepsilon$$

$$\leq V(x',t) + \varepsilon$$

By the arbitrariness of ε, it holds that $V(x,t) < V(x',t)$. In order to prove the concavity of the value function, let $x, x' > 0$, $(c_s)_{s=t,\ldots,T} \in \mathscr{C}_t^+(x)$, $(c'_s)_{s=t,\ldots,T} \in \mathscr{C}_t^+(x')$ and $\lambda \in (0,1)$ and define $x^\lambda := \lambda x + (1-\lambda)x'$. It is straightforward to check that the consumption process $(c_s^\lambda)_{s=t,\ldots,T}$ defined by $c_s^\lambda := \lambda c_s + (1-\lambda)c'_s$, for all $s = t,\ldots,T$, belongs to $\mathscr{C}_t^+(x^\lambda)$. Due to the concavity of the utility function u, it holds that

$$\lambda u(c_s) + (1-\lambda)u(c'_s) \leq u(c_s^\lambda),$$

for all $s = t,\ldots,T$, so that

$$\lambda\left(u(c_t) + \sum_{s=t+1}^{T} \delta^{s-t}\mathbb{E}[u(c_s)|\mathscr{F}_t]\right) + (1-\lambda)\left(u(c'_t) + \sum_{s=t+1}^{T} \delta^{s-t}\mathbb{E}[u(c'_s)|\mathscr{F}_t]\right)$$

$$\leq u(c_t^\lambda) + \sum_{s=t+1}^{T} \delta^{s-t}\mathbb{E}[u(c_s^\lambda)|\mathscr{F}_t].$$

By taking suprema, it then follows that

$$\lambda V(x,t) + (1-\lambda)V(x',t) \leq V(x^\lambda,t),$$

thus proving the concavity of $V(\cdot,t)$.

Solution of Exercise 6.4 The proposition can be proved by relying on the backward induction scheme explained after Proposition 6.3. Let us start at date $T-1$, with one period remaining until the terminal date T. Then, the optimality condition (6.20) implies that

$$\frac{1}{c_{T-1}^*} = \delta\mathbb{E}\left[\frac{1}{W_T^*}\left(\sum_{n=1}^{N} w_T^{*n}(r_T^n - r_f) + r_f\right)\Big|\mathscr{F}_{T-1}\right]$$

$$= \delta\mathbb{E}\left[\frac{1}{W_{T-1}^* - c_{T-1}^*}\Big|\mathscr{F}_{T-1}\right] = \frac{\delta}{W_{T-1}^* - c_{T-1}^*},$$

where the second equality follows from the self-financing condition (6.19) and the third equality from the adaptedness of the processes W^* and c^*. In turn, this implies that

$$c^*_{T-1} = \frac{W^*_{T-1}}{1+\delta}.$$

Having computed the optimal consumption at date $T-1$, we can now compute the value function $V(W^*_{T-1}, T-1)$ at date $T-1$. By the Bellman equation (6.11), it follows that, using the self-financing condition (6.19) and recalling that $W^*_T = c^*_T$, we get

$$V(W^*_{T-1}, T-1) = \log(c^*_{T-1}) + \delta \mathbb{E}\left[\log(c^*_T)|\mathscr{F}_{T-1}\right]$$

$$= \log\left(\frac{W^*_{T-1}}{1+\delta}\right)$$

$$+ \delta \mathbb{E}\left[\log\left(W^*_{T-1}\left(1 - \frac{1}{1+\delta}\right)\left(\sum_{n=1}^{N} w^{*n}_T(r^n_T - r_f) + r_f\right)\right)\bigg|\mathscr{F}_{T-1}\right]$$

$$= (1+\delta)\log(W^*_{T-1}) + \psi_{T-1},$$

with

$$\psi_{T-1} := -\log(1+\delta) + \delta \log\left(\frac{\delta}{1+\delta}\right) + \delta \mathbb{E}\left[\log\left(\sum_{n=1}^{N} w^{*n}_T(r^n_T - r_f) + r_f\right)\bigg|\mathscr{F}_{T-1}\right].$$

We have thus shown that (6.22) and (6.23) hold for $t = T-1$. Now we use backward induction. Hence, let us assume that the proposition holds at a generic date $t+1$, for $t = 0, 1, \ldots, T-1$. Then, equation (6.20) together with (6.22) implies that

$$\frac{1}{c^*_t} = u'(c^*_t) = \delta \mathbb{E}\left[V'(W^*_{t+1}, t+1)\left(\sum_{n=1}^{N} w^{*n}_{t+1}(r^n_{t+1} - r_f) + r_f\right)\bigg|\mathscr{F}_t\right]$$

$$= \delta \mathbb{E}\left[\frac{f(t+1)}{W^*_{t+1}}\left(\sum_{n=1}^{N} w^{*n}_{t+1}(r^n_{t+1} - r_f) + r_f\right)\bigg|\mathscr{F}_t\right]$$

$$= \delta f(t+1)\mathbb{E}\left[\frac{1}{W^*_t - c^*_t}\bigg|\mathscr{F}_t\right] = \delta\frac{f(t+1)}{W^*_t - c^*_t},$$

so that

$$c^*_t = \frac{W^*_t}{1 + \delta f(t+1)}.$$

The function $f : \{0, 1, \ldots, T\} \to \mathbb{R}$ satisfies $f(T) = 1$ and $f(t) = 1 + \delta f(t+1)$, for all $t = 0, 1, \ldots, T-1$, so that it is explicitly given by $f(t) = \frac{1-\delta^{T-t+1}}{1-\delta}$, for all

$t = 0, 1, \ldots, T$. In turn, due to condition (6.11), it holds that

$$
\begin{aligned}
V(W_t^*, t) &= u(c_t^*) + \delta \mathbb{E}[V(W_{t+1}^*, t+1)|\mathscr{F}_t] \\
&= u(c_t^*) + \delta \mathbb{E}\left[f(t+1)\log(W_{t+1}^*)|\mathscr{F}_t\right] + \delta \mathbb{E}[\psi_{t+1}|\mathscr{F}_t] \\
&= \log\left(\frac{W_t^*}{1 + \delta f(t+1)}\right) + \delta \mathbb{E}[\psi_{t+1}|\mathscr{F}_t] \\
&\quad + \delta f(t+1)\mathbb{E}\left[\log\left((W_t^* - c_t^*)\left(\sum_{n=1}^{N} w_{t+1}^{*n}(r_{t+1}^n - r_f) + r_f\right)\right)\bigg|\mathscr{F}_t\right] \\
&= (1 + \delta f(t+1))\log(W_t^*) - \log(1 + \delta f(t+1)) + \delta \mathbb{E}[\psi_{t+1}|\mathscr{F}_t] \\
&\quad + \delta f(t+1)\mathbb{E}\left[\log\left(\left(1 - \frac{1}{1 + \delta f(t+1)}\right)\left(\sum_{n=1}^{N} w_{t+1}^{*n}(r_{t+1}^n - r_f) + r_f\right)\right)\bigg|\mathscr{F}_t\right] \\
&= f(t)\log(W_t^*) + \psi_t,
\end{aligned}
$$

thus showing that equation (6.22) holds, with

$$
\begin{aligned}
\psi_t &= -\log(1 + \delta f(t+1)) + \delta \mathbb{E}[\psi_{t+1}|\mathscr{F}_t] \\
&\quad + \delta f(t+1)\mathbb{E}\left[\log\left(\left(1 - \frac{1}{1 + \delta f(t+1)}\right)\left(\sum_{n=1}^{N} w_{t+1}^{*n}(r_{t+1}^n - r_f) + r_f\right)\right)\bigg|\mathscr{F}_t\right] \\
&= \delta f(t+1)\left(\log\left(\frac{f(t)-1}{f(t)}\right)\right. \\
&\quad \left. -\frac{\log f(t)}{\delta f(t+1)} + \mathbb{E}\left[\log\left(\sum_{n=1}^{N} w_{t+1}^{*n}(r_{t+1}^n - r_f) + r_f\right)\bigg|\mathscr{F}_t\right]\right) \\
&\quad + \delta \mathbb{E}[\psi_{t+1}|\mathscr{F}_t].
\end{aligned}
$$

Finally, the optimal portfolio process $(w_t^*)_{t=1,\ldots,T}$ is characterized by equation (6.21), namely:

$$
\begin{aligned}
0 &= (W_t^* - c_t^*)\mathbb{E}\left[\frac{f(t+1)}{W_{t+1}^*}(r_{t+1}^n - r_f)\bigg|\mathscr{F}_t\right] \\
&= f(t+1)\mathbb{E}\left[\left(\sum_{n=1}^{N} w_{t+1}^{*n}(r_{t+1}^n - r_f) + r_f\right)^{-1}(r_{t+1}^n - r_f)\bigg|\mathscr{F}_t\right],
\end{aligned}
$$

for all $t = 1, \ldots, T-1$ and $n = 1, \ldots, N$, thus proving equation (6.24), since $f(t) > 0$ for all $t = 0, 1, \ldots, T$.

Solution of Exercise 6.5 Recall first that any self-financing trading-consumption strategy (θ, c) satisfying $\mathbb{P}(c_T > 0) = 1$ necessarily satisfies $W_t > c_t$ for all $t = 0, 1, \ldots, T - 1$. Indeed, by Lemma 6.19 (applied with $s = T$), if $c_T(A_T) > 0$ for all events $A_T \in \mathscr{F}_T$, then it holds that $c_t(A_t) > 0$ for all events $A_t \in \mathscr{F}_t$ and all dates $t = 0, 1, \ldots, T - 1$. Therefore, we can use the parametrization in terms of the portfolio process $(w_t)_{t=1,\ldots,T}$ defined in (6.18). The corollary can then be proved by means of the following computations:

$$
\mathbb{E}\left[\frac{W_{t+1}}{W^*_{t+1}}\Big|\mathscr{F}_t\right] = \frac{1}{f(t+1)}\mathbb{E}\left[W_{t+1} V'(W^*_{t+1}, t+1)|\mathscr{F}_t\right]
$$

$$
= \frac{W_t - c_t}{f(t+1)}\left(\mathbb{E}\left[V'(W^*_{t+1}, t+1)\left(\sum_{n=1}^{N} w^n_{t+1}(r^n_{t+1} - r_f) + r_f\right)\Big|\mathscr{F}_t\right]\right)
$$

$$
= \frac{W_t - c_t}{f(t+1)}\frac{1}{\delta c^*_t} = \frac{W_t - c_t}{(f(t)-1)c^*_t}
$$

$$
= \frac{W_t - c_t}{W^*_t - c^*_t},
$$

for all $t = 0, 1, \ldots, T - 1$, where the first equality follows from equation (6.22), the second equality from (6.19), the third equality from the optimality condition (6.20), the fourth equality from the fact that $f(t) = 1 + \delta f(t+1)$ (see Exercise 6.4) and, finally, the last equality from (6.23).

Solution of Exercise 6.6 The proposition can be proved by relying on the backward induction scheme explained after Proposition 6.3. Let us start at date $T-1$, with one period remaining until the terminal date T. Then, the optimality condition (6.20) implies that

$$
(c^*_{T-1})^{\gamma-1} = \delta\mathbb{E}\left[(W^*_T)^{\gamma-1}\left(\sum_{n=1}^{N} w^{*n}_T(r^n_T - r_f) + r_f\right)\Big|\mathscr{F}_{T-1}\right]
$$

$$
= \delta\mathbb{E}\left[(W^*_{T-1} - c^*_{T-1})^{\gamma-1}\left(\sum_{n=1}^{N} w^{*n}_T(r^n_T - r_f) + r_f\right)^{\gamma}\Big|\mathscr{F}_{T-1}\right],
$$

where the second equality follows from the self-financing condition (6.19). Hence, using the $\mathscr{F}_{T-1}$-measurability of W^*_{T-1} and c^*_{T-1}, it follows that

$$
c^*_{T-1} = \delta^{\frac{1}{\gamma-1}}(W^*_{T-1} - c^*_{T-1})\left(\mathbb{E}\left[\left(\sum_{n=1}^{N} w^{*n}_T(r^n_T - r_f) + r_f\right)^{\gamma}\Big|\mathscr{F}_{T-1}\right]\right)^{\frac{1}{\gamma-1}},
$$

which leads to

$$
c^*_{T-1} = a_{T-1}W^*_{T-1},
$$

with

$$
a_{T-1} = \frac{\delta^{\frac{1}{\gamma-1}} \left(\mathbb{E}\left[\left(\sum_{n=1}^{N} w_T^{*n}(r_T^n - r_f) + r_f \right)^{\gamma} \Big| \mathscr{F}_{T-1} \right] \right)^{\frac{1}{\gamma-1}}}{1 + \delta^{\frac{1}{\gamma-1}} \left(\mathbb{E}\left[\left(\sum_{n=1}^{N} w_T^{*n}(r_T^n - r_f) + r_f \right)^{\gamma} \Big| \mathscr{F}_{T-1} \right] \right)^{\frac{1}{\gamma-1}}}
$$

$$
= \left(1 + \left(\delta \mathbb{E}\left[\left(\sum_{n=1}^{N} w_T^{*n}(r_T^n - r_f) + r_f \right)^{\gamma} \Big| \mathscr{F}_{T-1} \right] \right)^{\frac{1}{1-\gamma}} \right)^{-1}.
$$

Having computed the optimal consumption at date $T - 1$, let us now compute the value function $V(W_{T-1}^*, T - 1)$ at date $T - 1$. By the Bellman equation (6.11), it follows that, using the self-financing condition (6.19) and recalling that $W_T^* = c_T^*$,

$$
V(W_{T-1}^*, T - 1) = \frac{(c_{T-1}^*)^{\gamma}}{\gamma} + \delta \mathbb{E}\left[\frac{(c_T^*)^{\gamma}}{\gamma} \Big| \mathscr{F}_{T-1} \right]
$$

$$
= \frac{(a_{T-1} W_{T-1}^*)^{\gamma}}{\gamma} + \delta \mathbb{E}\left[\frac{(W_{T-1}^*(1 - a_{T-1}))^{\gamma}}{\gamma} \left(\sum_{n=1}^{N} w_T^{*n}(r_T^n - r_f) + r_f \right)^{\gamma} \Big| \mathscr{F}_{T-1} \right]
$$

$$
= a_{T-1}^{\gamma-1} \frac{(W_{T-1}^*)^{\gamma}}{\gamma},
$$

where the third equality follows from the fact that, as can be checked by means of straightforward computations,

$$
a_{T-1} + \frac{\delta(1 - a_{T-1})^{\gamma}}{a_{T-1}^{\gamma-1}} \mathbb{E}\left[\left(\sum_{n=1}^{n} w_T^{*n}(r_T^n - r_f) + r_f \right)^{\gamma} \Big| \mathscr{F}_{T-1} \right] = 1.
$$

We have thus shown that (6.25) and (6.26) hold for $t = T - 1$. Now we use backward induction. Hence, let us assume that the proposition holds at a generic date $t + 1$, for $t = 0, 1, \dots, T - 1$. Then, equation (6.20) together with (6.25) implies that

$$
(c_t^*)^{\gamma-1} = u'(c_t^*) = \delta \mathbb{E}\left[V'(W_{t+1}^*, t + 1) \left(\sum_{n=1}^{N} w_{t+1}^{*n}(r_{t+1}^n - r_f) + r_f \right) \Big| \mathscr{F}_t \right]
$$

$$
= \delta \mathbb{E}\left[(W_t^* - c_t^*)^{\gamma-1} \left(\sum_{n=1}^{N} w_{t+1}^{*n}(r_{t+1}^n - r_f) + r_f \right)^{\gamma} a_{t+1}^{\gamma-1} \Big| \mathscr{F}_t \right]
$$

$$
= (W_t^* - c_t^*)^{\gamma-1} \delta \mathbb{E}\left[\left(\sum_{n=1}^{N} w_{t+1}^{*n}(r_{t+1}^n - r_f) + r_f \right)^{\gamma} a_{t+1}^{\gamma-1} \Big| \mathscr{F}_t \right],
$$

so that

$$c_t^* = a_t W_t^*,$$

with

$$a_t = \left(1 + \left(\delta \mathbb{E} \left[\left(\sum_{n=1}^{N} w_{t+1}^{*n} (r_{t+1}^n - r_f) + r_f \right)^{\gamma} a_{t+1}^{\gamma-1} \middle| \mathscr{F}_t \right] \right)^{\frac{1}{1-\gamma}} \right)^{-1}.$$

In turn, due to condition (6.11), it holds that

$$V(W_t^*, t) = u(c_t^*) + \delta \mathbb{E}[V(W_{t+1}^*, t+1) | \mathscr{F}_t]$$

$$= \frac{(a_t W_t^*)^{\gamma}}{\gamma} + \delta \mathbb{E} \left[a_{t+1}^{\gamma-1} \frac{(W_{t+1}^*)^{\gamma}}{\gamma} \middle| \mathscr{F}_t \right]$$

$$= \frac{(a_t W_t^*)^{\gamma}}{\gamma} + \delta \mathbb{E} \left[\frac{(W_t^*(1-a_t))^{\gamma}}{\gamma} \left(\sum_{n=1}^{N} w_{t+1}^{*n} (r_{t+1}^n - r_f) + r_f \right)^{\gamma} a_{t+1}^{\gamma-1} \middle| \mathscr{F}_t \right]$$

$$= a_t^{\gamma-1} \frac{(W_t^*)^{\gamma}}{\gamma} \left(a_t + \delta \frac{(1-a_t)^{\gamma}}{a_t^{\gamma-1}} \mathbb{E} \left[\left(\sum_{n=1}^{N} w_{t+1}^{*n} (r_{t+1}^n - r_f) + r_f \right)^{\gamma} a_{t+1}^{\gamma-1} \middle| \mathscr{F}_t \right] \right)$$

$$= a_t^{\gamma-1} \frac{(W_t^*)^{\gamma}}{\gamma},$$

thus showing that equation (6.25) holds. Finally, the optimal portfolio process $(w_t^*)_{t=1,\dots,T}$ is characterized by equation (6.21), namely:

$$0 = \mathbb{E} \left[a_{t+1}^{\gamma-1} (W_t^*)^{\gamma-1} (1-a_t)^{\gamma-1} \left(\sum_{n=1}^{N} w_{t+1}^{*n} (r_{t+1}^n - r_f) + r_f \right)^{\gamma-1} (r_{t+1}^n - r_f) \middle| \mathscr{F}_t \right]$$

for all $t = 1, \dots, T-1$ and $n = 1, \dots, N$, from which equation (6.27) follows, noting that the random variable $(W_t^*)^{\gamma-1}(1-a_t)^{\gamma-1}$ is strictly positive and $\mathscr{F}_t$-measurable.

Solution of Exercise 6.8 Starting from the terminal date T, if the $\mathscr{F}_{T-1}$-conditional distribution of the asset returns r_T^n, $n = 1, \dots, N$, is equal to the unconditional distribution of r_1^n, $n = 1, \dots, N$, then the optimal portfolio w_T^* which solves (6.27) is deterministic and can be characterized as the solution w^* to

$$\mathbb{E} \left[\left(\sum_{n=1}^{N} w^{*n} (r_1^n - r_f) + r_f \right)^{\gamma-1} (r_1^n - r_f) \right] = 0, \qquad \text{for all } n = 1, \dots, N.$$

$$(11.18)$$

In turn, this implies that a_{T-1} defined in Proposition 6.8 is also a deterministic quantity. By induction, for any $t = 1, \ldots, T - 1$, if the random variable a_{t+1} is deterministic, then the optimal portfolio w_{t+1}^* which solves (6.27) will not depend on the term $a_{t+1}^{\gamma-1}$ and, hence, is characterized by the deterministic vector w^* which solves (11.18). In particular, this implies that a_t is itself a deterministic quantity. This shows that the process $(a_t)_{t=0,1,\ldots,T}$ reduces to a deterministic function of time. As a consequence, the result of Corollary 6.9 holds true, in the more specific form (11.18). Note also that, due to Proposition 6.8, defining

$$b := \mathbb{E}\left[\left(\sum_{n=1}^{N} w^{*n}(r_1^n - r_f) + r_f\right)^{\gamma}\right],$$

the process $(a_t)_{t=0,1,\ldots,T}$ satisfies $a_T = 1$ and

$$\frac{1}{a_t} = 1 + (\delta b)^{\frac{1}{1-\gamma}} \frac{1}{a_{t+1}},$$

for all $t = 0, 1, \ldots, T - 1$, so that

$$\frac{1}{a_t} = \sum_{s=0}^{T-t} (\delta b)^{\frac{s}{1-\gamma}} = \frac{1 - (\delta b)^{\frac{T-t+1}{1-\gamma}}}{1 - (\delta b)^{\frac{1}{1-\gamma}}},$$

for all $t = 0, 1, \ldots, T$.

Solution of Exercise 6.10 Suppose that $q_{A_s|A_t}^* < q_{A_s}^*/q_{A_t}^*$ and consider the following strategy at the initial date $t = 0$: buy $q_{A_s}^*/q_{A_t}^*$ units of the Arrow security paying in correspondence of A_t and sell short one unit of the Arrow security paying in correspondence of A_s. The initial wealth needed to finance such a strategy is null, since $(q_{A_s}^*/q_{A_t}^*)q_{A_t}^* - q_{A_s}^* = 0$. At time t, if event A_t is realized, this strategy generates the payoff $q_{A_s}^*/q_{A_t}^*$ and we use part of it to buy one unit of the Arrow security paying in correspondence of A_s, at a price of $q_{A_s|A_t}^*$. This trade still leaves the strictly positive amount $q_{A_s}^*/q_{A_t}^* - q_{A_s|A_t}^*$. At the later date s, if event A_s is realized, then the cashflows related to the long and short positions in the Arrow security paying in A_s offset exactly, while if event A_s does not occur then nothing happens. Also, if at date t event A_t does not occur (and, hence, A_s does not occur as well), the strategy expires worthless. We have thus shown that this strategy has zero cost and generates a non-negative and non-null consumption plan, i.e., an arbitrage opportunity. If such a strategy would be possible, then every agent would invest in it in an unlimited way, thus contradicting the existence of an optimal portfolio. The case $q_{A_s|A_t}^* > q_{A_s}^*/q_{A_t}^*$ can be dealt with in an analogous way.

Solution of Exercise 6.11 The strict monotonicity of the utility functions implies that in equilibrium the available resources are fully allocated, i.e., $\sum_{i=1}^{I} \theta_t^{i*,n} = \sum_{i=1}^{I} \theta_0^{i,n} = 1$, for all $n = 1,\ldots,N$ and $t = 1,\ldots,T$. Moreover, for every $i = 1,\ldots,I$, the couple (θ^{i*}, c^{i*}) is a self-financing trading-consumption strategy, in the sense of Definition 6.1, so that condition (6.1) holds. Hence, summing condition (6.1) over i, we get

$$\sum_{i=1}^{I} c_t^{i*} = \sum_{i=1}^{I} \sum_{n=1}^{N} \left(\theta_t^{i*,n}(s_t^n + d_t^n) - \theta_{t+1}^{i*,n} s_t^n \right) = \sum_{n=1}^{N} d_t^n,$$

for all $t = 0,1,\ldots,T$, thus proving the claim.

Solution of Exercise 6.12 As a preliminary, let us introduce some notation. For a given consumption process $c = \{c_s(A_s); A_s \in \mathscr{F}_s, s = 0,1,\ldots,T\}$ and any date $t = 1,\ldots,T$ and event $A_{t-1} \in \mathscr{F}_{t-1}$, we denote by $C_t(A_{t-1})$ the $v_{t-1}(A_{t-1})$-dimensional vector of all possible values of the consumption plan c at date t in correspondence of the events $(A_t^1,\ldots,A_t^{v_{t-1}(A_{t-1})})$ that are contained in A_{t-1}.
$(i) \Rightarrow (iii)$: in view of Definition 6.14, this implication is trivial, since the dividend process of an Arrow security paying a unitary dividend at date t in correspondence of the event $A_t \in \mathscr{F}_t$ and zero otherwise can be represented as a consumption process $c = \{c_s(A_s); A_s \in \mathscr{F}_s, s = 0,1,\ldots,T\}$ such that $c_s(A_s) = \mathbf{1}_{A_t = A_s}$ if $s = t$ and zero otherwise.

$(iii) \Rightarrow (ii)$: for any $t = 1,\ldots,T$ and $A_t \in \mathscr{F}_t$, consider the Arrow security paying a unitary dividend in correspondence of event A_t at date t and zero otherwise. The dividend process of this Arrow security can be represented by means of a consumption process $c = \{c_s(A_s); A_s \in \mathscr{F}_s, s = 0,1,\ldots,T\}$ as above. By assumption, there exists a trading strategy $\theta = (\theta_s)_{s=0,1,\ldots,T}$ such that $(\theta,c) \in \mathscr{A}(x_0)$, for some $x_0 \in \mathbb{R}$. Since the process θ is predictable, it can be represented as $\theta = \{\theta_0, \theta_s(A_{s-1}); A_{s-1} \in \mathscr{F}_{s-1}, s = 1,\ldots,T\}$, with $\theta_s(A_{s-1}) \in \mathbb{R}^N$ denoting the position in the N securities over the time interval $[s-1,s]$ given the realization of the event A_{s-1} at date $s-1$. Moreover, since the process c is null after date t, we can assume without loss of generality that θ is null as well after date t. Therefore, in correspondence of date $t-1$ and any event $A_{t-1} \in \mathscr{F}_{t-1}$ such that $A_t \subseteq A_{t-1}$, condition (6.1) implies that

$$\mathbb{R}^{v_{t-1}(A_{t-1})} \ni (0 \cdots 0\, 1\, 0 \cdots 0)^\top = C_t(A_{t-1}) = \mathfrak{P}_t(A_{t-1})\theta_t(A_{t-1}),$$

with the element 1 being in correspondence of the event A_t. Equivalently, it holds that $(0 \cdots 010 \cdots 0)^\top \in I(\mathfrak{P}_t(A_{t-1}))$ and, since the event A_t is arbitrary, this implies that $I(\mathfrak{P}_t(A_{t-1})) = \mathbb{R}^{v_{t-1}(A_{t-1})}$, i.e., $\mathrm{rank}(\mathfrak{P}_t(A_{t-1})) = v_{t-1}(A_{t-1})$. Since this holds for any $t = 1,\ldots,T$, the implication $(iii) \Rightarrow (ii)$ is proved.

$(ii) \Rightarrow (i)$: let $c = \{c_t(A_t); A_t \in \mathscr{F}_t, t = 0,1,\ldots,T\}$ be an arbitrary consumption plan. At the terminal date T and for any event $A_{T-1} \in \mathscr{F}_{T-1}$, the assumption that $\mathrm{rank}(\mathfrak{P}_T(A_{T-1})) = v_{T-1}(A_{T-1})$ implies that there exists a $v_{T-1}(A_{T-1})$-dimensional vector $\theta_T(A_{T-1})$ such that

$$C_T(A_{T-1}) = \mathfrak{P}_T(A_{T-1})\theta_T(A_{T-1}).$$

Since the event A_{T-1} is arbitrary, we can then define an $\mathscr{F}_{T-1}$-measurable N-dimensional random vector θ_T such that $c_T = \theta_T^\top D_T$ holds with probability one (recall that $S_T = 0$). By backward induction, at any date $t = 1, \dots, T - 1$ and for any event $A_{t-1} \in \mathscr{F}_{t-1}$, the assumption that rank$(\mathfrak{P}_t(A_{t-1})) = \nu_{t-1}(A_{t-1})$ implies that there exists $\theta_t(A_{t-1}) \in \mathbb{R}^N$ such that

$$c_t(A_t) + \theta_{t+1}^\top(A_t)S_t(A_t) = \theta_t(A_{t-1})^\top \left(S_t(A_t) + D_t(A_t)\right) \tag{11.19}$$

holds true for all events $A_t \in \mathscr{F}_t$ such that $A_t \subseteq A_{t-1}$. In other words, letting $W_t(A_{t-1})$ be the $\mathbb{R}^{\nu_{t-1}(A_{t-1})}$-valued vector whose components are given by the left-hand side of (11.19), for all $A_t \in \mathscr{F}_t$ with $A_t \subseteq A_{t-1}$, it holds that

$$W_t(A_{t-1}) = \mathfrak{P}_t(A_{t-1})\theta_t(A_{t-1}).$$

Again, since the event A_{t-1} is arbitrary, we can define an $\mathscr{F}_{t-1}$-measurable N-dimensional random vector θ_t such that $W_t = c_t + \theta_{t+1}^\top S_t = \theta_t^\top (S_t + D_t)$ holds with probability one. In view of Definition 6.1, we have thus constructed a trading strategy θ which finances the consumption plan c, thus proving the claim. In view of (6.1), the initial wealth x needed to finance c is simply given by $x = c_0 + \theta_1^\top S_0$.

Solution of Exercise 6.13 The claim can be proved by relying on arguments analogous to those used in the proof of Proposition 4.4. For every $i = 1, \dots, I$, let the weight a^i be equal to the reciprocal of the Lagrange multiplier λ^i of agent i's optimal investment-consumption Problem (6.36) in correspondence of the optimal consumption plan c^{i*} and define the representative agent's utility function as in (6.48)–(6.49). The allocation $\{c^{i*}; i = 1, \dots, I\}$ corresponds to an equilibrium of the I agents economy and, hence, it is feasible. Moreover, in view of conditions (6.44)–(6.45), it holds that

$$\frac{u^{1\prime}\left(c_0^{1*}\right)}{\lambda^1} = \dots = \frac{u^{I\prime}\left(c_0^{I*}\right)}{\lambda^I} = q_0,$$

$$\frac{u^{1\prime}\left(c^{1*}(A_t)\right)}{\lambda^1} = \dots = \frac{u^{I\prime}\left(c^{I*}(A_t)\right)}{\lambda^I} = \frac{q_{A_t}}{\delta^t \pi_{A_t}},$$

for all $A_t \in \mathscr{F}_t$ and $t = 1, \dots, T$. In turn, in view of equation (6.49), the above conditions imply that the allocation $\{c^{i*}; i = 1, \dots, I\}$ defines the representative agent's utility function in correspondence of the endowment process e and weights $\{1/\lambda^i; i = 1, \dots, I\}$. This implies that the allocation $\{c^{i*}; i = 1, \dots, I\}$ and the prices $\{q_0, q_{A_t}; A_t \in \mathscr{F}_t, t = 1, \dots, T\}$ define a no-trade equilibrium in the representative agent economy.

Solution of Exercise 6.14 In correspondence of a no-trade equilibrium, the prices of the whole set of $\sum_{t=1}^T \nu_t$ Arrow securities are described by equation (6.50). Hence, since the endowment process can be expressed in terms of the Arrow securities (together with units of the consumption good at the initial date $t = 0$, the price of which is supposed to be normalized to $q_0 = 1$), the value at $t = 0$ of the aggregate endowment process e is given by

$$e_0 + \sum_{t=1}^{T} \sum_{A_t \in \mathscr{F}_t} q_{A_t} e(A_t) = e_0 + \sum_{t=1}^{T} \delta^t \sum_{A_t \in \mathscr{F}_t} \pi_{A_t} \frac{\mathbf{u}'(e(A_t))}{\mathbf{u}'(e_0)} e(A_t) = e_0 + \sum_{t=1}^{T} \delta^t \sum_{A_t \in \mathscr{F}_t} \pi_{A_t} e_0$$

$$= e_0 \sum_{t=0}^{T} \delta^t = e_0 \frac{1 - \delta^{T+1}}{1 - \delta},$$

since $\sum_{A_t \in \mathscr{F}_t} \pi_{A_t} = 1$ for all $t = 1, \ldots, T$.

Solution of Exercise 6.15 Let $c = (c_t)_{t=0,1,\ldots,T} \in \mathscr{C}_0^+(x)$ and $\tilde{c} = (\tilde{c}_t)_{t=0,1,\ldots,T} \in \mathscr{C}_0^+(\tilde{x})$ be two consumption processes satisfying the assumptions of the proposition and let $x := W_0(\theta)$ and $\tilde{x} := W_0(\tilde{\theta})$. Arguing by contradiction, suppose that $\tilde{x} > x$ and define the trading strategy $\hat{\theta} = (\hat{\theta}_t)_{t=0,1,\ldots,T}$ and the consumption process $\hat{c} = (\hat{c}_t)_{t=0,1,\ldots,T}$ by

$$\hat{\theta}_t := \begin{cases} 0, & \text{for } t = 0, \\ \theta_t - \tilde{\theta}_t, & \text{for } t = 1, \ldots, T; \end{cases} \quad \text{and} \quad \hat{c}_t := \begin{cases} \tilde{x} - x, & \text{for } t = 0, \\ c_t - \tilde{c}_t, & \text{for } t = 1, \ldots, T. \end{cases}$$

By the self-financing condition (6.1), since $(\theta, c) \in \mathscr{A}(x)$ and $(\tilde{\theta}, \tilde{c}) \in \mathscr{A}(\tilde{x})$, it is straightforward to check that $(\hat{\theta}, \hat{c}) \in \mathscr{A}(0)$, so that $\hat{c} \in \mathscr{C}_0^+(0)$. Moreover, the process $\hat{c}$ is identically null at all dates $t = 1, \ldots, T$ and satisfies $\hat{c}_0 = \tilde{x} - x > 0$ at the initial date $t = 0$. We have thus shown that $\hat{c}$ is an arbitrage opportunity of the second kind. The case $\tilde{x} < x$ can be treated in an analogous way.

Consider now the second part of Proposition 6.18. Let (S, D) be a couple of price-dividend processes not admitting arbitrage opportunities of the first kind and $c = (c_t)_{t=0,1,\ldots,T}$ and $\tilde{c} = (\tilde{c}_t)_{t=0,1,\ldots,T}$ two consumption processes financed by the strategies $\theta = (\theta_t)_{t=0,1,\ldots,T}$ and $\tilde{\theta} = (\tilde{\theta}_t)_{t=0,1,\ldots,T}$, respectively, and such that $\mathbb{P}(c_t = \tilde{c}_t) = 1$ for all $t = 0, 1, \ldots, T$. Arguing by contradiction, suppose that there exist a date $t \in \{1, \ldots, T\}$ and an event $A_t \in \mathscr{F}_t$ such that $W_t(\theta) < W_t(\tilde{\theta})$ holds on A_t. Similarly as above, define the trading strategy $\hat{\theta} = (\hat{\theta}_s)_{s=0,1,\ldots,T}$ and the consumption process $\hat{c} = (\hat{c}_s)_{s=0,1,\ldots,T}$ by

$$\hat{\theta}_s := \begin{cases} 0, & \text{for } s = 0, \ldots, t, \\ (\theta_s - \tilde{\theta}_s)\mathbf{1}_{A_t}, & \text{for } s = t + 1, \ldots, T; \end{cases}$$

$$\hat{c}_s := \begin{cases} 0, & \text{for } s = 0, \ldots, t - 1, \\ (W_t(\tilde{\theta}) - W_t(\theta))\mathbf{1}_{A_t}, & \text{for } s = t, \\ (c_s - \tilde{c}_s)\mathbf{1}_{A_t}, & \text{for } s = t + 1, \ldots, T. \end{cases}$$

By the self-financing condition (6.1), since $(\theta, c) \in \mathscr{A}(x_0)$ and $(\tilde{\theta}, \tilde{c}) \in \mathscr{A}(\tilde{x}_0)$, for some $x_0 > 0$ and $\tilde{x}_0 > 0$, it is straightforward to check that $(\hat{\theta}, \hat{c}) \in \mathscr{A}(0)$, so that $\hat{c} \in \mathscr{C}_0^+(0)$. Moreover, the consumption process $\hat{c}$ is identically null except at date t, when it takes the value $(W_t(\tilde{\theta}) - W_t(\theta))\mathbf{1}_{A_t} \geq 0$, with strict inequality holding in correspondence of A_t. We have thus shown that $\hat{c}$ is an arbitrage opportunity of the first kind. Note also that the strategy $(\hat{\theta}, \hat{c})$ is identically null if the event A_t does not occur at date t. The case $W_t(\tilde{\theta}) < W_t(\theta)$ can be treated in an analogous way.

Solution of Exercise 6.16 For $t = T$, the equality $W_T(\theta) = c_T$ holds by definition. By induction, for any $t \in \{0, 1, \ldots, T - 1\}$, suppose that $\mathbb{P}(W_s(\theta) \geq c_s) = 1$ holds for all $s = t + 1, \ldots, T$ and that $W_t(\theta) < c_t$ holds in correspondence of some event $A_t \in \mathscr{F}_t$. Define the consumption process $\hat{c} = (\hat{c}_s)_{s=0,1,\ldots,T}$ and the trading strategy $\hat{\theta} = (\hat{\theta}_s)_{s=0,1,\ldots,T}$ by

$$\hat{c}_s := \begin{cases} (c_t - W_t(\theta))\mathbf{1}_{A_t}, & \text{for } s = t, \\ W_{t+1}(\theta)\mathbf{1}_{A_t}, & \text{for } s = t + 1, \\ 0, & \text{otherwise;} \end{cases} \qquad \hat{\theta}_s := \begin{cases} \theta_{t+1}\mathbf{1}_{A_t}, & \text{for } s = t + 1, \\ 0, & \text{otherwise.} \end{cases}$$

Note that the process $\hat{c}$ is non-negative, since $\mathbb{P}(W_{t+1}(\theta) \geq c_{t+1} \geq 0) = 1$, and $\hat{c}_t$ is strictly positive on A_t. Moreover, it can be easily checked that $(\hat{\theta}, \hat{c}) \in \mathscr{A}(0)$. In fact, the self-financing condition (6.1) is satisfied for all $s \in \{0, \ldots, t - 1, t + 1, \ldots, T\}$, while at date t it holds that

$$\hat{\theta}_{t+1}^{\top} S_t = \mathbf{1}_{A_t} \theta_{t+1}^{\top} S_t = \mathbf{1}_{A_t}(W_t(\theta) - c_t) = -\hat{c}_t = W_t(\hat{\theta}) - \hat{c}_t.$$

Note that the strategy $(\hat{\theta}, \hat{c})$ is identically null if the event A_t does not occur at date t. Since $W_0(\hat{\theta}) = 0$, we have shown that $\hat{c} \in \mathscr{C}_0^+(0)$ is an arbitrage opportunity, thus proving the claim.

Consider now the second part of Lemma 6.19. Suppose that the price-dividend couple (S, D) does not admit arbitrage opportunities and let $(\theta, c) \in \mathscr{A}(x_0)$. Let $t \in \{0, \ldots, T - 1\}$ and suppose that there exists a date $s \in \{t + 1, \ldots, T\}$ and an event A_s such that $c_s(A_s) > 0$. Arguing as in the proof of Lemma 6.19, suppose that there exists an event $A_t \in \mathscr{F}_t$ such that $A_s \subseteq A_t$ and $W_t(\theta)(A_t) = c_t(A_t)$ and define the consumption process $\hat{c} = (\hat{c}_s)_{s=0,1,\ldots,T}$ and the trading strategy $\hat{\theta} = (\hat{\theta}_s)_{s=0,1,\ldots,T}$ by

$$\hat{c}_s := \begin{cases} 0, & \text{for } s = 0, \ldots, t, \\ c_s \mathbf{1}_{A_t}, & \text{for } s = t + 1, \ldots, T; \end{cases} \qquad \hat{\theta}_s := \begin{cases} 0, & \text{for } s = 0, \ldots, t, \\ \theta_s \mathbf{1}_{A_t}, & \text{for } s = t + 1, \ldots, T. \end{cases}$$

Observe that on the event A_t it holds that

$$\hat{\theta}_{t+1}^{\top} S_t = \theta_{t+1}^{\top} S_t = W_t(\theta) - c_t = 0 = W_t(\hat{\theta}) - \hat{c}_t,$$

so that the self-financing condition (6.1) is satisfied and $(\hat{\theta}, \hat{c}) \in \mathscr{A}(x_0)$, with $W_0(\hat{\theta}) = 0$. We have thus constructed an arbitrage opportunity (of the first kind), thus yielding a contradiction.

Solution of Exercise 6.17 In view of Theorem 6.23, the absence of arbitrage opportunities is equivalent to the existence of a risk neutral probability measure $\mathbb{P}^*$. Equation (6.54) of Proposition 6.22 implies that, for each $s \in \{0, 1, \ldots, T\}$,

$$W_s(\theta) = \mathbb{E}^* \left[\sum_{r=s}^{T} \frac{c_r}{r_f^{r-s}} \middle| \mathscr{F}_s \right] = c_s + \mathbb{E}^* \left[\sum_{r=s+1}^{T} \frac{c_r}{r_f^{r-s}} \middle| \mathscr{F}_s \right] \geq c_s,$$

thus proving the first statement of Lemma 6.19. The second statement immediately follows by noting that the last conditional expectation is strictly positive if there exist a date $r \in \{s+1, \ldots, T\}$ and an event $A_r \in \mathscr{F}_r$ such that $c_r > 0$ holds on A_r.

Solution of Exercise 6.18 $(i) \Rightarrow (ii)$: let $c \in \mathscr{C}_0^+(0)$ be an arbitrage opportunity for (S, D), in the sense of Definition 6.17. In view of Lemma 6.19 and of the following discussion, if there exist a date $t \in \{0, 1, \ldots, T\}$ and an event $A_t \in \mathscr{F}_t$ such that $W_t(\theta) < c_t$ holds on A_t, then there exists an arbitrage opportunity (in the sense of Definition 4.17) in the trading period $[t, t+1]$ and, hence, there is nothing to prove. Therefore, suppose that $\mathbb{P}(W_t(\theta) \geq c_t \geq 0) = 1$ for all $t = 0, 1, \ldots, T$ and let

$$\tau := \min \left\{ t \in \{0, 1, \ldots, T\} : \mathbb{P}\big(W_t(\theta) > 0\big) > 0 \right\}.$$

Since $c \in \mathscr{C}_0^+(0)$ and c is non-null, τ is well defined and takes values in $\{1, \ldots, T\}$. We now claim that the strategy $\bar{\theta} := \theta_\tau$ realizes an arbitrage opportunity in the trading period $[\tau - 1, \tau]$ with respect to the couple $(S_{\tau-1}, P_\tau)$. Indeed, it holds that $\bar{\theta}^\top S_{\tau-1} = \theta_\tau^\top S_{\tau-1} = W_{\tau-1}(\theta) - c_{\tau-1} = 0$ and $\bar{\theta}^\top P_\tau = W_\tau(\theta) \geq 0$ with strict inequality holding on some event with strictly positive probability.

$(ii) \Rightarrow (i)$: suppose that there exists a date $t = \{0, 1, \ldots, T-1\}$ such that the couple (S_t, P_{t+1}) admits an arbitrage opportunity, in the sense of Definition 4.17. This means that there exists a vector $\bar{\theta} \in \mathbb{R}^{N+1}$ such that $\bar{\theta}^\top S_t \leq 0$ and $\bar{\theta}^\top P_{t+1} \geq 0$, with at least one of the two inequalities being strict on some event with strictly positive probability. Define then the trading strategy $\theta = (\theta_s)_{s=0,1,\ldots,T}$ by $\theta_{t+1} := \bar{\theta}$ and $\theta_s := 0$ for all $s \neq t+1$. Define also the consumption process $c = (c_s)_{t=0,1,\ldots,T}$ by

$$c_s := \begin{cases} -\bar{\theta}^\top S_t, & \text{if } s = t, \\ \bar{\theta}^\top P_{t+1}, & \text{if } s = t+1, \\ 0, & \text{otherwise.} \end{cases}$$

It can be easily checked that the self-financing conditions (6.1)–(6.2) are satisfied, so that $(\theta, c) \in \mathscr{A}(0)$. Since $c \in \mathscr{C}_0^+(0)$ and the consumption process c is non-negative and non-null, we have thus constructed an arbitrage opportunity in the sense of Definition 6.17.

Solution of Exercise 6.19 Equation (6.53) easily follows from condition (6.52) in Definition 6.21. Indeed, it holds that

$$
s_t^n = \frac{1}{r_f} \mathbb{E}^* \left[s_{t+1}^n + d_{t+1}^n | \mathscr{F}_t \right] = \frac{1}{r_f} \mathbb{E}^* \left[\frac{1}{r_f} \mathbb{E}^* \left[s_{t+2}^n + d_{t+2}^n \Big| \mathscr{F}_{t+1} \right] + d_{t+1}^n \Big| \mathscr{F}_t \right]
$$

$$
= \dots = \mathbb{E}^* \left[\frac{s_T^n}{r_f^{T-t}} + \sum_{s=t+1}^{T} \frac{d_s^n}{r_f^{s-t}} \Big| \mathscr{F}_t \right] = \mathbb{E}^* \left[\sum_{s=t+1}^{T} \frac{d_s^n}{r_f^{s-t}} \Big| \mathscr{F}_t \right],
$$

where we have used the tower property of the conditional expectation (under the risk neutral probability measure $\mathbb{P}^*$) and the assumption that $s_T^n = 0$. In order to prove that the process $(s_t^n/r_f^t + \sum_{r=0}^t (d_r^n/r_f^r))_{t=0,1,\dots,T}$ is a martingale under the probability measure $\mathbb{P}^*$, let $s, t \in \{0, 1, \dots, T\}$ with $s < t$. Then, in view of equation (6.52) in Definition 6.21, it holds that

$$
\mathbb{E}^* \left[\frac{s_t^n}{r_f^t} + \sum_{r=0}^{t} \frac{d_r^n}{r_f^r} \Big| \mathscr{F}_s \right] = \frac{s_s^n}{r_f^s} + \sum_{r=0}^{s} \frac{d_r^n}{r_f^r} + \sum_{r=s+1}^{t} \mathbb{E}^* \left[\frac{d_r^n}{r_f^r} + \frac{s_r^n}{r_f^r} - \frac{s_{r-1}^n}{r_f^{r-1}} \Big| \mathscr{F}_s \right]
$$

$$
= \frac{s_s^n}{r_f^s} + \sum_{r=0}^{s} \frac{d_r^n}{r_f^r} + \sum_{r=s+1}^{t} \mathbb{E}^* \left[\mathbb{E}^* \left[\frac{d_r^n}{r_f^r} + \frac{s_r^n}{r_f^r} - \frac{s_{r-1}^n}{r_f^{r-1}} \Big| \mathscr{F}_{r-1} \right] \Big| \mathscr{F}_s \right]
$$

$$
= \frac{s_s^n}{r_f^s} + \sum_{r=0}^{s} \frac{d_r^n}{r_f^r},
$$

thus establishing the martingale property.
Finally, in order to prove equation (6.54), it suffices to prove that, for all $t = 0, 1, \dots, T-1$, it holds that

$$
\mathbb{E}^* \left[W_{t+1}(\theta) | \mathscr{F}_t \right] = r_f \big(W_t(\theta) - c_t \big).
$$

The latter condition easily follows from the self-financing condition (6.4) together with the risk neutral condition (6.52). In fact, recalling that θ_{t+1} is $\mathscr{F}_t$-measurable, for all $t = 0, 1, \dots, T-1$,

$$
\mathbb{E}^* \left[W_{t+1}(\theta) | \mathscr{F}_t \right] = W_t(\theta) + \theta_{t+1}^\top \big(\mathbb{E}^* \left[S_{t+1} + D_{t+1} | \mathscr{F}_t \right] - S_t \big) - c_t
$$

$$
= W_t(\theta) + (r_f - 1) \theta_{t+1}^\top S_t - c_t
$$

$$
= r_f \theta_{t+1}^\top S_t = r_f \left(W_t(\theta) - c_t \right),
$$

where the last two equalities follow from the self-financing condition (6.1).

Solution of Exercise 6.20   Using equation (6.61), it suffices to compute

$$
\begin{aligned}
\mathbb{P}^*(A_t|A_s) &= \frac{\mathbb{P}^*(A_t \cap A_s)}{\mathbb{P}^*(A_s)} = \frac{\mathbb{P}^*(A_t)}{\mathbb{P}^*(A_s)} = \frac{\sum_{\omega \in A_t} \pi_\omega^*}{\sum_{\omega \in A_s} \pi_\omega^*} \\[2mm]
&= \frac{\sum_{\omega \in A_t} u^{i'}\left(c_T^{i*}(\omega)\right)\pi_\omega}{\sum_{\omega \in A_s} u^{i'}\left(c_T^{i*}(\omega)\right)\pi_\omega} \\[2mm]
&= \frac{\pi_{A_t} u^{i'}\left(c_t^{i*}(A_t)\right) \sum_{\omega \in A_t} \frac{u^{i'}\left(c_T^{i*}(\omega)\right)}{u^{i'}\left(c_t^{i*}(A_t)\right)} \frac{\pi_\omega}{\pi_{A_t}}}{\pi_{A_s} u^{i'}\left(c_s^{i*}(A_s)\right) \sum_{\omega \in A_s} \frac{u^{i'}\left(c_T^{i*}(\omega)\right)}{u^{i'}\left(c_s^{i*}(A_s)\right)} \frac{\pi_\omega}{\pi_{A_s}}} \\[2mm]
&= \frac{\pi_{A_t} u^{i'}\left(c_t^{i*}(A_t)\right)(\delta r_f)^{t-T}}{\pi_{A_s} u^{i'}\left(c_s^{i*}(A_s)\right)(\delta r_f)^{s-T}} \\[2mm]
&= (\delta r_f)^{t-s} \frac{u^{i'}\left(c_t^{i*}(A_t)\right)}{u^{i'}\left(c_s^{i*}(A_s)\right)} \pi_{A_t|A_s}
\end{aligned}
$$

where the second last equality follows from the fact that

$$
u^{i'}(c_t^{i*}) = \delta r_f \mathbb{E}[u^{i'}(c_{t+1}^{i*})|\mathscr{F}_t],
$$

for all $t = 0, 1, \ldots, T - 1$ (see Corollary 6.5 and the following discussion).

Solution of Exercise 6.21   Let $\mathbb{P}^{*,1}$ and $\mathbb{P}^{*,2}$ be two risk neutral probability measures and let $c \in \mathscr{C}_0^+(x)$, for some $x \in \mathbb{R}_+$. The claim follows from the simple observation that

$$
c_0 + \sum_{t=1}^{T} \frac{1}{r_f^t}\mathbb{E}^{*,1}[c_t] = x = c_0 + \sum_{t=1}^{T} \frac{1}{r_f^t}\mathbb{E}^{*,2}[c_t].
$$

Solution of Exercise 6.22   It suffices to compute, using the predictability of the process $(\theta_t)_{t=1,\ldots,T}$,

$$
\sum_{n=1}^{N} \mathbb{E}^*\left[\theta_t^n g_t^n | \mathscr{F}_{t-1}\right] = \sum_{n=1}^{N} \theta_t^n \mathbb{E}^*\left[g_t^n | \mathscr{F}_{t-1}\right]
$$

$$
= \sum_{n=1}^{N} \theta_t^n \frac{1}{r_f^{t-1}} \mathbb{E}^*\left[\frac{1}{r_f}(d_t^n + s_t^n) - s_{t-1}^n \middle| \mathscr{F}_{t-1}\right] = 0,
$$

where the last equality follows from Definition 6.21. The remaining part of the exercise simply follows by summing over all dates and taking the risk neutral (unconditional) expectation.

Solution of Exercise 6.23 Suppose first that $c \in \mathscr{C}_0^+(x)$, so that there exists a trading strategy $\theta = (\theta_t)_{t=0,1,\dots,T}$ with $W_0(\theta) = x$ such that $(\theta, c) \in \mathscr{A}(x_0)$. In particular, recall that the process θ is predictable. Recalling that $W_T(\theta) = c_T$, Exercise 6.2 implies that

$$\bar{c}_T = \bar{W}_T(\theta) = x + \sum_{t=1}^{T} \theta_t^\top (\Delta \bar{S}_t + \bar{D}_t) - \sum_{t=0}^{T-1} \bar{c}_t.$$

Moreover, in view of equation (6.54) of Proposition 6.22, by taking the expectation with respect to the risk neutral probability measure $\mathbb{P}^*$ we get

$$x = W_0(\theta) = \sum_{t=0}^{T} \mathbb{E}^*[\bar{c}_t].$$

This shows that any process belonging to $\mathscr{C}_0^+(x)$, for some $x \in \mathbb{R}_+$, satisfies the requirements (*i*) and (*ii*). Let us now turn to the converse implication and suppose that there exists a predictable process $(\theta_t)_{t=0,1,\dots,T}$ satisfying condition (*i*). As can be easily verified, it holds that $\Delta \bar{s}_t^0 + \bar{d}_t^0 = 0$, for all $t = 1, \dots, T$. In turn, this means that the first component $(\theta_t^0)_{t=0,1,\dots,T}$ (representing the investment in the risk free security) does not play a role in requirement (*i*). Therefore, we can freely modify the process $(\theta_t^0)_{t=0,1,\dots,T}$ in order to construct a self-financing trading-consumption strategy which finances the consumption plan c. To this effect, let us define the process $\tilde{\theta} = (\tilde{\theta}_t)_{t=0,1,\dots,T}$ by $\tilde{\theta}_t^n := \theta_t^n$ for all $n = 1, \dots, N$ and $t = 0, 1, \dots, T$, while the process $(\tilde{\theta}_t^0)_{t=0,1,\dots,T}$ will be chosen in order to satisfy the self-financing condition (expressed in terms of discounted quantities, as in Exercise 6.2):

$$\tilde{\theta}_{t+1}^0 + \sum_{n=1}^{N} \tilde{\theta}_{t+1}^n \bar{s}_t^n = \tilde{\theta}_t^0 + \sum_{n=1}^{N} \tilde{\theta}_t^n \bar{s}_t^n - \bar{c}_t, \qquad \text{for all } t = 0, 1, \dots, T-1.$$

This implies that

$$\tilde{\theta}_0^0 := x - \sum_{n=1}^{N} \theta_0^n s_0^n \qquad \text{and} \qquad \tilde{\theta}_t^0 := \tilde{\theta}_{t-1}^0 - \sum_{n=1}^{N} (\theta_t^n - \theta_{t-1}^n) \bar{s}_{t-1}^n - \bar{c}_{t-1},$$

for all $t = 1, \dots, T$. By construction, the process $(\tilde{\theta}_t^0)_{t=0,1,\dots,T}$ is predictable and satisfies the self-financing condition. Since the self-financing condition is invariant with respect to discounting (see Exercise 6.2), this shows that $(\tilde{\theta}, c) \in \mathscr{A}(x_0)$. Moreover, since $W_0(\tilde{\theta}) = x$, it follows that $c \in \mathscr{C}_0^+(x)$, thus completing the argument.

Solution of Exercise 6.25 Due to Theorem 6.23, the absence of arbitrage opportunities is equivalent to the existence of a risk neutral probability measure $\mathbb{P}^*$ (note that, for the purposes of this exercise, the choice of the specific risk neutral probability measure is irrelevant and, actually, we can allow agents to exhibit heterogeneous beliefs as long as they all belong to the family of risk neutral probability measures). For each $i = 1, \ldots, I$, consider a risk neutral agent i with probability beliefs represented by the probability measure $\mathbb{P}^*$ and discount factor $\delta = 1/r_f$. For such an agent, the optimal investment-consumption Problem (6.6) can be represented as

$$\max \left(c_0^i + \sum_{t=1}^{T} \frac{1}{r_f^t} \mathbb{E}^*[c_t^i] \right),$$

where the maximization is with respect to all $(\theta^i, c^i) \in \mathscr{A}$ such that $\theta_0^i = \bar{\theta}_0^i$ or, equivalently, with respect to all consumption processes $c^i \in \mathscr{C}_0^+(x^i)$, with $x^i = (\bar{\theta}_0^i)^\top P_0$. However, for any $c^i \in \mathscr{C}_0^+(x^i)$, Proposition 6.22 implies that

$$c_0^i + \sum_{t=1}^{T} \frac{1}{r_f^t} \mathbb{E}^*[c_t^i] = x^i.$$

In particular, this implies that an optimal solution is simply given by consuming the dividends generated by the initial endowment of securities. By Definition 6.13, such a solution corresponds to a Radner equilibrium of the economy.

Solution of Exercise 6.27 Observe first that, due to Theorem 6.23, the absence of arbitrage opportunities is equivalent to the existence of a risk neutral probability measure $\mathbb{P}^*$. Observe then that, in view of Exercise 6.23, the maximization in Problem (6.6) can be equivalently formulated with respect to all processes belonging to $\mathscr{K}^+$ (recall that $\mathscr{K}^+$ is the set of all non-negative stochastic processes adapted to the information flow $\mathbb{F}$) which satisfy requirements $(i) - (ii)$ of Exercise 6.23. In particular, since the set

$$\left\{ c \in \mathscr{K}^+ : \bar{c}_T = x + \sum_{t=1}^{T} \theta_t^\top (\Delta \bar{S}_t + \bar{D}_t) - \sum_{t=0}^{T-1} \bar{c}_t, \text{ for some predictable } (\theta_t)_{t=0,1,\ldots,T} \right\}$$

is closed (due to the standing assumption of a finite probability space), the set of all processes in $\mathscr{K}^+$ satisfying requirements $(i) - (ii)$ of Exercise 6.23 is also closed. Moreover, the condition $\sum_{t=0}^{T} \mathbb{E}^*[\bar{c}_t] = x$ (together with the finiteness of the probability space) implies that the same set is also bounded and, hence, compact. The claim now follows from the fact that any continuous function admits a maximum over a compact set.

Solution of Exercise 6.28 It suffices to combine equations (6.69) and (6.71).

Solution of Exercise 6.29 Let $t \in \{1, \ldots, T\}$ and consider the problem of minimizing $\text{Corr}\left(r_{t+1}^w, \mathbf{u}'(e_{t+1})|\mathscr{F}_t\right)$ over all $\mathscr{F}_t$-measurable vectors $w_{t+1} \in \mathbb{R}^{N+1}$

satisfying $\sum_{n=0}^{N} w_{t+1}^n = 1$. Observe that

$$\text{Corr}\left(r_{t+1}^{\mathsf{T}} w_{t+1}, \mathbf{u}'(e_{t+1}) | \mathscr{F}_t\right)$$

$$= \frac{\mathbb{E}\left[\sum_{n=1}^{N} w_{t+1}^n (r_{t+1}^n - \mathbb{E}[r_{t+1}^n | \mathscr{F}_t]) \left(\mathbf{u}'(e_{t+1}) - \mathbb{E}[\mathbf{u}'(e_{t+1}) | \mathscr{F}_t]\right) | \mathscr{F}_t\right]}{\sqrt{\text{Var}\left(\sum_{n=1}^{N} w_{t+1}^n r_{t+1}^n | \mathscr{F}_t\right)} \sqrt{\text{Var}\left(\mathbf{u}'(e_{t+1}) | \mathscr{F}_t\right)}},$$

so that the minimization problem only depends on the last N components of the vector w, since security 0 is risk free. The first order conditions of the minimization problem give

$$\text{Cov}\left(r_{t+1}^n, \mathbf{u}'(e_{t+1}) | \mathscr{F}_t\right) \sqrt{\text{Var}\left(r_{t+1}^{\mathsf{T}} \hat{w}_{t+1} | \mathscr{F}_t\right)}$$

$$- \text{Cov}\left(r_{t+1}^{\mathsf{T}} \hat{w}_{t+1}, \mathbf{u}'(e_{t+1}) | \mathscr{F}_t\right) \left(\text{Var}\left(r_{t+1}^{\mathsf{T}} \hat{w}_{t+1} | \mathscr{F}_t\right)\right)^{-\frac{1}{2}} \text{Cov}\left(r_{t+1}^n, r_{t+1}^{\mathsf{T}} \hat{w}_{t+1} | \mathscr{F}_t\right) = 0,$$

for all $n = 1, \ldots, N$, thus proving relation (6.76).

Solution of Exercise 6.30   Since the couple $(\log m_t, \log r_t^n)$ conditionally on $\mathscr{F}_{t-1}$ is distributed according to a bivariate normal distribution with mean $\mu^n \in \mathbb{R}^2$ and covariance $\Sigma^n \in \mathbb{R}^{2\times 2}$, it holds that

$$\log\left(\mathbb{E}[m_{t+1} r_{t+1}^n | \mathscr{F}_t]\right) = \log\left(\mathbb{E}[e^{\log m_{t+1} + \log r_{t+1}^n} | \mathscr{F}_t]\right) = \mu_1^n + \mu_2^n + \frac{\Sigma_{11}^n + 2\Sigma_{12}^n + \Sigma_{22}^n}{2}.$$

Since $\mathbb{E}[m_{t+1} r_{t+1}^n | \mathscr{F}_t] = 1$, in view of equation (6.70), this implies that

$$\mathbb{E}[\log m_{t+1} | \mathscr{F}_t] + \mathbb{E}[\log r_{t+1}^n | \mathscr{F}_t] = \mu_1^n + \mu_2^n = -\frac{\Sigma_{11}^n + 2\Sigma_{12}^n + \Sigma_{22}^n}{2},$$

for all $n = 0, 1, \ldots, N$. In particular, taking the risk free security (i.e., $n = 0$), this gives

$$\log r_f = -\mathbb{E}[\log m_{t+1} | \mathscr{F}_t] - \frac{1}{2} \text{Var}\left(\log m_{t+1} | \mathscr{F}_t\right) = -\mu_1^n - \frac{\Sigma_{11}^n}{2}.$$

Formula (6.78) follows by combining the last two relations.

Solution of Exercise 6.31   Consider the following two strategies:

(A) at date t do nothing, at date $t + 1$ buy one unit of the zero-coupon bond with maturity $t + 2$, paying the price $B(t + 1, t + 2)$;

(B) at date t, buy one unit of the zero-coupon bond with maturity $t + 2$, paying the price $B(t, t + 2)$ and sell short $B(t, t + 2)/B(t, t + 1)$ units of the zero-coupon bond with maturity $t + 1$.

Both strategies yield the risk free payoff 1 at date $t + 2$, with a zero net investment at the initial date t. By the Law of One Price and comparing the payoff of strategies A and B at the intermediate date $t + 1$, it follows that

$$\frac{B(t, t+2)}{B(t, t+1)} = \mathbb{E}^*[B(t+1, t+2)|\mathscr{F}_t],$$

thus proving the claim.

Solution of Exercise 6.32 Equation (6.80) implies that

$$\log\left(1 + i(t, t + \tau)\right) = -\log\delta - \frac{1}{\tau}\log\left(\mathbb{E}\left[\left(\frac{e_{t+\tau}}{e_t}\right)^{-\gamma}\bigg|\mathscr{F}_t\right]\right)$$

$$= -\log\delta - \frac{1}{\tau}\log\mathbb{E}\left[g(t, \tau)^{-\tau\gamma}|\mathscr{F}_t\right]$$

$$\approx -\log\delta + \gamma\,\mathbb{E}[\log g(t, \tau)|\mathscr{F}_t],$$

where we have used the first order approximation

$$\log\mathbb{E}\left[g(t, \tau)^{-\tau\gamma}|\mathscr{F}_t\right] \approx -\tau\gamma\,\mathbb{E}[\log g(t, \tau)|\mathscr{F}_t].$$

Solution of Exercise 6.34 Equation (6.79) implies that, for any $t \in \mathbb{N}$,

$$B(t, t+1) = \mathbb{E}\left[\delta\frac{e_t}{e_{t+1}}\bigg|\mathscr{F}_t\right] = \mathbb{E}\left[\delta\frac{1}{\rho u_{t+1}}\bigg|\mathscr{F}_t\right] = \frac{\delta}{\rho}\mathbb{E}\left[\frac{1}{u_1}\right],$$

where we have used the assumption that the random variables $(u_t)_{t\in\mathbb{N}}$ are i.i.d. To prove the second equality, note that, due to the independence of the random variables composing the sequence $(u_t)_{t\in\mathbb{N}}$, the covariance term in equation (6.83) is null. Hence:

$$B(t, t+2) = B(t, t+1)\mathbb{E}[B(t+1, t+2)|\mathscr{F}_t] = \frac{\delta^2}{\rho^2}\left(\mathbb{E}\left[\frac{1}{u_1}\right]\right)^2.$$

Solution of Exercise 6.35 In view of equation (6.71), the equilibrium risk free interest rate $r_f(l)$ corresponding to the event $\{x_t = l\}$, for $t \in \mathbb{N}$, is given by

$$r_f(l) = \frac{1}{\delta\mathbb{E}[x_{t+1}^{-\alpha}|x_t = l]} = \frac{1}{\delta(\pi_{lh}h^{-\alpha} + \pi_{ll}l^{-\alpha})}$$

and, similarly, in the alternative state $\{x_t = h\}$, for $t \in \mathbb{N}$,

$$r_f(h) = \frac{1}{\delta\mathbb{E}[x_{t+1}^{-\alpha}|x_t = h]} = \frac{1}{\delta(\pi_{hl}l^{-\alpha} + \pi_{hh}h^{-\alpha})}.$$

To answer the second question, equation (6.84) with $\gamma = 1 - \alpha$ gives

$$\xi(l) = \delta\left(\pi_{lh}h^{1-\alpha}(1 + \xi(h)) + \pi_{ll}l^{1-\alpha}(1 + \xi(l))\right),$$
$$\xi(h) = \delta\left(\pi_{hh}h^{1-\alpha}(1 + \xi(h)) + \pi_{hl}l^{1-\alpha}(1 + \xi(l))\right),$$

where we recall that $\xi(k)$ denotes the value of the ratio s_t/e_t in correspondence of the event $\{x_t = k\}$, with $k \in \{l, h\}$ and for every $t \in \mathbb{N}$. The above linear system admits a unique solution $(\xi(l), \xi(h))$ which can be easily computed. The price at any date $t \in \mathbb{N}$ of the security delivering the aggregate endowment can be expressed as

$$s_t = e_t(\xi(l)\mathbf{1}_{\{s_t/e_t=l\}} + \xi(h)\mathbf{1}_{\{s_t/e_t=h\}}).$$

Finally, to answer the last question, the expected return of the security can be computed as, for each $t \in \mathbb{N}$,

$$\mathbb{E}[r_{t+1}|\mathscr{F}_t] = \mathbb{E}\left[\frac{s_{t+1} + e_{t+1}}{s_t}\middle|\mathscr{F}_t\right]$$

$$= \frac{1}{\xi(l)}\left(\pi_{lh}h(1 + \xi(h)) + \pi_{ll}l(1 + \xi(l))\right)\mathbf{1}_{\{x_t=l\}}$$

$$+ \frac{1}{\xi(h)}\left(\pi_{hl}l(1 + \xi(l)) + \pi_{hh}h(1 + \xi(h))\right)\mathbf{1}_{\{x_t=h\}}.$$

Solution of Exercise 6.36

(i): It suffices to use formula (6.86), noting that the risk neutrality of the representative agent implies that his marginal utility is constant and using the fact that $\sum_{s=1}^{\infty}\delta^s = \delta/(1-\delta)$.

(ii): The second claim can be shown by an analogous argument.

Solution of Exercise 6.37

(i) Note first that $\mathbb{E}[d_{t+1}|\mathscr{F}_t] = \bar{d} + \varrho(d_t - \bar{d})$ and, hence,

$$\mathbb{E}[d_{t+2}|\mathscr{F}_t] = \bar{d} + \varrho(\bar{d} + \varrho(d_t - \bar{d}) - \bar{d}) = (1 - \varrho^2)\bar{d} + \varrho^2 d_t.$$

Iterating the same argument and applying formula (6.86), we obtain that, for all $t \in \mathbb{N}$,

$$s_t^* = \sum_{s=1}^{\infty}\delta^s(1 - \varrho^s)\bar{d} + \sum_{s=1}^{\infty}\delta^s\varrho^s d_t = \left(\frac{\delta}{1-\delta} - \frac{\delta\varrho}{1-\delta\varrho}\right)\bar{d} + \frac{\delta\varrho}{1-\delta\varrho}d_t.$$

(ii): This can be shown by similar arguments, noting that

$$\sum_{s=1}^{\infty}\delta^s\beta^s = \delta\beta/(1 - \delta\beta).$$

(iii): Note first that, by the tower property of the conditional expectation,

$$\mathbb{E}[d_{t+s}|\mathscr{F}_t] = \mathbb{E}\big[\mathbb{E}[d_{t+s}|\mathscr{F}_{t+s-1}]|\mathscr{F}_t\big] = \mathbb{E}\big[e^{\mu + \log d_{t+s-1} + \frac{\sigma^2}{2}}|\mathscr{F}_t\big] = \ldots = d_t e^{s\left(\mu + \frac{\sigma^2}{2}\right)},$$

for all $t \in \mathbb{N}$ and $s \in \mathbb{N}$. The claim then follows by a simple application of formula (6.86).

Solution of Exercise 6.38 The first claim follows from the martingale property of the process $\big(\delta^t \mathbf{u}'(e_t)\beta_t\big)_{t \in \mathbb{N}}$ together with the fact that a bubble is always non-negative. Indeed, if $\beta_t = 0$ for some $t \in \mathbb{N}$, then

$$\mathbb{E}[\mathbf{u}'(e_{t+s})\beta_{t+s}|\mathscr{F}_t] = \delta^{-s}\mathbf{u}'(e_t)\beta_t = 0,$$

for all $s \in \mathbb{N}$. In turn, since the utility function $\mathbf{u}$ is assumed to be strictly increasing, this implies that $\beta_{t+s} = 0$, for all $s \in \mathbb{N}$.
To prove the second claim, note that equation (6.87) implies that, for each $t \in \mathbb{N}$,

$$\beta_t = \lim_{T\to\infty} \frac{\delta^T}{\mathbf{u}'(e_t)}\mathbb{E}[\mathbf{u}'(e_{t+T})s_{t+T}|\mathscr{F}_t] \le K \lim_{T\to\infty} \frac{\delta^T}{\mathbf{u}'(e_t)}\mathbb{E}[\mathbf{u}'(e_{t+T})|\mathscr{F}_t]$$

$$= K \lim_{T\to\infty} B(t, t+T) = 0,$$

where we have used equation (6.79). Since a rational bubble is always non-negative, this shows that $\beta_t = 0$, thus proving the claim.

Solution of Exercise 6.39 In view of Proposition 6.40, it suffices to verify that the process $(\delta^t \mathbf{u}'(e_t)\beta_t)_{t \in \mathbb{N}}$ is a martingale. This can be shown as follows:

$$\delta\mathbb{E}[\mathbf{u}'(e_{t+1})\beta_{t+1}|\mathscr{F}_t] = \delta c\mathbb{E}[d_{t+1}^{\lambda}|\mathscr{F}_t]$$

$$= \delta c\mathbb{E}[e^{\lambda(\mu + \log d_t + \varepsilon_{t+1})}|\mathscr{F}_t] = \delta c e^{\lambda(\mu + \log d_t) + \frac{\lambda^2 \sigma^2}{2}} = \mathbf{u}'(e_t)\beta_t,$$

for all $t \in \mathbb{N}$.

11.6 Exercises of Chap. 7

Solution of Exercise 7.1 Let us define $s_t^{\mathbb{F}} := \mathbb{E}[s_t^e|\mathscr{F}_t]$ and $s_t^{\mathbb{G}} := \mathbb{E}[s_t^e|\mathscr{G}_t]$, for all $t \in \mathbb{N}$. Observe also that, since $\mathscr{F}_t \subseteq \mathscr{G}_t$ and due to the tower property of the conditional expectation, it holds that $s_t^{\mathbb{F}} = \mathbb{E}[s_t^{\mathbb{G}}|\mathscr{F}_t]$. This implies that

$$\text{Cov}\left(s_t^{\mathbb{G}} - s_t^{\mathbb{F}}, s_t^{\mathbb{F}}\right) = \mathbb{E}\big[(s_t^{\mathbb{G}} - s_t^{\mathbb{F}})s_t^{\mathbb{F}}\big] - \mathbb{E}[s_t^{\mathbb{G}} - s_t^{\mathbb{F}}]\mathbb{E}[s_t^{\mathbb{F}}] = 0,$$

which in turn implies that

$$\text{Var}\left(s_t^G\right) = \text{Var}\left(s_t^G - s_t^F + s_t^F\right)$$

$$= \text{Var}\left(s_t^G - s_t^F\right) + \text{Var}\left(s_t^F\right) \geq \text{Var}\left(s_t^F\right),$$

thus showing inequality (7.8). The second inequality follows from (7.8) together with equations (7.6) and (7.7).

Solution of Exercise 7.2   Note first that $s_t^e - s_t^f = s_t^e - s_t^* + (s_t^* - s_t^f)$ and, since $s_t^e - s_t^* = u_t$ (see equation (7.3)) with $\mathbb{E}[u_t|\mathscr{F}_t] = 0$, for all $t \in \mathbb{N}$, it holds that $\mathbb{E}[(s_t^e - s_t^*)(s_t^* - s_t^f)|\mathscr{F}_t] = 0$, for all $t \in \mathbb{N}$. In turn, this implies that

$$\mathbb{E}[(s_t^e - s_t^f)^2|\mathscr{F}_t] = \mathbb{E}[(s_t^e - s_t^*)^2|\mathscr{F}_t] + \mathbb{E}[(s_t^* - s_t^f)^2|\mathscr{F}_t],$$

thus proving both inequalities in (7.11).

Solution of Exercise 7.3   Note that, as in decomposition (7.3), it holds that

$$s_t^F = s_t^e - \sum_{s=1}^{\infty} \frac{1}{r_f^s}\varepsilon_{t+s}^F \quad \text{and} \quad s_t^G = s_t^e - \sum_{s=1}^{\infty} \frac{1}{r_f^s}\varepsilon_{t+s}^G, \qquad \text{for all } s \in \mathbb{N}.$$

Hence,

$$\mathbb{E}\left[\left(\sum_{s=1}^{\infty} \frac{1}{r_f^s}\varepsilon_{t+s}^F\right)^2\right] = \mathbb{E}\left[(s_t^F - s_t^e)^2\right] = \mathbb{E}\left[(s_t^F + s_t^G - s_t^G - s_t^e)^2\right]$$

$$= \mathbb{E}\left[\left(s_t^F - s_t^G - \sum_{s=1}^{\infty} \frac{1}{r_f^s}\varepsilon_{t+s}^G\right)^2\right]$$

$$= \mathbb{E}\left[(s_t^F - s_t^G)^2\right] + \mathbb{E}\left[\left(\sum_{s=1}^{\infty} \frac{1}{r_f^s}\varepsilon_{t+s}^G\right)^2\right]$$

$$\geq \mathbb{E}\left[\left(\sum_{s=1}^{\infty} \frac{1}{r_f^s}\varepsilon_{t+s}^G\right)^2\right],$$

where we have used the fact that the quantities $s_t^F - s_t^G$ and $\sum_{s=1}^{\infty} \frac{1}{r_f^s}\varepsilon_{t+s}^G$ are uncorrelated, since $\mathbb{E}[\sum_{s=1}^{\infty} \frac{1}{r_f^s}\varepsilon_{t+s}^G|\mathscr{G}_t] = 0$, for all $t \in \mathbb{N}$. Then, noting that

$$\varepsilon_t^F = s_t^F + d_t - r_f s_{t-1}^F = \mathbb{E}[s_t^F + d_t|\mathscr{F}_t] - \mathbb{E}[s_t^F + d_t|\mathscr{F}_{t-1}]$$

$$= \mathbb{E}[s_t^e + d_t|\mathscr{F}_t] - \mathbb{E}[s_t^e + d_t|\mathscr{F}_{t-1}],$$

and similarly for ε_t^G, it follows that the sequences $(\varepsilon_{t+s}^F)_{s \in \mathbb{N}}$ and $(\varepsilon_{t+s}^G)_{s \in \mathbb{N}}$ are composed of uncorrelated random variables, so that

$$\frac{1}{r_f^2 - 1} \mathbb{E}\big[(\varepsilon_t^F)^2\big] = \mathbb{E}\left[\left(\sum_{s=1}^{\infty} \frac{1}{r_f^s} \varepsilon_{t+s}^F\right)^2\right]$$

$$\geq \mathbb{E}\left[\left(\sum_{s=1}^{\infty} \frac{1}{r_f^s} \varepsilon_{t+s}^G\right)^2\right] = \frac{1}{r_f^2 - 1} \mathbb{E}\big[(\varepsilon_t^G)^2\big],$$

thus proving inequality (7.12), under the standing assumption that the dividend process is adapted to both information flows $\mathbb{F}$ and $\mathbb{G}$.

Solution of Exercise 7.5 As a preliminary, equation (7.17) can be rewritten in terms of de-trended quantities as

$$\bar{d}_{t+1} = \bar{d}_t + \sum_{k=0}^{N} \lambda_k (\bar{E}_{t-k+1} - \bar{E}_{t-k}),$$

where $\lambda_k := \gamma_k / (1 + g)^{k+1}$. Then, the assumption that $E_t = r_f V_t$, for all $t \in \mathbb{N}$, implies that

$$\bar{d}_{t+1} = \bar{d}_t + r_f \sum_{k=0}^{N} \lambda_k (\bar{V}_{t-k+1} - \bar{V}_{t-k})$$

$$= \bar{d}_t + r_f \delta \sum_{k=0}^{N} \theta_k (\bar{V}_{t-k+1} - \bar{V}_{t-k}),$$

where we define $\bar{V}_t := V_t / (1 + g)^t$, for all $t \in \mathbb{N}$, and

$$\delta := \sum_{k=0}^{N} \lambda_k \quad \text{and} \quad \theta_k := \frac{\lambda_k}{\delta}, \quad \text{for all } k = 0, 1, \ldots, N.$$

By summing over t in the last relation and using the assumption that $d_0 = 0$, we obtain

$$\bar{d}_t = r_f \delta \sum_{k=0}^{N} \theta_k \bar{V}_{t-k}.$$

Then, the assumption that stocks are priced rationally (i.e., $s_t = V_t$) implies that

$$\bar{d}_t = r_f \delta \sum_{k=0}^{N} \theta_k \bar{s}_{t-k},$$

where $\bar{s}_t := s_t/(1+g)^t$ denotes the de-trended price, for all $t \in \mathbb{N}$. Hence, in view of equation (7.10) together with the assumption that $\bar{s}_T^e := \sum_{t=0}^{T-1} \bar{s}_t/T$, it holds that (in de-trended terms)

$$\hat{s}_t^e = \sum_{s=1}^{T-t-1} \frac{1}{r_f^s} \bar{d}_{t+s} + \frac{1}{Tr_f^{T-t}} \sum_{t=0}^{T-1} \bar{s}_t$$

$$= r_f \delta \sum_{s=1}^{T-t-1} \frac{1}{r_f^s} \sum_{k=0}^{N} \theta_k \bar{s}_{t+s-k} + \frac{1}{Tr_f^{T-t}} \sum_{t=0}^{T-1} \bar{s}_t$$

$$=: \sum_{k=-N}^{T-1} w_{tk} \bar{s}_k,$$

with $w_{tk} = 0$ for $k < t - N$.

Solution of Exercise 7.6 In order to prove part *(i)*, it suffices to note that (assuming without loss of generality that $k < t$)

$$\text{Cov}(r_{t+s}, r_t | \mathcal{K}_k) = \text{Cov}(z_{t+s}, z_t | \mathcal{K}_k) = \mathbb{E}[z_{t+s} z_t | \mathcal{K}_k] = \mathbb{E}[\mathbb{E}[z_{t+s}|\mathcal{K}_t] z_t | \mathcal{K}_k] = 0,$$

where the last equality follows from equation (7.15), since

$$\mathbb{E}[z_{t+s}|\mathcal{K}_t] = \mathbb{E}[\mathbb{E}[z_{t+s}|\mathcal{F}_t]|\mathcal{K}_t] = 0.$$

To prove part *(ii)*, note that the value based on the information set $\mathcal{K}_t$ of a $\mathcal{K}_t$-measurable trading strategy θ_{t+1} is given by (compare with equation (6.68) and recall the assumption of risk neutrality):

$$\frac{1}{r_f}\mathbb{E}[\theta_{t+1}^\top(S_{t+1} + D_{t+1})|\mathcal{K}_t] = \sum_{n=0}^{N} \frac{\theta_{t+1}^n}{r_f} \mathbb{E}[\mathbb{E}[s_{t+1}^n + d_{t+1}^n|\mathcal{F}_t]|\mathcal{K}_t]$$

$$= \sum_{n=0}^{N} \theta_{t+1}^n \mathbb{E}[s_t^n|\mathcal{K}_t] = \sum_{n=0}^{N} \theta_{t+1}^n s_t^n,$$

for all $t \in \mathbb{N}$, where we have used the fact that s_t^n is $\mathcal{K}_t$-measurable, for all $n = 0, 1, \ldots, N$ and $t \in \mathbb{N}$. Provided that $\theta_{t+1}^\top S_t > 0$, the corresponding return of the strategy is $r_{t+1}^{\theta_{t+1}} := \theta_{t+1}^\top(S_{t+1} + D_{t+1})/(\theta_{t+1}^\top S_t)$. Therefore, in view of the previous arguments, the $\mathscr{H}_t$-conditional expected return is equal to

$$\mathbb{E}[r_{t+1}^{\theta_{t+1}}|\mathscr{H}_t] = \frac{\theta_{t+1}^{\mathsf{T}}\mathbb{E}[\mathbb{E}[S_{t+1}+D_{t+1}|\mathscr{F}_t]|\mathscr{H}_t]}{\theta_{t+1}^{\mathsf{T}}S_t} = r_f\frac{\theta_{t+1}^{\mathsf{T}}\mathbb{E}[S_t|\mathscr{H}_t]}{\theta_{t+1}^{\mathsf{T}}S_t} = r_f,$$

where we have used the assumption that $S_t = (s_t^0, s_t^1, \ldots, s_t^N)^{\mathsf{T}}$ is $\mathscr{H}_t$-measurable.

Solution of Exercise 7.7   In equilibrium, the aggregate consumption coincides with the aggregate dividend, so that the key asset pricing relation (6.68) implies that the price process $(s_t)_{t\in\mathbb{N}}$ satisfies

$$s_t e_t^{\gamma} = \delta\mathbb{E}[s_{t+1}e_{t+1}^{\gamma}|\mathscr{F}_t] + \delta\mathbb{E}[e_{t+1}^{\gamma+1}|\mathscr{F}_t], \quad \text{for all } t \in \mathbb{N}.$$

Let us now replace a candidate solution of the form $s_t = \varrho(X_t)e_t$ into the above relation:

$$\varrho(X_t)e_t^{\gamma+1} = \delta\mathbb{E}\big[\big(\varrho(X_{t+1}) + 1\big)e_{t+1}^{\gamma+1}|\mathscr{F}_t\big].$$

Replacing into the last expression the explicit form of the endowment process, we get

$$\varrho(X_t) = \delta e^{\alpha_0(1+\gamma)+(1+\gamma)^2\sigma^2/2+\alpha_1(1+\gamma)X_t}\big(\mathbb{E}[\varrho(X_{t+1})|\mathscr{F}_t] + 1\big).$$

Now, since X_t can only assume the two possible values $\{0, 1\}$, the last equation yields a system of two linear equations with unknowns $\varrho(0)$ and $\varrho(1)$, which can be explicitly computed as stated in the exercise.

Solution of Exercise 7.8   Note first that, due to the time additivity of the utility function U, it holds that

$$\frac{\partial U}{\partial c_t} = \delta^t u_t'(c_t), \quad \text{for every } t = 0, 1, \ldots, T.$$

Moreover, by computing the total derivative,

$$d\log(c_t/c_s) = \frac{1}{c_t}dc_t - \frac{1}{c_s}dc_s, \quad \text{for every } s, t = 0, 1, \ldots, T,$$

and similarly

$$d\log\left(\frac{\partial U}{\partial c_s}\bigg/\frac{\partial U}{\partial c_t}\right) = \frac{u_s''(c_s)}{u_s'(c_s)}dc_s - \frac{u_t''(c_t)}{u_t'(c_t)}dc_t, \quad \text{for every } s, t = 0, 1, \ldots, T.$$

Hence, by the definition of $\eta_U(s, t)$:

$$\eta_U(s, t) = \frac{\frac{1}{c_t}\frac{\partial c_t}{\partial U} - \frac{1}{c_s}\frac{\partial c_s}{\partial U}}{\frac{u_s''(c_s)}{u_s'(c_s)}\frac{\partial c_s}{\partial U} - \frac{u_t''(c_t)}{u_t'(c_t)}\frac{\partial c_t}{\partial U}}$$

$$= \frac{\frac{1}{c_t\delta^t u_t'(c_t)} - \frac{1}{c_s\delta^s u_s'(c_s)}}{\frac{u_s''(c_s)}{\delta^s (u_s'(c_s))^2} - \frac{u_t''(c_t)}{\delta^t (u_t'(c_t))^2}}$$

$$= \frac{1 - \frac{c_t}{c_s}\frac{\delta^{t-s} u_t'(c_t)}{u_s'(c_s)}}{r_{u_t}^r(c_t) - \frac{c_t}{c_s}\frac{\delta^{t-s} u_t'(c_t)}{u_s'(c_s)} r_{u_s}^r(c_s)},$$

thus proving that (7.19) holds. The special case of CRRA utility functions immediately follows.

Solution of Exercise 7.9   As a preliminary, observe that the representative agent's stochastic discount factor process $(M_t)_{t \in \mathbb{N}}$ satisfies

$$\frac{M_{t+1}}{M_t} = m_{t+1} = e^{-\delta}\frac{e_{t+1}^{-\gamma}}{e_t^{-\gamma}} = \begin{cases} e^{-R}, & \text{if there is no disaster at } t+1, \\ e^{-R}B_{t+1}^{-\gamma}, & \text{otherwise,} \end{cases}$$

where m_{t+1} is the one-period stochastic discount factor between t and $t + 1$. Moreover, in view of the hypotheses of the model, we obtain

$$\mathbb{E}\left[m_{t+1}\frac{d_{t+1}}{d_t}\Big|\mathscr{F}_t\right]$$

$$= e^{-R+g_d}\bigg(\mathbb{E}[(1 + \varepsilon_{t+1})\mathbf{1}_{\{\text{no disaster at } t+1\}}|\mathscr{F}_t]$$

$$+ \mathbb{E}[(1 + \varepsilon_{t+1})B_{t+1}^{-\gamma}F_{t+1}\mathbf{1}_{\{\text{disaster at } t+1\}}|\mathscr{F}_t]\bigg)$$

$$= e^{-R+g_d}\left((1 - p_t) + p_t\mathbb{E}[B_{t+1}^{-\gamma}F_{t+1}|\{\text{there is a disaster at } t+1\} \vee \mathscr{F}_t]\right)$$

$$= e^{-R+g_d}(1 + H_t),$$

where H_t is the asset resilience introduced above. Note first that, due to the key asset pricing relation (6.68), the cum-dividend equilibrium price p_t of the security paying the dividend process $(d_t)_{t \in \mathbb{N}}$ satisfies

$$p_t = d_t + \mathbb{E}[m_{t+1}p_{t+1}|\mathscr{F}_t], \qquad \text{for all } t \in \mathbb{N}.$$

Let us conjecture a solution structure of the form $p_t = d_t\big(a + b(H_t - H^*)\big)$, for all $t \in \mathbb{N}$. By substitution into the last relation and using the hypotheses of the model, we can compute:

$$a + b(H_t - H^*) = 1 + \mathbb{E}\left[m_{t+1} \frac{d_{t+1}}{d_t}\big(a + b(H_{t+1} - H^*)\big) \Big| \mathscr{F}_t \right]$$

$$= 1 + e^{-R+gd}(1 + H_t)\left(a + be^{-\phi}\frac{1 + H^*}{1 + H_t}(H_t - H^*) \right)$$

$$= 1 + e^{-R+gd}\big(a(1 + H^* + (H_t - H^*)) + be^{-\phi}(1 + H^*)(H_t - H^*)\big),$$

for all $t \in \mathbb{N}$. Since the above equation has to hold for all $t \in \mathbb{N}$ and for all possible values of H_t, we can use a separation of variables argument and solve for a and b, thus yielding

$$a = \frac{1}{1 - e^{-r}} \quad \text{and} \quad b = \frac{e^{-R+gd}a}{1 - e^{-r-\phi}},$$

thus proving the representation (7.20).

11.7   Exercises of Chap. 8

Solution of Exercise 8.1   The claim simply follows from the fact that $U^*(\pi)$ is defined as the upper envelope of the family of linear functions $\pi \mapsto \sum_{s=1}^{S} \pi_s u(x_s)$, with $(x_1, \ldots, x_S) \in B$.

Solution of Exercise 8.2   Part *(i)* directly follows from the first order condition of the optimal saving problem (8.42), since it must hold that $u'(w_0 - \alpha_0^*) = u'(\alpha_0^* - \alpha_1^*)$, which in turn implies that $\alpha_1^* = 2\alpha_0^* - w_0$ because the function $u'(\cdot)$ is strictly decreasing. Let us then denote

$$H(\alpha_0) := 2u(w_0 - \alpha_0) + \mathbb{E}[u(2\alpha_0 - w_0 + \tilde{x})], \qquad (11.20)$$

which corresponds to (8.42) when consumption is smoothed over the first two dates and $\tilde{x}$ is observed only at $t = 2$. In order to prove part *(ii)*, observe first that, by a similar reasoning, it is optimal for the agent to smooth consumption between date $t = 1$ and date $t = 2$, so that the optimal saving problem (8.42) when $\tilde{x}$ is observed at $t = 1$ can be rewritten as

$$\max_{\alpha_0 \in \mathbb{R}} \left(u(w_0 - \alpha_0) + 2\mathbb{E}\left[u\left(\frac{\alpha_0 + \tilde{x}}{2} \right) \right] \right).$$

In this case, the first order optimality condition is given by (8.43). Since the function H introduced in (11.20) is strictly concave, in order to prove part *(iii)* it suffices to show that $H'(\alpha_0^{i*}) \geq 0$, i.e.,

$$\mathbb{E}[u'(2\alpha_0^{i*} - w_0 + \tilde{x})] \geq u'(w_0 - \alpha_0^{i*}). \tag{11.21}$$

Define

$$z := w_0 - \alpha_0^{i*} \quad \text{and} \quad \tilde{y} := \frac{3\alpha_0^{i*}}{2} - w_0 + \frac{\tilde{x}}{2}.$$

With this notation, condition (11.21) holds if and only if

$$\mathbb{E}[u'(z + 2\tilde{y})] - u'(z) \geq 0. \tag{11.22}$$

In turn, condition (8.43) can be equivalently rewritten as $\mathbb{E}[u'(z + \tilde{y})] - u'(z) = 0$. Since the map $k \mapsto \mathbb{E}[u'(z + k\tilde{y})]$ is convex, as a consequence of the assumption that $u''' > 0$, it follows that condition (11.22) necessarily holds, thus proving the claim.

Solution of Exercise 8.3 By definition, in order to show that the information structure (π^a, v^a) is finer than the information structure (π^b, v^b), we need to show that condition (8.5) holds, for any utility function $u : \mathbb{R}_+ \to \mathbb{R}$. Observe that

$$\sum_{j=1}^{J} \pi^b(y_j) U^*\left(v^b(y_j)\right) = \sum_{j=1}^{J} \pi^b(y_j) \max_{(x_1,\dots,x_S) \in B} \sum_{s=1}^{S} v_s^b(y_j) u(x_s)$$

$$= \sum_{j=1}^{J} \pi^b(y_j) \max_{(x_1,\dots,x_S) \in B} \sum_{s=1}^{S} \sum_{k=1}^{J} v_s^a(y_k) K_{kj} u(x_s)$$

$$\leq \sum_{j=1}^{J} \pi^b(y_j) \sum_{k=1}^{J} K_{kj} \max_{(x_1,\dots,x_S) \in B} \sum_{s=1}^{S} v_s^a(y_k) u(x_s)$$

$$= \sum_{k=1}^{J} \pi^a(y_k) \max_{(x_1,\dots,x_S) \in B} \sum_{s=1}^{S} v_s^a(y_k) u(x_s)$$

$$= \sum_{k=1}^{J} \pi^a(y_k) U^*\left(v^a(y_k)\right),$$

where the first and the last equalities follow from the definition of $U^*(v^\ell(y_j))$, for $\ell = a, b$, the second and the third equalities follow from condition (8.6) and the inequality in the middle uses the fact that the map $\pi \mapsto U^*(\pi)$ from Δ_S onto $\mathbb{R}$ is convex in the vector of probabilities (see Exercise 8.1). This shows that condition (8.5) holds, thus proving the claim.

Solution of Exercise 8.4 Let $\{x^i \in \mathbb{R}^S_+; i = 1, \ldots, I\}$ be an ex-ante Pareto optimal allocation and suppose that there exists a feasible allocation $\{x^{i'} \in \mathbb{R}^S_+; i = 1, \ldots, I\}$ such that

$$\sum_{s=1}^{S} v^i_s(y^i_j)u^i(x^{i'}_s) \geq \sum_{s=1}^{S} v^i_s(y^i_j)u^i(x^i_s), \qquad \text{for all } i = 1, \ldots, I \text{ and } j = 1, \ldots, J,$$

with strict inequality holding for some $i \in \{1, \ldots, I\}$ and $j \in \{1, \ldots, J\}$. Noting that $\sum_{j=1}^{J} \pi(y^i_j)v^i_s(y^i_j) = \pi_s$, for every $i = 1, \ldots, I$ and $s = 1, \ldots, S$, it holds that

$$\sum_{s=1}^{S} \pi_s u^i(x^{i'}_s) = \sum_{j=1}^{J}\sum_{s=1}^{S} \pi(y^i_j)v^i_s(y^i_j)u^i(x^{i'}_s) \geq \sum_{j=1}^{J}\sum_{s=1}^{S} \pi(y^i_j)v^i_s(y^i_j)u^i(x^i_s) = \sum_{s=1}^{S} \pi_s u^i(x^i_s)$$

and, assuming that $\pi(y^i_j) > 0$, for all $i = 1, \ldots, I$ and $j = 1, \ldots, J$, a strict inequality holds for some $i \in \{1, \ldots, I\}$, thus contradicting the ex-ante Pareto optimality of the allocation $\{x^i \in \mathbb{R}^S_+; i = 1, \ldots, I\}$. Hence, the allocation $\{x^{i'} \in \mathbb{R}^S_+; i = 1, \ldots, I\}$ cannot be an interim Pareto improvement of $\{x^i \in \mathbb{R}^S_+; i = 1, \ldots, I\}$. Assuming that all the elements of the matrix v of conditional probabilities are strictly positive, a similar argument allows to show that every interim Pareto optimal allocation is also ex-post Pareto optimal.

Solution of Exercise 8.5 In correspondence of a Green-Lucas equilibrium (see Definition 8.6), every agent maximizes

$$\max_{z^i \in \mathbb{R}} \mathbb{E}\big[(\tilde{d} - \phi^*(\tilde{y}))z^i | \phi^*(y), y^i\big], \qquad \text{for each } i = 1, \ldots, I.$$

As a consequence, in correspondence of the optimal choice z^{i*}, for each $i = 1, \ldots, I$, it holds that

$$\mathbb{E}\big[(\tilde{d} - \phi^*(\tilde{y}))z^{i*} | \phi^*(y), y^i\big] \geq 0.$$

By the law of iterated expectations together with the market clearing condition, it holds that

$$\sum_{i=1}^{I} \mathbb{E}\big[(\tilde{d} - \phi^*(\tilde{y}))z^{i*} | \phi^*(y)\big] = 0.$$

In turn, this implies that $\mathbb{E}[(\tilde{d} - \phi^*(\tilde{y}))z^{i*} | \phi^*(y), y^i] = 0$, for all $i = 1, \ldots, I$, i.e., the expected gain from trading is null. Clearly, letting $\hat{z}^{i*} = 0$, for all $i = 1, \ldots, I$, satisfies the market clearing condition appearing in Definition 8.6 and, hence, corresponds to another Green-Lucas equilibrium (without trade).

Solution of Exercise 8.6   Suppose first that $\tilde{z}$ is a sufficient statistic for the conditional density $f(y, z|x)$. Then, the Bayes rule implies that

$$
\begin{aligned}
f(x|y, z) &= \frac{g(x)f(y, z|x)}{\int_{-\infty}^{\infty} g(x)f(y, z|x)dx} \\[2mm]
&= \frac{g(x)g_1(y, z)g_2(z, x)}{\int_{-\infty}^{\infty} g(x)g_1(y, z)g_2(z, x)dx} \\[2mm]
&= \frac{g(x)g_2(z, x)}{\int_{-\infty}^{\infty} g(x)g_2(z, x)dx},
\end{aligned}
$$

where $g(\cdot)$ denotes the density function of the random variable $\tilde{x}$. This shows that the conditional distribution of $\tilde{x}$ given $\{\tilde{y} = y, \tilde{z} = z\}$ does not depend on y.

Conversely, if the conditional distribution of $\tilde{x}$ given $\{\tilde{y} = y, \tilde{z} = z\}$ does not depend on y, then it holds that

$$
f(y, z|x) = \frac{f(y, z, x)}{g(x)} = \frac{f(x|y, z)h(y, z)}{g(x)} = \frac{f(x|z)h(y, z)}{g(x)} = g_1(y, z)g_2(z, x),
$$

with $g_1(y, z) := h(y, z)$ and $g_2(z, x) := f(x|z)/g(x)$ and where $h(\cdot, \cdot)$ denotes the density function of the pair of random variables $(\tilde{y}, \tilde{z})$. We have thus shown that $\tilde{z}$ is a sufficient statistic for the conditional density $f(y, z|x)$.

Solution of Exercise 8.8   Let $p^{a*} : \mathcal{Y} \to \mathbb{R}^N$ be the equilibrium price functional of the artificial economy and suppose that it is a sufficient statistic for the conditional distribution of the random dividends of the N risky assets given the signal random vector $\tilde{y}$. We want to prove that letting $\phi^*(\cdot) := p^{a*}(\cdot)$ defines an equilibrium price functional for the original economy (in the sense of Definition 8.6) which is a sufficient statistic. In the setting of this exercise, the optimal choice problem of agent i in the original economy where he observes his private signal together with the vector of equilibrium prices can be formulated as follows (compare with Section (3.1)):

$$
\max_{w^i \in \mathbb{R}^N} \mathbb{E}\left[u^i\left(w_0^i r_f + \sum_{n=1}^{N} w_n^i(\tilde{r}_n - r_f) \right) \Bigg| \tilde{y}^i, \phi^*(\tilde{y}) \right],
$$

where w_0^i represents the initial wealth of agent i and $w^i \in \mathbb{R}^N$ denotes the amounts of wealth invested in the N risky assets, for $i = 1, \ldots, I$, and $\tilde{r}_n$ represents the random return of asset n, for $n = 1, \ldots, N$. Since $\phi^*(\tilde{y}) = p^{a*}(\tilde{y})$ and using the tower property of the conditional expectation, this problem can be rewritten as

$$
\max_{w^i \in \mathbb{R}^N} \mathbb{E}\left[\mathbb{E}\left[u^i\left(w_0^i r_f + \sum_{n=1}^{N} w_n^i(\tilde{r}_n - r_f) \right) \Bigg| \tilde{y}^i, \tilde{y}, p^{a*}(\tilde{y}) \right] \Bigg| \tilde{y}^i, p^{a*}(\tilde{y}) \right].
$$

Since $\tilde{y}^i$ is an element of $\tilde{y}$ and p^{a*} is a sufficient statistic for the conditional distribution of the dividends on the N risky assets (and, hence, for the conditional distribution of the returns as well), it holds that

$$\mathbb{E}\left[u^i\left(w_0^i r_f + \sum_{n=1}^{N} w_n^i(\tilde{r}_n - r_f)\right)\bigg|\, \tilde{y}^i, \tilde{y}, p^{a*}(\tilde{y})\right]$$

$$= \mathbb{E}\left[u^i\left(w_0^i r_f + \sum_{n=1}^{N} w_n^i(\tilde{r}_n - r_f)\right)\bigg|\, p^{a*}(\tilde{y})\right],$$

so that

$$\mathbb{E}\left[\mathbb{E}\left[u^i\left(w_0^i r_f + \sum_{n=1}^{N} w_n^i(\tilde{r}_n - r_f)\right)\bigg|\, \tilde{y}^i, \tilde{y}, p^{a*}(\tilde{y})\right]\bigg|\, \tilde{y}^i, p^{a*}(\tilde{y})\right]$$

$$= \mathbb{E}\left[u^i\left(w_0^i r_f + \sum_{n=1}^{N} w_n^i(\tilde{r}_n - r_f)\right)\bigg|\, p^{a*}(\tilde{y})\right]$$

$$= \mathbb{E}\left[u^i\left(w_0^i r_f + \sum_{n=1}^{N} w_n^i(\tilde{r}_n - r_f)\right)\bigg|\, \tilde{y}\right],$$

where the last equality uses again the fact that p^{a*} is a sufficient statistic. We have thus shown that

$$\max_{w^i \in \mathbb{R}^N} \mathbb{E}\left[u^i\left(w_0^i r_f + \sum_{n=1}^{N} w_n^i(\tilde{r}_n - r_f)\right)\bigg|\, \tilde{y}^i, \phi^*(\tilde{y})\right]$$

$$= \max_{w^i \in \mathbb{R}^N} \mathbb{E}\left[u^i\left(w_0^i r_f + \sum_{n=1}^{N} w_n^i(\tilde{r}_n - r_f)\right)\bigg|\, \tilde{y}\right].$$

Hence, for every agent $i = 1, \ldots, I$, the optimal choice problem in the artificial economy corresponds exactly to the optimal choice problem in the original economy. Since market clearing holds, $\phi^* := p^{a*}$ defines an equilibrium price functional which is a sufficient statistics. The converse implication can be proved by an analogous argument.

Solution of Exercise 8.10   For every $i = 1, \ldots, I$, the optimal choice problem of agent i can be formulated as

$$\max_{(x_1^i, \ldots, x_S^i) \in \mathbb{R}_+^S} \frac{1}{\alpha} \sum_{s=1}^{S} \pi_s (x_s^i)^\alpha =: U(x_1^{i*}, \ldots, x_S^{i*}; \pi),$$

subject to the budget constraint $\sum_{s=1}^{S} p_s^* e_s^i = \sum_{s=1}^{S} p_s^* x_s^i$. For this problem, the first order optimality conditions imply

$$\frac{p_s^*}{p_r^*} = \frac{\pi_s}{\pi_r} \left(\frac{x_s^{i*}}{x_r^{i*}}\right)^{\alpha-1}, \qquad \text{for all } r, s = 1, \ldots, S.$$

Since markets are complete, the equilibrium allocation is ex-ante Pareto optimal. In view of Proposition 4.5, there exists a sharing rule $\{y^i : \mathbb{R}_+ \to \mathbb{R}_+; i = 1, \ldots, I\}$ such that $x_s^{i*} = K_i e_s$, for some $K_i > 0$, for all $i = 1, \ldots, I$. Using the budget constraint together with the above optimality condition in correspondence of the aggregate endowment, it follows that

$$K_i \sum_{s=1}^{S} \pi_s e_s^\alpha = \sum_{s=1}^{S} \pi_s e_s^i e_s^{\alpha-1}, \qquad \text{for all } i = 1, \ldots, I,$$

so that

$$K_i = \frac{\sum_{s=1}^{S} \pi_s e_s^i e_s^{\alpha-1}}{\sum_{s=1}^{S} \pi_s e_s^\alpha}, \qquad \text{for all } i = 1, \ldots, I.$$

In turn, this implies that, for every $i = 1, \ldots, I$, the maximal expected utility of agent i can be written as

$$U(x_1^{i*}, \ldots, x_S^{i*}; \pi) = U(K_i e_1, \ldots, K_i e_S; \pi) = \frac{1}{\alpha} \sum_{s=1}^{S} \pi_s \left(\frac{\sum_{r=1}^{S} \pi_r e_r^i e_r^{\alpha-1}}{\sum_{r=1}^{S} \pi_r e_r^\alpha} e_s\right)^\alpha$$

$$= \frac{1}{\alpha} \left(\sum_{s=1}^{S} \pi_s e_s^i e_s^{\alpha-1}\right)^\alpha \left(\sum_{s=1}^{S} \pi_s e_s^\alpha\right)^{1-\alpha}.$$

Moreover, the function $\pi \mapsto U(x_1^{i*}, \ldots, x_S^{i*}; \pi)$ is concave in the vector of probabilities $\pi = (\pi_1, \ldots, \pi_S)$ (compare also with Exercise 8.1) and, clearly, is non-linear with respect to π if and only if the individual endowment e_s^i is not proportional to the aggregate endowment, for every $s = 1, \ldots, S$. The claim then follows by the arguments analogous to those used in Exercise 8.3.

Solution of Exercise 8.11 The claim simply follows by inserting the equilibrium price (8.20) into the optimal demand of an agent, thus yielding

$$w^{i*} = \frac{1}{a^i \sigma^2(\tilde{d}|y^i, \phi^*(y^1, \ldots, y^I))} \left(\mathbb{E}[\tilde{d}|y^i, \phi^*(y^1, \ldots, y^I)] - \phi^*(y^1, \ldots, y^I) r_f\right)$$

$$= \frac{1}{a^i} \left(\sum_{i=1}^{I} \frac{1}{a^i}\right)^{-1}, \qquad \text{for every } i = 1, \ldots, I.$$

Solution of Exercise 8.12 Suppose that $\phi^* : \{1,2\} \to \mathbb{R}$ is a fully revealing equilibrium price functional. Then, in correspondence of the equilibrium price $q^* = \phi^*(y)$, the optimal demand of the informed agent is given by

$$w^{1*,y} = \arg\max_{w \in \mathbb{R}} \mathbb{E}\left[-e^{-a(w(\tilde{d}-\phi^*(\tilde{y})))}\Big|\tilde{y} = y\right] = \frac{\mu_y - \phi^*(y)}{a\sigma^2}, \qquad \text{for } y = 1, 2.$$

Since ϕ^* is a fully revealing equilibrium price functional, it perfectly reveals to the uninformed agent the private information of the informed agent. Therefore, the optimal demand in equilibrium of the uninformed agent is given by

$$w^{2*,y} = \arg\max_{w \in \mathbb{R}} \mathbb{E}\left[-e^{-a(w(\tilde{d}-\phi^*(\tilde{y}))+k_{\tilde{y}}p)}\Big|\tilde{y} = y\right]$$

$$= \arg\max_{w \in \mathbb{R}} \mathbb{E}\left[-e^{-a(w(\tilde{d}-\phi^*(\tilde{y}))+k_{\tilde{y}}p)}\Big|\phi^*(\tilde{y}) = \phi^*(y)\right] = \frac{\mu_y - \phi^*(y)}{a\sigma^2} - k_y,$$

for $y = 1, 2$. In correspondence of the equilibrium, the future market must clear, so that

$$w^{1*,y} + w^{2*,y} = \frac{\mu_y - \phi^*(y)}{a\sigma^2} + \frac{\mu_y - \phi^*(y)}{a\sigma^2} - k_y = 0, \qquad \text{for } y = 1, 2.$$

The market clearing condition implies that the equilibrium price functional ϕ^* is given by

$$\phi^*(y) = \mu_y - \frac{\sigma^2 k_y}{2}, \qquad \text{for } y = 1, 2.$$

It is easy to see that, for the values chosen, it holds that $\phi^*(1) = \phi^*(2) = 3$, thus contradicting the assumption that the equilibrium price functional is fully revealing.

Solution of Exercise 8.13 Similarly as in the proof of Proposition 8.11, we conjecture that the equilibrium price functional ϕ^* is linear with respect to the aggregate supply $\tilde{u}$ and the average signal $\bar{\tilde{y}} := \sum_{i=1}^{I} \tilde{y}^i / I$ (and recall that $\bar{\tilde{y}}$ was the sufficient statistic in the Grossman [839] model presented at the end of Sect. 8.2). In other words, we start from the conjecture that ϕ^* admits the representation (8.23), for some coefficients α, η and γ to be determined. By exploiting the basic properties of the normal multivariate distribution, it can be easily seen that the conditional distribution of the random variable $\tilde{d}$ given $\{\tilde{y}^i = y^i, \tilde{e}^i = e^i, \phi^*(\bar{\tilde{y}}, \tilde{u}) = \phi\}$ is again normal with conditional mean and variance that can be explicitly computed. The optimal demand of each agent can then be determined in terms of the conditional mean and variance of $\tilde{d}$, similarly as in the proof of Proposition 8.11. Imposing the market clearing condition then leads to a system of three non-linear equations in three unknowns which can be explicitly solved, leading to the values reported after (8.23) for the coefficients α, η and γ (compare also with Huang & Litzenberger [971, Chapter 9]).

Solution of Exercise 8.14   For brevity, we denote by $p := \phi^*(d, u)$, for $d \in \mathbb{R}$ and $u \in \mathbb{R}$, a generic realization of the random variable $\phi^*(\tilde{d}, \tilde{u})$ representing the candidate price functional, conjectured to be linear with respect to $\tilde{d}$ and $\tilde{u}$. Recall that, by the properties of the normal multivariate distribution, the conditional expectations $\mathbb{E}[\tilde{d}|y^i, p]$ and $\mathbb{E}[\tilde{d}|p]$ are linear functions of (y^i, p) and p, respectively, and the conditional variances $\mathrm{Var}(\tilde{d}|y^i, p)$ and $\mathrm{Var}(\tilde{d}|p)$ do not depend on the realizations y^i and p. Note that $\mathbb{E}[\tilde{d}|y^i, p]$ denotes the conditional expectation of $\tilde{d}$ given the observation of the event $\{\tilde{y}^i = y^i, \phi^*(\tilde{d}, \tilde{u}) = p\}$ (and similarly for the conditional variance). Since the two classes of informed and uninformed agents are composed of homogeneous agents (i.e., with the same preferences and with the same type of information) and since the utility functions are negative exponential, the optimal demands of the informed and of the uninformed agents in correspondence of a price p and of a private signal y^i are of the form
vspace*-3pt

$$w^{\mathrm{inf}}(y^i, p) = \gamma\, y^i - c^{\mathrm{inf}}p + \hat{b}^{\mathrm{inf}},$$
$$w^{\mathrm{un}}(p) = -c^{\mathrm{un}}p + \hat{b}^{\mathrm{un}}, \tag{11.23}$$

for suitable coefficients γ, c^{inf}, c^{un}, $\hat{b}^{\mathrm{inf}}$ and $\hat{b}^{\mathrm{un}}$. The market clearing condition, together with the assumption that $\int_0^\mu \tilde{\varepsilon}^i di = 0$ almost surely, implies that

$$\mu(\gamma d - c^{\mathrm{inf}}p + \hat{b}^{\mathrm{inf}}) + (1-\mu)(-c^{\mathrm{un}}p + \hat{b}^{\mathrm{un}}) + u = 0,$$

for every realization $d \in \mathbb{R}$ and $u \in \mathbb{R}$. Hence, the market clearing condition implies that the equilibrium price functional ϕ^* necessarily satisfies

$$\phi^*(d, u) = \lambda(\mu\gamma d + u + \hat{b}), \qquad \text{for all } d \in \mathbb{R} \text{ and } u \in \mathbb{R},$$

where $\hat{b} := \mu\hat{b}^{\mathrm{inf}} + (1-\mu)\hat{b}^{\mathrm{un}}$ and $\lambda := 1/(\mu c^{\mathrm{inf}} + (1-\mu)c^{\mathrm{un}})$, provided that $\mu c^{\mathrm{inf}} + (1-\mu)c^{\mathrm{un}} \neq 0$. We now conjecture that the optimal demands are of the form, for $p = \phi^*(d, u)$,

$$w^{\mathrm{inf}}(y^i, p) = \gamma(y^i - p) - b^{\mathrm{inf}}(p - \bar{d}),$$
$$w^{\mathrm{un}}(p) = -b^{\mathrm{un}}(p - \bar{d}). \tag{11.24}$$

By comparing (11.23) with (11.24), it must hold that

$$c^{\mathrm{inf}} = \gamma + b^{\mathrm{inf}}, \quad \hat{b}^{\mathrm{inf}} = b^{\mathrm{inf}}\bar{d}, \quad c^{\mathrm{un}} = b^{\mathrm{un}} \quad \text{and} \quad \hat{b}^{\mathrm{un}} = b^{\mathrm{un}}\bar{d}.$$

With this parametrization, it is easy to check that $\mathbb{E}[\phi^*(\tilde{d}, \tilde{u})] = \mathbb{E}[\tilde{d}] = \bar{d}$, so that the conditional expectation $\mathbb{E}[\tilde{d}|p]$ and the conditional variance $\mathrm{Var}(\tilde{d}|p)$ are respectively given by

$$\mathbb{E}[\tilde{d}|p] = \bar{d} + \frac{1}{\lambda}\frac{\mu\gamma\sigma^2}{\mu^2\gamma^2\sigma^2 + \sigma_u^2}(p - \bar{d}) \quad \text{and} \quad \mathrm{Var}(\tilde{d}|p) = \sigma^2\frac{\sigma_u^2}{\mu^2\gamma^2\sigma^2 + \sigma_u^2}. \tag{11.25}$$

Considering first the case of uninformed agents, due to the assumption of negative exponential utility functions, their optimal demand necessarily satisfies

$$w^{\mathrm{un}}(p) = \frac{\mathbb{E}[\tilde{d}|p] - p}{a^{\mathrm{un}}\,\mathrm{Var}(\tilde{d}|p)} = -b^{\mathrm{un}}(p - \bar{d}). \tag{11.26}$$

By substituting (11.25) into (11.26), we obtain that the parameter b^{un} satisfies

$$b^{\mathrm{un}} = \frac{1}{a^{\mathrm{un}}}\left(\frac{1}{\sigma^2} + \frac{\mu^2\gamma^2}{\sigma_u^2} - \frac{\mu\gamma}{\lambda\sigma_u^2}\right).$$

Considering now the class of informed agents, we can perform a similar procedure. First, by relying on the properties of the multivariate normal distribution, we can compute the conditional expectation and the conditional variance of $\tilde{d}$:

$$\mathbb{E}[\tilde{d}|y^i, p] = \frac{1}{\frac{1}{\sigma_\varepsilon^2} + \frac{1}{\sigma^2} + \frac{\mu^2\gamma^2}{\sigma_u^2}}\left(\frac{y^i}{\sigma_\varepsilon^2} + \left(\frac{1}{\sigma^2} + \frac{\mu^2\gamma^2}{\sigma_u^2} - \frac{\mu\gamma\sigma_u^2}{\lambda}\right)\bar{d} + \frac{\mu\gamma\sigma_u^2}{\lambda}p\right),$$

$$\mathrm{Var}(\tilde{d}|y^i, p) = \left(\frac{1}{\sigma_\varepsilon^2} + \frac{1}{\sigma^2} + \frac{\mu^2\gamma^2}{\sigma_u^2}\right)^{-1}.$$

The optimal demand of each informed agent necessarily satisfies

$$w^{\mathrm{inf}}(p) = \frac{\mathbb{E}[\tilde{d}|y^i, p] - p}{a^{\mathrm{un}}\,\mathrm{Var}(\tilde{d}|y^i, p)} = \gamma(y^i - p) - b^{\mathrm{inf}}(p - \bar{d}).$$

By substitution, we obtain that the parameters γ and b^{inf} satisfy

$$\gamma = \frac{1}{a^{\mathrm{inf}}\sigma_\varepsilon^2}, \qquad b^{\mathrm{inf}} = \frac{1}{a^{\mathrm{inf}}}\left(\frac{1}{\sigma^2} + \frac{\mu^2\gamma^2}{\sigma_u^2} - \frac{\mu\gamma}{\lambda\sigma_u^2}\right)$$

$$\gamma + b^{\mathrm{inf}} = \frac{1}{a^{\mathrm{inf}}}\left(\frac{1}{\sigma_\varepsilon^2} + \frac{1}{\sigma^2} + \frac{\mu^2\gamma^2}{\sigma_u^2} - \frac{\mu\gamma}{\lambda\sigma_u^2}\right).$$

We can then solve for λ, obtaining

$$\lambda = \frac{1 + \mu\gamma\left(\frac{\mu}{a^{\mathrm{inf}}} + \frac{1-\mu}{a^{\mathrm{un}}}\right)\frac{1}{\sigma_u^2}}{\mu\gamma\left(\frac{\mu}{a^{\mathrm{inf}}} + \frac{1-\mu}{a^{\mathrm{un}}}\right)\left(\frac{1}{\sigma^2} + \frac{\mu^2\gamma^2}{\sigma_u^2}\right)} > 0,$$

which in turn allows to explicitly compute the coefficients b^{inf} and b^{un} as given in Proposition 8.12.

Solution of Exercise 8.15   As follows from Exercise 8.14, the optimal demands of the informed and of the uninformed agents in correspondence of the equilibrium price $p := \phi^*(d, u)$ are respectively given by

$$w^{\text{inf},*}(y^i, p) = \frac{1}{a^{\text{inf}}} \left(\frac{1}{\sigma_\varepsilon^2} + \frac{1}{\sigma^2} + \frac{\mu^2 \gamma^2}{\sigma_u^2} \right) \left(\mathbb{E}[\tilde{d}|y^i, p] - p \right),$$

$$w^{\text{un},*} = \frac{1}{a^{\text{un}}} \left(\frac{1}{\sigma^2} + \frac{\mu^2 \gamma^2}{\sigma_u^2} \right) \left(\mathbb{E}[\tilde{d}|p] - p \right),$$

for all $i \in [0, \mu]$ and for every realization $\{\tilde{y}^i = y^i, \phi^*(\tilde{d}, \tilde{u}) = p\}$. The market clearing condition (8.24) implies that

$$\phi^*(d, u) = p = \frac{\frac{1}{a^{\text{inf}}} \left(\frac{1}{\sigma_\varepsilon^2} + \frac{1}{\sigma^2} + \frac{\mu^2 \gamma^2}{\sigma_u^2} \right) \int_0^\mu \mathbb{E}[\tilde{d}|y^i, p] di + \frac{1-\mu}{a^{\text{un}}} \left(\frac{1}{\sigma^2} + \frac{\mu^2 \gamma^2}{\sigma_u^2} \right) \mathbb{E}[\tilde{d}|p] + u}{\frac{\mu}{a^{\text{inf}}} \left(\frac{1}{\sigma_\varepsilon^2} + \frac{1}{\sigma^2} + \frac{\mu^2 \gamma^2}{\sigma_u^2} \right) + \frac{1-\mu}{a^{\text{un}}} \left(\frac{1}{\sigma^2} + \frac{\mu^2 \gamma^2}{\sigma_u^2} \right)},$$

thus proving the claim. In turn, the last expression can be easily seen to imply that if $a^{\text{un}} \to 0$ then $\phi^*(d, u)$ converges to $\mathbb{E}[\tilde{d}|p]$.

Solution of Exercise 8.16   Following Grossman & Stiglitz [849, Appendix B], we start by computing the optimal expected utility of an informed agent who buys the information $\tilde{d}^o$ by paying the fixed cost c and can invest in the two traded securities, with the price of the risky security being $\phi_\lambda^*(\tilde{d}^o, \tilde{u})$:

$$\mathbb{E}\left[e^{-a\widetilde{W}_{1,\lambda}^{\text{inf},i,*}} \right] = \mathbb{E}\left[e^{-a\left((W_0^i - c)r_f + w^{\text{inf},*}(\tilde{d}^o, \phi_\lambda^*(\tilde{d}^o, \tilde{u}))(\tilde{d} - \phi_\lambda^*(\tilde{d}^o, \tilde{u})r_f) \right)} \right]$$

$$= e^{-a(w_0^i - c)r_f} \mathbb{E}\left[e^{-aw^{\text{inf},*}(\tilde{d}^o, \phi_\lambda^*(\tilde{d}^o, \tilde{u}))(\tilde{d} - \phi_\lambda^*(\tilde{d}^o, \tilde{u})r_f)} \right]$$

$$= e^{-a(w_0^i - c)r_f} \mathbb{E}\left[\mathbb{E}\left[e^{-aw^{\text{inf},*}(\tilde{d}^o, \phi_\lambda^*(\tilde{d}^o, \tilde{u}))(\tilde{d} - \phi_\lambda^*(\tilde{d}^o, \tilde{u})r_f)} \,\middle|\, \tilde{d}^o, \tilde{u} \right] \right],$$

where the last equality follows by iterated conditioning. Using the conditional moment generating function of a normal random variable, it follows that

$$\mathbb{E}\left[e^{-aw^{\text{inf},*}(\tilde{d}^o, \phi_\lambda^*(\tilde{d}^o, \tilde{u}))(\tilde{d} - \phi_\lambda^*(\tilde{d}^o, \tilde{u})r_f)} \,\middle|\, \tilde{d}^o, \tilde{u} \right]$$

$$= e^{-aw^{\text{inf},*}(\tilde{d}^o, \phi_\lambda^*(\tilde{d}^o, \tilde{u}))(\tilde{d}^o - \phi_\lambda^*(\tilde{d}^o, \tilde{u})r_f) + \frac{a^2}{2} (w^{\text{inf},*}(\tilde{d}^o, \phi_\lambda^*(\tilde{d}^o, \tilde{u})))^2 \sigma_\varepsilon^2}$$

$$= e^{-\frac{1}{2\sigma_\varepsilon^2} \left(\tilde{d}^o - \phi_\lambda^*(\tilde{d}^o, \tilde{u})r_f \right)^2} = e^{-\frac{h_\lambda}{2\sigma_\varepsilon^2} \left(\frac{\tilde{d}^o - \phi_\lambda^*(\tilde{d}^o, \tilde{u})r_f}{\sqrt{h_\lambda}} \right)^2},$$

where the second equality uses the explicit form of the optimal demand of an informed agent (see Proposition 8.13) and where for brevity we let

$$h_\lambda := \sigma^2 \left(\tilde{d}^o | \phi_\lambda^*(\tilde{d}^o, \tilde{u}) \right),$$

(recall also that $\phi_\lambda^*(\tilde{d}^o, \tilde{u})$ and $\eta_\lambda(\tilde{d}^o, \tilde{u})$ are informationally equivalent), so that

$$\mathbb{E}\left[e^{-a\widetilde{W}_{1,\lambda}^{\text{inf},i,*}}\right] = e^{-a(w_0^i - c)r_f}\mathbb{E}\left[e^{-\frac{h_\lambda}{2\sigma_\varepsilon^2}\left(\frac{\tilde{d}^o - \phi_\lambda^*(\tilde{d}^o,\tilde{u})r_f}{\sqrt{h_\lambda}}\right)^2}\right].$$

Recall now that, if a pair of random variables $(\tilde{z}, \tilde{y})$ follows a bivariate normal distribution with $\sigma^2(\tilde{z}|\tilde{y}) = 1$, then

$$\mathbb{E}\left[e^{-t\tilde{z}^2} \mid \tilde{y}\right] = \frac{1}{\sqrt{1 + 2t}}e^{-\frac{t}{1+2t}(\mathbb{E}[\tilde{z}|\tilde{y}])^2}.$$

By applying this result to the previous relation, we get

$$\mathbb{E}\left[e^{-\frac{h_\lambda}{2\sigma_\varepsilon^2}\left(\frac{\tilde{d}^o - \phi_\lambda^*(\tilde{d}^o,\tilde{u})r_f}{\sqrt{h_\lambda}}\right)^2} \mid \phi_\lambda^*(\tilde{d}^o, \tilde{u})\right] = \frac{\sqrt{\sigma_\varepsilon}}{\sqrt{h_\lambda + \sigma_\varepsilon^2}}e^{-\frac{1}{2(h_\lambda + \sigma_\varepsilon^2)}\left(\mathbb{E}[\tilde{d}|\phi_\lambda^*(\tilde{d}^o,\tilde{u})] - \phi_\lambda^*(\tilde{d}^o,\tilde{u})r_f\right)^2}$$

$$= \sqrt{\frac{\sigma^2(\tilde{d}|\tilde{d}^o)}{\sigma^2(\tilde{d}|\eta_\lambda(\tilde{d}^o, \tilde{u}))}}e^{-\frac{1}{2\sigma^2(\tilde{d}|\eta_\lambda(\tilde{d}^o,\tilde{u}))}\left(\mathbb{E}[\tilde{d}|\phi_\lambda^*(\tilde{d}^o,\tilde{u})] - \phi_\lambda^*(\tilde{d}^o,\tilde{u})r_f\right)^2},$$

thus proving that

$$\mathbb{E}\left[e^{-a\widetilde{W}_{1,\lambda}^{\text{inf},i,*}} \mid \eta_\lambda(\tilde{d}^o, \tilde{u})\right]$$

$$= e^{-a(w_0^i - c)r_f}\sqrt{\frac{\sigma^2(\tilde{d}|\tilde{d}^o)}{\sigma^2(\tilde{d}|\eta_\lambda(\tilde{d}^o, \tilde{u}))}}\,e^{-\frac{1}{2\sigma^2(\tilde{d}|\eta_\lambda(\tilde{d}^o,\tilde{u}))}\left(\mathbb{E}[\tilde{d}|\phi_\lambda^*(\tilde{d}^o,\tilde{u})] - \phi_\lambda^*(\tilde{d}^o,\tilde{u})r_f\right)^2}. \tag{11.27}$$

By performing analogous computations, it can be shown that the optimal expected utility of an uninformed agent, conditionally on $\eta_\lambda(\tilde{d}^o, \tilde{u})$ is given by

$$\mathbb{E}\left[e^{-a\widetilde{W}_{1,\lambda}^{\text{un},i,*}} \mid \eta_\lambda(\tilde{d}^o, \tilde{u})\right] = e^{-aw_0^i r_f}e^{-\frac{1}{2\sigma^2(\tilde{d}|\eta_\lambda(\tilde{d}^o,\tilde{u}))}\left(\mathbb{E}[\tilde{d}|\phi_\lambda^*(\tilde{d}^o,\tilde{u})] - \phi_\lambda^*(\tilde{d}^o,\tilde{u})r_f\right)^2}. \tag{11.28}$$

Hence, by combining (11.27) and (11.28) and using iterated expectations, it follows that

$$\gamma(\lambda) = \frac{\mathbb{E}\left[u(\widetilde{W}_{1,\lambda}^{\text{inf},i,*})\right]}{\mathbb{E}\left[u(\widetilde{W}_{1,\lambda}^{\text{un},i,*})\right]} = \frac{\mathbb{E}\left[\mathbb{E}[u(\widetilde{W}_{1,\lambda}^{\text{inf},i,*})|\eta_\lambda(\tilde{d}^o, \tilde{u})]\right]}{\mathbb{E}\left[\mathbb{E}[u(\widetilde{W}_{1,\lambda}^{\text{un},i,*})|\eta_\lambda(\tilde{d}^o, \tilde{u})]\right]} = e^{acr_f}\sqrt{\frac{\sigma^2(\tilde{d}|\tilde{d}^o)}{\sigma^2(\tilde{d}|\eta_\lambda(\tilde{d}^o, \tilde{u}))}},$$

thus proving the first representation in (8.34). The second representation given in (8.34) simply follows by using relations (8.33). It remains to show that the function $\lambda \mapsto \gamma(\lambda)$ is strictly increasing. This follows since the quantity $\sigma^2(\tilde{d}|\eta_\lambda(\tilde{d}^o, \tilde{u}))$ is strictly decreasing with respect to λ, as shown in (8.33).

Solution of Exercise 8.17   Noting that

$$\gamma(\lambda) = e^{acr_f} \sqrt{\frac{1+m}{1+m+mn}} = e^{acr_f} \left(1 + \frac{nm}{1+m} \right)^{-1/2}$$

and recalling that the function $\lambda \mapsto \gamma(\lambda)$ is increasing, it follows that λ^* increases if and only if the ratio $nm/(1+m)$ increases. Suppose that $\sigma^2(\tilde{d}^o)$ increases while the variance σ^2 remains constant (so that $\sigma_\varepsilon^2 = \sigma^2 - \sigma^2(\tilde{d}^o)$ decreases, i.e., the information accessible to the informed agents becomes more precise). Then, differentiating the ratio $nm/(1+m)$ with respect to σ_ε^2 and keeping constant σ^2 (meaning that $\partial\sigma^2(\tilde{d}^o) = -\partial\sigma_\varepsilon^2$) gives that

$$\frac{\partial}{\partial\sigma^2(\tilde{d}^o)} \frac{nm}{1+m} = \frac{1}{\sigma_\varepsilon^2} \frac{m}{1+m} - \frac{n}{(1+m)^2} \frac{m}{\sigma^2(\tilde{d}^o)} + \frac{n}{\sigma_\varepsilon^2} \frac{m}{1+m} - \frac{2n}{(1+m)^2} \frac{m}{\sigma_\varepsilon^2}.$$

Simplifying this expression, it can be shown that

$$\mathrm{sign}\left(\frac{\partial}{\partial\sigma^2(\tilde{d}^o)} \frac{nm}{1+m} \right) = \mathrm{sign}\left(m\left(1 + \frac{1}{n} \right) - 1 \right)$$

$$= \mathrm{sign}\left(\frac{e^{2acr_f} - 1}{n - e^{2acr_f} + 1} \frac{n+1}{n} - 1 \right),$$

where the second equality follows from (8.35). Part *(iv)* of Proposition 8.15 then follows by letting $\bar{n}$ be the value satisfying the equation

$$\frac{e^{2acr_f} - 1}{\bar{n} - e^{2acr_f} + 1} = \frac{\bar{n}}{\bar{n}+1}.$$

Solution of Exercise 8.18   Suppose that $\sigma_u \to 0$. Then, for any fixed $\lambda > 0$, the quantity m introduced in Proposition 8.14 converges to zero and, hence, $\gamma(\lambda) \to e^{acr_f} > 1$ for $\sigma_u^2 \to 0$. Hence, any value $\lambda > 0$ cannot be an equilibrium if $\sigma_u^2 \to 0$ and it must hold that $\lambda \to 0$ as well. The limit of the quantity in the right hand side of (8.34) for $\lambda \to 0$ and $\sigma_u^2 \to 0$ is easily seen to be $e^{acr_f}(1+n)^{-1/2}$. Hence, it follows that, if $e^{acr_f} > \sqrt{1+n}$ then $\lambda^* = 0$ defines an overall equilibrium. Furthermore, it follows from (8.35) that the informativeness of the equilibrium price remains unchanged. On the contrary, if $e^{acr_f} < \sqrt{1+n}$, then there does not exist an overall equilibrium, since $\gamma(0) < 1$ and the map $\lambda \mapsto \gamma(\lambda)$ is discontinuous at $\lambda = 0$.

It remains to prove part *(ix)* of the proposition. If $\sigma_\varepsilon^2 = 0$, then $nm = 0$, so that $\gamma(\lambda) = e^{acr_f} > 1$ for every $\lambda > 0$. Moreover, if $\sigma_\varepsilon^2 = 0$ then $n = \infty$, so that $\gamma(0) = e^{acr_f}(1+n)^{-1/2} = 0$. It then follows that no overall equilibrium can exist.

Solution of Exercise 8.19

(i): Note first that the terminal wealth of an early informed trader is given by

$$W^{\text{early}} := x_2(\theta, p_2)(\bar{f} + \theta + \varepsilon) - x_1(\theta, p_1)p_1 - (x_2(\theta, p_2) - x_1(\theta, p_1))p_2 + e_0$$

while the terminal wealth of a late informed trader is given by

$$W^{\text{late}} := y_2(\theta, p_2)(\bar{f} + \theta + \varepsilon) - y_1(p_1)p_1 - (y_2(\theta, p_2) - y_1(p_1))p_2 + e_0,$$

for every realization $\{\tilde{\theta} = \theta\}$ and $\{\tilde{\varepsilon} = \varepsilon\}$. Since $\widetilde{W}^{\text{early}}$ and $\widetilde{W}^{\text{late}}$, viewed as random variables, are normally distributed conditionally on the information set $\mathcal{G}_2$ and given the above assumptions of independence of the random variable $\tilde{\varepsilon}$ and of exponential utility functions, it can be easily shown that the optimal demand of early and late informed agents at date $t = 2$ (and recall that at $t = 2$ their information sets coincide) is given by

$$x_2(\theta, p_2) = y_2(\theta, p_2) = \frac{\bar{f} + \theta - p_2}{a\sigma_\varepsilon^2},$$

conditionally on a realization $\{\tilde{\theta} = \theta\}$, thus proving (8.47).

Let us now turn to the proof of (8.48), which requires more computations. We start by replacing equation (8.47) into the expression of the terminal wealth of an early informed agent, thus yielding

$$
\begin{aligned}
W^{\text{early}} &= \frac{\bar{f} + \theta - p_2}{a\sigma_\varepsilon^2}(\bar{f} + \theta + \varepsilon) - \frac{\bar{f} + \theta - p_2}{a\sigma_\varepsilon^2}p_2 - x_1(\theta, p_1)(p_1 - p_2) + e_0 \\
&= \frac{(\bar{f} + \theta - p_2)^2}{a\sigma_\varepsilon^2} + \frac{\bar{f} + \theta - p_2}{a\sigma_\varepsilon^2}\varepsilon - x_1(\theta, p_1)(p_1 - p_2) + e_0,
\end{aligned}
$$

for every realizations $\{\tilde{\theta} = \theta\}$ and $\{\tilde{\varepsilon} = \varepsilon\}$. By iterated conditioning, we have that

$$\mathbb{E}\big[e^{-a\widetilde{W}^{\text{early}}} | \mathcal{G}_1\big] = \mathbb{E}\Big[\mathbb{E}\big[e^{-a\widetilde{W}^{\text{early}}} | \mathcal{G}_2\big] | \mathcal{G}_1\Big]$$

and, noting that, conditionally on the information set $\mathcal{G}_2$, the only unknown random variable is $\tilde{\varepsilon}$, which is independent of the conditioning information, it holds that

$$\mathbb{E}\big[e^{-a\widetilde{W}^{\text{early}}} | \mathcal{G}_2\big] = e^{-a\left(e_0 - x_1(\theta, p_1)p_1 + x_1(\theta, p_2)p_2 + \frac{(\bar{f} + \theta - p_2)^2}{2a\sigma_\varepsilon^2}\right)},$$

so that

$$\mathbb{E}\big[e^{-a\widetilde{W}^{\text{yearly}}}\big|\mathscr{G}_1\big] = \mathbb{E}\left[e^{-a\left(e_0 - x_1(\theta,p_1)p_1 + x_1(\theta,p_2)p_2 + \frac{(\bar{f}+\bar{\theta}-p_2)^2}{2a\sigma_\varepsilon^2}\right)}\bigg|\mathscr{G}_1\right],$$

from which (8.48) follows by explicitly computing the last conditional expectation and optimizing with respect to $x_1(\theta, p_1)$.

Finally, referring to Hirshleifer et al. [947, Appendix] for full details, $y_1(p_1)$ can be derived, similarly as above, by computing iterated conditional expectations and showing that

$$y_1(p_1) = \frac{\mu_{p_2} - p_1}{aS_1} + \frac{\bar{f} + \mu_\theta - \mu_{p_2}}{a\sigma_\varepsilon^2}\frac{S_2 - S_1}{S_1}, \tag{11.29}$$

for suitable coefficients S_1 and S_2 and where μ_θ and μ_{p_2} denote the means of $\tilde{\theta}$ and p_2 conditionally on the observation of the equilibrium price p_1. However, equation (8.46) (by risk neutrality and perfect competition of the market makers) implies that in correspondence of the equilibrium it holds that

$$\mu_{p_2} = \bar{f} + \mathbb{E}\big[\mathbb{E}[\tilde{\theta}|D_1(\cdot), D_2(\cdot)]\big|D_1(\cdot)\big] = \bar{f} + \mathbb{E}[\tilde{\theta}|D_1(\cdot)] = p_1,$$

so that $\mu_{p_2} - p_1 = 0$. Moreover, it holds that

$$\bar{f} + \mu_\theta = \bar{f} + \mathbb{E}[\tilde{\theta}|D_1(\cdot)] = p_1 = \mu_{p_2}$$

and, hence, $\bar{f} + \mu_\theta - \mu_{p_2} = 0$. Equation (8.49) then follows from (11.29), thus showing that at equilibrium the late informed agents do not trade at date $t = 1$.

(ii): The two inequalities (8.50) follow from the linear representation given in part (i) of the equilibrium price functionals ϕ_1^* and ϕ_2^* together with the assumption that the random variables $\tilde{z}_1$ and $\tilde{z}_2$ are independent from $\tilde{\theta}$.

Solution of Exercise 8.20   Let us conjecture that the equilibrium price of the risky asset is indeed given by (8.51). Then, the excess return per share, denoted by $q_{t+1} := p_{t+1} + d_{t+1} - r_f p_t$, for $t \in \mathbb{N}$, can be written as

$$q_{t+1} = -(r_f - 1)\phi_0 + \phi_z(z_{t+1} - r_f z_t) + \frac{1}{r_f - \alpha_d}\tilde{y}_{t+1} + \frac{r_f}{r_f - \alpha_d}\tilde{\varepsilon}_{t+1}.$$

For each $t \in \mathbb{N}$, the distribution of the random variable q_{t+1} conditioned on the observation of the current price and dividend and of the signal $\tilde{y}_t$ is normal with conditional mean and variances given by

$$\mathbb{E}[q_{t+1}|p_t, d_t, y_t] = -(r_f - 1)(\phi_0 + \phi_z\bar{z}) + \phi_z(\alpha_z - r_f)\tilde{z}_t,$$

$$\text{Var}(q_{t+1}|p_t, d_t, y_t) = \sigma_f^2 + \phi_z^2\sigma_z^2 =: \sigma_q^2.$$

Since every agent is supposed to maximize a negative exponential utility function, the optimal demand of class A agents is given by

$$w_t^{A,*} = \frac{\mathbb{E}[q_{t+1}|p_t, d_t, y_t]}{a \operatorname{Var}(q_{t+1}|p_t, d_t, y_t)} = \frac{1}{a\sigma_q^2}\left(-(r_f - 1)(\phi_0 + \phi_z\bar{z}) + (\alpha_z - r_f)\phi_z\tilde{z}_t\right),$$

while the optimal demand of class B agents is given by

$$w_t^{B,*} = \frac{\mathbb{E}[q_{t+1}|p_t, d_t, y_t]}{b_t \operatorname{Var}(q_{t+1}|p_t, d_t, y_t)} = \frac{1}{b_t\sigma_q^2}\left(-(r_f - 1)(\phi_0 + \phi_z\bar{z}) + (\alpha_z - r_f)\phi_z\tilde{z}_t\right).$$

Market clearing requires that $\lambda w_t^{A,*} + (1 - \lambda)w_t^{B,*} = 1$, so that

$$\left(\frac{\lambda}{a} + \frac{1 - \lambda}{b_t}\right)\left(-(r_f - 1)(\phi_0 + \phi_z\bar{z}) + (\alpha_z - r_f)\phi_z\tilde{z}_t\right) = \sigma_q^2.$$

Since $\lambda/a + (1 - \lambda)/b_t = 1/z_t$, it holds that

$$(\alpha_z - r_f)\phi_z = \sigma_q^2 \qquad \text{and} \qquad -(r_f - 1)(\phi_0 + \phi_z\bar{z}) = \sigma_q^2\bar{z}.$$

Under the condition $\sigma_z^2 \leq \sigma^{2,*} := (r_f - \alpha_z)^2/(4\sigma_f^2)$, there are two real solutions ϕ_z to the second order equation

$$(\alpha_z - r_f)\phi_z = \sigma_f^2 + \phi_z^2\sigma_z^2,$$

given by

$$\phi_z = \frac{r_f - \alpha_z}{2\sigma_z^2}\left(-1 \pm \sqrt{1 - (\sigma_z^2/\sigma^{2,*})}\right).$$

Among the two solutions, we select the one associated with the positive sign, since it satisfies $\phi_z \to 0$ for $\sigma_f^2 \to 0$. The term ϕ_0 is then given by $\phi_0 = (1 - \alpha_z)\phi_z\bar{z}/(r_f - 1)$.

Solution of Exercise 8.21 By the same arguments used in the first part of the proof of Proposition 8.17, the optimal demand w^{i*} of each agent belonging to the first group, in correspondence of a price p at date $t = 0$, is given by

$$w^{i*} = \max\left(\frac{\bar{d}_i - p}{a}; 0\right), \qquad \text{for all } \bar{d}_i \in [\bar{d} - k, \bar{d} + k].$$

Concerning the second group of agents, their aggregate optimal demand is simply given by $(\bar{d} - p)/a_{\text{arb}}$, since they are allowed to short sell the risky asset. The market clearing condition then requires

$$\frac{1}{2k}\int_{\max(p, \bar{d} - k)}^{\bar{d} + k} \frac{\bar{d}_i - p}{a}\, d\bar{d}_i + \frac{\bar{d} - p}{a_{\text{arb}}} = u.$$

Representation (8.38) can then be obtained by solving the last equation with respect to p.

Solution of Exercise 8.22   Suppose first that condition (8.40) holds. Then, for every $a \in A$, it holds that

$$p_t \geq \delta \mathbb{E}^a \left[d_{t+1} + p_{t+1} | \mathscr{F}_t \right], \qquad \text{for all } t \in \mathbb{N}.$$

By iterating the above relation (and using the tower property of conditional expectation), it follows that

$$p_t \geq \mathbb{E}^a \left[\sum_{k=t+1}^{T} \delta^{k-t} d_k + \delta^{T-t} p_T \bigg| \mathscr{F}_t \right],$$

for every $a \in A$ and $T \geq t$. By maximizing with respect to all $a \in \{1, \ldots, A\}$ and $T \geq t$, this implies that condition (8.39) holds.
Conversely, suppose that condition (8.39) holds. Then, it holds that (choosing $T = t + 1$)

$$p_t \geq \max_{a \in \{1, \ldots, A\}} \delta \mathbb{E}^a \left[d_{t+1} + p_{t+1} | \mathscr{F}_t \right], \qquad \text{for all } t \in \mathbb{N}. \tag{11.30}$$

Arguing by contradiction, suppose that there exists an event $F_t \in \mathscr{F}_t$ such that a strict inequality holds in (11.30). Then, applying equation (8.39) at $t + 1$ and using iterated conditioning, it follows that on the event F_t

$$p_t > \max_{a \in \{1, \ldots, A\}} \mathbb{E}^a \left[\beta d_{t+1} + \beta p_{t+1} | \mathscr{F}_t \right]$$

$$\geq \max_{a \in \{1, \ldots, A\}} \sup_{T \geq t} \mathbb{E}^a \left[\sum_{k=t+1}^{T} \beta^{k-t} d_k + \beta^{T-t} p_T \bigg| \mathscr{F}_t \right],$$

thus leading to a contradiction with (8.39) and proving that (11.30) must hold as an equality.

11.8   Exercises of Chap. 9

Solution of Exercise 9.1   Consider two gambles $\tilde{x}_1$ and $\tilde{x}_2$ taking values in $[-1, 1]$ and denote by F_1 and F_2 their distribution functions, respectively. Note that, in the context of prospect theory, F_1 and F_2 are defined with respect to some given weighting function $w(\cdot)$ (the result of this exercise does not depend on the choice of the weighting function). For simplicity, we suppose that $F_1(-1) = F_2(-1) = 0$ and that F_1 and F_2 are continuous. Let $u(\cdot) : [-1, 1] \to \mathbb{R}$ be an arbitrary continuous and increasing value function, such that $u''(x) \geq 0$ for all $x \in [-1, 0]$ and $u''(x) \leq 0$ for all $x \in [0, 1]$. Define then

$$G(x, y) := \int_y^x \left(F_1(z) - F_2(z) \right) dz, \qquad \text{for all } -1 \leq y \leq 0 \leq x \leq 1.$$

Note that

$$\frac{\partial G(x, y)}{\partial x} = F_1(x) - F_2(x) \qquad \text{and} \qquad \frac{\partial G(x, y)}{\partial y} = -\left(F_1(y) - F_2(y) \right).$$

Similarly as in (2.13), using twice the integration by parts formula, we can compute

$$\begin{aligned}
U(\tilde{x}_1) - U(\tilde{x}_2) &= \int_{-1}^1 u(z) dF_1(z) - \int_{-1}^1 u(z) dF_2(z) \\
&= \int_{-1}^0 u(z) \left(dF_1(z) - dF_2(z) \right) + \int_0^1 u(z) \left(dF_1(z) - dF_2(z) \right) \\
&= -\int_{-1}^0 \left(F_1(z) - F_2(z) \right) u'(z) dz - \int_0^1 \left(F_1(z) - F_2(z) \right) u'(z) dz \\
&= -\int_{-1}^0 G(x, z) u''(z) dz + \int_0^1 G(z, y) u''(z) dz
\end{aligned}$$

for arbitrary $y \in [-1, 0]$ and $x \in [0, 1]$. Recalling that $u''(x) \leq 0$ for $x \geq 0$ and that $u''(x) \geq 0$ for $x \leq 0$, it follows that $G(x, y) \leq 0$ for all $-1 \leq y \leq 0 \leq x \leq 1$ implies that $U(\tilde{x}_1) \geq U(\tilde{x}_2)$. The converse implication follows by the same calculation by considering arbitrary value functions $u(\cdot)$.

Solution of Exercise 9.2 Note first that, by Proposition 2.8, the fact that $\tilde{x}_1 \succeq_{\text{FSD}} \tilde{x}_2$ implies

$$\sum_{k=1}^s \pi_s \leq \sum_{k=1}^s \pi_s', \qquad \text{for every } s = 1, \ldots, S.$$

By (9.4) and using the above property, we can compute

$$\begin{aligned}
U(\tilde{x}_1) &= \sum_{s=1}^S u(x_s) \left(w\left(\sum_{k=1}^s \pi_k \right) - w\left(\sum_{k=1}^{s-1} \pi_k \right) \right) \\
&= \sum_{s=1}^S u(x_s) w\left(\sum_{k=1}^s \pi_k \right) - \sum_{s=1}^{S-1} u(x_{s+1}) w\left(\sum_{k=1}^s \pi_k \right)
\end{aligned}$$

$$= \sum_{s=1}^{S-1} \big(u(x_s) - u(x_{s+1})\big) w\Big(\sum_{k=1}^{s} \pi_k\Big) + u(x_S) w(1)$$

$$\geq \sum_{s=1}^{S-1} \big(u(x_s) - u(x_{s+1})\big) w\Big(\sum_{k=1}^{s} \pi_k'\Big) + u(x_S) w(1) = U(\tilde{x}_2),$$

where we have used the fact that $x_s < x_{s+1}$, for every $s = 1, \ldots, S-1$, and that the functions $u(\cdot)$ and $w(\cdot)$ are increasing.

Solution of Exercise 9.3

(i): It suffices to compute, using the definition of the preference functional (9.7) together with the linearity of the function v with respect to its second argument and the definition of the function $\hat{v}$ given in Proposition 9.5,

$$\mathbb{E}\left[\sum_{t=0}^{\infty} \big(\delta^t u(c_t^* + \alpha c_t) + b_t \delta^{t+1} v(X_{t+1}, w_{t+1}^* + \alpha w_{t+1}, z_t)\big)\right]$$

$$- \mathbb{E}\left[\sum_{t=0}^{\infty} \big(\delta^t u(c_t^*) + b_t \delta^{t+1} v(X_{t+1}, w_{t+1}^*, z_t)\big)\right]$$

$$= \mathbb{E}\left[\sum_{t=0}^{\infty} \big(\delta^t (u(c_t^* + \alpha c_t) - u(c_t^*)) + b_t \alpha w_{t+1} \delta^{t+1} \hat{v}(r_{t+1}, z_t)\big)\right]$$

$$\leq \mathbb{E}\left[\sum_{t=0}^{\infty} \big(\delta^t u'(c_t^*) \alpha c_t + b_t \alpha w_{t+1} \delta^{t+1} \hat{v}(r_{t+1}, z_t)\big)\right],$$

where, in the last step, we have used the concavity of the function $x^{1-\gamma}/(1-\gamma)$.

(ii): Let $(W_t^*)_{t \in \mathbb{N}}$ be the self-financing wealth process associated to the strategy (w^*, c^*) and $(W_t)_{t \in \mathbb{N}}$ the self-financing wealth process associated to the strategy (w, c). Note that, by the self-financing condition, it holds that

$$\alpha W_{t+1} = W_{t+1}^{*,\alpha} - W_{t+1}^* = (\alpha W_t - \alpha c_t) r_f + \alpha w_{t+1}(r_{t+1} - r_f),$$

where $(W_t^{*,\alpha})_{t \in \mathbb{N}}$ is the wealth process associated to the self-financing strategy $(w^* + \alpha w, c^* + \alpha c)$. Making use of this last relation together with the Euler conditions (9.81)–(9.82), it then suffices to compute

$$\delta \mathbb{E}\left[u'(c_{t+1}^*) \alpha W_{t+1} | \mathscr{F}_t\right] = \delta \alpha W_t r_f \mathbb{E}\left[u'(c_{t+1}^*) | \mathscr{F}_t\right] - \delta \alpha c_t r_f \mathbb{E}\left[u'(c_{t+1}^*) | \mathscr{F}_t\right]$$

$$+ \delta \alpha w_{t+1} \mathbb{E}\left[u'(c_{t+1}^*)(r_{t+1} - r_f) | \mathscr{F}_t\right]$$

$$= \alpha W_t u'(c_t^*) - \alpha c_t u'(c_t^*) - \alpha w_{t+1} \delta b_t \mathbb{E}\left[\hat{v}(r_{t+1}, z_t) | \mathscr{F}_t\right],$$

from which (9.83) immediately follows.

(iii): It suffices to sum relation (9.83) over all $t \in \mathbb{N}$. For the proof of the remaining part of the claim, we refer the reader to the Appendix of Barberis et al. [150].

(iv): The Euler condition (9.81) together with the dynamics (9.6) satisfied by the aggregate consumption process $(\bar{c}_t)_{t \in \mathbb{N}}$ implies that

$$\frac{1}{r_f} = \delta \mathbb{E}\left[\frac{u'(\bar{c}_{t+1})}{u'(\bar{c}_t)}\middle|\mathscr{F}_t\right] = \delta \mathbb{E}\left[\left(\frac{\bar{c}_{t+1}}{\bar{c}_t}\right)^{-\gamma}\middle|\mathscr{F}_t\right] = \delta \mathbb{E}\left[e^{-\gamma g_c - \gamma \sigma_c \eta_{t+1}}|\mathscr{F}_t\right]$$

$$= \delta e^{-\gamma g_c + \frac{\gamma^2 \sigma_c^2}{2}},$$

thus proving the claim.

(v): Note first that, making use of relation (9.5),

$$r_{t+1} = \frac{p_{t+1}^* + d_{t+1}}{p_t^*} = \frac{1 + \frac{p_{t+1}^*}{d_{t+1}}}{\frac{p_t^*}{d_t}}\frac{d_{t+1}}{d_t} = \frac{1 + f(z_{t+1})}{f(z_t)}\frac{d_{t+1}}{d_t} = \frac{1 + f(z_{t+1})}{f(z_t)}e^{g_d + \sigma_d \varepsilon_{t+1}},$$

for all $t \in \mathbb{N}$. To prove the claim, it then suffices to compute, using basic properties of the bivariate normal distribution and noting that z_{t+1} is determined only by z_t and ε_{t+1},

$$\mathbb{E}\left[r_{t+1}\left(\frac{\bar{c}_{t+1}}{\bar{c}_t}\right)^{-\gamma}\middle|\mathscr{F}_t\right]$$

$$= \mathbb{E}\left[\frac{1 + f(z_{t+1})}{f(z_t)}e^{g_d + \sigma_d \varepsilon_{t+1} - \gamma(g_c + \sigma_c \eta_{t+1})}\middle|\mathscr{F}_t\right]$$

$$= e^{g_d - \gamma g_c}\mathbb{E}\left[\mathbb{E}\left[e^{-\gamma \sigma_c \eta_{t+1}}|\mathscr{F}_t \vee \sigma(\varepsilon_{t+1})\right]\frac{1 + f(z_{t+1})}{f(z_t)}e^{\sigma_d \varepsilon_{t+1}}\middle|\mathscr{F}_t\right]$$

$$= e^{g_d - \gamma g_c + \frac{\gamma^2 \sigma_c^2(1-\rho^2)}{2}}\mathbb{E}\left[\frac{1 + f(z_{t+1})}{f(z_t)}e^{(\sigma_d - \gamma \rho \sigma_c)\varepsilon_{t+1}}\middle|\mathscr{F}_t\right].$$

(vi): The Euler conditions (9.81)–(9.82) are necessary and sufficient for the optimality of a trading-consumption strategy $((w_t^*, c_t^*))_{t \in \mathbb{N}}$. Moreover, parts *(iv)* and *(v)* of the exercise show that the strategy consisting in consuming the aggregate consumption and holding the whole supply of the risky asset and the zero net supply of the risk free asset satisfies the Euler optimality conditions, from which the optimality of this strategy follows, thus completing the proof of Proposition 9.5.

Solution of Exercise 9.5

(i) Denote by $(W_t^*)_{t=0,1,\dots,T}$ and $(c_t^*)_{t=0,1,\dots,T}$ the optimal wealth and consumption processes, respectively, associated to the maximization of the recursive preferences (9.19). In view of Proposition 9.7, the value function V satisfies (9.21)–(9.22). In particular, similarly as in the proof of Proposition 6.4, the

first order condition of the maximization problem implies that, at each date $t \in \{0, 1, \ldots, T - 1\}$,

$$\frac{\partial}{\partial c} v\Big(c_t^*, \mu_t(V(W_{t+1}^*, t + 1))\Big) = \frac{\partial}{\partial y} v\Big(c_t^*, \mu_t(V(W_{t+1}^*, t + 1))\Big)\Big(\mu_t(V(W_{t+1}^*, t + 1))\Big)^{\alpha}$$

$$\times \mathbb{E}\left[V(W_{t+1}^*, t + 1)^{-\alpha} V'(W_{t+1}^*, t + 1) r_{t+1}^n \Big| \mathscr{F}_t \right]$$

$$= \frac{\partial}{\partial y} v\Big(c_t^*, \mu_t(V(W_{t+1}^*, t + 1))\Big)\Big(\mu_t(V(W_{t+1}^*, t + 1))\Big)^{\alpha}$$

$$\times \mathbb{E}\left[V(W_{t+1}^*, t + 1)^{-\alpha} \frac{\partial}{\partial c} v\Big(c_{t+1}^*, \mu_{t+1}(V(W_{t+2}^*, t + 2))\Big) r_{t+1}^n \Big| \mathscr{F}_t \right],$$

where the second equality follows from the envelope condition given in Proposition 9.7 and where $\frac{\partial}{\partial y} v$ denotes the first derivative of the function v with respect to its second argument. Note that (9.19) implies that

$$\frac{\partial}{\partial c} v(c, y) = v(c, y)^{\varrho}(1 - \delta)c^{-\varrho} \quad \text{and} \quad \frac{\partial}{\partial y} v(c, y) = v(c, y)^{\varrho} \delta y^{-\varrho}.$$

By substituting into the above relation, we obtain

$$1 = \delta \mathbb{E}\left[\left(\frac{V(W_{t+1}^*, t + 1)}{\mu_t(V(W_{t+1}^*, t + 1))} \right)^{\varrho - \alpha} \left(\frac{c_{t+1}^*}{c_t^*} \right)^{-\varrho} r_{t+1}^n \Big| \mathscr{F}_t \right],$$

for all $t = 0, 1, \ldots, T - 1$ and $n = 1, \ldots, N$, where we have also used relation (9.22). This proves the validity of the Euler condition (9.23). Furthermore, if there exists a risk free asset with a constant rate of return $r_f > 0$, then the Euler condition (9.24) follows from (9.23) similarly as in the proof of Corollary 6.5.

(ii): Note first that, as a consequence of the Euler condition (9.23) together with the self-financing condition (6.19), the optimal wealth process $(W_t^*)_{t=0,1,\ldots,T}$ satisfies the given backward stochastic difference equation. Let then $(X_t)_{t=0,1,\ldots,T}$ be an arbitrary solution to the backward stochastic difference equation. Note first that

$$X_T = \frac{V(W_T^*, T)}{u'(c_T^*)} = \frac{u(c_T^*)}{u'(c_T^*)} = c_T^* = W_T^*.$$

By induction, suppose that $X_{t+1} = W_{t+1}^*$, for some $t = 0, 1, \ldots, T - 1$. Then, the backward stochastic difference equation implies that

$$X_t = c_t^* + \mathbb{E}\left[\frac{\xi_{t+1}}{\xi_t} X_{t+1} \Big| \mathscr{F}_t \right] = c_t^* + \mathbb{E}\left[\frac{\xi_{t+1}}{\xi_t} W_{t+1}^* \Big| \mathscr{F}_t \right] = W_t^*,$$

thus showing the uniqueness of the solution to the backward difference equation.

(iii): Define the process $(X_t)_{t=0,1,\ldots,T}$ by

$$X_t := \frac{V(W_t^*, t)}{\frac{\partial}{\partial c} v\big(c_t^*, \mu_t(V(W_{t+1}^*, t+1))\big)} = V(W_t^*, t)^{1-\varrho} \frac{(c_t^*)^\varrho}{1-\delta}, \quad \text{for all } t = 0, 1, \ldots, T.$$

It can be checked that the process $(X_t)_{t=0,1,\ldots,T}$ is a solution to the backward stochastic difference equation considered in part (ii). Indeed, for all $t = 0, 1, \ldots, T-1$, it holds that

$$\mathbb{E}\left[\frac{\xi_{t+1}}{\xi_t} X_{t+1} \Big| \mathscr{F}_t\right] = \frac{\delta}{1-\delta} \mathbb{E}\left[V(W_{t+1}^*, t+1)^{1-\alpha} \mu_t(V(W_{t+1}^*, t+1))^{\alpha-\varrho} (c_t^*)^\varrho \big| \mathscr{F}_t\right]$$

$$= \frac{\delta}{1-\delta} (c_t^*)^\varrho \mu_t(V(W_{t+1}^*, t+1))^{\alpha-\varrho} \mu_t(V(W_{t+1}^*, t+1))^{1-\alpha}$$

$$= V(W_t^*, t)^{1-\varrho} \frac{(c_t^*)^\varrho}{1-\delta} - c_t^* = X_t - c_t^*,$$

where we have used the $\mathscr{F}_t$-measurability of $\mu_t(V(W_{t+1}^*, t+1))$ together with its definition. In part (ii), we have shown that the above backward stochastic difference equation admits a unique solution, thus proving the validity of representation (9.25). Relation (9.26) then follows by means of elementary computations, using the fact that

$$W_t^* - c_t^* = \mathbb{E}\left[\frac{\xi_{t+1}}{\xi_t} W_{t+1}^* \Big| \mathscr{F}_t\right] = \frac{\delta}{1-\delta} (c_t^*)^\varrho \mu_t\big(V(W_{t+1}^*, t+1)\big)^{1-\varrho},$$

for all $t = 0, 1, \ldots, T-1$.

(iv): This follows by simply substituting (9.26) in the Euler condition (9.24).

Solution of Exercise 9.6

(i) Using (9.29) and (9.30), relation (9.23) can be rewritten as follows:

$$1 = \mathbb{E}\left[e^{\gamma \log \delta + (\gamma-1) \log r_{t+1}^* - \gamma\varrho \log(c_{t+1}^*/c_t^*) + \log r_{t+1}^n} \Big| \mathscr{F}_t\right],$$

for all $t = 0, 1, \ldots, T-1$ and $n = 1, \ldots, N$. Since r_{t+1}^* can be written as a linear combination of the returns on the single assets, we can rewrite the above relation replacing r_{t+1}^n with r_{t+1}^*. Making use of the distributional assumptions of Proposition 9.9, the conditional expectation can then be explicitly computed as follows:

$$1 = e^{\gamma \log \delta + \gamma\mu_r - \gamma\varrho\mu_c + \frac{\gamma^2\sigma_r^2}{2} + \frac{\gamma^2\varrho^2\sigma_c^2}{2} - \varrho\gamma^2\sigma_{cr}},$$

from which part (i) of the exercise follows.

(ii) Making use of relation (9.23), written in terms of the optimal return process $(r_t^*)_{t=1,\dots,T}$ and with respect to the risk free asset with return r_f, we can compute, similarly as above,

$$1 = e^{\gamma \log \delta + (\gamma-1)\mu_r - \gamma\varrho\mu_c + \frac{(\gamma-1)^2\sigma_r^2}{2} + \frac{\gamma^2\varrho^2\sigma_c^2}{2} - (\gamma-1)\gamma\varrho\sigma_{cr} + \log r_f}.$$

Relation (9.33) follows directly by substituting μ_r with its expression computed in part *(i)* of the exercise.

(iii) Similarly as in part *(ii)* of the exercise, we can compute

$$1 = \mathbb{E}\left[e^{\gamma \log \delta + (\gamma-1)\log r_{t+1}^* - \gamma\varrho \log(c_{t+1}^*/c_t^*) + \log r_{t+1}^n}\,\Big|\,\mathscr{F}_t\right]$$

$$= e^{\gamma \log \delta + (\gamma-1)\mu_r + \frac{(\gamma-1)^2\sigma_r^2}{2} - \gamma\varrho\mu_c + \frac{\gamma^2\varrho^2\sigma_c^2}{2} + \mathbb{E}[\log r_{t+1}^n|\mathscr{F}_t] + \frac{\mathrm{Var}(\log r_{t+1}^n|\mathscr{F}_t)}{2}}$$

$$\times\, e^{-(\gamma-1)\gamma\varrho\sigma_{cr} + (\gamma-1)\,\mathrm{Cov}(\log r_{t+1}^*,\log r_{t+1}^n|\mathscr{F}_t) - \gamma\varrho\,\mathrm{Cov}(\log(c_{t+1}^*/c_t^*),\log r_{t+1}^n|\mathscr{F}_t)}.$$

Relation (9.34) then follows directly by relying on parts *(i)* and *(ii)* of the exercise.

Solution of Exercise 9.7

(i) As a first step, the relation $W_{t+1}^* = (W_t^* - c_t^*)r_{t+1}^*$ can be rewritten in logarithmic terms as

$$\Delta \log W_{t+1}^* = \log r_{t+1}^* + \log\left(1 - e^{\log c_t^* - \log W_t^*}\right).$$

Letting $\bar{x} := \mathbb{E}[\log c_t^* - \log W_t^*]$, the non-linear function $\log(1 - e^x)$ admits the following Taylor expansion up to the first order:

$$\log(1 - e^x) \approx \log(1 - e^{\bar{x}}) + \frac{e^{\bar{x}}}{e^{\bar{x}} - 1}(x - \bar{x}).$$

By combining the last two equations, we can then write

$$\Delta \log W_{t+1}^* \approx \log r_{t+1}^* + \log \lambda - \frac{1-\lambda}{\lambda}(\log c_t^* - \log W_t^* - \log(1 - \lambda)),$$

thus proving part *(i)* of the exercise.

(ii): Making use of part *(i)* of the exercise together with the approximate relation (9.85), it holds that

$$\Delta \log W_{t+1}^* \approx \theta_{t+1}^*(\log r_{t+1} - \log r_f) + \log r_f + \frac{\theta_{t+1}^*}{2}(1 - \theta_{t+1}^*)\,\mathrm{Var}(\log r_{t+1}|\mathscr{F}_t)$$

$$+ \left(1 - \frac{1}{\lambda}\right)(\log c_t^* - \log W_t^*) + K.$$

Let us then write the elementary identity

$$\Delta \log c^*_{t+1} = (\log c^*_{t+1} - \log W^*_{t+1}) - (\log c^*_t - \log W^*_t) + \Delta \log W^*_{t+1},$$

for all $t = 0, 1, \ldots, T - 1$. It holds that

$$\text{Cov}(\log r_{t+1}, \Delta \log c^*_{t+1} | \mathscr{F}_t) = \text{Cov}(\log r_{t+1}, \log c^*_{t+1} - \log W^*_{t+1} | \mathscr{F}_t)$$
$$+ \text{Cov}(\log r_{t+1}, \Delta \log W^*_{t+1} | \mathscr{F}_t).$$

Making use of the above relations, we get that

$$\text{Cov}(\log r_{t+1}, \log r^*_{t+1} | \mathscr{F}_t) \approx \theta^*_{t+1} \, \text{Var}(\log r_{t+1} | \mathscr{F}_t)$$

and

$$\text{Cov}(\log r_{t+1}, \log W^*_{t+1} | \mathscr{F}_t) \approx \theta^*_{t+1} \, \text{Var}(\log r_{t+1} | \mathscr{F}_t),$$

so that

$$\text{Cov}(\log r_{t+1}, \Delta \log c^*_{t+1}) \approx \text{Cov}(\log r_{t+1}, \log c^*_{t+1} - \log W^*_{t+1} | \mathscr{F}_t)$$
$$+ \theta^*_{t+1} \, \text{Var}(\log r_{t+1} | \mathscr{F}_t).$$

Making use of (9.85), relation (9.86) then follows directly.

(iii): To solve the final part of the exercise, it suffices to solve relation (9.86) with respect to θ^*_{t+1} and note that

$$\frac{1}{1 - \gamma + \varrho\gamma} = \frac{1}{\alpha} \quad \text{and} \quad \frac{\varrho\gamma}{1 - \gamma + \varrho\gamma} = \frac{\varrho}{\varrho - 1} \frac{\alpha - 1}{\alpha}.$$

Solution of Exercise 9.8

(i) For a given consumption process $c = (c_t)_{t=0,1,\ldots,T}$, let us denote

$$U(c) = \sum_{t=0}^{T} \delta^t u(c_t, z_t).$$

Similarly as in the proof of Proposition 6.4, at optimality it must hold that

$$\mathbb{E}\left[\frac{\frac{\partial}{\partial c_{t+1}} \mathbb{E}[U(c^*) | \mathscr{F}_{t+1}]}{\frac{\partial}{\partial c_t} \mathbb{E}[U(c^*) | \mathscr{F}_t]} (r^n_{t+1} - r_f) \middle| \mathscr{F}_t \right] = 0,$$

where $c^* = (c_t^*)_{t=0,1,\dots,T}$ denotes the optimal consumption process. Condition (9.39) then follows by computing

$$\frac{\partial}{\partial c_t} U(c^*) = \delta^t \partial_c u(c_t^*, z_t^*) + a \sum_{s=t+1}^{T} \delta^s \partial_z u(c_s^*, z_s^*) b^{s-t}$$

$$= \delta^t \left(\partial_c u(c_t^*, z_t^*) + a \sum_{s=1}^{T-t} \delta^s \partial_z u(c_{t+s}^*, z_{t+s}^*) b^s \right),$$

for all $t = 0, 1, \dots, T$, where we have used relation (9.36) and where we have denoted by $\partial_c u$ the partial first derivative of the function u with respect to the first argument and, analogously, by $\partial_z u$ the partial first derivative of the function u with respect to the second argument.

(ii): The second part of the exercise follows from a straightforward computation.

Solution of Exercise 9.9

(i) It suffices to compute

$$\frac{\partial}{\partial c_t} U(c) = \delta^t c_t^{-\alpha} z_t^{-(1-\alpha)} - \delta^{t+1} \beta \gamma c_{t+1}^{1-\alpha} c_t^{-\beta\gamma(1-\alpha)-1} c_{t-1}^{-\gamma(1-\beta)(1-\alpha)},$$

from which (9.87) follows by elementary computations.

(ii): Relation (9.88) follows directly from (9.87) together with the assumption of market equilibrium, using the notation introduced in the exercise. The Euler condition can be obtained as usual, making use of relation (9.88).

(iii): It suffices to observe that

$$\frac{d_{t+1} + p_{t+1}^*}{d_t} = \frac{d_{t+1}}{d_t} \left(1 + \frac{p_{t+1}^*}{d_{t+1}} \right) = x_{t+1} \left(1 + \frac{p_{t+1}^*}{d_{t+1}} \right)$$

and make use of the Euler condition obtained in part *(ii)* of the exercise.

Solution of Exercise 9.10 As a first step, note that the intertemporal marginal rate of substitution is given by

$$m_{t+1} = \delta \frac{\frac{\partial}{\partial c_{t+1}} u(c_{t+1}, z_{t+1})}{\frac{\partial}{\partial c_t} u(c_t, z_t)} = \delta \left(\frac{c_{t+1}}{c_t} \frac{s_{t+1}}{s_t} \right)^{-\alpha},$$

for all $t = 0, 1, \dots, T - 1$, from which (9.48) follows in view of (9.46)–(9.47). In order to prove (9.49), it suffices to compute

$$\mathbb{E}[m_{t+1}|\mathscr{F}_t] = \delta e^{-\alpha(g+(\varphi-1)(\log s_t - \bar{s}))} \mathbb{E}\left[e^{-\alpha(1+\lambda(s_t))v_{t+1}}|\mathscr{F}_t \right]$$

$$= \delta e^{-\alpha(g+(\varphi-1)(\log s_t - \bar{s}))+\alpha^2(1+\lambda(s_t))^2 \frac{\sigma^2}{2}}$$

and

$$\mathrm{Var}(m_{t+1}|\mathscr{F}_t) = \delta^2 e^{-2\alpha(g+(\varphi-1)(\log s_t - \bar{s}))} \, \mathrm{Var}(e^{-\alpha(1+\lambda(s_t))v_{t+1}}|\mathscr{F}_t)$$

$$= \delta^2 e^{-2\alpha(g+(\varphi-1)(\log s_t - \bar{s}))} e^{\alpha^2(1+\lambda(s_t))^2 \sigma^2} \left(e^{\alpha^2(1+\lambda(s_t))^2 \sigma^2} - 1 \right),$$

from which (9.49) directly follows. Finally, in order to determine the equilibrium risk free rate r_f, it suffices to compute

$$r_f = \frac{1}{\mathbb{E}[m_{t+1}|\mathscr{F}_t]} = \frac{1}{\delta} e^{\alpha(g+(\varphi-1)(\log s_t - \bar{s})) - \alpha^2(1+\lambda(s_t))^2 \frac{\sigma^2}{2}}.$$

Solution of Exercise 9.11 Suppose that we are at date t, for some arbitrary $t \in \mathbb{N}$, and denote the initial wealth at date t of an agent (rational or noise trader) by W_t. At date t, both the rational and the noise trader have to choose how much to invest in the risky asset in order to maximize the expected utility of wealth at the next period. Let us first consider the case of a rational trader. Denoting by θ^r_{t+1} the quantity of the risky asset held in the portfolio by a rational trader in the period $[t, t+1]$, it holds that

$$W_{t+1} = (W_t - \theta^r_{t+1} p_t)(1 + r_f) + \theta^r_{t+1}(r_f + p_{t+1})$$

$$= W_t(1 + r_f) + \theta^r_{t+1}\left(r_f + p_{t+1} - p_t(1 + r_f)\right).$$

The same relation also holds for a noise trader, replacing θ^r_{t+1} with θ^n_{t+1}, with the latter denoting the quantity of the risky asset held in the portfolio by a noise trader in the period $[t, t+1]$. Due to the assumption of an exponential utility function, the rational trader's optimal portfolio problem consists in choosing the optimal quantity $\theta^{r,*}_{t+1}$ which maximizes the expression

$$W_t(1 + r_f) + \theta^r_{t+1}\left(r_f + \mathbb{E}[p_{t+1}|\mathscr{F}_t] - p_t(1 + r_f)\right) - \gamma\left(\theta^r_{t+1}\right)^2 \mathrm{Var}(p_{t+1}|\mathscr{F}_t)$$

over all $\theta^r_{t+1} \in \mathbb{R}$. As can be easily verified, the optimal $\theta^{r,*}_{t+1}$ is given by

$$\theta^{r,*}_{t+1} = \frac{r_f + \mathbb{E}[p_{t+1}|\mathscr{F}_t] - (1 + r_f)p_t}{2\gamma \, \mathrm{Var}(p_{t+1}|\mathscr{F}_t)}.$$

We now perform a similar analysis in the case of a noise trader. Similarly as above, the optimal portfolio problem consists in determining the quantity $\theta^{n,*}_{t+1} \in \mathbb{R}$ which maximizes the expression

$$W_t(1 + r_f) + \theta^n_{t+1}\left(r_f + \mathbb{E}[p_{t+1}|\mathscr{F}_t] - p_t(1 + r_f)\right) - \gamma\left(\theta^n_{t+1}\right)^2 \mathrm{Var}(p_{t+1}|\mathscr{F}_t) + \theta^n_{t+1}\varrho_t,$$

where ϱ_t represents the noise trader's misperception of the next period's expected price of the risky asset. As can be easily verified, the optimal $\theta_{t+1}^{n,*}$ is given by

$$\theta_{t+1}^{n,*} = \frac{r_f + \mathbb{E}[p_{t+1}|\mathscr{F}_t] - (1 + r_f)p_t}{2\gamma \operatorname{Var}(p_{t+1}|\mathscr{F}_t)} + \frac{\varrho_t}{2\gamma \operatorname{Var}(p_{t+1}|\mathscr{F}_t)}.$$

In equilibrium, the aggregate demand of the risky asset by rational and noise traders must be equal to the aggregate supply, which is fixed at one. Hence, it must hold that

$$\mu\theta_{t+1}^{n,*} + (1 - \mu)\theta_{t+1}^{r,*} = 1,$$

for all $t \in \mathbb{N}$, leading to the condition

$$\theta_{t+1}^{r,*} + \mu\frac{\varrho_t}{2\gamma \operatorname{Var}(p_{t+1}|\mathscr{F}_t)} = 1.$$

Solving for p_t, this implies that the equilibrium price process $(p_t^*)_{t\in\mathbb{N}}$ must satisfy

$$p_t^* = \frac{1}{1 + r_f}\left(r_f + \mathbb{E}[p_{t+1}^*|\mathscr{F}_t] - 2\gamma \operatorname{Var}(p_{t+1}^*|\mathscr{F}_t) + \mu\varrho_t\right).$$

Note that this formula relates the equilibrium price p_t^* of the risky asset to the noise traders' misperception ϱ_t, to the dividend r_f on the risk free asset, to the coefficient of absolute risk aversion 2γ as well as to the $\mathscr{F}_t$-conditional distribution of the next period's equilibrium price p_{t+1}^*. Considering a steady-state equilibrium where the unconditional distribution of p_t^* is constant over time and following the arguments given in De Long et al. [549, Section I.B], the last relation can be solved recursively yielding

$$p_t^* = 1 + \frac{\mu(\varrho_t - \bar{\varrho})}{1 + r_f} + \frac{\mu\bar{\varrho}}{r_f} - \frac{2\gamma}{r_f} \operatorname{Var}(p_{t+1}^*|\mathscr{F}_t).$$

Observe that, in the right-hand side of the last relation, only the second term is variable, since γ, $\bar{\varrho}$ and r_f are constant and the conditional variance $\operatorname{Var}(p_{t+1}^*|\mathscr{F}_t)$ is a function of the constant variance of noise traders' misperception ϱ_t:

$$\operatorname{Var}(p_{t+1}^*|\mathscr{F}_t) = \operatorname{Var}(p_{t+1}^*) = \frac{\mu^2\sigma_\varrho^2}{(1 + r_f)^2}.$$

Formula (9.51) then follows directly from the last two relations.

Let us now prove part (ii) of the proposition, following the arguments given in De Long et al. [549, Section II.A]. In equilibrium, the difference in the returns on the optimal portfolios of noise and rational traders is given by

$$\Delta R_{t+1} := (\theta_{t+1}^{n,*} - \theta_{t+1}^{r,*})(r_f + p_{t+1}^* - p_t^*(1 + r_f)).$$

By relying on the results established in the first part of the exercise, it holds that

$$\theta_{t+1}^{n,*} - \theta_{t+1}^{r,*} = \frac{\varrho_t}{2\gamma \, \mathrm{Var}(p_{t+1}|\mathcal{F}_t)} = \frac{(1+r_f)^2 \varrho_t}{2\gamma \mu^2 \sigma_\varrho^2},$$

for all $t \in \mathbb{N}$. On the other hand, the $\mathcal{F}_t$-conditional expected value of the excess return on the risky asset is given by

$$\mathbb{E}\big[r_f + p_{t+1}^* - p_t^*(1+r_f)|\mathcal{F}_t\big] = 2\gamma \, \mathrm{Var}(p_{t+1}^*|\mathcal{F}_t) - \mu\varrho_t = \frac{2\gamma\mu^2\sigma_\varrho^2}{(1+r_f)^2} - \mu\varrho_t,$$

so that

$$\mathbb{E}[\Delta R_{t+1}|\mathcal{F}_t] = \varrho_t - \frac{(1+r_f)^2 \varrho_t^2}{2\gamma\mu\sigma_\varrho^2}.$$

By the law of iterated expectation, it then follows that

$$\mathbb{E}[\Delta R_{t+1}] = \mathbb{E}\big[\mathbb{E}[\Delta R_{t+1}|\mathcal{F}_t]\big] = \bar{\varrho} - \frac{(1+r_f)^2 \bar{\varrho}^2 + (1+r_f)^2 \sigma_\varrho^2}{2\gamma\mu\sigma_\varrho^2},$$

thus proving part *(ii)* of the proposition.

Solution of Exercise 9.12 Part *(i)* of the exercise easily follows from the assumption of exponential utility functions together with the assumption of a normal distribution for all the random variables appearing in the model. Indeed, it suffices to observe that, due to the assumptions on the sequence of random variables $(\tilde{\varepsilon}_t)_{t\in\mathbb{N}}$, it holds that

$$\mathbb{E}[r_f + \tilde{\varepsilon}_{t+1} + p_{t+1}|\mathcal{F}_t] = r_f + \mathbb{E}[p_{t+1}|\mathcal{F}_t]$$

and

$$\mathrm{Var}(r_f + \tilde{\varepsilon}_{t+1} + p_{t+1}|\mathcal{F}_t) = \sigma_\varepsilon^2 + \mathrm{Var}(p_{t+1}|\mathcal{F}_t),$$

for all $t \in \mathbb{N}$. The explicit expressions for the optimal demand of the risky asset by rational and noise investors then follow exactly as in Exercise 9.11.

In correspondence of a steady-state equilibrium of the economy, part *(ii)* of the exercise simply follows by noting that

$$\mathrm{Var}(\tilde{\varepsilon}_{t+1} + p_{t+1}^*|\mathcal{F}_t) = \sigma_\varepsilon^2 + \mathrm{Var}(p_{t+1}^*|\mathcal{F}_t) = \sigma_\varepsilon^2 + \mathrm{Var}(p_{t+1}^*) = \sigma_\varepsilon^2 + \frac{\mu^2\sigma_\varrho^2}{(1+r_f)^2}$$

and by following the derivation of part *(i)* of Proposition 9.11 as detailed in Exercise 9.11, replacing μ with μ_t in the limit as ξ converges to zero.

Finally, part *(iii)* of the exercise can be obtained by following the same steps as in the last part of Exercise 9.11 and making use of the results established above.

Solution of Exercise 9.13 We first prove that the equilibrium price of the risky security at dates $t \in \{1, 2, 3\}$ is given by (9.55). Since the informed agents are risk neutral, they set market prices equal to the future expected dividend of the security, conditionally on the available information at each date $t \in \{1, 2, 3\}$. It is trivial that $p_3^* = \tilde{d}$, while at the previous dates it holds that

$$p_1^* = \mathbb{E}[\tilde{d}|\tilde{d} + \tilde{\varepsilon}],$$

$$p_2^* = \mathbb{E}[\tilde{d}|\tilde{d} + \tilde{\varepsilon}, \tilde{d} + \tilde{\eta}].$$

Formula (9.55) then follows by standard properties of the multivariate normal distribution, making use of the independence among the random variables $\tilde{d}$, $\tilde{\varepsilon}$ and $\tilde{\eta}$.

We now compute the covariances stated in (9.56), by making use of formula (9.55) together with the distributional assumptions of the model (compare with Daniel et al. [516, Appendix B]):

$$\mathrm{Cov}(p_3^* - p_2^*, p_2^* - p_1^*) = \frac{\sigma_d^6 \sigma_c^2 \sigma_\eta^2 (\sigma_\varepsilon^2 - \sigma_c^2)}{(\sigma_d^2 + \sigma_c^2)(\sigma_d^2(\sigma_c^2 + \sigma_\eta^2) + \sigma_c^2 \sigma_\eta^2)^2},$$

which is strictly positive since $\sigma_\varepsilon^2 > \sigma_c^2$, due to the informed agents' overconfidence. Let us then compute

$$\mathrm{Cov}(p_2^* - p_1^*, p_1^* - p_0^*) = -\frac{\sigma_d^6 \sigma_c^2 (\sigma_\varepsilon^2 - \sigma_c^2)}{(\sigma_d^2 + \sigma_c^2)^2(\sigma_d^2(\sigma_c^2 + \sigma_\eta^2) + \sigma_c^2 \sigma_\eta^2)},$$

which is strictly negative always due to the assumption that $\sigma_\varepsilon^2 > \sigma_c^2$. Similarly, we can compute (noting that $p_0^* = \bar{d} = 0$)

$$\mathrm{Cov}(p_3^* - p_1^*, p_1^* - \bar{d}) = -\frac{\sigma_d^4(\sigma_\varepsilon^2 - \sigma_c^2)}{(\sigma_d^2 + \sigma_c^2)^2} < 0,$$

and

$$\mathrm{Cov}(p_3^* - p_2^*, p_1^* - \bar{d}) = -\frac{\sigma_d^4 \sigma_\eta^2 (\sigma_\varepsilon^2 - \sigma_c^2)}{(\sigma_d^2 + \sigma_c^2)(\sigma_c^2(\sigma_d^2 + \sigma_\eta^2) + \sigma_d^2 \sigma_\eta^2)} < 0.$$

Solution of Exercise 9.14 Similarly as in Exercise 9.13, it holds that

$$p_1^* = \mathbb{E}[\tilde{d}|\tilde{d} + \tilde{\varepsilon}] = \frac{\sigma_d^2}{\sigma_d^2 + \sigma_c^2}(\tilde{d} + \tilde{\varepsilon}).$$

Concerning the equilibrium price at date $t = 2$, if $\text{sign}(d + \varepsilon) \neq \text{sign}(s_2)$, then the (over)confidence remains constant and, since the public signal $\tilde{s}_2$ is uninformative regarding the fundamental value of the asset, the equilibrium price does not change at date $t = 2$, so that $p_2^* = p_1^*$. On the other hand, if $\text{sign}(d + \varepsilon) = \text{sign}(s_2)$, then the equilibrium price at date $t = 2$ must be computed by taking into account the increased degree of overconfidence of informed investors, so that

$$p_2^* = \mathbb{E}[\tilde{d}|\tilde{d} + \tilde{\varepsilon}] = \frac{\sigma_d^2}{\sigma_d^2 + \sigma_c^2 - k}(\tilde{d} + \tilde{\varepsilon}),$$

thus establishing (9.57). We now compute the covariances between the equilibrium prices at different dates stated in (9.58) (see also Daniel et al. [516, Appendix D]). Since the probability p is assumed to be exogenously given and the probability that the price change at date $t = 1$ is positive is $1/2$, the law of iterated expectations gives that

$$\text{Cov}(p_2^* - p_1^*, p_1^* - p_0^*) = \mathbb{E}\big[\mathbb{E}[(p_2^* - p_1^*)(p_1^* - p_0^*)|\tilde{s}_2]\big]$$

$$= \frac{k\sigma_d^4(\sigma_d^2 + \sigma_\varepsilon^2)}{2(\sigma_d^2 + \sigma_c^2)^2(\sigma_d^2 + \sigma_c^2 - k)} > 0.$$

Similarly, it holds that

$$\text{Cov}(p_3^* - p_2^*, p_2^* - p_1^*) = \mathbb{E}\big[\mathbb{E}[(p_3^* - p_2^*)(p_2^* - p_1^*)|\tilde{s}_2]\big]$$

$$= -\frac{\sigma_d^2(k^2\sigma_d^2 + k(\sigma_\varepsilon^2 - \sigma_c^2))}{2(\sigma_d^2 + \sigma_c^2)(\sigma_d^2 + \sigma_c^2 - k)^2} < 0.$$

Finally, we have that

$$\text{Cov}(p_3^* - p_1^*, p_1^* - p_0^*) = \mathbb{E}[(\tilde{d} - p_1^*)p_1^*] = -\frac{\sigma_d^4(\sigma_\varepsilon^2 - \sigma_c^2)}{(\sigma_d^2 + \sigma_c^2)^2} < 0,$$

thus completing the proof of the proposition.

Solution of Exercise 9.15 We start from (9.62). Solving recursively for s_t, we obtain

$$s_t = \sum_{s=1}^{\infty} \frac{1}{r_f^s}\mathbb{E}[d_{t+s}|\mathscr{F}_t] + \lim_{T \to \infty} \frac{\mathbb{E}[s_{t+T}|\mathscr{F}_t]}{r_f^T} = \sum_{s=1}^{\infty} \frac{1}{r_f^s}\mathbb{E}[d_{t+s}|\mathscr{F}_t] + \beta_t,$$

where $\beta_t := \lim_{T \to \infty} \mathbb{E}[s_{t+T}|\mathscr{F}_t]/r_f^T$, $t \in \mathbb{N}$, represents the bubble component (compare with the proof of Proposition 6.40). Making use of relation (9.63), solving recursively for d_t and relying on the independence of the random variables $(\tilde{\varepsilon}_t)_{t \in \mathbb{N}}$, it holds that

$$\mathbb{E}[d_{t+2}|\mathscr{F}_t] = \mathbb{E}[\alpha + \psi d_{t+1} + \tilde{\varepsilon}_{t+1}|\mathscr{F}_t] = \alpha + \psi d_{t+1},$$

where the last equality uses the fact that d_{t+1} is $\mathscr{F}_t$-measurable. Similarly, for $s \geq 3$, it holds that

$$\mathbb{E}[d_{t+s}|\mathscr{F}_t] = \alpha(1 + \ldots + \psi^{s-2}) + \psi^{s-1}d_{t+1}.$$

Therefore, we get that

$$s_t = \alpha \sum_{s=2}^{\infty} \frac{1}{r_f^s} \sum_{k=0}^{s-2} \psi^k + \frac{d_{t+1}}{\psi} \sum_{s=1}^{\infty} \frac{\psi^s}{r_f^s} + \beta_t,$$

Let us first compute

$$\frac{1}{\psi} \sum_{s=1}^{\infty} \frac{\psi^s}{r_f^s} = \frac{1}{\psi} \left(\frac{r_f}{r_f - \psi} - 1 \right) = \frac{1}{r_f - \psi}.$$

For the other summation, it holds that

$$\sum_{s=2}^{\infty} \frac{1}{r_f^s} \sum_{k=0}^{s-2} \psi^k = \sum_{s=2}^{\infty} \frac{1}{r_f^s} \frac{1 - \psi^{s-1}}{1 - \psi}$$

$$= \frac{1}{1 - \psi} \left(\frac{r_f}{r_f - 1} - 1 - \frac{1}{r_f} - \frac{1}{\psi} \left(\frac{r_f}{r_f - \psi} - 1 - \frac{\psi}{r_f} \right) \right)$$

$$= \frac{1}{(r_f - 1)(r_f - \psi)}.$$

Formula (9.64) then follows directly by combining the above results.

Solution of Exercise 9.16 Following the original arguments given in Weil [1650, Proposition 2], if $u_1''' > 0$, then the function u_1' is convex and Jensen's inequality implies that

$$\mathbb{E}[u_1'(\tilde{d}_1 + \tilde{e}_1^i)] = \mathbb{E}\big[\mathbb{E}[u_1'(\tilde{d}_1 + \tilde{e}_1^i)|\tilde{d}_1]\big]$$

$$\geq \mathbb{E}\big[u_1'(\tilde{d}_1 + \mathbb{E}[\tilde{e}_1^i|\tilde{d}_1])\big]$$

$$= \mathbb{E}\big[u_1'(\tilde{d}_1 + \bar{e}_1)\big],$$

where we have used the fact that the random variables $\tilde{d}_1$ and $\tilde{e}_1$ are independent, so that

$$\mathbb{E}[\tilde{e}_1|\tilde{d}_1] = \mathbb{E}[\tilde{e}_1] = \bar{e}_1.$$

In view of equations (9.78) and (9.80), this immediately implies that $\hat{r}_f^* \geq r_f^*$ if and only if $u_1''' < 0$. Similarly, in the case of the expected equilibrium return, it holds that

$$\mathbb{E}[\tilde{d}_1 u_1'(\tilde{d}_1 + \tilde{e}_1^i)] = \mathbb{E}[\mathbb{E}[\tilde{d}_1 u_1'(\tilde{d}_1 + \tilde{e}_1^i)|\tilde{d}_1]]$$
$$\geq \mathbb{E}[\tilde{d}_1 u_1'(\tilde{d}_1 + \mathbb{E}[\tilde{e}_1^i|\tilde{d}_1])]$$
$$= \mathbb{E}[\tilde{d}_1 u_1'(\tilde{d}_1 + \bar{e}_1)].$$

In view of equations (9.77) and (9.79), this completes the proof of the proposition.

11.9   Exercises of Chap. 10

Solution of Exercise 10.1   As a preliminary, note that the precisions τ_{inf} and τ_{un} are both bounded from below by the prior precision τ_d and bounded from above by the full information precision τ_F. Hence, there exist two constants $\varphi_{\text{inf}} \in [0, 1]$ and $\varphi_{\text{un}} \in [0, 1]$ such that (10.6) holds. In correspondence of a symmetric linear equilibrium of the form (10.4), the market clearing condition (10.1) implies that

$$N\mu_{\text{inf}} + N\beta\tilde{d} + \beta \sum_{n=1}^{N} \tilde{\varepsilon}_n - N\gamma_{\text{inf}}\tilde{p}^* + M(\mu_{\text{un}} - \gamma_{\text{un}}p^*) + \tilde{z} = 0.$$

Solving for $\tilde{p}^*$ yields

$$\tilde{p}^* = \lambda \left(N\beta\tilde{d} + \beta \sum_{n=1}^{N} \tilde{\varepsilon}_n + \tilde{z} + N\mu_{\text{inf}} + M\mu_{\text{un}} \right).$$

Observe that the random variable $\tilde{p}^*$ is informationally equivalent to the random variable $\tilde{h}$ defined by

$$\tilde{h} := \tilde{d} + \frac{1}{N} \sum_{n=1}^{N} \tilde{\varepsilon}_n + \frac{\tilde{z}}{N\beta},$$

while, for every $n = 1, \ldots, N$, the couple $(\tilde{p}^*, \tilde{y}_n)$ is informationally equivalent to the couple $(\tilde{h}_n, \tilde{y}_n)$, where

$$\tilde{h}_n := \tilde{d} + \frac{1}{N-1} \sum_{\substack{k=1 \\ k \neq n}}^{N} \tilde{\varepsilon}_k + \frac{\tilde{z}}{(N-1)\beta}.$$

Relations (10.7) and (10.8) then follow by standard computations, using the basic properties of the normal multivariate distribution (compare also with Kyle [1148,

Appendix A]) and computing

$$\mathbb{E}[\tilde{d}|\tilde{p}^*] = \mathbb{E}[\tilde{d}|\tilde{h}] \qquad \text{and} \qquad \mathbb{E}[\tilde{d}|\tilde{p}^*, \tilde{y}_n] = \mathbb{E}[\tilde{d}|\tilde{h}_n, \tilde{y}_n],$$

for every $n = 1, \ldots, N$. The fact that, if $\mu_{inf} = \mu_{un} = 0$, then $\mathbb{E}[\tilde{d}|\tilde{p}^*] = \tilde{p}^*$ holds if and only if $\varphi_{un}\tau_\varepsilon = \beta\lambda\tau_{un}$ simply follows by substituting $\mu_{inf} = \mu_{un} = 0$ into (10.8). Finally, it remains to show that (10.9) holds. This can be shown by making use of the definition of the parameters φ_{inf} and φ_{un}. Then, in order to show that

$$\tau_{inf} - \tau_{un} = (1 - \varphi_{un})(1 - \varphi_{inf})\tau_\varepsilon,$$

it suffices to substitute $N(\varphi_{inf} - \varphi_{un}) = -(1 - \varphi_{inf})\varphi_{un}$, which follows from the relations in (10.9), into

$$\tau_{inf} - \tau_{un} = \big(1 - \varphi_{inf} + N(\varphi_{inf} - \varphi_{un})\big)\tau_\varepsilon,$$

which is obtained from (10.6).

Solution of Exercise 10.2

(i): Given the conjectured linear structure of the equilibrium, the expected profit of the insider trader, conditionally on the observation of the dividend $\tilde{d}$, if he chooses to trade a quantity x of the asset can be written as

$$\mathbb{E}\big[(\tilde{d} - P(x + \tilde{z}))x|\tilde{d}\big] = (\tilde{d} - \mu - \lambda x)x,$$

where we have used the assumption that the random variables $\tilde{d}$ and $\tilde{z}$ are independent and that $\mathbb{E}[\tilde{z}] = 0$. Note also that we have implicitly used the assumption that the insider trader takes the price setting rule of the market maker as given and exploits it when deciding the quantity to trade. Since the insider trader is assumed to maximize his expected profits, he chooses the quantity x^* which maximizes the above expression, so that

$$x^* = \frac{\tilde{d} - \mu}{2\lambda},$$

thus showing that $\beta = 1/(2\lambda)$ and $\alpha = -\mu\beta$.

(ii): Given the conjectured linear structure of the measurable functions X and P, the market efficiency condition (10.11) can be rewritten as

$$\mu + \lambda y = \mathbb{E}[\tilde{d}|\alpha + \beta\tilde{d} + \tilde{z} = y] = \mathbb{E}[\tilde{d}|\beta\tilde{d} + \tilde{z} = y - \alpha],$$

for any $y \in \mathbb{R}$ representing an arbitrary realization of the aggregate demand. In view of the distributional assumptions on the random variables $\tilde{d}$ and $\tilde{z}$, the couple $(\tilde{d}, \beta\tilde{d} + \tilde{z})$ is distributed as a bivariate normal random variable with mean vector and covariance matrix respectively given by

$$\begin{pmatrix} \bar{d} \\ \beta\bar{d} \end{pmatrix} \qquad \text{and} \qquad \begin{pmatrix} \sigma_d^2 & \beta\sigma_d^2 \\ \beta\sigma_d^2 & \beta^2\sigma_d^2 + \sigma_z^2 \end{pmatrix}.$$

Hence, the above conditional expectation can be explicitly computed as

$$\mathbb{E}[\tilde{d}\,|\alpha + \beta\tilde{d} + \tilde{z} = y] = \bar{d} + \frac{\beta\sigma_d^2}{\beta^2\sigma_d^2 + \sigma_z^2}(y - \alpha - \beta\bar{d}).$$

In turn, this implies that

$$\lambda = \frac{\beta\sigma_d^2}{\beta^2\sigma_d^2 + \sigma_z^2} \qquad \text{and} \qquad \mu - \bar{d} = -\frac{\beta\sigma_d^2}{\beta^2\sigma_d^2 + \sigma_z^2}(\alpha + \beta\bar{d}).$$

Combining the above results and solving for λ and β lead directly to the result of Proposition 10.3.

Solution of Exercise 10.4 We follow the arguments given in the solution of Exercise 10.3. Observe first that, given the conjectured linear structure of the equilibrium, the optimal demand by an informed agent observing the realization y of the signal $\tilde{y}$ can be obtained by determining the quantity x^* which maximizes with respect to x the expected profits conditionally on the observation of $\{\tilde{y} = y\}$:

$$\mathbb{E}\big[(\tilde{d} - P(x + \tilde{z}))x|\tilde{y} = y\big] = \mathbb{E}\big[(\tilde{d} - \mu - \lambda(x + \tilde{z}))x|\tilde{y} = y\big]$$

$$= \left(\frac{\sigma_d^2}{\sigma_d^2 + \sigma_\varepsilon^2}(y - \bar{d}) - \mu - \lambda x\right)x.$$

It follows that the optimal demand x^* is explicitly given by

$$x^* = \frac{\frac{\sigma_d^2}{\sigma_d^2 + \sigma_\varepsilon^2}(y - \bar{d}) - \mu}{2\lambda},$$

so that $\alpha = -\mu/(2\lambda)$ and $\beta = \sigma_d^2/((\sigma_d^2 + \sigma_\varepsilon^2)2\lambda)$. The equilibrium price is set by the risk dealer according to the zero expected profit condition (10.11), given the observation of the total market demand by informed and noise traders. Hence, in equilibrium the following condition must hold, letting $\omega := x^* + z$ denote an arbitrary realization of the random variable $\tilde{x}^* + \tilde{z}$:

$$\mu + \lambda\omega = \mathbb{E}[\tilde{d}|\alpha + \beta\tilde{y} + \tilde{z} = \omega].$$

By computing explicitly the conditional expectation appearing on the right-hand side of the last equation, relying on the assumption of normal distribution, we have that

$$\mathbb{E}[\tilde{d}|\alpha + \beta\tilde{y} + \tilde{z} = \omega] = \frac{\beta\sigma_d^2}{\beta^2(\sigma_d^2 + \sigma_\varepsilon^2) + \sigma_z^2}(\omega - \alpha - \beta\bar{d}) + \bar{d}.$$

From the last expression, it follows that

$$\lambda = \frac{\beta\sigma_d^2}{\beta^2(\sigma_d^2 + \sigma_\varepsilon^2) + \sigma_z^2}.$$

Putting together the equations for β and λ yields the explicit values

$$\beta = \sqrt{\frac{\sigma_z^2}{\sigma_d^2 + \sigma_\varepsilon^2}} \qquad \text{and} \qquad \lambda = \frac{1}{2}\sqrt{\frac{\sigma_d^4}{(\sigma_d^2 + \sigma_\varepsilon^2)\sigma_z^2}},$$

thus proving the claim.

Solution of Exercise 10.5

(i): The expected utility maximization problem of each informed trader consists in determining the quantity x which maximizes the conditional expectation

$$\mathbb{E}\left[-e^{-a(\tilde{d}-\tilde{p})x}\big|\tilde{\varepsilon} + \tilde{\eta}\right] = \mathbb{E}\left[-e^{-a(\tilde{\varepsilon}-\lambda(x+(N-1)\beta(\tilde{\varepsilon}+\tilde{\eta})+\tilde{z}))x}\big|\tilde{\varepsilon} + \tilde{\eta}\right],$$

where we have used the fact that the total market demand, conditionally on the observation of the signal $\tilde{\varepsilon} + \tilde{\eta}$ and using the conjecture that each of the remaining $N - 1$ informed traders demands a quantity equal to $\beta(\tilde{\varepsilon} + \tilde{\eta})$, is given by

$$x + (N - 1)\beta(\tilde{\varepsilon} + \tilde{\eta}) + \tilde{z},$$

so that the price set by the market maker equals

$$\bar{d} + \lambda\big(x + (N - 1)\beta(\tilde{\varepsilon} + \tilde{\eta}) + \tilde{z}\big).$$

By the assumption of a normal distribution, the above maximization problem can be rewritten as the problem of finding the quantity $\tilde{x}^*$ which maximizes the mean-variance function

$$\mathbb{E}\big[\big(\tilde{\varepsilon} - \lambda(x + (N - 1)\beta(\tilde{\varepsilon} + \tilde{\eta}) + \tilde{z})\big)x\big|\tilde{\varepsilon} + \tilde{\eta}\big]$$
$$- \frac{a}{2}\,\mathrm{Var}\big[\big(\tilde{\varepsilon} - \lambda(x + (N - 1)\beta(\tilde{\varepsilon} + \tilde{\eta}) + \tilde{z})\big)x\big|\tilde{\varepsilon} + \tilde{\eta}\big].$$

By using the basic properties of the normal multivariate distribution, the conditional expectation and variance appearing above can be computed in closed form. The optimal demand $\tilde{x}^*$ can then be obtained by differentiating with respect to x, setting the derivative equal to zero and solving with respect to x. This yields the expression given in (10.31).

(ii): To prove the claim it suffices to set $\tilde{x} = \beta(\tilde{\varepsilon} + \tilde{\eta})$ in (10.31) and solve the resulting equation for β.

(iii): Note that the total market demand by noise traders and informed traders is given by

$$N\beta(\tilde{\varepsilon} + \tilde{\eta}) + \tilde{z}.$$

Hence, if the equilibrium price rule set by the market maker is linear with respect to the market demand, it holds that

$$\bar{d} + \lambda\big(N\beta(\tilde{\varepsilon} + \tilde{\eta}) + \tilde{z}\big) = \mathbb{E}\big[\tilde{d}|N\beta(\tilde{\varepsilon} + \tilde{\eta}) + \tilde{z}\big].$$

In particular, by the properties of the bivariate normal distribution, it holds that

$$\lambda = \frac{\mathrm{Cov}(\tilde{\varepsilon}, N\beta(\tilde{\varepsilon} + \tilde{z}) + \tilde{\eta})}{\mathrm{Var}(N\beta(\tilde{\varepsilon} + \tilde{\eta}) + \tilde{z})}.$$

The claim then follows due to the normality assumptions on the random variables $\tilde{z}$, $\tilde{\varepsilon}$ and $\tilde{\eta}$.

Solution of Exercise 10.7

(i): For any $k = 1, \ldots, k$, let x_k denote the demand submitted by the informed trader k. In correspondence of such a demand and conditionally on the observation of the signal $\tilde{y}_k = \tilde{\varepsilon} + \tilde{\eta}_k$, the expected profits of the informed trader are given by

$$\mathbb{E}\left[(\bar{d} + \tilde{\varepsilon} - \tilde{p})\, x_k | \tilde{y}_k\right] = \mathbb{E}\left[\left(\tilde{\varepsilon} - \lambda\left(x_k + \sum_{\substack{l=1 \\ l\neq k}}^{K} \beta(\tilde{\varepsilon} + \tilde{\eta}_l) + \sum_{n=1}^{N} \gamma\tilde{e}_n\right)\right) x_k \Big| \tilde{y}_k\right],$$

where we have used the conjecture on the market maker's price setting rule and on the demand submitted by the other market participants. By the distributional assumptions on the random variables appearing in the model, the above conditional expectation can be computed as follows:

$$
\mathbb{E}\left[\left(\tilde{\varepsilon} - \lambda\left(x_k + \sum_{\substack{l=1 \\ l \neq k}}^{K} \beta(\tilde{\varepsilon} + \tilde{\eta}_l) + \sum_{n=1}^{N} \gamma \tilde{e}_n\right)\right) x_k \middle| \tilde{\varepsilon} + \tilde{\eta}_k\right]
$$

$$
= \frac{\sigma_\varepsilon^2}{\sigma_\varepsilon^2 + \sigma_\eta^2}(\tilde{\varepsilon} + \tilde{\eta}_k)x_k - \lambda\beta(K-1)\frac{\sigma_\varepsilon^2}{\sigma_\varepsilon^2 + \sigma_\eta^2}(\tilde{\varepsilon} + \tilde{\eta}_k)x_k - \lambda x_k^2.
$$

Maximizing the last quantity with respect to x_k yields the optimal value

$$
\tilde{x}_k^* = \sigma_\varepsilon^2 \frac{1 - \lambda(K-1)\beta}{2\lambda(\sigma_\varepsilon^2 + \sigma_\eta^2)}(\tilde{\varepsilon} + \tilde{\eta}_k).
$$

Therefore, the conjecture

$$
\tilde{x}_k^* = X(\tilde{\varepsilon} + \tilde{\eta}_k) = \beta(\tilde{\varepsilon} + \tilde{\eta}_k)
$$

can hold in equilibrium if and only if the coefficient β satisfies condition (10.32).

(ii): Given the conjectured behavior of the other market participants, the profits of the uninformed agent $n \in \{1, \ldots, N\}$ if he demands a quantity w_n of the security are given by

$$
(\bar{d} + \tilde{\varepsilon})(\tilde{e}_n + w_n) - \left(\bar{d} + \lambda\left(K\beta\bar{d} + \beta\sum_{k=1}^{K}(\tilde{\varepsilon} + \tilde{\eta}_k) + w_n + \gamma\sum_{\substack{m=1 \\ m \neq n}}^{N}\tilde{e}_m\right)\right)w_n.
$$

Due to the assumption of a negative exponential utility function together with the multivariate normality assumption, the expected utility maximization problem of the uninformed trader n reduces to determining the quantity $\tilde{w}_n^*$ which maximizes the following mean-variance function with respect to w_n:

$$
\mathbb{E}\left[(\bar{d} + \tilde{\varepsilon})\tilde{e}_n + \tilde{\varepsilon}w_n - \lambda\left(K\beta\bar{d} + \beta\sum_{k=1}^{K}(\tilde{\varepsilon} + \tilde{\eta}_k) + w_n + \gamma\sum_{\substack{m=1 \\ m \neq n}}^{N}\tilde{e}_m\right)w_n \middle| \tilde{e}_n\right]
$$

$$
- \frac{a}{2}\operatorname{Var}\left((\bar{d} + \tilde{\varepsilon})\tilde{e}_n + \tilde{\varepsilon}w_n - \lambda\left(K\beta\bar{d} + \beta\sum_{k=1}^{K}(\tilde{\varepsilon} + \tilde{\eta}_k) + w_n + \gamma\sum_{\substack{m=1 \\ m \neq n}}^{N}\tilde{e}_m\right)w_n \middle| \tilde{e}_n\right).
$$

By relying on the multivariate normality assumption, the conditional expectation and the conditional variance appearing in the last expression can be explicitly calculated, similarly as above, and the optimal value $\tilde{w}_n^*$ can then be calculated by differentiating the resulting expression with respect to w_n and setting the derivative equal to zero. This yields the optimal value $\tilde{w}_n^*$ given in (10.33). Then, in equilibrium the conjecture

$$\tilde{w}_n^* = W(\tilde{e}_n) = \gamma \tilde{e}_n$$

is verified if and only if

$$\gamma \tilde{e}_n = -\frac{a\sigma_\varepsilon^2(1 - K\lambda\beta)\tilde{e}_n}{2\lambda + a\left(\sigma_\varepsilon^2(1 - K\lambda\beta)^2 + K\sigma_\eta^2\lambda^2\beta^2 + \lambda^2(N-1)\gamma^2\sigma_e^2\right)},$$

which is equivalent to the validity of equation (10.34).

(iii): The zero expected profits condition for the risk neutral market maker, together with the conjecture of a linear price formation rule, implies that a condition equivalent to the market efficiency condition (10.11) must be satisfied, i.e.,

$$\mathbb{E}\left[\tilde{d}\left|\sum_{k=1}^{K}\tilde{x}_k^* + \sum_{n=1}^{N}\tilde{w}_n^*\right.\right] = \bar{d} + \lambda\left(\sum_{k=1}^{K}\tilde{x}_k^* + \sum_{n=1}^{N}\tilde{w}_n^*\right)$$

or, equivalently,

$$\mathbb{E}\left[\tilde{\varepsilon}\left|\sum_{k=1}^{K}\tilde{x}_k^* + \sum_{n=1}^{N}\tilde{w}_n^*\right.\right] = \lambda\left(\sum_{k=1}^{K}\tilde{x}_k^* + \sum_{n=1}^{N}\tilde{w}_n^*\right).$$

By relying on the optimal demands of the informed and of the uninformed traders derived in parts *(i)* and *(ii)* of the exercise, the latter condition can be rewritten as

$$\mathbb{E}\left[\tilde{\varepsilon}\left|K\beta\tilde{\varepsilon} + \beta\sum_{k=1}^{K}\tilde{\eta}_k + \gamma\sum_{n=1}^{N}\tilde{e}_n\right.\right] = \lambda\left(K\beta\tilde{\varepsilon} + \beta\sum_{k=1}^{K}\tilde{\eta}_k + \gamma\sum_{n=1}^{N}\tilde{e}_n\right).$$

Due to the distributional assumptions of the model, the last equation reduces to

$$\lambda = \frac{K\beta\sigma_\varepsilon^2}{K^2\beta^2\sigma_\varepsilon^2 + K\beta^2\sigma_\eta^2 + \gamma^2 N\sigma_e^2}.$$

Recall that, from step *(i)* of the exercise, the coefficient β satisfies equation (10.32). By substitution, we obtain that λ and γ must satisfy equation (10.35). Moreover, condition (10.32) implies that $1 - K\lambda\beta > 0$, so that in turn equation (10.34) implies that we must choose the negative solution γ to equation (10.35), for a given λ, thus proving part *(iii)* of the exercise.

(iv): In order to complete the proof of Proposition 10.5, it suffices to replace the expression for the parameter γ obtained in part *(iii)* of the exercise into equation (10.34), which then yields a linear equation for γ with an explicit solution given as in the statement of Proposition 10.5. Substituting the explicit value of the parameter γ into the equation obtained in part *(iii)* of the exercise then yields the explicit value of λ. Finally, the equilibrium value of β can be deduced making use of relation (10.32).

Solution of Exercise 10.8

(i): The expected utility maximization problem of each informed trader $i \in [0, 1]$, conditionally on the observation of the market price $\tilde{p}$ and the private signal $\tilde{y}_i$, consists in determining the optimal demand $\tilde{x}_i^*$ which maximizes with respect to x_i the conditional expectation

$$\mathbb{E}\left[-e^{-a(\tilde{d}-\tilde{p})x_i}\,|\,p, \tilde{y}_i\right].$$

Due to the distributional assumptions of the model and the conjectured linearity of the equilibrium price, the distribution of the random variable $\tilde{d}$ conditionally on the observation of $(\tilde{p}, \tilde{y}_i)$ is normal. Hence, the above expected utility maximization problem can be reduced to the maximization over $x_i \in \mathbb{R}$ of the following mean-variance function:

$$\left(\mathbb{E}[\tilde{d}|\tilde{p}, \tilde{y}_i] - \tilde{p}\right)x_i - \frac{a}{2}\,\mathrm{Var}(\tilde{d}|\tilde{p}, \tilde{y}_i)x_i^2.$$

The optimal value $\tilde{x}_i^*$ is given by (see also Sect. 3.1)

$$\tilde{x}_i^* = X_i(p, y_i) = \frac{\mathbb{E}[\tilde{d}|p, y_i] - p}{a\,\mathrm{Var}(\tilde{d}|p, y_i)},$$

for every realization $\{\tilde{p} = p, \tilde{y}_i = y_i\}$, denoting $\mathbb{E}[\tilde{d}|p, y_i] := \mathbb{E}[\tilde{d}|\tilde{p} = p, \tilde{y}_i = y_i]$ (and similarly for the conditional variance).

(ii): Under the conjecture of linear demand schedules by informed traders, it holds that

$$L(\tilde{p}) = \alpha\tilde{d} + b\tilde{p} + c + \tilde{z}.$$

The market maker sets an equilibrium price $\tilde{p}$ as the solution of the equation

$$\tilde{p} = \mathbb{E}[\tilde{d}|L(\cdot)].$$

Note that, from the point of view of the market maker, the informative part of the aggregate demand schedule $L(\cdot)$ is given by $\alpha \tilde{d} + \tilde{z}$. Hence, the equilibrium price $\tilde{p}$ is determined by

$$\tilde{p} = \mathbb{E}[\tilde{d}|\alpha \tilde{d} + \tilde{z}].$$

By standard properties of the bivariate normal distribution, it can be easily computed that

$$\tilde{p} = \mathbb{E}[\tilde{d}|\alpha \tilde{d} + \tilde{z}] = \bar{d} + \frac{\alpha \sigma_d^2}{\alpha^2 \sigma_d^2 + \sigma_z^2}(\alpha(\tilde{d} - \bar{d}) + \tilde{z}),$$

from which the claim of part *(ii)* of the exercise follows by the definition of λ.

(iii): To complete the proof of Proposition 10.6, it remains to show that the demand schedule of the informed traders admits the representation (10.18). This easily follows by relying on the conditional distribution of a normal multivariate random variable in order to compute the terms $\mathbb{E}[\tilde{d}|\tilde{p}, \tilde{y}_i]$ and $\mathrm{Var}(\tilde{d}|\tilde{p}, \tilde{y}_i)$ appearing in the demand schedule computed in part *(i)* of the exercise:

$$\mathbb{E}[\tilde{d}|\tilde{p}, \tilde{y}_i] = \frac{\tau_\varepsilon \tilde{y}_i + (\alpha^2 \tau_z + \tau_d)\tilde{p}}{\tau_\varepsilon + \alpha^2 \tau_z + \tau_d} \quad \text{and} \quad \mathrm{Var}(\tilde{d}|\tilde{p}, \tilde{y}_i) = \frac{1}{\tau_\varepsilon + \alpha^2 \tau_z + \tau_d}.$$

It then follows that the coefficients α, b and c introduced in the conjectured linear representation in part *(ii)* of the exercise are given by

$$\alpha = \frac{\tau_\varepsilon}{a}, \qquad b = -\frac{\tau_\varepsilon}{a} \quad \text{and} \quad c = 0,$$

thus proving that representation (10.18) holds.

Solution of Exercise 10.9 Note first that, by the linearity of the expectation, it holds that

$$\mathbb{E}[\widetilde{W}'] = \mathbb{E}[\widetilde{W}^*] - x_n(\mathbb{E}[\tilde{d}_n] - \bar{p}_n),$$

where $\widetilde{W}^*$ denotes the optimal wealth which solves the expected utility maximization problem of the dealer at the initial date $t = 0$, x_n is the order arriving at the dealer and $\bar{p}_n = p_n^{\text{ask}} \mathbf{1}_{\{x_n > 0\}} + p_n^{\text{bid}} \mathbf{1}_{\{x_n < 0\}}$. Similarly, the variance of $\widetilde{W}'$ can be computed as

$$\mathrm{Var}(\widetilde{W}') = \mathrm{Var}(\widetilde{W}^*) + x_n^2 \sigma_n^2 - 2x_n \mathrm{Cov}(\widetilde{W}^*, \tilde{d}_n).$$

Hence, the mean-variance problem (10.20) can be rewritten as

$$-x_n(\mathbb{E}[\tilde{d}_n] - \bar{p}_n) - \frac{a}{2}\left(x_n^2 \sigma_n^2 - 2x_n \mathrm{Cov}(\widetilde{W}^*, \tilde{d}_n)\right) = 0.$$

Solving the last equation with respect to $\bar{p}_n$ gives

$$\bar{p}_n = \bar{d}_n + \frac{a}{2}x_n\sigma_n^2 - a\operatorname{Cov}(\widetilde{W}^*, \bar{d}_n),$$

from which the statement of Proposition 10.7 immediately follows.

Solution of Exercise 10.10

(*i*): It suffices to observe that, since the current ask price is part of the publicly available information (i.e., it is measurable with respect to the σ-algebra $\mathcal{D}_t \wedge \mathcal{F}_t$), it holds that

$$p_t^{\text{ask}} = \mathbb{E}_t[p_t^{\text{ask}}|C_t] = \mathbb{E}[\mathbb{E}[\tilde{v}|\mathcal{D}_t, C_t]|\mathcal{D}_t \wedge \mathcal{F}_t, C_t] = \mathbb{E}_t[\tilde{v}|C_t].$$

(*ii*): Again by the tower property of the conditional expectation, it holds that

$$\mathbb{E}_t[\tilde{v}|C_t] = \mathbb{E}_t\Big[\mathbb{E}\big[\mathbb{E}[\tilde{v}|\mathcal{K}_t \vee \sigma(\delta_t), C_t]|\mathcal{D}_t \wedge \mathcal{F}_t \vee \sigma(\delta_t), C_t\big]\Big|C_t\Big]$$

$$= \mathbb{E}_t\Big[\mathbb{E}\big[\mathbb{E}[\tilde{v}|\mathcal{K}_t \vee \sigma(\delta_t)]|\mathcal{D}_t \wedge \mathcal{F}_t \vee \sigma(\delta_t), C_t\big]\Big|C_t\Big]$$

$$= \mathbb{E}_t\Big[\mathbb{E}\big[\mathbb{E}[\tilde{v}|\mathcal{K}_t]|\mathcal{D}_t \wedge \mathcal{F}_t \vee \sigma(\delta_t), C_t\big]\Big|C_t\Big],$$

where the last equality follows from the fact that the knowledge of the random variable δ_t does not bring any useful information in order to forecast the fundamental value $\tilde{v}$ (i.e., condition (10.21) holds).

(*iii*): Observe first that

$$\mathbb{E}[X|X > a] = \frac{\mathbb{E}[X\mathbf{1}_{\{X>a\}}]}{\mathbb{P}(X > a)} = \frac{\mathbb{E}[(X - a + a)\mathbf{1}_{\{X>a\}}]}{\mathbb{P}(X > a)} = \frac{\mathbb{E}[(X - a)^+]}{\mathbb{P}(X > a)} + a$$

$$\geq \mathbb{E}[(X - a)^+] + a.$$

Moreover, the function $(\cdot)^+$ is a convex function and, hence, Jensen's inequality implies that $\mathbb{E}[(X - a)^+] \geq (\mathbb{E}[X] - a)^+$. This yields

$$\mathbb{E}[X|X > a] \geq a + \big(\mathbb{E}[X] - a\big)^+ \geq \mathbb{E}[X].$$

With the same reasoning, an analogous inequality can be obtained in the case of a conditional expectation of the form $\mathbb{E}[X|\mathcal{A}, X > a]$, for an arbitrary σ-algebra $\mathcal{A}$.

(*iv*): It holds that

$$\mathbb{E}\big[\mathbb{E}[\tilde{v}|\mathcal{K}_t]|\mathcal{D}_t \wedge \mathcal{F}_t \vee \sigma(\delta_t), C_t\big] \geq \mathbb{E}\big[\mathbb{E}[\tilde{v}|\mathcal{K}_t]|\mathcal{D}_t \wedge \mathcal{F}_t \vee \sigma(\delta_t)\big],$$

where we have used the result established in step *(iii)* applied to the random variable $X = \mathbb{E}[\tilde{v}|\mathcal{H}_t]$, with $a = p_t^{\text{ask}}/\delta_t$ and $\mathcal{A} = \mathcal{D}_t \wedge \mathcal{F}_t \vee \sigma(\delta_t)$. In view of the result of step *(ii)*, by taking the conditional expectation on both sides of the last inequality, it then follows that

$$
\mathbb{E}_t[\tilde{v}|C_t] = \mathbb{E}_t\Big[\mathbb{E}\big[\mathbb{E}[\tilde{v}|\mathcal{H}_t]|\mathcal{D}_t \wedge \mathcal{F}_t \vee \sigma(\delta_t), C_t\big]\big|C_t\Big]
$$

$$
\geq \mathbb{E}_t\Big[\mathbb{E}\big[\mathbb{E}[\tilde{v}|\mathcal{H}_t]|\mathcal{D}_t \wedge \mathcal{F}_t \vee \sigma(\delta_t)\big]\big|C_t\Big]
$$

$$
= \mathbb{E}_t\Big[\mathbb{E}\big[\mathbb{E}[\tilde{v}|\mathcal{H}_t]|\mathcal{D}_t \wedge \mathcal{F}_t\big]\big|C_t\Big]
$$

$$
= \mathbb{E}_t\Big[\mathbb{E}\big[\tilde{v}|\mathcal{D}_t \wedge \mathcal{F}_t\big]\big|C_t\Big] = \mathbb{E}_t[\tilde{v}],
$$

where in the second equality we have used condition (10.21) and again the tower property of the conditional expectation. This proves the claim.

Solution of Exercise 10.11   The exercise is a simple application of Bayes' rule. Indeed, note that

$$
p^{\text{ask}} = \overline{v}\mathbb{P}(\tilde{v} = \overline{v}|\text{buy order}) + \underline{v}\mathbb{P}(\tilde{v} = \underline{v}|\text{buy order}).
$$

Hence, in order to compute p^{ask}, it suffices to compute the two conditional probabilities $\mathbb{P}(\tilde{v} = \overline{v}|\text{buy order})$ and $\mathbb{P}(\tilde{v} = \underline{v}|\text{buy order})$. To this end:

$$
\mathbb{P}(\tilde{v} = \overline{v}|\text{buy order}) = \frac{\mathbb{P}(\tilde{v} = \overline{v}, \text{buy order})}{\mathbb{P}(\text{buy order})} = \frac{\mathbb{P}(\tilde{v} = \overline{v})\mathbb{P}(\text{buy order}|\tilde{v} = \overline{v})}{\mathbb{P}(\text{buy order})}
$$

$$
= \frac{\mathbb{P}(\tilde{v} = \overline{v})\mathbb{P}(\text{buy order}|\tilde{v} = \overline{v})}{\mathbb{P}(\tilde{v} = \overline{v})\mathbb{P}(\text{buy order}|\tilde{v} = \overline{v}) + \mathbb{P}(\tilde{v} = \underline{v})\mathbb{P}(\text{buy order}|\tilde{v} = \underline{v})}
$$

$$
= \frac{\pi(\alpha + \frac{1-\alpha}{2})}{\pi(\alpha + \frac{1-\alpha}{2}) + (1-\pi)\frac{1-\alpha}{2}}
$$

$$
= \frac{\pi(1+\alpha)}{1-\alpha+2\pi\alpha}.
$$

By means of similar computations, it holds that

$$
\mathbb{P}(\tilde{v} = \underline{v}|\text{buy order}) = \frac{(1-\pi)(1-\alpha)}{1-\alpha+2\pi\alpha}.
$$

Hence:

$$
p^{\text{ask}} = \overline{v}\frac{\pi(1+\alpha)}{1-\alpha+2\pi\alpha} + \underline{v}\frac{(1-\pi)(1-\alpha)}{1-\alpha+2\pi\alpha}.
$$

The computations in the case of the bid price are completely analogous.

Solution of Exercise 10.12

(i): Similarly as in Exercise 10.11, the equilibrium ask price $p^{\text{ask}}(1)$ can be computed as

$$p^{\text{ask}}(1) = \mathbb{E}[\tilde{v}|\text{buy size 1}],$$

where $\mathbb{E}[\tilde{v}|\text{buy size 1}]$ denotes the conditional expectation of the fundamental value of the asset given that a buy order of size 1 has been submitted. By the Bayes formula together with the assumptions of the present exercise, it holds that

$$\mathbb{E}[\tilde{v}|\text{buy size 1}] = \mathbb{P}(\tilde{v} = 1|\text{buy size 1})$$

$$= \frac{\mathbb{P}(\tilde{v} = 1)\mathbb{P}(\text{buy size 1}|\tilde{v} = 1)}{\mathbb{P}(\tilde{v} = 1)\mathbb{P}(\text{buy size 1}|\tilde{v} = 1) + \mathbb{P}(\tilde{v} = 0)\mathbb{P}(\text{buy size 1}|\tilde{v} = 0)}$$

$$= \frac{\alpha(1 - \mu) + \frac{(1-\alpha)(1-\beta)}{2}}{\alpha(1 - \mu) + (1 - \alpha)(1 - \beta)}.$$

The ask price $p^{\text{ask}}(2)$ associated to buy orders of size 2 can be computed in an analogous way.

(ii): By relying on the ask prices computed in the previous step of the exercise and using condition $2(1 - p^{\text{ask}}(2)) = 1 - p^{\text{ask}}(1)$, it holds that

$$2\left(1 - \frac{\alpha\mu + \frac{(1-\alpha)\beta}{2}}{\alpha\mu + (1-\alpha)\beta}\right) = 1 - \frac{\alpha(1-\mu) + \frac{(1-\alpha)(1-\beta)}{2}}{\alpha(1-\mu) + (1-\alpha)(1-\beta)}.$$

Solving for μ then gives the explicit value reported above. The values of the ask prices $p^{\text{ask}}(1)$ and $p^{\text{ask}}(2)$ can then be obtained by substituting the endogenously determined value of μ into the expressions computed in step (i) of the exercise.

(iii): By arguing similarly as in Exercise 10.11 and in the first part of this exercise, it holds that

$$p^{\text{ask}}(1) = \mathbb{P}(\tilde{v} = 1|\text{buy size 1})$$

$$= \frac{\mathbb{P}(\tilde{v} = 1)\mathbb{P}(\text{buy size 1}|\tilde{v} = 1)}{\mathbb{P}(\tilde{v} = 1)\mathbb{P}(\text{buy size 1}|\tilde{v} = 1) + \mathbb{P}(\tilde{v} = 0)\mathbb{P}(\text{buy size 1}|\tilde{v} = 0)}$$

$$= \frac{\frac{1}{2}\frac{(1-\alpha)(1-\beta)}{2}}{\frac{1}{2}\frac{(1-\alpha)(1-\beta)}{2} + \frac{1}{2}\frac{(1-\alpha)(1-\beta)}{2}} = \frac{1}{2}.$$

Similarly,

$$p^{\text{ask}}(2) = \mathbb{P}(\tilde{v} = 1|\text{buy size 2})$$

$$= \frac{\mathbb{P}(\tilde{v} = 1)\mathbb{P}(\text{buy size } 2|\tilde{v} = 1)}{\mathbb{P}(\tilde{v} = 1)\mathbb{P}(\text{buy size } 2|\tilde{v} = 1) + \mathbb{P}(\tilde{v} = 0)\mathbb{P}(\text{buy size } 2|\tilde{v} = 0)}$$

$$= \frac{\frac{1}{2}(\frac{(1-\alpha)\beta}{2} + \alpha)}{\frac{1}{2}(\frac{(1-\alpha)\beta}{2} + \alpha) + \frac{1}{2}\frac{(1-\alpha)\beta}{2}}$$

$$= \frac{\frac{(1-\alpha)\beta}{2} + \alpha}{(1-\alpha)\beta + \alpha}.$$

The necessary condition $\beta \geq \alpha/(1-\alpha)$ follows by substituting the explicit expression of $p^{\text{ask}}(1)$ and $p^{\text{ask}}(2)$ into the inequality

$$2(1 - p^{\text{ask}}(2)) \geq 1 - p^{\text{ask}}(1).$$

Solution of Exercise 10.13

(*i*): Due to the assumption of a normal distribution for the liquidation value of the asset together with the assumption of a negative exponential utility function, the expected utility maximization problem of each dealer $m \in \{1, \ldots, M\}$ corresponds to maximizing the following mean-variance function with respect to x_m, representing the quantity demanded of the asset, given a price $\tilde{p}$:

$$\mathbb{E}[(\tilde{d} - \tilde{p})x_m + \tilde{d}E_m] - \frac{a}{2}\text{Var}\left((\tilde{d} - \tilde{p})x_m + \tilde{d}E_m\right).$$

For each $m = 1, \ldots, M$, the optimal solution is given by

$$\frac{\bar{d} - \tilde{p}}{a\sigma_d^2} - E_m.$$

(*ii*): The market clearing condition implies that

$$M\left(\frac{\bar{d} - \tilde{p}}{a\sigma_d^2} - \bar{E}\right) + z = 0,$$

where z denotes an arbitrary realization of the random variable $\tilde{z}$. Solving for the equilibrium price $\tilde{p}$ directly yields the result.

(*iii*): It suffices to replace the explicit expression for the equilibrium price p into the optimal demand schedule computed in part (*i*) of the exercise, thus yielding

$$x_m^* = \bar{E} - E_m - \frac{z}{M},$$

for each $m = 1, \ldots, M$.

References

1. Abarbanell, J. and Bernard, V. (1992) Tests of analysts' overreaction/underreaction to earnings information as an explanation for anomalous stock price behavior. *Journal of Finance*, 47:1181–1207.
2. Abdellaoui, M., Bleichrodt, H. and Paraschiv, C. (2007) Loss aversion under prospect theory: a parameter-free measurement. *Management Science*, 57:1659–1674.
3. Abel, A. (1989) Asset prices under heterogeneous beliefs: implications for the equity premium puzzle. Mimeo.
4. Abel, A. (1990) Asset prices under habit formation and catching up with the Joneses. *The American Economic Review*, 80:38–42.
5. Abel, A. (1999) Risk premia and term premia in general equilibrium. *Journal of Monetary Economics*, 43:3–33.
6. Abel, A. (2001) The effects of investing social security funds in the stock market when fixed costs prevent some houses from holding stocks. *The American Economic Review*, 91:128–148.
7. Abel, A. (2002) An exploration of the effects of pessimism and doubt on asset returns. *Journal of Economic Dynamics and Control*, 26:1075–1092.
8. Abel, A., Mankiw, G., Summers, L. and Zeckhauser, R. (1989) Assessing dynamic efficiency: theory and evidence. *Review of Economic Studies*, 56:1–20.
9. Abreu, D. and Brunnermeier, M. (2002) Synchronization risk and delayed arbitrage. *Journal of Financial Economics*, 66:341–360.
10. Abreu, D. and Brunnermeier, M. (2003) Bubbles and crashes. *Econometrica*, 71:173–204.
11. Acharya, V. and Pedersen, L.H. (2005) Asset pricing with liquidity risk. *Journal of Financial Economics*, 77:375–410.
12. Ackert, R., Church, B. and Jayaraman, N. (2001) An experimental study of circuit breakers: The effects of mandated market closures and temporary halts on market behavior. *Journal of Financial Markets*, 4:185–208.
13. Ackert, R. and Smith, M. (1993) Stock price volatility, ordinary dividends, and other cash flows to shareholders. *Journal of Finance*, 48:1147–1165.
14. Adam, K. and Marcet, A. (2011) Internal rationality, imperfect market knowledge and asset prices. *Journal of Economic Theory*, 146:1224–1252.
15. Admati, A. (1985) A noisy rational expectations equilibrium for multi-asset securities markets. *Econometrica*, 53:629–657.
16. Admati, A. (1989) Information in financial markets: the rational expectations approach. In *Financial Markets and Incomplete Information*, S. Bhattacharya and G. Constantinindes (eds.), 139–152.

© Springer-Verlag London Ltd. 2017

E. Barucci, C. Fontana, *Financial Markets Theory*,
Springer Finance, DOI 10.1007/978-1-4471-7322-9

17. Admati, A. and Pfleiderer, P. (1985) Interpreting the factor risk premia in the arbitrage pricing theory. *Journal of Economic Theory*, 35:191–195.
18. Admati, A. and Pfleiderer, P. (1986) A monopolistic market for information. *Journal of Economic Theory*, 39:400–438.
19. Admati, A. and Pfleiderer, P. (1988) Selling and trading on information in financial markets. *The American Economic Review*, 78:96–103.
20. Admati, A. and Pfleiderer, P. (1988) A theory of intraday patterns. Volume and price variability. *Review of Financial Studies*, 1:3–40.
21. Admati, A. and Pfleiderer, P. (1989) Divide and conquer: a theory of intraday and day-of-the-week mean returns. *Review of Financial Studies*, 2:189–223.
22. Admati, A. and Pfleiderer, P. (1990) Direct and indirect sale of information. *Econometrica*, 58:901–928.
23. Admati, A. and Pfleiderer, P. (1991) Sunshine trading and financial market equilibrium. *Review of Financial Studies*, 4:443–482.
24. Admati, A. and Pfleiderer, P. (2000) Forcing firms to talk: financial disclosure regulation and externalities. *Review of Financial Studies*, 13:479–519.
25. Adrian, T. and Franzoni, F. (2009) Learning about beta: Time-varying factor loadings, expected returns, and the conditional CAPM. *Journal of Empirical Finance*, 16:537–556.
26. Affleck-Graves, J., Hegde, S. and Miller, R. (1994) Trading mechanisms and the components of the bid–ask spread. *Journal of Finance*, 49:1471–1488.
27. Aggarwal, R. and Wu, G. (2006) Stock market manipulations. *Journal of Business*, 79:1915–1953.
28. Ahn, D., Boudoukh, J., Richardson, M. and Whitelaw, R. (2002) Partial adjustment or stale prices? Implications from stock index and futures autocorrelations. *Review of Financial Studies*, 15:655–689.
29. Ahn, D., Conrad, J. and Dittmar, R. (2003) Risk adjustment and trading strategies. *Review of Financial Studies*, 16:459–485.
30. Ai, H. (2010) Information quality and long-run risk: asset pricing implications. *Journal of Finance* 65:1333–1367.
31. Ait-Sahalia, Y. and Brandt, M. (2001) Variable selection for portfolio choice. *Journal of Finance*, 56:1297–1351.
32. Ait-Sahalia, Y., Parker, J. and Yogo, M. (2004) Luxury goods and the equity premium. *Journal of Finance*, 59:2959–3004.
33. Aiyagari, R. (1994) Uninsured idiosyncratic risk and aggregate saving. *Quarterly Journal of Economics*, 109:659–684.
34. Aiyagari, R. and Gertler, M. (1991) Asset returns with transactions costs and uninsured individual risk. *Journal of Monetary Economics*, 27:311–331.
35. Aiyagari, R. and Gertler, M. (1999) Overreaction of asset prices in general equilibrium. *Review of Economic Dynamics*, 2:3–35.
36. Akgiray, V. (1989) Conditional heteroscedasticity in time series of stock returns: evidence and forecasts. *Journal of Business*, 62:55–80.
37. Aktas, N., de Bodt, E. and Van Oppens, E. (2008) Legal insider trading and market efficiency. *Journal of Banking and Finance*, 32:1379–1392.
38. Al-Najjar, N. (1999) On the robustness of factor structures to asset repackaging. *Journal of Mathematical Economics*, 31:309–320.
39. Alessie, R. and Lusardi, A. (1997) Consumption, saving and habit formation. *Economics Letters*, 55:103–108.
40. Aliber, R., Chowdhry, B. and Yan, S. (2003) Some evidence that a Tobin tax on foreign exchange transactions may increase volatility. *European Finance Review*, 7:481–510.
41. Allais, M. (1953) Le comportement de l'homme rationnel devant le risque, critique des postulates et axioms de l'école americaine. *Econometrica*, 21:503–546.
42. Allen, B. (1981) Generic existence of completely revealing equilibria for economies with uncertainty when prices convey information. *Econometrica*, 49:1173–1199.

43. Allen, B. (1982) Strict rational expectations equilibria with diffuseness. *Journal of Economic Theory*, 27:20–46.
44. Allen, F. and Gale, D. (1992) Stock price manipulation. *Review of Financial Studies*, 5:503–529.
45. Allen, F. and Gale, D. (1994) Limited market particitpation and volatility of asset prices. *The American Economic Review*, 84:933–955.
46. Allen, F. and Gale, D. (1994) *Financial Innovation and Risk Sharing*. MIT Press, Cambridge.
47. Allen, F. and Gorton, G. (1993) Stock price manipulation, market microstructure and asymmetric information. *European Economic Review*, 36:624–630.
48. Allen, F., Litov, L. and Mei, J. (2006) Large investors, price manipulation, and limits to arbitrage: an anatomy of market corners. *Review of Finance*, 10:645–693.
49. Allen, F., Morris, S. and Postlewaite, A. (1993) Finite bubbles with short sale constraints and asymmetric information. *Journal of Economic Theory*, 61:206–229.
50. Allen, F., Morris, S. and Shin, S. (2006) Beauty contests and iterated expectations in asset markets. *Review of Financial Studies*, 19:719–752.
51. Allingham, M. (1990) Existence theorems in the capital asset pricing model. *Econometrica*, 59:1169–1174.
52. Alon, S. and Schmeidler, D. (2014) Purely subjective maxmin expected utility. *Journal of Economic Theory*, 152:382–412.
53. Alvarez, F. and Jermann, U. (2000) Efficiency, equilibrium and asset pricing with default risk. *Econometrica*, 68:775–797.
54. Alvarez, F. and Jermann, U. (2001) Quantitative asset pricing implications of endogenous solvency constraints. *Review of Financial Studies*, 14:1117–1151.
55. Ameriks, J. and Zeldes, S. (2004) How do portfolio shares vary with age? Mimeo.
56. Amershi, A. and Stoeckenius, J. (1983) The theory of syndicates and linear sharing rules. *Econometrica*, 51:1407–1416.
57. Amihud, Y. (2002) Illiquidity and stock returns: cross–sectional and time series effects. *Journal of Financial Markets*, 5:31–56.
58. Amihud, Y. and Mendelson, H. (1980) Dealership market. *Journal of Financial Economics*, 31:31–53.
59. Amihud, Y. and Mendelson, H. (1986) Asset pricing and the bid–ask spread. *Journal of Financial Economics*, 17:223–249.
60. Amihud, Y., Mendelson, H. and Lauterbach, B. (1997) Market microstructure and securities values: evidence from the Tel Aviv Stock Exchange. *Journal of Financial Economics*, 45:365–390.
61. Anderson, E., Ghysels, E. and Juergens, J. (2005) Do heterogeneous beliefs matter for asset pricing? *Review of Financial Studies*, 18:875–924.
62. Anderson, E., Ghysels, E. and Juergens, J. (2009) The impact of risk and uncertainty on expected returns. *Journal of Financial Economics*, 94:233–263.
63. Anderson, E., Hansen, L.P. and Sargent, T. (2003) A quartet of semigroups for model specification, robustness, prices of risk, and model detection. *Journal of the European Economic Association*, 1:68–123.
64. Ang, A., Bekaert, G., and Liu, J. (2005) Why stocks may disappoint. *Journal of Financial Economics*, 76:471–508.
65. Ang, A. and Bekaert, G. (2001) Stock return predictability: is it there? *Review of Financial Studies*, 20:651–707.
66. Ang, A. and Chen, J. (2007) CAPM over the long run: 1926–2001. *Journal of Empirical Finance*, 14:1–40.
67. Ang, A., Chen, J. and Xing, Y. (2006) Downside risk. *Review of Financial Studies*, 19:1191–1239.
68. Ang, A., Hodrick, R., Xing, Y. and Zhang, X. (2009) High idiosyncratic volatility and low returns: International and further U.S. evidence. *Journal of Financial Economics*, 91:1–23.
69. Ang, A. and Liu J. (2004) How to discount cashflows with time-varying expected returns. *Journal of Finance*, 59:2745–2783.

70. Ang, A., Papanikolau, D. and Westerfield, M. (2014) Portfolio choice with illiquid assets. *Management Science*, 60:2737–2761.

71. Angeletos, G., Laibson, D., Repetto, A., Tobacman, J. and Weinberg, S. (2001) The hyperbolic consumption model: calibration, simulation, and empirical evaluation. *The Journal of Economic Perspectives*, 15:47–68.

72. Anscombe, F. and Aumann, R. (1963) A definition of subjective probability. *Annals of Mathematical Statistics*, 34:199–205.

73. Antonelli, F., Barucci, E. and Mancino, M.E. (2001) Asset pricing with endogenous aspirations. *Decisions in Economics and Finance*, 24:21–41.

74. Antonelli, F., Barucci, E. and Mancino, M.E. (2001) Asset pricing with a forward-backward stochastic differential utility. *Economics Letters*, 72:151–157.

75. Araujo, A., Páscoa, M.R. and Torres-Martínez, J. P. (2002) Collateral avoids Ponzi schemes in incomplete markets. *Econometrica*, 70:1613–1638.

76. Arrow, K. (1953) The role of securities in the optimal allocation of risk–bearing. *Review of Economic Studies*, 30:91–96.

77. Arrow, K. (1968) Economic equilibrium. *International Encyclopedia of the Social Sciences*. Crowell Collier & Macmillan, New York.

78. Arrow, K. (1970) *Essays, in the Theory of Risk Bearing*. North-Holland, Amsterdam.

79. Asness, C., Moskowitz, T. and Pedersen, L. (2013) Value and momentum everywhere. *Journal of Finance*, 68:929–985.

80. Attanasio, O. (1991) Risk, time–varying second moments and market efficiency. *Review of Economic Studies*, 58:479–494.

81. Attanasio, O., Banks, J. and Tanner, S. (2002) Asset holding and consumption volatility. *Journal of Political Economy*, 110:771–792.

82. Attanasio, O. and Browning, M. (1995) Consumption over the life cycle and over the business cycle. *The American Economic Review*, 85:1118–1137.

83. Attanasio, O. and Davis, S. (1996) Relative wage movements and the distribution of consumption. *Journal of Political Economy*, 104:1227–1250.

84. Attanasio, O., Goldberg, P. and Kyriazidou, E. (2008) Credit constraints in the market for consumer durables: evidence from micro data on car loans. *International Economic Review*, 49:401–436.

85. Attanasio, O. and Vissing-Jørgensen, A. (2003) Stock-market participation, intertemporal substitution, and risk-aversion. *The American Economic Review*, 93:383–391.

86. Attanasio, O. and Weber, G. (1989) Intertemporal substitution, risk aversion and the Euler equation for consumption. *The Economic Journal* 99:59–73.

87. Attanasio, O. and Weber, G. (1993) Consumption growth, the interest rate and the aggregation. *Review of Economic Studies*, 60:631–649.

88. Attanasio, O. and Weber, G. (1995) Is consumption growth consistent with intertemporal optimization? Evidence from the consumer expenditure survey. *Journal of Political Economy*, 103:1121–1157.

89. Ausubel, L. (1990) Insider trading in a rational expectations economy. *The American Economic Review*, 80:1022–1041.

90. Ausubel, L. (1990a) Partially revealing rational expectations equilibrium in a competitive economy. *Journal of Economic Theory*, 50:93–126.

91. Avramov, D. (2002) Stock–returns predictability and model uncertainty. *Journal of Financial Economics*, 64:423–458.

92. Avramov, D. (2004) Stock return predictability and asset pricing models. *Review of Financial Studies*, 17:699–738.

93. Avramov, D. and Chordia, T. (2006) Asset pricing models and financial market anomalies. *Review of Financial Studies* 19:1001–1040.

94. Avramov, D., Chordia, T. and Goyal, A. (2006) Liquidity and autocorrelations in individual stock returns. *Journal of Finance*, 61:2365–2394.

95. Bacchetta, P. and van Wincoop, E. (2008) Higher order expectations in asset pricing. *Journal of Money, Credit and Banking*, 40:837–866.

96. Bachelier, L. (1900) Theory of speculation. Reprinted in *The Random Character of Stock Market Prices*, Cootner ed. (1964). MIT Press, Cambridge.

97. Back, K. (1992) Insider trading in continuous time. *Review of Financial Studies*, 5:387–409.

98. Back, K. (2010) *Asset Pricing and Portfolio Choice Theory*. Oxford University Press.

99. Back, K. and Baruch, S. (2004) information in securities markets: Kyle meets Glosten and Milgrom. *Econometrica*, 72:433–465.

100. Back, K. and Baruch, S. (2007) Working orders in limit order markets and floor exchanges. *Journal of Finance*, 62:1589–1621.

101. Back, K., Cao, H. and Willard, G. (2000) Imperfect competition among informed traders. *Journal of Finance*, 55:2117–2155.

102. Back, K. and Pedersen, H. (1998) Long-lived information and intraday patterns. *Journal of Financial Markets*, 1:385–402.

103. Backus, D., Gregory, A. and Zin, S. (1989) Risk premium in the term structure. *Journal of Monetary Economics*, 23:371–399.

104. Badrinath, S., Kale, J. and Noe, T. (1995) Of shephelders, sheep, and the cross-autocorrelations in equity returns. *Review of Financial Studies*, 8:401–430.

105. Bagehot, W. (1971) The only game in town. *Financial Analysts Journal*, 27:12–14.

106. Bagnoli, M. and Khanna, N. (1992) Insider trading in financial signalling models. *Journal of Finance*, 47:1905–1934.

107. Bagnoli, M. and Lipman, B. (1996) Stock price manipulation through takeover bids. *Rand Journal of Economics*, 27:124–147.

108. Baker, M. and Stein, J. (2004) Market liquidity as a sentiment indicator. *Journal of Financial Markets*, 7:271–299.

109. Baker, M. and Wurgler, J. (2000) The equity share in new issues and aggregate stock returns. *Journal of Finance*, 55:2219–2257.

110. Baker, M. and Wurgler, J. (2006) Investor sentiment and the cross-section of stock returns. *Journal of Finance*, 61:1645–1680.

111. Baker, M. and Wurgler, J. (2007) Investor sentiment in the stock market. *Journal of Economic Perspectives*, 21:129–151.

112. Baker, M., Wurgler, J. and Yuan, Y. (2012) Global, local, and contagious investor sentiment. *Journal of Financial Economics*, 104:272–287.

113. Bakshi, G. and Chen, Z. (1996) The spirit of capitalism and stock-market prices. *The American Economic Review*, 86:133–157.

114. Balasko, Y. and Cass, D. (1989) The structure of financial equilibrium with exogenous yields: the case of incomplete markets. *Econometrica*, 57:135–162.

115. Balduzzi, P., Bertola, G. and Foresi, S. (1995) Asset price dynamics and infrequent feedback traders. *Journal of Finance*, 50:1747–1766.

116. Balduzzi, P., Foresi, S. and Hait, D. (1997) Price barriers and the dynamics of asset prices in equilibrium. *Journal of Financial and Quantitative Analysis*, 32:137–159.

117. Balduzzi, P. and Kallal, H. (1997) Risk premia and variance bounds. *Journal of Finance*, 52:1913–1950.

118. Balduzzi, P. and Lynch, A. (1999) Transaction costs and predictability: some utility cost calculations. *Journal of Financial Economics*, 52:47–78.

119. Balduzzi, P. and Robotti, C. (2001) Minimum-variance kernels, economic risk premia, and tests of multi-beta models. Mimeo.

120. Balduzzi, P. and Yao, T. (2001) Does heterogeneity matter for asset pricing? Mimeo.

121. Ball, R. (1978) Anomalies in relationships between securities' yields and yields-surrogates. *Journal of Financial Economics*, 6:103–126.

122. Ball, R. (1992) The earnings–price anomaly. *Journal of Accounting and Economics*, 15: 319–345.

123. Ball, R. and Brown, P. (1968) An empirical evaluation of accounting income numbers. *Journal of Accounting Research*, 6:159–177.

124. Ball, R. and Kothari, S. (1989) Nonstationary expected returns: implications for tests of market efficiency and serial correlation in returns. *Journal of Financial Economics*, 25:51–74.

125. Ball, R., Kothari, S. and Shanken, J. (1995) Problems in measuring portfolio performance. An application to contrarian investment strategies. *Journal of Financial Economics*, 38:79–107.
126. Balvers, R., Cosimano, T. and McDonald, B. (1990) Predicting stock returns in an efficient market. *Journal of Finance*, 45:1109–1128.
127. Balvers, R., Wu, Y. and Gilliland, E. (2000) Mean reversion across national stock markets and parametric contrarian investment strategies. *Journal of Finance*, 55:745–771.
128. Bamber, L., Barron, O. and Stober, T. (1999) Differential interpretations and trading volume. *Journal of Financial and Quantitative Analysis*, 34:369–386.
129. Banerjee, S. (2011) Learning from prices and the dispersion in beliefs. *Review of Financial Studies*, 24:3025–3068.
130. Banerjee, S. and Kremer, I. (2010) Disagreement and learning: dynamic patterns of trade. *Journal of Finance*, 55:1269–1302.
131. Banerjee, S., Kaniel, R. and Kremer, I. (2009) Price drift as an outcome of differences in higher-order beliefs. *Review of Financial Studies*, 22:3707–3743.
132. Banks, J., Blundell, R. and Brugiavini, A. (2001) Risk pooling, precautionary saving and consumption growth. *Review of Economic Studies*, 68:757–780.
133. Bansal, R. and Coleman II, J. (1996) A monetary explanation of the equity premium, term premium, and risk-free rate puzzles. *Journal of Political Economy*, 104:1135–1171.
134. Bansal, R., Dittmar, R. and Lundblad, C. (2005) Consumption, dividends, and the cross-section of equity returns. *Journal of Finance*, 60:1639–1672.
135. Bansal, R., Dittmar, R. and Kiku, D. (2009) Cointegration and consumption risk in equity returns. *Review of Financial Studies*, 22:1343–1375.
136. Bansal, R. and Lehman, B. (1997) Growth optimal portfolio restrictions on asset pricing models. *Macroeconomic Dynamics*, 1:333–354.
137. Bansal, R. and Lundblad, C. (2002) Market efficiency, fundamental values, and asset returns in global equity markets. *Journal of Econometrics*, 109:195–237.
138. Bansal, R. and Viswanathan, S. (1993) No-arbitrage and arbitrage pricing: A new approach. *Journal of Finance*, 48:1231–1262.
139. Bansal, R. and Yaron, A. (2004) Risks for the long run: a potential resolution of asset pricing puzzles. *Journal of Finance*, 59:1481–1509.
140. Banz, R. (1981) The relation between return and market value of common stocks. *Journal of Financial Economics*, 9:3–18.
141. Barber, B., Lehavy, R., McNichols, M. and Trueman, B. (2001) Can investors profit from the prophets? Security analyst recommendations and stock returns. *Journal of Finance*, 56:531–563.
142. Barber, B. and Lyon, J. (1997) Detecting long run abnormal stock returns: the empirical power and specification of test statistics. *Journal of Financial Economics*, 43:341–372.
143. Barber, B. and Odean, T. (2000) Trading is hazardous to your wealth: the common stock investment performance of individual investors. *Journal of Finance*, 55:773–805.
144. Barber, B. and Odean, T. (2001) Boys will be boys: gender, overconfidence, and common stock investment. *Quarterly Journal of Economics*, 116:261–292.
145. Barber, B. and Odean, T. (2002) Online investors: do the slow die first? *Review of Financial Studies*, 15:455–487.
146. Barberis, N. (2000) Investing for the long run when returns are predictable. *Journal of Finance*, 55:225–264.
147. Barberis, N., Greenwood, R.M., Jin, L. and Shleifer, A. (2015) X-CAPM: An extrapolative capital asset pricing model. *Journal of Financial Economics*, 115:1–24.
148. Barberis, N. and Huang, M. (2001) Mental accounting, loss aversion, and individual stock returns. *Journal of Finance*, 56:1247–1282.
149. Barberis, N. and Huang, M. (2008) Stocks as lotteries: the implications of probability weighting for security prices. *The American Economic Review*, 98:2066–2100.
150. Barberis, N., Huang, M. and Santos, T. (2001) Prospect theory and asset prices. *Quarterly Journal of Economics*, 116:1–53.

151. Barberis, N. and Shleifer, A. (2003) Style investing. *Journal of Financial Economics*, 68:161–199.
152. Barberis, N., Shleifer, A. and Wurgler, J. (2002) Comovement. *Journal of Financial Economics*, 75:-283–317.
153. Barberis, N., Shleifer, A. and Vishny, R. (1998) A model of investor sentiment. *Journal of Financial Economics*, 49:307–343.
154. Barberis, N. and Thaler, R. (2003) A survey of behavioral finance. In *Handbook of Economics and Finance*, G. Constantinides, M. Harris and R. Stulz (eds.), North-Holland, Amsterdam, 1051–1121.
155. Barberis, N. and Xiong, W. (2009) What drives the disposition effect? An analysis of a long-standing preference-based explanation. *Journal of Finance*, 64:751–784.
156. Barberis, N. and Xiong, W. (2012) Realization utility. *Journal of Financial Economics*, 104:251–271.
157. Barclay, M. (1997) Bid–ask spreads and the avoidance of odd-eighth quotes on Nasdaq: an examination of exchange listings. *Journal of Financial Economics*, 45:35–60.
158. Barclay, M., Christie, W., Harris, J., Kandel, E. and Schultz, P. (1999) Effects of market reform on the trading costs and depths of Nasdaq stocks. *Journal of Finance*, 44:1–34.
159. Barclay, M., Kandel, E. and Marx, L. (1999) The effects of transaction costs on stock prices and trading volume. *Journal of Financial Intermediation*, 7:130–150.
160. Barclay, M., Litzenberger, R. and Warner, J. (1990) Private information, trading volume, and stock–return variances. *Review of Financial Studies*, 3:233–253.
161. Barclay, M. and Warner, J. (1993) Stealth trading and volatility. *Journal of Financial Economics*, 34:281–305.
162. Barillas, F., Hansen, L. P. and Sargent, T. (2009) Doubts or variability. *Journal of Economic Theory*, 144:2388–2418.
163. Barlevy, G. and Veronesi, P. (2000) Information acquisition in financial markets. *Review of Economic Studies*, 67:79–90.
164. Barlevy, G. and Veronesi, P. (2003) Rational panics and stock market crashes. *Journal of Economic Theory*, 110:234–263.
165. Barner, M., Feri, F. and Plott, C. (2005) On the microstructure of price determination and information aggregation with sequential and asymmetric information arrival in an experimental asset market. *Annals of Finance*, 1:73–107.
166. Barro, R. (2006) Rare disasters and asset markets in the twentieth century. *The Quarterly Journal of Economics*, 121:823–866.
167. Barro, R. and Jin, T. (2011) On the size distribution of macroeconomic disasters. *Econometrica*, 79:1567–1589.
168. Barsky, R. and De Long, B. (1993) Why does the stock market fluctuate? *Quarterly Journal of Economics*, 108:291–311.
169. Barsky, R., Juster, T., Kimball, M. and Shapiro, M. (1997) Preference parameters and behavioral heterogeneity: an experimental approach in the health and retirement study. *Quarterly Journal of Economics*, 112:536–579.
170. Barucci, E. and Casna, M. (2014) On the market selection hypothesis in a mean reverting environment. *Computational Economics*, 44:101–126.
171. Barucci, E. and Landi, L. (1996) Speculative dynamics under bounded rationality. *European Journal of Operational Research*, 91:284–300.
172. Barucci, E. and Marazzina, D. (2012) Optimal investment, stochastic labor income and retirement. *Applied Mathematics and Computation*, 218:5588–5604.
173. Barucci, E., Monte R. and Renò, R. (2004) Asset price anomalies under bounded rationality. *Computational Economics*, 23:255–269.
174. Baruch, S. (2005) Who benefits from an open limit-order book? *Journal of Business*, 78:1267–1306.
175. Basak, S. (2002) A model of dynamic equilibrium asset pricing with heterogeneous beliefs and extraneous risk. *Journal of Economic Dynamics and Control*, 24:63–95.

176. Basak, S. and Croitoru, B. (2001) Nonlinear taxation, tax arbitrage and equilibrium asset prices. *Journal of Mathematical Economics*, 35:347–382.
177. Basak, S. and Croitoru, B. (2000) Equilibrium mispricing in a capital market with portfolio constraints. *Review of Financial Studies*, 13:715–748.
178. Basak, S. and Cuoco, D. (1998) An equilibrium model with restricted stock market particitpation. *Review of Financial Studies*, 11:309–341.
179. Basu, S. (1977) The investment performance of common stocks in relation to their prices to earning ratios: a test of the efficient market hypothesis. *Journal of Finance*, 32:663–682.
180. Bateman, I., Munro, A., Rhodes, B., Starmer, C. and Sugden, R. (1997) A test of the theory of reference-dependent preferences. *Quarterly Journal of Economics*, 112:479–505.
181. Bajeux-Besnainou, I., Jordan, J. and Portait, R. (2003) Dynamic asset allocation for stocks, bonds and cash. *The Journal of Business*, 76:263–287.
182. Bajeux-Besnainou, I., Jordan, J. and Portait, R. (2000) An asset allocation puzzle: a comment. *The American Economic Review*, 91:1170–1179.
183. Bajeux-Besnainou, I. and Rochet, J. (1996) Dynamic spanning: are options an appropriate instrument? *Mathematical Finance*, 6:1–16.
184. Bebchuk, L. and Fershtman, C. (1994) Insider trading and the managerial choice among risky projects. *Journal of Financial and Quantitative Analysis*, 29:1–14.
185. Bekaert, G., Campbell, R. and Lundblad, C. (2007) Liquidity and expected returns: lessons from emerging markets. *Review of Financial Studies*, 20:1783–1831.
186. Bekaert, G. and Hodrick, R. (1992) Characterizing predictable in excess returns on equity and foreign exchange markets. *Journal of Finance*, 47:467–507.
187. Bekaert, G., Hodrick, R. and Marshall, D. (1997) The implications of first–order risk aversion for asset market risk premiums. *Journal of Monetary Economics*, 40:3–39.
188. Bekaert, G. and Liu, J. (2004) Conditioning information and variance bounds on pricing kernels. *Review of Financial Studies*, 17:339–378.
189. Benabou, R. and Laroque, G. (1992) Using privileged information to manipulate market: insiders, gurus, and credibility. *Quarterly Journal of Economics*, 107:921–958.
190. Benartzi, S. (2001) Excessive extrapolation and the allocation of 401(k) accounts to company stock. *Journal of Finance*, 56:1747–1764.
191. Benartzi, S. and Thaler, R. (1995) Myopic loss aversion and the equity premium puzzle. *Quarterly Journal of Economics*, 110:73–92.
192. Benninga, S. and Protopapadakis, A. (1990) Time preference and the equity premium puzzle. *Journal of Monetary Economics*, 25:49–58.
193. Benos, A. (1998) Aggressiveness and survival of overconfident traders. *Journal of Financial Markets*, 1:353–383.
194. Benveniste, L., Marcus, A. and Wilhelm, W. (1992) What's special about the specialist? *Journal of Financial Economics*, 32:61–86.
195. Benzoni, L., Collin-Dufresne, P. and Goldstein, R. (2007) Portfolio choice over the life cycle when stock and labor markets are cointegrated. *Journal of Finance*, 62:2123–2167.
196. Bergman, Y. (1985) Time preference and capital asset pricing models. *Journal of Financial Economics*, 14:145–159.
197. Berk, J. (1995) A critique of size-related anomalies. *Review of Financial Studies*, 8:275–286.
198. Berk, J. (1997) The acquisition of information in a dynamic market. *Economic Theory*, 9:441–451.
199. Berk, J., Green, J. and Naik, V. (1999) Optimal investment, growth options, and security returns. *Journal of Finance*, 54:1553–1607.
200. Berk, J. and Uhlig, H. (1993) The timing of information in a general equilibrium framework. *Journal of Economic Theory*, 59:275–287.
201. Berkelaar, A., Kouwenberg, R. and Post, T. (2004) Optimal portfolio choice under loss aversion. *Review of Economics and Statistics*, 86:973–987.
202. Bernard, V. and Thomas, J. (1990) Evidence that stock prices do not fully reflect the implications of current earnings for future earnings. *Journal of Accounting Economics*, 13:305–340.

203. Bernardo, A. and Judd, K. (2000) Asset market equilibrium with general tastes, returns, and informational asymmetries. *Journal of Financial Markets*, 3:17–43.
204. Bernardo, A. and Ledoit, O. (2000) Gain, loss, and asset pricing. *Journal of Political Economy*, 108:144–172.
205. Bernhardt, D., Hollifield, B. and Hughson, E. (1995) Investment and insider trading. *Review of Financial Studies*, 8:501–543.
206. Bernhardt, D. and Hughson, E. (1997) Splitting orders. *Review of Financial Studies*, 10:69–101.
207. Berrada, T. (2009) Bounded rationality and asset pricing with intermediate consumption. *Review of Finance*, 13:693–672.
208. Berry, T. and Howe, K. (1994) Public information arrival. *Journal of Finance*, 49:1331–1346.
209. Bertaut, C. (1998) Stockholding behavior of US households: evidence from the 1983–1989 survey of consumer finances. *Review of Economics and Statistics*, 80:263–275.
210. Bertaut, C. and Starr-McCluer, M. (2001) Household portfolios in the United States. In *Household Portfolios*, L. Guiso, M. Haliassos and T. Jappelli (eds.), MIT Press, Cambridge, 181–217.
211. Bertola, G., Guiso, L. and Pistaferri, L. (2005) Uncertainty and consumer durables adjustment. *Review of Economic Studies*, 72:973–1007.
212. Bessembinder, H. (1997) The degree of price resolution and equity trading costs. *Journal of Financial Economics*, 45:9–34.
213. Bessembinder, H. (1999) Trade execution costs on Nasdaq and the NYSE: a post reform comparison. *Journal of Financial and Quantitative Analysis*, 34:387–407.
214. Bessembinder, H. (2003) Quote-based competition and trade execution costs in NYSE-listed stocks. *Journal of Financial Economics*, 70:385–422.
215. Bessembinder, H., Chan, K. and Seguin, P. (1996) An empirical examination of information, differences of opinion, and trading activity. *Journal of Financial Economics*, 40:105–134.
216. Bessembinder, H. and Kaufman, H. (1997) A comparison of trade execution costs for NYSE and NASDAQ–listed stocks. *Journal of Financial and Quantitative Analysis*, 32:287–310.
217. Bessembinder, H. and Venkataraman, K. (2004) Does an electronic stock exchange need an upstairs market? *Journal of Financial Economics*, 73:3–36.
218. Bettis, J., Coles, J., Lemmon, M. (2000) Corporate policies restricting trading by insiders. *Journal of Financial Economics*, 57:191–220.
219. Bewley, T. (1980) The optimum quantity of money. In *Models of Monetary Economies*, J. Kareken and N. Wallace (eds.), Federal Reserve Bank of Minneapolis.
220. Bhamra, H. and Uppal, U. (2014) Asset prices with heterogeneity in preferences and beliefs. *Review of Financial Studies*, 27:519–580.
221. Bhandari, L. (1988) Debt/equity ratio and expected common stock returns: empirical evidence. *Journal of Finance*, 43:507–528.
222. Bhattacharya, S. and Nicodano, G. (2001) Insider trading, investment, and liquidity: a welfare analysis. *Journal of Finance*, 56:1141–1156.
223. Bhattacharya, U. and Daouk, H. (2002) The world price of insider trading. *Journal of Finance*, 57:75–108.
224. Bhattacharya, U. and Daouk, H. (2009) When no law is better than a good law. *Review of Finance*, 13:577–627.
225. Bhattacharya, U., Daouk, H., Jorgenson, B. and Kehr, C. (2001) When an event is not an event: the curious case of an emerging market. *Journal of Financial Economics*, 55:69–102.
226. Bhattacharya, U. and Lipman, B. (1995) Ex ante versus interim rationality and the existence of bubbles. *Economic Theory*, 6:469–494.
227. Bhattacharya, U. and Spiegel, M. (1991) Insiders, outsiders, and market breakdowns. *Review of Financial Studies*, 4:255–282.
228. Bhattacharya, U. and Spiegel, M. (1998) Anatomy of a market failure: NYSE trading suspensions (1974–1988) *Journal of Business and Economic Statistics*, 16:216–226.

229. Bhojraj, S. and Swaminathan, B. (2001) Macromomentum: evidence of predictability in international equity markets. Mimeo.
230. Bhushan, R., Brown, D. and Mello, S. (1997) Do noise traders "create their own space?" *Journal of Financial and Quantitative Analysis*, 32:25–45.
231. Biais, B. (1993) Price formation and equilibrium liquidity in fragmented and centralized markets. *Journal of Finance*, 48:157–184.
232. Biais, B. and Bossaerts, P. (1998) Asset prices and trading volume in a beauty contest. *Review of Economic Studies*, 65:307–340.
233. Biais, B., Bossaerts, P. and Spatt, C. (2010) Equilibrium asset pricing and portfolio choice under asymmetric information. *Review of Financial Studies*, 23:1503–1543.
234. Biais, B., Foucault, T. and Salanié, F. (1998) Floors, dealer markets and limit order markets. *Journal of Financial Markets*, 1:253–284.
235. Biais, B. and Germain, L. (2002) Incentive compatible contracts for the sale of information. *Review of Financial Studies*, 15:987–1003.
236. Biais, B., Glosten, L. and Spatt, C. (1995) Market microstructure: a survey of microfoundations, empirical results, and policy implications. *Journal of Financial Markets*, 8:217–264.
237. Biais, B. and Hillion, P. (1994) Insider and liquidity trading in stock and option markets. *Review of Economic Studies*, 7:743–780.
238. Biais, B., Hillion, P., Mazurier, K. and Pouget, S. (2005) Judgemental overconfidence, self-monitoring, and trading performance in an experimental financial market. *Review of Economic Studies*, 72:287–312.
239. Biais, B., Hillion, P. and Spatt, C. (1995) An empirical analysis of the limit order book and the order flow in the Paris bourse. *Journal of Finance*, 50:1655–1689.
240. Biais, B., Martimort, D. and Rochet, J. (2000) Competing mechanisms in a common value environment. *Econometrica*, 68:799–837.
241. Biais, B and Shadur, R. (2000) Darwinian selection does not eliminate irrational traders. *European Economic Review*, 44:469–490.
242. Black, F. (1972) Capital market equilibrium with restricted borrowing. *Journal of Business*, 45:444–454.
243. Black, F. (1986) Noise. *Journal of Finance*, 41:529–541.
244. Black, F. (1993) Estimating expected return. *Financial Analyst Journal*, 49:36–38.
245. Black, F., Jensen, M. and Scholes, M. (1972) The capital asset pricing model: some empirical tests. In *Studies in the Theory of Capital Markets*, M.C. Jensen (ed.), Praeger, New York.
246. Blackwell, D. (1951) Comparison of experiments. In *Proceedings of the Second Berkeley Symposium on Mathematical Statistics and Probability*, J. Neyman (ed.), University of California Press, Berkeley, 93–102.
247. Blanchard, O. (1993) Movements in the equity premium. *Brooking papers on economic activity*, 2:75–138.
248. Blanchard, O. and Watson, M. (1982) Bubbles, rational expectations and financial markets. In *Crises in the Economic and Financial Structure*, P. Wachtel (ed.), Lexington Books.
249. Bloomfield, R. (1996) Quotes, prices, and estimates in a laboratory market. *Journal of Finance*, 51:1791–1808.
250. Bloomfield, R., Libby, R. and Nelson, M. (2000) Underreactions, overreactions and moderated confidence. *Journal of Financial Markets*, 3:113–137.
251. Bloomfield, R. and O'Hara, M. (1998) Does order preferencing matter? *Journal of Financial Economics*, 50:3–37.
252. Bloomfield, R. and O'Hara, M. (1999) Market transparency: Who wins and who loses? *Review of Financial Studies*, 12:5–35.
253. Bloomfield, R. and O'Hara, M. (2000) Can transparent markets survive? *Journal of Financial Economics*, 55:425–459.
254. Bloomfield, R., O'Hara, M. and Saar, G. (2005) The "make or take" decision in an electronic market: Evidence on the evolution of liquidity. *Journal of Financial Economics*, 75:165–199.
255. Blume, L. and Easley, D. (1992) Evolution and market behavior. *Journal of Economic Theory*, 58:9–40.

256. Blume, L. and Easley, D. (2006) If you're so smart, why aren't you rich? Belief selection in complete and incomplete markets. *Econometrica*, 74:929–966.
257. Blume, L., Easley, D. and O'Hara, M. (1994) Market statistics and technical analysis: the role of volume. *Journal of Finance*, 49:153–179.
258. Blume, M. and Friend I. (1973) A new look at the capital asset pricing model. *Journal of Finance*, 28:19–33.
259. Blundell, R., Pistaferri, L. and Preston, I. (2008) Consumption inequality and partial insurance. *The American Economic Review*, 98:1887–1921.
260. Bodie, Z., Merton R. and Samuelson, W. (1992) Labour supply flexibility and portfolio choice in a life cycle model. *Journal of Economic Dynamics and Control*, 16:427–449.
261. Bodurtha, J. and Mark, N. (1991) Testing the CAPM with time-varying risks and returns. *Journal of Finance*, 46:1485–1505.
262. Boehme, R., Danielsen, B. and Sorescu, S. (2006) Short-sale constraints, differences of opinion, and overvaluation. *Journal of Financial and Quantitative Analysis*, 41:455–487.
263. Boldrin, M., Christiano, L. and Fisher, J. (2001) Habit persistence, asset returns, and the business cycle. *The American Economic Review*, 91:149–166.
264. Bollerslev, T., Engle, R. and Wooldridge, J. (1988) A capital asset pricing model with time–varying covariances. *Journal of Finance*, 96:116–131.
265. Bollerslev, T., Tauchen, G. and Zhou, H. (2009) Expected stock returns and variance risk premia. *Review of Financial Studies*, 22:4463–4492.
266. Bonomo, M. and Garcia, R. (1996) Consumption and equilibrium asset pricing: an empirical assessment. *Journal of Empirical Finance*, 3:239–265.
267. Bonomo, M., Garcia, R., Meddahi, N. and Teodongap, R: (2011) Generalized disappointment aversion, long-run volatility risk, and asset prices. *Review of Financial Studies*, 24:82–122.
268. Boot, A. and Thakor, A. (2001) The many faces of information disclosure. *Review of Financial Studies*, 14:1021–1057.
269. Borch, K. (1962) Equilibrium in a reinsurance market. *Econometrica*, 30:424–444.
270. Bossaerts, P. and Green, R. (1990) A general equilibrium model of changing risk premia: theory and tests. *Review of Financial Studies*, 2:467–493.
271. Bossaerts, P., Frydman, C. and Ledyard, J. (2014) The speed of information revelation and eventual price quality in markets with insiders: comparing two theories. *Review of Finance*, 18:1–22.
272. Boswijk, H., Hommes, C. and Mazan, S. (2007), Behavioral heterogeneity in stock prices. *Journal of Economic Dynamics and Control*, 31:1938–1970.
273. Boudoukh, J., Richardson, M. and Whitelaw, R. (1994) A tale of three schools: insights on autocorrelations of short horizon stock returns. *Review of Financial Studies*, 7:539–573.
274. Bowman, D., Minehart, D. and Rabin, M. (1999) Loss aversion in a consumption-savings model. *Journal of Economic Behavior & Organization*, 38:155–178.
275. Boyer, B., Mitton, T. and Vorkink, K. (2010) Expected idiosyncratic skewness. *Review of Financial Studies*, 23:169–202.
276. Boyle, P., Garlappi, L., Uppal, R. and Wang, T. (2012) Keynes meets Markowitz: the trade-off between familiarity and diversification. *Management Science*, 58:253–272.
277. Branch, W. and Evans, G. (2010) Asset return dynamics and learning. *Review of Financial Studies*, 23:1651–1680.
278. Branch, W. and Evans, G. (2011) Learning about risk and return: a simple model of bubbles and crashes. *American Economic Journal: Macroeconomics*, 3:159–191.
279. Brandt, M. (1999) Estimating portfolio and consumption choice: a conditional Euler equations approach. *Journal of Finance*, 54:1609–1645.
280. Brandt, M. (2010) Portfolio choice problems. In *Handbook of Financial Econometrics*, Volume 1, Y. Ait-Sahalia and L.P. Hansen (eds.), North Holland, Amsterdam, 269–336.
281. Brandt, M., Zeng, Q. and Zhang, L. (2004) Equilibrium stock return dynamics under alternative rules of learning about hidden states. *Journal of Economic Dynamics and Control*, 28:1925–1954.

282. Braun, P., Constantinides, G. and Ferson, W. (1993) Time nonseparability in aggregate consumption. International evidence. *European Economic Review*, 37:897–920.
283. Brav, A., Constantinides, G. and Gezcy, C. (2002) Asset pricing with heterogeneous consumers and limited particitpation: empirical evidence. *Journal of Political Economy*, 110:793–824.
284. Brav, A. and Heaton, J. (2002) Competing theories of financial anomalies. *Review of Financial Studies*, 15:575–606.
285. Breeden, D. (1979). An intertemporal asset pricing model with stochastic consumption and investment opportunities. *Journal of Financial Economics*, 7:265–296.
286. Breeden, M., Gibbons, R. and Litzenberger, R. (1989) Empirical tests of the consumption-oriented CAPM. *Journal of Finance*, 46:231–262.
287. Breeden, D. and Litzenberger, R. (1978) Prices of state-contingent claims implicit in option prices. *Journal of Business*, 51:621–651.
288. Breen, W. and Korajczyk, R. (1995) On selection biases in book-to-market based tests of asset pricing models. Mimeo.
289. Brennan, M. (1970) Taxes, market valuation and corporate financial policy. *National Tax Journal*, 23:417–427.
290. Brennan, M. (1998) The role of learning in dynamic portfolio decisions. *European Finance Review*, 1:295–306.
291. Brennan, M. and Cao, H. (1997) International portfolio flows. *Journal of Finance*, 52:1851–1880.
292. Brennan, M. and Chordia, T. (1993) Brokerage commission schedules. *Journal of Finance*, 48:1379–1402.
293. Brennan, M., Chordia, T. and Subrahmanyam, A. (1998) Alternative factor specifications, security characteristics and the cross expected stock returns. *Journal of Financial Economics*, 49:345–373.
294. Brennan, M., Chordia, T., Subrahmanyam, A. and Tong, Q. (1998) Sell-order liquidity and the cross-section of expected stock returns. *Journal of Financial Economics*, 105:523–541.
295. Brennan, M., Jegadeesh, N. and Swaminathan, B. (1993) Investment analysis and the adjustment of stock prices to common information. *Review of Financial Studies*, 6:799–824.
296. Brennan, M., Schwartz, E. and Lagnado, R. (1997) Strategic asset allocation. *Journal of Economic Dynamics and Control*, 21:1377–1403.
297. Brennan, M. and Subrahmanyam, A. (1996) Market microstructure and asset pricing: on the compensation for adverse selection in stock returns. *Journal of Financial Economics*, 41:341–364.
298. Brennan, M. and Subrahmanyam, A. (1998) The determinants of average trade size. *Journal of Business*, 71:1–23.
299. Brennan, M., Wang, A. and Xia, Y. (2004) Estimation and test of a simple model of intertemporal capital asset pricing. *Journal of Finance* 59:1743–1775.
300. Brennan, M. and Xia, Y. (2000) Stochastic interest rates and the bond–stock mix. *European Finance Review*, 4:197–210.
301. Brennan, M. and Xia, Y. (2002) Dynamic asset allocation under inflation. *Journal of Finance*, 57: 1201–1238.
302. Brennan, M. and Xia, Y. (2001) Stock price volatility, learning, and the equity premium. *Journal of Monetary Economics*, 47: 249–283.
303. Brennan, M. and Xia, Y. (2001) Assessing asset pricing anomalies. *Review of Financial Studies*, 14:905–942.
304. Bris, A. (2005) Do insider trading laws work? *European Financial Management*, 11:267–312.
305. Brock, W. (1982) Asset prices in a production economy. In *The Economics of Information and Uncertainty*, J. McCall (ed.), University of Chicago Press, 1–46.
306. Brock, W. and Hommes, C. (1998) Heterogeneous beliefs and routes to chaos in a simple asset pricing model. *Journal of Economic Dynamics and Control*, 22:1235–1274.

307. Brock, W. and Kleidon, A. (1992) Periodic market closure and trading volume. *Journal of Economic Dynamics and Control*, 16:451–489.
308. Brock, W., Lakonishok, J. and LeBaron, B. (1992) Simple technical trading rules and the stochastic properties of stock returns. *Journal of Finance*, 47:1731–1764.
309. Brock, W. and LeBaron, B. (1996) A dynamic structural model for stock return volatility and trading volume. *Review of Economics and Statistics*, 78:94–110.
310. Brown, D. and Jennings, R. (1989) On technical analysis. *Review of Financial Studies*, 2:527–552.
311. Brown, D. and Ross, S. (1991) Spanning, valuation and options. *Economic Theory*, 1:3–12.
312. Brown, D. and Zhang, Z. (1997) Market orders and market efficiency. *Journal of Finance*, 52:277–306.
313. Brown, J., Ivkovic, Z., Smith, P. and Weisbenner, S. (2008) Neighbors matter: causal community effects and stock market participation. *Journal of Finance*, 63:1509–1531.
314. Brown, S. and Goetzmann, W. (1995) Performance persistence. *Journal of Finance*, 50:679–698.
315. Brown, S., Goetzmann, W. and Ross, S. (1995) Survival. *Journal of Finance*, 50:853–873.
316. Brown, S. and Weinstein, M. (1983) A new approach to testing asset pricing models: the bilinear paradigm. *Journal of Finance*, 38:711–743.
317. Browning, M. and Crossley, T. (2001) The life–cycle model of consumption and saving. *Journal of Economic Perspectives*, 15:3–22.
318. Brunnermeier, M. (2005) Information leakage and market efficiency. *Review of Financial Studies*, 18:417–457.
319. Brunnermeier, M. (2001) *Asset Pricing Under Asymmetric Information*. Oxford University Press.
320. Brunnermeier, M. and Nagel, S. (2008) Do wealth fluctuations generate time-varying risk aversion? Micro-evidence on individuals' asset allocation. *The American Economic Review*, 98:713–736.
321. Buckley, G. and Tonks, I. (1989) Are UK stock prices excessively volatile? Trading rules and variance bounds tests. *The Economic Journal*, 99:1083–1098.
322. Buckley, G. and Tonks, I. (1992) Trading rules and excess volatility. *Journal of Financial and Quantitative Analysis*, 27:365–377.
323. Burmeister, E. and McElroy, M. (1998) Joint estimation of factor sensitivities and risk premia for the arbitrage pricing theory. *Journal of Finance*, 43:721–733.
324. Caballé, J. and Krishnam, M. (1994) Imperfect competition in a multi–security market with risk neutrality. *Econometrica*, 62:695–704.
325. Caballero, R. (1990) Consumption puzzles and precautionary savings. *Journal of Monetary Economics*, 25:113–136.
326. Cabrales, A. and Hoshi, T. (1996) Heterogeneous beliefs, wealth accumulation, and asset price dynamics. *Journal of Economic Dynamics and Control*, 20:1073–1100.
327. Camerer, C. (1989) Bubbles and fads in asset pricing. *Journal of Economic Surveys*, 3:3–41.
328. Camerer, C. (1995) Individual decision making. In *The Handbook of Experimental Economics*, J. Hagel and A. Roth (eds.), Princeton University Press.
329. Camerer, C. and Weber, M. (1992) Recent developments in modeling preferences: uncertainty and ambiguity. *Journal of Risk and Uncertainty*, 5:325–370.
330. Campbell, J. (1987) Stock returns and the term structure. *Journal of Financial Economics*, 18:373–399.
331. Campbell, J. (1991) A variance decomposition for stock returns. *The Economic Journal*, 101:157–179.
332. Campbell, J. (1993) Intertemporal asset pricing without consumption data. *The American Economic Review*, 83:487–512.
333. Campbell, J. (1996) Understanding risk and return. *Journal of Political Economy*, 104:298–343.
334. Campbell, J. (1998) Asset prices, consumption, and the business cycle. In *Handbook of Macroeconomics*, J.B. Taylor and M. Woodford (eds.), North-Holland, Amsterdam.

335. Campbell, J. (2000) Asset pricing at the millennium. *Journal of Finance*, 55:1515–1567.

336. Campbell, J. (2003) Consumption-based asset pricing. In *Handbook of Economics and Finance*, G. Constantinides, M. Harris and R. Stulz (eds.), North-Holland, Amsterdam, 803–887.

337. Campbell, J. (2006) Household finance. *Journal of Finance*, 61:1553–1603.

338. Campbell, J. and Ammer, J. (1993) What moves the stock and bond markets? A variance decomposition for long-term asset returns. *Journal of Finance*, 48:3–36.

339. Campbell, J., Chan, L., Gomes, F. and Maenhout, P. (2001) Investing retirement wealth: a life cycle model. In *Risk Aspects of Investment Based Social Security Reform*, J.Y. Campbell and M. Feldstein (eds.), University of Chicago Press.

340. Campbell, J., Chan, L. and Viceira, L. (2004) A multivariate model of strategic asset allocation. *Journal of Financial Economics*, 67:41–80.

341. Campbell, J. and Cocco, J. (2005) How do house prices affect consumption? Evidence from micro data. *Journal of Financial Economics*, 54:591–621.

342. Campbell, J., Cocco, J., Gomes, F., Maenhout, P. and Viceira, L. (2001) Stock market mean reversion and the optimal equity allocation of a long lived investor. *European Finance Review*, 5:269–292.

343. Campbell, J. and Cochrane, J. (1999) By force of habit: a consumption-based explanation of aggregate stock market behavior. *Journal of Political Economy*, 107:205–251.

344. Campbell, J. and Cochrane, J. (2000) Explaining the poor performance of consumption-based asset pricing models. *Journal of Finance*, 55:2863–2878.

345. Campbell, J. and Deaton, A. (1989) Why is consumption so smooth? *Review of Economic Studies*, 56:357–373.

346. Campbell, J., Grossman, S. and Wang, J. (1993) Trading volume and serial correlation in stock returns. *Quarterly Journal of Economics*, 108:905–939.

347. Campbell, J. and Kyle, A. (1993) Smart money, noise trading and stock price behaviour. *Review of Economic Studies*, 60:1–34.

348. Campbell, J., Lo, A. and MacKinlay, C. (1997) *The Econometrics of Financial Markets*. Princeton University Press.

349. Campbell, J. and Mankiw, G. (1991) The response of consumption to income: a cross country investigation. *European Economic Review*, 35:715–721.

350. Campbell, J., Polk, C. and Vuolteenaho, T. (2009) Growth or glamour? Fundamentals and systematic risk in stock returns. *Review of Financial Studies*, 23:305–344.

351. Campbell, J. and Shiller, R. (1988) The dividend–price ratio and expectations of future dividends and discount factors. *Review of Financial Studies*, 1:195–228.

352. Campbell, J. and Shiller, R. (1988) Stock prices, earnings, and expected dividends. *Journal of Finance*, 43:661–676.

353. Campbell, J. and Shiller, R. (2001) Valuation ratios and the long-run stock market outlook: an update. *NBER working paper*, no. 8221.

354. Campbell, J. and Viceira, L. (1999) Consumption and portfolio decisions when expected returns are time varying. *Quarterly Journal of Economics*, 114:433–493.

355. Campbell, J. and Viceira, L. (2001) Who should buy long-term bonds? *The American Economic Review*, 91:99–127.

356. Campbell, J. and Viceira, L. (2002) *Strategic Asset Allocation: Portfolio Choice for Long-Term Investors*. Oxford University Press.

357. Campbell, J. and Thompson, B. (2008) Predicting excess stock returns out of sample: can anything beat the historical average? *Review of Financial Studies*, 21:1509–1531.

358. Campbell, J. and Vuolteenaho, T. (2004) Bad beta, good beta. *The American Economic Review*, 94:1249–1275.

359. Campbell, Y. and Yogo, M. (2006) Efficient tests of stock return predictability. *Journal of Financial Economics*, 81:27–60.

360. Canner, N., Mankiw, G. and Weil, D. (1997) An asset allocation puzzle. *The American Economic Review*, 87:181–191.

361. Cao, H. (1999) The effect of derivative assets on information acquisition and price behavior in a rational expectations equilibrium. *Review of Financial Studies*, 12:131–163.
362. Cao, H. and Ou-Yang, H. (2009) Differences of opinion of public information and speculative trading in stocks and options. *Review of Financial Studies*, 22:299–335.
363. Cao, H., Wang, T., Zhang, H. (2005) Model uncertainty, limited market participation, and asset prices. *Review of Financial Studies*, 18:1219–1251.
364. Caplin, A. and Lehay, J. (1994) Business as usual, market crashes, and wisdom after the fact. *Quarterly Journal of Economics*, 84:448–505.
365. Caplin, A. and Lehay, J. (2001) Psychological expected utility theory and anticipatory feelings. *Quarterly Journal of Economics*, 166:55–79.
366. Carceles-Poveda, E. and Giannitsarou, C. (2008) Asset pricing with adaptive learning. *Review of Economic Dynamics*, 11:629–651.
367. Carhart, M. (1997) On persistence in mutual fund performance. *Journal of Finance*, 52:57–82.
368. Carroll, C. (1994) How does future income affect current consumption? *Quarterly Journal of Economics*, 109:111–147.
369. Carroll, C. (1997) Buffer-stock saving and the life cycle/permanent income hypothesis. *Quarterly Journal of Economics*, 112:1–55.
370. Carroll, C. (2001) A theory of the consumption function, with and without liquidity constraints. *Journal of Economic Perspectives*, 15:23–45.
371. Carroll, C. (2002) Portfolios of the rich. In *Household Portfolios*, L. Guiso, M. Haliassos and T. Jappelli (eds.), MIT Press, Cambridge, 389–429.
372. Carroll, C., Overland, J. and Weil, D. (2000) Saving and growth with habit formation. *The American Economic Review*, 90:341–355.
373. Carroll, C. and Samwick, A. (1997) The nature of precautionary wealth. *Journal of Monetary Economics*, 40:41–71.
374. Carroll, C. and Samwick, A. (1998) How important is precautionary saving? *Review of Economics and Statistics*, 80:410–419.
375. Cass, D. (1984) Competitive equilibrium with incomplete financial markets. *CARESS working paper*, no.84–09, University of Pennsylvania.
376. Cass, D. and Stiglitz, J. (1970) The structure of investor preferences and assets returns, and separability in portfolio allocation: a contribution to the pure theory of mutual funds. *Journal of Economic Theory*, 2:122–160.
377. Cassano, M. (1999) Learning and mean reversion in asset returns. *Quarterly Review of Economics and Finance*, 39:529–545.
378. Cecchetti, S., Lam, P. and Mark, N. (1990) Mean reversion in equilibrium asset prices. *The American Economic Review*, 80:398–418.
379. Cecchetti, S., Lam, P. and Mark, N. (1993) The equity premium and the risk-free rate. *Journal of Monetary Economics*, 31:21–45.
380. Cecchetti, S., Lam, P. and Mark, N. (1994) Testing volatility restrictions on the intertemporal marginal rates of substitution implied by Euler equations and asset returns. *Journal of Finance*, 49:123–151.
381. Cecchetti, S., Lam, P. and Mark, N. (2000) Asset pricing with distorted beliefs: are equity returns too good to be true? *The American Economic Review*, 90:787–805.
382. Cecchetti, S. and Mark, N. (1990) Evaluating empirical tests of asset pricing models: alternative interpretations. *The American Economic Review*, 80:48–52.
383. Chabakauri, G. (2013) Dynamic equilibrium with two stocks, heterogeneous investors, and portfolio constraints. *Review of Financial Studies*, 26:3104–3141.
384. Chacko, G. and Viceira, L. (2005) Dynamic consumption and portfolio choice with stochastic volatility in incomplete markets. *Review of Financial Studies*, 18:1369–1402.
385. Chae, J. (2005) Trading volume, information asymmetry, and timing information. *Journal of Finance*, 60:413–442.
386. Chai, J., Horneff, W., Maurer, R. and Mitchell, O. (2014) Optimal portfolio choice with annuities and life insurance for retired couples. *Review of Finance*, 18:147–188.

387. Chacko, G. and Viceira, L. (2005) Dynamic consumption and portfolio choice with stochastic volatility in incomplete markets. *Review of Financial Studies*, 18:1369–1402.

388. Chakraborty, A. and Yilmaz, B. (2004) Manipulation in market order models. *Journal of Financial Markets*, 7:187–206

389. Chakraborty, A. and Yilmaz, B. (2004) Informed manipulation. *Journal of Economic Theory*, 114:132–152.

390. Chakravarty, S. (2001) Stealth-trading: which traders' trades move stock prices? *Journal of Financial Economics*, 61:289–307.

391. Chakravarty, S., and Holden, C. (1995) An integrated model of market and limit orders. *Journal of Financial Intermediation* 4:213–241.

392. Chakravarty, S. and McConnell, J. (1999) Does insider trading really move stock prices? *Journal of Financial and Quantitative Analysis*, 34:191–209.

393. Chalmers, J. and Kadlec, G. (1998) An empirical examination of amortized spreads. *Journal of Financial Economics*, 48:159–188.

394. Chamberlain, G. (1983) Funds, factors and diversification in the arbitrage pricing theory. *Econometrica*, 51:1305–1323.

395. Chamberlain, G. (1983) A characterization of the distributions that imply mean-variance utility functions. *Journal of Economic Theory*, 29:185–201.

396. Chamberlain, G. (1985) Asset pricing in multiperiod securities markets. *Econometrica*, 56:1283–1300.

397. Chamberlain, G. and Rothschild, M. (1983) Arbitrage, factor structure, and mean variance analysis on large asset markets. *Econometrica*, 51:1281–1304.

398. Chamley, C. (2008) On "acquisition of information in financial markets". *Review of Economic Studies*, 75:1081–1084.

399. Chan, K. (1988) On the contrarian investment strategy. *Journal of Business*, 61:147–163.

400. Chan, K. (1993) Imperfect information and cross–autocorrelation among stock prices. *Journal of Finance*, 48:1211–1230.

401. Chan, K. and Chen, N. (1991) Structural and return characteristics of small and large firms. *Journal of Finance*, 46:1467–1483.

402. Chan, K., Chen, N. and Hsieh, H. (1985) An exploratory investigation of the firm size effect. *Journal of Financial Economics*, 14:451–471.

403. Chan, K., Christie, W. and Schultz, P. (1995) Market structure and the intraday pattern of bid-ask spreads for NASADQ securities. *Journal of Business*, 68:35–60.

404. Chan, K. and Fong, W. (2000) Trade size, order imbalance, and the volatility-volume relation. *Journal of Financial Economics*, 57:247–273.

405. Chan, K., Hameed, A. and Tong, W. (2000) Profitability of momentum strategies in the international equity markets. *Journal of Financial and Quantitative Analysis*, 35:153–172.

406. Chan, L., Jegadeesh, N. and Lakonishok, J. (1995) Evaluating the performance of value versus glamour stocks: the impact of selection bias. *Journal of Financial Economics*, 38:269–296.

407. Chan, L., Jegadeesh, N. and Lakonishok, J. (1996) Momentum strategies. *Journal of Finance*, 51:1681–1711.

408. Chan, L., Karceski, J. and Lakonishok, J. (1998) The risk and return from factors. *Journal of Financial and Quantitative Analysis*, 33:159–185.

409. Chan, Y. and Kogan, L. (2002) Catching up with the Joneses: heterogeneous preferences and the dynamics of asset prices. *Journal of Political Economy*, 110:1235–1285.

410. Chapman D. (2002) Does intrinsic habit formation actually resolve the equity premium puzzle? *Review of Economic Dynamics*, 5:618–645.

411. Chateauneuf, A. (1991) On the use of capacities in modeling uncertainty aversion and risk aversion. *Journal of Mathematical Economics*, 20: 343–369.

412. Chateauneuf, A., Dana, R. and Tallon, J. (2000) Optimal risk-sharing rules and equilibria with Choquet-expected-utility. *Journal of Mathematical Economics*, 34:191–214.

413. Chateauneuf, A., Eichberger, J. and Grant, S. (2007) Choice under uncertainty with the best and worst in mind: neo-additive capacities. *Journal of Economic Theory*, 137:538–567.

414. Chau, M. and Vayanos, D. (2008) Strong-form efficiency with monopolistic insiders. *Review of Financial Studies*, 21:2275–2306.
415. Chen, H., Ju, N. and Miao, J. (2014) Dynamic asset allocation with ambiguous return predictability. *Review of Economic Dynamics*, 17:799–823.
416. Chen, J. (2001) Intertemporal Capm and the cross-section of stock returns. Mimeo.
417. Chen, J., Hong, H. and Stein, J. (2001) Forecasting crashes: trading volume, past returns and conditional skewness in stock prices. *Journal of Financial Economics*, 61:311–344.
418. Chen, J., Hong, H. and Stein, J. (2002) Breadth of ownership and stock returns. *Journal of Financial Economics*, 66:171–205.
419. Chen, L. and Zhang, L. (2010) A better three-factor model that explains more anomalies. *Journal of Finance*, 65:563–595.
420. Chen, L. and Ludvigson, S. (2009) Land of addicts? An empirical investigation of habit-based asset pricing models. *Journal of Applied Econometrics*, 24:1057–1093.
421. Chen, N. (1991) Financial investment opportunities and the macroeconomy. *Journal of Finance*, 46:529–544.
422. Chen, N. (1983) Some empirical tests of arbitrage pricing theory. *Journal of Finance*, 38:1393–1414.
423. Chen, N. and Ingersoll, J. (1983) Exact pricing in linear factor models with finitely many assets. *Journal of Finance*, 38: 985–988.
424. Chen, N., Roll, R. and Ross, S. (1986) Economic forces and the stock market. *Journal of Business*, 59:383–403.
425. Chen, N. and Zhang, F. (1998) Risk and return of value stocks. *Journal of Business*, 71:501–535.
426. Chen, Z. and Epstein, L. (2002) Ambiguity, risk and asset returns in continuous time. *Econometrica*, 70:1403–1443.
427. Cheng, S., Nagar, V. and Rajan, M. (2007) Insider trades and private information: the special case of delayed-disclosure trades. *Review of Financial Studies*, 20:1833–1864.
428. Chew, H., Karni, E. and Safra, Z. (1987) Risk aversion in the theory of expected utility with rank dependent probabilities. *Journal of Economic Theory*, 42:370–381.
429. Chiarella, C. and He, T. (2001) Asset price and wealth dynamics under heterogeneous expectations. *Quantitative Finance*, 1:509–526.
430. Cho, J., Shin, J. and Singh, R. (1999) Endogenous informed trading in the presence of trading costs: theory and evidence. *Journal of Financial Markets*, 2:273–305.
431. Chopra, N., Lakonishok, J. and Ritter, J. (1992) Measuring abnormal performance: do stock market overreact? *Journal of Financial Economics*, 31:235–268.
432. Chordia, T., Huh, S. and Subrahmanyam, A. (2007) The cross-section of expected trading activity. *Review of Financial Studies*, 20:709–740.
433. Chordia, T., Huh, S. and Subrahmanyam, A. (2009) Theory-based illiquidity and asset pricing. *Review of Financial Studies*, 22:3629–3668.
434. Chordia, T., Roll, R. and Subrahmanyan, A. (1995) Market liquidity and trading activity. *Journal of Finance*, 56:501–530.
435. Chordia, T., Roll, R. and Subrahmanyan, A. (2011) Recent trends in trading activity and market quality. *Journal of Financial Economics*, 101:243–263.
436. Chordia, T. and Shivakumar, L. (2002) Momentum, business cycle and time-varying expected returns. *Journal of Finance*, 57:985–1019.
437. Chordia, T. and Subrahmanyan, A. (1995) Market making, the tick size, and payment for order flow: theory and evidence. *Journal of Business*, 68:543–573.
438. Chordia, T. and Subrahmanyan, A. (2004) Order imbalance and individual stock returns: Theory and evidence. *Journal of Financial Economics*, 72:485–518.
439. Chordia, T., Subrahmanyan, A. and Anshuman, R. (2001) Trading activity and expected stock returns. *Journal of Financial Economics*, 59:3–32.
440. Chordia, T. and Swaminathan, B. (2000) Trading volume and cross-autocorrelations in stock returns. *Journal of Finance*, 55:913–935.

441. Chowdhry, B. and Nanda, V. (1991) Multimarket trading and market liquidity. *Review of Financial Studies*, 4:483–511.
442. Chowdhry, B. and Nanda, V. (1998) Leverage and market stability: the role of margin rules and price limits. *Journal of Business*, 71:179–210.
443. Christelis, D., Jappelli, T. and Padula, M. (2010) Cognitive abilities and portfolio choice. *European Economic Review*, 54:18–38.
444. Christiano, L., Eichenbaum, M. and Marshall, D. (1991) The permanent income hypothesis revisited. *Econometrica*, 59:371–396.
445. Christie, W., Harris, J. and Schultz, H. (1994) Why did NASDAQ market makers stop avoiding odd-eight quotes? *Journal of Finance*, 49:1841–1860.
446. Christie, W. and Schultz, H. (1994) Why do NASDAQ market makers avoid odd-eight quotes? *Journal of Finance*, 49:1813–1840.
447. Chuang, W. and Lee, B. (2006) An empirical evaluation of the overconfidence hypothesis. *Journal of Banking and Finance*, 30:2489–2515.
448. Chung, K. L. (1979) *Elementary Probability Theory with Stochastic Processes*. Springer, New York.
449. Chung, K., Chuwonganant, C. and McCormick, T. (2004) Order preferencing and market quality on NASDAQ before and after decimalization. *Journal of Financial Economics*, 71:581–612.
450. Chung, K., Van Ness, B. and Van Ness, R. (1999) Limit orders and the bid ask spread. *Journal of Financial Economics*, 53:255–287.
451. Chung, K., Van Ness, B. and Van Ness, R. (2001) Can the treatment of limit orders reconcile the differences in trading costs between NYSE and NASDAQ Issues? *Journal of Financial and Quantitative Analysis*, 36:266–287.
452. Chung, K. and Van Ness, R. (2001) Order handling rules, tick size, and the intraday pattern of bid-ask spreads for Nasdaq stocks. *Journal of Financial Economics*, 4:143–161.
453. Citanna, A. and Villanacci, A. (2000) Existence and regularity of partially revealing rational expectations equilibrium in finite economies. *Journal of Mathematical Economics*, 34:1–26.
454. Clark, A., Frijters, P. and Shields, M. (2008) Relative income, happiness, and utility: An explanation for the Easterlin paradox and other puzzles. *Journal of Economic Literature*, 46:95–144.
455. Cocco, J. (2005) Portfolio choice in the presence of housing. *Review of Financial Studies*, 18:535–567.
456. Cocco, J., Gomes, F. and Maenhout, P. (2005) Consumption and portfolio choice over the life cycle. *Review of Financial Studies*, 18:491–533.
457. Cochrane, J. (1988) How big is the random walk in GNP? *Journal of Political Economy*, 96:893–920.
458. Cochrane, J. (1991) Volatility tests and efficient markets. *Journal of Monetary Economics*, 27:463–485.
459. Cochrane, J. (1991) A simple test of consumption insurance. *Journal of Political Economy*, 99:957–976.
460. Cochrane, J. (1991) Production based asset pricing and the link between stock returns and economic fluctuations. *Journal of Finance*, 46:209–237.
461. Cochrane, J. (1992) Explaining the variance of price-dividend ratios. *Review of Financial Studies*, 5:243–280.
462. Cochrane, J. (1996) A cross-sectional test of an investment-based asset pricing model. *Journal of Political Economy*, 104:572–621.
463. Cochrane, J. (1997) Where is the market going? Uncertain facts and novel theories. *Economic Perspectives*, 21:3–37.
464. Cochrane, J. (1999) New facts in finance. *Economic Perspectives*, 23:36–58.
465. Cochrane, J. (2001) *Asset Pricing*. Princeton University Press.
466. Cochrane, J. (2008) The dog that did not bark: a defense of return predictability. *Review of Financial Studies*, 21:1533–1575.

467. Cochrane, J. and Hansen, L. (1992) Asset pricing explorations for macroeconomics. *NBER Annual Report*: 117–182.
468. Cochrane, J. and Saa'-Requejo, J. (2000) Beyod arbitrage: good deal asset price bounds in incomplete markets. *Journal of Political Economy*, 108:79–119.
469. Cogley, T. (2002) Idiosyncratic risk and the equity premium: evidence from the consumer expenditure survey. *Journal of Monetary Economics*, 49:309–334.
470. Cogley, T. and Sargent T. (2008) The market price of risk and the equity premium: A legacy of the Great Depression? *Journal of Monetary Economics*, 55:454–476.
471. Cohen, R., Gompers, P. and Vuolteenaho, T. (2002) Who underreacts to cash flow news? Evidence from trading between individuals and institutions. *Journal of Financial Economics*, 66:409–462.
472. Cohen, R., Polk, C. and Vuolteenaho, T. (2003) The value spread. *Journal of Finance*, 58:609–641.
473. Coles, J., Loewenstein, U. and Suay, J. (1995) On equilibrium pricing under parameter uncertainty. *Journal of Financial and Quantitative Analysis*, 30:347–364.
474. Conlon, J. R. (2004) Finite horizon bubbles robust to higher order knowledge. *Econometrica*, 72:927–936.
475. Connor, G. (1984) A unified beta pricing theory. *Journal of Economic Theory*, 34:13–31.
476. Connor, G. (1989) Notes on the arbitrage pricing theory. In *Theory of Valuation*, S. Bhattacharya and G. Constantinides (eds.), Rowman & Littlefield, New Jersey.
477. Connor, G. (1995) The three types of factor models: a comparison of their explanatory power. *Financial Analyst Journal*, 51:42–46.
478. Connor, G. and Korajczyk, R. (1988) Risk and return in an equilibrium APT: applications of a new methodology. *Journal of Financial Economics*, 21:255–289.
479. Connor, G. and Korajczyk, R. (1990) An intertemporal equilibrium beta pricing model. *Review of Financial Studies*, 2:373–392.
480. Connor, G. and Korajczyk, R. (1993) A test for the number of factors in an approximate factor model. *Journal of Finance*, 48:1263–1290.
481. Connor, G. and Korajczyk, R. (1995) The arbitrage pricing theory and multifactor models of asset returns. In *Handbooks in Operations Research and Management Science*, Vol. 9 (Finance), R.A. Jarrow, V. Maksimovic and W.T. Ziemba (eds.), North Holland, Amsterdam.
482. Conrad, J., Hameed, A. and Niden, C. (1994) Volume and autocovariances in short-horizon individual security returns. *Journal of Finance*, 49:1305–1329.
483. Conrad, J. and Kaul, G. (1988) Time-variation in expected returns. *Journal of Business*, 61:409–425.
484. Conrad, J. and Kaul, G. (1989) Mean reversion in short horizon expected returns. *Review of Financial Studies*, 2:225–240.
485. Conrad, J. and Kaul, G. (1993) Long-term market overreaction or biases in computed returns. *Journal of Finance*, 48:39–63.
486. Conrad, J. and Kaul, G. (1998) An anatomy of trading strategies. *Review of Financial Studies*, 11:489–519.
487. Conrad, J., Kaul, G. and Nimalendran, M. (1991) Components of short-horizon individual security returns. *Journal of Financial Economics*, 29:365–284.
488. Constantinides, G. (1982) Intertemporal asset pricing with heterogeneous consumers and without demand aggregation. *Journal of Business*, 55:253–267.
489. Constantinides, G. (1983) Capital market equilibrium with personal tax. *Econometrica*, 51:611–636.
490. Constantinides, G. (1984) Optimal stock trading with personal taxes. *Journal of Financial Economics* 13:65–89.
491. Constantinides, G. (1986) Capital market equilibrium with transaction costs. *Journal of Political Economy*, 94:842–862.
492. Constantinides, G. (1989) Theory of valuation: overview and recent developments. In *Theory of Valuation*, S. Bhattacharya and G. Constantinides (eds.), Rowman & Littlefield, New Jersey.

493. Constantinides, G. (1990) Habit formation: a resolution of the equity premium puzzle. *Journal of Political Economy*, 98:519–543.
494. Constantinides, G. (2002) Rational asset prices. *Journal of Finance*, 57:1567–1591.
495. Constantinides, G., Donaldson, G. and Mehra, R. (1998) Junior can't borrow: a new perspective on the equity premium puzzle. *Quarterly Journal of Economics*, 117:269–296.
496. Constantinides, G. and Duffie, D. (1996) Asset pricing with heterogeneous consumers. *Journal of Political Economy*, 104:219–240.
497. Constantinides, G. and Ghosh, A. (2011) Asset pricing tests with long-run risks in consumption growth. *Review of Asset Pricing Studies*, 1:96–136.
498. Copeland, T. and Galai, D. (1983) Information effects on the bid-ask spread. *Journal of Finance*, 38:1457–1468.
499. Cornell, B. and Sirri, E. (1992) The reaction of investors and stock prices to insider trading. *Journal of Finance*, 47:1031–1059.
500. Corwin, S. and Lipson, M. (2000) Order flow and liquidity around NYSE trading halts. *Journal of Finance*, 55:1771–1801.
501. Coval, J. and Moskowitz, T. (1999) Home bias at home: local equity preference in domestic portfolios. *Journal of Finance*, 54:145–166.
502. Coval, J. and Moskowitz, T. (2001) The geography of investment: informed trading and asset prices. *Journal of Political Economy*, 109:811–841.
503. Coval, J. and Shumway, T. (2001) Do behavioral biases affect asset prices? *Journal of Finance*, 60:1–34.
504. Cox, J. and Huang, C. (1989) Optimal consumption and portfolio policies when asset prices follow a diffusion process. *Journal of Economic Theory*, 49:33–83.
505. Cox, J. and Huang, C. (1991) A vartiational problem occurring in financial economics. *Journal of Mathematical Economics*, 20:465–487.
506. Cox, J., Ingersoll, J. and Ross, S. (1985) An intertemporal general equilibrium model of asset prices. *Econometrica*, 53:363–384.
507. Cox, J., Ross, S. and Rubinstein, M. (1979) Option pricing: a simplified approach. *Journal of Financial Economics*, 7:229–263.
508. Cuoco, D. and Liu, H. (2000) Optimal consumption of a divisible good. *Journal of Economic Dynamics and Control*, 24:561–613.
509. Cutler, D., Poterba, J. and Summers, L. (1990) Speculative dynamics and the role of feedback traders. *The American Economic Review*, 80:63–68.
510. Cutler, D., Poterba, J. and Summers, L. (1991) Speculative dynamics. *Review of Economic Studies*, 58:529–546.
511. Cvitanić, J. and Zapatero, F. (2004) *Introduction to the Economics and Mathematics of Financial Markets*. MIT Press, Cambridge.
512. Damodaran, A. and Liu, C. (1993) Insider trading as a signal of private information. *Review of Financial Studies*, 6:79–119.
513. Dana R. (2004) Ambiguity, uncertainty aversion and equilibrium welfare. *Economic Theory*, 23:569–587.
514. Dana, R. and Jeanblanc-Picqué, M. (2003) *Financial Markets in Continuous Time*. Springer, Berlin/Heidelberg.
515. Daniel, K. (2001) The power and size of mean reversion tests. *Journal of Empirical Finance*, 8:493–535.
516. Daniel, K., Hirshleifer, D. and Subrahmanyam, A. (1998) Investor psychology and security market under- and overreactions. *Journal of Finance*, 53:1839–1883.
517. Daniel, K., Hirshleifer, D. and Subrahmanyam, A. (2001) Overconfidence, arbitrage, and equilibrium asset pricing. *Journal of Finance*, 56:921–965.
518. Daniel, K., Hirshleifer, D. and Teoh, S. (2002) Investor psychology in capital markets: evidence and policy implications. *Journal of Monetary Economics*, 49:139–209.
519. Daniel, K. and Marshall, D. (1997) The equity premium puzzle and the risk-free rate puzzle at long horizons. *Macroeconomic Dynamics*, 1:452–484.

520. Daniel, K. and Titman, S. (1997) Evidence of the characteristics of cross-sectional variation in stock returns. *Journal of Finance*, 52:1–33.

521. Daniel, K. and Titman, S. (2006) Market reactions to tangible and intangible information. *Journal of Finance*, 61:1605–1643.

522. Daniel, K. and Titman, S. (2012) Testing factor-models explanations of market anomalies. *Critical Finance Review*, 1:103–139.

523. Daniel, K., Titman, S. and Wei, J. (2001) Explaining the cros-section of stock returns in Japan: factors or characteristics. *Journal of Finance*, 56:743–766.

524. Danthine, J. and Donaldson, J. (2002) Labour relations and asset returns. *Review of Economic Studies*, 69:41–64.

525. Datar, V., Naik, N. and Radcliffe, R. (1998) Liquidity and stock returns: an alternative test. *Journal of Financial Markets* 1:203–219.

526. Davis, J. (1994) The cross-section of realized stock returns: the pre-COMPUSTAT evidence. *Journal of Finance*, 49:1579–1593.

527. Davis, J., Fama, E. and French, K. (2000) Characteristics, covariances, and average returns: 1929–1997. *Journal of Finance*, 55:389–406.

528. Davis, S., Kubler, F. and Willen, P. (2006) Borrowing costs and the demand for equity over the life cycle. *Review of Economics and Statistics*, 88:348–362.

529. Deaton, A. (1991) Saving and liquidity constraints. *Econometrica*, 59:1221–1248.

530. Deaton, A. (1992) *Understanding Consumption*. Clarendon Press, Oxford.

531. Deaves, R., Luders, E. and Luo, G. (2009) An experimental test of the impact of overconfidence and gender on trading activity. *Review of Finance*, 13:555–575.

532. De Bondt, W. and Thaler, R. (1985) Does the stock market overreact? *Journal of Finance*, 40:793–805.

533. De Bondt, W. and Thaler, R. (1987) Further evidence on investor overreaction and stock market seasonality. *Journal of Finance*, 42:557–582.

534. De Bondt, W. and Thaler, R. (1990) Do security analysts overreact? *The American Economic Review*, 80:52–57.

535. Debreu, G. (1959) *Theory of Value*. Yale University Press, New Haven.

536. Dechow, P. and Sloan, R. (1997) Returns to contrarian investment strategies: tests of naive expectations hypothesis. *Journal of Financial Economics*, 43:3–27.

537. De Finetti, B. (1952) Sulla preferibilità. *Giornale degli Economisti and Annali di Economia*, 11:685–709.

538. De Finetti, B. (1970) *Teoria della Probabilità*. Einaudi, Torino.

539. De Jong, F., Nijman, T. and Roell, A. (1996) Price effects of trading and components of bid-ask spreads on the Paris Bourse. *Journal of Empirical Finance*, 3:193–213.

540. De Jong, F. and Rindi, B. (2009) *The Microstructure of Financial Markets*. Cambridge University Press.

541. DeJong, D. and Whiteman, C. (1992) The temporal stability of dividends and stock prices: evidence from the likelihood function. *The American Economic Review*, 81:600–617.

542. DeJong, D. and Ripoll, M. (2007) Do self-control preferences help explain the puzzling behavior of asset prices? *Journal of Monetary Economics*, 54:1035–1050.

543. Dekel, E. (1989) Asset demand without the independence axiom. *Econometrica*, 57:163–169.

544. Delbaen, F. and Schachermayer, W. (1994) A general version of the fundamental theorem of asset pricing. *Mathematische Annalen*, 300:463–520.

545. Delbaen, F. and Schachermayer, W. (1998) The fundamental theorem of asset pricing for unbounded stochastic processes. *Mathematische Annalen*, 312:215–250.

546. De Long, B. and Magin, K. (2009) The U.S. equity return premium: past, present, and future. *The Journal of Economic Perspectives*, 23:193–208.

547. De Long, B., Shleifer, A., Summers, L. and Waldmann, R. (1989) The size and incidence of the losses from noise trading. *Journal of Finance*, 44:681–696.

548. De Long, B., Shleifer, A., Summers, L. and Waldmann, R. (1990) Positive feedback investment strategies and destabilizing rational speculation. *Journal of Finance*, 45:379–395.

549. De Long, B., Shleifer, A., Summers, L. and Waldmann, R. (1990) Noise trader risk in financial markets. *Journal of Political Economy*, 98:703–737.

550. De Long, B., Shleifer, A., Summers, L. and Waldmann, R. (1991) The survival of noise traders in financial markets. *Journal of Business*, 64:1–19.

551. Demange, G. and Laroque, G. (2006) *Finance and the Economics of Uncertainty*. Blackwell Publishing.

552. DeMarzo, P., Fishman, M. and Hagerty, K. (1998) The optimal enforcement of insider trading regulations. *Journal of Political Economy*, 106:602–632.

553. DeMarzo, P., Kaniel, R. and Kremer, I. (2004) Diversification as a public good: community effects in portfolio choice. *Journal of Finance*, 59:1677–1716.

554. DeMarzo, P., Kaniel, R. and Kremer, I. (2008) Relative wealth concerns and financial bubbles. *Review of Financial Studies*, 21:19–50.

555. DeMarzo, P. and Skiadas, C. (1998) Aggregation, determinacy, and informational efficiency for a class of economies with asymmetric information. *Journal of Economic Theory*, 80:123–152.

556. DeMarzo, P. and Skiadas, C. (1999) On the uniqueness of fully informative rational expectations equilibria. *Economic Theory*, 13:1–24.

557. De Miguel, V., Nogales, F. and Uppal, R. (2014) Stock return serial dependence and out-of-sample portfolio performance. *Review of Financial Studies*, 27:1031–1073.

558. De Nardi, M., French, E. and Jones, J. (2010) Why do the elderly save? The role of medical expenses. *Journal of Political Economy*, 118:39–75.

559. Dennert, J. (1993) Price competition between market makers. *Review of Economic Studies*, 60:735–751.

560. De Santis, G. and Gerard, B. (1997) International asset pricing and portfolio diversification with time varying risk. *Journal of Finance*, 52:1881–1912.

561. Detemple, J. (1986) Asset pricing in an economy with incomplete information. *Journal of Finance*, 61:383–392.

562. Detemple, J. and Giannikos, C. (1996) Asset and commodity prices with multi-attribute durable goods. *Journal of Economic Dynamics and Control*, 20:1451–1504.

563. Detemple, J. and Gottardi, P. (1998) Aggregation, efficiency and mutual fund separation in incomplete markets, *Economic Theory*, 11:443–455.

564. Detemple, J. and Murthy, S. (1994) Intertemporal asset pricing with heterogeneous beliefs. *Journal of Economic Theory*, 62:294–320.

565. Detemple, J. and Murthy, S. (1997) Equilibrium asset prices and no-arbitrage with portfolio constraints. *Review of Financial Studies*, 10:1133–1174.

566. Detemple, J. and Serrat, A. (2003) Dynamic equilibrium with liquidity constraints. *Review of Financial Studies*, 16:597–629.

567. Detemple, J. and Zapatero, F. (1991) Asset prices in an exchange economy with habit formation. *Econometrica*, 59:1633–1657.

568. Diamond, D. (1985) Optimal release of information by firms. *Journal of Finance*, 40:1071–1094.

569. Diamond, D. and Verrecchia, R. (1981) Information aggregation in a noisy rational expectations economy. *Journal of Financial Economics*, 9:221–235.

570. Diamond, D. and Verrecchia, R. (1991) Disclosure, liquidity, and the cost of capital. *Journal of Finance*, 46:1325–1360.

571. Diamond, P. (1967) The role of a stock market in a general equilibrium model with technological uncertainty. *The American Economic Review*, 57:759–776.

572. Díaz, A., Pijoan-Mas, J. and Ríos-Rull, J. (2003) Precautionary savings and wealth distribution under habit formation preferences. *Journal of Monetary Economics*, 50:1257–1291.

573. Diba, B. and Grossman, H. (1987) On the inception of rational bubbles. *Quarterly Journal of Economics*, 102:697–700.
574. Diba, B. and Grossman, H. (1988) Explosive rational bubbles in stock prices? *The American Economic Review*, 78:520–530.
575. Diba, B. and Grossman, H. (1988) The theory of rational bubbles in stock prices. *The Economic Journal*, 98:746–754.
576. Diether, K., Lee, K. and Werner, I. (2008) Short-sale strategies and return predictability. *Review of Financial Studies*, 22:575–607.
577. Diether, K., Malloy, C. and Scherbina, A. (2002) Differences of opinion and the cross section of stock returns. *Journal of Finance*, 57:2113–2141.
578. Dimson, E. and Marsh, P. (1999) Murphy's law and market anomalies. *Journal of Portfolio Management*, 23:53–69.
579. Dittmar, R. (2002) Nonlinear pricing kernels, kurtosis preference, and evidence from the cross-section of equity returns. *Journal of Finance*, 57:369–403.
580. Doidge, C., Karolyi, A., Lins, K., Miller, D. and Stulz, R. (2009) Private benefits of control, ownership, and the cross-listing decision. *Journal of Finance*, 64: 425–466.
581. Dothan, M. (1990) *Prices in Financial Markets*. Oxford University Press.
582. Dothan, U. and Feldman, D. (1986) Equilibrium interest rates and multiperiod bonds in a partially observable economy. *Journal of Finance*, 61:369–382.
583. Doukas, J., Kim, C. and Pantzalis, C. (2006) Divergence of opinion and equity returns. *Journal of Financial and Quantitative Analysis*, 41:573–606.
584. Dow, J. and Gorton, G. (1994) Arbitrage chains. *Journal of Finance*, 49:819–849.
585. Dow, J. and Gorton, G. (1995) Profitable informed trading in a simple general equilibrium model of asset pricing. *Journal of Economic Theory*, 67:327–369.
586. Dow, J. and Rahi, R. (2003) Informed trading, investment, and welfare. *Journal of Business*, 76:439–454.
587. Dow, J. and Rahi, R. (2000) Should speculators be taxed? *Journal of Business*, 73:89–107.
588. Dow, J. and Werlang, S. (1992) Uncertainty aversion, risk aversion, and the optimal choice of portfolio. *Econometrica*, 60:197–204.
589. Dow, J. and Werlang, S. (1992) Excess volatility of stock prices and Knightian uncertainty. *European Economic Review*, 36:631–638.
590. Drees, B. and Eckwert, B. (1997) Intrinsic bubbles and asset price volatility. *Economic Theory*, 9:499–510.
591. Duarte, J. and Young, L. (2009) Why is PIN priced? *Journal of Financial Economics*, 91:119–138.
592. Duffie, D. (1987) Stochastic equilibria with incomplete financial markets. *Journal of Economic Theory*, 41:405–416.
593. Duffie, D. (1996) *Dynamic Asset Pricing Theory*. Princeton University Press.
594. Duffie, D. and Epstein, L. (1992) Stochastic differential utility. *Econometrica*, 60:353–394.
595. Duffie, D. and Epstein, L. (1992) Asset pricing with stochastic differential utility. *Review of Financial Studies*, 5:411–436.
596. Duffie, D. and Huang, C. (1985) Implementing Arrow-Debreu equilibria by continuous trading of few long-lived securities. *Econometrica*, 53:1337–1356.
597. Duffie, D. and Rahi, R. (1995) Financial market innovation and security design: an introduction. *Journal of Economic Theory*, 65:1–42.
598. Duffie, D. and Shafer, W. (1985) equilibrium in incomplete markets I. A basic model of generic existence. *Journal of Mathematical Economics*, 14:285–300.
599. Duffie, D. and Shafer, W. (1986) Equilibrium in incomplete markets II. Generic existence in stochastic economies. *Journal of Mathematical Economics*, 15:199–216.
600. Duffie, D. and Zame, W. (1989) The consumption-based capital asset pricing model. *Econometrica*, 57:1279–1297.
601. Dumas, B., Kurshev, A. and Uppal, R. (2009) Equilibrium portfolio strategies in the presence of sentiment risk and excess volatility. *Journal of Finance*, 64:579–629.

602. Dunn, K. and Singleton, K. (1986) Modeling the term structure of interest rates under non-separability of preferences and durability of goods. *Journal of Financial Economics*, 17:27–55.

603. Dutta, P. and Madhavan, A. (1995) Price continuity rules and insider trading. *Journal of Financial and Quantitative Analysis*, 30:199–223.

604. Dutta, P. and Madhavan, A. (1997) Competition and collusion in dealer markets. *Journal of Finance*, 52:245–276.

605. Dybvig, P. (1983) An explicit bound on individual assets' deviations from Apt pricing in a finite economy. *Journal of Financial Economics*, 12:483–496.

606. Dybvig, P. (1984) Short sales restrictions and kinks on the mean–variance frontier. *Journal of Finance*, 39:239–244.

607. Dybvig, P. and Ingersoll, J. (1982) Mean-variance theory in complete markets. *Journal of Business*, 55:233–251.

608. Dybvig, P. and Liu, H. (2010) Lifetime consumption and investment: Retirement and constrained borrowing. *Journal of Economic Theory*, 145:885–907.

609. Dybvig, P. and Ross, S. (1985) Yes, the APT is testable. *Journal of Finance*, 40:1173–1187.

610. Dybvig, P. and Ross, S. (1986) Tax clienteles and asset pricing. *Journal of Finance*, 41:751–763.

611. Dybvig, P. and Ross, S. (1992) Arbitrage. In *The New Palgrave Dictionary of Money and Finance*, J. Eatwell, M. Milgate and P. Newman (eds.), Palgrave Macmillan, London.

612. Dynan, K. (1993) How prudent are consumers? *Journal of Political Economy*, 101:1104–1113.

613. Dynan, K. (2000) Habit formation in consumer preferences: evidence from panel data. *The American Economic Review*, 90:391–406.

614. Dynan, K., Skinner, J. and Zeldes, S. (2004) Do the rich save more? *Journal of Political Economy*, 112:397–444.

615. Easley, D., Hvidkjaer, S. and O'Hara, M. (2002) Is information risk a determinant of asset returns? *Journal of Finance*, 57:2185–2221.

616. Easley, D., Kiefer, N. and O'Hara, M. (1997) The information content of the trading process. *Journal of Empirical Finance*, 4:159–186.

617. Easley, D., Kiefer, N. and O'Hara, M. (1997) One day in the life of a very common stock. *Review of Financial Studies*, 10:805–835.

618. Easley, D., Kiefer, N., O'Hara, M. and Paperman, J. (1996) Liquidity, information, and infrequently traded stocks. *Journal of Finance*, 51:1405–1432.

619. Easley, D. and O'Hara, M. (1987) Price, trade size, and information in securities markets. *Journal of Financial Economics*, 19:69–90.

620. Easley, D. and O'Hara, M. (2004) Information and the cost of capital. *Journal of Finance*, 59:1553–1583.

621. Easley, D. and O'Hara, M. (2009) Ambiguity and nonparticipation: the role of regulation. *Review of Financial Studies*, 22:1817–1843.

622. Easley, D. and O'Hara, M. (2010) Liquidity and valuation in an uncertain world. *Journal of Financial Economics*, 97:1–11.

623. Easley, D. and O'Hara, M. (2010) Microstructure and ambiguity. *Journal of Finance*, 65:1817–1846.

624. Eckbo, E. and Smith, D. (1998) The conditional performance of insider trades. *Journal of Finance*, 53:467–498.

625. Eckwert, B. and Zilcha, I. (2001) The value of information in production economies. *Journal of Economic Theory*, 100:172–186.

626. Eckwert, B. and Zilcha, I. (2003) Incomplete risk sharing arrangements and the value of information. *Economic Theory*, 21:43–58.

627. Edelen, R. and Gervais, S. (2003) The role of trading halts in monitoring a specialist market. *Review of Financial Studies*, 16:263–300.

628. Ederington, L. and Lee, H. (1993) How markets process information: news releases and volatility. *Journal of Finance*, 48:1161–1191.

629. Edwards, R. (2008) Health risk and portfolio choices. *Journal of Business & Economic Statistics*, 26:472–485.
630. Eeckhoudt, L. and Gollier, C. (2000) The effects of change in risk on risk taking: A survey. Mimeo.
631. Eeckhoudt, L., Gollier, C. and Schlesinger, H. (2005) *Economic and Financial Decisions under Risk*. Princeton University Press.
632. Eeckhoudt, L. and Kimball, M. (1992) Background risk, prudence and the demand for insurance. In *Contributions to Insurance Economics*, G. Dionne (ed.), Kluwer Academic Publishers.
633. Eichenbaum, M. and Hansen, L. (1990) Estimating models with intertemporal substitution using aggregate time-series data. *Journal of Business and Economic Statistics*, 8: 53–69.
634. Eichenbaum, M., Hansen L. and Singleton, K. (1988) A time series analysis of representative agent models of consumption and leisure. *Quarterly Journal of Economics*, 103:51–78.
635. Eleswarapu, V. (1997) Cost of transacting and expected returns in the NASDAQ market. *Journal of Finance*, 52:2113–2127.
636. Ellsberg, D. (1961) Risk, ambiguity, and the Savage axioms. *Quarterly Journal of Economics*, 75:647–679.
637. Elmendorf, D. and Kimball, M. (2000) Taxation of labor income and the demand for risky assets. *International Economic Review*, 41:801–833.
638. Elton, E., Gruber, M. and Blake, C. (1995) Fundamental economic variables, expected returns, and bond fund performance. *Journal of Finance*, 50:1229–1257.
639. Elton, E., Gruber, M. and Busse, J. (1998) Do investors care about sentiment? *Journal of Business*, 71:477–499.
640. Epstein, L. (1999) A definition of uncertainty aversion. *Review of Economic Studies*, 66:579–608.
641. Epstein, L. (2001) Sharing ambiguity. *The American Economic Review*, 91:45–50.
642. Epstein, L. and Melino, A. (1995) A revealed preference analysis of asset pricing under recursive utility. *Review of Economic Studies*, 62:597–618.
643. Epstein, L. and Miao, J. (2003) A two-person dynamic equilibrium under ambiguity. *Journal of Economic Dynamics and Control*, 27:1253–1288.
644. Epstein, L. and Schneider, M. (2007) Learning under ambiguity. *Review of Economic Studies*, 74:1275–1303.
645. Epstein, I. and Schneider, M. (2008) Ambiguity, information quality, and asset pricing. *Journal of Finance*, 63:197–228.
646. Epstein, L. and Wang, T. (1994) Intertemporal asset pricing under Knightian uncertainty. *Econometrica*, 62:283–322.
647. Epstein, L. and Wang, T. (1995) Uncertainty, risk-neutral measures and security price booms and crashes. *Journal of Economic Theory*, 67:40–82.
648. Epstein, L. and Zhang, J. (2001) Subjective probabilities on subjectively unambiguous events. *Econometrica*, 69: 265–306.
649. Epstein, L. and Zin, S. (1989) Substitution, risk aversion, and the temporal behavior of consumption and asset returns I: a theoretical framework. *Econometrica*, 57:937–969.
650. Epstein, L. and Zin, S. (1990) 'First-order' risk aversion and the equity premium puzzle. *Journal of Monetary Economics*, 26:387–407.
651. Epstein, L. and Zin, S. (1991) Substitution, risk aversion, and the temporal behavior of consumption and asset returns II: an empirical analysis. *Journal of Political Economy*, 57:937–969.
652. Epstein, L. and Zin, S. (2001) The independence axiom and asset returns. *Journal of Empirical Finance*, 6:537–572.
653. Etner, J., Jeleva, M. and Tallon, J.-M. (2012) Decision theory under ambiguity. *Journal of Economic Surveys*, 26:234–270.
654. Evans, G. (1986) Selection criteria for models with non-uniqueness. *Journal of Monetary Economics*, 18:147–157.

655. Evans, G. (1989) The fragility of sunspots and bubbles. *Journal of Monetary Economics*, 23:297–317.
656. Evans, G. (1991) Pitfalls in testing for explosive bubbles in asset prices. *The American Economic Review*, 81:922–930.
657. Evans, G. and Honkapohja, S. (1998) Learning dynamics. In *Handbook of Macroeconomics*, J. Taylor and M. Woodford (eds.), North Holland, Amsterdam.
658. Evans, M. (1994) Expected returns, time-varying risk, and risk premia. *Journal of Finance*, 49:654–679.
659. Evans, M. (1998) Dividend variability and stock market swings. *Review of Economic Studies*, 65:711–740.
660. Fama, E. (1965) The behaviour of stock market prices. *Journal of Business*, 38:34–105.
661. Fama, E. (1970) Efficient capital markets: a review of theory and empirical work. *Journal of Finance*, 25:383–417.
662. Fama, E. (1990) Stock returns, expected returns, and real activity. *Journal of Finance*, 45:1089–1107.
663. Fama, E. (1991) Efficient capital markets: II. *Journal of Finance*, 46:1575–1618.
664. Fama, E. (1998) Market efficiency, long-term returns, and behavioural finance. *Journal of Financial Economics*, 49:283–306.
665. Fama, E. and Blume, M. (1966) Filter rules and stock market trading profits. *Journal of Business*, 39:226–241.
666. Fama, E. and French, K. (1988) Permanent and temporary components of stock prices. *Journal of Political Economy*, 96:246–273.
667. Fama, E. and French, K. (1988) Dividend yields and expected stock returns. *Journal of Financial Economics*, 22:3–27.
668. Fama, E. and French, K. (1989) Business conditions and expected returns on stocks and bonds. *Journal of Financial Economics*, 25:23–49.
669. Fama, E. and French, K. (1992) The cross-section of expected stock returns. *Journal of Finance*, 47:427–465.
670. Fama, E. and French, K. (1993) Common risk factors in the returns of stocks and bonds. *Journal of Financial Economics*, 33:3–56.
671. Fama, E. and French, K. (1995) Size and book to market factors in earnings and returns. *Journal of Finance*, 50:131–155.
672. Fama, E. and French, K. (1996) The CAPM is wanted, dead or alive. *Journal of Finance*, 51:1947–1958.
673. Fama, E. and French, K. (1996) Multifactor explanations of asset pricing anomalies. *Journal of Finance*, 51:55–85.
674. Fama, E. and French, K. (1998) Value versus growth: the international evidence. *Journal of Finance*, 53:1975–1999.
675. Fama, E. and French, K. (2002) The equity premium. *Journal of Finance*, 57:637–660.
676. Fama, E. and French, K. (2006) The Value Premium and the CAPM. *Journal of Finance*, 61:2163–2186.
677. Fama, E. and French, K. (2012) Size, value, and momentum in international stock returns. *Journal of Financial Economics*, 105: 457–472.
678. Fama, E. and MacBeth, J. (1973) Risk, return and equilibrium: empirical tests. *Journal of Political Economy*, 81:607–636.
679. Fama, E. and Schwert, G. (1977) Asset returns and inflation. *Journal of Financial Economics*, 5:115–146.
680. Farhi, E. and Panageas, S. (2007) Saving and investing for early retirement: A theoretical analysis. *Journal of Financial Economics*, 83: 87–121.
681. Faure-Grimaud, A. and Gromb, D. (2004) Public trading and private incentives. *Review of Financial Studies*, 17:985–1014.
682. Fernandes, N. and Ferreira, M. (2009) Insider trading laws and stock price informativeness. *Review of Financial Studies*, 22:1845–1887.
683. Fernholz, R. (2002) *Stochastic Portfolio Theory*. Springer, New York.

684. Ferris, S., Haugen, R. and Makhija, A. (1988) Predicting contemporary volume with historic volume at differential price levels: evidence supporting the disposition effect. *Journal of Finance*, 43:677–697.

685. Ferson, W. (1989) Changes in expected security returns, risk, and the level of interest rates. *Journal of Finance*, 44:1191–1217.

686. Ferson, W. (1990) Are the latent variables in time varying expected returns compensation for consumption risk? *Journal of Finance*, 45:397–427.

687. Ferson, W. (1995) Theory and empirical testing of asset pricing models. In *Handbooks in Operations Research and Management Science*, Vol. 9 (Finance), R.A. Jarrow, V. Maksimovic and W.T. Ziemba (eds.), North Holland, Amsterdam.

688. Ferson, W. (2003) Tests of multifactor pricing models, volatility bounds and portfolio performance. In *Handbook of the Economics of Finance*, Vol. 21, G. Constantinides, M. Harris and R. Stulz (eds.), Elsevier Science Publisher, 743–802.

689. Ferson, W. and Constantinides, G. (1991) Habit persistence and durability. *Journal of Financial Economics*, 29:199–240.

690. Ferson, W., Foerster, S. and Keim, D. (1993) General tests of latent variable models and mean variance spanning. *Journal of Finance*, 48:131–156.

691. Ferson, W. and Harvey, C. (1991) The variation of economic risk premiums. *Journal of Political Economy*, 99:385–415.

692. Ferson, W. and Harvey, C. (1992) Seasonality and consumption-based asset pricing. *Journal of Finance*, 47:511–551.

693. Ferson, W. and Harvey, C. (1999) Conditioning variables and the cross-section of stock returns. *Journal of Finance*, 54:1325–1360.

694. Ferson, W., Kandel, S. and Stambaugh, R. (1987) Tests of asset pricing with time-varying expected risk premiums and market betas. *Journal of Finance*, 42:201–220.

695. Ferson, W. and Korajczyk, R. (1995) Do arbitrage pricing models explain the predictability of stock returns? *Journal of Business*, 68:309–349.

696. Ferson, W. and Jagannathan, R. (1996) Econometric evaluation of asset pricing models. In *Handbook of Statistics* (Vol. 14), G.S. Maddala and C.R. Rao (eds.), Elsevier Science Publisher.

697. Ferson, W., Sarkissian, S. and Simin, T. (1999) The alpha factor asset pricing model: a parable. *Journal of Financial Markets*, 2:49–68.

698. Ferson, W. and Siegel, A. (2001) The efficient use of conditioning information in portfolios. *Journal of Finance*, 56:967–982.

699. Ferson, W. and Siegel, A. (2003) Stochastic discount factor bounds with conditioning information. *Review of Financial Studies*, 16:567–595.

700. Fielding, D. and Stracca, L. (2007) Myopic loss aversion, disappointment aversion, and the equity premium puzzle. *Journal of Economic Behavior and Organization*, 64:250–268.

701. Figlewski, S. (1978) Market "efficiency" in a market with heterogeneous information. *Journal of Political Economy*, 86:581–97.

702. Figlewski, S. (1979) Subjective information and market efficiency in a betting market. *Journal of Political Economy*, 87:75–88.

703. Figlewski, S. (1981) The informational effects of restrictions on short sales: some empirical evidence. *Journal of Financial and Quantitative Analysis* 16:463–476.

704. Fischer, P. (1992) Optimal contracting and insider trading restrictions. *Journal of Finance*, 47:673–694.

705. Fischer, P. and Verrecchia, R. (1999) Public information and heuristic trade. *Journal of Accounting and Economics*, 27:89–124.

706. Fishburn, P. (1970) *Utility Theory for Decision Making*. Wiley, New York.

707. Fishburn, P. and Porter, B. (1976) Optimal portfolios with one safe and one risky asset: effects of changes in rate of return and risk. *Management Science*, 22:1064–1073.

708. Fishman, M. and Hagerty, K. (1989) Disclosure decisions by firms and the competition for price efficiency. *Journal of Finance*, 46:633–646.

709. Fishman, M. and Hagerty, K. (1992) Insider trading and the efficiency of stock prices. *Rand Journal of Economics*, 23:106–122.
710. Fishman, M. and Hagerty, K. (1995) The mandatory disclosure of trades and market liquidity. *Review of Financial Studies*, 8:637–676.
711. Fishman, M. and Hagerty, K. (1995) The incentive to sell financial market information. *Journal of Financial Intermediation*, 4:95–115.
712. Fishman, M. and Longstaff, F. (1992) Dual trading in futures markets. *Journal of Finance*, 47:643–671.
713. Flannery, M. and Protopapadakis, A. (2002) Macroeconomic factors do influence aggregate stock returns. *Review of Financial Studies*, 15:751–782.
714. Flavin, M. (1981) The adjustment of consumption to changing expectations about future income. *Journal of Political Economy*, 89:974–1009.
715. Flavin, M. (1983) Excess volatility in financial markets: a reassessment of the empirical evidence. *Journal of Political Economy*, 91:929–956.
716. Flavin, M. and Nakagawa, S. (2008) A model of housing in the presence of adjustment costs: a structural interpretation of habit persistence. *The American Economic Review*, 98:474–495.
717. Flood, M., Huisman, R., Koedijk, K. and Mahieu, R. (1999) Quote disclosure and price discovery in multiple-dealer financial markets. *Review of Financial Studies*, 12:37–59.
718. Flood, R. and Hodrick, R. (1986) Asset price volatility, bubbles, and prices switching. *Journal of Finance*, 41:831–842.
719. Flood, R. and Hodrick, R. (1990) On testing for speculative bubbles. *Journal of Economic Perspective*, 4:85–101.
720. Flood, R., Hodrick, R. and Kaplan, P. (1986) An evaluation of recent evidence on stock market bubbles. *NBER working paper*, no.1971.
721. Föllmer, H. and Schied, A. (2011) *Stochastic Finance - An Introduction in Discrete Time.* 3rd edition. Walter de Gruyter, Berlin.
722. Föllmer, H. and Schweizer, M. (1991) Hedging of contingent claims under incomplete information. In *Applied Stochastic Analysis*, M.H.A. Davis and R.J. Elliott (eds.), Gordon & Breach, London/New York, 389–414.
723. Föllmer, H. and Sondermann, D. (1986) Hedging of non-redundant contingent claims. In *Contributions to Mathematical Economics*, W. Hildenbrand and A. Mas-Colell (eds.), North Holland, Amsterdam, 205–223.
724. Fontana, C. (2015) Weak and strong no-arbitrage conditions for continuous financial markets. *International Journal of Theoretical and Applied Finance*, 18:1550005.
725. Fontana, C. and Schweizer, M. (2012) Simplified mean-variance portfolio optimisation. *Mathematics and Financial Economics*, 6:125–153.
726. Fornari, F. (2002) The size of the equity premium. Temi di Discussione no.447, Banca d'Italia.
727. Forster, M. and George, T. (1992) Anonymity in securities markets. *Journal of Financial Intermediation*, 2:168–206.
728. Forsythe, R. and Lundholm, R. (1990) Information aggregation in an experimental market. *Econometrica*, 58:309–347.
729. Foster, D. and Viswanathan, S. (1990) A theory of intraday variations in volume, variance and trading costs in securities markets. *Review of Financial Studies*, 3:593–624.
730. Foster, D. and Viswanathan, S. (1993) The effect of public information and competition on trading volume and price volatility. *Review of Financial Studies*, 6:23–56.
731. Foster, D. and Viswanathan, S. (1993) Variations in trading volume, returns volatility, and trading costs: evidence of recent price formation models. *Journal of Finance*, 48:187–211.
732. Foster, D. and Viswanathan, S. (1996) Strategic trading when agents forecast the forecasts of others. *Journal of Finance*, 51:1437–1478.
733. Foucault, T. (1999) Order flow composition and trading costs in a dynamic limit order market. *Journal of Financial Markets*, 2:99–134.
734. Foucault, T., Kadan, O. and Kandel, E. (2005) Limit order book as a market for liquidity. *Review of Financial Studies*, 18:1171–1217.

735. Foucault, T., Pagano, M. and Röell, A. (2013) *Market Liquidity: Theory, Evidence and Policy.* Oxford University Press.
736. Frederick, S., Loewenstein, G. and O'Donoghue, T. (2002) Time discounting and time preference: a critical review. *Journal of Economic Literature*, 40:351–401.
737. French, E. (2005) The effects of health, wealth, and wages on labour supply and retirement behaviour. *Review of Economic Studies*, 72:395–427.
738. French, E. and Jones, J.B. (2011) The effects of health and self-insurance on retirement behavior. *Econometrica*, 79:693–732.
739. French, K. (1980) Stock returns and the weekend effect. *Journal of Financial Economics*, 8:55–69.
740. French, K. and Poterba, J. (1991) Investor diversification and international equity markets. *The American Economic Review*, 81:222–236.
741. French, K. and Roll, R. (1986) Stock return and variances: the arrival of information and reaction of traders. *Journal of Finance*, 17:5–26.
742. French, K., Schwert, G. and Stambaugh, R. (1987) Expected stock returns and volatility. *Journal of Financial Economics*, 19:3–29.
743. Friedman, M. (1953) *Essays in Positive Economics*, University of Chicago Press.
744. Friedman, M. and Savage, L. (1948) The utility analysis of choice involving risk. *Journal of Political Economy*, 56:279–304.
745. Froot, K. and Obstfeld, M. (1991) Intrinsic bubbles: the case of stock prices. *The American Economic Review*, 81:1189–1213.
746. Fuydenberg, D. and Tirole, J. (1991) *Game Theory*. MIT Press, Cambridge.
747. Gabaix, X. (2012) Variable rare disasters: an exactly solved framework for ten puzzles in macro-finance. *Quarterly Journal fo Economics*, 127:645–700.
748. Gabaix, X. and Laibson, D. (2001) The 6D bias and the equity premium puzzle. *NBER Macroeconomics Annual 2001*, 257–328.
749. Gali, J. (1994) Keeping up with the Joneses: consumption externalities, portfolio choice, and asset prices. *Journal of Money, Credit, and Banking*, 26:1–9.
750. Gallant, R., Hansen, L. and Tauchen, G. (1990) Using conditional moments of asset payoffs to infer the volatility of intertemporal marginal rates of substitution. *Journal of Econometrics*, 45:141–179.
751. Gallant, R., Rossi, P. and Tauchen, G. (1992) Stock prices and volume. *Review of Financial Studies*, 5:199–242.
752. Gallant, R., Rossi, P. and Tauchen, G. (1993) Nonlinear dynamic structures. *Econometrica*, 61:871–907.
753. Gallant, R. and Tauchen, G. (1989) Seminonparametric estimation of conditionally constrained heterogeneous processes: asset pricing applications. *Econometrica*, 57:1091–1120.
754. Ganguli, J. V. and Yang, L. (2009) Complementarities, multiplicity, and supply information. *Journal of the European Economic Association*, 7:90–115.
755. Garcia, D. and Sangiorgi, F. (2011) Information sales and strategic trading. *Review of Financial Studies*, 24:3069–3104.
756. Garcia, D. and Strobl, G. (2011) Relative wealth concerns and complementarities in information acquisition. *Review of Financial Studies*, 24:169–207.
757. Garcia, D. and Vanden, J. (2009) Information acquisition and mutual funds. *Journal of Economic Theory*, 144:1965–1995.
758. Garfinkel, J. and Nimalendran, M. (2003) Market structure and trader anonymity: an analysis of insider trading. *Journal of Financial and Quantitative Analysis*, 38:591–610.
759. Garlappi, L., Uppal, R. and Wang, T. (2007) Portfolio selection with parameter and model uncertainty: a multi-prior approach. *Review of Financial Studies*, 20:41–81.
760. Garleanu, N. (2009) Portfolio choice and pricing in illiquid markets. *Journal of Economic Theory*, 144:532–564.
761. Garleanu, N. and Pedersen, L.H. (2013) Dynamic trading with predictable returns and transaction costs. *Journal of Finance*, 68:2309–2340.

762. Gaunersdorfer, A. (2000) Endogenous fluctuations in a simple asset pricing model with heterogeneous agents. *Journal of Economic Dynamics and Control*, 24:799–831.
763. Geanakoplos, J. (1990) An introduction to general equilibrium with incomplete markets. *Journal of Mathematical Economics*, 19:1–38.
764. Geanakoplos, J. and Mas-Colell, A. (1989) Real indeterminacy with financial assets. *Journal of Economic Theory*, 47:22–38.
765. Geanakoplos, J. and Polemarchakis, H. (1986) Existence, regularity and constrained suboptimality of competitive allocations when the asset market is incomplete. In *Essays in Honor of K. Arrow*, W. Heller, R. Starr and D. Starrett (eds.), Cambridge University Press, 65–95.
766. Gemmill, G. (1994) Transparency and liquidity: a study of block trades on the London Stock exchange under different publication rules. *Journal of Finance*, 51:1765–1790.
767. Gennotte, G. (1986) Optimal portfolio choice under incomplete information. *Journal of Finance*, 61:733–749.
768. George, T., Kaul, G. and Nimalendran, M. (1991) Estimation of the bid-ask spread and its components: a new approach. *Review of Financial Studies*, 4:623–656.
769. George, T., Kaul, G. and Nimalendran, M. (1994) Trading volume and transaction costs in specialist markets. *Journal of Finance*, 49:1489–1505.
770. Gerard, B. and Nanda, V. (1993) Trading and manipulation around seasoned equity offerings. *Journal of Finance*, 48:213–245.
771. Gerety, M. and Mulherin, J.H. (1992) Trading halts and market activity: an analysis of volume at the open and the close. *Journal of Finance*, 47:1765–1784.
772. Gerety, M. and Mulherin, J.H. (1994) Price formation on stock exchanges: the evolution trading within the day. *Review of Financial Studies*, 6:609–629.
773. Gervais, S. and Odean, T. (2001) Learning to be overconfident. *Review of Financial Studies*, 14:1–27.
774. Ghirardato, P. and Marinacci, M. (2002) Ambiguity made precise: A comparative foundation. *Journal of Economic Theory*, 102:251–289.
775. Gibbons, M. (1982) Multivariate tests of financial models: a new approach. *Journal of Financial Economics*, 10:3–27.
776. Gibbons, M. and Ferson, W. (1985) Testing asset pricing models with changing expectations and an unobservable market portfolio. *Journal of Financial Economics*, 14:217–236.
777. Gibbons, M., Ross, S. and Shanken, J. (1989) A test of the efficiency of a given portfolio. *Econometrica*, 57:1121–1152.
778. Gilboa, I. (1987) Expected utility with purely subjective non-additive probabilities. *Journal of Mathematical Economics*, 16:65–88.
779. Gilboa, I. and Marinacci, M. (2013) Ambiguity and the Bayesian paradigm. In *Advances in Economics and Econometrics*, D. Acemoglu, M. Arellano and E. Dekel (eds.), Cambridge University Press.
780. Gilboa, I. and Schmeidler, D. (1989) Maxmin expected utility with nonunique prior. *Journal of Mathematical Economics*, 18:141–153.
781. Gilles, C. and LeRoy, S. (1991) On the arbitrage pricing theory. *Economic Theory*, 1:213–229.
782. Gilles, C. and LeRoy, S. (1991) Econometric aspects of variance-bounds tests: a survey. *Review of Financial Studies*, 4:753–791.
783. Gilles, C. and LeRoy, S. (1992) Bubbles and charges. *International Economic Review*, 33:323–339.
784. Gilles, C. and LeRoy, S. (1997) Bubbles as payoffs at infinity. *Economic Theory*, 9:261–281.
785. Giovannini, A. and Weil, P. (1989) Risk aversion and intertemporal substitution in the capital asset pricing model. *NBER working paper* no.2824.
786. Glaser, M. and Weber, M. (2007) Overconfidence and trading volume. *The Geneva Risk and Insurance Review*, 32:1–36.
787. Glosten, L. (1989) Insider trading, liquidity, and the role of the monopolist specialist. *Journal of Business*, 62:211–235.

788. Glosten, L. (1994) Is the electronic open limit order book inevitable? *Journal of Finance*, 49:1127–1162.
789. Glosten, L. and Harris, L. (1988) Estimating the components of the bid-ask spread. *Journal of Financial Economics*, 21:123–142.
790. Glosten, L. and Milgrom, P. (1985) Bid, ask and transaction prices in a specialist market with heterogeneously informed traders. *Journal of Financial Economics*, 14:71–100.
791. Gneezy, U., Kapteyn, A. and Potters, J. (2003) Evaluation periods and asset prices in a market experience. *Journal of Finance*, 58:821–838.
792. Gneezy, U. and Potters, J. (1997) An experiment on risk taking and evaluation periods. *Quarterly Journal of Economics*, 112:631–645.
793. Godek, P. (1996) Why Nasdaq market makers avoid odd-eight quotes. *Journal of Financial Economics*, 41:465–474.
794. Goettler, R., Parlour, C. and Rajan, U. (2005) Equilibrium in a dynamic limit order market. *Journal of Finance*, 60:2149–2192.
795. Goetzmann, W. and Jorion, P. (1993) Testing the predictive power of dividend yields. *Journal of Finance*, 48:663–679.
796. Goetzmann, W. and Jorion, P. (1995) A longer look at dividend yields. *Journal of Business*, 68:482–508.
797. Goetzmann, W. and Jorion, P. (1999) Global stock markets in the twentieth century. *Journal of Finance*, 54:953–980.
798. Goldstein, M. and Kavajecz, K. (2000) Eighths, sixteenths, and market depth: changes in tick size and liquidity provision on the NYSE. *Journal of Financial Economics*, 56:125–149.
799. Gollier, C. (1995) The comparative statics of changes in risk revisited. *Journal of Economic Theory*, 66:522–535.
800. Gollier, C. (2001) *The Economics of Risk and Time*. MIT Press, Cambridge.
801. Gollier, C. (2001) Wealth inequality and asset pricing. *Review of Economic Studies*, 68:181–203.
802. Gollier, C. (2009) Portfolio choices and asset prices. The comparative statics of ambiguity aversion. *Review of Economic Studies*, 78:1329–1344.
803. Gollier, C. and Pratt, J. (1996) Risk vulnerability and tempering effect of background risk. *Econometrica*, 64:1109–1123.
804. Gollier, C. and Schlesinger, H. (2002) Changes in risk and asset prices. *Journal of Monetary Economics*, 49:747–760.
805. Gomes, F. (2005) Portfolio choice and trading volume with loss-averse investors. *Journal of Business*, 78:675–706.
806. Gomes, F., Kotlikoff, L. and Viceira, L. (2008) Optimal life-cycle investing with flexible labor supply: a welfare analysis of life-cycle funds. *The American Economic Review: Papers and Proceedings*, 98:297–303.
807. Gomes, F. and Michaelides, A. (2003) Portfolio choice with internal habit formation: a life-cycle model with uninsurable labor income risk. *Review of Economic Dynamics*, 6:729–766.
808. Gomes, F. and Michaelides, A. (2005) Optimal life cycle asset allocation: understanding the empirical evidence. *Journal of Finance*, 60:869–904.
809. Gomes, F. and Michaelides, A. (2008) Asset pricing with limited risk sharing and heterogeneous agents. *Review of Financial Studies*, 21:415–448.
810. Gomes, J., Kogan, L. and Zhang, L. (2003) Equilibrium cross-section of returns. *Journal of Political Economy*, 111:693–732.
811. Gomes, J., Yaron, A. and Zhang, L. (2006) Asset pricing implications of firms' financing constraints. *Review of Financial Studies*, 19:1321–1356.
812. Goodhart, C. and O'Hara, M. (1997) High frequency data in financial markets: issues and applications. *Journal of Empirical Finance*, 4:73–114.
813. Gottardi, P. and Rahi, R. (2014) Value of information in competitive economies with incomplete markets. *International Economic Review*, 55:1468–2354.
814. Gould, J. and Verrecchia, R. (1985) The information content of specialist pricing. *Journal of Political Economy*, 93:66–83.

815. Gourinchas, P. and Parker, J. (2002) Consumption over the life cycle. *Econometrica*, 70:47–89.
816. Gourio, F. (2007) Disasters and recoveries. *The American Economic Review* 98:68–73.
817. Goyal, A. and Welch, I. (2008) A comprehensive look at the empirical performance of equity premium prediction. *Review of Financial Studies*, 21:1455–1508.
818. Graham, B. and Dodd, D. (1934) *Security Analysis*. McGraw-Hill, New York.
819. Graham, J., Harvey, C. and Huang, H. (1996) Investor competence, trading frequency, and home bias. *Management Science*, 55:1094–1106.
820. Green, J. (1973) Information, efficiency and equilibrium. Discussion paper n.284. Harvard University.
821. Green, J. (1981) Value of Information with Sequential Futures Markets *Econometrica*, 49:335–358.
822. Green, J. and Jarrow, R. (1987) Spanning and completeness in markets with contingent claims. *Journal of Economic Theory*, 41:202–210.
823. Greene, J. and Smart, S. (1999) Liquidity provision and noise trading: evidence from the investment dartboard column. *Journal of Finance*, 54:1885–1899.
824. Greenwald, B. and Stein, J. (1991) Transactional risk, market crashes, and the role of circuit breakers. *Journal of Business*, 64:443–462.
825. Griffin, J. (2002) Are the Fama and French factors global or country specific? *Review of Financial Studies*, 15:775–806.
826. Griffin, J. and Lemmon, M. (2002) Book-to-market equity, distress risk, and stock returns. *Journal fo Finance*, 57:2317–2336.
827. Grinblatt, M. and Han, B. (2005) Prospect theory, mental accounting, and momentum. *Journal of Financial Economics*, 78:311–339.
828. Griffin, J., Harris, J., Shu, T. and Topaloglu, S. (2011) Who drove and burst the tech bubble? *Journal of Finance*, 66:1251–1290.
829. Grinblatt, M. and Keloharju, M. (2001) What makes investors trade? *Journal of Finance*, 56:589–616.
830. Grinblatt, M. and Keloharju, M. (2004) Tax-loss trading and wash sales. *Journal of Financial Economics*, 71:51–76.
831. Grinblatt, M. and Keloharju, M. (2009) Sensation seeking, overconfidence, and trading activity. *Journal of Finance*, 69:549–578.
832. Grinblatt, M., Keloharju, M. and Linnainmaa, J. (2011) IQ and stock market participation, *Journal of Finance*, 66:2121–2164.
833. Grinblatt, M. and Moskowitz T. (2004) Predicting stock price movements from past returns: the role of consistency and tax-loss selling. *Journal of Financial Economics* 71:541–579.
834. Grinblatt, M. and Titman, S. (1983) Factor pricing in a finite economy. *Journal of Financial Economics*, 12:497–507.
835. Grinblatt, M. and Titman, S. (1985) Approximate factor structures. Interpretations and implications for empirical tests. *Journal of Finance*, 40:1367–1373.
836. Grinblatt, M. and Titman, S. (1987) The relation between mean-variance efficiency and arbitrage pricing. *Journal of Business*, 60:97–112.
837. Gromb, D. and Vayanos, D. (2002) Equilibrium and welfare in markets with financially constrained arbitrageurs. *Journal of Financial Economics*, 66:361–407.
838. Gross, D. and Souleles, N. (2001) Do liquidity constraints and interest rates matter for consumer behavior? Evidence from credit card data. *Quarterly Journal of Economics*, 117:149–186.
839. Grossman, S. (1976) On the efficiency of competitive stock markets where traders have diverse information. *Journal of Finance*, 31:573–585.
840. Grossman, S. (1978) Further results on the informational efficiency of competitive stock markets. *Journal of Economic Theory*, 18:81–101.
841. Grossman, S. (1981) An introduction to the theory of rational expectations under asymmetric information. *Review of Economic Studies*, 48:541–559.
842. Grossman, S. (1989) *The Informational Role of Prices*. MIT Press, Cambridge. ·

843. Grossman, S. (1992) The informational role of upstairs and downstairs trading. *Journal of Business*, 65:509–530.

844. Grossman, S. and Laroque, G. (1990) Asset pricing and optimal portfolio choice in the presence of illiquid durable consumption goods. *Econometrica*, 58:25–51.

845. Grossman, S., Melino A. and Shiller, R. (1987) Estimating the continuous time consumption based asset pricing model. *Journal of Business and Economic Statistics*, 5:315–327.

846. Grossman, S. and Miller, M. (1988) Liquidity and market structure. *Journal of Finance*, 43:617–637.

847. Grossman, S., Miller, M., Cone, K., Fischel, D. and Ross, D. (1997) Clustering and competition in dealer markets. *Journal of Law and Economics*, 40:23–60.

848. Grossman, S. and Shiller, R. (1981) The determinants of the variability of stock markets prices. *The American Economic Review*, 71:222–227.

849. Grossman, S. and Stiglitz, J. (1980) On the impossibility of informationally efficient markets. *The American Economic Review*, 70:393–408.

850. Grundy, B. and Martin, S. (2001) Understanding the nature of the risks and the source of the rewards to momentum investing. *Review of Financial Studies*, 14:29–78.

851. Grundy, B. and Martin, S. (2001) Trade and revelation of information through prices and direct disclosure. *Review of Financial Studies*, 2:495–526.

852. Grundy, B. and Kim, Y. (2002) Stock market volatility in a heterogeneous information economy. *Journal of Financial and Quantitative Analysis*, 37:1–27.

853. Guesnerie, R. and Jaffray, J. Y. (1974) Optimality of equilibrium of plans, prices and price expectations. In *Allocation under uncertainty: equilibrium and optimality*, J. Drèze (ed.), Palgrave Macmillan, London.

854. Guidolin, M.(2006) Pessimistic beliefs under rational learning: Quantitative implications for the equity premium puzzle. *Journal of Economics and Business*, 58:85–118.

855. Guidolin, M. and Timmermann, A. (2007) Properties of equilibrium asset prices under alternative learning schemes. *Journal of Economic Dynamics and Control*, 31:161–217.

856. Guidolin, M. and Rinaldi, F. (2013) Ambiguity in asset pricing and portfolio choice: a review of the literature. *Theory and Decisions*, 74:183–217.

857. Guiso, L., Haliassos, M. and Jappelli, T. eds. (2001) *Household Portfolios*. MIT Press, Cambridge.

858. Guiso, L. and Jappelli, T. (2005) Awareness and stock market participation. *Review of Finance*, 9:537–567.

859. Guiso, L., Jappelli, T. and Terlizzese, D. (1992) Earnings uncertainty and precautionary saving. *Journal of Monetary Economics*, 30:307–332.

860. Guiso, L., Jappelli, T. and Terlizzese, D. (1996) Income risk, borrowing constraints, and portfolio choice. *The American Economic Review*, 86:158–172.

861. Guiso, L., Sapienza, P. and Zingales, L. (2008) Trusting the stock market. *Journal of Finance*, 63:2557–2600.

862. Guiso, L. and Sodini, C. (2012) Household finance. An emerging field. Mimeo.

863. Gul, F. (1991) A theory of disappointment aversion. *Econometrica*, 59:667–686.

864. Gul, F. and Pesendorfer, W. (2001) Temptation and self-control. *Econometrica*, 69:1403–1435.

865. Gul, F. and Pesendorfer, W. (2004) Self-control and the theory of consumption. *Econometrica*, 72:119–158.

866. Gurkaynak, R. (2008) Econometric tests of asset price bubbles: taking stock. *Journal of Economic Surveys*, 22:166–186.

867. Guvenen, F. (2006) Reconciling conflicting evidence on the elasticity of intertemporal substitution: A macroeconomic perspective. *Journal of Monetary Economics*, 53:1451–1472.

868. Guvenen, F. (2009) A parsimonious macroeconomic model for asset pricing. *Econometrica*, 77:1711–1750.

869. Hadar, J. and Russell, W. (1971) Stochastic dominance and diversification. *Journal of Economic Theory*, 3:288–305.

870. Hadar, J. and Russell, W. (1974) Diversification of interdependent prospects. *Journal of Economic Theory*, 7:231–240.
871. Hadar, J., Russell, W. and Seo, T. (1977) Gains from diversification. *Review of Economic Studies*, 44:363–368.
872. Hadar, J. and Seo, T. (1990) The effects of shifts in a return distribution on optimal portfolios. *International Economic Review*, 31:721–736.
873. Hagerty, K. (1991) Equilibrium bid-ask spreads in markets with multiple assets. *Review of Economic Studies*, 58:237–257.
874. Hahn, F. (1973) *On the Notion of Equilibrium in Economics*. Cambridge University Press.
875. Hakansson, N. (1970) Optimal investment and consumption under risk for a class of utility functions. *Econometrica*, 38:587–607.
876. Haliassos, M. and Bertaut, C. (1995) Why does so few hold stocks? *The Economic Journal*, 105:1110–1129.
877. Haliassos, M. and Michaelides, A. (2003) Portfolio choice and liquidity constraints. *International Economic Review*, 44:143–177.
878. Hall, R. (1978) Stochastic implications of the life cycle-permanent income hypothesis: theory and evidence. *Journal of Political Economy*, 86:971–987.
879. Hall, R. and Mishkin, F. (1982) The sensitivity of consumption to transitory: estimates from panel data households. *Econometrica*, 50:461–481.
880. Hall, R. (1988) Intertemporal substitution in consumption. *Journal of Political Economy*, 96:339–357.
881. Hamilton, J. (1994) *Time Series Analysis*. Princeton University Press.
882. Hamilton, J. and Whiteman, C. (1985) The observable implications of self-fulfilling expectations. *Journal of Monetary Economics*, 16:353–373.
883. Han, Y. (2006) Asset allocation with a high dimensional latent factor stochastic volatility model. *Review of Financial Studies*, 19:237–271.
884. Han, Y., Ke, Y. and Zhou G. (2013) A new anomaly: the cross-sectional profitability of technical analysis. *Journal of Financial and Quantitative Analysis*, 48:1433–1461.
885. Handa, P., Kothari, S. and Wasley, C. (1993) Sensitivity of multivariate tests of the capital asset pricing model to the returns measurement interval. *Journal of Finance*, 48:1543–1551.
886. Handa, P. and Linn, S. (1993) Arbitrage pricing with estimation risk. *Journal of Financial and Quantitative Analysis*, 28:81–100.
887. Handa, P. and Tiwari, A. (2006) Does stock return predictability imply improved asset allocation and performance? Evidence from the U.S. stock market (1954–2002). *The Journal of Business*, 79:2423–2468.
888. Hankansson, N., Kunkel, G. and Ohlson, J. (1982) Sufficient and necessary conditions for information to have social value in pure exchange. *Journal of Finance*, 37:1169–1181.
889. Hansen, L., Heaton, J. and Li, N. (2008) Consumption strikes back? Measuring long run risk. *Journal of Political Economy*, 116:260–302.
890. Hansen, L., Heaton, J. and Lutmer, E. (1995) Econometric evaluation of asset pricing models. *Review of Financial Studies*, 8:237–274.
891. Hansen, L. and Jagannathan, R. (1991) Implications of security market data for models of dynamic economies. *Journal of Political Economy*, 91:249–265.
892. Hansen, L. and Jagannathan, R. (1997) Assessing specification errors in stochastic discount factors models. *Journal of Finance*, 52:557–590.
893. Hansen, L. and Sargent, T. (2001) Robust control and model uncertainty. *The American Economic Review*, 91:60–66.
894. Hansen, L., Sargent, T. and Tallarini, T. (1999) Robust permanent income and pricing. *Review of Economic Studies*, 66:873–907.
895. Hansen, L. and Singleton, K. (1982) Generalized instrumental variables estimation of nonlinear rational expectations models. *Econometrica*, 50:1269–1286.
896. Hansen, L. and Singleton, K. (1983) Stochastic consumption, risk aversion, and the temporal behavior of asset returns. *Journal of Political Economy*, 91:249–265.

897. Harris, C. and Laibson, D. (2001) Dynamic choices of hyperbolic consumers. *Econometrica*, 69:935–957.
898. Harris, L. (1986) A transaction data study of weekly and intradaily patterns in stock returns. *Journal of Financial Economics*, 16:99–117.
899. Harris, L. (1994) Minimum price variations, discrete bid-ask spreads, and quotation sizes. *Review of Financial Studies*, 7:149–178.
900. Harris, L. and Hasbrouck, J. (1996) Market vs. limit orders: The SuperDOT evidence on order submission strategy. *Journal of Financial and Quantitative Analysis*, 31:213–231.
901. Harris, M. and Raviv, A. (1993) Differences of opinion make a horse race. *Review of Financial Studies*, 6:473–506.
902. Harris, M. and Venkatesh, P. (2005) The information content of the limit order book: evidence from NYSE specialist trading decisions. *Journal of Financial Markets*, 8:25–67.
903. Harrison, M. and Kreps, D. (1978) Speculative investor behavior in a stock market with heterogeneous expectations. *Quarterly Journal of Economics*, 92:323–336.
904. Harrison, M. and Kreps, D. (1979) Martingales and arbitrage in multiperiod securities markets. *Journal of Economic Theory*, 28:183–91.
905. Harrison, M. and Pliska, S. (1981) Martingales and stochastic integrals in the theory of continuous trading. *Stochastic Processes and their Applications*, 11:215–260.
906. Hart, O. (1975) On the optimality of equilibrium when the market structure is incomplete. *Journal of Economic Theory*, 11:418–443.
907. Hart, O. and Kreps, D. (1986) Price destabilizing speculation. *Journal of Political Economy*, 94:927–952.
908. Harvey, C. (1989) Time varying conditional covariances in tests of asset pricing models. *Journal of Financial Economics*, 24:289–317.
909. Harvey, C. (1991) The world price of covariance risk. *Journal of Finance*, 46:111–157.
910. Harvey, C. and Siddique, A. (2000) Conditional skewness in asset pricing tests. *Journal of Finance*, 55:1263–1296.
911. Hasbrouck, J. (1988) Trades, quotes, inventories, and information. *Journal of Financial Economics*, 22:229–252.
912. Hasbrouck, J. (1991) Measuring the information content of stock trades. *Journal of Finance*, 46:179–207.
913. Hasbrouck, J. and Sofianos, G. (1993) The trades of market makers: an empirical analysis of NYSE specialists. *Journal of Finance*, 48:1565–1595.
914. Hasbrouck, J. and Schwartz, R. (1988) Liquidity and execution costs in equity markets. *Journal of Portfolio Management*, 14:10–16.
915. Haugen, R. and Baker, N. (1996) Commonality in the determinants of expected stock returns. *Journal of Financial Economics*, 41:401–439.
916. Hawawini, G. and Keim, D. (1995) On the predictability of common stock returns: worldwide evidence. In *Handbooks in Operations Research and Management Science*, Vol. 9 (Finance), R.A. Jarrow, V. Maksimovic and W.T. Ziemba (eds.), North Holland, Amsterdam.
917. Hayashi, F. (1985) The effect of liquidity constraints on consumption: a cross-sectional analysis. *Quarterly Journal of Economics*, 100:183–206.
918. He, H. and Modest, D. (1995) Market frictions and consumption-based asset pricing. *Journal of Political Economy*, 103:94–117.
919. He, H. and Pearson, N.D. (1991) Consumtpion and portfolio choices with incomplete markets and short-sale constraints: the infinite dimensional case. *Journal of Economic Theory*, 54:259–304.
920. He, H. and Wang, J. (1995) Differential information and dynamic behavior of stock trading volume. *Review of Financial Studies*, 8:919–972.
921. He, J., Kan, R., Ng, L. and Zhang, C. (1996) Tests of the relations among marketwide factors, firm-specific variables, and stock returns using a conditional asset pricing model. *Journal of Finance*, 51:1891–1908.

922. Healy, P. and Palepu, K. (2001) Information asymmetry, corporate disclosure, and the capital markets: a review of the empirical disclosure literature. *Journal of Accounting and Economics*, 31:405–440.
923. Heaton, J. (1993) The interaction between time-nonseparable preferences and time aggregation. *Econometrica*, 61:353–385.
924. Heaton, J. (1995) An empirical investigation of asset pricing with temporally dependent preference specifications. *Econometrica*, 63:681–717.
925. Heaton, J. and Lucas, D. (1992) The effects of incomplete insurance markets and trading costs in a consumption-based asset pricing model. *Journal of Economic Dynamics and Control*, 16:601–620.
926. Heaton, J. and Lucas, D. (1995) The importance of investor heterogeneity and financial market imperfections for the behavior of asset prices. *Carnegie-Rochester Conference series on Public policy*, 42:1–32.
927. Heaton, J. and Lucas, D. (1996) Evaluating the effects of incomplete markets on risk sharing and asset pricing. *Journal of Political Economy*, 104:443–487.
928. Heaton, J. and Lucas, D. (1997) Market frictions, saving behavior and portfolio choice. *Macroeconomic Dynamics*, 1:76–101.
929. Heaton, J. and Lucas, D. (2000) Portfolio choice and asset prices: the importance of entrepreneurial risk. *Journal of Finance*, 55:1163–1197.
930. Hellwig, C. and Lorenzoni, G. (2009) Bubbles and self-enforcing debt. *Econometrica*, 77:1137–1164.
931. Hellwig, M. (1980) On the aggregation of information in competitive markets. *Journal of Economic Theory*, 22:477–498.
932. Hellwig, M. (1982) Rational expectations equilibrium with conditioning on past prices: a mean variance example. *Journal of Economic Theory*, 26:279–312.
933. Hendershott, T. and Mendelson, H. (2000) Crossing networks and dealer markets: competition and performance. *Journal of Finance*, 55:2071–2115.
934. Hendershott, T. and Seasholes, M. (2007) Market maker inventories and stock prices. *The American Economic Review*, 97:210–214.
935. Henderson V. (2005) Explicit solutions to an optimal portfolio choice problem with stochastic income. *Journal of Economic Dynamics and Control*, 29:1237–1266.
936. Henderson, V. and Hobson, D. (2008) Utility indifference pricing: an overview. In *Indifference Pricing*, R. Carmona (ed.), Princeton University Press, 44–74.
937. Hens, T. and Pilgrim, B. (2003) *General Equilibrium Foundations of Finance: Structure of Incomplete Markets Models*, Kluwer Academic Publishers.
938. Hens, T. and Rieger M.O. (2010) *Financial Economics*, Springer, Berlin/Heidelberg.
939. Hernandez, A., and Santos, M. (1996) Competitive equilibria for infinite-horizon economies with incomplete markets. *Journal of Economic Theory*, 71:102–130.
940. Heston, S., Loewenstein, M. and Willard G. (2007) Options and Bubbles. *Review of Financial Studies*, 20:359–390.
941. Heston, S. and Sadka, R. (2008) Seasonality in the cross-section of stock returns. *Journal of Financial Economics*, 87:418–445.
942. Hindy, A. (1995) Viable prices in financial markets with solvency constraints. *Journal of Mathematical Economics*, 24:105–135.
943. Hindy, A. and Huang, C. (1993) Optimal consumption and portfolio rules with durability and local substitution. *Econometrica*, 61:85–121.
944. Hindy, A., Huang, C. and Zhu, S. (1997) Optimal consumption and portfolio rules with durability and habit formation. *Journal of Economic Dynamics and Control*, 21:525–550.
945. Hirshleifer, D. (2001) Investor psychology and asset pricing. *Journal of Finance*, 56:1533–1597.
946. Hirshleifer, D. and Luo, G. (2001) On the survival of overconfident traders in a competitive securities market. *Journal of Financial Markets*, 4:73–84.

947. Hirshleifer, D., Subrahmanyam, A. and Titman, S. (1994) Security analysis and trading patterns when some investors receive information before others. *Journal of Finance*, 49:1664–1698.
948. Hirshleifer, J. (1971) The private and social value of information and the reward to inventive activity. *The American Economic Review*, 61:561–574.
949. Ho, T. and Stoll, H. (1981) Optimal dealer pricing under transactions and return uncertainty. *Journal of Financial Economics*, 9:47–73.
950. Ho, T. and Stoll, H. (1983) The dynamics of dealer markets under competition. *Journal of Finance*, 38:1053–1074.
951. Hodrick, R. (1992) Dividend yields and expected stock returns: alternative procedures for inference and measurement. *Review of Financial Studies*, 5:357–386.
952. Hodrick, R. and Zhang, X. (2001) Evaluating the specification errors of asset pricing models. *Journal of Financial Economics*, 62:327–376.
953. Holden, C. and Subrahmanyan, A. (1992) Long-lived private information and imperfect competition. *Journal of Finance*, 47:247–270.
954. Holden, C. and Subrahmanyan, A. (1994) Risk aversion, imperfect competition and long-lived information. *Economics Letters*, 44:181–190.
955. Hollifield, B., Miller, R. and Sandas, P. (2004) Empirical analysis of limit order markets. *Review of Economic Studies*, 71:1027–1063.
956. Holmstrom, B. and Myerson, R.B. (1983) Efficient and durable decision rules with incomplete information. *Econometrica*, 51:1799–1819.
957. Hong, H., Kubik, J. and Stein, J. (2004) Social interaction and stock-market participation. *Journal of Finance*, 59:137–163.
958. Hong, H., Lim, T. and Stein, J. (2000) Bad news travels slowly: size, analyst coverage and the profitability of momentum strategies. *Journal of Finance*, 55:-265–295.
959. Hong, H. and Rady, S. (2002) Strategic trading and learning about liquidity. *Journal of Financial Markets*, 5:419–450.
960. Hong, H., Scheinkman, J. and Xiong, W. (2006) Asset float and speculative bubbles. *Journal of Finance*, 61:1073–1117.
961. Hong, H. and Stein, J. (1999) A unified theory of underreaction, momentum trading and overreaction in asset markets. *Journal of Finance*, 44:2143–2184.
962. Hong, H. and Stein, J. (2003) Differences of opinion, rational arbitrage and market crashes. *Review of Financial Studies*, 16:487–525.
963. Hong, H. and Stein, J. (2007) Disagreement and the stock market. *Journal of Economic Perspectives*, 21:109–128.
964. Hong, H. and Stein, J. (2007) Simple forecasts and paradigm shifts. *Journal of Finance*, 62:1207–1242.
965. Hong, H. and Wang, J. (2000) Tradinng and returns under periodic market closures. *Journal of Finance*, 55:297–320.
966. Horneff, W., Maurer, R. and Stamos, M. (2008) Life-cycle asset allocation with annuity markets. *Journal of Economic Dynamics and Control* 32:3590–3612.
967. Hou, K. (2007) Industry information diffusion and the lead-lag effect in stock returns. *Review of Financial Studies*, 20:1113–1138.
968. Hou, K. and Moskowitz, T. (2005) Market frictions, price delay, and the cross-section of expected returns. *Review of Financial Studies*, 18:981–1020.
969. Hou, K. and Robinson, D. (2006) Industry concentration and average stock returns. *The Journal fo Finance*, 61:1927–1955.
970. Hsieh, D. and Miller, M. (1990) Margin regulations and stock market volatility. *Journal of Finance*, 45:3–30.
971. Huang, C. and Litzenberger, R. (1988) *Foundations for Financial Economics*. North-Holland, Amsterdam.
972. Huang, K. and Werner, J. (2000) Asset price bubbles in Arrow-Debreu and sequential equilibrium. *Economic Theory*, 15:253–278.

973. Huang, K. and Werner, J. (2004) Implementing Arrow-Debreu equilibria by trading infinitely-lived securities. *Economic Theory*, 19:35–64.

974. Huang, K., Liu, Z. and Zhu, J. (2015) Temptation and self-control: some evidence and applications. *Journal of Money Credit and Banking*, 47:581–615.

975. Huang, M. (2003) Liquidity shocks and equilibrium liquidity premia. *Journal of Economic Theory*, 109:104–129.

976. Huang, R. and Stoll, H. (1994) Market microstructure and stock return predictions. *Review of Financial Studies*, 7:179–213.

977. Huang, R. and Stoll, H. (1996) Dealer versus auction markets: a paired comparison of execution costs on NASDAQ and the NYSE. *Journal of Financial Economics*, 41:313–357.

978. Huang, R. and Stoll, H. (1997) The components of the bid-ask spread: a general approach. *Review of Financial Studies*, 10:995–1034.

979. Huberman, G. (1982) A simple approach to arbitrage pricing theory. *Journal of Economic Theory*, 28:183–91.

980. Huberman, G. (1992) A simple approach to arbitrage pricing theory. In *The New Palgrave Dictionary of Money and Finance*, J. Eatwell, M. Milgate and P. Newman (eds.), Palgrave Macmillan, London.

981. Huberman, G. (2001) Familiarity breeds investment. *Review of Financial Studies*, 14:659–680.

982. Huberman, G. and Kandel, S. (1987) Mean-variance spanning. *Journal of Finance*, 52:873–888.

983. Huberman, G. and Kandel, S. (1990) Market efficiency and Value Line's record. *Journal of Business*, 63:187–215.

984. Huberman, G., Kandel, S. and Stambaugh. R. (1987) Mimicking portfolios and exact arbitrage pricing. *Journal of Finance*, 52:1–9.

985. Huberman, G. and Ross, S. (1983) Portfolio turnpike theorems, risk aversion, and regularly varying utility functions. *Econometrica*, 51:1345–1361.

986. Huberman, G. and Schwert, W. (1985) Information aggregation, inflation, and the pricing of indexed bonds. *Journal of Political Economy*, 93:92–114.

987. Huddart, S., Hughes, J. and Brunnermeier, M. (1999) Disclosure requirements and stock exchange listing choice in an intertemporal context. *Journal of Accounting Economics*, 26:237–269.

988. Huddart, S., Hughes, J. and Levine, C. (1999) Public disclosure of trades by corporate insiders in financial markets and tacit coordination. Mimeo.

989. Huddart, S., Hughes, J. and Levine, C. (2001) Public disclosure and dissimulation of insider trades. *Econometrica*, 69:665–682.

990. Huddart, S. and Ke, B. (2007) Information asymmetry and cross-sectional determinants of insider trading. *Contemporary Accounting Research*, 24:195–232.

991. Huddart, S., Ke, B. and Shi, C. (2007) Jeopardy, non-public information, and insider trading around SEC 10-K and 10-Q filings. *Journal of Accounting and Economics*, 43:3–36.

992. Huggett, M. (1993) The risk-free rate in heterogeneous-agent incomplete-insurance economies. *Journal of Economic Dynamics and Control*, 17:953–969.

993. Huggett, M. (2001) Aggregate precautionary savings: when is the third derivative irrelevant? *Journal of Monetary Economics*, 48:373–396.

994. Hussman, J. (1992) Market efficiency and inefficiency in rational expectations equilibria. *Journal of Economic Dynamics and Control*, 16:655–680.

995. Hvidkjaer, S. (2006) A trade-based analysis of momentum. *Review of Financial Studies*, 19:458–491.

996. Ibbotson, R. and Ritter, J. (1995) Initial public offerings. In *Handbooks in Operations Research and Management Science*, Vol. 9 (Finance), R.A. Jarrow, V. Maksimovic and W.T. Ziemba (eds.), North Holland, Amsterdam.

997. Ikeda, S. and Shibata, A. (1992) Fundamentals-dependent bubbles in stock prices. *Journal of Monetary Economics*, 30:143–168.

998. Ikenberry, D., Lakonishok, J. and Vermaelen, T. (1995) Market underreaction to open market share repurchases. *Journal of Financial Economics*, 39:181–208.
999. Ingersoll, J. (1984) Some results in the theory of arbitrage pricing. *Journal of Finance*, 39:1021–1039.
1000. Ingersoll, J. (1987) *Theory of Financial Decision Making*. Rowman & Littlefield, New Jersey.
1001. Ingrao, B. and Israel, G. (1987) *La Mano Invisibile*. Laterza, Roma.
1002. Jackson, M. (1991) Equilibrium, price formation, and the value of private information. *Review of Financial Studies*, 4:1–16.
1003. Jackson, M. (1994) A proof of the existence of speculative equilibria. *Journal of Economic Theory*, 64:221–233.
1004. Jackson, M. and Peck, J. (1991) Speculation and price fluctuations with private, extrinsic signals. *Journal of Economic Theory*, 55:274–295.
1005. Jacobs, F. (1999) Incomplete markets and security prices: do asset-pricing puzzles result from aggregation problems? *Journal of Finance*, 46:123–163.
1006. Jacobs, K. and Wang, K. (2004) Idiosyncratic consumption risk and the cross section of asset returns. *Journal of Finance*, 59:2211–2252.
1007. Jaffe, J. (1974) Special information and insider trading. *Journal of Business*, 47:410–428.
1008. Jaffe, J. and Westerfield, R. (1999) The week-end effect in common stock returns: the international evidence. *Journal of Finance*, 40:433–454.
1009. Jagannathan, R. and Kocherlakota, N. (1996) Why should older people invest less in stocks than younger people? *Federal Reserve Bank of Minneapolis Quarterly Review*, 20:11–23.
1010. Jagannathan, R., Kubita, K. and Takehara, H. (1998) Relationship between labor-income risk and average returns: empirical evidence from the Japanese stock market. *Journal of Business*, 71:319–347.
1011. Jagannathan, R., McGrattan, E. and Scherbina, A. (2001) The declining U.S. equity premium. *NBER working paper*, no.8172.
1012. Jagannathan, R. and Wang, Z. (1996) The conditional CAPM and the cross-section of expected returns. *Journal of Finance*, 51:3–55.
1013. Jagannathan, R. and Wang, Z. (2007) Lazy investors, discretionary consumption, and the cross-section of stock returns. *Journal of Finance*, 62:1623–1661.
1014. Jang, B., Keun Koo, H., Liu, H. and Loewenstein, M. (2007) Liquidity premia and transaction costs. *Journal of Finance*, 62:2329–2366.
1015. Jappelli, T. (1990) Who is credit constrained in the US economy? *Quarterly Journal of Economics*, 105:219–234.
1016. Jappelli, T. and Pagano, M. (1994) Saving, growth, and liquidity constraints. *Quarterly Journal of Economics*, 109:83–109.
1017. Jappelli, T., Pischke, J. and Souleles, N. (1998) Testing for liquidity constraints in Euler equations with complementary data sources. *The Review of Economics and Statistics*, 80:251–262.
1018. Jarrow, R. (1986) The relationship between arbitrage and first order stochastic dominance. *Journal of Finance*, 41:159–172.
1019. Jarrow, R. (1988) Preferences, continuity, and the arbitrage pricing theory. *Review of Financial Studies*, 1:159–172.
1020. Jarrow, R. (1992) Market manipulation, bubbles, corners, and short squeezes. *Journal of Financial and Quantitative Analysis*, 27:311–336.
1021. Jarrow, R. (1994) Derivative security markets, market manipulation, and option pricing theory. *Journal of Financial and Quantitative Analysis*, 29:241–261.
1022. Jarrow, R. and Zhao, F. (2006) Downside loss aversion and portfolio management. *Management Science*, 52:558–566.
1023. Jegadeesh, N. (1990) Evidence of predictable behavior of security returns. *Journal of Finance*, 45:881–898.
1024. Jegadeesh, N. (1991) Seasonality in stock price mean reversion: evidence from the US and UK. *Journal of Finance*, 46:1427–1444.

1025. Jegadeesh, N. (1992) Does market risk really explain the size effect. *Journal of Financial and Quantitative Analysis*, 27:337–352.
1026. Jegadeesh, N. and Titman, S. (1993) Returns to buying winners and selling losers: implications for stock market efficiency. *Journal of Finance*, 48:65–91.
1027. Jegadeesh, N. and Titman, S. (1995) Overreaction, delayed reaction, and contrarian profits. *Review of Financial Studies*, 8:973–993.
1028. Jegadeesh, N. and Titman S. (2001) Profitability of momentum strategies: an evaluation of alternative explanations. *Journal of Finance*, 56:699–720.
1029. Jegadeesh, N. and Titman S. (2002) Cross-sectional and time-series determinants of momentum returns. *Review of Financial Studies*, 15:143–158.
1030. Jeng, L., Metrick, A. and Zeckauser, R. (2003) The profits to insider trading: a performance-evaluation perspective. *Review of Economics and Statistics*, 85:453–471.
1031. Jensen, M. (1978) Some anomalous evidence regarding market efficiency. *Journal of Financial Economics*, 6:95–101.
1032. Jermann, U. (1998) Asset pricing in production economies. *Journal of Monetary Economics*, 41:257–275.
1033. Jobson, J. and Korkie, B. (1982) Potential performance and tests of portfolio efficiency. *Journal of Financial Economics*, 10:433–466.
1034. John, K. and Narayanan, R. (1997) Market manipulation and the role of insider trading regulations. *Journal of Business*, 70:217–247.
1035. John, K. and Lang, L. (1991) Insider trading around dividend announcements: theory and evidence. *Journal of Finance*, 46:1361–1389.
1036. Johnson, T. (2004) Forecast dispersion and the cross section of expected returns. *Journal of Finance*, 59:1957–1978.
1037. Johnson, T. (2008) Volume, liquidity, and liquidity risk. *Journal of Financial Economics*, 87:388–417.
1038. Jones, C. (2001) Extracting factors from heteroskedastic asset returns. *Journal of Financial Economics*, 62:293–325.
1039. Jones, C. (2001) A century of stock market liquidity and trading costs. Mimeo.
1040. Jones, C., Kaul, G. and Lipson, M. (1994) Information, trading, and volatility. *Journal of Financial Economics*, 36:127–154.
1041. Jones, C., Kaul, G. and Lipson, M. (1994) Transactions, volume and volatility. *Review of Financial Studies*, 7:631–651.
1042. Jones, C. and Lamont, O. (2002) Short sale constraints and stock returns. *Journal of Financial Economics*, 66:207–239.
1043. Jones, C. and Lipson, M. (2001) Sixteenths: direct evidence on institutional execution costs. *Journal of Financial Economics*, 59:253–278.
1044. Jones, C. and Seguin, P. (1997) Transaction costs and price volatility: evidence from commission deregulation. *The American Economic Review*, 87:728–737.
1045. Jordan, J. (1982) The generic existence of rational expectations equilibrium in the higher dimensional case. *Journal of Economic Theory*, 26:2224–2243.
1046. Jordan, J. (1983) On the efficient markets hypothesis. *Econometrica*, 51:1325–1343.
1047. Jordan, J. and Radner, R. (1982) Rational expectations in microeconomic models: an overview. *Journal of Economic Theory*, 26:201–223.
1048. Jorion, P. and Giovannini, A. (1993) Time-series tests of a non-expected-utility model of asset pricing. *European Economic Review*, 37:1083–1100.
1049. Jouini, E. and Kallal, H. (1995) Martingales and arbitrage in securities markets with transaction costs. *Journal of Economic Theory*, 66:178–197.
1050. Jouini, E. and Kallal, H. (1995) Arbitrage in securities markets with short sales constraints. *Mathematical Finance*, 5:197–232.
1051. Jouini, E. and Napp, C. (2006) Heterogeneous beliefs and asset pricing in discrete time. *Journal of Economic Dynamics and Control* 30:1233–1266.
1052. Jouini, E. and Napp, C. (2007) Consensus consumer and intertemporal asset pricing with heterogeneous beliefs. *Review of Economic Studies*, 74:1149–1174.

1053. Ju, N. and Miao, J. (2012) Ambiguity, learning, and asset returns. *Econometrica*, 80:559–591.

1054. Julliard, C. and Ghosh, A. (2012) Can rare events explain the equity premium puzzle? *Review of Financial Studies*, 25:3037–3076.

1055. Jurek, J. and Viceira, L. (2011) Optimal value and growth tilts in long-horizon portfolios *Review of Finance*, 15:29–74.

1056. Kabanov, Y. and Safarian, M. (2009) *Markets with Transaction Costs - Mathematical Theory*. Springer, Berlin/Heidelberg.

1057. Kahle, K. (2000) Insider trading and the long run performance of new security issues. *Journal of Corporate Finance*, 6:25–53.

1058. Kahn, C. and Winton, A. (1998) Ownership structure, speculation, and shareholder intervention. *Journal of Finance*, 53:99–129.

1059. Kahn, J. (1990) Moral hazard, imperfect risk-sharing, and the behavior of asset returns. *Journal of Monetary Economics*, 26:27–44.

1060. Kahneman, D. and Tversky, A. (1979) Prospect theory: an analysis of decision under risk. *Econometrica*, 47:263–291.

1061. Kahneman, D. and Tversky, A. (1982) Intuitive prediction: biases and corrective procedures. In *Judgement under Uncertainty: Heuristics and Biases*, D. Kahneman, P. Slovic and A. Tversky (eds.), Cambridge University Press.

1062. Kamihigashi, T. (1998) Uniqueness of asset prices in an exchange economy with unbounded utility. *Economic Theory*, 12:103–122.

1063. Kan, R. and Zhang, C. (1997) Two pass tests of asset pricing models with useless factors. *Journal of Finance*, 54:204–235.

1064. Kandel, E. and Marx, L. (1997) Nasdaq market structure and spread patterns. *Journal of Financial Economics*, 45:61–89.

1065. Kandel, E. and Marx, L. (1999) Payments for order flow on Nasdaq. *Journal of Finance*, 54:35–66.

1066. Kandel, S. and Pearson, N. (1995) Differential interpretation of public signals and trade in speculative markets. *Journal of Political Economy*, 103:831–872.

1067. Kandel, S. and Stambaugh, R. (1987) On the correlations and inferences about mean variance efficiency. *Journal of Financial Economics*, 18:61–90.

1068. Kandel, S. and Stambaugh, R. (1990) Expectations and volatility of consumption and asset returns. *Review of Financial Studies*, 3:207–232.

1069. Kandel, S. and Stambaugh, R. (1991) Asset returns and intertemporal preferences. *Journal of Monetary Economics*, 27:39–71.

1070. Kandel, S. and Stambaugh, R. (1995) Portfolio inefficiency and the cross-section of expected returns. *Journal of Finance*, 50:157–184.

1071. Kandel, S. and Stambaugh, R. (1996) On the predictability of stock returns: an asset allocation perspective. *Journal of Finance*, 51:385–424.

1072. Kang, J. and Stulz, R. (1997) Why is there a home bias? An analysis of foreign portfolio equity ownership in Japan. *Journal of Financial Economics*, 46:3–28.

1073. Karatzas, I. and Shreve, S. (1998) *Methods of Mathematical Finance*. Springer, New York.

1074. Karolyi, A. (2006) The world of cross-listings and cross-listings of the world: challenging conventional wisdom. *Review of Finance*, 10:99–152.

1075. Karpoff, J. (1987) The relation between price changes and trading volume: a survey. *Journal of Financial and Quantitative Analysis*, 22:109–126.

1076. Kaul, G. (1996) Predictable components in stock returns. In *Handbook of Statistics* (Vol.14), G.S. Maddala and C.R. Rao (eds.), Elsevier Science Publisher.

1077. Ke, B., Huddart, S. and Petroni, K. (2003) What insiders know about future earnings and how they use it: evidence from insider traders. *Journal of Accounting and Economics*, 35:315–346.

1078. Kehoe, T. and Levine, D. (1993) Debt constrained asset markets. *Review of Economic Studies*, 60:865–888.

1079. Keim, D. (1983) Size-related anomalies and stock return seasonality: further empirical evidence. *Journal of Financial Economics*, 12:13–32.
1080. Keim, D. (1989) Trading patterns, bid-ask spreads, and estimated security returns: the case of common stock at calendar turning points. *Journal of Financial Economics* 25:75–97.
1081. Keim, D. and Madhavan, A. (1996) The upstairs market for large-block transactions: analysis and measurement of price effects. *Review of Financial Studies*, 9:1–36.
1082. Keim, D. and Stambaugh, R. (1984) A further invetsigation of the weekend effect in stock returns. *Journal of Finance*, 39:819–835.
1083. Keim, D. and Stambaugh, R. (1986) Predicting returns in the stock and bond markets. *Journal of Financial Economics*, 17:357–390.
1084. Kelly, B. and Ljungqvist, A. (2012) Testing asymmetric-information asset pricing models. *Review of Financial Studies*, 25:1366–1413.
1085. Kelly, M. (1997) Do noise traders influence stock prices? *Journal of Money Credit and Banking*, 29:251–363.
1086. Kelsey, D. and Milne, F. (1995) The arbitrage pricing theorem with non-expected utility preferences. *Journal of Economic Theory*, 65:557–574.
1087. Keynes, J.M. (1936) *The General Theory of Unemployment, Interest and Money*. Harcourt, Brace & World.
1088. Khanna, N., Slezak, S. and Bradley, M. (1994) Insider trading, outside search, and resource allocation: why firms and society may disagree on insider trading restrictions. *Review of Financial Studies*, 7:575–608.
1089. Kihlstrom, R. and Mirman, L. (1974) Risk aversion with many commodities. *Journal of Economic Theory*, 8:361–368.
1090. Kihlstrom, R. and Mirman, L. (1981) Constant, increasing, decreasing risk aversion with many commodities. *Review of Economic Studies*, 48:271–280.
1091. Kim, D. (1995) The errors in the variables problem in the cross-section of expected stock returns. *Journal of Finance*, 50:1605–1635.
1092. Kim, D. (1997) A Reexamination of firm size, book-to-market, and earnings price in the cross-section of expected returns. *Journal of Financial and Quantitative Analysis*, 32:463–491.
1093. Kim, J., Nelson, C. and Startz, R. (1991) Mean reversion in stock prices? A reappraisal of the empirical evidence. *Review of Economic Studies*, 58:515–528.
1094. Kim, K. (2001) Price limits and stock market volatility. *Economics Letters*, 71:131–136.
1095. Kim, O. and Verrecchia, R. (1991) Market reaction to anticipated announcements. *Journal of Financial Economics*, 30:273–309.
1096. Kim, O. and Verrecchia, R. (1991) Trading volume and price reactions to public announcements. *Journal of Accounting Research*, 29:302–321.
1097. Kim, O. and Verrecchia, R. (1994) Market liquidity and volume around earnings announcements. *Journal of Accounting and Economics*, 17:41–67.
1098. Kim, S., Lin, J. and Slovin, M. (1997) Market structure, informed trading, and analysts' recommendations. *Journal of Financial and Quantitative Analysis*, 32:507–532.
1099. Kim, T. and Omberg, E. (1996) Dynamic nonmyopic portfolio behavior. *Review of Financial Studies*, 9:141–161.
1100. Kimball, M. (1990) Precautionary saving in the small and in the large. *Econometrica*, 58:53–73.
1101. Kimball, M. (1993) Standard risk aversion. *Econometrica*, 61:589–611.
1102. Kirby, C. (1997) Measuring the predictable variation in stock and bond returns. *Review of Financial Studies*, 10:579–630.
1103. Kirby, C. (1998) The restrictions on predictability implied by rational asset pricing models. *Review of Financial Studies*, 11:343–382.
1104. Kleidon, A. (1986) Variance bounds tests and stock price valuation models. *Journal of Political Economy*, 94:953–1001.
1105. Klibanoff, P., Marinacci, M. and Mukerji, S. (2005) A smooth model of decision making under ambiguity. *Econometrica*, 73:1849–1892.

1106. Knez, P. and Ready, M. (1996) Estimating the profits from trading strategies. *Review of Financial Studies*, 9:1121–1163.
1107. Knez, P. and Ready, M. (1997) On the robustness of size and book-to-market in cross-sectional regressions. *Journal of Finance*, 52:1355–1382.
1108. Knight, F. (1921) *Risk, Uncertainty, and Profit*. Houghton Mifflin, Boston.
1109. Kocherlakota, N. (1990) On tests of representative consumer asset pricing models. *Journal of Monetary Economics*, 26:285–304.
1110. Kocherlakota, N. (1990) Disentangling the coefficient of relative risk aversion from the elasticity of intertemporal substitution: an irrelevance result. *Journal of Finance*, 45:175–190.
1111. Kocherlakota, N. (1992) Bubbles and constraints on debt accumulation. *Journal of Economic Theory*, 57:245–256.
1112. Kocherlakota, N. (1996) The equity premium: it's still a puzzle. *Journal of Economic Literature*, 34:42–71.
1113. Kocherlakota, N. (1998) The effects of moral hazard on asset prices when financial markets are complete. *Journal of Monetary Economics*, 41:39–56.
1114. Kocherlakota, N. (2008) Injecting rational bubbles. *Journal of Economic Theory*, 142:218–232.
1115. Kogan, L., Makarov, I. and Uppal, R. (2007) The equity risk premium and the riskfree rate in an economy with borrowing constraints. *Mathematics and Financial Economics*, 1:1–19.
1116. Kogan, L., Ross, S., Wang, J. and Westerfield, M. (2006) The price impact and survival of irrational traders. *Journal of Finance* 61:195–229.
1117. Kogan, L. and Wang, T. (2003) A simple theory of aset pricing under model uncertainty. *MIT working paper*.
1118. Kondor, P. (2012) The more we know about the fundamental, the less we agree on the price. *Review of Economic Studies*, 79:1175–1207.
1119. Koo, H. (1999) Consumption and portfolio selection with labor income: a discrete time approach. *Mathematical Methods of Operations Research*, 50:219–243.
1120. Korajczyk, R. A. and Sadka, R. (2004) Are momentum profits robust to trading costs? *Journal of Finance*, 59:1039–1082.
1121. Koszegi, B. and Rabin, M. (2006) A model of reference-dependent preferences. *Quarterly Journal of Economics*, 121:1133–1165.
1122. Koski, J. and Michaely, R. (2000) Prices, liquidity, and the information content of trades. *Review of Financial Studies*, 13:659–696.
1123. Kothari, S. (2001) Capital markets research in accounting. *Journal of Accounting and Economics*, 31:105–231.
1124. Kothari, S. and Shanken, J. (1992) Stock return variation and expected dividends. *Journal of Financial Economics*, 31:177–210.
1125. Kothari, S. and Shanken, J. (1997) Book-to-market, dividend yield, and expected market returns: a time series analysis. *Journal of Financial Economics*, 44:169–203.
1126. Kothari, S., Shanken, J. and Sloan, R. (1995) Another look at the cross-section of expected stock returns. *Journal of Finance*, 50:185–224.
1127. Kothari, S. and Warner, J. (1997) Measuring long-horizon security price performance. *Journal of Financial Economics*, 43:301–339.
1128. Kramkov, D. and Schachermayer, W. (1999) The asymptotic elasticity of utility functions and optimal investment in incomplete markets. *Annals of Applied Probability*, 9:904–950.
1129. Kramkov, D. and Schachermayer, W. (2003) Necessary and sufficient conditions in the problem of optimal investment in incomplete markets. *Annals of Applied Probability*, 13:1504–1516.
1130. Krasa, S. (1987) Existence of competitive equilibrium for option markets. *Journal of Economic Theory*, 47:413–431.
1131. Krasa, S. and Werner, J. (1991) Equilibrium with options: existence and indeterminacy. *Journal of Economic Theory*, 54:305–320.

1132. Krebs, T. (2004) Testable implications of consumption-based asset pricing models with incomplete markets. *Journal of Mathematical Economics*, 40:191–206.
1133. Krebs, T. (2007) Rational expectations equilibrium and the strategic choice of costly information. *Journal of Mathematical Economics*, 43:532–548.
1134. Kreps, D. (1977) A note on fulfilled expectations equilibria. *Journal of Economic Theory*, 14:32–44.
1135. Kreps, D. (1982) Multiperiod securities and the efficient allocation of risk: a comment on the Black and Scholes option pricing model. In *The Economics of Uncertainty and Information*, J. McCall (ed.), University of Chicago Press.
1136. Kreps, D. (1988) *Notes on the Theory of Choice*. Westview Press, Boulder.
1137. Kroll, Y., Levy, H. and Rapoport, A. (1988) Experimental tests of the separation theorem and the capital asset pricing model. *The American Economic Review*, 78:500–519.
1138. Krueger, D. and Lustig, H. (2010) When is market incompleteness irrelevant for the price of aggregate risk (and when is it not)? *Journal of Economic Theory*, 145:1–41.
1139. Krusell, P., Kuruscu, B. and Smith, A. (2002) Time orientation and asset prices. *Journal of Monetary Economics*, 49:107–135.
1140. Krusell, P. and Smith, A. (1997) Income and wealth heterogeneity, portfolio choice, and equilibrium asset returns. *Macroeconomic Dynamics*, 1:387–422.
1141. Kuhn, (1970) *The Structure of Scientific Revolutions*. University of Chicago Press.
1142. Kumar, P. and Seppi, D. (1992) Futures manipulation with "cash settlement". *Journal of Finance*, 47:1485–1502.
1143. Kupiec, P. (1997) Margin requirements, volatility, and market integrity: what have we learned since the crash? Mimeo.
1144. Kupiec, P. and Sharpe, S. (1991) Animal spirits, margin requirements, and stock price volatility. *Journal of Finance*, 46:717–731.
1145. Kurz, M. and Motolese, M. (2001) Endogenous uncertainty and market volatility. *Economic Theory*, 17:497–544.
1146. Kyle, A. (1984) A theory of futures market manipulation. In *The Industrial Organization of Futures Markets*, R. Anderson (ed.), Lexington Books, 141–175.
1147. Kyle, A. (1985) Continuous auctions and insider trading. *Econometrica*, 53:1315–1334.
1148. Kyle, A. (1989) Informed speculation with imperfect competition. *Review of Economic Studies*, 56:317–356.
1149. Kyle, A. and Viswanathan, S. (2008) How to define illegal price manipulation. *The American Economic Review*, 98:274–79.
1150. Kyle, A. and Wang, A. (1997) Speculation duopoly with agreement to disagree: can overconfidence survive the market test? *Journal of Finance*, 52:2073–2090.
1151. Kyle, A. and Xiong, W. (2001) Contagion as a wealth effect. *Journal of Finance*, 56:1401–1440.
1152. Labadie, P. (1989) Stochastic inflation and the equity premium. *Journal of Monetary Economics*, 24:277–298.
1153. Laffont, J. (1985) On the welfare analysis of rational expectations equilibria with asymmetric information. *Econometrica*, 53:1–30.
1154. Laffont, J. (1989) *The Economics of Uncertainty and Information*. MIT Press, Cambridge.
1155. Laffont, J. and Maskin, E. (1990) The efficient market hypothesis and insider trading on the stock market. *Journal of Political Economy*, 98:70–89.
1156. Laibson, D. (1997) Golden eggs and hyperbolic discounting. *The Quarterly Journal of Economics*, 112:443–477.
1157. Laibson, D., Repetto, A. and Tobacman, J. (1998) Self-control and saving for retirement. *Brookings Papers on Economic Activity*, 1:91–196.
1158. Lakonishok, J. and Lee, D. (2001) Are insiders' trades informative? *Review of Financial Studies*, 14:79–111.
1159. Lakonishok, J., Shleifer, A. and Vishny, R. (1994) Contrarian investment, extrapolation and risk. *Journal of Finance*, 49:1541–1578.

1160. Lakonishok, J. and Smidt, A. (1988) Are seasonal anomalies real? A 90 years perspective. *Review of Financial Studies*, 1:403–425.

1161. Lambert, R., Leuz, C. and Verrecchia, R. (2007) Accounting information, disclosure, and the cost of capital. *Journal of Accounting Research*, 45:385–420.

1162. Lamont, O. (1998) Earnings and expected returns. *Journal of Finance*, 53:1563–1587.

1163. Lamont, O., Polk, C. and Saa-Requejo, J. (2001) Financial constraints and stock returns. *Review of Financial Studies*, 14:529–554.

1164. Lamoureux, C. and Schnitzlen, C. (1997) When it's not the only game in town: the effect of bilateral search on the quality of a dealer market. *Journal of Finance*, 52:683–712.

1165. Lamoureux, C. and Zhou, G. (1996) Temporary components of stock returns: what do the data tell us? *Review of Financial Studies*, 9:1033–1059.

1166. Lang, L., Litzenberger, R. and Madrigal, V. (1992) Testing financial market equilibrium under asymmetric information. *Journal of Political Economy*, 100:317–348.

1167. La Porta, R. (1996) Expectations and the cross-section of stock returns. *Journal of Finance*, 51:1715–1742.

1168. La Porta, R., Lakonishok, J., Shleifer, A. and Vishny, R. (1997) Good news for value stocks: further evidence on market efficiency. *Journal of Finance*, 52:859–874.

1169. Lauterbach, B. and Ben-Zion, U. (1993) Stock market crashes and the performance of circuit breakers: empirical evidence. *Journal of Finance*, 48:1909–1920.

1170. Leach, C. and Madhavan, A. (1992) Intertemporal price discovery by market makers: active versus passive learning. *Journal of Financial Intermediation*, 2:207–235.

1171. Leach, J. (1991) Rational speculation. *Journal of Political Economy*, 99:131–144.

1172. LeBaron, B., Arthur, B. and Palmer, R. (1999) Time series properties of an artificial stock market. *Journal of Economic Dynamics and Control*, 23:1487–1516.

1173. Lee, C., Ready, M. and Seguin, P. (1994) Volume, volatility, and New York Stock Exchange trading halts. *Journal of Finance*, 49:183–213.

1174. Lee, C., Shleifer, A. and Thaler, R. (1991) Investor sentiment and closed-end fund puzzle. *Journal of Finance*, 46:75–109.

1175. Lee, C. and Swaminathan, B. (2000) Price momentum and trading volume. *Journal of Finance*, 55:2017–2069.

1176. Lee, W., Jiang, C. and Indro, D. (2002) Stock market volatility, excess returns, and the role of investor sentiment. *Journal of Banking and Finance*, 26:2277–2299.

1177. Lehman, B. (1990) Fads, martingales, and market efficiency. *Quarterly Journal of Economics*, 105:1–28.

1178. Lehmann, B. and Modest, D. (1988) The empirical foundations of the arbitrage pricing theory. *Journal of Financial Economics*, 21:213–254.

1179. Leippold, M., Trojani, F. and Vanini, P. (2008) Learning and asset prices under ambiguous information. *Review of Financial Studies*, 21:2565–2597.

1180. Leland, H. (1972) On the existence of optimal policies under uncertainty. *Journal of Economic Theory*, 4:35–44.

1181. Leland, H. (1992) Insider trading: should it be prohibited? *Journal of Political Economy*, 100:859–887.

1182. Lengwiler, Y. (2004) *Microfoundations of Financial Economics*. Princeton University Press.

1183. LeRoy, S. (1973) Risk aversion and the martingale property of stock prices. *International Economic Review*, 14:436–446.

1184. LeRoy, S. (1989) Efficient capital markets and martingales. *Journal of Economic Literature*, 27:1583–1621.

1185. LeRoy, S. (2004) Rational exuberance. *Journal of Economic Literature*, 42: 783–804.

1186. LeRoy, S. and Gilles, C. (1992) Asset price bubbles. In *The New Palgrave Dictionary of Money and Finance*, J. Eatwell, M. Milgate and P. Newman (eds.), Palgrave Macmillan, London.

1187. LeRoy, S. and LaCivita, C. (1981) Risk aversion and the dispersion of asset prices. *Journal of Business*, 54:535–547.

1188. LeRoy, S. and Parke, W. (1992) Stock price volatility; tests based on the geometric random walk. *The American Economic Review*, 82:981–992.
1189. LeRoy, S. and Porter, R. (1981) The present-value relation: tests based on implied variance bounds. *Econometrica* 49:555–574.
1190. LeRoy, S. and Steigerwald, D. (1995) Volatility. In *Handbooks in Operations Research and Management Science*, Vol. 9 (Finance), R.A. Jarrow, V. Maksimovic and W.T. Ziemba (eds.), North Holland, Amsterdam.
1191. LeRoy, S. and Werner, J. (2001) *Principles of Financial Economics*, Cambridge University Press.
1192. Letendre, M. and Smith, G. (2001) Precautionary saving and portfolio allocation: DP and GMM. *Journal of Monetary Economics*, 48:197–215.
1193. Lettau, M. (2002) Idiosyncratic risk and volatility bounds, or, can models with idiosyncratic risk solve the equity premium puzzle? *Review of Economics and Statistics*, 84:376–380.
1194. Lettau, M. and Ludvigson, S. (2001) Consumption, aggregate wealth, and expected stock returns. *Journal of Finance*, 56:815–850.
1195. Lettau, M. and Ludvigson, S. (2001) Resurrecting the (C)CAPM: a cross-sectional test when risk premia are time varying. *Journal of Political Economy*, 109:1238–1287.
1196. Lettau, M. and Ludvigson, S. (2005) Expected returns and expected dividend growth. *Journal of Financial Economics*, 76:583–626.
1197. Lettau, M., Ludvigson, S. and Wachter J. (2008) The declining equity premium: what role does macroeconomic risk play? *Review of Financial Studies*, 21:1653–1687.
1198. Lettau, M. and Uhlig, H. (2000) Can habit formation be reconciled with business cycle facts? *Review of Economic Dynamics*, 3:79–99.
1199. Lettau, M. and Uhlig, H. (2000) Sharpe ratios and preferences an analytical approach. *Macroeconomic Dynamics*, 6:242–265.
1200. Lettau, M. and Van Nieuwerburgh, S. (2008) Reconciling the return predictability evidence. *Review of Financial Studies*, 21:1607–1652.
1201. Lettau, M. and Wachter, J. (2007) Why is long-horizon equity les risky? A duration based explanation of the value premium. *Journal of Finance*, 62:55–92.
1202. Leuz, C. and Verrecchia, R. (2000) The economic consequences of increased disclosure. *Journal of Accounting Research*, 38:91–124.
1203. Levine, D. and Zame, W. (2002) Does market incompleteness matter? *Econometrica*, 70:1805–1839.
1204. Levy, H. (1992) Stochastic dominance and expected utility: survey and analysis. *Management Science*, 38:555–593.
1205. Levy, H. (2016) *Stochastic Dominance: Investment Decision Making under Uncertainty*, 3rd edition. Springer, Cham.
1206. Levy, H. and Levy, M. (2002) Prospect theory: much ado about nothing? *Management Science*, 48:1334–1349.
1207. Levy, H. and Levy, M. (2004) Prospect theory and mean-variance analysis. *Review of Financial Studies*, 17:1015–1041.
1208. Levy, H. and Wiener, Z. (1998) Stochastic dominance and prospect dominance with subjective weighting functions. *Journal of Risk and Uncertainty*, 12:147–163.
1209. Lewellen, J. (1999) The time series relations among expected return, risk, and book-to-market. *Journal of Financial Economics*, 54:5–43.
1210. Lewellen, J. (2002) Momentum and autocorrelation in stock returns. *Review of Financial Studies*, 15:533–564.
1211. Lewellen, J. (2004) Predicting returns with financial ratios. *Journal of Financial Economics*, 74:209–235.
1212. Lewellen, J. and Nagel, S. (2006) The conditional CAPM does not explain asset-pricing anomalies. *Journal of Financial Economics*, 82:289–314.
1213. Lewellen, J., Nagel, S. and Shanken, J. (2010) A skeptical appraisal of asset pricing tests. *Journal of Financial Economics* 96:175–194.

1214. Lewellen, J. and Shanken, J. (2002) Learning, asset-pricing tests, and market efficiency. *Journal of Finance*, 57:1113–1145.

1215. Lewis, K. (1996) What can explain the apparent lack of international consumption risk sharing? *Journal of Political Economy*, 104:267–297.

1216. Lewis, K. (1999) Trying yo explain home bias in equities and consumption. *Journal of Economic Literature*, 37:571–608.

1217. Li, Y. (2001) Expected returns and habit persistence. *Review of Financial Studies*, 14:861–899.

1218. Liew, J. and Vassalou, M. (2000) Can book to market, size, and momentum be risk factors that predict economic growth? *Journal of Financial Economics*, 57:221–245.

1219. Lin, J., Sanger, G. and Booth, G. (1995) Trade size and components of the bid-ask spread. *Review of Financial Studies*, 8:1153–1183.

1220. Lintner, J. (1956) Distribution of incomes of corporations among dividends, retained earnings, and taxes. *The American Economic Review*, 61:97–113.

1221. Lintner, J. (1965) Security prices, risk and maximal gains from diversification. *Journal of Finance*, 20:587–615.

1222. Litzenberger, R. and Ramaswamy, K. (1979) On distributional restrictions for two fund separation. *TIMS Studies in the Management Science*, 11:99–107.

1223. Liu, H. (2011) Dynamic portfolio choice under ambiguity and regime switching mean returns. *Journal of Economic Dynamics and Control*, 35:623–640.

1224. Liu, H. and Loewenstein, M. (2002) Optimal portfolio selection with transaction costs and finite horizons. *Review of Financial Studies*, 15:805–835.

1225. Liu, H. and Miao, J. (2015) Growth uncertainty, generalized disappointment aversion and production-based asset pricing. *Journal of Monetary Economics*, 69:70–89.

1226. Liu, J. (2006) Portfolio selection in stochastic environments. *Review of Financial Studies*, 20:1–39.

1227. Liu, J., Pan, J. and Wang, T. (2006) An equilibrium model of rare-event premia and its implication for option smirks. *Review of Financial Studies*, 18:131–164.

1228. Liu, L., Whited, T. and Zhang, L. (2009) Investment-based expected stock returns. *Journal of Political Economy*, 117:1105–1139.

1229. Liu, W. (2006) A liquidity-augmented capital asset pricing model. *Journal of Financial Economics*, 82:631–671.

1230. Liu, J. and Longstaff, F. (2004) Losing money on arbitrage: optimal dynamic portfolio choice in markets with arbitrage opportunities. *Review of Financial Studies*, 17:611–641.

1231. Ljungqvist, L. and Sargent, T. (2004) *Recursive Macroeconomic Theory*, 2nd edition. MIT Press, Cambridge.

1232. Llorente, G., Michaely, R., Saar, G. and Wang, J. (2002) Dynamic volume-return relation of individual stocks. *Review of Financial Studies*, 15:1005–1047.

1233. Lo, A. and MacKinlay, C. (1988) Stock market prices do not follow random walks: evidence from a simple specification test. *Review of Financial Studies*, 1:41–66.

1234. Lo, A. and MacKinlay, C. (1989) The size and power of the variance ratio test in finite samples. A Monte Carlo investigation. *Journal of Econometrics*, 40:203–238.

1235. Lo, A. and MacKinlay, C. (1990) Data-snooping biases in tests of financial asset pricing models. *Review of Financial Studies*, 3:431–467.

1236. Lo, A. and MacKinlay, C. (1990) An econometric analysis of nonsynchronous trading. *Journal of Econometrics*, 45:181–211.

1237. Lo, A. and MacKinlay, C. (1990) When are contrarian profits due to stock market overreaction? *Review of Financial Studies*, 3:175–206.

1238. Lo, A. Mamaysky H. and Wang, J. (2001) Foundations of technical analysis: computational algorithms, statistical inference, and empirical implementation. *Journal of Finance*, 55:1705–1756.

1239. Lo, A. Mamaysky H. and Wang, J. (2001) Asset prices and trading volume under fixed transactions costs. *NBER working paper*, no.8311.

1240. Lo, A. and Wang, J. (2000) Trading volume: definitions, data analysis and implications of portfolio theory. *Review of Financial Studies*, 13:257–300.

1241. Locke, P., Sarkar, A. and Wu, L. (1999) Market liquidity and trader welfare in multiple dealer markets: evidence from dual trading restrictions. *Journal of Financial and Quantitative Analysis*, 34:35–47.

1242. Loewenstein, M. and Willard, G. (2000) Rational equilibrium asset-pricing bubbles in continuous trading models. *Journal of Economic Theory*, 91:17–58.

1243. Loewenstein, M. and Willard, G. (2000) Local martingales, arbitrage, and viability. *Economic Theory*, 16:135–161.

1244. Longstaff, F. (2001) Optimal portfolio choice and the valuation of illiquid securities. *Review of Financial Studies*, 14:407–431.

1245. Longstaff, F. (2009) Portfolio claustrophobia: asset pricing in markets with illiquid assets. *The American Economic Review*, 99:1119–1144.

1246. Loughran, T. (1997) Book to market across firm size, exchange, and seasonality: is there an effect? *Journal of Financial and Quantitative Analysis*, 32:249–268.

1247. Loughran, T. and Ritter, J. (1996) Long-term market overreaction: the effect of low priced stocks. *Journal of Finance*, 51:1959–1970.

1248. Loewenstein, G. and Prelec, D. (1992) Anomalies in intertemporal choice: evidence and an interpretation. *The Quarterly Journal of Economics*, 107:573–597.

1249. Loewenstein, M. and Willard, G. (2006) The limits of investor behavior. *Journal of Finance*, 61:231–258.

1250. Lucas, D. (1994) Asset pricing with undiversifiable income risk and short sales constraints deepening the equity premium puzzle. *Journal of Monetary Economics*, 34:325–341.

1251. Lucas, R. (1972) Expectations and the neutrality of money. *Journal of Economic Theory*, 2:103–104.

1252. Lucas, R. (1978) Asset pricing in an exchange economy. *Econometrica*, 46:1429–1445.

1253. Ludvigson, S. and Michaelides, A. (2001) Does buffer stock saving explain the smoothness and excess sensitivity of consumption? *The American Economic Review*, 91:631–647.

1254. Luo, G. (1998) Market efficiency and natural selection in a commodity futures market. *Review of Financial Studies*, 11:647–674.

1255. Luttmer, E. (1996) Asset pricing in economies with frictions. *Econometrica*, 64:1439–1467.

1256. Luttmer, E. (1999) What level of fixed costs can reconcile consumption and stock returns? *Journal of Political Economy*, 107:969–997.

1257. Lynch, A. (1996) Decision frequency and synchronization across agents: implications for aggregate consumption and equity returns. *Journal of Finance*, 51:1479–1497.

1258. Lynch, A. (2001) Portfolio choice and equity characteristics: characterizing the hedging demands induced by returns predictability. *Journal of Financial Economics*, 62:67–130.

1259. Lynch, A. and Balduzzi, P. (2000) Predictability and transaction costs: the impact on rebalancing rules and behavior. *Journal of Finance*, 55:2285–2309.

1260. Lynch, A. and Tan, S. (2011) Labor income dynamics at business cycle frequencies: implications for portfolio choice. *Journal of Financial Economics*, 101:333–359.

1261. Lyons, R. (1996) Optimal transparency in a dealer market with an application to foreign exchange. *Journal of Financial Intermediation*, 5:225–254.

1262. Maccheroni, F., Marinacci, M. and Rustichini, A. (2006) Ambiguity aversion, robustness, and the variational representation of preferences. *Econometrica*, 74:1447–1498.

1263. Maccheroni, F., Marinacci, M. and Rustichini, A. (2006) Dynamic variational preferences. *Journal of Economic Theory*, 128:4–44.

1264. Maccheroni, F., Marinacci, M., Rustichini, A. and Taboga, M. (2009) Portfolio selection with monotone mean-variance preferences. *Mathematical Finance*, 19:487–521.

1265. Mace, B. (1991) Full insurance in the presence of aggregate uncertainty. *Journal of Political Economy*, 99:928–957.

1266. Machina, M. (1982) "Expected utility" analysis without the independence axiom. *Econometrica*, 50:277–323.

1267. Machina, M. (1987) Choice under uncertainty: problems solved and unsolved. *Journal of Economic Perspectives*, 1:121–159.
1268. Machina, M. (1989) Dynamic consistency and non-expected utility models of choice under uncertainty. *Journal of Economic Literature*, 27:1622–1668.
1269. MacKinlay, C. (1995) Multifactor models do not explain deviations from the CAPM. *Journal of Financial Economics*, 38:3–28.
1270. MacKinlay, C. (1997) Event studies in economics and finance. *Journal of Economic Literature*, 35:13–40.
1271. MacKinlay, C. and Richardson, M. (1991) Using generalized methods of moments to test mean-variance efficiency. *Journal of Finance*, 46:511–527.
1272. Madhavan, A. (1992) Trading mechanisms in securities markets. *Journal of Finance*, 47:607–641.
1273. Madhavan, A. (1995) Consolidation, fragmentation, and the disclosure of trading information. *Review of Financial Studies*, 8:579–603.
1274. Madhavan, A. (1996) Security prices and market transparency. *Journal of Financial Intermediation*, 5:255–283.
1275. Madhavan, A. (2000) Market microstructure: a survey. *Journal of Financial Markets*, 3:205–258.
1276. Madhavan, A. and Cheng, M. (1997) In search of liquidity: block trades in the upstairs and downstairs markets. *Review of Financial Studies*, 10:175–203.
1277. Madhavan, A., Porter, D. and Weaver, D. (2005) Should securities markets be transparent? *Journal of Financial Markets*, 8:265–287.
1278. Madhavan, A., Richardosn, M. and Roomans, M. (1997) Why do security prices change? A transaction-level analysis of NYSE stocks. *Review of Financial Studies*, 10:1035–1064.
1279. Madhavan, A. and Smidt, S. (1991) A Bayesian model of intraday specialist pricing. *Journal of Financial Economics*, 30:99–134.
1280. Madhavan, A. and Smidt, S. (1993) An analysis of changes in specialist inventories and quotations. *Journal of Finance*, 48:1595–1628.
1281. Madhavan, A. and Sofianos, G. (1998) An empirical analysis of NYSE specialist trading. *Journal of Financial Economics*, 48:189–210.
1282. Madrigal, V. and Scheinkman, J. (1997) Price crashes, information aggregation, and market-making. *Journal of Economic Theory*, 75:16–63.
1283. Madrigal, V. and Smith, S. (1995) On fully revealing prices when markets are incomplete. *The American Economic Review*, 85:1152–1159.
1284. Maenhout, P. (2004) Robust portfolio rules and asset pricing. *Review of Financial Studies*, 17:951–983.
1285. Magill, M. and Quinzii, M. (1996) Incomplete markets over an infinite horizon: long-lived securities and speculative bubbles. *Journal of Mathematical Economics*, 26:133–170.
1286. Magill, M. and Quinzii, M. (1998) *Theory of Incomplete Markets*. MIT Press, Cambridge.
1287. Magill, M. and Shafer, W. (1985) Characterization of generically complete real asset structures. *Journal of Mathematical Economics*, 19:167–194.
1288. Magill, M. and Shafer, W. (1991) Incomplete markets. In *Handbook of Mathematical Economics* (Vol.4), W. Hildenbrand and H. Sonnenschein (eds.), Elsevier Science Publisher.
1289. Makarov, I. and Rytchkov, O. (2012) Forecasting the forecasts of others: implications for asset pricing. *Journal of Economic Theory*, 147:941–966.
1290. Malkiel, B. (2003) The efficient market hypothesis and its critics. *Journal of Economic Perspectives*, 17:59–82.
1291. Malloy, C., Moskowitz, T. and Vissing-Jorgensen, A. (2009) Long-run stockholder consumption risk and asset returns. *Journal of Finance*, 64:2427–2479.
1292. Manaster, S. and Mann, S. (1996) Life in the pits: competitive market making and inventory control. *Review of Financial Studies*, 9:953–975.
1293. Mankiw, G. (1986) The equity premium and the concentration of aggregate shocks. *Journal of Financial Economics*, 17:211–219.

1294. Mankiw, G., Romer, D. and Shapiro, M. (1985) An unbiased reexamination of stock market volatility. *Journal of Finance*, 40:677–689.

1295. Mankiw, G., Romer, D. and Shapiro, M. (1991) Stock market forecastability and volatility: a statistical appraisal. *Review of Economic Studies*, 38:455–477.

1296. Mankiw, G., Rotemberg, J. and Summers, L. (1985) Intertemporal substitution in macroeconomics. *Quarterly Journal of Economics*, 100:225–252.

1297. Mankiw, G. and Shapiro, M. (1986) Risk and return: consumption beta versus market beta. *Review of Economics and Statistics*, 48:452–459.

1298. Mankiw, G. and Zeldes, S. (1991) The consumption of stockholders and nonstockholders. *Journal of Financial Economics* 29:97–112.

1299. Manne, H. (1966) *Insider Trading: Regulation, Enforcement and Prevention*. Free Press, New York.

1300. Manove, M. (1989) The harm from insider trading and informed speculation. *Quarterly Journal of Economics*, 104:823–846.

1301. Manzano, C. and Vives, A. (2011) Public and private learning from prices, strategic substitutability and complementarity, and equilibrium multiplicity. *Journal of Mathematical Economics*, 47:346–369.

1302. Marcet, A. and Singleton, K. (1999) Equilibrium asset prices and savings of heterogeneous agents in the presence of incomplete markets and portfolio constraints. *Macroeconomic Dynamics*, 3:243–277.

1303. Markovitz, H. (1952) Portfolio selection. *Journal of Finance*, 7:77–91.

1304. Marin, J. and Rahi, R. (1999) Speculative securities. *Economic Theory*, 14:653–668.

1305. Marin, J. and Rahi, R. (2000) Information revelation and market incompleteness. *Review of Economic Studies*, 67:563–589.

1306. Marsh, T. and Merton, R. (1986) Dividend variability and variance bounds tests for the rationality of stock market prices. *The American Economic Review*, 76:483–498.

1307. Marsh, T. and Merton, R. (1987) Dividend behavior for the aggregate stock market. *Journal of Business*, 60:1–40.

1308. Marshall, D. (1992) Inflation and asset returns in a monetary economy. *Journal of Finance*, 47:1315–1342.

1309. Marshall, D. and Parekh, N. (1999) Can costs of consumption adjustment explain asset pricing puzzles? *Journal of Finance*, 54:623–654.

1310. Mas-Colell, A., Whinston, M. and Green, J. (1995) *Microeconomic Theory*. Oxford University Press.

1311. Massa, M. and Simonov, A. (2006) Hedging, familiarity and portfolio choice. *Review of Financial Studies*, 19:633–685.

1312. Maug, E. (1998) Large shareholders as monitors: is there a trade-off between liquidity and control? *Journal of Finance*, 53:65–98.

1313. Mayers, D. (1972) Nonmarketable assets and capital market equilibrium under uncertainty. In *Studies in the Theory of Capital Markets*, M. Jensen (ed.), Praeger, New York.

1314. McKelvey R. and Page, T. (1990) Public and private information: an experimental study of information pooling. *Econometrica*, 58:1321–1339.

1315. McQueen, G., Pinegar, M. and Thorley, S. (1996) Delayed reaction to good news and the cross-autocorrelation of portfolio returns. *Journal of Finance*, 51:889–918.

1316. Mech, T. (1993) Portfolio return autocorrelation. *Journal of Financial Economics*, 34:307–344.

1317. Medrano, L.A. and Vives, X. (2004) Regulating insider trading when investment matters. *Review of Finance*, 8:199–277.

1318. Meghir, C. and Weber, G. (1996) Intertemporal nonseparability or borrowing restrictions? A disaggregate analysis using a U.S. consumption panel. *Econometrica*, 64:1151–1181.

1319. Mehra, R. and Prescott, E. (1985) The equity premium: a puzzle. *Journal of Monetary Economics*, 15:145–161.

1320. Mehra, R. and Prescott, E. (2003) The equity premium in retrospect. In *Handbook of the Economics of Finance* (Vol. 2B), G. Constantinides, M. Harris and R. Stulz, Elsevier Science Publisher, 889–938.

1321. Mehar, R. and Sah, R. (2002) Mood fluctuations, projection bias, and volatility of equity prices. *Journal of Economic Dynamics and Control*, 26:869–897.
1322. Mei, J. (1993) A semiautoregression approach to the arbitrage pricing theory. *Journal of Finance*, 48:599–619.
1323. Mendelson, H. (1987) Consolidation, fragmentation, and market performance. *Journal of Financial and Quantitative Analysis*, 22:189–207.
1324. Mendelson, H. and Tunca, T. (2004) Strategic trading, liquidity, and information acquisition. *Review of Financial Studies*, 17:295–337.
1325. Menzly, L., Santos, T. and Veronesi, P. (2004) Understanding predictability. *Journal of Political Economy*, 112:1–47.
1326. Merrick, J., Naik, N. and Yadav, P. (2005) Strategic trading behavior and price distortion in a manipulated market: anatomy of a squeeze. *Journal of Financial Economics*, 77:171–218.
1327. Merton, R. (1969) Lifetime portfolio selection: the continuous time case. *Review of Economics and Statistics*, 51:247–257.
1328. Merton, R. (1971) Optimal consumption and portfolio rules in a continuous time model. *Journal of Economic Theory*, 3:373–413.
1329. Merton, R. (1972) An analytic derivation of the efficient portfolio frontier. *Journal of Financial and Quantitative Analysis*, 7:1851–1872.
1330. Merton, R. (1973) An intertemporal capital asset pricing model. *Econometrica*, 41:867–887.
1331. Merton, R. (1974) On the pricing of corporate debt: the risk structure of interest rates. *Journal of Finance*, 29:449–470.
1332. Merton, R. (1987) A simple model of capital market equilibrium with incomplete information. *Journal of Finance*, 42:483–510.
1333. Meulbroek, L. (1992) An empirical analysis of illegal insider trading. *Journal of Finance*, 47:1661–1699.
1334. Meyer, J. (1987) Two-moment decision models and expected utility maximization. *The American Economic Review*, 77:421–430.
1335. Michaelides, A. (2001) Portfolio choice, liquidity constraints and stock market mean reversion. Mimeo.
1336. Michaely, R., Thaler, R. and Womack, K. (1995) Price reactions to dividend initiations and omissions: overreaction or drift? *Journal of Finance*, 50:573–607.
1337. Michaely, R. and Vila, J. (1996) Trading volume with private valuation: evidence from the ex-dividend day. *Review of Financial Studies*, 9:471–509.
1338. Michaely, R., Vila, J. and Wang, J. (1996) A model of trading volume with tax-induced heterogeneous valuation and transaction costs. *Journal of Financial Intermediation*, 5:340–371.
1339. Michaely, R. and Womack, K. (1999) Conflict of interest and the credibility of underwriter analysts recommendations. *Review of Financial Studies*, 12:573–608.
1340. Milevsky, M. and Young, V. (2007) Annuitization and asset allocation. *Journal of Economic Dynamics and Control*, 31:3138–3177.
1341. Milgrom, P. and Stokey, N. (1982) Information, trade and common knowledge. *Journal of Economic Theory*, 26:17–27.
1342. Miller, E. (1977) Risk, uncertainty, and divergence of opinion. *Journal of Finance* 32:1151–1168.
1343. Miller, M. and Scholes, M. (1972) Rates of return in relation to risk: a re-examination of some recent findings. In *Studies in the Theory of Capital Markets*, M. Jensen (ed.), Praeger, New York.
1344. Milne, F. (1979) Consumer preferences, linear demand functions and aggregation in competitive asset markets. *Review of Economic Studies*, 50:407–417.
1345. Milne, F. (1988) Arbitrage and diversification in a general equilibrium asset economy. *Econometrica*, 46:815–840.
1346. Milne, F. (1995) *Finance Theory and Asset Pricing*. Clarendon Press, Oxford.
1347. Mitchell, M. and Mulherin, H. (1994) The impact of public information on the stock market. *Journal of Finance*, 49:923–950.

1348. Modigliani, F. and Miller, M. (1958) The cost of capital, corporate finance, and the theory of investment. *The American Economic Review*, 48:261–297.

1349. Modigliani, F. and Miller, M. (1963) Corporate income taxes and the cost of capital: a correction. *The American Economic Review*, 53:433–443.

1350. Moinas, S. and Pouget, S. (2013) The bubble game: an experimental study of speculation. *Econometrica*, 81:1507–1539.

1351. Montrucchio, L. and Privileggi, F. (2001) On the fragility of bubbles in equilibrium asset pricing models. *Journal of Economic Theory*, 101:158–188.

1352. Morris, S. (1994) Trade with heterogeneous prior beliefs and asymmetric information. *Econometrica*, 62:1327–1347.

1353. Morris, S. (1996) Speculative investor behavior and learning. *Quarterly Journal of Economics*, 111:1111–1131.

1354. Moskowitz, T. (2003) An analysis of risk and pricing anomalies. *Review of Financial Studies*, 16:417–457.

1355. Moskowitz, T. and Grinblatt, M. (1999) Do industries explain momentum? *Journal of Finance*, 44:1249–1290.

1356. Moskowitz, T. and Vissing-Jorgensen, A. (2002) The returns to entrepreneurial investment: a private equity premium puzzle? *The American Economic Review*, 92:745–778.

1357. Mossin, J. (1966) Equilibrium in a capital asset market. *Econometrica*, 34:261–276.

1358. Mossin, J. (1968) Aspects of rational insurance purchasing. *Journal of Political Economy*, 79:553–568.

1359. Muendler, M.A. (2007) The possibility of informationally efficient markets. *Journal of Economic Theory*, 133:467–483.

1360. Muhle-Karbe, J. and Guasoni, P. (2013) Portfolio choice with transaction costs: a user's guide. *Paris-Princeton Lectures on Mathematical Finance 2013*, V. Henderson and R. Sircar (eds.), Springer, Cham.

1361. Mukerji, S. and Tallon, J. (2001) Ambiguity aversion and incompleteness of financial markets. *Review of Economic Studies*, 68:883–904.

1362. Mukerji, S. and Tallon, J. (2004) Ambiguity aversion and the absence of indexed debt. *Economic Theory*, 24:665–685.

1363. Munk, C. (2013) *Financial Asset Pricing Theory*. Oxford University Press.

1364. Musiela, M. and Rutkowski, M. (1997) *Martingale Methods in Financial Modelling*. Springer, Berlin/Heidelberg.

1365. Nachman, D. (1989) Spanning and completeness with options. *Review of Financial Studies*, 1:311–328.

1366. Naik, N. (1997) On the aggregation of information in competitive markets: the dynamic case. *Journal of Economic Dynamics and Control*, 21:1199–1227.

1367. Naik, N. (1997) Multi-period information markets. *Journal of Economic Dynamics and Control*, 21:1229–1258.

1368. Naik, N., Neuberger, A. and Viswanathan, S. (1999) Trade disclosure regulation in markets with negotiated trades. *Review of Financial Studies*, 12:873–900.

1369. Naik, V. (1995) Finite state securities market models and arbitrage. In *Handbooks in Operations Research and Management Science*, Vol. 9 (Finance), R.A. Jarrow, V. Maksimovic and W.T. Ziemba (eds.), North Holland, Amsterdam.

1370. Naranjo, A., Nimalendran, M. and Ryngaert, M. (1998) Stock returns, dividend yields, and taxes. *Journal of Finance*, 53:2029–2057.

1371. Narayanan, R. (2000) Insider trading and the voluntary disclosure of information by firms. *Journal of Banking and Finance*, 24:395–425.

1372. Nawalkha, M. (1997) A multibeta representation theorem for linear asset pricing Theories. *Journal of Financial Economics*, 46:357–381.

1373. Nelson, C. and Kim, M. (1993) Predictable stock returns: the role of small sample bias. *Journal of Finance*, 48:641–661.

1374. Nguyen, P. and Portait, R. (2002) Dynamic asset allocation with mean variance preferences and a solvency constraint. *Journal of Economic Dynamics and Control*, 26:11–32.

1375. Nielsen, L. (1990) Existence of equilibrium in CAPM. *Journal of Economic Theory*, 32:223–229.
1376. Nielsen, L. (1987) Portfolio selection in the mean-variance model: a note. *Journal of Finance*, 42:1371–1376.
1377. Nielsen, L. (1992) Positive prices in CAPM. *Journal of Finance*, 47:791–808.
1378. Noe, C. (1999) Voluntary disclosures and insider transactions. *Journal of Accounting Economics*, 27:305–326.
1379. Odean, T. (1998) Volume, volatility, price, and profit when all traders are above the average. *Journal of Finance*, 53:1887–1934.
1380. Odean, T. (1998) Are investors reluctant to realize their losses? *Journal of Finance*, 53:1775–1798.
1381. Odean, T. (1999) Do investors trade too much? *The American Economic Review*, 89:1279–1298.
1382. Ofek, E. and Richardson M. (2003) DotComMania: the rise and fall of internet stock prices. *Journal of Finance*, 58:1113–1137.
1383. Ogaki, M. and Zhang, Q. (2001) Decreasing relative risk aversion and tests of risk sharing. *Econometrica*, 69:515–526.
1384. O'Hara, M. (1995) *Market Microstructure Theory*. Blackwell Publishers.
1385. O'Hara, M. and Oldfield, G. (1986) The microeconomics of market making. *Journal of Financial and Quantitative Analysis*, 21:361–376.
1386. Orosel, G. (1996) Informational efficiency and welfare in the stock market. *European Economic Review*, 40:1379–1411.
1387. Orosel, G. (1998) Participation costs, trend chasing, and volatility of stock prices. *Review of Financial Studies*, 11:521–557.
1388. Otrok, C., Ravikumar, B. and Whiteman, C. (2002) Habit formation: a resolution of the equity premium puzzle? *Journal of Monetary Economics* 49:1261–1288.
1389. Owen, J. and Rabinovitch, R. (1983) On the class of elliptical distributions and their applications to the theory of portfolio choice. *Journal of Finance*, 37:745–752.
1390. Ozguz, A. (2009) Good times or bad times? Investors' uncertainty and stock returns. *Review of Financial Studies*, 22:4377–4422.
1391. Pagan, A. (1996) The econometrics of financial markets. *Journal of Empirical Finance*, 3:15–102.
1392. Pagan, A. and Schwert, W. (1990) Alternative models for conditional stock volatility. *Journal of Econometrics*, 45:267–290.
1393. Pagano, M. (1989) Endogenous market thinness and stock price volatility. *Review of Economic Studies*, 56:269–288.
1394. Pagano, M. (1989) Trading volume and asset liquidity. *Quarterly Journal of Economics*, 104:255–274.
1395. Pagano, M. and Roell, A. (1996) Transparency and liquidity: a comparison of auction and dealer markets with informed trading. *Journal of Finance*, 51:123–145.
1396. Paiella, M. (2004) Heterogeneity in financial market participation: appraising its implications for the C-CAPM. *Review of Finance*, 8:445–480.
1397. Palfrey, T. and Wang, S. (2012) Speculative overpricing in asset markets with information flows. *Econometrica*, 80:1937–1976.
1398. Palomino, F. (1996) Noise trading in small markets. *Journal of Finance*, 51:1537–1550.
1399. Palumbo, M. (1999) Uncertain medical expenses and precautionary saving near the end of the life cycle. *Review of Economic Studies*, 66:395–421.
1400. Park, C. and Irwin, S. (2007) What do we know about the profitability of technical analysis? *Journal of Economic Surveys*, 21:786–826.
1401. Parker, J. and Julliard, C. (2005) Consumption risk and the cross section of expected returns. *Journal of Political Economy*, 113:185–222.
1402. Parlour, C. (1998) Price dynamics in limit order markets. *Review of Financial Studies*, 11:789–816.

1403. Parlour, C. and Seppi, D. (2003) Liquidity-based competition for order flow. *Review of Financial Studies*, 16:301–343.
1404. Pascucci, A. and Runggaldier, W. (2012) *Financial Mathematics: Theory and Problems for Multi-period Models*. Springer, Milan.
1405. Pástor, L. (2000) Portfolio selection and asset pricing models. *Journal of Finance*, 55:179–223.
1406. Pástor, L. and Stambaugh, R. (2000) The equity premium and structural breaks. *Journal of Finance*, 56:1207–1239.
1407. Pástor, L. and Stambaugh, R. (2003) Liquidity risk and expected stock returns. *Journal of Political Economy*, 111:642–685.
1408. Pástor, L. and Veronesi, P. (2003) Stock valuation and learning about profitability. *Journal of Finance*, 58:1749–1789.
1409. Pástor, L. and Veronesi, P. (2006) Was there a Nasdaq bubble in the late 1990s? *Journal of Financial Economics*, 81:61–100.
1410. Patelis, A. (1997) Stock return predictability and the role of monetary policy. *Journal of Finance*, 52:1951–1972.
1411. Pennacchi, G. (2008) *Theory of Asset Pricing*. Pearson.
1412. Peress, J. (2004) Wealth, information acquisition and portfolio choice. *Review of Financial Studies*, 17:879–914.
1413. Peress, J. (2011) Erratum to "Wealth, information acquisition and portfolio choice". *Review of Financial Studies*, 24:3187–3195.
1414. Pesaran, H. (1989) *The Limits to Rational Expectations*. Blackwell Publishers.
1415. Pesaran, H. and Timmermann, A. (1995) Predictability of stock returns: robustness and economic significance. *Journal of Finance*, 50:1201–1227.
1416. Petkova, R. (2006) Do the Fama-French factors proxy for innovations in predictive variables? *Journal of Finance*, 61:581–612.
1417. Petkova, R. and Zhang, L. (2005) Is value riskier than growth? *Journal of Financial Economics*, 78:187–202.
1418. Pham, H. (2000) On quadratic hedging in continuous time. *Mathematical Methods of Operations Research*, 51:315–339.
1419. Pietra, T. and Siconolfi, P. (1998) Fully revealing equilibria in sequential economies with asset markets. *Journal of Mathematical Economics*, 29:211–223.
1420. Platen, E. and Heath, D. (2006) *A Benchmark Approach to Quantitative Finance*. Springer, Berlin/Heidelberg.
1421. Pliska, S. (1997) *Introduction to Mathematical Finance*. Blackwell Publishers.
1422. Plott, C. and Sunder, S. (1988) Rational expectations and the aggregation of diverse information in laboratory security markets. *Econometrica*, 56:1085–1118.
1423. Po-Hsuan, H. and Chung-Ming, K. (2005) Reexamining the profitability of technical analysis with data snooping checks. *Journal of Financial Econometrics*, 3:606–628.
1424. Polemarchakis, H. and Ku, B. (1990) Options and equilibrium. *Journal of Mathematical Economics*, 19:107–112.
1425. Polemarchakis, H. and Siconolfi, P. (1993) Asset markets and the information revealed by prices. *Economic Theory*, 3:645–661.
1426. Polkovnichenko, V. (2007) Life-cycle portfolio choice with additive habit formation preferences and uninsurable labor income risk. *Review of Financial Studies*, 20:83–124.
1427. Pontiff, J. and Schall, L. (1998) Book-to-market ratios as predictors of market returns. *Journal of Financial Economics*, 49:141–160.
1428. Porter, D. and Weaver, D. (1998) Post-trade transparency on Nasdaq's national market system. *Journal of Financial Economics*, 50:231–252.
1429. Poterba, J. and Samwick, A. (1999) Household portfolio allocation over the life-cycle. *NBER working paper*, no. 6185.
1430. Poterba, J. and Summers, L. (1988) Mean reversion in stock prices: evidence and implications. *Journal of Financial Economics*, 22:27–60.

1431. Poterba, J. and Weisbenner, S. (2001) Capital gains tax rules, tax loss trading, and turn-of-the-year returns. *Journal of Finance* 56:353–368.
1432. Pratt, J. (1964) Risk aversion in the small and in the large. *Econometrica*, 32:122–136.
1433. Pratt, J. and Zeckauser, R. (1987) Proper risk aversion. *Econometrica*, 55:143–154.
1434. Presidential Task Force on Market Mechanism (1988) Report of the Presidential Task Force on Market Mechanism. United States Government, Washington.
1435. Protter, P. (2013) A mathematical theory of financial bubbles. *Paris-Princeton Lectures on Mathematical Finance*, Henderson, V. and Sircar, R. (eds.), Springer, Cham, 1–108.
1436. Quiggin, J. (1982) A theory of anticipated utility. *Journal of Economic Behavior and Organization*, 3:323–343.
1437. Radner, R. (1972) Existence of equilibrium of plans, prices, and price expectations in a sequence of markets. *Econometrica*, 40:289–303.
1438. Radner, R. (1979) Rational expectations equilibrium: generic existence and the information revealed by price. *Econometrica*, 47:655–678.
1439. Radner, R. (1982) Equilibrium under uncertainty. In *Handbook of Mathematical Economics* (Vol.I), K. Arrow and M. Intriligator (eds.), Elsevier Science Publisher.
1440. Rahi, R. (1985) Partially revealing rational expectations equilibria with nominal assets. *Journal of Mathematical Economics*, 24:137–146.
1441. Reinganum, M. (1981) The arbitrage pricing theory: some empirical results. *Journal of Finance*, 36:313–322.
1442. Reinganum, M. (1981) A new empirical perspective on the CAPM. *Journal of Financial and Quantitative Analysis*, 16:439–462.
1443. Reinganum, M. (1981) Misspecification of capital asset pricing : Empirical anomalies based on earnings' yields and market values. *Journal of Financial Economics*, 9:19–46.
1444. Reinganum, M. (1983) The anomalous stock market behavior of small firms in January: Empirical tests for tax-loss selling effects. *Journal of Financial Economics*, 12:89–104.
1445. Reisman, H. (1992) Reference variables, factor structure, and the approximate multibeta representation. *Journal of Finance*, 47:1303–1314.
1446. Repullo, R. (1999) Some remarks on Leland's model of insider trading. *Economica*, 66:359–374.
1447. Richards, A. (1997) Winner-loser reversals in national stock market indices: can they be explained? *Journal of Finance*, 52:2129–2144.
1448. Richardson, M. (1993) Temporary components of stock prices: a skeptic view. *Journal of Business and Economic Statistics*, 11:199–207.
1449. Richardson, M. and Smith, T. (1994) A unified approach to testing for serial correlation in stock returns. *Journal of Business*, 67:371–399.
1450. Richardson, M. and Stock, J. (1989) Drawing inferences from statistics based on multi-year asset returns. *Journal of Financial Economics*, 25:323–348.
1451. Rietz, T. (1988) The equity risk premium: a solution. *Journal of Monetary Economics*, 22:117–131.
1452. Rigotti, L. and Shannon, C. (2005) Uncertainty and risk in financial markets. *Econometrica*, 73:203–243.
1453. Rigotti, L., Strzalecki, T. and Shannon, C. (2008) Subjective beliefs and ex-ante trade. *Econometrica*, 76:1167–1190.
1454. Rinaldi, F. (2009) Endogenous incompleteness of financial markets: the role of ambiguity and ambiguity aversion. *Journal of Mathematical Economics*, 45:872–893.
1455. Rochet, J. and Vila, J. (1994) Insider trading without normality. *Review of Economic Studies*, 61:131–152.
1456. Roell, A. (1990) Dual capacity and the quality of the market. *Journal of Financial Intermediation*, 1:105–124.
1457. Roll, R. (1977) A critique of the asset's pricing theory's test: part I. *Journal of Financial Economics*, 4:129–176.
1458. Roll, R. (1983) Vas ist das? The turn-of-the-year effect and the return premia of small firms. *Journal of Portfolio Management* 9:18–28.

1459. Roll, R. (1984) Orange juice and weather. *The American Economic Review*, 74:861–880.
1460. Roll, R. (1988) R^2. *Journal of Finance*, 43:541–566.
1461. Roll, R. and Ross, S. (1980) An empirical investigation of the arbitrage pricing theory. *Journal of Finance*, 35:1073–1103.
1462. Roll, R. and Ross, S. (1994) On the cross-sectional relation between expected returns and betas. *Journal of Finance*, 49:101–121.
1463. Romer, D. (1993) Rational asset-price movements without news. *The American Economic Review*, 83:1112–1130.
1464. Ross, S. (1976) The arbitrage pricing theory of capital asset pricing. *Journal of Economic Theory*, 13:341–360.
1465. Ross, S. (1976) Options and efficiency. *Quarterly Journal of Economics*, 90:75–89.
1466. Ross, S. (1977) Return, risk and arbitrage. In *Risk and Return in Finance*, I. Friend and J. Bicksler (eds.), Ballinger, Cambridge.
1467. Ross, S. (1978) Mutual fund separation in financial theory: the separating distributions. *Journal of Economic Theory*, 17:254–286.
1468. Ross, S. (1981) Some stronger measures of risk aversion in the small and large with applications. *Econometrica*, 49:621–638.
1469. Ross, S. (1987) Arbitrage and martingales with taxation. *Journal of Political Economy*, 95:371–393.
1470. Ross, S. (1989) Information and volatility: the no arbitrage martingale approach to timing and resolution irrelevance. *Journal of Finance*, 44:1–18.
1471. Rosu, I. (2009) A dynamic model of the limit order book. *Review of Financial Studies*, 22:4601–4641.
1472. Rothschild, M. and Stiglitz, J. (1970) Increasing risk I: a definition. *Journal of Economic Theory*, 2:225–243.
1473. Rothschild, M. and Stiglitz, J. (1971) Increasing risk II: its economic consequences. *Journal of Economic Theory*, 3:66–84.
1474. Routledge, B. (1999) Adaptive learning financial markets. *Review of Financial Studies*, 12:1165–1202.
1475. Routledge, B. and Zin, S. (2009) Model uncertainty and liquidity. *Review of Economic Dynamics* 12:543–566.
1476. Routledge, B. and Zin, S. (2010) Generalized disappointment aversion and asset prices. *Journal of Finance*, 65:1303–1332.
1477. Rouwenhorst, G. (1995) Asset pricing implications of equilibrium business cycle models. In *Frontiers of Business Cycle Research*, T. Cooley (ed.), Princeton University Press, 294–330.
1478. Rouwenhorst, G. (1998) International momentum strategies. *Journal of Finance*, 53:267–284.
1479. Rouwenhorst, G. (1999) Local return factors and turnover in emerging stock markets. *Journal of Finance*, 54:1439–1464.
1480. Rozeff, M. (1984) Dividend yields are equity risky premiums. *Journal of Portfolio Management*, 11:68–75.
1481. Rozeff, M. and Kinney, W. (1976) Capital market seasonality: the case of stock returns. *Journal of Financial Economics*, 3:57–63.
1482. Rozeff, M. and Zaman, M. (1988) Market efficiency and insider trading: new evidence. *Journal of Business*, 61:25–44.
1483. Rozeff, M. and Zaman, M. (1998) Overreaction and insider trading: evidence from growth and value portfolios. *Journal of Finance*, 53:701–716.
1484. Rozen, K. (2010) Foundations of intrinsic habit formation. *Econometrica*, 78:1341–1373.
1485. Rubinstein, M. (1974) An aggregation theorem for securities markets. *Journal of Financial Economics*, 1:225–244.
1486. Rubinstein, M. (1975) Securities market efficiency in an Arrow-Debreu economy. *The American Economic Review*, 65:812–824.
1487. Rubinstein, M. (1976) The valuation of uncertain income streams and the pricing of options. *Bell Journal of Economics*, 7:407–425.

1488. Runkle, D. (1991) Liquidity constraints and the permanent income hypothesis: evidence from panel data. *Journal of Monetary Economics*, 27:73–98.
1489. Sadka, R. and Scherbina, A. (2007) Analyst disagreement, mispricing, and liquidity. *Journal of Finance*, 62: 2367–2403.
1490. Salyer, K. (1998) Crash states and the equity premium: solving one puzzle raises another. *Journal of Economic Dynamics and Control*, 22:955–965.
1491. Samuelson, P. (1965) Proof that properly anticipated prices fluctuate randomly. *Industrial Management Review*, 6:41–49.
1492. Samuelson, P. (1967) General proof that diversification pays. *Journal of Financial and Quantitative Analysis*, 2:1–13.
1493. Samuelson, P. (1969) Lifetime portfolio selection by dynamic stochastic programming. *Review of Economics and Statistics*, 51:239–246.
1494. Sandas, P. (2001) Adverse selection and competitive market making: empirical evidence from a limit order market. *Review of Financial Studies*, 14:705–734.
1495. Sandroni, A. (2000) Do markets favor agents able to make accurate predictions? *Econometrica*, 68:1303–1341.
1496. Santos, M. and Woodford, M. (1997) Rational asset pricing bubbles. *Econometrica*, 65:19–58.
1497. Santos, T. and Veronesi, P. (2006) Labor income and predictable stock returns. *Review of Financial Studies*, 19:1–44.
1498. Santos, T. and Veronesi, P. (2010) Habit formation, the cross section of stock returns and the cash flow risk puzzle. *Journal of Financial Economics*, 98:385–413.
1499. Savage, L. (1954) *The Foundations of Statistics*. Wiley, New York.
1500. Schachermayer, W. (2004) The fundamental theorem of asset pricing under proportional transaction costs in finite discrete time. *Mathematical Finance*, 14:19–48.
1501. Schachermayer, W. (2010) The fundamental theorem of asset pricing. In *Encyclopedia of Quantitative Finance*, Vol. 2, R. Cont (ed.), Wiley, Chichester, 792–801.
1502. Scheinkman, J. and Xiong, W. (2003) Overconfidence and speculative bubbles. *Journal of Political Economy*, 111:1183–1219.
1503. Scheinkman, J. and Xiong, W. (2004) Heterogeneous beliefs, speculation and trading in financial markets. *Paris-Princeton Lectures on Mathematical Finance 2003*, Lecture Notes in Mathematics Volume 1847, 217–250.
1504. Schlee, E. (2001) The value of information in efficient risk-sharing arrangements *The American Economic Review*, 91:509–524.
1505. Schmeidler, D. (1989) Subjective probability and expected utility without additivity. *Econometrica*, 57:571–587.
1506. Schmeidler, D. (1989) A rational expectations equilibrium with informative trading volume. *Journal of Finance*, 64:2783–2805.
1507. Schnitzlen, C. (1996) Call and continuous trading mechanisms under asymmetric information: an experimental investigation. *Journal of Finance*, 51:613–636.
1508. Schweizer, M. (2000) A guided tour through quadratic hedging approaches. In *Option Pricing, Interest Rates and Risk Management*, E. Jouini, J. Cvitanic and M. Musiela (eds.), Cambridge University Press, 538–574.
1509. Schwert, W. (1989) Why does stock market volatility change over time? *Journal of Finance*, 44:1115–1153.
1510. Schwert, W. (2001) Anomalies and market efficiency. In *Handbook of the Economics of Finance*, Vol. 2, G. Constantinides, M. Harris and R. Stulz (eds.), Elsevier Science Publisher, 939–974.
1511. Schwert, W. and Seguin, P. (1990) Heteroskedasticity in stock returns. *Journal of Finance*, 45:1129–1155.
1512. Schwert, W. and Seguin, P. (1993) Securities transaction taxes: an overview of costs, benefits and unresolved questions. *Financial Analysts Journal*, 49:27–35.
1513. Scott, L. (1985) The present value model of stock prices: regression tests and Monte Carlo results. *The Review of Economics and Statistics*, 67:599–605.

1514. Segal, U. and Spivak, A. (1990) First order versus second order risk aversion. *Journal of Economic Theory*, 51:111–125.
1515. Sentana, E. and Wadhwani, S. (1992) Feedback traders and stock return autocorrelations: evidence from a century daily data. *The Economic Journal*, 102:415–425.
1516. Seo, K. (2009) Ambiguity and second-order belief. *Econometrica*, 77:1575–1605.
1517. Seppi, D. (1990) Equilibrium block trading and asymmetric information. *Journal of Finance*, 45:73–94.
1518. Seppi, D. (1992) Block trading and information revelation around quarterly earnings announcements. *Review of Financial Studies*, 5:281–305.
1519. Seppi, D. (1997) Liquidity provision with limit orders and strategic specialist. *Review of Financial Studies*, 10:103–150.
1520. Seyhun, H. (1986) Outsiders' profits, costs of trading, and the markets efficiency. *Journal of Financial Economics*, 16:189–212.
1521. Seyhun, H. (1998) *Investment Intelligence from Insider trading*. MIT Press, Cambridge.
1522. Shalen, C. (1993) Volume, volatility, and the dispersion of beliefs. *Review of Financial Studies*, 6:405–34.
1523. Shanken, J. (1982) The arbitrage pricing theory: Is it testable? *Journal of Finance*, 37:1129–1140.
1524. Shanken, J. (1985) Multi-beta CAPM or equilibrium-APT? A reply. *Journal of Finance*, 40:1189–1196.
1525. Shanken, J. (1987) Multivariate proxies and asset pricing relations: living with the Roll critique. *Journal of Financial Economics*, 18:91–110.
1526. Shanken, J. (1992) On the estimation of beta-pricing models. *Review of Financial Studies*, 5:1–55.
1527. Shanken, J. (1996) Statistical methods in tests of portfolio efficiency: a synthesis. In *Handbook of Statistics*, Vol.14, G.S. Maddala and C.R. Rao (eds.), North-Holland, Amsterdam.
1528. Shanken, J. and Zhou, G. (2007) Estimating and testing beta pricing models: alternative methods and their performance in simulations. *Journal of Financial Economics*, 84:40–86.
1529. Sharpe, W. (1964) Capital asset prices: a theory of market equilibrium under conditions of risks. *Journal of Finance*, 19:425–442.
1530. Sharpe, W. (1991) Capital asset prices with and without negative holdings. *Journal of Finance*, 46:489–509.
1531. Shefrin, H. and Statman, M. (1985) The disposition to sell winners too early and ride losers too long. *Journal of Finance*, 40:777–790.
1532. Shefrin, H. and Statman, M. (1994) Behavioral capital asset pricing. *Journal of Financial and Quantitative Analysis*, 29:323–349.
1533. Shiller, R. (1981) Do stock prices move too much to be justified by subsequent changes in dividends? *The American Economic Review*, 71:421–435.
1534. Shiller, R. (1981) The use of volatility measures in assessing market efficiency. *Journal of Finance*, 36:291–304.
1535. Shiller, R. (1984) Stock prices and social dynamics. *Brookings Papers on Economic Activity*, 2:457–498.
1536. Shiller, R. (1986) The Marsh-Merton model of managers smoothing of dividends. *The American Economic Review*, 76:499–503.
1537. Shiller, R. (1988) The probability of gross violations of a present value variance inequality. *Journal of Political Economy*, 96:1089–1092.
1538. Shiller, R. (1989) *Market Volatility*. MIT Press, Cambridge.
1539. Shin, J. (1996) The optimal regulation of insider trading. *Journal of Financial Intermediation*, 5:49–73.
1540. Shleifer, A. and Summers, L. (1990) The noise trader approach to finance. *Journal of Economic Perspectives*, 4:19–33.
1541. Shleifer, A. and Vishny, R. (1997) The limits of arbitrage. *Journal of Finance*, 52:35–55.
1542. Siegel, J. (1988) *Stocks for the Long Run*. McGraw-Hill.

1543. Siegel, J. (1999) The shrinking equity premium. *Journal of Portfolio Management*, 26:10–17.

1544. Simaan, Y., Weaver, D. and Whitcomb, D. (2003) Market maker quotation behavior and pretrade transparency. *Journal of Finance*, 58:1247–1268.

1545. Simon H. (1955) A behavioral model of rational choice. *Quarterly Journal of Economics*, 69:99–118.

1546. Simon, H. (1976) From substantive to procedural rationality. In *Method and Appraisal in Economics*, S. Latsis (ed.), Cambridge University Press.

1547. Simsek, A. (2013) Belief disagreements and collateral constraints. *Econometrica*, 81:1–53.

1548. Singleton, K. (1987) Asset prices in a time series model with disparately informed competitive traders. In *New Approaches to Monetary Economics*, W. Barnett and K. Signleton (eds.), North-Holland, Amsterdam.

1549. Singleton, K. (1990) Specification and estimation of intertemporal asset pricing models. In *Handbook of Monetary Economics*, Vol.1, B. Friedman and F. Hahn (eds.), North-Holland, Amsterdam.

1550. Skinner, J. (1988) Risky income, life cycle consumption, and precautionary saving. *Journal of Monetary Economics*, 22:237–255.

1551. Smith, W. (2001) How does the spirit of capitalism affect stock market prices? *Review of Financial Studies*, 14:1215–1232.

1552. Smith, V., Suchanek, G. and Williams, W. (1988) Bubbles, crashes and endogenous expectations in experimental spot asset markets. *Econometrica*, 56:1119–1151.

1553. Snow, K. (1991) Diagnosis asset pricing models using the distribution of asset returns. *Journal of Finance*, 46:955–983.

1554. Sogner, L. and Mitlohner, H. (2002) Consistent expectations equilibria and learning in a stock market. *Journal of Economic Dynamics and Control*, 26:171–185.

1555. Solnik, (1974) An equilibrium model of the international capital market. *Journal of Economic Theory* 8:500–524.

1556. Solow, R. (1956) A contribution to the theory of economic growth. *Quarterly Journal of Economics*, 70:65–94.

1557. Song, F. and Zhang, J. (2005) Securities transaction tax and market volatility. *The Economic Journal*, 115:1103–1120.

1558. Spiegel, M. and Subrahmanyam, A. (1992) Informed speculation and hedging in a noncompetitive securities market. *Review of Financial Studies*, 5:307–329.

1559. Spiegel, M. and Subrahmanyam, A. (2000) Asymmetric information and news disclosure rules. *Journal of Financial Intermediation*, 9:363–403.

1560. Sriboonchitta, S., Wong, W.-K., Dhompongsa, S. and Nguyen, H.T. (2010) *Stochastic Dominance and Applications to Finance, Risk and Economics*. CRC Press, Boca Raton.

1561. Stambaugh, R. (1982) On the exclusion of assets from tests of the two parameter model. *Journal of Financial Economics*, 10:235–268.

1562. Stambaugh, R. (1983) Arbitrage pricing with information. *Journal of Financial Economics*, 12:357–369.

1563. Stambaugh, R., Yu, J. and Yuan, Y. (2012) The short of it: investor sentiment and anomalies. *Journal of Financial Economics*, 104:288–302.

1564. Starmer, C. (2000) Developments in non-expected utility theory: the hunt for a descriptive theory of choice under risk. *Journal of Economic Literature*, 38:332–382.

1565. Statman, M., Thorley, S. and Vorkink, K. (2006) Investor overconfidence and trading volume. *Review of Financial Studies*, 19:1531–1565.

1566. Stattman, D. (1980) Book values and stock returns. *The Chicago MBA: A Journal of Selected Papers*, 4:25–45.

1567. Stein, J. (1987) Informational externalities and welfare-reducing speculation. *Journal of Political Economy*, 95:1123–1145.

1568. Stiglitz, J. (1982) The inefficiency of the stock market equilibrium. *Review of Economic Studies*, 49:241–261.

1569. Stiglitz, J. (1989) Using tax policy to curb speculative short term trading. *Journal of Financial Services Research*, 3:101–115.

1570. Stoll, H. (1978) The supply of dealer services in securities markets. *Journal of Finance*, 33:1133–1151.

1571. Stoll, H. and Whaley, R. (1990) Stock market structure and volatility. *Review of Financial Studies*, 3:37–71.

1572. Storesletten, K., Telmer, C. and Yaron, A. (2004) Consumption and risk sharing over the life cycle. *Journal of Monetary Economics*, 51:609–633.

1573. Storesletten, K., Telmer, C. and Yaron, A. (2004) Asset pricing with idiosyncratic risk and overlapping generations. *Review of Economic Dynamics*, 10:519–548.

1574. Stulz, R. (1981) A model of international asset pricing. *Journal of Financial Economics*, 9:383–406.

1575. Subrahmanyam, A. (1991) Risk aversion, market liquidity, and price efficiency. *Review of Financial Studies*, 4:417–441.

1576. Subrahmanyam, A. (1994) Circuit breakers and market volatility: a theoretical perspective. *Journal of Finance*, 49:237–252.

1577. Subrahmanyam, A. (1998) Transaction taxes and financial market equilibrium. *Journal of Business*, 71:81–118.

1578. Summers, L. (1986) Does the stock market rationality reflect fundamental values? *Journal of Finance*, 41:591–601.

1579. Summers, L. and Summers, V. (1989) When financial markets work too well: a cautious case for a securities transactions tax. *Journal of Financial Services Research*, 3:261–268.

1580. Sundaresan, S. (1989) Intertemporally dependent preferences and the volatility of consumption and wealth. *Review of Financial Studies*, 2:73–88.

1581. Sunder, S. (1992) Market for information: experimental evidence. *Econometrica*, 60:667–695.

1582. Tallarini, T. and Zhang, H. (2005) External habit and the cyclicality of expected stock returns. *Journal of Business*, 78:1023–1048.

1583. Telmer, C. (1993) Asset pricing puzzles and incomplete markets. *Journal of Finance*, 48:1803–1832.

1584. Teoh, S. and Hwang, C. (1991) Nondisclosure and adverse disclosure as signals of firm value. *Review of Financial Studies*, 4:283–313.

1585. Tepla, L. (2000) Optimal hedging and valuation of nontraded assets. *European Finance Review* 4:231–251.

1586. Tetlock, P. (2010) Does public financial news resolve asymmetric information. *Review of Financial Studies*, 23:3520–3557.

1587. Thaler, R. (1999) Mental accounting matters. *Journal of Behavioral Decision Making*, 12:183–206.

1588. Thaler, R., Tversky, A., Kahneman, D. and Schwartz, A. (1997) The effect of myopia and loss aversion on risk taking: an experimental test. *Quarterly Journal of Economics*, 112:647–661.

1589. Thorisson, H. (1995) Coupling methods in probability theory. *Scandinavian Journal of Statistics*, 22:159–182.

1590. Timmermann, A. (1993) How learning in financial markets generates excess volatility and predictability in stock prices. *Quarterly Journal of Economics*, 108:1135–1145.

1591. Timmermann, A. (1994) Present value models with feedback. *Journal of Economic Dynamics and Control*, 18:1093–1119.

1592. Timmermann, A. (1994) Can agents learn to form rational expectations? Some results on convergence and stability of learning in the UK stock market. *The Economic Journal*, 104:777–797.

1593. Timmermann, A. (1994) Excess volatility and predictability of stock prices in autoregressive dividend models with learning. *Review of Economic Studies*, 63:523–557.

1594. Tirole, J. (1982) On the possibility of speculation under rational expectations. *Econometrica*, 50:1163–1181.

1595. Tirole, J. (1985) Asset bubbles and overlapping generations. *Econometrica*, 53:1499–1528.
1596. Tobin, J. (1958) Liquidity preferences as behavior towards risk. *Review of Economic Studies*, 25:65–86.
1597. Tobin, J. (1984) On the efficiency of the financial system. *Lloyds Bank Review*, 153:1–25.
1598. Townsend, R. (1983) Forecasting the forecasts of others. *Journal of Political Economy*, 91:546–588.
1599. Townsend, R. (1994) Risk and insurance in village India. *Econometrica*, 62:539–591.
1600. Trojani, F. and Vanini, P. (2004) Robustness and ambiguity aversion in general equilibrium. *Review of Finance*, 8:279–324.
1601. Tsay, R. (2010) *Analysis of Financial Time Series*, 3rd ed. Wiley, Chichester.
1602. Tuckman, B. and Vila, J. (1992) Arbitrage with holding costs: a utility based approach. *Journal of Finance*, 47:1283–1302.
1603. Tversky, A. and Kahneman, D. (1992) Advances in prospect theory: cumulative representation of uncertainty. *Journal of Risk and Uncertainty*, 5:297–323.
1604. Umlauf, S. (1993) Transaction taxes and the behavior of the Swedish stock market. *Journal of Financial Economics*, 33:227–240.
1605. Uppal, R. and Wang, T. (2003) Model misspecification and under diversification. *Journal of Finance*, 58:2465–2486.
1606. Van Ness, B., Van Ness, R. and Pruitt, S. (2000) The impact of the reduction in tick increments in major US markets on spreads, depth, and volatility. *Review of Quantitative Finance and Accounting*, 15:153–167.
1607. Van Nieuwerburgh, S. and Veldkamp, L. (2009) Information immobility and the home bias puzzle. *Journal of Finance* 64:1187–1215.
1608. Van Nieuwerburgh, S. and Veldkamp, L. (2010) Information acquisition and under-diversification. *Review of Economic Studies*, 77:779–805.
1609. van Rooij, M., Lusardi, A. and Alessie, R. (2011) Financial literacy and stock market participation. *Journal of Financial Economics*, 101:449–472.
1610. Varian, H. (1984) *Microeconomic Analysis*. Norton, New York.
1611. Varian, H. (1989) Differences in opinion in financial markets. In *Financial Risk: Theory, Evidence and Implications*, C. Stone (ed.), Kluwer Academic Publishers.
1612. Vassalou, M. (2003) News related to future GDP growth as a risk factor in equity returns. *Journal of Financial Economics*, 68:47–73.
1613. Vassalou, M. and Xing, Y. (2004) Default risk in equity returns. *Journal of Finance*, 59:831–868.
1614. Vayanos, D. (1998) Transaction costs and asset prices: a dynamic equilibrium model. *Review of Financial Studies*, 11:1–58.
1615. Veldkamp, L. (2006) Information markets and the comovement of asset prices. *Review of Economic Studies*, 73:823–845.
1616. Veronesi, P. (1999) Stock market overreaction to bad news in good times: a rational expectations equilibrium model. *Review of Financial Studies*, 12:975–1007.
1617. Veronesi, P. (2004) The Peso problem hypothesis and stock market returns. *Journal of Economic Dynamics and Control*, 28:707–725.
1618. Verrecchia, R. (1982) Information acquisition in a noisy rational expectations economy. *Econometrica*, 50:1415–1430.
1619. Verrecchia, R. (1993) Insider trading and the voluntary disclosure of information by firms. *Journal of Banking and Finance*, 24:395–425.
1620. Verrecchia, R. (1993) Discretionary disclosure. *Journal of Accounting and Economics*, 5:179–194.
1621. Verrecchia, R. (2001) Essays on disclosure. *Journal of Accounting and Economics*, 32:97–180.
1622. Viceira, L. (2001) Optimal portfolio choice for long horizon investors with nontradable labor income. *Journal of Finance*, 56:433–470.
1623. Vila, J. (1989) Simple games of market manipulation. *Economics Letters*, 29:21–26.

1624. Villaverde, J.F. and Krueger, D. (2011) Consumption and saving over the life cycle: how important are consumer durables? *Macroeconomic Dynamics*, 15:725–770.

1625. Vissing-Jorgensen, A. (2000) Towards an explanation of household portfolio choice heterogeneity: nonfinancial income and participation cost structures. *NBER working paper*, no.8884.

1626. Vissing-Jorgensen, A. (2002) Limited asset market participation and the elasticity of intertemporal substitution. *Journal of Political Economy*, 110:825–853.

1627. Viswanathan, S. and Wang, J. (1999) Market architecture: limit order books versus dealership markets. *Journal of Financial Markets*, 5:127–167.

1628. Vives, X. (1995) Short-term investment and the informational efficiency of the market. *Review of Financial Studies*, 8:125–160.

1629. Vives, X. (1995) The speed of information revelation in a financial market mechanism. *Journal of Economic Theory*, 67:178–204.

1630. Vives, X. (2008) *Information and Learning in Markets*. Princeton University Press.

1631. Vives, X. (2014) On the possibility of informationally efficient markets. *Journal of the European Economic Association*, 12:1200–1239.

1632. Von Neumann, J. and Morgenstern, O. (1953) *Theory of Games and Economic Behaviour*. Princeton University Press.

1633. Vuolteenaho, T. (2002) What drives firm level stock returns? *Journal of Finance*, 57:233–264.

1634. Wachter, J. (1999) Risk aversion and allocation to long-term bonds. *Journal of Economic Theory*, 112:325–333.

1635. Wachter, J. (2006) A consumption-based model of the term structure of interest rates. *Journal of Financial Economics*, 79:365–399.

1636. Wachter, J. (2001) Portfolio and consumption decisions under mean-reverting returns: an exact solution for complete markets. *Journal of Financial and Quantitative Analysis*, 37:63–91.

1637. Wachter, J. (2013) Can time-varying risk of rare disasters explain aggregate stock market volatility? *Journal of Finance*, 68:987–1035.

1638. Wachter, J. and Warusawitharana, M. (2009) Predictable returns and asset allocation: Should a skeptical investor time the market? *Journal of Econometrics*, 148:162–178.

1639. Wachter, J. and Yogo, M. (2010) Why do household portfolio shares rise in wealth? *Review of Financial Studies*, 23:3929–3965.

1640. Wahal, S. (1997) Entry, exit, market makers, and the bid-ask spread. *Review of Financial Studies*, 10:871–901.

1641. Wang, A. (2001) Overconfidence, investor sentiment and evolution. *Journal of Financial Intermediation*, 10:138–170.

1642. Wang, F. (1998) Strategic trading, asymmetric information and heterogeneous beliefs. *Journal of Financial Markets*, 1:321–352.

1643. Wang, J. (1993) A model of intertemporal asset prices under asymmetric information. *Review of Economic Studies*, 60:249–282.

1644. Wang, J. (1994) A model of competitive stock trading volume. *Journal of Political Economy*, 102:127–168.

1645. Wang, S. (1993) The local recoverability of risk aversion and intertemporal substitution. *Journal of Economic Theory*, 59:333–363.

1646. Watanabe, M. (2008) Price volatility and investor behavior in an overlapping generations model with information asymmetry. *Journal of Finance*, 63:229–272.

1647. Weil, P. (1987) Confidence and the real value of money in overlapping generation models. *Quarterly Journal of Economics*, 102:1–22.

1648. Weil, P. (1989) The equity premium puzzle and the risk-free rate puzzle. *Journal of Monetary Economics*, 24:401–21.

1649. Weil, P. (1990) On the possibility of price decreasing bubbles. *Econometrica*, 58:1467–1474.

1650. Weil, P. (1992) Equilibrium asset prices with undiversifiable labor income risk. *Journal of Economic Dynamics and Control*, 16:769–790.

1651. Weitzman, M. (2007) Subjective expectations and asset-return puzzles. *The American Economic Review*, 97:1102–1130.
1652. Werner, I. and Kleidon, A. (1996) U.K. and U.S. trading of British cross-listed stocks: an intraday analysis of market integration. *Review of Financial Studies*, 9:619–664.
1653. Werner, J. (1985) Equilibrium in economies with incomplete financial markets. *Journal of Economic Theory*, 36:110–119.
1654. Werner, J. (2014) Rational asset pricing bubbles and debt constraints. *Journal of Mathematical Economics*, 53:145–152.
1655. West, D. (1987) A specification test for speculative bubbles. *Quarterly Journal of Economics*, 102:553–580.
1656. West, D. (1988) Bubbles, fads, and stock price volatility tests: a partial evaluation. *Journal of Finance*, 43:639–656.
1657. West, D. (1988) Dividend innovations and stock price volatility. *Econometrica*, 56:37–61.
1658. Weston, J. (2000) Competition on the NASDAQ and the impact of recent market reforms. *Journal of Finance*, 55:2565–2598.
1659. Wheatley, S. (1988) Some tests of the consumption-based asset pricing model. *Journal of Monetary Economics*, 22:193–215.
1660. Willard, G. and Dybvig, P. (1999) Empty promises and arbitrage. *Review of Financial Studies*, 12:807–834.
1661. Wilson, R. (1968) The theory of syndacates. *Econometrica*, 36:119–132.
1662. Womack, K. (1996) Do brokerage analysts' recommendations have investment value? *Journal of Finance*, 51:137–167.
1663. Yan, H. (2008) Natural selection in financial markets: Does it work? *Management Science*, 54:1935–1950.
1664. Yao, R. and Zhang, H. (2005) Optimal consumption and portfolio choices with risky housing and borrowing constraints. *Review of Financial Studies*, 18:197–239.
1665. Yogo, M. (2006) A consumption-based explanation of expected stock returns. *Journal of Finance*, 61:539–580.
1666. Yogo, M. (2008) Asset prices under habit formation and reference-dependent preferences. *Journal of Business and Economic Statistics*, 26:131–143.
1667. Yogo, M. (2016) Portfolio choice in retirement: health risk and the demand for annuities, housing, and risky assets. *Journal of Monetary Economics*, 80:17–34.
1668. Yu, J. (2011) Disagreement and return predictability of stock portfolios. *Journal of Financial Economics*, 99:162–183.
1669. Xia, Y. (2001) Learning about predictability: the effects of parameter uncertainty on dynamic asset allocation. *Journal of Finance*, 56:205–246.
1670. Xiong, W. (2001) Convergence trading with wealth effects: an amplification mechanism in financial markets. *Journal of Financial Economics*, 62:247–292.
1671. Xiong, W. and Hongjun, Y. (2010) Heterogeneous expectations and bond markets. *Review of Financial Studies*, 23:1433–1466.
1672. Zapatero, F. (1998) Effects of financial innovations on market volatility when beliefs are heterogeneous. *Journal of Economic Dynamics and Control*, 22:597–626.
1673. Zarowin, P. (1989) Does the stock market overreact to corporate earnings information? *Journal of Finance*, 54:1385–1399.
1674. Zarowin, P. (1990) Size, seasonality, and stock market overreaction. *Journal of Financial and Quantitative Analysis*, 25:113–125.
1675. Zeldes, S. (1989) Consumption and liquidity constraints: an empirical investigation. *Journal of Political Economy*, 97:305–346.
1676. Zeldes, S. (1989) Optimal consumption with stochastic income: deviations from certainty equivalence. *Quarterly Journal of Economics*, 104:275–297.
1677. Zhang, F. (2006) Information uncertainty and stock returns. *Journal of Finance*, 61:105–137.
1678. Zhang, H. (1997) Endogenous borrowing constraints with incomplete markets. *Journal of Finance*, 52:2187–2209.

1679. Zhang, H. (2000) Explaining bond returns in heterogeneous agent models: The importance of higher-order moments. *Journal of Economic Dynamics and Control*, 24:1381–1404.
1680. Zhang, L. (2005) The value premium. *Journal of Finance*, 60:67–103.
1681. Zhou, C. (1998) Dynamic portfolio choice and asset pricing with differential information. *Journal of Economic Dynamics and Control*, 22:1027–1051.
1682. Zhou, C. (1999) Informational asymmetry and market imperfections: another solution to the equity premium puzzle. *Journal of Financial and Quantitative Analysis*, 34:445–464.

Index

σ-algebra, 256

absolute risk aversion, 28
actuarially fair, 24
adapted stochastic process, 258
adverse selection, 398, 422, 433, 447, 450,
        594, 601, 605, 607, 617, 619, 637,
        638
aggregate demand, 7
aggregate risk, 129
aggregation property, 12, 157, 159, 191, 291
Allais paradox, 48, 489
ambiguity aversion, 486
ambiguity premium, 488
anonymity of trading, 640
anticipation, 500, 567
anxiety, 500
arbitrage opportunity, 148, 165, 175, 184, 209,
        231, 293, 304
    asymptotic, 234
    of the first kind, 165, 293
    of the second kind, 165, 293
arbitrage pricing theory, 230–245, 319, 381,
        572
    empirical analysis, 241
    book-market value ratio, 244
    equilibrium models, 249
    macroeconomic factors, 244
    Shanken critique, 241
    size, 244
Arrow security, 102, 143, 150, 282
Arrow-Debreu equilibrium, 139
artificial economy, 415
asset allocation puzzle, 556, 572

asset return, 56
asymmetric information, 398
auction market, 594, 638

background risk, 108, 112, 114
backward recursion, 264, 266
bank account, 56
batch auction market, 594
Bayes rule, 258
behavioral finance, 358, 360, 370, 372,
        520–541
Bellman equation, 266
biased self-attribution, 529, 531
bid-ask spread, 560, 561, 607, 616, 619, 625,
        628, 638, 641
    information ambiguity, 485
bifurcation index, 287
binomial model, 183
    multi-period, 308
blackout period, 632
Blackwell effect, 407, 421, 422
bond, 183
Borch condition, 131, 136
borrowing constraints, 562
bounded rationality, 541–548
breadth of ownership, 454
bubble, 330, 336, 468, 538
    overconfidence, 533
budget constraint, 4
buffer stock behavior, 556, 561
business cycle, 375

Call option, 159
capacity, 484

© Springer-Verlag London Ltd. 2017
E. Barucci, C. Fontana, *Financial Markets Theory*,
Springer Finance, DOI 10.1007/978-1-4471-7322-9

Printed in the United States
By Bookmasters